GEOPHYSICS AND GEOCHEMISTRY IN THE SEARCH FOR METALLIC ORES

EXPLORATION 77

©Minister of Supply and Services Canada 1979

Available in Canada through

authorized bookstore agents
and other bookstores

or by mail from

Canadian Government Publishing Centre
Supply and Services Canada
Hull, Québec, Canada K1A 0S9

and from

Geological Survey of Canada
601 Booth Street
Ottawa, Canada K1A 0E8

A deposit copy of this publication is also available
for reference in public libraries across Canada

Cat. No. M43-31/1979   Canada: $35.00
ISBN —   0-660-10425-3   Other countries: $42.50

Price subject to change without notice

Geological Survey of Canada
Economic Geology Report 31

GEOPHYSICS AND GEOCHEMISTRY IN THE SEARCH FOR METALLIC ORES

edited by
Peter J. Hood

Proceedings of Exploration 77
— an international symposium held in
Ottawa, Canada in October 1977

1979

Preface

Exploration, the prerequisite to the development of any mining industry, provides the means whereby a country determines its mineral wealth. During 1977 the non-fuel sector of the mining industry in Canada produced more than $7 billion dollars of new wealth, a significant fraction of the country's gross national product. Mining also generates a demand for a wide range of goods and services, and expenditures by the mining industry have a multiplier effect throughout the economy. Mines demand an infrastructure of new roads, electric power and communications. This can have a major effect on the growth of a region or even a whole country. Thus establishment of a mining industry is often the first step in the development of a country.

Mineral exploration today is highly dependent on the application of geophysics and geochemistry. The relevant knowledge and technology is expanding rapidly, becoming more and more specialized, and expensive to apply. The ever increasing costs must be paid for by society as a whole in the price paid for mineral products. Thus the public interest is served by providing those involved in exploration with up-to-date reviews of the state-of-the-art to assist them in their task. In 1970, the Geological Survey of Canada published the Proceedings of the 1967 Conference on Mining and Groundwater Geophysics; the demand for this volume proved to be so great that three reprintings were required. It is therefore appropriate that the Geological Survey of Canada publish the successor to this volume not only as a contribution to increasing the efficiency and effectiveness of exploration but also as a measure of the advances made during the past decade.

Ottawa, March 1979

D.J. McLaren
Director General
Geological Survey of Canada

FOREWORD

This volume contains the proceedings of an International Symposium on Geophysics and Geochemistry applied to the search for Metallic Ores, held in Ottawa in October 1977.

The symposium was planned and organized by a committee of the Canadian Geoscience Council, with the support of the Geological Survey of Canada and numerous industrial benefactors. It was designed to be a successor to the Canadian Centennial Conference on Mining and Groundwater Geophysics held at Niagara Falls, Ontario, ten years previously. During the intervening decade the National Research Council Associate Committees had been dissolved and succeeded, in the case of the earth sciences, by the Canadian Geoscience Council. This council is composed of representatives of all scientific and professional societies concerned with the diverse aspects of earth science in Canada. Mineral exploration, particularly metalliferous mineral exploration is an activity for which Canada has established a world-wide reputation. Much of this reputation stems from the development of, and experience in using modern technology to explore for concealed mineralization. Geophysical techniques have been progressively developed and applied for this purpose over the past 30 years, and within the last 10 years geochemical techniques have achieved equal importance. For these reasons the Canadian Geoscience Council fully supported the recommendation of the Symposium Organizing Committee that Exploration '77 should give equal weight to geophysics and geochemistry in its review of modern exploration techniques. So in this respect the 1977 symposium differed from that held in 1967.

The symposium was opened by the President of the Canadian Geoscience Council, P.J. Savage. It was attended by 770 registrants from 36 countries including an official delegation from the People's Republic of China. Forty papers were presented over a 4 day period. Concurrently with the symposium an exhibition of exploration equipment and results was held, which provided a forum for 33 exhibitors. Overall every effort was made to provide a balanced picture of metalliferous mineral exploration methods as they exist in the late 1970's. Thanks are due to all those who contributed their time and energy to the planning, preparation and execution of the symposium, and to the authors who spent many long hours before and after preparing their material for presentation and publication. Especial thanks are due to Peter Hood for his painstaking and indefatigable efforts in shepherding papers through the critical readers into their final form.

The following were members of the Organizing Committee for Exploration '77:

Stanley W. Holmes	Honorary Treasurer
Bernard E. Manistre	Secretary/Treasurer
Peter J. Hood	Chairman, Program Committee
Harold O. Seigel	Convener Geophysics
J. Alan Coope	Convener Geochemistry
Roger M. Pemberton	Co-convener, Case Histories
Kenneth A. Morgan	Co-convener, Case Histories
Pauline Moyd	Adviser
John Needham	Chairman, Exhibition Committee

Arthur G. Darnley
General Chairman

INTRODUCTION

In planning the technical program for the Exploration '77 Symposium, the Program Committee decided to retain a format similar to that used for the 1967 Centennial Conference to enable the proceedings volume to be essentially a textbook describing recent advances in geophysical and geochemical techniques applied to the search for metallic ores. Accordingly it was decided to have general papers presenting overviews of economic geology, geophysics and geochemistry followed by papers describing the state-of-the-art for each of the main subdisciplines of geophysics and geochemistry. Papers describing the application of these techniques in finding mines were then to be presented by explorationists active in the profession; for this purpose two co-conveners from mining companies were selected from the Program Committee to solicit these case-histories with as great a geographic distribution and type of deposit as possible. Thus the Program Committee has attempted to make the Exploration '77 Symposium as international in scope as could be achieved albeit with some Canadian bias. The result was that 42 per cent of the papers were by authors resident outside of Canada and nine of the sixteen case history papers dealt with exploration in foreign countries.

The production of the proceedings volume has been facilitated by the contributions of many individuals both within the Geological Survey of Canada and in the universities and mining and survey companies both in Canada and overseas. Each paper was critically read either by one or two reviewers whose efforts have helped greatly to improve the quality of the papers. The critical readers were:

CRITICAL READERS

State-of-the-Art

A. Becker	A.G. Darnley	P.G. Killeen	A.W. Rose
R.A. Bosschart	P.H. Dodd	K. Kosanke	W.J. Scott
P.M.D. Bradshaw	A.V. Dyck	J. Lynch	H.O. Seigel
K.B.S. Burke	I.L. Elliott	G.W. Mannard	W.W. Shilts
E.M. Cameron	W.A. Finney	S.S. Nargolwalla	A.J. Sinclair
R.H. Carpenter	E.H.S. Gaucher	I. Nichol	V.R. Slaney
J.M. Carson	R.L. Grasty	P. Norgaard	D.W. Strangway
L.G. Closs	A.F. Gregory	N.R. Paterson	D.W. Wagg
L.S. Collett	P.G. Hallof	A.R. Rattew	S.H. Ward
J.G. Conaway	M.T. Holroyd	M.S. Reford	H.V. Warren
J.A. Coope	R.D. Hutchinson	K.A. Richardson	G.F. West

Case Histories

J.G. Baird	J.D. Crone	E.H.W. Hornbrook	R.S. Middleton
E.J. Ballantyne	T. Flanagan	F.L. Jagodits	K.A. Morgan
R.W. Boyle	D.C. Fraser	J.J. Lajoie	R.H. Pemberton
E.M. Cameron	R.J. Henderson	J. McAdam	L.E. Reed
J.D. Corbett			

This report was printed from copy prepared in the Word Processing Unit of the Geological Survey of Canada, Ottawa by:

Debby Busby	Susan Gagnon	Suzanne Lalonde
Judy Coté	Janet Gilliland	Sharon Parnham

Production editing and layout:

Leona R. Mahoney	Lorna A. Firth	Michael J. Kiel

Peter J. Hood
Editor and
Chairman, Program Committee

CONTENTS

Page

ix   FOREWORD
        A.G. Darnley

xi   INTRODUCTION
        Peter J. Hood

1   GEOLOGY AND ORE DEPOSITS
        Duncan R. Derry (1)*

7   AN OVERVIEW OF MINING GEOPHYSICS
        Harold O. Seigel (2)

25   GEOCHEMISTRY OVERVIEW
        R.W. Boyle (3)

33   AIRBORNE ELECTROMAGNETIC METHODS
        A. Becker (4)

45   GROUND ELECTROMAGNETIC METHODS AND BASE METALS
        Stanley H. Ward (5)

63   RECENT DEVELOPMENTS IN THE USE AND INTERPRETATION
        OF DIRECT-CURRENT RESISTIVITY SURVEYS
        George V. Keller (6)

77   MAGNETIC METHODS APPLIED TO BASE METAL EXPLORATION
        Peter J. Hood, M.T. Holroyd and P.H. McGrath (7)

105   GRAVITY METHOD APPLIED TO BASE METAL EXPLORATION
        J.G. Tanner and R.A. Gibb (8)

123   THE INDUCED-POLARIZATION EXPLORATION METHOD
        John S. Sumner (9)

135   GAMMA RAY SPECTROMETRIC METHODS IN URANIUM EXPLORATION –
        AIRBORNE INSTRUMENTATION
        Q. Bristow (10A)

147   GAMMA RAY SPECTROMETRIC METHODS IN URANIUM EXPLORATION –
        THEORY AND OPERATIONAL PROCEDURES
        R.L. Grasty (10B)

163   GAMMA RAY SPECTROMETRIC METHODS IN URANIUM EXPLORATION –
        APPLICATION AND INTERPRETATION
        P.G. Killeen (10C)

231   MODERN TRENDS IN MINING GEOPHYSICS AND NUCLEAR BOREHOLE LOGGING
        METHODS FOR MINERAL EXPLORATION
        Jan A. Czubek (11)

*Bracketed numbers refer to the Paper numbers

273   BOREHOLE LOGGING TECHNIQUES APPLIED TO BASE METAL ORE DEPOSITS
      W.E. Glenn and P.H. Nelson (12)

295   FOCUS ON THE USE OF SOILS FOR GEOCHEMICAL EXPLORATION
      IN GLACIATED TERRANE
      B. Bølviken and C.F. Gleeson (13)

327   THE APPLICATION OF SOIL SAMPLING TO GEOCHEMICAL EXPLORATION
      IN NONGLACIATED REGIONS OF THE WORLD
      P.M.D. Bradshaw and I. Thomson (14)

339   LITHOGEOCHEMISTRY IN MINERAL EXPLORATION
      G.J.S. Govett and Ian Nichol (15)

363   THE APPLICATION OF ATMOSPHERIC PARTICULATE GEOCHEMISTRY
      IN MINERAL EXPLORATION
      A.R. Barringer (16)

365   ANALYTICAL METHODOLOGY IN THE SEARCH FOR METALLIC ORES
      F.N. Ward and W.F. Bondar (17)

385   ADVANCES IN BOTANICAL METHODS OF PROSPECTING FOR MINERALS
      PART I – ADVANCES IN GEOBOTANICAL METHODS
      Helen L. Cannon (18A)

397   ADVANCES IN BOTANICAL METHODS OF PROSPECTING FOR MINERALS
      PART II – ADVANCES IN BIOGEOCHEMICAL METHODS OF PROSPECTING
      Robert R. Brooks (18B)

411   STREAM SEDIMENT GEOCHEMISTRY
      W.T. Meyer, P.K. Theobald, Jr., and H. Bloom (19)

435   LAKE SEDIMENT GEOCHEMISTRY APPLIED TO MINERAL EXPLORATION
      W.B. Coker, E.H.W. Hornbrook, and E.M. Cameron (20)

479   APPLICATION OF HYDROGEOCHEMISTRY TO THE SEARCH FOR BASE METALS
      W.R. Miller (21A)

489   APPLICATION OF HYDROGEOCHEMISTRY TO THE SEARCH FOR URANIUM
      W. Dyck (21B)

511   REMOTE SENSING IN THE SEARCH FOR METALLIC ORES:
      A REVIEW OF CURRENT PRACTICE AND FUTURE POTENTIAL
      Alan F. Gregory (22)

527   COMPUTER COMPILATION AND INTERPRETATION OF GEOPHYSICAL DATA
      Allan Spector and Wilf Parker (23)

545   COMPUTER-BASED TECHNIQUES IN THE COMPILATION, MAPPING
      AND INTERPRETATION OF EXPLORATION GEOCHEMICAL DATA
      R.J. Howarth and L. Martin (24)

575   SOME ASPECTS OF INTEGRATED EXPLORATION
      J.A. Coope and M.J. Davidson (25)

593   EXPLORATION DISCOVERIES, NORANDA DISTRICT, QUEBEC
        J. Boldy (26)

605   EXPLORATION CASE HISTORIES OF THE ISO AND NEW INSCO OREBODIES
        W.M. Telford and Alex Becker (27)

631   THE DISCOVERY AND DEFINITION OF THE LESSARD BASE METAL DEPOSIT, QUEBEC
        Laurie E. Reed (28)

641   IZOK LAKE DEPOSIT, NORTHWEST TERRITORIES, CANADA:
        A GEOPHYSICAL CASE HISTORY
        George Podolsky and John Slankis (29)

653   GEOPHYSICAL EXPLORATION AT THE PINE POINT MINES LTD. ZINC-LEAD PROPERTY,
        NORTHWEST TERRITORIES, CANADA
        Jules J. Lajoie and Jan Klein (30)

665   ON THE APPLICATION OF GEOPHYSICS IN THE INDIRECT EXPLORATION
        FOR COPPER SULPHIDE ORES IN FINLAND
        M. Ketola (31)

685   GEOPHYSICAL AND GEOCHEMICAL METHODS USED IN THE DISCOVERY
        OF THE ISLAND COPPER DEPOSIT, VANCOUVER ISLAND, BRITISH COLUMBIA
        K.E. Witherly (32)

697   GEOPHYSICAL AND GEOCHEMICAL CASE HISTORY OF THE QUE RIVER DEPOSIT,
        TASMANIA, AUSTRALIA
        S.S. Webster and E.H. Skey (33)

721   GEOPHYSICS AND GEOCHEMISTRY IN THE DISCOVERY AND DEVELOPMENT
        OF THE LA CARIDAD PORPHYRY COPPER DEPOSIT, SONORA, MEXICO
        D.F. Coolbaugh (34)

727   EXPLORATION OF THE REAL DE ANGELES SILVER-LEAD-ZINC SULPHIDE DEPOSIT,
        ZACATECAS, MEXICO
        Lee R. Stoiser and José Bravo Nieto (35)

735   APPLICATION OF X-RAY DIFFRACTION ALTERATION AND GEOCHEMICAL TECHNIQUES
        AT SAN MANUEL, ARIZONA
        D.M. Hausen (36)

745   EXPLORATION FOR MASSIVE SULPHIDES IN DESERT AREAS USING
        THE GROUND PULSE ELECTROMAGNETIC METHOD
        Duncan Crone (37)

757   THE APPLICATION OF AIRBORNE AND GROUND GEOPHYSICAL TECHNIQUES
        TO THE SEARCH FOR MAGNETITE-QUARTZITE ASSOCIATED BASE-METAL
        DEPOSITS IN SOUTHERN AFRICA
        Geoff Campbell and R. Mason (38)

779   GEOPHYSICAL AND GEOCHEMICAL METHODS FOR MAPPING
        GOLD-BEARING STRUCTURES IN NICARAGUA
        R.S. Middleton and E.E. Campbell (39)

799   AN OUTLINE OF MINING GEOPHYSICS AND GEOCHEMISTRY IN CHINA
        Hsia Kuo-chih et. al. (40)

811   AUTHOR INDEX

xii   PLATES 1-4

Plate 1

Residual magnetic anomalies of central Canada south of Hudson Bay; WB – Wabigoon volcanic belt, CSB – Boundary between Churchill and Superior geological provinces, KL – Kapuskasing Lineament, and QS – Quetico structural zone. Red > +200 γ, yellow 0 to +200 γ, green 0 to -200 γ, blue < -200 γ; (see p. 101) (from Magnetic Anomaly Map of Canada, McGrath et al., 1977).

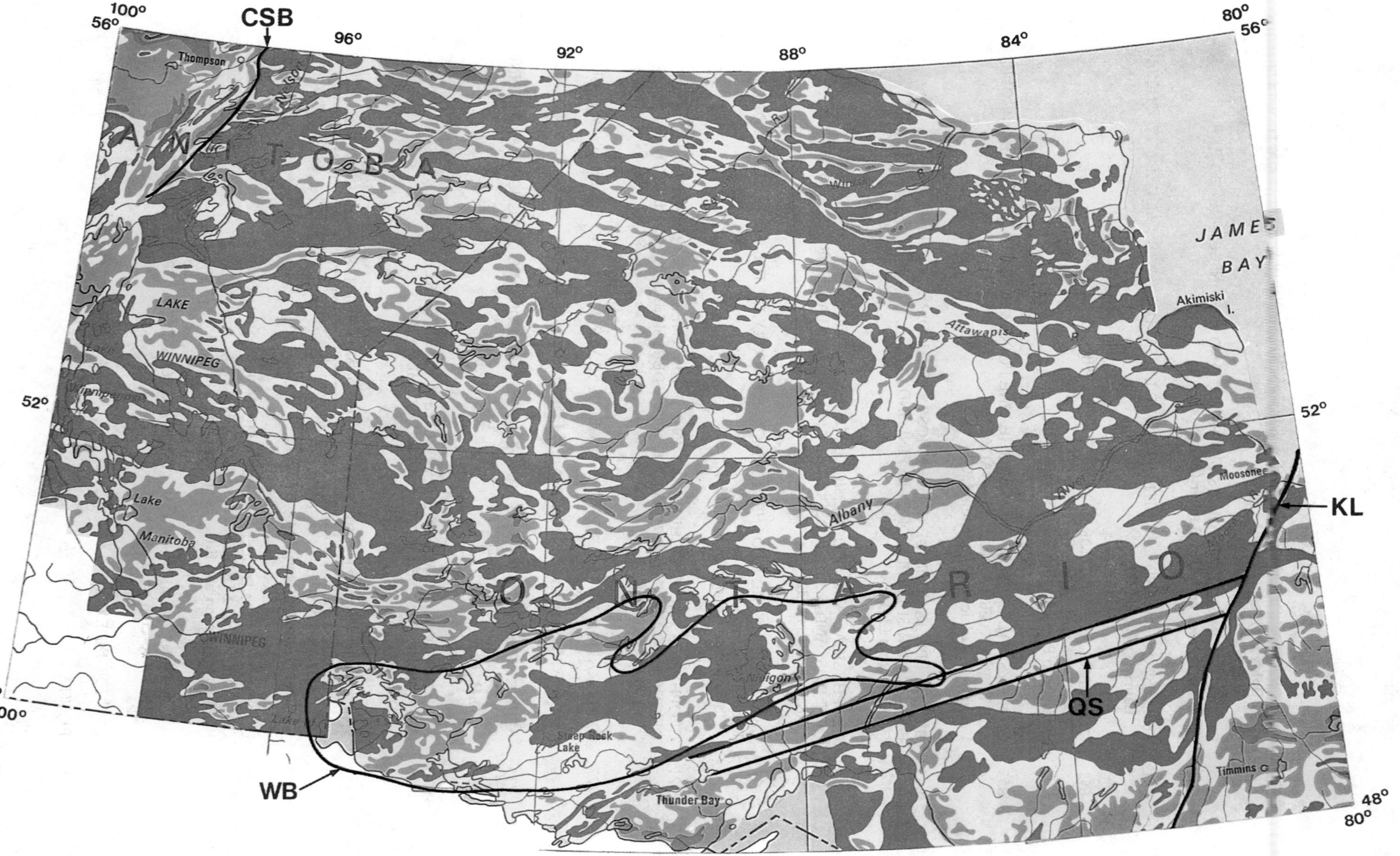

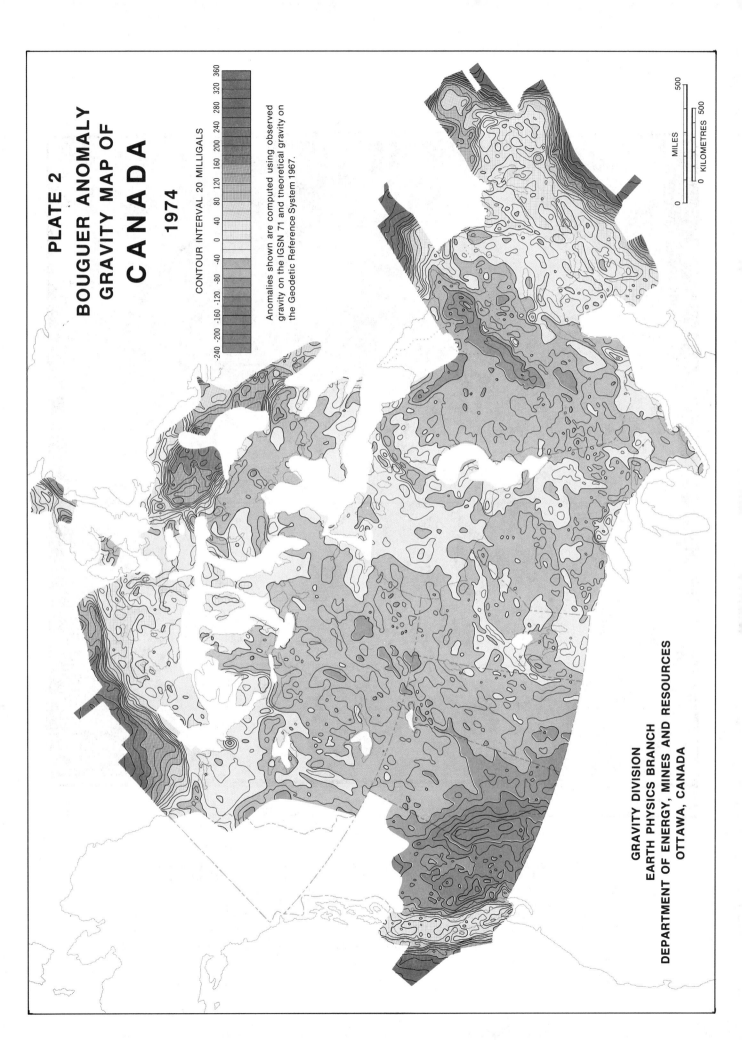

Plate 3

False colour Landsat image of southwestern Bolivia and northeastern Chile. Scale: approximately 1:1 million. Mesozoic to Recent volcanic and sedimentary rocks of the Cordillera Occidental and Atacama Desert. Note Tertiary volcanoes, Cretaceous granitic intrusions, the Salar de Atacama, a variety of geological structures and the alluvial cover. The Chuquicamata porphyry copper deposit can be seen about 20 km north of the large vegetated (red) area in the northwest corner of the image (see p. 518).

Digitally processed image courtesy of M.J. Abrams, Jet Propulsion Laboratory, California Institute of Technology.

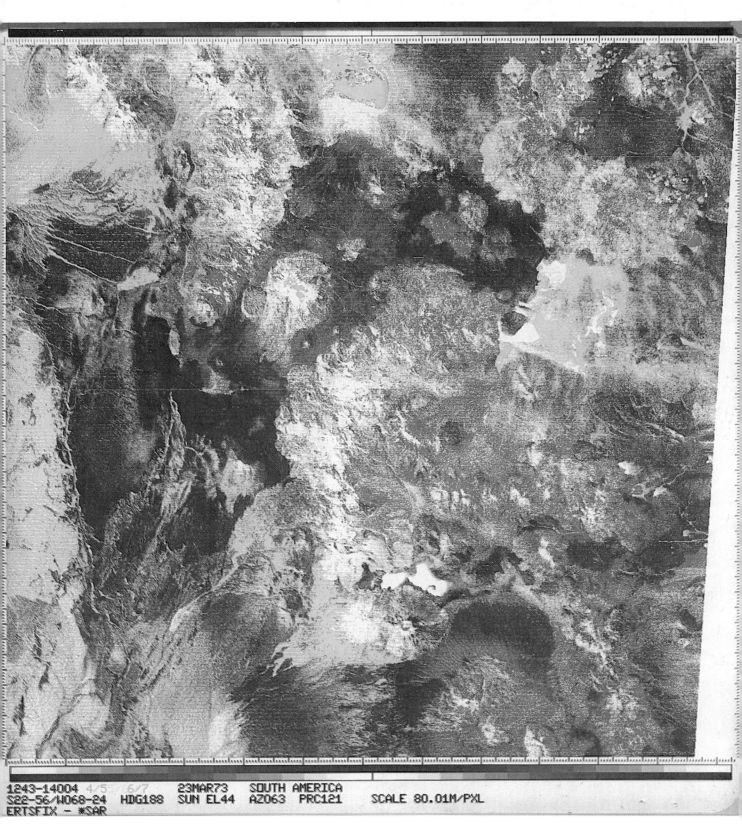

Plate 4

Landsat ratio composite of northeast quarter of Plate 3. Scale approximately 1:450 000. Ratios of bands 4/5, 5/6 and 6/7 are displayed as blue, green and red respectively. The reddish orange areas represent limonitic, hydrothermal alteration. Hot springs, gypsum and sulphur deposits typical of a young volcanic environment occur on the ground. Note that ratioing is specific for selected features, e.g., orange-red for limonitic alteration; however, the same ratios group different features in another colour, e.g., medium blue includes vegetation, clouds, alluvium and unaltered volcanic rocks which have quite different reflectances in Plate 3. (see p. 521).

Digitally processed image courtesy of M.J. Abrams, Jet Propulsion Laboratory, California Institute of Technology.

GEOLOGY AND ORE DEPOSITS

Duncan R. Derry

Derry, Michener & Booth, Toronto, Canada

Derry, Duncan R., Geology and Ore Deposits; in Geophysics and Geochemistry in the search for Metallic Ores; Peter J. Hood, editor; Geological Survey of Canada, Economic Geology Report 31, p. 1-6, 1979.

Abstract

The plate tectonic concept has had implications for almost all aspects of Earth Science over the past ten years including the distribution and genesis of ore deposits. This has coincided with:

(a) Increased emphasis on the conditions surrounding the formation of the rocks that enclose an ore deposit rather than emphasis on later metallogenic processes.

(b) The increased diversity and sensitivity of geophysical and geochemical methods used in the search for new ore deposits.

The relationships of many metallic deposits to plate boundaries has become increasingly clear, the best example perhaps being that of porphyry copper deposits. Other plate boundary relationships range through various metallic deposits to concentrations of evaporites and petroleum.

The fact that few if any new discoveries can be attributed directly to the application of the plate tectonic concept must be accepted. It has shown convincing reasons why the deposits were there after they have been discovered. The actual discovery has usually resulted from the application of geophysics, geochemistry or conventional prospecting to areas selected by empirical reasoning, e.g. the observed association of porphyry copper deposits with mountain systems bordering the edge of continents. The concept of volcanogenic base metal deposits related to volcanic centres can claim more discovery credit but even here acid lavas were empirically used as favourable indications with success before the present concepts were established.

In future geologists must aim to be more directly constructive in using the plate tectonic concept and its various implications in the concentration on more specific areas for exploration that could not be accurately selected by purely empirical reasoning. This may turn them more towards the search for deeply-penetrating crustal fractures which may not be visible at surface. Such may be forecast by plate tectonic studies and identified by deeply-penetrating geophysical techniques or by broad but sensitive geochemical surveys.

Résumé

La notion de tectonique des plaques est utilisée dans presque tous les domaines des sciences de la terre depuis dix ans y compris l'étude de la répartition et de la genèse des gisements métallifères.

En même temps, on a davantage prêté attention:

(a) aux conditions dans lesquelles se forment les roches encaissantes qu'aux processus métallogéniques ultérieures,

(b) à la diversité et à la sensibilité des méthodes géophysiques et géochimiques utilisées dans la recherche de nouveaux gisements métallifères.

Les relations entre un grand nombre de gisements métallifères et les limites des plaques sont devenues de plus en plus claires, le meilleur exemple étant probablement celui des gisements porphyriques cuprifères. Les autres cas de gîtes liés aux limites de plaques vont des gîtes métallifères aux accumulations d'évaporites et d'hydrocarbures.

Cependant, on doit reconnaître que très peu de découvertes peuvent être attribuées directement à l'application de la notion de tectonique des plaques. Ce concept a davantage démontré, une fois que les gisements ont été découverts, à quoi était due leur présence. Les découvertes actuelles résultent normalement de l'application de la prospection géophysique, géochimique, ou conventionnelle dans des régions choisies par un raisonnement empirique; on a par exemple observé que les gisements porphyriques de cuivre étaient associés aux systèmes montagneux qui bordent les continents. Le concept de l'association de gisements de métaux de base avec des centres volcaniques, a sans doute favorisé un plus grand nombre de découvertes, mais, dans ce cas, aussi, c'est la présence de laves acides qui nous a guidés, avant de faire appel aux notions actuelles.

A l'avenir, les géologues doivent apprendre à utiliser la notion de tectonique des plaques et ses diverses implications, et chercher à explorer plus spécifiquement des secteurs dont le choix ne peut être fait de fa on purement empirique. Ceci pourrait orienter les recherches vers l'exploration des fractures profondes de la croûte terrestre, et qui restent cachées en surface. On peut les déceler par l'étude de la tectonique des plaques, et les identifier par des méthodes d'exploration géophysique à grande profondeur, ou par des levés géochimiques larges, mais très précis.

INTRODUCTION

The principal objective of this symposium volume is to review the state of the art of the application of geophysics and geochemistry to the search for new mineral deposits. By means of the case histories and comparative studies contributed to provide this review the co-ordination of geology, geophysics and geochemistry and the co-operation of those specializing in each should also be improved.

This paper provides a geological framework of the changing concepts that have coincided with the refinements and new techniques in geophysics and geochemistry that will be discussed in the remaining papers presented at the Symposium.

Two major concepts developed over the past decade and a half, the contributions these have made to mineral exploration, but also the limitations in their practical applications, and how these might be overcome in the future are reviewed briefly.

CONTEMPORANEITY OF ORE AND ENCLOSING ROCKS

The first concept deals directly with ore genesis and applies to widely differing groups of metals. Its basic trend, however, has been to relate mineralization more closely to the conditions prevailing at the time of the formation of the enclosing rocks rather than to later geological events. An obvious example is the massive sulphide deposit typified by those of the Noranda area which were once thought to have originated by hydrothermal replacement that took place millions — perhaps hundreds of millions — of year after the volcanic or sedimentary host rock was formed. Today most accept the thesis that the initial deposition of such orebodies was approximately synchronous with the enclosing rocks although recognizing that much later remobilization has played a critical role. In many cases this remobilization has resulted in changing the final grade and mineral form in directions favourable to economic exploitation.

Another example of the basic change in thinking towards contemporaneity of deposition with at least the hanging wall and towards formation at, or close to, the then surface is that of uranium in the so-called "vein class" which I prefer to term the "unconformity class" since the form of the orebodies is only incidentally in fracture fillings. In this case the concept that supergene concentration played at least a major part in the present ore grades has replaced, in the minds of many geologists, the former assumption of much later hydrothermal replacement.

Both examples are typical of the general swing towards accepting contemporaneity between ore and wallrock and towards surface or near surface origin for many metallic deposits. It is a general concept, that has a direct application in the selection of areas for concentrated exploration and it has already contributed significantly to the discovery of individual orebodies. Without any clear understanding of their meaning it is, of course, not always easy to separate sharply the conscious application of the working hypothesis from the empirical use of observed relationships. For instance, long before the volcanogenic origin of Archean sulphide orebodies was suggested many geologists had noticed that such orebodies commonly appeared to have a spatial relationship with acid lavas or pyroclastics. The name "mill rock" was applied by one geologist to the coarse volcanic fragmental that he observed was nearly always within sight or sound of an operating mill. In the exploration field I remember when I was with RioCanex that in 1958 we laid out an airborne EM survey over an area of northern Quebec that was selected partly on the presence of a belt of acid volcanics because we had noticed that copper-zinc mineralization elsewhere had this association. The program resulted in finding a modest orebody that made the Poirier mine. In this case empirical information was used without any specific theory to account for the association. The volcanogenic thesis has since been effectively used in selecting areas for the detailed application of geophysics and geochemistry and it was a program based on this theory, started in New Brunswick, that eventually led Texasgulf to the Kidd Creek orebody in northern Ontario.

PLATE TECTONICS AND METALLOGENY

The second major conceptual change that has influenced mining exploration over the past 15 or 20 years is, of course, that of Plate Tectonics. Early in its development — even in the writings of Wegener, Du Toit and Holmes — the implications of Continental Drift on concentrations of ore or petroleum were noted. But it was not until the modern concept of Plate Tectonics was developed that the significance of the overlapping of crustal plates, i.e. subduction zones, came to be considered in the genesis of orebodies as opposed only to their location relative to continental margins. Here again we were probably using ancient plate margins as a guide to exploration in a purely empirical way. The relationship of orebodies to geochronological boundaries had been noted by various people in the geological or mining exploration field for some years before the present Plate Tectonic concept course. I read a paper, as the Presidential Address to the Society of Economic Geologists in 1960, pointing out that if we had restricted our search in the Canadian Precambrian Shield to within 100 miles of any Archean-Proterozoic boundary we would have found 80 per cent of all the known ore deposits in the Shield. However I was not ingenious enough to come up with a reason for the relationship.

In May 1974, a symposium was sponsored in St. John's, Newfoundland by the Geological and Mineralogical Associations of Canada which resulted in a volume of which Canadian earth scientists generally may be justifiably proud. Edited by David Strong (1976) it contains 32 papers by authorities from Britain, Europe, the United States, South Africa and Australia, as well as from Canada from coast to coast. It is the most comprehensive report to date on the relationship of Plate Tectonics to metallogeny and I believe it will acquire increasing respect in years ahead. Examples of particular interest include contributions by A.H.G. Mitchell on the relation of mineral deposits to subduction margins; A.H. Clarke and Associates from Queen's University describing excellent studies on two selected cross-sections in the Andes and another paper by R.H. Sillitoe of Imperial College, London on the same general region; D.F. Sangster and F.J. Sawkins in separate studies on volcanogenic massive sulphide deposits; Takeo Sato on the Japanese sulphide deposits; T.P. Thayer on mineralization related to ophiolites; P. Laznika on the global distribution of lead deposits; A.Y. Glickson of Australia on Proterozoic structures related to Plate Tectonics; W. Walker on global orogeny/unconformity relationships with ore genesis. There have been, of course, many valuable articles and textbooks published before and since this publication dealing with the same field but for sheer, factual information on a worldwide basis it is doubtful if it can be equalled. Despite this it is a sobering conclusion that I do not know of any significant metallic discovery that can be unequivocally credited to the application of the Plate Tectonics concept. All the articles referred to above, and many others in relevant publications, present impressive data showing the relationships of various types of ore deposits to plate margins, etc. and may explain the reasons for their presence. The actual discoveries, however, mainly by drilling on geochemical or geophysical anomalies, have been in areas selected by empirical reasoning. A few examples follow.

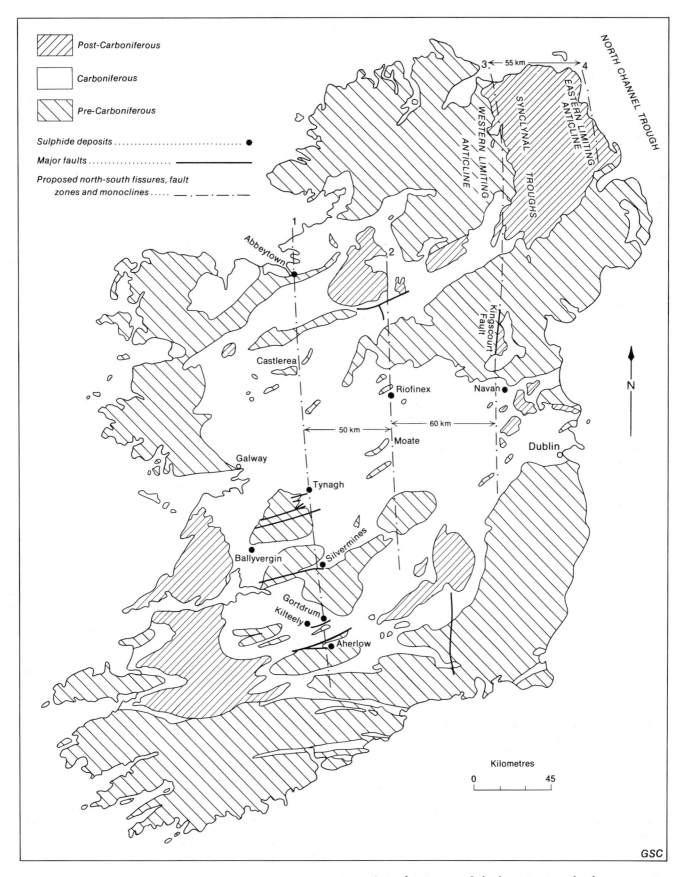

Figure 1.1. Geology of Ireland with postulated north-south geofractures and the larger copper-lead-zinc-deposits in Dinantian (lower Carboniferous) rocks (after Russell, 1968).

The most consistent relationship with plate margins is that of porphyry copper deposits with subduction zones. But many years before the present concept of Plate Tectonics, the pattern of porphyry coppers paralleling the trace of continental margin orogenies had been noted and used in long-term exploration strategies by major base metal companies. Such empirical reasoning indicated favourable zones measurable in thousands of linear kilometres. I do not believe, much as I would like to, that the admirable and formidable accumulation of research as summarized in the GAC Volume referred to above has really enabled us to concentrate our efforts in, say, 15 or 20 per cent of the linear extent of a potential porphyry copper belt over a subduction zone while eliminating the rest as being relatively unfavourable. Much the same must be said for the relationship of Plate Tectonics to lead/zinc deposits which tend to form belts parallel to the porphyry coppers but farther from the plate margins — we can explain the controls logically but can we apply the knowledge or theory, with any conviction, in the selection of areas on which to concentrate our exploration expenditures and geochemical or geophysical tools? Our selection of areas is still largely empirical based on observations of petrological or sedimentological associations.

The recognition of ophiolites as material originating from the ocean floor was of particular importance in the development of the Plate Tectonics concept. Linear zones of discontinuous ophiolite occurrences are probably the most reliable indicators of sutures marking the lines of collision between tectonic plates, e.g. the Tseng-Po – Indus valleys line in Asia, marks a major, if not the main, line of impact of the Indian and Eurasian plates. More recent studies in Newfoundland have identified ophiolites as indicating the sutures of former plate collisions in Appalachian-Caledonian times.

From the economic standpoint the studies of the Cyprus pyrite and copper deposits in ophiolites by Hutchinson (1965) and others, which led to conclusions of a dominantly syngenetic rather than replacement origin, predated the general acceptance of the Plate Tectonic concept. The more recent discovery, or rather re-discovery of deposits worked between 500 and 2500 years ago, in the most extensive ophiolite area in the world, the Sultanate of Oman, was made by Charles Huston of Toronto and his associates, not on the basis of Plate Tectonic reasoning but as the result of reading a book on the archaeology of the area and following this with geological and geophysical reconnaissance which defined drilling targets (Crone, 1979).

Substantial reserves of chromite are known in ophiolites in various parts of the world and there is no doubt that future searches for this strategic metal will be concentrated in ophiolite zones already known and, possibly, in some not yet known which may be found as a result of Plate Tectonic reasoning. Up to the present (1977), however, it is questionable if any new production or substantial reserves can be credited directly to Plate Tectonic reasoning.

The Kuroko and related deposits of Japan have been restudied by many Japanese geologists in the light of Plate Tectonic concepts and the relationship of this type of deposit to island arc stages of plate boundaries has been convincingly demonstrated. As far as I am aware, however, any new discoveries made of this type of deposit have been the result of drilling based on detailed studies of the volcanic stratigraphy and the configuration of the underlying basement, followed by the application of various geophysical techniques, and cannot be attributed even indirectly to Plate Tectonic reasoning. Nevertheless such reasoning may instigate new exploration in the volcanic sequences of other island arcs such as the North Island of New Zealand and other parts of the South Pacific.

I have mentioned a few types of deposits that seem to have particularly close relationships with Plate Tectonics. However it is difficult to show convincing evidence that the acceptance of these principles has directly aided exploration. It is to be hoped that the next stage of study and increased knowledge in the Plate Tectonics field may bring out new concepts or working hypotheses that could help concentrate exploration efforts more precisely than on, for example, the general trend of plate boundaries. The following are merely speculations in the directions these might take.

Metallographic Selection

The concept, or perhaps more correctly the observation, of metallographic areas long predates Plate Tectonics and was discussed by de Launay in the first quarter of this century. Nevertheless it is still almost as much a mystery as it was 75 years ago why certain areas of the world seemed to produce an anomalous number of deposits of a particular metal or association of metals spread over several geological events. Perhaps tin is the most intriguing and frustrating in that about 59 per cent of the world's known reserves of this metal (from Commodity Data Summaries, 1977) are in an area, including Thailand, Malaysia, western Indonesia and the southwest part of the Peoples Republic of China, amounting to about 3 million km^2 or less than 0.6 per cent of the world's surface. This tin mineralization, moreover, is associated with granites intruded in at least three periods between 300 and 48 million years ago. An additional 12 per cent of global tin reserves are in a relatively small area of Bolivia and are distributed over at least four orogenic periods, the last of which (about 20 million years ago but extending up to a million years or so ago) is responsible for the largest proportion. Both the Far Eastern and the Andean examples are within plate boundary zones but occupy a frustratingly small proportion of their extent. If we could understand the reasons for these two concentrations we might be able to apply the knowledge to finding new tin metallographic areas (probably smaller than the two mentioned) which would reduce our concern on future global supplies of what is probably the most critical of all base metals.

A highly respected geologist and author, Pierre Routhier (1976), has presented some thoughts on metallographic zones. His suggested relationships of tin areas of the world are, to me, unconvincing but some of his more positive observations in central and southern Europe, in which he is an acknowledged authority, are of considerable interest. He points, for example, to the broad belt of mineralization that contains the famous Rio Tinto mines in Spain. He noted that, within this easterly-trending belt of Paleozoic sediments and volcanics, zones of copper and pyrite mineralization alternate with those of manganese and lie at an oblique angle to the mineralized belt as a whole. He also pointed to belts of lead-zinc mineralization in western Europe that cut across geological boundaries and include deposits of varying ages and type (Routhier, 1976). The first of these extends from the Ardennes (Belgium) to Upper Harz (West Germany) and contains metallic deposits dating from mid-Devonian to Triassic including both vein fillings and tabular-bedded bodies. A second belt, less firmly established, extends from the north of Spain in a northeasterly direction across the Pyrenees and along the southern edge of the Alps. This also includes deposits that seem to disregard age boundaries and types of deposits but the dominant metals are lead and zinc. It should be noted that the Trepca deposits in Yugoslavia, the most important source of lead and zinc in continental Europe, are not within either of the above zones but may fall into a separate, less extensive one.

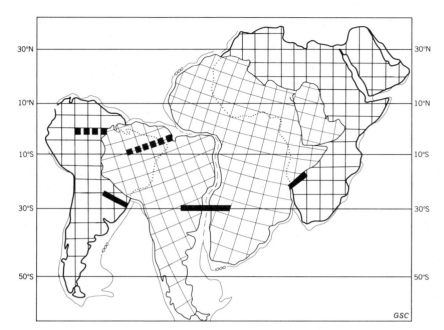

Figure 1.2. Orientation of geofractures in Africa and South America pre- and post-continental drifting (after Kutina, 1976).

Routhier 1976 also drew attention to a belt of predominantly copper deposits, some being significant producers and others of mainly scientific interest, that extends in an arc-shape oriented roughly north-south for 550 km through Bulgaria, eastern Yugoslavia and Romania in host rocks varying from Upper Mesozoic to mid-Tertiary age.

What is the controlling feature in each of these examples? No observable fracture zones exist coinciding with the metallographic belts but one is driven towards the conclusion that some crustal fracture systems predating present host rocks must exist. Routhier reached no definite conclusion but considered that "the Zn-Pb belts can, therefore, be considered as revealing former structural directions partly hidden by more recent phenomena". He continued further "that is the fundamental question: were the metallogenic provinces grafted on to the pre-formed geochemical provinces, and how?"

Primordial Crustal Fractures

The concern with fractures, invisible at the surface, that individually seem to coincide with several ore deposits or occurrences on separate structural belts, has occupied the minds of geologists in several parts of the world. E.A. Noble has studied these phenomena for many years and speculated on their origin. E.B. Brock, working mainly in South Africa but applying his ideas to global tectonics generally, included the relationship of crustal fractures to concentrations of mineral deposits.

One approach to the possible significance of crosscutting "geofractures" that cannot be recognized on surface is that of Russell (1968) studying especially the lead-zinc deposits in Ireland (Fig. 1.2) and also those in parts of Britain and Greenland. He pointed to the fact that of twelve deposits in Ireland, each containing over 10 000 tons of lead-zinc (or the equivalent value in other metals), five lie on a line trending 8° west of north, these being Abbeytown, Tynagh, Silvermines, Gortdrum and Aherlow (the last two copper-silver mineralization). Russell postulated other north-south geofractures at roughly equidistant intervals of between 45 and 65 km. It may be noted that the biggest zinc-lead deposit at Navan cannot be "matched" on a north-south line with any other known deposits, although it is on a line passing through the Kingscourt inlier in an expected, roughly equidistant position.

I.S. Thompson, who worked with me in Ireland some years ago, independently noted the roughly equidistant north-south "lines" on which mineral concentrations appeared to lie. These observations were not published, for reasons of exploration strategy, but were used empirically in the selection of areas for geochemical programs with some encouraging results although not as yet of economic significance.

Russell (1976) suggested that these vertical zones of weakness formed in the upper crust are a result of east-west relative tension and high pore fluid pressures. These may have permitted hot waters that were freed at depth and migrated towards tensile stress concentrations at the bottom of the geofractures there to be transferred towards surface by convection flow carrying metals leached from geosynclinal rocks. Where there is adequate sulphur, the metals are precipitated over the intersections of the geofractures with crosscutting favourable structures – in this case mainly normal faults trending east-northeasterly.

The primary cause of such geofractures, if they exist, is a matter of speculation but Russell has suggested that they might be related to the initiation of the major fracture that resulted in the mid-Atlantic ridge.

Another writer who has been impressed with the influence of crosscutting, and sometimes equidistant fractures and veins, is J. Kutina. His first studies were in the Bohemian Massif in Czechoslovakia but subsequently in other parts of the world including North and South America. In western Czechoslovakia he showed the roughly equidistant pattern of north-south faults and vein systems that crosscut the boundaries of age and structural units. In a more recent paper Kutina (1976), compared the source of patterns shown in Kutina (1968) with the spacing of fracture zones of deep ocean floors and suggested a possible relationship between mineralization and "hot spots" in the mantle such as proposed by Sillitoe (1974).

Kutina pointed out that geofractures that extended across the boundary of two formerly-joined continents do not necessarily have common orientation today (Fig. 1.2). For example, it is known that Africa, when it was attached to South America over 140 Ma ago, had its long axis tilted in a southwesterly direction 30° from its present north-south one. Similarly South America had its present north-south axis tilted 15° in a southeasterly direction. Accordingly if an east-west fracture system crossed the boundary prior to the parting of the continents the two halves of the fracture system would now differ in orientation by 45°.

Observations on deep and ancient structures that may have influenced metallogenesis have been made by a number of writers in the U.S.S.R. including Favorskaya (1976) who attempted to correlate the metallogenic generalizations resulting from Plate Tectonic concepts with the observed regularities in the distribution of large deposits when considering them as major geochemical anomalies. The depth of the lineaments is shown by their through-going character, cutting across not only boundaries of different crustal structures but also the boundaries of continents and oceans. The conclusion reached is that these ore-concentrating structures are remarkable for their longevity and variety of sources and are "apparently connected with deeper processes than those that occur at the boundaries of plates".

DISCUSSION

The difficulty with the above observations and speculative conclusions is just that — they are speculative. On small-scale maps it is rather easy to start drawing lines that appear to go through a number of deposits. If, however, on closer and more detailed study the conclusions are found to be too highly based on small-scale assumptions there is a tendency for the earth science fraternity to discard the whole idea. Geophysicists tend to give particularly short shrift to such ideas, finding little if any physical evidence to support them. However geophysicists, who justifiably can claim the major role in the development of the Plate Tectonics concept, should remember, as an exercise in humility, that the fiercest opposition to the ideas of Wegener, Du Toit and Holmes came from geophysicists who claimed that the movement of continents in any form was physically and mathematically impossible. Geologists, a fair proportion of whom (including the writer) must be classified as mathematical morons, accepted such informed criticism rather too easily.

I would suggest that a greater concentration of effort is justified in assembling the evidence of such "geofractures" that may be associated with metallographic belts or alignments of mineral concentrations on a succession of parallel structural belts. The next stage would be to find methods for identifying such geofractures or fracture systems not visible at the Earth's surface by means of seismic or other geophysical applications. I suggest that this is the biggest challenge for geophysicists concerned with mineral deposits and one that should logically be undertaken by government. There are some arguments, in a free enterprise system, against governments carrying out surveys that may specifically pinpoint mineral deposits since this would or could be done by industry at no direct cost to the taxpayer. The sort of project suggested here, however, could only be done properly by a central co-ordinating agency.

In addition to any geophysical approach, since any such fracture systems by their very nature are assumed to extend to a great depth and, perhaps to the mantle, they should be identifiable by means of broad regional geochemical surveys. In this regard the Russian experience should be particularly valuable since the Ministry of Geology of the U.S.S.R. has for many years been carrying out geochemical surveys in mineral exploration over vast areas. This line of research in our own country, and elsewhere where mineral exploration might be carried out, should logically be done by government organizations or by government sponsorship so that the results could be co-ordinated by a single agency.

CONCLUSIONS

I don't want to leave with you the impression that I am denigrating the fundamental relationships between ore genesis and Plate Tectonics. Far from it. I think we will find increasingly close controls as we learn more about the history and processes involved. What I am saying is that there has been a tendency, particularly in papers written for non-specialist earth scientists or for those in other sciences, to give the impression that a new tool has been presented to the exploration geoscientist that has fundamentally changed and simplified his task in the selection of areas. This is not yet the case. Most recent successful exploration for any minerals, whether metals, non-metals or petroleum, has been in areas selected empirically on the basis of observed geological or geographical relationships which fit well into the Plate Tectonic concept but were not directly derived from it. As we learn more of the detailed tectonic mechanisms involved I believe we will be able to use greater precision in the selection of areas for exploration. The identification of the more detailed structures and mechanisms of plate movement will depend heavily on regional geophysics and geochemistry and the resulting follow-up programs will require still greater penetration and sensitivity in defining drilling targets.

REFERENCES

Commodity Data Summaries
   1977: U.S. Bur. Mines.

Crone, J.D.
   1979: Exploration for massive sulphides in desert areas; in Geophysics and Geochemistry in the search for metallic ores; Geol. Surv. Can., Econ. Geol. Rep. 31, rep. 37.

Favorskaya, M.A.
   1976: Metallogeny of Deep Lineaments; 25th Int. Geol. Cong. Abs., Sydney, p. 736-737.

Hutchinson, R.W.
   1965: Genesis of Canadian massive sulphides reconsidered by comparison to Cyprus deposits; Can. Inst. Min. Met. Trans., v. 68, p. 286-300.

Kutina, J.
   1968: On the application of the principles of equidistances in the search for ore veins; Proc. 23rd Int. Geol. Cong., v. 7, p. 99-110.

   1976: Metallogenesis and the motion of lithospheric plates; 25th Int. Geol. Cong. Abs., Sydney, p. 741-742.

Routhier, P.
   1976: A new approach to metallogenic provinces: the example of Europe; Econ. Geol., v. 71, p. 803-811.

Russell, M.J.
   1968: Structural controls of base metal mineralization in Ireland in relation to continental drift; Inst. Min. Met. Trans., London, v. B77, p. 117-128.

   1971: North-South geofractures in Scotland and Ireland; Scottish J. Geol., (1), p. 75-84.

   1976: Incipient plate separation and possible related mineralization in lands bordering the North Atlantic; Geol. Assoc. Can., Spec. Paper 14, p. 339-349.

Sillitoe, R.H.
   1974: Tin mineralization above mantle hot spots; Nature, v. 248, p. 497-499.

Strong, D.F. (Editor)
   1976: Metallogeny and Plate Tectonics; Geol. Assoc. Can., Spec. Paper 14, 660 p.

AN OVERVIEW OF MINING GEOPHYSICS

Harold O. Seigel
Scintrex Limited, Concord, Ontario

Seigel, Harold O., An overview of mining geophysics; in Geophysics and Geochemistry in the Search for Metallic Ores; Peter J. Hood, editor; Geological Survey of Canada, Economic Geology Report 31, p. 7-23, 1979.

Abstract

The technology of mining geophysics is advancing rapidly in methodology, instrumentation and interpretation techniques. A surprising diversity of approach exists in a number of methods, because of differences in the local geological, geophysical, topographic or even socio-political setting.

A general trend is towards the simultaneous gathering of increasingly large amounts of independent geophysical data in airborne, ground or borehole surveys. The resultant challenge is to extract the maximum amount of useful geological information from the mass of field data generated by these surveys. Increasingly, computers perform the task of routine interpretation, using either direct inversion programs or automatic selection from a family of theoretical models.

Minicomputers are routinely employed in the control of multisensor airborne systems and in the recording of their output in digital form on magnetic tape. Some simple data manipulation is even being done by such minicomputers in real time prior to recording.

Microprocessors are being increasingly incorporated into field portable geophysical instruments for surface and borehole surveys.

Résumé

L'application de la géophysique à l'extraction minière connaît une évolution rapide tant au niveau de la méthodologie et de l'instrumentation que des techniques d'interprétation. Il existe des approches étonnamment variées à certaines méthodes à cause des différences sur les plans géologique, géophysique, topographique ou même socio-politique au niveau local.

De façon générale, la géophysique tend à s'orienter vers la collecte simultanée d'une quantité de plus en plus grande de données géophysiques indépendantes recueillies par des levés aériens, au sol ou par sondage. Le défi est alors d'extraire le plus grand nombre possible de données géologiques utiles de la masse de données obtenues par des levés sur le terrain. On fait de plus en plus appel aux ordinateurs pour les tâches d'interprétation courantes, en utilisant soit des programmes d'inversion directe ou de choix automatique à partir d'une famille de modèles théoriques.

On emploie couramment des mini-ordinateurs pour le contrôle des systèmes aéroportés à détecteurs multiples et pour l'enregistrement des résultats obtenus sous forme numérique sur des rubans magnétiques. On utilise même quelquefois ces mini-ordinateurs en temps réel lorsqu'il s'agit de manipulation simple de données, avant l'enregistrement.

De plus en plus, on incorpore les microprocesseurs aux appareils géophysiques transportables pour les levés en surface et par sondage.

INTRODUCTION

The science of geophysics applied to mineral exploration includes a kaleidoscope of methods and techniques, many of which are in a state of rapid flux. At least eighteen fundamentally different methods are in major or minor use in mining geophysics around the world, some being employed in up to twenty variants.

There is a zestful difference of opinion on the relative merits of methods, techniques and instrumental approaches, not only between the two main centres of development in North America and the U.S.S.R., but also among the scientists in each centre. These differences of opinion have, through scientific or commercial competition, been instrumental in stimulating advances in the art in both centres.

In my opinion, the most significant factor at present in mining geophysics is the data explosion. Through the miracles of 1977 microelectronics we can now endow a man or an aircraft with far more geophysical data gathering capability than would have been dreamed of only a few years ago. For example, airborne radiometric systems record 256 to 1024 channels of data simultaneously, whereas only a year or two ago four channels were standard and ten years ago only a single channel. Airborne electromagnetic systems today rarely gather fewer than six channels of data and many more channels are contemplated. Ground induced-polarization receivers commonly produce six channels of data and one even makes one thousand channels available. Computer controlled data acquisition systems are increasingly being used in airborne surveying as are microprocessors in ground based equipment, to facilitate the gathering and manipulation of multichannel data.

As a typical example of an integrated airborne geophysical system designed for multi-resource mapping at low level, we may find six audio frequency electromagnetic channels, two VLF EM channels, four gamma-ray spectrometer channels and a magnetometer channel. The thirteen independent geophysical channels in all are recorded digitally on magnetic tape every half second, or every thirty metres on the ground. Figure 2.1 shows a computer printout of a section of the digital tape from such a survey, with the various geophysical data streams, the altimeter output and the fiducial numbers identified. Eight sets of plans of stacked profiles or contours are normally produced by computer from these various data. This great volume of data strains the human capability to correlate and fully utilize its potential at the present time.

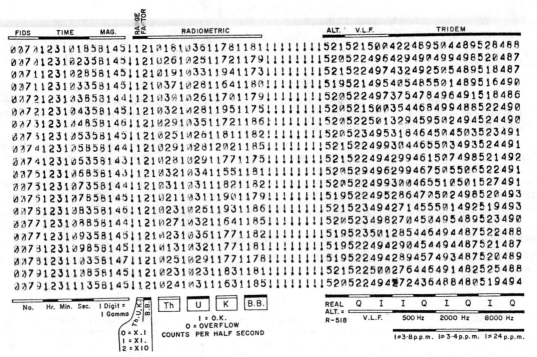

Figure 2.1. Computer printout of typical multimethod airborne geophysical magnetic tape. (Courtesy Scintrex Limited)

In fact, the greatest challenge now facing mining geophysicists is the distillation of the maximum amount of useful geological information from the mass of independent data currently arising from surveys embodying single or multiple methods. If this is to be done on a routine basis, it will clearly have to be carried out by the computer.

Computer processing and machine presentations of data have been already, of necessity, taken over from human hands and minds in the case of complex airborne-generated data of the type shown in Figure 2.1. Total "hands off" processing of such data is still not practiced in most low-level surveys, however, because the flight path is still recovered using tracking cameras due to the shortcomings of present radio navigational devices (such as Doppler). Navigational devices which are moderate in cost, light enough to be used in the relatively small aircraft employed in mining geophysical surveys and still give the high positioning accuracy required (± 25 m) are not yet available. Transponder systems (e.g. Motorola, Raydist and Decca), using fixed stations, have acceptable accuracy and weight but their cost and inconvenience precludes their general use.

The use of the computer has permitted some fundamental advances to be made in the quantitative interpretation of many types of geophysical data, particularly of a multichannel nature. The direct inversion of data for purposes of interpretation is now practical in resistivity, IP (e.g. Pelton et al., 1977) and electromagnetic surveys (e.g. Sternberg, 1977). Where direct inversion is too demanding of computer time, the computer still serves as a fast and economical means of selecting the best fit from a family of simplified models for which it has derived the forward solution.

In order to provide a broad perspective on this complex field and to place the various methods in their proper context, I propose to structure this discussion by geological objective. The broader features of each more important method will be considered as it is encountered in its major geological application.

REGIONAL GEOLOGICAL MAPPING

In the field of basic geological mapping, the magnetometer remains the primary airborne geophysical tool, indicating the distribution of the various rock types and minerals through their variation in magnetic susceptibility and remanent magnetization. However, the gamma-ray spectrometer is now almost a constant companion of the magnetometer for regional airborne geophysical surveys, providing a classification of rock or soil types based on their content of potassium, uranium and thorium.

A third mapping parameter, namely conductivity, utilizing one of several airborne electromagnetic devices is now available. For this application, multiple frequency measurements or multichannel transient measurements are commonly made in order to cover a broad range of earth conductivities.

Radio transmission measurements using VLF and broadcast band stations as source and measuring the ratio of electric and magnetic field components, can also map near-surface conductivities. These are quantitatively discriminating only in a somewhat lower range of conductivities than those mentioned above since their basic operating frequencies are high by normal airborne electromagnetic standards.

The Airborne Magnetometer

The proton magnetometer is the most commonly used airborne magnetometer. In a stable, well-compensated installation, the best of these can now achieve a useful sensitivity of about 0.1 gamma (at one reading per second). Some high sensitivity (cesium, rubidium or helium) optically-pumped magnetometers are also being used for identification of subtle features in areas of low magnetic activity and as vertical gradiometers.

Surprisingly enough, many total field self-orienting fluxgate magnetometers are still in operation, mostly in the East, but also in the West because the smooth analog profile data is still preferred by many interpreters.

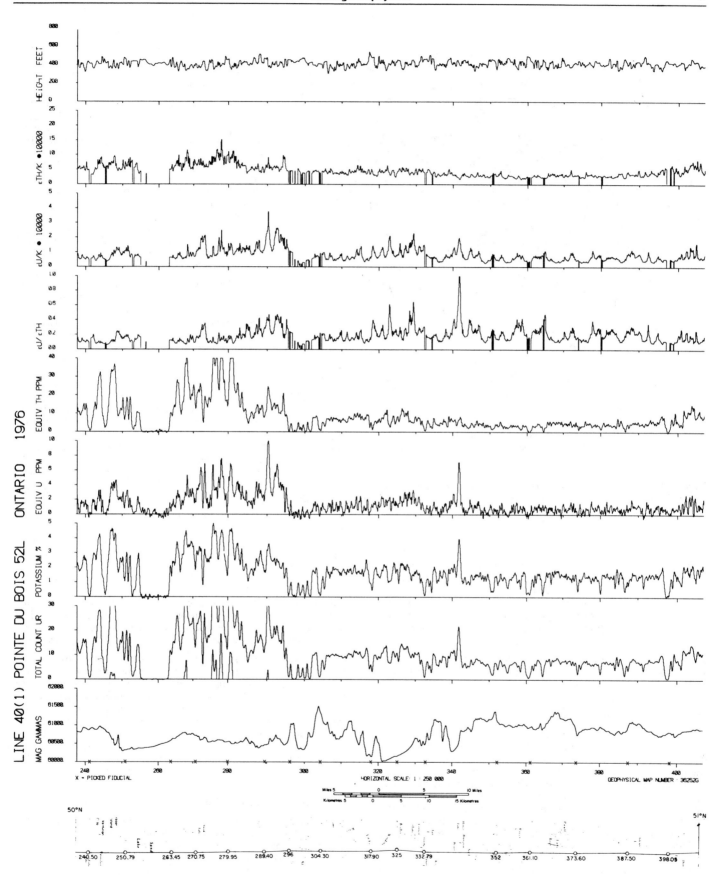

Figure 2.2. Computer plot of stacked airborne radiometric data in four channel profiles and elemental ratios. Pointe du Bois, Ontario, Canada. (Courtesy of Ontario Ministry of Natural Resources and Department of Energy, Mines and Resources, Ottawa).

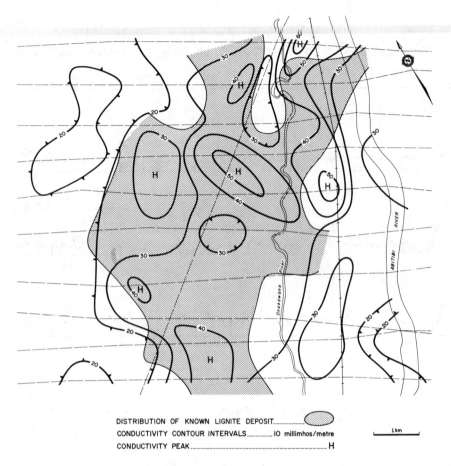

Figure 2.3. Computer interpretation of lower layer conductivity derived from Tridem airborne electromagnetic data. Onakawana lignite deposit, Ontario. (Courtesy Ontario Ministry of Natural Resources)

Regional aeromagnetic surveys for mapping purposes are normally flown at a mean terrain clearance of about 150 to 300 m (terrain permitting) and with line spacings depending on the ultimate plotting scale. For example, if the plotting scale is to be about 1:50 000 the line spacing may be 500 m (or 0.25 mile) and for 1:100 000 the line spacing may be 1 km (or 0.5 mile) etc. Surveys for specific exploration targets will have line spacings dependent on the expected target dimensions.

Compilation of the resultant aeromagnetic data, most of which are now digitally recorded, is now mostly carried out using computer techniques.

The Gamma-Ray Spectrometer

Until recently, sodium iodide crystal volumes for aerial gamma-ray spectrometry rarely exceeded 1000 cu. in. (16.4 L) and at most four channels or windows of gamma radiation were measured. Now it is common to employ up to 3000 cu. in. (49.2 L) for first order aerial mapping. The specifications for such high sensitivity surveys commonly call for about 200 cc of crystal per km/h of ground speed of the aircraft (or 20 cu. in. per mile per hour). With such large crystals, yielding much higher count rates than before, it has become meaningful to break the spectrum into more than four channels. For example, in the current Canadian or U.S. government-type specifications, the natural spectrum from 0 to 3 MeV is recorded in 256 channels on magnetic tape. In addition, cosmic-ray activity in the 3 to 6 MeV range is monitored in 256 channels. The local level of atmospheric (radon) and cosmic radiation is separately measured by an upward-looking crystal which is lead-shielded from gamma radiation of terrestrial origin. Five hundred and twelve channels of gamma radiation may be similarly recorded from this upward-looking crystal.

At least theoretically, the recording of the resulting 1024 channels of radiometric data permits corrections for the effect of atmospheric radon and cosmic radiation into the natural terrestrial spectrum, rock-type identification through absolute levels of uranium, thorium and potassium, as well as their ratios, and an independent check on the energy calibration of the gamma-ray spectrometer at all times. Some attempts are being made to recognize spectrum changes due to overburden attenuation and soil moisture changes, but with as yet uncertain success (e.g. Geodata, 1978).

The copious results of such multichannel surveys cannot possibly be compiled and presented in their entirety. In practice, they are corrected for terrain clearance variations, cosmic and atmospheric background and then grouped into four channels, one being "total count" and the other three centred about the primary gamma peak for each of K, U and Th. Corrections are made for Compton interference of the elements so that three "stripped" elemental channels are presented, one for each of K, U and Th, as well as total count. These elemental channel values may be presented directly, or as ratios, in profile or contour form.

Figure 2.2 shows typical stacked profiles of total count, K, U and Th. Each has been corrected for altitude (to 120 m) and background variations and each has been averaged over three adjacent one-second sampling periods. The K, U and Th are expressed in equivalent per cent or ppm of each element, assuming an infinite source area, having been "stripped" of Compton interference effects. Also presented in stacked profile form are the ratios eU/eTh, eU/K and eTh/K, as well as the altimeter and magnetometer profiles. In order to avoid statistically meaningless ratios the ratio values are based on running averages of successive readings over intervals where the numerator and denominator each exceed 100 counts.

Both types of profiles (elemental and ratio) may be meaningful in rock-type identification and in mineral exploration. The latter application is commonly restricted to prospecting for deposits of uranium, thorium, potash and phosphates (uranium rich) and those minerals associated with alkali complexes (e.g. columbium and tantalum), at least in the West. In the U.S.S.R., the ratio values are also used for the recognition of rock alteration associated with a variety of deposits of non-radioactive elements including bauxite, molybdenum and tin, etc. (Zietz et al., 1976).

Airborne Conductivity Mapping

Details of airborne electromagnetic systems for conductivity mapping will be found in the section of this paper on base metal exploration, for which they are more commonly employed. As an example of such conductivity mapping for other applications, consider Figure 2.3. This shows the mapping, by a multifrequency in-phase/quadrature system, of a lignite deposit which is buried under 10 m to 40 m of glacial clays and tills. In this case, the lignite and its associated fire clay horizon has a significantly higher conductivity (25 to 60 millimhos/m) than that of the overlying materials (about 5 to 10 millimhos/m). The conductivities of this contour plan were derived by computer interpretation using techniques described more fully in Seigel and Pitcher (1978).

The optimum terrain clearance may be different for each type of mapping survey. As a rule, prospecting-type radiometric surveys should be flown as close to the ground as possible (75 m or less) for improved detection of small targets. Aeromagnetic surveys may be flown somewhat higher (e.g. 150 m). Electromagnetic surveys must also be flown at a minimum terrain clearance. However, low terrain clearance means more severe turbulence, more difficult flight path recovery and higher unit costs. Thus, combined magnetic-radiometric mapping surveys are often flown at terrain clearances which are in excess of those desirable for the most effective radiometric prospecting purposes.

Similarly, fixed-wing airborne electromagnetic systems in which the receiver is mounted in a towed bird must fly (for safety reasons) at 115-130 m terrain clearance. This renders them only marginally suitable for those simultaneous radiometric surveys where uranium prospecting is a major objective. Totally on-board airborne electromagnetic systems are normally flown at very low terrain clearances and are therefore relatively compatible with radiometric surveys for uranium prospecting.

BASE METAL EXPLORATION

Mining geophysicists have been highly successful in the discovery of buried base metal orebodies of many types and in many environments. The most successful primary methods have been the electromagnetic induction technique for massive sulphide bodies and induced polarization for disseminated sulphide bodies.

Massive sulphide bodies commonly have electrical conductivities (1000-10 000 millimhos/m) which are two to four orders of magnitude higher than those of their host rocks. Thus, in most geological environments, the orebodies can be resolved from their host rocks by electromagnetic induction methods. Where large areas of favourable rock types are to be explored, it is now the accepted practice to employ a suitable airborne electromagnetic method for reconnaissance detection, followed by ground electromagnetic investigations and other ground-based methods for detail.

Airborne electromagnetic methods have been particularly cost-effective where they can be applied, because of their low unit cost, which is still of the order of $15-$30 per km surveyed. This is not to imply that the probability of success with these methods has been high. The ratio of base metal orebodies to total "conductors" has been found to be rather low, usually of the order of 1/500 to 1/1000. Despite this, the low unit conductor discovery cost and the high average orebody return have made this type of exploration approach highly rewarding in the search for stratabound copper-zinc and lead-zinc deposits, particularly in Precambrian environments. Similar success has been recorded in the exploration for nickeliferous pyrrhotite deposits associated with ultrabasic intrusives.

The high quality of modern airborne electromagnetic and magnetic data leaves little additional information to be gleaned from their counterpart ground surveys except greater precision of location, which is desirable for drilling purposes. Useful complementary information may be afforded from geological mapping and geochemical sampling of the conductor location, depending on the nature of the soil cover in the survey area. In the Canadian Shield, where outcrops are few and the soils are of transported glacial origin, the geophysical data alone may have to be relied upon for a drilling decision.

Gravimeter traverses are often useful in such areas to resolve conductors due to graphite from those due to massive sulphides. Both induced polarization and gravity (to a lesser extent) have proven useful in resolving conductors of ionic origin (bedrock troughs and shear zones) from those of electronic origin (sulphides or graphite).

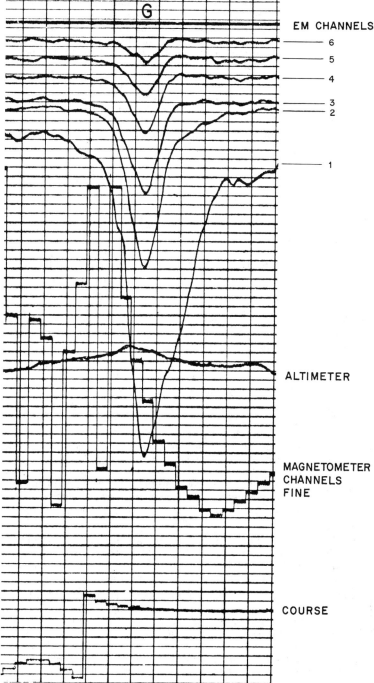

Figure 2.4. Input AEM traces over a bedrock conductor under 85 m of overburden. Sheraton Township, Ontario, Canada. (Courtesy of Questor Surveys Limited)

The IP method has been particularly useful as a ground follow-up technique in areas of tropical weathering, particularly in semiarid environments (e.g. Australia).

For a fuller discussion of ground follow-up philosophy and methods, the reader is referred to Seigel (1972).

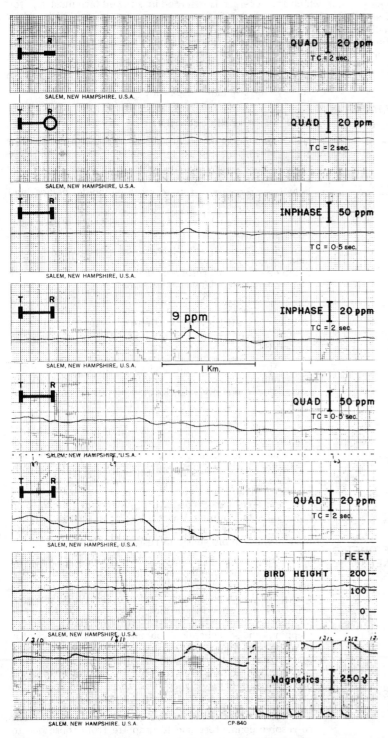

Figure 2.5. *Multichannel single-frequency Dighem helicopter electromagnetic discovery profile over nickel/copper deposit. Montcalm Township, Ontario, Canada. (Courtesy of Geophysical Engineering Ltd.)*

The Airborne Electromagnetic Method

At present, about seven basically different airborne electromagnetic (AEM) methods are in active use throughout the world. Most of these were first developed and utilized in Canada, because of various technical and historical reasons, although increasingly these are now being manufactured abroad, for example, in Sweden, Finland, U.S.S.R., India and China. A fuller description of AEM systems may be found in Ward (1970).

Recent airborne electromagnetic developments have been directed towards expanding the data-gathering capabilities of the systems, leading to the detection and resolution of a broader range of geological conductors, as well as to improvement in signal to noise ratios. The latter objective, when achieved, will automatically result in an increased depth of exploration as well as higher spatial resolution for near surface targets through a reduction of time constants.

It is common today for an airborne electromagnetic system to produce between 6 and 9 simultaneous channels of data (perhaps not all of it truly independent) for different (transient) times, different frequencies, or different coil configurations. The interpretative possibilities of this wealth of data are large. Some computer interpretations based on simple geological models have been developed by the operators of these systems and are now being applied on a routine basis.

Perhaps the most widely-used AEM system at the present time is the transient system known as Input (e.g. Lazenby, 1973). It measures six slices of the electromagnetic transient decay at various times out to $2.3\ s^{-3}$ following a $1.1\ s^{-3}$ primary pulse. In this way, it is able to detect a wide range of geological conductors.

Figure 2.4 shows an Input AEM test profile flown across a bedrock conductor consisting of a mixed graphite/sulphide body in Precambrian rocks, under 85 m of relatively poorly conducting overburden. The figure illustrates the exploration depth capability of the Input AEM system under the existing conditions of this test.

Multi-frequency continuous-wave AEM systems, measuring in-phase and quadrature disturbances simultaneously at several frequencies (e.g. Tridem, operating at 500, 2000 and 8000 Hz; see Seigel and Pitcher, 1978), have come into use for the same reasons.

Some AEM systems employ several transmitter and receiver coil configurations in order to obtain information about geological conductors having a variety of geometries relative to the survey line. For example, the Dighem helicopter system (see Fraser, 1978) employs three receivers, one coaxial and two orthogonal to the transmitter coil, with an operating frequency of 918 Hz and a coil separation of 9 m.

Figure 2.5 shows an actual Dighem discovery traverse over a body of pyrrhotite, pentlandite and chalcopyrite in a Precambrian gabbro intrusive, lying under about 20 m of glacially derived soil cover.

The struggle for greater useful depth of penetration has been frustrating. Because of the immutable laws of physics, the signal from a conducting body drops off, at best, as the inverse third power of the elevation of the airborne electromagnetic system and at worst as the inverse fifth power, depending on the relative parameters of the conductor and the system employed. Noise due to the aircraft being a conductor cannot be totally eliminated and moreover, the geological noise associated with undesirable earth conductors cannot be eliminated at all. These factors

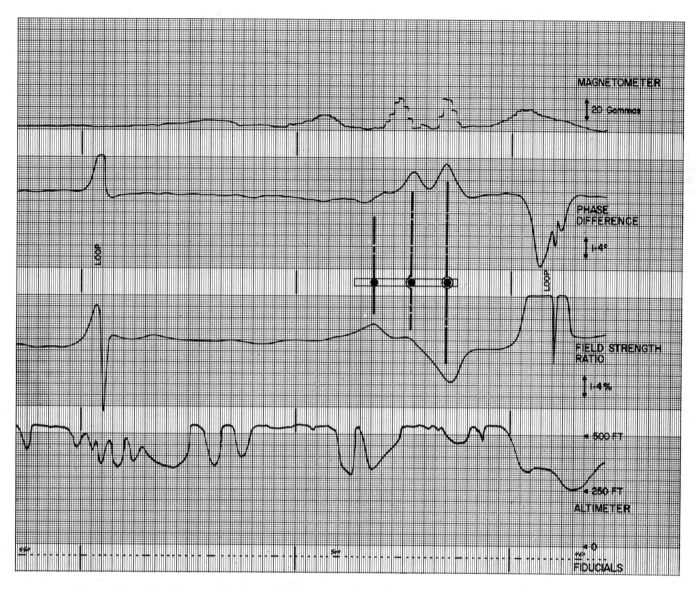

Figure 2.6. *Turair semiairborne profile over conductor in mountainous area. Operating frequency 400 Hz. Western Canada. (Courtesy of Scintrex Limited)*

put practical limits on the effective depth penetration of present day fully-airborne electromagnetic systems of between 100 and 125 m under most circumstances, despite the exceptional case which may be documented from time to time.

For this reason, semiairborne systems have been developed in the U.S.S.R. and Canada, to achieve much greater potential depth of penetration. These systems employ large, ground-based energizing loops or grounded cables, usually 4-10 km in dimension, as primary field sources. The mechanics of laying such large field sources by helicopter have been achieved in Canada. The use of these semiairborne electromagnetic systems has been especially rewarding in very mountainous areas, where it is difficult to maintain the relatively small terrain clearance necessary for wholly airborne electromagnetic surveys and yet the geological noise is often small. Such systems have demonstrated detection of conductors under as much as 200 m of cover under these conditions.

Figure 2.6 shows a profile generated by the Turair semiairborne, helicopter-supported system, in a particularly rugged section of the Canadian Cordillera. A conductor zone was revealed by this profile despite the fact that the helicopter, at the time, was flying at a height greater than 150 m above the ground surface. Since there was over 1000 m of local topographic relief, these conditions precluded the use of fixed wing aircraft and restricted the effectiveness of helicopter-borne totally-airborne electromagnetic systems. The figure shows also the effect of the loop sides, where the field gradients are too rapid for useful measurements to be made.

To utilize the airborne electromagnetic data from a particular system quantitatively, it is necessary for its calibration and zero levels to be well established. This has not always been sufficiently tied down by the operators. At present, however, much greater attention is being given to these factors, to the obvious benefit of both the systems and the users. As a result, the potential use of AEM systems has been broadened from the simple role of base metal anomaly hunting into the quantitative field of conductivity mapping.

One interesting aspect of recent airborne electromagnetic experience in Canada clearly illustrates the large element of chance (or statistical probability), involved in this type of exploration approach. The known base metal mining camps in eastern Canada have been flown and reflown by airborne electromagnetic systems since the first introduction of such surveys in Canada in 1951. Nevertheless, new base metal mines continue to be found by new surveys in these areas. Still we cannot say that these fresh discoveries occur at greater depths than were possible using the older systems, or were achieved because they did not show up earlier through some technical limitation of the older (and therefore presumably inferior) systems. Of five relatively recent discoveries, two had been missed by earlier surveys because they had a very short strike length and probably lay between the lines of the earlier survey. One had been previously found and drilled inadequately so that its nature had not been properly investigated. The other two were previously known to be conductors but had not been followed up because of inadequate geophysical-geological reasoning. Their combination of electrical and magnetic properties was such as to relegate them to the rank of uninteresting conductors. It remained for new prospectors with better geological reasoning (or was it less reliance on conventional geophysical reasoning?) to take the trouble to investigate these conductors and to reveal their ore potential.

As Slichter (1955) and others have pointed out, geophysical surveying is a statistical procedure and there is a relationship between the line spacing, the probable orebody length and the probability of discovery of the orebody. On the basis of such arguments, one may select the line spacing so as to optimize the cost effectiveness or prospecting profit ratio of the geophysical program. Ground investigation costs, including line cutting, geophysics and drilling, have escalated far more rapidly than those of airborne surveying in the past decade. As a result, the cost-effectiveness of an integrated exploration program, which includes airborne and ground phases, will be improved by flying more closely spaced lines today than was done ten years ago. As an example of this trend to closer line spacing, our company (Scintrex) has recently flown large blocks of mineralogically favourable country in Saskatchewan at 120 m (400 ft.) line separation with an integrated airborne system and has computer plotted the results on the scale of 1:5000. These line separations and plotting scales were formerly reserved for ground surveys.

It is interesting that plotting on such large scales clearly reveals any tiny uncertainty in the flight path recovery or in the mosaic preparation. A positioning error equivalent to only one half second in time produces the most remarkable "herringboning" in the contours.

The success ratio of airborne electromagnetic surveying around the world has been highly variable. In temperate and arctic areas the geophysical conditions are generally relatively good, with little oxidation and only moderately conducting soils. The Canadian Shield and particularly the Baltic Shield provide good geophysical and physiographic environments for this exploration technique and many major stratabound base metal bodies, usually of the copper-zinc types, as well as copper-nickel sulphide bodies, have been found using AEM techniques.

The application of these techniques in areas of tropical weathering, particularly in arid or semiarid conditions (e.g. Australia and the southwestern United States), has been far less rewarding to date because of a number of fundamental, adverse factors. Such weathering may easily oxidize a massive sulphide body to 50 m depth, whereas conductors of graphitic, serpentine and saline origin may persist through to the ground surface. In addition, semiarid and tropical soils are often highly conducting. In other words, the geological noise level is usually increased and the orebody signal level decreased in electromagnetic prospecting under conditions of tropical weathering and arid and semiarid desert terrain. This is not to say that airborne electromagnetic methods are of no value in such areas. In fact, there are successful case histories which prove the contrary. It just means that the odds for discovery are less under these conditions and one has to be more cautious in the use of AEM in areas of tropical weathering and arid and semiarid terrain.

The Induced Polarization Method

In ground exploration for base metal deposits, the induced polarization method is the primary electrical exploration tool for porphyry coppers, for contact metamorphic copper deposits and even for stratabound copper or lead-zinc deposits. It is often used as the preferred exploration tool in the search for massive copper-zinc or nickeliferous sulphide deposits, in areas with highly conducting surficial deposits, such as Australia, South Africa and in many wet tropical countries. As was mentioned earlier, the method is being used, with good reason, as a ground follow-up tool for airborne electromagnetic surveys in such areas. A volume by Sumner (1976) summarized much of the development that has occurred in IP in recent years.

The IP method, as presently practiced, exists in a number of possible methods of measurement and quantities measured. These fall into two main categories, viz. the frequency-domain (continuous wave) and time-domain (or transient) systems.

Most commonly, the electric fields associated with polarization current flows in the ground are measured between two potential electrodes (EIP). More recently (Seigel, 1974) there has developed an approach (MIP or Magnetic Induced Polarization) wherein the magnetic fields associated with polarization current flow are measured by a sensitivity magnetometer. The nature of EIP and MIP responses are very different in theory and practice and two methods appear to be complementary (i.e. advantageous for different problems) rather than competitive.

Within the frequency-domain systems, there are a number of possible methods of measurements and quantities measured. Traditionally, the percentage change of apparent resistivity with frequency (PFE) has been measured as the IP characteristic. More recently, the phase shift between the measured voltage and the primary current waveforms has been used to yield equivalent information (e.g. Hallof, 1974). Some workers now measure complex resistivities and their change with frequency to obtain IP information (e.g. Zonge and Wynn, 1975).

Both time-domain and frequency-domain IP systems are in use and the formerly heated conflict between proponents of the two different approaches has now died away. It is generally appreciated that, in IP as in EM, the time-domain or transient method has the advantages of higher potential sensitivity. Since transient measurements are made after the inducing current pulse has been cut off, they are "absolute" measurements. Thus, one may improve the signal-to-noise of such measurements by increasing the transmitter power, thereby also increasing the sensitivity of the measurement for low IP responses. In addition, many channels of data which are obtained simultaneously may give useful information relating to a range of polarizable materials. The low inherent frequencies commonly employed in time-domain surveys generally result in lower electromagnetic induction problems (at least at later decay times) although the earlier decay times may be markedly distorted by EM coupling effects.

The frequency-domain approach has the advantage of better signal-to-noise, or lower primary power, if you prefer. This can be very significant in areas of high magnetotelluric

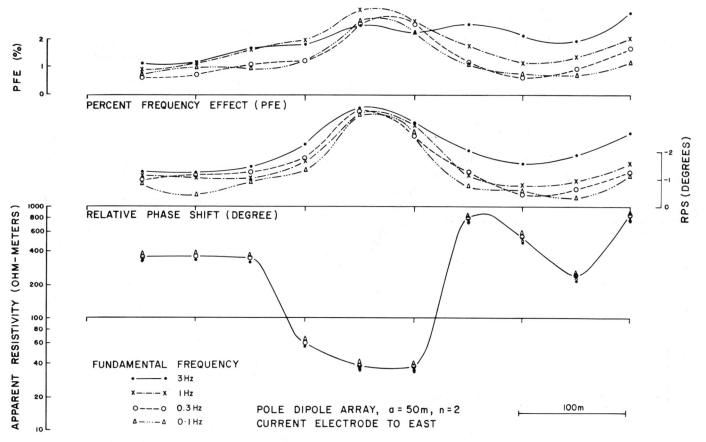

Figure 2.7. Scintrex IPRF-2 complex resistivity pole-dipole traverses at fundamental frequencies 0.1, 0.3, 1 and 3 Hz. State of Goias, Brazil. (Courtesy of Departamento Nacional Da Producao Mineral, Brazil)

or industrial noise activity. Being a "relative" method, its sensitivity is usually restricted by the stability of the transmitted wave form, regardless of the power transmitted.

Over the past few years the trend in IP measurements has been towards the gathering of more information on a routine basis. In the time-domain a number of channels, usually 3 to 6, of transient data, are measured simultaneously, in order to obtain the shape as well as the amplitude of the transient curve. In the frequency domain, complex resistivities may be measured at a number of frequencies in the range of 0.01 to 100 Hz.

Figure 2.7 shows the results of multi-frequency, frequency-domain EIP complex resistivity measurements in a tropical environment. Pole-dipole traverses were run over the same line with a comparator type of receiver which automatically compares the resistivity amplitudes (PFE) and phase shifts (RPS) at the fundamental and third harmonic resulting from a single transmitted square-wave current form. Base frequencies of 0.1, 0.3, 1 and 3 Hz were employed. It is noteworthy that the high polarization/low resistivity zone in the central part of the profile shows up more clearly as the operating frequency is decreased. In addition, the RPS, which has a measure of inherent EM suppression, has a better geological signal/noise than the PFE at the higher frequencies employed.

Figure 2.8 presents three of six transient traces measured by a six-channel time-domain receiver in Western Australia. A typical interrupted square-wave current pulse of two-seconds on-and-off time was employed. Each channel represents a time integral of the transient (IP) signal over 260 ms. The survey in this instance is actually an MIP survey over a nickeliferous pyrrhotite body in ultrabasic rocks, under about 30 m of oxidation. The various channels have already been normalized with respect to the "Newmont" universal decay curve, which is a transient IP decay curve form obtained by averaging a large number of such curves from many geological environments. Among other significant features of this profile is the divergence in decay curve form between various stations. The interior current flow appears to be characterized by an abnormally long time constant, which is consistent with the rather "massive" nature of the sulphide lenses in this deposit.

There is an awareness that in the frequency domain the phase angle measurement allows a simple, first-order removal of electromagnetic induction effects and that it is even possible to make a relative phase shift measurement which is automatically stripped of electromagnetic effects to this approximation (Seigel, 1974; Hallof, 1974). This is based on the fact, which has been theoretically predicted and experimentally observed in the field, that the phase shift in low frequency IP measurements varies very slowly with frequency, possibly less than 50 per cent per decade of frequency. At the same low frequencies, however, any phase shifts due to EM inductive effects increase almost proportionately with the frequency. Thus, the measurement of phase angles at two frequencies provides a simple means of correcting for EM effects.

Armed with the new tools that can obtain response curve shape information in the time or frequency domains, many workers are attempting to derive criteria for distinguishing one type of IP source from another. The

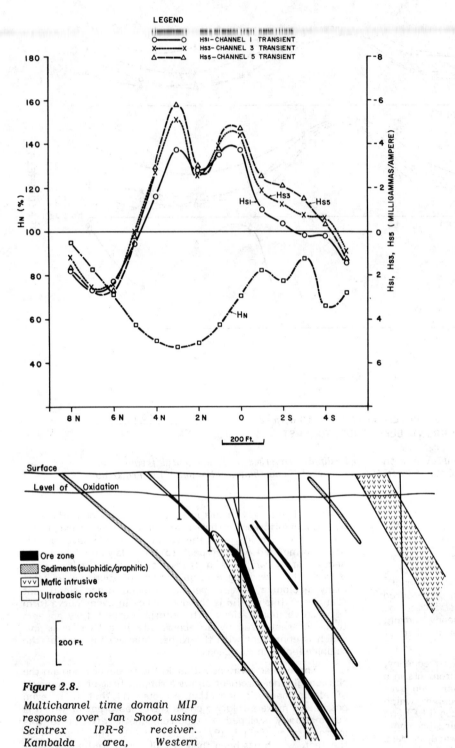

Figure 2.8.
Multichannel time domain MIP response over Jan Shoot using Scintrex IPR-8 receiver. Kambalda area, Western Australia. (Courtesy of Western Mines Limited)

dependence of the IP response curve shape on the size distribution of the metallic particles has been known for 25 years. This permits the differentiation of sources of small average particle size e.g. graphitic shales, from sources of larger average particle size, e.g. sulphide grains. The former commonly give rise to relatively short time-constant transients, whereas the latter give rise to relatively long time-constant transients.

Some workers (e.g. Zonge and Wynn, 1975) have claimed ability to differentiate one metallic sulphide from another through differences in their IP response curves. These claims have become somewhat muted in time as particular field cases arise which do not conform to the original hypothesis. In addition, the results of other investigations tend to refute this claim (e.g. Angoran and Madden, 1977).

In my own experience, IP curve-shape information can be valuable in resolving source ambiguities in a particular area, but only after the appropriate local experience has been obtained. Generalizations from one area to another, or from one geological environment to another may be dangerous.

Microprocessor-based receivers have started to appear. These can be programmed to record complete complex waveforms in a large number of channels (e.g. 1000) and to process these waveforms to derive factors of IP significance in either the time or frequency domains. The cost/benefit of these new receivers remains to be established.

The MIP method has been mostly used to date in Australia. This method can detect an induced polarization source through a conducting overburden layer which would normally seriously reduce the electric field IP effects of the source. In addition, it permits IP measurements to be made in areas of loose sand or permafrost, where electrical contact with the ground is difficult or even impossible.

For those IP receivers producing 10 or more channels of simultaneous information, there is a trend to having their data fed into a cassette-type magnetic tape recorder in the field for storage and later processing, by computer.

Ground Electromagnetic Systems

There are three basic types of continuous-wave (CW) ground electromagnetic systems in active use in the field today, viz. the vertical-loop or tilt system, the slingram or horizontal loop and the Turam, or fixed source gradient systems. All of these have been used for over 25 years and still remain in use because each fulfills its own useful functions. The vertical-loop system is primarily used for ground follow-up of airborne electromagnetic indications, particularly in forested areas and areas of high topographic relief, because it can operate without cutting lines. The horizontal-loop method has been improved in recent years with the addition of multiple frequencies — up to five in one unit, and greater intercoil separation — up to 240 m. The horizontal-loop technique is probably the preferred ground electromagnetic method for the majority of conducting base metal deposits, except for the very deep ones (in excess of 100 m) and except for areas of rugged topography.

The Turam method finds its best areas of application in the search for deeply buried conductors, or in rugged topography. Detection of sulphide bodies under more than 150 m of cover by the Turam method has been achieved in practice. The number of operating frequencies available

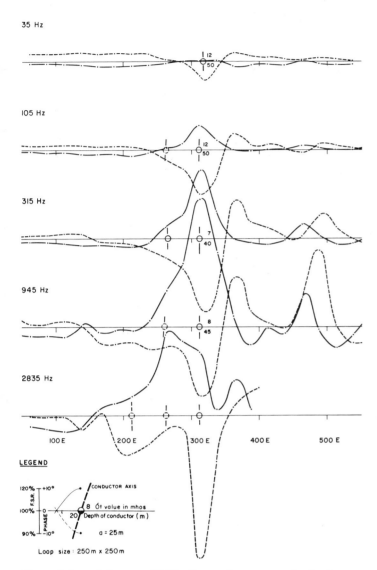

Figure 2.9. *Scintrex SE-77 Turam-type ground electromagnetic response over buried conductor. State of Goias, Brazil. (Courtesy of Departamento Nacional Da Producao Mineral, Brazil)*

today in standard production units has increased to five and the range of frequencies increased as well to cover both poorly and highly conducting environments.

Figure 2.9 shows the results of a five-frequency, Turam-type system, employed on ground follow-up of airborne electromagnetic surveys, in a tropical environment. The operating frequencies increase by factors of three, from 35 Hz to 2835 Hz. An examination of these results indicates that whereas a single, well-defined conductor axis is apparent on the lowest frequency employed, the response pattern is distorted by flanking, weaker conductors on the highest frequency.

Transient ground electromagnetic prospecting, long an exclusively Russian domain, has spread to the west and two models of multichannel transient systems are now in use in North America. Their increasing popularity is due to the broad range of conductors which can be detected and resolved simultaneously, as well as their lack of sensitivity to topographic relief. Of course there is the usual requirement for higher power or longer signal averaging times than in comparable CW systems. Both horizontal-loop and Turam configurations are being used, the latter requiring but a single horizontal receiving coil.

More recently, EM systems are being developed which may combine the broad conductor response of the transient systems and the lower power requirements of the CW systems. These newer systems employ complex waveforms which are repetitive but include a number of components of differing frequencies, often harmonically related. These include the square wave, the saw-toothed waveform of UTEM (ref. Lamontagne and West, 1973) and the pseudo-random waveform (Edwards, 1976). Coherent detection is employed using synchronized crystal clocks or radio links for time reference. Microprocessors may be employed for control of the receiver. With such systems the measurements may be made in terms of the earth impedance vs. frequency spectrum or as its step-function or impulse function response. The multichannel data are usually stored in digital form on magnetic tape cassettes. These data may be transformed, for greater ease of interpretation, from the frequency to the time domain, or vice versa, by computer or even by programmable calculator. Thus, in due course, the advent of these complex waveform EM systems will remove the distinction between time-domain and frequency-domain systems that presently exists.

Figure 2.10 presents the measurements made by an eight-channel, time-domain ground electromagnetic system, over a copper-zinc-gold sulphide orebody in a semiarid environment. Dual vertical-axis transmitter and receiver coils were employed with centres 50 m apart. A current pulse with an on/off time of 10.8 ms was employed. The eight channels of transient electromagnetic decay signals measured range in a logarithmic fashion from 0.15 ms (channel 1) to 8 ms (channel 8) in delay times. The orebody is in Cretaceous andesites and basalts, under 30-40 m of oxidation and soil cover. It is noteworthy that the orebody shows up best on the intermediate channels. The shortest time channel (1) does not distinguish the orebody response from geological noise. The longest time channel (8) gives only a low order orebody response.

The long wave VLF radio transmissions, primarily in the 15-30 kHz region, are employed as sources for one-man electromagnetic prospecting units. These units usually measure field amplitudes or inclinations. They are attractive from the standpoint of cost, weight and speed of measurement. They suffer from a number of limitations, the chief of which is their lack of conductor discrimination because of the high operating frequencies and distant source geometry. Nevertheless, VLF-measuring devices are in use in those areas where geologic noise, particularly overburden is minimal.

The magnetotelluric method, first proposed by Cagniard for deep sounding for petroleum exploration, has now been adapted for shallow sounding in mining geophysics, as well as permafrost problems, etc. In order to respond to shallow structures, in the first few hundred metres of the ground surface, frequencies in the audio or near audio range (1-10 000 Hz) are used. The resultant method is sometimes termed Audio Frequency Magnetotellurics or AMT (see Strangway et al., 1973).

Figure 2.11 is an AMT profile which presents the resistivities calculated from crossed E and H measurements at four irregularly-spaced frequencies in the range of 1-3000 Hz. The profile is over the Cluff Lake, Saskatchewan uranium orebody which is associated with a fracture zone in Precambrian rocks. Apparently this zone is highly conducting and extends almost to the present ground surface, for the resistivity decreases progressively as the frequency increases.

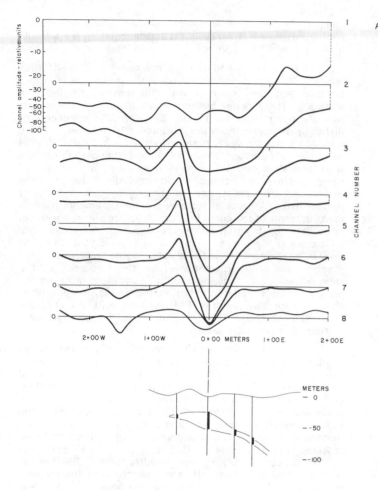

Figure 2.10. Crone PEM 8 channel time domain ground electromagnetic profile over Cu, Zn, Au orebody. Sultanate of Oman. (Courtesy of Sultanate of Oman)

Almost four orders of magnitude resistivity contrast is to be seen between the high frequency derived resistivity over the zone and the more resistive rocks to the west.

Drillhole Logging Methods

There is currently a minor upsurge of interest in drillhole electromagnetic methods for base metal exploration. At least three different systems are in use in North America, including a fixed-source time-domain unit, a fixed-source CW multifrequency unit and a single frequency moving, coaxial, transmitter-receiver system.

Figure 2.12 shows the results of a five-frequency CW electromagnetic log of an exploration hole which passes within about 50-60 m of a nickel-copper massive pyrrhotite body of the typical Sudbury (Canada) type. In this system the field source is a large, closed loop laid out on the ground, primarily on one side of the collar of the hole. A down-hole coil, coaxial with the hole, compares the magnetic field in that direction, in phase and amplitude, with that measured by a coil which remains fixed in a location near the collar. The field strength ratio in this example has been normalized for the normal field geometric changes through division by the lowest frequency field strength data, smoothed for the obvious local disturbance near 1000-foot depth in the hole.

From these results, it is clear that the range of detection of sulphide targets through the use of electromagnetic logs in exploratory boreholes is similar to that

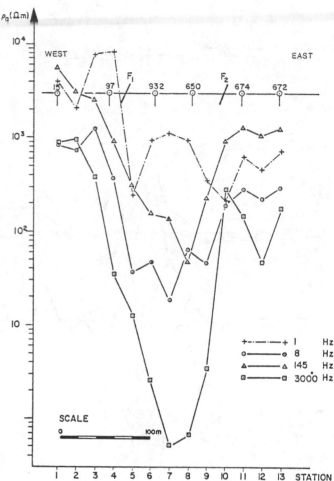

Figure 2.11. Audio frequency magnetotelluric traverse over uranium deposit. Cluff Lake, Saskatchewan, Canada. (Courtesy of Mineral Exploration Research Institute, Montreal and Amok Limited)

achieved through the use of the equivalent surface technique. The latter usually suffers from somewhat greater geological noise, due to surface conduction problems.

Despite such favourable results, it still cannot be said that it has become standard practice in the West to log by EM (or any other technique) all base metal exploration holes, even after at least two decades of educational effort by the geophysical community. In the U.S.S.R. and China, it does appear to be commonplace to log pairs of base metal exploration holes with the high frequency (0.15 to 40 MHz) "radio shadow" technique (Zietz et al., 1976). For example, in the Tongling, China copper area, an excellent discovery of a contact metamorphic copper deposit at 300 m depth was made with help from this technique, using a 1 MHz transmitter (pers. comm.).

A number of other geophysical logging techniques suitable for base metal exploration are used sporadically as well, including induced polarization, resistivity, mise à la masse, and three component magnetometer, usually to solve a specific local geological problem. All of these logging techniques are being used for remote detection purposes and are adaptations and extensions of techniques and apparatus developed for surface exploration.

The mining industry, unlike the petroleum industry, is accustomed to core drilling its exploration holes. Thus the mining industry is, in general, disinterested in the short range, or "at hole" physical property information provided by

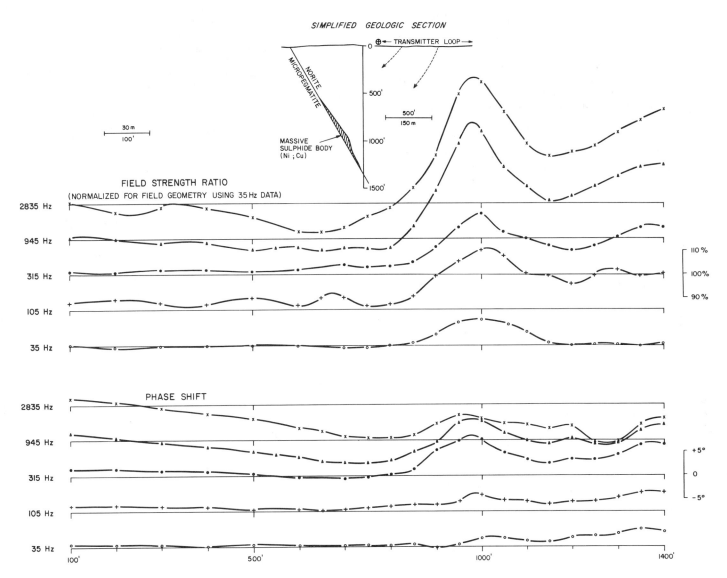

Figure 2.12. Scintrex DHEM-5 drillhole electromagnetic log showing nearby Ni/Cu deposit. Sudbury, Ontario, Canada. (Courtesy of Geological Survey of Canada and International Nickel Company of Canada Limited)

the broad array of logging tools (density, resistivity, SP, caliper, etc.) developed for the oil industry. For certain types of base metal exploration and in uranium exploration (see next section), specific short range logging tools are, however, being employed, because they can yield quantitative grade information with a resultant saving in overall exploration cost and time.

In situ grade information may be obtained through a variety of nuclear techniques including: simple natural gamma-ray spectral measurements in the case of K, U and Th, radioisotope source XRF measurements for a broad range of elements, neutron activation (delayed and prompt gamma measurements) for a variety of elements, gamma-neutron measurements for beryllium and delayed fission neutron measurements for uranium.

All of these techniques attempt to determine quantitatively the mean grade of specific elements in the vicinity of a borehole. They share a common problem in attempting to achieve acceptance by the mining geologist in that in order to establish their validity, their results are compared with chemical assays of cores or chip samples from the same borehole sections. Depending on the degree of heterogeneity of the elemental distribution in the specific deposit, there may be little correlation between these two sets of data (e.g. Czubek, 1976), particularly since they may present the analysis of far different volumes of rock.

Portable XRF analyzers using a radioisotope source for excitation and balanced "Compton edge" filters for energy selection have been in use in the West for about ten years (e.g. Clayton, 1976). They are employed for surface analysis and, to a lesser degree, for borehole analysis of a range of metals of economic interest. For borehole use they require a dry, clean hole because of the low penetrating power of the primary and secondary X-radiation employed.

One nuclear logging system of sufficiently high penetrating power to give a "bulk sample" analysis (500 kg or more per assay) is that employing a neutron source (e.g. ^{252}Cf) and measuring prompt gamma radiation resulting from neutron capture in the nuclei of the elements round the hole (Nargolwalla and Seigel, 1977). This system called Metalog by the manufacturers, Scintrex Limited, has been employed to determine the grade of Ni, Fe and Si in the lateritic nickel environment, copper in the porphyry copper environment and S and ash content of coal, through in situ borehole measurements.

Figure 2.13 presents experimental correlation charts between the Metalog values for Ni and Fe and bulk sample grades in a lateritic nickel test.

Figure 2.14 shows similar experimental correlation charts for S and ash in a coal field test. In this instance, the comparative chemical analysis was obtained from core samples, as bulk sampling was not feasible.

Other Ground Methods

I have briefly reviewed the primary methods currently in use in base metal exploration. Other methods are also in use, in the East or West, to a lesser extent and for the solution of special problems. The seismic method, for example, finds indirect application in situations where the definition of geological structure will be of value in the location of ore deposits. The method is also employed in the U.S.S.R., using two boreholes and is called the "acoustic shadow" method. It locates inhomogeneities lying between the two boreholes (ref. Zietz et al., 1976).

A piezoelectric method has been developed and used in the U.S.S.R. for exploring for quartz veins and quartz-rich pegmatites (ref. Zietz et al., 1976).

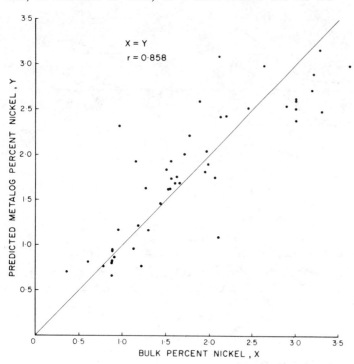

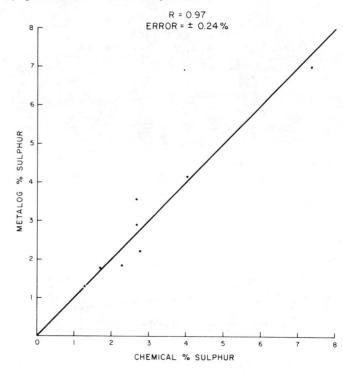

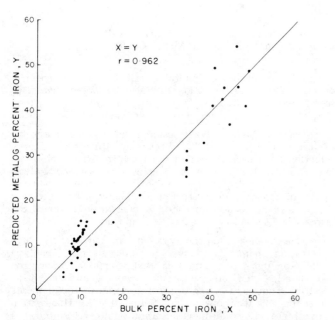

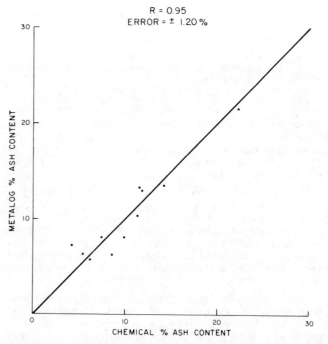

Figure 2.13. Metalog neutron-prompt gamma bulk sample comparison charts for Ni and Fe in a lateritic nickel deposit. (Courtesy of Scintrex Limited)

Figure 2.14. Metalog neutron-prompt gamma core sample comparison charts for S and ash in coal seams. (Courtesy of Shell Research and British Coal Board)

URANIUM EXPLORATION

Uranium exploration, presently a very important area of involvement of mining geophysicists around the world, involves airborne, ground, and borehole activities. In this search, geophysicists have the unique advantage of having at their disposal a powerful tool which is so direct that it can yield actual U grades and yet is rapid and inexpensive in execution. This is, of course, gamma-ray spectroscopy. Using adequate volumes of NaI(Tl) crystals as detectors, with properly calibrated and stabilized electronic pulse-height analyzers, we can quantitatively determine the average equivalent grade of uranium in rocks or soils. There is some question as to whether or not this method is properly denoted as a geophysical or geochemical one, but, since the devices are most often operated by technicians who are neither geophysicists nor geochemists, the controversy in fact should not arise.

The 2.62 MeV energy gamma-ray peak uniquely denotes the presence and amount of Th. The 1.76 MeV gamma-ray peak, after suitable correction for the Compton continuum of

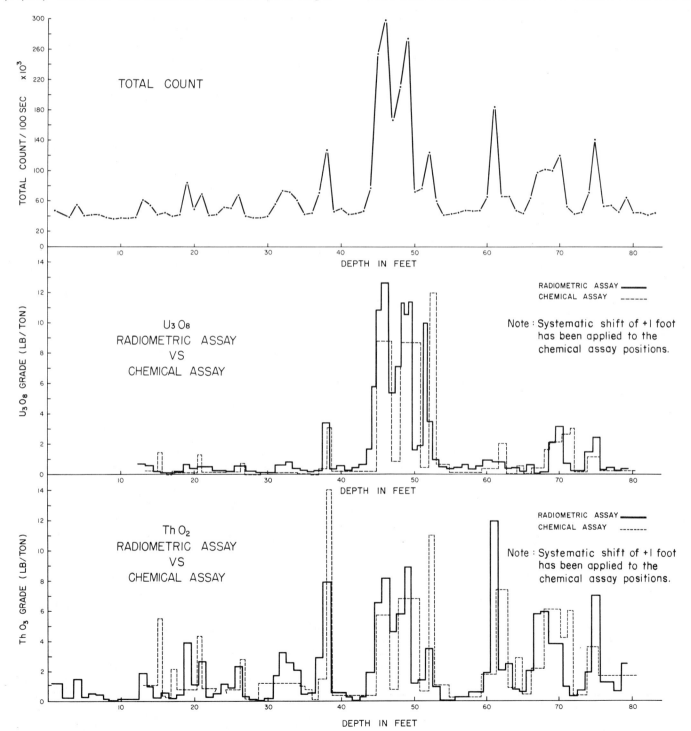

Figure 2.15. Comparison of total count and stripped Scintrex GAD-6 gamma-ray spectrometric assay logs with chemical assays of core for U and Th. Lake Agnew, Ontario, Canada. (Courtesy of Agnew Lake Mines)

Th, is specific for the presence and amount of ^{214}Bi, which is a daughter product of ^{238}U. Under those conditions where the ^{238}U is in equilibrium with this daughter product, then the 1.76 MeV peak (after correction) is specific for ^{238}U. Similarly the 1.46 MeV peak, after suitable Compton corrections, is related quantitatively to the ^{40}K radioisotope.

The current trend in gamma-ray spectrometric measurements is to tightly control the calibration and stabilization of the measuring devices so that quantitative eK, eU and eTh grade values may be deduced from such measurements, whether they are derived from airborne, ground or borehole surveys. The assumption of equilibrium, so necessary for U determinations, is, unfortunately, often invalid, particularly in porous or highly weathered rocks. Even fresh-looking, low-porosity Precambrian sediments and intrusives, scoured clean by glacial action only 10 000 years ago, may be in U disequilibrium in the first 50 cm or so, from where almost all the gamma radiation detectable at surface arises. Ignoring, for the moment, the disequilibrium problem, one can apply corrections to the observed data for the outcrop geometry (if known) and for the gamma radiation due to radon in the atmosphere and cosmic ray activity, to obtain eK, eU and eTh.

I have already briefly reviewed the state-of-the-art in airborne radiometric exploration (Fig. 2.2). The airborne techniques and instrumentation are not very different whether the geological objective is radiochemical mapping or the search for uranium deposits. The line spacing and survey height tend to be smaller in the uranium search. Of course, four-channel systems, with smaller crystal volumes, are also being used to good effect in light aircraft and helicopters, at elevations of 30-60 m, in the search for uranium. A good review of airborne radiometric considerations is to be found in a paper by Darnley (1973).

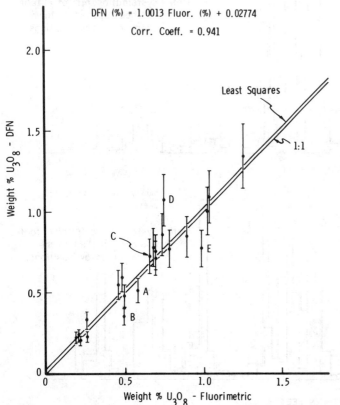

Figure 2.16. Comparison chart of delayed fission neutron assays and fluorimetric core assays for uranium. (Courtesy of Geophysics; after Givens et al., 1976)

For ground uranium prospecting, a broad range of portable gamma-ray detectors are employed, from simple, nondiscriminating scintillation counters to four-channel differential spectrometers with automatic Compton stripping and spectral stabilization, etc. Some of these are excellent examples of compact modern electronic engineering. Crystal volumes on ground instruments may range from 30 cm^3 for rough anomaly hunting, to over 1200 cm^3, for radiochemical mapping. The best of these spectrometers enable determinations to be made of the grade of U and Th in outcrops and rock samples, providing that radioactive equilibrium exists and that proper corrections are made for the outcrop or sample geometry.

In borehole logging for uranium, the natural gamma scintillation counters or spectrometers are mainly used, yielding at least semiquantitative results. With careful calibration against chemically-established grades of U and Th in test boreholes (either natural or man-made, e.g. Geological Survey of Canada pads outside of Ottawa or Department of Energy facilities near Grand Junction, Colorado) and with radioisotopic stabilization, for example using a ^{133}Ba source, U and Th grades of satisfactory accuracy may be determined. This assumes equilibrium conditions for U, which is normally valid in tight Precambrian rocks beneath the weathered zone.

Figure 2.15 is an example of a stabilized gamma-spectrometric log of an underground borehole which has been drilled to provide guidance for underground mining in an established uranium mine. Both U and Th are present in this mine, in highly variable relative proportions, in Precambrian sediments. The results of the radiometric log are presented as total count eU and eTh. Chemical assays for U and Th of core samples are shown for comparison purposes. The degree of correlation between the radiometric and chemical assays is deemed to be good, despite the limitations, mentioned above, pertaining to such correlations.

In areas where disequilibrium of U is known to be a significant problem, for example in the Colorado Plateau area of the Western United States, the natural gamma spectrum can be misleading and a more direct tool, such as prompt or delayed fission neutron (DFN) logging, is employed for more quantitative results. A comprehensive description of the latter variant, now being used on an operational scale in the West, is given by Givens et al. (1976). In DFN logging the rock around the hole is bombarded with 14 MeV neutrons from a pulsed neutron generator, producing 2 pulses per second. Since the half life of the DFN due to U is about 4 seconds, there will be a build up with time of the DFN levels. For quantitative U-grade determinations, correction to the DFN measurements are made for the effects of fluid salinity, strong thermal absorbers, borehole size and formation density.

Figure 2.16 shows a direct comparison of DFN assays and fluorimetric assays for U made on the core. The correlation coefficient is high (0.94).

Radon soil gas measurements based on alpha detection is a popular uranium exploration method, used in the hope of expanding the range of detection of buried deposits. Both short term and long term (or integrated) sampling techniques are employed (e.g. see Warren, 1977). The former use either an ionization chamber or zinc sulphide scintillator for measurement. The latter use fission track or solid-state alpha detectors, left in a hole in the ground for up to several weeks. The detection of buried uranium deposits by means of soil radon gas measurements is a well accepted technique. There is, however, some controversy about the effective depth of detection of this approach, since the short half life (3.8 days) of radon and its expected diffusion rates are incompatible with the 100 m or so depth detection claimed. Nondiffusion transport mechanisms for radon or its precursors appear necessary to account for the observed field results.

Because of its diversity of methods, mining geophysics gives great scope for creative originality on the part of its practitioners. I believe that is one of the secrets of its fascination for those of us who have made it our life's work. It also makes the task of compressing a comprehensive review of mining geophysics into a few pages a most difficult one.

ACKNOWLEDGMENTS

The author is indebted to a number of sources for permission to publish the various figures of this paper. An acknowledgment to the appropriate source has been made in respect of each figure, in its caption.

REFERENCES

Angoran, Y. and Madden, T.R.
    1977: Induced polarization: a preliminary study of its chemical basis; Geophysics, v. 42, p. 788-803.

Clayton, C.G.
    1976: Some experience with the use of nuclear techniques in mineral exploration and mining; Nuclear Techniques in Geochemistry and Geophysics, IAEA, Vienna.

Czubek, J.A.
    1976: Comparison of nuclear well logging data with the results of core analysis; Nuclear Techniques in Geochemistry and Geophysics, IAEA, Vienna.

Darnley, A.G.
    1973: Airborne gamma-ray survey techniques — present and future; Uranium Exploration Methods, IAEA, Vienna.

Edwards, R.N.
    1976: Electrical methods for the study of regional crustal conductivity anomalies; Acta Geodaet., Geophys. et Montanist. Acad. Sci. Hung., Tomus 11 (3-4), p. 399-425.

Fraser, D.C.
    1978: Geology of the Montcalm Township copper-nickel discovery; Can. Min. Metall. Bull., v. 71, January. 1978.

Geodata
    1978: Evaluation of soil moisture measurement using gamma-ray spectroscopy; Geodata Inter., Inc., Dallas, Texas, U.S. Dep. Energy, Open File Rep. GJBX-49(78).

Givens, W.W., Mills, W.R., Dennis, C.L., and Caldwell, R.L.
    1976: Uranium assay logging using a pulsed 14 MeV neutron source and detection of delayed fission neutrons; Geophysics, v. 41, p. 468-490.

Hallof, P.G.
    1974: The IP phase measurement and inductive coupling; Geophysics, v. 39, p. 650-665.

Lamontagne, Y. and West, G.F.
    1973: A wide-band, time-domain ground EM system; from Symposium on electromagnetic exploration methods, University of Toronto, May 1973.

Lazenby, P.G.
    1973: New developments in the Input airborne EM system; Can. Min. Metall. Bull., v. 66, no. 732, April 1973, p. 96-104.

Nargolwalla, S. and Seigel, H.O.
    1977: In-situ mineral deposit evaluation with the Scintrex metalog system; Can. Min. J., v. 98 (4), April 1977.

Pelton, W.H., Rijo, L., and Swift, C.M.
    1977: Inversion of two dimensional resistivity and induced polarization data; from Induced polarization for exploration geologists and geophysicists, Dep. Geosciences, Univ. Arizona, March 1977.

Seigel, H.O.
    1972: Ground investigation of airborne electromagnetic indications; Proc. 24th Int. Geol. Cong., Montreal, Sec. 9, p. 98-109.

    1974: The magnetic induced polarization (MIP) method; Geophysics, v. 39, p. 321-339.

Seigel, H.O. and Pitcher, D.H.
    1978: Mapping earth conductivities using a multi-frequency airborne electromagnetic system; Geophysics, v. 43, no. 3, p. 563-575.

Slichter, Louis B.
    1955: Geophysics applied to prospecting for ores; Econ. Geol., 50th Anniversary Volume; Lancaster Press Inc.

Sternberg, B.K.
    1977: Interpretation of electromagnetic sounding data. From Numerical Methods and Computer Program Listings, Univ. Wisc. Rep. 77-1, May 1977.

Strangway, D.W., Swift, C.M., and Holmer, R.C.
    1973: The application of audio frequency magneto-tellurics (AMT) to mineral exploration; Geophysics, v. 38, p. 1159-1175.

Sumner, J.S.
    1976: Principles of induced polarization for geophysical exploration; Elsevier Publ. Co., New York.

Ward, S.H.
    1970: Airborne electromagnetic methods. Mining and Groundwater Geophysics, 1967; Geol. Surv. Can., Econ. Geol. Rept. 26, p. 81-108.

Warren, R.K.
    1977: Recent advances in uranium exploration with electronic alpha cups; Geophysics, v. 42, p. 982-989.

Zietz, I., Gordon, P.E., Frischrecht, F.C., Kane, M.F., and Moss, C.K.
    1976: A western view of mining geophysics in the U.S.S.R.; Geophysics, v. 41, p. 310-323.

Zonge, K.L. and Wynn, J.C.
    1975: Recent advances and applications in complex resistivity measurements; Geophysics, v. 40, p. 851-864.

GEOCHEMISTRY OVERVIEW

R.W. Boyle
Geological Survey of Canada

Boyle, R.W., Geochemistry overview; in Geophysics and Geochemistry in the Search for Metallic Ores; Peter J. Hood, editor; Geological Survey of Canada, Economic Geology Report 31, p. 25-31, 1979.

Abstract

In retrospect, geochemical prospecting has made major advances in the last 40 years as a result of the development of rapid and accurate analytical methods for the determination of most of the elements of the Periodic Table, a circumstance that has facilitated the elucidation of the nature of primary and secondary dispersion halos associated with most mineral deposits and accumulations of hydrocarbons. Application of this knowledge has resulted in the accelerated discovery of numerous large mineral deposits.

In prospect, it is certain that geochemical prospecting will offer an efficient and economical way of prospecting both for mineral deposits and accumulations of hydrocarbons. Future research should be focused on techniques for discovering deeply buried deposits and on aspects such as heavy mineral surveys, specific elemental and mineral indicators of mineralization, the use of isotopes, mathematical interpretative procedures, and the fundamental factors controlling the migration of the elements in endogene and exogene environments.

Résumé

La prospection géochimique a effectué des progrès considérables ces 40 dernières années, par suite du développement de méthodes analytiques rapides et précises, permettant d'identifier la plupart des éléments du Tableau périodique; ceci nous a permis de mieux comprendre la nature des auréoles de dispersion primaires et secondaires associées à la plupart des gîtes minéraux et d'hydrocarbures. La mise en application de ces connaissances explique l'intensification des découvertes de gîtes d'importance.

La prospection géochimique représentera sans aucun doute à l'avenir une méthode efficace et rentable d'exploration des gîtes minéraux et d'hydrocarbures. Désormais, les recherches devraient surtout porter sur la mise au point de techniques d'investigation des gîtes profonds, les relevés géochimiques des concentrations en minéraux lourds, les minéraux ou éléments indicateurs des minéralisations, l'utilisation des isotopes, les procédés d'interprétation mathématique et les facteurs fondamentaux qui régissent la migration des éléments dans les milieux endogènes et exogènes.

INTRODUCTION

Briefly stated "geochemical prospecting" is the application of geochemical and biogeochemical principles and data in the search for economic deposits of minerals, petroleum, and natural gases. In this paper a review of some general concepts is given followed by a summary of advances that have been made in recent years; the paper concludes with some thoughts for the future.

Some of the techniques used in geochemical prospecting have been employed for centuries in the search for mineral deposits. One is reminded of the descriptions in ancient Greek and Roman treatises of the search for gold deposits by panning the streams of an area and of the detailed instructions in Georgius Agricola's De re metallica of the best methods for discovering mineral veins by analyzing the water of springs and observing the deleterious or toxic effects wrought on vegetation by the presence of metallic deposits. One is also reminded of the first successful geochemical prospector, Giovanni de Castro, a Genoese gentleman of the mid-fifteenth century. It is recorded that he was engaged in manufacturing alum from alunite at the mines near Edessa in Syria. While there he observed a particular type of holly plant that characteristically grew near the veins. On returning to Italy he found similar plants growing in the hills at Tolfa, near Rome. Prospecting in this area he soon found alunite float and later the veins from which it came. This led to an important alum industry at Tolfa, and for his efforts Pope Pius II granted Castro an annuity for life and erected a statue in his honour.

In the present century the techniques of geochemical prospecting had their origin mainly in the Soviet Union and in the Scandinavian countries, where much research on methods was carried out in the late 1930s. After World War II, the various methods were introduced into the United States, Canada, Great Britain, and other countries where they have since been used extensively in mineral- and petroleum-exploration programs by both mining companies and government agencies.

RETROSPECT AND PROSPECT

The Earth is characterized by five distinct spheres — lithosphere, pedosphere, hydrosphere, atmosphere, and biosphere — the various materials of which provide sampling media for geochemical surveys carried out in the search for mineral deposits, coal, oil, and natural gas. Geochemical surveys are, therefore, commonly classified as lithogeochemical, pedogeochemical, hydrogeochemical, atmogeochemical, and biogeochemical.

Within a specific geological terrane a general background or abundance for the various individual elements of the Periodic Table prevails in the five geospheres. Near mineral deposits or accumulations of hydrocarbons the abundance of their constituent elements generally exhibits an increase in the materials of the various spheres producing anomalies. These anomalies are manifest by primary and/or secondary elemental dispersion halos, trains, or fans in the rocks, soils, glacial deposits, hydrologic system, stream and lake sediments, atmosphere, plants and animals; also under certain conditions by the presence of indicator plants (and animals) or by elemental stimulative or deleterious (chlorotic) effects on the plants (and animals). The nature of the halos, trains, and fans are complex: in most cases there is a consistent increase in the content of a specific element or elements toward a mineral deposit or hydrocarbon accumulation (positive or addition halos); in other cases the reverse may be true (negative or subtraction halos). These various halos, trains, and fans provide a means of tracing and locating the focal point or source from which the ore elements, gangue elements, or hydrocarbons were dispersed. They constitute the fundamental basis for all geochemical surveys.

During geochemical prospecting surveys sampling is conducted on the grid principle, on the directed walk principle, or on the random walk principle. In the first, samples are obtained at standard intervals in two or three dimensions; in the second, samples are obtained at standard intervals along some topographic or geologic feature, e.g. along a stream net or along a specific stratum; in the third samples are selected here and there with no particular pattern. The last method is not generally employed except where the geochemical survey is ancillary to other types of surveys, and the geologist observes something that excites his interest. Surveys may be shoeborne, muleborne, camelborne, shipborne, carborne, or airborne, depending on the facilities available, the elements sought in the surveys, the terrain, economics, and other factors. When the samples from the various surveys are analyzed and the results plotted, anomalies are revealed by contouring or by an inspection of increases or decreases in metal or hydrocarbon content toward some point, hopefully an economic mineral deposit or concentration of hydrocarbons.

The modern methods of geochemical exploration owe their rapid development in the 20th century to a number of circumstances which can be briefly stated as follows:

1. Elucidation of the nature of primary and secondary dispersion halos and trains that are associated with all mineral deposits and accumulations of hydrocarbons. Halos and trains have been recognized by geologists for centuries. Their importance, however, in a geochemical sense was first emphasized by A.E. Fersman and his co-workers in U.S.S.R. in the early 1930s.

2. Development of accurate and rapid analytical methods utilizing the spectrograph and the various specific sensitive colorimetric reagents, especially dithizone. G.R. Kirchhoff and R.W. von Bunsen founded the science of optical spectroscopic analysis in 1859, and Assar Hadding first employed X-ray spectrography in chemical analysis in 1922. Both methods have given immeasurable service in geochemical prospecting. Dithizone was first prepared in 1878 by Emil Fischer, who noted that its reactions with heavy metals gave brilliantly coloured products. No analytical use, however, was made of it until 1925, when Hellmut Fischer demonstrated its particular use in estimating the amounts of various trace metals in substances. Since then, dithizone and many other similar organic compounds have been widely used in geochemistry and geochemical prospecting. Actually, dithizone methods approach the limits obtainable by spectrography and in some cases they surpass them.

3. Development of polyethylene laboratory ware of all types. This permitted greater freedom in sampling and field analysis and reduced the incidence of contamination. The introduction of resins for the production of metal-free water for use in trace analysis requires no comment.

4. Development of atomic absorption spectroscopy, a particularly rapid and accurate method of determining many of the elements in earth materials in the parts per million range. For this advance we owe a great debt to the sustained research in instrumentation and methods by A. Walsh at C.S.I.R.O. in Australia. One can say without exaggeration that atomic absorption has been the most significant development in analytical geochemistry in recent years.

5. Development of fluorimetric methods of uranium analysis in the early 1950s. These methods are based on the discovery by E. Nichols and M.K. Slattery in 1926 that a trace of uranium fused in a sodium fluoride bead gives an intense yellow-green fluorescence, the magnitude of which is proportional to the amount of uranium present. More recent developments by Scintrex Ltd. of Concord, Ontario include a laser-induced fluorescence method for uranium in solution which is said to have a sensitivity of 0.05 ppb uranium in natural waters.

6. Development of sensitive methods utilizing specific ion electrodes particularly for elements such as fluorine and chlorine.

7. Development of gas, partition, paper, adsorption, and ion exchange chromatography and other precise methods of trace analysis of hydrocarbon compounds and various gaseous inorganic substances. These are probably the most significant developments as regards the rapid analysis of traces of hydrocarbons and associated gaseous inorganic compounds in petroleum prospecting surveys utilizing rocks, soils, waters, and lake and marine sediments.

8. Development of radioactive methods of detection of the radioelements, particularly uranium and thorium and their daughter elements. Following the discovery of the phenomenon of radioactivity by A.H. Becquerel in 1896, Marie Sklodowska Curie, the first radiochemist, began an extended study of natural radioactivity resulting in the discovery of polonium and radium in pitchblende from Jáchymov, Czechoslovakia. This was followed by extensive studies of the natural radioactivity of rocks, minerals, natural waters, and the atmosphere by many chemists, geologists, and physicists. The instruments used in the early work were the photographic plate, electroscope, electrometer, ionization chamber, and spinthariscope, the last a most valuable instrument developed for nuclear research by W. Crookes. In 1908 the Geiger tube was developed and modified in 1928 as the Geiger-Müller tube; the latter was adapted to field use by G.M. Shrum and R. Smith (1934) at the University of British Columbia in 1932. Prior to this (about 1930) H.V. Ellsworth (1932) of the Geological Survey of Canada used an improved version of the electroscope in the field surveying of uranium- and thorium-bearing pegmatites and other radioactive bodies. Electronic scintillation counters based on the principal of the spinthariscope, but employing sensitive photomultiplier tubes, were developed by S.C. Curran and W.R. Baker in 1944. This was followed in succeeding years by much research on large volume crystals for counting beta-particles and gamma radiation, culminating in the development by R. Hofstader in 1948 of thallium-activated sodium-iodide crystals for the detection of gamma radiation. These crystals have remained for many years the most important detector medium for gamma-ray scintillation spectrometry. Scintillation counters were first adapted to field use in 1949 by G.M. Brownell (1950) of the University of Manitoba. Today many of the airborne spectrometers have reached a high degree of sophistication employing as many as 1024 channels for total gamma counting and specific quantitative analyses of elements such as potassium (^{40}K), uranium (as ^{214}Bi), and thorium (as ^{208}Tl). Early radiometric surveys were mainly shoeborne; today many of the surveys are carborne, airborne, or shipborne.

9. Development of methods of radioactivation analysis: In a modern sense these methods, utilizing certain nuclear properties of the isotopes of the elements sought in the sample, were developed after the Second World War. Most methods utilize thermal neutrons as the bombarding particles (neutron activation analysis). Such methods have an extremely high sensitivity in the parts per billion range for elements such as gold, platinoids, uranium, and so on. They have been used extensively in basic geochemical research for many years and are now finding increasing use in geochemical prospecting.

10. Development of methods for estimating radioactive elements such as uranium, thorium, and radon by fission and alpha-track analyses. These methods have had a long history in nuclear research. Within the last five years a number of variations of the method have been adapted for the field and laboratory analysis of radon by consulting geochemical companies. A novel in-field method for the determination of radon by alpha-sensitive dielectric film under the trade name "Track Etch" has been developed by Terradex Corporation of Walnut Creek, California. This method utilizes cups that are placed in a shallow hole in the ground and left for a period of 3 weeks or more; they are then recovered and the film returned to a central laboratory for analysis. Another novel radon gas detector developed by Alpha Nuclear of Caledon East, Ontario utilizes a solid state detector (silicon diffused junction) coupled to an electronic integrating readout metering device. This reusable instrument is placed in a hole in the ground and left for one or two days, after which the radon concentration and the number of hours of integration can be read from a visual numeric display.

11. Development of rapid and precise methods of analyses of various volatile elements such as mercury, sulphur compounds, helium, and radon in rocks, soil gases, waters, and in the atmosphere. The instruments adapted for such work include the gas chromatograph, atomic absorption spectrometer, mass spectrometer, and a number of variations of alpha counters.

12. Utilization of mass spectrometric methods, originally developed by F.W. Aston in 1919, for the determination of the isotopic abundances of lead, sulphur, hydrogen, and so on. Some of the abundance data may indicate the presence of certain deposits (e.g. radiogenic lead derived from thorium and uranium deposits); other systematic abundance data may be the basis for vectoring in on mineral deposits or accumulations of hydrocarbons.

13. Development of rapid and precise methods of analysis of various types of both organic and inorganic particulates in the atmosphere. O. Weiss (1971) in South Africa and A.R. Barringer (1977) in Canada have pioneered these techniques utilizing aircraft as sampling vehicles. The methods provide a broad scanning technique of a terrane and under favourable circumstances may indicate the location of mineral deposits and oil fields.

14. Refinement of field techniques in carrying out reconnaissance and detailed surface geochemical surveys of all types but especially those based on stream sediments, lake-bottom sediments, heavy minerals, stream and lake waters, groundwaters, springs and their precipitates, and biological materials. In recent years the use of helicopters has revolutionized sample collection in reconnaissance surveys in practically all terrains.

15. Development and refinement of methods of detailed geochemical prospecting using overburden drilling techniques in permafrost terranes, glacial terranes, and in deeply weathered and lateritized terranes.

16. Development of methods for discovering ore shoots or their extensions based on primary halos and leakage halos.

17. Development and refinement of methods using ore boulder, gangue boulder, and heavy mineral trains and fans for prospecting in glacial terranes and in deeply weathered and lateritized terranes.

18. Research and development of efficient methods in the processing and assessment of geochemical prospecting data by statistical and computer techniques.

19. Improved understanding of the geochemistry of mineral deposits and concentrations of hydrocarbons, particularly the chemistry of their genesis.

Geochemical prospecting is now being employed with increasing emphasis in all terrains of the world from tundra to tropical belts. Among its successes can be claimed the discovery of a considerable number of large, low grade deposits that yield gold, copper, uranium, nickel, lead, and zinc. Here we may mention a few of these with brief comments: Carlin-type gold orebodies (Cortez deposit) in Nevada related to anomalies discovered by the United States Geological Survey using gold, arsenic, and other indicators in oxidized bedrock and residual soil surveys (U.S.G.S., 1968); the enormous auriferous Muruntau deposit in Uzbek S.S.R., discovered by a U.S.S.R. prospecting syndicate using arsenic as an indicator in soils and weathered residuum (Khamrabaev, 1969); the great Panguna (Bougainville) copper porphyry in the Solomons chain located by stream sediments and soil sampling (Baumer and Fraser, 1975); the extensive Beltana and Aroona willemite deposits in South Australia discovered by stream sediment sampling (Muller and Donovan, 1971); the McArthur River lead-zinc deposits, Northern Territory, Australia by soil sampling (Murray, 1975); the large Lady Loretta lead-zinc deposit in Queensland by grid soil sampling (Cox and Curtis, 1977); the large Woodlawn copper-lead-zinc deposit in New South Wales by roadside reconnaissance stream sediment and soil sampling (Malone et al., 1975); the Casino porphyry-copper-molybdenum deposit in Yukon by stream sediment, water, bedrock, and soil sampling (Archer and Main, 1971); a number of prospects and deposits, particularly the Husky lead-zinc-silver deposit in the Keno Hill area, Yukon by overburden drilling and geochemical sampling methods (Van Tassell, 1969); the Howards Pass zinc-lead belt in Yukon by stream sediment and soil surveys; the Island Copper porphyry deposit in Vancouver Island, British Columbia by soil sampling (Young and Rugg, 1971); the Sam Goosly copper-silver-molybdenum deposit in central British Columbia by stream sediment and soil sampling (Ney et al., 1972); the Brandywine gold-silver-copper-lead-zinc deposit in western British Columbia by stream sediment and soil sampling (Anon, 1974); the Coed-y-Brenin porphyry copper deposit in North Wales by soil surveys (Mehrtens et al., 1973; Rice and Sharp, 1976); and the Tynagh and a number of other lead-zinc-copper-silver deposits in Eire mainly by soil sampling (Schultz, 1971). This by no means exhausts the recent successes by geochemical methods. In a recent paper Glukhov (1974) estimates that some 80 000 surface (metallic) halos and anomalies have been identified by geochemical prospecting methods in U.S.S.R. in the last 20 years. Investigation of these has led to the discovery of some 220 deposits (now being mined or at the development stage) of a variety of elements and some 900 prospects. The percentage ratio of deposits to anomalies is, therefore, a not very impressive, 0.3 per cent. The percentage is no better in other countries. The present writer estimates that of some 100 000 anomalies located in Africa, Europe, Asia, North America, South America, and Australia in the last twenty years by geochemical methods, some 150 orebodies (now being mined or at the development stage) have been found associated with these anomalies, giving about the same order of magnitude for the ratio. In both U.S.S.R. and other countries many of these anomalies have not been adequately tested, and more discoveries of orebodies associated with perhaps about 500 anomalies may be expected in each case.

It is interesting to note which methods have been most successful. Glukhov (1974) attributes nearly all of the successes in U.S.S.R. to surface lithogeochemical methods (soils, stream sediments, heavy minerals, and rocks). Only a few discoveries have been made by hydrogeochemical (water) methods, and none by biogeochemical (plants, animals) and atmogeochemical (gases) methods. Glukhov (op. cit.) also mentions that the discovery rate and the potential for discovery of near-surface deposits is declining rapidly in U.S.S.R. and advocates greater attention to depth methods of geochemical prospecting. In other countries the story is about the same. Nearly all discoveries are attributable to stream sediment, pedogeochemical, and lithogeochemical work and very few to hydrogeochemical (water), biogeochemical (plants), and atmogeochemical work. With respect to airborne radioactivity surveys for thorium and uranium the recent report by the International Atomic Energy Agency, Vienna (1973) on uranium exploration methods is of interest if considered with caution because of different definitions of deposits, prospects, etc. Of some 11 284 anomalies outlined by all means in 21 countries (excluding Australia, Canada, France, South Africa, U.S.A. and U.S.S.R.) some 256 deposits were discovered. Of the total anomalies 3402 were outlined by aerial surveys resulting in 37 deposits. In the 17 conterminous Western States of U.S.A. some 100 000 radioactive anomalies were outlined resulting in some 700 deposits. Of the total anomalies some 4000 were discovered by aerial surveys resulting in 120 deposits. Again it is apparent that surface methods are the most effective. Without going into details it is also apparent from the data that airborne methods are most effective in arid, low vegetation, and desert terrains.

There is little public information available on the discovery of oil and gas fields by geochemical methods and one cannot draw any logical conclusions. In recent years some oil fields have been located by detailed studies of hydrocarbons in groundwaters and overlying soils and glacial deposits (Heroy, 1969). Even the hydrocarbon content of the sediments of the sea and of the overlying seawater has been employed in searching for submarine oil pools (Hitchon, 1977), but no case histories on the success or failure of this type of work appears to have been published.

The writer would now like to turn the reader's attention to desirable future research in geochemical prospecting and discuss some aspects of the subject that should receive immediate attention. Some of these are general whereas others are of a detailed nature and demand action if sufficient minerals and energy materials are to be available to maintain our industrial society as it now exists let alone possible future expansion in the world community.

The first general problem concerns the delineation of geochemical provinces and their relationship to metallogenic provinces. Many metallogenic provinces are known and were clearly outlined in many parts of the world long before geochemistry came into vogue. Others remain to be discovered and defined not only in remote and inaccessible regions but also in those that have been extensively explored. Numerous extensions of known metallogenic provinces under extensive drift or younger rocks, likewise, remain to be delineated. It is now clear that surficial and biological materials such as soils, stream sediments, lake sediments, groundwaters, and vegetation reflect the presence of certain types of exposed metallogenic provinces, particularly those mineralized with copper, lead, zinc, nickel, and uranium. But what about the other 50 or more important elements of the Periodic Table? Here we have very little precise knowledge. Furthermore, we do not yet know for certain whether or not fresh (unmineralized) rock samples (or lithogeochemistry) reflect the presence and type of metallogenic provinces in a region. It would appear that the cases for uranium (e.g. Bear Province in Canada's Northwest Territories; Lake Athabasca region, Saskatchewan; Bancroft area, Ontario) and copper (e.g. Kupferschiefer; Michigan copper belt, Seal Lake copper belt, Labrador) are positive, but in the writer's experience the cases for gold, silver, lead, zinc, and many other elements are equivocal in most regions. We need to know much more about the trace and major element chemistry of rocks and their geochemical relationship to metallogenic belts especially when drilling through deep overburden or younger overlying rocks. The targets in such types of mineral exploration are small, even the primary halos, and enlargement of the target utilizing favourable indicator rocks is highly desirable.

The second most important field of research in geochemical methods involves the development of methods for discovering mineral deposits and concentrations of hydrocarbons deeply buried in the rocks. We tend to think and work within a two-dimensional framework and generally have the mentality of areal explorers, paying practically no attention to the third or depth dimension. Mining geologists are forced to think in three dimensions if their respective mines are to remain in business, and they have developed a number of sophisticated methods in their search for ore. If these techniques were combined with those of geochemistry, basing the search for ore on the principles of primary halos, leakage halos, and favourable primary and secondary ore localizing features, it is certain that many orebodies would be discovered within the confines of known mineral belts at minimal cost. We need to know much more about the three-dimensional morphology of primary halos, particularly leakage halos, and their relationship to orebodies, subjects that have received very little attention in the last few years. In this respect it is well worth mentioning that lithogeochemical techniques are potentially powerful exploration tools which offer depth penetration well beyond the limits of any foreseeable conventional geophysical method. The nature of the groundwater systems in mineral belts has also received only cursory attention, yet sampling of this medium can be profitable in ore search as outlined by Boyle et al. (1971). Concerning concentrations of hydrocarbons it can be said without exaggeration that we have little precise knowledge about the nature, morphology, and chemistry of the rock and ground (stratal) water halos associated with accumulations of hydrocarbons, a circumstance that is ludicrous considering the enormous amount of drilling done for oil and gas in the last 75 years. Until we know the exact characteristics of both the elemental, salt, and hydrocarbon halos associated with oil and gas fields we cannot use geochemical prospecting to discover such fields. The writer predicts that when we seriously begin to use geochemical prospecting in petroleum and natural gas exploration we shall find throughout the world as many oil and gas pools as are now known. The writer made a similar prediction with respect to mineral deposits in a talk in 1960; that prediction is rapidly proving true.

Large sections of metalliferous regions in many countries containing potential orebodies are covered by deep surficial deposits; in countries like Canada and U.S.S.R. these deposits constitute various glacial materials; in Australia and parts of United States, Central America, South America, and Africa the deposits are laterites; and in various countries the surficial materials are recent sediments or volcaniclastics. After considerable research in many countries it is now evident that the only viable geochemical approach to mineral exploration in these deeply covered terrains is the overburden drill and chemical analyses of materials from favourable horizons, commonly the basal horizon. Some research has been done on defining the character (hydromorphic and/or mechanical), morphology, and relationship to mineralization of dispersion fans, trains, and halos buried deeply in or by surficial deposits. Much more research, however, must be done on the mechanical and chemical dispersion characteristics of the elements in glaciated, lateritized, and sediment-volcaniclastic covered terranes to provide a sound basis for the interpretation of the dispersion patterns found in such terranes.

During geochemical surveys it is common to find numerous anomalies that vary in elemental nature, size and intensity. At our present state of knowledge it is not generally possible to differentiate anomalies related to mineralization and mineral deposits from those that represent enrichments from the country rocks due to an infinite variety of geochemical factors. In other words we lack definitive screening techniques. In this respect much more work is required on specific indicators of particular types of mineralization and mineral deposits, on the relationship of the ratios between elements to mineralization, and on the intensity relationship of elemental dispersion with distance from the focus of mineralization. Concerning the same subject we should concentrate considerable research on the elucidation and formulation of interpretative techniques to relate the size and intensity of trace element halos and dispersion trains to the estimated size and grade of mineral deposits.

Biogeochemical, geobotanical, and atmogeochemical methods have received considerable research in recent years but have not as yet found extensive use in mineral and hydrocarbon exploration. The reason for this is not clear. Despite the fact that anomalies found during biogeochemical and atmogeochemical surveys are of a second-order nature and thus twice removed from primary mineralization, the dispersion patterns are commonly more distinct and often possess greater contrast than those in associated soils, weathered residuum, stream sediments, and waters. There is also the added factor in biogeochemical surveys that trees and shrubs draw metals from depth, in some cases a hundred feet or more in arid regions. This should be a great advantage in Australia and in many countries in Africa, Asia, and South America. A novel approach is the use of sap analysis where the appropriate tree coverage is present. This method has already proved of interest in defining auriferous zones and those enriched in elements such as copper, zinc, niobium, uranium, etc. Another novel method, the use of dogs as "ore sniffers" deserves much more attention than it has so far received.

Increased research in the use of geochemistry in locating viable thermal zones in the earth for harnessing geothermal power should be assiduously pursued. Much research of this kind has been conducted in the North Island of New Zealand and in the Larderello Valley of Italy where much electric power is generated from thermal waters (steam). Work in these and other countries has shown that thermal areas are indicated by elevated contents of elements such as mercury, arsenic, antimony, thallium, boron, and silica in spring and groundwaters. In addition there is considerable evidence that the thermal zones are indicated by increased amounts of the alkalis, including Na, K, and Rb, and by increases in the Na/K and other ratios in the waters.

For many years economic geologists have discussed the various mineralizers including B, CO_2, S, Se, Cl, Br, I, and F, elements and compounds that provide the ligands for the facile transport of the elements in both epigenetic and syngenetic processes. One or more of these elements are universal in practically all types of mineral deposits, yet we have not employed them to full advantage in geochemical prospecting methods. Two of these elements, B and F, are universal indicators of practically all types of epigenetic deposits; where one is not present the other is, and in many deposits both occur.

Many elements are specific indicators of certain types of deposits. Among these may be mentioned gold, beryllium, radium, antimony, bismuth, tellurium, and the platinoids. All have very low abundances in all geochemical spheres of the earth; when found in detectable concentrations (1 ppm or more for gold and 1.4×10^{-6} ppm or more for radium), one can be certain that deposits hosting these indicators (e.g. copper deposits, silver deposits, nickel deposits, uranium deposits, etc.) occur in the environment. Excepting gold and perhaps radium we have little precise knowledge of the distribution and geochemistry of these elements in mineral deposits and their dispersion halos, trains, and fans. This is a fertile field for research, and one that could prove most useful in various geochemical prospecting methods.

Much more attention should be paid to heavy mineral surveys in all terrains. The writer and his colleagues have observed that in the mountainous regions of Yukon Territory and in the Appalachians of Canada, elemental dispersion trains in the stream and river systems are caused as much by transport of particulate matter as by hydrologic processes. Traditionally, heavy mineral surveys have been used to define areas enriched in metals having a low degree of chemical mobility in the supergene cycle. Our studies indicate, however, that analyses of heavy mineral concentrates for the chemically more mobile elements can outline metalliferous zones. In this respect geochemical and mineralogical analyses of heavy mineral concentrates from the drainage net can complement normal stream sediment surveys and provide additional information that can aid in the interpretation of stream sediment data. Furthermore, particulate dispersion trains of hypogene and supergene minerals may be more extensive than hydromorphic trains especially in carbonate terranes where chemical mobility of a number of elements may be restricted due to the relatively high pH.

The writer would like to make a plea for the greater use of bogs (muskegs) as sampling media in northern terrains, especially those that have been glaciated. Many of these bogs represent ancient lakes infilled with gyttja and peat during the long interval since the last glaciation. Sampling of the bogs, particularly the lower horizons (which in actual fact represent ancient lake sediments), on a regional basis could be an effective method of locating mineralized belts and zones that are heavily drift-covered and not generally amenable to other methods of geochemical prospecting.

Studies of the isotopic distribution of a number of elements in mineral deposits and accumulations of hydrocarbons began in earnest some twenty-five years ago yet few of the results obtained have been utilized in geochemical prospecting. A number of papers have suggested the use of isotopes in geochemical prospecting, but the writer has seen no detailed accounts of case histories using the techniques available. The most suitable isotopes for prospecting purposes would appear to be sulphur, lead, uranium, carbon, and hydrogen although many others may eventually prove useful. Isotopes are a large subject, but a few simple examples of their possible use in prospecting may be given. High amounts of radiogenic lead isotopes in heavy minerals such as pyrite and galena in soils and stream sediments may indicate the presence of uranium and/or thorium deposits, and excess contents of ^{234}U manifest in higher ^{234}U/^{238}U abundance ratios in natural waters may reflect the presence of concentrations of uranium minerals when compared with background abundances produced by disseminations of uraniferous minerals in a terrane. Systematic variations in the sulphur isotopic ratios in the metamorphic zones and alteration halos associated with sulphide deposits may indicate approach to orebodies. Similarly, systematic changes in the carbon and hydrogen isotopic ratios in sedimentary rocks and stratal waters may signal approach to petroleum and natural gas reservoirs.

Countries should now give thought to the production of standardized regional geochemical maps on a national grid based on analyses for the principal elements in the Periodic Table in materials of the lithosphere (rocks), pedosphere (soils), hydrosphere (waters, stream sediments, lake sediments), and biosphere (vegetation, bogs, etc.). One sample per square kilometre would appear advisable; in mineral belts a much higher density is desirable. For most countries such a program may require a concentrated effort lasting a century or more. Such maps will provide invaluable data for mineral exploration and a host of environmental studies concerned with forestry, fishing, agriculture, recreation, etc.

Considerable thought and research should be directed to the problem of the proper timing of the use of geochemical methods in the overall exploration program. Having carefully observed for more than 25 years the procedures of many companies and other organizations in many countries pursuing exploration, the writer is convinced that this subject is one that requires much more attention than it has up to now received. In my opinion the proper sequence of events in exploration is: (1) Geological assessment, (2) Geochemical reconnaissance utilizing one or more of hydrogeochemical and stream and lake sediment methods, pedogeochemical methods, lithogeochemical methods, biogeochemical methods, airborne radioactive methods, and atmospheric particulate methods, (3) Utilization of geophysical and/or detailed geochemical methods to obtain a sharp focus on the sites of mineralization indicated by the anomalous patterns revealed by the reconnaissance geochemical surveys. Only in terranes where geochemical prospecting is found unsuitable, and they are few in this writer's experience, should step 2 be omitted.

All aspects of mathematical interpretation of geochemical data require concerted research in the future. Particular attention should be paid to the methods of vector analysis in both two and three dimensions. The simplest vector is a stream with increasing concentration of elements in the water or stream sediments upstream to the site of a mineralized zone or deposit. It seems to this writer that considering the large amount of data now available one should now be able to quantify vectors obtained from surface surveys and also those obtained in three dimensions. In many cases we have at our disposal vectors with direction and magnitude (including concentrations or ratios between various elements) from which we should be able to predict precisely which kinds of vectors are likely to indicate orebodies.

Finally, the writer would urge that we devote more attention and research funds to a better understanding of the factors controlling the migration of the elements in endogene and exogene environments. In this respect the universities, and especially the geological faculties, have an obligation that in the writer's opinion has been neglected in many parts of the world. An examination of the curricula and research projects of many universities discloses a sad omission of what could be called "useful geochemistry", namely the study of the elements as we find them in nature. In the recent geological literature there is a plethora of erudite studies dealing with the great manifestations of the earth but few dealing with the fundamental geochemical factors that are useful in interpreting the results of geochemical surveys.

A number of recent textbooks and reviews are now available on the subject of Geochemical Prospecting; some of these are included in the Selected Bibliography.

SELECTED BIBLIOGRAPHY

Anon
1974: How geochemical prospecting found the Brandywine deposit in B.C.; Western Miner, v. 47, no. 9, p. 56-58.

Archer, A.R. and Main, C.A.
1971: Casino Yukon — A geochemical discovery of an unglaciated Arizona-type porphyry; p. 67-77, in Geochemical Exploration, Can. Inst. Min. Met., Spec. Vol. 11, Editors R.W. Boyle and J.I. McGerrigle.

Barringer, A.R.
1977: Airtrace — An airborne geochemical exploration technique; in Proc. First William T. Pecora Mem. Symp., Oct. 1975, U.S. Geol. Surv., Prof. Paper 1015, p. 231-251.

Baumer, A. and Fraser, R.B.
1975: Panguna porphyry copper deposit, Bougainville; in Economic geology of Australia and Papua, New Guinea; Editor, C.L. Knight, Austr. Inst. Min. Met. Monogr. Ser. 5, p. 855-866.

Beus, A.A. and Grigorian, S.V.
1977: Geochemical exploration methods for mineral deposits; Editor A.A. Levinson, Translator R. Teteruk-Schneider, Applied Publishing Ltd., Wilmette, Illinois, 287 p.

Boyle, R.W.
1976: Geochemical prospecting; in McGraw-Hill Encyclopedia of Science and Technology, 4th ed., v. 6, p. 124-129.

Boyle, R.W. and Garrett, R.G.
1967: Geochemical prospecting — a review of its status and future; Earth Sci. Rev., v. 6, p. 51-75.

Boyle, R.W., Hornbrook, E.H.W., Allan, R.J., Dyck, W., and Smith, A.Y.
1971: Hydrogeochemical methods — Application in the Canadian Shield; Can. Inst. Min. Met. Bull., v. 64, no. 715, p. 60-71.

Boyle, R.W. and McGerrigle, J.I. (Editors)
1971: Geochemical exploration; Can. Inst. Min. Met., Spec. Vol. 11, 594 p.

Brooks, R.R.
1972: Geobotany and biogeochemistry in mineral exploration; Harper and Row, Publishers, New York, 290 p.

Brownell, G.M.
1950: Radiation surveys with a scintillation counter; Econ. Geol., v. 45, p. 167-174.

Butt, C.R.M. and Wilding, I.G.P. (Editors)
1977: Geochemical exploration 1976; J. Geochem. Explor., v. 8, no. 1/2, 494 p.

Cameron, E.M. (Editor)
1967: Proceedings, Symposium on Geochemical Prospecting, Ottawa, April, 1966; Geol. Surv. Can., Paper 66-54, 282 p.

Canney, F.C. (Editor)
1969: Proceedings Second International Geochemical Exploration Symposium; Colo. Sch. Mines Q., v. 64, no. 1, 520 p.

Cox, R. and Curtis, R.
1977: The discovery of the Lady Loretta zinc-lead-silver deposit, northwest Queensland, Australia — a geochemical exploration case history; J. Geochem. Explor., v. 8, no. 1/2, p. 189-202.

Elliott, I.L. and Fletcher, W.K. (Editors)
1975: Geochemical exploration 1974; Elsevier Publishing Co., Amsterdam, 720 p.

Ellsworth, H.V.
1932: Rare-element minerals of Canada; Geol. Surv. Can., Econ. Geol. Series 11, 272 p.

Ginzburg, I.I.
1960: Principles of geochemical prospecting; Pergamon Press, London, 311 p.

Glukhov, V.A.
1974: Effectiveness of geochemical methods of prospecting; in Theoretical aspects of the migration of elemental indicators of ore deposits and mathematical methods of processing geochemical information, Part II, Examples of the discovery of mineralization and deposits by means of geochemical methods; Akad. Nauk. U.S.S.R. Ministry of Geology, Institute of Geology, Armenian S.S.R.

Hawkes, H.E. and Webb, J.S.
    1962: Geochemistry in mineral exploration; Harper and Row, New York, 415 p.

Heroy, W.B. (Editor)
    1969: Unconventional methods in exploration for petroleum and natural gas; Southern Methodist University, Dallas, Texas.

Hitchon, B. (Editor)
    1977: Application of geochemistry to the search for crude oil and natural gas; J. Geoch. Explor., v. 7, no. 2, Special Issue.

International Atomic Energy Agency
    1973: Uranium exploration methods; Vienna, 320 p.

Jones, M.J. (Editor)
    1973: Geochemical exploration 1972; Inst. Min. Met. (London), 458 p.

    1975: Prospecting in areas of glaciated terrain, 1975; Inst. Min. Met. (London), 154 p.

Kartsev, A.A., Tabasaranskii, Z.A., Subbota, M.I., and Mogilevskii, G.A.
    1959: Geochemical methods of prospecting and exploration for petroleum and natural gas; Univ. California Press, Berkeley and Los Angeles, 349 p.

Khamrabaev, I. Kh. (Editor)
    1969: The auriferous formations and principal features of the metallogeny of gold in Uzbekistan; Acad. Sci. Uzbek S.S.R., "Fan" Publishing House, Tashkent, 396 p. (in Russian)

Kvalheim, A. (Editor)
    1967: Geochemical prospecting in Fennoscandia; Interscience Publishers, New York, 350 p.

Levinson, A.A.
    1974: Introduction to exploration geochemistry; Applied Publishing Ltd., Calgary, Alberta, 612 p.

Malone, E.J., Olgers, F., Cucchi, F.G., Nicholas, T., and McKay, W.J.
    1975: Woodlawn copper-lead-zinc deposit; in Economic geology of Australia and Papua, New Guinea; Editor, C.L. Knight, Austr. Inst. Min. Met., Monogr. Ser. 5, p. 701-710.

Malyuga, D.P.
    1964: Biogeochemical methods of prospecting; Consultants Bureau, New York, N.Y., 205 p.

Mehrtens, M.B., Tooms, J.S., and Troup, A.C.
    1973: Some aspects of geochemical dispersion from base-metal mineralization within glaciated terrain in Norway, North Wales and British Columbia, Canada; p. 105-115, in Geochemical Exploration 1972, Editor, M.J. Jones, Inst. Min. Met. (London).

Muller, D.W. and Donovan, P.R.
    1971: Stream-sediment reconnaissance for zinc silicate (willemite) in the Flinders Ranges, South Australia; p. 231-234, in Geochemical Exploration, Can. Inst. Min. Met., Spec. Vol. 11, Editors R.W. Boyle and J.I. McGerrigle.

Murray, W.T.
    1975: McArthur River H.Y.C. lead-zinc and related deposits; N.T.; in Economic geology of Australia and Papua, New Guinea; Editor, C.L. Knight, Austr. Inst. Min. Met., Monogr. Ser. 5, p. 329-339.

Ney, C.S., Anderson, J.M., and Panteleyev, A.
    1972: Discovery, geologic setting and style of mineralization, Sam Goosly deposit, B.C.; Can. Inst. Min. Met. Bull., v. 65, no. 723, p. 53-64.

Rice, R. and Sharp, G.J.
    1976: Copper mineralization in the forest of Coed-y-Brenin, North Wales; Inst. Min. Met. Trans., Sec. B, v. 85, p. 1-13.

Schultz, R.W.
    1971: Mineral exploration practice in Ireland; Inst. Min. Met. Trans., v. 80, Sec. B, p. B238-B258.

Shima, M.
    1970: Geochemical prospecting methods (Chikyu Kagaku Tanko-ho); 2nd ed.; Kyoritsu Publishing Co. Ltd. Tokyo (in Japanese), 273 p.

Shrum, G.M. and Smith, R.
    1934: A portable Geiger-Müller tube counter as a detector for radioactive ores; Can. J. Res., v. 11, no. 5, p. 652-657.

Siegel, F.R.
    1974: Applied geochemistry; John Wiley and Sons, New York, 353 p.

United States Geological Survey
    1968: U.S. Geological Survey heavy metals program progress report, 1966 and 1967; U.S. Geol. Surv., Circ. 560, p. 1.

Van Tassell, R.E.
    1969: Exploration by overburden drilling at Keno Hill Mines Limited; Colo. Sch. Mines Q., v. 64, no. 1, p. 457-478.

Weiss, O.
    1971: Airborne geochemical prospecting; in Geochemical Exploration, Can. Inst. Min. Met., Spec. Vol. 11, Editors R.W. Boyle and J.I. McGerrigle, p. 502-514.

Young, M.J. and Rugg, E.S.
    1971: Geology and mineralization of the Island copper deposit; Western Miner, v. 44, no. 2, p. 31-40.

AIRBORNE ELECTROMAGNETIC METHODS

A. Becker
Ecole Polytechnique, Montreal, Quebec

Becker, A., Airborne electromagnetic methods; in Geophysics and Geochemistry in the Search for Metallic Ores; Peter J. Hood, editor; Geological Survey of Canada, Economic Geology Report 31, p. 33-43, 1979.

Abstract

The past decade has witnessed a number of major technical advances in the art of airborne electromagnetic (AEM) surveying. Although in most cases these took the form of improvements to AEM systems available in their basic form prior to 1967, they resulted in a substantial upgrading of data quality. In this context the Canadian airborne survey industry has introduced multifrequency and multicomponent operation, has reduced noise and drift levels and has increased the recording bandwidth of AEM equipment. It is only natural that the introduction of digital data recording accompanied these changes so that more sophisticated methods of data reduction and interpretation could be implemented.

Refined AEM equipment and superior EM data quality has now given practical meaning to the concept of electromagnetic mapping. The AEM method is thus no longer restricted to prospecting applications, but can serve as an essential component of a flying laboratory used for the indirect search for mineral deposits or for the resolution of problems in geological engineering. Thus, much progress has been made in recognition and automated interpretation of overburden response.

It is expected that these trends will continue so that reliable theoretical tools will soon be available for translating high quality airborne electromagnetic data into corresponding geological sections of the underlying terrane.

Résumé

Nous avons été témoins au cours de la dernière décennie de progrès techniques importants dans le domaine des levés électromagnétiques aéroportés. Bien que, dans la plupart des cas, ces progrès aient été des améliorations apportées à des systèmes de base existant déjà avant 1967, ils ont constitué un enrichissement important de la qualité des données. C'est ainsi que l'industrie canadienne des levés aéroportés a introduit des travaux à fréquences et à composantes multiples, a réduit les niveaux de bruit et de dérive et a augmenté la largeur de bande d'enregistrement. Il est bien normal que l'introduction de la technique d'enregistrement numérique des données ait accompagné ces changements; ainsi des méthodes plus perfectionnées de réduction et de décodage des données pouvaient être mises en oeuvre.

Du matériel amélioré et une qualité supérieure des données ont ainsi accordé une signification pratique au concept de la cartographie électromagnétique. Les levés aéromagnétiques aéroportés ne sont alors plus limités à la prospection: ils peuvent constituer un élément essentiel d'un laboratoire volant servant à la recherche indirecte de ressources ou à la solution de problèmes d'étude géologique. Ainsi, d'importants progrès ont été réalisés dans les domaines de l'identification et du décodage automatisé de réponses des morts-terrains.

Il est à espérer qu'à l'avenir ces tendances se poursuivent de sorte que l'on dispose bientôt d'outils théoriques véritables pour la traduction de données électromagnétiques aéroportées de haute qualité en des sections géologiques correspondantes du sol sous-jacent.

INTRODUCTION

In the decade that has elapsed since the last review of the state of the art in Airborne Electromagnetic (AEM) surveying by S.H. Ward in 1967 (Ward, 1970), the survey industry has concentrated on the refinement and modification of existing AEM systems rather than on the development of new systems based on new concepts. Most of the recent development efforts have been aimed at improving AEM system sensitivity and penetration through a reduction in noise level, at improving the system capability for definition of conductor geometry through the introduction of multicomponent receivers, and at improving AEM system apertures for definition of conductor quality through the introduction of secondary field measurements at a number of simultaneously transmitted frequencies.

As they occurred, these innovations were reported by P.J. Hood in his annual review "*Mineral exploration trends and developments*" which appears each year in an early number of the Canadian Mining Journal (e.g. see Hood, 1978). The major capability in AEM systems is still located in Canada. For this reason, and because information on new AEM developments in Scandinavia and the Eastern World is difficult to obtain, the present review will concern itself principally with the North American scene.

Although no radical instrumentation changes were reported, the scope of application of the AEM technique has been considerably enlarged. In fact, the simple prospecting tool of ten years ago has evolved into a sophisticated mapping system whose multichannel information is recorded digitally for further computer processing on the ground. Our understanding of the electromagnetic induction process in a variety of geological environments has been considerably improved because high quality multi-frequency and/or multicomponent data are available. Thus, in addition to their main application in mineral exploration programs, airborne electromagnetic measurements are now also frequently used for the definition of geological parameters either as an aid to structural mapping or, in an engineering context, for mapping overburden. Finally, AEM systems are beginning to see service as part of a flying laboratory destined for the indirect detection of mineral deposits such as uranium mineralization which itself does not show an AEM response, but is sometimes associated with conductive marker horizons.

The progress made in improving AEM instrumentation quality and extent of applications was also accompanied by progress in the development of EM data interpretation methods. Thus the partial catalogues of scale model EM curves presented by Ward ten years ago were updated and completed for virtually all current AEM systems by West, Ghosh and Palacky (Ghosh and West, 1971; Palacky, 1978). Information is now available upon which a rational choice can be made for the deployment of a particular AEM system in a given environment. In parallel with this effort, new theoretical AEM tools were evolved for the automatic mapping of earth resistivity. It is expected that the trend towards the upgrading of EM interpretation methods will continue. As methods of numerical analysis for the calculation of the response of an arbitrarily shaped conductor to a dipole source become more accurate and economical, one can look forward to the replacement of the current techniques, which involve the use of master charts, by the direct inversion of the measured secondary field parameters into geological models.

TECHNICAL DEVELOPMENTS (1967-1977) IN AEM SYSTEMS

In spite of the fact that no radically new AEM system became operational during the 1968-1977 decade, the capability of the 1967 systems was considerably enlarged either through an updating and improvement of the electronics or through the addition of auxiliary data acquisition capability to the original system. Thus, today, as in 1967 (Ward, 1970), we still have a number of versions of the three basic AEM system types, namely rigid boom, towed bird, and VLF (Very Low Frequency-remote transmitter).

The most common system in the rigid boom category is the helicopter-towed coaxial configuration. In this type of AEM system the transmitter and receiver are rigidly supported in a plastic structure with a seven to ten metre separation between the two elements. The whole assembly is then suspended beneath the carrying platform on a cable which brings the secondary field information to the inphase and quadrature detectors and the recording apparatus located in the helicopter. If one includes in the above category, the new three-frequency coaxial Kenting-Scintrex Tridem (Bosschart and Seigel, 1974; Hood, 1978) system which is rigidly attached to the Canso aircraft that carries it, then it appears that this type of system is rapidly replacing the fixed-wing coplanar equipment which has seen much service since its inception more than twenty years ago. The current popularity of the helicopter AEM systems is probably related to the recent technological improvements that resulted in noise levels of the order of 1 ppm and a true "button-on" capability which is now available from a number of Canadian manufacturers. Although these AEM systems usually operate at a single frequency in the vicinity of 1kHz, at least two manufacturers offer a two-frequency version of this type of apparatus.

A further refinement to helicopter-borne, rigid-boom AEM systems was the introduction of the Dighem system which has a three-component receiver coil system (Fraser, 1972a). As illustrated in Figure 4.1, this type of system provides useful information on the strike and dip of small- and medium-sized conductors. The Dighem system has recently been further modified to contain two totally independent systems (one coaxial, one coplanar) within the same housing. This arrangement (Fraser, 1978a) facilitates recognition of conductor geometry.

Towed-bird AEM systems have also undergone considerable technological change during the past decade. In particular, as shown in Figure 4.2, the Barringer Input EM system (Lazenby, 1973) has been improved by its operators by increasing the transmitter power by 30 per cent for the Mark VI model. This simple change yielded a much improved signal-to-noise ratio, and permitted a decrease in the recording time constant, which resulted in an accompanying increase in anomaly resolution and depth of penetration. The new Mark VI installation has also been equipped with a digital recording system to take full advantage of improved data quality. In the frequency domain the McPhar F-400 two-frequency quadrature AEM system (Seiberl, 1975) has been modified to respond to a wider range of conductors by extending the capability of that system to five-frequency operation. The improved system, called Quadrem, which can be adapted to either fixed wing or helicopter operation operates at 95, 285, 855, 2565, and 7695 Hz. Typical data recorded by this system are shown in Figure 4.3. Another example of the current high level of technological achievement in towed-bird AEM systems is the new Hudson Bay EM-30 system constructed by Geonics Ltd. With this system it is possible to obtain inphase and quadrature data at a large coil separation in a compensated towed-bird configuration with a noise level lower than 200 ppm (Anonymous, 1977).

The gap between the moving source methods described above and the fixed remote-source VLF EM methods is bridged by the Turair EM system in which Scintrex Ltd. has introduced a semiairborne EM system in order to attain the

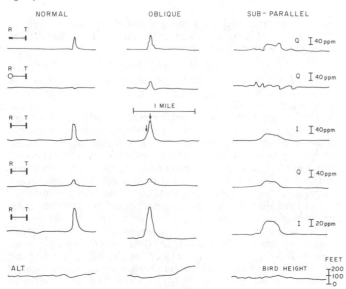

Figure 4.1. Variation of anomaly shape with conductor strike relative to flight direction; Dighem helicopter-borne EM system (after Fraser, 1972a).

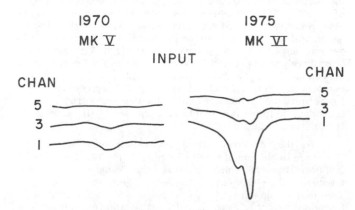

Figure 4.2. Improvement in signal quality over a given anomaly due to technological improvements; Input EM system (courtesy of Questor Surveys).

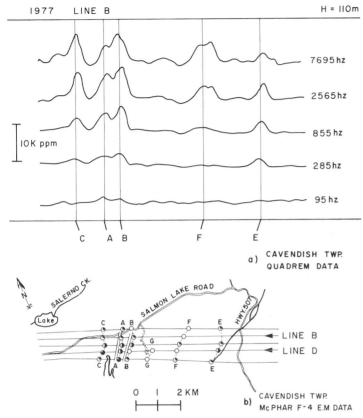

Figure 4.3. Five-frequency quadrature EM data, McPhar Quadrem and F-400 AEM systems (courtesy of McPhar Geophysics).

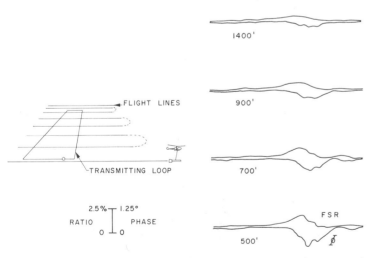

Figure 4.4. Variation of Turair EM response as a function of altitude (courtesy of Scintrex Ltd.).

depth of exploration capability of large-loop uniform primary-field ground systems such as the Turam system, while maintaining the efficient coverage of conventional AEM systems. The mode of operation of this equipment as shown in Figure 4.4 is entirely analogous to the conventional ground Turam system with the all important exception that the receiver coils, which can be set to measure the gradient of either the horizontal or the vertical EM field, are traversed across the survey area by helicopter on a short (3-10 m) rigid boom. The system operates in the audio frequency range.

The introduction of the first airborne VLF-EM system by Barringer in 1967 (Barringer, 1970) was closely followed by the development of a number of other airborne VLF-EM systems during the past decade. All the recent systems (Hood, 1974) resemble the AFMAG method (Ward, 1970) in that use is made of a remote VLF transmitter and conductors are detected by the presence of an anomalous vertical component of the secondary field. The Geonics and Scintrex VLF-EM systems measure directly the inphase and quadrature components of this quantity, with respect to the total horizontal EM field. The McPhar VLF-EM system, however, measures the tilt angle of the total field and the amplitude of the horizontal EM component.

Finally, it is appropriate to mention the trend in the use of integrated airborne geophysical survey systems. Although magnetic field data were always obtained simultaneously with EM data in airborne surveys, it is only recently that multisensor systems have begun to be fully utilized. An example of such an integrated aerogeophysical survey system is shown in Figure 4.5. The aircraft, a TU 206D Cessna, is equipped with a McPhar two-frequency F-400 EM system, a VLF-EM system, a magnetometer, and a gamma-ray spectrometer. For efficient processing, the large amounts of data generated by such an installation must be digitally recorded. Figure 4.6 shows data obtained with the Scintrex Tridem installation to illustrate this point. As our concept of base metal exploration evolves from one of anomaly hunting to that of geological mapping, installations such as these will become more and more common.

AEM INTERPRETATION

The electromagnetic scale model (Grant and West, 1965) still remains the most practical source of interpretation data for all AEM systems. As a result of a major research program at the geophysical laboratory of the University of Toronto (Ghosh and West, 1971), sufficient EM data were collected to establish the system response to a thin sheet model conductor for most Canadian airborne EM systems under a variety of flight conditions. While the degree to which the thin sheet model simulates a typical Canadian Shield massive sulphide deposit varies from case to case (Ghosh, 1972), it appears that this type of model is satisfactory for interpreting a fair percentage of the airborne EM anomalies encountered in practice. As shown by Ward (1970), individual scale model EM results may be summarized in the form of master charts which are used to translate the anomaly parameters (e.g. amplitude and phase) into conductor parameters (i.e. thickness-conductivity product and depth of burial).

Model studies are also extremely valuable as they permit the determination of the effects of special conditions such as the superposition of conductive overburden over the target conductor. This particular effect is illustrated in Figure 4.7 which shows the change in response of a vertical sheet conductor to a standard helicopter EM system as a function of overburden conductance. In this case, providing the overburden is not overly thick, its effect is mainly observed as a phase rotation so that the detected conductor appears to be of a better quality and at a somewhat greater depth than it actually is. At the cost of considerable additional effort, scale model EM experiments can also provide information on the effects of conductive host rock. Finally, with the same tool one can also investigate the effects of deviations of conductor geometry from the ideal "infinite" thin sheet model in current use.

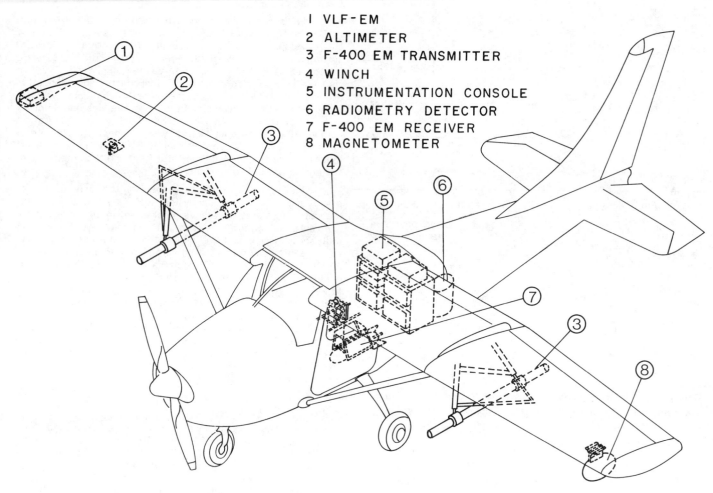

Figure 4.5. Integrated airborne geophysical survey system (courtesy of McPhar Geophysics).

In recent years, however, improved multifrequency AEM data and/or data obtained with two different AEM systems over a given conductor have revealed that the classical thin sheet model anomalies can differ appreciably from those actually observed in the field. Consider two particular cases which illustrate this fact. The first concerns the Sturgeon Lake orebody in Ontario and was first reported by Ghosh (1972). As shown in Figure 4.8, the anomaly recorded by a small scale helicopter EM system indicates a high quality conductor while the anomaly obtained with a large scale two-frequency quadrature AEM system indicates a medium quality conductor. The other case relates to the anomaly recorded by a fixed-wing, three-frequency Tridem AEM system over the New Insco deposit in the Noranda area of Quebec. When the AEM data (Fig. 4.9) is plotted on the appropriate interpretation chart (Fig. 4.10), it becomes evident that the apparent quality and depth of the conductor is roughly inversely proportional to the frequency of measurement as both quantities increase with decreasing frequency. Palacky (1978) has reported similar effects over a number of Canadian ore deposits only to confirm the occasional inappropriateness of the thin sheet model.

Although scale model EM experiments provide the basis for a qualitative understanding of the differences between the observed AEM field data and the laboratory infinite thin sheet model, it is unlikely that sufficient scale model data will ever be generated for the proper analysis of observable field anomalies. It is more probable that development of airborne EM data interpretation in future years will evolve through computer modelling. Computer programs for modelling two-dimensional VLF EM anomalies are already available and in reasonably common use (Vozoff, 1971; Telford et al., 1977). In addition, complete theoretical model VLF EM data sets have been generated and are available to the user (Vozoff and Madden, 1971). Computer anomaly modelling for a dipole source such as is used in AEM exploration, however, is still far from perfected. While recent efforts in this direction (e.g. Meyer, 1976) show much promise, the available programs lack accuracy and, above all, are very costly to execute (Anonymous, 1978).

AEM SYSTEM EVALUATION

An evaluation of AEM systems has been carried out by N.R. Paterson (Paterson, 1971) who found that the overall quality of performance for a given AEM system as an anomaly detector could be determined as a function of six major factors, namely: penetration, sensitivity, discrimination, resolution, aperture, and lateral coverage. Space does not allow a full discussion of all of these factors, but at least two factors, penetration and sensitivity, should be considered. Both are defined with reference to the signal-to-noise ratio variation with depth of burial in free space of a high quality conductor. Penetration is the depth at which this ratio is

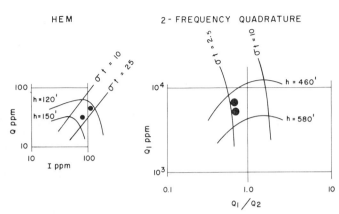

Figure 4.8. Variation of "apparent" conductor quality with AEM system scale; Sturgeon Lake orebody, Ontario (after Ghosh, 1972).

four, while sensitivity is defined as the actual value of this ratio for a conductor at surface. These quantities are graphically illustrated in Figure 4.11 which was compiled from scale model EM data. It should be noted that the effective penetration of an AEM system is not an absolute quantity and should be checked during the progress of a survey to establish its effectiveness. This is so because the noise of the AEM system can vary with survey conditions, especially if geological noise is taken into consideration. From a practical point of view, one can get a better appreciation between these two system attributes by comparing actual field data flown with different AEM systems over the same conductor. For instance, Figure 4.12 contrasts data obtained with two different McPhar AEM systems over the Cavendish test range in southern Ontario. The data clearly indicate the superior sensitivity of the KEM apparatus over the F-400 AEM system which is known to have better penetration.

APPLICATIONS

Today, as in the past, the principal application of AEM methods is to discover discrete conductive mineralized zones in the underlying bedrock. The geographical arena for the proper deployment of AEM systems has, however, been considerably enlarged as a result of much improved equipment. Thus AEM surveys in areas of deep overburden or in areas of moderately deep but quite conductive overburden, which were previously considered unsuitable for AEM, are now being successfully carried out. In view of this, the industry's remarkably consistent discovery rate of about two orebodies per year is not unduly surprising.

In addition to the expanding role of AEM in the direct discovery of conductive deposits, the last decade witnessed the firm establishment of the practice of airborne electromagnetic mapping. This practice can be divided into two distinct processes. The first involves the delineation of geological features by detecting the conductive material which they contain. An example of this type of application of the AFMAG system was given by Sutherland (1970). Similar information can be obtained with VLF-EM systems such as the Barringer Radiophase, for which typical data from the Noranda area of Quebec is shown in Figure 4.13. Finally, good use can also be made of active AEM systems for this purpose. As an illustration of such an application, Figure 4.14 indicates the association of Input EM anomalies with a fault system in the Lake Wanipigow area of southeastern Manitoba because of the presence of conductive serpentinite (Dyck et al., 1975).

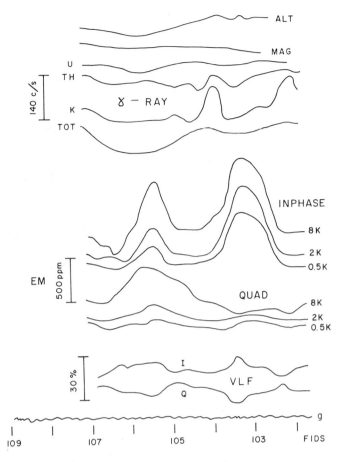

Figure 4.6. Typical data output for an integrated airborne geophysical mapping system (courtesy of Scintrex Ltd.).

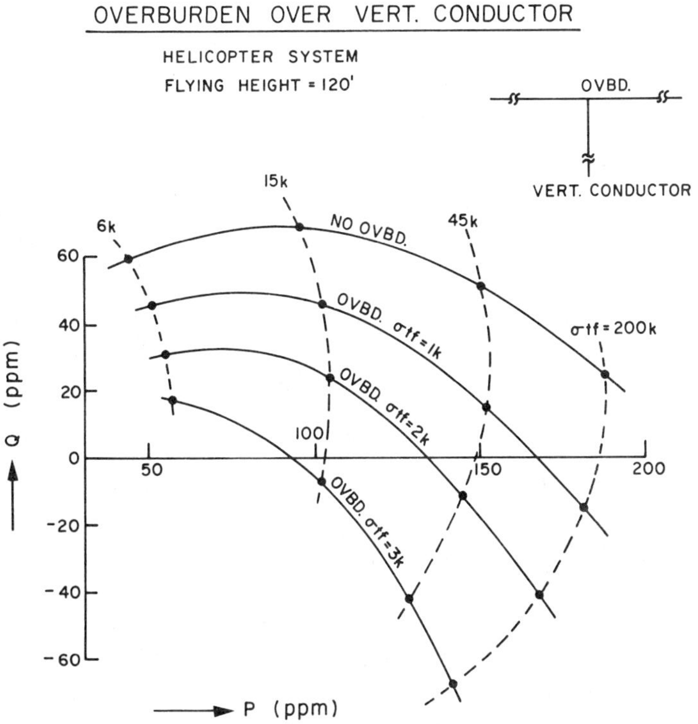

Figure 4.7. Change in response of a vertical sheet conductor to a standard helicopter coaxial EM system as a function of overburden conductance; P is inphase response and Q is quadrature response (after Ghosh, 1972).

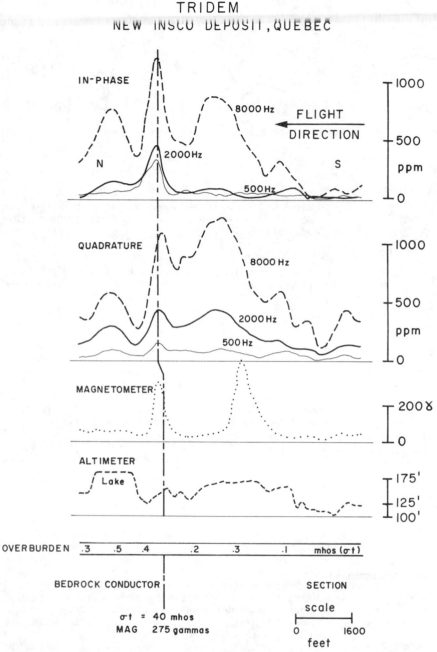

Figure 4.9. Three-frequency Tridem EM data over New Insco Deposit, Quebec (courtesy of Scintrex Ltd.).

The second AEM mapping application relates to the establishment of overburden parameters from the continuously recorded EM data. As pointed out by Seigel and Pitcher (1978) and Fraser (1978b) this procedure can be very useful for

i) Definition of bedrock conductors in mineral exploration surveys;

ii) Location of nonmetallic mineral deposits (e.g. lignite and kimberlite);

iii) Mapping and differentiation of surficial materials for civil engineering purposes.

In the first case, quantitative estimates of overburden parameters facilitate the interpreter's task of choosing between surficial conductivity anomalies and those likely to be caused by bedrock conductors. In the last case, the derived overburden parameters can be used in the selection of road sites and pipeline routes.

The basic techniques utilized in the automatic interpretation of AEM data to calculate overburden parameters involve the digital recording of the field data, the definition of suitable field data parameters and the calculation of ground parameters for a particular assumed model. The last step is done most efficiently through the use of table look-up routines on discrete data sets which are precalculated and stored in the computer. This was the approach taken by Dyck et al. (1974), Becker and Roy (unpubl. rep.) and most recently by Fraser (1978b) and Seigel and Pitcher (1978).

The EM interpretation procedures summarized above can be best illustrated with reference to some actual field data. The examples will be taken from surveys sponsored by the Geological Survey of Canada which were made with the Input (Dyck et al., 1974) and the Tridem (Seigel and Pitcher, 1978) AEM systems. Both surveys were carried out in the Hawkesbury area near Ottawa, Ontario where geological knowledge allowed the assumption of a single layer of conductive overburden (mainly clay) over a resistive half-space composed of limestone.

Typical 11-channel, vertical-axis receiver, Input EM data obtained over one line in the Hawkesbury area is shown in Figure 4.15. At each point along the survey line the observed data are summarized by defining the average amplitude and decay rate of the observed secondary field transient. These two parameters are then entered into an interpretation chart of the type shown in Figure 4.16 that is stored in a computer memory. The output is a reliable estimate of the conductivity-thickness product of the overburden and, in cases where the clay is exposed at surface, reliable estimates of the upper layer resistivity. Typical output for the data of Figure 4.15 is shown in Figure 4.17, where the airborne EM data are compared with DC resistivity data obtained independently. The DC resistivity spreads were spaced at about 2 km intervals. Thus, the airborne EM data, which are continuous, show much greater detail. As the overburden mapped in this area consists mainly of material whose resistivity was found to be about 3 ohm metres, the thickness of overburden here can, in places, extend to depths of the order of 75 m.

A similar procedure can be used to interpret overburden properties from data obtained with the Tridem system (Becker and Roy, 1977). Once again, Figure 4.18 shows a typical profile obtained along line 208 in the Hawkesbury area. Here the original digitally-recorded AEM data was replotted to show the variation of secondary field amplitude and phase at each of the three operating frequencies which are 0.5, 2 and 8 kHz. Figure 4.19 shows the type of chart used for automatic interpretation of Tridem data. In this

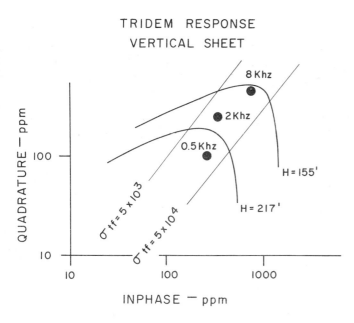

Figure 4.10. Interpretation of three-frequency Tridem data.

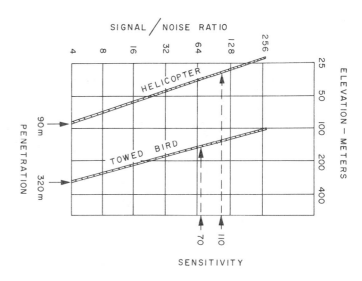

Figure 4.11. Signal-to-noise ratio as a function of elevation for helicopter-borne and fixed-wing towed-bird AEM systems (after Paterson, 1971).

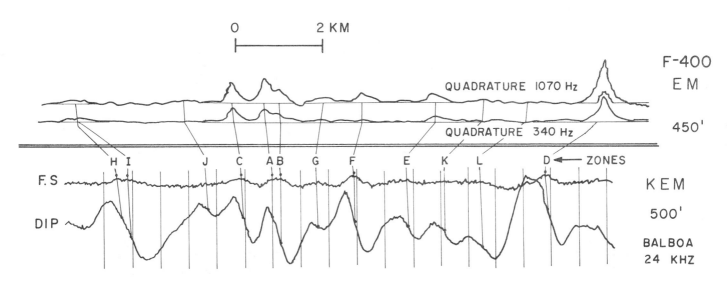

Figure 4.12. Comparative data for VLF-EM and F-400 AEM systems, Line D, Cavendish Test Range, Ontario (courtesy of McPhar Geophysics).

case, the thickness of the conductive layer is determined by the difference in phase angle of the secondary field between the extreme frequencies. The quality of the conductor is defined by the phase angle at the central frequency and its depth of burial by the amplitude data. The results of such an interpretation are shown in Figure 4.20, which is an automatic plot of the overburden conductance along Line 208 in the Hawkesbury survey where it is less than about 10 mhos, and a plot of the ground conductivity as well as the thickness of the resistive drift cover (if any) where the overburden appears as a homogeneous half-space. By compiling a series of such profiles, one can construct a terrane map for the Hawkesbury survey area as shown in Figure 4.21. Here the shaded area, which is bounded by the conductivity contour of 0.03 mhos/m, corresponds to resistive sands, tills, and limestone outcrop. The hatched area bounded by the 10 mho conductance contour corresponds to a thick (30 m) layer of conductive clay. The unshaded zone represents the area of transition between these two conductivity values. It is interesting to note that similar results for the Tridem interpretation were obtained by Seigel and Pitcher (1978) using a technique which differed in many details from the one described above.

FUTURE DEVELOPMENTS IN AEM METHODS

Future research in AEM technology will aim at developing systems that will more readily discover mineralized zones under the thick conductive cover found in many tropical areas. As the quest for increased efficiency in difficult survey conditions is pursued, AEM instrumentation will become more sophisticated and complex. One example of a step in this direction is the unique single-coil EM system invented by H.F. Morrison and W. Dolan which is now under

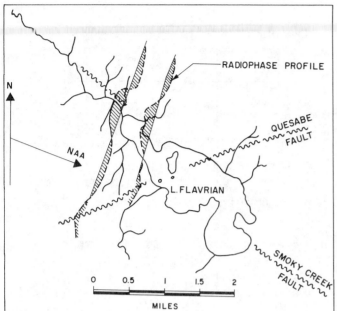

Figure 4.13. Example of Radiophase AEM data; Noranda area, Quebec (courtesy of Quebec Department of Natural Resources).

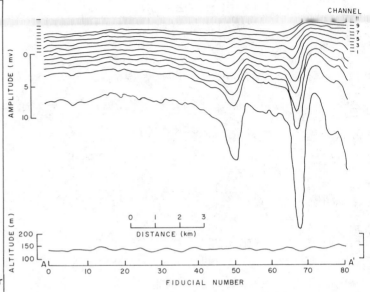

Figure 4.15. Typical 11-channel, vertical-axis receiver Input EM response over a conductive overburden layer; Hawkesbury area, eastern southeastern Ontario (after Dyck et al., 1974).

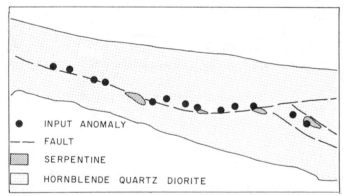

Figure 4.14. Structural mapping with the Input EM system; Lake Wanipigow area, southeastern Manitoba (after Dyck et al., 1975).

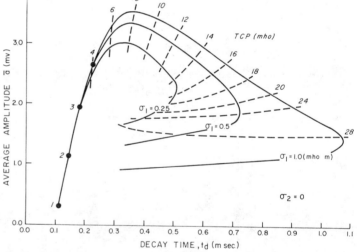

Figure 4.16. Overburden parameter interpretation chart (after Dyck et al., 1974).

construction at the Berkeley campus of the University of California (Morrison et al., 1976). This system (Fig. 4.22) detects anomalies by measuring minute changes in resistance of the single transmitter/receiver immersed in liquid helium in an especially constructed cryostat so that the coil winding is superconducting. The principal advantage of this type of apparatus is its ability to operate at extremely low (40 Hz) frequencies and thus to detect deep conductive zones covered by a highly conductive overburden. Another ongoing endeavour is the development of the Cotran AEM system by Barringer Research Ltd. (Barringer, 1976). Although the present concept for Cotran does not involve any departure from the conventional fixed-wing transmitter-receiver embodiment of an AEM system, the design differs radically from standard practice as the detector electronics are replaced by a microcomputer which continuously correlates the received time-domain EM signal with a prestored set of expected secondary field transients. It is probable that such sophisticated signal detection will result in improved system performance.

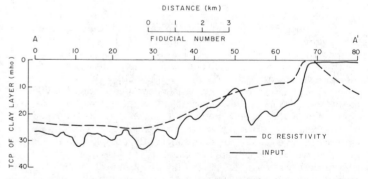

Figure 4.17. Comparison of interpreted thickness — conductivity products for surficial layer obtained by Input EM and DC resistivity surveys; Hawkesbury area, southeastern Ontario (after Dyck et al., 1974).

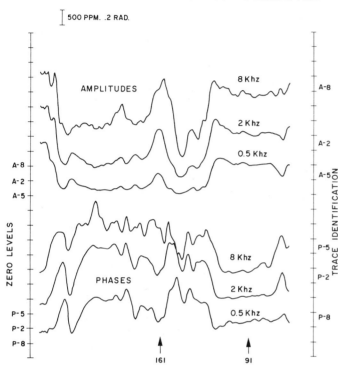

Figure 4.18. Typical Tridem response over a conductive overburden layer, Line 208, Hawkesbury area, southeastern Ontario.

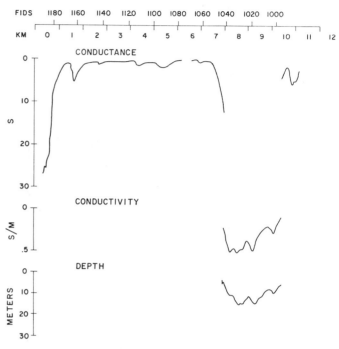

Figure 4.20. Computer-derived interpretation of overburden parameters; from Tridem data; Line 208, Hawkesbury area, southeastern Ontario (after Becker and Roy, 1977).

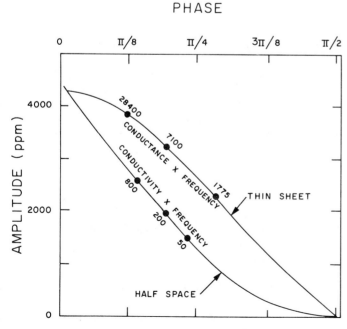

Figure 4.19. Interpretation chart for the calculation of overburden parameters from Tridem AEM data (after Becker and Roy, 1977).

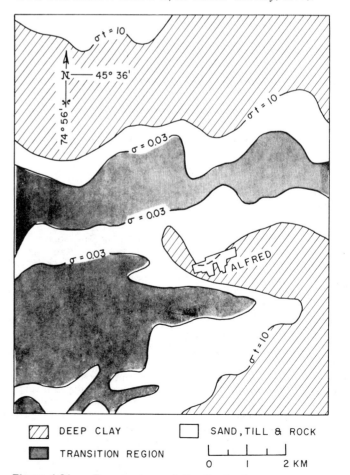

Figure 4.21. Terrane map of the Hawkesbury survey area, southeastern Ontario (after Becker and Roy, 1977).

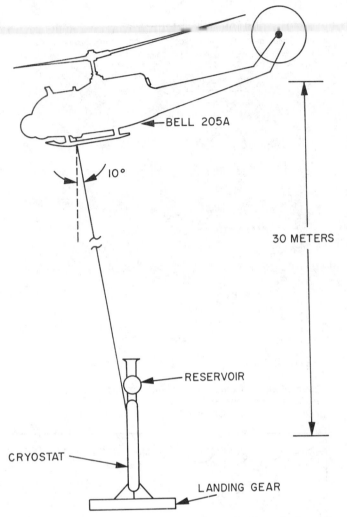

Figure 4.22. Synoptic representation of the single-coil cryogenic EM system.

Finally, it now appears, as a result of work in a number of laboratories, that electromagnetic systems can interact with nonconductive but magnetic minerals (Olhoeft and Strangway, 1974). Although the influence of magnetic susceptibility on the inphase component of the secondary field has been clearly demonstrated (Fraser, 1972b), it has only recently become feasible to measure the absorption of electromagnetic energy by magnetic minerals by detecting the subsequent influence of this process on the quadrature component of the secondary EM field. It is expected that as airborne EM measurements are made at increasingly lower frequencies, the magnetic interaction effects will allow some degree of differentiation between different minerals associated with the observed anomalies.

CONCLUSION

Because of time and space limitations only the highlights made in the technology and practice of airborne electromagnetic surveying during the past decade have been mentioned in this review. As the technical quality and scope of the AEM instrumentation improved, new methods of interpretation were developed so as to take full advantage of the increased amounts of data generated. It is expected that this trend to improve AEM interpretation techniques will continue and intensify in the future. Thus, it is conceivable that the final result of future airborne electromagnetic surveys will no longer be an "anomaly map", but rather a detailed geological section of the subsurface formations.

ACKNOWLEDGMENT

I am much indebted to the many individuals, companies, and organizations who have supplied me with the data that have made this review possible. Part of the work reported was done under National Research Council Grant A7472.

REFERENCES

Anonymous
   1977: HBED's unique EM-30 system; Can. Min. J., v. 98 (8), p. 41.
   1978: Workshop on modelling of electrical and electromagnetic methods, May 1978, Earth Science Division, Lawrence Berkeley Laboratory, University of California, 203 p.

Barringer, A.R.
   1970: Regional reconnaissance techniques applied to exploration; in Mining and groundwater geophysics, L.W. Morley, ed., Geol. Surv. Can., Econ. Geol. Rept. no. 26, p. 202-210.
   1976: Geophysical prospecting method utilizing correlation of received waveforms with stored reference waveforms, U.S. Patent 3,950,695.

Becker, A. and Roy, J.
   1979: Descriptive interpretation of Tridem data relating to the Hawkesbury area, Ontario by Scintrex Limited; Geol. Surv. Can., Open File 605, Pt. 2.

Bosschart, R.A. and Seigel, H.O.
   1974: The Tridem three-frequency airborne electromagnetic system; Can. Min. J., v. 95, April, p. 68-69.

Dyck, A.V., Becker, A., and Collett, L.S.
   1974: Surficial conductivity mapping with the airborne Input system; Can. Min. Metall. Bull., v. 67 (744), p. 104-109.
   1975: Input AEM results from Project Pioneer, Manitoba; Can. J. Earth Sci., v. 12 (6), p. 971-981.

Fraser, D.C.
   1972a: New multicoil aerial electromagnetic prospecting system; Geophysics, v. 37 (3), p. 518-537.
   1972b: Magnetite ore tonnage estimates from an aerial electromagnetic survey; Geoexploration, v. 11, p. 97-105.
   1978a: Geophysics of the Montcalm Township copper-nickel discovery; Can. Min. Metall. Bull., v. 71 (789), p. 99-104.
   1978b: Resistivity mapping with an airborne multicoil electromagnetic system; Geophysics, v. 43, no. 1, p. 144-172.

Ghosh, M.K.
   1972: Interpretation of airborne EM measurements based on thin sheet models; unpubl. Ph.D. thesis, University of Toronto, 195 p.

Ghosh, M.K. and West, G.F.
   1971: AEM analogue model studies; Norman Paterson & Assoc. Ltd., Toronto.

Grant, F.S. and West, G.W.
   1965: Interpretation theory in Applied Geophysics; McGraw-Hill, New York.

Hood, P.
   1974: Mineral exploration trends and developments in 1973; Can. Min. J., v. 95 (2), p. 163-214.
   1978: Mineral exploration trends and developments in 1977; Can. Min. J., v. 99 (1), p. 163-214.

Lazenby, P.G.
- 1973: New developments in the Input airborne EM system; Can. Min. Metall. Bull., v. 66, p. 96-104.

Meyer, W.H.
- 1976: Computer modelling of electromagnetic prospecting methods; unpubl. Ph.D. thesis, University of California, Berkeley, 207 p.

Morrison, H.F., Dolan, W., and Dey, A.
- 1976: Earth conductivity determinations employing a single superconducting coil; Geophysics, v. 41 (6A), p. 1184-1206.

Olhoeft, G. and Strangway, D.W.
- 1974: Magnetic relaxation and the electromagnetic response parameter; Geophysics, v. 39 (3), p. 302-311.

Palacky, G.J.
- 1978: Selection of a suitable model for quantitative interpretation of towed-bird AEM measurements; Geophysics, v. 43 (3), p. 576-587.

Paterson, N.R.
- 1971: Airborne electromagnetic methods as applied to the search for sulphide deposits; Can. Min. Metall. Bull., v. 64, p. 29-38.

Seiberl, W.A.
- 1975: The F-400 series quadrature component airborne electromagnetic system; Geoexploration, v. 13, p. 99-115.

Seigal, H.O. and Pitcher, D.H.
- 1978: Mapping earth conductivities using a multi-frequency airborne electromagnetic system; Geophysics, v. 43, no. 2, p. 563-575.

Sutherland, D.B.
- 1970: AFMAG for electromagnetic mapping; in Mining and Groundwater Geophysics/1967; L.W. Morley, ed., Geol. Surv. Can., Econ. Geol., Rept. no. 26, p. 228-237.

Telford, W.M., King, W.F., and Becker, A.
- 1977: VLF mapping of geological structure; Geol. Surv. Can., Paper 76-25, 13 p.

Vozoff, K.
- 1971: The effect of overburden on vertical component anomalies in AFMAG and VLF exploration; Geophysics, v. 36 (1), p. 53-57.

Vozoff, K. and Madden, T.R.
- 1971: VLF Model suite; Private edition, Lexington, Mass., 2000 p.

Ward, S.H.
- 1970: Airborne electromagnetic methods; in Mining and Groundwater Geophysics/1967; L.W. Morley, ed., Geol. Surv. Can., Econ. Geol. Rept. no. 26, p. 81-108.

GROUND ELECTROMAGNETIC METHODS AND BASE METALS

Stanley H. Ward
Department of Geology and Geophysics, University of Utah, Utah, U.S.A.

Ward, Stanley H., Ground electromagnetic methods and base metals; in Geophysics and Geochemistry in the Search for Metallic Ores; Peter J. Hood, editor; Geological Survey of Canada, Economic Geology Report 31, p. 45-62, 1979.

Abstract

The electromagnetic method of prospecting for base metals has been used successfully since the early 1920s. Its application has been marked by spectacular success and by dismal failure. Despite three decades of intensive research on the electromagnetic method, it still suffers from severe limitations of interpretability. These have gradually diminished over the last decade and probably will diminish much more rapidly over the next decade as the method encompasses more frequencies, more transmitter-receiver configurations, hardware of greater flexibility, in-field data processing, plus exceptionally sophisticated methods of data interpretation. We should expect the electromagnetic method to contribute to our knowledge of the three-dimensional distribution of electrical resistivity in the subsurface but we should not expect it to offer means for discriminating between specific minerals. Accordingly, we are concentrating on means for separating the responses of the various elements in the geoelectric section and on determining the geometries of these elements. On the other hand, a combination of electromagnetic and induced polarization methods could be mineral specific. The past, the present, and the future of the ground electromagnetic method are all discussed within this framework. This contribution is concerned only with active *ground electromagnetic systems. The outlook is optimistic.*

Résumé

La prospection par méthode électromagnétique en ce qui a trait aux métaux non précieux est utilisée avec succès depuis le début des années 1920. Son usage a été marqué de succès spectaculaires et d'échecs lamentables. Malgré trente années de recherches poussées sur la méthode électromagnétique, cette dernière est encore affligée de sérieuses limites de décodage. Ces limites ont diminué graduellement au cours de la dernière décennie et elles continueront probablement à décroître encore plus rapidement au cours des dix prochaines années à mesure que la méthode comporte des fréquences plus nombreuses, davantage de dispositions d'émission — réception, du matériel d'une plus grande souplesse d'emploi, un traitement sur place des données, de même que des méthodes exceptionnellement perfectionnées de décodage des données. Nous devrions nous attendre à ce que la méthode électromagnétique contribue à nous faire connaître la répartition tridimensionnelle de la résistance électrique dans le sous-sol, et non pas à ce qu'elle offre des moyens de distinguer des minéraux précis. En conséquence, nous portons notre attention sur différents moyens de séparer les réponses des divers éléments de la section géoélectrique et d'établir la géométrie de ces éléments. De plus, une utilisation combinée des méthodes électromagnétiques et de polarisation induite pourrait se prêter à la distinction des minéraux. Le passé, le présent et l'avenir de la méthode électromagnétique terrestre sont tous analysés à l'intérieur de ce cadre. L'étude s'intéresse ici seulement aux systèmes électromagnétiques terrestres "actifs". La perspective est optimiste.

INTRODUCTION

In applying the active ground electromagnetic method to the search for massive sulphide deposits of base metals, one can only expect to detect the ore if its signature is distinctive from those of the halo of disseminated non-economic sulphide mineralization, the weathered and fresh host rock, adjacent faults, shears, and graphitic structures, overburden and surface and buried topography. Figure 5.1 depicts the sources of signatures in a typical exploration problem.

In the last decade we have learned that the separation of anomaly (signature of ore deposit) from geological noise (combined signatures of disseminated halo, host rock, faults and shears, graphitic structures, overburden, and topography) is not easy. To attempt this separation we utilize a broad band of frequencies, we employ several transmitter-receiver separations and we employ several transmitter-receiver configurations.

If the various elements of the earth, depicted in Figure 5.1, responded independently of one another, the problem of separation of anomaly from geological noise would be much simplified. Unfortunately the elements react with one another to complicate the problem. For example, any source of electromagnetic waves will induce currents in the host rock. When a massive sulphide deposit is emplaced in this host, it causes the currents to be deflected into it on account of its lower resistivity. Current gathering by an inhomogeneity thereby occurs. Current gathering usually increases the anomaly due to the massive sulphide but it also modifies its response to make it appear deeper and of higher resistivity. Quantitative interpretation of electromagnetic data is, therefore, usually seriously hampered. What can be done about this and related problems? In this paper an attempt is made to expose both the optimum approach and the current practical approach to these problems. In the process I hope to indicate current and future advantages and limitations of the ground electromagnetic method.

Although not an objective in the past, the principal current objective of the electromagnetic method is to develop an ability to detect each element of the geoelectric section so that the resistivity environment surrounding the assumed ore may be assessed. In this fashion, for example, the massive sulphide ore can be distinguished from disseminated noneconomic mineralization in a volcanogenic environment.

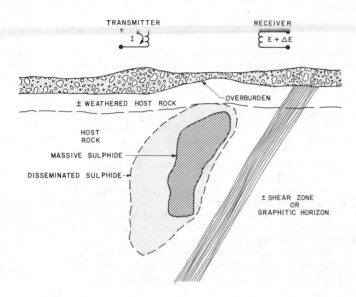

Figure 5.1. The elements, of a generalized three-dimensional earth, which contribute to the signature obtained with an electromagnetic prospecting system.

Recent papers which address the problems encountered by the electromagnetic method when faced with a real earth in which overburden, host rock, surface topography, buried topography, a disseminated halo around a massive sulphide body, faults, shears, and carbonaceous or graphitic structures all might be included are: Lowrie and West (1965), Roy (1970), Sarma and Maru (1971), Gaur, Verma, and Gupta (1972), Gaur and Verma (1973), Scott and Fraser (1973), Ward et al. (1974a), Ward et al. (1974b), Hohmann (1975), Lamontagne (1975), Palacky (1975), Verma and Gaur (1975), Spies (1976), Lajoie and West (1976), Lajoie and West (1977), Lodha (1977), Ward et al. (1977), and Braham et al. (1978). We refer the reader to them for details but summarize their conclusions in Table 5.1.

DATA ACQUISITION

Domain of Acquisition

Electromagnetic data are always collected as a time series describing an electromagnetic field at a point P and a time t as in Figure 5.2. The resulting data may be processed and interpreted in the frequency domain (fD) or in the time domain (tD). In (fD) the spectrum of the waveform is viewed through some frequency window (passband) as in Figure 5.2a while in (tD), the transient decay of an impulsive waveform is viewed through some time window (passband) as in Figure 5.2b. Observations at discrete frequencies or at discrete times are most commonly made. There is no

Table 5.1

Summary of effects of extraneous features in electromagnetic search for massive sulphides

Feature	Effect	Interpretation problem
Overburden	rotates phase	depth estimates invalid
	decreases amplitude	σt estimates invalid
Host rock	rotates phase	depth estimates invalid
	increases amplitude for shallow conductors	σt estimates invalid
		dip estimates invalid
	increases or decreases amplitude for deep conductors	
	changes shape of profiles	
	fall-off laws change	
Surface and buried topography	introduces geological noise	depth estimates invalid
		σt estimates invalid
		dip estimates invalid
		may obscure sulphide anomalies
Halo	rotates phase	depth estimates invalid
	increases amplitude	dip estimates invalid
		σt estimates invalid
Weathered host rock	introduces geological noise	obscures sulphide anomalies
		may invalidate all quantitative interpretation
Faults, shears, graphitic structures	introduces geological noise	obscures sulphide anomalies
		may invalidate all quantitative interpretation

NOTE: σt — conductivity-thickness product of massive sulphide body.

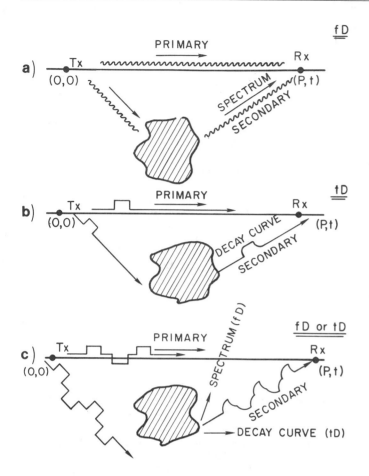

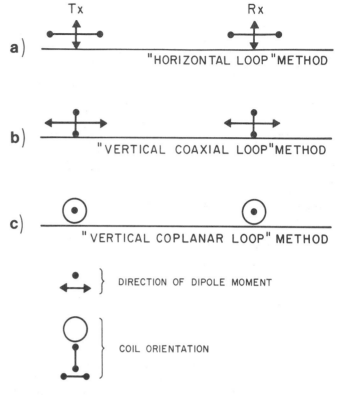

Figure 5.3. Coil configurations employed in (a) horizontal loop, (b) vertical coaxial loop, and (c) vertical coplanar loop methods of electromagnetic prospecting.

Figure 5.2. Schematic of (a) frequency domain, (b) time domain, and (c) combined time domain/frequency domain processing of a signature from a base metal deposit by an electromagnetic system.

particular reason why one transmitted waveform cannot be processed and interpreted in the frequency domain, in the time domain, or in both domains simultaneously as in Figure 5.2c.

Attempts to design optimum waveforms have been made in recent years. Thus, the UTEM system (Lamontagne and West, 1973) was designed to transmit a triangular waveform and receive its derivative, a square waveform. Pseudo random noise generators (Quincy et al., 1976) and sweep-frequency generators have also been proposed. Enhancement of signal-to-noise ratio is the reason for selecting these latter waveforms.

Regardless of the method of data processing and interpretation, the signal sensed at the receiver (R) is a superposition of a primary field (including the effects of a homogeneous earth) and a secondary field (reflected from a subsurface inhomogeneity).

Coil Configurations

Roving Coil Pairs — Fixed Orientations

The transmitting coil (T) may be oriented with its axis vertical (called a vertical magnetic dipole or a horizontal loop) or with its axis horizontal (called a horizontal magnetic dipole or a vertical loop). If a horizontal magnetic dipole is used its axis may be pointed at the receiver (R) or be orthogonal to this direction. Three orthogonal transmitting coil orientations are thus possible. Similarly, three orthogonal receiving coil orientations are possible.

In the so-called horizontal loop EM method both the receiving and transmitting coils have their axes vertical and the coil pair is transported, with constant coil separation, in-line in a direction normal to strike as in Figure 5.3a. Commercially available equipment which utilizes this configuration is listed in Table 5.2 (after Hood, 1977).

In the vertical coaxial loop EM method (Fig. 5.3b) both the receiving and transmitting coils have their axes horizontal, in-line, and the pair is transported, with constant coil separation, a) in-line in a direction normal to strike, or b) broadside in a direction normal to strike. Any one of the coil systems listed in Table 5.2 could be used in this configuration but this is not routinely done.

In the vertical coplanar loop EM method, the receiving and transmitting coils have their planes vertical and common and the pair is transported in-line with constant coil separation, in a direction 45 degrees to strike as in Figure 5.3c. This latter configuration is not used commonly but any one of the systems of Table 5.2 could be used this way. The need to traverse at 45 degrees to strike in order to obtain significant response usually mitigates against use of this configuration.

Roving Coil Pairs — Rotatable Orientations

In the Crone shootback method (Crone, 1966), the receiving and transmitting coils are interchangeable in the sense that each is used both as a receiver and as a transmitter. The remarkable advantage of the method is that the effects of elevation differences between transmitter and receiver are eliminated. Two variations of it, the Crone horizontal and vertical shootback methods are used (Crone, 1966, 1973). With each, the orientation of the transmitting coil is fixed with its axis at some angle to the horizontal or to

Table 5.2
Commercially available horizontal loop ground electromagnetic equipment

Manufacturer	Model designation	Frequency of operation in Hertz	Coil separation in metres	Dipole moment in amp turns sq. mtr (in m²)	Component measured I/P--in Phase O/P--Out of Phase	Readout device	Range of readings	Read-ability	Weight (kg)	Power source
ABEM (Sweden)	Demigun	880 & 2640	30,60,90, 150&180	50/880Hz 20/2640Hz	I/P & O/P	2 Dials	0-160%I/P ±80%O/P	±0.5%	23.2	D or Nicad cells
Apex Parametrics (C)	Max Min II	222,444, 888,1777 & 3555	30,60,90, 120,180, & 240	150/222 & 444Hz 75/888Hz 50/1777Hz	I/P & O/P	2 Dials +Tiltmeter	±100%I/P ±100%O/P	±0.5%		3 x 6V Cells
Geonics (C)	EM 17	1600	30,60,90 & 120	24	I/P & O/P	Meter (self-indicating)	±100%I/P	±0.5%	12.61	C cells --Rx
	EM17L	817	50,100, 150 & 200	24(reduced) 48(normal)			±50%O/P	±0.25%	13.4	D cells --Tx
McPhar (C)	VHEM	600 & 2400	30,60,90, or 40,80	60/600Hz 18/2400Hz	I/P & O/P	Dial & Headset	±100%I/P ±100%O/P	±0.5%	8.2	9v--Rx D cells --Tx
Scintrex (C)	SE-600	1600	60 or 90	27	I/P & O/P	Dial & Headset	±100%I/P ±50%O/P	±0.5%	15	6 & 13.5v cells

the vertical. The receiving coil, however, is rotated about an axis normal to the traverse line until a minimum signal is obtained at which time the tilt of the plane of the receiving coil, from the horizontal or from the vertical, is recorded. Two readings, taken at each observation stop with first one coil as transmitter then the other, are averaged. Elevational effects in the tilt angle reading are thus eliminated. The reader is referred to the literature for specific operational details. Figure 5.4 illustrates the two variations.

In the vertical loop broadside method, the transmitter is transported along one traverse line while the receiver is moved in unison along an adjacent traverse line. At each point of observation the transmitting coil is placed in that vertical plane which contains the location of the receiver as in Figure 5.5. The receiving coil is then rotated about an axis normal to the traverse line until a minimum signal is obtained, at which time the tilt of the plane of the receiving coil from the horizontal is recorded.

Fixed Transmitter, Roving Receiver

With the rotating vertical loop method the transmitting coil (T) is erected at a fixed position within the survey area (Fig. 5.6a) while the receiving coil (R) is carried systematically on traverse lines adjacent to it and oriented normal to strike. The plane of the vertical loop is oriented for each observation so as to contain the point of observation. Commercially-available equipment which utilizes this configuration is listed in Table 5.3 (after Hood, 1977).

For the coaxial vertical loop method, illustrated in Figure 5.6b, the axis of the transmitting coil is oriented normal to strike and the receiving coil is carried along the line of the axis of the transmitting coil. This is not a standard technique but has been used with success where tried (Ward et al., 1974b; Pridmore, 1978).

A fixed horizontal loop (or coil with its axis vertical) is used where large transmitting coil moments are required. When the loops are tens of metres in diameter or less, the transmitting loop is fixed at the end of a traverse line and the receiver traversed in increments along that line. Then the transmitting coil is moved to an adjacent line and the process repeated. Commercially available systems which use this technique are listed in Table 5.4 (derived from the data of Hood, 1977 and Buselli and O'Neill, 1977).

With the frequency-domain Turam method a large rectangular transmitting coil, hundreds or even thousands of metres to a side, is laid out on the ground and the field strength ratio and phase difference are recorded between a pair of receiving coils 30 m to 100 m apart. With the time-domain Russian MPP01 and the Australian Sirotem system, a single rectangular loop 50 m to 200 m to a side, is used first as a receiver, and at appropriate time delays, then as a receiver. Sirotem also offers the opportunity for use of separate transmitter and receiver loops separated by 100 m to 200 m.

The Crone Pulse EM (PEM) method uses a small, i.e. tens of metres, transmitting coil separated from the receiving coil by 30 m to 100 m. Typically, the transmitter

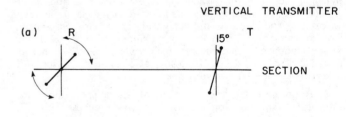

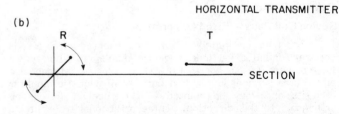

Figure 5.4. *Crone Shootback configurations.*

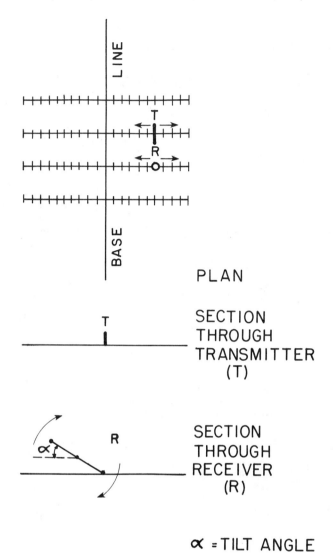

Figure 5.5. *Coil configuration for vertical-loop broadside EM method.*

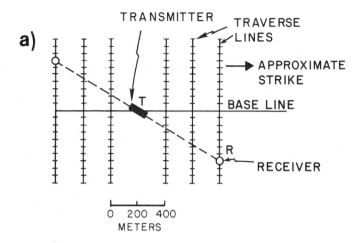

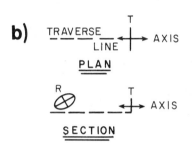

Figure 5.6. *(a) Survey procedure for vertical rotating-loop EM method, (b) orientation of projection of ellipse of polarization on vertical plane passing through axis of transmitter.*

and receiver are moved in unison along a traverse line so that this method could be listed under the section on roving coil pairs.

The Newmont EMP system, not available commercially, uses a large fixed rectangular transmitting loop, hundreds of metres to a side. The received signals are the three orthogonal components of magnetic field recorded at 32 discrete time channels after termination of each transmitted current pulse (Nabighian, 1977).

It can be argued that the larger the loop the larger the potential for exploring to greater depths. The basis for this argument is that in the small dimensional limit the loop is a dipole, whereas in the large dimensional limit the loop becomes four line sources. Fields from a dipole fall off as $1/r^3$ whereas fields from a line source fall off as $1/r$. Coupled with the attractiveness of using larger loops for greater depth of exploration is an opposing factor which is that a loop couples best with a body of its own dimensions and hence only very large targets would be excited optimally by very large loops.

The theoretical computations of Lajoie and West (1976) confirm this analysis in a general way but indicate that the optimum source dimension depends upon the overburden and host rock resistivities as well as upon the dimensions of the inhomogeneity. Figure 5.7a, b, c (after Lajoie and West, 1976) show two models and their respective phasor diagrams. The in-phase (IP) and quadrature (Q) amplitudes are normalized with respect to the vertical component of the primary magnetic field intensity on the surface, directly over the plate. Note that the largest percentage anomaly occurs for a 1000 m to 2000 m loop for model 5 while it occurs from a 250 m x 500 m loop for model 6.

Decades of Frequency Required to Separate Geological Signal from Geological Noise

If the earth responded as a single body, the anomaly to which it gave rise with any electromagnetic system would, as a function of frequency, trend from a low frequency asymptote to a high frequency asymptote in about three decades of frequency. Zone B at the Cavendish Test Site in southern Ontario, seems to respond this way to a vertical axial coil operating over the frequency range 10^2 to 10^5 Hz. Figure 5.8 contains plots of a) the tilt of the major axis of the projection of the ellipse of magnetic field polarization on a vertical plane passing through the axis of the transmitting coil, and b) the ratio of the minor to major axis of this projection of the ellipse of magnetic field polarization. These two quantities are referred to as tilt angle and ellipticity, respectively. Three decades of frequency also encompass the low and high frequency asymptotes for horizontal loop excitation as Figure 5.8 reveals. All curves are both smoothly varying and slowly varying. If the response of a sulphide body was always so simple and predictable, then only two or three frequencies spread over a decade or two, would be needed to learn all there is to know, electrically, about the sulphide body. Indeed, until ten years ago this was the common assumption.

Table 5.3

Commercially available dip-angle ground electromagnetic equipment

Manufacturer	Model designation	Frequency of operation in Hertz	Maximum coil separation in metres	Dipole moment in amp. turns sq. m. & weight (kg)	Transmitter power source	Weight of receiver (kg)	Readability of clinometer	Bandwidth of receiver system	Remarks
Crone Geophysics	CEM (shootback)	390, 1830 & 5010	200	45inm²/390Hz 30inm²/1830Hz 18inm²/5010Hz 10kg with batteries	3 x 6 volt lantern batteries	11	±0.5°	5Hz/390Hz 15Hz/1830Hz 30Hz/5010Hz	Transreceiver Units for Shootback & vertical-loop
	VEM	390 & 1830	700	1100inm²/390Hz 900inm²/1830Hz 20kg with battery	12 volt/ 24 amp hr. Gel battery	6	±0.5°	5Hz/390Hz 15Hz/1830Hz	Hi powered Vertical loop
McPhar	REM	1000 & 5000	200	60inm²/1000Hz 15inm²/5000Hz 4.5kg	Hg cells	2.4	±0.5°	20Hz/1000Hz	
	VHEM	600 & 2400	200	60inm²/600Hz 18inm²/2400Hz 4.1kg	300 cells	3.8	±0.5°	13Hz/600Hz	Horizontal loop also
Scintrex	SE-600	1600	300	27inm² 8 kg	2 x 6 volt cells	5.5	±0.5°	10 Hz?	Horizontal loop also

Table 5.4

Commercially available fixed horizontal loop electromagnetic equipment

Domain	Manufacturer	Model designation	Parameters measured	Frequency of operation in Hertz	Power Output	Power source (mg=motor generator)	Readout device	Weight (kg)	Remarks
Time	Crone Geophysics	Pulse EM (PEM)	8 samples of secondary field 1 sample of ramp voltage	Equivalent 18 to 1060 Hz	Max 450W	2 x 12v 20 amp.hr Gel batt.	Meter	21	Moving Tx-loop system
	Geonics	UTEM	Vert. magnetic & Hor. electric fields	10 time slots -base freq. 7 to 90 Hz		1.5kw Mg	Meter or digital tape deck	60	Large loop
	Geoex.	SIROTEM	Vert. Mag. field	12 to 32 time slots 0.25 ms to 180 ms	Max 176W	22v 10 amp.hr Nicads	Printer Output	20	Single or Double Loop Configuration Moving Loops Choice of Sizes
Frequency	Geoprobe	Maxi-probe EM 16	Vert. & Hor. Mag. field Hor. elec. fields	2^n(n=0 to 9), 1k, 2k, 4.1k, 8.2k, 16.4k, 32.8k, 41k	2100W	6Hp Mg	Meter	116	
	Scintrex	SE-77/ TSQ-2M (Turam)	Field Strength Ratio & Phase diff.	35, 105, 315, 945 & 2835	500W	3 HP Mg	Automatic Meter Display	42	Reads harmonics of transmitted square wave
	Geotronics	EMR-1/ GT20A	32f	10 to 10 000 Hz	20Kw	Mg	Meter	R.10 Kg	Coherent super without carried reference

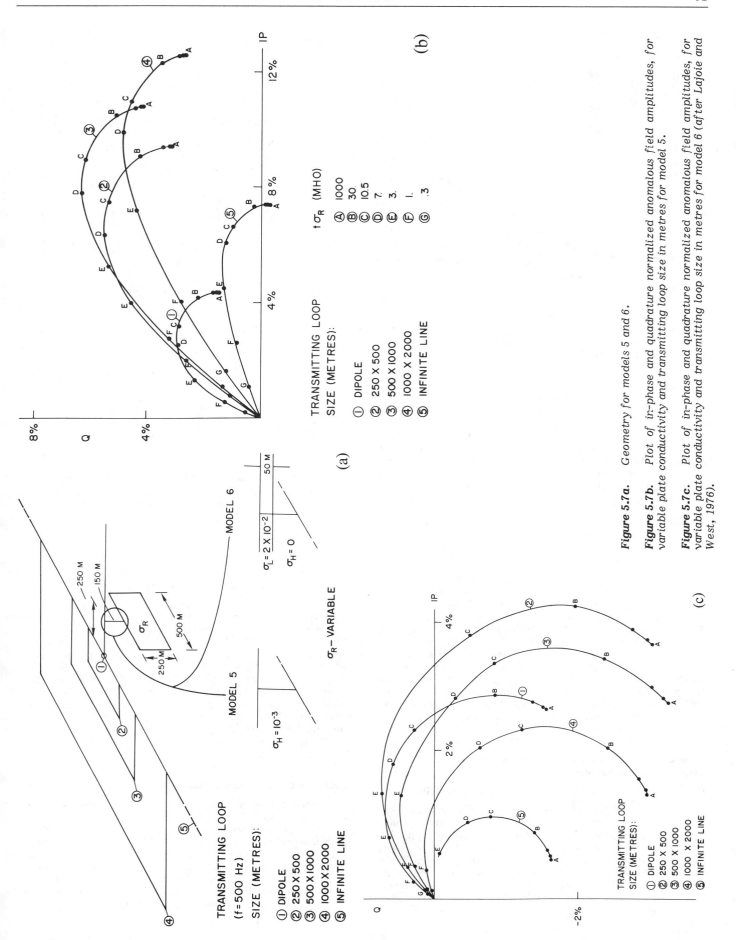

Figure 5.7a. Geometry for models 5 and 6.

Figure 5.7b. Plot of in-phase and quadrature normalized anomalous field amplitudes, for variable plate conductivity and transmitting loop size in metres for model 5.

Figure 5.7c. Plot of in-phase and quadrature normalized anomalous field amplitudes, for variable plate conductivity and transmitting loop size in metres for model 6 (after Lajoie and West, 1976).

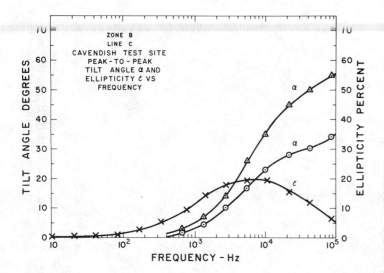

Figure 5.8. Peak-to-peak tilt angle ⊙ and ellipticity X for the vertical axial coil; and peak tilt angle ▲ for the horizontal coil. All are plotted against frequency for Zone B, Cavendish Test Site (after Ward et al., 1974b).

However, the earth can behave in a most unusual manner as a function of frequency when not only the sulphide body, but the host rock, the overburden and the other elements of the general earth also contribute in the frequency window used. This is illustrated in Figure 5.9 in which plots of ellipticity and tilt angle over Zone A of the Cavendish Test Site are contained. With vertical axial coil excitation, the tilt angle trends from asymptote to asymptote over four decades of frequency while the ellipticity at 10^5 Hz is at the high frequency asymptote, but never does reach a low frequency asymptote four decades below this. With horizontal coil and vertical rotating coil excitations the tilt angle at 10Hz is at the low frequency asymptote but the high frequency asymptote is never reached. The latter two curves are not at all smoothly varying or even slowly varying.

It should be evident from the previous illustration that at least four decades and preferably five decades of frequency (1Hz to 10^5Hz) are required to understand the earth. The commercially available systems listed in Tables 5.2 and 5.3 at most use 1+ decades and hence are not suited to use over complex earths. On the other hand, the systems of Table 5.4 use two to four decades of frequency or of spectrum and hence are to be preferred for complex earths. About four frequency or time samples per decade are necessary to assure delineation of rapid changes of response with frequency. The Sirotem system has a capability for about 10 time samples per decade but this high sampling rate may not be necessary.

Combined Resistivity, Induced Polarization, and Electromagnetic Surveys

Conventional and Inductive Measurement of Resistivity

The distribution of true resistivity of a homogeneous, layered, generally inhomogeneous, or layered earth in which local inhomogeneities exist may be estimated from apparent resistivity data acquired during routine surveys with electrode arrays such as the Schlumberger or dipole-dipole arrays. As discussed later, the data may be interpreted in terms of one-, two-, and three-dimensional earth models. References on this subject include Al'pin et al. (1966), Keller

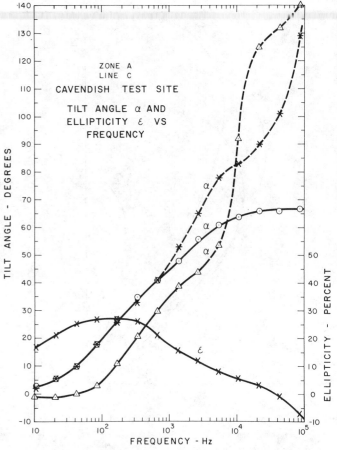

Figure 5.9. Peak-to-peak tilt angle ⊙ and ellipticity X for the vertical axial coil; peak-to-peak tilt angle * for the vertical rotating coil; and peak tilt angle ▲ for the horizontal coil. All are plotted against frequency for Zone A, Cavendish Test Site (after Ward et al., 1974b).

and Frischknecht (1966), Kunetz (1966), Van Nostrand and Cook (1966), Bhattacharyya and Patra (1968), Madden (1971), Roy and Apparao (1971), and Telford et al. (1976).

Apparent resistivity can be measured readily with an inductive electromagnetic system as has been noted by Keller and Frischknecht (1966), Frischknecht (1967), Vanyan (1967), Dey and Ward (1970), Ryu et al. (1970) and others. However, the feasibility of inductive measurement of the distribution of true resistivity in a two- or three-dimensional earth has only recently been recognized (Parry, 1969; Hohmann, 1971; Swift, 1971; Hohmann, 1975; Lamontagne, 1975; Lajoie and West, 1976; Dey and Morrison, 1977; Pelton, 1977; Rijo, 1977; Pridmore, 1978; and others). Pridmore (1978) demonstrates how the resistivity distribution of a homogeneous or layered earth may be recognized despite the presence of one or more inhomogeneities.

Inductive Induced Polarization (IIP)

Bhattacharyya (1964) computed the transient response of a loop antenna placed on a polarizable half-space but drew no inferences concerning the possibility of measuring induced polarization inductively. Dias (1968) indicated that detection of the induced polarization phenomenon was possible via broadband inductive electromagnetic methods. Hohmann et al. (1970) presented evidence that suggested that IIP was not feasible. Ward (1971a) argued that the tests of Hohmann et al. were not conclusive. Lamontagne (1975) presented field experimental evidence that IIP might indeed be possible. Pridmore (1978) has examined the problem both theoretically and by field experiments but is unable to reach a firm conclusion pro or con. As of early 1978 we can only conclude that IIP is theoretically and experimentally possible but we cannot yet assert that its use can be made probable let alone assure that it can be made economically feasible.

Extracting Resistivity, Induced Polarization, and Electromagnetic Data from a Broadband Dipole-Dipole Resistivity Survey

In conventional surveys designed to measure resistivity and induced polarization simultaneously, an unwanted contribution arises from inductive electromagnetic coupling between transmitting and receiving dipoles (see, for example: Sunde, 1949; Millet, 1967; Hohmann, 1973; Zonge and Wynn, 1975; Summer, 1976; Pelton, 1977). Most practitioners of the induced polarization method attempt to remove the inductive electromagnetic contribution from the data base. My colleagues and I, however, note that there is information about the earth in this data base and hence we advocate its use.

Epilogue

There seems to be information about resistivity and induced polarization in broadband electromagnetic surveys while there also appears to be inductive electromagnetic data inherent in broadband resistivity/induced polarization surveys. A merging of previously disparate technologies is now resulting from this awareness. Coherent detection systems utilizing in-field digital processors, combined with sophisticated and expensive computer modelling, may permit exploitation of this merging. Alternatively, successive application of the electromagnetic method and the induced polarization method, regardless of specific array geometries, may permit three-dimensional mapping of resistivity by electromagnetic methods (e.g. Hohmann, 1975) and specific mineral identification by linear or nonlinear induced polarization methods (e.g. Pelton, 1977; Klein and Shuey, 1978).

DATA PROCESSING

Until the last several years, processing of electromagnetic prospecting data has been entirely analog. Zonge (1973), Lamontagne and West (1973), Jain and Morrison (1976), Snyder (1975, 1976), Nabighian (1977), Buselli and O'Neill (1977), Hohmann et al. (1977) have presented descriptions of in-field digital processors which are being or might be applied to ground electromagnetic receiving systems. The in-field microprocessor permits a range of pre-programmed software applications including stacking, filtering, coherent harmonic detection, spectral storage, and spectral weighting. Cassette-recording systems rather than conventional tape decks, as recording media, provide the convenience of in-field or field camp same-day processing and even first-cut interpretation of data using processors such as the Hewlett-Packard 9825A. The costs of such processors and of computers in general will probably decrease rapidly during the next several years.

With these relatively inexpensive luxuries we are now dynamically changing the design of field surveys as the accumulated data warrant. No longer need we send a crew back to the field to obtain a missing data point or data profile or to repeat inadequate or noisy data. Now we know in the field or in the camp whether or not our data is adequate to meet our objectives.

This greater flexibility in the field is gradually leading to simultaneous collection of multichannel data such as n frequencies, m time samples, or r receiver locations. While simultaneous multiple receiver locations are now used commonly with advanced resistivity/induced polarization systems, they have yet to appear with ground electromagnetic systems. They will!

If we are to probe deeper for metallic sulphides via electromagnetic methods we must be able to utilize a broad spectrum, to employ more than one mode of excitation, to enhance the electromagnetic signal relative to natural or artificial electromagnetic noise, and to facilitate in-field processing and preliminary interpretation. Only by using in-field digital processors can we hope to accommodate all of these functions simultaneously.

The basic elements of a microprocessor-controlled electromagnetic receiver might include:

1. a multichannel analog front-end with gain and filtering automatically or manually controlled using the microprocessor,

2. an analog-to-digital converter,

3. an ultra-stable precision interval timer and synchronous clock locked to an identical clock which drives the transmitter,

4. a microcomputer, and

5. a keyboard and display unit.

The transmitter used most commonly with these receivers produces alternate cycles of reversed square wave although programmable waveform processors are available for complete generality of transmitted waveform. The microcomputer can be programmed to process the received signal in the time domain via stacking, in the frequency domain via synchronous demodulation, or both. The important point here is that software algorithms control the particular transmitted waveform and received signal processing scheme and these can be changed at will.

DATA INTERPRETATION

Introduction

If the earth is horizontally plane-layered it is referred to as a one-dimensional (1D) earth. If it exhibits extended strike length, locally, it is referred to as a two-dimensional (2D) earth. Both of these simplistic models are used today. Realistically, however, any base metal environment is three-dimensional (3D) in that any electrical parameter varies significantly over tens or hundreds of metres in all three dimensions. Interpretation proceeds by establishing models of the earth, calculating or finding in a catalogue the electromagnetic signatures of the models, selecting a model whose signature best approximates that observed, and then implying that this selected model is a reasonable representation of the real earth. Further, 3D models employed to interpret field data ideally must include all of the elements of the real earth, i.e. massive sulphide mineralization, halo of disseminated mineralization, fresh host rock, weathered host rock, overburden, faults, shears, graphitic zones plus surface and buried topography as in Figure 5.1. This task of modelling is a formidable one, the absence of which has led to the drilling of overburden anomalies (Fountain, 1972; Scott and Fraser, 1973, for example).

Interpretation of electromagnetic data can be accomplished by either forward or inverse methods. With the forward method, a catalogue of signatures is prepared with which observed signatures are compared by visual inspection or are matched with computed signatures by trial and error. The catalogues may be developed by scaled physical modelling, by analytic solution, or by numerical approximation. With the inverse method, observed signatures are automatically compared with numerically derived signatures via computer; the difference between the two is minimized in a least-squares sense and the ambiguity of interpretation is assessed statistically.

The Forward Method of Interpretation
Scaled Physical Modelling

For years we have used metallic sheets to simulate thin ore veins, metallic spheres to simulate equidimensional ore deposits, or slabs of carbon/graphite to represent tabular base metal deposits in scaled physical models. The variety of geometries of such models is great. Until about ten years ago, we modelled the earth by placing sheets, spheres or slabs in air and totally ignored all of the other elements of the real earth (Ward, 1971b). That we were successful with application of the electromagnetic method when using such crude models of interpretation is surprising. Ward et al. (1974a) reviewed some scaled physical modelling of the early 1970s that clearly indicated the need to take account of host rock and overburden. Many data from model tank measurements are now available in the literature.

The difficulty of physically modelling a real earth with all its complexities has deterred many from pursuing this method of forward modelling. Ward (1967) discussed the problems faced by the scaled physical modeller of that time. These problems led many of us, including me, to emphasize numerical solutions and to use scaled physical modelling only for checks on theoretical solutions. Perhaps it is time, given new materials, new perspectives, and greater imagination, to reconsider the use of scaled physical models not only as stand-alone data sets but also as data sets against which numerical models may be checked. The difficulty, computer limitations, and costs of analytic and numerical forward solutions may force us in this direction.

Analytic Solutions

Analytic solutions for a variety of simple earth models are available. These include:

1. electric or magnetic dipoles over a homogeneous earth (Wolf, 1946; Wait, 1953, 1955; Quon, 1963; Pridmore, 1978),
2. electric or magnetic dipoles over a layered earth (Wolf, 1946; Wait, 1958; Quon, 1963; Frischknecht, 1967; Ward, 1967),
3. a uniform alternating magnetic field incident upon a sphere or cylinder in free space (Wait, 1951, 1952; Negi, 1962; Ward, 1967),
4. magnetic dipoles near a spherical body in free space (Nabighian, 1971; Lodha and West, 1976; Best and Shammas, 1978), and
5. magnetic dipoles near a sphere in a conductive half-space (Singh, 1973).

These solutions are important in themselves because they serve as basic guides for interpretation. More importantly, of late, they serve as devices for checking the validity of the more flexible and more general numerical solutions which have entered the scene since the early 1970s.

A fundamental result using an analytic solution was obtained by Pridmore (1978). He demonstrated that reduction of magnetic field amplitude was more dependent upon geometric fall-off from a finite source than it was upon attenuation in a lossy medium.

Numerical Solutions

Hohmann (1977), in a publication of limited distribution, made an excellent analysis of the state-of-the-art of interpretation of resistivity, induced polarization, and electromagnetic data. In the next few paragraphs I shall draw heavily from that document.

In recent years solutions have been found to previously intractable electromagnetic boundary value problems (Coggon, 1971; Hohmann, 1971; Parry and Ward, 1971; Swift, 1971; Vozoff, 1971; Dey and Morrison, 1973; Lee, 1974; Hohmann, 1975; Lajoie and West, 1976; Stoyer and Greenfield, 1976; Rijo, 1977; and Pridmore, 1978). The earliest of these articles dealt with two-dimensional inhomogeneities in the fields of line sources; this is a true 2D problem. The article by Stoyer and Greenfield (1976) describes a two-dimensional inhomogeneity in the field of a three-dimensional source (the 2D-3D problem) while the articles by Lee (1974), Hohmann (1975), Lajoie and West (1976), and Pridmore (1978) described a three-dimensional inhomogeneity in the field of a three-dimensional source (the full 3D problem).

Four methods have been used to achieve numerical solutions to two- and three-dimensional problems:

1. finite difference,
2. network analogy,
3. finite element,
4. integral equation.

These four methods all use the method of moments (Harrington, 1968) to solve an equation of the form

$$L(f) = g \quad (1)$$

where L is a linear operator, g is a known source term, and f is an unknown field function. The unknown f is approximated by

$$f = \Sigma_n \alpha_n f_n \quad (2)$$

in which the α_n are constants and the f_n are referred to as basis functions. If (2) is to be an exact solution then its right hand side must be an infinite summation and the f_n form a complete set of basis functions. For approximate solutions, (2) is usually a finite summation based on an incomplete set of basis functions. If we substitute (2) in (1) we obtain

$$\Sigma_n \alpha_n L(f_n) = g \quad (3)$$

An inner product $<f,g>$ and a set of vector weighting functions W_m, are defined for the problem. If we then take the inner product of (3) with each W_m we obtain

$$\Sigma_n \alpha_n <W_m, L(f_n)> = <W_m, g> \quad (4)$$

This set of equations can be written in matrix form as (Harrington, 1968)

$$[\ell_{mn}] [\alpha_n] = [g_m] \quad (5)$$

where

$$\ell_{mn} = \begin{bmatrix} <W_1, L(f_1)>, & <W_1, L(f_2)> & \ldots \\ <W_2, L(f_1)>, & <W_2, L(f_2)> & \ldots \\ \ldots & \ldots & \end{bmatrix} \quad (6)$$

$$[\alpha_n] = \begin{bmatrix} \alpha_1 \\ \alpha_2 \\ \cdot \\ \cdot \\ \cdot \end{bmatrix} \quad (7)$$

and

$$[g_m] = \begin{bmatrix} <W_1,g> \\ <W_2,g> \\ \cdot \\ \cdot \\ \cdot \end{bmatrix} \quad (8)$$

When the matrix (1) is nonsingular its inverse $[\ell^{-1}]$ exists and under this circumstance the α_n are given by

$$[\alpha_n] = [\ell^{-1}_{mn}][g_m] \quad (9)$$

When the α_n are substituted in (2), the unknown field function f is obtained.

For methods (1), (2), and (3), L is a second-order differential operator, while for method (4) it is an integral operator. For each method one must choose effective basis and weighting functions. The articles referenced earlier illustrate how these choices are made. The particular choice $W_n = f_n$ is known as Galerkin's method.

For purposes of comparison, the four methods may be divided into two groups: differential equation (1, 2, 3) and integral equation (4). The differential equation solutions are simple to construct and result in large banded matrices. Because the fields in the entire earth are modelled on a grid or mesh with the differential equation methods, these methods are preferable for modelling complex earths. This is in contrast to the integral equation method in which unknown fields are required only in the anomalous parts of the earth. The net result is that integral equation solutions are less expensive per unit of accuracy (than differential methods) where the earth contains only a few regions of differing physical properties; the resulting matrices, though full, are smaller than those for the differential methods. Of the differential methods, the finite difference and network methods are easier to program, while the finite element method can handle more complex geometries. This short discourse on the over-riding method of moments and on comparisons between finite difference, network analogy, finite element, and integral equation are made a) to illustrate the similarities between each of the four methods and b) to point out that differences exist, on a cost per unit accuracy basis, between them. Rijo (1977) showed that the network analogy and finite element methods led to identical matrices but, facility to handle complex earths was quite different with the two methods.

The state-of-the-art in forward solutions for the electromagnetic scattering problem is shown in Table 5.5 opposite.

Table 5.5

Current progress in forward solutions by numerical methods

	2D	2D-3D	3D
Finite difference	x	[x]	[x]
Network analogy	x		
Finite element	x		x
Integral equation	x	[x]	x

[ ] denotes an unchecked algorithm.

Integral-equation 3D algorithms are inexpensive but are applicable to simple earths involving only one or two inhomogeneities. Finite element solutions are expensive but are applicable to complex earths. To illustrate the two methods, calculations of field quantities for the model of Figure 5.10 have been made with each of them. Figure 5.11 (after Hohmann, 1975) shows a horizontal-loop EM profile of real and imaginary components for background conductivities of 0.01 mhos/m and 0.0033 mhos/m. The effect of current gathering in the inhomogeneity is greater for the background of highest conductivity as expected.

Figure 5.12 (after Pridmore, 1978) shows tilt angle and ellipticity profiles measured with a roving receiver and a fixed transmitter. The background conductivity for this example was 0.033 mhos/m, so high that the ellipticity anomaly due to the inhomogeneity is very small. The tilt angle anomaly is not terribly diagnostic of the inhomogeneity because of the high background conductivity.

Figures 5.11 and 5.12 establish that full three-dimensional problems can be solved numerically. Each integral equation profile cost about US $30.00 while each finite element profile cost about ten times that much. On the other hand, a much more complex earth model could be treated by the finite element method at essentially the same cost. Note that each of Figures 5.11 and 5.12 were produced for a single frequency only. The costs for modelling broadband data obviously become excessive. The only economical means of treating such problems is via use of a dedicated computer, such as a DEC11/60, which is allowed to

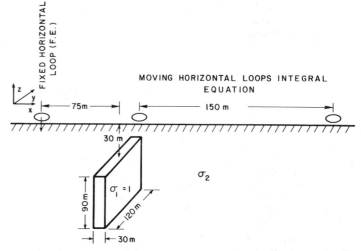

Figure 5.10. Geometry of the three dimensional model used in producing Figures 5.9 and 5.12.

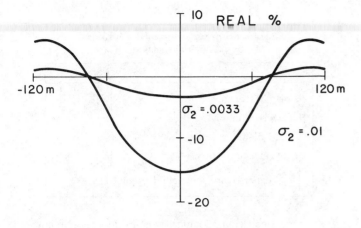

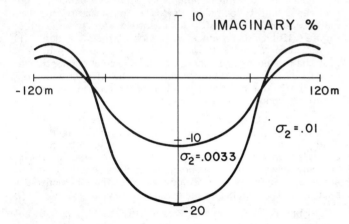

Figure 5.11. Horizontal loop EM profiles for the model used in Figure 5.10; frequency = 1000 Hz.

grind away at the problem at small operating cost. A special computer might be designed specifically to treat these problems at an even lower cost.

The algorithms of Hohmann and Pridmore cross check very well provided the conductivity contrast between inhomogeneity and host is not too large. For high contrasts and shallow depths to the top of the inhomogeneities, convergence problems force either error in computations or astronomical costs, again unless a dedicated computer is used.

The Inverse Method of Interpretation

While numerous algorithms exist for the inversion of resistivity and magnetotelluric data in terms of a layered earth, such is not the case for active source electromagnetic methods. Glenn et al. (1973), Glenn and Ward (1976), Ward et al. (1976), and Tripp et al. (1978) are the only published papers on the subject at the time of writing. Pridmore (1978) has applied this algorithm to estimate the mean conductivity of a half-space adjacent to a massive sulphide base metal deposit. Because of its limitation to a 1D earth, this particular algorithm is expected to be of minimal applicability in base metal exploration.

So far, no 2D or 3D inverse algorithms for active electromagnetic methods have been produced. The cost of applying them, if developed, may be prohibitive. However, some form of computer interactive inversion will be required if we are to derive the maximum possible information from EM data.

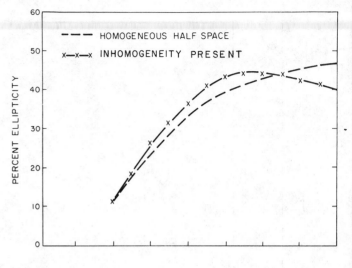

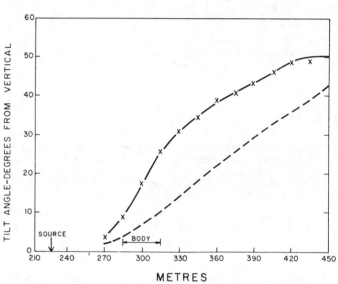

Figure 5.12. Tilt angle and ellipticity for fixed horizontal-loop EM transmitter for the model in Figure 5.10; frequency = 1000 Hz.

RECENT RESULTS WITH BROADBAND EM SYSTEMS

A few examples of the application of broadband electromagnetic methods follow. No attempt at completeness in terms of specific methods, specific geological problems, or specific geographical areas is attempted.

Cavendish Test Site, Frequency Domain

Ward et al. (1974b) obtained broadband frequency domain electromagnetic data over the Cavendish Test Site in Ontario. Figure 5.13 shows some of their data. The transmitter-receiver configuration used to obtain the data was the vertical coaxial coil method referred to earlier. The contours in frequency-distance space clearly reveal the existence of Zones A and B and also reveal that Zone A has a higher conductivity-thickness product than Zone B because the response commences at lower frequencies. The effects of overburden, host rock, and disseminated halo are not obvious in this form of data presentation, for this particular case history. However, Figure 5.9 clearly reveals that either the host rock or the overburden is contributing to the response at frequencies in excess of 10^4 Hz.

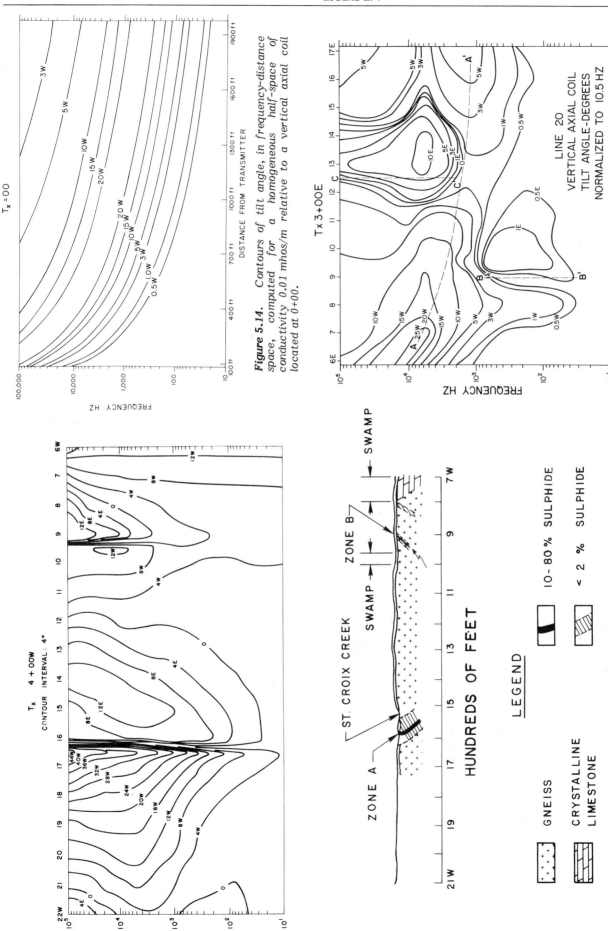

Figure 5.14. Contours of tilt angle, in frequency-distance space, computed for a homogeneous half-space of conductivity 0.01 mhos/m relative to a vertical axial coil located at 0+00.

Figure 5.15. Contours of tilt angle, in frequency-distance space, observed at a site in California where massive sulphides, disseminated sulphides, and a conductive host rock are present. A-A', B-B' and C-C' indicate trends of half-space, massive sulphide, and disseminated sulphide responses, respectively.

Figure 5.13. Contours of tilt angle, in frequency-distance space, for measurements near a vertical axial coil located at 4+00W on Line C at the Cavendish Test Site, Ontario, Canada.

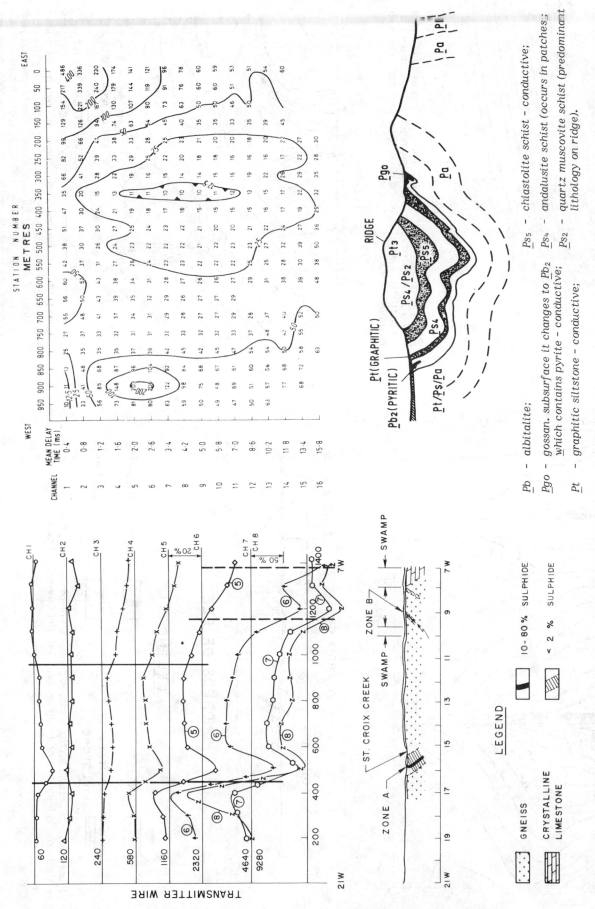

Figure 5.16. Profiles of seven channel UTEM time-domain response on Line C of Cavendish Test Site, Ontario, Canada (after Lamontagne and West, 1973).

Figure 5.17. Apparent resistivity plotted in time-distance space as derived from data obtained over the Willyama complex, South Australia, using the Sirotem system (after Geoex Pty. Ltd., 1977).

Pb — albitalite;

Pgo — gossan, subsurface it changes to Pb_2 which contains pyrite — conductive;

Pt — graphitic siltstone — conductive;

Ps_5 — chiastolite schist — conductive;

Ps_4 — andalusite schist (occurs in patches);

Ps_2 — quartz muscovite schist (predominant lithology on ridge).

The ability of a broadband system to evaluate the conductivity — thickness products of two adjacent zones in a resistive environment is brought out by this case history. The location of each conductor is indicated by a change of sign of tilt angle, in frequency-distance space, of vertical contours.

Theoretical Response of a Homogeneous Half-Space, Frequency Domain

Figure 5.14 (from Pridmore, 1978) shows the computed contours of tilt angle in frequency-distance space for a coaxial vertical-loop method. Broad, sweeping, semihorizontal contours of tilt angle in frequency-distance space are diagnostic of a homogeneous earth (or of a horizontally-layered earth, Pridmore, 1978).

California Test Site, Frequency Domain

When conductive overburden, a weathered layer, and conductive host rocks are encountered, tilt angle response combining the diagnostic features of Figures 5.13 and 5.14 is to be expected. Thus Figure 5.15 illustrated the half-space or layered half-space response (A-A') superimposed upon the response of an inhomogeneity of relatively high conductivity-thickness product (B-B') plus the response of an inhomogeneity of relatively low conductivity-thickness product (C-C'). The latter two responses arise in massive economic sulphides and in disseminated non-economic sulphides, respectively (Pridmore, 1978).

Cavendish Test Site, Time Domain

Figure 5.16 contains the early time-domain results obtained by Lamontagne and West (1973) over the Cavendish Test Site.

This time domain data clearly illustrate that Zone A is of higher conductivity-thickness product than Zone B, in accordance with the frequency domain results presented earlier. The transmitter was a large horizontal loop.

Willyama Complex, South Australia

Figure 5.17 contains a plot of apparent resistivity in time-distance space as obtained with the Sirotem system. The transmitting and receiving loops were coincident and were 100 m square. The loop was placed at 50 m intervals to obtain the data points of Figure 5.17. The boundaries of the conductive section of the syncline are indicated at 150W and 850W. The zone is most conductive under 350W which is the position of the conductive sediments at the depth of maximum response of the loop configuration. Near surface variations in resistivity are reflected in the variations at short time delays such as the low resistivities to the far west over thick conductive soil cover (Geoex Pty. Ltd., 1977). The ability of a broadband electromagnetic system to indicate the complexity of a geoelectric section is brought out by this example. Later results with Sirotem, illustrating penetration through a 90 m conductive cover to detect a massive sulphide body, are given in McCracken and Buselli (1978).

CONCLUSION

I have attempted to illustrate in this article what is being done and what will be done with electromagnetic systems designed to search for base metals. The newer systems currently in use, and especially those of the future, will obtain information on the parameters of the real three-dimensional earth. Failure to evaluate geological "noise" as well as geological "signal" will deter our attempts to search deeper for smaller base metal deposits. The variables to be employed are many and interrelated. We must be exceptionally clever if we are to make significant strides in our quest.

A new hardware item to look for in industrial use in the near future is a three-component cryogenic magnetometer which currently exhibits flat response from D.C. to 10^2 Hz at a sensitivity of 10^{-5} γ/Hz. Extension of the passband to 10^3 Hz is in the prototype stage, while extension to 10^4 Hz is on the drawing board. These exceptionally sensitive receivers permit three-component, and gradient of three-component, magnetic field measurement in an extremely compact package of about 12 kg weight. The accompanying in-field digital processing hardware is about 10 kg weight. If this in-field instrumental capability can be combined with a 3D interpretative capability via digital computer, then the electromagnetic method shall have achieved a viable status.

ACKNOWLEDGMENTS

I am grateful to K.G. McCracken, E. Burnside, D.F. Pridmore, and G. Hopkins for supplying data specifically requested for this article. The numerous other sources of data upon which I have drawn are indicated in the list of references. G.W. Hohmann made valuable editorial comments and contributed to the design of the content of the paper.

This report has been made possible by Grant AER76-11155 from the RANN program of the U.S. National Science Foundation.

REFERENCES

Al'pin, L.M., Berdichevskii, M.N., Vedrintsev, G.A., and Zagarmistr, A.M.
    1966: Dipole methods for measuring earth conductivity; selected and translated by G.V. Keller, Consultants Bureau, New York, 302 p.

Best, M.E. and Shammas, B.R.
    1978: A general solution for a spherical conductor in a magnetic dipole field; preprint Shell Canada Resources Ltd.

Bhattacharyya, B.K.
    1964: Electromagnetic fields of a small loop antenna on the surface of a polarizable medium; Geophysics, v. 29 (5), p. 814-831.

Bhattacharyya, B.K. and Patra, H.P.
    1968: Direct current geoelectric sounding; Elsevier Pub. Co., New York, 135 p.

Braham, B., Haren, R., Lappi, D., Lemaire, H., Payne, D., Raiche, A., Spies, B., and Vozoff, K.
    1978: Lecture notes from the US-Australia electromagnetic workshop; Bull. Aust. Soc. Explor. Geophys., v. 9 (1), p. 2-33.

Buselli, G. and O'Neill, B.
    1977: Sirotem: A new portable instrument for multichannel transient electromagnetic measurements; Bull. Aust. Soc. Explor. Geophys., v. 8 (3), p. 1-6.

Coggon, J.H.
    1971: Electromagnetic and electrical modelling by the finite element method; Geophysics, v. 36 (1), p. 132-155.

Crone, J.D.
    1966: The development of a new ground EM method for use as a reconnaissance tool; Mining Geophysics, v. I, Soc. Explor. Geophys., Tulsa.

    1973: Model studies with the Shootback method; in Symposium on electromagnetic exploration methods, Univ. Toronto.

Dey, A. and Morrison, H.F.
1973: Electromagnetic coupling in frequency- and time-domain induced polarization surveys over a multilayered earth; Geophysics, v. 38 (2), p. 380 405.

1973: Electromagnetic response of two-dimensional inhomogeneities in a dissipative half-space for Turam interpretation; Geophys. Prosp., v. 21, p. 340-365.

1977: Resistivity modelling for arbitrarily shaped three-dimensional structures; Lawrence Berkeley Lab., Preprint LBL-7010.

Dey, A. and Ward, S.H.
1970: Inductive sounding of a layered earth with a horizontal magnetic dipole; Geophysics, v. 35 (4), p. 660-703.

Dias, C.A.
1968: A non-grounded method for measuring induced electrical polarization and conductivity; Ph.D. thesis, Univ. California, Berkeley.

Fountain, D.K.
1972: Geophysical case history of disseminated sulfide deposits in British Columbia; Geophysics, v. 32 (1), p. 142-159.

Frischknecht, F.C.
1967: Fields about an oscillating magnetic dipole over a two-layer earth, and application to ground and airborne electromagnetic surveys; Colo. Sch. Mines Q., v. 62 (1), 326 p.

Gaur, V.K., Verma, O.P., and Gupta, C.P.
1972: Enhancement of electromagnetic anomalies by a conducting overburden; Geophys. Prosp., v. 20 (3), p. 580-604.

Gaur, V.K. and Verma, O.P.
1973: Enhancement of electromagnetic anomalies by a conducting overburden II; Geophys. Prosp., v. 21 (1), p. 159-184.

Geoex Pty. Ltd.
1977: Apparent resistivity time section, Willyama Complex, South Australia; Advertising brochure, Case-Study Series, No. 3.

Glenn, W.E., Ryu, J., Ward, S.H., Peeples, W.J., and Phillips, R.J.
1973: Inversion of vertical magnetic dipole data over a layered structure; Geophysics, v. 38 (6), p. 1109-1129.

Glenn, W.E. and Ward, S.H.
1976: Statistical evaluation of electrical sounding methods. Part I: Experiment design; Geophysics, v. 41 (6A), p. 1207-1221.

Harrington, R.F.
1968: Field computation by moment methods; Macmillan Co., New York, 229 p.

Hohmann, G.W.
1971: Electromagnetic scattering by conductors in the earth near a line source of current; Geophysics, v. 36 (1), p. 101-131.

1973: Electromagnetic coupling between grounded wires at the surface of a two-layer earth; Geophysics, v. 38 (5), p. 854-863.

1975: Three-dimensional induced polarization and electromagnetic modelling; Geophysics, v. 40 (2), p. 309-324.

Hohmann, G.W. (cont'd.)
1977: Modelling team report, Workshop on electrical methods in geothermal exploration; U.S. Geological Survey Grant 14-08-0001-G-359, Univ. Utah, January, 1977.

Hohmann, G.W., Kintzinger, P.R., Van Voorhis, G.D., and Ward, S.H.
1970: Evaluation of the measurement of induced polarization with an inductive system; Geophysics, v. 35 (5), p. 901-915.

Hohmann, G.W., Nelson, P.H., and Van Voorhis, G.D.
1977: A vector EM system and its field applications; Geophysics, vol. 43, no. 7, p. 1418-1440.

Hood, Peter
1977: Mineral exploration trends and developments in 1976; Can. Min. J., v. 98 (1), p. 8-47.

Jain, B. and Morrison, H.F.
1976: Inductive resistivity survey in Grass Valley, Nevada; Lawrence Berkeley Lab., Progress Report.

Keller, G.V. and Frischknecht, F.C.
1966: Electrical methods in geophysical prospecting; Pergamon Press, New York, 517 p.

Klein, J.D. and Shuey, R.T.
1978: Nonlinear impedance of mineral-electrolyte interfaces; Parts I and II, Geophysics, vol. 43, no. 6, p. 1222-1249.

Kunetz, G.
1966: Principles of direct current resistivity prospecting; Geoexplor. Mon. Ser. 1, No. 2, Gebruder Borntraeger, Berlin.

Lajoie, J.J. and West, G.F.
1976: The electromagnetic response of a conductive inhomogeneity in a layered earth; Geophysics, v. 41 (6A), p. 1133-1156.

1977: Two selected field examples of EM anomalies in a conductive environment; Geophysics, v. 42 (3), p. 655-660.

Lamontagne, Y.
1975: Applications of wide-band, time-domain EM measurements in mineral exploration; unpubl. Ph.D. thesis, Univ. Toronto.

Lamontagne, Y. and West, G.F.
1973: A wide-band, time-domain, ground EM system; in Proc., Symposium on electromagnetic exploration methods, Univ. Toronto, p. 2-1 to 2-5.

Lee, T.
1974: Transient electromagnetic response of a sphere in a layered medium; Geophys. Prosp., v. 23, p. 492-512.

Lodha, G.S. and West, G.F.
1976: The electromagnetic response of a conductive inhomogeneity in a layered earth; Geophysics, v. 41 (6A), p. 1133-1156.

Lodha, G.S.
1977: Time domain and multifrequency electromagnetic responses in mineral properties; unpubl. Ph.D. thesis, Univ. Toronto, 183 p.

Lowrie, W. and West, G.F.
1965: The effect of a conducting overburden on electromagnetic prospecting measurements; Geophysics, v. 30 (4), p. 624-632.

Madden, T.R.
1971: The resolving power of geoelectric measurements for delineating resistive zones within the crust; in J.G. Heacock (Editor), The structure and physical properties of the Earth's crust, Am. Geophys. Union, p. 95-105.

McCracken, K.G. and Buselli, G.
1978: Australian exploration geophysics — current performance and future prospects; Second Circum-Pacific energy and minerals resources conference, Honolulu.

Millett, F.B., Jr.
1967: Electromagnetic coupling of colinear dipoles on a uniform half-space; Mining Geophysics; Soc. Explor. Geophys., Tulsa, 708 p.

Nabighian, M.N.
1971: Quasi-static transient response of a conducting permeable two-layer sphere in a dipolar field; Geophysics, v. 36 (1), p. 25-37.

1977: The Newmont EMP methods; in Geophysics applied to detection and delineation of non-energy, non-renewable resources, Report on Grant AER76-80802, Nat. Sci. Foundation; Dep. Geol. Geophys., Univ. Utah.

Negi, J.G.
1962: Inhomogeneous cylindrical ore body in presence of a time varying magnetic field; Geophysics, v. 27 (3), p. 386-392.

Palacky, G.J.
1975: Interpretation of INPUT AEM measurements in areas of conductive overburden; Geophysics, v. 40 (3), p. 490-502.

Parry, J.R.
1969: Integral equation formulations of scattering from two-dimensional inhomogeneities in a conductive earth; unpubl. Ph.D. thesis, Univ. Calif., Berkeley.

Parry, J.R. and Ward, S.H.
1971: Electromagnetic scattering from cylinders of arbitrary cross-section in a conductive half-space; Geophysics, v. 36 (1), p. 67-100.

Pelton, W.H.
1977: Interpretation of induced polarization and resistivity data; unpubl. Ph.D. thesis, Univ. Utah.

Pridmore, D.F.
1978: Electromagnetic scattering of three-dimensional fields by three-dimensional earths; unpubl. Ph.D. thesis, Univ. Utah.

Quincy, E.H., Davenport, W.H., and Moore, D.F.
1976: Three-dimensional response maps for a new side-band induction system; IEEE Transactions on Geoscience Electronics, v. GE-14 (4), p. 261-269.

Quon, C.
1963: Electromagnetic fields of elevated dipoles on a two-layer earth; unpubl. M.Sc. thesis, Univ. Alberta.

Rijo, J.
1977: Modelling of electric and electromagnetic data; unpubl. Ph.D. thesis, Univ. Utah.

Roy, A.
1970: On the effect of overburden on EM anomalies, a review; Geophysics, v. 35 (4), p. 646-659.

Roy, A. and Apparao, A.
1971: Depth of investigation in direct current methods; Geophysics, v. 36 (5), p. 943-959.

Ryu, J., Morrison, H.F., and Ward, S.H.
1970: Electromagnetic fields about a loop source of current; Geophysics, v. 35 (5), p. 862-896.

Sarma, D.G. and Maru, W.M.
1971: A study of some effects of a conducting host rock with a new modelling apparatus; Geophysics, v. 36 (1), p. 166-183.

Scott, W.J. and Fraser, D.C.
1973: Drilling of EM anomalies caused by overburden; Can. Inst. Min. Met. Bull., v. 66 (735), p. 72-77.

Singh, S.K.
1973: Electromagnetic transient response of a conducting sphere embedded in a conductive medium; Geophysics, v. 38 (5), p. 864-893.

Snyder, D.D.
1975: A programmable digital electrical receiver; 45th Annual Meeting, Soc. Explor. Geophys., Denver.

1976: Field tests of a microprocessor-controlled electrical receiver; 46th Annual Meeting, Soc. Explor. Geophys., Houston.

Spies, B.S.
1976: The transient electromagnetic method in Australia; BMR J. Aust. Geol. Geophys., v. 1, p. 23-32.

Stoyer, C.H. and Greenfield, R.J.
1976: Numerical solutions of the response of a two-dimensional earth to an oscillating magnetic dipole source; Geophysics, v. 41 (3), p. 519-520.

Summer, J.S.
1976: Principles of induced polarization for geophysical exploration; Elsevier Publ. Co., New York, 277 p.

Sunde, E.D.
1949: Earth conduction effects in transmission systems; D. Van Nostrand Co., Inc., New York, 373 p.

Swift, C.M., Jr.
1971: Theoretical magnetotelluric and Turam response from two-dimensional inhomogeneities; Geophysics, v. 36 (1), p. 38-52.

Telford, W.M., Geldart, L.B., Sheriff, R.E., and Keys, D.A.
1976: Applied Geophysics; Cambridge Univ. Press, Cambridge, 860 p.

Tripp, A.C., Ward, S.H., Sill, W.R., Swift, C.M., Jr., and Petrick, W.R.
1978: Electromagnetic and Schlumberger resistivity sounding in the Roosevelt Hot Springs KGRA, Geophysics, v. 43 (7), p. 1450-1469.

Vanyan, L.L.
1967: Electromagnetic depth sounding; selected and translated by G.V. Keller, Consultants Bureau, New York, 312 p.

Van Nostrand, R.G. and Cook, K.L.
1966: Interpretation of resistivity data; U.S. Geol. Surv., Prof. Paper 499, 310 p.

Verma, O.P. and Gaur, V.K.
1975: Transformation of electromagnetic anomalies brought about by a conducting host rock; Geophysics, v. 40 (3), p. 473-489.

Vozoff, K.
1971: The effect of overburden on vertical component anomalies in AFMAG and VLF exploration: a computer model study; Geophysics, v. 36 (1), p. 53-57.

Wait, J.R.
- 1951: A conducting sphere in a time varying magnetic field; Geophysics, v. 16 (5), p. 666-672.
- 1952: The cylindrical ore body in the presence of a cable carrying an oscillating current; Geophysics, v. 17 (2), p. 378-386.
- 1953: Induction by a horizontal magnetic dipole over a conducting homogeneous earth; Trans. Am. Geophys. Union, v. 34, p. 185-189.
- 1955: Mutual electromagnetic coupling of loops over a homogeneous ground; Geophysics, v. 20 (3), p. 630-637.
- 1958: Induction by an oscillating magnetic dipole over a two layer ground; Appl. Sci. Res., Sec. B, v. 7, p. 73-80.

Ward, S.H.
- 1967: The electromagnetic method; in Mining Geophysics, v. 2, Soc. Explor. Geophys., Tulsa, p. 224-372.
- 1971a: Discussion on "Evaluation of the measurement of induced electrical polarization with an inductive system"; Geophysics, v. 36 (2), p. 427-429.
- 1971b: Foreword and Introduction; in Special Issue on electromagnetic scattering, Geophysics, v. 36 (1), p. 1-8.

Ward, S.H., Ryu, J., Glenn, W.E., Hohmann, G.W., Dey, A., and Smith, B.D.
- 1974a: Electromagnetic methods in conductive terranes; Geoexploration, v. 12, p. 121-183.

Ward, S.H., Pridmore, D.F., Rijo, L., and Glenn, W.E.
- 1974b: Multispectral electromagnetic exploration for sulfides; Geophysics, v. 39 (5), p. 666-682.

Ward, S.H., Smith, B.D., Glenn, W.E., Rijo, L., and Inman, J.R., Jr.
- 1976: Statistical evaluation of electrical sounding methods. Part II: Applied electromagnetic depth sounding; Geophysics, v. 41 (6A), p. 1222-1235.

Ward, S.H., Campbell, R., Corbett, J.D., Hohmann, G.W., Moss, C.K., and Wright, P.M.
- 1977: Geophysics applied to detection and delineation of non-energy non-renewable resources; report on grant AER76-80802, Nat. Sci. Foundation, Dep. Geol. Geophys., Univ. Utah, Salt Lake City.

Wolf, A.
- 1946: Electric field of an oscillating dipole over the surface of a two layer earth; Geophysics, v. 2 (4), p. 518-834.

Zonge, K.L.
- 1973: Minicomputer used in mineral exploration, or back-packing a box full of bits into the bush; 11th Symposium on computer applications in the mineral industry; Univ. Arizona, Tucson.

Zonge, K.L. and Wynn, J.C.
- 1975: EM coupling, its intrinsic value, its removal, and the cultural coupling problem; Geophysics, v. 40 (5), p. 831-850.

RECENT DEVELOPMENTS IN THE USE AND INTERPRETATION OF DIRECT-CURRENT RESISTIVITY SURVEYS

George V. Keller
Colorado School of Mines, Golden, Colorado

Keller, George V., Recent developments in the use and interpretation of direct-current resistivity surveys; in Geophysics and Geochemistry in the Search for Metallic Ores; Peter J. Hood, editor, Geological Survey of Canada, Economic Geology Report 31, p. 63-75, 1979.

Abstract

The direct-current resistivity survey method technique has developed slowly since the early part of this century, with applications being primarily for engineering problems and in the search for groundwater. Over the past two decades, extensive resistivity surveys have been carried out in mining exploration, but mainly as an integral part of induced polarization surveys. In the last few years, use of direct-current resistivity surveys has expanded rapidly because of their value in geothermal exploration. In part, geothermal surveys are carried out using conventional Schlumberger and dipole-dipole techniques, but in part, geothermal exploration is being done using arrays which previously had not been greatly utilized. In particular, the dipole mapping array (or, as it is also known, the bipole-dipole array), has been used extensively to map two- and three-dimensional resistivity structures. This expanded use of the DC resistivity method has led to increased concern with interpretation procedures. Numerical methods for the interpretation of layered structures are well developed, but the application of such methods to two- and three-dimensional structures is still at an early stage.

Résumé

La technologie des levés géophysiques par les méthodes de résistivité en courant continu s'est développée peu à peu à partir du début de ce siècle; les premières applications concernaient surtout les problèmes liés à l'ingéniérie et à la recherche des eaux souterraines. Au cours des deux dernières décennies, des levés de résistivité assez étendus ont été exécutés dans le cadre de campagnes de prospection minière, mais ces levés faisaient le plus souvent partie intégrante de levés de polarisation provoquée. Ces dernières années, l'emploi des levés de résistivité en courant continu s'est répandu rapidement, en raison des possibilités qu'ils offrent pour la prospection des ressources géothermiques. Les levés géothermiques se font en partie à l'aide des techniques classiques comme la technique Schlumberger et celle du dipôle-dipôle, mais en partie aussi en utilisant des dispositifs qu'on n'avait que peu utilisés auparavant. En particulier, le dispositif dipôle de levé de surface aussi connu sous le non de bipôle-dipôle a été largement utilisé pour établir la carte ou le bloc-diagramme de certaines structures de résistivité. Du fait de ce regain de la méthode de résistivité en courant continu, il a fallu apporter encore plus d'attention aux méthodes d'interpértation. Les méthodes numériques d'interprétation des structures stratifiées sont parfaitement au point, mais l'application de ces méthodes à des structures représentées en deux ou en trois dimensions n'en est encore qu'à ses débuts.

INTRODUCTION

Traditionally, electrical prospecting methods based on the use of direct current have been considered separately from those based on the use of alternating currents. Often, the term "resistivity method" is used to indicate a method based on the assumption of direct current flow in the earth. The use of this definition is no longer satisfactory inasmuch as earth resistivity is now being measured using alternating currents as well as direct currents. In this paper, I will discuss those methods for measuring earth resistivity in which (the assumption is made that) direct current is used to energize the earth. In actual fact a low-frequency alternating-current is most frequently used in field practice. The assumption of direct current merely requires that the frequency be low enough that magnetic coupling between current flow lines can be neglected, and the flow of current in the earth can be described adequately by Laplace's equation.

Roman (1960) cited the earliest uses of electrical resistivity methods as being due to Fox, prior to 1830, and to Barus, who in 1883 reported on the use of electrical methods in mapping extensions of the Comstock Lode in Nevada. Electrical resistivity methods developed slowly from these early beginnings and as Roman pointed out, although the results have never been spectacular, successful resistivity measurements have been made by many people for many purposes.

The induced polarization method (IP) is the most popular and the most effective of the electrical exploration methods used in the search for metallic ore deposits. It is impractical in most cases to do an IP survey without simultaneously measuring earth resistivity. Often, the earth resistivity data so obtained are useful in characterizing the nature of a rock mass which gives rise to an induced polarization effect. However, for this paper, the application of resistivity methods in parallel with induced polarization will not be considered.

In the last decade, the independent use of electrical resistivity methods has grown very rapidly in comparison to the extent of use in previous decades. Principal applications have been in the solution of engineering problems and in the search for groundwater. However, perhaps the greatest growth has occurred since electrical resistivity surveys have come into use in exploration for geothermal resources. Here, the methods are the primary approach to locating geothermal resources. Because of this, significant changes have been taking place both in the way in which resistivity surveys are carried out in the field and the way in which the data are interpreted to yield information about geologic structure.

Two trends can be recognized in the increased application of electrical resistivity methods. First, a wider variety of field techniques is being tried, and some of these new techniques are becoming adopted in general use. Second, significant changes in the way in which resistivity surveys are used have come about as a consequence of theoretical developments. Computational methods and analytical procedures are now available which permit a far more rigorous interpretation of field observations than has been possible in the past. Interpretation of resistivity data is less subjective and interpretations can now be made with greater detail than was previously possible. An important added feature of some of the new data processing methods is that the reliability of the interpretations can be estimated.

In the paper which follows, the development of new field techniques, and particularly the application of new interpretive techniques, will be reviewed.

FIELD TECHNIQUES

Most direct current methods for measuring resistivity employ an array of four electrodes in contact with the ground. Two electrodes are used to provide current to the ground, while the other two are used to measure the voltage developed by this current flow in the ground. The geometry of electrode layout is used to control the sensing area and depth of investigation of the resistivity measurement. While it is quite possible to compute an apparent resistivity value for measurements made with an array of completely arbitrarily located electrodes, in practice, a few standardized arrays are used much more commonly than others. These include the Wenner array, the Schlumberger array, the dipole-dipole array, and more recently, the general dipole array. All but the last have been thoroughly described in textbooks on resistivity surveying, including Keller and Frischknecht (1966), Orellano (1972), Dobrin (1976), and Telford et al. (1976).

The general dipole or bipole-dipole method was first described in detail by Alpin (1966), who described the use of a dipole current source and a dipole receiver arbitrarily positioned with respect to each other as a means for measuring earth resistivity. The advantages for the dipole method were operational ease in that the lengths of cable that had to be laid out were relatively short, and a relatively great sensitivity to lateral changes in resistivity in comparison with that of other electrode arrays.

Alpin employed a short separation between the current electrodes so that the field would behave as though it originated from a dipole source of zero length. In the subsequent development of bipole-dipole methods, it has become more common to use a source whose length is too great to permit the dipolar approximation. One of the first bipole-dipole surveys is described by Stefanescu and Tanasescu (1965) and by Doicin et al. (1965). In a survey of potentially oil-bearing strata in Transylvania, a bipole source which was actually an out-of-service power line several tens of kilometres in length was used to energize the area under study. The electric field distribution about this bipole source was then mapped over an area of nearly 1000 km^2. The results were said by the authors to provide an excellent representation of the structure of the sedimentary basin within the area surveyed.

In 1967 and 1968, a similar bipole-dipole resistivity survey was carried out in the vicinity of the Broadlands geothermal prospect in the central part of the north island of New Zealand (Risk et al., 1970). Two bipole current sources oriented more or less at right angles to one another were used sequentially to energize the area of the survey. The electric field was mapped in detail over an area of approximately 50 km^2. The survey outlined a region of low resistivity which is closely associated with a geothermal reservoir at depths of one kilometre and more.

Furgerson (1970) reports the use of a bipole-dipole resistivity survey in the Darwin Hills area in southeast California to map the geometry of a small intrusive associated with the Darwin Hills ore deposits. Again, the method appeared to be effective in locating the boundaries between major lithologic units over an area of several tens of square kilometres.

The essential features of the bipole-dipole electrode array are shown in Figure 6.1. The ground is energized by passing current through a relatively long wire grounded at points "A" and "B". The length of the source may range from a kilometre to several tens of kilometres. The amount of current provided to the source must be great enough to produce easily detectable signals over the area to be mapped. The current waveform is usually that of a square wave or an asymmetric square wave with low frequency content so that direct current behavior can be assumed. The amplitude of these square waves may range from a few amperes in a small scale survey to 500 amperes or more in a large scale survey. In order to obtain the required high current levels, it is necessary to expend considerable effort to obtain low resistance ground contacts, particularly in areas where the ground resistivity is high. Existing metal structures, such as well casings and highway culverts, may sometimes be used. In other cases, it may be necessary to drill holes in which to install electrodes, or dig ditches to the water table.

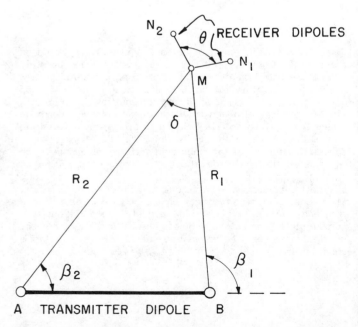

Figure 6.1. Definition of electrode layout used in bipole-dipole surveys (Arestad, J., 1977).

A heavy generator with a capacity ranging from ten kilowatts to several hundred kilowatts is required to provide the large currents mentioned above. Such equipment is heavy, and the current source cannot readily be located away from roads.

The electric field generated by this current is mapped over the area of interest. Usually, the electric field is detected by laying out two short receiver dipoles more or less at right angles to measure two components of electrical field. Measurements are made at many locations with the number ranging from 50 to 200 over the area of interest using a single bipole source for excitation. Measured electric field values are converted to apparent resistivity values as is done with other electrode arrays. When this is done, one recognizes that the field technique provides redundant data; that is, more quantities are measured than are necessary to compute a single value of apparent resistivity at each observation point. Sufficient data are recorded in the field to permit computation of two independent values of apparent resistivity. For example, Alpin (1966) has given expressions for computing apparent resistivity values for the components of the electric field parallel to and perpendicular to the direction of the bipole source.

Furgerson (1970) computed both parallel and perpendicular apparent resistivity values for his survey of the Darwin Hills. These two values of apparent resistivity have the unusual feature that negative values of resistivity can appear in the data. These are usually characteristic of particular types of lateral change in the actual resistivity in the earth. They may be useful in interpretation, but the theoretical results presented by Furgerson indicate that the complexity of the patterns is so great that interpretation would be difficult. An example of the behavior of parallel and perpendicular resistivities for a simple two-layer structure is shown in Figure 6.2.

Another way of making use of the redundant field observations is to compute an apparent resistivity value from the magnitude of the electric field vector, and consider the direction of the electric field vector as an independent piece of data. At present, this is the mode of presentation most commonly used in presenting bipole-dipole surveys. The expression for computing total field apparent resistivity, as it is called, is as follows

$$\rho_a = \frac{2\pi R_1^2}{\left[1+\left(\frac{R_1}{R_2}\right)^4 - 2\left(\frac{R_1}{R_2}\right)^2 \cos\delta\right]^{1/2}} \frac{|E_T|}{I} \quad \ldots 1$$

where R_1 and R_2 are the distance from an observation point to the two ends of the bipole source, δ is the angle between these two lines, E_T is the magnitude of the electric field vector at the observation point, and I is the amplitude of the current pulse provided to the ground. Separately, the direction of the electric field vectors can be presented as shown in Figure 6.3. These vectors will be deviated from their normal direction in a manner which will be indicative of the locations of resistivity changes away from the locations where measurements are actually made.

Still another way of using the redundant field data has been suggested by Zohdy (1970). Zohdy suggests treating the bipole source as though it were two independent single pole sources of current. Each pole of the source will contribute one electric field vector at a receiver station. The observed electric field would be the sum of these two vectors. Unless measurements are made along the equatorial axis of the bipole source, the field data are adequate to compute two separate values of apparent resistivity valid for the pole-dipole array, which is the same as the Schlumberger array.

Still another way of presenting bipole-dipole results is in the form of apparent conductance. The definition of apparent resistivity is based on the assumption that the earth is completely uniform. Another model which is sometimes more realistic is one in which the earth consists of a thin layer with a given conductance (the ratio of thickness to resistivity) covering an insulating substratum. In such a case, the value of apparent resistivity increases linearly with separation between bipole source and dipole receiver. This consistent increase can mask the changes in apparent resistivity caused by lateral effects. The lateral effects can be accentuated in these cases by computing an apparent conductance using the formula

$$S_a = \frac{2\pi R_1}{\left[1+\left(\frac{R_1}{R_2}\right)^2 - 2\left(\frac{R_1}{R_2}\right) \cos\delta\right]^{1/2}} \frac{|E_T|}{I} \quad \ldots 2$$

where the parameters are the same as defined earlier.

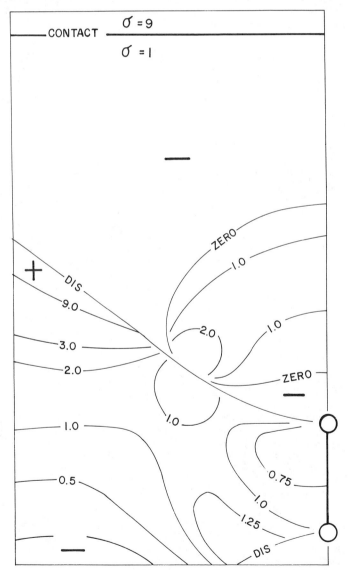

Figure 6.2. Map of apparent resistivity computed for the component of electric field perpendicular to the boundary, in a medium with a single vertical boundary separating regions with unit resistivity and with a resistivity of nine units. The line marked "DIS" represents the locus of points where the component of electric field parallel to the source is zero, and a discontinuity exists in the expression for apparent resistivity (from Furgerson, 1970). d = 4c; k = 0.8.

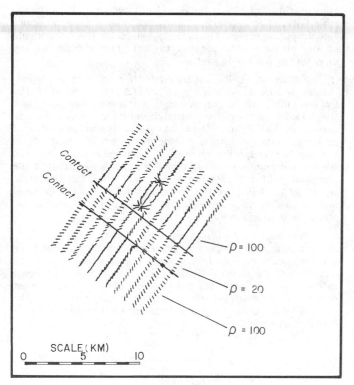

Figure 6.3. Computed directions for electric field vectors for the case of a vertical conductive slab (Arestad, 1977).

Both field studies and theoretical studies (Keller et al., 1975) indicate that under some circumstances, the dipole mapping method or bipole-dipole method provides excellent definition of the boundaries of zones with unique resistivity contrasts. In other cases, the method appears to provide confusing or even misleading results (Beyer, 1977). The best results appear to be obtained when the bipole source is located outside the target area when the target has relatively low resistivity compared to the rest of the terrane, or when only one terminal of the bipole source is located within a zone of low resistivity. Poor results are obtained if both ends of the bipole source are located within a relatively conductive feature. In order to avoid such problems, either the general nature of the resistivity structure must be known before the bipole source is located, or several surveys must be carried out with overlapping coverage to assure that at least one bipole source is partially located outside of the conductive features being mapped.

An extension of the bipole-dipole method which provides even more highly redundant data is the crossed dipole method. This was first suggested by Vedrintsev (1966), who proposed using the differences in character of the equatorial and polar-dipole sounding surveys. In field practice, two bipole sources oriented more or less at right angles to one another are established. Then, at a receiver station, two sets of electric field measurements are made, one for excitation of each of the two dipole sources. The results can be treated as though two separate bipole-dipole surveys were being carried out, but much of the advantage of this method lies in the comparison of the two electric fields. Morris (1975) described the application of the crossed-dipole method to the mapping of structure in a geothermal prospect in the Black Rock Desert area of northwestern Nevada. In the compilation of the data, he not only computed the apparent resistivities for the two separate sources, but combined the electric field vectors from the two sources in varying proportions so as to simulate apparent resistivity values for current flow in all possible directions at a given receiver station. An example of his field data along with a theoretical model which was used to interpret the field measurements is shown in Figure 6.4A, B.

According to Morris, when resistivity values are computed as a function of the direction of the electrical field vector at a receiver station, an elliptical pattern is generated which he called an ellipse of anisotropy. It is characterized by a maximum and minimum value of apparent resistivity, as well as a direction in which the maximum value of apparent resistivity is determined. He defined a coefficient of anisotropy as being the square root of the ratio of maximum to minimum apparent resistivity. There parameters are defined in Figure 6.5. According to simple theoretical models developed by Morris, plots of the direction of the resistivity ellipse axis, of the coefficient of anisotropy, and of the average of the maximum and minimum apparent resistivities all provide relatively simpler patterns than do the total field resistivities obtained from single dipole coverage.

Doicin (1976) has also carried out a theoretical evaluation of measurements made with two dipole sources oriented at right angles. According to Doicin, the most useful parameter to be obtained from the field measurements is the vector cross product of the two electric fields that are measured. This parameter behaves in much the same way as the average resistivity parameter defined by Morris. A tensor apparent resistivity has been defined by Bibby (1977).

With bipole-dipole mapping, the apparent resistivity computed for a given receiver station usually varies widely for different source locations. When overlapping coverage is provided from several sources to assure that proper excitation is obtained for any conductive features in the area being surveyed, a problem arises when one attempts to present the results on a common map. Multiple values measured at each receiver station will not agree. One solution is to consider each map separately, and obtain compatible interpretations in areas where coverage from several sources overlap. Another approach is to provide sufficiently redundant coverage that averaging of values computed from various sources is possible.

An example of a survey in which averaging appears to work well is shown in Figure 6.6B. Here, the results of a bipole-dipole survey of the Imperial Valley in southern California are shown. Fifty bipole sources were used, placed in pairs at intervals of approximately 10 km, as shown in Figure 6.6A. Approximately 5000 individual values of apparent resistivity were determined over the area of survey. To compile Figure 6.6B, a grid with a spacing of 1.6 km was placed over the surveyed area, and the geometric means for all apparent resistivity values measured within a radius of 1.6 km about each node on the grid were contoured. The presentation provides a coherent picture of major resistivity changes in the Imperial Valley. Arestad (1977) also gives an example of spatial averaging to present the results of many overlapping bipole-dipole surveys.

A technique which is closely related to the bipole-dipole method is the mise-à-la-masse method. In this, a conductive mass of rock is energized by placing one current electrode within it, with the current return being placed at a considerable distance. The electric field behavior is mapped to detect a sudden increase in field strength as current leaves the conductive mass to enter more resistant rock around the outside. The significant difference between the mise-à-la-masse method and the bipole-dipole method as described here is that in the mise-à-la-masse method, commonly the electrode placed in the conductive rock mass is situated in a borehole. An example of the use of the mise-à-la-masse method in recent years is reported by Ketola (1972). The problem of field behavior with the mise-à-la-masse method has been considered in some detail by Merkel (1971), by Snyder and Merkel (1973), and by Merkel and Alexander (1971). They have considered both the case in which the electrode is placed within the conductive mass, and the case in which the electrode is placed close to the outer boundary of the conductive mass in a borehole. In either case, considerable advantage is gained in being able to map the boundaries of the conductive zone.

In summary, the recent increase in field utilization of resistivity methods particularly for exploration in geothermal areas has led to the extensive use of electrode arrays in which electric field behavior is mapped over a planar surface, rather than at a few specified points. This greater amount of information about the structure of the electric field from a bipole source may ultimately lead to the better definition of the electrical structure of the earth in areas where surveys are being carried out.

THEORETICAL DEVELOPMENTS

One-Dimensional Forward Problem

By convention, the computation of the behavior of electric potential when the resistivity structure has been specified is called the forward problem. The inverse of this problem — determining the resistivity structure from observations of the potential field behavior — is the one of interest in interpretation. Sor far, it has been necessary to be able to solve the appropriate forward problem before the inverse interpretation can be carried out. Forward problems are of varying degrees of complexity, ranging from the relatively simple case of a one-dimensional variation of resistivity to the highly complex case of a three-dimensional variation. The one-dimensional forward problem has been studied for a very long period of time, and is still the object of many theoretical analyses. The most common one-dimensional problem considered in earth resistivity studies is the one in which resistivity varies only with depth in the earth. However, a problem in which the resistivity varies only with one of the horizontal directions is equally amenable to solution.

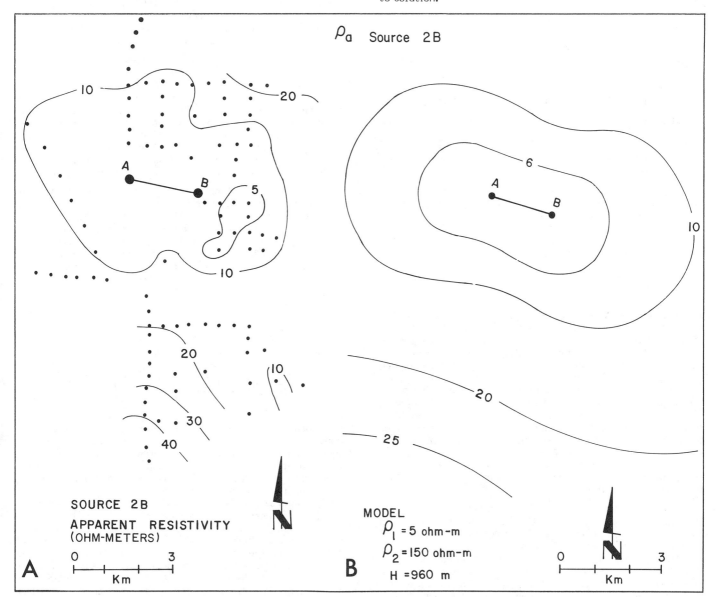

Figure 6.4A. Apparent resistivity map (total field) for a survey done in northwestern Nevada. Values in ohm-m.

Figure 6.4B. Apparent resistivity map computed for a two-layer model to match the data in Figure 6.4A (from Morris, 1975).

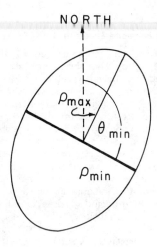

ELLIPTICITY = $\sqrt{\dfrac{\rho_{max}}{\rho_{min}}}$

AVERAGE RESISTIVITY $\rho_{ave} = \dfrac{(\rho_{max}+\rho_{min})}{2}$

BEARING OF MINOR AXIS, θ_{min}, MEASURED CLOCKWISE

Figure 6.5. Definition of ellipse of resistivity measured with the quadripole resistivity method (Morris, 1975).

The standard solution for the one-dimensional variation of resistivity with depth which is in use today was apparently first published by Stefanescu et al. (1930). It is based on a solution of Laplace's equation in cylindrical coordinates which leads to an expression for the potential on the surface of the earth in the following form:

$$U = \dfrac{\rho_1 I}{2\pi}\left[\dfrac{1}{r} + 2\int_0^\infty \Theta(\lambda) J_0(\lambda r)d\lambda\right] \qquad \ldots 3$$

where U is the potential, r is the distance from a single pole source of current to a single measurement point, ρ^1 is the resistivity at the surface of the earth, λ is a dummy variable of integration which enters in the solution of the differential equation, J_0 is a Bessel function of the first kind of order 0, and Θ is a kernel function (Slichter, 1933) which depends on a function of the resistivity structure of the earth. According to Slichter, kernel functions can readily be found for a variety of resistivity distribution functions, including those which are piecewise uniform (a series of layers with fixed resistivities), or a variety of functional relationships between resistivity and depth including an exponential function of depth, a power function of depth, a hyperbolic function of depth, or a trigonometric function of depth. The kernel functions which correspond to each of these variations have been tabulated by Meinardus (1967).

This expression has been evaluated numerous times for the piecewise continuous variation of resistivity with depth that characterizes a sequence of uniform layers. Numerical results have also been obtained for more complex changes in resistivity with depth, including the case of a linear transitional layer in which the resistivity varies linearly from one level to another as reported by Jain (1972), a power law variation within a single layer (Niwas and Upadhyay, 1974), and a layer with an exponential variation of resistivity with depth as reported by Stoyer and Wait (1977). Naidu (1970) has developed an expression for the potential over an earth in which resistivity is a random function of depth characterized by a standard deviation and an autocorrelation function. Lee (1977) has extended this analysis, presented some numerical results, and considered the case of a sequence of random layers resting upon a deterministic lower medium.

The form of the forward solution for the one-dimensional problem as given in equation 3 prevented any extensive numerical computations until high-speed computers became available. The Bessel function within the integral is oscillatory, so that numerical methods of evaluating this integral could not be used prior to the development of high-speed computers. The only cases for which the integral could be evaluated were those in which the integral could be expressed in a closed form (Koefoed, 1966, 1968). The kernel function can be expanded in a Taylor series, with each term being in a form which can be evaluated in a closed form. However, this can be done only when the layers have integer thicknesses and uniform resistivities (Mooney et al., 1966). Expansion of the integral in terms of a series for the kernel function leads to a simple series which has been called a series of images by Roman (1959, 1960, 1963). Even with the image approach for evaluating the integral, computations were tedious so that with hand calculations solutions could be obtained only for a few layers.

With the availability of high-speed computers, it became possible to carry out direct numerical evaluations of the integral in equation 3. Meinardus (1967, 1970) describes a method based on numerical quadrature for evaluating the integral for an arbitrarily large number of layers. While feasible, the method is expensive in terms of computer time. Later, Crous (1971) used a method in which the kernel function is approximated by a spline interpolation formula and the integral in equation 3 is broken into a series of definite integrals which could be evaluated with less effort.

The method which is most commonly used today in evaluating the Hankel transform integral in equation 3 is the convolution method first described by Strakhov (1968) and by Ghosh (1971a, b) and subsequently expanded in papers by Anderson, 1973; Koefoed, 1976a, b, c; Verma and Koefoed, 1973; Das and Ghosh, 1973, 1974; and Das et al., 1974. In this approach the Hankel Transform Integral is converted to the form of a convolution integral by making the substitutions

$$x = \ln\lambda$$
$$y = \ln\left(\dfrac{1}{r}\right) \qquad \ldots 4$$

in equation 3 to yield the result

$$U = \dfrac{\rho_1 I}{2\pi}\left[y+2\int_0^\infty T(x)e^x J_0(e^{x-y})dx\right] \qquad \ldots 5$$

where T(x) is the kernel function expressed in terms of the transformed variable, x. This integral has the form of a convolution, which is a rapid numerical procedure on a high-speed computer, and efficient programs making use of linear filtering are available (Anderson, 1973). The filter representing the Bessel function is determined by considering the Hankel transform of any analytical kernel for which an analytical answer is available, such as $e^{-a\lambda}$ and determining the filter numerically by division of the Fourier transforms. The exact nature of the filter operator will depend on the particular combination of Bessel functions in the Hankel transform, on the sampling density, and on the length of the operator. Sampling densities ranging from three points per decade of λ to ten points per decade have been used (Anderson, 1973; Daniels, 1974). The length of the filter operator may vary from nine points (Strakhov, 1968) to forty points (Anderson, 1973) or longer. The convolution approach

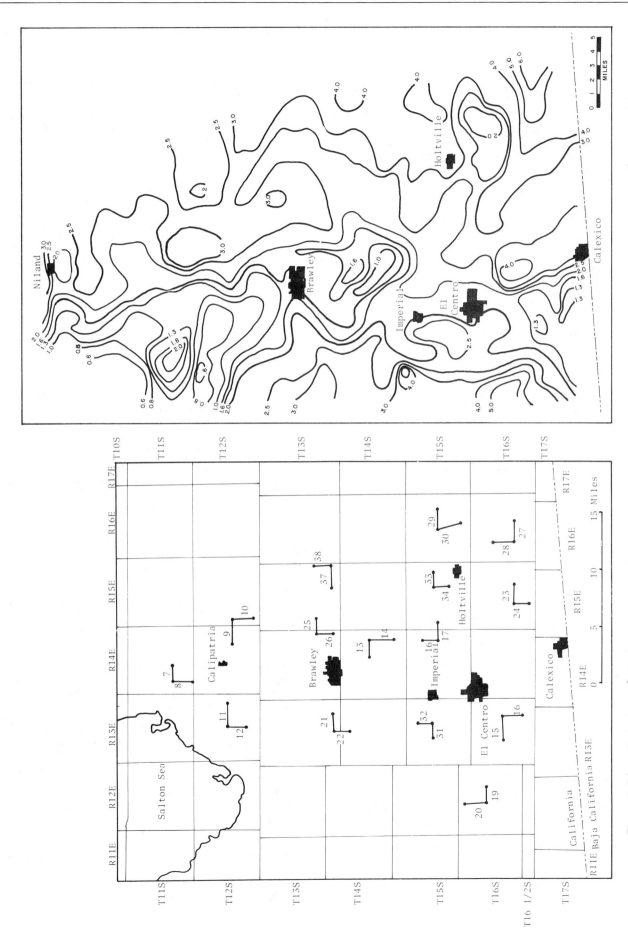

Figure 6.6B. Spatially averaged total-field resistivity map of the Imperial Valley, California. Contours in ohm-m.

Figure 6.6A. Locations of bipole sources used in a survey of the Imperial Valley, California.

has made evaluation of potential functions over a multilayer earth on any one dimensional earth which can be represented by a Hankel transform a straight-forward matter even with small computers.

Three-Dimensional Forward Problem

Three-dimensional resistivity distributions are more difficult to handle for a variety of reasons. The primary reason is the large number of parameters which are necessary to describe a three-dimensional structure. If the structure is simple, as in the case of a dyke, a buried sphere or figure of revolution, straightforward analytical solutions can often be obtained. For example, Jain (1974) has presented curves for potential field behavior over simple dykes intersecting the earth's surface. Singh and Espindola (1976) and Scurta (1972) have examined the case of a single sphere buried in the earth. Bibby and Risk (1973) have given results for an ellipsoid of revolution which is bisected by the earth's surface. Each of these cases represents a body whose shape can be fitted to an equipotential surface in an orthogonal system of coordinates. Tabulation of such bodies and the appropriate coordinate systems has been given by Van Nostrand and Cook (1966). Another set of geometries which is amenable to direct solution is that of planar-perpendicular boundaries (Keller et al., 1975). Examples of the boundaries which may be considered are shown in Figure 6.7.

While many shapes of arbitrary bodies can be simulated with figures of revolution, the restrictions of the analytical approach leave one with the feeling that a large class of bodies cannot be handled (Hohman, 1975). Therefore, considerable effort has been spent in recent years in the solution of the problem of the potential distribution when current flows around and through a body of completely arbitrary shape. Basically, two approaches to the solution of this problem have appeared; the network approach, and the surface integration approach.

In the network approach, sometimes also known as the finite-element method or the finite-difference method, the earth is divided into a series of cells. Sometimes the cells are rectangular in shape, while in other cases the faces of the cells are triangular or cylindrical. At each point in the mesh representing that portion of the earth through which appreciable current flows, some relationship between current, potential, and structure of the earth is evaluated. For example, in the finite-difference method, the Laplacian equation is converted to a finite-difference equation where the second derivatives are calculated in terms of the differences in potential between adjacent points in the mesh. Then, at each point in the mesh, Laplace's equation can be evaluated in terms of the potentials at adjacent points in the mesh. A set of simultaneous equations which in number equals the number of mesh points nearly is developed. Solution of this system of equations yields values for the potential throughout the medium as well as at the surface where it is to be measured.

Other equivalent approaches may also be used. For example, the cells into which the earth is divided can be considered as resistance networks. The size of the resistor used to represent each cell is simply related to the size of the cell and its resistivity. At each point where the resistors are connected in the network representing the earth, continuity of current must be preserved. This provides the equation valid at that point. Simultaneous solution of equations for all points representing the medium again provides a numerical solution in terms of the voltage distribution throughout the model and at the surface. This approach has been used by Pires (1975) to provide an interpretation of bipole-dipole surveys carried out in northwestern Nevada and described in a previous part of this paper.

Still another condition which may be used to generate equations at each of the points in a mesh-represented earth is one used by Coggon (1971, 1973). He derived an expression for the energy loss for each element in a mesh representing the earth. He then applied the condition that the energy dissipated should be minimum for the actual flow of current in the earth. This provides a set of equations for solution which yields the potential at each point through the medium and on the surface.

Methods based on the use of meshes to represent the earth have not proved to be entirely satisfactory. They work best if the earth can be represented in two dimensions; that is, if the resistivity varies in one horizontal and one vertical direction. Then the cross section in apparent resistivity can be represented with reasonable detail by a small enough number of mesh points that a solution can be obtained on a reasonably large size computer. Mufti (1976) reduces the complexity of the two-dimensional problem by assuming infinite line electrodes, so that the problem can be specified in terms of a single cross section. Use of point electrodes leads to the need for three-dimensional representation of the medium, even though the earth structure is only two-dimensional. The complexity of this problem is reduced by recognizing that the field along the structure can be Fourier transformed so that parallel cross-sections of the earth can be treated sequentially rather than simultaneously (Stoyer, 1974; Coggon, 1971, 1973; Pelton et al., 1978). Axially symmetric bodies have been done by two-dimensional finite elements and Fourier transform by Bibby (1978).

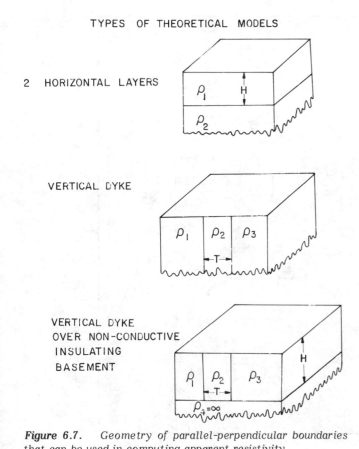

Figure 6.7. Geometry of parallel-perpendicular boundaries that can be used in computing apparent resistivity.

When three-dimensional bodies are considered, the number of mesh points required to represent the body can be quite large. Bibby (1978) has shown how to handle axially symmetric bodies. However, to get a good representation of a general body, the number of equations which must be solved simultaneously exceeds the capability of even the large-scale computers currently available.

The surface integration method (Alfano, 1959; Keller and Frischknecht, 1966; Morozova, 1967; Dieter et al., 1969; Barnett, 1972; Snyder, 1976) is capable of representing a homogeneous three-dimensional body also. The surface integration method is based on the solution of the fundamental relationship between current and voltage in a resistive earth. Combination of the definition of electrostatic potential, U, in terms of electric field, E:

$$\vec{E} = -\text{grad } U$$

with the divergence condition for current flow, J

$$\text{div } \vec{J} = 0 \quad \ldots 6$$

and a relationship between electric field and current density

$$\vec{E} = \rho \vec{J} \quad \ldots 7$$

(where ρ is resistivity)

leads to an expression in the form of Poisson's equation representing the behavior of potential in the earth

$$\nabla^2 U = -\rho \left[\nabla \cdot \vec{J}_o + \nabla U \cdot \nabla(\frac{1}{\rho}) \right] \quad \ldots 8$$

where J_o is the source current system, which has potential U_o in a uniform region. The solution to Poisson's equation is well known:

$$U = U_o - \frac{1}{2\pi} \int_V \frac{\rho \nabla U \cdot \nabla(\frac{1}{\rho})}{|R|} \, dV \quad \ldots 9$$

The first term in equation 9 represents the normal potential due to current flowing in a uniform earth with resistivity ρ. The second term represents the disturbing potential due to changes in resistivity within that earth. If changes take place at discrete boundaries, the second term is nonzero only at the boundaries. It can be considered conceptually as representing the effect of a series of current sources distributed over the surface of the arbitrary body. The current sources are of such strength as to match the boundary condition for current flow through that surface. The strength required for each of the elementary current sources over a surface is a function of the potential from all other current sources throughout the medium, both real and fictitious. Thus, a surface can be represented by some number, n, of current sources, and their strengths can be determined by a set of simultaneous equations. An example of the apparent resistivity curves over a buried body computed using this approach is shown in Figure 6.8.

Inversion

In the preceding paragraphs, the various approaches which are available for determining the potential on the surface of the earth for virtually any conceivable structure of resistivity in the subsurface have been enunciated. With this capability, the concept of analytical inversion has appeared, and the technique has proved to be progressively more and more feasible. Various approaches to the inversion of field data to find the most likely resistivity distribution giving rise to those data have appeared (Parker, 1977).

The concept of linearization and solution of the inverse problem can be developed as follows. Following Crous (1971), the most frequently used criterion of how well two sets of points agree is the least squares criterion, defined as

$$E = \sum_{i=1}^{n} (\rho(x_i) - \hat{\rho}(x_i, P))^2 \quad \ldots 10$$

where $\rho(x_i) = $ observed apparent resistivities at points x_i

and $\hat{\rho}(x_i, P) = $ theoretical model apparent resistivities at points x_i when the model parameters are $P = (p_j)$, i j k

when the model parameters are changed from $\underline{P}$ to $\underline{P} + \Delta$

where $\Delta = (\delta_i)$

the theoretical apparent resistivities can be represented by a Taylor series

$$\hat{\rho}(x_i, \underline{P} + \Delta) = \hat{\rho}(x_i, \underline{P}) + \sum_{j=1}^{k} \left(\frac{\partial \rho}{\partial p_j}\right)_{x_i, \underline{P}} \delta_j + \ldots \quad \ldots 11$$

where $\hat{\rho}(x_i, \hat{P} + \Delta)$ is the resistivity at points x; for a set of parameters, $\hat{P}$, which describe the resistivity profile (layer thicknesses and resistivities) for the minimum error, and δ are displacements of each parameter from its ideal value

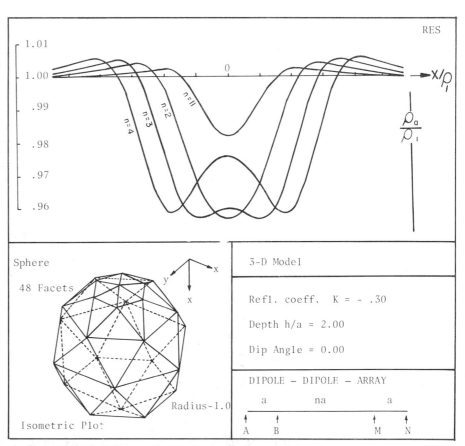

Figure 6.8. *Apparent resistivity profiles computed using surface integration for a buried, nearly spherical body (from Barnett, 1972).*

for the initial model. Only the first derivative from the Taylor's expansion is given in equation 11, this corresponds to linearizing the problem. The mean-square error expressed in terms of equation 11 is minimized at each observation point. This provides a set of normal equations which must be solved to find the values for the parameters describing the earth that give the minimum error.

This set of equations can be solved only when each of the parameters of the earth model is uniquely related to the error. In many cases, this is not the case. Ambiguity may result from two or more parameters describing the resistivity in the earth affecting the error in a like manner or when the model has more parameters than there are independent observations. Therefore, before a solution can be obtained, the dependence of the error on each one of the parameters must be established to see that the parameters are mutually independent. If they are not, the parameters may be lumped in groups which are independent, or the problem may be modified to make the parameters independent. Details of the mathematical methods used in obtaining a solution to the set of normal equations have been described by Marquardt (1963), Backus and Gilbert (1967), Crous (1971), Inman et al. (1973), Inman (1975), Parker (1977), and Oldenburg (1978). Some examples of interpretation are given in Rijo et al. (1977) and Petrick et al. (1977).

Linearization is not necessary in finding the model which gives minimum error in interpretation. Another approach which is commonly used is that of the Fibinachi search. Here, some initial estimate of the interpretation is made. The error is computed in the least squares sense. Then, each of the parameters describing the resistivity structure of the earth is perturbed to find if the error is reduced. In this way, progressively better models are obtained to simulate the earth. If the perturbations follow a Fibinachi progression, it can be shown that the minimum is found with the least number of evaluations of the function, providing that the shape of the error function is at least convex. The assumption of linearity of the error is not required. However, if the initial guess is at all reasonable, the linearization can lead to a very rapid convergence on the minimum.

Still another approach to interpretation is sometimes called pseudo-inversion. Because the problems involved in inversion relate to the nonindependence of the parameters describing the model, difficulty can be avoided by fixing part of the parameters. One way of doing this is to assume that earth is made up of a sequence of layers of fixed thickness, usually thicknesses which increase in a geometric progression (Marsden, 1973). Then, solutions are sought only in terms of the resistivities for these preassigned layer thicknesses. Because ambiguity arises primarily from the interdependence of error between a resistivity and a thickness, most of the ambiguity in interpretation is removed by fixing the thickness.

Inversion techniques have proved highly effective for cases in which one-dimensional models are useful. An example of the comparison between field data and the expected curve for an interpretation obtained by inversion is shown in Figure 6.9. With a good set of field data, the RMS error between the curve calculated for the interpretation and

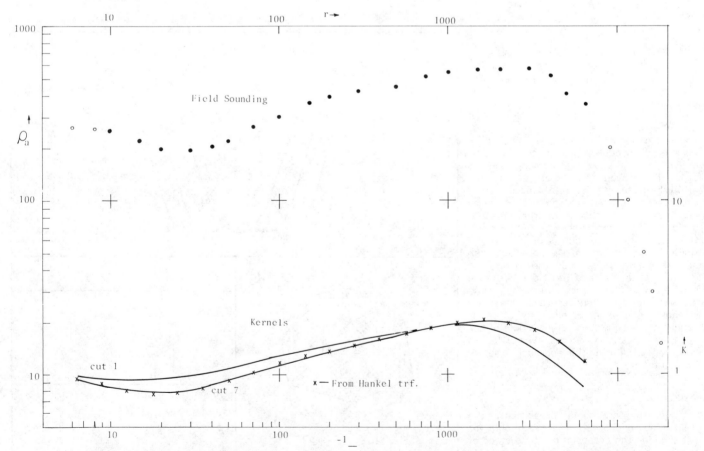

Figure 6.9. Example of the inversion of a Schlumberger sounding curve. Cut 1 represents the initial estimate, while cut 7 reflects the fit to date after seven interations of an inversion scheme (Crous, 1971).

of field observations can be expected to be less than one percent. To date, no convincing three-dimensional inversions have been reported in the literature. The difficulty with three-dimensional inversions arises because of the number of parameters that is necessary to describe the earth. Often, the number of parameters exceeds the number of observations so that the least squares process becomes indeterminate. Considerable research is being done at the present time on three-dimensional inversion schemes, so that some results may be expected in the near future.

SUMMARY AND CONCLUSIONS

While advances have been made in both the acquisition of field data in resistivity surveys and in the interpretation of those data, it appears that the capabilities of interpretive schemes has gone forward more rapidly than the capability for acquiring field data. It is to be expected that the capability to interpret field data will bring about some modifications in the way in which the field data are acquired to take advantage of these capabilities. It is my opinion that we can look forward to continued modification of field techniques, particularly in the direction of obtaining larger data sets from quantities of data to represent one particular survey to utilize in interpretation.

REFERENCES

Alfano, L.
    1959: Introduction to the interpretation of resistivity measurements for complicated structural conditions; Geophys. Prospect., v. 7, p. 311-368.

Alpin, L.M.
    1966: The theory of dipole sounding; in Dipole methods for measuring earth conductivity: Consultants Bureau, New York, p. 1-60.

Anderson, W.L.
    1973: Fortran IV programs for the determination of the transient tangential electric field and vertical magnetic field about a vertical magnetic dipole for an M-layered stratified earth by numerical integration and digital linear filtering; U.S. Geol. Surv. Pub. PB-226 240/5, Denver, Colorado.

Arestad, J.F.
    1977: Resistivity studies in the upper Arkansas Valley and northern San Luis Valley, Colorado; M.Sc. thesis 1934, Colo. Sch. Mines, Golden, Colorado, 129 p.

Backus, G.E. and Gilbert, J.F.
    1967: Numerical application of a formalism for geophysical inverse problems; Geophys. J. Roy. Astron. Soc., v. 13, p. 247-276.

Barnett, C.T.
    1972: Theoretical modeling of induced polarization effects due to arbitrarily shaped bodies; D.Sc. thesis T-1453, Colo. Sch. Mines, Golden, Colorado.

Beyer, J.H.
    1977: Telluric and DC resistivity techniques applied to the geophysical investigation of Basin and Range geothermal systems; Ph.D. thesis, Univ. of Calif., Berkeley, Rept LBL-6325.

Bibby, H.M.
    1977: The apparent resistivity tensor; Geophysics, v. 42 (6), p. 1258-1261.

    1978: Direct current resistivity modeling for axially symmetric bodies using the finite element method; Geophysics, v. 43 (3), p. 550-562.

Bibby, H.M. and Risk, G.F.
    1973: Interpretation of dipole-dipole resistivity surveys using a hemispheroidal model; Geophysics, v. 38 (4), p. 719-736.

Coggon, J.H.
    1971: Electromagnetic and electrical modelling by the finite-element model; Geophysics, v. 36, p. 132-155.

    1973: A comparison of IP electrode arrays; Geophysics, v. 38 (4), p. 737-761.

Crous, C.M.
    1971: Computer-assisted interpretation of electrical soundings; M.Sc. thesis 1363, Colo. Sch. Mines, Golden, Colorado, 108 p.

Daniels, J.J.
    1974: Interpretation of electromagnetic soundings using a layered earth model; Ph.D. thesis 1627, Colo. Sch. Mines, Golden, Colorado, 85 p.

Das, U.C. and Ghosh, D.P.
    1973: A study on the direct interpretation of dipole sounding resistivity measurements over a layered earth; Geophys. Prospect., v. 21 (2), p. 379-400.

    1974: The determination of filter coefficients for the computation of standard curves for dipole resistivity sounding over layered earth by linear digital filtering; Geophys. Prospect., v. 22 (4), p. 765-780.

Das, U.C., Ghosh, D.P., and Biewinga, D.T.
    1974: Transformation of dipole resistivity sounding measurements over a layered earth by linearly digital filtering; Geophys. Prospect., v. 22 (3), p. 476-489.

Dieter, K., Paterson, N.R., and Grant, F.S.
    1969: IP and resistivity type curves for three-dimensional bodies; Geophysics, v. 34 (4), p. 615-632.

Dobrin, M.B.
    1976: Introduction to geophysical prospecting; New York, McGraw-Hill, 630 p.

Doicin, D.
    1976: Quadripole-quadripole arrays for direct current measurements – model studies; Geophysics, v. 41 (1), p. 79-95.

Doicin, D., Ionescu, D., Direa, O., Trimbitos, I., and Mihalathe, I.
    1965: Etude des Massifs de sol en Transylvanie a l'aide des méthodes électriques; Revue Acad. Rep. Soc. Roumanie, Ser. Géophysique, v. 9, p. 101-117.

Furgerson, R.L.
    1970: A controlled source telluric current technique and its application to structural investigations; M.Sc. thesis 1813, Colo. Sch. Mines, Golden, Colorado, 123 p.

Ghosh, D.P.
    1971a: The application of linear filter theory to the direct interpretation of geoelectrical resistivity sounding measurements; Geophys. Prospect., v. 19, p. 192-217.

    1971b: Inverse filter coefficients for the computation of apparent resistivity standard curves for a horizontally stratified earth; Geophys. Prospect., v. 19 (4), p. 769-775.

Hohmann, G.W.
　1975: Three dimensional induced polarization and electromagnetic modeling; Geophysics, v. 40 (2), p. 309-324.

Inman, J.R.
　1975: Resistivity inversion with ridge regression; Geophysics, v. 40 (5), p. 798-817.

Inman, J.R., Jr., Ryu, J., and Ward, S.H.
　1973: Resistivity inversion; Geophysics, v. 38 (6), p. 1088-1108.

Jain, S.C.
　1972: Resistivity sounding on a three-layer transitional model; Geophys. Prospect., v. 20 (2), p. 283-292.

　1974: Theoretical broadside resistivity profiles over an outcropping dyke; Geophys. Prospect., v. 22 (3), p. 445-457.

Keller, G.V. and Frischknecht, F.
　1966: Electrical methods in geophysical prospecting: Pergamon Press, Oxford, 527 p.

Keller, G.V., Furgerson, R., Lee, C.Y., Harthill, N., and Jacobson, J.J.
　1975: The dipole mapping method; Geophysics, v. 40 (3), p. 451-472.

Ketola, M.
　1972: Some points of view concerning mise-à-la-masse measurements; Geoexploration, v. 10 (1), p. 1-22.

Koefoed, O.
　1966: The direct interpretation of resistivity observations made with the Wenner electrode configuration; Geophys. Prospect., v. 14 (1), p. 71-79.

　1968: The application of the kernel functions in interpreting geoelectrical measurements; Geoexploration Mon., Ser. 1(2), Gebruder Borntraeger, Stuttgart.

　1976a: Recent developments in the direct interpretation of resistivity soundings; Geoexploration, v. 14 (3/4), p. 243-270.

　1976b: Error propagation and uncertainty in the interpretation of resistivity sounding data; Geophys. Prospect., v. 24 (1), p. 31-48.

　1976c: An approximate method of resistivity sounding interpretation; Geophys. Prospect, v. 24 (4), p. 617-632.

Lee, C.Y.
　1977: Behavior of electric potential fields over randomly layered earth models; Ph.D. thesis 1950, Colo. Sch. Mines, Golden, Colorado, 112 p.

Marquardt, D.W.
　1963: An algorithm for least-squares estimation of non-linear parameters; J. Soc. Indust. Applied Math, v. 11, p. 431-441.

Marsden, D.
　1973: The automatic fitting of a resistivity sounding by a geometrical progression of depths; Geophys. Prospect., v. 21 (2), p. 266-280.

Meinardus, H.A.
　1967: The kernel function in direct-current resistivity sounding; D.Sc. thesis 1103, Colo. Sch. Mines, Golden, Colorado, 151 p.

　1970: Numerical interpretation of resistivity soundings over horizontal beds; Geophys. Prospect., v. 18, p. 415-433.

Merkel, R.H.
　1971: Resistivity analysis for plane-layer half-space models with buried current sources; Geophys. Prospect., v. 19 (4), p. 626-639.

Merkel, R.H. and Alexander, S.S.
　1971: Resistivity analysis for models of a sphere in a half-space with buried current sources; Geophys. Prospect., v. 19 (4), p. 640-651.

Mooney, H.M., Orellana, E., Picket, H., and Tornheim, L.
　1966: A resistivity computation method for layered earth models; Geophysics, v. 31 (1), p. 192-203.

Morozova, G.M.
　1967: Primeneniya metoda integral'nikh uravnenii pri reshenii zadach teorii electrorazvedki postroyannim tokom; Geologiya i Geofizika, n. 11, p. 104-111.

Morris, Drew,
　1975: Quadripole mapping near the Fly Ranch geothermal prospect, northwest Nevada; M.Sc. thesis 1699, Colo. Sch. Mines, Golden, Colorado, 100 p.

Mufti, I.R.
　1976: Finite difference resistivity modeling for arbitrarily shaped two-dimensional structures; Geophysics, v. 41 (1), p. 62-78.

Naidu, P.S.
　1970: A response of a randomly layered earth to an electric point or dipole source; Geophys. Prospect., v. 22 (2), p. 279-296.

Niwas, S. and Upadhyay, S.K.
　1974: Theoretical resistivity sounding results over a transition layer model; Geophys. Prospect., v. 22 (2), p. 279-296.

Oldenburg, D.W.
　1978: The interpretation of DC resistivity measurements; Geophysics, v. 43 (3), p. 610-625.

Orellano, Ernesto
　1972: Prospecion geoelectrica en corriente continua; Madrid, Paraninfo, 523 p.

Parker, R.L.
　1977: Understanding inverse theory; in Ann. Rev. Earth Planet. Sci., v. 5, p. 35-64.

Pelton, W.H., Rijo, L., and Swift, C.M.
　1978: Inversion of two-dimensional resistivity and induced polarization data; Geophysics, v. 43 (4), p. 788-803.

Petrick, W.R., Pelton, W.H., and Ward, S.H.
　1977: Ridge regression inversion applied to crustal resistivity sounding data from South Africa; Geophysics, v. 42 (5), p. 995-1005.

Pires, A.B.C.
　1975: Resistor network modeling in electric dipole mapping; D.Sc. thesis 1746, Colo. Sch. Mines, Golden, Colorado, 135 p.

Rijo, L., Pelton, W.H., Fertosa, E.C., and Ward, S.H.
　1977: Interpretation of apparent resistivity data from Apodi Valley, Rio Grande do Norte, Brazil; Geophysics, v. 42 (4), p. 811-822.

Risk, G.F. Macdonald, W.J.P., and Dawson, G.B.
　1970: DC resistivity surveys of the Broadlands geothermal region, New Zealand; Geothermics, Special Issue 2 (2), p. 287-294.

Roman, Irwin:
- 1959: An image analysis of multiple layer resistivity problems; Geophysics, v. 24, p. 485-509.
- 1960: Apparent resistivity of a single uniform overburden; U.S. Geol. Surv., Prof. Paper 365, 99 p.
- 1963: The kernel function in the surface potential for a horizontally stratified earth; Geophysics, v. 28 (2), p. 232-249.

Sampaio, E.S.
- 1976: Electrical sounding of a half-space whose resistivity or its inverse function varies linearly with depth; Geophys. Prospect., v. 24 (1), p. 112-122.

Scurtu, E.F.
- 1972: Computer calculation of resistivity pseudo sections of a buried spherical conductor body; Geophys. Prospect., v. 20 (3), p. 605-625.

Singh, S.K. and Espindola, J.M.
- 1976: Apparent resistivity of a perfectly conducting sphere buried in a half-space; Geophysics, v. 31 (4), p. 742-751.

Slichter, L.B.
- 1933: The interpretation of the resistivity prospecting method for horizontal structures; Physics, v. 4, p. 307-322.

Snyder, D.D.
- 1976: A method for modeling the resistivity and IP response of two-dimensional bodies; Geophysics, v. 41 (5), p. 995-1015.

Snyder, D.D. and Merkel, R.M.
- 1973: Analytic models for the interpretation of electrical surveys using buried current electrodes; Geophysics, v. 38 (3), p. 513-529.

Stefanescu, S., Schlumberger, C., and Schlumberger, M.
- 1930: Sur la distribution électrique potentielle autour d'une prise de terre ponctuelle dans un terrain à couches horizontales homogènes; Le Journal de Physique et le Radium, Series 7, v. 1, p. 132-140.

Stefanescu, S.S. and Tanasescu, P.
- 1965: Sur la prospection électrique par la méthode des émetteurs croisés; Carpatho-Balkan Geol. Assoc., VII Congress, Sofia, Part VI, Proc., p. 199-205.

Stoyer, C.H.
- 1974: Numerical solutions of the response of a two-dimensional earth to an oscillating magnetic dipole source with application to a groundwater field study; Ph.D. thesis, Penn. State Univ., Pennsylvania.

Stoyer, C.H. and Wait, J.R.
- 1977: Resistivity probing of an "exponential" earth with a homogeneous overburden; Geoexploration, v. 15 (1), p. 11-18.

Strakhov, V.N.
- 1968: O reshenii obratnoi zadachi v metode vertikal'nikh elektricheskikh zondirovanii; Fizika Zemli, no. 4.

Telford, W.M., Geldart, L.P., Sheriff, R.E., and Keys, D.A.
- 1976: Applied Geophysics; Cambridge, Cambridge Univ. Press, 860 p.

Van Nostrand, R.G. and Cook K.L.
- 1966: Interpretation of resistivity data; U.S. Geol. Surv., Prof. Paper 499, 310 p.

Vedrintsev, G.A.
- 1966: Theory of electric sounding in a medium with lateral discontinuities; in Dipole methods for measuring earth conductivity; New York, Consultants Bureau.

Verma, R.K. and Koefoed, O.
- 1973: A note on the linear filter method of computing electromagnetic sounding curves; Geophys. Prospect., v. 21 (1), p. 70-76.

Zohdy, A.A.R.
- 1970: Geometric factors of bipole-dipole arrays; U.S. Geol. Surv., Bull 1313-B.

MAGNETIC METHODS APPLIED TO BASE METAL EXPLORATION

Peter J. Hood, M.T. Holroyd, and P.H. McGrath
Geological Survey of Canada

Hood, Peter J., Holroyd, M.T., and McGrath, P.H., Magnetic methods applied to base metal exploration; in Geophysics and Geochemistry in the Search for Metallic Ores; Peter J. Hood, editor; Geological Survey of Canada, Economic Geology Report 31, p. 77-104, 1979.

Abstract

During the past decade, both ground and airborne magnetometers have been improved considerably by miniaturization and made more reliable by the extensive use of integrated circuit devices. For aeromagnetic surveys, proton precession magnetometers are now in common use for standard-sensitivity surveys in inboard installations and their sensitivity is commonly 1.0 gammas or better. An inboard vertical gradiometer has been fabricated by the Geological Survey of Canada using single-cell optical absorption magnetometers; results from test surveys clearly demonstrate the superior resolution of the gradiometer compared to total field results, although the line spacing for such surveys must consequently be somewhat closer. Presently a new generation of magnetometer is under active development which utilizes superconducting quantum interference detectors (Squid). If successful, this development will result in a short-base three-component gradiometer system with a much higher sensitivity than is possible with optical-absorption magnetometers.

Digital recording of aeromagnetic survey data is now standard procedure and except for the flight path recovery process, the compilation of aeromagnetic maps is now a completely automated process. Computer contouring is more objective than hand contouring, and far less costly. Furthermore, the aesthetics of machine-produced contour maps are now almost as good as the manual product. The grid data generated for the machine contouring process also facilitate the production, efficiently and at low cost, of a variety of derived and processed maps, to any scale or map projection for use in the subsequent interpretation of the aeromagnetic data.

Quantitative interpretation methods of magnetic survey data have also been improved considerably over the past decade. Several automated techniques for aeromagnetic profile data have been introduced which produce continuous depth-to-causative body profiles although the results of such routine procedures must be modified by the interpreter to take account of strike direction and the finite lengths of the causative bodies. Several three-dimensional "best-fit" interpretation techniques for individual anomalies have been published in recent years which are quite general with regard to the geometry and magnetization vector of the causative body. These techniques may also be programmed so that the results may be viewed on an interactive graphics terminal so that the interpreter can constrain various parameters to conform to the known geological realities.

Résumé

Au cours de la dernière décennie, on a amélioré de façon considérable, grâce à la miniaturisation, les magnétomètres aéroportés et terrestres; de même, on les a rendu plus fiables grâce à l'utilisation poussée de dispositifs à circuit intégré. En ce qui concerne les levés aéromagnétiques, les magnétomètres à protons sont maintenant d'utilisation courante pour les levés à sensibilité normale à bord des avions et leur niveau de sensibilité est habituellement de 1,0 gamma ou d'un niveau meilleur. Un gradiomètre vertical embarqué a été construit par la Commission géologique du Canada: il utilise des magnétomètres à absorption optique à cellule simple. Les résultats des levés démontrent clairement que le gradiomètre a une résolution supérieure à l'ensemble des résultats sur le terrain, quoique l'espacement linéaire des levés soit nécessairement un peu plus serré. A l'heure actuelle, on est à mettre au point une nouvelle génération de magnétomètres utilisant des détecteurs supraconducteurs à interférence quantique (Squid). S'ils s'avèrent un succès, il en résultera un système à base courte de gradiomètres à trois éléments, d'une sensibilité plus élevée qu'il est possible d'atteindre avec les magnétomètres à absorption optique.

L'enregistrement numérique des données aéromagnétiques est maintenant chose courante, à l'exception du procédé de correction de la trajectoire de vol, l'établissement des cartes aéromagnétiques est entièrement automatisé. Le traçage des courbes par ordinateur est plus objectif que le traçage à la main, et de loin plus économique. Bien plus, la présentation des cartes en courbes produites mécaniquement est maintenant presque aussi bonne que celle des cartes produites à la main. Les données de quadrillage fournies par le procédé mécanique de traçage des courbes facilitent également la production, de façon plus efficace et à un coût moindre, d'une gamme de cartes dérivées et traitées, à n'importe quelle échelle ou projection cartographique devant être utilisées lors de l'interprétation ultérieure des données aéromagnétiques.

Les méthodes d'interprétation quantitative des données magnétiques ont aussi été considérablement améliorées au cours de la dernière décennie. Plusieurs techniques automatisées de profilage aéromagnétique ont été introduites; elles permettent de produire des profils continus de profondeur des masses, bien que les résultats de tels procédés de routine doivent être modifiés par l'interprète afin de tenir compte de la direction de trace horizontale et des longueurs limites des masses. Plusieurs techniques tridimensionnelles d'interprétation "idéales" d'anomalies individuelles ont été publiées au cours des dernières années; elles sont quelque peu générales du point de vue de la géométrie et du vecteur de magnétisation de la masse. Ces techniques peuvent également être programmées de façon à ce que les résultats puissent être visionnés sur un terminal de dialogue des graphiques; l'interprète peut ainsi rendre certains paramètres conformes aux réalités géologiques connues.

INTRODUCTION

More than 500 articles dealing with the magnetic survey technique have been published since the first symposium in 1967 so it is possible to give more than a condensed overview of the significant progress that has been made in the past decade in this review; we will, therefore, present what appear to us to be the highlights with a consequent but unavoidable omission of much important work.

There have been some important reviews published in the last decade. A comprehensive review of aeromagnetic survey instrumentation and techniques was published by Hood and Ward (1969). For general reviews of data processing and interpretation, the reader is referred to the articles by Steenland (1970) and by Grant (1972).

Units Used in Magnetic Surveying

In September 1973, the International Association of Geomagnetism and Aeronomy recommended the adoption of Systeme International (SI) units in the field of geomagnetism. Of main interest to exploration geophysicists are the SI units for magnetic field (T), intensity of magnetization (J) and volume susceptibility (k) and the changes are briefly reviewed in the following discussion. For further elaboration the reader is referred to Hahn (1978) and to IAGA (1973); an understanding of the units used in geomagnetism appears to have been somewhat obfuscated by the adoption of SI units, and what follows, therefore, is mainly intended for the guidance of exploration geophysicists.

Magnetic Field (T)

In practice, the SI and cgs units of magnetic field are the tesla and oersted respectively and their relationship is

$$1 \text{ tesla} = 10\ 000 \text{ oersteds}$$

which is, however, a rather large unit for magnetic survey use. The unit in common usage by exploration geophysicists for at least 50 years is the gamma (symbol γ) and

$$1 \text{ gamma} = 10^{-5} \text{ oersted}$$

so that $1 \text{ gamma} = 10^{-9}$ teslas = 1 nanotesla (symbol nT)
$= \dfrac{10^{-2}}{4\pi} \text{ Am}^{-1} \simeq 8 \times 10^{-4} \text{ Am}^{-1}$

In the IAGA resolution adopted at the 1973 meeting in Kyoto, Japan, it was perhaps anticipated that common usage would keep the gamma alive for the following statement was made: "If it is desired to express values in gammas, a note should be added stating that 'one gamma is equal to one nanotesla'".

Intensity of Magnetization (J)

In the cgs system, the units of magnetization were usually expressed in emu units. In SI, the units are ampere per metre and the relationship is $1 \text{ emu} = 10^3 \text{ Am}^{-1}$ so that

$$J_{emu} \times 10^{-6} \text{ emu} = J_{SI} \times 10^{-3} \text{ Am}^{-1} = J_{SI} \text{mAm}^{-1}$$

i.e. milliamp per metre.

Susceptibility (k)

Volume susceptibility was usually expressed in dimensionless cgs units as $\times 10^{-6}$ emu/cc or cgsu. For SI units, 1 cgsu of susceptibility = $1/4\pi$ SI unit of susceptibility, i.e.

$$k_{SI} = 4\pi k_{cgs}$$

and common usage is to express volume susceptibility in $\times 10^{-3}$ SI units so that 1000×10^{-6} emu/cc = 12.6×10^{-3} SI units.

Because volume susceptibility is the ratio between the induced intensity of magnetization and applied magnetic field which are both expressed in Am^{-1}, it results in the necessity of expressing susceptibility in the somewhat meaningless terms of SI units. Actually susceptibility (or rather permeability) is a fundamental physical constant and should therefore have its own units. Accordingly for exploration geophysicists it would be convenient if the $\times 10^{-3}$ SI susceptibility unit were called a kappa; then for the induced magnetization case

$$k_{SI} = \dfrac{J_{SI} \text{ in mAm}^{-1}}{T \text{ in gammas}} \times \dfrac{4\pi}{10}$$

so that k_{SI} in kappas $= \dfrac{J_{SI} \text{ in mAm}^{-1}}{T \text{ in gammas}} \times 400\pi$

Thus, a rock with 1 per cent magnetite by volume which usually has a susceptibility of about 3000×10^{-6} emu/cc, would have a susceptibility of 37.7 kappas, which is a convenient size number to deal with.

MAGNETIC SURVEY INSTRUMENTATION

During the past decade, both ground and airborne magnetometers have been improved considerably by miniaturization and made more reliable by the extensive utilization of integrated circuit devices. The theory of operation of the various types of magnetometer and much additional background material was published in the proceedings volume for the 1967 symposium held in Niagara Falls (Hood, 1970) and will not be repeated here.

Ground Survey Magnetometers

For ground magnetometers, the mechanical balance type has been almost completely replaced in survey use throughout the world by electronic magnetometers because of the higher rate of survey production that can be achieved with the latter and the fact that electronic magnetometers have a higher sensitivity. There are two main types of ground magnetometer in common survey use for mineral exploration, namely the fluxgate and proton free-precession magnetometer (Table 7.1).

Ground Fluxgate Magnetometers

The typical fluxgate magnetometer weighs from 1.5 to 4 kg (that is 3 to 9 lb), measures the vertical component of the earth's magnetic field and in survey use is carried by a strap worn around the operator's neck (see Fig. 7.1). The range of a typical fluxgate instrument is $\pm 100\ 000\ \gamma$ which is more than adequate for worldwide use except over very strongly magnetized iron formation. The majority of ground fluxgate magnetometers have a meter as the readout device with the highest sensitivity selectable usually being 10 or 20 γ per small division. However, Scintrex manufacture a one-gamma digital fluxgate magnetometer, the MFD-4, in which the vertical component is read from a five-digit light-emitting diode display.

In 1977, Littlemore Scientific Engineering of Oxford, U.K. produced a fluxgate gradiometer which has a differential resolution of 0.5 γ. The 4.6 kg Elsec 781 fluxgate gradiometer (Fig. 7.2) consists of two fluxgate elements mounted coaxially 1 m apart at the ends of a Pyrex glass tube which has a low coefficient of thermal expansion and therefore negligible temperature effect. Thus the instrument measures the vertical gradient of the vertical component of the earth's magnetic field with a resolution of 0.5 γ/m. The

Table 7.1

Commercially available electronic magnetometers for ground prospecting. All magnetometers are hand held and operate from rechargeable batteries or cells (from Hood, 1978)

Type	Manu-facturer	Model designation	Component measured T or V	Readout display	Sens in gammas	Range in gammas (K=1000)	Weight in field use	Power Source
Fluxgate	Adams Marine (Canada)	Sable Mark 2	V	Meter	20q/div	100K	4 kg	9 VDC transistor cell
	Littlemore (UK)	Elsec 781	V & vert. grad.	LCD Bar	0.5γ	±5000q/m	kg	18 V Pb-acid battery
	McPhar (Canada)	M700	V	Meter	20q/div	±100K	3.8 kg	18 VDC 9V or C cells
	Phoenix (Canada	MV-1	V	Meter	20q/div	±100K	1.7 kg	2x6 V gel-cell batteries
	Scintrex (Canada)	MF2-100 MFD-4	V V	Meter 5 digit LED	2q/div 1	±200K ± 99 999	2.9 kg 2.1 kg	24 VDC cells 4 D cells or 6V Pb-acid battery
Proton precession	Austral (Australia)	PPM-1	T	5 digit	1γ	20 to 100K	6.8 kg	12-15 VDC D cells or Pb acid battery
	Barringer (Canada	GM-122	T	5 digit LED	1γ	20 to 100K	5.1 kg	18 VDC 12 D cells
	Geometrics (US)	G816	T	5 digit	1γ	20 to 100K	2.8 kg	18 VDC D cells
		G836	T	4 digit	10γ		2 kg	12 VDC
	Littlemore (UK)	Elsec 595	T	5 digit Nixie	1γ	24 to 72K	6.7 kg	17 VDC Nicad battery
		Elsec 770	T	LCD	0.25γ	20 to 90K	2.5 kg	18 VDC D cells
	Scintrex (Canada)	MP-4	T	5 digit LED	1γ	20 to 100K	3.7 kg	12 VDC 8 D cells
Optical absorption	Varian (Canada)	V1W 2302	T & gradient	LED Audio	0.1q and 0.1γ/m	20 to 100K	12.5 kg	30 VDC 5 gel-cell batteries

electronic circuitry utilizes Cmos components and a novel readout method. The output of the differential fluxgates is in analog form which can be used for any external recorder system; the output signal is digitized and displayed on a liquid-crystal two-level bar readout. The bar readout combines the analog aspect of travelling in a positive or negative direction but at the same time giving a precise digital indication. Two parallel bar displays are provided, one having 10 times the resolution of the other. Besides this visual readout, an audio signal is provided when the field change exceeds a front panel preset value. The Elsec 781 fluxgate gradiometer is intended for archeological investigations and should be of particular value in the search for ferrous objects, e.g. steel pipes, munitions, located some distance below the ground which are beyond the limited range of mine detector-type devices.

Ground Proton Precession Magnetometers

Ground proton precession magnetometers have been improved considerably over the past decade particularly with respect to their weight in field use, their ability to function properly in high magnetic field gradients and lower power consumption. A typical ground proton precession magnetometer measures the total intensity of the earth's magnetic field which is read on a LED display with a sensitivity of one gamma or better, weighs from 2.5 to 7 kg (5 to 15 lb) and operates over the range 20 000 to 100 000 γ using Nicad or lead-acid batteries. Figure 7.3 shows the Scintrex MP-4 instrument which uses an optimized noise-cancelling dual sensor with a gradient tolerance of 5000 γ/m. An indicator light warns of excessive gradient, ambient noise or electronic failure and a digital readout of the battery voltage indicates whether the batteries should be replaced. Many of the ground proton precession magnetometers can be used without modification as diurnal base station monitors with a suitable digital or analog chart recorder.

Optical Absorption Magnetometers

In 1975, Varian Associates of Georgetown, Ontario introduced a portable cesium-vapour magnetometer, the V1W-2302, which was intended to be used to search for ferromagnetic objects such as pipelines or in archeological investigations. In 1977, Varian introduced a gradiometer version of the same instrument which was designated the V1W-2302C1 (Fig. 7.4). The V1W-2302C1 is an instrument that uses two cesium magnetometers to determine the

Figure 7.1. McPhar M700 fluxgate magnetometer with strap around operator's neck, being operated on the Cavendish geophysical test range, southwestern Ontario; McPhar Instrument Corp., Willowdale, Ontario. (GSC 202228-G)

Figure 7.2. Elsec 781 digital fluxgate gradiometer; Littlemore Scientific Engineering Co., Oxford, U.K. (GSC 203492-O)

magnitude of the magnetic field intensity at two points spaced 2 m apart. By subtracting the two measured values, a gradient of the total field intensity can be determined. The sensitivity of the gradient measurement is $\pm 0.1 \, \gamma/2$ m and the range is $\pm 9000 \, \gamma/2$ m. The instrument is also designed to work in the so-called variometric mode whereby one Cs sensor is stationary while the other, connected via a single coaxial cable, is used as a mobile sensor. In this mode a very detailed survey of an area may be conducted with the results being independent of time-varying components of the earth's magnetic field, i.e. diurnal changes and micropulsations. Each of the two Varian cesium sensors can also be used for the total field measurements. In this mode of operation, the magnitude of the magnetic field intensity can be measured over the range of 20 000 to 100 000 γ, which permits the instrument to be used for magnetic surveys in any part of the world. The sensitivity of the measurements is $\pm 0.1 \, \gamma$ and the absolute accuracy is $\pm 0.5 \, \gamma$. In all three modes of operation (gradiometer/variometer/total field magnetometer) the results are displayed in the form of a 6-digit number by a 7 segment incandescent display on the 2.1 kg console. The display is updated 2 times per second. In addition to the visual display, an audio output is available whose frequency is 7 Hz/γ in the total field mode and 7 Hz per 1 $\gamma/2$ m in the gradiometric mode. The audio display permits a rapid survey for localized magnetic anomalies due to buried artifacts. The visual display update time is limited by the human response. The instrument takes measurements at the rate of 11 times per second and each of the measurements takes 0.045 seconds. This information is available in the digital (BCD) form or analog form (potentiometric or galvanometric) for the recording version of the instrument. The Varian V1W-2302C1 portable gradiometer is powered by 5 gel rechargeable batteries connected in series (providing 30 V) for periods of up to 6 hours before the batteries have to be recharged; the temperature operating range is 0° to 50°C.

Instrumentation for Magnetic Property Measurements

It is often useful to investigate the magnetic properties of a rock formation to ascertain whether it has sufficient magnetization to produce a given anomaly or whether the direction of the magnetization vector differs markedly from that of the present earth's field e.g. is reversed. The magnetization (J) of a given rock formation is due to two components, namely the induced and the remanent magnetization (R). These are vectors and the induced magnetization is accurately aligned in the direction of the earth's magnetic field (T).

Magnetic Methods

Figure 7.3. Scintrex MP-4 direct-reading proton precession magnetometer; Scintrex Ltd., Toronto, Ontario. (GSC 202228-E)

Figure 7.4. V1W-2302C1 cesium gradiometer; Varian Associates, Georgetown, Ontario. (GSC 203492-T)

Thus $\quad \bar{J} = k\bar{T} + \bar{R}$

where k is the susceptibility of the rock formation. Thus in order to ascertain the total magnetization of a rock formation it is necessary to measure both the susceptibility and the remanent magnetization vector.

Susceptibility Meters

The preferred technique in measuring the susceptibility of a given rock formation is to carry out in situ measurements because this physical parameter is somewhat variable and a representative determination cannot normally be made from a few hand specimens.

Table 7.2 lists the susceptibility meters which are presently available. Several portable susceptibility meters have appeared in the last decade. One example is the Elliot PP-2A instrument (Fig. 7.5) which weighs 0.5 kg and may be used to measure the susceptibility of hand and drill core samples or for in situ measurements on outcropping rock formations. The direct digital readout of susceptibility to two significant figures over the range 100×10^{-6} to $99\,000 \times 10^{-6}$ cgsu (1244×10^{-3} SIu) may be obtained with a resolution of 100×10^{-6} cgsu (0.13×10^{-3} SIu) by pushing a button. The instrument may also be calibrated in percentage magnetite equivalents for the direct logging of drill core for magnetite content.

Figure 7.5. Elliot PP-2A susceptibility meter; Elliot Geophysical Co., Tucson, Arizona. (GSC 203492-S)

Table 7.2

Commercially available susceptibility meters (from Hood, 1978)

Manu- facturer	Model Designation	Type	Frequency of operation in Hertz	Readout device	Range (K=1000) x10⁻⁶cgsu	Accuracy	Weights in field use (kg)	Power Source
Bison Instruments (US)	3101 A	in situ and lab		Digital dial & meter	1-100K	Absolute 5% Relative 1%	4.5	2x5.4V Hg cells
Elliot Geophysical (US)	PP 2A	in situ and lab		2 digit	100-99K	100x10⁻⁶ cgsu	0.5	
Institute of Applied Geophysics (Czecho- slovakia)	KT-3	in situ and lab	10 kHz	Meter	300-40K	Absolute 5%	1.25	6 D cells
	Kappabridge KLY-1[a]	Lab	970 Hz	Meters	16-7960	± 0.05% of range	60	115/220 VAC
Scintrex (Canada)	CTU-2[b]	in situ and lab	5 kHz	Meter	100-40K ± 10% Relative ± 2%	Absolute	5.5	9 D cells

a. Also measures anisotropy.
b. Also measures IP and conductivity.

In 1969, the Institute of Applied Geophysics in Brno, Czechoslovakia introduced the KT-2 Kappameter which was a robust susceptibility meter for in situ measurements. It operated on a frequency of 10 kHz and had a measuring range of 0.2×10^{-3} to 2000×10^{-3} SIu, with a sensitivity of 0.05×10^{-3} SIu. An accurate reading took only 15 seconds. The KT-2 could also be used for determination of susceptibilities on drill cores of any diameter without the need for special core holders, and for laboratory testing of samples. It was battery operated and was designed primarily for use in the field before and during magnetic surveying. In 1970, an improved model, the KT-3 Kappameter (Fig. 7.6) was introduced which had improved sensitivity (0.02×10^{-3} SI units) and six measuring ranges from 4×10^{-3} to 500×10^{-3} SI units.

In 1970, the Institute of Applied Geophysics in Brno, Czechoslovakia also introduced a high sensitivity laboratory bridge, the KLY-1 Kappabridge, for the measurement of susceptibility of rocks in low intensity (80 A/m) alternating (970 Hz) magnetic field which had a magnetic anisotropy measurement capability. Measurement of the anisotropy of susceptibility makes it possible to determine for example the bedding planes of nonlayered sediments, the fluid structure of igneous rocks, and the magnetic lineation and foliation of metamorphic rocks. The sensitivity of the instrument for a 8 cm³ sample is 4×10^{-8} SIu and five measuring ranges were incorporated being 0.2, 1, 4, 20 and 100×10^{-3} SIu.

Remanent Magnetometers

In 1969, the Institute of Applied Geophysics in Brno, Czechoslovakia completed the development of two laboratory remanent magnetometers, the LAM-1 astatic magnetometer and the JR-2 spinner magnetometer. The JR-2 spinner magnetometer was designed to measure remanent magnetic polarization even when specimens were only slightly magnetized. It consisted of three units, a spinner with anti- magnetic shielding in which the sample rotated at 80 rev/min, a power supply unit, and an amplifier-demodulator measuring unit. Measurements of six different sample orientations, whereby each of the three components was determined four

Figure 7.6. KT-3 Kappameter, being used to measure the volume susceptibility of a rock formation; Institute of Applied Geophysics, Brno, Czechoslovakia. (GSC 203492-V)

times, took only 10 minutes. The JR-2 had a sensitivity of 4 $\times 10^{-6}$ Am^{-1}. It did not require any special laboratory conditions. The LAM-1 astatic magnetometer was capable of measuring the remanence, susceptibility, and susceptibility anisotropy using irregular and large sized rock samples. The sensitivity of the LAM-1 was about 1.6×10^{-5} Am^{-1}, i.e. 2×10^{-7} Oe. In 1973, the range of LAM astatic magnetometers was increased by the introduction of three new models. The basic, simplest version LAM-2 measures the astatic system deflection by a light beam spot on a graduated scale with sensitivity 4×10^{-15} to 6×10^{-6} Am^{-1} scale division (5×10^{-7} to 5×10^{-8} Oe/scale division). The LAM-3 incorporates an electronic negative feedback loop and system deflection is displayed on an analog indicator. This gives a shorter setting time, stabilizes the sensitivity and allows simple range selection. A digital output device converts the LAM-3 to a LAM-3D which permits automatic data recording for computer handling. The LAM-4 has automatic ranging which further shortens the measuring time. The LAM-3 and LAM-4 have 7 sensitivity ranges from 8×10^{-4} to 8×10^{-1} Am^{-1} (1×10^{-5} to 1×10^{-2} Oe/fsd). Each LAM is provided with an orthogonal manipulator that will take irregular samples of up to 10 cm diameter; a microlamp for small samples is also provided so that measurements are quickly carried out.

In 1974, the Institute of Applied Geophysics in Brno, Czechoslovakia commenced delivery of a digital version of their JR series (JR-2 and the subsequent JR-3) of spinner magnetometers. The new instrument, designated UGF-JR4 has digital indication of two simultaneous components of remanent magnetic polarization and parallel BCD output for direct computer processing of measured data. Samples may be cubes $20 \times 20 \times 20$ mm or cylinders 25.4 mm diameter by 22 mm length. Cylinders are measurable with axis vertical or horizontal. The major improvements incorporated in the JR-4 rock magnetometer include increased sensitivity, a wider measuring range, simpler measuring procedure, automatic compensation for magnetic moment of sample holder, higher accuracy, and shorter measuring time. Some of the features of the new instrument are automatic compensation for the magnetic moment of the sample holder, use of photo transistors instead of permanent magnets to obtain the reference signal, and an automatic stopping of the sample rotation at the end of integration time. The most sensitive measuring range covers 0-9.99 picoteslas with a sensitivity of 1 pT for an integration time of 100 seconds.

A portable instrument designated the PSM-1 (Fig. 7.7) has been developed by Schonstedt Instrument Company of Reston, Virginia, in which the remanent magnetization of irregularly-shaped rock specimens may be measured. Full-scale ranges of 10^{-4} to 1 emu (10^{-1} to 10^{3} Am^{-1}) permit the measurement of virtually all igneous rocks and many types of sediments with an accuracy of ±10% of moment and ±5% direction angle. Schonstedt also manufactures a range of laboratory spinner magnetometers for more accurate remanent magnetism measurements.

Superconducting quantum interference device (Squid) sensors have also been developed for use in rock magnetism measurements. The interested reader is referred to an excellent review of the topic by Goree and Fuller (1976) for descriptions of the theory and design of the Squid instrumentation.

Drillhole Logging Instrumentation

For iron-ore prospecting, the measurement of the magnetic parameters of rock formations through which drillholes pass are carried out using susceptibility logging equipment or by the use of three-component fluxgate magnetometers. In 1974, Zablocki described the magnetic susceptibility system being used by the U.S. Steel Corporation at the Minntac open-pit taconite mine in northern Minnesota for short blast holes 40-60 feet (12-18 m) in depth with a diameter of 12.25 in. (31 cm). Figure 7.8 shows a schematic diagram of the system which uses an air-cored multi-turn induction coil connected to one arm of an inductance bridge and designed using an operational amplifier so that there was constant current through the induction coil. The system had a capability of measuring the Fe content in the drillholes with a standard error of ±1.2 weight per cent Fe, and was considered to be especially useful in establishing cut-off boundaries between ore and waste. In situ bulk measurements are much preferable to laboratory measurements of much smaller volumes using drill core.

In recent years much effort has been devoted in Sweden to the development of a reliable drillhole fluxgate magnetometer which would be useful for the detection of iron-ore deposits. In 1969, ABEM introduced a new version of the Swedish three-component drillhole magnetometer for EX (36 mm) holes designated the HBM 4 (Fig. 7.9). This transistorized instrument measures three orthogonal magnetic components as well as dip angle down to 1200 m (3600 ft.). The measurements have an accuracy of 50 γ within the range ±100 000 γ and 150 γ in the range ±300 000 γ. The dip of the hole is measurable from +90° (horizontal) to -40° with an accuracy of 0.5°. The major difference from the previous model is the electronic switching of fluxgate probes instead of servomotor changeover.

Lantto (1973) has presented a general interpretation procedure for magnetic logging data called the characteristic curve method. This method is based on the characteristic points of a profile where either the horizontal or vertical field components are zero. Standard curves are given in the paper for models which can be represented by two parallel infinite line poles (long tubular bodies) or dipoles (rod-shaped bodies). The standard curves were applied to determining the depth of the bottom of an iron ore deposit at the Otanmaki mine in Finland using characteristic points located on the surface and in two boreholes.

Airborne Magnetometers

Table 7.3 shows the airborne magnetometers which are commercially available at the time of writing. Direct-reading proton precession magnetometers are now in common use for standard sensitivity aeromagnetic surveys in inboard

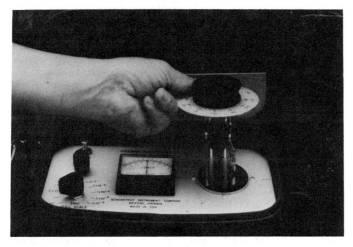

Figure 7.7. Schonstedt PSM-1 remanent magnetometer for field measurements; Schonstedt Instrument Co., Reston, Virginia. (GSC 202228-F)

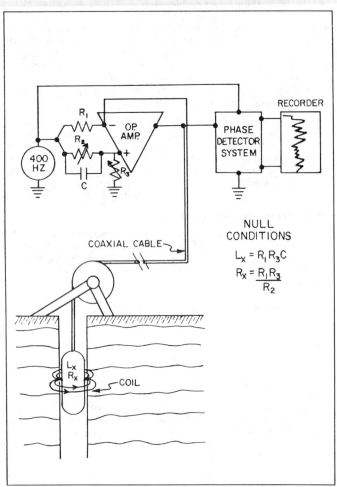

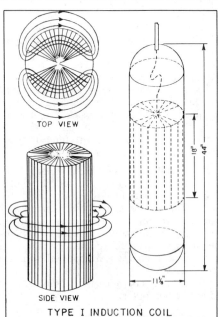

Figure 7.8. Magnetic susceptibility drillhole logging system (Zablocki, 1974).

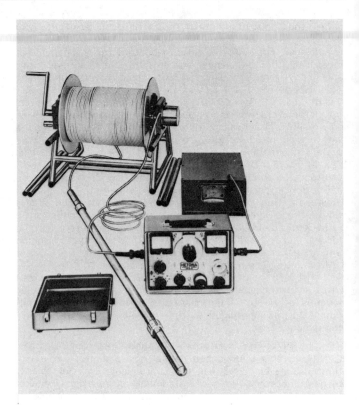

Figure 7.9. ABEM three-component fluxgate magnetometer for drillhole logging; Atlas Copco ABEM AB, Stockholm, Sweden. (GSC 203492-U)

installations and their sensitivity is mostly 1.0 gamma or better. However, the use of fluxgate magnetometers persists perhaps because they produce a continuous analog output instead of a stepped one and this is preferred by many magnetic interpreters who use graphical interpretation techniques. The range of airborne proton precession magnetometers is typically 20 000 to 100 000 γ and this permits their use worldwide except over the most strongly magnetized iron formations.

During 1969, Geometrics introduced their Model G-803 direct-reading airborne proton precession magnetometer (Fig. 7.10). The complete 22.3 kg aeromagnetic survey system consists of a 6.4 kg magnetometer console in a standard 19 inch rack, a 5.5 kg 13 cm chart recorder, and a bird and towing-system weighing 10.5 kg. The range of the instrument is 20 000 to 100 000 γ in 10 manually-selected overlapping ranges. The sensitivity of the standard instrument is 1 γ at a sampling rate of 1 second controlled either by external or internal automatic triggers or manually. Other options are available such as a 0.5 γ magnetometer with an 0.8 second sampling rate. The total field output of the instrument may be recorded digitally and/or potentiometrically or galvanometric analog recording employed.

Gam Service Ltd. of Grenoble, France, which is part of the French Atomic Energy Commission, is manufacturing an Overhauser proton precession magnetometer, the MPPE101, which has a 0.01 γ sensitivity at a 1 second sampling interval. The range of the 27 kg magnetometer is 20 000 to 80 000 γ. Both analog and BCD digital outputs are provided. The instrument has been used for aeromagnetic surveys and as a differential magnetometer for archeological investigations (Collin et al., 1973); it is readily apparent that it would be well suited for use in a vertical gradiometer aeromagnetic survey system.

Table 7.3

Airborne magnetometers available for purchase. All magnetometers measure the total intensity of the earth's magnetic field, have visual displays, and are direct reading

Magnetometer and type	Manufacturer	Model No.	Max sens in gammas	Range in gammas (K=1000)	Sampling rate for maximum sensitivity	Readout A=Analog D=Digital	Power Requirements	Weight (kg)
Proton precession- direct reading	Barringer (Canada)	AM 123	1γ	20-100K	1 sec	A & D	12-30VDC 5A polarize	9.1 (incl. chart recorder)
	Geometrics (US)	G 803	1γ	20-100K	1 sec	A & D	22-32VDC 150W average	6.4
	Littlemore (UK)	Elsec 5951	1γ	24-72K	1.2 sec	A & D	20-28VDC	25
		Elsec 7702	1γ	20-90K	2 sec	A	20-28VDC	7
	Sander (Canada)	NPM-5	0.1γ	20-100K	1 sec	A & D	28VDC 1A + 7A polarize	5 (console)
	Scintrex (Canada)	MAP-4	1γ	20-100K	0.5 sec	A & D	24-30VDC 3.5A max.	6 (console) 3 (sensor)
	Sonotek (Canada)	IGSS/SDS systems	0.1γ	15-100K	1 sec	A & D	28VDC	
	Varian (Canada	V-85	0.1γ	20-100K	1 sec	A & D	28VDC 4A polarize	7.7
Overhauser	Gam Service (France)	MPPE101	0.01γ	20-80K		A & D	12&28VDC 2.5A	27

Figure 7.10. *Geometrics G803 proton precession airborne magnetometer; Geometrics Inc., Sunnydale, California (GSC 203492-R)*

Aeromagnetic Gradiometers

The use of optical absorption magnetometers for aeromagnetic surveying appears to have been mostly confined to petroleum exploration with the exception of the experimental airborne system installed inboard on a Beechcraft B80 aircraft of the Geological Survey of Canada. During the middle 70's the system was modified by the addition of a second boom (Fig. 7.11) so that the vertical gradient of the earth's total field could be measured directly (Sawatzky and Hood, 1975). The successful development of this system was only made possible by the use of two active magnetic compensation systems which are manufactured by Canadian Aviation Electronics of Montreal. The vertical separation of the single-cell self-orienting optical absorption magnetometers in the two booms is approximately 2 metres and the system is designed for gradiometer surveys of Precambrian Shield areas rather than for petroleum exploration.

With regard to the desirable sensitivity for magnetometers to be used in an inboard vertical gradiometer system, two main factors enter into the consideration: 1) the basic sensitivity of the magnetometers themselves, and 2) their vertical separation. It is clear from the results of the surveys flown at 500 feet (152 m) over the Canadian Precambrian Shield by the Geological Survey of Canada that a contour interval of at least 0.025 γ/m should be utilized in delineating magnetic gradient anomalies. Anomalies of this amplitude commonly extend across several flight lines (see Fig. 2 in Coope, 1979, where the spacing of the N-S flight lines is 1000 ft. (305 m)) and therefore reflect the presence of underlying geological features having a low magnetization contrast with respect to the surrounding rocks. The ratio between the basic sensitivity of the survey magnetometer and

Figure 7.11. *Vertical gradiometer system installed on the Beechcraft B80 Queenair aircraft of the Geological Survey of Canada. (GSC 202036-K)*

the contour interval used in the compilation of the resultant aeromagnetic maps is usually between 5:1 and 10:1 depending upon the quality of the recorded data which is affected by various factors such as the compensation figure-of-merit of the survey aircraft. For instance, using a well-compensated one-gamma magnetometer, it is quite feasible to compile the final aeromagnetic maps using a 5-gamma contour interval. Consequently an effective (Precambrian Shield) gradiometer would have to have a sensitivity of 0.005 γ/m if the resultant maps were to be compiled using a 0.025 γ/m interval. For light twin-engine aircraft which are now commonly utilized for aeromagnetic surveys, a sensor separation around 2 m is feasible for an inboard gradiometer system. In the case of the GSC Queenair B80 gradiometer (Fig. 7.11), the sensor separation is actually 2.06 m. Thus magnetometers with a minimum sensitivity of 0.01 γ are required for an inboard vertical gradiometer system to be used in aeromagnetic surveys of Precambrian terrane.

Squid Magnetometers

For total field magnetometers, the useful sensitivity that can be utilized in airborne surveys is about 0.01 γ and there would not be much point in developing more sensitive magnetometers than now exist except as gradiometers. Thus for some years now a new generation of more sensitive cryogenic gradiometer has been under active development which utilizes the Josephson junction effect; the device is usually referred to as a Squid which stands for superconducting quantum interference device. In 1969, an aeromagnetic gradiometer was developed in the Laboratory of Electronics of the University of Oulu in Finland (Otala, 1969), which had a gradient resolution of 8×10^{-3} γ/m but the project appears to be dormant at the present time. In Canada, CTF Systems Ltd. of North Burnaby, British Columbia are presently (1978) developing a six-component gradiometer system for airborne use which will permit gradient measurements in three orthogonal directions. The potential advantage of such Squid devices is that they possess a sufficiently high sensitivity that the gradient can be measured over a distance of 25 cm or less so that a gradiometer sensor can be installed in a single aircraft stinger with a realizable sensitivity of 10^{-5} gammas per metre. Sarwinski (1977) has described some of the practical considerations in the design of a Squid gradiometer.

Figure 7.12 shows a block diagram of a typical RF-driven Squid magnetometer (Goree and Fuller, 1976). The sensor is in a cryogenic environment, namely liquid helium at a temperature of 4.2°K. The superconducting drive coil is resonated close to 30 MHz by using a fixed capacitor near the sensor. The coil is connected to an RF amplifier and detector, to the audiomodulator, and to the feedback circuit. The sensor and drive coil are placed in a tightly fitting tube

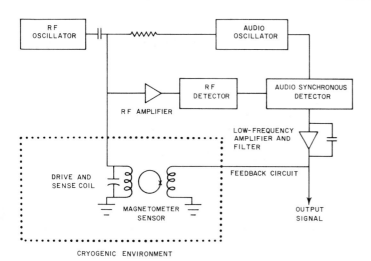

Figure 7.12. Block diagram of the circuitry for a RF-driven Squid magnetometer (from Goree and Fuller, 1976).

on which the superconducting field coil is wound. This assembly is then placed in a superconducting shield to isolate the sensor from all magnetic field changes except that from the field coil. The field coil of the standard sensor requires a current of about 0.3 µA to change the flux linking the Squid by one flux quantum, which is a field change of 1 γ for a 1.6-mm diameter sensor. Typical sensors are capable of a resolution of one part in 1500 of a flux quantum. Thus the peak-to-peak current resolution is 0.2 nA, which gives a field sensitivity at the sensor of 7×10^{-4} γ.

Digital Data Acquisition Systems

Digital recording on magnetic tape of the total field and doppler and altitude information is now the standard data acquisition technique for aeromagnetic survey data, and this permits the subsequent compilation of the digital data to be carried out using the computer.

For a description of the various digital data acquisition systems in current use in the industry, the reader is referred to the annual reviews published by Hood during the review period (see for instance Hood, 1972, 1973, 1976, 1977).

DIGITAL COMPILATION OF AEROMAGNETIC SURVEY DATA

Great progress has been made in the last decade in digital compilation techniques and all the major airborne geophysical survey companies now possess their own in-house computer hardware and software capability. This has come about partly as a result of the introduction of optical absorption magnetometers which required digital-recording systems to avoid the dynamic range recording problem with high resolution data, and of a general realization of the fact that digital compilation techniques would produce more objective results and would permit much greater versatility in data processing operations.

Fifteen years ago with the fluxgate and proton precession magnetometers then in use, a paper strip chart about 25 cm in width provided adequate resolution for recording the aeromagnetic field variations. However, with such a chart calibrated to allow the trace to be read to the precision of the instrument, the chart datum would be automatically reset perhaps 20 times in covering a very large anomaly, making the chart difficult to read in high gradient areas.

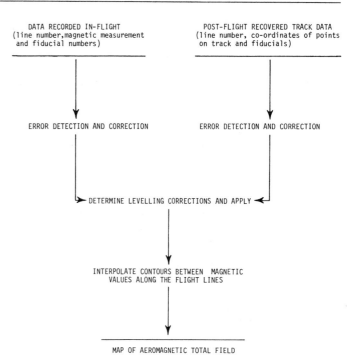

Figure 7.13. Basic processes of aeromagnetic compilation.

With the advent of the high sensitivity optical absorption magnetometer, the use of paper recording charts became altogether impractical. Were a strip chart recorder to be calibrated so as to be readable at the full sensitivity of an optical absorption magnetometer, then over a large anomaly the datum on the chart would be reset as much as 1000 times, producing an unusable record.

It therefore became necessary to employ a recording medium whose sensitivity matched that of the new magnetometer, and one advantageous for the less sensitive instruments. A digital recording system has, for all practical purposes, an unlimited sensitivity. This phenomenon of a need to change from analog to digital methods was not, of course, restricted to aeromagnetic work. It was widespread throughout the whole of the science and technology sector. The reason was that, although the early digital systems were more expensive than analog ones, each increase in sensitivity by an order of magnitude required a small fixed cost for a digital system ("add one more digit") whereas the increase for an analog system may require a corresponding order of magnitude increase in cost ("make the chart paper and recorder ten times wider"). A further potential advantage was offered by the lower weight, and fewer moving parts of principally electronic devices. Thus direct inflight digital recording began to be adopted by the aerogeophysical survey industry and the need for digital compilation systems arose.

Figure 7.13 summarizes the main processes of aeromagnetic compilation whether manual or digital. The levelling process removes the nongeological datum variations from line to line caused by inexact flight altitude, diurnal variation of the earth's magnetic field and other causes.

In the manual compilation process, magnetic values at the intersection points of a "levelling network" formed by the control and the main traverse lines were picked off the analog chart. Levelling adjustments were determined manually and the adjustment values marked on the analog chart at the intersections of the control and traverse lines. These points were then joined by straight lines drawn on the chart to establish a corrected datum. The magnetic profiles were then intercepted at the map contour interval required, usually 10 γ, using the datum. These intercepted values were then transcribed onto a flight path manuscript and hand drawn contours visually interpolated between the flight lines. With the advent of digital recording, computer programs were developed to carry out those manual operations previously performed on the chart. A minimal software system would require a routine to retrieve the value of the magnetic field at the manually determined levelling network points, a routine to make a linear interpolation between these points to produce the computer equivalent of the chart datum, and a routine to output corrected values as the difference between the originally measured value and the datum at the required contour interval. These values could then be transcribed onto the flightpath manuscript and contoured manually as before.

Once having been required to employ the computer, the advantages and indeed necessities of developing beyond this minimal software system became obvious. An early requirement was the need to be able to determine errors and aberrations in the digitally recorded data. Compared with the highly visible trace on the strip chart, errors in the digitally recorded data were less easy to detect.

The most direct and familiar way to present the data for inspection being the form in which it had previously been recorded, namely, as an analog profile, development moved into the field of computer graphics. This allowed the digital data to be output in analog form and inspected for obvious errors in exactly the same manner that the strip chart had been treated previously.

Access to computer graphics having been gained, the obvious next step was automation of the laborious transcription process. This, however, would have resulted in an awkward hybrid process unless the base map of the flight path could also be produced by computer.

Another necessity then presented itself, namely the need to employ digitizing facilities. Although recovered by manual methods, flight path data once digitized, can be submitted to computer automated processes and the transcription of the data to the flight path base map made completely automatic.

Up to this point the production of computer automated compilation software had been a relatively straightforward process, the level of software sophistication required being relatively low. At this point however the required complexity of the software increased significantly.

The acquisition sequence of the in-flight data, and therefore the order of the data on the magnetic tape, is determined by logistical and economic considerations. Lines are not flown all in the same direction and in strict geographical order across the survey area as this would require a prohibitively high proportion of nonproductive flying merely to position the aircraft for each survey line. The sequence of digitization of recovered flight path data is likewise governed by practical considerations. Unless the map area of the survey at the recovery scale is small enough to fit onto the digitizing table, then flight lines must be digitized map sheet by map sheet. Hence a line which exists on the in-flight data tape as a continuous unbroken sequence of records, will exist on the digitized flight path data tape as a series of map sheet-sized segments separated by segments of other lines. In order to bring the two data sets together, it is necessary to submit them, either or both, to a complex "sort-merge" process. Once software for this purpose has been implemented, then automatic transcription of in-flight data onto flight path base maps can be carried out.

All the previously mentioned processes, though complex in some cases, are eminently suited for computer automation, consisting as they do of a sequence of well defined and repetitive processes. Such is not the case with the levelling and contouring stages.

Although the basics of the levelling process can be well defined many further factors dependent upon the skill and experience of the map compiler enter into the process as it is executed. The compiler will seek sources other than the data immediately at hand to explain and resolve dubious levelling corrections. Certain results will not "feel right" to the experienced compiler and special attention would be paid to these cases. The computer is not, of course, capable of such beneficial digressions from the specified task.

Gridding and Contouring

Similarly in the case of contouring, although the computer is capable of doing an excellent job with properly sampled data, if the data are undersampled (e.g. if the flight line spacing is too wide for the chosen flight altitude) then the computer does its best to produce a map which best represents the data as sampled — namely a poor map. With hand contouring the draughtsman would decide upon an appropriate trend and string the contours along this trend as though they were smoothly and well defined along their length. The resulting product has a pleasing, albeit potentially deceiving, appearance.

Figures 7.14a, b, c demonstrate just how appearances can be deceiving. All three maps are computer-contoured segments of an aeromagnetic gradiometer survey flown at 500 ft. (150 m) altitude with 500 ft. (150 m) line spacing. They all cover the same area of the survey but Figure 7.14a was contoured from all the original data. Figure 7.14b was contoured from every second flight line only and Figure 7.14c was contoured from every fourth flight line only. Hence Figures 7.14b and 7.14c represent the end product as it would appear had the survey been flown at 1000 (305 m) and 2000 ft. (610 m) line spacings respectively.

It must be noted that the gridding and contouring programs employed ensure that the contour positions are exact to within 0.01 cm where the contours pass over the flight lines. The contour positions between flight lines are somewhat conjectural whether contoured by man or machine, but with undersampled data, manual contours appear more reasonable. Figure 7.14c clearly exhibits the typical, undesirable features of machine-contoured, undersampled data. Namely isolated "potato" anomalies (or magnetic boudinage) elongated at right angles to the flight lines and the tendency for contours which cross several flight lines to undulate between the flight lines rather than link up the flight line contour intersection points with a smooth continuous curve, as is the case with manual contours.

Figures 7.14a and 7.14b, however, do not clearly exhibit the symptoms of undersampling. In both cases, contours cross the flight lines as smooth continuous curves and elongated trends are shown at sharp angles to the flight line direction — symptomatic of good sampling.

The interesting consequence is, however, that the 500 ft. (150 m) line spacing, which we will regard as the accurate depiction of the magnetic field variations, shows a series of four-parallel "ridges" separated by "troughs" running approximately 15° east of north, whereas the less well sampled data in Figure 7.14b exhibits only one such, but stronger, feature with a very emphatic cross-cutting trough oriented about 20° west of north. This feature is clearly spurious, but so well defined by the data as sampled that both man and machine would confidently depict it as such.

This example should leave no doubt as to the need for adequate sampling in aeromagnetic surveys. The presence of such a cross-cutting feature could be very misleading to the interpreter with potentially costly results, if, for example, it occurred in an area where economic mineralization was associated with cross trend faulting and it became the subject of exploratory drilling.

A) 500 ft. (150 m)
B) 1000 ft. (305 m)
C) 2000 ft. (610 m)

Figure 7.14. *Computer-contoured segments of an aeromagnetic gradiometer survey flown at 500 ft. (150 m) altitude in southeastern Ontario.*

Hence, the data must be well sampled if such pitfalls are to be avoided, and if the data are well sampled, machine contouring will produce the most objective and accurate results from the recorded data. It should be borne in mind that the term 'well sampled' not only refers to the flight line spacing but also includes the orientation of the flight lines with respect to the strike of the magnetic anomalies.

Contours are merely a means of depicting in two dimensions, a surface which has three dimensions. For aeromagnetic surveys, the "surface" is an analogous representation of the amplitude of the earth's magnetic field in a given horizontal plane above the earth's surface. The compiled aeromagnetic data may be considered as continuous in one dimension only — along the flight lines. It is therefore necessary to generate a surface from this data before contours can be traced. This initial stage is in fact, the most critical. Once such a surface exists the positions of the (as yet untraced) contours are firmly fixed. The actual contouring stage is mostly concerned with cosmetics (e.g. labelling contours and marking depressions).

The surface could be represented as an analytical algebraic expression and if so, the contours could be defined also as algebraic expressions by substitution of a constant value for the vertical (Z) variable. The contours could then be generated by solution and evaluation of this contour equation. Though theoretically possible, such a method is ruled out in practice, as a typical aeromagnetic map would require an astronomically high order polynomial. A surface can be adequately defined however, by a set of points upon it if the points are sufficiently closely spaced. Such a "numerical surface" is a viable alternative to the impractical algebraic surface.

Holroyd and Bhattacharyya (1970) employed a hybrid method which used a numerical surface to define the coarse features of the data. Individual algebraic surfaces were then fitted within each cell of the numeric surface for the definition of fine detail.

The potential advantage of this approach is a reduction in the amount of data required to define the surface. Only the coefficients of the polynomial defining the surface in each coarse cell need be sorted. When a pair of points on the surface are needed to fix the position of a contour, the polynomial is evaluated at these points only, thus obviating the storage of redundant surface points which will not be required to define a contour position. Although this method was originally intended for application to aeromagnetic data, subsequent evaluation showed that it was in fact particularly unsuited to aeromagnetic contouring due to the following reasons:

i) In order to realize its potential advantage, the coarse grid of the numerical surface must be fairly large compared to the fine grid that is eventually evaluated to define the contours.

ii) The coarse grid cells must be rectangular and as nearly square as possible.

iii) Aeromagnetic data are not sampled at points falling upon a regular grid. They are sampled along approximately equispaced subparallel lines at intervals five to fifteen times smaller than the line spacing.

Thus the coarse grid points will not normally fall upon a sampled point and although the grid size may approximate to the line spacing, it will have to be significantly larger than the sample interval along the lines. This means that a two stage interpolation process is required to create the contourable surface — firstly to interpolate values at the coarse grid points; secondly to interpolate the fine grid values. As a result the contours as traced generally will deviate substantially from their correct flight line intercept point and furthermore, much fine detail will be lost.

Much more effective gridding systems have been subsequently devised. The methods used by the Geological Survey of Canada and those of the Canadian aerogeophysical survey industry are essentially the same.

In general the method involves fitting smooth, continuous interpolation functions along parallel lines normal to the flight line direction. Each function has values equal to the magnetic measurements on the flight lines at the points where the two sets of lines intersect. The interpolation function lines are spaced apart at an interval equal to the ultimately desired fine grid interval and the functions are evaluated along these lines at points separated by the same interval, thus producing the fine grid directly.

By this method the contours when traced pass within 0.01 cm of their true flight line intercept position and all fine detail is preserved. The desirability of such a result far outweighs the minor disadvantage of having to calculate values at every fine grid point even though many of such values may not be required to fix a contour position. Even this necessity becomes an advantage if further digital processing, such as digital filtering, is to be applied as such processes require values at all grid points.

As noted, once a numerical surface defined by a closely spaced grid has been created, contour positions are fixed. The Geological Survey of Canada and the Canadian aerogeophysical industry have adopted a general standard square grid interval of 0.25 cm measured on the contour map.

Thereafter the actual "contouring" programs are mainly concerned with matters such as contour labels, line weights and feathering of lower order contours in high gradient areas. These cosmetic processes however should not be underestimated as, in the better programs, the complexity and sophistication of the algorithms required exceed that of many other stages of the compilation and cartography processes.

Most aeromagnetic maps currently produced in Canada for the federal government by aerogeophysical exploration companies or by the Federal Government itself are now machine contoured. This is feasible because the survey specifications ensure adequate sampling of the magnetic field and hence allow a machine-contouring package to do its work with little difficulty.

Figure 7.15 shows a flow chart of a digital aeromagnetic compilation system similar to the one in use by the Geological Survey of Canada.

Several stages are annotated as being optional. The digital filter stage for example, need not be applied to aeromagnetic total field data except to remove high frequency sampling noise (without distortion of the geological anomalies) if such is present. The extraction of the levelling data set and the application of level adjustments, are not necessary for aeromagnetic gradiometric data. Such data are presently levelled using a digital filter which removes any long-wavelength variation from the profiles. Thereafter the data, as they are not subject to diurnal variation, should be contourable without further levelling adjustment.

The derivation of alternative data forms is also an optional process. Some of the alternative forms that can be produced are for example regional or residual maps, vertically continued maps, and derivative maps.

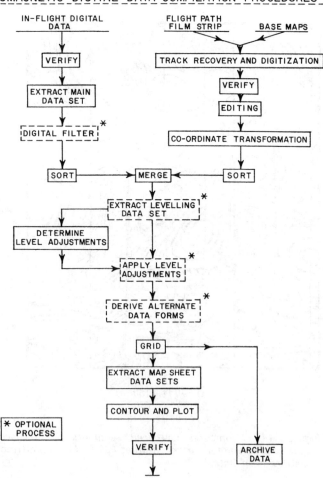

Figure 7.15. Flow chart for digital aeromagnetic compilation system.

Levelling methods however are still not totally automated. Some Canadian companies and the Federal Government use entirely automatic methods, others rely upon manual methods and yet others on a hybrid technique. Current trends in development in computer software and automated methods however are making it easier to decide between fully automated and manual methods.

The trend referred to is the increasing use of interactive systems rather than the previous almost universally used batch system. With an interactive terminal (preferably a graphics terminal) on line to a computer service bureau or as a peripheral of an in-house computer system, those parts of the work best done by the computer can be automated, and those parts requiring human interaction can be presented to the map compiler via the terminal. Such a system obviates the time consuming need to switch from a manual to a batch process and back again with the attendant problems of interfacing between the man and the machine.

With such a system, applied to levelling for example, the clearly defined processes of the levelling can be carried out on the computer at its own very high speed. When a problem arises that cannot be resolved by the computer algorithm the computer can then present the information to the compiler at the terminal for his inspection and decision as to the best course of action. After the decision is made the computer can continue until another similar decision point is reached.

Work currently in progress at the Geological Survey of Canada includes investigation of methods which will eventually allow the interactive process to be applied to contouring. The data will be contoured and the contours presented to a compiler at a graphics terminal. If the data appear to have been adequately sampled and the automatically produced contours are acceptable, then no action by the compiler will be required. If however the data are undersampled and trends, which are visible to the compiler, are inadequately depicted by the machine contouring, then the compiler, by use of a graphics input device, will be able to indicate to the machine the linkages to be made between spuriously isolated anomalies. After this the machine will recontour the data enforcing the appropriate trending of contours across the map. After every session of automatic contouring, the data will be presented to the compiler until final acceptability has been reached. After this the contours can be output in final form on a digital plotter.

Error Detection

The most time-consuming stages of the compilation process are the verification of the in-flight data and the track data. Identification and removal of data errors and abberations are vital if time-consuming reprocessing is to be avoided.

Both the in-flight and the recovered track data set are subject to a great variety of types, and degree of severity, of man and machine errors. If, for example, either the in-flight instrument operator or the digitizer operator miskeys a line number, then a lack of correspondence between the two data sets will arise at the sort-merge stage. An even worse situation is when line numbers are somehow transposed. All lines of in-flight data would have a corresponding line of track data and thus the error could remain hidden up to the levelling or even final contouring stage. As well as such gross logical errors, many physical errors are encountered. With flight path data, a recovered point may be inaccurately positioned on the recovery map or its position may be inaccurately digitized. Means exist, however, to detect automatically severe cases of such errors (and also the logical corollary, where a track point is correctly positioned but mislabelled).

The principal error detection method is known as a "speed check", and was also employed in manual compilation; the technique has been adapted with much improved sensitivity to digital compilation. The time at which each recovered track point was flown over is known. From this information and the positional co-ordinates, the speed-check program calculates the apparent speed of the aircraft between each pair of track points. The actual aircraft speed will vary but this variation should be smooth i.e. no sudden significant changes in speed. If a track point is misplaced, for example, some distance along the direction of flight then the track segment before this point will be lengthened and the one after shortened. As the time of passage over these points will not have changed this causes an apparent increase in speed before the point and a corresponding decrease after it. If the positional error is significant, the apparent change in speed at the misplaced point is readily detectable and the error clearly indicated. Without such a check, the contours would be distorted in the vicinity of the misplaced point.

Physical errors in the in-flight data set stem from two principal sources: malfunction of either the sensing or the encoding-multiplex-recording system and to such effects as miscompensation of the airborne magnetometer system to aircraft motions. Figure 7.16a shows the four ways in which these errors manifest themselves in the data. The first three, spikes, steps or hash are high frequency features; the last, drift, is a medium to low frequency feature. At any given distance from a magnetic body, there is a calculable minimum width to the anomaly it causes. This minimum width increases as the distance from the body. At the usual flight altitudes and with the usual measurement frequency, the minimum anomaly size is several measurement intervals wide since the measurements are normally spaced about 75 m (250 ft.) apart on the ground for aeromagnetic surveys.

Thus features in the magnetic field record defined by a single point or several single point features in succession — such as spikes, steps or hash — are clearly erroneous and can be detected and corrected at the earliest stage of compilation.

Low or medium frequency drift, however, possesses the same frequency characteristics as genuine anomalies and usually remains undetected until the levelling or even contouring stage is reached. Even though detectable, very little can be done about it as the drift and the genuine anomalies upon which it is superimposed, are inseparable. The best that can be done is to discard that section of the data containing it.

The high frequency errors present a much more tractable problem. High frequency noise detection programs were recently added to the ADAM system (Holroyd, 1974) which is employed by the Geological Survey of Canada for the autocompilation of aeromagnetic and airborne gradiometric data. These programs are designed to find and remove spikes, the most common form of error, and to indicate the presence of steps and hash. The kernel of the process is a routine to recognize general disturbances and certain specific patterns in the fourth difference of the data values.

As the values are equispaced the fourth differences, about the ith data element, d_i, is given by the expression (see Fig. 7.16b):

$$d_{i+2} - 4d_{i+1} + 6d_i - 4d_{i-1} + d_{i-2}$$

As can be seen, the expression involves the five consecutive data values centred upon the ith value and the error in d_i would be amplified by a factor of 6 with the same sign; adjacent fourth differences would have error values four times that of d_i but with the opposite sign. The fourth

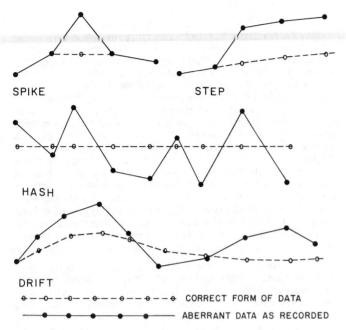

Figure 7.16a. Commonly encountered types of error in digitally-recorded aeromagnetic survey data.

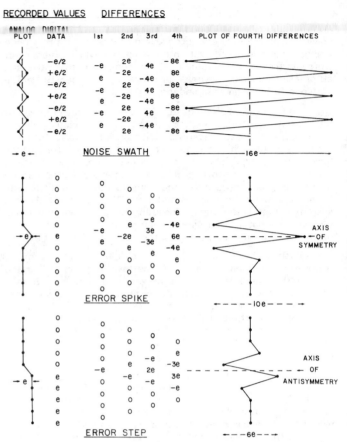

Figure 7.16b. Amplification of errors by calculating fourth differences and the resultant waveforms of the plotted fourth differences.

difference is employed specifically because a true spike needs five consecutive points to define it. The first difference (slope) can be large on the sides of a spike but it could be even steeper on a genuine anomaly. The second difference can be large on the point of a spike but can also be large naturally, similarly for the third difference. When the fourth difference is reached, its value over all the correct and smooth parts of the data tends to zero, but the value over a spike becomes very large.

With "hash" every point can be considered as a spike and consequently the fourth difference tends to multiply the hash swath by a factor of 2^4, i.e. 16 (see Fig. 7.16b). The actual fourth difference pattern would not be significant if the intent was merely to locate aberrant data of any of the high frequency types. The intent, however, is to separately identify spikes. This is because spikes are the most common of the three types of high frequency aberrations, and unlike the other two, are simple to correct automatically. To this end, the program seeks not only for high values of the fourth difference but also for the special patterns of variations within the fourth difference which typify the spikes and step. For the spike, this pattern is five consecutive fourth differences whose values are in the ratio +1:-4:+6:-4:+1 and the waveform is symmetrical about the error point (see Fig. 7.16b). To eliminate the spike in the original data using a fourth-difference table, the two fourth-difference (4 x error e) values on either side of the symmetrical peak are first averaged and then added to the fourth-difference peak (6 x error e) value to give a resultant which is ten times the spike error e in the original data. The appropriate value in the original data corresponding to the position of the peak in the fourth-difference values is then corrected using one tenth of the calculated 10 e value. For the step the pattern is four consecutive values in the ratio +1:-3:+3:-1 (see Fig. 7.16b) so that the resultant waveform is antisymmetrical about the error point. This test is extremely sensitive and in use it has revealed genuine spikes of such a low amplitude that their presence in the data was previously unsuspected. Due to the precision and smoothness of high sensitivity data, it has been possible to detect spikes of 0.1 γ in data with very high gradients and high frequency anomalies of over 10 000 γ amplitude.

The advent of interactive processing systems has greatly speeded up quality control of the work. The practice of verification on an interactive system varies significantly from that on a batch system. With a batch system where no human interaction takes place, it is necessary to pre-define the boundary between acceptable and unacceptable data. As it is almost impossible to predict in advance exactly the types of error to be encountered, it is necessary to make the error-detection routine "over zealous". That is, it is necessary to accept misidentification of good data as bad, rather than allow bad data to be accepted as good. This means that a batch system will not only correctly identify genuinely erroneous values, but it will also identify a significant quantity of valid data as invalid. This then requires detailed, usually manual, work to separate the true errors from the spuriously identified ones.

The nature of interactive systems allows a much more effective method to be applied. Namely, three classes of data are defined. Firstly the data which are clearly wrong, lastly the data which are clearly correct, and in between the two extremes a grey area is defined where the data are suspect but not clearly right or wrong. A batch-processing program can be used to separate the data into these three types. The clearly erroneous data may be processed automatically, the clearly good data may be passed on without any further attention, but the data falling into the grey area between the two may be reserved by the system for interactive inspection. Thereafter a compiler seated at a graphic or other type of terminal can make the decisions regarding acceptability or non-acceptability of the data, a task which the human being can do far more effectively than the computer.

Such a system, incorporating the above mentioned "spike finder" has been implemented at the Geological Survey of Canada. A batch program detects disturbances in the

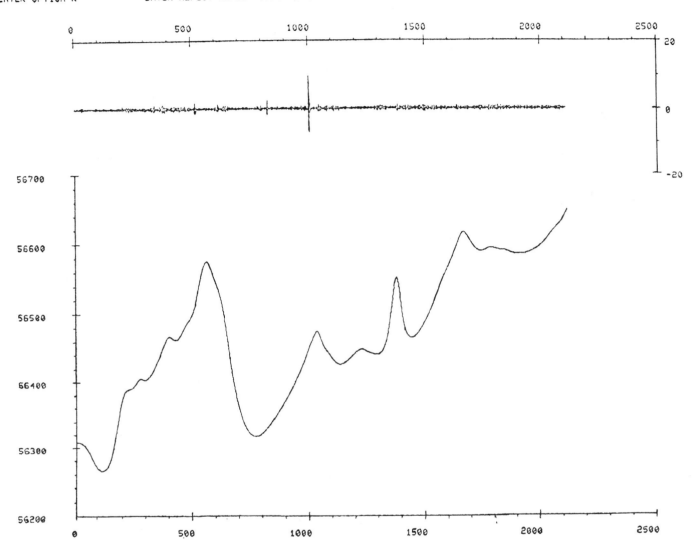

Figure 7.17. Data inspection by interactive graphics.

fourth difference of the data, clearly defined spikes are automatically removed, and those segments of the data containing disturbances, whether spikes or other types, are reserved on a disc file for later access by the interactive program.

Figures 7.17 and 7.18 show examples of interactive viewing of data set aside by the batch program. The graphs, axes and fine print text are a photograph of the actual graphics terminal hard copy unit output, i.e. what was actually displayed on the screen. Figure 7.17 shows the initial display for a segment of data together with a graph of the fourth difference values above. The y axes units are those of the data or fourth difference values, the x axes units are simply the number of the data point with respect to the first data point in the displayed sequence. The data profile has a range of over 400 γ on the y axis and contains 2109 data points. It is as smooth as the resolution of the graphics screen permits. The fourth difference profile however shows several spikes, the largest of which according to its position on the x axis, lies somewhere in the vicinity of the 1000th point in the sequence. After plotting the graphs, the compiler may select that part of the data containing the spike and display it on the screen.

Figure 7.18 shows the resultant display in which 10 points on either side of 1000 have been selected. The specified segment is replotted and as it covers only 20 points with a y axis range of only 17 γ, the 1.6 γ spike which caused the original disturbance in the fourth difference becomes clearly visible. It should also be noted that the amplitude of the spike can be calculated from the upper trace in Figure 7.17 by dividing the peak-to-peak fourth difference signature by 10. It should be noted that the basic noise level of the aeromagnetic survey system can be calculated from the upper trace of Figure 7.17, by dividing its width by 16. In this example the noise swath is estimated to be approximately 0.1 γ.

As to future development, Dutton and Nisen (1978) noted that during the 1960's the emphasis was on hardware. This for the aerogeophysical survey industry meant the period during which the new magnetometers and digital recording equipment plus the digitizing tables and computer graphics devices came into use. The writers stated that during the 1970's the emphasis was on software. This for the aerogeophysical industry was the development of the digital compilation systems which are now almost universally in use throughout the industry. Finally, the prediction for the 1980's

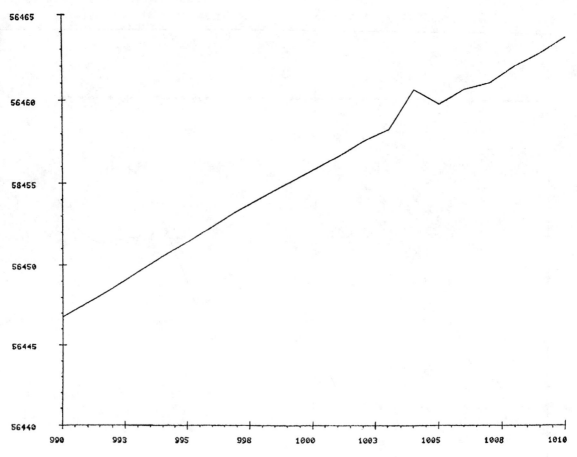

Figure 7.18. Amplified replot of that part of the aeromagnetic profile presented in Figure 7.17 in which an error spike occurs.

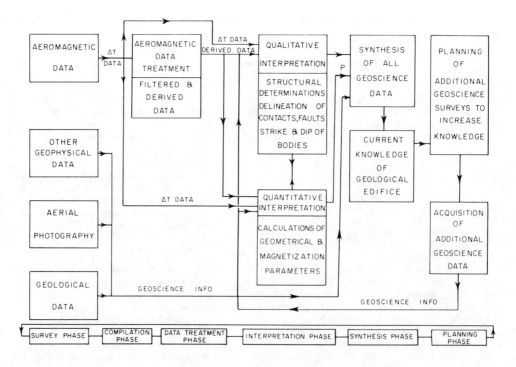

Figure 7.19. General scheme for the interpretation of aeromagnetic data.

was that the emphasis will be on data bases. One could interpret this with respect to the aerogeophysical industry as meaning that now the data are digitally recorded and processed by mainly automated means, the question remains as to what to do with the ever increasing quantity of digital information on hand. The aeromagnetic digital data bank of the Geological Survey of Canada already costs several thousand dollars per annum merely for rental and storage of the digital tapes upon which it resides. It is evident that, as the quantity of data continues to increase and more and more methods are developed for further usage of this valuable resource, then more sophisticated means will have to be developed to retrieve and further utilize this data.

DATA PROCESSING

Advances in data processing techniques during the past decade have mostly involved the application of digital data filtering techniques and power spectrum analysis (see Spector and Grant, 1970) in order to emphasize the higher frequency components of the crustal field and remove the regional gradient of the earth's magnetic field and the longer wavelengths of the crustal field. Because of space restrictions, the topic is left to the paper in this volume by Spector and Parker (1979) for a more comprehensive review.

Several techniques for determining the magnetization or effective susceptibility of underlying rock formations were developed during the past decade all of which may be considered to be a form of filtering. Bott and Hutton (1970) published a technique for use on profiles in which the depths to the top of the magnetic formations had to be known. O'Brien (1970) has reported on an inverse filtering technique which combined pole reduction with downward continuation in order to transform the anomaly due to a magnetic dipole into a narrow spike. It was claimed to be effective in locating contacts and in outlining areas of higher magnetization contrasts.

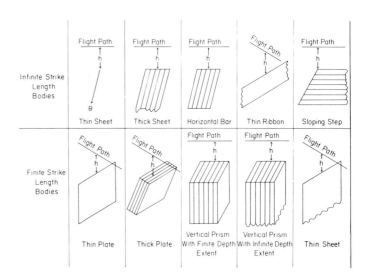

Figure 7.21. *Geometrical models that may be generated from the thin plate model by numerical integration (from McGrath and Hood, 1973).*

In 1973, Paterson, Grant and Watson Ltd. of Toronto developed a new technique for the interpretation of aeromagnetic maps in which the magnetic susceptibilities of the underlying rock formations are calculated (Grant, 1974). Magnetic susceptibility mapping assumes that the bedrock is composed of a large number of homogeneous, vertical rectangular prisms whose upper surfaces follow the bedrock topography. By generating a horizontal grid of total intensity values from the flight-line data, the magnetic susceptibilities of the prisms may be calculated directly from the gridded data through an inversion. Tandem software provides for the rectification of drape-flown surveys, for regional-residual (deep versus shallow) separation, and for final adjustment to flight-line observations if desired. The resultant magnetic susceptibility contour map is a useful tool for determining geological boundaries. It is most effective when sufficient outcrop information is available to identify the lithology of the various magnetic rock units. In such circumstances it may be used with appropriate care and caution to extend geological mapping into unmapped areas or beneath nonmagnetic cover. In the absence of such controls, the magnetic susceptibility map may be viewed as a contour presentation of the volume concentration of magnetite (or magnetite equivalent) in the bedrock. Such a map can be useful in exploration studies for making preliminary geological interpretations, and for solving special problems such as locating felsic/mafic contacts, outlining the metamorphic zones in an intrusion, etc.

AEROMAGNETIC INTERPRETATION

It is axiomatic that the ultimate value of a given aeromagnetic survey lies in the geological information that can be derived from an interpretation of the resultant data. Figure 7.19 is a block diagram showing ideally how aeromagnetic data should be interpreted to maximize their value and this can only be done effectively by using the other kinds of geoscience information available. On the left hand side of Figure 7.19 are the information inputs into the system, namely aeromagnetic data, other kinds of geophysical data, aerial photography and the geological information known about the area. First the aeromagnetic data may be treated in some fashion to improve the resolving power of the technique. Such treatment will result in some form of filtered map being produced, for instance, a second vertical derivative or downward continuation map, which emphasizes

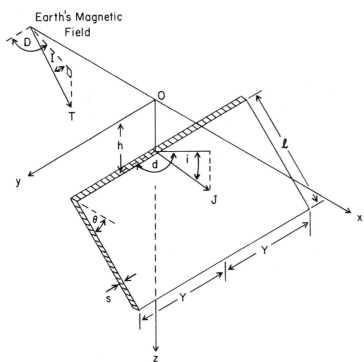

Figure 7.20. *Oblique view of thin plate showing nomenclature used (from McGrath and Hood, 1973).*

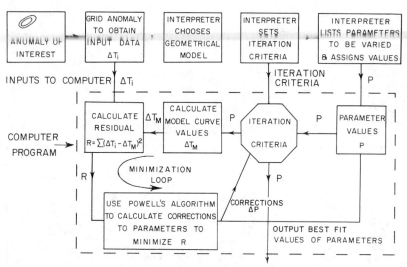

Figure 7.22. Automated least-squares multi-model method of magnetic interpretation.

the magnetic effects of the near-surface geology and removes the gradient effects of the main earth's field. Then the interpretation proper of the aeromagnetic data is carried out and this important step may be divided into two phases. The first phase is a <u>qualitative</u> interpretation in which areas underlain by a common rock type are delineated and structural features such as faults are recognized. Then a <u>quantitative</u> interpretation is carried out in which the geometrical shape and position of each causative body producing an individual anomaly is calculated together with its intensity of magnetization; the latter being, of course, the diagnostic physical parameter for the magnetic survey technique. The value obtained for the intensity of magnetization will be a guide to the lithology of the causative body. However, as a control on the quantitative interpretation, and to set permissible limits for the upper and lower values acceptable for the intensity of magnetization, it is recommended procedure to obtain representative remanent magnetization and susceptibility measurements of the main formations in the survey area. The results emanating from the qualitative and quantitative interpretation should, of course, be compatible, which is the reason for the arrows between these two blocks on Figure 7.19. Now, at the synthesis phase all types of geoscience information including that resulting from the interpretation of the aeromagnetic data are combined in order to deduce the most probable geological edifice for the area. At this stage it may become apparent that knowledge of the geological edifice is weak in some particular aspect and so the interpreter may recommend that additional geoscience surveys be carried out to fill in these gaps. There is a point that we particularly wish to emphasize, namely that there should be a feedback loop in the interpretation sequence. After additional geoscience information is acquired and subsequently fed back into the synthesis phase, a further interpretation of the aeromagnetic data is often worthwhile. Thus the first interpretation of the aeromagnetic data should not necessarily be considered the final one. There is always additional information which can be gleaned from the data and the acquisition of other types of geoscience information may indicate what additional facets of information can be derived from the aeromagnetic data. Thus the point of this discussion is to bring out the interrelationship between all the kinds of geoscience information useful in the interpretation, and the fact that the results from a particular geoscience discipline should not be interpreted in isolation.

Quantitative Interpretation of Aeromagnetic Survey Data

We have counted 130 articles on quantitative magnetic interpretation which have been published during the past decade so it is impossible to comprehensively review the interpretation literature in the space available for a general review. We will therefore only attempt to summarize and highlight what appear to be some of the more significant advances since 1967:

In general, the articles have been concerned with four main types of interpretation method:

1) the more classical interpretation techniques in which special points on the anomalies, such as the maxima, half-maxima, minima, inflection points etc. are utilized together with a set of charts to calculate the geometrical parameters;

2) computer depth-profiling techniques in which depth estimates are made along digitized magnetic profiles using a single or multigeometry causative body model;

3) computer curve-fitting techniques in which a three-dimensional geometrical model is derived by an optimization technique whose calculated anomaly best fits a set of gridded data in a horizontal plane; and

4) interactive computer graphics techniques utilizing curve-fitting methods in which the interpreter may communicate interactively with the computer through a CRT display and light pencil to modify the interpretation.

An example of the first type of interpretation method which does not utilize the computer, concerned the dipping-dyke case (Koulomzine et al., 1970). Their algorithm involves splitting the anomaly curve into symmetrical and antisymmetrical parts by folding but more importantly presents a quantitative technique for first ascertaining the position of the origin for the anomaly so that an estimate of the background datum level is not required. The technique then utilizes maximum, three-quarter and half maximum abscissal distances and the maximum amplitudes of the symmetrical and antisymmetrical curves together with a single master curve.

Another notable article was that by Am (1972) which presented a mathematical review of the magnetic anomalies of thin and thick dipping dykes and derives the characteristic points which can be utilized for depth estimation and presents a set of curves by which the parameters of the causative body may be calculated.

In the second group, one of the first notable articles which dealt with a computer depth-profiling technique was published by Hartman et al. (1971). They applied the so-called Werner deconvolution method to vertical gradient, horizontal gradient, and total field data in which the elemental models utilized were a single dipping sheet for a body with a thickness less than its depth or a stack of such sheets to simulate the wide block or contact case. Because the Werner equation has four unknowns and in order to reduce the effects of interference, a set of six or seven equally-spaced points were taken in order to produce a set of simultaneous equations which are solved by the computer. The entire sequence of data points was then advanced by one point along the profile and the calculation repeated. A correction had to be made for each anomaly determination if the profile did not cross the strike of the causative body at right angles or if the strike of the body was not effectively infinite.

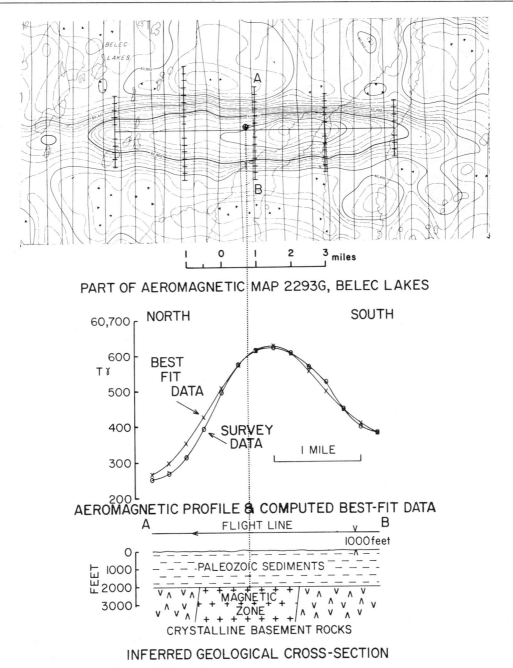

Figure 7.23. Aeromagnetic and geological data for the Belec Lake anomaly caused by a magnetic zone in the basement rocks of the Hudson Bay Lowlands, northern Ontario (from McGrath and Hood, 1973).

Computer curve-fitting techniques for profiles have undergone a great deal of development during the past decade and techniques have been published by Johnson (1969) for two-dimensional structures, by McGrath and Hood (1970), and by Rao et al. (1973) for the dipping dyke. These techniques differ from the Werner deconvolution technique in that a single solution for the causative body is obtained from the whole anomaly whereas in the computer curve-scanning technique a set of solutions is obtained as the computer scans along the anomaly taking a relatively small number of data points for each individual calculation.

Naudy (1971) also described a computer depth-profiling technique which utilized the dipping dyke and lens models in which an individual anomaly was split into its symmetrical and antisymmetrical components. The final result was a series of symbols placed on the chart by the computer which correspond to the depths to the tops of the corresponding model.

O'Brien (1971, 1972) developed a computer depth profiling technique, called Compudepth, for determining the edges of and depths to two-dimensional prismatic bodies. The computer algorithm employs the spatial equivalent of auto-regression in frequency space but the accuracy of the technique depends greatly upon the system noise level and magnetization contrasts.

For the third group, several articles have been published which describe a computer algorithm for the derivation of three-dimensional models whose calculated anomalies best fit

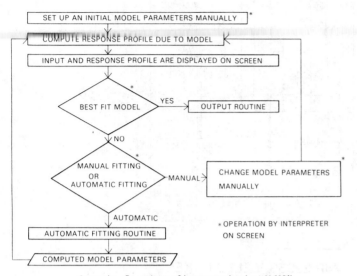

Operation flow chart of interpretation by "IMIS".

Figure 7.24. Interactive computer graphics interpretation technique (from Ogawa and Tsu, 1976).

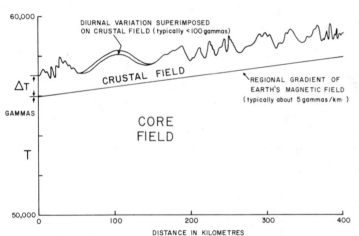

Figure 7.25. Typical aeromagnetic profile across the earth's surface in Canada showing the contribution due to the earth's core, the crustal rocks and diurnal variation.

the observed anomaly on a contour map. In 1973, McGrath and Hood published such a computerized curve-matching technique which utilizes a numerical integration of the equation for the magnetic anomaly due to a thin rectangular plate (Fig. 7.20) to generate an anomaly for one of the geometrical shapes shown in Figure 7.21. Figure 7.22 is a block diagram for this automated least-squares interpretation method. After selecting the anomaly of interest, the anomaly is gridded to obtain the input data. The interpreter then chooses the geometrical model that he thinks best fits the causative body based on the shape of the anomaly and other geoscience information that is available for the area. By selecting a geometrical model, the interpreter in essence selects the mathematical equation to which the input data are to be fitted by a least-squares iteration technique, and this is of course a subjective judgment. The mathematical equation for the model used contains the geometrical parameters such as the depth, width, vertical thickness and length of the causative body and also its magnetization contrast with the surrounding country rock; the magnetization is a vector, that is, it has both magnitude and direction. To obtain the values for the foregoing parameters,

the computer program varies each of these parameters in turn until the best fit criterion is obtained. Thus the interpreter has first to decide how the iteration procedure should be stopped, for instance this decision may simply be an arbitrary number of iterations. It is also necessary for the interpreter to decide the parameters that should be varied and then assign initial values which are usually obtained by simple graphical methods. These initial parameter values are fed to the computer, inserted in the equation and the resultant model curve values, which correspond with each of the input data values, is generated. Then the sum of the least-square differences between the anomaly and model curve values is calculated and an algorithm devised by Powell is used to calculate corrections to the parameters in order to minimize this least-square difference residual value. These corrected parameter values are then fed back into the equation and the process continues until the iteration criterion is reached, when the computer outputs the best-fit values obtained for the parameters. This means, of course, that the anomaly values have been fitted as closely as possible with those generated by the geometrical model using the calculated parameters.

The example selected to illustrate the use of the computerized interpretation algorithm is located in the Moose River basin of the Hudson Bay Lowlands in northern Ontario, which is underlain by Paleozoic sedimentary rocks resting on the Precambrian crystalline basement. The upper part of Figure 7.23 shows the Belec Lakes magnetic anomaly which appears on GSC Aeromagnetic Map 2293G. The contour interval is 10 γ and the survey flight lines are spaced 0.5 mile (0.8 km) apart; survey elevation was 1000 feet (305 m). The amplitude of the anomaly is approximately 380 gammas. The second set of thicker solid lines shows the position of the sampling grid which was placed over the anomaly. These consist of 5 profiles such as AB along each of which 9 to 17 sampling points at 0.2 mile (322 m) intervals were placed; the distance between each profile is ten sampling intervals. Thus there are 63 sampling points used in the interpretation of this anomaly. The thick dipping plate model was chosen, and it was found that the least-squares fit was insensitive to the vertical thickness of the model thus indicating that the depth extent of the causative body is at least five times the depth to its upper surface. The computer determined that the best-fit width of the causative body was 5670 feet (1728 m) and its strike length was 47 000 feet (14 325 m). The calculated depth to the top of the causative body was 2960 feet (900 m) which indicates a thickness of 1960 feet (600 m) for the Paleozoic sedimentary formations and the dip of the causative body was 82° to the north. The best-fit computer values for the total intensity of magnetization vector were a dip of 64° and a declination of 107° east of magnetic north. The dip of the earth's magnetic field in this area is 79°, which means that the causative body must possess a significant remanent component of magnetization. The effective susceptibility contrast which includes the remanent component was calculated to be 2300×10^{-6} cgsu (29 kappas) which is a typical value for igneous rocks. We think that the automatic computer interpretation methods will prove of great value to the industry geophysicist if applied judiciously and enable him to interpret such anomalies more objectively than he has been able to do in the past.

As an extension of the computer curve-fitting technique, it is now possible to adapt the method using an interactive graphics terminal. Ogawa and Tsu (1976) have described such a system in which a combination of either manual or automatic curve-matching procedures may be used in interpreting profiles or contour maps (Fig. 7.24). Their automatic adjustment is based on a least-square minimizing technique which utilizes a modified form of the Marquardt algorithm.

Magnetic Methods

A useful aid to interpretation was published by Andreasen and Zietz (1969) who produced a catalog of 925 theoretical anomalies for a 4 x 6 depth unit vertical prism with vertical thicknesses ranging from 0.1 depth unit to infinity. The anomalies were computed for inclinations of the earth's field ranging from 0° to 90° in steps of 15°. For each model to indicate the effects of remanent magnetism, the azimuth with respect to magnetic north and the inclination of the polarization vector were systematically varied from 0° to 90° and from 0° to 150° respectively. The resultant changes in the shape and skewness of the total field anomalies are most instructive in carrying out qualitative interpretation.

AEROMAGNETIC GRADIOMETER SURVEYS

The vertical gradiometer has many advantages over the single sensor instrument. The first advantage is that the gradiometer records essentially only the magnetic field produced by the rock formations in the upper part of the earth's crust. This is, of course, the required signal and the noise (see Fig. 7.25) which consists of a time-varying component that is the effect of diurnal variation is automatically removed, together with the main earth's field produced by the earth's core and most of the long-wavelength anomalies due to deep-seated bodies in the crust.

However, the most impressive feature is the greatly improved resolution of the gradiometer compared to the total field instrument due to the fact that vertical gradient anomalies are always narrower than their associated total field anomalies. This fact is illustrated in Figure 7.26 which shows the theoretical total field (ΔT) and vertical gradient ($\frac{\partial \Delta T}{\partial Z}$) anomalies due to a thin vertical dyke model where the dip of the earth's field is 90°. The ordinate values have been normalized for amplitude and the abscissal values are expressed in depth units. The nomenclature used is that of Hood (1971) where D is the half-width to depth ratio.

Anomaly 1 is the anomaly produced by a single thin vertical dyke and it can be seen that the vertical gradient anomaly is much narrower than the total field anomaly. Anomalies 2 and 3 are for two thin dykes whose distance apart is at the limit of resolution of the combined total field and vertical gradient anomalies respectively. The limit of resolution is here defined as being the maximum separation of the dykes where the top of the anomaly is still flat without breaking downward at the top to show that two anomalies are present. For the total field case, the separation has to be at least 1.15 D for the two dykes to be resolved but only 0.85 D in the case of vertical gradient measurements. Thus vertical gradient measurements have a 30 per cent better resolving capability for the thin dyke case at high magnetic latitudes than for total field measurements.

The superior resolution of vertical gradient data is, of course, also readily apparent from a comparison of actual total field and vertical gradient profiles. Figure 7.27 shows such a comparison from the survey flown in the Kasmere Lake area of Manitoba (GSC Open File 528, 1978). The anomalous zone on the left hand end of the profile can from the vertical gradient profile be seen to consist of at least 9 recognizable anomalies which have been labelled from A to I in both the total field and vertical gradient profiles. For comparison purposes amplitudes of the anomalies have been normalized with respect to anomaly F. Two anomalies, J and K, occur on the flank of the anomalous zone and L is an isolated anomaly. It can be seen that composite anomalies CD, EFG and HI are below the resolution of the total field instrument whereas for the vertical gradiometer each of the anomalies has been clearly resolved from its neighbour. Note also that the relative amplitudes of the vertical gradient anomalies differ from the total field ones (although this comparison is complicated by the interference of adjacent anomalies except for the case of anomaly L). However vertical gradient anomaly H has clearly been amplified in comparison with anomalies B or C, and this is due to the shorter wavelength of anomaly H. The shorter wavelength features will in general be produced by near-surface causative bodies, so the vertical gradiometer tends to emphasize the magnetic effects of the near-surface geology over deep-seated features. Thus it is a better geological mapping tool than the single sensor instrument.

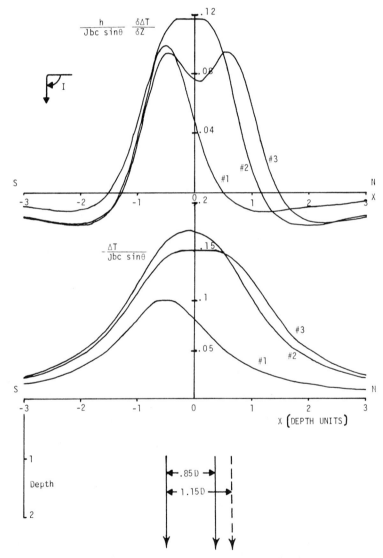

Figure 7.26. Normalized theoretical total field and vertical gradient profiles due to:

#1 - a single thin vertical dyke;

#2 - two thin dykes whose separation is 0.85 times their depth of burial — the limit of resolution for the total field anomalies;

#3 - two thin dykes whose separation is 1.15 times their depth of burial — the limit of separation for the vertical gradient anomalies.

Inclination of the earth's field is 90°.

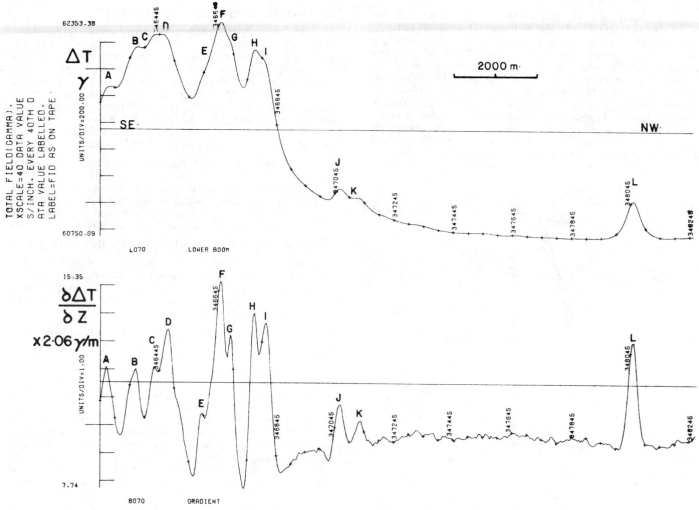

Figure 7.27. *Aeromagnetic profiles demonstrating the superior resolution of vertical gradient over total field data, Kasmere Lake, Manitoba (GSC Open File 528, 1978).*

In Figures 25.2 and 25.3 of Paper 25 of this volume, Coope and Davidson (1979) have presented the vertical gradient and total field maps of a survey flown over the White Lake granite pluton in southeastern Ontario. Figure 25.1 in the same paper is part of GSC geological map 1046A (Quinn, 1952) covering the survey area. The airborne results were also issued as Open File Report 339 by the Geological Survey of Canada (Hood et al., 1976). It is readily apparent from a comparison of the maps that a greater amount of geological information of a more precise nature may be derived from vertical gradient maps. For instance, the two diabase dykes that cut the granite pluton are much narrower on the vertical gradient map. It should be noted in those maps that there is evidence of zoning within the granite pluton itself in the vertical gradient map and several fault zones are evident. In fact, fault zones are much more readily apparent on vertical gradient than on total field maps. Thus vertical gradient aeromagnetic surveys would appear to be of particular value in porphyry copper exploration programs. It should however be pointed out that a somewhat closer flight line spacing must be used in aeromagnetic gradiometer surveys because of the higher resolution of the technique in order to avoid coherency problems. We conclude that vertical gradient aeromagnetic surveys are a viable alternative to total or vertical field ground magnetic surveys.

Another important property of vertical gradient data is its ability to delineate geological contacts because Precambrian Shield areas usually consist of rock formations of relatively large areal extent separated from one another by steeply dipping contacts. Figure 7.28 shows the resultant vertical gradient (solid line) and total field (dashed line) curves over a wide elongated rock formation with sloping sides which is located at a high magnetic latitude such as Canada. For this particular geometric model, actually a dyke, the half width to depth ratio is 5. For comparison purposes, the amplitude of the two curves has been made approximately equal. It can be seen that there are two crossovers from positive to negative values for the vertical gradient profile with the zero gradient values occurring close to either of the contacts. In contrast, the total field anomaly is much less definitive in delineating the edges of the causative rock formation. Actually if the dip of the earth's field and the contacts were vertical, then the zero gradient value would accurately coincide with the position of the contact. However, it can be shown mathematically that the line joining the maximum and minimum gradient values crosses the vertical gradient profile itself at the point where the contact is located. This important property has been illustrated in Figure 7.28. Thus on vertical gradient aeromagnetic maps flown at high magnetic latitudes, the zero gradient contours will delineate the contacts of major rock formations having some measurable magnetization contrast with adjacent formations in a reasonably accurate fashion.

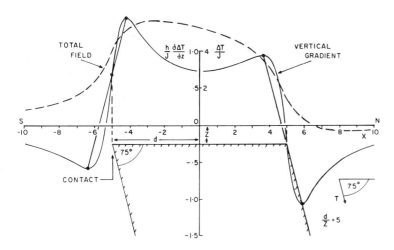

Figure 7.28. Vertical gradient and total field profiles over a wide dipping dyke; inclination of the earth's magnetic field = 75°.

It follows therefore that vertical gradient aeromagnetic survey results are in fact a better tool for geological mapping programs than the more conventional total field results. Moreover, because many orebodies are located at or near contacts, the vertical gradient technique should prove invaluable in tracing such contacts using airborne techniques.

There is a great deal of aeromagnetic survey data available throughout the world and it is possible to derive the first vertical gradient data from the total field results. Although there is some loss of accuracy when compared with the measured vertical gradient data (Hood et al., 1976), nevertheless the resultant end product reflects the underlying geology much better than the total field map. McGrath et al. (1976) has devised a compilation procedure for the purpose which uses a two-dimensional vertical gradient operator (McGrath, 1975). It is necessary, however, that the total field data be of high quality without significant levelling errors because these tend to be emphasized in the resultant map.

Thus to summarize, the main advantages of the aeromagnetic gradiometer as a geological mapping tool compared to the single sensor (total field) technique are:

— direct delineation of vertical contacts by the zero gradient contour value, i.e. vertical contact mapper;
— superior resolution of anomalies produced by closely-spaced geological formations;
— anomalies produced by near-surface features are emphasized with respect to those resulting from more deeply-buried rock formations; and
— regional gradient of the earth's magnetic field and diurnal variation are automatically removed.

REGIONAL MAGNETIC ANOMALY MAPS

Recognition of the value of regional magnetic compilations has grown appreciably in the last decade since the first magnetic anomaly map of Canada, and probably the first such map of its kind in the world, was presented at the 1967 symposium in Niagara Falls by Morley et al. (1968). A third edition of the Magnetic Anomaly Map of Canada was published by the Geological Survey of Canada (McGrath et al., 1977) in time for the Exploration 77 Symposium. The map was derived from approximately 8250 1:63 360 total-field aeromagnetic maps and was published at the scale of 1:5 000 000 using a colour contour interval of 200 γ. Sea magnetometer data were also included. For the compilation, the main geomagnetic field which has its origin in the earth's core was subtracted from the map data by a graphical separation technique (McGrath et al., 1978). Thus the residual magnetic features shown in this map are related mainly to sources within the crust of the earth. Because of the small scale of the map, magnetic features less than 8 km wide do not show up. Hence the map can be regarded as reflecting large-scale near-surface and major deep-seated crustal features. The main uses of a map of this type are threefold. Firstly, as an index map it presents an overview of the aeromagnetic survey coverage of Canada. Secondly, the map presents the major patterns produced by the continental rocks in the Canadian landmass and may be used both as an aid in the interpretation of regional geological features in the basement rocks as well as to help in the planning of more detailed investigations. Lastly and probably the most important use of the map is as a vehicle or medium to initially stimulate comparisons of magnetic features with other types of geoscientific compilations.

Plate 1 shows part of the area extending from James Bay on the east to the Province of Manitoba on the west from the Magnetic Anomaly Map of Canada. The large magnetic high areas (red) correlate well with terrain underlain by rocks of granitic composition. Also there is an obvious correspondence between greenstone belts and their metamorphic equivalents with the broad magnetic low areas (green). For example, the Wabigoon volcanic belt (WB) is associated with a magnetic low. Thus it is within the broad magnetic low areas in which most of the economic ore deposits occur. Locally magnetic highs in greenstone belts are associated with iron formation, iron-rich tholeiitic basalts and iron-rich basic intrusive rocks. Volcanic rocks of calc-alkaline and komatiitic affinities are generally nonmagnetic.

The continuity of the magnetic patterns over large areas is particularly evident and contrast with the variation in the patterns from one area to another. Some of the magnetic anomaly pattern changes occur at geological province boundaries. For example, the trace of the change from the predominantly east-west trending anomalies in the magnetic high area over northwestern Ontario as compared to the magnetic low northeast-southwest trends to the northwest in Manitoba delineates the edge of the Churchill-Superior province (CSB).

Other pattern changes occur within a given geological province. The Kapuskasing Lineament (KL) extending from James Bay to the eastern end of Lake Superior is a structural zone within a geological province in which both the magnetic base level and the magnetic anomaly patterns change. On the other hand, the Quetico structural zone (QS) is associated with a linear magnetic low which transgresses through the magnetic anomaly pattern associated with the western Superior province.

It seems apparent from the examples presented in the foregoing discussion that in general a great deal of correlation exists between the magnetic and geologic compilations and that the production of such national magnetic anomaly maps is worthwhile and should be encouraged by other nations.

A number of interesting regional magnetic studies have been carried out by workers in various countries and among the more interesting are those by McGrath and Hall, 1969; Hall and Dagley, 1970; Kornik, 1971; Zietz et al., 1971; Krutikhovskaya et al., 1973; Hood and Tyl, 1977.

International Geomagnetic Reference Field

As aeromagnetic coverage of large continental areas was completed during the early sixties and the results were compiled into regional magnetic anomaly maps, the need to eliminate the dominating effect of the earth's main field became apparent. In a country such as Canada the earth's magnetic field varies by an amount of 7000 γ or so in going from the east and west coasts to the centre of the country. The amplitude of anomalies due to crustal rocks except in the case of magnetic iron formation is less than 1000 γ. Consequently if the main field which is due to the magnetic properties of the earth's core were not removed, the resultant map would be dominated by a series of parallel contours due to the main earth's field and there is a consequent distortion of the crustal anomalies e.g. the peaks are displaced. If the main earth's field is removed, it is possible to produce a coloured map utilizing four or five colours each representing a 200 γ contour interval to represent the crustal magnetic field. Accordingly a mathematical model of the main earth's field called the International Geomagnetic Reference Field (IGRF 1965.0) was adopted at the Symposium on the Description of the Earth's Magnetic Field held in Washington from October 22-25, 1968. The 1965.0 IGRF was defined by a set of spherical harmonic coefficients to degree and order 8, which requires 80 coefficients, and a linear secular variation correction was incorporated (Fabiano and Peddie, 1969; Cain and Cain, 1971). It has transpired that the secular variation corrections chosen were not particularly accurate and this has resulted in errors of the IGRF of up to a few hundred gammas in five years (Regan and Cain, 1975). At the Grenoble Assembly of the IUGG in 1975 it was resolved that IGRF 1965.0 be replaced by IGRF 1975.0 to be used until at least 1980. This new model consists of IGRF 1965.0 brought up to epoch 1975.0 for its main field coefficients plus new coefficients for secular change (Barraclough and Fabiano, 1977). It is hoped that with the launching of U.S. magnetic field satellite (Magsat) in September 1979, that an analysis of the resultant data will result in a more accurate mathematical model of the earth's magnetic field (Cain, 1971; Langel et al., 1977).

SUMMARY AND CONCLUSIONS

During the past decade both ground and airborne magnetometers have been improved considerably by miniaturization and made more reliable by the extensive use of integrated circuit devices. Digital-recording of aeromagnetic survey data is now standard procedure and except for the flight recovery process, the compilation of aeromagnetic maps is now a completely automated process. The resultant gridded data generated for the machine contouring process also facilitate the production of a variety of derived and processed maps to any scale or map projection for use in the subsequent interpretation of the aeromagnetic data. Aeromagnetic gradiometry has also been proven to be a useful tool in the mapping of Precambrian terrane.

Quantitative interpretation methods have seen noticeable advances during the past decade particularly those using the computer to match observed and calculated anomalies by a best-fit process.

Thus after more than a quarter century the aeromagnetic survey technique is still being improved in many new and novel ways and it remains the most utilized airborne survey technique in terms of the line kilometrage flown each year throughout the world. This continued popularity is due in part to a variety of reasons. Of all the airborne geophysical survey techniques, the aeromagnetic method has by far the greatest depth penetration being able to detect features down to the Curie point geotherm some 20 km or so beneath the earth's surface. Moreover in contrast to other airborne techniques, the aeromagnetic method is unaffected by the presence of surficial material such as overburden and tropical weathering or by the presence of lakes and swamps. Aeromagnetic surveys also provide a continuity of information at low cost that is impossible to achieve in ground geophysical or geological surveys, and one of the outstanding advantages of aeromagnetic surveys becomes apparent when large areas are surveyed because large regional geological features are often discovered. These may not be recognizable on the ground because they are so large or are perhaps obscured by sedimentary formations. Clearly further significant developments in the aeromagnetic survey technique can be expected in the next decade especially in the application of digital recording and processing techniques.

REFERENCES

Am, K.
1972: The arbitrarily magnetized dyke: interpretation by characteristics; Geophys. Explor., v. 10 (2), p. 63-90.

Andreasen, G.E. and Zietz, I.
1969: Magnetic fields for a 4 x 6 prismatic model; U.S. Geol. Surv., Prof. Paper 666, 219 p.

Barraclough, D.R. and Fabiano, E.B.
1977: Grid values and charts for the IGRF 1975.0; Int. Assoc. Geomag. Aeron., Bull. 38, 134 p.

Bott, M.H.P. and Hutton, M.A.
1970: A matrix method for interpreting oceanic magnetic anomalies; Geophys. J. v. 20 (2), p. 149 157.

Cain, J.C.
1971: Geomagnetic models from satellite surveys; Rev. Geophys. Space Phys., v. 9, p. 259-273.

Cain, J.C. and Cain, S.J.
1971: Derivation of the International Geomagnetic Reference Field; NASA, Tech. Note D-6237 (IGRF 10/68).

Collin, C.R., Salvi, A., Lemercier, D., Lemercier, P., and Robach, F.
1973: Magnétomètre différentiel à haute sensibilité; Geophys. Prosp., v. 21 (4), p. 704-715.

Coope, J.A. and Davidson, M.J.
1979: Some aspects of integrated exploration; in Geophysics and Geochemistry in the Search for Metallic Ores, Geol. Surv. Can., Econ. Geol. Rep. 31, Paper 25.

Dutton, D.H. and Nisen, W.G.
1978: The expanding realm of computer cartography; Datamation, p. 134-142, June 1978.

Fabiano, E.B. and Peddie, N.W.
1969: Grid values of total magnetic intensity IGRF-1965; U.S. Env. Sci. Serv. Admin., Tech. Rept. 38, 55 p.

Geological Survey of Canada
1978: Aeromagnetic gradiometer survey, Kasmere Lake area, Manitoba; Geol. Surv. Can., Open File 528.

Goree, W.S. and Fuller, M.
1976: Magnetometers using RF-driven Squids and their applications in rock magnetism and paleomagnetism; Rev. Geophys. Space Phys., v. 14 (4), p. 591-608.

Grant, F.S.
1972: Review of data processing and interpretation methods in gravity and magnetics, 1964-1971; Geophysics, v. 37 (4), p. 647-661.

1974: Timmins magnetic susceptibility map; Geol. Surv. Can., Open File 229.

Hahn, A.
    1978: Die Einheiten des internationalen Systems in der Geomagnetik; J. Geophys., v. 44, p. 189-202.

Hall, D.H. and Dagley, P.
    1970: Regional magnetic anomalies: an analysis of the smoothed aeromagnetic map of Great Britain and Northern Ireland; Inst. Geol. Sci., UK, Rep. 70/10, 8 p.

Hartman, R.R., Teskey, D.J., and Friedberg, J.L.
    1971: A system for rapid digital aeromagnetic interpretation; Geophys., v. 36 (5), p. 891-918.

Holroyd, M.T.
    1974: The aeromagnetic data automatic mapping (ADAM) system; in Report of Activities, Part B, Geol. Surv. Can., Paper 74-1B, p. 79-82.

Holroyd, M.T. and Bhattacharyya, B.K.
    1970: Automatic contouring of geophysical data using bicubic spline interpolation; Geol. Surv. Can., Paper 70-55, 40 p.

Hood, P.
    1970: Magnetic surveying instrumentation; a review of recent advances; in Mining and Groundwater Geophysics 1967 (L.W. Morley, Ed.), Geol. Surv. Can., Econ. Geol. Rep. 26, p. 3-31.

    1971: Geophysical applications of high resolution magnetometers; in Encyclopedia of Physics, Springer-Verlag, Berlin, v. 49 (3), p. 422-460.

    1972: Mineral exploration: trends and developments in 1972; Can. Min. J., v. 93 (2), p. 175-199.

    1973: Mineral exploration: trends and developments in 1972; Can. Min. J., v. 94 (2), p. 167-182.

    1976: Mineral exploration: trends and developments in 1975; Can. Min. J., v. 97 (2), p. 163-197.

    1977: Mineral exploration: trends and developments in 1976; Can. Min. J., v. 98 (1), p. 8-47.

    1978: Mineral exploration: trends and developments in 1977; Can. Min. J., v. 99 (1), p. 8-53.

Hood, P., Sawatzky, P., Kornik, L.J., and McGrath, P.H.
    1976: Aeromagnetic gradiometer survey, White Lake, Ontario; Geol. Surv. Can., Open File 339.

Hood, P. and Tyl, I.
    1977: Residual magnetic anomaly map of Guyana and its regional geological interpretation; Mem. Second Latin Amer. Cong., Caracas, v. 3, p. 2219-2235.

Hood, P. and Ward, S.H.
    1969: Airborne geophysical methods; in Advances in Geophysics, Academic Press, New York, v. 13, p. 1-112.

IAGA
    1973: Adoption of SI units in geomagnetism; Trans. 2nd Gen. Sci. Assembly, Int. Assoc. Geomag. Aeronomy, Kyoto, Japan, 1973, Bull. 35, p. 148 151.

Johnson, W.W.
    1969: A least-squares method of interpreting magnetic anomalies caused by two-dimensional structures; Geophysics, v. 34 (1), p. 65-74.

Kornik, L.J.
    1971: Magnetic subdivision of Precambrian rocks in Manitoba; Geol. Assoc. Can., Spec. Paper 9, p. 51 60.

Koulomzine, T., Lamontagne, Y., and Nadeau, A.
    1970: New methods for the direct interpretation of magnetic anomalies caused by inclined dikes of infinite length; Geophysics, v. 35 (5), p. 812-830.

Krutikhovskaya, Z.A., Pashkevich, I.K., and Simonenko, T.N.
    1973: Magnetic anomalies of Precambrian Shields and some problems of their geological interpretation; Can. J. Earth Sci., v. 10 (5), p. 629-636.

Langel, R.A., Regan, R.D., and Murphy, J.P.
    1977: Magsat: a satellite for measuring near earth magnetic fields; Goddard Space Flight Center, Rep. X-922-77-199.

Lantto, V.
    1973: Characteristic curves for interpretation of highly magnetic anomalies in borehole measurements; Geoexpl., v. 11 (2), p. 75-85.

McGrath, P.H.
    1975: A two-dimensional first vertical derivative operator; in Report of Activities, Part A, Geol. Surv. Can., Paper 75-1A, p. 107-108.

McGrath, P.H., Haley, E.L., Reveler, D.A., and Letourneau, C.P.
    1978: Compilation techniques employed in constructing the Magnetic Anomaly Map of Canada; in Current Research, Part A, Geol. Surv. Can., Paper 78-1A, p. 509-515.

McGrath, P.H. and Hall, D.H.
    1969: Crustal structure in northwestern Ontario: regional magnetic anomalies; Can. J. Earth Sci., v. 6 (1), p. 101-107.

McGrath, P.H. and Hood, P.J.
    1970: The dipping dike case: a computer curve-matching method of magnetic interpretation; Geophysics, v. 35 (5), p. 831-848.

    1973: An automatic least-squares multi-model method for magnetic interpretation; Geophysics, v. 38 (2), p. 349-358.

McGrath, P.H., Hood, P.J., and Darnley, A.G.
    1977: Magnetic anomaly map of Canada; Geol. Surv. Can., Map 1255A (3rd Ed.).

McGrath, P.H., Kornik, L.J., and Dods, S.D.
    1976: A method for the compilation of high quality calculated first vertical derivative aeromagnetic maps; in Report of Activities, Part C, Geol. Surv. Can., Paper 76-1C, p. 9-17.

Morley, L.W., MacLaren, A.S., and Charbonneau, B.W.
    1968: Magnetic anomaly map of Canada; Geol. Surv. Can., Map 1255A (1st Ed.).

Naudy, H.
    1971: Automatic determination of depth on aeromagnetic profiles; Geophysics, v. 36 (4), p. 717-722.

O'Brien, D.P.
    1970: Two-dimensional inverse convolution filters for analysis of magnetic field; 40th Ann. Int. Meet., Soc. Explor. Geophys., New Orleans (Abstr.).

    1971: An automated method for magnetic anomaly resolution and depth-to-source computation; Proc. Symp. on Treatment and Interpretation of Aeromagnetic Data, Berkeley, California.

    1972: Compudepth: a new method for depth-to-basement computation; 42nd Ann. Mtg., Soc. Explor. Geophys., Anaheim, California (Preprint).

Ogawa, K. and Tsu, H.
  1976: Magnetic interpretation using interactive computer graphics; Japan Pet. Dev. Corp., Tech. Res. Center, Rep. 3, p. 19-39.

Otala, M.
  1969: The theory and construction of a proposed superconducting aeromagnetic gradiometer; Acta Polytech. Scand., Helsinki, Elect. Eng. Ser. 21, 55 p.

Quinn, H.A.
  1952: Renfrew map area, Renfrew and Lanark Counties, Ontario; Geol. Surv. Can., Paper 51-27, 79 p.

Rao, B.S.R., Murthy, I.V.R., and Rao, C.V.
  1973: Two methods for computer interpretation of magnetic anomalies of dikes; Geophysics, v. 38 (4), p. 710-718.

Regan, R.D. and Cain, J.C.
  1975: The use of geomagnetic field models in magnetic surveys; Geophysics, v. 40 (4), p. 621-629.

Sarwinski, R.E.
  1977: Superconducting instruments; Cryogenics, p. 671 679, Dec.

Sawatzky, P. and Hood, P.J.
  1975: Fabrication of an inboard digital-recording vertical gradiometer system for aeromagnetic surveying: a progress report; in Report of Activities, Part A, Geol. Surv. Can., Paper 75-1A, p. 139-140.

Spector, A. and Grant, F.S.
  1970: Statistical models for interpreting aeromagnetic data; Geophysics, v. 35 (2), p. 293-302.

Spector, A. and Parker, W.
  1979: Computer compilation and interpretation of geophysical data; in Geophysics and Geochemistry in the Search for Metallic Ores, Geol. Surv. Can., Econ. Geol. Rep. 31, Paper 23.

Steenland, N.C.
  1970: Recent developments in aeromagnetic methods; Geoexplor., v. 8 (3/4), p. 185-204.

Zablocki, C.J.
  1974: Magnetite assays from magnetic susceptibility measurements in taconite production blast holes; Geophysics, v. 39 (2), p. 174-189.

Zietz, I., Hearn, B.C., Higgins, M.W., Robinson, G.D., and Swanson, D.A.
  1971: Interpretation of an aeromagnetic strip across the northwestern United States; Geol. Soc. Am. Bull., v. 82 (12), p. 3347-3372.

GRAVITY METHOD APPLIED TO BASE METAL EXPLORATION*

J.G. Tanner and R.A. Gibb
Earth Physics Branch
Department of Energy, Mines and Resources

Tanner, J.G., and Gibb, R.A., Gravity method applied to base metal exploration; in Geophysics and Geochemistry in the Search of Metallic Ores; Peter J. Hood, editor; Geological Survey of Canada, Economic Geology Report 31, p. 105-122, 1979.

Abstract

During the past decade there have been a number of significant advances in the field of gravity investigations. One important step has been the adoption of a worldwide absolute gravity standard at the 1971 General Assembly of the International Union of Geodesy and Geophysics in Moscow. The reference system, known as the International Gravity Standardization Net 1971, has greatly facilitated the collection and compilation of gravity data on a uniform worldwide basis and hence gravity-based studies of global geological features.

Improvements to gravity instrumentation include the transportable absolute gravity apparatus and the development of the microgravimeter. The absolute apparatus, in particular, has permitted much greater flexibility in establishing and maintaining gravity standards and the possibility of studying secular changes in gravity. The microgravimeter is a highly sensitive version of the standard spring gravimeter capable of measuring changes in gravity of the order of parts in a billion. This improved instrumentation with its attendant increased precision has already proved useful for the detection of chrome nodules in Colombia.

The computer has brought a number of benefits to gravity studies. Perhaps the principal among these stems from its power as a device to store, retrieve and display gravity data. In Canada the ease and convenience with which data can be retrieved has been one of the main factors leading to an increased and more effective use of the gravity method. The process of gravity interpretation has also been greatly facilitated through the use of the computer. As a result it is now possible to derive complex geological models to satisfy any given gravity anomaly by a computer-automated iterative process. This advance has led to the development of statistical and other analytical procedures which place realistic limits on the models.

Gravimetrists are more and more constraining their interpretations within the context of dynamic models, particularly plate tectonic models. Thus modern plate tectonic processes are being used as analogues of paleotectonic developments in the Canadian Shield, which, coupled with geochemical survey data, may result in a better insight into the distribution of minerals within a given structural province. Such studies are becoming increasingly multidisciplinary, an approach that is necessary for the most effective use of gravity data. Thus gravity has played a significant role in the development of statistical methods for base metal exploration by both government and university laboratories. The relationship between gravity anomalies and copper porphyry deposits in British Columbia and sulphide bodies in the Northwest Territories and Yukon Territory, and nickel deposits in Manitoba illustrate some more direct applications of the gravity method to base metal exploration.

Résumé

Il y a eu, au cours des dix dernières années, des progrès importants réalisés dans le domaine des recherches sur la gravité. Une étape majeure a été l'adoption d'une norme mondiale de gravité absolue à l'assemblée générale de 1971 de l'Union de géodésie et de géophysique internationale, tenue à Moscou. Le système de référence, connu sous le nom de Réseau international de normalisation gravimétrique de 1971, a grandement facilité la cueillette et la compilation de données gravimétriques de façon uniforme à l'échelle mondiale, de même que la réalisation d'études gravimétriques de l'ensemble des particularités géologiques.

Les améliorations apportées aux instruments de mesures gravimétriques concernent, notamment le dispositif portatif de gravité absolue et la mise au point du microgravimètre. L'appareil de mesure absolue a permis, en particulier, une plus grande flexibilité dans l'élaboration et le maintien de normes gravimétriques, ainsi que la possibilité d'étudier des variations séculaires de la gravité. Le microgravimètre est une version très sensible du gravimètre classique à ressort qui peut mesurer des variations de la gravité de l'ordre du milliardième. Ce dispositif amélioré, d'une précision accrue, s'est déjà révélé utile dans la détection de nodules de chrome en Colombie.

L'ordinateur a également contribué à améliorer la qualité des études gravimétriques. Son principal avantage, parmi tant d'autres, est peut-être de pouvoir stocker, rechercher et afficher les données gravimétriques. Au Canada, la facilité et la commodité avec lesquelles les données peuvent être obtenues ont été parmi les principaux facteurs qui ont conduit à l'utilisation accrue

---

* Contribution of the Earth Physics Branch No. 740.

et plus efficace des méthodes gravimétriques. L'analyse des données gravimétriques a aussi été grandement facilitée par l'utilisation de l'ordinateur, avec pour résultat qu'il est aujourd'hui possible de puiser dans des modèles géologiques complexes pour expliquer toute anomalie gravimétrique donnée au moyen d'un procédé interactif automatisé. Ce progrès a conduit à la mise au point de procédés statistiques et analytiques qui délimitent de façon réaliste les modèles.

Les experts en gravimétrie limitent de plus en plus leurs interprétations au contexte des modèles dynamiques, principalement ceux de la tectonique des plaques. Etant donné que les analyses modernes de la technique des plaques sont utilisées comme des facteurs analogue à des formations paléotectoniques dans le Bouclier canadien, alliées à des données provenant de levés géochimiques, il peut en résulter un meilleur aperçu de la distribution des minéraux dans une province structurale donnée. De telles études sont en train de revêtir un caractère des plus multidisciplinaires, ce qui constitue une approche nécessaire à l'utilisation plus efficace des données gravimétriques. C'est ainsi que la gravité a joué un rôle important dans la mise au point, par les laboratoires universitaires et gouvernementaux, de méthodes statistiques pour l'exploration des métaux communs. La relation entre les anomalies gravimétriques et les gisements de cuivre porphyrique en Colombie-Britannique, les inclusions de sulfure dans les Territoires du Nord-Ouest et le Yukon et les gisements de nickel au Manitoba illustrent quelques autres applications plus directes de la méthode gravimétrique à la recherche des métaux communs.

INTRODUCTION

As the government agency responsible for operating the Gravity Service of Canada, the Gravity and Geodynamics Division of the Earth Physics Branch is not directly involved in exploration for base metals. The objectives of the Gravity Service are rather to ensure the availability of data and information describing the gravity field in Canada; to provide a uniform reference standard for gravity measurements in Canada; to provide data and information on the structure and figure of the Canadian landmass from studies of gravity anomalies; to contribute to knowledge of the earth's evolutionary processes for the benefit of the resource development industry, earth scientists, standards laboratories and other government agencies; to participate in international gravity studies; and to provide scientific and technical advice and services to the public and private sectors.

In this paper we first review significant developments in the gravity method that have occurred within the last decade. Major advances have occurred in gravity standards and in instrumentation and the rapid development of computer technology has led to greatly improved methods of data base management, data reduction, and gravity interpretation. We next review progress over the last ten years in mapping the Canadian landmass. We then discuss the role that gravity has played in developing a hypothesis for the formation of the Canadian Shield by plate tectonic processes. According to this hypothesis the Shield is constructed of continental parts of ancient plates joined together at suture lines, the sites of ancient subduction. Such a model may be useful in predicting the occurrence of ordered metalliferous zones analogous to examples found in modern orogens forming above active subduction zones. We describe briefly two other indirect applications of the gravity method to base metal exploration involving a statistical approach. Finally, we briefly describe several recent examples of how the gravity method can contribute in more direct ways to base metal exploration.

RECENT ADVANCES IN GRAVITY INVESTIGATIONS

Gravity Standards

One of the most significant advances in gravity investigations in Canada within the past decade has been the conversion of the reference standard for relative gravity observations to the new world system of absolute gravity values adopted by the International Union of Geodesy and Geophysics (McConnell and Tanner, 1974) which replaces the Potsdam Gravity System. The new system, known as the International Gravity Standardization Net 1971 (IGSN71) consists of 1854 stations around the world (Morelli et al., 1974). The datum for IGSN71 is provided by absolute measurements, the scale is controlled by both absolute and pendulum measurements, and the internal structure is provided by some 24 000 gravimeter observations.

In Canada all gravity measurements made by the Gravity Service are tied to the National Gravity Net (Fig. 8.1), which comprises approximately 3400 control stations having an absolute accuracy of ±0.1 mgal, and a relative accuracy of ±0.05 mgal (McConnell and Tanner, 1974). Most of the control stations have been established with LaCoste and Romberg gravimeters, although a limited number of older connections were made with Worden and North American gravimeters. The network is tied to the 20 stations of IGSN71 that are spread throughout Canada and provide datum and scale for the adjustment of the Canadian net to the new absolute standard. The Geodetic Reference System 1967 has also been adopted in Canada to replace the International Ellipsoid of 1930 as the reference surface for the computation of theoretical gravity.

Not only geophysics but also geodesy and metrology require accurate values of gravity over the earth's surface. The new homogeneous worldwide reference system permits standardization of gravity measurements on land and at sea with obvious benefits to national and international studies. The new system also gives datum and scale with an accuracy compatible with modern instrumental capability (Morelli et al., 1974).

Gravity Instrumentation

Concomitant with the improvement in gravity standards have been major improvements in gravity instrumentation. A decade ago the first generation of absolute gravity equipment was successfully developed in standards laboratories at several locales around the world. These instruments were generally not transportable and were capable of measuring gravity to about ±0.1 mgal. Since that time several types of transportable apparatus have been developed and tested. One of these, the so-called French-Italian apparatus (Sakuma, 1971) is currently operational and a series of measurements is either in progress or planned. Another has been developed in the United States and it is planned to have it operational some time in 1978. In addition to their transportability, the current generation of instruments have an accuracy that is improved by almost an order of magnitude (±0.02 mgal).

Aside from engineering or technical improvements in the design and construction of the components, the principle of operation is standard. An object (usually a corner cube) is either tossed and allowed to travel an up-and-down path or dropped and the time and distance it travels over a particular path are carefully measured. The distance is usually measured by laser interferometer. Any measurement at a location comprises a number of tosses or drops, usually about one hundred, with the time for a complete set of measurements being about 5 days including setting-up and breaking down.

The French-Italian transportable gravity apparatus shown in Figure 8.2 was brought to Canada recently to make a measurement at the national gravity reference site in Ottawa. It is a tossed-object type apparatus designed to provide an absolute gravity measurement according to the formula

$$S = 1/2gt^2$$

where S is the distance between the apex of the trajectory and some reference point and

t is the duration of the trajectory

The distance is measured by a Michelson interferometer in which the interference fringes of a laser beam reflecting upon a corner cube are counted digitally. Time is measured by a high stability digital clock. Both the launching chamber and the interferometer are mounted on damping devices to ensure the corner cube and the laser beam operate as closely as possible in the same reference frame. At unusually noisy sites the apparatus is designed so that the damping units may be controlled by a long period vertical seismometer.

The theory and operation of the various types of absolute apparatus have been well described. For a full discussion of error bounds and corrections the reader is referred to Preston-Thomas et al. (1960), Cook (1965), Faller (1965), and Sakuma (1971).

As dramatic as have been the improvements in absolute gravity measurements, they have not kept pace with developments in gravimeters. Perhaps the single most important development of practical significance to the mining industry is the so-called microgravimeter. This instrument is capable, under ideal conditions, of detecting changes in gravity of ±0.002 mgal. Like most spring-type gravimeters, the performance of the microgravimeter is largely governed by the degree of exposure to external

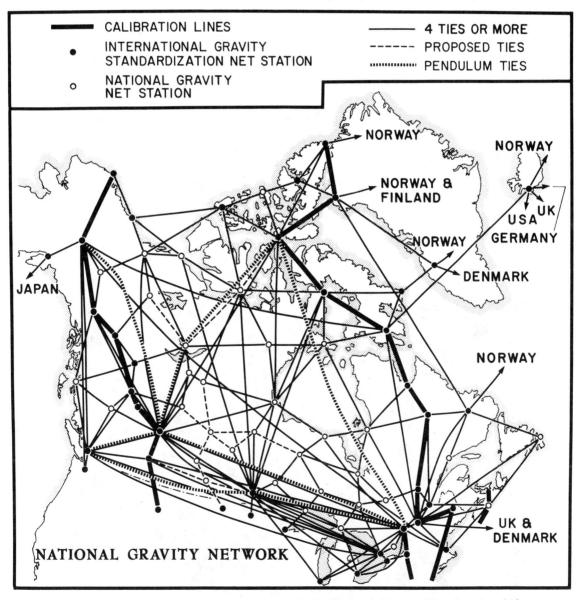

Figure 8.1. The principal control stations of the National Gravity Net, Canada. (After Gibb and Thomas, 1977a; reproduced by permission of Earth Physics Branch.)

sources of vibration during transport. Figure 8.3 shows typical results that might be expected under various transportation modes with the LaCoste and Romberg microgravimeter. Improvements in this performance, in rougher transportation modes, probably can be realized by better designed carrying devices.

Figure 8.4 shows a typical example of the way in which the observations must be carried out if the best results are to be obtained. Theoretically, an ideal network should be established in which every site is connected to every other site, but in practice it is usually sufficient to connect each to three or four other sites. When operating at the limits of accuracy of these instruments it is also usual to observe each leg in the network at least four times. Networks such as that shown in Figure 8.4 are adjusted rigorously by least squares to produce the final gravity values.

Examples of the application of the microgravimeter are comparatively few in the literature. From the standpoint of the mining industry, however, it seems apparent that the microgravimeter would be useful in obtaining accurate estimates of ore tonnages and perhaps in studying the details of the shape of an orebody. The latter application would clearly depend on drillhole information and good density sampling. One example of the use of highly precise gravity measurements in mining has been drawn to our attention by T. Feininger, Escuela Politecnica Nacional, Quito (pers. comm.). In Colombia, chrome nodules occurring in glacial tills have in the past been located by trial and error drilling. Recently, however, the company has adopted the practice of conducting microgravity surveys over the tills and digging for the chrome under the gravity "highs". This method has proved highly successful and consequently has replaced the more cumbersome hit-and-miss approach of drilling.

Another exciting development underway in the field of gravity instrumentation is the airborne vertical gradiometer. Thus far only bench models exist (Heller, 1977; Metzger and Jircitano, 1977; Trageser, 1977) but those engaged in research in this field seem confident that an operational airborne gradiometer accurate to between one and ten Eötvös units is conceivable within the next decade given adequate support. To understand the degree of precision involved, one Eötvös unit, equal to 10^{-9} gal/cm, is equivalent to the pressure exerted by part of the leg of a mosquito on a man's spine, as

Figure 8.2. The French-Italian transportable gravity apparatus. (GSC 203492)

it alights on his shoulder. Any successful development of an airborne gravity gradiometer will have an immediate application to the mining industry because of its potential for rapid, efficient and highly accurate geophysical prospecting surveys.

Computer Technology

Data Base Management

Although employed initially as a data reduction tool in Canada (Tanner and Buck, 1964), the rapid development of the computer soon led to its use for the storage and retrieval of gravity data (Buck and Tanner, 1972; McConnell, 1977). Gravity and related data collected by the Earth Physics Branch are stored in a National Gravity Data Base which presently contains over 300 000 gravity records. Significant contributions to the data base have been made by various government agencies, universities, and mineral and petroleum companies. A notable contribution of 51 000 dynamic gravimeter observations has been made by the Atlantic Geoscience Centre, Dartmouth – a division of the Geological Survey of Canada. The National Gravity Data Base consists of five files (Fig. 8.5), two of which (Instrument Data and Control Station Data) reside on random access devices and are directly accessible by applications software systems (Fig. 8.6) (McConnell, 1977). The two remaining digital files (Anomaly Data and Network Observation Data) due to their large size reside in binary form on magnetic tape. Applications software programs which require data from these files must first search the file sequentially and prepare a subfile upon which subsequent operations are performed. The Control Station Description File is maintained in the form of hard copy reproduced by photo offset or Xerox printing.

Information may be retrieved from the data base in several formats – listings, punched cards, magnetic tapes, and plots depending on the customer's preference. Plots are available at any specified scale and a variety of map projections. Customers who use this service include exploration companies and consultants to the petroleum and mineral industry, provincial and federal agencies responsible for mapping and resource inventories, research geophysicists, geodesists and geologists, and the international scientific community.

The Gravity Service publishes the results of its surveys in a series of maps, generally at a scale of 1:500 000, known as the Gravity Map Series. These are usually accompanied by a report describing the gravity surveys, the gravity anomalies, and their correlation with geology. Data are also released through the open file system of the Earth Physics Branch.

Data Reduction

Although procedures to observe and reduce gravity data have remained comparatively standard for the most part, the increasingly widespread use of the computer and its greatly enlarged capacity and computing power have led to some significant changes in the reduction of gravity observations in the past ten years. Perhaps the most important change has been the gradual trend away from the simple but rigid concept of a twofold system of gravity surveys (viz. the establishment of a permanent system of control stations from which gravity traverses can be run to provide the more detailed coverage needed for a particular investigation) to a more flexible system better suited to the improved performance of modern gravimeters. The mathematical model for the reduction of observations made in such a fashion is necessarily more complex, often involving a least squares process, but easily within the capacity of present day computers. Side benefits include the capability to process the data in a single pass (if desired) and the availability of much more statistical information from which the results of a survey can be evaluated. Nowhere is this trend more in evidence than for marine gravimetry where the volume of data is enormous, the quality of the individual observations distinctly lower than for land data and port ties few and often far between. Given that properly evaluated data sets are a major goal of any gravity survey, a flexible technique capable of giving reproducible results is needed to provide an unbiased adjustment of the data and at the same time provide reliable

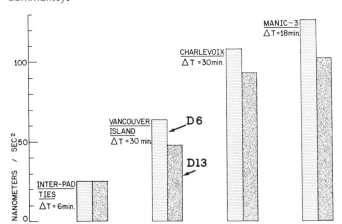

Figure 8.3. Variability of standard deviation of gravity ties for LaCoste and Romberg microgravimeters D6 and D13 due to different modes of transportation as follows: inter-pad ties, hand carried; Vancouver Island, vehicle on paved roads; Charlevoix, vehicle on paved and unpaved roads; Manic -3, helicopter. The major cause of the variation is exposure to different degrees of vibration. ΔT denotes the average time interval for a gravity tie. (After Lambert et al., 1977; reproduced by permission of Earth Physics Branch.)

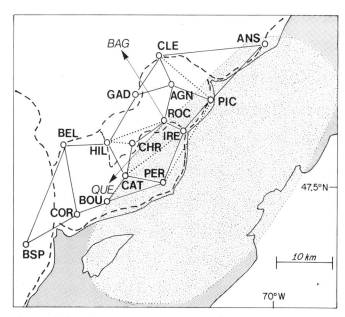

Figure 8.4. Precise gravity network (solid and dotted lines) and first order levels (dashed lines) at Charlevoix, Quebec. Shading denotes area of seismic activity. Each connection in the gravity network comprises an average of eight to ten gravity ties. Arrows are connections to airport gravity stations at Quebec City and Bagotville. (After Lambert et al., 1977; reproduced by permission of Earth Physics Branch.)

estimates of the quality of the observations. An appropriate observational technique in the case of marine gravimetry is the use of frequent crossovers of the cruise lines, ties to at least two different ports preferably at the extremes of the range of gravity and, if possible, the use of a signature line for the calibration of the dynamic gravimeter during the course of the marine survey. The resulting data set can be adjusted by least squares to obtain the most likely values of the crossover points which provide control for the concurrent or subsequent reduction of the intermediate points along the cruise lines. A heavy demand is placed on the software system to edit and compile the data for the least squares adjustment but the results are well worth the effort since the operator gets an immediate overview of the data set from the adjustment, can see immediately where to concentrate his efforts on improving it and gets a feel for the overall quality from the statistics of the adjustment. One example of such a software system is the ASSOB (Adjustment of Sea Surface Observations) system which has been developed at the Earth Physics Branch in Ottawa in response to a clearly indicated need for improved procedures to observe and reduce marine gravity data.

Clearly there is an analogue of the marine case on land where temporarily relocatable stations can be used as repeat points at specified times and locations during a traverse with the result that each traverse is linked to other traverses by an interconnected ad hoc network of repeated sites which plays the same role as crossovers in the case of marine observations. Provided that the data set is tied to primary reference stations for calibration control and an absolute gravity reference (at least in the case of regional surveys), there is no major need to develop an extensive network of control stations in the traditional sense. Modern gravimeters are easily adaptable to this mode of operation without any loss of quality and with the decided advantage that a unified, well understood data set can be produced with a minimum of time and effort on the part of the gravimetrist. If gravity surveys are being carried out to increase the density or upgrade the results of previous surveys more complex reduction models may be the only way of producing a comprehensible data set, other than adopting the somewhat unsatisfactory procedure of adjusting datum and scale arbitrarily to amalgamate the different data sets.

Gravity Interpretation

As with other aspects of the gravity method the computer has brought about many desirable improvements in the procedure of interpreting gravity data in terms of models of earth structure. These improvements stem mainly from the increased power and memory capacity of the computer which permits the use of more flexible, more complex, and more effective interpretation procedures. The end result has been a significant change in the standards for gravity interpretation. No longer is the objective simply to provide a gravity model that is consistent with the geological constraints and other geophysical data, but rather the question must involve a consideration of the tectonic setting and of the possible processes leading to the development of

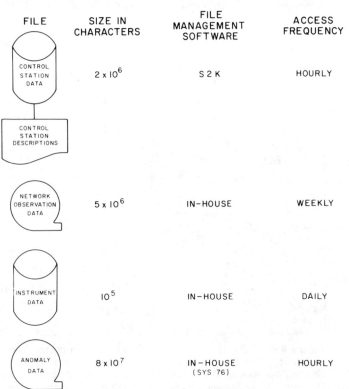

Figure 8.5. *The four digital files and one manuscript file that comprise the National Gravity Data Base. (Modified from McConnell, 1977; reproduced by permission of Earth Physics Branch.)*

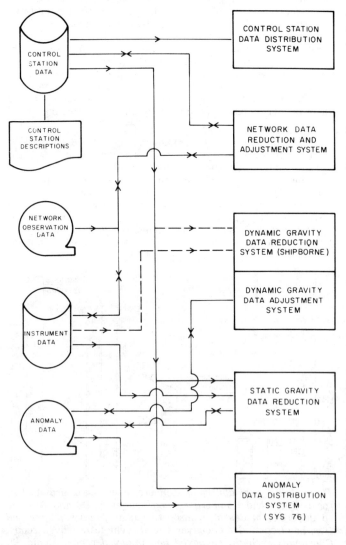

Figure 8.6. *Interaction between the National Gravity Data Base and its associated applications software. (Modified from McConnell, 1977; reproduced by permission of Earth Physics Branch.)*

the structure, preferably in quantitative terms. It is also likely that statistical or analytical methods of evaluating the reasonableness of the assumptions made during the interpretation will soon become a standard part of gravity interpretations. Some progress in this direction has already been made. Thus Miller (1977) has tried to put statistical limits on the information that can be gained from gravity data and Parker (1977) has attempted to place limits on the possible density distribution that can cause a particular gravity anomaly in special cases. The latter is a particularly important aspect of gravity interpretation that needs pursuing because, while samples can be collected to provide estimates of the densities of various rock types exposed in a given region, the spread of results is all too often so great that an interpreter must make a subjective judgment on the density contrast. Efforts on the part of interpreters to seek improved methods of arriving at density contrasts or determining limits for them should be both encouraged and applauded.

Interpretation procedures themselves have been improved greatly by the computer with the result that a wide variety of methods to deduce models to satisfy a given gravity anomaly have been developed in the last decade. Several authors have recently reviewed the role of the computer in gravity interpretation procedures. The reader is referred to reviews by Grant (1972), Talwani (1973) and Nettleton (1976, chapters 6 to 8). An excellent summary of inverse methods i.e. numerical methods for determining density distributions directly from gravity anomalies has been given by Bott (1973). He has succinctly reviewed linear inverse methods in which the shape of the body is specified and the problem is to determine the density distribution (e.g. Bott, 1967; Kanasewich and Agarwal, 1970) and nonlinear inverse methods in which the density is known and the shape of the body is to be determined (e.g. Tanner, 1967; Cordell and Henderson, 1968; Al-Chalabi, 1970; 1972). Recent papers on inverse methods include those by Parker (1972), Oldenburg (1974) and Lee (1977).

STATUS OF CANADA'S REGIONAL GRAVITY MAPPING PROGRAM

Canada has an area of almost 10 million square kilometres; the adjacent shelf seas cover an additional 3.8 million square kilometres. Within this vast region considerable variations in climate (semiarid to Arctic), in terrain (ancient peneplains to Cordilleran peaks), in vegetation (dense bush to treeless barren lands), and in the means of communication (roads restricted to the southern populated regions), have necessitated a variety of approaches to the gathering of gravity data, both in regard to transportation and to instrumentation. Four main categories of gravity surveys are conducted by the Gravity Service: land surveys; ice-surface surveys; underwater surveys; and sea-surface surveys. The planned coverage by each type of survey is indicated in Figure 8.7. The progress to January, 1977 in surveying these regions is also indicated (the 500 m bathymetric contour is a purely arbitrary limit in Figure 8.7 as many surveys have gone and will continue to go beyond this limit). Approximately 80 per cent of the country has been covered by 183 000 discrete (static gravimeter) stations and 134 000 shipborne (dynamic gravimeter) stations as shown schematically in Figure 8.8.

The first gravity measurement in Canada was made at Winter Harbour, Melville Island in 1820 (Sabine, 1821) during a voyage in search of the Northwest passage. Truly systematic surveys, however, did not commence until 1944 when the gravity meter supplanted the pendulum for routine work. Figure 8.9 is a histogram showing the number of gravity stations established per year from 1945 to 1977 by the Gravity Service. The introduction of helicopters in the early 1960s resulted in a sevenfold increase in annual gravity station production. This increase coincided fortunately with the advent of electronic computers and the Gravity Service rapidly adopted computer methods for gravity data processing (Tanner and Buck, 1964). Perhaps the most significant change in recent years is the increase in systematic sea-surface gravity surveys undertaken by the Gravity Service (Fig. 8.9).

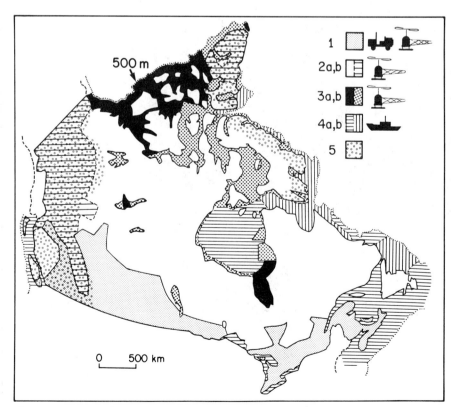

Figure 8.7

Gravity coverage in Canada according to type of survey (to January, 1978). Surveyed and unsurveyed areas beyond the 500 m bathymetric contour are not shown. The distribution of data in oceanic areas is shown in Figure 8.8.

1 - *road covered area, 100% surveyed;*
2a - *hinterland area, 84% surveyed;*
2b - *unsurveyed area 1 315 000 km^2;*
3a - *ice-covered area, 35% surveyed;*
3b - *unsurveyed area 941 000 km^2;*
4a - *water-covered area, 75% surveyed;*
4b - *unsurveyed area 526 000 km^2;*
5 - *mountainous terrain*
Total unsurveyed area 2 782 000 km^2.

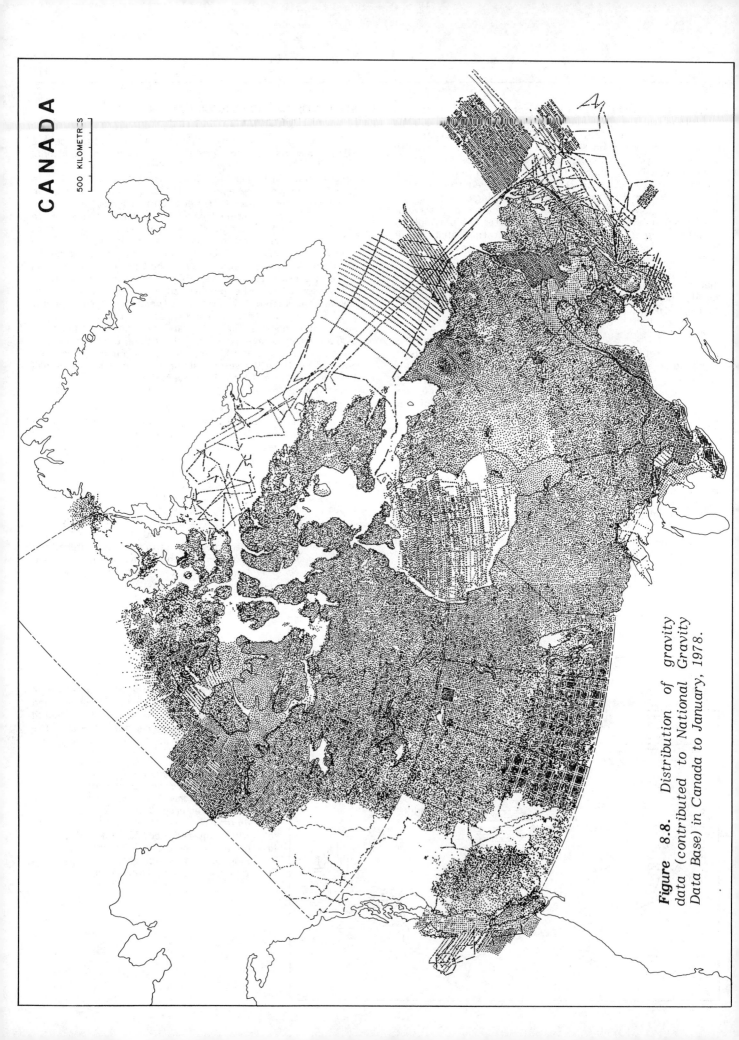

Figure 8.8. Distribution of gravity data (contributed to National Gravity Data Base) in Canada to January, 1978.

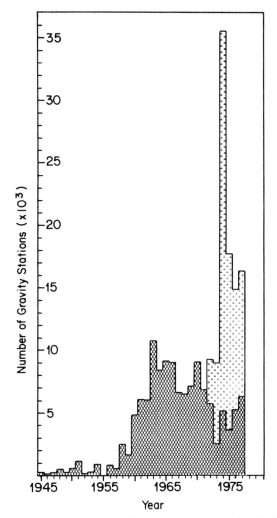

Figure 8.9. Histogram of number of static (dark shading) and dynamic (light shading) gravity stations per year measured by the Gravity Service, Earth Physics Branch for the period 1945 to 1977.

This trend is likely to continue for several years. Figure 8.10 shows the cumulative total number of stations obtained by the Service (graph a) and from all sources (graph b) for the same period 1945-1977. The data are graphically illustrated by the most recent edition of the Bouguer Gravity Map of Canada (Plate 2, p. xiii). The methods used to prepare the map have been described by Nagy (1977).

The remaining unsurveyed areas of Canada have a common requirement for nonroutine survey techniques due to difficulties of terrain or other hostile environments. On land the main unsurveyed areas are the western Cordillera of British Columbia and the Yukon, and the mountainous northern Arctic Islands. There progress will depend largely on the availability of monumented stations with known elevations and sufficiently detailed topographic maps. At sea the remaining areas include parts of the Atlantic and Pacific continental shelves, and parts of Hudson Bay, Hudson Strait, Davis Strait and Baffin Bay. Ice-covered regions include the Canadian sector of the Arctic Ocean, the inter-island channels of the Arctic Islands, Foxe Basin, and some of the large inland lakes. The sea and ice surveys are usually undertaken as co-operative efforts with other mapping agencies such as the Canadian Hydrographic Service, the Polar Continental Shelf Project, and the Geological Survey of Canada. The rate of progress will therefore depend not only on the availability of ships and navigation aids but also on the

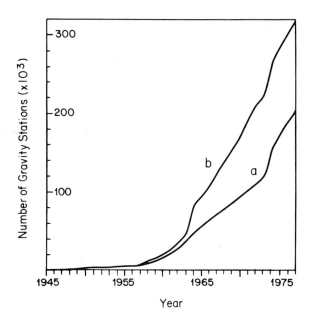

Figure 8.10. Cumulative total gravity stations contributed to the National Gravity Data Base from all sources for the period of 1945 to 1977 by the Gravity Service (graph a), all sources (graph b).

concerns, priorities and continued co-operation of these agencies. At the present rate of coverage the reconnaissance gravity mapping program is likely to continue for 15 to 20 years. In addition to the environmental constraints, the increased cost of navigation and transportation in the remaining areas compared to the rest of Canada will mean that the area mapped each year will decrease if operations are maintained at about the present level. Various aspects of the national gravity mapping program have been described in more detail by Gibb and Thomas (1977a) and McConnell (1977).

SOME APPLICATIONS OF THE GRAVITY METHOD TO BASE METAL EXPLORATION

Although the Gravity Service is not directly involved in exploration for base metals, gravity has played a role in developing a hypothesis for the formation of the Canadian Shield which may provide a regional framework for base metal exploration. We first discuss this hypothesis and other indirect applications of gravity to base metal exploration and we then describe several examples of more direct applications of the method.

Indirect Applications

The theory of plate tectonics has revolutionized concepts of ore genesis. There is growing agreement among economic geologists that there is a fundamental relationship between volcanic processes associated with midocean ridges and zones of subduction and the formation of certain types of ores. Recent studies have confirmed that large deposits of base metal sulphides are formed initially at or near midocean ridges. Such deposits may be recoverable directly from the ocean floors or from obducted sheets of ophiolites preserved (often in orogenic belts) on the continents. Most of these ores, however, appear to have undergone further alteration and concentration in volcanic processes associated with partial melting of oceanic lithosphere subducted beneath island arcs or continental margins. The ore minerals and magmas thus generated rise through the crust and appear to be spatially and temporally related to the subduction zone and its evolution e.g. the type of ore and the composition of magma vary with distance from the trench. Mitchell (1976) has examined the relationship between magma composition and tectonic setting in Cenozoic subduction zones. He showed that characteristic mineral deposits are related to different magmas in different tectonic settings. Such a pattern of mineral occurrences has been described in the Andes (Sillitoe, 1976). From west to east, major longitudinal mineral zones have been identified as follows: Fe, Cu-(Mo-Au), Cu-Pb-Zn-Ag, Cu-Fe and Sn-(W-Ag-Bi) (Fig. 8.11). According to Sillitoe (1976), magmatism and mineralization in the Fe and Cu belts migrated progressively eastwards from early Jurassic to mid-Tertiary time but about 15 Ma ago the Cu-Pb-Zn-Ag belt was formed by a sudden expansion of magmatism to the east.

Mineral zonations related to paleosubduction zones have been recognized in the Mesozoic, Paleozoic, and even the Proterozoic (e.g. Mitchell, 1976). Recent studies by Gibb and Thomas (1976) have suggested that gravity signatures across structural province boundaries in the Canadian Shield may originate from essentially identical structures formed in response to plate tectonic processes operating during Proterozoic time. They have suggested that plate convergence, cratonic collision, and suturing have been instrumental in forming the Shield as we know it today. Three geosutures have so far been proposed, partly on the basis of gravity studies, at or near structural province boundaries. The suture peripheral to the Superior Province is the most easily recognized (Gibb and Walcott, 1971); it is 3200 km in length and extends from the Manitoba Nickel Belt (Nelson front), across the Hudson Bay Lowlands (Gibb, 1975), to eastern Hudson Bay and thence to the Cape Smith foldbelt (Thomas and Gibb, 1977), and Labrador Trough (Kearey, 1976). A second suture has been suggested within the Grenville Province near the Grenville front (Thomas and Tanner, 1975) which separates the Superior and Grenville provinces. This location based on gravity interpretation is but one of several other suggested locations within or bordering the Grenville Province based on paleomagnetic and geological evidence. A third suture has been postulated in the vicinity of the Thelon front, the boundary between the Slave and Churchill provinces in the Northwest Territories (Gibb and Halliday, 1974; Gibb and Thomas, 1977b).

Bouguer anomaly profiles numbered 1 to 5 across five structural province boundaries of the Shield (Fig. 8.12) are shown on a common datum in Figure 8.13a. Apart from short wavelength anomalies attributable to local geological features, all five profiles are very similar. The smoothed gravity signature, called the type anomaly by Gibb and Thomas (1976), was derived by averaging the profiles and is shown in Figure 8.13b within an envelope which varies according to the standard deviation calculated at intervals of 5 km along the profiles. A type crustal model derived from the gravity signature and constrained by seismic results, rock densities and geological information is shown in Figure 8.13c. The correspondence between the computed anomaly and the smoothed gravity signature is shown in Figure 8.13b. The crustal blocks are in approximate isostatic equilibrium.

The model indicates that the younger crustal block is consistently thicker and slightly denser than the older. This consistency is surprising because not only do the structures have a wide geographic distribution but they are also of vastly different ages, suggesting that similar processes have operated throughout much of the Proterozoic time. The density discontinuity of the type model penetrates the whole crust and separates cratons of different density, thickness, age, and internal structure; it was interpreted as a vestigial suture between collided continental blocks. The model with slight modifications to crustal density and thickness applies to all five boundary zones and may apply to other examples in Canada and elsewhere.

Paleosutures have been identified or proposed by several authors at other sites in the Canadian Shield. Talbot (1973), Goodwin and West (1974) and Langford and Morin (1976) among others have suggested that some form of primitive plate tectonic processes played a role in the formation of the Archean crust of the Superior province. Within the Churchill Province possible Proterozoic paleosutures have been recognized in the Fond du Lac area (Walcott and Boyd, 1971; Gibb and Halliday, 1974; Cavanaugh and Seyfert, 1977), in the Wollaston foldbelt (Weber in Donaldson et al., 1976; Camfield and Gough, 1977), in the Flin Flon area (Stauffer, 1974), and in the Foxe foldbelt of Melville Peninsula (Henderson in Donaldson et al., 1976).

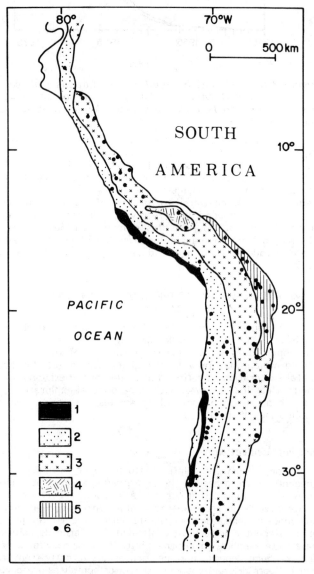

1. Fe belt;
2. Cu-(Mo-Au) belt;
3. Cu-Pb-Zn-Ag belt;
4. Cu-Fe belt;
5. Sn-(W-Ag-Bi) belt;
6. ore deposits.

Figure 8.11. Metallogenic belts of the Central Andes. (Redrawn after Sillitoe, 1976; reproduced by permission of Geological Association of Canada.)

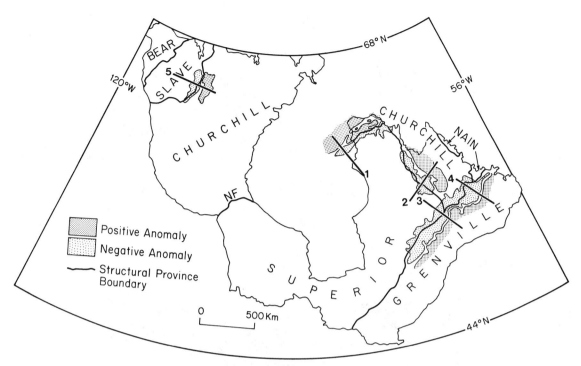

1. Superior – Churchill boundary (offshore extension of Cape Smith foldbelt);
2. Superior – Churchill boundary (Labrador Trough);
3. Superior – Grenville boundary (Grenville front);
4. Churchill – Grenville boundary (Grenville front);
5. Slave–Churchill boundary (Thelon front).

Figure 8.12. Bouguer gravity anomalies at five structural province boundaries in the Canadian Shield. Gravity profiles are shown in Figure 8.13a. NF–Nelson front. (After Gibb and Thomas, 1976; reproduced by permission of Macmillan Press).

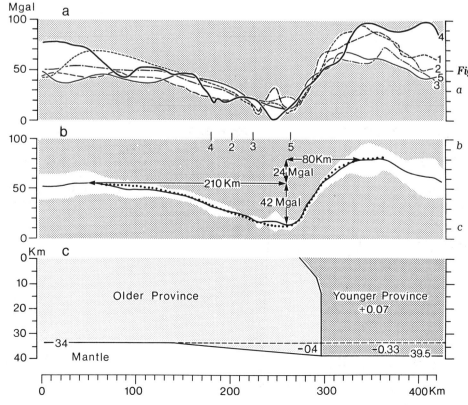

Figure 8.13

a Bouguer gravity anomaly signatures along profiles 1 to 5 of Figure 8.12. Vertical lines numbered 2 to 5 show positions of corresponding inter-province boundaries.

b Gravity signature (type anomaly) derived by averaging profiles 1 to 5. Unshaded envelope is described by standard deviation calculated at 5 km intervals along profiles. Dotted curve is the gravity effect of type model shown in c.

c Type crustal structure derived from type anomaly. Density contrasts in g/cm^3; depths in km. (After Gibb and Thomas, 1976; reproduced by permission of Macmillan Press.)

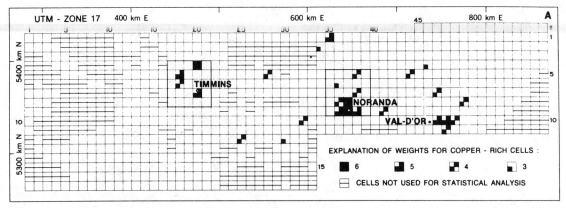

a Control cells for copper;

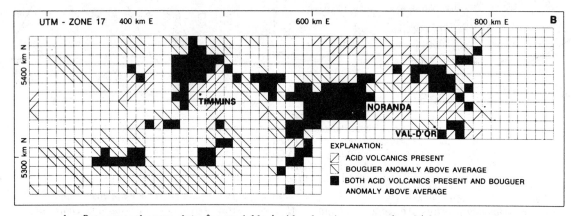

b Presence-absence data for variable 'acid volcanics present' and 'above average Bouguer anomaly' and their combination.

Figure 8.14. (After Agterberg et al., 1972.)

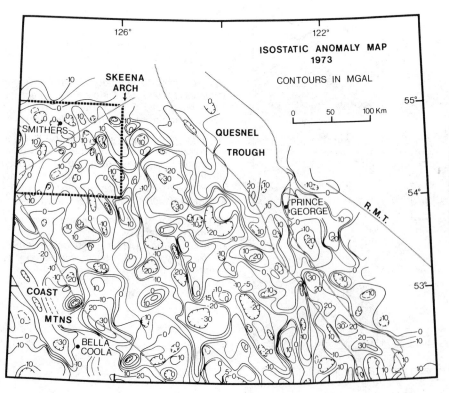

Figure 8.15

Airy-Heiskanen isostatic anomalies for $T = 30$ km and $\Delta\rho = 0.4$ g/cm^3 and major geological units of the Parsnip River map sheet (N.T.S. 93). The Smithers-Houston region is outlined in the northwest part of the map. (After Stacey, 1976; reproduced by permission of Geological Association of Canada.)

Hoffman (1973), Sutton and Watson (1974) and Burke et al. (1977) have interpreted the Coronation orogen as a product of Andean-type orogeny or as a product of the Wilson cycle of ocean opening and closing. Van Schmus (1976) has interpreted rocks of the Southern Province involved in the Penokean Orogeny as products of processes similar to modern plate tectonics. Possible suture sites have also been proposed along or near the northern boundary of the Grenville Province (Vine and Hess, 1968; Krogh and Davis, 1971; Irving et al., 1972), along the buried southern boundary (Dewey and Burke, 1973; Baer, 1976) and within the Grenville Province (Chesworth, 1972; Brown et al., 1975). Identification of paleosutures in the Shield and their true polarity may prove to be a useful guide to the distribution of base metals. By analogy with modern examples, equivalent mineral zonations should occur in the hinterland of paleosutures in the Shield. On the assumption that erosion has exposed the deeper levels of large areas of the Shield, we would expect to find minerals formed at levels of the crust deeper than those of modern examples. Evidence that this may indeed be the case is forthcoming from recent geochemical studies in the Shield (Badham, 1976; Allan, 1978).

Agterberg et al. (1972) have also employed gravity in an indirect way to assist with the prediction of copper and zinc potential in a test area of the Abitibi belt in the Superior Province. This study provided the first quantitative estimates in terms of statistical probability made by the Geological Survey of Canada of the mineral potential of an area by analysis of certain geological and geophysical parameters. The test area was divided into cells for analysis and the average Bouguer anomaly value per cell was one of ten basic geological and geophysical variables used to express the probability of occurrence of mineral deposits. Bouguer values in each cell were averaged and mean values for cells with no gravity stations were computed by interpolation from surrounding cells. Associations of the basic variables were used to produce 45 new variables. It is of interest to note that for copper in Abitibi, the variable with the largest correlation coefficient was the combination of 'acid volcanic rocks present' and 'above average Bouguer anomaly' (Fig. 8.14). The calculated probability index was also converted to a tonnage of copper or zinc expected from a given region.

In a similar statistical study in the Abitibi area, Favini and Assad (1974) relied exclusively on variables derived from aeromagnetic and gravity maps to predict massive sulphide potential. The parameters used were 'average values' and 'logarithm of the variance', both over a four mile traverse interval, and 'gradient'. Stepwise discriminant analysis was used to derive discriminant functions which could best characterize the geophysical variables in a control area in relation to known sulphide potential. The derived probabilistic sulphide potential profile conformed reasonably well to the known profile emphasizing the usefulness of such methods in the exploration for massive sulphides.

Direct Applications

In British Columbia, Triassic and Cenozoic copper porphyry deposits are associated with granitic plutons ranging in size from large batholiths to small plugs. In the Smithers-Houston area (Fig. 8.15), Stacey (1976) has used the gravity method in a more direct way to formulate a predictive model for copper porphyry exploration. He noted a relationship between local negative gravity anomalies (of about 30 mgal amplitude) and exposed quartz monzonite plutons and suggested that similar but lower amplitude anomalies over the volcanics of the Skeena arch are related to similar but buried plutons. A typical exposed pluton has a diameter of about 10 km and a gravity anomaly of -30 mgal. The anomaly can be explained by a density difference of 0.2 g/cm^3 between quartz monzonite and surrounding quartz diorite and a body 10 km wide extending to a depth of 5 km.

Anomalies associated with the plutons tend to be elongated parallel to faults suggesting that their emplacement was fault-controlled. Figure 8.16a from Stacey (1976) shows the anomaly over a rectangular body similar to that in the Dean River Valley. Figure 8.16b shows the anomaly derived from a body at a depth of 3 km with a thickness of 2 km above a uniform granitic terrain and surrounded by volcanics 0.2 g/cm^3 denser. Figure 8.16c shows small stocks occurring preferentially at the intersection of faults bounding the larger intrusion. Stacey suggested

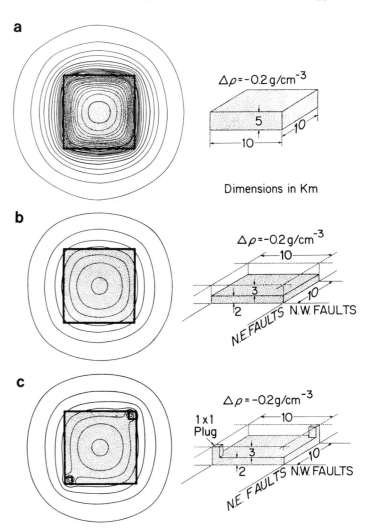

a quartz monzonite body similar to that exposed in the Dean River valley (amplitude of the gravity anomaly is -25 mgal);

b unexposed quartz monzonite body similar to those believed to lie below the Skeena arch (amplitude of the gravity anomaly is -7 mgal);

c unexposed body similar to the above with small stocks reaching the surface at fault plane intersections (amplitude of the gravity anomaly is -7 mgal).

Figure 8.16. Computed gravity anomalies contoured at 1 mgal intervals. (After Stacey, 1976; reproduced by permission of Geological Association of Canada.)

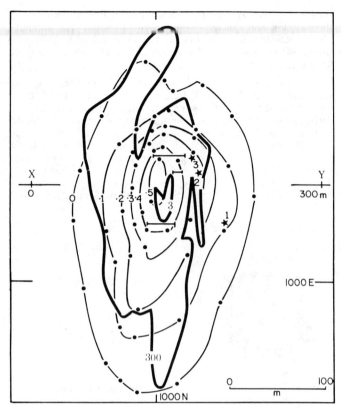

Figure 8.17. Residual gravity anomaly associated with the Agricola Lake geochemical anomaly Northwest Territories. Contour interval 0.1 mgal. Heavy lines represent VLF EM apparent resistivity contours from Scott, 1975. Stars indicate drillhole positions. Solid bars indicate plan views of sulphides as proved by drilling at that time. (After Boyd et al., 1975.)

that Cu mineralization associated with these small stocks came from the surrounding Mesozoic volcanics because the exposed monzonites such as the Dean River Valley example are barren. Using this model, gravity surveys in the vicinity of porphyry stocks can give an estimate of the thickness of overlying Mesozoic volcanics and detailed surveys at the intersection points of regional faults may lead to the discovery of new porphyry stocks in the region.

In a similar application of the gravity method Ager et al. (1973) derived a three-dimensional model of the Guichon Creek batholith from gravity data in British Columbia. They concluded that large low grade copper porphyry deposits of higher than average economic importance are spatially related to the surface projection of the core of the batholith and suggested that precise regional gravity studies could be used to delineate other low grade copper deposits related to batholiths.

In another recent direct application of the gravity method Boyd et al. (1975) mapped the gravity anomaly associated with a massive sulphide body in the Northwest Territories. The target was one of several located by follow-up studies (Cameron and Durham, 1974a,b) of a prominent Cu-Zn geochemical anomaly outlined by a regional lake sediment survey in 1972 (Allan et al., 1972). The body occurs in intermediate to acid volcanics of the Archean Beechey Lake sedimentary-volcanic belt. The target was selected as a test area for multidisciplinary study using a variety of geophysical and geochemical methods.

A Bouguer anomaly map of the area shows anomalies accurate to 0.1 mgal (Boyd et al., 1975) and the residual gravity high separated from the regional field corresponds in

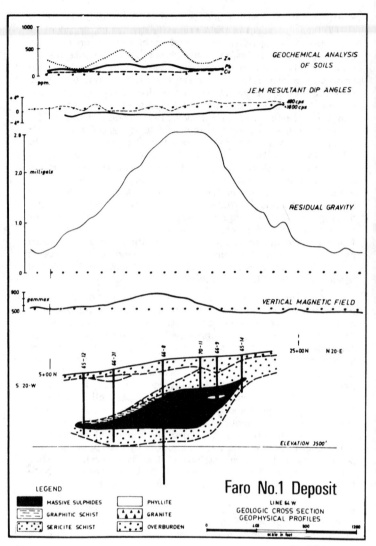

Figure 8.18. Geological cross-section of Faro deposit and geophysical profiles. (After Brock, 1973; reproduced by permission of Canadian Institute of Mining and Metallurgy.)

position with the prominent VLF Radiohm electrical resistivity anomaly over the mineralized zone (Scott, 1975) (Fig. 8.17). The residual anomaly is elliptical in plan and is 320 m by 180 m and has a peak amplitude of 0.5 mgal. The close association between gravity and resistivity anomalies taken together with other geophysical and geochemical survey results strongly suggested the presence of a sulphide body. A massive Zn-Cu-Pb-Ag-Au-bearing sulphide body was confirmed by drilling and further detailed gravity surveys by the YAVA syndicate (Cameron, 1977).

Gravity played an important role in the integrated airborne and ground geophysical exploration program that led to the discovery of the Faro Pb-Zn sulphide deposit of the Yukon Territory (Brock, 1973). Initially gravity was used to further define coincident magnetic and geochemical anomalies. It was later replaced as a primary tool by EM because of relatively high costs, immobility and ambiguity of interpretation. Gravity surveys were finally reserved for follow-up work following discovery of sulphide indications by rotary drilling and proved to be an excellent guide for subsequent diamond drilling programs. The Faro No. 1 deposit was best outlined by a gravity survey (Brock, 1973); a 2.8 mgal anomaly coincides with the thickest section of the

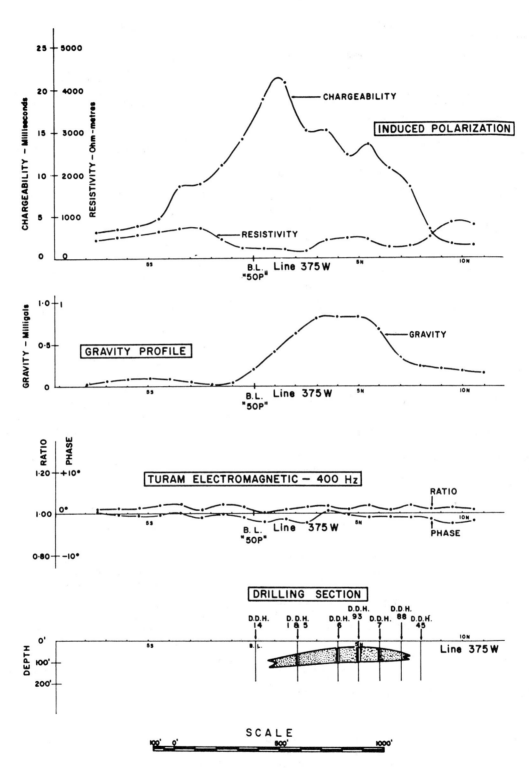

Figure 8.19. Geological cross-section of Pyramid No. 1 orebody and geophysical profiles. (*After Seigel et al., 1968; reproduced by permission of Geophysics.*)

No. 1 zone (Fig. 8.18). Preliminary calculations using the gravity results suggested a mass of 44 million tons. The surface area of the anomaly was later proven to contain about 46 million tons covering the No. 1 zone.

At Pine Point gravity was also used primarily to expedite development drilling after discovery of Pb-Zn orebodies by the induced polarization method (Seigel et al., 1968). The gravity anomalies showed an extremely good correlation with the distribution of Pb-Zn mineralization in the orebodies (Fig. 8.19) and permitted optimal selection of drillhole locations. The gravity results were also used successfully to estimate the total tonnage of the orebodies using the classical method (Hammer, 1945).

In Manitoba, the exposed portion of the Nickel Belt underlies a pronounced gravity minimum and orebodies lie scattered along its axis. This correlation, first noted by Innes (1960) was recently used by Roth (1975) to assist in determining the concealed southerly extension of the Nickel Belt although AFMAG, EM, IP and magnetic methods were the primary tools of investigation. Detailed gravity surveys were also used in this investigation to help outline serpentinite zones discovered by other methods. The correlation between nickel belts and linear regional negative gravity anomalies was also noted by Eckstrand (1976) who compared features of the Manitoba Nickel Belt and the Kotalahti Nickel Belt of Finland.

REFERENCES

Ager, C.A., Ulrych, T.J., and McMillan, W.J.
1973: A gravity model for the Guichon Creek batholith, south-central British Columbia; Can. J. Earth Sci., v. 10, p. 920-935.

Agterberg, F.P., Chung, C.F., Fabbri, A.G., Kelly, A.M., and Springer, J.S.
1972: Geomathematical evaluation of copper and zinc potential of the Abitibi area, Ontario and Quebec; Geol. Surv. Can., Paper 71-41.

Al-Chalabi, M.
1970: Interpretation of two-dimensional magnetic profiles by non-linear optimisation; Boll. Geofis. Teor. Appl., v. 12, p. 3-20.

1972: Interpretation of gravity anomalies by non-linear optimisation; Geophys. Prospect, v. 20, p. 1-16.

Allan, R.J.
1978: Regional geochemical anomalies related to plate tectonic models for the northwestern Canadian Shield; J. Geochem. Explor., v. 10, p. 203-218.

Allan, R.J., Cameron, E.M., and Durham, C.C.
1972: Lake geochemistry: a low sample density technique for reconnaissance geochemical exploration and mapping of the Canadian Shield; in Geochemical Exploration (ed. M.J. Jones), Inst. Min. Metall., London, p. 131-160.

Badham, J.P.N.
1976: Orogenesis and metallogenesis with reference to the silver-nickel, cobalt arsenide ore association; in Metallogeny and Plate Tectonics (ed. D.F. Strong), Geol. Assoc. Can., Spec. Paper 14, p. 559-571.

Baer, A.J.
1976: The Grenville Province in Helikian times: a possible model of evolution; Phil. Trans. Roy. Soc. Lond., Ser. A, v. 280, p. 499-515.

Bott, M.H.P.
1967: Solution of the linear inverse problem in magnetic interpretation with application to oceanic magnetic anomalies; Geophys. J. Roy. Astr. Soc., v. 13, p. 313-323.

Bott, M.H.P. (cont'd.)
1973: Inverse methods in the interpretation of magnetic and gravity anomalies; in Methods in Computational Physics (ed. Bruce A. Bolt), v. 13, p. 133-162, Academic Press, New York.

Boyd, J.B., Gibb, R.A., and Thomas, M.D.
1975: A gravity investigation within the Agricola Lake geochemical anomaly, District of Mackenzie; in Report of Activities, Part A, Geol. Surv. Can., Paper 75-1A, p. 193-198.

Brock, J.S.
1973: Geophysical exploration leading to the discovery of the Faro deposit; Can. Min. Metall. Bull., v. 66, no. 738, p. 97-116.

Brown, R.L., Chappell, J.F., Moore, J.M. Jr., and Thompson, P.H.
1975: An ensimatic island arc and ocean closure in the Grenville province of southeastern Ontario, Canada; Geosci. Can., v. 2(3), p. 141-144.

Buck, R.J. and Tanner, J.G.
1972: Storage and retrieval of gravity data; Bull. Geod., v. 103, p. 63-84.

Burke, K., Dewey, J.F., and Kidd, W.S.F.
1977: World distribution of sutures – the sites of former oceans; Tectonophysics, v. 40, p. 69-99.

Cameron, E.M.
1977: Geochemical dispersion in mineralized soils of a permafrost environment; J. Geochem. Explor., v. 7, p. 301-326.

Cameron, E.M. and Durham, C.C.
1974a: Follow-up investigations on the Bear-Slave geochemical operation; in Report of Activities, Part A, Paper 74-1A, p. 53-60.

1974b: Geochemical studies in the eastern part of the Slave structural province, 1973; Geol. Surv. Can., Paper 74-27.

Camfield, P.A. and Gough, D.I.
1977: A possible Proterozoic plate boundary in North America; Can. J. Earth Sci., v. 14, p. 1229-1238.

Cavanaugh, M.D. and Seyfert, C.K.
1977: Apparent polar wander paths and the joining of the Superior and Slave Provinces during Proterozoic time; Geology, v. 5, p. 207-211.

Chesworth, W.
1972: Possible plate contact in the Precambrian of eastern Canada; Nature (Phys. Sci.), v. 237, p. 11-12.

Cook, A.H.
1965: The absolute determination of the acceleration due to gravity; Metrologia, v. 1(3), p. 84-114.

Cordell, L. and Henderson, R.G.
1968: Iterative three-dimensional solution of gravity anomaly data using a digital computer; Geophysics, v. 33, p. 596-601.

Dewey, J.F. and Burke, K.C.A.
1973: Tibetan, Variscan and Precambrian basement reactivation: products of continental collision; J. Geol., v. 81, p. 683-692.

Donaldson, J.A., Irving, E., Tanner, J., and McGlynn, J.
1976: Stockwell symposium on the Hudsonian orogeny and plate tectonics; Geosci. Can., v. 3(4), p. 285-291.

Eckstrand, O.R.
1976: Striking similarities and some differences are evident from comparison of Finnish and Canadian (nickel) deposits; Northern Miner, March 4.

Faller, J.E.
  1965: Results of an absolute determination of the acceleration of gravity; J. Geophys. Res., v. 70, p. 4035-4038.

Favini, G. and Assad, R.
  1974: Statistical aeromagnetic and gravimetric criteria for sulphide districts in greenstone areas of Quebec and Ontario; Can. Min. Metall. Bull., v. 67, no. 752, p. 58-63.

Gibb, R.A.
  1975: Collision tectonics in the Canadian Shield?; Earth Planet. Sci. Lett., v. 27, p. 378-382.

Gibb, R.A. and Halliday, D.W.
  1974: Gravity measurements in southern District of Keewatin and southeastern District of Mackenzie, N.W.T.; Earth Phys. Branch, Ottawa, Gravity Map Ser. Nos. 124-131.

Gibb, R.A. and Thomas, M.D.
  1976: Gravity signature of fossil plate boundaries in the Canadian Shield; Nature, v. 262, p. 199-200.

  1977a: Gravity mapping in Canada; in Geophysics in the Americas (eds. J.G. Tanner and M.R. Dence), Publ. Earth Phys. Branch, Ottawa, v. 46, p. 48-57.

  1977b: The Thelon Front: a cryptic suture in the Canadian Shield; Tectonophysics, v. 38, p. 211-222.

Gibb, R.A. and Walcott, R.I.
  1971: A Precambrian suture in the Canadian Shield; Earth Planet. Sci. Lett., v. 10, p. 417-422.

Goodwin, A.M. and West, G.F.
  1974: The Superior geotraverse project; Geosci. Can., v. 1(3), p. 21-29.

Grant, F.S.
  1972: Review of data processing and interpretation methods in gravity and magnetics, 1964-71; Geophysics, v. 37, p. 647-661.

Hammer, S.
  1945: Estimating ore masses in gravity prospecting; Geophysics, v. 10, p. 50-62.

Heller, W.G.
  1977: Vertical deflection recovery by a gradiometer-aided inertial system; Proc. 1st International Symposium on Inertial Technology for Surveying and Geodesy; Can. Inst. Surv., Ottawa, p. 343-350.

Hoffman, P.F.
  1973: Evolution of an early Proterozoic continental margin: the Coronation geosyncline and associated aulacogens, northwest Canadian Shield; Phil. Trans. Roy. Soc. Lond., Ser. A, v. 273, p. 547-581.

Innes, M.J.S.
  1960: Gravity and isostasy in Manitoba and northern Ontario; Publ. Dominion Obs., Ottawa, v. 21, p. 265-338.

Irving, E., Park, J.K., and Roy, J.L.
  1972: Paleomagnetism and the origin of the Grenville Front; Nature, v. 263, p. 344-346.

Kanasewich, E.R. and Agarwal, R.G.
  1975: Analysis of combined gravity and magnetic fields in wave number domains; J. Geophys. Res., v. 75, p. 5702-5712.

Kearey, P.
  1976: A regional structural model of the Labrador Trough, northern Quebec, from gravity studies and its relevance to continental collision in the Precambrian; Earth Planet. Sci. Lett., v. 28, p. 371-378.

Krogh, T.E. and Davis, G.L.
  1971: The Grenville Front interpreted as an ancient plate boundary; Carnegie Inst. Wash., Year Book 70, p. 239-240.

Lambert, A., Liard, J., and Dragert, H.
  1977: Canadian precise gravity networks for crustal movement studies: an instrument evaluation; Presented at International Symposium on Recent Crustal Movements, Pao Alta, California, July, 1977.

Langford, F.F. and Morin, J.A.
  1976: The development of the Superior province of northwestern Ontario by merging island arcs; Amer. J. Sci., v. 276, p. 1023-1034.

Lee, S.K.J.
  1977: Multilayer gravity inversion using Fourier transforms; unpubl. M.Sc. thesis, Univ. Alberta, Edmonton.

McConnell, R.K.
  1977: The management of the Canadian national gravity data base; in Geophysics in the Americas (eds. J.G. Tanner and M.R. Dence), Publ. Earth Phys. Branch, Ottawa, v. 46, p. 107-112.

McConnell, R.K. and Tanner, J.G.
  1974: Gravity standards in Canada; in Proc. 1973 Can. Soc. Explor. Geophys. National Convention (eds. A.E. Wren and R.B. Cruz), p. 235-237.

Metzger, E.H. and Jircitano, A.
  1977: Application analysis of gravity gradiometers for mapping of earth gravity anomalies and derivation of the density distribution of the earth crust; Proc. 1st International Symposium on Inertial Technology for Surveying and Geodesy; Can. Inst. Surv., Ottawa, p. 334-342.

Miller, H.G.
  1977: Statistical decision making applied to geophysical modelling; Geol. Assoc. Can., Program with Abstracts, v. 2, p. 36.

Mitchell, A.H.G.
  1976: Tectonic settings for emplacement of subduction-related magmas and associated mineral deposits; in Metallogeny and Plate Tectonics (ed. D.F. Strong), Geol. Assoc. Can., Spec. Paper 14, p. 3-21.

Morelli, C., Gantar, C., Honkasalo, T., McConnell, R.K., Tanner, J.G., Szabo, B., Uotila, U., and Whalen, C.T.
  1974: The international gravity standardization net 1971 (IGSN71); Int. Assoc. Geod., Spec. Publ. 4.

Nagy, D.
  1977: Bouguer anomaly map of Canada; Can. Cartog., v. 14, p. 59-66.

Nettleton, L.L.
  1976: Gravity and magnetics in oil prospecting; McGraw-Hill Book Co., New York, 464 p.

Oldenburg, D.G.
  1974: The inversion and interpretation of gravity anomalies; Geophysics, v. 39, p. 526-536.

Parker, R.L.
  1972: The rapid calculation of potential anomalies; Geophys. J. Roy. Astr. Soc., v. 31, p. 447-455.

  1977: Understanding inverse theory; Ann. Rev. Earth Planet. Sci., v. 5, p. 35-64.

Preston-Thomas, H., Turnbull, L.G., Green, E., Dauphinee, T.M., and Kalra, G.N.
  1960: An absolute measurement of the acceleration due to gravity at Ottawa; Can. J. Phys., v. 38, p. 824-852.

Roth, J.
  1975: Exploration of the southern extension of the Manitoba Nickel Belt; Can. Min. Metall. Bull., v. 68, no. 761, p. 73-80.

Sabine, E.
  1821: An account of experiments to determine the acceleration of the pendulum in different latitudes; Phil. Trans. Roy. Soc. Lond., v. 111, p. 163-190.

Sakuma, A.
  1971: Recent developments in the absolute measurement of gravitational acceleration; in Proc. International Conference on Precision Measurement and Fundamental Constants (eds. D.N. Laugenberg and B.N. Taylor), Nat. Bur. Stand., Spec. Publ., 343, p. 447-456.

Scott, W.J.
  1975: VLF resistivity (radiohm) survey, Agricola Lake area, District of Mackenzie; in Report of Activities, Part A, Geol. Surv. Can., Paper 75-1A, p. 223-225.

Seigel, H.O., Hill, H.L., and Baird, J.G.
  1968: Discovery case history of the Pyramid ore bodies, Pine Point, Northwest Territories, Canada; Geophysics, v. 33, p. 645-656.

Sillitoe, R.H.
  1976: Andean mineralization: a model for the metallogeny of convergent plate margins; in Metallogeny and Plate Tectonics (ed. D.F. Strong), Geol. Assoc. Can. Spec. Paper 14, p. 59-100.

Stacey, R.A.
  1976: Deep structure of porphyry ore deposits in the Canadian Cordillera; in Metallogeny and Plate Tectonics (ed. D.F. Strong), Geol. Assoc. Can. Spec. Paper 14, p. 391-412.

Stauffer, M.R.
  1974: Geology of the Flin Flon area; Geosci. Can., v. 1(3), p. 30-35.

Sutton, J. and Watson, J.V.
  1974: Tectonic evolution of continents in early Proterozoic times; Nature, v. 247, p. 433-435.

Talbot, C.J.
  1973: A plate tectonic model for the Archean crust; Phil. Trans. Roy. Soc. Lond., Ser. A, v. 273, p. 413-427.

Talwani, M.
  1973: Computer usage in the computation of gravity anomalies; in Methods in Computational Physics (ed. Bruce A. Bolt), v. 13, p. 344-389, Academic Press, New York.

Tanner, J.G.
  1967: An automated method of gravity interpretation; Geophys. J. Roy. Astr. Soc., v. 13, p. 339-347.

Tanner, J.G. and Buck, R.J.
  1964: A computer oriented system for the reduction of gravity data; Publ. Dominion Obs., Ottawa, v. 31, p. 53-65.

Thomas, M.D. and Gibb, R.A.
  1977: Gravity anomalies and deep structure of the Cape Smith foldbelt, northern Ungava, Quebec; Geology, v. 5, p. 169-172.

Thomas, M.D. and Tanner, J.G.
  1975: Cryptic suture in the eastern Grenville Province; Nature, v. 256, p. 392-394.

Trageser, M.B.
  1977: The floated gravity gradiometer in geodesy; Proc. 1st International Symposium on Inertial Technology for Surveying and Geodesy; Can. Inst. Surv., Ottawa, p. 322-333.

Van Schmus, W.R.
  1976: Early and Middle Proterozoic history of the Great Lakes area, North America; Phil. Trans. Roy. Soc. Lond., Ser. A, v. 280, p. 605-628.

Vine, F.J. and Hess, H.H.
  1968: Sea-floor spreading; in The Sea 4(2) (ed. A.E. Maxwell), Wiley Interscience, New York, p. 587-622.

Walcott, R.I. and Boyd, J.B.
  1971: The gravity field of northern Alberta and part of the Northwest Territories and Saskatchewan; Earth Phys. Branch, Ottawa, Gravity Map Ser. Nos. 103-111.

THE INDUCED-POLARIZATION EXPLORATION METHOD

John S. Sumner
Department of Geosciences, University of Arizona, Tucson, U.S.A.

Sumner, John S., The induced-polarization exploration method; in Geophysics and Geochemistry in the Search for Metallic Ores; Peter J. Hood, editor; Geological Survey of Canada, Economic Geology Report 31, p. 123-133, 1979.

Abstract

Induced polarization is a current-stimulated electrical phenomenon observed as a delayed voltage response in earth materials. It is important as a method of exploration for buried metallic mineral deposits. With recent improvements in electrical instrumentation and computer analysis techniques, the method has become well developed and is now the most widely used of the ground geophysical exploration methods.

Induced-polarization measurements are made in the time-domain as a voltage decay curve, in the frequency-domain as a voltage difference with variation in frequency, and in the phase-domain as a phase lag angle. The complex-resistivity method requires a time link between the current transmitter and the voltage receiver to obtain the real and imaginary components of the earth's resistivity. A phase-coupled spectral IP response obtained in this way can be analyzed and interpreted to improve signal-to-noise ratio and remove electromagnetic coupling effects from the field data. The electromagnetic coupling between transmitter and receiver can also be interpreted to give an independent structural picture of the earth in the survey area.

Induced-polarization measurements are routinely made down drillholes both to log the near-hole properties and to probe deeper into the earth. Underground surveys are also made. Induced polarization data are used in mining areas to estimate the grade of metallic minerals and to seek a direction toward better mineralization. Also, there is increasing encouragement that it may be possible to make a distinction between the electrode polarization of metallic minerals and the membrane polarization phenomenon of clays.

Research in IP includes investigation into discrimination between metallic mineral species, removal of electromagnetic coupling effects, improvement of signal-to-noise ratio, and measurement of magnetic induced polarization effects. Mathematical modeling of different geometric shapes of polarizable bodies is proving to be an effective way of interpreting IP results, as has been the simulation of the subsurface using analog models. Research in IP instrumentation has been directed toward large-scale integrated circuits and computers used in the field to process, analyze, and interpret data.

The future for IP surveying appears to be favorable, and the method continues to be the best geophysical means for locating small volume percentages of metallic minerals in concealed mineral deposits.

Résumé

La polarisation induite est un phénomène électrique favorisé par le passage d'un courant; une fois que le courant est interrompu, on constate l'existence d'un potentiel transitoire dans le sol. Cette méthode d'exploration est importante pour la recherche des minerais métalliques enfouis. Grâce aux récents perfectionnements de l'appareillage électrique et des techniques d'analyse par ordinateur, cette méthode a pris un développement important, et actuellement, elle est celle que l'on utilise le plus pour l'exploration géophysique au sol.

Les mesures de polarisation induites sont, dans le domaine des temps, une courbe de décroissance du potentiel; dans le domaine des fréquences, la différence caractérisant le potentiel lorsqu'on fait varier la fréquence, et dans le domaine des phases, l'angle de retard de phase. La méthode de résistivité complexe exige que l'on établisse une relation de temps, entre l'émetteur de courant et le récepteur de tension, afin d'obtenir les composantes imaginaires et réelles de la résistivité terrestre. Ainsi, on obtient une réponse IP (de polarisation induite) spectrale, avec couplage de phase, que l'on peut analyser et interpréter afin d'améliorer le rapport signal-bruit, et d'éliminer des données obtenues sur le terrain les effets du couplage électromagnétique. Le couplage électromagnétique entre l'émetteur et le récepteur peut aussi être interprété de manière à donner dans la région étudiée une image structurale indépendante.

On effectue couramment des relevés de polarisation induite dans les trous de forage, à la fois pour établir un log des propriétés du terrain à proximité du trou de forage, et explorer le sol à plus grande profondeur. On effectue aussi des levés souterrains. On utilise les résultats de la polarisation induite dans les zones minières pour évaluer la teneur des minerais métalliques, et chercher les zones les mieux minéralisées. Et de plus en plus, il semble que l'on pourra établir une distinction entre la polarisation d'électrode des minéraux métalliques, et la polarisation de membrane des argiles.

En polarisation induite, la recherche vise aussi à nous permettre de mieux distinguer les unes des autres les espèces minérales métalliques, d'éliminer les effets de couplage électromagnétique, d'améliorer le rapport signal-bruit, et de mesurer les effets magnétiques de la polarisation induite. La modélisation mathématique de diverses formes géométriques des corps polarisables s'avère comme une méthode efficace d'interprétation des résultats IP, de même que la simulation des zones proches de la surface à l'aide de modèles analogiques. La recherche relative à l'appareillage IP s'est orientée vers l'étude de circuits intégrés de grandes dimensions et d'ordinateurs, que l'on pourrait utiliser sur le terrain pour le traitement, l'analyse et l'interprétation des données. Il semble que l'avenir soit prometteur pour les méthodes de levés IP, et cette méthode géophysique semble être la plus appropriée pour localiser des concentrations peu volumineuses de minéraux métalliques, dans les gîtes minéraux dissimulés.

This report is an update review of the induced-polarization (IP) method of geophysical exploration. Since its rediscovery and first extensive field use three decades ago, the method has enjoyed increasing popularity until it is now the most widely used ground geophysical surveying technique employed in exploration for metallic-luster minerals.

METHODS OF IP MEASUREMENT

Whereas 10 years ago there were only two different commonly used ways of measuring the IP phenomenon, more recently the phase and the complex-resistivity systems have gained favour. The measuring method is mainly a matter of desired sensitivity and the availability of field equipment, as will be discussed later. The standardization of units of IP measurements is presently being debated. This writer believes that the results of field measurements should be compatible with those of laboratory measurements and that units relevant to measurements should be used.

Time-domain IP

Prior to 1950 all IP measurements were of the time-domain type using the waveforms shown (Fig. 9.1). A simple on-off step-function current was impressed in the earth by grounded contacts and the analysis of the voltage waveform gave a measure of the IP effect. A satisfactory theoretical explanation of the IP phenomenon was developed by Seigel (1959), who formulated that since the polarization ($\vec{P}$) was stimulated by electric current the dimensionless chargeability response of V_s/V_p or M, must be directly proportional to polarization and inversely proportional to the current density ($\vec{J}$), or

$$M = -\vec{P}/\vec{J} \qquad (1)$$

The minus sign indicates the opposing vector relationship between polarization and current density. The rather unusual feature of equation (1) is that $\vec{P}$ and $\vec{J}$ must have similar dimensional units, which means that the polarization can be physically interpreted either as a blocking resistivity or as the generation in polarizable ground of opposing electrical currents. In any event, induced polarization basically is quite different from the charge separation phenomenon of dielectric materials.

Wait (1959a) pointed out that IP has a linear behavior in materials, at least at low current densities. Induced-polarization linearity is important because it means that IP measurements are repeatable under different current conditions. The IP phenomenon becomes nonlinear at higher currency densities, a fact that may prove useful in identifying the causes of polarization. Also, the depolarization or decay curve is generally logarithmic in shape, although contributing components with different time constants can usually be identified.

Frequency-domain IP

Since the blocking resistivity is a time-dependent behavior, it must also have a frequency dependence because time and frequency values can be related one to the other. Thus polarizable materials can be viewed as having an impedance which is frequency dependent, leading to frequency-domain measurements. Madden et al. (1957) of MIT and Wait (1959b) of Newmont developed frequency-domain equipment and accompanying electrical theory. The frequency method current and waveform patterns are shown on Figure 9.2. The measured parameter, per cent frequency effect (PFE), used now by most field workers is given by

$$PFE = 100 \times \frac{\rho_{dc} - \rho_{ac}}{\rho_{ac}} \left[\log \left(\frac{f_{ac}}{f_{dc}} \right) \right]^{-1} \qquad (2)$$

where ρ_{dc} and ρ_{ac} and are the low- and high-frequency apparent resistivities. Early frequency-domain field equipment lacked a capability to measure low frequency voltages accurately, hence the ρ_{ac} rather than the ρ_{dc} in the denominator of equation (2). If ρ_{dc} were used rather than ρ_{ac} there would be an exact proportionality between chargeability and frequency effect. However, these two IP response parameters are nearly proportional at fairly low polarization values.

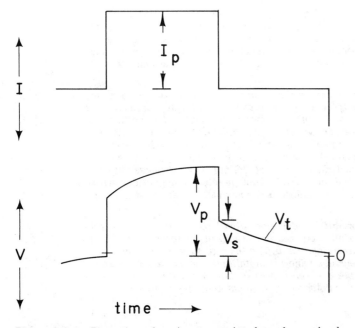

Figure 9.1. The time-domain transmitted and received voltage waveforms, showing the inducing primary current I_p being detected as a maximum primary voltage V_p. When current is turned off, voltage drops to a secondary level V_s and the transient voltage V_t decays with time.

Groups using the frequency-domain IP method have advocated the use of a ratio factor of the apparent frequency effect with the apparent resistivity, resulting in a parameter dubbed "metal factor". To put the metal factor (MF) into the range of commonly used numbers it is multiplied by 2000,

$$MF = (PFE/\rho_a) \times 2000 \qquad (3)$$

where ρ_a is in ohm metres. Controversies have arisen regarding the merits and significance of the metal factor — with inconclusive results. Table 9.1 is an attempt to summarize the pros and cons of the metal factor argument. There seem to be areas where the metal factor is a useful parameter in estimating the amount of mineralization and in putting a priority on anomaly patterns. Recently Snyder and Merkel (1977) have used a working relationship

$$\% \text{ wt. sulphides} = (100 \times \sigma)^{1/3} \qquad (4)$$

where σ is the quadrature conductivity (metal factor) in millimhos per metre. However, the metal factor can be misleading if used without regard to threshold IP response and possible low resistivities. Figure 9.3 indicates a simple nomograph relationship between these three factors.

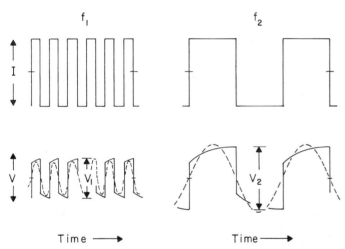

Figure 9.2. Frequency-method waveforms, showing a controlled constant inducing current I at frequencies f_{ac} and f_{dc} being detected as voltages V_{ac} and V_{dc} where $V_{ac} < V_{dc}$. The dashed line is the sinusoidal filtered voltage.

Table 9.1

The metal factor debate

For	Against
1. Increases resolution	1a. Emphasizes low resistivities
	1b. Emphasizes EM coupling errors
2. Useful in mineralization estimation	2. No physical basis for sulphide grade estimation
3. Physically related to the dielectric constant	3. No precise physical interpretation
4. Useful as a correlation factor between polarization and low resistivity	4. Exaggerates resistivity anisotropy

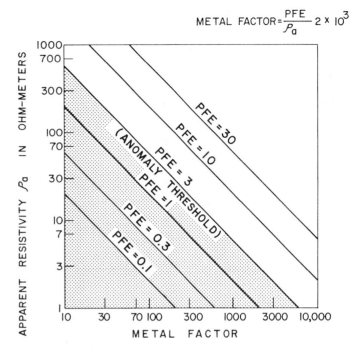

EXAMPLE: IF $\rho_a = 30$ AND PFE = 10
THEN METAL FACTOR $\approx$ 670

Figure 9.3. Per cent frequency effect (PFE) plotted as a function of metal factor and resistivity.

Phase IP measurements

If the current waveform transmitting system and the voltage receiving system are temporally time linked, the phase difference between transmitted and received signals can be measured and this difference, determined either in time or as an angle, gives the polarization of the intervening earth. Significantly, the ratio interval between time and total time (or angular difference and angle) remains generally constant over the measuring range (Fig. 9.4), although other patterns and variations do exist. One advantage in making phase measurements is that only a single waveform need be used, so speedy measurements are possible. However, the necessary time correlation between transmitter and receiver has posed problems in instrumentation. Many reference methods have been proposed and used, including an electrical cable, a precise clock reference, a ground signal, comparison of harmonics, and a radio link.

Phase-angle measurements are usually made by taking a ratio of out-of-phase and in-phase components and then finding the tangent of the defined angle. The rotating phase diagram is illustrated in Figure 9.5. It can be demonstrated that phase angles are closely related to chargeability (Fig. 9.6) relating a phase diagram to the polarization quantities of equation (1). The quadrature polarization vector P_Q leads the resistive component J_R of the total inducing current vector J by $\pi/2$, then

$$\tan \beta = -P_Q/J_R \qquad (5)$$
$$\tan \beta = -M \qquad (6)$$

so that at small phase angles

$$\beta = -M \qquad (7)$$

The approximate relationship between PFE, chargeability, and phase angle can be summarized as follows

	Domain	
Frequency (Per cent over one decade)	Time (Milliseconds) relative to M_{331}	Phase (Milliradians)
1.0	6.6	−5.6
0.15	1.0	−0.83
0.18	1.2	−1.0

The measured IP response voltage lags behind the inducing current, so normal IP phase angles are negative.

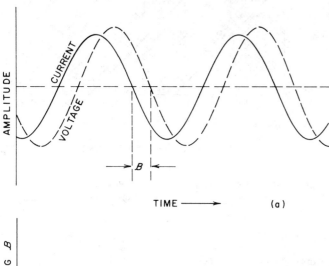

Figure 9.4. Phase determinations: (a) phase lag angle β between input (solid) and output (dashed) sinusoidal waveforms and (b) an ideal IP phase spectrum diagram.

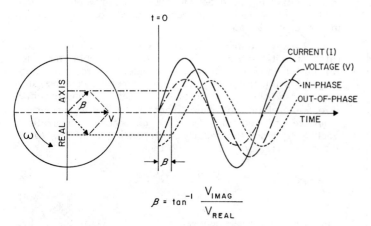

Figure 9.5. Rotating vector components of a phase diagram, showing the phase lag angle.

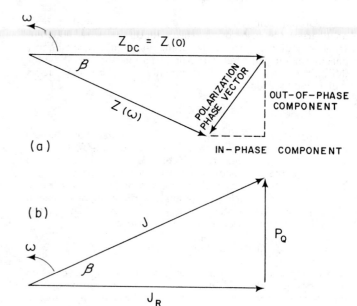

Figure 9.6. Phasor diagrams showing (a) components of the polarization vector and (b) components of the rotating current vector.

Complex Resistivity

Even over a continuous range of determinations, ordinary time- and frequency-domain IP measurements measure only absolute values without regard to in-phase and out-of-phase, or vector, components. But in order to measure all effects truly and to be able to transform back and forth from time to frequency the in-phase and out-of-phase IP components over a wide range must be taken into account. The in-phase and out-of-phase components are sometimes plotted in the complex plane of rotating vectors, which leads to the concept of complex resistivity measurements.

Complex impedance measurements of materials have been made at least since 1941 (Cole and Cole, 1941; Grant, 1958) in studies of dielectric phenomena. Van Voorhis et al. (1973) and more recently Snyder (1976) and Zonge (1976) have used the capabilities of minicomputers to observe IP phase component responses at multiple frequencies. One of the ways of graphically plotting complex-resistivity data, bringing out any existing phase and amplitude differences over the spectrum of observing frequencies, is the Cole-Cole plot shown on Figure 9.7. Note that this type of diagram can also be used to relate conventional PFE and phase angle to the complex-resistivity spectrum.

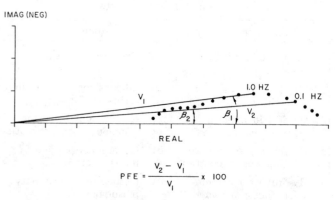

Figure 9.7. A Cole-Cole plot in the complex plane.

IP Measurement Units

Induced-polarization units are, by the unusual nature of the phenomenon, a bit unconventional. Because of the nearly dc frequencies used, most field workers do not use Maxwell's equations and electromagnetic parameters for measurement. However, theoretical electromagnetic relationships can be developed, and some laboratory groups (Olhoeft, 1975) advocate their adoption.

Before changing to other IP units it is probably better that we understand IP theory and that we research the phenomenon more thoroughly and come to closer agreement on reasons for the past units.

FIELD PROCEDURES AND METHODS

Over the years, IP field methods have not changed much. It is still a task and a chore to assemble the equipment, check it out, transport it in good condition to the field site, and conduct field operations in an efficient manner. Miniaturization and simplification of key components, such as the transmitter and receiver, have been a real boon, but upkeep of ancillary equipment, especially on a deep-search survey, remains troublesome.

Surface Arrays

Despite attendant signal-to-noise problems, there has been a continuing trend toward the use of the dipole-dipole array for IP field work. The main arguments for this layout scheme seem to be the advantages in anomaly resolution, depth of exploration, and the flexibility in survey procedure. For shallow exploration the three-electrode or pole-dipole array and gradient (Schlumberger n>10) array are feasible. Figure 9.8 is a summation of the features of the various array geometries. Whiteley (1973) has carefully analyzed the relative advantages and disadvantages of all reasonable arrays, and Table 9.2 summarizes the concepts along these lines.

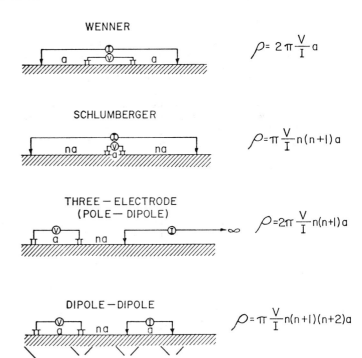

Figure 9.8. Geometric array factors for the commonly used IP exploration array configurations.

Table 9.2
Summary of features of IP arrays

	Dipole-Dipole	Pole-Dipole	Gradient	Schlumberger and Wenner
Response amplitude	Good	Fair	Poor	Poor
Dip of structure	Poor	Poor	Good	Fair
Depth of exploration	Good	Good	Fair	Poor
Resolution of mineralization	Good	Good	Fair	Poor
Freedom from EM coupling	Fair	Fair	Poor	Poor
Interpretability of layering	Poor	Poor	Fair	Good
Depth estimates	Fair	Fair	Fair	Fair
Signal-to-noise ratio	Poor	Fair	Good	Good
Labor needed	Poor	Fair	Good	Good
Susceptibility to noise	Fair	Fair	Good	Good

Drillhole IP Methods

The two categories of IP drillhole surveying are: in-hole surveys (near-hole surveys, including hole logging) and downhole surveys (exploration IP surveys), employing one current or potential electrode down the drillhole. Figure 9.9 is an illustration of the in-hole normal array and the downhole azimuthal array. One of the enigmas of IP surveying is the negative response that is often observed in subsurface work, particularly in drillhole surveying. In every analyzed circumstance, the IP response has been explained by the interactive geometric relationship between the orientation of the inducing currents with the polarizable body and its associated secondary electric fields.

While the mise-à-la-masse drillhole resistivity method has often been successful in finding the direction toward better mineralization, mise-à-la-masse polarization effects can be quite peculiar. The reason for this peculiarity is that in most observed instances IP is a dipolar phenomenon; that is, current is passed through a polarizable body and induced secondary currents flow in and around the body between induced current sources and sinks. However, when an inducing electrode is in contact with the body, nonlinear polarization effects occur in regions of high current density and also the secondary current fields can more readily flow in an opposite sense to the inducing currents and this opposing flow is then observed as a negative polarization. Thus, IP mise-à-la-masse interpretation must be approached with some caution.

INSTRUMENTATION CURRENTLY EMPLOYED FOR IP SURVEYS

Manufacturers of geophysical field equipment have been quick to take maximum advantage of the use of semiconductor devices in innovating new field instruments. Also integrated circuits and microprocessors are in prominent use. Time- and frequency-domain instruments continue to be improved by their competitive manufacturers, and the inquisitive reader is referred to Hood's (1977), Mineral Exploration Trends and Development article (and those in previous years) for particulars on specific instruments. Research in time-domain equipment appears to be focused on noise elimination by stacking and on decay curve shape determination by observing successive selectable voltage windows. Newer frequency-domain IP receivers are capable of removing first- and second-order EM coupling effects by quadratic curve synthesis using phase measurements.

Phase IP Instruments

Nilssen (1971) has described a phase-measuring IP instrument, which is in popular use by the Boliden Company of Sweden. Parasnis (1973) has mentioned complex measurements in Europe, which because a single frequency is used would be classed here as phase measurements.

Phoenix Geophysics Ltd. employs synchronized crystal clocks at the transmitter and receiver for IP phase measurements. The system can also detect and eliminate most normally encountered EM coupling. Scintrex Ltd. of Canada has developed an interesting single-frequency ground-coupled phase-measuring instrument. The relative phase shift is determined by a comparison of the time or "phase" shift of the fundamental and third harmonic components, so the method does not require a radio link or synchronized crystal clocks.

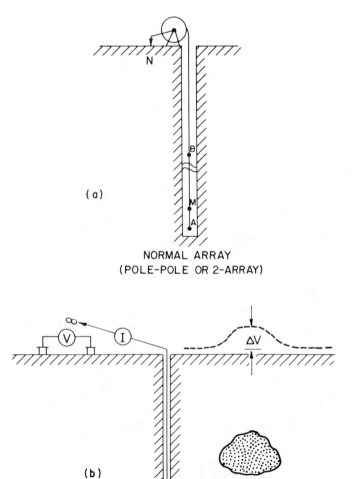

Figure 9.9. Drillhole IP arrays: (a) the normal in-hole array and (b) the downhole azimuthal array.

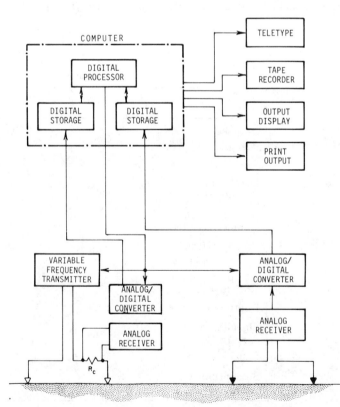

Figure 9.10. Block diagram of the components of a complex-resistivity system in the dipole-dipole array.

A major advantage of IP phase instruments is the effective suppression of unwanted noise voltages. This is brought about because a phase shift angle rather than a signal amplitude is measured, and this is a more basic kind of electrical measurement.

Complex-resistivity Instruments

The first complete multispectral phase-coupled IP field system was developed at Kennecott Copper Corporation prior to 1972 by Van Voorhis, Nelson, and Drake (1973). Multispectral time- and frequency-domain measurements have previously been studied in the laboratory by many researchers (Collett, 1959; Katsube and Collett, 1973; Fraser et al., 1964; Madden et al., 1957; Zonge, 1972) in the hope that mineralized rocks would have a unique spectral signature or that particular minerals or ions could be identified. Van Voohris et al. (1973) did not find any significant variation using their early equipment, but the method could be used to virtually eliminate bothersome EM coupling effects.

More recently, Miller et al. (1975), Snyder (1976) and Zonge (1976) have used modern microprocessor technology in the field to obtain phase-coupled spectral IP, or complex-resistivity, data. All these systems utilize a computer-controlled variable-frequency current transmitter and a linked voltage receiving system permitting the transmitted and received signal to be compared in amplitude and phase. In order to accomplish this, it is usually convenient to Fourier transform signals from one domain to the other and then deconvolve them. Figure 9.10 is a block diagram of a typical complex-resistivity field system, and Figure 9.11 is a program flow diagram (after Zonge, 1976) showing the computer processing of data. The inputted field data proceed from the control point through a decoder, variable selection step, and command operation step to be printed out finally in a selectable number of forms. Once in digital form the processed data can be displayed in several different ways: as time-domain chargeabilities, frequency-domain per cent frequency effect, phase-domain milliradians, as polar or other types of graphical plots, or in tabular form.

IP INTERPRETATION TECHNIQUES

The methods of interpreting processed IP data have been advanced considerably in the past decade by innovative computer modeling methods. In general, there are two main computer approaches to model interpretation, which can be called the forward solution and the inverse solution. Of course, the interpreter must always be guided by geologically reasonable boundary conditions, and intuition and experience remain important factors.

Forward Solutions in IP Interpretation

Forward solutions are the precomputed models of specified electric potentials over subsurface structures. These can be calculated in any one of several different ways, depending mainly on economic limitations and the subsurface geometry of resistivity and IP contrasts. The four numerical computing techniques have been finite difference, finite element, network analogy, and integral equation. The one-dimensional problem involving a change in electrical property with depth has been extensively treated by Nabighian and Elliot (1974) and a fairly complete thirteen-volume library involving all commonly used surface arrays is available from Elliot Geophysical Company of Tucson. For irregularly shaped, two- and three-dimensional subsurface bodies, Hohmann (1977) has summarized the forward interpretation theory and presents several interesting examples, two of which are shown as Figures 9.12 and 9.13.

Inversion Techniques in IP Interpretation

Comprehensive programs can be written to calculate a model based on the field data using the inversion method. Thus far the method has not been too successful beyond the one-dimensional problem, but even so this can greatly simplify the interpreter's task. It appears likely that with computers now being a part of some IP field equipment, the field geophysicist may be able to interact with a simple inversion model to at least narrow down the large number of possible subsurface conditions.

Complex-resistivity Data Interpretation

It is appropriate to mention some of the interpretational features of the complex-resistivity (CR) method, even though some aspects of the data are not yet well understood. The general frequency trends of spectral response have been called types A, B, and C (Fig. 9.14), and apparently these spectral types are related to alteration and mineralization. The type A response, in which the out-of-phase component decreases with increasing frequency, is usually associated with strong alteration and sulphide mineralization, usually including pyrite. Type B has a constant out-of-phase component over the frequency range, similar to the Drake model of Figure 9.4. Type C has an increasing out-of-phase component with increasing frequency and is not often associated with pyritic sulphide mineralization. At higher CR frequencies, electromagnetic effects become stronger and these are interpretable, giving resistivity contrast ratio and depth to resistivity contrast of up to five or six dipole lengths.

PROBLEM AREAS IN IP

As geophysical exploration techniques continue to mature, old problems tend to be solved, only to have new ones appear — and so it is with induced polarization.

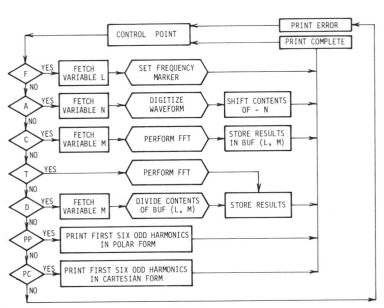

Figure 9.11. Flow diagram of a complex-resistivity computer program. After Zonge (1976).

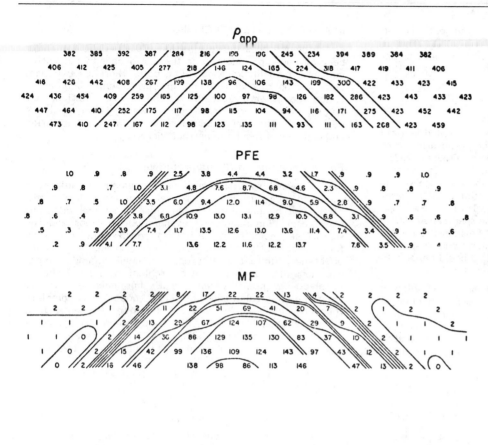

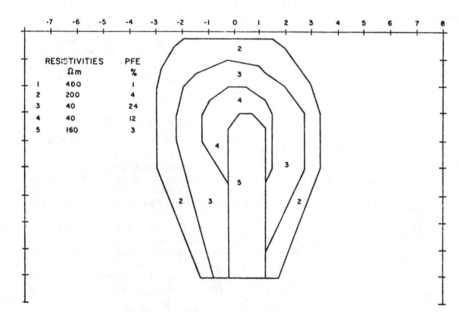

Figure 9.12. Finite element results of the dipole-dipole array for a two-dimensional porphyry copper model. From Hohmann (1977).

EM Coupling Problems

With the availability of lower frequencies and the phase-measuring IP method, electromagnetic coupling is becoming much less of a problem. However, there is a trade-off in combating the uncertainty of whether an anomaly is due to coupling. Additional data must be obtained and interpreted, and the field equipment is more sophisticated and therefore more expensive and complex. Also, the field geophysicist must be trained and experienced in solving these problems. On the brighter side of the EM coupling problem, if complete coupling removal is indeed the fact that it appears to be, then the isolated and interpreted EM effects can constitute an additional facet to the total interpretation of subsurface conditions.

Masking by Resistivity Contrast

The difficult physical environment posed by near-surface low-resistivity layers continues to be bothersome to IP surveys. Of course, this condition must first be recognized to exist and then the thickness and extent of low-resistivity material must be determined in order finally to overcome the masking effects. The problem therefore involves much more than the IP measurements alone and includes making and interpreting resistivity and possibly EM measurements. The interpretation problem due to near-surface, low-resistivity layers would probably be more severe in Australia than in the southwestern United States. As yet complex-resistivity measurements have not been reported from the Australian environment. The magnetic induced polarization (MIP) method when developed to a routine technique, should reduce the masking effect of conductive overburden conditions.

Nonmetallic IP Response

Fine grained clays and fibrous minerals such as serpentinite can produce a moderately strong IP effect, and in the past many exploration holes have been drilled on nonmetallic polarization sources. In many areas over certain of the offending rock types, the clay response can be recognized, even using the polarization signatures obtained with conventional time-domain and spectral frequency-domain instruments. In general, nonmetallic zones have an increasing response with increasing frequencies or shorter times, which is the complex-resistivity response type C of Figure 9.14. In a few problem areas, even the sophisticated spectral patterns obtained by CR equipment give ambiguous results, particularly if small amounts of metallic minerals are present in the nonmetallic zone. Of course, it must be mentioned that the response from nonmetallic sources is generally much lower in amplitude than from metallic-luster minerals.

Metallic Mineralized Zones with a Negligible IP Response

Occasionally the author hears of a mineralized body that does not yield an IP response to an exploration survey. However, on detailed inquiry, it is found that the body was below the depth of exploration or so small that the response

Induced Polarization

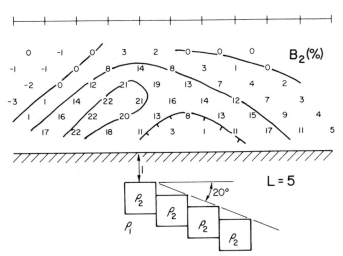

Figure 9.13. Induced-polarization response for the dipole-dipole array over a dipping conductive body composed of square cells, $\rho_2/\rho_1 = 0.2$. The B_2 (%) on the pseudo-section gives the percentage of the intrinsic IP value of the dipping body; L is the ratio of length to depth of the body. From Hohmann (1977).

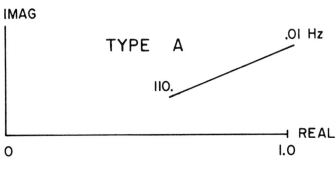

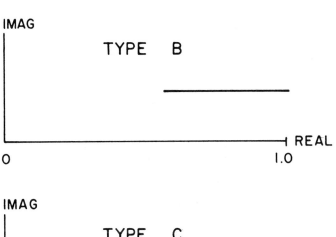

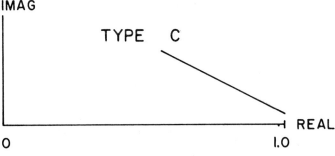

Figure 9.14. Complex-resistivity spectral response types A, B, and C. As on Figure 9.7, the lowest frequencies are on the right. After Zonge and Wynn (1975).

was diluted below an anomaly threshold level. In general the presence of pyrite greatly improves the IP response, possibly because of the porous nature of most pyrite in mineral deposits and its high electrochemical activity. Conversely, deposits low in pyrite, as a general rule have a lower IP response.

RECENT ADVANCES, TRENDS, AND NEEDS IN IP SURVEYING

There is little doubt that the complex-resistivity method of IP surveying is an important trend of research activity. Also numerical modeling and particularly inversion techniques deserve attention for future development. Mineral discrimination using spectral frequency-domain equipment has recently been reported by Pelton et al. (1977); Figure 9.15 shows the results of some of their findings.

Magnetic Induced Polarization

Magnetic induced polarization (MIP) was announced by Seigel (1974) in a theoretical analysis, and since then equipment and survey techniques have been developed. The method measures the weak magnetic fields associated with electrical depolarization currents employing a very sensitive magnetometer.

Development of the MIP technique is continuing, but it has yet to be accepted by the exploration fraternity as a routine search scheme. Figure 9.16 shows the concept of the method. The transmitted current I_t encounters a polarizable body and effectively creates subsurface charges, which in turn produce a secondary electrical field at E_s;

which can be measured by the relationship of Ampere's law $\nabla \times \vec{H} = \sigma \vec{E}$. Grounded contact receiver wires are not required, so airborne IP is conceivable. Model studies of the MIP response by Hohmann (1977) indicate that the MIP response is relatively smaller than conventional IP, at least for bodies in an electrically homogeneous earth. However, the method has advantages where elongated polarizable bodies are present, and where conductive overburden is a major problem.

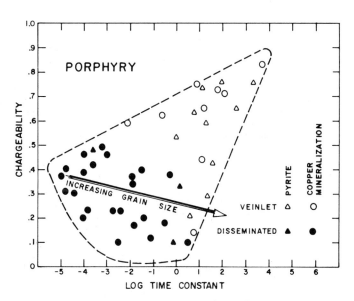

Figure 9.15. Induced-polarization data from western porphyry copper deposits, showing a grouping of veinlet mineralization (open circles) versus discretely disseminated mineralization (closed circles). From Pelton et al. (1977).

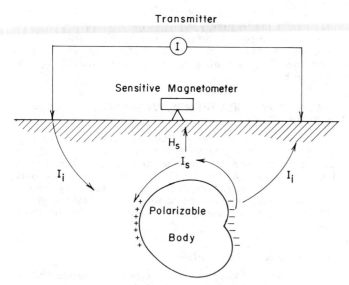

Figure 9.16. Schematic illustration of magnetic induced polarization.

Electrochemistry and IP Theory

One of the more important areas for future IP research in this writer's opinion is in electrochemistry. Laboratories of the Geological Survey of Canada (Katsube, 1977), the U.S. Geological Survey (Olhoeft, 1975), and the University of Utah (Klein and Shuey, 1975) have provided useful information about the basic phenomena, but much more remains to be accomplished before the IP mechanism is completely understood.

A stronger foundation of electrochemical principles will assist in establishing better theory for IP. As matters now stand, there are several proposed electrical theories (Seigel, 1959; Patella, 1972; Nilssen, 1971), none of which has a firm modern foundation based on the underlying electrochemical nature of polarizable materials. Also, electrochemistry holds the clues to the reality and meaning of mineral discrimination as gained from spectral IP data.

Prediction of Telluric Noise

As a result of a long-term correlation of interplanetary magnetic field polarity, solar wind speed, and geomagnetic disturbance index during the declining phase of the recent sunspot cycle (1973-1975) it has become apparent that telluric noise patterns can be predicted. This predictability is due to recognition of the source mechanism for low-frequency geomagnetic disturbances and telluric noise. Heretofore sunspots and solar storms in general were thought to be the source of the telluric noise that so plagues IP surveys and low signal conditions caused by wide electrode separations and low resistivities. Now the exact disturbance source and mechanism has been found thanks to Kitt Peak National Observatory research by Sheeley et al. (1976).

The solar wind sources for low-frequency telluric noise are solar surface structures known as "coronal holes", which are associated with the dying phases of sunspots. A coronal hole is a vortex disturbance pattern that is not readily visible on the sun's surface, and it is essentially a north or south magnetic pole. Coronal holes are the origin of most high-speed solar wind streams, and these charged particles are projected into space like high-pressure water from a fire hose. Since coronal holes are detectable from earth several days before they are carried across the central meridian by solar rotation and since two to three days remain before the wind from the hole will reach the earth, we should be able to predict the arrival of the high-speed streams and their associated magnetic effects approximately a week in advance. The accurate prediction of noisy conditions can assist in scheduling IP surveys in problem areas. The correlations by Sheeley et al. (1976) indicate relationships of magnetic field polarity, solar wind intensity, and the magnitude of recorded geomagnetic disturbances.

Of course, the results of this solar research should also be useful for signal prediction for magnetotelluric exploration. Information on solar magnetic distrubances is available from Space Environment Services Center, Space Environment Laboratory, ERL, National Oceanic and Atmospheric Administration, Boulder, Colorado 80302.

FUTURE TRENDS AND DEVELOPMENTS IN IP SURVEYING

Although IP is the newest of the mining geophysical exploration methods, it has progressed to the extent that it has become the most popular despite the fact that costs for IP surveys are comparatively high. A cost reduction would be a major achievement. Innovations in the IP method are still being perfected, but airborne IP surveys do not appear to be feasible at this time.

There is still debate as to the significance of the metal factor in IP interpretation. There is also an uncertainty about the capability of the IP method to discriminate between different polarizing materials. However, the effectiveness of an IP survey is high, and it is readily possible to obtain resistivity, self-potential, and electromagnetic data as well as complex-resistivity information. Indeed it seems likely that as data are compared and correlated at least a limited amount of mineral discrimination will be possible.

The digital computer is becoming a necessary addition to the IP equipment list, and the use of the microprocessor is substantially transforming routine IP surveys. With continuing decreases in cost of computing devices, their inclusion as integral parts of future IP systems can be foreseen.

REFERENCES

Cole, K.S. and Cole, R.H.
    1941: Dispersion and absorption in dielectrics. I. Alternating current fields; J. Chem. Phys., v. 9, p. 341.

Collett, L.S.
    1959: Laboratory investigation of overvoltage; in J.R. Wait, ed., Overvoltage research and geophysical applications: London, Pergamon Press, p. 50-70.

Fraser, D.C., Keevil, N.B., and Ward, S.H.
    1964: Conductivity spectra of rocks from the Craigmont ore environment; Geophysics, v. 29, p. 832-847.

Grant, F.S.
    1958: Use of complex conductivity in the representation of dielectric phenomena; J. App. Phys., v. 29, p. 76-80.

Hohmann, G.W.
    1977: Numerical IP modeling; in Induced polarization for exploration geologists and geophysicists: Tucson, Dep. Geosciences, Univ. Arizona, p. 15-44.

Hood, P.
1977: Mineral Exploration: trends and developments in 1976; Can. Min. J., v. 98, p. 8-47.

Katsube, T.J.
1977: Electrical properties of rocks; in Induced polarization for exploration geologists and geophysicists: Tucson, Dep. Geosciences, Univ. Arizona, p. 15-44.

Katsube, T.J. and Collett, L.S.
1973: Electrical characteristic differentiation of sulfide minerals by laboratory techniques; 43rd Ann. Int. Meeting, Soc. Explor. Geophys. and 5th Meeting, Asoc. Mexicana Geofis. Explor., Mexico City, 1973, Abstracts, p. 54.

Klein, J.D. and Shuey, R.T.
1975: A laboratory investigation on non-linear impedance of mineral-electrolyte interfaces; 45th Ann. Int. Meeting, Soc. Explor. Geophys., Tulsa, Abstracts.

Madden, T.R., Fahlquist, D.A., and Neves, A.S.
1957: Background effects in the induced polarization method of geophysical exploration; U.S. AEC Rept. RME-3150.

Miller, D., Chapman, W., and Dunster, D.
1975: Mark II – A multichannel IP system with minicomputer control and processing; 45th Ann. Int. Meeting, Soc. Explor. Geophys., Tulsa, Abstracts.

Nabighian, M.N. and Elliot, C.L.
1974: Unusual induced polarization effects from a horizontally three-layered earth; 44th Ann. Int. Meeting, Soc. Explor. Geophys., Dallas, Texas, 1974, Abstracts, p. 52-53.

Nilssen, B.
1971: A new combined resistivity and induced polarization-instrument and a new theory of the induced polarization phenomenon; Geoexploration, v. 9, p. 35-54.

Olhoeft, G.R.
1975: The electrical properties of permafrost; unpubl. Ph.D. thesis, Univ. Toronto.

Parasnis, D.S.
1973: Mining geophysics; ed. 2: Amsterdam, Elsevier Sci. Publ. Co., 356 p.

Patella, D.
1972: An interpretation theory for induced polarization vertical soundings (time-domain); Geophys. Prosp., v. 20, p. 561-579.

Pelton, W.H., Ward, S.H., Hallof, P.G., Sill, W.R., and Nelson, P.H.
1977: Mineral discrimination and removal of inductive coupling with multifrequency IP; in Induced polarization for exploration geologists and geophysicists: Tucson, Dep. Geosciences, Univ. Arizona, p. 285-354.

Seigel, H.O.
1959: A theory for induced polarization effects (for stepfunction excitation); in J.R. Wait, ed., Overvoltage research and geophysical applications; London, Pergamon Press, p. 4-21.

1974: The magnetic induced polarization (MIP) method; Geophysics, v. 39, p. 321-339.

Sheeley, N.R., Jr., Harvey, J.W., and Feldman, W.C.
1976: Coronal holes, solar wind streams, and recurrent geomagnetic disturbances; 1973-1976 Skylab/Naval Res. Lab. preprint; Solar Physics, v. 49, p. 271-278.

Snyder, D.D.
1976: Field tests of a microprocessor-controlled electrical receiver; 46th Ann. Int. Meeting, Soc. Explor. Geophys., Tulsa, Oklahoma, Abstracts.

Snyder, D.D. and Merkel, R.H.
1977: Induced polarization measurements in and around boreholes; in Induced polarization for exploration geologists and geophysicists: Tucson, Dep. Geosciences, Univ. Arizona, p. 161-220.

Van Voorhis, G.D., Nelson, P.H., and Drake, T.L.
1973: Complex resistivity spectra of porphyry copper mineralization; Geophysics, v. 38, p. 49-60.

Wait, J.R.
1959a: A phenomenological theory of overvoltage for metallic particles; in J.R. Wait, ed., Overvoltage research and geophysical applications: London, Pergamon Press, p. 22-28.

1959b: The variable-frequency method; in J.R. Wait, ed., Overvoltage research and geophysical applications: London, Pergamon Press, p. 29-49.

Whiteley, R.J.
1973: Electrode arrays in resistivity and IP prospecting: a review; Aust. Soc. Explor. Geophys. Bull., v. 4, p. 1-29.

Zonge, K.L.
1972: Electrical parameters of rocks as applied to geophysics; Ph.D. dissertation, Univ. Arizona, Tucson. Ann Arbor, Michigan, University Microfilms.

1976: Method using induced polarization for ore discrimination in disseminated earth deposits; U.S. Patent Office, Patent No. 3,967,190, 13 p.

Zonge, K.L. and Wynn, W.C.
1975: Recent advances and applications in complex resistivity measurements; Geophysics, v. 40, p. 851-864.

GAMMA RAY SPECTROMETRIC METHODS IN URANIUM EXPLORATION — AIRBORNE INSTRUMENTATION

Q. Bristow
Geological Survey of Canada

Bristow, Q., Gamma ray spectrometric methods in uranium exploration: airborne instrumentation; in Geophysics and Geochemistry in the Search for Metallic Ores; Peter J. Hood, editor; Geological Survey of Canada, Economic Geology Report 31, p. 135-146, 1979.

Abstract

Gamma ray spectrometry is used in uranium exploration to determine as precisely as possible the crustal distributions of potassium, uranium and thorium by quantitative measurements of the gamma radiation from the natural radioisotopes associated with these elements. This applies to airborne and ground surveys and to borehole logging.

In recent years and particularly since 1972, the commercially available instrumentation for such surveys has achieved a level of sophistication found only in research laboratories. Technical specifications generated by various national agencies involved in uranium resource assessment programs have hastened this process in the case of airborne surveys by requiring multichannel analysis using analogue-to-digital converters, complete sectral recording, atmospheric radon monitoring and by setting sensitivity and overall detector performance standards for the equipment.

The advent of 4 x 4 x 16 inch (100 x 100 x 406 mm) prismatic-shaped sodium iodide detectors, with a volume equivalent to a conventional 9 x 4 inch (230 x 100 mm) detector, but with only one photomultiplier tube, has greatly simplified the packaging of multidetector arrays and associated electronics, allowing lighter and more compact spectrometers with lower power consumption. It is now possible, for example, to operate a system with a 2000 cubic inch (32.8 L) volume, complete with full spectral recording using a helicopter.

Most airborne gamma ray spectrometry instrumentation is now designed as one component of multisensor integrated data-acquisition systems based on microprocessors or minicomputers. This new found flexibility has encouraged the development of a variety of features such as real-time on-line corrections for Compton scatter, background radiation, detector gain shift and verification of taped data by read-back-and-compare, all of which can be implemented by software.

Electromechanical methods of mass storage such as magnetic tapes and discs are the most failure-prone components of any field system. Recent advances in memory technology may bring solid state equivalents such as bubble memories, into use as practical and economically viable replacements before 1985.

Résumé

Dans la recherche de l'uranium, la spectrométrie de detection de rayons gamma permet de déterminer de fa on aussi précise que possible la répartition de potassium, d'uranium et de thorium dans la croûte terrestre par des mesures quantitatives de la radiation gamma émise par les radio isotopes naturels de ces éléments. Cette technique s'applique aux levés aériens et terrestres ainsi qu'à la diagraphie par trou de sonde.

Au cours des dernières années, particulièrement depuis cinq ans, les instruments que l'on trouve sur le marché pour de tels levés ont atteint des niveaux de perfectionnement que l'on retrouve seulement dans les laboratoires de recherche. Les spécifications techniques exigées par divers organismes nationaux qui s'occupent de programme d'évaluation des ressources en uranium ont accéléré ce processus dans le cas des levés aériens, en exigeant l'utilisation d'analyses multi canal faites au moyen de convertisseurs numériques-analogiques, des techniques complètes d'enregistrement spectral, le contrôle du radon atmosphérique et l'établissement des normes de sensibilité et de rendement global des détecteurs pour le matériel.

La mise au point de détecteurs prismatiques à iodure de sodium, de 4 x 4 x 16 po (100 x 100 x 406 mm), d'un volume équivalent à celui d'un détecteur classique de 9 x 4 po (230 x 100 mm), mais avec un seul tube multiplicateur, a grandement simplifié l'assemblage d'ensembles de multi détecteurs et des composantes connexes, permettant ainsi l'utilisation de spectromètres plus légers et plus compacts, d'une consommation énergétique moindre. Il est maintenant possible, par exemple, de faire fonctionner un réseau complet de 2 000 po cubes (32,8 litres), avec tout l'enregistrement spectral, à bord d'un hélicoptère.

La plupart des instruments aéroportés de spectrométrie à rayons gamma se retrouvent maintenant sous forme de composantes de systèmes intégrés de saisie de données à multicapteurs basés sur des microprocesseurs ou des miniordinateurs. Ce nouvel apport de flexibilité a favorisé l'élaboration d'un ensemble de particularités comme les corrections en direct et en temps réel pour la diffusion Compton, le rayonnement de la zone de fond, le décalage de rendement du détecteur et la vérification des données sur bandes par relecture/comparaison, opérations qui peuvent toutes être effectuées par logiciel.

Les méthodes électromécaniques de mémoires de grandes capacités comme les bandes et les disques magnétiques constituent les composantes les plus sujettes aux défaillances de tout système d'exploration sur place. Des progrès récents dans les techniques de mise en mémoire peuvent favoriser l'utilisation, avant 1985, d'equivalents transistorisés comme les mémoires à bulles, comme moyen de remplacement pratique et économiquement rentable.

HISTORICAL DEVELOPMENT OF AIRBORNE GAMMA RAY SPECTROMETRY

There are three radioelements in the terrestrial crust which are readily detectable by the gamma radiation emitted by the natural radioisotopes associated with them, potassium, uranium and thorium. However, until the early 1950s there was no method of distinguishing between different gamma radiation energies. The standard radiation detector, the Geiger-Müller counter, provided only a total count indication. Thus it was not possible to distinguish between potassium, uranium and thorium. The advent of the thallium-activated sodium iodide scintillation detector was a major advance for two reasons, it had a stopping power (i.e. detection efficiency) several hundred times that of the Geiger-Müller counter for the energy range of interest in this application and the output pulses it produced had amplitudes proportional to the energy of the gamma radiation which caused them. By adding electronic circuitry capable of sorting and counting pulses according to their amplitudes it was possible to separate the contributions from the three radioelements according to their energies, which is by definition gamma ray spectrometry.

The essential details of a scintillation counter and its method of operation can be understood by reference to Figure 10A.1. Gamma rays which interact in the crystal cause tiny flashes of light (scintillations) the intensities of which are proportional to the energy deposited in the crystal by the gamma rays. The photomultiplier tube which is optically coupled to the crystal, usually with transparent grease, converts the scintillations to corresponding electrical signals which can be amplified and sorted in the subsequent electronic circuitry.

The technique of sorting pulses according to their amplitudes is known as pulse height analysis. The most elementary method is to use a single discriminator circuit which allows all pulses above a certain preset amplitude (i.e. energy) to be counted and rejects all others, or vice versa. A slightly more sophisticated arrangement uses two such discriminators which can be preset to different levels. Pulses having amplitudes which fall between the two levels are counted while all others are rejected. This arrangement is known as a Single Channel Analyser (SCA) and is shown in Figure 10A.2.

Until the mid 1960s airborne gamma radiation surveys were largely confined to "total count" measurements. The first gamma ray spectrometry systems for airborne use which became available at that time consisted broadly speaking of the following items:-

- Scintillation detector, usually one 150 x 100 mm (6 x 4 inch).
- 4 counting channels, each with a single channel analyzer and analogue rate meter
- 4 channel strip chart recorder

The block diagram is shown in Figure 10A.3. Three of the single channel analyzers were set to cover the pulse amplitudes (i.e. energy bands) corresponding to those of potassium, uranium and thorium, while the fourth was set to cover an energy range encompassing all three, the "total count".

Various systems based on the above concept were produced in the years following and they found wide acceptance in the exploration industry. A lack of standards, and in many cases a lack of understanding of the principles of the new technique, caused difficulties in the collection and interpretation of spectrometric data. It also became evident that a gap had developed between the level of instrumental technology in the equipment used for uranium prospecting and similar equipment which was available in the laboratory at that time.

Since then and particularly during the last five years the situation has changed dramatically with the dissemination of more information and the development of more sophisticated techniques and instrumentation, the latter being covered in some detail below.

No attempt is made in this article to describe available commercial instrumentation. For this information the reader is referred to an excellent annual review which appears in the Canadian Mining Journal (P.J. Hood, 1967, et seq). The purpose here is to acquaint the reader with currently used techniques, their advantages and their limitations.

FUNDAMENTAL CONSIDERATIONS

While the radiation physics aspects of gamma ray spectrometry are covered in detail elsewhere in this three part article, the spectrum from a standard calibration pad (Fig. 10A.4) is included here as a convenient aid for discussing the instrumental aspects of the technique.

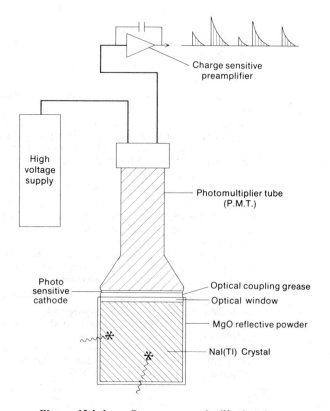

Figure 10A.1.   *Gamma ray scintillation detector.*

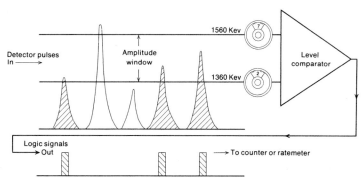

Figure 10A.2. Single channel analyzer.

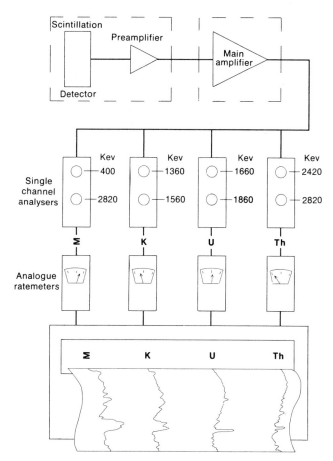

Figure 10A.3. Simple four-channel spectrometer, widely used for airborne surveys until about 1975.

From the strictly instrumental point of view the purpose of a radiometric survey is to record accurately the radiation intensities (count rates) at the detector in predetermined energy bands or "windows". Instrumental factors which degrade or cause errors in this process can be summarized as follows: —

— Inadequate counting statistics for reliable quantitative data.
— Inadequate detector resolution (i.e. inability to resolve energy peaks sufficiently well for satisfactory pulse height analysis) due to poor quality crystals or defective assembly.
— Detector gain drift, i.e. systematic variations in the pulse height corresponding to scintillations generated by a given energy.

— Inaccurate or varying window width settings.
— Excessive noise in the pulse amplifying and shaping chain causing an additional statistical variation in pulse heights.
— Spectrum distortion at high counting rates.
— Pulse height distortion due to detector overload by high energy events from cosmic sources.

It should be noted that even if these pitfalls are avoided and the objective indicated above is achieved, the measurements still require correction for the following factors: —

— Spectral interferences; due to the physical processes involved in the interaction of gamma radiation with the detector material and to the introduction of statistical noise by the photomultiplier tube, there is a mutual overspill or crosstalk between the closely adjacent potassium (^{40}K) and uranium (^{214}Bi) windows and a contribution to both of them from the thorium (^{208}Tl) window. "Stripping ratios" are experimentally determined and applied to recorded data to correct for this.

— Background radiation due to sources other than the crustal radioelement content. These include atmospheric radon, cosmic sources and radioactive material in either the equipment itself or the aircraft used to carry it.

— Atmospheric attenuation.

The theory behind these corrections and the methods of applying them are dealt with in more detail elsewhere in this article.

CURRENT LEVEL OF TECHNOLOGY FOR GAMMA RAY SPECTROMETERS

The instrumental techniques and hardware which are now coming into use in modern exploration gamma ray spectrometers are discussed here with reference to the fundamental considerations outlined above For convenience they are separated into groups as follows: —

— Detectors
— Signal-conditioning electronics
— Pulse height analysis
— Spectrum stabilization
— Data display and recording

Detectors

The scintillation detector is still the most widely used type for airborne and ground follow-up equipment and appears likely to remain so for some time. Figure 10A.1 shows the crystal material labelled as NaI(Tl), which is an abbreviation for sodium iodide (thallium-activated). While this is the material most commonly used in scintillation detectors designed for uranium exploration applications, it is by no means the only one. Other inorganic crystals sometimes used in borehole logging applications are cesium iodide, activated with either sodium or thallium [CsI(Na), CsI(Tl)] and bismuth germinate ($Bi_4Ge_3O_{12}$). The cesium iodide crystals have a density which is approximately 25 per cent greater than that of sodium iodide with a correspondingly higher efficiency or stopping power for sensing gamma radiation; however the ability to resolve spectral peaks (resolution) is somewhat less than for sodium iodide. Bismuth germinate has a density very nearly double that of sodium iodide, but the resolution of detectors made from this material is barely adequate at present for spectrometry. An entirely different class of scintillators is that based on the use of organic plastics. These materials have low atomic numbers and hence very low efficiencies for the complete absorbtion of gamma rays in the

0.5-3.0 MeV region which is the useful range in uranium exploration applications. While they are widely used in laboratory work as active shields and in related applications where the requirement is for detecting and timing radiation events as opposed to energy measurements of such events, their very limited ability to resolve spectral peaks coupled with their poor efficiency renders them unsuitable for use in uranium survey work. Detectors made from plastic scintillators are much less costly than their inorganic crystal counterparts and some workers have conducted experiments to evaluate them as a less expensive alternative for airborne gamma radiation surveys (see e.g. Duval et al., 1972).

Experiments and studies have been conducted to evaluate the potential for airborne work of intrinsic germanium solid state detectors which have a resolution far superior to that of scintillation detectors. These indicate that the advantage of high resolution does not offset the cost of an array of sufficient volume to match the detection sensitivity of currently used scintillation detectors. The requirement for cryogenic operation with liquid nitrogen also presents an operational problem with solid state detectors. It seems probable however that solid state detectors will be used in borehole gamma ray spectrometry. In this application the detector volume is limited by the borehole tool diameter which in turn severely limits detector stopping power or efficiency for energies much above 1.0 Mev. Consequently it would be an advantage to have a detector capable of resolving the closely spaced lower energy peaks also associated with uranium and thorium.

For portable spectrometers single scintillation detectors with one photomultiplier tube (PMT) and a 76 x 76 mm (3x3 inch) crystal are still the standard unit. In airborne systems arrays of large detectors are in common use, and usually these are offered as add-on-modules of 16.4 L (1000 cu. in.) pre-packaged with suitable thermal insulation and shock mounting. Until recently the individual detectors have been of the traditional cylindrical type 100 to 130 mm thick by 150, 179, 203, 229 or 279 mm in diameter (4 or 5 inches thick by 6, 7, 8, 9 or 11 inches in diameter). The number of PMTs per crystal vary from three on the smaller ones to four on the 9 x 4 inch to seven on the 11 x 4 inch crystal. The reason for having more than one PMT on large crystals is to ensure more nearly constant light collection no matter where in the crystal a scintillation of given intensity occurs. This requires that each PMT be fitted with a separate gain control and that they be equalized by experiment to optimize the detector resolution, a tedious and time consuming task.

During the last three years a detector has become available in a prismatic configuration with a 100 x 100 mm (4 x 4 inch) square cross-section by 406 mm (16 inch) long with a single PMT mounted on one end. This has become very popular for airborne systems as it lends itself to compact packaging with an array of four or six side by side forming a "slab". Figure 10A.5 is a photograph showing such an arrangement.

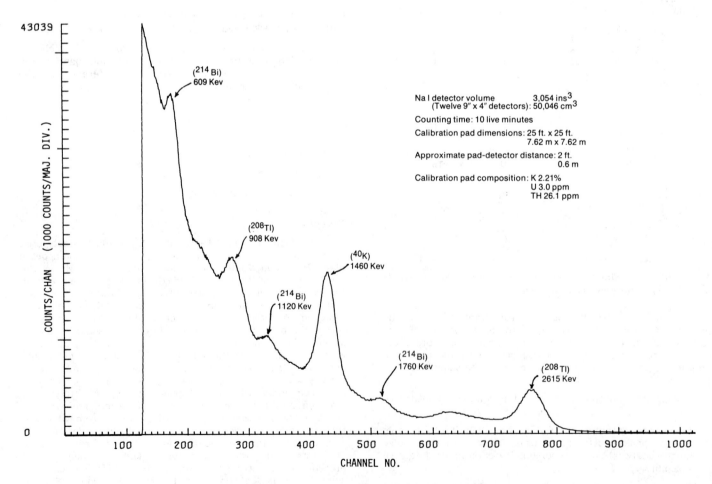

Figure 10A.4. *Typical natural radioactivity spectrum.*

Figure 10A.5. Typical array of six 100 x 100 x 406 mm (4 x 4 x 16 inch) NaI detectors installed in a container designed to provide both thermal and electrical shielding. GSC 203492-K

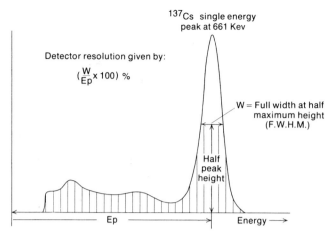

Figure 10A.6. Standard method of measuring detector resolution.

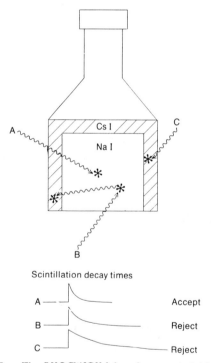

Figure 10A.7. The PHOSWICH (phosphor-sandwich) Detector. *Different scintillation decay times in CsI and NaI enable only gamma rays which are absorbed in the NaI core (A in figure) to be recognized and accepted on basis of pulse shape. This greatly reduces the Compton Scatter interference normally generated by analyzing and counting partially absorbed gamma rays.*

More recently an actually slab-shaped detector has appeared measuring 279 x 279 mm by 100 mm (11 x 11 x 4 inches) thick with four PMTs mounted on the upper surface. It remains to be seen how widely this type will be used.

One of the standard measures of performance of a scintillation detector is the resolution of the 661KeV single energy peak of the isotope ^{137}Cs. This is a measure of the sharpness of the photo peak and is by implication a measure of the ability of the detector to resolve two closely spaced peaks. Figure 10A.6 illustrates the way that detector resolution is calculated and specified as a percentage. Individual detectors used in an airborne array such as the 100 x 100 x 406 mm (4 x 4 x 16 inch) size and others of comparable volume should have resolutions of 9.5 per cent or better (i.e. less).

A gamma ray in the 0.5 – 3.0 MeV range interacts with sodium iodide by a series of interactions with electrons in the material, during which most of the energy is converted to a series of virtually simultaneous scintillations. If all of the energy of an incident gamma ray is absorbed by the detector in this way then the resulting composite scintillation will produce a pulse height corresponding to the full energy. However if after a number of energy absorbing collisions the gamma ray, now reduced in energy, escapes from the detector, then the pulse height produced by the event will be identified as being of a lower energy and will not be counted in the proper energy window. The process is analagous to dropping a bead of mercury into a beaker; it shatters into a large number of small droplets which if all remain in the beaker will add up to the mass of the original bead. If some are "scattered" over the top then the mass will be less than that of the original bead.

The partially absorbed photons which appear in lower energy windows after the pulse sorting process, are reduced in a new type of detector known as the "Phoswich" detector, the name being a mnemonic for "phosphor-sandwich". The principle is illustrated in Figure 10A.7. The crystal is in a form similar to that of an iced cake with the "cake" being made from sodium iodide and the "icing" from cesium iodide. The decay time for scintillations in the two materials is different by about 4:1, so that it is possible to differentiate between photons which have been completely absorbed in the sodium iodide central portion, (the active part of the detector) and those which have been absorbed by both parts, i.e. those which have been partially absorbed and scattered.

Pulse shape discrimination circuitry is used in the subsequent electronics to reject all events which have occurred either wholly or partly in the caesium iodide "icing" and accept only those which have been completely absorbed in the sodium iodide. The "Phoswich" detector is still in the experimental stage and it remains to be seen whether the advantages will offset the additional complication and expense.

A new type of photomultiplier tube, presently at an early stage of development, makes use of a silicon diode to replace the dynodes used in the conventional photo multiplier tube. Scintillations liberate photoelectrons from a photocathode deposited on glass in the usual way. These are then accelerated by a potential of the order of 10 kilovolts and focused onto a silicon photodiode which produces charge at an effective rate of approximately one electron per 3.3 eV (electron volts) of energy of each of the accelerated electrons from the photocathode. Since each of these has an energy of about 10 keV, approximately 3000 times as much charge is produced by the silicon diode as was liberated from the photocathode. This effective multiplication, while nowhere near as large as with a conventional multidynode arrangement, is sufficient and is produced in one step compared to the ten or more steps using a conventional tube. Consequently the statistical variation in the number of electrons produced per photoelectron liberated from the photocathode is less, offering the potential of improved energy resolution. However a high-gain low-noise charge preamplifier is then required to boost the signal to the level normally produced by a conventional photomuliplier tube. Early indications are that this technique is capable of providing a useful improvement in scintillation detector resolution over what can now be obtained with ordinary photomultiplier tubes. Additional advantages are a reduced dependence of the overall detector gain on high voltage and ambient temperature variations.

Signal-Conditioning Electronics Circuitry

This is the term generally used to describe the process of extracting, filtering or amplifying analogue signals so that the information they contain can either be digitized or fed directly to some form of visual or audio device with the minimum of unwanted noise. Examples would be the demodulation of stereo or colour TV signals, or the removal of d.c. levels from strain gauge signals to obtain an output suitable for driving a strip chart recorder. In this case the object is to convert the tiny charge signals which appear at the anode of the PMT of a scintillation detector into voltage pulses whose amplitudes are proportional to the energies of the gamma ray events which caused them.

Well-designed systems have a charge sensitive preamplifier for each detector, usually located in the detector package, which converts the charge pulses to voltage pulses capable of driving a reasonable length of line. This is followed by a main amplifier containing special filter circuits which shape the pulses to give an optimum signal to noise ratio. Figure 10A.8 shows the arrangement and how the zero level or base line between pulses following the preamplifier is poorly defined and noisy, but is considerably improved after the filtering and pulse shaping in the main amplifier. There are a number of excellent main amplifiers available in the popular "NIM" (Nuclear Instrument Module) package from laboratory instrumentation manufacturers. The generally preferred kind are those which produce Gaussian-shaped pulses. Most main amplifiers designed for laboratory work in the NIM package are more than adequate for exploration spectrometers and have provision for "base line restoration", an important feature when moderately high counting rates (greater than 20 000 c/s) are likely to be involved.

The need for base line restoration can be understood by reference to Figure 10A.9. At some point in the signal conditioning chain the pulse train will be passed through a capacitor to remove a d.c. level, following which the average d.c. level will always be zero. This means that at high count rates with a large number of pulses causing a net positive shift, the base line between pulses settles down to a slightly negative level to maintain the overall average at zero. The subsequent pulse-height analysis circuitry however does not recognize the negative shift of the base line, but sees all the pulses apparently reduced in amplitude and sorts them into lower energy slots than they should be. In other words at high count rates the entire spectrum is shifted to the left causing peaks to move out of their windows by an amount which is dependent on the count rate – a very undesirable effect indeed.

This effect can be compensated for in a variety of ways which are collectively termed base line restoration and as indicated earlier most good quality main amplifiers now have this feature incorporated.

The multidetector arrays used for airborne surveys require that the pulse signals from each of the individual detectors be adjusted so that the pulse amplitudes for a given energy are the same from each one. The outputs from all the detectors are then summed together at some point between the preamplifiers and the main amplifier. The circuitry for this is normally in a separate module with separate gain controls and switches for each detector to facilitate gain matching.

A modern airborne survey system with a detector volume of 50 L (3000 cu. in.) should have an overall resolution at the ^{137}Cs energy of at least 12 per cent (FWHM); although with good quality detectors and carefully designed signal conditioning electronics much better resolution is achievable.

Pulse-Height Analysis

The simplest form of pulse height measurement device, the Single Channel Analyzer (SCA) described earlier, accepts detector pulses from the main amplifier and puts out a logic signal whenever a detector pulse falls between amplitude limits set on two front panel controls. The most serious problem with SCAs is the difficulty in determining what energy limits the calibrated front panel controls actually correspond to. In the case of portable four-channel spectrometers for exploration work, a fifth channel is usually incorporated preset to cover the energy peak from an isotope such as the 661 KeV peak of ^{137}Cs, to allow for manual calibration. The other window limits also have preset controls not accessible on the instrument front panel. Initial settings of these is virtually a factory adjustment, requiring a reliable nuclear pulse generator with calibrated controls and provision for varying the pulse shape to match that of the detector being used. Most SCA discriminator circuits are sensitive to pulse shape to some degree as well as pulse height, so that initial pulse generator calibrations by the manufacturer can be in error if this point is not appreciated.

Once the initial calibration is done the accuracy of the subsequent measurements by the user (perhaps over many seasons), depends entirely on all of these preset controls maintaining their relative positions corresponding to the required window limits.

For many years so called "multichannel analyzers" have been available for making much more accurate pulse height measurements. These require an analogue-to-digital converter (ADC) which is similar to a very high speed digital voltmeter but with no visual display of the numbers it produces. The amplitude of each incoming detector pulse is

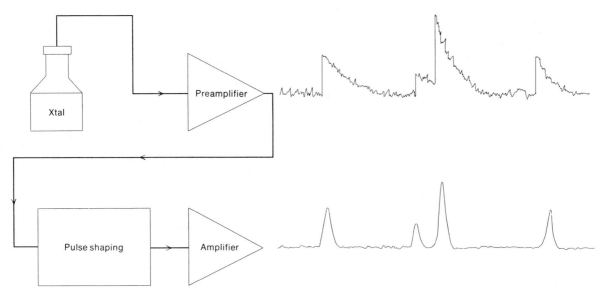

Figure 10 A.8. Signal conditioning.
Well-designed signal conditioning electronics are essential for accurate pulse-height analysis.

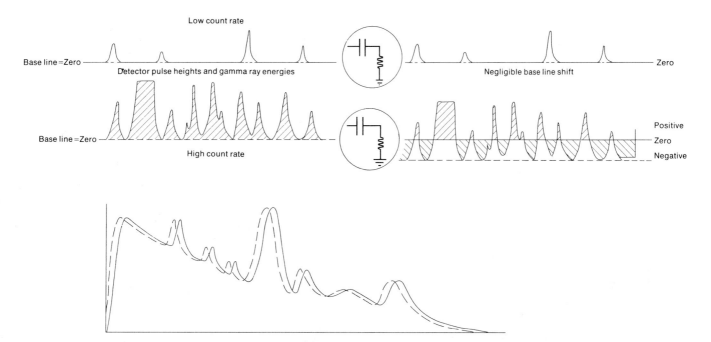

Figure 10 A.9. Baseline shift.
If the detector pulse train is passed through a resistor-capacitor network (centre), the baseline shifts downward to maintain equal areas above and below the true zero level. Since pulse heights are measured from the true zero, this shift can cause serious errors at high count rates causing the entire spectrum to be shifted to the dashed line position (bottom).

measured by the ADC which assigns it a number between 1 and some maximum. In laboratory systems this could be as high as 8192 if high resolution solid-state detectors are being used; in airborne survey systems 256 is usually the maximum. The ADC thus has the capability of grading or sorting incoming detector pulses according to amplitude (i.e. energy), into one of 256 or more slots or channels. This is equivalent to 256 Single Channel Analysers in one box, except that to make use of all this information we now need 256 counters to keep track of the number of detector pulses that are accumulating in each channel. From this point there are two alternative methods for handling these data as indicated in Figure 10A.10.

- Use 256 locations of a minicomputer memory or equivalent device and increment each location when its number is "called" by the ADC
- Use digital circuitry that will increment a single counter for all numbers generated by the ADC between certain preset limits corresponding to a desired energy window.

The first method produces a representation of the gamma ray energy spectrum in the memory which is simply a histogram of energy divided into 256 discrete steps, showing the number of gamma photons which fell into each channel during the counting period. This can be displayed on a CRT, or plotted on hard copy and provides complete information on the sources of radiation which were detected (within the limits of the detector resolution).

The second method is fairly straightforward and can be implemented without any memory, with the necessary circuitry being duplicated to produce as many "digital windows" as required. The advantage over the equivalent Single Channel Analyser arrangement is that all the window limits are set by specifying channel numbers generated by the ADC, and these have a fixed and predictable relationship with each other, since each channel spans a known energy increment.

As to the ADCs themselves there are two distinct principles of operation for converting the pulse height into a number proportional to it. One is known as the "Wilkinson Ramp" method and the other as the method of "Successive Approximations". Since there has been some controversy as to which is most suitable, their relative merits are worth some discussion.

The Wilkinson ramp principle is illustrated in Figure 10A.11 and is analogous to the "hour glass" principle. Imagine an hour glass being filled with sand to a level which is equal to the height of a detector pulse. If the hour glass is then inverted, the time the sand takes to run out is then proportional to the pulse height. The number of seconds of elapsed time could be interpreted as the channel number for that pulse height (gamma photon energy). Note that if a second pulse arrives during the time the hour glass is running for the first one, the second one is ignored and neither analyzed nor counted. This is the ADC "dead time" and in the Wilkinson ramp type it is proportional to the height of the pulse being analyzed, plus a fixed time for some logical operations in the ADC

In electrical terms a capacitor is charged to the height of each detector pulse and then discharged linearly to zero while pulses from a high frequency oscillator (50 to 200 MHz) are counted. The number in the counter at the end of this process is then taken as the channel number of that detector pulse.

The successive approximations method can be understood by reference to Figure 10A.12. Essentially it works by comparing the incoming detector pulse height to a series of yardsticks of varying lengths on a trial-and-error basis starting with the longest yardstick. If the pulse amplitude is greater than this then an amount equal to this yardstick is subtracted and the next shortest one is compared to the remainder; if this is too long the next shortest is tried. The process continues until a match is found. The circuitry then infers from the particular yardsticks which were used to subtract pieces from the incoming pulse, what the channel number is.

By way of example suppose that it was only required to sort the incoming detector pulse into one of sixteen channels rather than 256. Suppose further the particular pulse being analyzed had an amplitude of 13 on this scale. The ADC requires only four yardsticks to sort any pulse into one of sixteen channels, their lengths (in arbitrary units) being 1, 2, 4 and 8 units. Beginning with the longest (8 units) it would find that the pulse was larger and subtract the equivalent of 8 units. It would then compare the remainder with the 4 unit yardstick and still find a mismatch. Having subtracted off 4 more units it would test the remainder against the 2 yardstick which would be too large. Finally the 1 yardstick would be used to find a match. The height of the pulse would thus be computed as 8 + 4 + 1 = 13 units.

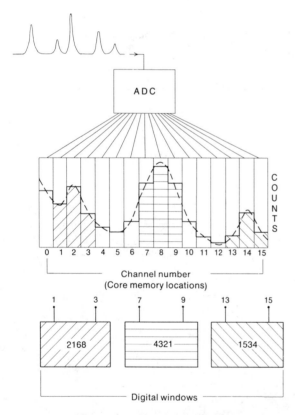

ADC measures height of each incoming pulse, in this case on a scale of 16 and increments contents of corresponding memory locations by one to build up a 16 channel spectrum or histogram. By using more channels a smoother curve can be obtained corresponding to the dotted line. If only counting windows are required then counters can be connected to the ADC via a simple interface and arranged to count all pulses in preset ranges of channel numbers – no computer or memory is required.

Figure 10A.10. *Advantages of an Analogue/Digital Converter (ADC).*

This is a much quicker process than the clock pulse counting required by the Wilkinson ramp type. For example in order to sort a pulse into one of 256 slots, (including zero) only 8 yardsticks are required at most only 8 comparisons. The difficulty arises in making accurate comparisons and subtractions of the yardsticks, (voltages developed across precision resistors in fact) with and from the incoming detector pulses.

In the case of a 256 channel successive approximations ADC, a detector pulse having a channel number of 128 will be analyzed after applying the very first yardstick, although the remaining seven comparisons will be made anyway because the logic is simplified by so doing, but one having a value of 127 will require the successive subtraction of voltages corresponding to all of the 8 yardsticks before it is determined. This places very stringent conditions on the accuracy of both the yardsticks and the subtraction/comparison circuitry used if the effective widths in terms of energy increments for the pairs of channels which occur at "yardstick change" points are to be equal.

The advantages and disadvantages of the two types of analogue-to-digital converter can be summarized as follows: –

Successive Approximations Type

— rapid conversion time which is the same for all detector pulses regardless of amplitude
— high probability of sharp discontinuities in adjacent channel widths particularly at 1/4, 1/2 and 3/4 of full scale due to the nature of the conversion process

Wilkinson Ramp Type

— Conversion time which is proportional to the amplitude of the pulse being analyzed plus a small constant and which is of the order of twice as long as for the successive approximations ADC at full scale.
— Differential nonlinearity, i.e. variation in channel widths across the range, far less than for the successive approximations type due to the inherently linear "hour glass" principle of conversion.
— Universally used by nuclear laboratories where high precision is required and where the constraints of high count rate and limited available counting time are not a problem.
— As a result of many years of refinement, well-designed ADCs of this type are available off-the-shelf in one or two width NIM (Nuclear Instrument Module) packages at relatively modest cost from a number of sources.

So long as the scintillation detector with its comparatively poor resolution is being used with only 256 channels to cover the spectral region of interest, then there is probably little to choose between the two types unless the data are to be used for very precise spectral analysis. However if solid-state detectors come into vogue as they probably will for borehole logging, requiring 2048 or 4096 channels for adequate spectral information, then the successive approximations ADC will require considerably more development to achieve the necessary differential linearity.

ADC "dead time" is a factor which requires some form of correction no matter which type is used. The dead time is the interval during which the ADC is making a conversion and is unable to accept any other detector pulses. In the case of the successive approximations type the dead time is a fixed length, whereas it is proportional to the amplitude of the pulse being analyzed for the Wilkinson-ramp type. A correction can be made in the former case by multiplying the total number of pulses counted in say one second by the known conversion time (a few microseconds) and increasing the contents of all channels pro rata to compensate for the lost time. A hardware method applicable to both fixed and variable dead time ADCs involves two time counters, one of which is interrupted for all intervals when the ADC is busy (i.e. "dead"). At the end of each sample time the difference between the contents of the two counters is a measure of the dead time for that record and can be used to make an on-line correction to the data just acquired, or simply recorded on tape along with other data for use in off-line data processing.

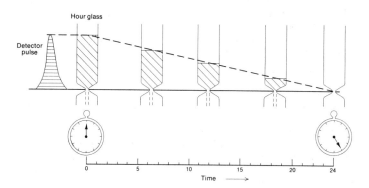

Imagine an hour glass filled to the height of an incoming detector pulse. The time for the sand to run out would then be proportional to the energy of the corresponding gamma ray. A measurement of the time in appropriate units (microseconds in practice) is then a digital representation of the energy.

Figure 10A.11. *Principle of the Wilkinson Ramp type ADC.*

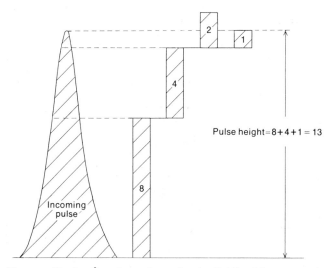

The amplitude of an incoming pulse is digitized by comparing it with a series of precision voltage "yardsticks". In the case shown four comparisons only are required to determine the amplitude as a number on a scale of 1-15. A more precise determination on a scale of 0-255 would require eight yardsticks with relative values 1:2:4:8:16:32:64:128.

Figure 10A.12. *Principle of Successive Approximation ADC.*

Spectrum Stabilization

All spectrometers are vulnerable to spectrum drift, i.e. the possibility that the detector pulse height corresponding to a given energy changes over a period of some hours. This manifests itself as a uniform scale change on the horizontal axis of the spectrum. Figure 10A.13 shows the effect of a reduction in pulse height causing the spectral peaks to shift to the left and off the centres of the counting windows.

The major sources of spectrum drift are the gains of the PMTs which are notoriously temperature dependent and not particularly stable even at constant temperature, and changes in the high voltage supply. A change of 0.1 per cent in the high voltage applied to the PMTs changes the pulse heights by about 1 per cent. Modern high voltage supplies designed specifically for this purpose are now more than adequate to reduce this source of spectrum drift to negligible proportions. The PMT gain variation can be reduced to manageable limits by packaging the detectors in a thermally controlled environment. Virtually all airborne spectrometers currently on the market have heated detector packages for this reason.

There are some spectrometers available which incorporate active spectrum stabilization. This involves monitoring detector pulses which are known to be due to scintillations of a certain energy generated in the detector either by a specially implanted radioisotope, or an artificial light source. The principle is illustrated in Figure 10A.14. Two adjacent counting windows are set up to centre on the spectral peak generated by the implanted source. The circuitry is so arranged that when their counting rates are equal no output is obtained, but when they differ a suitable analogue error signal is generated, positive say if the peak is off centre to the left and negative if it is off centre to the right. This error signal can then be used to change the gain of the system somewhere in the chain to restore the proper relation between spectral peaks and counting windows. The usual technique for applying corrective action is through the high voltage supply, although changing the conversion gain of the ADC or the gain of the main amplifier (more difficult) would achieve the same result.

Spectrum stablization is not difficult to implement and many variations on the general technique have been utilized at one time or another, particularly in portable spectrometers. The problem is to avoid interfering with the spectral information which is being sought in the first place. This constraint means that the implanted source should not generate a peak in the region of interest i.e. approx. 0.5-3.0 MeV. In the case of an implanted light source such as a PIN diode, it can be pulsed on during predetermined short intervals when data acquisition is disabled to avoid any such interference. However a radio-isotope cannot be so controlled other than by a mechanical shutter which is cumbersome.

To avoid interference with the data and the uncertainties of the PIN diode light source (still in the experimental stage), some manufacturers use low energy sources such as ^{133}Ba which generates a peak at 356 KeV. At this low energy even a small base-line shift at high count rates equivalent to say 20 KeV, would be interpreted by a spectrum stabilized using ^{133}Ba as a reduction in gain of 20/356 or about 6 per cent. It would then call for an overall gain increase of 6 per cent to correct for a minor offset which would not of itself have caused a significant error in the results. The 6 per cent gain increase would however in this case result in a very significant error.

The problem can be circumvented in airborne systems with arrays of large detectors by using the potassium ^{40}K peak itself as the reference, since it is a reliably prominent one under most conditions. In a mini-computer-based experimental system designed by the author

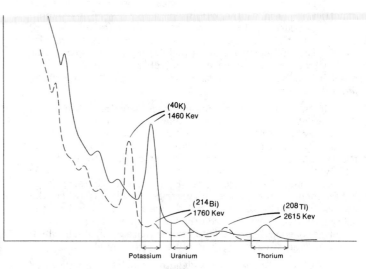

Figure 10A.13. *The dashed line shows what happens to the spectrum if the detector gain or subsequent amplifier gains drop. In this extreme case the ^{214}Bi peak has moved into the ^{40}K window. Unless drifts of this sort are quickly spotted by the equipment operator, useless and misleading data may be used in good faith to complete radiometric maps or make exploraton decisions.*

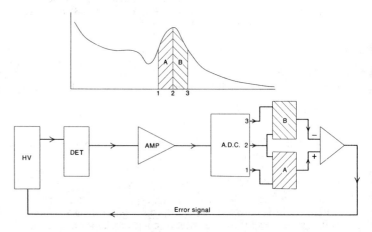

When count rates in windows A and B are equal the error signal is zero. When the count rates differ a positive or negative error signal is generated according as A>B or A<B which alters the high voltage in such a way as to move the peak back and restore symmetry. A stabilizing peak must always be present for such an arrangement and can be generated by a pulsed light source or radioisotope implanted in the detector, or by using an external source.

Figure 10A.14. *Spectrum Stabilization.*

(description to be published), the ^{40}K peak position is monitored and the counting window positions are recomputed periodically to correct for any drift.

Data Acquisition: Display and Recording

More often than not the gamma ray spectrometer is but one component of a multiparameter system in airborne applications and some complexity is necessary in handling the incoming data in order to display and or record it in an orderly manner. Until recently this function was performed by hardwired logic controllers with operator entries being made via thumbwheel switches.

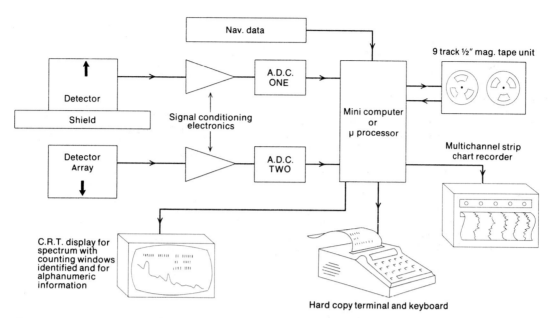

Such a system will have full spectral recording, means for atmospheric radon correction, a spectrum display with windows identified and a strip chart recorder on which a variety of profiles can be plotted. Such a system would normally be minicomputer or microprocessor based.

Figure 10A.15. *Modern airborne gamma-ray spectrometer*

Most equipment now incorporates programmable hardware based on minicomputers or their microprocessor equivalents. The advent of this technology in a form suitable for field and airborne instrumentation has suddenly opened up the possibility for sophisticated on-line data correction, CRT displays, keyboard entry of commands etc., to an extent limited only by the imagination of the designers.

Manufacturers have been understandably cautious in exploiting these possibilities, however most have the capability for storing complete gamma ray spectra and most either have or are adding CRTs for displaying the full spectrum with provision for identifying the counting windows. This is a well nigh indispensable feature in the author's view for detector gain matching and other performance checks. Window counts and other alphanumeric data are displayed either on CRT terminals or LED readouts; some manufacturers offer hard copy terminals rather than CRT displays.

Multichannel strip chart recorders are still universally used to display inflight profiles of window counts and other data such as ratios and radar altimeter readings. Some systems also use the same recorder for generating plots of complete spectra for system performance records.

Digital recording of airborne survey data was for many years done on 0.5 inch magnetic tape at a density of 200 or 300 bits per inch (bpi), using incremental seven-track tape transports. The years 1976 and 1977 saw an almost universal change over to the newer nine-track tape transports which are compatible with data centres all over the world and with 16 bit minicomputer systems. They also offer much higher recording rates than the older seven-track systems and more economical information storage as the bit density is 800 bpi, about three times that of the older standard. One disadvantage is that this high bit density makes factors such as tape condition, cleanliness of the recording heads and capstan drive etc., very much more critical than they were with the older systems, so that more attention must be paid to tape storage and handling practices and preventive maintenance schedules for the tape transports during survey operations.

The higher recording rates now available make it possible to record complete gamma ray spectra at the end of each counting period if they are stored in a suitable memory by the ADC as described earlier. Several survey specifications drawn up by national agencies now call for complete spectral recording from two separate detector arrays, one with a downward field of view and a smaller one shielded from direct terrestial radiation which is used to monitor atmospheric radon. The recorded spectra extend to 6.0 MeV, allowing cosmic background to be monitored so that corrections can be made continuously for this contribution to the potassium, uranium and thorium derived signals. The arrangement is fairly elaborate and requires two ADCs and buffer memories, so that one pair of spectra can be acquired while the previous ones are being written to tape. Figure 10A.15 shows the block diagram of a typical modern airborne spectrometer.

FUTURE TRENDS IN AIRBORNE GAMMA RAY SPECTROMETRY

One thing which has bedevilled electronic data processing in general and field recording systems in particular has been the need for electromechanical mass storage systems such as magnetic discs, drums, and tapes large and small. It now seems likely that "magnetic bubble" memories and similar technology will replace these devices in the forseeable future. Bubble memories can be considered (very broadly indeed) as the solid state analogue of the magnetic disc memory. Instead of the disc physically rotating past a sensing head, the tiny magnetic anomalies (bubbles) in a piece of otherwise uniform material are moved in the material itself. Storage and readout are not yet as fast as with core or solid state memory devices, but even now are orders of magnitude faster than for a disc. There seems little doubt that by 1985 or even sooner, the cost and the development will have reached the point where bubble memories could replace magnetic tape for in-flight storage of survey data.

As microprocessors become faster and more sophisticated there will be no reason not to do on-line data processing, including the deconvolution of spectra to correct for the known and imperfect response of detectors to the gamma radiation they receive.

Analogue-to-digital converters are already available in single width Nuclear Instrument Modules (e.g. Nuclear Data model 575). As large scale integration becomes more sophisticated, their sizes will shrink to the point where it would be possible to consider having one ADC for each detector in an airborne array. If at the same time the PIN, diode light source or an implanted alpha emitter such as ^{241}Am becomes a reliable method of providing a reference source, then the necessary signal conditioning electronics circuitry an ADC and a high voltage power supply could be designed to be integral with the detector as a single package. The detector "output" would then be a digital address rather than an analogue pulse height, with the energy and address relation internally stabilized by the reference scintillation source using appropriate windows on the ADC as described earlier. This would avoid the analogue problems which develop at high count rates since the detector "outputs" would be "summed" digitally. The superimposition of analogue pulses from different detectors would also be avoided by pushing the digital processing one stage farther back in this way.

Detectors with internally stabilized digital outputs could then be assembled in very large arrays indeed without any loss in overall resolution, thereby allowing higher survey speeds.

REFERENCES

The literature cited below is a selected bibliography where additional discussions of the instrumental aspects of gamma ray spectrometry can be found. These are not necessarily in the context of uranium exploration as this represents a fairly limited and specialized application of the science. For references to earlier work in this field the reader is referred to the proceedings of the Symposium on Mining and Groundwater Geophysics held at Niagara Falls, Ontario 1967 (see below for citation).

Breiner, S., Lindow, J.T., and Kaldenbach, R.J.
   1976: Gamma-ray measurement and data reduction considerations for airborne radiometric surveys; in I.A.E.A. Proc., International Symposium Exploration of Uranium ore deposits, Vienna 1976.

Bristow, Q.
   1974: Gamma-Ray; sensors; in geoscience instrumentation; E.A. Wolff, E.P. Mercanti, ed., John Wiley & Sons, New York.
   1975: Gamma-ray spectrometry instrumentation; in Report of Activities, Part C, Geol. Surv. Can., Paper 75-1C, p. 213-219.
   1977: A system for the offline processing of borehole gamma-ray spectrometry data on a NOVA minicomputer; in Report of Activities, Part A, Geol. Surv. Can., Paper 77-1A, p. 87-89.
   1978: The application of airborne gamma-ray spectrometry in the search for radioactive debris from the Russian satellite COSMOS 954 (Operation "Morning Light"); in Current Research, Part B, Geol. Surv. Can., Paper 78-1B, p. 151-162.

Chase, R.L.
   1961: Nuclear pulse spectrometry, McGraw Hill, New York, 221 p.

Crouthamel, C.E.
   1960: Applied gamma-ray spectrometry, Pergamon Press, New York, 443 p.

Duval, J.S., Worden, J.M., Clarke, R.B., and Adams, J.A.S.
   1972: Experimental comparison of Na(Tl) and solid organic scintillation detectors for use in remote sensing of terrestrial gamma-rays; Geophysics, v. 37(5), p. 879-888.

Fairstein, E. and Hahn, J.
   1965/
   1966: Nuclear pulse amplifiers – fundamentals and design practice, Parts 1-5, Nucleonics: v. 23(7), p. 56 (July 1965); v. 23(9), p. 81, (Sept. 1965) v. 23(11), p. 50 (Nov. 1965); v. 24(1), p. 54 (Jan. 1966); v. 24(3), p. 68 (Mar. 1966).

Foote, R.S.
   1970: Radioactive methods in mineral exploration; in Mining and Groundwater Geophysics, ed., L.W. Morley; Geol. Surv. Can., Econ. Geol. Rept. No. 26, p. 177-190.

I.A.E.A.
   1974: Recommended instrumentation for uranium and thorium exploration; I.A.E.A. Technical Rept. No. 158, Vienna.

Killeen, P.G. and Bristow, Q.
   1975: Uranium exploration by borehole gamma-ray spectrometry using off-the-shelf instrumentation; I.A.E.A. Proc. International Symposium on Exploration of Uranium Ore Deposits, Vienna.

Nicholson, P.W.
   1974: Nuclear electronics; John Wiley & Sons, London.

Reed, J.H. and Reynolds, G.M.
   1977: Gamma-ray spectrum enhancement; U.S., E.R.D.A. Rept. No. GJBX-25(77).

Orphan, V., Polichar, R., and Ginaven, R.
   1978: A "solid state" photomultiplier tube for improved gamma counting techniques; Uranium Exploration Technology Ann. Amer. Nucl. Soc. Meet., San Diego, Calif, June, 1978. (Abs.)

Pemberton, R.H.
   1970: Airborne radiometric surveying of mineral deposits; in Mining and Groundwater Geophysics, ed., L.W. Morley; Geol. Surv. Can., Econ. Geol. Rept. No. 26, p. 416-424.

Schneid, E.J., Swanson, F.R., Kamykowski, E.A., and Mendelsohn, A.
   1977: Airborne detector improvement; U.S. E.R.D.A. report No. GJBX-40(77).

GAMMA RAY SPECTROMETRIC METHODS IN URANIUM EXPLORATION — THEORY AND OPERATIONAL PROCEDURES

R.L. Grasty
Geological Survey of Canada, Ottawa

Grasty, R.L., Gamma ray spectrometric methods in uranium exploration — theory and operational procedures; in Geophysics and Geochemistry in the Search for Metallic Ores; Peter J. Hood, editor; Geological Survey of Canada, Economic Geology Report 31, p. 147-161, 1979.

Abstract

Many of the instrumental and operational problems in airborne gamma ray surveys have been solved and reliable data can now be provided. Multi-channel recording is an integral part of many survey operations and can be used to minimize energy calibration problems due to spectral drift and can also be used to increase sensitivity for uranium. In areas where suitable lakes cannot be found, upward looking crystals have proven essential for monitoring variations of atmospheric background from decay products of radon. To correct for variations of cosmic radiation due to changes in topographic relief, an energy window above that of the natural gamma ray emissions from the ground is often used.

Construction of concrete calibration pads and the utilization of calibration strips has greatly facilitated the standardization of airborne data from different detector configurations. The energy and angular distribution of the natural gamma-radiation field over uniformly radioactive ground can now be calculated reliably. By incorporating the detector response, the sensitivity of a particular system to each of the radioelements can be evaluated.

Résumé

De nombreux problèmes reliés aux instruments et aux travaux de levés aériens à rayons gamma ont été résolus; il est maintenant possible d'en obtenir des données sûres. L'enregistrement multicanal fait partie intégrante de nombreux travaux de levés et peut servir à réduire au minimum les problèmes d'étalonnage énergétique dus à la dérive spectrale; il peut aussi être utilisé pour accroître la sensibilité à l'uranium. Dans des régions où il n'est pas possible de trouver des lacs convenables, des cristaux d'orientation ascendante se sont avérés essentiels pour le contrôle des variations de la zone de fond atmosphérique à partir de la famille radioactive du radon. On utilise souvent, pour corriger les variations de radiation cosmique dues aux changements du relief topographique, une fenêtre énergétique au-dessus de celle des émissions naturelles de rayons gamma provenant du sol.

La construction de blocs d'étalonnage en béton et l'utilisation de bandes d'étalonnage ont grandement facilité la normalisation des données aériennes à partir des diverses configurations décelées par le détecteur. L'énergie et la distribution angulaire du champ naturel de rayonnement gamma au-dessus d'un terrain radioactif uniforme peuvent maintenant être calculées avec justesse. En joignant les données du détecteur, on peut évaluer la sensibilité d'un système à chacun des radioéléments.

THE NATURAL GAMMA-RADIATION FIELD

Basic Considerations

While studying the phosphorescence of various materials Becquerel discovered that an invisible radiation was emitted by several uranium salts that was capable of traversing thin layers of opaque material and fogging a photographic plate (Becquerel, 1896a, b). Soon afterwards Schmidt (1898) and Curie (1898) independently observed that a similar radiation was emitted by compounds of thorium. Through the work of Villard (1900), Rutherford (1903), and Strutt (1903) it was shown that three characteristic types of radiation were emitted, alpha, beta, and gamma radiation. Potassium was found by Campbell and Wood to emit beta radiation in 1906 although it was not until 1927 that it was observed by Kolhorster to emit gamma radiation (Campbell and Wood, 1906; Campbell, 1907; Kolhorster, 1928).

Alpha rays or alpha particles are doubly positively-charged helium nuclei and are absorbed by a few centimetres of air. Beta particles are electrons carrying unit negative charge, are more penetrating and can travel up to a metre or so. Gamma radiation, an electromagnetic radiation similar in nature to X-rays is strongly penetrating and was found from measurements by Wulf (1910) on the Eiffel Tower to be capable of ionizing air at heights of 300 m.

The absorption of gamma radiation takes place in three distinct ways, by the photoelectric effect, by scattering, and by pair production.

In the photoelectric effect the energy of the gamma ray is completely absorbed through the emission of an electron. The scattering process known as the Compton effect takes place when a gamma ray photon collides with an electron, imparts part of its energy to the electron, and is scattered at an angle to the original direction of the incident photon. This process predominates for moderate gamma ray energies in a wide range of materials. The third process, pair production can only take place if the incident photon has an energy greater than 1.02 MeV, since 1.02 MeV is necessary for the creation of an electron-positron pair. This interaction predominates at high energies particularly in materials of high atomic number. Because most materials (rocks, air and water) encountered in airborne radioactivity measurements have a low atomic number and because most natural gamma rays have moderate to low energies (less than 2.62 MeV) Compton scattering is the predominant absorption process occurring between the source of the radioactivity and the detector.

If a collimated beam of radiation of intensity I is incident upon an absorbing layer of thickness dx, the amount of radiation absorbed dI is proportional both to dx and to I so that:

$$dI = -\mu I dx$$

The proportional factor μ is a characteristic property of the medium known as the linear attenuation coefficient and is a function of the gamma ray energy. If the intensity I has the value I_o when no absorbing material is present then it follows that

$$I = I_o e^{-\mu x}$$

The thickness of absorbing material that reduces the intensity to half its original value is called the half-value thickness ($x_{1/2}$). It follows from the previous equation that

$$x_{1/2} = \frac{\log_e 2}{\mu} = \frac{0.693}{\mu}$$

Table 10B.1 shows the calculated half-thicknesses and mass attenuation coefficients at various energies, for water, air and rock (Hubbell and Berger, 1968). The mass attenuation coefficient is the linear attenuation coefficient divided by the density of the material. At aircraft altitudes of 100 m or more the intensity of gamma rays below 0.10 MeV emitted by rocks and soils in the ground will be considerably reduced and dominated by Compton-scattered high-energy gamma radiation. It is apparent that the measurement of natural radioactivity must be carried out within a few hundred metres of the ground and only gamma rays originating from a few tens of centimetres below the surface of the ground can be detected.

All rocks and soils are radioactive and emit gamma radiation. The three major sources are:

1. Potassium-40, which is 0.12 per cent of the total potassium and emits gamma ray photons of energy 1.46 MeV.
2. Decay products in the uranium-238 decay series.
3. Decay products in the thorium-232 decay series.

The gamma ray spectrum from the uranium and thorium decay series is extremely complex. Table 10B.2 and 10B.3 show the principal gamma rays over 100 KeV that are emitted by uranium-238 and thorium-232 in equilibrium with their decay products as tabulated by Smith and Wollenberg (1972). Their relative abundance, measured as photons per disintegration is also indicated. Tables 10B.4 and 10B.5 show the two radioactive series together with the principal emissions and half lives of the decay products.

Table 10B.1

Mass attenuation coefficients and half thickness for various gamma ray energies in air, water and concrete

Photon Energy MeV	Mass Attenuation Coefficient (cm2/g)			Half Thickness[a]		
	Air[b]	Water	Rock[c]	Air[d] (m)	Water (cm)	Rock[e] (cm)
0.01	4.82	4.99	26.5	1.11	0.139	0.01
0.10	0.151	0.168	0.171	35.5	4.13	1.62
0.15	0.134	0.149	0.140	40.0	4.65	1.98
0.20	0.123	0.136	0.125	43.6	5.10	2.22
0.30	0.106	0.118	0.107	50.6	5.87	2.59
0.40	0.0954	0.106	0.0957	56.2	6.54	2.90
0.50	0.0868	0.0966	0.0873	61.8	7.18	3.18
0.60	0.0804	0.0894	0.0807	66.7	7.75	3.43
0.80	0.0706	0.0785	0.0708	75.9	8.83	3.92
1.0	0.0635	0.0706	0.0637	84.4	9.82	4.35
1.46	0.0526	0.0585	0.0528	102	11.8	5.25
1.5	0.0517	0.0575	0.0519	104	12.1	5.34
1.76	0.0479	0.0532	0.0482	112	13.0	5.75
2.0	0.0444	0.0493	0.0447	121	14.1	6.20
2.62	0.0391	0.0433	0.0396	137	16.0	7.00
3.0	0.0358	0.0396	0.0365	150	17.5	7.60

a — The thickness of material which reduced the intensity of the beam to half its initial value.
b — 75.5% N, 23.2% O, 1.3% Ar by weight.
c — Composition of typical concrete, see Hubbell and Berger (1968).
d — For air at 0°C and 76 cm of Hg with a density of 0.001293 g/cm^3.
e — Density of concrete is 2.5 g/cm^3.

Table 10B.2

Principal[a] gamma rays over 100 KeV emitted by uranium in equilibrium with its decay products

Isotope[b]	γ-Energy (KeV)	Intensity[c]	Isotope[b]	γ-Energy (KeV)	Intensity[c]
Th-234	115	0.42	Bi-214	786	0.29
U-235	144	0.48	Bi-214	806	1.10
Ra-223	144	0.14	Bi-214	821	0.16
Ra-223	154	0.24	Bi-214	826	0.13
U-235	163	0.22	Pb-211	832	0.14
U-235	186	2.52	Bi-214	839	0.59
Ra-226	186	3.90	Pb-214	904	0.59
U-235	205	0.22	Bi-214	934	3.10
Th-227	236	0.51	Bi-214	964	0.37
Pb-214	242	7.60	Pa-234M	1001	0.83
Th-227	256	0.28	Bi-214	1052	0.33
Pb-214	259	0.80	Bi-214	1070	0.26
Ra-223	269	0.61	Bi-214	1104	0.16
Rn-219	271	0.45	Bi-214	1120	15.0
Pb-214	275	0.70	Bi-214	1134	0.25
Pb-214	295	18.9	Bi-214	1155	1.70
Ra-223	324	0.16	Bi-214	1208	0.47
Th-227	330	0.13	Bi-214	1238	6.10
Ra-223	338	0.12	Bi-214	1281	1.50
Bi-211	351	0.60	Bi-214	1304	0.11
Pb-214	352	36.3	Bi-214	1378	4.30
Bi-214	387	0.31	Bi-214	1385	0.80
Bi-214	389	0.37	Bi-214	1402	1.50
Rn-219	402	0.29	Bi-214	1408	2.60
Pb-211	405	0.18	Bi-214	1509	2.20
Bi-214	406	0.15	Bi-214	1539	0.53
Bi-214	427	0.10	Bi-214	1543	0.34
Bi-214	455	0.28	Bi-214	1583	0.73
Pb-214	462	0.17	Bi-214	1595	0.30
Pb-214	481	0.34	Bi-214	1600	0.34
Pb-214	487	0.33	Bi-214	1661	1.16
Rn-222	511	0.10	Bi-214	1684	0.24
Pb-214	534	0.17	Bi-214	1730	3.20
Bi-214	544	0.10	Bi-214	1765	16.7
Pb-214	580	0.36	Bi-214	1839	0.37
Bi-214	609	42.8	Bi-214	1848	2.30
Bi-214	666	14.0	Bi-214	1873	0.22
Bi-214	703	0.47	Bi-214	1890	0.10
Bi-214	720	0.38	Bi-214	1897	0.18
Bi-214	753	0.11	Bi-214	2110	0.10
Pa-234M	766	0.31	Bi-214	2119	1.30
Bi-214	768	4.80	Bi-214	2204	5.30
Pb-214	786	0.86	Bi-214	2294	0.33
			Bi-214	2448	1.65

a — Photons with an intensity greater than 0.1%.
b — Decaying isotope.
c — Decays per 100 decays of the longest lived parent.

Table 10B.3

Principal[a] gamma rays over 100 KeV emitted by thorium in equilibrium with its decay products

Isotope[b]	γ-Energy (KeV)	Intensity[c]	Isotope[b]	γ-Energy (KeV)	Intensity[c]
Pb-212	115	0.61	Bi-212	727	6.66
Ac-228	129	3.03	Ac-228	755	1.14
Th-228	132	0.26	Tl-208	763	0.61
Ac-228	146	0.23	Ac-228	772	1.68
Ac-228	154	1.02	Ac-228	782	0.56
Ac-228	185	0.11	Bi-212	785	1.11
Ac-228	192	0.13	Ac-228	795	5.01
Ac-228	200	0.36	Ac-228	830	0.64
Ac-228	204	0.18	Ac-228	836	1.88
Ac-228	209	4.71	Ac-228	840	10.2
Th-228	217	0.27	Tl-208	860	4.32
Tl-208	234	0.12	Bi-212	893	0.37
Pb-212	239	44.6	Ac-228	904	0.90
Ra-224	241	3.70	Ac-228	911	30.0
Tl-208	253	0.25	Ac-228	944	0.11
Ac-228	270	3.90	Ac-228	948	0.13
Tl-208	277	2.34	Bi-212	952	0.18
Ac-228	279	0.24	Ac-228	959	0.33
Bi-212	288	0.34	Ac-228	965	5.64
Pb-212	300	3.42	Ac-228	969	18.1
Ac-228	322	0.26	Ac-228	988	0.20
Bi-212	328	0.14	Ac-228	1033	0.23
Ac-228	328	3.48	Ac-228	1065	0.15
Ac-228	332	0.49	Bi-212	1079	0.54
Ac-228	338	12.4	Tl-208	1094	0.14
Ac-228	341	0.44	Ac-228	1096	0.14
Ac-228	409	2.31	Ac-228	1111	0.36
Ac-228	440	0.15	Ac-228	1154	0.17
Bi-212	453	0.37	Ac-228	1247	0.59
Ac-228	463	4.80	Ac-228	1288	0.12
Ac-228	478	0.25	Ac-228	1459	1.08
Ac-228	504	0.22	Ac-228	1496	1.09
Ac-228	510	0.51	Ac-228	1502	0.60
Tl-208	511	8.10	Bi-212	1513	0.31
Ac-228	523	0.13	Ac-228	1557	0.21
Ac-228	546	0.23	Ac-228	1580	0.74
Ac-228	562	1.02	Ac-228	1588	3.84
Ac-228	571	0.19	Bi-212	1621	1.51
Ac-228	572	0.17	Ac-228	1625	0.33
Tl-208	583	31.0	Ac-228	1630	2.02
Ac-228	651	0.11	Ac-228	1638	0.56
Ac-228	675	0.11	Ac-228	1666	0.22
Ac-228	702	0.20	Ac-228	1686	0.11
Ac-228	707	0.16	Bi-212	1806	0.11
Ac-228	727	0.83	Ac-228	1887	0.11
			Tl-208	2614	36.0

a — With more than 0.1 decays per 100 decays of the longest lived parent.
b — Decaying isotope.
c — Decays per 100 decays of the longest lived parent.

Table 10B.4

The U-238 series decay chain

Isotope	Radiation	Half Life
U^{238} ↓	α	4.507×10^9 y
Th^{234} ↓	β	24.1 d
Pa^{234} ↓	β	1.18 m
U^{234} ↓	α	2.48×10^5 y
Th^{230} ↓	α	7.52×10^4 y
Ra^{226} ↓	α	1600 y
Rn^{222} ↓	α	3.825 d
Po^{218} ↓	α	3.05 m
Pb^{214} ↓	β	26.8 m
Bi^{214} ↓	β	19.7 m
Po^{214} ↓	α	1.58×10^{-4} s
Pb^{210} ↓	β	22.3 y
Bi^{210} ↓	β	5.02 d
Po^{210} ↓	α	138.4 d
Pb^{206}		stable

Isotopes constituting less than 0.2 per cent of the decay products are omitted.

Table 10B.5

The Th-232 decay series

Isotope	Radiation	Half Life
Th^{232} ↓	α	1.39×10^{10} y
Ra^{228} ↓	$β^-$	6.7 y
Ac^{228} ↓	$β^-$	6.13 h
Th^{228} ↓	α	1.91 y
Ra^{224} ↓	α	3.64 d
Rn^{220} ↓	α	55.3 s
Po^{216} ↓	α	0.158 s
Pb^{212} ↓	$β^-$	10.64 h
Bi^{212} 36% 64%	$β^-$ (64%) α (36%)	60.5 m
Po^{212} ↓	α	3.04×10^{-7} s
Tl^{208} ↓	$β^-$	3.1 m
Pb^{208}		stable

Characteristics of Gamma Radiation

In order to monitor variations of the three radioelements in the ground by airborne gamma ray spectrometry, it is necessary to understand the behaviour of the gamma radiation field. The first theoretical work on the variation in intensity of the natural gamma radiation field with elevation above the surface of the earth was carried out by Eve (1911). He evaluated the intensity of the gamma radiation, measured in terms of the number (n) of ions produced per second per cubic centimetre of air and showed that

$$n = \frac{2\pi Q n_0}{\mu} \int_0^1 e^{-\lambda h/z} dz \qquad (1)$$

where

Q is the mean radium content of the rocks

n_0 is the number of ions produced per cubic centimetre per second in air at normal temperature and pressure, one centimetre from one curie of radium,

λ is the linear attenuation coefficient of gamma rays in air,

μ is the linear attenuation coefficient of gamma rays in the ground, and

z is sinθ, (π/2-θ) being the angle subtended at a detector a distance h above the surface by an elementary ring below the surface.

Gockel (1910) carried out the first airborne experiments to measure the ionizing effect of the gamma radiation from the ground using an electroscope mounted in a balloon and found an erratic variation with altitude, probably because of fluctuations in the concentration of radon daughters in the atmosphere.

Hess (1911, 1912) was the first to obtain definite results using a balloon and showed that while the ionization decreased slightly up to a distance of 1000 m, above 2000 m it began to increase and at 5000 m was two to three times the value found at ground level. These results can be explained if the radon daughter concentration decreases initially with altitude and at the higher elevation the ionization is predominated by cosmic radiation.

Substituting x = 1/z in Equation (1) we arrive at the commonly used expression

$$N = N_0 \int_1^\infty \frac{e^{-\lambda h x}}{x^2} dx = N_0 E_2(\lambda h) \qquad (2)$$

where

N_0 is the count rate with a gamma ray detector at ground level, and

N is the count rate at an altitude h.

The E_2 function is known as the exponential integral of the second kind and in Russian literature is often referred to as the King function. King (1912) generalized Equation (2) and derived the variation of gamma ray intensity, N, with altitude above a circular disc of thickness d subtending an angle 2φ at the point of measurement. N is given by

$$N = \frac{2\pi Q n_0}{\mu} \left[E_2(\lambda h) - E_2(\lambda h + \mu d) - \cos\phi \left\{ E_2\left(\frac{\lambda h}{\cos\phi}\right) - E_2\left(\frac{\lambda h + \mu d}{\cos\phi}\right) \right\} \right] \qquad (3)$$

where Q, n_0, μ, h, and λ are the same parameters as in Equation (1).

Table 10B.6

Percentage of detected gamma radiation originated from circular areas beneath the point of detection (altitude 120 m)

Percentage of Infinite Source	Diameter of Circle (m)		
	Potassium[a] (μ = 0.00680/m)	Uranium[a] (μ = 0.00619/m)	Thorium[a] (μ = 0.00506/m)
10	74.5	76.2	79.6
20	110.9	113.3	100.0
30	143.4	146.8	154.1
40	176.1	180.6	190.1
50	211.5	217.2	229.5
60	252.2	259.5	275.5
70	302.7	312.3	333.5
80	372.7	385.8	415.2
90	493.3	513.4	559.5

[a] – Linear attenuation coefficient taken from Table 1 for air at 0°C and 76 cm Hg.

It should be pointed out that this expression has been recalculated on many occasions e.g. Godby et al. (1952), Darnley et al. (1968), Duval et al. (1971), Kellogg (1971), and in some instances mathematical or typographical errors have occurred. From this equation it can readily be calculated that the percentage of the total radiation detected, P, originating from a circular area subtended an angle 2ϕ is given by

$$P = 100 \times \frac{\left[E_2(\lambda h) - \cos\phi\ E_2(\frac{\lambda h}{\cos\phi})\right]}{E_2(\lambda h)} \quad (4)$$

These results are tabulated in Table 10B.6 for potassium, uranium and thorium gamma ray energies of 1.46, 1.76 and 2.62 MeV and are also illustrated in Figure 10 B.1, for potassium and thorium. However, they are only valid for unscattered mono-energetic gamma radiation, since gamma rays can be Compton scattered and still contribute to the ionization or be detected by a gamma ray detector.

The complete solution of this gamma ray transport problem is extremely complex since several hundred gamma ray energies are involved, each with different attenuation coefficients and with multiple scattering occurring both in the ground and in the air. However with the advent of high-speed computers the energy and angular distribution of both the direct and scattered gamma ray component can now be evaluated. This has been carried out by Beck and his co-workers at the health and Safety Laboratory in New York (Beck and de Planque, 1968) for the purpose of evaluating the exposure rate from natural gamma radiation and fallout from nuclear weapons tests. Independently Kirkegaard (1972) has carried out similar calculations to aid in the interpretation of gamma ray surveys for exploration and arrived at similar solutions. Both calculation procedures solve the Boltzmann transport equation for two semi-infinite homogeneous media, one being the ground with a uniform distribution of gamma ray emitters and the other being the air. For this particular geometry the mathematics is considerably simplified because of symmetry. They both derive separately the scattered and uncollided gamma ray fluxes.

The continuous nature of the scattered component of the gamma ray flux is illustrated in Figure 10B.2 as calculated by Kirkegaard and Løvborg (1974), for an infinite homogeneous source of 1 per cent potassium, covered with a layer of 20 cm of water which is equivalent to approximately 200 m of air. The scattered component can be seen to increase significantly at energies below about 500 KeV. The calculated results for an infinite source of 1 per cent uranium and 1 per cent thorium are also shown in Figures 10B.3 and 10B.4. Figure 10B.5 shows the energy distribution of the gamma ray flux and its "skyshine" contribution 1 m above a typical granite containing 3.4 per cent potassium, 3 ppm uranium in equilibrium, and 12 ppm thorium (Løvborg et al., 1976). The "skyshine" contribution arises from gamma rays from the ground that have been scattered back towards the ground by the air from heights above 1 m. The "skyshine" flux is predominantly of low energy and contributes approximately 50 per cent of the total flux at energies less than about 200 KeV. At these low energies the radiation is virtually isotropic having a uniform angular distribution, since the gamma rays have suffered multiple collisions and in

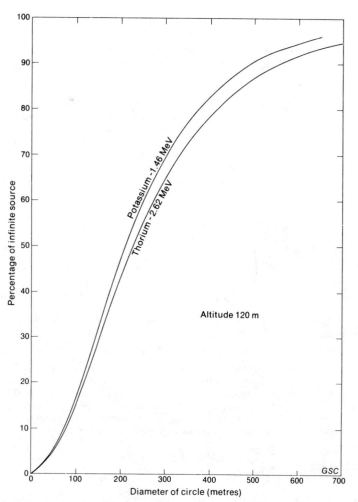

Figure 10B.1. *Percentage of total detected gamma radiation from circular areas beneath the detector.*

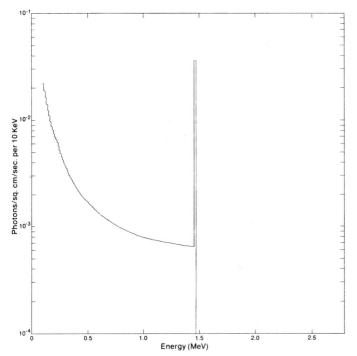

Figure 10B.2. *Energy distribution of the photon flux produced by potassium in sand at a water depth of 20 cm.*

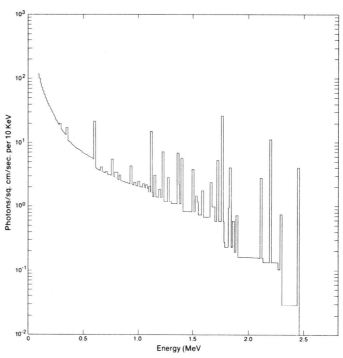

Figure 10B.3. *Energy distribution of the photon flux produced by uranium in sand at a water depth of 20 cm.*

essence lost all knowledge of their original direction. In Figure 10B.6 the angular distribution of a typical gamma ray flux at 1 m illustrates the fact that very little flux comes from sources directly beneath the detector and that the majority of the radiation comes from between 60 and 80° from the vertical (Beck, 1972).

From a knowledge of the energy distribution of the gamma ray flux above the ground containing the different radioelements it is a relatively simple matter to calculate the exposure rate above the ground. Table 10B.7 shows the contribution from potassium, uranium, and thorium to the exposure rate 1 m above the ground. The agreement between the results of Beck et al. (1972) and Løvborg and Kirkegaard (1974) is a good indication that the energy distribution of the gamma ray flux can be derived reliably.

In interpreting airborne gamma ray spectrometry data it is not sufficient to know the gamma ray flux distributions from potassium, uranium, and thorium, since the detector modifies the spectrum considerably. Incorporating the detector response is an extremely complex problem and in general can only be evaluated satisfactorily through a combination of experiment and Monte Carlo simulations. Løvborg and Kirkegaard (1974) have incorporated the detector response of a 7.6 x 7.6 cm (3 x 3 inch) sodium iodide detector and obtained excellent agreement between their theoretical and experimental work.

For a 7.6 x 7.6 cm (3 x 3 inch) detector the energy deposited in the crystal has little dependence on the angle the gamma ray photon strikes the crystal. This is not the case for large diameter crystals commonly used in airborne surveys. In this case it is necessary to derive a detector response which varies with angle. Løvborg et al. (1977) have attempted to do this by making some simplifying assumptions and modifying their theoretical results based on experimental data. From their work they have been able to make estimates of the detector sensitivities and various calibration constants for a variety of cylindrical sodium iodide detectors which are 10.2 cm (4 inches) thick.

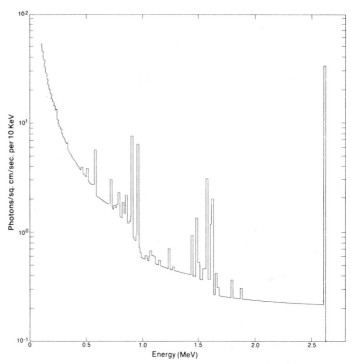

Figure 10B.4. *Energy distribution of the photon flux produced by thorium in sand at a water depth of 20 cm.*

Grasty and Holman (1974) measured the angular sensitivity variation at 2.62 MeV for a variety of detectors commonly employed in airborne survey operation. Grasty (1976a) incorporated these experimental results into the theoretical calculations (Equation (4)) and showed how the percentage contribution of circular areas beneath the aircraft to the total radiation detected varied with detector. These results are presented in Figure 10B.7 for a 29.2 x 10.2 cm (11.5 x 4 inch) and a 12.7 x 12.7 cm (5 x 5 inch) detector. The

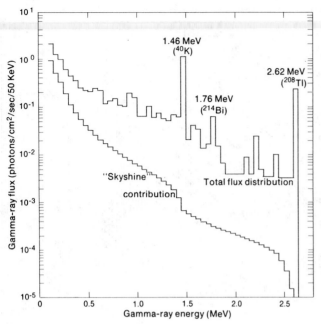

Figure 10B.5. *Energy distribution of the total photon flux and the skyshine component 1 m above a typical granite.*

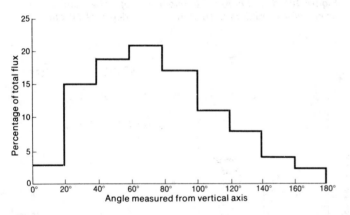

Figure 10B.6. *Angular distribution of the gamma ray flux at 1 m.*

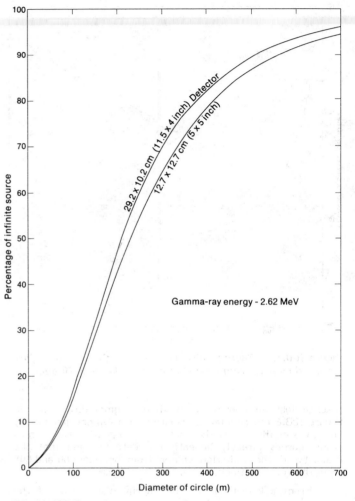

Figure 10B.7. *Percentage of total detected gamma radiation from circular areas beneath two different detectors.*

Table 10B.7

Calculated contributions from potassium, thorium, and uranium to the exposure rate 1 m above soil

	Exposure Rate (µR/h)	
	Løvborg and Kirkegaard (1974)	Beck et al. (1972)
1% K	1.52	1.49
1 ppm Th	0.31	0.31
1 ppm U	0.63	0.62

29.2 x 10.2 cm detector has the greatest angular sensitivity variation whereas the 12.7 x 12.7 cm detector shows little angular sensitivity variation. The results show that at these high energies the assumption of a spherical detector is a reasonable approximation even for a 29.2 x 10.2 cm detector. This is because at large angles from the vertical air absorption is significantly more important than the reduced sensitivity of the detector.

OPERATIONAL PROCEDURES

The Airborne System

In aerial measurements of natural radioactivity for geological mapping or uranium exploration, large volume cylindrical sodium iodide detectors are commonly employed. Due to the physical characteristics of the photomultiplier and detector assembly the discrete nature of the unscattered gamma ray photon flux as illustrated in Figures 10B.2 to 10B.4 cannot be observed, and it is necessary to select energy windows which are best representative of the particular radioelements concerned. A typical gamma ray spectrum taken at 120 m is shown in Figure 10B.8 (Foote, 1968). The peaks at 2.62, 1.76, and 1.46 MeV representing thallium-208 in the thorium decay series, bismuth-214 in the uranium decay series, and potassium-40, can be readily distinguished. These particular gamma ray photons have been generally accepted as being most suitable for the measurement of uranium and thorium because they are relatively abundant and being high in energy are not appreciably absorbed in the air. They can also be readily discriminated from other gamma rays in the spectrum. Typical gamma ray energy windows for monitoring these particular gamma rays are shown in Table 10B.8. According to McSharry (1973) these particular radioelement windows

Gamma-Ray Surveying

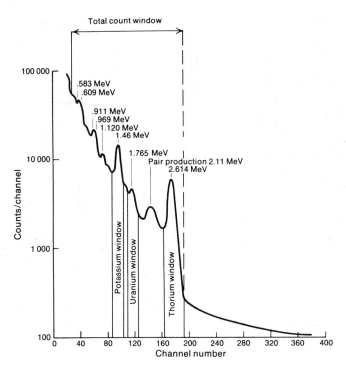

Figure 10B.8. *Typical gamma ray spectrum at 120 m.*

Table 10B.8

Spectral window widths

Element Analyzed	Isotope Used	Gamma Ray Energy (MeV)	Energy Window (MeV)
Potassium	K-40	1.46	1.37-1.57
Uranium	Bi-214	1.76	1.66-1.86
Thorium	Tl-208	2.62	2.41-2.81
Total Count			0.41-2.81

give the most reliable estimates of the individual radioelements. A total count window is also almost always used since the total count reflects general lithological variation and is therefore useful in geological mapping.

In the last few years, full energy spectral recording of up to 512 channels is a common requirement in government contracts for uranium reconnaissance programs and will also certainly become common practice in the future for all airborne surveys. Full spectral recording on magnetic tape has the advantage that subsequent to the survey operations the spectrum can be accurately calibrated from the prominent positions of the potassium and thorium peaks at 1.46 and 2.62 MeV respectively. However until more sophisticated data processing procedures are developed the particular windows shown in Table 10B.8 will be those generally used to convert the airborne data to ground concentrations.

In order to relate the airborne count rates from the three windows to ground concentrations, four particular data processing steps are necessary. These are:

1. the removal of background radiation,
2. a spectral stripping procedure,
3. an altitude correction, and
4. the conversion of the corrected data to ground concentrations.

Background Radiation

In any airborne radioactivity survey three sources of background radiation exist:

1. the radioactivity of the aircraft and its equipment,
2. cosmic radiation, and
3. airborne radioactivity arising from daughter products of radon gas in the uranium decay series.

The radioactivity of the aircraft and its equipment is found to remain constant and is due to the presence of small quantities of natural radioactive nuclides in the detector system and in the airframe. Particularly large contributors can arise from luminous watches and the radium dials on the instrument panels which must be removed from the aircraft.

The cosmic ray background is caused primarily by photons generated by cosmic ray interactions with nuclei present in the air, aircraft or in the detection system itself. The cosmic ray contribution increases with aircraft altitude but shows little variation on a day-to-day basis (Dahl and Odegaard, 1970; Grasty, 1973). Small variations are observed with latitude and with the eleven-year solar cycle and will also vary somewhat with the size of the aircraft. Figure 10B.9 shows a cosmic generated gamma ray spectrum obtained by subtracting over water spectra from two different altitudes (Burson, 1973). The prominent peak near 0.5 MeV is due to the annihilation of positrons created predominantly by pair production from high energy gamma ray photons in the aircraft structure or detector assembly. These positrons annihilate into two gamma ray photons of 0.511 MeV. The cosmic ray contribution in each radioelement window can be removed by monitoring a high-energy window from 3-6 MeV which will be unaffected by natural variations in the ground (Burson, 1973).

By far the most difficult background radiation correction arises from the decay products of radon. Radon, being a gas, can diffuse out of the ground. Furthermore it has a half life of 3.8 days. The rate of diffusion will depend on such factors as air pressure, soil moisture, ground cover, wind, and temperature. The decay products, lead-214 and bismuth-214, are attached to airborne aerosols and consequently their distribution is dependent to a large extent on wind patterns. Under early morning still-air conditions, there can be measurable differences in atmospheric radioactivity at sites a few miles apart. As the day progresses increasing air turbulence tends to mix the air to a greater extent and reduce the atmospheric background close to the ground. Figure 10B.10 shows the morning and afternoon radon concentrations in Cincinnati over a four-year period, taken from the results of Gold et al. (1964). Large annual variations probably arise from the trapping of the radon in the frozen ground during the winter. Darnley and Grasty (1970) reported that on the average 70 per cent of the photons detected in the uranium window arise from radon daughters occurring in the air. Figure 10B.11 shows some typical over water background measurements taken by the Geological Survey of Canada's high sensitivity gamma ray spectrometer while carrying out a large airborne survey in the Northwest Territories of Canada. Typical uranium channel count rates from the ground are around 20 counts/second. It is apparent that the variation of count rate in the total count and potassium window are essentially due to variations of bismuth-214.

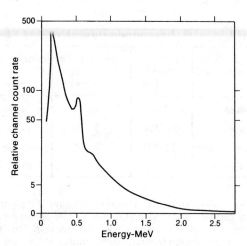

Figure 10B.9. Cosmic-generated spectrum.

Problems in measuring the uranium background also arise because of temperature inversions. Figure 10B.12 shows measurements on the same day at different altitudes over Lake Ontario. The thorium channel shows a relatively smooth and expected exponential increase with altitude. The uranium channel shows a general decrease from ground level to 1000 m as has been commonly observed (Hess, 1911, 1912; Burson et al., 1972) but at a temperature inversion at approximately 2000 m the uranium channel increases significantly. Above 2300 m the uranium count rate increases in a similar manner to the thorium channel due to the increasing cosmic ray contribution. From 150 to 300 m the uranium count rate is found to increase since at low altitudes the air cannot be considered as an infinite source (Cook, 1952; Burson et al., 1972).

Since accurate measurements of the count rate in the uranium windows are of prime importance in locating possible uranium targets, it is essential to measure the uranium background as accurately as possible. The technique adopted by the Geological Survey of Canada has been to fly over a lake before the commencement of a survey flight. Since the concentrations of radioactive nuclides in the water are several orders of magnitude lower than that of normal crustal material, the activity measured will be the total background contribution from all three sources. Fortunately in most of Canada lakes are abundant, and the background values can be updated frequently during the course of the survey. Many experimenters have found this method satisfactory when large lakes are present and homogeneous mixing of the radioactive decay products has occurred. An alternative approach when large lakes are not available has been to sample the air by the use of filters (Burson, 1973). Reasonable estimates of the radioactivity of the air can be made from the beta or gamma activity of the dust collected on the filter papers. Foote (1968) used a detector shielded from ground radiation by 10 cm of lead to monitor atmospheric radiation. This procedure has also been employed in Russia, Iran and in the United States (Purvis and Buckmeier, 1969), however the extra detectors and shielding use up valuable space and weight. It is also a complex procedure since gamma rays from the ground can be scattered in the air or in the lead shield and still be detected. Unless the shield is well designed, direct radiation from the ground can also be detected. Unfortunately there is very little documentation on the reliability of this procedure, although in areas with no lakes this may prove to be the only possible technique to use.

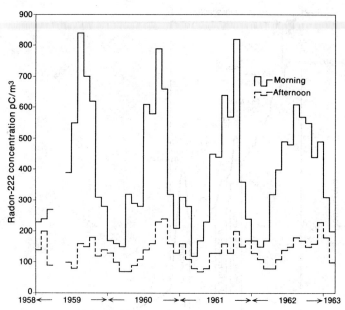

Figure 10B.10. Monthly average morning (0800 hrs.) and afternoon (1500 hrs.) radon 222 concentrations at Cincinnati.

Spectral Stripping

Due to Compton scattering in the ground and in the air of 2.62 MeV thallium-208 photons some counts will be recorded in the lower energy potassium and uranium windows from a pure thorium source. Counts in the lower energy windows may also arise from the incomplete absorption of 2.62 MeV photons in the detector or from other lower energy gamma ray photons in the thorium decay series. Similarly counts will be recorded in the lower energy potassium windows from a pure uranium source and can also appear in the high energy thorium window due to high energy gamma ray photons of bismuth-214 in the uranium decay series. Due to the poor resolution of sodium iodide detectors, counts can also be recorded in the uranium channel from a pure potassium source. The ratio of the counts in a lower energy window to those in a high energy window for a pure uranium or thorium source is termed a stripping ratio or spectral stripping coefficient and have generally been called alpha, beta, and gamma where

> alpha is the uranium counts per thorium count from a pure thorium source,
>
> beta is the potassium count per thorium count from a pure thorium source, and
>
> gamma is the potassium count per uranium count from a pure uranium source.

Grasty (1977) has adopted the terminology a, b and g for the reverse stripping ratios where

> a is the reverse stripping ratio, uranium into thorium,
>
> b is the reverse stripping ratio, potassium into thorium, and
>
> g is the reverse stripping ratio, potassium into uranium.

Unless multiple thorium windows are used, b will have a value of zero.

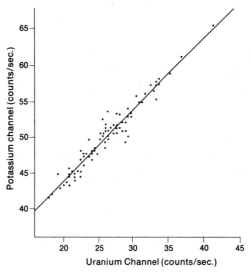

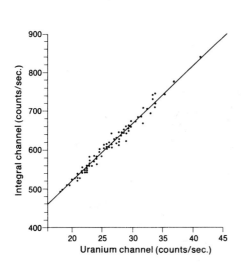

Figure 10B.11. Variation of over water backgrounds for potassium, uranium and total count.

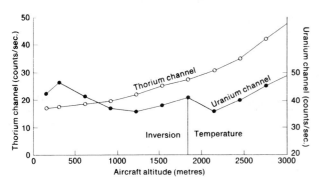

Figure 10B.12. Variation of uranium and thorium channel count rates over Lake Ontario.

In order to relate the airborne count rates in the three windows to ground concentrations it is first necessary to determine these six calibration constants. This is generally achieved through the use of large radioactive concrete calibration sources or pads. Sets of these calibration sources have been constructed in Ottawa by the Geological Survey of Canada (Grasty and Darnley, 1971), at Grand Junction for the U.S. Energy Research and Development Agency (Ward, 1978) and also in Iran. The concentration of the Ottawa and Grand Junction pads are given by Killeen (1979).

From measurements on these calibration pads the observed count rates in the three energy windows are all linear combinations of the radioelement compositions of the individual pads. As shown by Løvborg et al. (1972), a general matrix equation can be formulated to relate the observed count rates $N_{2.62}$, $N_{1.76}$ and $N_{1.46}$ to the radioelement concentrations Th_{ppm}, U_{ppm} and K_{pct} of each pad

$$\begin{bmatrix} N_{2.62} \\ N_{1.76} \\ N_{1.46} \end{bmatrix} = \begin{bmatrix} A_{11} & A_{12} & A_{13} \\ A_{21} & A_{22} & A_{23} \\ A_{31} & A_{32} & A_{33} \end{bmatrix} \begin{bmatrix} Th_{ppm} \\ U_{ppm} \\ K_{pct} \end{bmatrix}$$

The matrix coefficients A_{11}, A_{22}, and A_{33} are the sensitivities measured as counts in the thorium, uranium and potassium windows per unit concentration of thorium, uranium and potassium. Expressing this matrix equation as three separate equations (and incorporating background count rates) we obtain for each pad

$$N_{2.62} = A_{11} Th_{ppm} + A_{12} U_{ppm} + A_{13} K_{pct} + B_T \quad (5)$$

$$N_{1.76} = A_{21} Th_{ppm} + A_{22} U_{ppm} + A_{23} K_{pct} + B_U \quad (6)$$

$$N_{1.46} = A_{31} Th_{ppm} + A_{32} U_{ppm} + A_{33} K_{pct} + B_K \quad (7)$$

where B_K, B_U and B_T are the background count rates arising from the radioactivity of the ground surrounding the pads, the radioactivity of the aircraft and equipment, plus the contribution from cosmic radiation and the radioactivity of the air. The calibration constants α, β, and γ, a, b, and g are then related to the various A_{IJ}'s by the equations

$\alpha = A_{21}/A_{11}$

$\beta = A_{31}/A_{11}$

$\gamma = A_{32}/A_{22}$

a $= A_{12}/A_{22}$

b $= A_{13}/A_{33}$

g $= A_{23}/A_{33}$

Each of these equations (5, 6 and 7) have four unknowns and consequently from measurements on all five calibration pads the unknowns can be evaluated. Grasty and Darnley (1971) used a standard least squares procedure but have simplified these equations and assumed that there is no interference in the uranium and thorium windows from pure potassium and that any uranium present has little influence on the thorium window i.e. the values of a, b and g are 0. In a more recent paper, a is given a value of 0.05 (Grasty, 1975a). Stromswold and Kosanke (1977) used combinations of the pads to arrive at several unique solutions. They then take a weighted average of these results, the weights depending on the estimated accuracy of the individually calculated calibration constants.

It is interesting to note that from the results of background flights over lakes (Fig. 10B.11), the increase in the potassium channel per unit increase in the uranium channel provides a good estimate of the value of the calibration constant γ. Grasty (1975b) has also shown how the value of α can be estimated in areas where the thorium/uranium ratio is high.

The calibration constants derived from the use of these pads are for infinite sources at ground level. Due to Compton scattering in the air the uranium stripping ratio alpha will increase with altitude. Grasty (1975a) derived an analytical solution for the increase of the uranium stripping ratio with altitude. A similar increase with aircraft altitude has been calculated by Løvborg et al. (1977). In the range of altitudes from 50 to 300 m this increase can be approximated by a straight line (Grasty, 1976b).

Altitude Correction

Corrections have to be made to the detector count rates depending on the altitude (h) of the aircraft above the ground. In the range of altitudes normally encountered in airborne survey operations the count rates in each window can be adequately represented by a simple exponential expression in the form

$$N = Ae^{-\mu h} \qquad (8)$$

where A and $\bar{\mu}$ are constants (Darnley et al., 1968; Kogan et al., 1971; Burson, 1973). Figure 10B.13 shows the stripped and background corrected potassium count rate variation with aircraft altitude and the two curves given by Equation (8) and the theoretical expression Equation (2) (Grasty, 1976b).

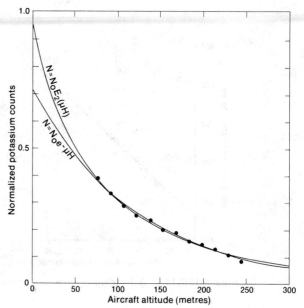

Figure 10B.13. *Potassium variation with aircraft altitude.*

Table 10B.9

Estimated calibration constants for a variety of sodium iodide detectors

Crystal Dimensions (mm)	Stripping Ratio								
	Thorium-into-Uranium (α)			Thorium-into-Potassium (β)			Uranium-into-Potassium (γ)		
	0 m	50 m	125 m	0 m	50 m	125 m	0 m	50 m	125 m
102x102 (4x4 inch)	0.47	0.50	0.53	0.65	0.68	0.73	0.99	1.02	1.07
152x102 (6x4 inch)	0.41	0.44	0.47	0.57	0.59	0.63	0.94	0.97	1.01
229x102 (9x4 inch)	0.39	0.41	0.44	0.52	0.55	0.58	0.90	0.93	0.97
292x102 (11.5x4 inch)	0.37	0.40	0.43	0.50	0.52	0.56	0.88	0.91	0.95

Spectral sensitivities per detector

Crystal Dimensions (mm)	Sensitivity in Counts/Sec Per Radioelement Concentration Unit								
	Potassium Window 1 pct K			Uranium Window 1 ppm eU			Thorium Window 1 ppm eTh		
	0 m	50 m	125 m	0 m	50 m	125 m	0 m	50 m	125 m
102x102 (4x4 inch)	5.7	3.1	1.4	0.52	0.30	0.15	0.26	0.16	0.086
152x102 (6x4 inch)	13	7.1	3.4	1.2	0.70	0.36	0.62	0.39	0.21
229x102 (9x4 inch)	29	16	7.9	2.7	1.6	0.83	1.4	0.90	0.50
292x102 (11.5x4 inch)	47	27	13	4.4	2.6	1.4	2.3	1.5	0.84

Exponential attenuation coefficients for survey heights of 50m and 125m

Crystal Dimensions (mm)	Height Attenuation Coefficients Per Metre x 10^2							
	Potassium Window		Uranium Window		Thorium Window		Total-Count Window	
	50 m	125 m	50 m	125 m	50 m	125 m	50 m	125 m
102x102 (4x4 inch)	1.22	0.97	1.09	0.86	0.96	0.76	0.91	0.73
152x102 (6x4 inch)	1.19	0.95	1.06	0.85	0.94	0.75	0.89	0.72
229x102 (9x4 inch)	1.15	0.94	1.03	0.83	0.90	0.74	0.86	0.71
292x102 (11.5x4 inch)	1.12	0.93	1.01	0.82	0.88	0.73	0.84	0.70

Conversion to Ground Concentration

From flights over a test strip of known ground concentration the sensitivity of the spectrometer in terms of counts per unit concentration per unit time can be readily obtained. The United States Department of Energy has selected a calibration strip near Las Vegas for the purpose of calibrating systems involved in the U.S. National Uranium Reconnaissance program (Geodata International Inc., 1977). The mean concentrations of this strip are 2.4 per cent potassium, 2.8 ppm uranium and 11.6 ppm thorium. This calibration strip suffers from the fact that the ground concentration is not uniform and different concentrations must be assigned to the strip depending on the aircraft altitude and particular detector configuration. The Geological Survey of Canada test strip, a few miles from Ottawa, has concentrations of 2.0 per cent potassium, 0.9 ppm uranium, and 7.7 ppm thorium (Grasty, 1975c; Grasty and Charbonneau, 1974). Because of the low uranium concentration, this strip is not ideal for the accurate calibration of the uranium channel.

Løvborg et al. (1977) have calculated the sensitivities, exponential height correction parameters and stripping ratios, α, β, and γ for four different cylindrical sodium iodide detectors which are 10 cm thick. These results are presented in Table 10B.9 and serve as a useful guide in the design of airborne systems. When the calibration constants for a particular system are known, it is a relatively simple matter to convert the airborne count rates to ground concentrations (Grasty, 1977).

Recommendations for Future Work

Probably one of the most difficult problems to overcome in providing reliable and consistent airborne gamma ray spectrometry data is due to the presence of radon and its decay products in the air. This is particularly true in areas where lakes cannot be found. A systematic study of the use of upward-looking crystals could provide valuable information on the best way of utilizing this particular technique.

In areas with large variations in topographic relief it is common practice to remove the effect of the varying cosmic ray component by monitoring a cosmic ray window from 3-6 MeV. A problem in utilizing this procedure arises because of the low count rate observed in this particular window. A possible procedure which would allow more frequent and accurate updates of the cosmic ray component could be to relate it directly to a barometric altimeter.

Considerable effort is now being spent in uranium reconnaissance programs and it is essential to use systems which are properly calibrated in order that the results from the different systems can be compared. There is considerable difficulty in finding suitable calibration strips which are readily accessible. A possible solution which warrants further attention is to utilize the calibration pads and simulate the absorption effects of the air by covering them with material such as plywood sheet. This technique could well prove to be the most reliable and accurate technique for evaluating the calibration constants for an airborne detection system.

REFERENCES

Beck, H.
    1972: The physics of environmental gamma radiation fields; Proceedings of the Second International Symposium on the Natural Radiation Environment, Houston, Texas. J.A.S. Adams, W.M. Lowder, and T.F. Gesell (eds.).

Beck, H., DeCampo, J., and Gogolak, C.
    1972: In situ Ge(Li) and NaI(Tl) gamma-ray spectrometry; Rept. HASL-258, U.S. Atomic Energy Comm.

Beck, H. and de Planque, G.
    1968: The radiation field in air due to distributed gamma-ray sources in the ground; Rept. HASL-195, U.S. Atomic Energy Comm.

Becquerel, H.
    1896a: Sur les radiations invisibles émises par les corps phosphorescents; C.R. Acad. Sci., Paris, v. 122, p. 500-503.

    1896b: Sur les radiations invisibles émises par les sels d'uranium; C.R. Acad. Sci., Paris, v. 122, p. 689-694.

Burson, Z.G.
    1973: Airborne surveys of terrestrial gamma radiation in environmental research; IEEE Transactions on Nuclear Science, v. NS-21 (1).

Burson, Z.G., Boyns, P.K., and Fritzsche, A.E.
    1972: Technical procedures for characterizing the terrestrial gamma radiation environment by aerial surveys; Proceedings of the Second International Symposium on the Natural Radiation Environment, Houston, Texas. J.A.S. Adams, W.M. Lowder, and T.F. Gesell (eds.).

Campbell, N.R.
    1907: The β rays from potassium; Proc. Camb. Phil. Soc., v. 14, p. 211.

Campbell, N.R. and Wood, A.
    1906: The radioactivity of the alkali metals; Proc. Camb. Phil. Soc., v. 14, p. 15.

Cook, J.C.
    1952: An analysis of airborne surveying for surface radioactivity; Geophysics, v. 17 (4), p. 687-706.

Curie, S.
    1898: Rayons emis par les composés de l'uranium et du thorium; C.R. Acad. Sci., Paris, v. 126, p. 1101-1103.

Dahl, J.B. and Ødegaard, H.
    1970: Areal measurements of water equivalent of snow deposits by means of natural radioactivity in the ground; in Isotope Hydrology, IAEA Vienna, Austria, p. 191-210.

Darnley, A.G., Bristow, Q., and Donhoffer, D.K.
    1968: Airborne gamma-ray spectrometer experiments over the Canadian Shield; in Nuclear Techniques and Mineral Resources (International Atomic Agency, Vienna), p. 163-186.

Darnley, A.G. and Grasty, R.L.
    1970: Mapping from the air by gamma-ray spectrometry; Proc. Third International Geochemical Symposium, Toronto, Can. Inst. Min. Met. Spec. Vol. 11, p. 485-500.

Duval, J.S., Jr., Cook, B., and Adams, J.A.S.
    1971: Circle of investigation of an airborne gamma-ray spectrometer; J. Geophys. Res., v. 76, p. 8466.

Eve, A.S.
    1911: On the ionization of the atmosphere due to radioactive matter; Phil. Mag., v. 21, p. 26.

Foote, R.S.
1968: Improvement in airborne gamma radiation data analyses for anomalous radiation by removal of environmental and pedologic radiation changes; in Symposium on the Use of Nuclear Techniques in the Prospecting and Development of Mineral Resources, International Atomic Energy Meeting, Buenos Aires.

Geodata International Inc.
1977: Lake Mead dynamic test range for calibration of airborne gamma radiation measuring systems; ERDA Report GJBX46(77).

Gockel, V.A.
1910: Luftelektrische beobachtungen bei einer Ballonfahrt; Phys. Zeit., v. 11, p. 280-282.

Godby, E.A., Connock, S.H.G., Steljes, J.F., Cowper, G., and Carmichael, H.
1952: Aerial prospecting for radioactive materials; Nat. Res. Coun. Lab. Joint Rept., MR-17 CRR-495, 90 p.

Gold, S., Barkham, H.W., Shlien, B., and Kahn, B.
1964: Measurement of naturally occurring radionuclides in air; The Natural Radiation Environment, Univ. Chicago Press, p. 369-382.

Grasty, R.L.
1973: Snow-water equivalent measurement using natural gamma emission; Nord. Hydr., v. 4, p. 1-16.

1975a: Uranium measurement by airborne gamma-ray spectrometry; Geophysics, v. 40 (3), p. 503-519.

1975b: Uranium stripping determination au naturel for airborne gamma-ray spectrometry; in Report of Activities, Part A, Geol. Surv. Can., Paper 75-1A, p. 87.

1975c: Atmospheric absorption of 2.62 MeV gamma-ray photons emitted from the ground; Geophysics, v. 40 (6), p. 1058-1065.

1976a: The circle of investigation of airborne gamma-ray spectrometers; in Report of Activities, Part B, Geol. Surv. Can., Paper 76-1B, p. 77-79.

1976b: A calibration procedure for an airborne gamma-ray spectrometer; Geol. Surv. Can., Paper 76-16, p. 1-9.

1977: A general calibration procedure for airborne gamma-ray spectrometers; in Report of Activities, Part C, Geol. Surv. Can., Paper 77-1C, p. 61-62.

Grasty, R.L. and Charbonneau, B.W.
1974: Gamma-ray spectrometer calibration facilities; in Report of Activities, Part B, Geol. Surv. Can., Paper 74-1B, p. 69-71.

Grasty, R.L. and Darnley, A.G.
1971: The calibration of gamma-ray spectrometers for ground and airborne use; Geol. Surv. Can., Paper 71-17, 27 p.

Grasty, R.L. and Holman, P.B.
1974: Optimum detector sizes for airborne gamma-ray surveys; in Report of Activities, Part B, Geol. Surv. Can., Paper 74-1B, p. 72-74.

Hess, V.F.
1911: Uber die Absorption der γ-strahlen in der Atmosphare; Phys. Zeit., v. 12, p. 998-1001.

1912: Uber beobachtungen der durchdringenden Strahlung bei sieben Freiballonfahrten; Phys. Zeit., v. 13, p. 1084-1091.

Hubbell, J.H. and Berger, M.J.
1968: Attenuation coefficients, energy absorption coefficients, and related quantities; Engineering Compendium on Radiation Shielding, v. 1, Springer Verlag, Berlin.

Kellogg, W.C.
1971: Calculation of airborne radioactivity survey responses: theory, method, and field test; Annual Meeting of A.I.M.E., New York.

Killeen, P.G.
1979: Gamma ray spectrometric methods in uranium exploration: Application and interpretation; in Geophysics and Geochemistry in the Search for Metallic Ores; Geol. Surv. Can., Econ. Geol. Rept. 31, Paper 10C.

King, L.V.
1912: Absorption problems in radioactivity; Phil. Mag., v. 23, p. 242.

Kirkegaard, P.
1972: Double-P_1 calculation of gamma-ray transport in semi-infinite media; Risø-M-1460.

Kirkegaard, P. and Løvborg, L.
1974: Computer modelling of terrestrial gamma-radiation fields; Risø Report No. 303.

Kogan, R.M., Nazarov, I.M., and Fridman, Sh.D.
1971: Gamma spectrometry of natural environments and formations — theory of the method, applications to geology and geophysics; Israel Program for Scientific Translations, 5778, Jerusalem.

Kolhorster, W.
1928: Gammastrahlen an Kaliumsalzen; Naturwiss, v. 16 (2), p. 28.

Løvborg, L., Grasty, R.L., and Kirkegaard, P.
1977: A guide to the calibration constants for aerial gamma-ray surveys in geoexploration; American Nuclear Society Symposium on Aerial Techniques for Environmental Monitoring, Las Vegas.

Løvborg, L. and Kirkegaard, P.
1974: Response of 3"x3" NaI(Tl) detectors to terrestrial gamma radiation; Nuclear Instruments and Methods 121, p. 239-251, North-Holland Publishing Co.

Løvborg, L., Kirkegaard, P., and Mose Christiansen, E.
1976: The design of NaI(Tl) scintillation detectors for use in gamma-ray surveys of geological sources; IAEA Symposium on Exploration of Uranium Ore Deposits, Vienna.

Løvborg, L., Kirkegaard, P., and Rose-Hansen, J.
1972: Quantitative interpretation of the gamma-ray spectra from geologic formations; Proc. of the Second Int. Symp. on the Natural Radiation Environment, Houston, Texas, J.A.S. Adams, W.M. Lowder, and T.F. Gesell (eds.).

McSharry, P.J.
1973: Reducing errors in airborne gamma ray spectrometry; Austr. Soc. Explor. Geophys. Bull., v. 4 (1), p. 31.

Purvis, A.E. and Buckmeier, F.J.
1969: Comparison of airborne spectral gamma radiation data with field verification measurements; Proc. Sixth Int. Symp. on Remote Sensing of Environment, v. 1.

Rutherford, E.
1903: The magnetic and electric deviation of the easily absorbed rays from radium; Phil. Mag., v. 5, p. 177.

Schmidt, G.C.
  1898: Über die von den Thorverbindungen and einigen anderen substanzen ausgehende Strahlung; Ann. d. Phys. u. Chem., v. 65, p. 141.

Smith, A.R. and Wollenberg, H.A.
  1972: High-resolution gamma ray spectrometry for laboratory analysis of the uranium and thorium decay series; Proc. 2nd Int. Symp. on the Natural Radiation Environment, Houston, Texas. J.A.S. Adams, W.M. Lowder, and T.F. Gesell (eds.). U.S. Dept. of Commerce, Springfield, Virginia, p. 181-231.

Stromswold, D.C. and Kosanke, K.L.
  1977: Calibration and error analysis for spectral radiation detectors; IEEE Nuclear Science Symp., San Francisco, California.

Strutt, R.J.
  1903: On the intensely penetrating rays of radium; Proc. Roy. Soc., v. 72, p. 208.

Villard, M.P.
  1900: Sur le rayonnement du radium; Comptes Rendus, v. 130, p. 1178.

Ward, D.L.
  1978: Construction of calibration facility Walker Field, Grand Junction, Colorado; ERDA Open File GJBX-37(78).

Wulf, V.T.
  1910: Beobachtungen uber die Strahlung hoher durchdringungsfahigkeit auf dem Eiffelturm; Phys. Zeit, v. 11, p. 810.

GAMMA RAY SPECTROMETRIC METHODS IN URANIUM EXPLORATION – APPLICATION AND INTERPRETATION

P.G. Killeen
Geological Survey of Canada

Killeen, P.G., Gamma ray spectrometric methods in uranium exploration – application and interpretation; in Geophysics and Geochemistry in the Search for Metallic Ores; Peter J. Hood, editor; Geological Survey of Canada, Economic Geology Report 31, p. 163-229, 1979.

Abstract

One of the most significant advances in uranium exploration in recent years has been the development of gamma ray spectrometric techniques. A brief review of the planning and stages of a uranium exploration program using gamma ray spectrometry, and disposition of costs is given as a background to the more technical considerations.

To obtain the fullest advantage of the use of a gamma ray spectrometer for uranium exploration the factors influencing gamma ray measurements must be understood. This paper reviews and discusses such factors as radioactive disequilibrium in the uranium decay series and its ramifications; geometry of the measurements; counting statistics and their effect on accuracy of results and required counting times; background radiation; and calibration of all types of gamma ray spectrometers, including information on calibration facilities.

Reviews and discussions of procedures, including examples of various types of results, presentation formats, and interpretations are given, for gamma ray spectrometric surveys in airborne, surface, underwater and borehole environments.

Résumé

Un des progrès les plus importants dans l'exploration de l'uranium des dernières années a été la mise au point de techniques de spectrométrie à rayons gamma. Une analyse succincte de la planification et des étapes d'un programme d'exploration de l'uranium utilisant la spectrométrie à rayons gamma, et la ventilation des coûts sont données comme point de départ pour une analyse des considérations techniques en cause.

Il faut bien comprendre les facteurs influant sur les mesures faites aux rayons gamma, afin d'obtenir les meilleurs résultats de l'emploi du spectromètre à rayons gamma pour l'exploration de l'uranium. Ce rapport analyse les facteurs comme le déséquilibre dans la famille radioactive de l'uranium et ses ramifications; la géométrie des mesures; les statistiques des calculs et leur effet sur la précision des résultats et sur les temps de calcul; la radiation de fond et l'étalonnage de tous les genres de spectromètres à rayons gamma, y compris l'information sur les installations d'étalonnage.

Le texte procède à l'analyse de marche à suivre, y compris des exemples de divers types de résultats, de modèles de présentation et des interprétations, pour les levés spectrométriques à rayons gamma effectués par voie d'air, en surface, sous l'eau et dans les sondages.

INTRODUCTION

Gamma ray spectrometric surveys of all types form only part of a complex series of interrelated investigations which are referred to as 'uranium exploration'. Before the application and interpretation of gamma ray spectrometry can be discussed, some of the stages of exploration should be considered in order to appreciate where each type of survey may be utilized most profitably.

Exploration Stages

The International Atomic Energy Agency in a report of a panel (IAEA, 1973a) summarized diagrammatically the various exploration methods and programs leading to the discovery of uranium. Stage I includes the selection of the region based on geologic considerations, and the collection of all available data such as regional geological maps, air photos, and topographic maps. Other considerations in the preliminary decision making of stage I are whether the surface exposure is large or small, and whether a multi-element exploration program requiring a geochemical survey is desired. If the surface exposure is large and uranium is the only target, a radiometric survey is warranted as a first step. Stage II, the progressive reduction of search areas, may be based on either a geochemical or radiometric approach. If the ground accessibility is poor or the area large, an airborne survey may be the best approach for this stage. The progressive reduction of search areas based on airborne radiometric surveys is shown in Figure 10C.1 (IAEA, 1973a). This figure indicates the input to the decision making process which results in the selection of an appropriate type of airborne radiometric survey. The method selected may vary from detailed total count (scintillometer) surveys flown over small areas with closely spaced flight lines, to large scale, reconnaissance, high-sensitivity gamma ray spectrometric surveys flown with widely spaced flight lines to provide regional coverage. The desired result of the survey is usually the identification of some form of anomaly or anomalous area which must then be further evaluated. This leads to the surface surveys and ground investigations of stage III. The ground investigations to evaluate these anomalous areas are summarized in Figure 10C.2 (after IAEA, 1973a). Again depending on the size of the area and ease of access, the appropriate type of vehicle-borne gamma ray surveys or foot traverses may be chosen. The final stage of evaluation includes sampling, trenching, drilling, and borehole gamma ray logging.

Other such "stages" in the exploration for minerals of various types have been proposed (e.g. Holmes, 1978), and they all vary in detail. However, the various stages mentioned above relate specifically to uranium exploration. The application and interpretation of the various types of gamma ray surveys (both total count and spectrometric) are interrelated. For example a discussion of the problem of 'geometry' with respect to surface gamma ray spectrometric measurements is also instructive for those interested in

Table 10C.1

Cost comparisons for various radiometric exploration surveys, on both a distance and time basis. Actual costs have increased through inflation but the data are still a good indication of relative costs. (Data from IAEA, 1973b)

		Instrument Cost and Cost Per Unit Distance And Area Covered		Distance or Area Covered Per Unit Time		Total Cost of an Average Minimum Program	Effective Coverage of Area: Comments
		Normal Low	Normal High	Normal Low	Normal High		
Ground surveys by portable GM and scintillation counters	Instrument km km²	GM: $200 Scint: $355 $5/km $50/km²	GM: $600 Scint: $1400 $10/km $400/km²	— 5 km/d 0.25 km²/d	— 10 km/d 1 km²/d	$3000 for instruments plus labour and transportation. Minimum program: 10–20/km²	100 gm-cm rock and soil. 1–5 m each side of traverse line.
Gross-count (total-count) carborne surveys	Instrument km km²	$4000 $0.50/km $5.00/km²	$8000 $3.00/km $30.00/km²	— 50 km/d 5 km²/d	— 300 km/d 30 km²/d	$10 000 for instruments. Minimum program: 3000–5000 km or 500–1000 km²	100 gm-cm rock and soil. 5–10 m each side of traverse line.
Gross-count (total-count) airborne surveys	Instrument km	$12 000 $3.00/km	$60 000 $6.00/km	— 600 km/d	— 1200 km/d	$50 000	150-m-wide belt.
Portable and carborne gamma-ray spectrometer surveys	Instrument Point Stations	$3000 $3.00/station	$15 000 $6.00/station	— 10 min/station	— 20 min/station	$8000 for instruments and one month operation	5–10 m each side of station.
Airborne gamma-ray spectrometer surveys	Instrument km	$30 000 $7.50/km	$400 000 $25.00/km	— 600 km/d	— 1200 km/d	$200 000	150-m-wide belt.
Radon measurement surveys in soil and sub-soil	Instrument Samples	$1500 $1.00/sample	$6000 $15.00/sample	— 10–20 samples/d	— 100–300 samples/d	$3000 for equipment $300–$350/km²	1 m diameter–often less in clay or wet soil. Upward diffusion of radon in soil probably limited to 5–10 m.
Exploration drilling (including logging) (a) Rotary (b) Diamond drill (8-h shift)	Metres Metres	$1.50/m $15.00/m	$6.00/m $60.00/m	150 m/d 7.5 m/d	300 m/d 25 m/d	$50 000 $100 000	Depth and close or wide coverage of areas depending on hole spacing.

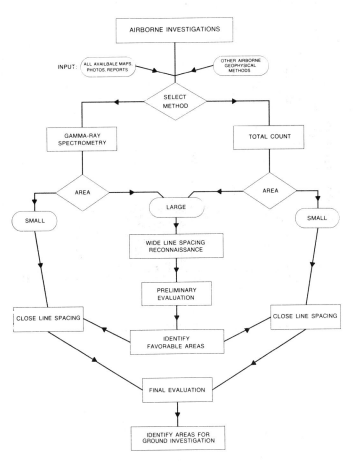

Figure 10C.1. The selection of an airborne radiometric survey for a uranium exploration program. This leads to identification of anomalous areas to be followed up by detailed investigation (after IAEA, 1973a).

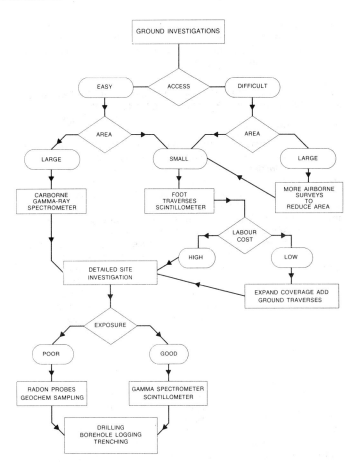

Figure 10C.2. The selection of a radiometric ground follow-up technique for detailed investigation of anomalous areas (after IAEA, 1973a).

either airborne or subsurface gamma ray spectrometry. The following is intended to tie together most of the present state-of-the-art information regarding the application and interpretation of gamma ray spectrometric methods in uranium exploration.

Exploration Costs

One additional consideration in the application of any of these methods is cost. The ideal sequence of radiometric exploration techniques may prove to be very costly, and a compromise must be reached in order to balance the size of the area to be searched against the thoroughness of the search. Since costs fluctuate widely from area to area, and with time, it is impossible to associate concrete cost figures with any given method. A useful compromise is the data presented in Table 10C.1, modified after a table given by the IAEA (1973b). The figures refer to average costs in 1972, primarily in North America. However inflation has increased these figures by about 75 per cent. It is the relative costs that are useful here, allowing cost per unit distance (or area) covered, and distance (or area) covered per unit time to be compared, for the various techniques. Low and high values are presented to account for variations in location and other conditions. Costs for radon measurements in soil are included because such surveys are also radiometric. A good breakdown of uranium exploration costs for several different areas has been given by Barnes (1972). His examples of the distribution of uranium exploration expenditures demonstrate some of the modifying factors and also indicate the percentage of the total exploration budget taken up by geophysics and geochemistry. This percentage increases as exploration moves to more remote areas with poor accessibility.

Prologue

The foregoing brief description of costs and stages in a uranium exploration program provides the background necessary to determine the sequence of different types of radiometric surveys for a successful search. The following five sections cover the application and interpretation of gamma ray spectrometric methods.

The common constraints and parameters for each technique are discussed in the next section. These include consideration of radioactive disequilibrium, geometry of the measurements, counting statistics, background, calibration and the parameters affecting calibration, and a description of existing calibration facilities.

The last four sections are specifically directed at reviews of the application and interpretation of: 1) airborne gamma ray spectrometric surveys; 2) surface gamma ray spectrometric surveys including portable (man carried) spectrometric surveys, carborne surveys, and snowmobile surveys; 3) underwater gamma ray spectrometric surveys, and finally 4) borehole gamma ray spectrometric logging. In these last four sections those problems which are unique to the survey mode, vehicle, environment or instrument of each type of gamma ray spectrometric survey are reviewed. This method of organization of the material should present the reader with a relatively efficient means of obtaining an overview of the state of the art in the application and interpretation of any given gamma ray spectrometric survey method for uranium exploration.

FACTORS INFLUENCING GAMMA RAY SPECTROMETRIC MEASUREMENTS IN URANIUM EXPLORATION

Introduction

It is difficult to make meaningful gamma ray spectrometric measurements, much less interpret them, without a thorough understanding of the factors which affect the measurements. Many of these influencing factors have been investigated in detail whereas others require additional evaluation. The meaning of 'radioactive equilibrium' and its importance as a basic assumption of gamma ray spectrometry will be described first. This is followed by discussion of the subject of geometry of the measurements; the effective sample volume; counting statistics; dead time and sum peaks; and background radiation. Calibration of gamma ray spectrometric equipment is described in detail including a review of the "why, how, when, and where" of calibration. References to the available literature regarding the design and construction of calibration facilities are included.

Radioactive Equilibrium

Radioactive equilibrium or disequilibrium is an important consideration in all gamma ray spectrometric measurements. Gamma ray spectrometry can be used to determine the concentrations of uranium, thorium, and potassium in a rock because gamma rays of specific energies are associated with each radioelement. By looking at peaks in the energy spectrum of gamma rays being emitted by the source, the radioelement content of the source can be inferred. The method involves the counting of gamma ray photons with specified energies, most conveniently those emitted by daughter products, bismuth-214 in the ^{238}U decay series and thallium-208 in the ^{232}Th decay series (see Figure 10C.3). The gamma ray count rate can then be related to the amount of parent, by assuming there is a direct relation between the amount of daughter and parent. This assumption is valid when the radioactive decay series is in a state of secular equilibrium.

A radioactive decay series such as that of ^{238}U is said to be in a state of secular equilibrium when the number of atoms of each daughter being produced in the series is equal to the number of atoms of that daughter being lost by radioactive decay.

The rate of loss by decay is proportional to the amount of radioelement present, for example:

$$\frac{dN_1}{dt} = -\lambda_1 N_1 \qquad (1)$$

where N_1 = the amount of element 1 and λ_1 = the decay constant for the element 1.

In a radioactive decay series, N_1, is decaying into N_2 at the above rate while at the same time N_2 is decaying with the decay constant λ_2 into N_3 and so on. If the parent has a relatively long half life, after a long period of time the amount of any given daughter becomes constant. The rate of production from its parent is equal to its rate of decay. The series is then in a state of secular equilibrium.

For a radioactive decay series, secular equilibrium implies that

$$\lambda_1 N_1 = \lambda_2 N_2 = \lambda_3 N_3 = \ldots\ldots\ldots \lambda_n N_n \qquad (2)$$

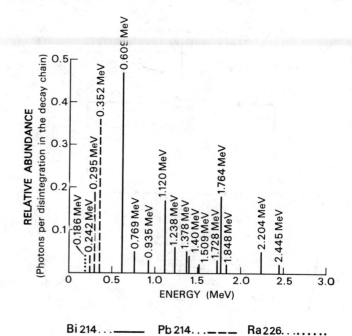

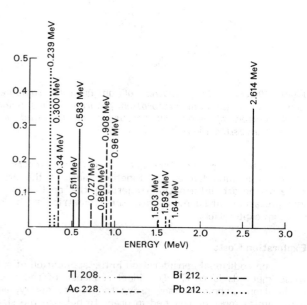

Figure 10C.3. Theoretical energy spectrum of principal gamma rays emitted by the ^{238}U decay series (top) and by the ^{232}Th decay series (bottom).

When this condition is obtained it is possible to determine the amount of the parent of the decay series by measuring the radioactivity from any daughter element.

The question then is whether the assumption of secular equilibrium, required for analysis by gamma ray spectrometric techniques, is valid for the geologic material being analyzed for its uranium content.

If one or more of the daughter products is being lost by any process other than radioactive decay, or if the parent was not deposited too long ago, equation (2) is not satisfied. Since each daughter product is an element with its own characteristic physical and chemical properties it may behave differently within a given environment. For example, in

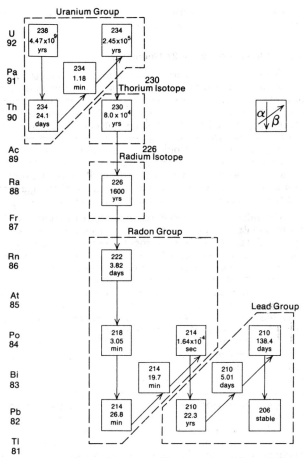

Figure 10C.4. Classification of natural radioactivity of the ^{238}U decay series into groups of isotopes, with respect to their state of radioactive equilibrium (after Rosholt, 1959).

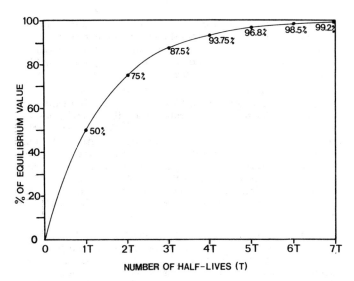

Figure 10C.5. Length of time (in half-lives; T) taken by a radioactive isotope to reach its radioactive equilibrium value with its parent.

the ^{238}U decay system there is a gas, radon-222, with a 3.85 day half life. The solubilities of radium, uranium and thorium differ, and preferential leaching of elements may occur. One of the most complete references on this subject is by Rosholt (1959) who subdivided the uranium decay series into five separate groups, as shown in Figure 10C.4. The elements within each group tend to remain in equilibrium with the parent of the group, although the parent of the group may not itself be in equilibrium with the parent of the decay series, ^{238}U.

How long does it take for secular equilibrium to become established? This is dependent on the half life of the longest-lived daughter in the decay chain below the parent. For example if uranium in solution moves into a chemically reducing environment such as a swamp and accumulates in substantial amounts over a short time, it will be relatively undetectable by gamma ray counting for considerable time. The daughter products must be given time to build up into their equilibrium proportions. If it is assumed (and it reasonably can be) that U isotopes travel in approximately equilibrium proportions then the length of time is controlled by the ^{230}Th with a half life of 80 000 years. The daughters below ^{230}Th have shorter half lives, and will remain in equilibrium with the ^{230}Th as it builds up from decay of the ^{234}U. Fifty per cent of the equilibrium amount of ^{230}Th will be attained in 80 000 years. Another half life will contribute 1/4 of the equilibrium amount, making a total of 75 per cent. A third half life contributes 1/8 for a total of 87.5 per cent and so on. This is illustrated in Figure 10C.5.

It can be seen from this figure that after seven half lives (about 560 000 years in this case) the daughter products will reach over 99 per cent of the equilibrium value. For most practical purposes gamma ray spectrometry can be used to determine the amount of parent ^{238}U by counting 1.76 MeV gamma rays from the daughter ^{214}Bi after about 300 000 years (four half lives). ^{238}U itself does not emit gamma rays and so cannot be detected directly. ^{214}Bi is usually chosen because some of its gamma rays are of high energy, and have less interference from gamma rays of similar energies than other daughter isotopes. High resolution solid state detectors make it possible to separate and measure gamma ray peaks which are indistinguishable using sodium iodide or other scintillators. In this way daughter isotopes higher in the decay series can be utilized (see e.g. Tanner et al., 1977a).

Another example of the need to know the length of time to reach equilibrium is the case of rock samples which are crushed for analysis by laboratory gamma ray spectrometry. The crushing operation releases radon and the sample must be placed in a sealed container and allowed to return to equilibrium. In this case it is the radon group as shown in Figure 10C.4 which is out of equilibrium with the rest of the decay series. The ^{210}Pb (half life = 22 years) below the radon group will not be affected by the missing radon if only a short time elapses before the sample is sealed. Again, in about seven half lives of the longest lived member of the radon group (3.85 days; ^{222}Rn) equilibrium will be re-established, i.e. in less than 28 days. In practice it is highly unlikely that <u>all</u> the radon in the sample was lost during crushing. Thus equilibrium could be established in 5 or 6 half lives, or even less. One method to determine this is to analyze the sample, wait a week or so, and re-analyze the sample. If the gamma ray spectrometric analysis is higher the second time, equilibrium was not established at the time of the first analysis. In fact with appropriate calculations, from two such analyses, the final analysis at 100 per cent equilibrium could be computed, without actually waiting for the full 28 days. These calculations were described in detail by Scott and Dodd (1960).

The thorium decay series can generally be assumed to be in radioactive equilibrium. The longest lived daughter in the thorium series is ^{228}Ra with a half life of 6.7 years. Seven half lives, totaling less than 50 years, is a geologic 'instant', and any redistribution of thorium would be followed by a relatively rapid re-establishment of secular equilibrium. Thus in practically all geological samples the amount of parent ^{232}Th can be computed by measuring the 2.62 MeV gamma ray activity of the daughter ^{208}Tl.

In conventional gamma ray spectrometry, the actual elements being measured are ^{214}Bi and ^{208}Tl. If results are expressed in terms of count rates, then it is sometimes the practice to label them as the ^{214}Bi and ^{208}Tl cps. This latter terminology has been used in radiometric results of the U.S. NURE program. However it is preferable that the results are expressed in terms of <u>equivalent</u> uranium and <u>equivalent</u> thorium concentrations, respectively (IAEA, 1976). For this reason determinations of U and Th by gamma ray spectrometry are denoted by a prefixed 'e' (e.g. eU, eTh, and the U/Th ratio becomes the eU/eTh ratio).

Radioactive disequilibrium is accepted as the general case in roll front or sandstone-type uranium deposits. The reason is that uranium is mobile within the sandstone and daughter product formation lags behind. This leads to a distribution of radioelements wherein the daughter products (e.g. ^{214}Bi) are left behind, creating a daughter-excess or parent-deficiency state, with strong gamma ray activity, while at some nearby location there is a (relatively) weakly radioactive uraniferous zone with a daughter-deficiency. For most other rocks, not a great deal is known about the state of radioactive equilibrium in general, although studies of specific areas have been done (e.g. Richardson, 1964). Disequilibrium investigations of the Elliot Lake uranium mining area by Ostrihansky (1976) showed that disequilibrium can occur on a small scale along joints or fractures. Killeen and Carmichael (1976) pointed out that the problem of radioactive disequilibrium is minimized by large sample volumes. Whereas a hand specimen taken from an outcrop might show radioactive disequilibrium, an in situ assay by portable gamma ray spectrometer on the same outcrop comprises such a large sample it may be effectively in equilibrium. i.e. the parents and daughters may have moved apart on the scale of a hand specimen, but not on the scale of a cubic metre of rock. Similarly both parent and daughter nuclides, even if separated on a small scale, could be included in the large surface area "seen" by an airborne gamma ray spectrometer making it more likely that the equilibrium assumption is valid.

Geometry of the Radiometric Measurement

Since the amount of radiation which can be detected is related to the size and shape of the radiometric source as well as its intensity, the so-called "geometry" must be taken into consideration. Generally the angular measurement or solid angle which the source subtends at the detector is used as a reference, where 4π steradians is equivalent to complete enclosure of the detector by the source. For example, in a laboratory the radiation may be collimated by lead bricks or other absorbers such that the source only subtends an angle of a few degrees at the detector. A detector placed above an 'infinite' planar source, would be an example of 2π geometry. Some considerations of geometry of measurements were discussed by Gregory and Horwood (1961, 1963) with respect to laboratory measurements.

The calibration of any gamma ray spectrometer is for a specific geometry. For this reason it is important to know the geometry of the measurements. A hand specimen of 1 per cent U_3O_8 ore will produce a much lower count rate than a rock outcrop averaging 1 per cent U_3O_8, even if the detector is located at the same distance away from the source. Therefore, to make some sense out of recorded count rates, the geometry of the measurement must be noted. A few examples are given below to illustrate situations in which the geometry of a measurement can change to produce a false anomaly, or to mask a real anomaly.

Surface Measurements — 2π Geometry

Generally radiometric measurements made above the surface of the earth by either hand-held portable gamma ray detectors, vehicle-mounted detectors, or airborne detectors are considered to be measurements in a 2π geometry. The instruments should be calibrated on flat calibration sources of effectively infinite diameter, such as the concrete pads discussed in the section on calibration facilities. A measurement made in a trench or near a cliff effectively changes the geometry such that the calibration factors are invalid. A correction factor may be applied if the geometry is measurable, however this is most often not the case.

Consider the portable gamma ray spectrometer field measurements made with the geometry shown in Figure 10C.6a. Assuming the spectrometer was calibrated in 2π geometry, the results will be valid and an in situ assay of the surface can be calculated from the measured count rates. In Figure 10C.6b, the detector is located near a cliff face of the same material. This 3π geometry would produce a count rate 50 per cent higher than the 2π geometry because of the additional radiation "seen" by the detector. Doig (1968) reported measuring the expected 50 per cent increase in count rate in moving the detector from a 2π (flat) to 3π (cliff) geometry. Wormold and Clayton (1976) have carried out an extensive study of the effects of geometry on measurements by a portable gamma ray spectrometer, with a view to applying corrections for measurable geometries differing from the calibration geometry. One such situation occurs when making measurements along the benches of an open-pit mine. Here the geometry is fairly well known. A worse case is shown in Figure 10C.6c where the detector is placed in a trench or gully and the geometry is approaching 4π. The geometry shown in Figure 10C.6b or c can occur during a carborne survey, if the vehicle passes through a road cut. Figure 10C.6d shows a fourth geometrical configuration of less than 2π steradians such as that found in measurements

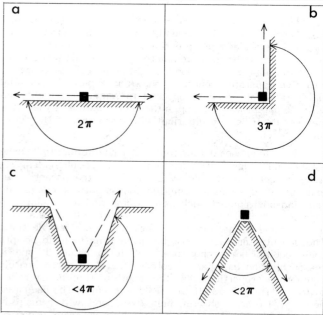

a) 2π geometry   c) approximately 4π geometry
b) 3π geometry   d) less than 2π geometry

Figure 10C.6. *Four different source-detector geometries encountered in surface gamma ray spectrometric measurements in the field.*

Gamma-Ray Interpretation

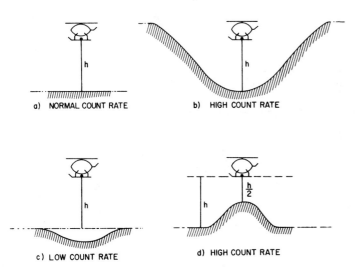

Figure 10C.7. The effect of source geometry and height upon count-rate for airborne gamma ray spectrometric surveys (after Grasty, 1976a). Note how a topographic depression can cause either a high count rate as in (b) or a low count rate as in (c), depending on whether the source-detector geometry has effectively changed.

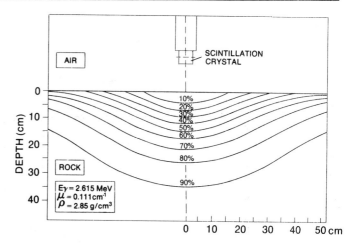

Figure 10C.8. The sample volume for an in situ assay by portable gamma ray spectrometer as computed by Løvborg et al. (1969). Curved lines represent nearly-hemispherical shells contributing 10 per cent of the detected gamma radiation for the case of the 2.62 MeV gamma rays from the ^{232}Th decay series. (Values computed using attenuation $\mu = 0.111 \text{ cm}^{-1}$, $\rho = 2.85 \text{ g/cm}^3$; typical of the Illimaussaq intrusion in Greenland.)

made on a hill or ridge. False anomalies may be recorded if the operator does not make a note of the changes in geometry which cause increased count rates. A good operator can save considerable time and expense in later follow-up investigations if proper field notes are kept.

A solution to problems caused by geometry is the use of a lead shield to control the geometry. Mahdavi (1964) described the use of shielding to measure radioelement concentrations in Texas coast beach sands, and Løvborg et al. (1969) employed a lead-shielded detector for a gamma ray spectrometer used for in situ surface analysis of the rough terrain of the Illimausaq intrusion in Greenland.

Lead shielding to control geometry in a carborne survey is highly impractical, however it may be used beneath the detector to shield it from the road bed if the latter is known to contain non-locally derived material which may be the main source of radioactivity being detected.

Airborne Measurements — 2π Geometry

Generally speaking there is not much variation from 2π geometry for an airborne survey of the reconnaissance type if the terrain is not too rugged. Large fixed-wing aircraft carrying large detector arrays for reconnaissance work however are less manoeuverable than the smaller aircraft or helicopters usually used in detailed follow-up surveys. If the terrain becomes rugged, maintaining a constant elevation is difficult and may become impossible with a fixed wing aircraft. Aircraft elevation corrections may be applied, but geometry corrections may usually be applied only in a qualitative sense or at best semi-quantitatively. Figure 10C.7 (after Grasty, 1976a) illustrates the effect of source geometry and height upon count rate. As shown in this figure the aircraft is able to maintain constant elevation (h) over flat ground (10C.7a) or over broad topographic changes (10C.7b), but over sharp hills (10C.7d) or narrow valleys (10C.7c) the height will be less than h and more than h respectively (Fig. 10C.7). The height correction factors can compensate for this, but not for the changes in geometry. In these last two cases, the geometrical configuration is less than 2π over hills and greater than 2π in valleys, and geometry considerations similar to those for portable gamma ray spectrometer measurements will be applicable.

Borehole Measurements — 4π Geometry

It can easily be seen that the "best" possible geometry is obtained when the source completely surrounds the detector, as in the case of a gamma ray probe inside a borehole. The source effectively surrounds the detector except for a narrow solid angle above and below it.

In an air-filled borehole, the geometry will still be 4π even if a large cave-in or wash-out is encountered, because the rock surrounds the detector. However in a water-filled hole the attenuation of the additional water in the enlarged portion of the hole must be taken into consideration. This is discussed in the section on borehole logging.

The Effective Sample Volume

The size of the sample being analyzed in a field measurement has been discussed by several authors. Dodd and Eschliman (1972) considered the borehole measurement case for total count logging surveys, Løvborg et al. (1971) did considerable work with respect to portable gamma ray spectrometers and Grasty et al. (in press) considered the sample volume of airborne radiometric measurements. Several factors can affect the sample volume, such as the energy of gamma radiation being measured, the density of the source material (rock, overburden etc.), the absorption coefficient of the material and whether the detector is moving or stationary. The definition of sample volume is also important. For example it could be arbitrarily defined as the volume within which 90 per cent of the detected gamma rays originated.

The sample volume for an in situ assay by portable gamma ray spectrometer is shown in Figure 10C.8 (Løvborg et al., 1969). The sample volume for a total count gamma ray log is illustrated in Figure 10C.9, where the distribution of gamma ray sources detected during any given sample interval is shown. This is intended to demonstrate the difficulty of defining the sample 'volume'. In addition the volume includes a cylinder of length L if the detector moves a distance L during the time of measurement. In an airborne gamma ray spectrometric survey, the volume of the sample is the product of the surface area "seen" by the airborne detector, and the thickness of the source material. The sample volume

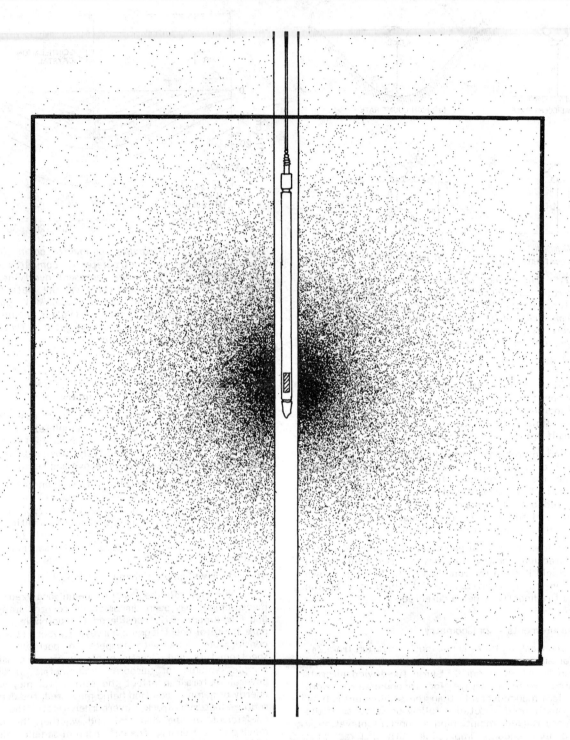

Figure 10C.9. Sample volume for a total count gamma ray borehole log as shown by the distribution of gamma ray sources detected during any given sample interval Δt by a detector located in a borehole penetrating a homogeneous radioactive zone. The difficulty in defining the boundaries of the 'sample volume' is apparent from this figure (after Conaway and Killeen, 1979).

for an airborne measurement will depend to a large extent on the height of the aircraft above the ground. This affects the area of the surface which is analyzed. The area of the surface has been called the field of view (F.O.V.), instantaneous field of view (I.F.O.V.), circle of investigation, and area of influence, by various authors. The size of the sample area is a moot point at present, and there is some dispute as to whether the width of the area supplying a given percentage of the radiation detected can be considered the same for measurements made with the aircraft moving or stationary (see Fig. 10C.10, after Grasty et al., in press). Usually the computations are made on the basis of a stationary detector at a fixed height (e.g. see Fig. 10C.11, after Duval et al., 1971). Grasty (1979) has also shown that the diameter of the circular area beneath the aircraft contributing a given percentage of the gamma radiation

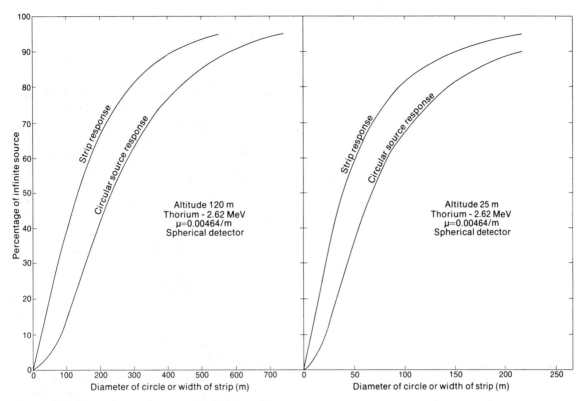

Figure 10C.10. *Area of coverage for an airborne gamma ray spectrometric survey for a stationary detector and a moving detector at altitudes of 120 m and 25 m (after Grasty et al., 1979). For a given percentage of infinite source detected, the width of the strip (moving detector) is less than the diameter of circle (stationary detector). Calculations of amount of surface area covered for a given airborne survey, based on the stationary detector model, will produce over-estimates, since the survey is actually done with a moving detector.*

varied with detector size. This is due to the differing sensitivity of the detectors in different directions, and the data of Figure 10B.7 in Grasty (1979) show the change in diameter can be in the order of 10 per cent variation.

The use of the concept of the "field of view" in computing the percentage coverage of an area in a given survey is important in choosing the desired flight line spacing for an airborne survey. Figure 10C.12 shows the thickness of material penetrated for various gamma ray energies computed from the mass absorption data given in Table 10B.3 of Grasty (1979). Figure 10C.12b shows the results using a density $\rho = 2.67$ g/cm^3 (average rock), Figure 10C.12a shows a low density case of $\rho = 2.0$ g/cm^3 and Figure 10C.12c, shows a high density case $\rho = 3.0$ g/cm^3. From these data an appreciation of the sample volume (thickness) can be obtained.

Counting Statistics

One of the factors often quoted as being an important requirement for a survey is "good counting statistics". This is a rather broad term and will be elaborated upon somewhat below, but basically it means that the gamma ray count rate, or total number of gamma rays counted in a given measurement must be large enough to be considered a statistically reliable measurement. This is related to the desired accuracy of a measurement.

Definitions and Effects of Varying Survey Parameters

Radioactive decay is a random process, and one standard deviation (σ) in a counting measurement equals the

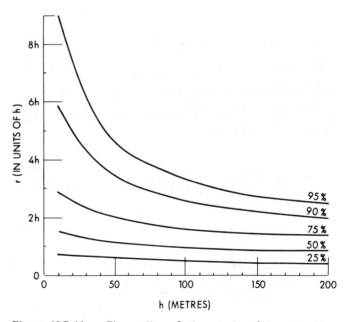

Figure 10C.11. *The radius of the circle of investigation versus altitude for a given percentage of infinite source yield (after Duval et al., 1971).*

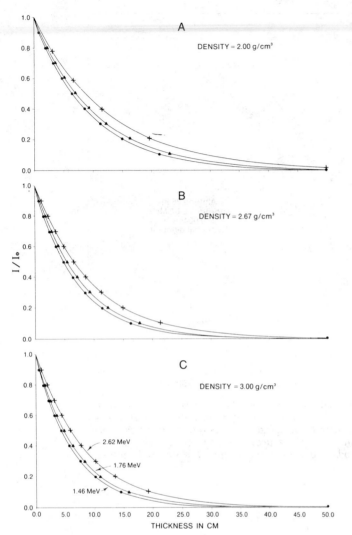

Figure 10C.12. Attenuation of gamma rays versus thickness for the three gamma ray energies associated with K(1.46 MeV), U(1.76 MeV), and Th(2.62 MeV) for three rock densities (ρ). In a strict sense these curves refer to collimated beams penetrating rock absorbers. However the relative effects on sample volume of both energy and rock density can be appreciated from the figure. (I/I_o = (intensity)/(initial intensity)).

square root of the count (n) obtained. This can then be expressed as a percentage of the count or percentage error. One standard deviation is:

$$\sigma = \pm \sqrt{n} \qquad (3)$$

Thus the standard deviation in a measurement where n = 100 counts is ±10 counts, or a percentage error of ±10%. A count of 1000 would yield ±31.6 counts, about ±3% error.

The standard deviation in the counting rate is

$$\sigma_{RATE} = \sqrt{n}/t = \sqrt{RATE/t} \qquad (4)$$

where the counting rate is defined as RATE = n/t.

The error in the final result will be increased when a number of corrections are applied such as background subtraction and spectral stripping, each of which includes a counting measurement with its associated errors.

In subtracting background for example the error in the net count will be

$$\pm \sigma_{NET} = \sqrt{\sigma_R^2 + \sigma_B^2} \qquad (5)$$

where σ_R = the standard deviation in the RAW count, and σ_B = the standard deviation in the BACKGROUND count.

Errors are always <u>additive</u>, even though the operation is <u>subtraction</u> of background.

When spectral stripping is included, the error calculations are quite ponderous. The following error analysis is given by Darnley et al. (1969):

"Assuming that all count rates are given in counts per minute, the standard deviation for the corrected count-rates are as follows:

$$\sigma_{Th} = \left[\frac{N_{Th}}{t} + \frac{BGD_{Th}}{T} \right]^{1/2} \qquad (6)$$

$$\sigma_U = \left[\frac{N_U}{t} + \frac{BGD_U}{T} + \alpha^2 \sigma^2_{Th} \right]^{1/2} \qquad (7)$$

$$\sigma_K = \left[\frac{N_K}{t} + \frac{BGD_K}{T} + \beta^2 \sigma^2_{Th} + \gamma^2 \sigma^2_U \right]^{1/2} \qquad (8)$$

where N_{Th}, N_U, N_K are the observed count rates in the Th, U and K channels respectively and BGD_{Th}, BGD_U, and BGD_K are the background count rates in the same channels, t is the measuring time for the interval under consideration, T is the time for which the background was observed. The corresponding relative standard deviations are given by

$$\sigma_{Th}/N_{Th}\text{ corr.}, \quad \sigma_U/N_U\text{ corr.}, \quad \text{and} \quad \sigma_K/N_K\text{ corr.}$$

where the subscripts refer to the corrected channel count rates.

As an example of the statistical error which arises from the count-rates measured and the application of the various corrections the data obtained from one Elliot Lake test run at 150 m have been used. Mean count-rates obtained by three 12.5 cm x 12.5 cm NaI(Tl) detectors in a 15-sec measuring period were calculated, background was subtracted and Compton scattering corrections were made. The 15-sec counting time at a helicopter speed of 40 km/hr represented a forward travel of about 170 m. The mean counts per minute for 15-sec counting intervals are as follows:

1.35-1.58 MeV, 'K' window 740 ± 11%

1.65-1.88 MeV, 'U' window 190 ± 22%

2.42-2.82 MeV, 'Th' window 230 ± 16%

The quoted uncertainty is based on one standard deviation."

To obtain a lower error the count must be increased; this can be accomplished by: 1) counting for a longer time, 2) increasing the detector size, or 3) moving the detector closer to the source. These three alternatives are relatively easy to accomplish in a laboratory, but in the field, changing these parameters could have far-reaching effects. Increasing the counting time for a portable (man-carried) gamma ray spectrometer reduces the number of readings that can be made each day. In a mobile survey, increasing the counting time means that each measurement is taken over a longer distance, and therefore each is representative of a larger volume of rock. This smoothing effect may not be acceptable, and the vehicle speed may have to be reduced by a direct ratio to the increase in sample time. In the case of an airborne survey, this could reduce the aircraft speed to the point where the selection of a rotary wing aircraft is required, increasing the cost.

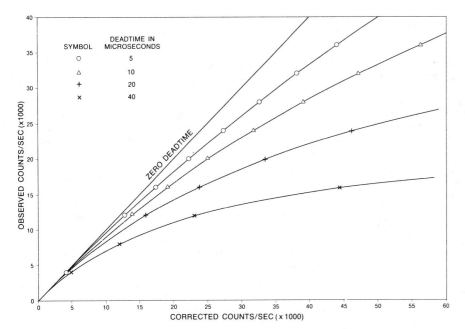

Figure 10C.13. *Deadtime correction curves are based on equation (10): $N = n/(1 - nT)$ (see text). Corrected count rate for any given measured count rate can be determined if the appropriate deadtime curve is known. Because the equation is an approximation, the corrections become increasingly erroneous as the percentage correction increases.*

Alternative (2), increasing the size of an airborne detector package, may require a larger, and consequently more costly type of aircraft. For a portable or a carborne survey, this may be a more economical solution. Alternative (3) moving closer to the source, may be impossible, as in the case of a carborne survey, or it may involve flying at unsafe altitudes for airborne surveys. Under alternative (3) the sample volume will decrease appreciably and it may be necessary to make more closely spaced measurements for the same ground coverage. This of course, also represents an increase in costs.

The relationship between sample time, aircraft speed, and altitude has been considered by Killeen et al. (1975) and Killeen et al. (1971). They suggested that once the exploration target size (or spatial wavelength) is selected, the aircraft altitude is optimized at twice this wavelength. It is then possible to choose from several combinations of speed and sampling time such that:

$$\Delta t = h/4V \quad (9)$$

where

Δt = sample time in seconds; V = aircraft velocity in metres/sec.; and h = the previously selected altitude in metres.

Once this has been done, and the aircraft selected, the detector size must be chosen to yield good counting statistics.

Dead Time and 'Sum' Peaks

The dead time of a gamma ray spectrometer is effectively the time taken by the equipment to analyze a single gamma ray which has been detected; during this period the equipment is busy and cannot analyze any other gamma rays. This dead time (sometimes called resolving time) is only a few microseconds per recorded count, but it becomes an appreciable percentage of the counting time at high count rates, and must be taken into consideration. The true count N can be computed from the measured count n if the dead time T is known, using the following approximation:

$$N = n/(1 - nT) \quad (10)$$

Since this correction is an approximation it introduces error when the percentage losses due to dead time are high (Chase and Rabinowitz, 1968). Dead time should be kept at less than 50 per cent in the worst instance and preferably below 10 per cent, by suitably adjusting detector sizes, source-detector spacing, or other variables.

One of the most common methods of measuring dead time is the method of paired sources. Two sources of activities R_1 and R_2 are selected such that, individually counted there are no significant dead time losses, but when counted together (R_{12}) the percentage loss becomes significant. The dead time has been given by Chase and Rabinowitz (1968) as follows:

$$T = \frac{R_1 + R_2 - R_{12}}{2R_1R_2} \quad (11)$$

where R_1, R_2 and R_{12} are activities that have had the background subtracted. This equation is also an approximation but is more convenient to use than the complex exact equation. Several other approximate relations for computing dead time have been given by Chase and Rabinowitz (1968).

Another method of determining the dead time of a borehole gamma ray logging system has been proposed by Crew and Berkoff (1970). This technique involves measurements in the infinitely thick, homogeneous ore zones in two model holes, which contain uranium ore grades differing by about a factor of 10 or more. They stated that the following approximation will yield dead time values generally within 1 microsecond of the rigorous solution

$$T = \frac{N_{LOW} - G_{RAT} N_{HIGH}}{N_{LOW} N_{HIGH} (1 - G_{RAT})} \quad (12)$$

where

T = apparent dead time in seconds

N_{LOW} = apparent measured peak count, low model

N_{HIGH} = apparent measured peak count, high model

G_{RAT} = (% eU_3O_8, low model)/(% eU_3O_8, high model) i.e. grade ratio

Having determined the dead time of a system it is important to know approximately the size of the dead time correction at various count rates. Using equation (10), the corrected counts are plotted against the measured counts for several values of dead time in Figure 10C.13. It can be seen from the graph that at a dead time of 20 microseconds, for example, a measured count rate of 25 000 cps would be corrected to 50 000 counts or 100 per cent dead time correction which is unacceptably high. With the same dead time a correction of 50 per cent occurs at 17 500 cps measured, corrected to about 26 250 cps. Thus the upper limit is about 17 500 cps for a system with 20 microsecond dead time and, in fact, it would be preferable to choose equipment which would operate at lower count rates than this in the radioactive zones of interest in the field.

Besides dead time an additional consideration at high count rates is sum peaks or coincidence peaks. These are especially important in the case of gamma ray spectrometry. When two gamma rays strike the detector at the same time, the two scintillations produced are "seen" as one by the photomultiplier tube and only one gamma ray count is registered, at an energy approximately equal to the sum of the energies of the two gamma rays which struck the detector. This is usually only important for a large detector with the source very close to it, as in a laboratory gamma ray spectrometer.

For example the 0.58 MeV and 2.62 MeV gamma rays from the thorium decay series are both of high intensity and the sum peak at 3.20 MeV becomes significant for measurements of samples with high thorium concentrations. Sum peaks can cause gamma rays of low energy to be counted in higher energy windows of a gamma ray spectrometer, introducing the possibility of erroneous results.

Background Radiation

The literature describes two values for background: the "local" background (i.e. the mean count rates), and the "over water" background. Only the latter permits comparison of data from different survey areas.

Background radiation is any radiation detected by the gamma ray spectrometer not originating from the source which is being analyzed; in this case, the lithosphere (IAEA, 1976). In the case of a laboratory it includes radiation coming from or through the walls, ceiling and floor, and the lead counting chamber or shield. In the field it includes the radiation from the vehicle, be it a man carrying the spectrometer, a truck, or an aircraft. In addition there is a cosmic ray component (most important for airborne surveys) and radioactivity in the atmosphere caused by radon and its daughter products, and products from nuclear fallout. To the extent possible the background should be minimized. Flight instrument dials or emergency exit signs etc. in aircraft, which are luminized with radium, should be replaced or removed. Similarly radium dial wrist watches should not be worn by personnel carrying out surface surveys. Calibration sources should be removed completely or at least shielded if they must be carried on the survey. Any remaining background should be measured accurately to enable its subtraction. The greatest problem of background is the variable or unknown component. This is primarily caused by the radioactivity of particles in the air. Grasty (1979) reviewed this in some detail especially with reference to airborne gamma ray surveys. The subject of background radiation in airborne measurements is not included in this section because it has been discussed in Paper 10B (Grasty, 1979).

Background Radiation in Surface Measurements

There are several ways in which the background can be determined for both carborne and portable gamma ray spectrometers. Ideally, a measurement can be made over water in the survey area using a boat or by driving the vehicle onto a bridge (not stone or concrete) over a wide river or onto the ice of a lake (Bowie et al., 1955). This will yield a background value for the equipment itself, the vehicle, cosmic rays and radioactivity in the atmosphere. Bowie et al. (1955) described a method of determining both the cosmic ray component and the vehicle component if it is possible to take readings as follows:

R_o = 2π reading taken on a road crossing a given rock type
  (include vehicle + cosmic ray component)

R_T = 4π reading in a tunnel through that same rock type
  (includes vehicle component)

R_1 = 2π reading on lake ice (vehicle + cosmic components only)

Then the cosmic ray component (C) is:

$$C = 2R_o - R_T - R_1 \qquad (13)$$

If the vehicle and equipment are unchanged during the survey and periodic background measurements differ, the difference will be due primarily to fluctuations in the radioactivity of the air. Background can fluctuate significantly both with time and location. Changes with time can be observed by keeping a base-station which is measured every morning and night during the survey. The base station should give the same count rate at all times (within counting measurement errors) except for differences in atmospheric radioactivity or cosmic radiation. Using a base station also provides a check on the condition of the instrument and detector and will permit early detection of electronic problems, a cracked crystal, or other problems.

Another method to determine the background radiation level is to extrapolate to zero concentration the count rates measured on calibration sources such as the ten calibration pads of the Geological Survey of Canada in Ottawa. The pads must be considered to be infinite 2π sources, and this will only hold for detectors placed directly on the pad near the centre.

Background Radiation in Borehole Measurements

Generally the background radiation in a borehole gamma ray log is considered to be zero since the detector is surrounded by the source. Cosmic ray effects are generally negligible, and the radioactivity of the air has no influence in a liquid-filled hole and negligible influence in an air-filled hole due to the small volume of air. Two situations, however, are possible in which the background becomes significant. The first occurs when drilling muds rich in potassium compounds, such as potassium chloride (KCl) and potassium bicarbonate ($KHCO_3$) have been used. Cox and Raymer (1976) discussed the effect of potassium muds on gamma ray logs and concluded that the contribution of mud radioactivity to the gamma ray measurement can be large, that the effect is greater in larger holes than small, and is directly proportional to the concentration of the potassium in the mud. Their correction factors presented in graphical form for various mud weights are instructive, but unnecessary for most hard rock mining applications which utilize small diameter boreholes and water as a drilling fluid.

The second situation in which the background in a borehole becomes significant occurs when the borehole is logged after a considerable time has elapsed since drilling. In this case it is possible that radon will migrate from radioactive zones into the borehole, creating a broader anomaly on the gamma ray log than if the hole had been logged shortly after drilling. This has been reported as a problem in air-filled model boreholes used for calibration but the effect can be reduced to an insignificant amount by keeping water in model holes (IAEA, 1976). The characteristics of radon emanation and corrections for it, have been discussed in detail by Scott and Dodd (1960), Austin (1975) and Barretto (1975).

In the field, it is difficult to know if a radon background problem exists in a borehole. Tanner et al. (1977b) have utilized high resolution solid state germanium detectors (both intrinsic Ge and Ge(Li)) to overcome the problem in areas of known radioactive disequilibrium. It has also been suggested by Lyubavin and Orchinnikov (1961) that disequilibrium (e.g. radon excesses) could be detected by spectral logging methods. Dodd et al. (1969) indicated that although

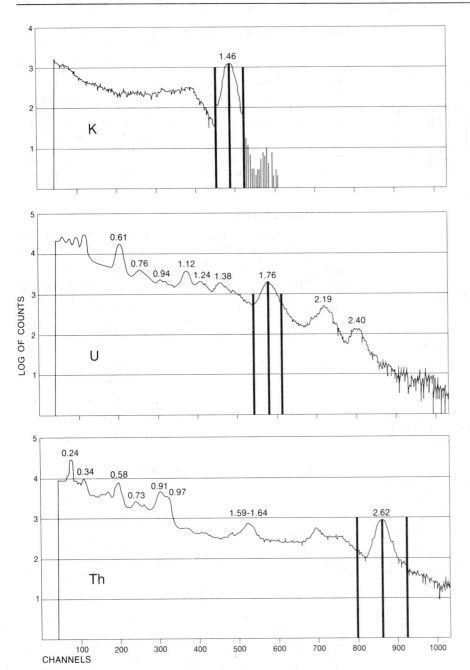

Figure 10C.14. Measured energy spectrum of gamma rays emitted by ^{40}K (top), the ^{238}U decay series (middle), and the ^{232}Th decay series (bottom). Measurements were made with a 75 x 75 mm NaI(Tl) detector. Energy axis (horizontal) is approximately 3.0 keV/channel. Locations of commonly used windows for gamma ray spectrometry are indicated at 1.46 MeV, 1.76 MeV (^{214}Bi) and 2.62 MeV (^{208}Tl). Window widths are 200 keV, 200 keV, and 400 keV respectively for K, U, and Th (after Killeen and Bristow, 1976).

preliminary investigations confirmed the feasibility of the idea, the complexity and stability of instrumentation required at that time (1967) exceeded the state of the art for routine application.

Background Radiation in Submarine Measurements

Background measurements with a 75 x 200 mm NaI(Tl) detector for underwater gamma ray spectrometry in lake bottoms in Saskatchewan have been reported by Stolz and Standing (1977) as 100 to 500 cps for total count, 5 to 15 cps for potassium and 0 to 10 cps for uranium and thorium (energies not given). Their detector was towed along the bottom in an 'eel', 7 m long by 17 cm diameter which gouged a trough of semi-circular cross-section in the lake bottom sediments. This background is the non-anomalous or regional radioactive component, rather than the 'true' background in the sense of radiation originating from other than the lake-bottom. "True" background could be obtained by suspending the detector half way between the lake bottom and water surface. Noakes et al. (1974a) used a detector consisting of a sled housing four 75 x 75 mm NaI(Tl) detectors towed along the sea floor sediments seaward of Amelia Island, Florida, at a speed of from three to five knots. They reported background levels on the sea floor for a two channel spectrometer as 50 to 75 counts per second for channel B (less than 1.0 MeV) and 10 to 20 cps for channel A (greater than 1.0 MeV). Gaucher et al. (1974) reported background measurements for sea water which they determined using a 75 x 75 mm detector on a sled. With a 500 second counting time they obtained counts of 144, 19, and 27 for their 'K', 'Th' and 'Ra' windows respectively (energies not given) on a 100 channel spectrometer. Clayton et al. (1976) reported approximately 7 counts per second for background with the probe surrounded by seawater with a detector of 75 x 125 mm. They give typical count-rates for the spectrometer channels (no energies given) of 1500 to 12 000 counts/minute (total count), 90 to 300 counts/minute (K), 10 to 50 counts/minute (U), and 7 to 30 counts/minute (Th) depending on rock type.

Calibration

For any physical measurement to be meaningful, the measuring device must be calibrated with respect to some standard. This is true whether the measurement concerns weight measurements with a spring balance, distance measurements with a ruler, or radioelement measurements with a gamma ray spectrometer. Confusion can result when comparing gamma ray counting measurements made with different instruments, sometimes even when the instruments are of the same manufacture. In addition, different instruments may display the data in different units, such as counts per second, counts per minute, micro roentgens per hour or millivolts. Darnley (1977) pointed out the advantages of standardizing radiometric exploration measurements. The International Atomic Energy Agency (IAEA, 1976) formulated recommendations for standardizing measurements in uranium exploration by calibration, and presenting results in units of concentration (e.g. %K, ppm eU, ppm eTh).

The basic equation relating the detected radioactivity to the radioelement concentration in the source of the radiation is:

Radioelement content = (a constant)(gamma ray intensity)

= (1/sensitivity)(gamma-ray intensity)   (14)

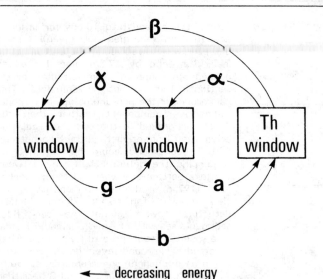

Figure 10C.15. *Schematic representation of the interaction between the K, U, and Th energy windows indicates the stripping factors which are used to remove the interference denoted by the arrows. Commonly used stripping factors are α, β, and γ. The 'upward' stripping factors a, b, and g are generally small or zero and are often ignored.*

The problem of calibration relates to the determination of this constant which is the "K factor" or the reciprocal of the sensitivity. First, however, the gamma ray intensity must be corrected for background, and for interference from gamma rays related to radioelements other than the one being determined.

The gamma rays emitted by the daughter nuclides of uranium and thorium have been tabulated by Smith and Wollenberg (1972), and a summary of their table of gamma ray energies and intensities is given by Grasty (1979). The principle gamma rays of the uranium and thorium series have also been tabulated in an appendix of Adams and Lowder (1964). Most compilations like these are derived from Lederer and Shirley (1978), Lederer et al. (1968), or Hyde et al. (1964). The principle gamma rays produce the energy spectra shown in Figure 10C.3. The potassium spectrum is a single gamma ray peak at 1.46 MeV. The relative activities or intensities are those obtained under conditions of radioactive equilibrium. Disequilibrium may introduce differences in the relative peak heights depending on whether any given daughter emitter has been removed or deposited in the sample.

Due to the inherently imperfect energy measuring capabilities of scintillation detectors and scattering of gamma rays and degradation of gamma ray energies in the source rock, the measured spectra for uranium and thorium decay series will look more like those shown in Figure 10C.14 (centre) and 10C.14 (bottom) respectively. The measured potassium spectrum will resemble that shown in Figure 10C.14 (top).

Gamma ray spectra for numerous isotopes useful for calibration such as Cs-137, Co-60, Y-88 and the uranium and thorium series have been published (e.g. Heath, 1964; Crouthamel, 1960; Adams and Dams, 1970).

Figure 10C.14 also indicates commonly chosen K, U, and Th energy windows for a differential spectrometer. These are 1.36-1.56 MeV (K window), 1.66-1.86 MeV (U window), and 2.4-2.8 MeV (Th window). Sometimes the upper limit of the K window coincides with the lower limit of the U window, and some workers raise the upper limit of the U window to 2.4 MeV. For a threshold spectrometer, similar lower limits would be used for the three windows, and the upper limit would be at 3.0 MeV or higher. In any case it can be seen that there is interference between the three spectra; some gamma rays originating from the Th decay series will be counted in the U and K windows, some gamma rays originating from the U decay series will be counted in the K window and to a small extent in the Th window, and finally some small portion of the K gamma rays may be counted in the U window. To determine the "K factor" or sensitivity of equation (14) for the individual radioelements K, U and Th, the gamma ray counts due to each individual radioelement must be determined. This is accomplished by the procedure known as spectral stripping.

The Stripping Factors

The determination and various definitions of the stripping factors (often called stripping ratios) have been discussed in numerous papers (Adams and Fryer, 1964; Wollenberg and Smith, 1964; Doig, 1968; Killeen and Carmichael, 1970; Grasty and Darnley, 1971; Grasty, 1976b, 1977a; Killeen and Cameron, 1977; Grasty, 1979). Figure 10C.15 is a schematic representation of the interplay between the three radioelement windows K, U, and Th and identifying the stripping factors, as listed below:

Stripping Factor	Used to Strip Off
α	Th gamma rays in U window
β	Th gamma rays in K window
γ	U gamma rays in K window
a	U gamma rays in Th window (usually small)
b	K gamma rays in Th window (zero)
g	K gamma rays in U window (approx. zero)

For many purposes only the first three stripping factors are used and a, b and g are assumed to be zero. For high uranium concentrations the upward stripping factor 'a' is necessary. It has a value of approximately 0.05 (Grasty, 1975), but the actual value is dependent on factors such as detector size and resolution, window widths and energy settings.

These stripping factors are determined by making measurements on calibration sources, as described below in greater detail. The count rates obtained, and the radioelementanalysis (least squares) computer program which solves for the stripping factors and the sensitivities or K factors.

These sensitivity factors have units of count rate per unit of radioelement concentration. The rigorous solution for the calibration equations is given by Grasty (1979). Since the simplified solution is used in most cases, the method of computing radioelement concentrations from gamma ray counts as set forth by Killeen and Cameron (1977) is reproduced below.

Assume the counts measured in (cpm) in the three spectrometer windows are Kc (potassium), Uc (uranium) and Thc (thorium).

A. To calculate eTh in ppm (parts per million)
   1. Obtain the net thorium count by subtracting thorium background from the measured thorium count:

$$\text{Th net} = \text{Thc} - \text{Thb} \quad (15)$$

   2. Obtain eTh ppm by dividing the net thorium count by the thorium sensitivity (Ths)

$$\text{eTh ppm} = (\text{Th net})/\text{Ths} \quad (16)$$

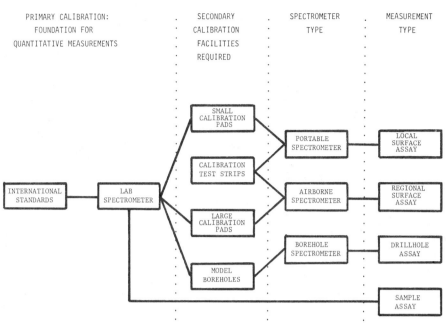

Figure 10C.16. *The relationship between primary and secondary calibration facilities for various types of gamma ray spectrometer. International standard samples are used to calibrate lab spectrometers. These in turn analyze calibration pads and model boreholes which can then be used as secondary calibration facilities for field spectrometers. The calibrated field spectrometers can then carry out assays in situ.*

B. To calculate eU in ppm:
1. Determine the thorium contribution to the uranium count by multiplying the net thorium count by the factor α:

$$(\alpha)(\text{Th net})$$

2. Obtain the net uranium count by subtracting the uranium background and the thorium contribution from the measured uranium count:

$$U\text{ net} = U_c - U_b - (\alpha)(\text{Th net}) \qquad (17)$$

3. Obtain eU ppm by dividing the net uranium count by the uranium sensitivity (U_s):

$$eU\text{ ppm} = (U\text{ net})/U_s \qquad (18)$$

C. To calculate potassium in per cent:
1. Determine the thorium contribution to the potassium count, by multiplying the net thorium count by the factor β.

$$(\beta)(\text{Th net})$$

2. Determine the uranium contribution to the potassium count, by multiplying the net uranium count by the factor γ.

$$(\gamma)(U\text{ net})$$

3. Obtain the net potassium count by subtracting the potassium background and the thorium and uranium contributions from the measured potassium count:

$$K\text{ net} = K_c - K_b - (\beta)(\text{Th net}) - (\gamma)(U\text{ net}) \qquad (19)$$

4. Obtain the per cent potassium by dividing the net potassium count by the potassium sensitivity (K_s).

$$\%K = (K\text{ net})/K_s \qquad (20)$$

Killeen and Cameron (1977) also presented a numerical example for the case of a portable 4 channel gamma ray spectrometer with a 75 x 75 mm NaI(Tl) detector.

In 1976, the International Atomic Energy Agency recommended a new unit, the 'ur', or 'unit of radioelement concentration', for total count gamma ray spectrometer or scintillometer results (IAEA, 1976). The unit is described as follows: "A geological source with 1 unit of radioelement concentration produces the same instrument response (e.g. count rate) as an identical source containing only 1 part per million uranium in radioactive equilibrium".

In other words a "geological source", which probably contains some potassium and thorium as well as uranium, produces a certain instrument response (e.g. count rate). This is expressed in units of the amount of equivalent uranium which alone would give the same instrument response. Thus the use of the 'ur' unit replaces the use of the term 'equivalent uranium' which was used in the above sense in many early publications on total count radiometric work. The present-day usage of "equivalent uranium" however refers only to the single element <u>uranium</u> but measured indirectly by counting techniques (see the earlier section on equilibrium).

Thus, 1 unit of radioelement concentration is equivalent to 1 part per million uranium in equilibrium, or

$$1\text{ ur} = 1\text{ eU} \qquad (21)$$

The use of the ur unit for calibration of total count gamma ray surveys has been discussed by Grasty (1977b). Any natural radioelement calibration source can be used to calibrate in ur units by converting the radioelement concentrations of the source to ur units with the following relations:

$$1\% \text{ K} = 2.6\text{ ur} \qquad (22)$$
$$1\text{ ppm U} = 1.0\text{ ur} \qquad (23)$$
$$1\text{ ppm Th} = 0.477\text{ ur} \qquad (24)$$

These relationships are strictly valid only for measurements of gamma rays above 0.4 MeV. "A total count scintillation detector operated with an energy threshold of 0.4 MeV will enable the determination of radioelement concentration to be made in ur units almost independently of changing proportions of the radioelement" (IAEA, 1976).

Calibration Procedures

In this section the procedures for calibration of gamma ray spectrometer survey equipment are considered. A great deal of this information originates from unpublished experience in gamma ray spectrometer surveys, and from specifications for contracted surveys based on this experience at the Geological Survey of Canada (Bristow et al., 1977).

Calibration or standardization of gamma ray spectrometer equipment depends on some known, primary reference standard. Secondary calibration standards are then derived from the primary reference. This connection between primary and secondary calibration facilities is illustrated in Figure 10C.16. The types of spectrometer and relationships of their measurements to the different secondary calibration facilities are indicated. From the Figure 10C.16 it can be seen that ultimately all measurements depend on calibration

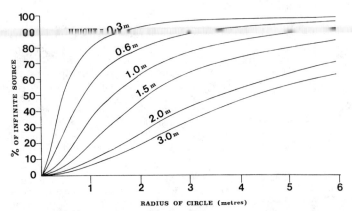

Figure 10C.17. Radius of the circle of investigation for a 75 x 75 mm NaI(Tl) detector placed at various heights showing the percentage of infinite source area (2.62 MeV gamma rays) for a given radius of circle.

of a laboratory gamma ray spectrometer using standard samples. The laboratory spectrometer analyzes samples taken during construction of concrete calibration pads, and model holes, and samples taken from test strips. These concrete models become secondary calibration sources for field gamma ray spectrometers. The calibration of an airborne gamma ray spectrometer requires in addition to the concrete calibration pads, a calibration test strip over which test flights may be made. The analysis of this test strip is done partly by in situ assays (or traverses) on a grid with a calibrated portable (or carborne) gamma ray spectrometer, and partly by laboratory assay of samples from the test strip.

The questions of when to calibrate, and how often are also important. Once an instrument is calibrated, it is not calibrated forever, as its characteristics may change with time and it should thus be re-calibrated periodically. It is essential that the user have information about the normal values of the calibration factors for a given instrument. Any large changes upon calibration would probably indicate one or more of the following:

1) a malfunction in the spectrometer,
2) an error in recording the counts,
3) the counting time was too short to provide good counting statistics,
4) an error in entering the data into the computer (compare the computer printout of raw data to your original numbers),
5) calibration just after a rainfall when all the radon daughters in the air are washed out increasing background on the pads,
6) drifting of energy windows due to change in temperature of the detector during calibration,
7) improper setting of the window locations or window widths,
8) low battery power,
9) cracked crystal in the detector package,
10) improper energy calibration.

Thus, calibration, in addition to permitting quantitative measurements also maintains a check on the performance of the system.

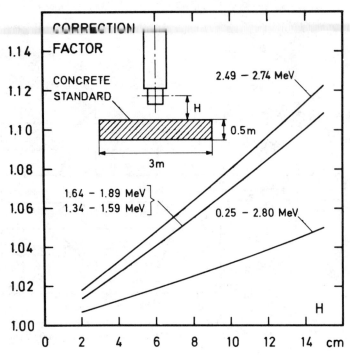

Figure 10C.18. Correction factors for calibrating portable gamma ray spectrometers on calibration pads of less than infinite diameter (after Løvborg et al., 1972). Factors are for a height H above a pad of 3 m diameter as shown in the inset.

Temporary calibration facilities can be set up in the field using the calibrated system. This field calibration facility (a base station, test road, test strip of land) can be monitored on a daily basis to ensure the system remains in calibration.

Calibration of Surface Systems

A surface gamma ray spectrometer system may be either carried by a vehicle or placed on the ground surface during the measurement. It is most important to simulate the field geometry during the calibration. The sample time or counting time may be as long as required to provide good counting statistics for the calibration measurement. For a portable gamma ray spectrometer the measurements are usually made by placing the detectors directly on the ground surface. The complete calibration can be carried out by making measurements on a set of calibration pads such as those suggested by the IAEA (1976).

If the detector is to be located at a raised elevation with respect to the ground, such as when mounted on a backpack or a vehicle, the system should first be calibrated with the detector on or near the surface of the calibration pad to ensure that the pad represents an infinite 2π geometry source. The field measurement, with detector mounted will be representative of a larger sample volume.

Figure 10C.17 gives the radius of the circle of investigation for a 75 x 75 mm detector placed at various heights for thorium series 2.62 MeV gamma rays and several different percentages of infinite source area. For example 90 per cent of the radiation from an infinite source area is achieved at a radius of about 5 m for a detector in a backpack carried at 1 m elevation. The 90 per cent radius is 15.8 m for a detector elevation of 3 m, and the 90 per cent

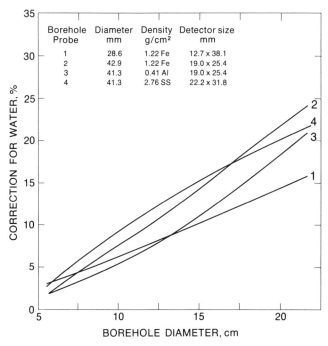

Figure 10C.19. Water correction factors for gamma ray logging in different diameter boreholes (after Dodd and Eschliman, 1972). Probe housings of iron (Fe), aluminum (Al) and stainless steel (SS) produce different gamma ray attenuations according to the different density of probe shell.

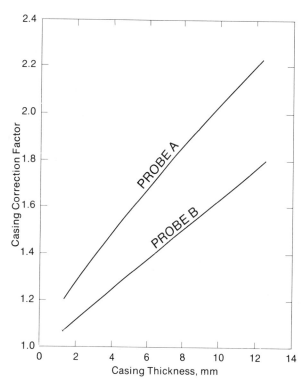

Figure 10C.20. Casing correction factors for gamma ray logging in different thickness of casing (after Dodd and Eschliman, 1972).

radius is 29 m for a detector at 6 m elevation (These last two values are not shown in Fig. 10C.17). Similar curves for larger detectors and higher elevations (for airborne surveys) have been given by Grasty (1976b) and by Adams and Clark (1972).

When taking measurements on calibration pads there is always a possibility of some inhomogeneity in the concrete mixture, and in the case of a small detector, several measurements at slightly different locations should be averaged. These measurements (count rates) are then combined with the analyses of the calibration pads in a regression computer program to solve for the calibration factors. It may be possible in the case of a carborne survey, if the detector elevation is not too high, to accomplish the calibration without removing the detector from the system by calibrating on a set of large calibration pads. For example if the detector is about 60 cm above the surface, the pad diameter should be greater than 7 m. It can be judged from Figure 10C.17 whether this is feasible for other detector elevations. Figure 10C.18 (after Løvborg et al., 1972) give correction factors for calibration made at various heights above pads. Matolin (1973) described the problems associated with artificial calibration standards.

Calibration of Airborne Systems

The calibration of an airborne gamma-ray spectrometer system requires both pads and an airborne test strip. The aircraft can be towed or taxied under its own power to a set of concrete calibration pads such as those built by the Geological Survey of Canada at Ottawa, or those built by the U.S. Department of Energy at Grand Junction. The gamma ray spectral measurements made on each pad are combined by a regression analysis program with the analyses of the calibration pads, and a set of stripping factors are derived. The stripping factors, are assumed valid for airborne work, except for the thorium into uranium stripping factor, α, which varies significantly with altitude (Grasty, 1976b). For example Løvborg et al. (1978a) have shown that for a 292 x 102 mm (11.5 x 4 inch) airborne detector, the value of α can increase by over 16 per cent when moving from elevation of 0 to 125 m. Under the same circumstances β (thorium into potassium) and γ (uranium into potassium) increased by 12 per cent and 8 per cent respectively. These increases are not as great for larger detectors.

A test strip is necessary for determination of the height correction parameters. The IAEA (1976) recommends for example that for surveys to be flown at 120 m, that altitude tests be flown between 50 and 250 m spanning the range of normal survey altitudes. Preferably at least five different altitudes should be tested to derive the height attenuation parameters.

The sensitivities for an airborne system are obtained from flights over a radiometric test range or test strip. The stripped count rates (computed using the stripping factors obtained on the calibration pads but corrected to survey altitude using attenuation coefficients derived on the test strip) are combined with the assay data from the test strip to provide the sensitivities.

The desired characteristics for a test strip are sometimes difficult to obtain. The IAEA (1976) made the following recommendations for a test strip:

1) it should be flat (low topographic relief),

2) uniform radioactivity (tested by surface traverses and sampling),

3) minimum dimensions of 1.0 km x 3.0 km,

4) close to a large body of water for background measurements,

5) easy repeatable navigability.

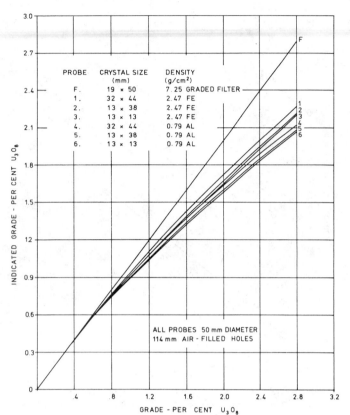

Figure 10C.21. Non-linear increase of indicated ore grade (computed from count rate) with actual ore grade due to Z effect at high ore grades (after IAEA, 1976). The filtered probe 'F', by attenuating all gamma rays of original energies less than 400 keV produces a linear relation. Note the departure from linearity starts at about 0.5% U_3O_8. At higher grades the count rate (if it includes gamma rays of energies below 400 keV) will always cause an under estimate when computing ore grade.

The Breckenridge test strip used by the Geological Survey of Canada is located on the Ottawa River flood plain which is flat, and uniform in radioactivity (Charbonneau & Darnley, 1970a; Grasty and Charbonneau, 1974), 7 km long, and close to the river which is about 1 km wide. A railroad track along the test strip is a convenient aid to navigation and facilitates reproducible results. Both the over-railway and over-water test strips are flown at altitudes of 60, 90, 120, 150, 180, 210 and 240 m. Typical values for sensitivities, stripping ratios, and height attenuation coefficients are given by Grasty (1979) for various detector sizes.

To ensure the continuing calibration of airborne gamma ray spectrometer during a field survey, the system can be monitored with a series of tests which are carried out daily to see that point sources give unchanging results.

Calibration of Borehole Systems

The calibration of gamma ray spectral logging systems is affected by many parameters, and some of these effects still remain to be determined. However a great deal of information is available on total count gamma ray logging, and calibration for quantitative measurement (Scott et al., 1961; Scott, 1963; Rhodes and Mott, 1966; Conaway and Killeen, 1978a).

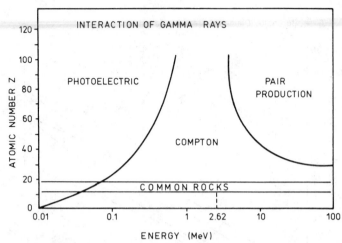

Figure 10C.22. The most likely type of gamma ray interaction for a range of gamma ray energies and equivalent atomic number (Zeq). The range of Zeq for common rocks is also indicated (after Dodd and Eschliman, 1972). When Zeq of a rock increases due to high uranium content, the photoelectric effect becomes a more dominant reaction at low energies. This causes a non-linear relation between count rate and ore grades unless only energies above 0.4 MeV are considered.

The calibration of a total count gamma-ray logging system requires the determination of the K-factor (or sensitivity) which is the constant of proportionality between the grade thickness product (GxT) of a radioactive zone, and the area (A) under the curve of the gamma ray log, i.e.

$$GT = KA \qquad (25)$$

In the case of an 'infinitely' thick homogeneous zone and <u>only</u> in that case is the grade directly proportional to the peak height or intensity (I) of the gamma ray log anomaly, i.e.

$$G = KI \qquad (26)$$

The determination of K is thus simplified by making measurements in model holes with "infinitely" thick ore zones (approximately 1 m thick or greater).

A set of appropriate model holes can also be used to provide water-correction factors, casing correction factors, and dead time measurements (Dodd and Eschliman, 1972). For these measurements, model holes of different diameters and casing or drill pipes of different thicknesses must be available. A graph of water correction factors for different borehole diameters is shown in Figure 10C.19 and typical casing correction factors are shown in Figure 10C.20 for various thicknesses of casing (after Dodd and Eschliman, 1972). These authors also pointed out the nonlinear response of total gamma ray intensity caused by the increasing equivalent atomic number (Zeq) of the rock as the ore grade increases. This effect is shown in Figure 10C.21. Gamma rays of low energies are strongly attenuated when the Zeq increases above that for common rocks.

An important diagram illustrating the relationship between equivalent atomic number (Zeq) and gamma ray energy with respect to the most probable gamma ray interaction is shown in Figure 10C.22. This figure has been used in various forms by Evans (1955), Siegbahn (1968), Adams and Gasparini (1970), and Dodd and Eschliman (1972). It can be seen that as Zeq increases, the proportion of photoelectric effect interactions increases for low energy gamma rays. However, for gamma rays above about 400 keV there is little change. Thus there should be a linear relation between gamma ray count and ore grade when a suitable filter or shield is used on the detector to eliminate these low-energy gamma rays. Larger detectors can then be used to increase the count rate which was diminished by the filter. In gamma ray spectral logging, a total count channel with a 400 keV low level threshold will provide count rates which vary nearly linearly with ore grade. It has been suggested by Czubek (1968) that this Z effect might be used to advantage by a gamma ray spectral logging system. He suggested the ratio of high energy to low energy gamma ray count rates could be used as a parameter which varies inversely with ore grade (i.e. Zeq). The method has been tested to some extent as mentioned by Dodd and Eschliman (1972), although it has not become commonly used. This may be partly due to the fact that radioactive disequilibrium may produce the same effect as a change in Zeq.

Gamma ray spectral logging systems must be calibrated in model holes containing "ore zones" of thorium, and of potassium, in addition to the usual uranium ore zone found in model holes used to calibrate total count logging systems. The calibration procedure is similar to that described for calibration of surface systems. A statistically adequate count rate must be obtained for each ore zone. These count rates and the radioelement analysis (grade) of the ore zone material are the input data to a regression analysis computer program which solves for the values of the sensitivities and stripping factors as described in earlier sections.

Instead of one K-factor as for the total count gamma ray log, there are three K-factors which relate to the area under the curve of the gamma ray log for each radioelement. Thus:

$$G_K T = K_K A_{K(NET)} \quad (27)$$

$$G_U T = K_U A_{U(NET)} \quad (28)$$

$$G_{Th} T = K_{Th} A_{Th(NET)} \quad (29)$$

where the subscripts denote the particular radioelement. The areas "$A_{(NET)}$" are the areas under the curve of the 'stripped' gamma ray spectral window logs. For example continuing in the simplified fashion of the section describing sensitivities or K-factors for in situ assaying the net counts are:

$$A_{Th(NET)} = A_{Th(RAW)} - B_{Th} \quad (30)$$

$$A_{U(NET)} = A_{U(RAW)} - \alpha X A_{Th(NET)} - B_U \quad (31)$$

$$A_{K(NET)} = A_{K(RAW)} - \beta X A_{Th(NET)} - \gamma X A_{U(NET)} - B_K \quad (32)$$

where the B_{Th}, B_U and B_K are the backgrounds (if any).

For the reduction of data from gamma ray logs recorded in the field, two methods are available. An iterative technique has been described by Scott (1963) and Scott et al. (1961), and a deconvolution or inverse filtering technique has been described by Conaway and Killeen (1978a). Both techniques require that the shape of the anomaly produced by an ore zone in a model hole be accurately determined. This shape defines the response of the gamma ray logging system and is used as the basis for the iteration or the inverse filtering (Conaway and Killeen, 1978b; Czubek, 1971, 1972).

The iterative technique is essentially the fitting of a synthetic anomaly (based on the system response function) to the measured anomaly. The synthetic anomaly is made by summing individual anomalies produced by hypothetical layers. During each iteration the grades of the hypothetical layers are adjusted, the synthetic anomaly is recomputed and its goodness of fit is tested. In the deconvolution technique an inverse filter is developed which can remove the effects of the system response. This inverse filter can be used in nearly real time to produce a deconvolved gamma ray log as suggested by Killeen et al. (1978) wherein the gamma ray logscale has been converted to grade, rather than count rate.

Calibration Facilities

Calibration facilities for total count gamma ray logging were first made available in the USA in the early 1960s, with an additional model hole each for K, U and Th for gamma ray spectral logging being built in 1974 (Knapp and Bush, 1975). The first large calibration pads for calibrating airborne and ground gamma ray spectrometer systems were built in Ottawa in 1968 (Darnley, 1970; Grasty and Darnley, 1971); five pads were built in Grand Junction, Colorado in 1976 (NURE, 1976). Four calibration pads built specifically for portable gamma ray spectrometers were constructed at Risø in Denmark in 1971 (Løvborg et al., 1972; Løvborg and Kirkegaard, 1974; Løvborg et al., 1978a, b). Radiometric test ranges (test strips) were established at Breckenridge near Ottawa in 1970 (Charbonneau and Darnley, 1970a; Grasty and Charbonneau, 1974), and near Lake Mead, Arizona (NURE, 1976). Extensive calibration facilities for borehole gamma ray spectral logging (3 radioelements, 9 models, 27 holes), and for surface portable gamma ray spectrometry (10 pads) were constructed in Ottawa in 1977 (Killeen, 1978; Killeen and Conaway, 1978). In addition seven pads were constructed in Calgary, and 2 model holes in Fredericton. Calibration facilities in South Africa at Pelindaba have been described by Toens et al. (1973). Recently a set of five large calibration pads has been constructed in Iran for airborne gamma ray spectrometry, and a set of large pads is in preparation in Brazil. Additional calibration facilities are being constructed in many countries, as their desirability is being recognized for uranium exploration programs. Some of the existing facilities are described in greater detail below.

Canada

Five concrete pads for calibration of airborne gamma ray spectrometer systems are located at Uplands airport in Ottawa. They are located adjacent to, and level with, a concrete parking area. Aircraft may taxi or be towed onto or off of the slabs. The aircraft must be able to back off each

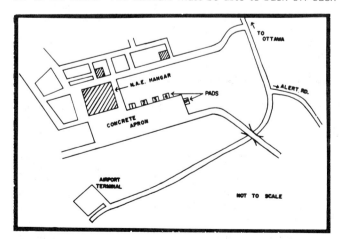

Figure 10C.23. *Geological Survey of Canada calibration pads for airborne gamma ray spectrometric systems at Ottawa, Ontario (after Grasty and Darnley, 1971). Radioelement concentrations are given in Table 10C.2.*

Table 10C.2

Radioelement concentrations of calibration facilities for airborne gamma ray spectrometer systems

Location	Pad Number	K% ± 1σ	eU ppm ± 1σ	eTh ppm ± 1σ
Canada[1]	1	1.70 ± 0.08	2.4 ± 0.2	8.9 ± 0.6
	2	2.27 ± 0.10	7.3 ± 0.2	12.6 ± 0.7
(Uplands Airport,	3	2.21 ± 0.08	3.0 ± 0.3	26.1 ± 0.9
Ottawa)	4	2.21 ± 0.01	2.9 ± 0.3	40.8 ± 1.9
	5	2.33 ± 0.09	11.7 ± 0.3	13.2 ± 0.7
		K% ± 2σ	eU ppm ± 2s	eTh ppm ± 2σ
United States[2]	1	1.45 ± 0.01	2.2 ± 0.1	6.3 ± 0.1
	2	5.14 ± 0.09	5.1 ± 0.3	8.5 ± 0.3
(Walker Field,	3	2.01 ± 0.04	5.1 ± 0.2	45.3 ± 0.7
Grand Junction)	4	2.03 ± 0.05	30.3 ± 1.6	9.2 ± 0.3
	5	4.11 ± 0.06	20.4 ± 1.3	17.5 ± 0.3
Iran[3]	1	2.31	1.80	7.91
	2	2.23	18.64	8.79
(Tehran)	3	2.17	2.80	46.0
	4	2.30	9.71	19.1
	5	2.09	1.93	9.11
Lake Mead Test Range[4] (U.S.A.)	—	2.5	2.6	11.6
Breckenridge Test Range[5] (Canada)	—	2.0	0.9	7.7

[1] from Grasty and Darnley (1971).
[2] from Stromswold (1978).
[3] from D. Blohm, pers. comm. (1978).
[4] from Foote (1978).
[5] from Grasty and Charbonneau (1974).

pad under its own power or be towed onto the concrete apron to reposition the aircraft on the next pad. The pad dimensions are 7.6 m x 7.6 m x 46 cm thick and they are spaced 15.2 m apart. A diagram of their relative locations is shown in Figure 10C.23. The concentrations of the radioelements in the calibration pads are given in Table 10C.2.

The five pads at Uplands Airport, which had also previously been used for calibration of portable instruments, have a limited range of radioelement concentrations because they were designed for calibrating airborne gamma ray spectrometers with large detector volumes. In 1977, a calibration facility consisting of ten calibration pads and 9 test columns with 3 model boreholes in each column was constructed at Bells Corners, approximately 10 km west of Ottawa. The calibration pads are concrete cylinders, 60 cm thick by 3 m in diameter, making effectively infinite sources if the detector of the portable gamma ray spectrometer is located near the centre of, and within a few centimetres of, the upper surface of the pad. Three of the pads contain different potassium concentrations, three contain different uranium concentrations and three contain different thorium concentrations. A tenth pad, referred to as the blank pad, was constructed with no radioelement additives. The radioelement concentrations in these 10 calibration pads are given in Table 10C.3.

For calibration of borehole gamma ray spectrometers under controlled conditions, nine concrete test columns have been constructed along the wall of an abandoned rock quarry at the same location (Fig. 10C.24). Each of these columns is 3.9 m in height with a simulated ore zone 1.5 m thick sandwiched between upper and lower barren zones (Fig. 10C.25). Each test column contains 3 boreholes of nominal diameters 48 mm (size A), 60 mm (size B), and 75 mm (size N) intersecting the ore zones. Three of the test columns contain ore zones of different concentrations for potassium, three for thorium, and three for uranium. The radioelement concentrations in these 9 ore zones are given in Table 10C.4.

Seven additional calibration pads for portable gamma ray spectrometers have been constructed in Calgary with the same physical dimensions as those at Ottawa. The radioelement concentrations are given in Table 10C.3. Two model boreholes constructed in Fredericton have ore zones of uranium, 1.5 m thick and of approximate grades 100 and 1000 ppm. These may be used to calibrate total count gamma ray logging equipment. Additional calibration facilities are planned at other Canadian locations.

The radioelement concentrations measured on the Breckenridge test range for airborne systems (described earlier) are given in Table 10C.2.

The U.S.A.

The U.S. Department of Energy has five concrete calibration pads, each 9.1 m by 12.2 m and 46 cm thick, located at Walker airfield, Grand Junction, Colorado. The pads are arranged in a line as shown in Figure 10C.26, with a turn-around at the end of the line of pads making it possible for an aircraft to taxi in a forward direction from pad to pad and return. The radioelement concentrations are given in Table 10C.2.

Table 10C.3
Radioelement concentrations of calibration facilities for
portable gamma ray spectrometer systems

Location	Pad Number	K%	eU ppm	eTh ppm
Canada[1]	PK-1-OT	0.8	—	—
	PK-2-OT	1.5	—	—
	PK-3-OT	3.0	—	—
	PK-4-OT	—	10	—
Ottawa	PU-5-OT	—	45	—
(Bell's Corners)	PU-6-OT	—	450	—
	PT-7-OT	—	—	8
	PT-8-OT	—	—	60
	PT-9-OT	—	—	300
	PB-10-OT	0.2	0.2	1
Canada[2]	PK-1-C	1.4	—	—
	PK-2-C	2.4	—	—
	PU-3-C	—	30	—
Calgary	PU-4-C	—	300	—
	PT-5-C	—	—	45
	PT-6-C	—	—	355
	PB-7-C	0.5	1	2
Denmark[3]	0	1.0	0.8	2.4
	1	7.0	4.2	2.7
Risø	2	0.8	6.3	151.
	3	1.0	198.	8.
South Africa[4]	Uranium - Pad 1	0.29	3731	290
	Uranium - Pad 2	.33	2078	202
	Uranium - Pad 3	.29	1255	114
	Uranium - Pad 4	.37	458	44
	Uranium - Pad 5	.33	12	<0.9
Pelindaba	Thorium - Pad 1	±1.66	280	12570
	Thorium - Pad 2	±1.66	102	3870
	Thorium - Pad 3	±1.66	68	3080
	Potassium Pad	10.1	0.4	0.9
	Mixed Pad	±1.2	763	4395
	Background Pad	0.33	0.5	0.9
South Africa[4] Beaufort West	Uranium Pad	.12	1153	75
United States[5]	Background	2	5	10
	Potassium	7.8	5	10
Grand Junction	Uranium	2	500	10
and Field Sites	Thorium	2	30	700
(Proposed)	Mix	4	350	250

[1] from preliminary data of Killeen and Conaway (1978).
[2] from preliminary data of Richardson and Killeen (1979).
[3] from Løvborg et al. (1978).
[4] from Corner and Toens (1979).
[5] from Evans (1978).

For calibration of portable (total count) scintillometers, four circular concrete pads containing uranium ore of dimensions 43 cm deep by 107 cm diameter are located at Grand Junction. Uranium grades are 80, 260, 1220, and 2830 ppm eU. In addition, two calibration pads are located at each of three field locations

Casper, Wyoming (265 and 1250 ppm eU)

Grants, New Mexico (270 and 1230 ppm eU)

George West, Texas (270 and 1250 ppm eU)

Additional facilities at other field locations are planned.

For calibration of portable gamma ray spectrometers, sets of five calibration pads, 60 cm thick by 1.2 m in diameter, are planned at all sites. The proposed composition of the pads is given in Table 10C.3.

Calibration facilities at Grand Junction for total count gamma ray logging include four uranium models, and a water factor model. These models have the following uranium concentrations: 17 400, 9075, 3765, 2020 and 2710 ppm eU. For gamma ray spectral logging, three model boreholes with ore zones 1.5 m thick are available with radioelement concentrations as given in Table 10C.4. In addition, a spectral or 'KUT' water-factor model was recently completed with proposed radioelement concentrations of 4 per cent K, 350 ppm eU, 250 ppm eTh.

At the three field locations, there are two uranium model boreholes, with additional sets planned at other field locations. The uranium concentrations in the existing models are:

Casper, Wyoming: 16 090 and 2790 ppm eU

Grants, New Mexico: 14 750 and 2630 ppm eU

George West, Texas: 14 400 and 2290 ppm eU.

The U.S. Department of Energy also plans additional spectral logging calibration models at all sites.

Recently, two new models have been constructed in Grand Junction containing thin dipping ore zones (Fig. 10C.27). The beds are 5 cm thick with ore grade of 1780 ppm eU. The models are 1.2 m in diameter, and the four beds or ore zones are at angles of 30, 45, 60 and 90 degrees (i.e. perpendicular) with respect to the borehole.

Besides the above-mentioned manmade calibration facilities, a radiometric test range is available for airborne gamma ray spectrometers. Referred to as the "Lake Mead Dynamic Test Range", it is located about 48 miles due east of Las Vegas, Nevada. Radioelement concentrations are given in Table 10C.2. An additional test range is planned.

Denmark

Four concrete calibration pads were constructed at the Research Establishment, Risø, near Roskilde, with dimensions 3 m diameter by 0.5 m thick. Their radioelement concentrations are given in Table 10C.3. The construction and monitoring of these pads has been described by Løvborg et al. (1978a).

Iran

A set of large calibration pads has been constructed in Iran at an airport in Tehran. These pads are circular with a 30 m diameter, and are spaced approximately 50 m apart in

Figure 10C.24. Geological Survey of Canada concrete test columns containing three model boreholes each, penetrating "ore zones". Bells Corners, Ottawa, Ontario (after Killeen, 1978). Three grades for each radioelement (K, U, and Th) are available as described in the text. Concentrations are given in Table 10C.4. (GSC 203254-O)

such a fashion as to allow an aircraft to taxi down the row of pads and back onto the tarmac. The radioelement concentrations of the pads are given in Table 10C.2.

South Africa

Calibration facilities were established at the national Nuclear Research Centre in Pelindaba in 1972 (Toens et al., 1973). These facilities now have eleven concrete pads 2 m in diameter by 0.3 m thick (Corner and Toens, in press) including five uranium pads, three thorium pads, one potassium pad, one mixed, and one background pad. The two model boreholes at the facility also contain single uranium and thorium 'ore' zones respectively. The models are 2.2 m long with 'ore' zones 0.95 m thick and a 12 cm diameter borehole. An additional calibration pad and model borehole were constructed in the uranium exploration area of the southern Karoo in Beaufort West. The radioelement concentrations of all of these calibration models are given in Tables 10C.3, 10C.4 (after Corner and Toens, in press). An 8 km long calibration strip for airborne gamma ray spectrometers is also available, located on Ecca Series sediments. Radioelement concentrations in the test strip average 2.44 ppm U, 3.09 ppm Th and 0.43% K_2O (D. Richards, pers. comm., 1978).

The use of the model holes was briefly mentioned by Corner and de Beer (1976).

Other Locations

New calibration pads and model holes are either planned or already under construction in many parts of the world as nations implement the recommendations of the IAEA (1976), (e.g. in Brazil, Spain, Australia). Pads for calibration of airborne gamma ray spectrometers have recently been constructed in Sweden using assemblies of precast concrete blocks fitted together to form large pads. Construction of model boreholes is also planned in Sweden (A. Hesselbom, pers. comm., 1978). A set of four pads, each 8 m square, has been constructed recently at an airport near Helsinki, Finland (Peltoniemi, pers. comm., 1978). Expansion of existing facilities in Canada and the U.S.A. is either planned or underway.

Some Recommendations Regarding Construction of Calibration Facilities

The IAEA has summarized the main points regarding construction of calibration facilities in Technical Report Number 174 (1976). Some data on the details of construction of the calibration pads in Denmark have been presented by Løvborg et al. (1978a, b; 1972). The construction of the large calibration pads for airborne systems was described by Grasty and Darnley (1971) for the pads at Ottawa, and by Ward (1978) for the pads at Grand Junction. A good description of the construction of the three K, U and Th model holes at Grand Junction is given by Knapp and Bush (1975). A description of the detailed construction specifications for the 10 calibration pads and test columns with model holes at Ottawa has been given by Killeen (1977).

National and International 'Standards'

Recalling Figure 10C.16, it can be seen that all calibration relies on some internationally agreed upon 'standards'. These are usually specially selected samples prepared in bulk, analyzed by several methods by several laboratories, and distributed by some recognized laboratory which maintains a large quantity of the standard for distribution, for a fee. In the United States and Canada, the samples generally used for standards are those available from the Standards and Reference Materials Section, U.S. Department of Energy, New Brunswick Laboratory, Building D350, 9800 South Cass Ave., Argonne, Illinois, 60439. These consist of 100 gram

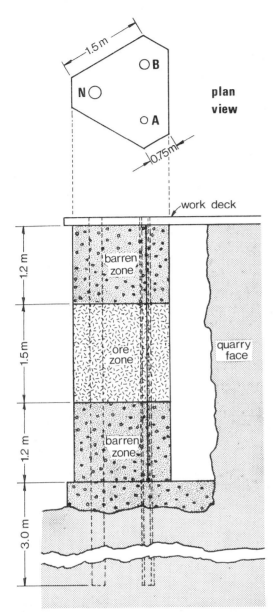

Figure 10C.25. Construction schematic showing one of the Geological Survey of Canada test columns at Bells Corners, Ottawa, Ontario (after Killeen and Conaway, 1978).

Table 10C.4

Radioelement concentrations of calibration facilities for borehole gamma ray spectrometer systems

Location	Borehole Model	K%	eU ppm	eTh ppm
CANADA[1]	BK-1-OT		0.7	—
	BK-2-OT	1.1	—	—
	BK-3-OT	3.0	—	—
Ottawa	BU-4-OT	—	15	—
(Bell's Corners)	BU-5-OT	—	100	—
	BU-6-OT	—	950	—
	BT-7-OT	—	—	8
	BT-8-OT	—	—	35
	BT-9-OT	—	—	350
SOUTH AFRICA[2]	Uranium	0.33	1221	114
Pelindaba	Thorium	0.51	19	890
SOUTH AFRICA[2]	Uranium	0.22	127	14
Beaufort West				
UNITED STATES[3]	K	6.30	2.9	2.5
(Grand Junction)	U	0.95	522	18.7
	T	1.36	26.1	508

[1] from preliminary values of Killeen and Conaway (1978).
[2] from Corner and Toens (1979).
[3] from Mathews et al. (1978).

AIRBORNE GAMMA RAY SPECTROMETRIC SURVEYS

Introduction

The application and interpretation of airborne gamma ray spectrometry rests on the foundation laid by earlier workers in total count scintillometry. Many of the techniques developed for airborne scintillometer surveys with respect to logistics and planning, data reduction and presentation, interpretation and correlation with geology, are applicable to airborne gamma ray spectrometer surveys. It is instructive to read the earlier literature on the subject (see e.g. Stead, 1950; Peirson and Franklin, 1951; Cook, 1952; Cowper, 1954; Agocs, 1955; Gregory, 1955, 1956, 1960; Bowie et al., 1958; Moxham, 1958, 1960; Guillou and Schmidt, 1960; Gregory and Horwood, 1963; Pitkin et al., 1964; Pitkin, 1968).

By 1966 some of the first papers on airborne gamma ray spectrometric surveying were being published (e.g. Pemberton and Seigel, 1966; Hartman, 1967) as well as the first descriptions of the data acquisition systems and processing of airborne gamma ray spectral survey data (see e.g. Bristow and Thompson, 1968; Foote, 1969; Grasty, 1972). The many aspects of data acquisition systems have been reviewed by Bristow (1979) and the theory and operational procedures have been reviewed by Grasty (1979) and Breiner et al. (1976). The application and interpretation of airborne gamma ray spectrometry requires a background knowledge of these two aspects of the subject as well as an integrated knowledge of the geology of uranium and thorium.

Selecting Survey Parameters

Airborne gamma ray spectrometric applications to uranium exploration fall into two main categories: reconnaissance surveys and detailed surveys.

The design of the survey must be aimed at answering the question: "How do the results of such a survey relate to potential uranium deposits?" The literature on reconnaissance airborne surveys and on detailed surveys is

bottles of crushed uranium ore and are available in concentrations of 10 ppm to 4 per cent U. Thorium samples (monazite sand mixtures) are also available from 10 ppm Th to 1 per cent Th.

It is important that the samples be in radioactive equilibrium or that the state of disequilibrium be known since the samples will be used to calibrate a radiometric method of analysis rather than a chemical method. Another source of standards is the Department of Energy, Mines and Resources, Canadian Certified Reference Materials Project, Mineral Sciences Laboratories, CANMET (Canada Centre for Mining and Energy Technology), 555 Booth St., Ottawa, Ontario, K1A 0G1. Samples available include Beaverlodge, Saskatchewan pitchblende uranium ore in concentrations from 220 ppm U to about 6 per cent U and Elliot Lake uranium ore containing both U and Th up to 1000 ppm.

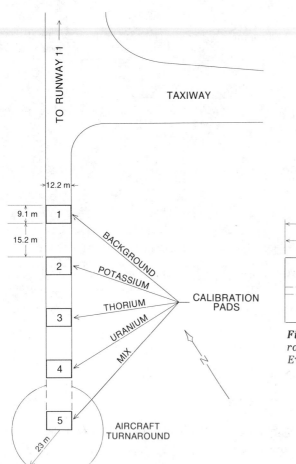

Figure 10C.26. Arrangement of U.S. Dept. of Energy (U.S. D.O.E.) calibration pads at Walker Field, Grand Junction, Colorado for calibration of airborne gamma ray spectrometers (after Ward, 1976). Radioelement concentrations are given in Table 10C.2.

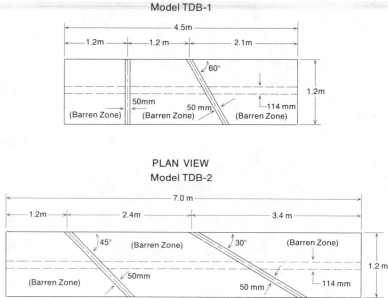

Figure 10C.27. Thin dipping ore zone models for borehole gamma ray logging at Grand Junction, Colorado (U.S. D.O.E.) (after Evans, 1978).

often intermingled; the latter is frequently a logical followup to the former. The design of a survey should consider line spacing, terrain clearance, detector volume, aircraft speed and sampling time.

A rather general paper on airborne geophysics by Willox and Tipper (1969) included examples of surveys by gamma ray spectrometry and Tipper (1969) described the airborne gamma ray spectrometer system used and the survey operation. Tipper pointed out that the line spacing (commonly 400 to 800 m) is generally a compromise between cost, the required detail, and the size of the survey area. The size of radioactive targets being sought and the possibility of subsequent more detailed flying must also be taken into account. The ground clearance which is generally between 75 and 150 m should be related to line spacing and aircraft safety. For detailed surveys the aircraft should be flown at a constant terrain clearance which is as low as possible. However, navigation then becomes more difficult, and high quality aerial photography becomes essential for flight path recovery.

A comparison of parameters of both a fixed wing and a helicopter gamma ray spectrometer was given by Darnley (1970) along with some measurements of the correlation between ground and airborne measurements.

A paper by Darnley (1972) includes three appendices on:

1. cost effectiveness of airborne radiometric surveys;
2. sample specifications for high sensitivity gamma ray spectrometer surveys; and
3. common causes of unsatisfactory airborne radiometric surveys.

An early application of Fourier analysis to airborne gamma ray spectrometric data was given by Killeen et al. (1971, 1975). They considered the target to cause an anomaly of a certain spatial 'wavelength', and from the desired target size derived some relations about the effects of altitude and sampling rate. Tipper and Lawrence (1972) presented gamma ray spectrometer profiles across the Nabarlek orebody in Australia flown at four different altitudes (120 to 300 m) as examples in a case history.

Calculations of the volume of material viewed by an airborne gamma ray spectrometer were presented by Duval et al. (1971) and by Grasty (1976a, b, and Grasty et al., in press). Figure 10C.11 shows the computed radius of the circle of investigation plotted as a function of altitude for a given percentage of infinite source yield, after Duval et al. (1971). Further work on the same type of computations was presented by Clark et al. (1972) and by Adams and Clark (1972).

Bowie (1973) discussed airborne radiometric survey requirements as well as other surveys in general. Nininger (1973) in reviewing exploration costs, considered relative costs of reconnaissance and detailed surveys. An IAEA Panel (1973a) summarized the application of airborne gamma ray spectrometer surveys incorporating Bowie's and Nininger's contributions in the following general statement:

"In practice two rather distinct approaches to airborne gamma-ray spectrometry have been developed to meet somewhat different objectives. One method uses a combination of gross-count and a minimal spectral capability to

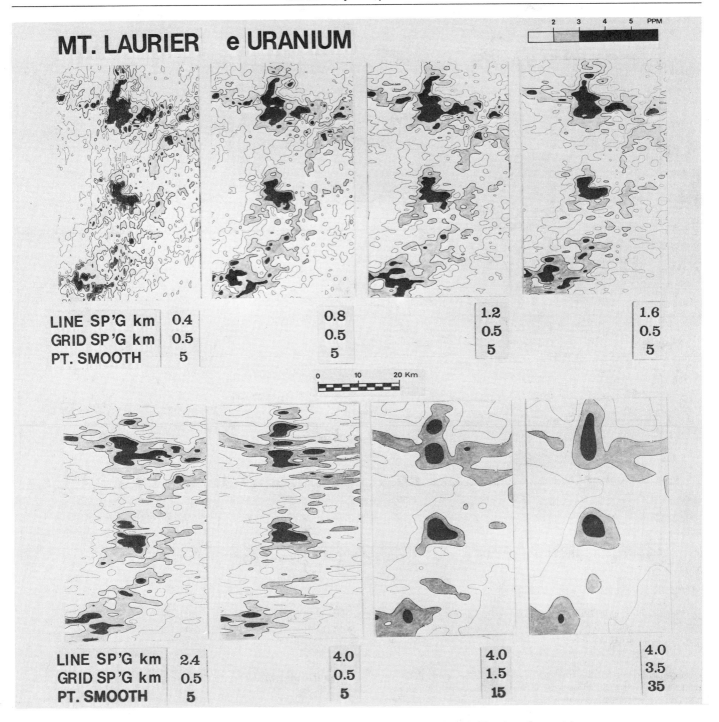

Figure 10C.28. *Comparison of the effect of flight line spacing on contoured eU values from airborne gamma ray spectrometric data (after Cameron et al., 1976). By appropriately deleting flight lines from a detailed (0.4 km spacing) survey, line spacings of 0.8 km, 1.2 km, 1.6 km, 2.4 km and 4.0 km are obtained and contoured with a grid spacing of 0.5 km and a 5 point smooth. A 15 point smooth appears to be optimum to avoid stretching of contours in the 4.0 km spacing case. The regional eU pattern evident from the 0.4 km spacing data is also readily apparent in the map produced from data along flight lines with 4 km spacing.*

achieve limited objectives. The other utilizes a high-sensitivity spectrometric capability to provide a sophisticated multi-parameter geochemical-statistical evaluation. The minimum system depends on gross-count for rapid regional search for anomalies and rudimentary geological mapping; spectral measurement of anomalies may identify the primary isotope, U, Th or K. The high-sensitivity spectral method is used for regional and area geochemical-geological mapping and to detect and map variations in lithology and anomalous radioelement ratios. Continuous ratio mapping is used to discover subtle anomalies often not indicated by gamma intensity alone. Sensitive spectrometry is most efficiently used to evaluate favourability of broad areas and for preliminary geological mapping".

Specifications for airborne gamma ray spectrometric surveys in Finland were arrived at by Ketola et al. (1975) on the basis of the results of a helicopter-borne test survey.

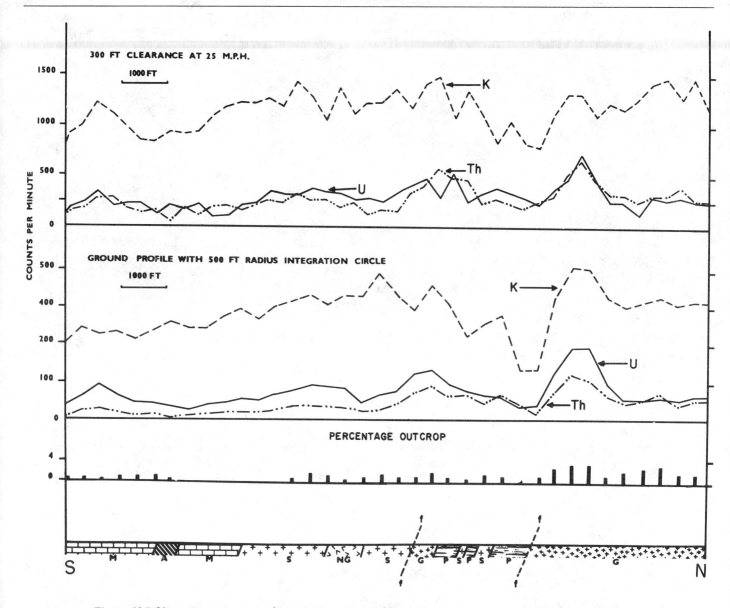

Figure 10C.29. Count-rate profiles in the Bancroft area measured by an airborne gamma ray spectrometer system at 90 m altitude and by a portable gamma ray spectrometer on the ground (after Darnley and Fleet, 1968).

They concluded that flight speed, altitude and detector volume were the most essential parameters affecting survey results. They also considered high quality data processing a dominant factor in determining the final result.

The effects of flight line spacing on contoured airborne gamma ray spectrometric data were investigated by Cameron et al. (1976). Using a survey of the Mont Laurier area with 400 m flight line spacing they selectively deleted alternate flight lines to simulate the same area flown with line spacings of 0.4, 0.8, 1.2, 1.6, 2.4, and 4.0 km. The resulting contour maps are shown in Figure 10C.28. They repeated this for the Elliot Lake area obtaining comparisons of flight line spacings of 0.5, 1.0, 1.5, 2.0, 3.0 and 5.0 km. Data from two coincident surveys in the Uranium City area with perpendicular flight line directions were also compared. They were able to demonstrate that surveys with 5 km flight line spacing were adequate to provide data for contoured regional radioelement distribution patterns over the Canadian Precambrian Shield.

Significance of Airborne Spectrometer Measurements

Darnley and Fleet (1968) described ground and airborne gamma ray spectrometry measurements over the Bancroft and Elliot Lake uranium mining areas of the Canadian Shield. A comparison of count-rate profiles measured at 90 m terrain clearance and on the ground is shown in Figure 10C.29 for the Bancroft area. The ground data were obtained from the integration of measurements within a 150 m radius circle moved along the flight line. The ground measurements were made on a grid with 60 m spacing. The airborne measurements were made with three 125 x 125 mm NaI(Tl) detectors in a helicopter flown at about 40 km/h. The similarity between the profiles is clear, indicating that these airborne measurements do indeed reflect the radioelement content of the ground. Darnley and Fleet (1968) presented histograms of count-rates obtained over various rock types, pointing out the potential for airborne gamma ray spectrometry as an aid to geological mapping. This paper drew attention to the importance of the ratio of uranium to thorium and uranium to potassium as indicators of possibly significant uranium mineralization.

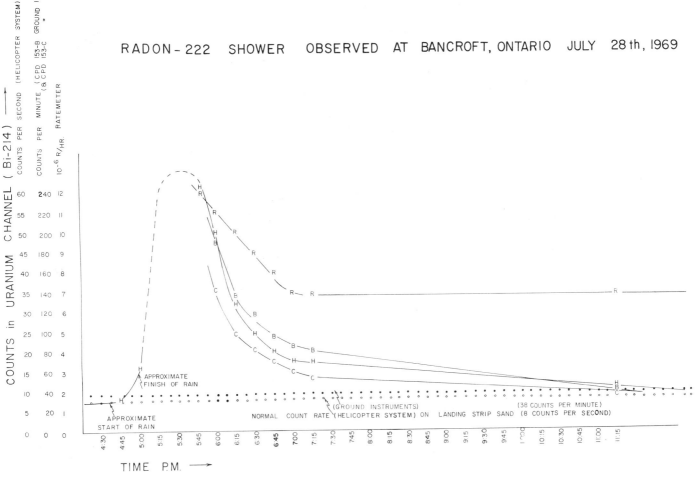

Figure 10C.30. Count-rates recorded by a helicopter gamma ray system parked on the ground, 2 portable spectrometers, and a ratemeter, from the onset of a rain shower until 7 hours later (after Charbonneau and Darnley, 1970b).

Experiences with helicopter-borne gamma ray equipment were described by Adams (1969). Profiles were presented for flights across beach sands, and granite jetties in the water off the Texas coast. A 3.5 m wide jetty was easily detected at a speed of 60 km/hr, at 15 m above the water using a 0.5 sec time constant and one 125 × 125 mm NaI(Tl) detector. However, the apparent width of the jetty determined from the anomaly was 58 m. Cook et al. (1971) presented further results using two 125 × 125 mm NaI(Tl) detectors with the same system.

For interpretation of the data other factors must be considered such as vegetation cover, percentage outcrop, and soil thickness and origin. For uncorrected data the variations in terrain clearance when the aircraft passes over valleys or hilltops may produce anomalies. Tipper (1969) indicated that the shape of anomalies on a contour map of total count may help to determine the contribution of the terrain effect. He also recommended that anomalies be interpreted on the basis of the ratios of the amplitudes in the different channels, and an understanding of the interference of the uranium, thorium, and potassium spectra in the energy windows.

Tipper further recommended that a short fixed traverse be reflown at least twice daily at the chosen survey altitude to monitor fluctuations in gamma ray attenuation caused by variations in humidity and soil dampness. These variations

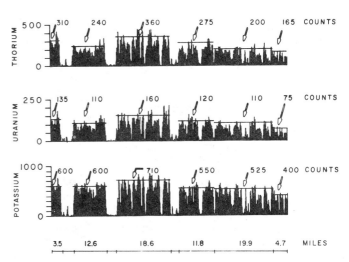

Figure 10C.31. Illustration of the method of determining the mean count rate levels from gamma ray spectrometric profiles in the Hardisty Lake area (after Richardson et al., 1972) (see text).

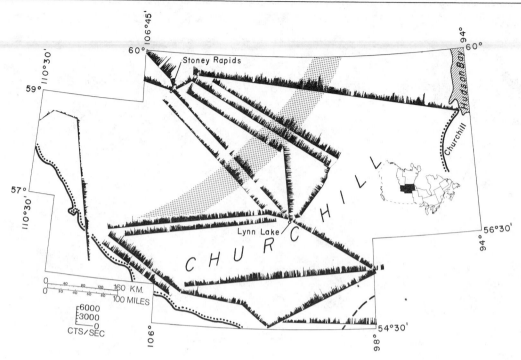

Figure 10C.32. A set of cross-country profiles illustrating a belt of anomalous radioactivity 70 km wide and 480 km long in the Churchill Province of the Canadian Shield (after Richardson et al., 1972).

are most significant if the timing of the survey is not carefully related to climate. Unless proper corrections are applied to each profile, spurious linear features parallel to the flight lines may occur on the resultant contour map.

Some measurements of the effect of rain on spectrometer measurements were presented by Charbonneau and Darnley (1970b). They plotted count rate changes with time during and after a heavy July rainfall which lasted for a 20 minute period. Four instruments were used: a helicopter spectrometer (parked on the ground), two portable gamma ray spectrometers, and a scintillometer. Figure 10C.30 shows a plot of the count rates recorded from the time of start of rain at 4:30 p.m. until about 7 hours later. The background radioactivity increased by a factor of 6 to a maximum about one hour after the rain started and thereafter decreased with about 30 minute half-life. Thus it is clear that gamma ray spectrometric surveys should not be carried out during or shortly after heavy rainstorms. Foote (1964) reported similar observations.

Flanigan (1972) described results of an airborne gamma-ray spectrometric survey with two 230 x 75 mm NaI(Tl) detectors in western Saudi Arabia. The area covered was flown with 1 km flight line spacing, and 90 m terrain clearance at 160 km/h. Data were recorded on the tape every 0.5 seconds. The author presented the results in the form of contoured count rates of the three radioelement windows and the total count channel. Count rate ranges were related to rock-type in order to extrapolate lithologies.

An interesting approach to comparison of radioelement concentrations measured on the ground and by airborne gamma ray spectrometry was described by Richardson et al. (1972) for parts of the Canadian Shield. In order to compare results from 14 000 rock specimen assays to the airborne data, they determined average count rates along profiles for each radioelement. The authors described the procedure as follows: They obtained a visual estimate for each of the radioelements from appropriate profile cross sections of different portions of the Shield. They did this by:

(1) neglecting narrow zones of high count rate (which constitute a small fraction of the distance along the flight path), (2) ignoring low levels of radioactivity over water and swampy areas, and (3) visually estimating a mean count rate level from the upper surface of the profiles. This procedure is illustrated in Figure 10C.31 which shows thorium, uranium, and potassium profiles along 130 km of cross country reconnaissance flight line over the Hardisty Lake area in the Northwest Territories. Estimated radioelement concentrations were obtained by converting weighted averages of these count rates using the calibration facilities. Using this method of analysis, and compilation of cross country profiles, a number of radioactive belts were defined and their radioelement contents were evaluated. Figure 10C.32 illustrates a set of cross country profiles, and the belt of anomalous radioactivity (70 km wide, 480 km long) defined by Richardson et al. (1972). They gave the mean radioelement content of the belt as 4.8% K, 6.2 ppm eU, and 31.1 ppm eTh. This trend runs along the Wollaston Lake Fold Belt and the parallel gneissic zone to the east. In the same paper, computer contoured K, U, and Th distribution maps of the Fort Smith area, Northwest Territories are presented. Correlation with magnetic data in these areas is also discussed in some detail. The authors suggest that the coincidence of distinct magnetic properties and radiometric properties in certain zones of the crust indicate both have been controlled by the same large scale fundamental processes.

The Canadian Uranium Reconnaissance Program was described by Darnley et al. (1975) who stated that the overall objectives of the program were: "to provide industry with high quality reconnaissance exploration data to indicate those areas of the country where there is the greatest probability of finding new uranium deposits, and to provide government with nationally systematic data to serve as a base for uranium resource appraisal".

The Canadian program involved high sensitivity airborne gamma ray spectrometry over areas of low relief and geochemistry in mountainous terrain and in areas with extensive overburden. The airborne gamma ray spectrometry

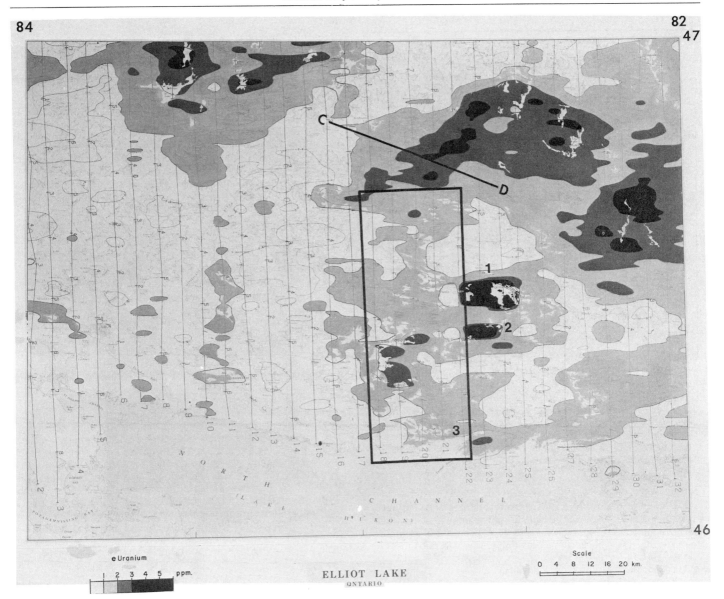

Figure 10C.33. Contoured eU distribution for the Blind River map sheet, Ontario compiled from the GSC Skyvan data (after Darnley et al., 1977). Note the broad regional high to the north of the Elliot Lake uranium mining area. The principal ore deposits adjoin 1 (Quirke, Denison), 2 (Nordic), and 3 (Pronto). The outlined area, and the profile C-D are discussed in the above-mentioned paper.

part of the program was carried out with 5 km flight line spacing, 135 m terrain clearance, 200 km/h aircraft speed and 50 000 cm^3 of NaI(Tl) detectors. The basis for the program was given by Darnley et al. (1975) in the following words:

"The program rests upon the concept that most uranium deposits occur within or marginal to regions of the crust containing higher than average amounts of uranium. Uranium may be found to be weakly concentrated in granitic rocks especially those late in an orogenic cycle. It may be found concentrated in high temperature pegmatites or in lower temperature vein deposits. These are all components of a primary source area which through erosion and redistribution can provide the material to form secondary deposits in any suitable adjacent geochemical trap. The reconnaissance program is designed primarily to identify all zones of primary enrichment within the country, and secondly to indicate, if possible, the limits of areas where secondary processes have operated. It is important not to dismiss anomalous areas as simply being low-grade igneous rocks of no economic importance. Such areas may have considerable potential as source areas, and geological knowledge must be brought into play to determine where the eroded material from these source areas has been deposited. It is the first objective of the Uranium Reconnaissance Program to delineate as rapidly as possible the major areas of uranium enrichment in Canada".

The concept above was illustrated with survey data from the Elliot Lake uranium area, the Johan Beetz (Havre St. Pierre) area of Quebec, the Beaverlodge uranium area of northern Saskatchewan, the Bancroft uranium area, and the Mont Laurier area of Quebec (Darnley et al., 1975) and other areas Darnley et al. (1977). Figure 10C.33, reproduced from Darnley et al. (1975), is one of these examples showing the uranium distribution pattern in the Blind River-Elliot Lake area. This map is described by the authors as showing a possible source-area to the northwest of the Huronian sedimentary rocks which contain the uranium deposits of the Elliot Lake area. Further descriptions of this particular

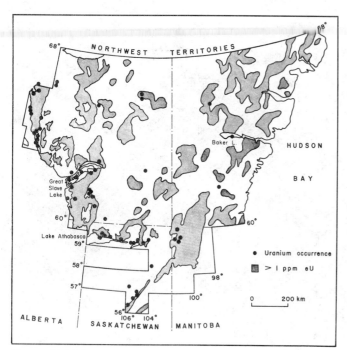

Figure 10C.34. *The relation between uranium occurrences and regional distribution of uranium (after Richardson and Carson, 1976). Based on airborne gamma ray spectrometric data, surface concentrations greater than 1 ppm eU have been contoured and shaded in this compilation from northern Manitoba, northern Saskatchewan and the Northwest Territories. Most occurrences lie within or near areas of regional uranium enrichment.*

survey were given by Richardson et al. (1975). Richardson and Carson (1976) presented a compilation map showing both the location of known uranium occurrences and areas determined to have an average surface eU content >1 ppm as measured by airborne gamma ray spectrometry. Figure 10C.34 shows the striking relationship between occurrences and regional enrichment.

Morris (1969) described the use of airborne gamma ray spectrometers in the search for uranium and gave as examples spectrometer records obtained at 150 m above ground with two 150 x 100 mm NaI(Tl) detectors flown over areas in southern England.

Darnley et al. (1969) presented further experimental results of airborne gamma ray spectrometer tests over the Canadian Shield, describing a high sensitivity spectrometer system used by the Geological Survey of Canada. A comparison of mean count rates for arrays of twelve 230 x 100 mm NaI(Tl) detectors and three 125 x 125 mm NaI(Tl) detectors was presented from flights over the Bancroft area.

Darnley et al. (1970) presented example profiles for five different areas of Canada showing variation of total count, K, eU, eTh, and eU/eTh ratio. An example of the interpretation of one of these profiles is given in Figure 10C.35. The authors stated that the figure is a south to north profile from the Bancroft area. The Grey Hawk uranium property is marked by the distinct equivalent uranium peak at 17 miles which is not accompanied by any increase in potassium or equivalent thorium. About two miles to the north the Faraday granite area is marked by another equivalent uranium peak associated with anomalous equivalent thorium and potassium but no eU/eTh ratio anomaly. Further north still there is a localized potassium concentration, without a matching increase in equivalent thorium or uranium. The eU/eTh ratio in addition to clearly distinguishing the Grey Hawk uranium occurrences shows two other weak anomalies which could have been easily overlooked on the equivalent uranium profile alone.

Darnley (1972) has several illustrations of data that demonstrate the use of gamma ray spectrometric data. One of these illustrations, Figure 10C.36, is a profile from the Uranium City area of Saskatchewan demonstrating how a significant equivalent uranium anomaly can be lost in a total count gamma survey because it coincides with low equivalent thorium and potassium values. The existence of the uranium anomaly is clearly indicated by the spectrometric data but not by the integral or total count profile. Figure 10C.37, also adapted from Darnley (1972), is an illustration from the Bancroft area of a small uranium occurrence at 19 km which is readily distinguished by a spectrometric survey, but which would not be found by a total radiation survey. (These last two classic examples have also been used by Grasty (1976c) to illustrate airborne gamma ray spectrometric data.)

Darnley and Grasty (1971) discussed results of surveys in the Bancroft area in detail, presenting seven contour maps showing total count, the three radioelements, and their ratios eU/eTh, eU/K and eTh/K, each superimposed on the geologic map of the area. The authors concluded that use of the ratio maps (e.g. eU/K and eU/eTh) can result in an order of magnitude reduction in the area to be ground searched, as compared with the total count map. Darnley (1973), in reviewing developments in airborne gamma ray survey techniques, stated that districts containing uranium mineralization generally fall within or on the margins of regions containing above-average radioelement abundances. Profiles from the Elliot Lake area, Mont Laurier area, and Fort Smith area in Canada were presented to illustrate this point. Figure 10C.38 is a reproduction of a profile from the Fort Smith, N.W.T. survey, adapted from Darnley (1973). Note the broad regional high in the total count channel, predominantly caused by thorium enrichment as indicated by the equivalent thorium channel. Uranium anomalies are located on the margin of the radioactive high as denoted by the equivalent uranium channel profile and the eU/eTh and eU/K ratios.

A comparison of reconnaissance techniques, made by Allan and Richardson (1974) for the northwestern Canadian Shield, concluded that in that area airborne gamma ray spectrometry and lake-sediment geochemistry can produce similar regional distribution maps for uranium at a similar cost. The comparison involved the high sensitivity airborne system of the Geological Survey of Canada flown at 5 km line spacing, and lake-sediment sampling by helicopter with one sample per 25 km². In 1975, Darnley (1975) presented a short review of geophysics in uranium exploration, emphasizing airborne gamma ray spectrometry. The concept of regional uranium enrichment and its relation to source areas was discussed. Several examples illustrating how gamma ray spectrometry can explain the cause of a total count anomaly were given. Some of these included: a classic profile across a thorium anomaly caused by the St. Andrews East carbonatite, a profile across the South March uranium occurrence near Ottawa, and a profile over a potassium anomaly.

Data Presentation Techniques

The form of presentation of results was a major consideration of a paper by Darnley (1972) and several methods of data presentation were compared. The description of the methods and their advantages and disadvantages are reproduced below.

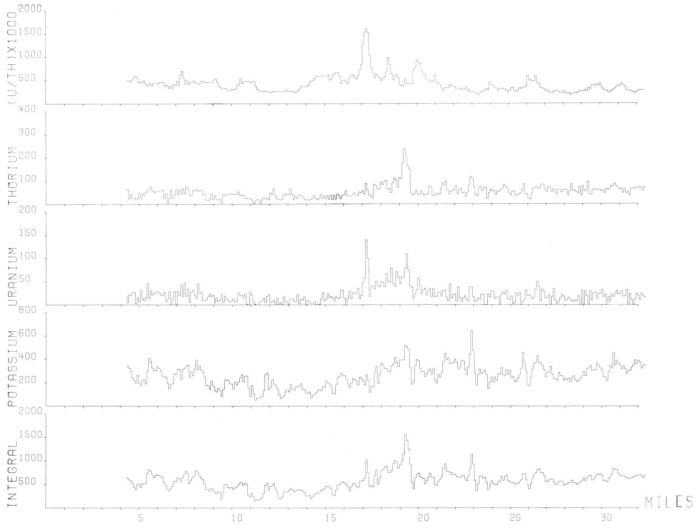

Figure 10C.35. A radiometric profile from south to north across the Greyhawk uranium property in the Bancroft area; see text (after Darnley et al., 1970).

The output of the first commercially available gamma ray spectrometers consisted of an analog chart recording with no corrections whatsoever, making only simple data presentations possible. Contouring was either not attempted or was limited to thorium only. A common form of data presentation was the use of a symbol superimposed on a flight line plan to indicate an anomaly exceeding an arbitrary background value by some given factor. Figure 10C.39 is an illustration of this type of uranium 'anomaly map'. This type of presentation although inexpensive is unsatisfactory according to Darnley from several points of view: (1) it is somewhat subjective; (2) since there are no Compton scattering corrections many uranium anomalies could be caused by high thorium values; (3) the background used is the average overland radiation level and therefore the information which the overland radiation base level can provide about the general geochemical environment is ignored; (4) since no terrain clearance correction was applied some anomalies may be caused by topographic highs.

In Figure 10C.40 the uranium and thorium count rates in the anomaly peak are shown along-side the uranium and thorium backgrounds (used in the same sense as the first example). This improves the amount of information available on the map.

A further elaboration is to contour thorium content and add this to the display of anomalies as shown in Figure 10C.41. Since the thorium count is usually more reproducible and statistically more significant than the uranium count it can be used as an aid in interpreting the geology.

Figure 10C.42 shows another type of data presentation, a profile map; also called offset profiles. Here the strip chart data have been plotted on a flight line map alongside the flight lines. In theory this shows the relationship in radiometric pattern from line to line and similar features can be joined. In this particular example, no allowance was made for the lag in plotted positions due to the time constant employed in the survey. Thus anomalies on adjoining lines are laterally displaced relative to one another because adjoining lines were flown in opposite directions.

Darnley (1972) also pointed out that elaborate presentation of data is not warranted if the counting statistics of the measurements are inadequate.

Recently, more effort has been put into displaying the data in forms which make the interpretation easier. The use of statistical treatments to enhance the data, filtering,

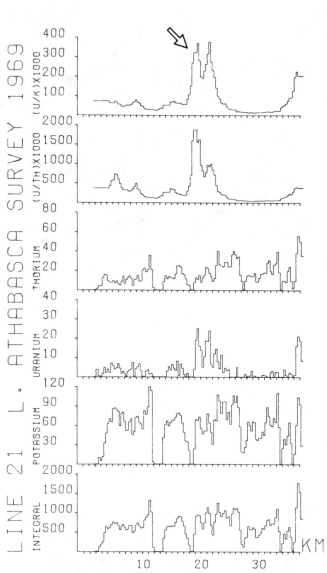

Figure 10C.36. A classic example from the Uranium City area of Saskatchewan demonstrating how a significant eU anomaly can be lost in a total-count gamma ray survey because it is coincident with decreased eTh and potassium values (after Darnley, 1972).

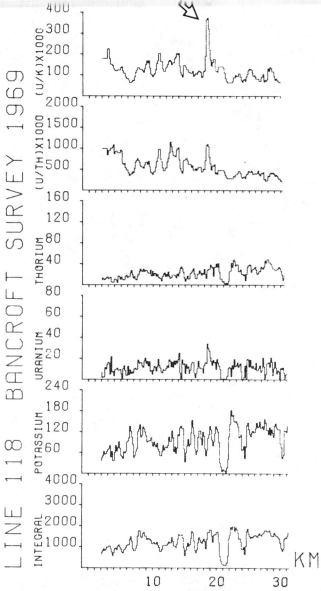

Figure 10C.37. A small uranium occurrence in the Bancroft area (at 19 km) which does not show on the total count profile, but which is easily seen by a spectrometric survey (after Darnley, 1972).

factor analysis, and the use of colour to combine information from all three radioelements in one map have been demonstrated and are discussed below.

A standard deviation map has become commonly used (see for example Geodata International, 1975a, b, c) as a form of data presentation by the U.S. Department of Energy in their National Uranium Resource Evaluation (NURE) program. Richardson and Carson (1976) utilized this type of presentation to display data from the Athabasca Formation in northern Saskatchewan which produced a rather uniform level of radioactivity and which was consequently difficult to contour. To produce these standard deviation maps (or anomaly maps) the mean value of equivalent uranium for each flight line was calculated and data points that exceed the mean by 1, 2, 3 or more standard deviations are indicated by 1, 2, 3 or more stars plotted above the flight line. An example from Richardson and Carson (1976) (Fig. 10C.43) shows the geology map for an area and the equivalent uranium anomaly map. "The maps show prominent anomalies on several flight lines near the contact between the quartz monzonitic gneiss (unit 1) and the Virgin River Schist Group (unit 3a). A few young uranium anomalies also occur on the southeastern edge of the quartz monzonitic gneiss, near its contact with biotite-garnet (unit 2) and diorite gneisses and schists (unit 3)" (Richardson and Carson, 1976). The authors suggested that these anomalous zones may be geologically favourable for Key Lake-type uranium deposits.

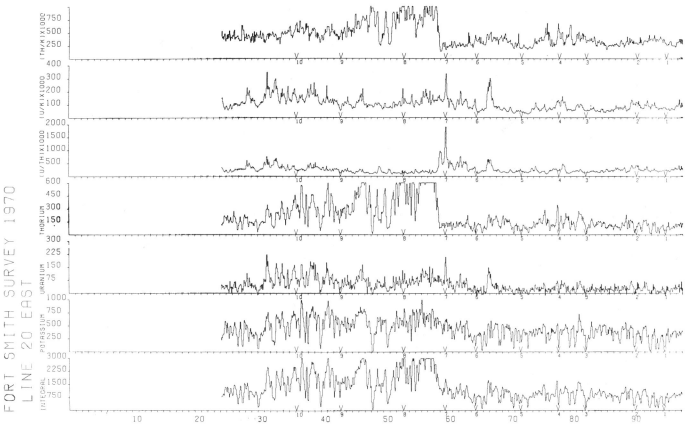

Figure 10C.38. Airborne gamma ray spectrometric profile from the Fort Smith area, Northwest Territories flown with a high sensitivity system (50 000 cm^3 NaI(Tl), after Darnley, 1972). Particularly interesting is the broad high eTh area of the Fort Smith belt (30 to 60 miles) with a narrow eU anomaly on the flank (at 60 miles), accentuated by the U/Th and U/K ratios.

In areas of well known geology the mean values for all data measured over each rock unit may be treated as above as has been done for the U.S. NURE program (Foote and Humphrey, 1976; Saunders and Potts, 1976). In this way the standard deviations correspond to a given rock unit rather than to the mean values for a given flight line. Potts (1976) presented a contour map of these standard deviations from the mean for equivalent uranium, calling it a "significance factor map". The significance factors are fractional multiples of the standard deviation above or below the mean, and can be considered the "degree of rarity" of a measurement. Figures 10C.44 and 10C.45 show the uranium countrate contour map and the significance factor contour map of uranium data for a survey area in South America (after Potts, 1976). The difference between the two maps is surprising, indicating the area with the greatest number of standard deviations from the mean value is nearly 10 km from the area with the highest uranium count rates.

A colour presentation of airborne gamma ray spectrometric data (Linden, 1976) shows the gamma radiation related to K, eU and eTh in the form of coloured columns plotted on the flight lines. The surveys used a 250 x 125 mm NaI(Tl) detector, flown at 200 m flight line spacing, at a height of 30 m. Digital recording of data occurs every 0.4 seconds, representing about 40 m of flight line. Every measurement is taken to represent gamma radiation from an area of 40 m x 200 m, and is depicted on the 1:50 000 map as an area of 0.8 mm x 4.0 mm. Linden (1976) described the process further:

"Within this area are plotted three centre-orientated columns. The lengths of the columns are proportional to the radiation that relates to each element. Potassium is represented by yellow, uranium by red and thorium by blue. Each group of three columns is separated from the next group by a white field equivalent to the width of one column. There is sufficient space to specify 20 different levels of radiation intensity. In areas of abnormally high or low contents of K, eU or eTh it is possible to improve resolution by increasing the contrast between element intensities".

The three component colour map technique has been used with success in Sweden for several years, and has been applied to geological mapping as well as uranium exploration.

Tammenmaa et al. (1976) applied digital time series analysis techniques to airborne gamma ray spectrometric data to derive suitable filters to improve ground resolution and reduce distortion of radiometric anomalies.

Gunn (1978) discussed the deconvolution of airborne gamma ray spectrometric data, and the possibility of utilizing "downward continuation" as it is often applied to gravity and magnetic data. Tammenmaa and Grasty (1977) demonstrated upward and downward continuation of gamma radiation fields. Richards (1977) also applied digital filtering to airborne gamma ray spectrometric data to remove statistical noise.

The technique known as factor analysis was applied to airborne gamma ray spectrometric data by Duval (1976, 1977). Basically the technique can be considered as a method of sorting the K, eU, and eTh data into groupings with similar

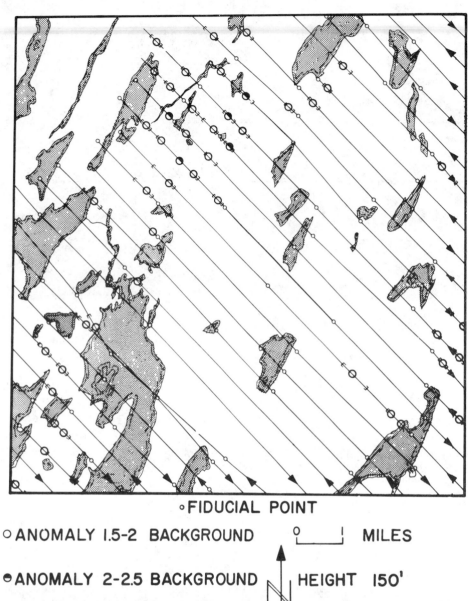

Figure 10C.39. *Data presentation example 1: symbols superimposed on a flight line map indicating anomalies exceeding some arbitrary background value by some given factor (after Darnley, 1972).*

Ziegler (1976) discussed some trial applications of geostatistics to airborne gamma ray spectrometric data. Some of the techniques tried included locating maximum variance segments of the data, "robust" techniques, cluster analysis, discrimination analysis, and data display by histograms and three-dimensional maps of frequency distribution by flight line for each given geological formation.

Foote (1976) reviewed the data presentation techniques currently in use by the U.S. NURE program including the following data presentation formats:

"1. Flight line profiles of intensity of radiation from uranium, potassium, thorium, and their ratios.

2. Histograms showing data distribution.

3. Radiation data by surface geologic unit.

4. Radiation data by flight line showing statistical variation from a mean value.

5. Data by flight line superimposed on surface geologic map."

Several other commercially used data presentation techniques are described below. The "zoning technique" first used in 1968 is described as follows by Hogg (pers. comm., 1977).

First, the equivalent thorium, equivalent uranium and potassium channels as presented in an analog airborne record or a computer-compiled multichannel profile are analyzed. The boundaries of recognizable changes in signal amplitude, ratio or character are marked on the profile. These subunits are described semi-quantitatively by an alpha-numeric code. K, U and T refer to the potassium, equivalent uranium and equivalent thorium channels, respectively, and the signal strength indicated by symbols ((+) strong, ( ) average and (-) weak) which relate to preselected levels (Fig. 10C.46). A computer/plotter may then be used to produce a profile map of fully corrected total count profiles with the zone boundaries and related coding annotated (Fig. 10C.47). These maps may then be coloured using different intensities for count rate amplitude and a different colour for each radioelement K, eU, and eTh. The advantage of this technique is that all of the information is presented on one map. The main disadvantage, however, is the amount of time involved in the initial steps of the processing.

Another data presentation method (Hogg, pers. comm., 1977) consists of a computer line-printer listing of anomalies meeting specified criteria. Thus for each anomaly, a set of co-ordinates, where the three co-ordinate axes are K, eU, and eTh. A similar approach was described by Killeen (1976a). The areas in which the data fall into groupings with similar co-ordinates are then coloured or shaded on maps. This technique has particular potential as a geological mapping aid. Newton and Slaney (1978) developed a classification system for airborne gamma ray spectrometric data in a survey area based on test flight lines which were studied in detail by photogeology to assign radiometric signatures to each rock type. Once rock classes were identified they were used to classify the entire survey area. The authors stated that zones of anomalously high radioactivity often cross lithological boundaries and may be considered useful indicators for uranium exploration. Further results of these investigations have been presented by Slaney (1978).

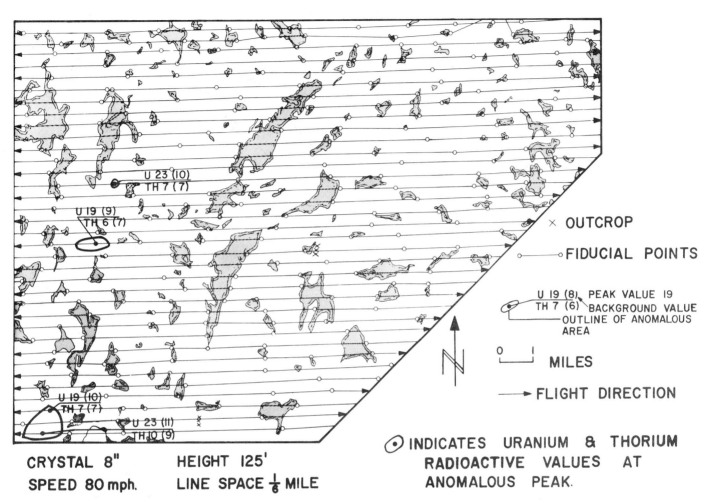

Figure 10C.40. Data presentation example 2: uranium count-rate in the anomaly peak is given beside the uranium background, and the thorium count-rate in the anomaly is given beside the thorium background (after Darnley, 1972).

statistics is printed including flight line number, fiducial numbers nearest the anomaly, anomaly amplitude, half width (left), area (left side), half width (right), area (right side), total half width, total area as shown in Figure 10C.48. Postscripts 1 and 2 refer to parameters calculated from stripped, smoothed, altitude-corrected equivalent uranium profile with subtraction of either: (1) atmospheric background or (2) local background plus atmospheric background. A line printer plot may also be produced to illustrate roughly the shape of the anomaly. Below the printer plot are given the count rates and ratios calculated at the uranium anomaly peak.

A useful anomaly classification technique has been described by D.B. Morris (pers. comm., 1977). The count rates in the three radioelement channels are expressed as a percentage of the count rates in the total count channel by a normalization process. Then the position of the anomaly is located on a ternary diagram as shown in Figure 10C.49. The diagram has 100% K, 100% eU, and 100% eTh as the three points of the triangle. It can be seen that any anomaly can be located on the diagram by the relative percentage contributions to the total count from channel 2 (K), channel 3 (U) and channel 4 (Th). The ternary diagram is divided into 9 fields.

An example of the application of this anomaly classification system is given in the anomaly analyses presented in Table 10C.5. Three example anomalies have been analyzed by this process and they fall into fields S, K, and L respectively.

Locating Favourable Areas for Uranium Exploration

In the past, most interpretation efforts were directed toward explaining, classifying, and setting priorities on gamma ray spectrometric anomalies to aid exploration. However, locating a favourable area in the first place is another problem. Reconnaissance surveys will outline large regions with above average radioelement content (geochemical provinces) which may be considered favourable areas, or in the vicinity of favourable areas (Darnley et al., 1977). Dodd (1976) described two suggested techniques to outline favourable areas. One approach consists of first producing the uranium anomaly map as described earlier, in terms of positive and negative standard deviations above or below the mean for each geologic unit. Anomalies were defined as data points exceeding one standard deviation either positive or negative. Areas of major and minor clusters of positive anomalous eU records were marked on the map, followed by the contouring of the ratios of the number of positive to the number of negative anomalies within each

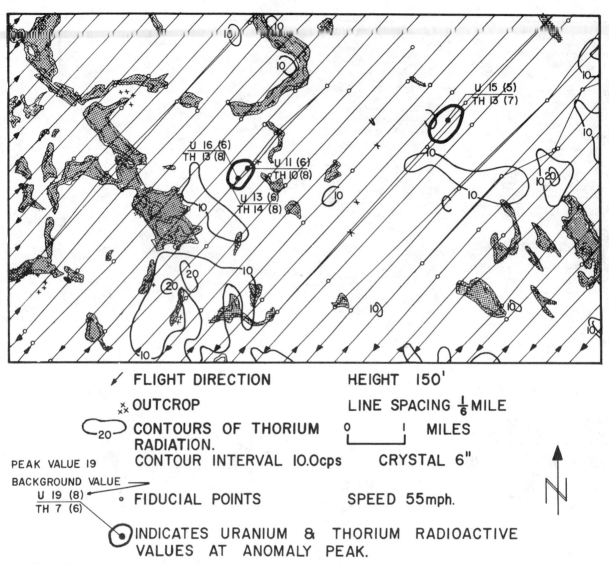

Figure 10C.41. Data presentation example 3: similar to Figure 10C.40 but with eTh content contoured in the vicinity of the anomalies (after Darnley, 1972).

fifteen minute quadrangle. This procedure was applied to an area of 37 300 km² in Wyoming. The resultant favourable areas (Fig. 10C.50) show good correlation with the Gas Hills, Crooks Gap-Green Mountain, Shirley Basin and Copper Mountain mining districts.

The second method described by Dodd (1976) is illustrated in Figure 10C.51. Here average values of the radioelement concentrations and their ratios are computed for each flight line in a survey with 8 km line spacing. These values are plotted as a profile spanning 410 km along the strike of the Goliad Formation on the Texas Gulf Coast. There is a regional change of K and eU and all known uranium occurrences are restricted to the area covered by lines 1 to 31. This area is the region of the Goliad formation containing the higher mean values on the profile.

Saunders and Potts (1978), attempted to determine a general "uranium favourability index" by plotting histograms of various possible indices such as the eU/eTh ratio for about 30 different areas where existing mines and occurrences were known to be favourable. They concluded that the median values of aerial gamma ray spectrometer parameters for geologic map units could be used as a guide to identify uraniferous provinces, reasoning that where the crustal abundance of uranium is high it is available to be chemically concentrated in economic deposits. They further reasoned that geochemical processes must have concentrated a part of the uranium in deposits. Removal of uranium from "average" rocks, separating it from thorium and potassium, results in low eU/Th and eU/K median values. They found the following parameters <u>decrease</u> with increasing uranium potential (where M denotes mean value): M (eU/eTh), M (eU/K), MeU/MeTh, and MeU/K.

Parameters which generally <u>increase</u> with increasing uranium potential are: MeU, MeTh, $\overline{MK}$, RSD eU, RSD (eU/eTh), and RSD (eU/K)

where RSD = relative standard deviation. i.e. (standard deviation/mean). Saunders and Potts (1978) thus arrived at a uranium favourability index, U_1, given by the equation:

$$U_1 = \frac{(MeU + MeTh + MK) \cdot RSD\ eU \cdot RSD(eU/eTh) \cdot RSD(eU/K)}{M(eU/eTh) \cdot M(eU/K)}$$

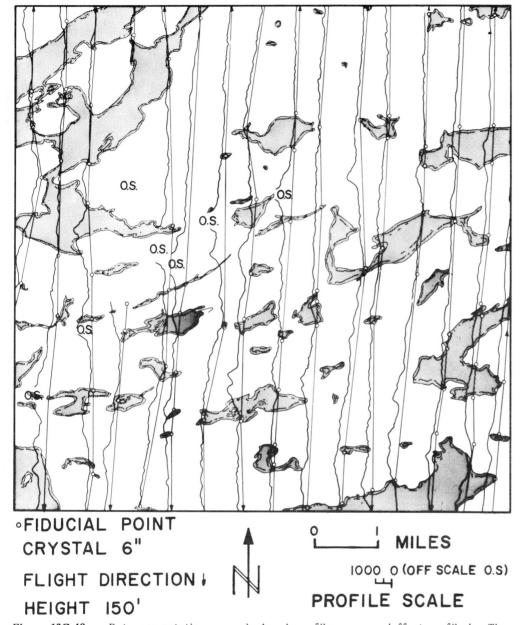

Figure 10C.42. Data presentation example 4: A profile map, or 'offset profiles'. The radiometric record is plotted beside the flight lines (after Darnley, 1972).

Deriving a uranium favourability index for a region is a complex problem. The parameters reflecting high favourability for a sandstone-type deposit such as those discussed above may not be applicable to other types of deposits.

SURFACE GAMMA RAY SPECTROMETRIC SURVEYS

Introduction

Portable gamma ray spectrometers are versatile instruments and have been used in carborne surveys, underwater surveys, airborne surveys and borehole logging. This section will deal principally with hand-carried spectrometers, i.e. foot-traverse surveys and detailed ground investigations of anomalies detected by other radiometric surveys, such as airborne or carborne. A number of instruments are available from different manufacturers ranging from simple single-channel instruments to four-channel instruments, with analog and/or digital outputs, audible alarms and stabilization. Some instruments can be programmed to present results corrected for spectral scattering and reduced to counts per second. Count rate displays are by LED, LCD or rate-meter needle, and a variety of detector sizes are available. The choice of instrument depends on many factors such as the skill or training of the operator, the objective of the survey, the locality (desert area, tropical jungle, cold northern areas), the cost, and possible future usage of the instrument. For example the immediate requirement may be for a simple model, but if a foreseeable survey requirement includes analog chart recording then the instrument chosen should have an analog output. Size and weight of the instrument or use in aircraft or lab also play an important role. Many aspects of surface gamma ray spectrometry such as energy windows, calibration, in situ assay, sample volume, counting statistics, geometry, deadtime and background are considered in other sections of this paper.

A simpler uranium index which avoids the use of RSD values was also suggested by Saunders and Potts (1978). Based on the observation that high mean uranium content indicates there is sufficient uranium for possible geochemical concentrating processes to work, and that low mean eU/eTh and eU/K values indicate these processes took place, they derived the index U_2:

$$U_2 = \frac{MeU}{\frac{MeU}{MeTh} \cdot \frac{MeU}{MK}} = \frac{MeTh \cdot MK}{MeU} \quad (34)$$

The histogram in Figure 10C.52 illustrates how the uranium index indicates the high uranium potential of the Casper and Delta areas, and ranks the favourability of 27 other quadrangles below them. The Casper area contains three major mining districts, and the Delta area contains one mine and numerous uranium occurrences.

Portable Gamma Ray Spectrometers

The earlier reports on the use of portable instruments in the field were primarily concerned with scintillometers (e.g. Russell and Scherbatskoy (1951), Gross (1952), Russell (1955)) or interpretation of laboratory gamma ray spectrometer measurements of rocks (e.g. Hurley, 1956; Horwood, 1960;

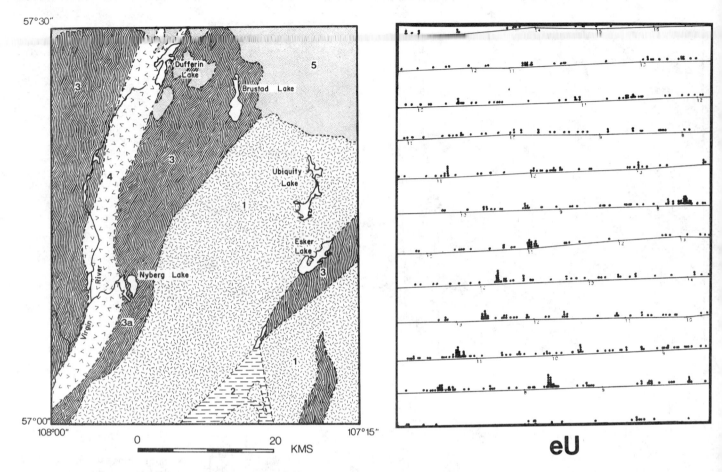

Figure 10C.43.   Comparison of geology map and eU anomaly map for the Nyberg Lakes – Brustad River area (after Richardson and Carson, 1976). The statistical treatment of the data (described in the text) produces a map which relates to the geology of the area, whereas the simple eU contour map did not.

Mero, 1960; Bunker and Bush, 1966). Gregory and Horwood (1961) carried out some fundamental research on the shape of gamma ray spectra with variation of source thickness. They demonstrated the feasibility of field gamma ray spectrometry. Adams and Fryer (1964) described the first portable gamma ray spectrometer in use in the United States, and Mahdavi (1964) described its application to the study of K, U, and Th concentrations in beach sands of the Gulf of Mexico coast. This spectrometer utilized a lead shield for collimation of the gamma rays. In Canada the first portable gamma ray spectrometer was constructed (Doig, 1964, 1968) under the auspices of the Geological Survey of Canada. Killeen (1966) and Killeen and Carmichael (1972) described the application of that spectrometer to uranium exploration in the Elliot Lake area. Darnley and Fleet (1968) produced gamma ray spectrometer maps of test areas at Bancroft and Elliot Lake. The first commercially available portable gamma ray spectrometers in Canada were described by Pemberton (1968). About the same time in the United States, airborne gamma ray spectrometers, which had been developed potentially for tracking the effluent of nuclear submarines, were becoming de-classified by the military. Foote (1969) reported on both surface and airborne applications of spectrometry to mineral exploration. Almost simultaneously with the development of portable gamma ray spectrometers in Canada and the U.S.A., Løvborg et al. (1969) developed a field portable unit in Denmark. They applied it to exploration for uranium and thorium deposits of South Greenland. Due to the rugged terrain a detector with lead collimator was used to control the geometry. They also described the calibration procedure, which consisted of comparison of field count rates to laboratory analyses of samples. Killeen and Carmichael (1970) described a similar method of calibration for an uncollimated portable gamma ray spectrometer. Adams and Gasparini (1970) reviewed all aspects of gamma ray spectrometry in a text book on the subject. Kogan et al. (1969) also provided extensive mathematical background on the subject. Miller and Loosemore (1972) described a portable gamma ray spectrometer with automatic gain stabilization, and more recently Clayton et al. (1976) described the updated version. Puibaraud (1972) summarized the generally desirable characteristics for a variety of portable radiometric equipment, including gamma ray spectrometers.

Løvborg et al. (1971) presented a detailed description of their use of the portable gamma ray spectrometer to evaluate the Ilimaussaq Alkaline intrusion in South Greenland. Making measurements on grids of 1 m spacing with a collimated detector, and taking as many as 100 measurements per day, it was possible to contour the resulting data. The same paper reported on calibration with a new set of concrete calibration pads. The computation of the "effective" sample volume was described. This is different from the 90% source volume commonly used to define the sample (i.e. the volume from which 90% of the detected radiation originates). The "effective" sample volume is defined as the "rock body in which the variance of a particular radioelement is equal to the estimation variance of a gamma ray spectrometric

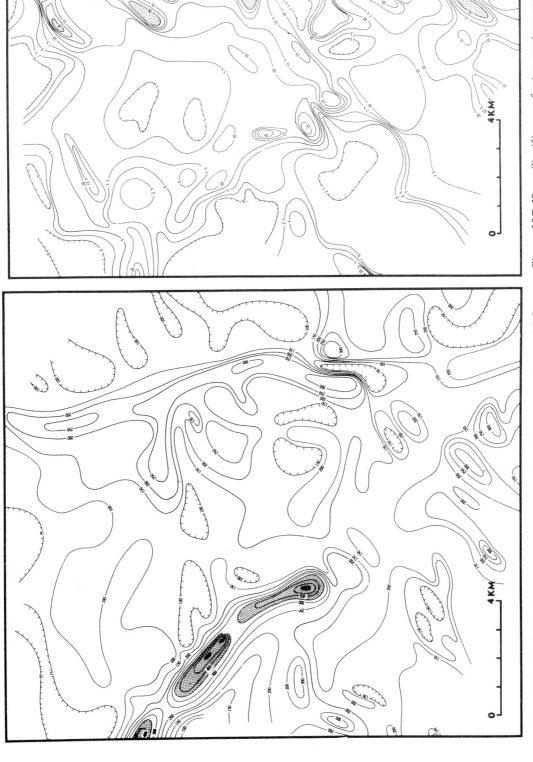

Figure 10C.45. Significance factor contour map (described in the text) for the area in South America shown in Figure 10C.44 (after Potts, 1976).

Figure 10C.44. Uranium count-rate contour map (counts per 2 seconds) for the area in South America shown in Figure 10C.45 (after Potts, 1976) based on airborne gamma ray spectrometry.

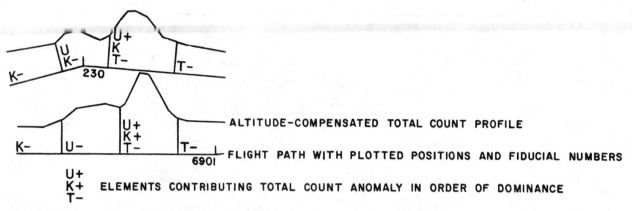

Figure 10C.46. The "zoning" technique applied to a section of radiometric total count profile as described in the text (after Hogg, pers. comm., 1977).

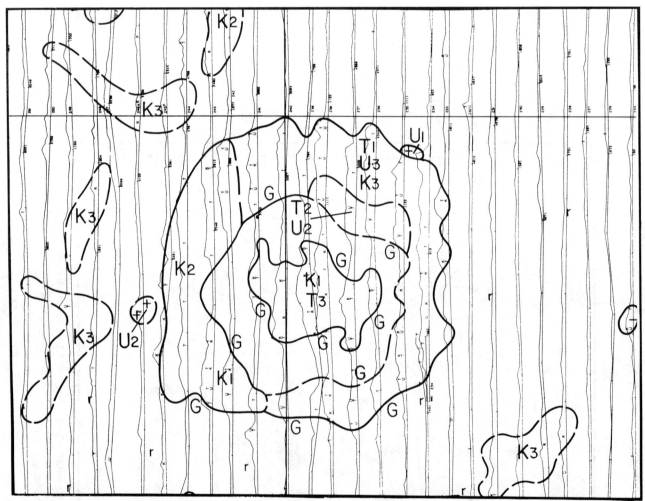

Figure 10C.47. The "zoned" map based on total count profiles processed as in Figure 10C.46, with the zone boundaries joined (after Hogg, pers. comm., 1977).

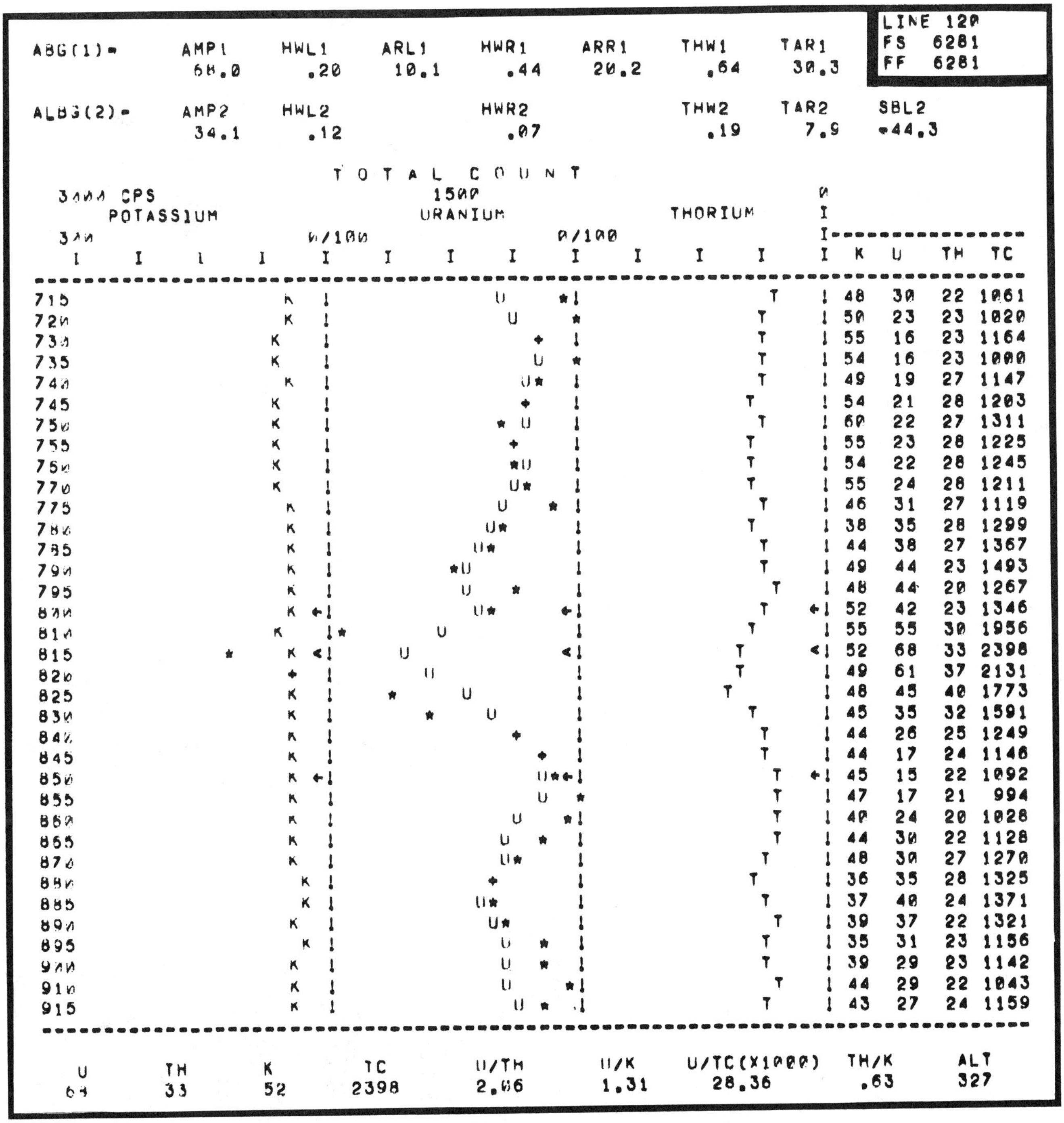

Figure 10C.48. A 'printer-plot' anomaly description (see text) used as an aid in classifying anomalies for symbol presentation on maps (after Hogg, pers. comm., 1977).

determination of this radioelement" (Løvborg et al., 1971). The "effective" sample contributes only about 60 per cent of the detected gamma rays. They stated that "for an isotropic detector the volume of the 60% effective sample is about one-tenth the volume of its 90% counterpart". A good discussion on accuracy and precision of field gamma ray spectrometric measurements is given. Løvborg (1972) reviewed the applications of gamma ray spectrometry and the portable spectrometer developed in Denmark was also described.

Løvborg et al. (1972) described a mathematical approach to the computation of theoretical gamma ray energy spectra. From this they were able to derive correction factors to compensate for the finite dimensions of concrete calibration pads which are assumed to be of infinite extent (2π geometry) when calibrating a portable spectrometer. For example, using their graph of correction factors (see Fig. 10C.18) count rate corrections of +8% in the 2.62 MeV channel, +7% in the 1.76 and 1.46 MeV channels and +3% in the total count channel would be applied when the detector is 10 cm above the pads. These studies were expanded (Løvborg et al., 1976) to include airborne spectrometers and the effects of sodium iodide's inherent characteristics in geophysical work.

A comprehensive table of gamma ray energies for the ^{238}U, ^{232}Th, and ^{235}U radioactive decay series, compiled by Smith and Wollenberg (1972), is a valuable aid to the interpretation of gamma ray spectra. Løvborg (1973) considered the future of gamma ray spectrometry, reviewing the available equipment, the precision of measurement, counting statistics and calibration standards. A modeling study of the response of 75 x 75 mm NaI(Tl) detectors was done by Løvborg and Kirkegaard (1974) to improve their computation of theoretical gamma ray spectra and aid in interpretation of field recorded spectra.

Sibbald (1975) described in considerable detail the ground follow-up program to investigate anomalies located by airborne gamma ray spectrometric surveys in northern Saskatchewan. Airborne radiometric anomalies were first marked on airphotos and then outlined on the ground by pace and compass traversing, with a scintillometer. Readings were taken continuously at hip level, and in contact with the surface at interesting locations. Samples were collected, and the geology was recorded. Of 21 anomalies, five were selected for more detailed investigation, and an integrated geological-geophysical-geochemical program was carried out. On the basis of geology, anomalies were divided into type 1 (relating to granites) and type 2 (relating to granite pegmatites). Pegmatitic anomalies were investigated in more detail than granitic anomalies. All outcrops were located and mapped, and checked with scintillometers in the detailed investigations. A grid was established at either 7.5 or 15 m spacings and portable four-channel gamma ray spectrometer readings were taken at waist height using 10 second counting intervals. The nature of the terrain was recorded at each station. The geology was mapped from these grid results alone in the less detailed investigations. Sibbald (1975) indicated that type 1 anomalies are commonly well exposed and occupy high land areas. The anomalies are typically recorded as spectacular and often broad highs in radiometric profiles. Type 2 anomalies occur in deeply eroded valleys and are less well exposed.

Another ground follow-up investigation in northern Saskatchewan between Wollaston Lake and Reindeer Lake was described by Munday (1975). The anomalies had been located by the GSC Skyvan survey flown in 1974 on flight lines at 1600 m spacing at an altitude of 120 m, and speed of

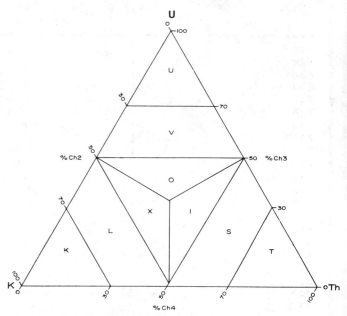

Figure 10C.49. Ternary diagram illustrating a process for classifying airborne gamma ray spectrometric anomalies into nine different fields (after Morris, pers. comm., 1977).

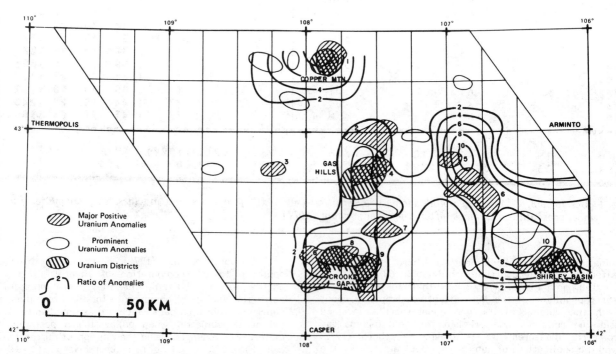

Figure 10C.50. Favourable areas for uranium exploration in Wyoming delineated by clusters of positive anomalies and high values of the contoured ratios of the number of positive to the number of negative anomalies (after Dodd, 1976).

Table 10C.5

Example of three anomalies classified as type S, K, and L
using the fields shown in Figure 10C.49 (after Morris, pers. comm., 1977).

						Corrected Counts					
					Peak	B/ground	Anomaly	%			
Anomaly No. 6	Channel 1										
	Peak time	6478	Secs						U/Th ratio		0.0
	1/2 width time 1	6476	Secs	Channel 1	200.	124.	76.		Th/K ratio		1.9
	1/2 width time 2	6482	Secs	Channel 2	31.	21.	10.	35	Anomaly type		S
	Peak Altitude	426	Feet	Channel 3	11.	10.	2.	0			
	Peak raw counts	707.	CPS	Channel 4	19.	8.	11.	65	Source		Th+U/K
	Ch 3 peak time	6475	Secs								
Anomaly No. 7	Channel 1										
					Peak	B/ground	Anomaly	%			
	Peak time	6578	Secs						U/Th ratio		.3
	1/2 width time 1	6575	Secs	Channel 1	131.	59.	72.		Th/K ratio		.9
	1/2 width time 2	6605	Secs	Channel 2	29.	13.	17.	45	Anomaly type		X
	Peak Altitude	471	Feet	Channel 3	15.	9.	6.	14			
	Peak raw counts	514.	CPS	Channel 4	16.	6.	9.	41	Source		Mixed(K)
	Ch 3 peak time	6580	Secs								
Anomaly No. 8	Channel 1										
					Peak	B/ground	Anomaly	%			
	Peak width time 1	6600	Secs						U/Th ratio		.1
	1/2 width time 1	6575	Secs	Channel 1	116.	50.	67.		Th/K ratio		.6
	1/2 width time 2	6605	Secs	Channel 2	31.	10.	21.	61	Anomaly type		L
	Peak Altitude	468	Feet	Channel 3	13.	9.	4.	4			
	Peak raw counts	599.	CPS	Channel 4	15.	6.	9.	35	Source		K+U/Th
	Ch 3 peak time	6597	Secs								

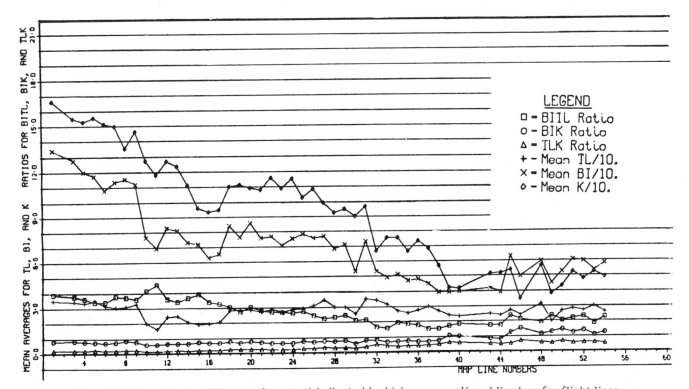

Figure 10C.51. Favourable areas (see text) indicated by high average eU and K values for flight lines 1 to 31 over the Goliad formation of Texas (after Dodd, 1976).

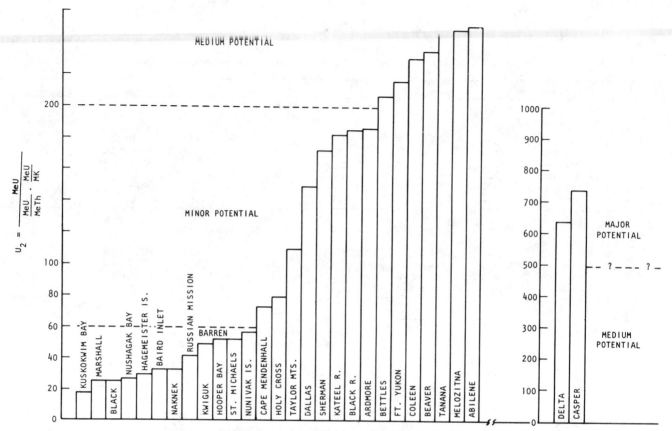

Figure 10C.52. Histogram of "uranium favourability indexes" as described in the text (U_2 values) for whole quadrangles. A high value indicates high potential for uranium (after Saunders and Potts, 1978).

190 km/hr with 50 000 cm³ of NaI(Tl) detector. The following priorities were adopted for investigating the anomalies:

Priority 1 Anomaly:
high total count with high eU/eTh ratio

Priority 2 Anomaly:
low total count with high eU/eTh ratio.

Munday (1975) indicated that areas of abnormal radioactivity were outlined by traversing in a direction roughly perpendicular to the glacial strike. Traverses were made at 1000 m spacing with scintillometer readings every 15 m at ground and waist level. Readings at 7 m and 3.5 m intervals were tested but the increased detail did not warrant the extra time. Also they found that waist height readings gave a better indication of general radioactivity, especially after a 3 point running average filter was applied to the data. Munday (1975) described the boulder count technique to determine the shape of the boulder fan and thus locate the source as follows:

"When anomalies had been delimited, boulder counts were taken every 150 m along traverse lines. Only boulders of 25 cm or more, diameter were measured and the scintillometer crystal was centered on the boulder. At least 100 radioactivity readings were established for each station and plotted as a histogram. Background glacial till yields a log-normal distribution, but as anomalous populations invade the system, bimodal distributions appear. Finally, when an anomalous population dominates the background, a second log-normal distribution can be plotted with higher mode value. The anomalous population is readily attributable to a

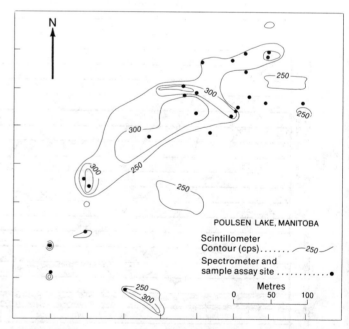

Figure 10C.53. Contoured scintillometer data, Poulsen Lake uranium occurrence, Manitoba (after Whitworth et al., 1977). The anomaly size and shape are typical of a boulder train, and as such is useful in estimating target size for an airborne survey.

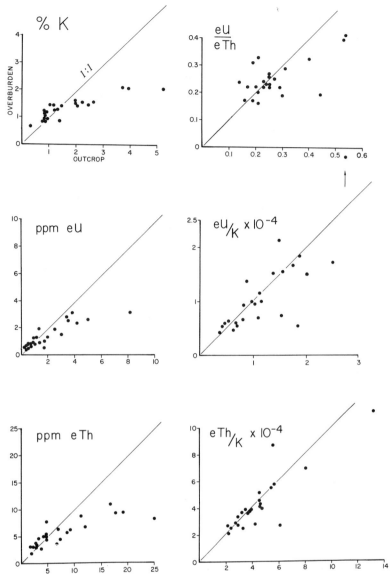

Figure 10C.54. *Mean radioelement values in outcrop versus radioelement values in the associated overburden for 24 different rock types in Canada (after Charbonneau et al., 1976). Radioelement ratios in outcrop and overburden cluster about the 1:1 line, whereas the radioelement concentrations in overburden do not increase in 1:1 with increasing concentration in outcrop.*

source in the field, and in this area was invariably related to a pink leucogranite. By systematically mapping the areal distribution of the anomaly, a fan was established and source area determined".

Several ground follow-up investigations in northern Manitoba were described by Soonawala (1977), Garber and Soonawala (1977), Whitworth et al. (1977), and Smith et al. (1977). These were investigations of anomalies located by airborne gamma ray spectrometric surveys of the Canadian Uranium Reconnaissance Program. The systematic exploration sequence began with a helicopter-borne scintillometer survey at 250 m flight line spacing, 40 m terrain clearance, and speed of 110 km/hr. The detector was a 150 x 100 mm NaI(Tl) (1850 cm^3). This was followed by ground scintillometer surveys (with NaI(Tl) detectors having a volume of 43 cm^3) in many cases on grid lines at 50 m spacing with stations 5 m apart. Anomalies located by this survey were evaluated by a combination of in situ assay by portable gamma ray spectrometer, and laboratory sample analyses. The scintillometer survey results were contoured, and the in situ and laboratory assays were marked on the contour map. The contoured scintillometer data obtained at the Poulsen Lake occurrence (Whitworth et al., 1977), reproduced in Figure 10C.53, delineates the shape of a boulder train. In situ spectrometric assays at several locations along the boulder train indicated equivalent uranium concentrations in boulders of up to 870 ppm, with several discrete groupings of boulders in the 250 to 500 ppm range.

Other descriptions of ground follow-up investigations with portable gamma ray spectrometers and scintillometers have been given by Charbonneau and Ford (1977, 1978) and Charbonneau et al. (1975). Charbonneau et al. (1976) compiled portable gamma ray spectrometer data consisting of over 2500 in situ assays from 24 sites across Canada. These results established two relationships:

1. The sympathetic relationship between radioelement contents of glacial overburden and the underlying bedrock.

2. The relationship between "average surface" radioelement concentrations measured by airborne gamma ray spectrometry, and concentrations measured on the ground.

Figure 10C.54 is a plot of the mean radioelement values in outcrop versus the radioelement values in the associated overburden, for 24 different rock types (after Charbonneau et al., 1976). Note that the radioelement ratios (eU/eTh, eU/K, eTh/K) show relatively little difference between the overburden and bedrock. Figure 10C.55 compares ground and airborne measurements of eU and eTh in the Elliot Lake and Mont Laurier areas. The values along the horizontal axis are the contour levels from the airborne spectrometric maps; values along the vertical axis are averages of ground-measured concentrations for each of the airborne contour levels. Radioelement concentrations in glacial drift measured on the ground are slightly higher than indicated by contoured radioelement values measured by the airborne survey in the same area. This is believed to be the result of the presence of surface waters within the area of investigation by the airborne system, which will reduce the measured airborne values. Radioelement concentrations in bedrock measured in situ are considerably higher than corresponding airborne measurements indicate, and this difference between bedrock and airborne survey values increases at higher radioelement concentrations. Similarly, outcrop radioelement contents become increasingly greater than the associated overburden radioelement contents, as the radioelement concentration increases.

In the areas discussed by Charbonneau et al. (1976) the airborne survey radioelement contour maps are primarily a measure of overburden radioelement content, but the airborne results do give an indication of the bedrock radioelement content. For example, an airborne measurement of 3 to 4 ppm eU in a drift covered area probably relates to an overburden content of 4 to 5 ppm eU, and a concentration of about 8 to 10 ppm eU in the underlying bedrock. Airborne contour maps of the radioelement ratios give values that are similar to the ratios determined by ground measurements on overburden and outcrop.

It has been suggested that portable gamma ray spectrometer surveys be carried out with solid state detectors rather than sodium iodide. The main advantage of the solid state detectors (e.g. lithium drifted germanium (Ge(Li)), or

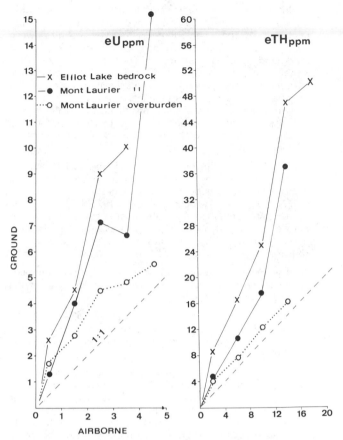

Figure 10C.55. Comparison of ground measurements and airborne measurements by gamma ray spectrometer in the Elliot Lake and Mont Laurier areas of Canada (after Charbonneau et al., 1976). Ground measurements are average values for all ground stations within the contour intervals indicated on a map from the airborne measurements. The ground measurements are higher than the corresponding airborne measurement since the latter is an 'average surface measurement' and includes water, swamp, outcrop and overburden.

hyperpure germanium) is energy resolution which permits distinction between peaks in the gamma ray energy spectrum which are indistinguishable with sodium iodide detectors. The main disadvantages of solid state detectors are the requirement to operate at liquid nitrogen temperature and the increased counting time necessitated by the relatively small detector size. To take advantage of the high resolution, a multi-channel spectrometer (preferably 4096 channels) is required. Field spectrometer systems using solid state detectors have been described by Anspaugh et al. (1972), Ragaini et al. (1974), Dickson et al. (1976), and Finck et al. (1976). The high resolution of solid state detectors presents the possibility of relatively direct determination of uranium concentration by spectrometry as well as indirectly by detection of the daughter ^{214}Bi. This has been investigated by Moxham and Tanner (1977) and their results indicated that at least a semi-quantitative measure of the state of equilibrium can be obtained in the field.

Carborne Gamma Ray Spectrometric Surveys

Most of the literature presently available pertaining to carborne surveys concern total-count rather than spectrometric surveys. Many of the principles involved and the field procedures will be illustrated with examples based on total-count surveys, but which apply to spectrometric surveys.

Calibration of Carborne Systems

Calibration of a carborne gamma ray spectrometer system has been mentioned earlier in the discussion of calibration of surface systems. It is difficult to do quantitative work with a carborne system since the geometry generally varies along the road traverses. However, accurate determination of the stripping factors on a set of calibration pads makes it possible to produce stripped counts and therefore radioelement ratios can be utilized. This is advantageous because the ratios of the radioelements are not seriously affected by geometry changes or changes in vegetation or moisture content (Charbonneau et al., 1976).

Detector Type and Volume

Some consideration has been given to the volume of the detector and its location in the vehicle by Berbezier et al. (1958). They also compared different types of detectors including Geiger-Muller (GM) tubes, sodium iodide crystals, and plastic scintillators. They found that a volume of 232 cm^3 of NaI(Tl) gave about the same results as 3750 cm^3 of plastic scintillator (tetraphenylbutadiene), but they considered the higher cost and fragility of NaI(Tl) was a disadvantage. However they pointed out that NaI(Tl) crystals make it possible to do gamma ray spectrometry and therefore determine which radioelements were present in the source. The GM counters were much less sensitive, produced lower count rates and consequently required a longer time constant than the NaI(Tl), making it difficult to detect narrow anomalous sources. Bowie et al. (1955) also compared GM counters with a sodium iodide scintillation counter and Kamiyama et al. (1973) reported on usage of two 75 x 125 mm detectors by carborne survey teams in Japan.

Nelson (1953) presented a thorough discussion of carborne radiometric surveying with a GM counter. With today's more modern electronics and the present availability of sodium iodide crystals, the other types of detectors mentioned above may be considered outmoded. For gamma ray spectrometry, scintillation crystals are mandatory. Irrespective of the type of detector, the determination of the required detector volume will depend on the desired reproducibility of measurement. Nelson (1953) considered an acceptable standard deviation to be ±10%. Recalling the earlier section on counting statistics, this requires N to be at least 100 counts in a given counting period. The next consideration is the minimum target or anomaly width which it is desired to detect. The anomaly could be considered to be of a certain 'wavelength' (Killeen et al., 1975). By sampling theory the <u>minimum</u> wavelength detectable is given by

$$\lambda_m = 2V \Delta t \qquad (35)$$

where

V   is the velocity of the vehicle in m/sec.

Δt is the sampling time in seconds.

Thus, for example, at a speed of 15 km/h (i.e. 4 m/sec) and a sampling time of 1 second, then $\lambda_m = 8$ m. Note that this anomaly of wavelength 8 m would only be sampled twice when crossing it at a velocity 4 m/sec. To define it more accurately a larger number of samples, at least four, is desirable. The important point is that once the target size is chosen, the speed and sampling time are essentially also determined. For example, if the target size is 4 m, it is desirable to make a measurement every metre; with a velocity of 15 km/hr, then the value of Δt is 0.25 seconds. This means for a desired 10% standard deviation per measurement, the count must be 100 counts in 0.25 seconds or at least 400 counts per second in areas of background

Figure 10C.56. Carborne gamma ray spectrometric survey installation in Mexico (Instituto Nacional de Energia Nuclear) with roof mounted detector. (GSC 203254-I)

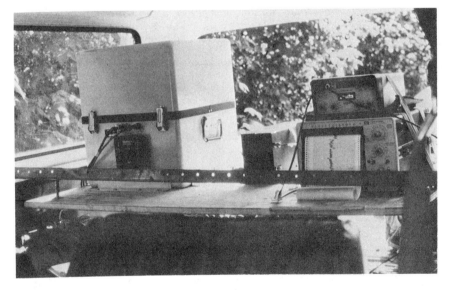

Figure 10C.57. Carborne gamma ray spectrometric survey installation in Canada (after Killeen et al., 1976); detector mounted inside the vehicle (left), portable gamma ray spectrometer and single pen strip chart recorder (right). (GSC 202941-E)

the material at the roadside and beyond, and decreases the percentage contribution from the road itself. Some discussion of the effective sample volume can be found earlier in this paper. Figure 10C.17 illustrates the effect of increasing the height of the detector in increasing the sample volume and minimizing road effects. Generally, however, the detector height is limited by bridges, tunnels and overhead wires. Goso et al. (1976) reported a carborne radiometric survey in Uraguay in which a 1230 cm^3 NaI(Tl) detector was mounted in a tower on the vehicle at an elevation of 3.50 m. The installation of a 1850 cm^3 detector on the roof of a 4 wheel drive jeep used by the Instituto Nacional de Energia Nuclear (INEN) of Mexico for gamma ray spectrometric work is shown in Figure 10C.56. The installation of 4200 cm^3 detector inside a four wheel drive vehicle used for a carborne gamma ray spectrometric survey on Prince Edward Island, Canada (Killeen et al., 1976) is shown in Figure 10C.57. Moxham et al. (1965) utilized a 6800 cm^3 NaI(Tl) detector on a tripod at 2.5 m height to make stationary measurements of gamma ray spectra with a 400 channel analyzer in a panel truck.

The orientation of the detector is not usually considered, but can be of some significance, especially if the thickness and diameter of the detector are very different. Nelson (1953) considered this problem with respect to the orientation of GM tubes. He arranged the detectors so as to present the largest detector surface area to gamma rays emitted from roadside rocks. The smallest area was then oriented to minimize gamma rays detected from road material below and cosmic rays above. If a prismatic (100 x 100 x 400 mm) sodium iodide detector were to be used for a carborne survey, it should be oriented with its long axis vertical.

Shielding by the Vehicle

In addition to the above considerations, the shielding effect of the vehicle may be used to advantage, especially if it is known that the road material is not locally derived. In some cases the radioactivity of the road material presents considerable interference, especially if the road material changes frequently, introducing man-made anomalies. A lead shield beneath the detector may be necessary in addition to the shielding by the vehicle (Bowie et al., 1955). In areas of unimproved dirt roads the vehicle shielding may be a hindrance, and the detector may be suspended behind the vehicle or over a hole in the floor. Shideler and Hinze (1971) in a carborne radiometric survey relating to petroleum exploration of glaciated regions encased the detectors in "lead containers" (not described) to minimize the effects of cosmic rays.

Pre-Survey Performance Checks

It is important to run some pre-survey tests under known conditions in order to be able to recognize anomalies when traversing new territory. In addition it is instructive to evaluate the effect of various parameters on the performance of the carborne system.

radioactivity. From this information the detector volume can be determined if the count rate per cubic centimetre of detector is known approximately for that background radioactivity from some preliminary measurement.

The detector volume required for a gamma ray spectrometric survey would be considerably larger than for a total-count survey. The IAEA (1973a) stated that "This type of survey should not be attempted unless a large volume of crystal can be provided". No specific volume was suggested.

Detector Location and the Sample Volume

The detector should be located as high as possible in the vehicle. This increases the diameter of the circle of investigation, and the percentage of radiation contributed by

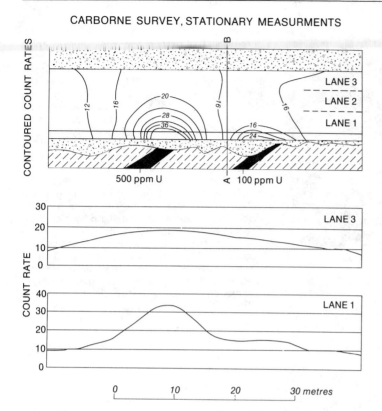

Figure 10C.58. Radioactive zones of 500 and 100 ppm uranium as a test area for carborne gamma ray survey (after Nelson, 1953). Traverses in the "near" and "far lane" made with stationary measurements are shown. A cross-section A-B of the highway is shown in Figure 10C.59.

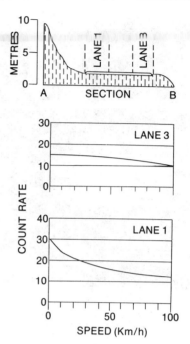

Figure 10C.59. Effect of vehicle speed on the maximum amplitude of the anomaly detected both in the "near" and "far" lane (after Nelson, 1953). The cross-section A-B (top) shows the source-detector geometry in Figure 10C.58.

A good example of this type of pre-survey performance study was given by Nelson (1953). For his test area Nelson chose a road cut which intersected two radioactive zones containing approximately 500 ppm and 100 ppm uranium respectively as shown in Figure 10C.58. A series of stationary total-count measurements were made past the radioactive zones with the carborne system and these are plotted for two highway lanes. It can be seen from the bottom of Figure 10C.58 that both the maximum amplitude of the anomaly and its sharpness are decreased considerably for measurements made in the "far lane". Figure 10C.59 (also from Nelson, 1953) illustrates the effect of vehicle speed on the maximum amplitude of the anomaly for both the "near lane" and "far lane". For these tests, an analog ratemeter with a time constant of 2 seconds was used. Similar graphs were constructed by Bowie et al. (1955) from data recorded with a carborne system traversing across an artificial vein 5.6 m long, 0.3 wide, and 0.3 m deep, filled with a homogeneous mixture containing 0.5 per cent U_3O_8. A profile of stationary measurements made across the vein showed an interesting unexpected asymmetry in the anomaly, caused by the detector having more absorbing material near one end than the other.

Field Procedure

Some of the first considerations in designing a carborne system are the installation of the survey equipment in the field vehicle, the source of power for the equipment, and the desired form of data presentation and hence the recording method to be used. For long-term surveys it is generally recommended that the source of power be furnished from the vehicle power supply (generator or alternator, and battery). This may require replacing the generator and regulator with equipment which can furnish additional power at low speeds (Berbezier et al., 1958; Bowie et al., 1955). For shorter term use, power may be supplied from a spare set of car batteries which can be recharged daily or as required (Killeen et al., 1976; Chandra and Leveille, 1977).

The addition of an adjustable threshold audible alarm has been found useful in many carborne surveys (Nelson, 1953; Bowie et al., 1955; Berbezier et al., 1958), especially if the chosen survey procedure is to investigate anomalies as soon as they are detected, rather than to return later after inspecting the recordings made during the road traverses. For this purpose it is useful to have an additional portable detector available in the vehicle. If the carborne survey is purely a total-count survey, a portable gamma ray spectrometer would enable the operator to determine whether the cause of a given anomaly was uranium, thorium or potassium by stopping to inspect it.

Killeen et al. (1976) utilized a four-channel portable gamma ray spectrometer with a single-pen strip chart recorder. Traversing was carried out while recording the output of the total-count channel. Anomalies were checked later by re-traversing while recording the output of the differential windows on the strip chart. The portable spectrometer could also be removed from the vehicle and attached to a 75 x 75 mm NaI(Tl) detector for in situ assaying. Chandra and Leveille (1977) recorded all four channels of a portable gamma ray spectrometer with a four-pen recorder in a carborne survey, returning to the interesting anomalies to inspect them on foot with the same spectrometer connected to a small detector. They also re-traversed anomalous sections of road while recording stripped count rates on the chart.

The methods of recording carborne radiometric data range from simply indicating anomalies on the road map with an X as they are spotted, to recording on strip charts anywhere from one to four channels of information plus fiducials or event markers. In any case, all of the roads to be traversed in the area of the survey are first marked on the field map. The beginning and end of each traverse are pre-marked with numbers such that each traverse has its own characteristic identifying numbers. Some workers prefer to pre-mark, with numbers on the map, any special land mark which can be used for navigation and location recovery such as road crossing or bridges (Killeen et al., 1976). Because anomalies may be smeared by effects of analog time constants the direction of the traverse should also be indicated on the map. Long traverses may be broken into segments, also with identifying codes (Chandra and Leveille, 1977). When a strip chart recording is made, event or fiducial marks are valuable. Thus, when a landmark is passed a manually operated event marker can be activated or the operator must make these marks by hand. Some carborne surveys have been equipped with automatic fiducials which are activated by an interconnection to the speedometer of the vehicle (e.g. Berbezier et al., 1958).

As with other gamma ray spectrometers, carborne systems require proper calibration. Pre-survey calibration should be carried out with the use of calibration pads as described above with respect to portable gamma ray spectrometers. Usually a calibration source such as ^{137}Cs or Th is used for the initial survey calibration. This calibration ensures that the energy windows chosen to represent K, U, and Th are in their proper positions. Because of instrument drift with temperature changes some spectrometric systems require periodic checking, while others have automatic stabilization or warning indicators for out-of-calibration conditions. The calibration source, if carried in the vehicle, must always be stored in the same location (producing a constant contribution to the background) and should be shielded (to reduce the contribution to background). It may be possible to use the battery of the vehicle as a shield. All systems should be shock mounted and have thermally insulated detectors to minimize drift. The thermostatically controlled, heated, insulated detector packages often used in airborne spectrometer systems are usually unnecessary for carborne surveys since frequent calibration checks are easier to make.

Most gamma ray spectrometer systems in use today have available a digital readout after a preset counting time, of the counts accumulated in the four windows (TC, K, U, Th). This feature is used to calibrate the scales on the strip chart recorder. The strip chart is allowed to run while the vehicle is stationary at the beginning of a road traverse, and counts are accumulated for a preset time. Then the results are marked on the chart, and used to determine and check the chart scale factor. This type of check should be carried out at the beginning and end of each traverse, and also at a base station occupied at the beginning and end of each day. This latter check ensures the reproducibility of results, is a check on background, and indicates instrumental problems. In addition, background measurements should be made periodically over water, if possible, during the survey (see the earlier section regarding background in surface measurements).

Carborne radiometric surveys have been used to advantage in France (Berbezier et al., 1958), in West Africa, South Africa and Norway (Bowie et al., 1955), in the United States (Nelson, 1953; Shideler and Hinze, 1971), in Canada (Killeen et al., 1976; Chandra and Leveille, 1977), and in Japan (Kamiyama et al., 1973). The success of the survey depends primarily on the nature and extent of the road network available.

The method of data presentation does not differ greatly from that used in airborne surveys. Data from each road traverse may be plotted in the same way as data from a flight line, as profiles. If road traverses are closely spaced the data may be contoured. It is therefore instructive when planning a carborne spectrometric survey to read the more abundant and generally more up to date literature on airborne radiometric surveying.

Snowmobile Surveys

Radiometric surveys over snow can be successful, even though the snow attenuates gamma radiation emitted by the underlying rock or overburden. This application is based on the assumption that the snow cover absorbs most of the radioactive emitted from the ground and only very highly radioactive occurrences would be detected. About 7 cm of water will attenuate the gamma radiation by 50 per cent. This could represent from 14 to 70 cm of snow, depending on its density. Minor changes in radioactivity would be undetectable beneath the snow-blanket and the method could be considered as "prospecting for hot spots". In many areas the only time of year when there is good accessibility is during winter when there is snow cover. The target of uranium rich boulders in glacial terrain seems well suited to this technique (Ketola and Sarikkola, 1973).

Field Procedure

The snowmobile survey technique has been used with some degree of success in Scandinavia. The detector assembly, which is fragile and must not be subjected to rapid temperature changes, was pre-chilled slowly in a thermostatically controlled refrigerator before the survey. Once it had reached outdoor field survey temperature, the detector assembly was always left outdoors where the temperature was relatively constant.

The survey procedure was relatively simple, consisting of traversing the area with closely spaced parallel lines. The traverse lines had to be kept close together since attenuation by snow off to the sides of the traverse line was greater than directly below the snowmobile. Anomalies were investigated as they were detected, by stopping and digging a hole in the snow (R. Sinding-Larsen, pers. comm., 1976). A pole with a

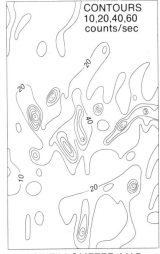

Figure 10C.60. Comparison of summer and winter total count ground gamma ray survey over a boulder train in Finland (after Ketola and Sarikkola, 1973, see text).

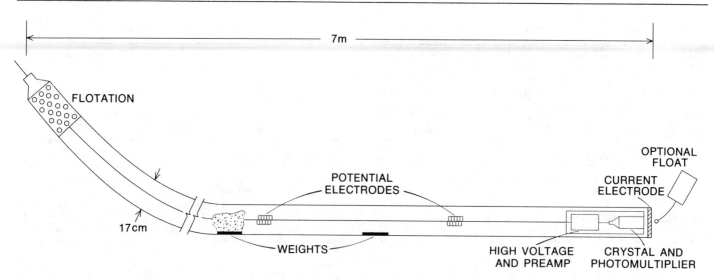

Figure 10C.61. Diagram of 'eel' assembly used for underwater gamma ray spectrometry in northern Saskatchewan (after Stolz and Standing, 1977). Detector dimensions are 75 x 200 mm NaI(Tl) (see text).

flag was erected at the location, and the approximate location was marked on a field map so that it could be found again in the spring after the snow melted.

In another type of winter radiometric survey, stationary measurements are made by pushing the detector into the snow. An example comparison of summer and winter total count gamma ray surveys over a boulder train in Finland carried out by Ketola and Sarikkola (1973) is shown in Figure 10C.60 to illustrate the validity of winter surveys. The survey employed a 50 m separation between traverses and a 10 m spacing between stations. Discrepancies are explained by two factors:

1. In winter the detector of the scintillometer is pushed into the snow to the bottom layers while in summer measurements are made at about 0.5 m above the surface (a geometry problem).

2. Radon gas may be trapped and concentrated below the impermeable frozen layers of snow producing more anomalies (a background problem).

UNDERWATER GAMMA RAY SPECTROMETRIC SURVEYS

Introduction

The earliest underwater gamma ray surveys, using total count scintillometers, were for applications other than uranium exploration. For example Summerhayes et al. (1970) used a conventional NaI(Tl) detector in a sealed container for stationary sea floor measurements to locate phosphorite by detecting radiation from its high uranium content. The application of radiometric techniques to locating offshore mineral deposits has increased, and much of the experience gained is useful for uranium exploration (see for example Noakes and Harding, 1971; Noakes et al., 1974a, b, 1975).

Offshore Sea-Bottom

One of the first reports of the use of a gamma ray spectrometer for sea-bottom surveying was given by Bowie and Clayton (1972). They described a prototype system consisting of a 75 x 75 mm NaI(Tl) detector mounted in a stainless steel cylinder 125 mm in diameter, fixed at the end of a reinforced rubber hose of the same diameter to avoid the possibility of the probe being caught on the bottom. This "eel", 25 m in length, was towed on a double armoured coaxial cable at a speed of 3 to 4 knots in up to 200 m water depth. The system employed a weak link and marker buoy to facilitate recovery in case of a snag on the bottom. The spectrometer was a 4 channel portable model. Miller and Symons (1973) presented total count, eU and eTh profiles, from a survey with this system off the Yorkshire Coast of England, showing the correlation with the geology of the seabed material. Some of the problems they encountered which could explain some discrepancies in the correlation were:

a) limited accuracy in position fixing.

b) lack of detailed knowledge of the seabed geological succession, and

c) the inconstent geometry of the detector with respect to the sea floor i.e. the variable depth of furrow cut by the 'eel' which could, in fact, be zero sometimes if the 'eel' left the sea floor.

The latest version of this system with a stabilized spectrometer was described by Clayton et al. (1976). The stainless steel probe contains a 75 x 125 mm NaI(Tl) detector. The 'eel' is reinforced P.V.C. of diameter 17.5 cm and length 30 m. Towing speed is normally 4 to 5 knots, with a maximum limit of 7 knots. The system is designed primarily for geological mapping of the seabed, and gives a spatial resolution of approximately 25 m at 4 knots with the total count channel, and 1 km at 4 knots with the K, U, and Th channels. The data are recorded on paper tape, and subsequently computer processed to produce contour maps. The system could be modified for use on smaller vessels for uranium exploration on lake-beds. Miller et al. (1977) reported on the use of this equipment for surveys of the continental shelf.

Gaucher et al. (1974) described a sea-floor system mounted in a sled. Experiments with two NaI(Tl) detectors (75 x 75 mm and 150 x 100 mm) towed at 1 knot produced reasonable results with 500 second counting times for the smaller detector and 300 seconds for the larger detector. Results of a survey off the Mediterranean Coast near Banyuls, at the eastern extension of the Pyrenees mountains were presented as contour maps of the potassium and thorium channel count rates which related to the sea-bottom unconsolidated materials.

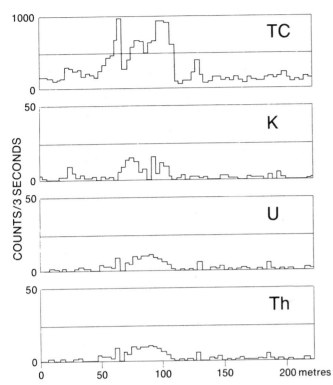

Figure 10C.62. Anomaly caused by sand, gravel and boulders from granite and pegmatite measured in lake bed gamma ray spectrometry survey in northern Saskatchewan (after Stolz and Standing, 1977).

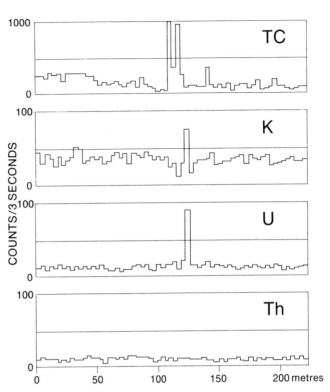

Figure 10C.63. Anomaly caused by radioactive boulders in lake bed gamma ray spectrometry survey in northern Saskatchewan (after Stolz and Standing, 1977).

Lake-Bottom Surveys

In regions where a significant percentage of the land is covered by lakes, and if the lakes cover geologically favourable areas for uranium exploration, a lake-bottom gamma ray spectrometer survey may be warranted. Such is the case in northern Saskatchewan where about 40 per cent of the land surface is covered by lakes and rivers. The lakes and rivers are often expressions of fractures, faults and other lineaments which have potential for uranium mineralization.

Hoeve (1975) and Beck et al. (1977) reported on the 'St. Louis Fault Project', the evaluation of a lake bed gamma ray survey in the Beaverlodge area. The survey was carried out in Alces Lake, 40 km northeast of Uranium City. The lake is situated on the postulated extension of the St. Louis Fault on which two uranium mines are located. The lake is elongate, about 8 km long by 0.8 km wide. The field procedure consisted of collecting lake-bottom samples and making lake-bottom scintillometer readings at locations about 20 m apart along grid lines at 200 m spacing. About 80 stations were sampled per day. The survey was carried out from an inflatable rubber boat with two canoes lashed alongside catamaran-style, on which a large wooden deck surface was constructed. For sample locations, distances were measured along a nylon rope tied to markers on either shore. The scintillometer used in this work (Goldak, 1975) comprised a waterproof detector package which was lowered to the lake-bottom while the readout electronics remained in the boat. At each station an echo sounder was used to provide depth information.

In 1976 the same field method was used in Seahorse lake, a part of which overlies the Key Lake uranium deposit, and Prince Lake on the postulated extension of the St. Louis Fault (Parslow and Stolz, 1976). Grid lines were located closer together (50 m) than in the Alces lake survey.

In addition to the above-mentioned stationary lake-bottom measurements, a gamma ray spectral logger, or continuous measurement system has also been evaluated (Stolz, 1976). The system included the measurement of apparent electrical resistivity of the lake-bottom, and a "scrape" microphone to assure the operator that the probe was scraping along the bottom. A diagram of the 'eel' assembly is shown in Figure 10C.61 (after Stolz and Standing, 1977).

It is expected that the system can detect differences of about 10 ppm uranium. Typical background count rates for the system have been given in the earlier section on backgrounds for submarine systems.

Because of attenuation of gamma radiation by the water, the width of the swath contributing gamma rays to the detector is quite narrow (e.g. 7 cm of water attenuates 50% of the gamma radiation). Thus, if the objective is to map uraniferous boulder trains as in exploration in glaciated regions, close line spacings are necessary. The gouging of the trench by the 'eel' increases the count rate by improving the geometry, but further narrows the width of the effective coverage thus hindering the detection of boulders. The angle between the cable and horizontal is kept small (less than 30°; Stolz, 1976) to prevent the eel from lifting off the bottom. This means the eel is a long way out and its precise location is not known. This makes follow-up investigation of the anomalies quite difficult.

For follow-up of anomalies found with the above system, hand held underwater single channel spectrometers containing 75 x 75 mm NaI(Tl) detectors were developed (Stolz, 1976; Stolz and Drevor, 1977). Counts are displayed on a four digit LED display and when the count rate exceeds a preset level, a flashing light turns on.

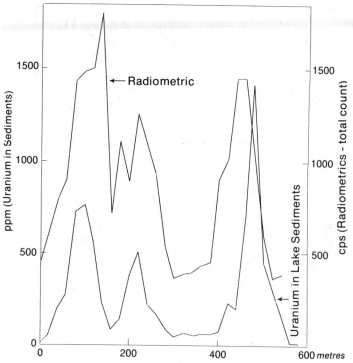

Figure 10C.64. Profile across Seahorse Lake, Saskatchewan comparing lake sediment sample assays at 20 m intervals with gamma ray profile (after Stolz and Standing, 1977).

Radioactive boulders produce short-wavelength, high amplitude anomalies as shown in Figure 10C.63 (after Stolz and Standing, 1977). The lake bottom logger results were in good agreement with bottom sediment uranium contents. The agreement between the geochemical and radiometric data is clearly shown in Figure 10C.64, a profile across Seahorse Lake where lake sediments were collected at 20 m intervals. The authors concluded that underwater radiometric surveying is both effective and economical for detecting and mapping radioactive occurrences and boulders. They further indicated that since the underwater survey requires very close line spacing it is more suited to detailed surveying of areas with high potential rather than reconnaissance.

BOREHOLE GAMMA RAY SPECTRAL LOGGING

Introduction

A number of specific borehole logging parameters common to other types of gamma ray spectrometric surveys such as the question of geometry, dead time, background, calibration, the effects of borehole diameter and casing, and of the equivalent atomic number of the rock have been covered in earlier sections. Many of the parameters are energy-dependent, and experimental data are sparse or unavailable with respect to variation with energy. However, as in the case of the previously discussed modes of gamma ray spectral surveying, a great deal can be learned from the experience gained in total count gamma ray work.

Previous Reviews

An early review by Russell (1955) included a good description of gamma ray logging, and discussion of the effects of some of the logging parameters. An idea of the potential usefulness of gamma ray logging as a lithological tool can be obtained by observing the mean radioelement concentrations of a number of different rock types given in Table 10C.6. With gamma ray spectrometry there is the additional possibility of determining ranges of radioelement ratios as identifiers of rock-types. Gamma ray logging specifically for uranium was reviewed by Stead (1956). Technical details of the logging equipment in use at that time were also described.

Beckerley (1960) presented a good review of all nuclear methods for subsurface prospecting, including the state of the art on gamma ray spectral logging. The calibration facilities established by the American Petroleum Institute at the University of Houston in 1959 were described. The use of these model holes was intended to improve the intercomparison of gamma logs. This would mean that differences between measurements by different service companies would be real and not just calibration differences. Gamma ray spectral logging was described as having great potential. It is interesting that Beckerley (1960) predicted that the problems involved with gamma energy detection techniques made it likely that spectral logging would remain scarce unless there was a real break-through.

A very well organized review of the application of nuclear techniques in oil and mineral boreholes was given by Clayton (1967). Natural gamma and gamma ray spectral logging were represented by about 35 of the 187 references given, and therefore the text is heavily weighted towards neutron and related logging. Dodd et al. (1969) reviewed borehole logging methods for uranium exploration covering the principles of calibration and analysis, the effects of many borehole parameters on the gamma ray log, and presented the state of the art in gamma ray spectral logging. They stated that "the advantages of downhole spectral measurements largely remain to be developed and demonstrated". The use of the energy region above 1 MeV for spectral logging was

The follow-up procedure consisted of prospecting the bottom in the vicinity of the anomaly with the underwater spectrometer in total count mode (thresholds at 0.30 MeV to 4.00 MeV). Background readings were typically 100 cps, with the anomaly in Seahorse Lake, Saskatchewan reading 1100 cps. Stolz (1976) reported that although underwater visibility was good, when the divers disturbed the bottom sediment, visibility became poor. The divers could read the LED display on the instrument but couldn't orient themselves. To avoid the problem and to locate anomalies more exactly, the anomalies were re-traversed until the count-rate reached a maximum over the anomaly and the boat was stopped. The divers then followed the cable down to the detector which was very close to the source.

Stolz and Drevor (1977) reported on additions and modifications to the above equipment and procedures. The major changes included recording all parameters separately on a six channel recorder, and the addition of an acoustic navigation system. Stolz and Drevor (1977) reported on the use of the system in Black Lake and near Brochet Island in Lake Athabasca. They concluded that the geophysical portion of the system worked well and several significant radioactive anomalies were discovered, but no detectable halos were found around uraniferous boulders. They recommended very close line spacing (<20 m) to map boulders on lake bottoms. The acoustic navigation system did not perform well in shallow, confined inland water. Stolz and Standing (1977) indicated the effective range of the navigation system is generally less than 1 km. They also report additions to the system such as a sub-bottom profiler (seismic) for added bottom sediment information. Thickness of up to 10 m of soft sediment on bedrock can be measured.

Figure 10C.62 shows an example from Stolz and Standing (1977) of the broad low amplitude anomaly characteristic of sands, gravels, and boulders from mechanical weathering of granite and pegmatite.

Table 10C.6

Radioelement concentrations in different classes of rocks[1]

Rock Class	U (ppm)		Th (ppm)		K (%)	
	Mean	Range	Mean	Range	Mean	Range
Acid Extrusives	4.1	0.8 - 16.4	11.9	1.1 - 41.0	3.1	1.0 -6.2
Acid Intrusives	4.5	0.1 - 30.0	25.7	0.1 -253.1	3.4	0.1 -7.6
Intermediate Extrusives	1.1	0.2 - 2.6	2.4	0.4 - 6.4	1.1	0.01-2.5
Intermediate Intrusives	3.2	0.1 - 23.4	12.2	0.4 -106.0	2.1	0.1 -6.2
Basic Extrusives	0.8	0.03- 3.3	2.2	0.05- 8.8	0.7	0.06-2.4
Basic Intrusives	0.8	0.01- 5.7	2.3	0.03- 15.0	0.8	0.01-2.6
Ultrabasic	0.3	0 - 1.6	1.4	0 - 7.5	0.3	0 -0.8
Alkali Feldspathoidal Intermediate Extrusives	29.7	1.9 - 62.0	133.9	9.5 -265.0	6.5	2.0 -9.0
Alkali Feldspathoidal Intermediate Intrusives	55.8	0.3 -720.0	132.6	0.4 -880.0	4.2	1.0 -9.9
Alkali Feldspathoidal Basic Extrusives	2.4	0.5 - 12.0	8.2	2.1 - 60.0	1.9	0.2 -6.9
Alkali Feldspathoidal Basic Intrusives	2.3	0.4 - 5.4	8.4	2.8 - 19.6	1.8	0.3 -4.8
Chemical Sedimentary Rocks*	3.6	0.03- 26.7	14.9	0.03-132.0	0.6	0.02-8.4
Carbonates	2.0	0.03- 18.0	1.3	0.03- 10.8	0.3	0.01-3.5
Detrital Sedimentary Rocks	4.8	0.1 - 80.0	12.4	0.2 -362.0	1.5	0.01-9.7
Metamorphosed Igneous Rocks	4.0	0.1 -148.5	14.8	0.1 -104.2	2.5	0.1 -6.1
Metamorphosed Sedimentary Rocks	3.0	0.1 - 53.4	12.0	0.1 - 91.4	2.1	0.01-5.3

*Includes carbonates

[1] compiled from English language literature by Wollenberg, pers. comm. (1978).

advocated since photopeaks in the lower energy region are obscured by scattering. The possibility of detecting radioactive disequilibrium was mentioned, but the required instrumental stability was cited as a real problem. Dodd and Eschliman (1972) expanded on the previous review indicating that many of the problems of spectral logging had been solved by the latest generation of instrumentation, but most of the review was concerned with total count gamma ray logging. Czubek et al. (1972) in a review of nuclear techniques in geophysics in Poland discussed natural gamma ray logging briefly, especially considering the effect of the equivalent atomic number of the rock. Scott and Tibbetts (1974) reviewed borehole logging techniques for mineral deposit evaluation, in which 67 references in English and 109 references in other languages were cited. The review is organized according to the different metals (e.g. uranium) and all the available techniques which can be used to evaluate the associated mineral deposits for each metal are given. References to gamma ray logging figure prominently in the section on uranium evaluation methods.

In a technical report entitled 'Recommended Instrumentation for Uranium and Thorium Exploration' the IAEA (1974) set forth a list of the advantages of gamma ray logging methods, along with some of the limitations. Often many of these are overlooked or disregarded, and proper importance is not given to gamma ray total count or gamma ray spectral logging in an exploration program. For this reason these lists are reproduced here from the IAEA (1974). The advantages of total count gamma ray logging include:

"1) High-cost coring can be largely replaced with less expensive non-core drilling.

2) Logs provide information lost by poor core recovery.

3) Data can be obtained from holes drilled previously for other purposes.

4) The volume 'sampled' is generally larger, virtually undisturbed and hence more representative than most cores or cutting samples.

5) Delays and costs of sampling and laboratory analysis are reduced.

6) The continuous log permits 'resampling' for additional statistical and economic studies.

7) Logs are objective and unbiased by personal observation and experience.

8) Logs require minimal space to store the information."

The IAEA (1974) report goes on to list the limitations of total count gamma logging in general:

"1) Inability to identify or separately measure the specific radio-isotopes which are the source(s) of the gamma radiation. This precludes independent analyses for K, U, and Th at normal to slightly anomalous concentrations in the rock.

2) The components of mixed ores of U and Th cannot be readily evaluated.

3) Disequilibrium within the uranium (and thorium) decay series may introduce locally significant errors in the quantitative analysis for U (or Th).

4) Variation from standard conditions of calibration, e.g. borehole fluid and diameter, formation moisture and composition (Z equivalent), casing etc., will influence the response; additional logs may be required to obtain reliable correction values".

The first three of these limitations may be overcome through the use of gamma ray spectral logging, although the third may require the use of solid state detectors. Gamma ray spectral logging includes all of the advantages of total count gamma logging and eliminates many of its limitations.

Killeen (1975) reviewed nuclear techniques for borehole logging in mineral exploration, briefly discussing the 'passive' systems of gamma ray logging and gamma ray spectral logging in addition to the 'active' systems which employ radiation sources for their measurements. Dodd (1976) discussed gamma ray spectral logging in some detail in a review of uranium exploration technology. An example was presented to illustrate the advantage of gamma ray spectral logging over total count gamma logging in an environment with significant concentrations of the three radioelements K, U, and Th. Dodd (1976) stated that experience at the U.S. Department of Energy (formerly ERDA) indicates that for reliable counting statistics, detector size and logging speed must be matched for the concentration levels being measured. Dodd estimated that about 100 to 200 cm^3 of NaI(Tl) could adequately measure typical rock concentrations (Clarke values) in one minute in a borehole. He suggested that assaying a sample length (thickness) of 1.5 m was possible at a logging speed of 1.5 m/min. Smaller detectors may be used for higher radioelement concentrations (e.g. 500 ppm eU or 1000 ppm eTh). A dual detector probe was being developed and tested by the U.S. Department of Energy to cover both high and low radioelement ranges (Dodd, 1976). In addition to the above mentioned reviews, useful information on gamma ray logging applications can be found in Faul and Tittle (1951) and Fons (1969); the advantages of digital logging were discussed by Burgen and Evans (1975) and by Moseley (1976); slim tool systems were described by Reeves (1976); some of the advantages of correlating gamma ray log information with data obtained by other logging techniques especially in uranium roll front exploration were discussed by Daniels et al. (1977).

Total Count Gamma Ray Logging

Sometimes referred to as 'gross count' gamma logging, total count gamma ray logging became firmly established as a quantitative method of measuring uranium concentrations with the publication of papers by Scott et al. (1961) on quantitative interpretation of gamma ray logs. Scott and Dodd (1960) discussed corrections for disequilibrium based on a knowledge of the ratio of chemical assays to radiometric assays in the area in question. Scott (1962) described the computer program 'GAMLOG' developed to carry out the quantitative interpretation of the gamma ray logs. Pirson (1963) described the gamma ray log, calibration in API units, and presented example logs. Scott (1963) gave further information on the use of the 'GAMLOG' program, describing in detail the iterative process of analyzing the logs. Carlier (1964) described work in France on quantitative measurements by gamma ray logging, pointing out some of the problems encountered when other than "text book ore zones" were evaluated. Dodd (1966) updated the earlier reports, presenting information for quantifying some of the necessary correction factors to the gamma ray log. Edwards et al. (1967) considered the application of gamma ray logging to quantitative evaluation of potash deposits, in a similar fashion to the work on evaluation of uranium deposits.

Hawkins and Gearhart (1968, 1969) discussed uranium prospecting with gamma ray logging, including information on many practical details, often omitted by other authors, such as logging practices and cable types. Spectral logging was mentioned briefly, and a gamma ray spectrum from a sample of monazite thorium sand was shown, as obtained by a detector with 470 m of standard 4 conductor 5 mm logging cable. Some practical 'rule-of-thumb' types of information were given by Hallenburg (1973) in a discussion of the interpretation of gamma ray logs. Sprecher and Rybach (1974) described a total count logging probe (25 mm outside diameter) designed for slim hole exploration in Switzerland. A good discussion of the determination of uranium grade in boreholes in South Africa by gamma ray logging was presented by Corner and de Beer (1976). Example logs from numerous boreholes were given, and disequilibrium problems were discussed. They found that uranium grades could be calculated to an accuracy of better than 10 per cent since the equipment was calibrated in the model holes located at Pelindaba, provided the ore was in equilibrium and thorium-free.

Corner and de Beer (1976) found disequilibrium was prevalent in the Karoo, primarily consisting of uranium depletion relative to its daughter products in the holes which they logged. This disequilibrium was mostly confined to the zone above the water table (i.e. in the air filled holes). They concluded that the high radiometric background levels observed over extended distances in some boreholes in Karoo-type occurrences were indicative of radon-gas buildup. Corner and de Beer (1976) indicated that radiometric borehole logging, to a great extent, could replace chemical assays for determining uranium grade for ore-reserve calculations. However they suggested that chemical checks for disequilibrium be made.

Pochet (1976) described the present practice in France for quantitative gamma ray logging. Restricted to very slim holes, a number of logging tools (probes) with very narrow diameters have been developed. The standard gamma ray probe is only 22 mm outside diameter, and contains a GM (geiger mueller) tube section (two GM tubes 1.5 x 4 cm each) for high grade ore evaluation, and a scintillation section (either a 1.2 cm x 2.5 or 1.2 x 5.0 cm NaI(Tl) detector). This solves the problem of a wide range of radioelement concentrations, but ambiguity is still present when mixed U and Th ores are encountered.

Gamma Ray Spectral Logging

The earliest pulse height analyzers utilized in gamma ray spectral logging tests were of the photographic type, and were not easily adaptable to quantitative measurements (Brannon and Osoba, 1956). Caldwell et al. (1963) reported gamma ray spectral measurements made through 1500 m of logging cable with a 64 x 64 mm NaI(Tl) detector and a pulse height analyzer of the type in use today. Most of the paper, however, was devoted to the study of gamma ray spectra resulting from bombardment of the rock by a neutron source. Rhodes and Mott (1966) presented a series of curves computed from theory which were designed to provide correction factors for gamma ray spectral logs. Correction factor curves for a range of gamma ray energies (up to 8 MeV) are given for the effects of casing, mud density, eccentricity of the detector, bed thickness and borehole diameter. Much of the data is for oil-well situations (e.g. large diameter holes, heavy drilling mud etc.) but some would be applicable to uranium exploration boreholes. Corroboration of the theoretical curves by empirical measurements is also necessary. Czubek (1969) considered the effect of borehole parameters (size, fluid, casing, etc.) on the spectral log using a different theory of absorption of gamma rays: the

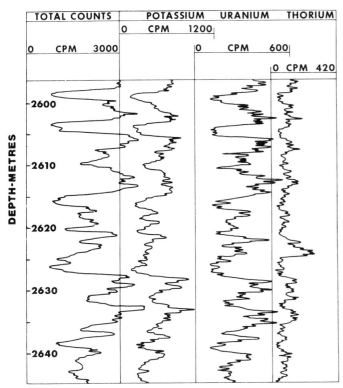

Figure 10C.65. A gamma ray spectral log showing a thorium marker anomaly just below 2620 m not shown on the total count log (after Lock and Hoyer, 1971).

so-called transmission factor (T). Czubek (1969) claimed this gave improved results and closer agreement between theory and experiment than other treatments of the problem. Czubek and Lenda (1969) considered the problem of the choice of units in which gamma ray logging measurements should be expressed by studying gamma ray energy distributions. They concluded that in any case measurements can only be standardized if the low level discriminator is set at 200 KeV, or for rocks with a high Zeq (equivalent atomic number) set at 400 KeV. This agrees with the more recent discussions concerning the so-called 'ur' unit used for total count scintillometry. Lock and Hoyer (1971) described a gamma ray spectral logging system and gave an example spectrum recorded through 6700 m of cable. They stated that "experience has shown that the potassium peak is always strong enough to provide a dependable reference for monitoring gain during the logging operation". They indicated that the gamma ray spectral log proved very useful in recognizing a distinctive thorium-rich bed that was used as a marker bed, but which did not appear in the total count log. Figure 10C.65 (after Lock and Hoyer, 1971) illustrates a gamma ray spectral log, and the thorium marker anomaly just below 2620 m on the log. (No spectral stripping has been performed on the logs shown in Figure 10C.65.)

Løvborg et al. (1972) compared data from a laboratory drill-core-scanning gamma ray spectrometer and total count gamma ray borehole logs. They found consistency between U and Th contents determined by scanning the drill core, and the total count gamma ray borehole log, with the exception of an apparent downward displacement of the gamma-log peaks by about 1.5 per cent which was attributed to cable stretching. Gamma ray spectral logging incorporating stripping was described by Wichmann et al. (1975). The output of a 512 channel analyzer was fed in groups of channels to three single channel analyzers with associated ratemeters. Window energies were 1.37 to 1.55 MeV for K(^{40}K), 1.58 to 1.95 MeV for U(^{214}Bi), and 2.40 to 2.85 MeV for Th(^{208}Tl). One of the main features was the use of a set of four "calibrators" each consisting of a cylindrical source constructed of plaster of paris. These contained K, U, Th, and a mixture of the three. These sources were placed on the detector by sliding the probe inside a hole along the axis of the source. They were then used to derive the stripping factors. The API gamma ray test hole in Houston (4%K, 13 ppm U, 24 ppm Th) was analyzed with this logging system and very close agreement was reported. The detector size was unspecified, but the probe dimensions were 92 mm in diameter by 2.1 m long. Wichmann et al. (1975) recommended logging speeds of less than about 4 m per minute. Hassan et al. (1976) referred to the gamma ray spectral log as the differential gamma ray log. They mention the problem of low count rates at high gamma ray energies and suggest several improvements such as increasing detector size, reducing logging speed, or increasing the number of energy windows. They suggested adding the ^{228}Ac peak at 0.91 MeV and the ^{214}Bi peak at 1.12 MeV to improve the Th and U window count rates respectively. Marett et al. (1976) attempted to incorporate all the counts from all the windows (channels of a multi-channel spectrometer) to improve counting statistics. The standard logging speed was quoted as 4.6 m/min. and the detector was a 5.1 cm diameter by 30.5 cm long NaI(Tl) crystal. A gamma-reference source was used for stabilization of the gain of the photomultiplier. Data was transmitted in digital form, multiplexed on a single conductor. The technique was used in the North Sea to identify micaceous sandstones, utilizing crossplots of the radioelements to help in the identification. Some examples of the use of this system in crystalline Precambrian basement rocks in northern New Mexico were presented by West and Laughlin (1976). The authors were able to recognize biotite-rich granitic or granodioritic gneiss and felsic dykes. Fracture zones with increased uranium concentration were interpreted as either sealed or open depending on whether or not Th and K peaks in the log were associated with the U peaks.

The use of portable borehole gamma ray spectral loggers for uranium exploration was reported briefly by Killeen (1976b), and in more detail by Killeen and Bristow (1976). They evaluated two commercially available types of spectral loggers in boreholes in uranium mining areas. Results using three different detector sizes were compared. Probes were slim (outside diameters 32 mm and 38 mm). Detectors were NaI(Tl) of dimensions 19 x 51 mm, 19 x 76 mm and 25 x 76 mm. Gamma ray spectra obtained with the three detectors were presented, both for K, U, and Th sources, and for in-hole measurements. Some example gamma ray spectral logs recorded in the same hole (Total Count, K, U and Th) were given for comparison. Recording was by single pen analog strip chart. Suggestions for improvement of portable borehole gamma ray spectral logging systems were given. Killeen et al. (1978) reported on an improved portable gamma ray spectral logging system which incorporated digital recording on cassette tape. This facilitated data processing by computer. The system was designed for Canadian uranium exploration conditions where access to boreholes by vehicles is often extremely difficult. The entire battery operated system weighed 73 kg including spectrometer, chart recorder, tape recorder, winch, cable, probe, and other accessories. The offline processing of data recorded on cassette tape by this system was accomplished by a mini-computer as described by Bristow (1977). This 'mini' was part of a larger truck mounted digital gamma ray spectral logging system referred to as the Geological Survey of Canada 'DIGI-PROBE' logger (Killeen et al., 1978). In situ assaying in the boreholes was also discussed, and a list of recommendations were given for a new generation of portable

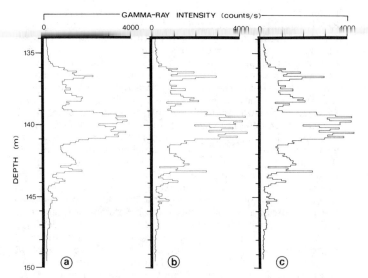

Figure 10C.66. A comparison of (a) a raw gamma ray log and the same log processed by: (b) iteration, and by (c) inverse filtering (after Conaway and Killeen, 1978b).

gamma ray spectral logger. The suggested system could produce a Radiometric Assay log (RA-log) directly in real time by deconvolving the raw gamma ray logs using a microprocessor. A detailed description of the deconvolution technique (also called inverse filtering) was presented by Conaway and Killeen (1978a). This inverse filtering technique is based on the determination of the response function of a gamma ray detector using data obtained in model boreholes such as those available in Ottawa, Canada or in Grand Junction, U.S.A. From the measured response an inverse operator is derived, to be used as a filter on the raw gamma ray log, removing the deleterious effects of the logging system response function. The method is illustrated with numerous theoretical examples of the effects of the processing technique on thin beds, thick beds, widely separated and closely spaced beds and a bed with linearly increasing radioelement contents across its width. An example of data recorded in a model borehole at the GSC calibration facilities is also given, processed to produce the RA-log. Conaway and Killeen (1978b) compared the inverse filter technique with the iterative technique that forms the basis of the GAMLOG computer program (Scott, 1963) which is commonly used to process gamma ray logs. They determined that the iterative technique and the inverse filter technique approach theoretical equivalence as the number of iterations increases. Two advantages of the inverse filter technique are the reduction in computing time (by a factor of over 20), and the possibility of processing data in real time by minicomputer or microprocessor with very little core storage required. Figure 10C.66 illustrates a comparison of a raw gamma ray log (digitally recorded at Δz = 10 cm intervals) and two processed logs, one by iteration and one by inverse filtering (after Conaway and Killeen, 1978b). The similarity is evident. With shorter sampling interval improved resolution is possible. An example comparison similar to the above, but for Δz = 3.3 cm is also shown by Conaway and Killeen (1978b). For the shorter sampling interval a smoothing operator is required by both techniques; GAMLOG at present has no facility for smoothing.

A detailed description of a truck-mounted borehole gamma ray spectral logging system developed by the U.S. Department of Energy was presented by George et al. (1978). The system includes a dual detector probe (small NaI(Tl) = 115 cm^3, large NaI(Tl) = 500 cm^3) to cover a wide range of radioelement concentrations. Three single channel analyzers and a lower level discriminator provide the K, U, Th, and Total Count outputs. The count rates from each window are recorded digitally on magnetic tape cartridges, and also displayed on an analog strip chart. Data are collected on a depth basis rather than on the more commonly used time basis. The large detector (5.1 x 25.4 cm) is used except when count rates exceed 20 000 cps. The smaller detector (4.4 x 7.5 cm) is switch selected by the operator in that case. A stabilization source of Mn-54 is used for each detector, providing a peak at 835 keV. Typical logging speed is 1.5 m/min. Counts are accumulated for 10 seconds during each measurement, representing about a 25 cm interval. A discussion of the calibration of the system, and some example applications are included in the report. Bristow and Killeen (1978) also presented a detailed report on the construction and operation of the G.S.C. DIGI-PROBE logging system, which records up to 1024 channels of gamma ray spectral logging data on 9-track tape as often as every 0.25 seconds, displaying the reduced K, U, and Th and/or any radioelement ratio data on a 6 pen strip chart recorder via digital to analog converters. The whole system is built around a 16-bit minicomputer operated interactively via a keyboard and a CRT display with alphanumeric and graphic capabilities. Commonly the DIGI-PROBE system utilizes slim hole probes with 25 x 76 mm CsI(Na) or NaI(Tl) detectors (or smaller) inside 38 mm outside diameter probes (or smaller), at logging speeds of 0.6 m/min. to 6.0 m/min., recording 256 channels of data with sample times of 1.0 second to 0.2 seconds respectively, representing sample intervals of 1 or 2 cm.

Gamma Ray Spectral Logging with Solid State Detectors

The main advantage of the solid state detector is its high energy resolution compared to sodium iodide detectors. This makes it possible to detect daughter products other than ^{214}Bi for uranium estimation, thus avoiding the problem caused by radioactive disequilibrium. The main problem in the application of solid state detectors such as lithium drifted germanium (or Ge(Li)) is their low operating temperature of below minus 150°C. The detector must be cooled by liquid nitrogen or some other equivalent coolant at all times, or it is rendered useless, losing its detection properties. Also large detectors are difficult to manufacture and are therefore very costly. Lauber and Landstrom (1972) reported on the use of a Ge(Li) borehole probe for gamma ray spectral logging in a uranium mine in Sweden. Their cryostat kept the probe cool for ten hours under working conditions, after which time the liquid nitrogen had to be replenished. The natural gamma ray spectrum recorded in the Ranstad uranium mine was illustrated to show the possibilities of the method. This is reproduced in Figure 10C.67. The counting times however were fairly long due to the small size of the detector (22 cm^3) and the need for a larger number of channels to utilize the high resolution of the detector. The authors suggested a 4 to 6 channel analyzer with its channels centred on peaks of interest should be a viable arrangement. Gorbatyuk et al. (1973) suggested the use of a borehole Ge(Li) detector to determine the uranium content of ore from the size of the 186 keV gamma ray peak which is a combination of the 185.7 and 186.2 keV gamma rays from ^{235}U and ^{226}Ra respectively. They also tried the 1.001 MeV peak of ^{234}Pa. This is a low-count peak, but is high energy and relatively free of interference and is as high as possible in the decay series. Boynton (1975) described a simplification of the detector cooling problem. It consists of using canisters or cartridges of solid propane 3.7 cm diameter by 57 cm long instead of liquid nitrogen. The solid propane converted to a liquid during the cooling, without much volume change, unlike the liquid nitrogen which converts to a gas, increasing in volume drastically requiring venting to be incorporated in the probe design. Landstrom (1976) described the interesting

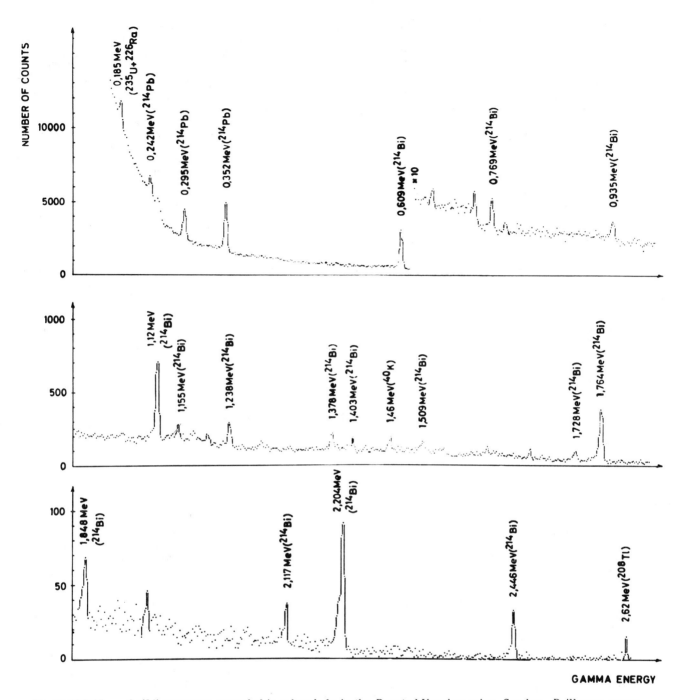

Figure 10C.67. Ge(Li) spectrum recorded in a borehole in the Ranstad Uranium mine, Sweden. Drill core assay; 400 ppm U, 10 ppm Th, 4% K; counting time: 10 min. (after Lauber and Landstrom, 1972).

possibilities of identifying elements in boreholes by X-ray fluorescence using natural gamma radiation as the source of excitation. Evaluation of the source itself must be carried out by gamma ray spectral logging. Christell et al. (1976) reviewed nuclear geophysics in Sweden, describing borehole gamma ray spectral logging measurements with both NaI(Tl) and Ge(Li) detectors. Senftle et al. (1976) described the use of <u>intrinsic</u> germanium (also called hyperpure Ge) in borehole probes used for uranium exploration. The intrinsic Ge has the advantage of only requiring cooling to operate, but not during storage or transportation as is the case with Ge(Li) detectors.

They discussed several gamma ray peaks which may be utilized for analysis of uranium such as the 63.3 keV peak of ^{234}Th, first daughter of ^{238}U. Tanner et al. (1977a) described the measurement of disequilibrium by using a solid state detector in a borehole probe. They utilized two probes, one with a Ge(Li) detector, the other with an intrinsic Ge detector. The latter is suitable for low energy gamma ray measurement, whereas the former, being of larger volume (45 cm^3) is used for high energy gamma ray measurements. Their procedure is to first delineate zones for detailed investigation by logging continuously at about 1.0 m/min.

The interesting zones are then analyzed with 10 minute counting times to determine their state of radioactive equilibrium or disequilibrium. In situ assaying is based on the 63.3 keV gamma ray of ^{234}Th and also the 1001.4 keV gamma ray of ^{234}Pa. These are in equilibrium with the parent ^{238}U. In total six isotopes or groups of isotopes are evaluated in a single measurement of disequilibrium. Tanner et al. (1977a) showed comparisons of scintillation detector logs and solid state detector logs. A series of holes drilled through a roll front uranium deposit were logged, and the state of equilibrium was displayed as an equilibrium ratio. Sensitivity of the method is about 80 ppm U_3O_8, for the 10 minute counting time.

It is apparent that the use of the solid state detector has a number of advantages over scintillation detectors but which can be obtained at present only with some difficulty. The recent rapid improvements in the application of solid state borehole probes in only a few years indicates that it won't be long before it is a commercially viable technique and will be offered by logging service companies.

CONCLUSIONS

It is evident that gamma ray spectrometry has made considerable advances in the last 10 years. Its applications have expanded into many different environments as the equipment has become more refined and the effects of these environments on gamma radiation have become better understood. It is likely that the next 10 years will see further improvements which will permit the extraction of more information from the gamma ray spectral measurements than is presently possible in routine surveys.

In the future, microprocessor technology will permit the presentation of the data, completely corrected and processed in the field for increased on-site decision-making in exploration as well as recording the data for later enhancement.

REFERENCES

Adams, J.A.S.
1969: Total and spectrometric gamma-ray surveys from helicopters and vehicles; in Nuclear Techniques and Mineral Resources, Proc. Series, IAEA, Vienna, p. 147-162.

Adams, J.A.S. and Clark, R.B.
1972: Computer modeling and experimental calibration of airborne gamma spectrometer systems; in the Natural Radiation Environment II, Adams, J.A.S., Lowden, W.M., and Gesell, T.F. (ed.), U.S. Dep. of Commerce, Springfield, Virginia, p. 641-648.

Adams, F. and Dams, R.
1970: Applied gamma-ray spectrometry; Pergamon Press, Toronto, 753 p.

Adams, J.A.S. and Fryer, G.E.
1964: Portable γ-ray spectrometer for field determination of thorium, uranium and potassium; in The Natural Radiation Environment, Adams, J.A.S. and Lowder, W.M., (eds.), Univ. of Chicago Press, Chicago, p. 577-596.

Adams, J.A.S. and Gasparini, P.
1970: Gamma-ray Spectrometry of Rocks; Elsevier Publishing Company, New York, 295 p.

Adams, J.A.S. and Lowder, W.M. (eds.)
1964: The Natural Radiation Environment; University of Chicago Press, Chicago, 1069 p.

Agocs, W.B.
1975: Airborne scintillation counter surveys; Trans. Can. Inst. Mining Met., v. 58, p. 59-61.

Allan, R.J. and Richardson, K.A.
1974: Uranium distribution by lake sediment geochemistry and airborne gamma-ray spectrometry; a comparison of reconnaissance techniques; Can. Min. Metall. Bull., v. 67, no. 746, p. 109-120.

Anspaugh, L.R., Phelps, P.L., Gudiksen, P.H., Lindeken, C.L., and Huckabay, G.W.
1972: The in-situ measurement of radionuclides in the environment with a Ge(Li) spectrometer; in The Natural Radiation Environment II, Adams, J.A.S., Lowder, W.M., and Gesell, T.F. (eds.), U.S. Dep. of Commerce, Springfield, Virginia, p. 279-304.

Austin, S.R.
1975: A laboratory study of radon emanation from domestic uranium ores; Radon in Uranium Mining (Proc. Panel, Washington, 1973), IAEA, Vienna, p. 151.

Barnes, F.Q.
1972: Uranium exploration costs; in Uranium Prospecting Handbook, S.H.U. Bowie et al. (eds.), Inst. Min. Met., London, p. 79-94.

Barretto, P.M.C.
1975: Radon-222 emanation characteristics of rocks and minerals; Radon in Uranium Mining (Proc. Panel, Washington, 1973), IAEA, Vienna, p. 129.

Beck, L.S., Parslow, G.R., and Hoeve, J.
1977: Evaluation of the uranium potential of areas covered by lake waters, using geophysical, geochemical and radiometric techniques; in Recognition and Evaluation of Uraniferous Areas, Proc. Series IAEA, Vienna, p. 261-280.

Beckerley, J.G.
1960: Nuclear methods for subsurface prospecting; in Annual Review of Nuclear Science, E. Segre, G. Friedlander, and W. Meyerhof (eds.), v. 10, p. 425-460.

Berbezier, J., Blangy, B., Guitton, J., and Lallemant, C.
1958: Methods of car-borne and air-borne prospecting: The technique of Radiation Prospecting by Energy Discrimination; Proc. Second U.N. Int. Conf. on Peaceful Uses of Atomic Energy, v. 2, p. 799-814.

Bowie, S.H.U.
1973: Methods, trends and requirements in uranium exploration; in Uranium Exploration Methods, Proc. Series, IAEA, Vienna, p. 57-65.

Bowie, S.H.U. and Clayton, C.G.
1972: Gamma spectrometer for sea- or lake-bottom surveying; Trans. Inst. Min. Metall., Sect. B, Appl. Earth Sci., v. 81, p. B251-256.

Bowie, S.H.U., Davis, M., and Ostle, D.
1972: Uranium Prospecting Handbook; Inst. Min. Metall., London, 346 p.

Bowie, S.H.U., Hale, F.H., Ostle, D., and Beer, K.E.
1955: Radiometric surveying with a car-borne counter; Bull. Geol. Surv. G. Brit., v. 10, p. 1-23.

Bowie, S.H.U., Miller, J.M., Pickup, J., and Williams, D.
1958: Airborne radiometric survey of Cornwall; Proc. of 2nd U.N. Conf. on Peaceful Uses of Atomic Energy, Geneva, Paper P/43, p. 787-798.

Boynton, G.R.
1975: Canister cryogenic system for cooling germanium semiconductor detectors in borehole and marine probes; Nuclear Instruments and Methods, v. 123, p. 599-603.

Brannon, H.R., Jr., and Osoba, J.S.
  1956: Spectral gamma-ray logging; J. Pet. Tech., v. 8, p. 30-35.

Breiner, S., Lindow, J.T., and Kaldenbach, R.J.
  1976: Gamma-ray measurement and data reduction considerations for airborne radiometric surveys; in Exploration for Uranium Ore Deposits, Proc. Series, IAEA, Vienna, p. 93-106.

Bristow, Q.
  1977: A system for the offline processing of borehole gamma-ray spectrometry data on a NOVA minicomputer; in Report of Activities, Part A, Geol. Surv. Can., Paper 77-1A, p. 87-89.

Bristow, Q. and Killeen, P.G.
  1978: A new computer-based gamma-ray spectral logging system; Society of Exploration Geophysicists, 48th Annual International Meeting Abstracts, p. 117-118.

Bristow, Q. and Thompson, C.J.
  1968: A computer P.H.A. system for real time off line analysis of spectra from an aerial survey of radioactive materials; IEEE Transactions on Nuclear Science, N.S. 15, No. 1, p. 150-156.

Bristow, Q., Carson, J.M., Darnley, A.G., Holroyd, M.T., and Richardson, K.A.
  1977: Specifications for federal-provincial uranium reconnaissance program 1976-1980 airborne radioactivity surveys; Geol. Surv. Can., Open File No. 335.

Bristow, Q.
  1979: Gamma ray spectrometric methods in uranium exploration: Airborne instrumentation; in Geophysics and Geochemistry in the Search for Metallic Ores, Geol. Surv. Can., Econ. Geol. Rep. 31, Paper 10A.

Bunker, C.M. and Bush, C.A.
  1966: Uranium, thorium, and radium analyses by gamma-ray spectrometry (0.184-0.352 million electron volts); U.S. Geol. Surv., Prof. Paper 550B, p. B176-B181.

Burgen, J.G. and Evans, H.B.
  1975: Direct digital laserlogging; Society of Petroleum Engineers of the A.I.M.E., Paper Number SPE 5506.

Caldwell, R.L., Baldwin, W.F., Bargainer, J.D., Berry, J.E., Salaita, G.N., and Sloan, R.W.
  1963: Gamma-ray spectroscopy in well logging; Geophysics, v. 28, no. 4, p. 617-632.

Cameron, G.W., Elliott, B.E., and Richardson, K.A.
  1976: Effects of line spacing on contoured airborne gamma-ray spectrometry data; in Exploration for Uranium Ore Deposits, proc. series, I.A.E.A., Vienna, p. 81-92.

Carlier, A.
  1964: Contribution aux methodes d'estimation des gisements d'uranium; Commissariat a l'energie atomique, Report CEA-R2332.

Chandra, J.J. and Leveille, J.
  1977: Ground mobile gamma-ray radiometric survey 1977, preliminary report; Mineral Resources Branch, Dep. Nat. Resourc., New Brunswick, Open File Report 78-1, 18 p.

Charbonneau, B.W. and Darnley, A.G.
  1970a: A test strip for calibration of airborne gamma-ray spectrometers; in Report of Activities, Geol. Surv. Can., Paper 70-1, pt. B, p. 27-32.

  1970b: Radioactive precipitation and its significance to high-sensitivity gamma-ray spectrometer surveys; in Report of Activities, Part B, Geol. Surv. Can., Paper 70-1, pt. B, p. 32-36.

Charbonneau, B.W. and Ford, K.L.
  1977: Ground radiometric investigations Kennetcook area, Nova Scotia; in Geol. Surv. Can., Open File 467.

  1978: Uranium mineralization at the base of the Windsor Group, South Maitland, Nova Scotia; in Current Research, Part A, Geol. Surv. Can., Paper 78-1A, p. 419-425.

Charbonneau, B.W., Jonasson, I.R., and Ford, K.L.
  1975: Cu-U mineralization in the March Formation Paleozoic Rocks of the Ottawa-St. Lawrence Lowlands; in Report of Activities, Part A, Geol. Surv. Can., Paper 75-1A, p. 229-233.

Charbonneau, B.W., Killeen, P.G., Carson, J.M., Cameron, G.W., and Richardson, K.A.
  1976: The significance of radioelement concentration measurements made by airborne gamma-ray spectrometry over the Canadian Shield; in Proceedings of International Symposium on Exploration for Uranium Deposits, proc. series, IAEA, Vienna, p. 35-54.

Chase, G.D. and Rabinowitz, J.L.
  1968: Principles of radioisotope methodology; Burgess Pub. Co., Minneapolis, 633 p.

Christell, R., Ljunggren, K., and Landstrom, O.
  1976: Brief review of developments in nuclear geophysics in Sweden; Nuclear Techniques in Geochemistry and Geophysics, Proc. Series, IAEA, Vienna, p. 21-46.

Clark, R.B., Duval, J.S., and Adams, J.A.S.
  1972: Computer simulation of an airborne gamma-ray spectrometer; J. Geophys. Res., v. 77, no. 17, p. 3021-3031.

Clayton, C.G.
  1967: A survey of the application of radiation techniques in oil and mineral boreholes; United Kingdom Atomic Energy Authority Report A.E.R.E.—R 5368.

Clayton, C.G., Cole, H.A., Munnock, W.C.T., Ostle, D., and Symons, G.D.
  1976: New instruments for uranium prospecting in exploration for uranium ore deposits; Proc. Series, IAEA, Vienna, p. 173-184.

Conaway, J.G. and Killeen, P.G.
  1978a: Quantitative uranium determinations from gamma-ray logs by application of digital time series analysis; Geophysics, v. 43, no. 6, p. 1204-1221.

  1978b: Computer processing of gamma-ray logs: iteration and inverse filtering; in Current Research, Part B, Geol. Surv. Can., Paper 78-1B, p. 83-88.

  Gamma-ray spectral logging for uranium; Can. Inst. Min. Metall. (in press)

Cook, J.C.
  1952: An analysis of airborne surveying for surface radioactivity, Geophysics, v. 17, no. 4, p. 687-706.

Cook, B., Duval, J., and Adams, J.A.S.
exploration; in Geochemical Exploration, Can. Inst. Min. Metall., Special Volume 11, p. 480-484.

Corner, B. and de Beer, G.P.
1976: The use of radiometric logging techniques to determine the uranium grade in certain mineralized Karoo boreholes; Atomic Energy Board, Republic of South Africa, Pelindaba PEL-252.

Corner, B. and Toens, P.D.
The Pelindaba facility for calibrating radiometric field instruments; Nuclear Active, Atomic Energy Board, Republic of South Africa. (in press)

Cowper, G.
1954: Aerial prospecting with scintillation counters; Nucleonics, v. 12, p. 29-32.

Cox, J.W. and Raymer, L.L.
1976: The effect of potassium-salt muds on gamma-ray and spontaneous potential measurements; Seventeenth Annual Logging Symposium Transactions, Society of Professional Well Log Analysts, Paper II, 19 p.

Crew, M.E. and Berkoff, E.W.
1970: TWOPIT, a different approach to calibration of gamma-ray logging equipment; The Log Analyst, v. 11, no. 6, p. 26-32.

Crouthamel, C.E., editor
1960: Applied gamma-ray spectrometry, Pergamon Press, London, 443 p.

Czubek, J.
1968: Natural selective gamma-logging, a new log of direct uranium determination, Nukleonika, v. 13, no. 1.

1969: Influence of borehole construction on the results of spectral gamma logging; Nuclear Techniques and Mineral Resources, proc. series, IAEA, Vienna, p. 37-53.

1971: Differential interpretation of gamma-ray logs: I. Case of the static gamma-ray curve; Report No. 760/1, Nuclear Energy Information Center, Polish Government Commissioner for Use of Nuclear Energy, Warsaw, Poland.

1972: Differential interpretation of gamma-ray logs: II. Case of the dynamic gamma-ray curve; Report No. 793/I, Nuclear Energy Information Center, Polish Government Commissioner for Use of Nuclear Energy, Warsaw, Poland.

Czubek, J.A. and Lenda, A.
1969: Energy distribution of scattered gamma-rays in natural gamma-logging; Nuclear Techniques and Mineral Resources, proc. series, IAEA, Vienna, p. 105-116.

Czubek, J.A., Florkowski, T., Niewodniczanski, J., and Przewlocki, K.
1972: Progress in the application of nuclear tectoniques in geophysics, mining and hydrology in Poland; Proc. 4th Int. PUAE conf., IAEA, v. 14, p. 123-143.

Daniels, J.J., Scott, J.H., Blackmon, P.D., and Starkey, H.S.
1977: Borehole geophysical investigations in the South Texas uranium district; J. Research, U.S. Geol. Survey, v. 5, no. 3, p. 343-357.

Darnley, A.G.
1970: Airborne gamma-ray spectrometry; Can. Min. Metall. Bull., v. 63, p. 145-154.

1972: Airborne gamma-ray survey techniques; in Uranium Prospecting Handbook (S.H.U. Bowie et al., ed.), Instit. Min. Met., London, p. 174-211.

1973: Airborne gamma-ray techniques — present and future; Uranium exploration methods, Proc. Series, IAEA, Vienna, p. 67-108.

1975: Geophysics in uranium exploration; in Uranium Exploration '75, Geol. Surv. Can., Paper 75-26, p. 21-31.

1977: The advantages of standardizing radiometric exploration measurements, and how to do it; Can. Min. Metall. Bull., v. 71, p. 91-95.

Darnley, A.G. and Fleet, M.
1968: Evaluation of airborne gamma-ray spectrometry in the Bancroft and Elliot Lake areas of Ontario, Canada; Proc. 5th Symposium on Remote Sensing of Environment, University of Michigan, Ann Arbor, p. 833-853.

Darnley, A.G. and Grasty, R.L.
1971: Mapping from the air by gamma-ray spectrometry; Can. Inst. Min. Metall., Special Volume II, Proc. Third International Geochemical Symposium, Toronto, p. 485-500.

Darnley, A.G., Bristow, Q., and Donhoffer, D.K.
1969: Airborne gamma ray spectrometer experiments over the Canadian Shield; in Nuclear Techniques and Mineral Resources, proc. series, IAEA, Vienna, p. 163-185.

Darnley, A.G., Cameron, E.M., and Richardson, K.A.
1975: The federal-provincial uranium reconnaissance program; in Uranium Exploration '75, Geol. Surv. Can., Paper 75-26, p. 49-63.

Darnley, A.G., Charbonneau, B.W., and Richardson, K.A.
1977: Distribution of uranium in rocks as a guide to the recognition of uraniferous regions; in Recognition and Evaluation of Uraniferous Areas, Proc. Series, IAEA, Vienna, p. 55-86.

Darnley, A.G., Grasty, R.L., and Charbonneau, B.W.
1969a: Airborne gamma-ray spectrometry and ground support operations; in Report of Activities, Part B, Geol. Surv. Can., Paper 69-1, pt. B, p. 10.

1970: Highlights of G.S.C. airborne gamma spectrometry in 1969; Can. Min. J., v. 91, p. 98-101.

Dickson, H.W., Kerr, G.D., Perdue, P.T., and Abdullah, S.A.
1976: Environmental gamma-ray measurements using in situ and core sampling techniques; Health Physics, v. 30, p. 221-227.

Dodd, P.H.
1966: Quantitative logging and interpretation systems to evaluate uranium deposits; Society of Professional Well Log Analysts 7th Ann. Logging Symp., Paper P.

1976: Uranium exploration technology; in Geology, Mining and Extractive Processing of Uranium, Inst. Min. Metall., London, p. 158-171.

Dodd, P.H. and Eschliman, D.H.
1972: Borehole logging techniques for uranium exploration and evaluation; in Uranium Prospecting Handbook, S.H.U. Bowie et al. (ed.), Inst. Min. Metall., London, p. 244-276.

Dodd, P.H., Droullard, R.F., and Lathan, C.P.
  1969: Borehole logging methods for exploration and evaluation of uranium deposits; in Mining and Groundwater geophysics/1967; Geol. Surv. Can., Econ. Geol. Rep. 26, p. 401-415.

Doig, R.
  1964: A portable gamma-ray spectrometer and its geological application; Ph.D. Thesis, McGill Univ., Dep. Geol., April.

  1968: The natural gamma-ray flux. In situ analysis. Geophysics, v. 33, no. 2, p. 311-328.

Duval, J.S.
  1976: Statistical interpretation of airborne gamma-ray spectrometric data using factor analysis; in Exploration for Uranium Deposits, Proc. Series, IAEA, Vienna, p. 71-80.

  1977: High sensitivity gamma-ray spectrometry — state of the art and trial application of factor analysis; Geophysics, v. 42, no. 3, p. 549-559.

Duval, J.S., Jr., Cook, B., and Adams, J.A.S.
  1971: Circle of investigation of an air-borne gamma-ray spectrometer; J. Geophys. Resear., v. 76, p. 8466-8470.

Edwards, J.M., Ottinger, N.H., and Haskell, R.E.
  1967: Nuclear log evaluation of potash deposits; Society of Professional Well Log Analysts, 8th Ann. Symp., Paper L.

Evans, R.D.
  1955: The atomic nucleus; McGraw-Hill, New York, 972 p.

Evans, H.B.
  1978: Review of U.S. D.O.E. calibration facilities; NEA/IAEA Workshop on Borehole Logging for Uranium, Grand Junction, Feb. 14-16.

Faul, H. and Tittle, C.W.
  1951: Logging of drill holes by the neutron, gamma method, and gamma-ray scattering; Geophysics, v. 16, no. 2, p. 260.

Finck, R.R., Liden, K., and Persson, R.B.R.
  1976: In situ measurements of environmental gamma radiation by the use of a Ge(Li) spectrometer; Nuclear Instruments and Methods, v. 135, p. 559-567.

Flanigan, V.J.
  1972: Gamma radiation; an aid to geologic mapping on the Arabian Shield, Kingdom of Saudi Arabia; Proc. 2nd Int. Symp. Natural Radiation Environment, Houston, p. 667-697.

Fons, L.
  1969: Geological applications of well logs; Society of Professional Well Log Analysts, 10th Ann. Log. Symp., Paper AA.

Foote, R.S.
  1964: Time variation of terrestrial gamma radiation; in The Natural Radiation Environment, J.A.S. Adams and W.M. Lowder (eds.), University of Chicago Press, Chicago, p. 757-766.

Foote, R.S.
  1969: Radioactive methods in mineral exploration; in Mining and Groundwater Geophysics/1967, L.W. Morley (ed.), Geol. Surv. Can., Econ. Geol. Rep. 26, p. 177-190.

  1976: Radiometric data presentation; in Uranium Geophysical Technology Symposium Sept. 1976, Summaries and Visual Presentations, U.S. Energy Research and Development Administration, Grand Junction Office, Colorado, p. 53-62.

  1978: Development of a U.S. ERDA calibration range for airborne gamma radiation surveys; Proc. Symp. on Aerial Techniques for Environmental Monitoring, American Nuclear Society, p. 158.

Foote, R.S. and Humphrey, N.B.
  1976: Airborne radiometric techniques and applications to uranium exploration; in Exploration for Uranium Ore Deposits, IAEA, Vienna, p. 17-34.

Garber, R.J. and Soonawala, N.M.
  1977: Koona Lake; in Report of Field Activities, 1977, Manit. Dep. Mines, Resourc. Environ. Management, p. 173-177.

Gaucher, J.C., Got, H., Labeyrie, J., Lalou, C., and Lansiart, A.
  1974: Application de la spectrometrie q "in situ" a la cartographie granulometrique sous-marine; Bulletin du BRGM (deuxieme serie) Section IV, No. 4-1974, p. 231-241.

Geodata International Inc.
  1975a: Aerial radiometric and magnetic survey of Lubbock and Plainview National topographic maps, NW Texas; Doc. No. GJO-1654, U.S. Energy Research and Development Administration Contract AT (05-1)-1654, Grand Junction, Colo.

  1975b: Aerial radiometric and magnetic survey of central Appalachian Basin — parts of Virginia and the Carolinas; Doc. No. GJO-1644, U.S. Energy Research and Development Administration Contract No. AT (05-1)-1644, Grand Junction, Colo.

  1975c: Aerial radiometric and magnetic survey of Greenville, Augusta, Florence, Georgetown, Athens, Savannah and Spartanburg National topographic maps, North and South Carolina areas; Doc. No. GJO-1663-1 (7 separate reports), U.S. Energy Research and Development Administration Contract No. E (05-1)-1663, Grand Junction, Colo.

George, D.C., Evans, H.B., Allen, J.W., Key, B.N., Ward, D.L., and Mathews, M.A.
  1978: A borehole gamma-ray spectrometer for uranium exploration; U.S. Dep. of Energy, Grand Junction Office, Report GJBX-82(78).

Goldak, G.R.
  1975: Underwater radiometry proving useful tool to locate uranium; Northern Miner, March 6.

Gorbatyuk, O.V., Kadisov, E.M., Miller, V.V., and Troitskii, S.G.
  1973: Possibilities of determining uranium and radium content of ores by measuring gamma radiation in a borehole using a spectrometer with a Ge(Li) detector; Translated from Atomnaya Energiya, v. 35, no. 5, p. 355-357, Nov. 1973: Consultants Bureau, Plenum Pub. Co., New York.

Goso, H., Spoturno, J., and Preciozzi, G.
  1976: Una methodologia de prospeccion autoportada — Primeros resultados obtenidos en la Cuenca del Nordeste (Uruguay); Exploration for Uranium Ore Deposits, Proc. Series, IAEA, Vienna, p. 531-544.

Grasty, R.L.
- 1972: Airborne gamma-ray spectrometry data processing manual; Geol. Surv. Can., Open File 109.
- 1975: Uranium measurement by airborne gamma-ray spectrometry; Geophysics, v. 40, p. 503-519.
- 1976a: The "field of view" of gamma-ray detectors — a discussion; in Report of Activities, Part B, Geol. Surv. Can., Paper 76-1B, p. 81-82.
- 1976b: A calibration procedure for an airborne gamma-ray spectrometer; Geol. Surv. Can., Paper 76-16, p. 1-9.
- 1976c: Applications of gamma radiation in remote sensing; in Remote Sensing for Environmental Sciences, E. Schanda (ed.), Springer-Verlag, New York.
- 1977a: A general calibration procedure for airborne gamma-ray spectrometers; in Report of Activities, Part C, Geol. Surv. Can., Paper 77-1C, p. 61-62.
- 1977b: Calibration for total count gamma-ray surveys; in Report of Activities, Part B, Geol. Surv. Can., Paper 77-1B, p. 81-84.
- 1979: Gamma ray spectrometric methods in uranium exploration: Theory and operational procedures; in Geophysics and Geochemistry in the Search for Metallic Ores, Geol. Surv. Can., Econ. Geol. Rep. 31, Paper 10B.

Grasty, R.L. and Charbonneau, B.W.
- 1974: Gamma-ray spectrometer calibration facilities; in Report of Activities, Part B, Geol. Surv. Can., Paper 74-1B, p. 69-71.

Grasty, R.L. and Darnley, A.G.
- 1971: The calibration of gamma-ray spectrometers for ground and airborne use; Geol. Surv. Can., Paper 71-17, 27 p.

Grasty, R.L., Kosanke, K.L., and Foote, R.S.
- Fields of view of airborne gamma-ray spectrometers; Geophysics. (in press)

Gregory, A.F.
- 1955: Aerial detection of radioactive mineral deposits, Trans. Can. Inst. Min. Met., v. 58, p. 261-267.
- 1956: Analysis of radiometric sources in aeroradiometric surveys over oilfields; Bull. Am. Assoc. Petrol. Geol., v. 40, p. 2457-2474.
- 1960: Geological interpretation of aeroradiometric data; Geol. Surv. Can., Bull. 66.

Gregory, A.F. and Horwood, J.L.
- 1961: A laboratory study of gamma-ray spectra at the surface of rocks; Dep. Energy, Mines and Resources, Ottawa, Mines Br., Res. Rep. R 85, 52 p.
- 1963: A spectrometric study of the attenuation in air of gamma rays from mineral sources, Dep. of Mines and Technical Surveys, Ottawa, Mines Branch Research Report R 110, 110 p.

Gross, W.H.
- 1952: Radioactivity as a guide to ore; Econ. Geol., v. 47, p. 722-741.

Guillou, R.B. and Schmidt, R.G.
- 1960: Correlation of aeroradioactivity data and areal geology; in U.S. Geol. Surv. Prof. Paper 400-B, p. B119-B121.

Gunn, P.J.
- 1978: Inversion of airborne radiometric data; Geophysics, v. 43, no. 1, p. 133-143.

Hallenburg, J.K.
- 1973: Interpretation of gamma ray logs; Log Analyst, v. 14, p. 3-15.

Hartman, R.R.
- 1967: Airborne gamma-ray spectrometry; Aero Service Report, Aug. 24.

Hassan, M., Hossin, A., and Combay, A.
- 1976: Fundamentals of the differential gamma ray logs; in Transactions of Society of Prof. Well Log Analysts 17th Annual Symposium, Paper H, 18 p.

Hawkins, W.K. and Gearhart, M.
- 1968: Use of logging in uranium prospecting; Society of Professional Well Log Analysts, 9th Ann. Log. Symp., Paper T.
- 1969: Gamma-ray logging in uranium prospecting; in Nuclear Techniques and Mineral Resources, Proc. Series, IAEA, Vienna, p. 213-222.

Heath, R.L.
- 1964: Scintillation spectrometry, gamma-ray spectrum catalogue 2nd ed., Vol. 1 and 2, U.S.A.E.C. Research and Development Report IDO-16880-1, Physics T.I.D.-4500 (31st ed.).

Hoeve, J.
- 1975: The St. Louis Fault Project; in Summary of Investigations by the Saskatchewan Geological Survey 1975, edited by Christopher, J.E. and Macdonald, R., p. 120-122.

Holmes, S.W.
- 1978: Methodology a must for efficient exploration; The Northern Miner, March 2, page C16.

Horwood, J.L.
- 1960: A graphical determination of uranium and thorium ores from their gamma-ray spectra; Int. J. Appl. Radiation Isotopes, v. 9, p. 16-26.

Hurley, P.M.
- 1956: Direct radiometric measurement by gamma-ray scintillation spectrometer: Parts I and II; Bull. Geol. Soc. Am., v. 67, p. 395-412.

Hyde, E.K., Perlman, I., and Seaborg, G.T.
- 1964: The nuclear properties of the heavy elements; Vol. 2, Detailed Radioactivity Properties, Prentice Hall, New Jersey, 1107 p.

IAEA
- 1973a: Panel report no. 2: Survey of present methods; in Uranium Exploration Methods, Proc. Series, IAEA, Vienna, p. 257-291.
- 1973b: Panel report no. 4: Exploration costs; in Uranium Exploration Methods, Proc. Series, IAEA, Vienna, p. 297-299.
- 1974: Recommended instrumentation for uranium and thorium exploration; Technical Report 158, IAEA, Vienna, 93 p.
- 1976: Radiometric reporting methods and calibration in uranium exploration; Technical Report 174, IAEA, Vienna, 57 p.

Kamiyama, T., Okada, S., and Shimayaki, Y.
- 1973: Exploration of uranium deposits in Tertiary conglomerates and sandstones in Japan; Uranium Exploration Methods, Proc. Series, IAEA, Vienna, p. 45-54.

Ketola, M. and Sarikkola, R.
　1973: Some aspects concerning the feasibility of radiometric methods for uranium exploration in Finland; in Uranium Exploration Methods, proc. series, IAEA, Vienna, p. 31-43.

Ketola, M., Piiroinen, E., and Sarikkola, R.
　1975: On feasibility of airborne radiometric surveys for uranium exploration in Finland; Geol. Surv. Fin., Rep. Investig., No. 7, 43 p.

Killeen, P.G.
　1966: A gamma-ray spectrometric study of the radioelement distribution of the Quirke Lake syncline, Blind River, Ontario; Unpublished M.Sc. Thesis, Univ. Western Ontario, London, Ontario, 154 p.

　1975: Nuclear techniques for borehole logging in mineral exploration, in Borehole Geophysics Applied to Metallic Mineral Prospecting — A review, A.V. Dyck (ed.), Geol. Surv. Can., Paper 75-31, p. 39-52.

　1976a: Discussion of a paper by Duval (1976); in Exploration for Uranium Ore Deposits, Proc. Series, IAEA, Vienna, p. 80.

　1976b: Portable borehole gamma-ray spectrometer tests; in Report of Activities, Part A, Geol. Surv. Can., Paper 76-1A, p. 487-489.

　1977: Specifications for G.S.C. calibration facilities; Energy, Mines and Resources, Pilot Plant Bells Corners, Ottawa, Ontario, Feb. 1977.

　1978: Gamma-ray spectrometric calibration facilities — a preliminary report; in Current Research, Part A, Geol. Surv. Can., Paper 78-1A, p. 243-247.

Killeen, P.G. and Bristow, Q.
　1976: Uranium exploration by borehole gamma-ray spectrometry using off-the-shelf instrumentation; in Exploration for Uranium Ore Deposits, Proc. Series, IAEA, Vienna, p. 393-414.

Killeen, P.G. and Cameron, G.W.
　1977: Computation of in situ potassium, uranium and thorium concentration from portable gamma-ray spectrometer data; in Report of Activities, Part A, Geol. Surv. Can., Paper 77-1A, p. 91-92.

Killeen, P.G. and Carmichael, C.M.
　1970: Gamma-ray spectrometer calibration for field analysis of thorium, uranium and potassium; Can. J. Earth Sci., v. 7, no. 4, p. 1093-1098.

　1972: Case History — an experimental survey with a portable gamma-ray spectrometer in the Blind River area, Ontario; in Uranium Prospecting Handbook, S.H.U. Bowie, M. Davis, and D. Ostle (eds.), Inst. Min. Met., London, p. 306-312.

　1976: Radioactive disequilibrium determinations, Part 1: Determination of radioactive disequilibrium in uranium ores by alpha-spectrometry; Geol. Surv. Can., Paper 75-38, p. 1-18.

Killeen, P.G. and Conaway, J.G.
　1978: New facilities for calibrating gamma-ray spectrometric logging and surface exploration equipment; Can. Inst. Mining Metall. Bull., v. 71, no. 793, p. 84-87.

Killeen, P.G., Bernius, G.R., and Hall, N.
　1976: Carborne gamma-ray survey, Prince Edward Island; in Report of Activities, Part C, Geol. Surv. Can., Paper 76-1C, p. 269-271.

Killeen, P.G., Carson, J.M., and Hunter, J.A.
　1975: Optimizing some parameters for airborne gamma-ray spectrometric surveying; Geoexploration, v. 13, p. 1-12.

Killeen, P.G., Conaway, J.G., and Bristow, Q.
　1978: A gamma-ray spectral logging system including digital playback, with recommendations for a new generation system; in Current Research, Part A, Geol. Surv. Can., Paper 78-1A, p. 235-241.

Killeen, P.G., Hunter, J.A., and Carson, J.M.
　1971: Some effects of altitude and sampling rate in airborne gamma-ray spectrometric surveying; Geoexploration, v. 9, no. 4, p. 231-234.

Knapp, K.E. and Bush, W.E.
　1975: Construction KUT test pits; Internal Report by Lucius Pitkin Inc., Grand Junction, Colorado.

Kogan, R.M., Nazarov, I.M., and Fridman, Sh.D.
　1969: Gamma spectrometry of natural environments and formations; Trans. 1971 by Israel Program for Scientific Translations Ltd. No. 5778, available from the U.S. Dept. of Commerce, Nat. Tech. Inf. Ser., Springfield, Va., 22151, 337 p.

Landstrom, O.
　1976: Analysis of elements in boreholes by means of naturally occurring X-ray fluorescence radiation; in Nuclear Techniques in Geochemistry and Geophysics, Proc. Series, IAEA, Vienna, p. 47-52.

Lauber, A. and Landstrom, O.
　1972: A Ge(Li) borehole probe for in situ gamma ray spectrometry; Geophys. Prospect., v. 20, p. 800-813.

Lederer, C.M. and Shirley, V.S. (ed.)
　1978: Tables of Isotopes; seventh edition, John Wiley and Sons, New York.

Lederer, C.M., Hollander, J.M., and Perlman, I.
　1968: Table of Isotopes; Sixth Edition, John Wiley and Sons, Inc., New York, N.Y., 594 p.

Linden, A.H.
　1976: Method of detecting small or indistinct radioactive sources by airborne gamma-ray spectrometry; in Geology, Mining and Extractive Processing of Uranium, Inst. Min. Metall., London, p. 113-120.

Lock, G.A. and Hoyer, W.A.
　1971: Natural gamma-ray spectral logging; The Log Analyst, v. 12, no. 5, p. 3-9.

Løvborg, L.
　1972: Assessment of uranium by gamma-ray spectrometry; in Uranium Prospecting Handbook, S.H.U. Bowie et al. (ed.), Instit. Min. Met., London, p. 157-173.

　1973: Future development in the use of gamma-ray spectrometry for uranium prospecting on the ground; in Uranium Exploration Methods, Proc. Series, IAEA, Vienna, p. 141-153.

Løvborg, L. and Kirkegaard, P.
　1974: Response of 3"x3" NaI(Tl) detectors to terrestrial gamma radiation; Nuclear Instruments and Methodo., v. 121, p. 239-251.

Løvborg, L., Bøtter-Jensen, L., and Kirkegaard, P.
　1978a: Experiences with concrete calibration sources for radiometric field instruments; Geophysics, v. 43, no. 3, p. 543-549.

Løvborg, L., Grasty, R.L., and Kirkegaard, P.
1978b: A guide to the calibration constants for aerial gamma-ray surveys in geoexploration; Proceedings of American Nuclear Society Symposium on Aerial Techniques for Environmental Monitoring, March 1977, Las Vegas.

Løvborg, L., Kirkegaard, P., and Christiansen, E.M.
1976: Design of NaI(Tl) scintillation detectors for use in gamma-ray surveys of geological sources; in Exploration for Uranium Ore Deposits, Proc. Series, IAEA, Vienna, p. 127-148.

Løvborg, L., Kirkegaard, P., and Rose-Hansen, J.
1972: Quantitative interpretation of the gamma ray spectra from geologic formations; in The Natural Radiation Environment II, Adams, J.A.S., Lowder, W.M., and Gesell, T.F. (eds.), U.S. Dept. of Commerce, Springfield, Va., p. 155-180.

Løvborg, L., Kunzendorf, H., and Hansen, J.
1969: Use of field gamma spectrometry in the exploration of uranium and thorium deposits in South Greenland; Nuclear Techniques and Mineral Resources, Proc. Series, IAEA, Vienna, p. 197-211.

Løvborg, L., Wollenberg, H., Rose-Hansen, J., and Nielsen, B.L.
1972: Drill-core scanning for radioelements by gamma-ray spectrometry; Geophysics, v. 37, no. 4, p. 675-693.

Løvborg, L., Wollenberg, H., Sørensen, P., and Hansen, J.
1971: Field determination of uranium and thorium, exemplified by measurements in the Ilimaussaq alkaline intrusion, South Greenland; Econ. Geol., v. 66, p. 368-384.

Lyubavin, Yu. P. and Orchinnikov, A.K.
1961: Gamma radiation of uranium and its daughter products in radioactive ore bodies. Vopr. rudn. geofiz. (problems of mining geophysics); Ministry of Geology and Conservation of Natural Resources, U.S.S.R., no. 3 (AEC Trans. 6830), p. 87-94.

Mahdavi, A.
1964: The thorium, uranium, and potassium contents of Atlantic and Gulf Coast beach sands; in The Natural Radiation Environment, J.A.S. Adams and W.M. Lowder (eds.), Chicago, University of Chicago Press, p. 87-114.

Marett, G., Chevalier, P., Souhaite, P., and Suau, J.
1976: Shaly sand evaluation using gamma-ray spectrometry, applied to the North Sea Jurassic; Trans., Society of Professional Well Log Analysts, Seventeenth Annual Logging Symposium, Denver, Colorado.

Matolin, M.
1973: Artificial standards for calibration of airborne, field portable, and logging gamma spectrometers; in Uranium Exploration Methods, Proc. Series, IAEA, Vienna, p. 125-139.

Mathews, M.A., Koizumi, C.J., and Evans, H.B.
1978: D.O.E. — Grand Junction logging model data synopsis; U.S. Dept. of Energy, Grand Junction Office, Report GJBX-76(78).

Mero, J.L.
1960: Uses of the gamma-ray spectrometer in mineral exploration; Geophysics, v. 25, p. 1054-1076.

Miller, J.M. and Loosemore, W.R.
1972: Instrumental techniques for uranium prospecting; in Uranium Prospecting Handbook, S.H.U. Bowie et al. (eds.), Instit. Min. Met., London, p. 135-148.

Miller, J.M., Roberts, P.D., Symons, G.D., Merrill, N.N., and Wormald, M.R.
1977: A towed sea-bed gamma-ray spectrometer for continental shelf surveys; Int. Symp. Nuc. Tech. Expl., Extr., and Process. Min. Res. IAEA/SM/216/62, Vienna.

Miller, J.M. and Symons, G.D.
1973: Radiometric traverse of the seabed off the Yorkshire Coast; Nature, G.B., v. 242, no. 5394, p. 184-186.

Morris, D.
1969: New airborne exploration methods; Hunting Group Review, No. 70, p. 4-7.

Moseley, L.M.
1976: Field evaluation of direct digital well logging; in Transactions of Society of Professional Well Log Analysts 17th Annual Symposium, Paper NN.

Moxham, R.M.
1958: Geologic evaluation of airborne radioactivity survey data; Proc. of the 2nd Int. Conf. on the Peaceful Uses of Atomic Energy, Geneva, v. 2, p. 815-819.

1960: Airborne radioactivity surveys in geologic exploration; Geophysics, v. 25, no. 2, p. 408-432.

Moxham, R.M. and Tanner, A.B.
1977: High resolution gamma-ray spectrometry in uranium exploration; U.S. Geol. Survey, Jour. Resear., v. 5, no. 6, p. 783-795.

Moxham, R.M., Foote, R.S., and Bunker, C.M.
1965: Gamma-ray spectrometer studies of hydrothermally altered rocks; Econ. Geol., v. 60, no. 4, p. 653-671.

Munday, R.J.
1975: Investigation of radiometric anomalies within crown reserves 617, 618, 619 and CBS 3548; in Summary of Investigations 1975, Sask. Geol. Surv., p. 114-119.

Nelson, J.M.
1953: Prospecting for uranium with car-mounted equipment: U.S. Geol. Surv., Bull. 988-I, p. 211-221.

Newton, A.R. and Slaney, V.R.
1978: Geological interpretation of an airborne gamma-ray spectrometer survey of the Hearne Lake area, Northwest Territories; Geol. Surv. Can., Paper 77-32, 14 p.

Nininger, R.D.
1973: Uranium exploration policy, economics and future prospects; in Uranium Exploration Methods, panel proc. series, IAEA, Vienna, p. 3-17.

Noakes, J.E. and Harding, J.L.
1971: New Techniques in seafloor mineral exploration; J. Marine Tech., v. 5, no. 6, p. 41-44.

Noakes, J.E., Harding, J.L., and Spaulding, J.O.
1974a: Locating offshore mineral deposits by natural radioactive measurements; Marine Technol. Soc. J., v. 8, no. 5, p. 36-39.

1974b: Californium 252 as a new oceanographic tool; in Marine Technology Society 8th Annual Conference Preprints, p. 415-427.

Noakes, J.E., Harding, J.L., Spaulding, J.D., and Fridge, D.S.
1975: Surveillance system for subsea survey and mineral exploration; Offshore Technology Conference, Dallas, Paper No. OCT 2239, p. 909-914.

NURE
  1976: National Uranium Resource Evaluation Annual NURE Report, U.S. Energy Research and Development Administration, Grand Junction Office Report GJBX-11(77).

Ostrihansky, L.
  1976: Radioactive disequilibrium determinations, Part 2: Radioactive disequilibrium investigations, Elliot Lake area, Ontario; Geol. Surv. Can., Paper 75-38, p. 19-48.

Parslow, G.R. and Stolz, H.
  1976: Evaluation of techniques for assessing the uranium potential of lake covered areas: Parts I and II; in Summary of investigations by the Saskatchewan Geological Survey 1976, edited by Christopher, J.E. and Macdonald, R., p. 128-143.

Pemberton, R.
  1968: Radiometric exploration: modern tools in the search for uranium; Mining in Canada, May, p. 34-42.

Pemberton, R.H. and Seigel, H.O.
  1966: Airborne radioactivity tests, Elliot Lake area, Ontario; Can. Min. J. v. 87, no. 10, p. 81-87.

Pierson, D.H. and Franklin, E.
  1951: Aerial prospecting for radioactive minerals; Br. J. Appl. Pys., v. 2, p. 281-291.

Pirson, S.J.
  1963: Handbook of Well Log Analysis for Oil and Gas Formation Evaluation; Prentice-Hall Inc.

Pitkin, J.A.
  1968: Airborne measurements of terrestrial radioactivity as an aid to geologic mapping: U.S. Geol. Surv., Prof. Paper 516-F, 29 p.

Pitkin, J.A., Neuschel, S.K., and Butes, R.G.
  1964: Aeroradioactivity surveys and geologic mapping; in The Natural Radiation Environment, Adams, J.A.S. and Lowder, W.M. (eds.), Univ. of Chicago Press, p. 723-736.

Pochet, F.R.
  1976: Le developpement des diagraphies dans les forages des recherche et d'exploitation minière au Commissariat A L'Energie Atomique; in Exploration for Uranium Ore Deposits, Proc. Series, IAEA, p. 353-366.

Potts, M.J.
  1976: Computer methods for geologic analysis of radiometric data; in Exploration for Uranium Ore Deposits, Proc. Series, IAEA, Vienna, p. 55-69.

Puibaraud, Y.
  1972: Portable equipment for uranium prospecting; in Uranium Prospecting Handbook S.H.U. Bowie et al. (ed.), Instit. Min. Met., London, p. 149-156.

Purvis, A.E. and Foote, R.S.
  1964: Atmospheric attenuation of gamma radiation; in Natural Radiation Environment, J.A.S. Adams and W.M. Lowden (eds.), Univ. of Chicago Press, Chicago, p. 747-756.

Ragaini, R.C., Jones, D.E., Huckaboy, G.W., and Todachine, T.
  1974: Terrestrial gamma-ray surveys at preoperational nuclear power plants using an in situ Ge(Li) spectrometer; Proceedings IEEE Nuclear Science Symposium, Washington, D.C., Dec. 11-14.

Reeves, D.R.
  1976: Development of slimline logging systems for coal and mineral exploration; Transactions of Society of Professional Well Log Analysts 17th Annual Logging Symposium, Paper KK.

Rhodes, D.F. and Mott, W.E.
  1966: Quantitative interpretation of gamma-ray spectral logs; Geophysics, v. 31, no. 2, p. 410-418.

Richards, D.J.
  1977: Karoo uranium occurrences — modelling of the airborne gamma-ray spectrometer response for digital cross-correlation filtering of observed profiles and removal of statistical noise; Geological Survey, Republic of South Africa, Report No. GL 2275.

Richardson, K.A.
  1964: Thorium, uranium, and potassium in the Conway granite, New Hampshire, U.S.A.; in The Natural Radiation Environment, J.A.S. Adams and W.M. Lowder (eds.), Univ. Chicago Press, p. 39-50.

Richardson, K.A. and Carson, J.M.
  1976: Regional uranium distribution in northern Saskatchewan; in Symposium on Uranium in Saskatchewan, Sask. Geol. Soc., Spec. Pub. 3, p. 27-50.

Richardson, K.A., Darnley, A.G., and Charbonneau, B.W.
  1972: Airborne gamma-ray spectrometric measurements over the Canadian Shield; Proc. 2nd Int. Symp. of the Natural Radiation Environment, U.S. Dept. of Commerce, Springfield, Virginia, p. 681-704..

Richardson, K.A. and Killeen, P.G.
  1979: Calibration facilities for gamma-ray spectrometers made available by Geological Survey of Canada; The Northern Miner, March 8, p. C2.

Richardson, K.A., Killeen, P.G., and Charbonneau, B.W.
  1975: Results of a reconnaissance type airborne gamma-ray spectrometer survey of the Blind River-Elliot Lake area, Ontario; in Report of Activities, Part A, Geol. Surv. Can., Paper 75-1A, p. 133-135.

Rosholt, J.N., Jr.
  1959: Natural radioactive disequilibrium of the uranium series; U.S. Geol. Surv. Bull. 1084-A, 30 p.

Russell, W.L.
  1955: The use of gamma-ray measurements in prospecting; Econ. Geol., 50th Ann. Vol., p. 835-866.

Russell, W.L. and Scherbatskoy, S.A.
  1951: The use of sensitive gamma ray detectors in prospecting; Econ. Geol., v. 40, p. 432.

Saunders, D.F. and Potts, M.J.
  1976: Interpretation and application of high-sensitivity airborne gamma-ray spectrometer data; in Exploration for Uranium Ore Deposits, Proc. Series, IAEA, Vienna, p. 107-125.

  1978: Manual for the application of NURE 1974-1977 aerial gamma-ray spectrometer data; U.S. Dept. of Energy, Grand Junction Office, Report GJBX-13(78).

Scott, J.H.
  1962: The GAMLOG computer program; U.S.A.E.C. Report RME-143, Grand Junction, Colorado.

  1963: Computer analysis of gamma-ray logs; Geophysics, v. 28, no. 3, p. 457-465.

Scott, J.H. and Dodd, P.H.
  1960: Gamma-only assaying for disequilibrium corrections; U.S. Atomic Energy Comm. RME-135, p. 1-20.

Scott, J.H. and Tibbetts, B.L.
1974: Well-logging techniques for mineral deposit evaluation: A review; United States Department of the Interior, Bureau of Mines, Information Circular 8627.

Scott, J.H., Dodd, P.H., Droullard, R.F., and Mudra, P.J.
1961: Quantitative interpretation of gamma-ray logs; Geophysics, v. 26, no. 2, p. 182-191.

Senftle, F.E., Moxham, R.M., Tanner, A.B., Boynton, G.R., Philbin, P.W., and Baicker, J.A.
1976: Intrinsic germanium detector used in borehole sonde for uranium exploration; Nuclear Instruments and Methods, v. 138, p. 371-380.

Shideler, G.L. and Hinze, W.J.
1971: The utility of carborne radiometric surveys in petroleum exploration of glaciated regions; Geophys. Prospect., v. 19, p. 568-585.

Sibbald, T.I.
1975: Investigation of certain radiometric anomalies in Crown Reserve 621; in Summary of Investigations 1975, Sask. Geol. Surv., p. 123-129.

Siegbahn, K.
1968: Alpha-, Beta-, and Gamma-ray Spectroscopy; North Holland Pub. Co. Amsterdam, 1742 p.

Slaney, V.R.
1978: A study of airborne gamma-ray spectrometry for geological mapping; Proc. 12th Int. Symp. on Remote Sensing of Environment, Ann Arbor, Mich., p. 1995-2009.

Smith, A.R. and Wollenberg, H.A.
1972: High-resolution gamma ray spectrometry for laboratory analysis of the uranium and thorium decay series; Proceedings of the Second International Symposium on the Natural Radiation Environment, Houston, Texas, J.A.S. Adams, W.M. Lowder, and T.F. Gesell (eds.), U.S. Dept. of Commerce, Springfield, Virginia, p. 181-231.

Smith, B.R., Whitworth, R.A., and Soonawala, N.M.
1977: Red Sucker Lake; in Report of Field Activities 1977, Manit. Dep. Mines, Resources and Environmental Management, p. 187-189.

Soonawala, N.M.
1977: Uranium exploration; in Report of Field Activities 1977, Manit. Dep. Mines, Resources and Environmental Management, p. 171.

Sprecher, C. and Rybach, L.
1974: Design and field test of a scintillation probe for γ-logging of small diameter boreholes; Pure Appl. Geophys., v. 112, p. 563-570.

Stead, F.W.
1950: Airborne radioactivity surveying speeds uranium prospecting; Engin. Min. J., v. 151, no. 9, p. 74-77.

1956: Subsurface radiometric techniques; in Exploration for Nuclear Raw Materials, R.D. Nininger (ed.), D. Van Nostrand Pub. Co., Princeton, New Jersey.

Stolz, H.
1976: Evaluation of techniques for assessing the uranium potential of lake covered areas, Part II: radiometric techniques; in Summary of Investigations 1976, Sask. Geol. Surv., p. 134-141.

Stolz, H. and Drevor, G.
1977: Evaluation of techniques for assessing the uranium potential of lake covered areas: underwater radiometry; in Summary of Investigations 1977, Sask. Geol. Surv., p. 72-77.

Stolz, H. and Standing, K.F.
1977: Underwater prospecting for radioactive minerals; Preprint presented at 47th Annual Meeting, Society of Exploration Geophysicists, Calgary.

Stromswold, D.C.
1978: Monitoring of the airport calibration pads at Walker Field, Grand Junction, Colorado for long term radiation variations; U.S. Dept. of Energy, Grand Junction Office Report No. GJBX-99(78).

Summerhayes, C.P., Hazelhoff-Roelfzema, B.H., Tooms, J.S., and Smith, D.B.
1970: Phosphorite prospecting using a submersible scintillation-counter; Econ. Geol., v. 65, p. 718-723.

Tammenmaa, J.K. and Grasty, R.L.
1977: Upward and downward continuation of gamma-radiation fields; Society of Exploration Geophysicists, Abstracts of Papers, 47th Annual International Meeting, Calgary, Alberta, p. 61-62.

Tammenmaa, J.K., Grasty, R.L., and Peltoniemi, M.
1976: The reduction of statistical noise in airborne radiometric data; Can. J. Earth Sci., v. 13, no. 10, p. 1351-1357.

Tanner, A.B., Moxham, R.M., and Senftle, F.E.
1977a: Assay for uranium and determination of disequilibrium by means of in situ high resolution gamma-ray spectrometry; U.S. Geol. Survey Open File Report 77-571.

1977b: Assay for uranium and measurement of disequilibrium by means of high-resolution gamma-ray borehole sondes; in Campbell, J.A., ed., Short papers of the U.S. Geological Survey uranium-thorium symposium, U.S. Geol. Surv., Circ. 753, p. 56-57.

Tipper, D.B.
1969: Airborne gamma-ray spectrometry: Australian Mining, April, p. 42-44.

Tipper, D.B. and Lawrence, G.
1972: The Narbarlek area, Arnhemland Australia: a case history; in Uranium Prospecting Handbook, S.H.U. Bowie, M. Davis, and D. Ostle (eds.), Instit. Min. Metall., London, p. 301-305.

Toens, P.D., van As, D., and Vleggaar, C.M.
1973: J. S. Afr. Inst. Min. Metall., v. 73; A facility at the national nuclear research centre, Pelindaba for the calibration of gamma-survey meters used in uranium prospecting operations, p. 428.

Ward, D.L.
1976: Development of airport calibration pads; in Uranium Geophysical Technology Symposium, Sept. 1976, Summaries and Visual Presentations, U.S. Energy Research and Development Administration, Grand Junction Office, Colorado, p. 99-104.

1978: Construction of the calibration pads facility, Walker Field, Grand Junction, Colorado; U.S. Dept. of Energy, Grand Junction Office Report GJBX-37(78).

West, F.G. and Laughlin, A.W.
1976: Spectral gamma logging in crystalline basement rocks; Geology, v. 4, p. 617-618.

Whitworth, R.A., Garber, R.J., and Soonawala, N.M.
1977: Kasmere-Munroe regional follow-up; in Report of Field Activities 1977, Manit. Dep. Mines, Resources and Environmental Management, p. 178-186.

Wichmann, P.A., McWhirter, V.C., and Hopkinson, C.E.
   1975: Field results of the natural gamma-ray spectralog: Trans., Society of Professional Well Log Analysts Sixteenth Annual Logging Symposium, June 4-7, New Orleans, Louisiana.

Willox, W.A. and Tipper, D.B.
   1969: Aerial techniques in mineral exploration; Mining Technology Magazine, May.

Wollenberg, H.A. and Smith, A.R.
   1964: Studies in terrestrial gamma radiation; in The Natural Radiation Environment. J.A.S. Adams and W.M. Lowder (eds.), p. 513-566.

Wormold, M.R. and Clayton, C.G.
   1976: Observations on the accuracy of gamma spectrometry in uranium prospecting; in Exploration for Uranium Ore Deposits, Proc. Series, IAEA, Vienna, p. 149-172.

Zeigler, R.K.
   1976: Radiometric data analysis and graphical display; in Uranium Geophysical Technology Symposium, Sept. 1976, Summaries and Visual Presentations; U.S. Energy Research and Development Administration, Grand Junction Office, Colorado, p. 121-132.

MODERN TRENDS IN MINING GEOPHYSICS AND NUCLEAR BOREHOLE LOGGING METHODS FOR MINERAL EXPLORATION

Jan A. Czubek
Institute of Nuclear Physics, Kraków, Poland

Czubek, Jan A., Modern Trends in Mining Geophysics and Nuclear Borehole Logging Methods for Mineral Exploration; in Geophysics and Geochemistry in the Search for Metallic Ores; Peter J. Hood, editor; Geological Survey of Canada, Economic Geological Report 31, p. 231-272, 1979.

Abstract

This paper reviews recent developments in basic research and in the field practice of nuclear logging for mineral exploration, which have been published during 1974-1977, in about 200 papers from geophysical laboratories mostly in eastern Europe.

The main achievements in theory and experimentation in the following logging methods are presented: gamma-ray logging, XRF logging, Mössbauer effect, nuclear gamma resonance, neutron-neutron (resonance, epithermal and thermal) logging, spectrometric neutron-gamma logging, photon-neutron logging, die-away logging with pulsed neutron sources, and activation logging. Some problems of geostatistics applied to nuclear borehole logging also are described. For each logging method the fields of application reported during the last four years are given.

In the second part of the paper, the combined application of different nuclear logging methods for different groups of deposits is reviewed. The deposits discussed are: Iron-bauxite group – Fe, Mn, Al; Base metal group – Cu, Zn, Pb, Hg, Ba; Tin-rare metal group Sn, W, Be, Mo; Ultrabasic group: Cr, Ni; Gold group: Au, Sb, U; Sediments, evaporites and other types of deposit: S, K, B, phosphorite, apatite, fluorite, alunite. Some new possibilities in the future development of nuclear logging methods are also described.

The general conclusion is that the period 1974-1977 was characterized by a moderate development of theoretical research for different kinds of mineral nuclear logging (except for gamma and gamma-gamma methods). The practical application of XRF logging has been rather broad, the neutron methods are starting to be very promising, especially in the spectrometric version (both for radiative capture and activation). New neutron methods for uranium detection have been tried. The practical application of borehole neutron generators is still very limited, and has been confined to the search for mercury deposits. The first applications of the pulsed borehole photon generators for density logging have been reported. For each nuclear logging method the accuracy of the grade determination was at least comparable to that obtained by the usual chemical assay of cores.

Résumé

Dans cet article, l'auteur examine les derniers développements de la technologie des diagraphies nucléaires en prospection minérale (recherche fondamentale et applications sur le terrain), en se basant sur les résultats publiés de 1974 à 1977, dans près de 200 rapports provenant de laboratoires de géophysique dont la majeure partie se trouvent dans l'Europe de l'Est.

On y trouvera exposés les résultats les plus importants, en ce qui concerne la théorie et l'expérimentation, obtenus dans les méthodes de diagraphie suivantes: rayons gamma naturels, fluorescence X, effet Mössbauer, fluorescence nucléaire résonante, neutron-neutron (n. de résonance, n. épithermique et n. thermique), photoneutrons, spectrométrie du rayonnement gamma de capture, temps de relaxation neutronique (avec une source de neutrons pulsés), activation neutronique. On y aborde aussi quelques problèmes de géostatistique appliquée aux diagraphies nucléaires. Pour chaque méthode, les domaines d'application sont indiqués, tels qu'ils ressortent des travaux publiés pendant les quatre dernières années.

En deuxième partie, l'auteur donne des exemples d'application combinée des différentes diagraphies nucléaires dans différents groupes de gisements. Il s'agit des gisements suivants: le groupe fer-bauxite –Fe, Mn, Al; le groupe des métaux de base – Cu, Zn, Pb, Hg, Ba; le groupe étain-métaux rares – Sn, W, Be, Mo; le groupe ultrabasique – Cr, Ni; le groupe de l'or – Au, Sb, U; les gisements sédimentaires, d'évaporites et autres – S, K, B, phosphorite, apatite, fluorine, alunite. L'auteur décrit aussi quelques nouvelles possibilités dans le développement futur des diagraphies nucléaires.

On peut conclure en disant que la période de 1974 à 1977 se caractérise par une évolution modérée dans la recherche théorique des différents types des diagraphies nucléaires dans le domaine minéral (à l'exception des diagraphies de rayons gamma naturels et gamma-gamma). L'application pratique de la diagraphie de fluorescence X est assez répandue; les diagraphies de neutrons commencent à être assez prometteuses, surtout dans la version spectrométrique (spectrométrie du rayonnement gamma de capture et diagraphie d'activation neutronique). Pour la recherche de l'uranium, on a essayé de nouvelles diagraphies de neutrons. L'application pratique des générateurs de neutrons dans les sondages reste encore très limitée: en prospection minérale ils n'ont guère été utilisés que pour la recherche des gisements de mercure. On signale les premières applications des générateurs de photons à impulsions pour les diagraphies de densité. La détermination de la richesse d'un minerai est au moins aussi précise par les méthodes de diagraphies nucléaires que par l'analyse chimique des carottes.

INTRODUCTION

Nuclear logging methods are very sensitive to the presence of a given element in the multielemental system for this reason they are as convenient as the more common drillhole logging techniques for mineral exploration and mining. The three main groups of radiation sources used in practical applications are:

— Sources naturally existing in rocks, and those artificially introduced into drillholes:

— Gamma-ray sources,

— Neutron sources.

The gamma-ray sources due to the natural radioactivity of rocks are the most important in the first group.

The second group is composed of isotopic gamma-ray sources with a broad range of primary energies. Some accelerator-type gamma-ray sources are also included. Very soft gamma-ray energy (in the X-ray region) is used for X-ray fluorescence (XRF) logging for heavy elements and for the Mössbauer effect for tin ores. Gamma-ray energies in the region 100 to 300 keV are used in the so-called selective gamma-gamma method for detecting the presence of heavy elements (without distinguishing between them) in ores. Higher energies are used for density logging, the nuclear gamma resonance method (for Cu or Ni), and for the photon-neutron method (for Be mainly).

Among the neutron sources, the steady state and pulse operated ones have to be distinguished. The steady state neutron sources are usually of the α-Be type although the application of ^{252}Cf sources is becoming popular.

The pulsed neutron sources are usually those generated by the D, T reaction and are 14 MeV in energy. However, the application of these sources is not yet in common use because of the high cost of borehole neutron generators and the utilization of large diameter tools (except for the oil industry which is outside of the scope of this paper).

SUMMARY OF METHODS

Usually the development of any nuclear technique follows this sequence:

1. physical idea and feasibility of measurement,
2. laboratory experiments,
3. design and development of field equipment,
4. qualitative application,
5. establishing the theory for a given method,
6. solution of the so-called inverse problem to obtain the algorithms for quantitative interpretation,
7. quantitative applications.

The particular steps of this development can sometimes be changed into another sequence if new advances in apparatus design so dictate. Various nuclear methods used for mineral exploration and mining are classified at different stages in this development. Here we shall try to give a very short and simple description of the actual state-of-the-art for each method.

Steady state methods

In general the problem of quantitative interpretation and even the applicability of the method is linked with the solution of the equation:

$$R(\vec{r}) = \int_V G(\vec{r},\vec{r}^1) \cdot P(\vec{r}^1) \cdot d\vec{r}^1 \qquad (1)$$

where $R(\vec{r})$ is the logging probe response at the point $\vec{r}$ due to a given set of values of geological parameters P at the points $\vec{r}^1$. $G(\vec{r},\vec{r}^1)$ is the effect (for a given particular nuclear method) caused at the point $\vec{r}$ by the presence of a given geological parameter P at the point $\vec{r}^1$. The integration is over the whole space V at which the parameter P is different from zero. The goal is to find the set of values $P(\vec{r}^1)$ from a knowledge of the experimental values of R_{ex} at $\vec{r}$:

$$R_{ex}(\vec{r}) = R(\vec{r}) + \delta_R \qquad (2)$$

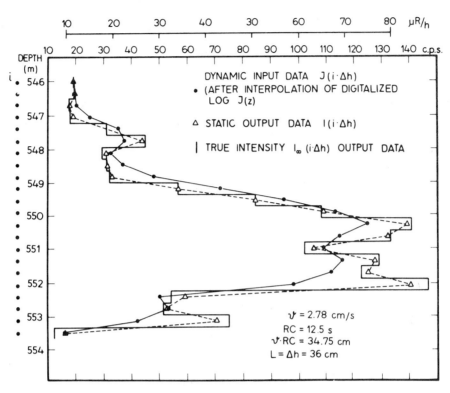

Figure 11.1. Example of a gamma-ray log interpretation of a uranium deposit. Uncased borehole, diameter 132 mm, filled with drilling mud. Linear absorption coefficient taken for interpretation: $\mu = 0.09$ cm^{-1}, $p = 4$, (after Czubek and Zorski, 1976).

where δ_R is a random error of the experimental value of the probe response R. The functions $G(\vec{r},\vec{r}^1)$, different for different nuclear methods, usually are not known exactly. Sometimes only a very rough approximation of their form can be postulated.

In view of these remarks the solution $P(\vec{r}^1)$ of Equation 1, when the values $R_{ex}(\vec{r})$ are known, belongs to the so-called "improper" problems of mathematics (cf. Tikhonov and Arsenin, 1976).

Natural gamma-ray logging

The natural gamma-ray logging technique which has undergone development for many years especially for uranium and potassium ores, has the best quantitative achievements among all the nuclear logging techniques. But even for that technique, the problem of the exact theory of the method is not entirely solved. The state-of-the-art in this subject was recently reviewed by Czubek and Zorski (1976). Under some simplifying conditions about the function $G(\vec{r},\vec{r}^1)$ in Equation 1, the depth distribution of the radioactive ore grade $q(z) \equiv P(\vec{r}^1)$ along the borehole axis z with a given depth resolution Δh can be obtained. The simplifying condition mainly involves the nature of the build-up factor for scattered radiation. A unique, equivalent attenuation coefficient μ is postulated for the whole observed gamma-ray spectrum in the rock. The value of μ can depend upon the detectors and the gamma-ray windows used in the logging tool. Novikov et al. (1974, 1976) have found the values:

$\mu = 0.032 \pm 0.034$ cm^2/g for uranium series

$\mu = 0.028 \pm 0.032$ cm^2/g for thorium series

$\mu = 0.05$ cm^2/g for potassium

which are essentially the same as were found earlier for the gross count of the whole scattered spectrum.

The final results of interpretation of a gamma-ray log can be presented as a step-wise function as shown in Figure 11.1. Here, for the elementary layer of thickness Δh, the gamma-ray intensity I_∞ can be determined. This intensity, free of the influence of measuring parameters (such as logging speed, ratemeter time constant, detector dead time, borehole diameter, borehole fluid, activity of the neighbour layers, etc) is directly proportional to the grade (per unit weight of the natural wet rock) of radioactive material. This natural radioactivity can be related either to the uranium, thorium or potassium ore grade, or to the grade of some other mineral, whose concentration is sometimes correlated with the natural radioactivity of the rock. Such a correlation has been reported, for example, for phosphorites by Rudyk et al. (1974) where the correlation coefficient was about 0.95, or for alunites (Muravev and Yakubson, 1975).

The problems occurring in the quantitative interpretation of gamma-ray logs have not been fully investigated and solved. The existing systems of interpretation need to be compared in detail to define the most accurate. The problem of accuracy of interpretation was discussed by Varga (1975) for the interpretation procedure proposed earlier by Rösler. His procedure is based on the solution of an infinite set of linear equations using the matrix method. Varga found that for this method of interpretation the statistical error of the grade determination could be three times higher than the statistical error of the input data.

The gamma-ray logger, especially the spectrometric versions, is an excellent tool for potassium determination; however it also has great disadvantages when used for uranium grade determination. Uranium which is commonly in radioactive disequilibrium with its decay products, cannot be accurately determined without a good knowledge of the disequilibrium factor. This is the main reason why there has been an increased effort in recent years to find other possibilities of measuring uranium content.

Other theoretical problems related to the contribution of different radioactive series to the whole gamma-ray spectrum as recorded by the gamma-ray logger have been discussed by Arakcheev and Bondar (1975) in order to distinguish different lithologies.

Methods of using gamma-ray sources

Very broad and successful application of gamma-ray sources to exploration and mining control was recently reviewed in an excellent monograph by Ochkur (1976). The main progress in recent years has been in the practical application of XRF logging of multicomponent ores with up to four elements being simultaneously determined (Bolotova and Leman, 1976).

XRF logging

In the X-ray fluorescence assay, the measurements are performed either on outcrops or in boreholes. The principal method is the so-called "spectral ratio" with an internal self-standardization of background as described by Medvedev et al. (1973). This method avoids the influence of the XRF and primary scattered radiation of other elements on the registered energy line $I(E)$ of the investigated element. When one takes the ratio

$$\eta = \frac{I(E)}{I(E_s)} \qquad (3)$$

where $I(E_s)$ is the intensity of scattered radiation at energy E_s, it is possible to find a value of E_s, for which the η value for a given, fixed energy E of the XRF line is independent of the grade of the interfering elements. An example is given in Figure 11.2. Here the source used was ^{147}Pm ($E_\gamma = 38$ keV) and the detector was a xenon proportional counter with a resolution of 16 per cent for the 22.5 keV line. The value of η for the antimony K_α line ($E = 26$ keV) does not depend upon the admixtures of Fe, Ba or Pb in the alabaster when the E_s energy is chosen in the region of 37 to 38 keV.

There are a number of different XRF probe constructions. The one used by Zgardovskiy et al. (1974) is shown in Figure 11.3. Other constructions have been reported by Leman et al. (1975) and by Yanshevskiy et al. (1976a, 1976b) and were designed for variable borehole diameters. The logs can be recorded either in dry or in water-filled holes (Krasnoperov and Zvuykovskiy, 1976). Other designs have been reported by Meyer and Filippov (1974), Meyer and Rozuvanov (1974), Baldin et al. (1976), Christell et al. (1976), Landström (1976) and many others. Landström has measured Pb ore grade using the spectral ratio but without any artificial gamma-ray source. The K_α line of lead was excited by the natural rock radioactivity.

Some details of XRF apparatus, filters, etc. have been described by Chernyavskaya et al. (1975). The optimum parameters for XRF loggers have been discussed by Nakhabtsev (1977) and Yanshevskiy et al. (1976a, 1976b). The application of solid-state detector to this kind of logger has been presented by Baldin et al. (1975, 1976).

In order to indicate how the system with background self-standarization works for the simultaneous determination of four elements, the results of Bolotova and Leman (1976a) are presented in Figures 11.4 and 11.5. The relative accuracy of the XRF logger is not very high being within the limits of ± 25 per cent for any particular point assay. Such a low accuracy is

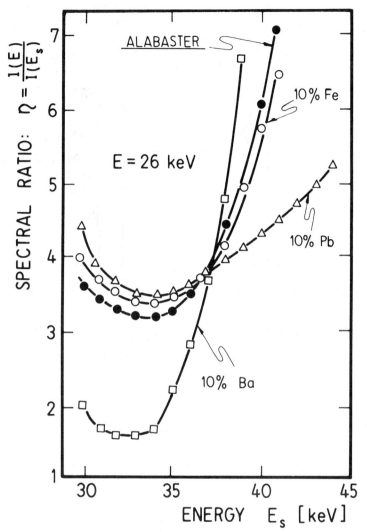

Figure 11.2. Determination of the self-standardized background in XRF logging for antimony ores (in alabaster) (after Medvedev et al., 1973).

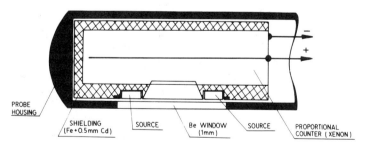

Figure 11.3. Measuring head of the XRF probe SRPD (after Zgardovskiy et al., 1974).

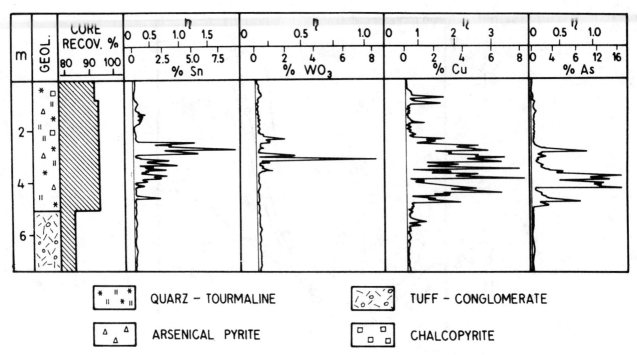

Figure 11.4. Separate determination of Sn, W, Cu, and As by XRF logging (after Bolotova and Leman, 1976a).

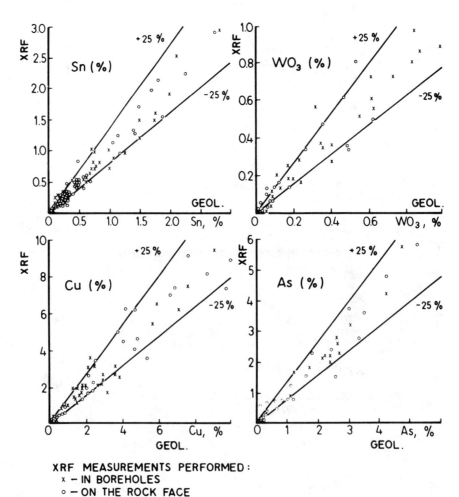

Figure 11.5. Comparison of XRF assaying with the chemical sampling results, (after Bolotova and Leman, 1976a).

due mainly to the very short range of penetration of the XRF method and variable ore grain size. The problem of the proper averaging or filtering of the raw data together with the physical principles of penetration of gamma radiation through heterogeneous rock media remain, therefore, a main research subject in many geophysical laboratories. In spite of rather low accuracy, the limit of detection of XRF methods is quite good. It depends, of course, upon the particular logging equipment being used and upon the investigated element. Bolotova and Leman (1976b), for example, give the following limits of detection: 0.05% for Sn, 0.1% for Cu, 0.1% for WO_3, 0.01% for Mo, 0.1% for As with the final accuracies being comparable with those obtained by chemical analysis.

The theoretical problems of XRF logging concerned with the cross-sections for scattered radiation have been discussed by Pshenichnyy and Meyer (1974, 1975) and Nakhabtsev (1974, 1975).

Mössbauer effect

Methodological problems similar to those of the XRF method are experienced in the practical application of the Mössbauer effect to geophysics. This effect is applied in the field and in logging practice mainly to tin ores (Ochkur, 1976; "Nuclear Geophys. Assay...", 1976; Goldanskiy et al., 1974). Three count rates are measured in this method:

I(O) — scattered radiation when the source velocity is $\nu = 0$

I(∞) — scattered radiation when the source velocity is $\nu = \infty$
in practice for tin ores, this velocity is $\nu = 2$ mm/s,

I_b — background.

The parameter

$$\varepsilon_s = \frac{I(O) - I(\infty)}{I(\infty) - I_b} \qquad (4)$$

depends upon the tin grade of the ore.

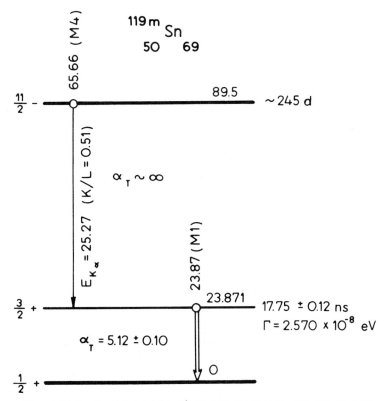

Figure 11.6. Decay scheme of Sn-119 m isotope, Energy in keV (after Stevens and Stevens, 1975).

The source of the resonance radiation is $^{119m}SnO_2$ of the activity of the order of 10^8 gammas/s. The decay scheme of this isotope is given in Figure 11.6. To eliminate the 65.66 keV line, which is almost completely converted, which gives the K_α line of Sn of energy 25.27 keV, the source is covered by a palladium sheet of thickness of 100 to 120 mg/cm^2. The source is vibrated by a bimorphous piezoelement using the polarized ceramic system of $PbZrTiO_3$. For detection purposes, a special resonance scintillation detector is utilized. Logging is performed step by step in 5 cm intervals. The range of investigation is of the order of 100 mg/cm^2.

Gamma-gamma methods

When a gamma-ray source is imbedded in a rock medium, the scattered gamma-ray spectrum reflects different properties of the rock surrounding the borehole. The whole spectrum is usually a function of the geometry of measurement such as borehole radius, type of shielding, presence or absence of drilling fluid, etc. In the energy region below 100 keV, the intensity of this spectrum is strongly influenced by the X-ray fluorescence of the rock and probe elements.

The measurement of scattered gamma radiation above the 100 keV energy level is the principal goal of the so-called gamma-gamma methods. They are used in geophysics in two distinct versions: for the measurement of density and in selective gamma-gamma methods. The first application is based on the measurement of the scattered gamma radiation from a relatively high energy gamma-ray source usually ^{137}Cs or ^{60}Co in the energy region above 200 or 300 keV. For this energy region the amount of scattered radiation is closely related to the bulk density of the scatterer, i.e. the rock. When ore bulk density is well correlated with ore grade, this type of measurement can be used to estimate ore grade. Sometimes ore density is used as one of the parameters in the ore grade determination by means of the multivariational correlation analysis.

The low energy region of the scattered gamma-ray spectrum, between 100 and about 300 keV, depends upon the photoelectric absorption properties of the rock. Therefore, this part of the spectrum is used in the second application of the method i.e. the selective gamma-gamma method. It is sensitive to the presence of heavy elements in the rock, or to the change in the equivalent atomic number Z_{eq} of the rock (Czubek, 1966; 1971). When the ores are of the monoelemental type, this log may be used to determine ore grade. To negate the influence of noncorrelated density variations on the selective gamma-gamma log results, the so-called P_z technique was introduced (Czubek, 1966). P_z is simply the ratio of the high energy part to the low energy part of the scattered spectrum. The P_z parameter is a strong function of Z_{eq} and is relatively insensitive to variations in rock bulk density. In dry boreholes, the shape of the scattered spectrum near the source energy is a strong function of the borehole diameter. By taking the ratios of intensities in the energy region above 400 or 500 keV Aylmer et al., (1978); Charbucinski et al., (1977); and Eisler et al., (1976) were able to correct the P_z and density values for the influence of borehole diameter.

The density method has been used for iron ore grade determination by Szymborski (1975) with an accuracy of 3.4 per cent of iron and by Ochkur et al. (1976). This logging technique was also used by Koshelev et al. (1976) to distinguish apatite-nepheline ores from spheno-apatite ores, and P_2O_5 was determined in apatite-nepheline ores by Startsev et al. (1975a) by the correlation with density. In a qualitative way this logging technique has been used by Kozlov et al. (1975a, 1975c) on copper deposits. The lithological differentiation of the bauxite ores by means of density logging was reported by Shishakin et al. (1974). Density logging has been used together with other kinds of gamma-gamma logs by Eisler et al. (1976), Charbucinski et al. (1977) and Aylmer et al. (1978) to determine iron content very precisely in West Australian deposits.

The selective log was used by Gera (1974) for the localization of quartz veins in gold deposits using backscattering radiation from a ^{170}Tm source (52 and 84 keV) with a source-detector distance of the order of 50 mm. The same method, but using a higher energy ^{137}Cs source (the registered window was 130 to 150 keV) was used for lead and barium assays by Shmonin et al. (1976a) as well as for the analysis of a lead-zinc deposit by Shmonin et al. (1976b). Tungsten grades higher than 0.2 per cent were investigated with this method by Kuchurin et al. (1976). It was also possible to determine the iron content of skarns with an accuracy of 2.5 per cent (Senko and Zorin, 1975).

The majority of applications of the gamma-gamma methods were more or less purely experimental and without a deep physical knowledge of the problems from the point of view of the gamma-ray transport in heterogeneous rock media. The real need for such knowledge stimulated research work in this subject. Gulin (1975) has published a monograph on gamma-gamma methods where the most important physical features of the method are given. His results have been obtained mainly by a theoretical approach using the Monte-Carlo technique for calculation of the space-angle-energy distributions of scattered photons in actual tool-borehole-rock systems. The other problems concerned with angular distribution of scattered radiation, together with the question of how to distinguish thin layers, were considered by Popov et al. (1974), Popov and Vishnyakov (1974), Utkin et al. (1974, 1975, 1976), Lukhminskiy and Galimbekov (1975), Galimbekov et al. (1976), Galimbekov (1975), and Utkin and Ermakov (1975).

In order to characterize the ores of minerals using the gamma-gamma method, their equivalent atomic number Z_{eq} and their heterogeneity have to be known. The Z_{eq} defined by Czubek (1966) has been investigated experimentally by Artsybashev and Ivanyukovich (1974). They obtained the same results as Czubek obtained theoretically for the gamma-ray energy range from 30 to 2500 keV. By the Monte-Carlo technique, the problem of Z_{eq} was also treated by Lukhminskiy and Galimbekov (1975) with similar results.

Rock heterogeneity problems were treated usually by the Monte-Carlo technique, by Leman et al. (1975b), Lukhminskiy (1975), Umiastowski et al. (1976) and Umiastowski and Buniak (1977). For a given grade of heavy mineral, the effect of its heterogeneity has the same influence on scattered radiation as the decrease in grade for homogeneous ore. For some range of gamma-ray energies, the concurrent, reciprocal influence of grade and grain size on scattered radiation becomes so strong that grade determination is impossible unless the grain size, or heterogeneity, is fixed or known. These theoretical results, confirmed by experiments, were used by Charbucinski and Umiastowski (1977) to develop experimental tools for selective gamma-gamma measurements on lead-zinc ores.

In order to calculate the probe response for a given ore grade in the selective gamma-gamma method, Galimbekov and Soboleva (1976) have established a special computer program called MOK-22.

Simultaneously with the development of theory and experimental research, progress in the design of new tools was made. Utkin and Burdin (1975) have discussed the necessary specifications for digital recording in order to obtain a given accuracy in the density measurements. New small diameter borehole tools were designed for the applications of both selective and density versions of the method (Voskoboynikov et al., 1975a, 1975b; Utkin, 1975). The most important requirement for each tool is good spectral stability. This problem was solved using either the property of the stable shape of the scattered gamma-ray spectrum (Utkin, 1975) or by using a light emitting diode as the energy mark for the spectrometer (Bakhterev et al., 1975b).

Parallel to an improvement in the performance of the "classic" gamma-gamma tools, an effort was made to construct a borehole gamma-ray generator. This project was carried out in the All-Union Science and Research Institute of Nuclear Geophysics and Geochemistry (VNIIYaGG) in Moscow (Belkin and Kolesov, 1976; Belkin et al., 1975; Grumbkov et al., 1975, 1976). It consists of a pulsed X-ray tube with an average energy of photons equal to 250 keV and with the maximum energy between 550 and 600 keV. The total photon output (for photons of energy above 200 keV) is of the order of 10^{12} photons per second. The sensitivity of the density measurement with this generator was doubled due to the increase in the source-detector spacing compared to the usual density tools.

Nuclear gamma resonance method

The nuclear gamma resonance technique (Sowerby and Ellis, 1974; Sowerby, 1974) also belongs to this group of logging methods. As opposed to the Mössbauer effect, the energy levels of nuclei (at the level of about 1 MeV) are excited by gamma rays of the same energy as of the nuclear level itself. The energy of the excitation source is increased by increasing the thermal motion of the atoms. The difference in energy, ΔE, needed to observe this resonance

$$\Delta E = \frac{E_o^2}{M.c^2} \qquad (5)$$

where E_o is the energy of the nuclear level, M is the atomic mass and c is the velocity of light, is obtained by irradiation of the surrounding rock with the gamma-ray source being in the vapour state. This method was tested in Australia under laboratory conditions on copper (with 5 Ci* of the ^{65}Zn source) and nickel (with 0.7 Ci of the ^{60}Co source) ores. This method needs very active gamma-ray sources of the order of few curies which have to be transformed by heating them into the gaseous state. Probably for this reason no field application has been reported as yet (Sowerby et al., 1977)

Photo-neutron method

The photo-neutron method is usually used for beryllium exploration because of its low energy threshold for this reaction (Fig. 11.7). This energy threshold is achieved using a ^{124}Sb gamma-ray source. The other isotopes, deuterium and ^{13}C are interesting for geohydrology and petroleum geology applications but are outside the scope of this paper. The possible detection of other heavy elements that have relatively low energy thresholds for this reaction (in the vicinity of 5 MeV) will be possible when borehole generators of gamma rays with sufficiently high photon energy are available.

---

* 1 Ci = 37 GBq

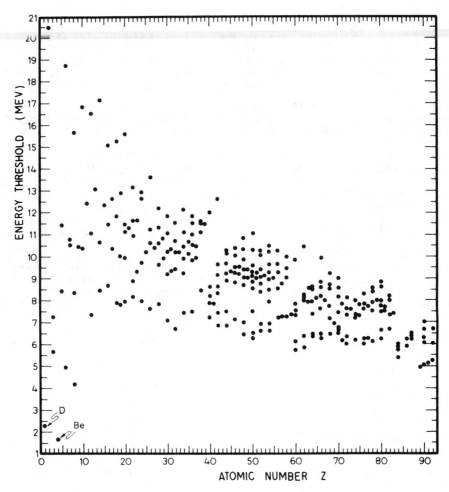

Figure 11.7. Energy thresholds for γ, n reactions for natural isotopes.

The theoretical approach to the photo-neutron method was made by Krapivskiy and Saltsevich (1975) using analytical methods for photon and neutron transport (age, diffusion and age-diffusion approximations). Gorev et al. (1975a) have used the Monte-Carlo technique of calculation for the same purpose and obtained a much better agreement with experiment than Krapivskiy did. Berzin et al. (1975), have nvestigated the pulsed photo-neutron method on a layered model using the betatron as a source of gamma rays of energy between 5 and 30 MeV.

It is very difficult to follow all the applications of the gamma source methods in different countries. Just to give some idea of how large the application of the XRF assay technique is in the exploration and mining of minerals, we can present some figures reported by Osmonbetov et al. (1976) for the Kirgiz Republic, USSR: In 1975, more than 75 000 assays were performed for Sn, W, Sb, and since 1970 more than 400 000 were made. The cost of one XRF assay is between 2 and 4 times cheaper than a chemical assay which saves a lot of money in exploration and mining. No similar figures have been published for other methods.

Neutron methods

The main development in the neutron methods of borehole logging has been carried out by the petroleum industry. These achievements are of some use for mineral exploration, but some new effects should also be considered. The very sophisticated methods developed in the petroleum industry for the calculations of neutron flux in rock can be utilized in mineral exploration, especially when the rock medium is weakly absorbing. On the other hand, some new nuclear reactions should also be taken into account. The possible presence of strongly absorbing elements requires some different approaches to the problems of neutron transport. The heterogeneous structure of the rock media, when the absorbing elements are concentrated in some grains whereas others are free of them, presents new difficulties for theoretical considerations. If all of these problems were solved, a good knowledge of the physics of a given logging device could be obtained which would permit the optimization of the logging tools.

In the design of logging tools, it is not only a question of using different borehole diameters or borehole fluids from those used in oil fields, but very often the neutron sources, detectors, source-detector spacings, filters, etc. should be specially chosen for a given type of neutron method to be applied on a given mineral deposit.

In spite of the foregoing remarks, the fundamental understanding of the physics of most of the phenomena listed above and already achieved in the petroleum industry is of great importance and help in the development of neutron methods for solid mineral exploration.

It is not possible to present a very detailed and distinct classification of neutron methods. Each element that occurs in a given type of deposit and in given borehole conditions requires its own specific discussion. In general the neutron methods can be divided into two groups: the first utilizing steady state sources, and the second utilizing pulsed neutron sources. In the first group, the measurement of a given product of neutron interaction with the rock material is observed at a given distance from the source along the borehole axis. In the second group, almost the same interaction products are observed in some time sequence related to the time sequence of the neutron source pulsing. This kind of measurement permits the measurement of some physical phenomena which are not obtainable using the steady state methods.

When the steady state source is used the following phenomena, which can be used as logging parameters, can be observed in the borehole at a distance z from the source:

1. Epithermal neutron flux of slowed-down neutrons.
2. Resonance neutron flux of slowed-down neutrons.
3. Thermal neutron flux from the thermalized slowed-down neutrons.
4. Gamma radiation from the radiative capture of resonance neutrons.
5. Gamma radiation from the radiative capture of thermal neutrons.
6. Gamma radiation from activation by fast neutrons.
7. Gamma radiation from activation by thermal neutrons.
8. Fission neutrons (prompt or delayed) when the uranium or thorium series are present.

The possibility of detection of gamma radiation from the inelastic scattering of fast neutrons or from activation by resonance neutrons is of less importance here and gives rather an increase in the background than the measurable effect. When induced gamma radiation is detected with the spectrometric tool, especially when a solid-state detector is utilized, the possibility of detection of separate elements increases considerably.

With the pulse neutron source, the phenomena accessible for observation are:

1. Die-away curve of epithermal neutrons.
2. Die-away curve of thermal neutrons.
3. Die-away curve of photons from radiative capture of resonance neutrons.
4. Die-away curve of photons from radiative capture of thermal neutrons.
5. Photons from inelastic scattering of fast neutrons.
6. Photons from activation by fast neutrons.
7. Photons from activation by thermal neutrons.
8. Fission neutrons (prompt or delayed) when the uranium or thorium series are present.

The detection of the phenomena listed above for pulsed neutron sources is always performed in the time windows which are correlated with the pulsing of the neutron beam.

Let the tool response for all phenomena listed above be R. The R value is always observed inside the borehole of a given geometry. In this case the tool response R can be presented in the form:

$$R(z) = R_\infty(r=z, P) \cdot \left[1 + S(r_B, z, P, P_B) \right] \qquad (6)$$

where $R_\infty(r=z,P)$ is the tool response when the borehole radius is $r_B = 0$ at a distance $r=z$ from the point neutron source. The rock medium is characterized by a set of geological parameters P (chemical composition of the rock, ore grade, density, etc.); $S(r_B,z,P,P_B)$ is the borehole influence function; P_B is the set of the physical and chemical parameters for the borehole. The main objective in designing a given logging tool for a given method is to minimize the function $S(...)$ and to get it as close as possible to zero. When this is achieved the tool response R is a strong function of the geological parameters P which are the object of investigation using a given method.

The behaviour of the function $S(...)$ will not be discussed because of its complexity. Suffice it to say that for each method it can be minimized; attention will be focused on the principal value of the tool response, namely the $R_\infty(r=z,P) = R_\infty(r,P)$ functions.

Very sophisticated methods of description of the neutron and gamma-ray transport in the rock media are used when neutron and neutron-gamma logs are considered. For neutrons, the most common approximations of the Boltzmann transport equation used are: age and multigroup diffusion approximations, Greuling-Goertzel approximation, Monte-Carlo techniques

(Shimelevich et al., 1976; Postelnikov et al., 1975; Drozhzhinov et al., 1975), P_2 approximation (Kozachok 1975; Kozachok and Riznik 1977), whereas for gamma rays, radiative and build-up factor approximation (Shimelevich et al., 1976), and diffusion approximation (Davydov, 1975a; Davydov et al., 1976) are used.

Steady state neutron methods

The steady state neutron methods are used for two different purposes:

1. as lithology and porosity logging in the petroleum industry,
2. as a logging tool which gives information about ore grade.

In the latter case, the grade is determined either from the general decrease of the epithermal resonance or thermal neutron flux when the grade of the absorbing elements is increased, or from the increase of a given gamma-ray line from radiative capture or activation connected with the increase in grade of a given element.

When the neutron flux $\emptyset$ is measured i.e. the $R_\infty(r,P)$ function in Equation 6, its general behaviour is:

$$\emptyset(r,P) = \frac{Q}{\Sigma_a} \cdot F_1(r, L_s, L) \qquad (7)$$

where Q is the source neutron output, Σ_a is the total absorption cross-section of the rock and the $F_1(...)$ function depends upon the source-detector spacing r, slowing-down (L_s) and diffusion (L) lengths for neutrons in the rock medium, which translated into geological language means that it depends upon the lithology, bulk density and porosity of the rock. These latter factors being constant, the neutron flux $\emptyset$ becomes a function of Σ_a, which in the case of only one anomalously absorbing element "x" present in the rock is:

$$\Sigma_a = \rho \cdot N_0 \cdot \left[(1-p) \cdot (\sigma_a/A)_R + p (\sigma_a/A)_x \right] \qquad (8)$$

where ρ is the bulk density, N_0 is the Avogadro number, p is the grade of a strongly absorbing element x with microscopic absorption cross-section σ_a and atomic weight A. $(\sigma_a/A)_R$ is the average σ_a/A ratio for the barren rock. Finally the relative effect of the presence of an absorbing element observed in neutron flux is:

$$\frac{\emptyset(r,p)}{\emptyset(r,p=0)} \approx \frac{1}{1 + p \left[(\sigma_a/A)_x / (\sigma_a/A)_R - 1 \right]} \qquad (9)$$

which can be considered as the first approximation for the calibration curve. When $(\sigma_a/A)_x \gg (\sigma_a/A)_R$, the calibration curve very quickly reaches the value zero, even for a very low value of the grade p. For common rocks, for which the macroscopic absorption cross-section is in the range 4.10^{-3} cm$^{-1} \leq \Sigma_a \leq 20.10^3$ cm^{-1}, the $(\sigma_a/A)_R$ value is in the limits 2 mb $\leq (\sigma_a/A)_R \leq$ 10 mb (milibarn); for example for boron $(\sigma_a/A)_B = 73.47$ barns $= 73.47 \cdot 10^3$ mb. To avoid any difficulty with a lack of probe sensitivity when logging higher ore grades, registration of higher energy neutrons is recommended because for the higher neutron energies the ratio $(\sigma_a/A)_x/(\sigma_a/A)_R$ becomes lower. This was the reason for the application of the resonance neutron-neutron method for boron deposits (Vakhtin et al., 1973, 1975) where it was possible to extend the measuring range up to 16 per cent of B_2O_3, instead of 4 per cent for the thermal region. An example of such a log is given in Figure 11.8, where the detector was the sandwich-type resonance detector described by Vakhtin et al. (1972).

This method was also used in the epithermal version by Grigoryan (1975) on copper deposits to distinguish the homogeneous or heterogeneous quality of ores. Fatkhutdinov and Urmanov (1975), and Fatkhutdinov et al. (1976) used it on a Hg-Sb deposit using Cd-In filters. For the multielemental gold deposit in East Zabaykal, Kuchurin et al. (1976b) have used the thermal version of this neutron-neutron logging tool. Ochkur et al. (1976) report the application of both the thermal and epithermal versions on chromite and manganese deposits. Krapivskiy (1976) has used a combination of the thermal neutron method with the gamma-neutron method for the determination of lithium in boreholes. The accuracy of this assay was comparable to that of chemical assays.

When the neutron-gamma log of radiative capture or activation is taken into account, the neutron flux given by Equation 7 at the point $\vec{r}^1$ interacts with the rock matter at a rate given by the value of the macroscopic cross-section Σ_r for that reaction (radiative capture on a given isotope or activation, etc.). The gamma photon of a given energy originates at the elementary volume $d\vec{r}^1$ around the point $\vec{r}^1$ and it has a probability $G_2(|\vec{r} - \vec{r}^1|)$ of reaching the point $\vec{r}$ at

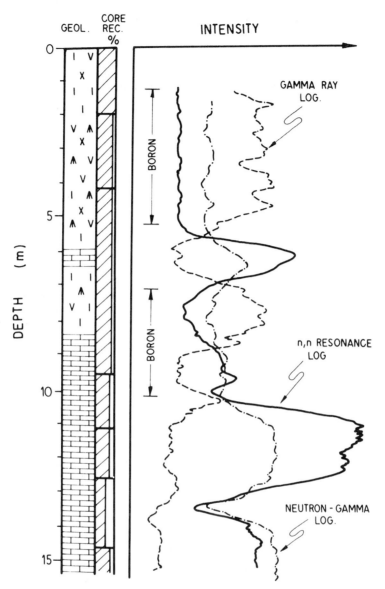

Figure 11.8. Boron determination in a magmatic limestone lithology by neutron resonance logging. Po-Be source: 4.3×10^6 n/s. Cd-Rh-In-Tu-Ta-Ag filter (after Vakhtin et al., 1973).

which the detector is situated. This being valid for the elementary volume $\vec{dr}^1$, it can now be integrated over the whole space giving the response N_γ of the gamma-ray detector:

$$N_\gamma(\vec{r},P) = Q \cdot \frac{\Sigma_r}{\Sigma_a} \cdot F_2(\vec{r}, L_s, L, \mu) \qquad (10)$$

Here again the function $F_2(...)$ is a weak function of the ore grade and a rather strong function of the moisture content, bulk density and lithology. μ is the gamma-ray absorption coefficient for a given rock. For the same type of ore, the functions $F_1(\vec{r},L_s,L)$ and $F_2(\vec{r},L_s,L,)$ are very similar in form. Therefore, when one needs to eliminate the influence of the variable porosity or bulk density on the logging data, the ratio

$$\frac{N_\gamma(\vec{r},P)}{\emptyset(\vec{r},P)} \approx \text{const.} \, \Sigma_r \qquad (11)$$

is recommended.

The reaction cross-section on the element "x" is given as:

$$\Sigma_r = p \cdot a_x \cdot (\sigma^r/A)_x \qquad (12)$$

where a_x is the abundance of the isotope of element x on which the reaction with the cross-section σ^r is going on. Similarly to Equation 9, the net calibration curve i.e. when the background gamma radiation is subtracted, for the neutron-gamma methods is:

$$N_\gamma(\vec{r},p) \approx \text{const} \cdot \frac{p}{\dfrac{(\sigma/A)_R}{a_x \cdot (\sigma^r/A)_x} + p \cdot \left\{1 - \dfrac{(\sigma/A)_R}{a_x \cdot (\sigma^r/A)_x}\right\}} \qquad (13)$$

Here again, when

$$a_x \cdot (\sigma^r/A)_x \gg (\sigma/A)_R$$

the calibration curve given by Equation 13 is saturated for a low p value and the method is not convenient in this case for use in the quantitative application. The normalization given by Equation 11 does not help too much in this case, because the background is usually too high in comparison with the net effect.

Bakhterev and Senko-Bulatnyy (1975) have used the neutron-gamma spectrometric log on nickel deposits. The calibration factor in Equation 13 was influenced by the iron content, thus the second energy window for measurement of the iron line served as the correction for the calibration factor.

Chrusciel (1976) has performed some laboratory experiments with the solid state detector to determine the detection limits of W, Ti, Ni, Mn, Cu and S in the borehole geometry for radiative capture. His results were next applied by Niewodniczanski and Palka (1976) and Niewodniczanski et al. (1977) for the determination of the sulphur content in boreholes using the spectral ratio from the radiative capture of thermal neutrons and the NaI(Tl) detector. Similar work was done by Blinova et al. (1974). Much more sophisticated neutron-gamma spectrometry was used by Eisler et al. (1977) for iron determination in blast holes in Australian deposits. Egorov et al. (1974), Sokolov et al. (1975), Afanasev et al. (1974) and Balakshin and Kravchenko (1976) have used neutron-gamma spectrometry in different combinations for mercury determination in boreholes with a precision between 0.02 to 0.08 per cent of mercury. To determine chromite in boreholes, Ochkur et al. (1976) have used a combination of neutron-gamma spectrometry with density and neutron (thermal) logs.

Fundamental theoretical and experimental research in the application of neutron methods to absorbing media i.e. for ores have been done by Krapivskiy and Brem (1975a, 1975b), Postelnikov et al. (1975), Drozhzhinov et al. (1975), Egorov et al. (1975), Fatkhutdinov and Urmanov (1975), Kozachok (1975), and Kozachok and Riznik (1977).

Pulsed neutron methods

Pulsed neutron methods using the borehole neutron generator have no wide application, as yet, in mineral exploration. One reason is the difficulty in the availability of small diameter tools needed for the small diameter boreholes usually drilled in mineral exploration. The other reason is the high cost of neutron generators.

The time sequence of different radiations occurring in the rock space when a fast neutron burst is injected is given in Figure 11.9. Almost instantly with the neutron burst, the photons resulting from the excited states of nuclei by the inelastic scattering of fast neutrons can be observed. This technique, used sometimes in the petroleum industry, and very sensitive, especially for heavy elements, to the presence of different elements, has been not yet reported in any mineral exploration application.

Following the neutron burst is a transition time zone, t_s, where the neutron flux from the burst is influenced by the proximity of the borehole and by the properties of the tool, and is not in equilibrium. During this period the epithermal neutron flux disappears. Following the transition time zone is the die-away time zone, where the thermal neutron flux from the slowed down neutrons becomes the most important component. During this period (of the few neutron lifetimes τ, with $\tau \gg t_s$), the thermal neutron flux is the carrier of information about the environment absorption cross-section Σ_a. The logarithmic time decrement λ of the neutron flux $\phi_{th}(t)$ is directly related to the rock absorption cross-section $\Sigma_{a,R}$:

$$\lambda = \frac{d \ln \phi_{th}(t)}{dt} \approx \sim \Sigma_{a,R} \qquad (14)$$

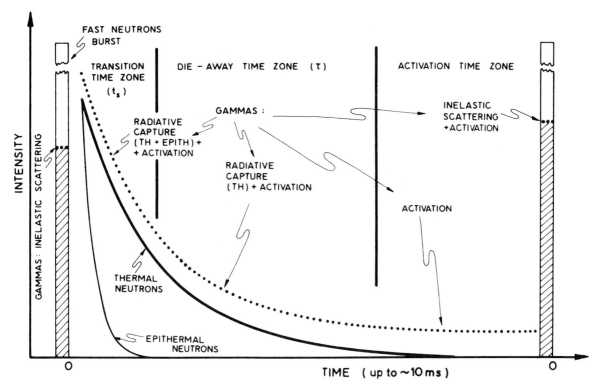

Figure 11.9. Behaviour of different radiations in the pulsed neutron methods during one cycle.

This is only true, however, when the absorption cross-section $\Sigma_{a,R}$ of the rock is lower than the absorption cross-section $\Sigma_{a,B}$ of the borehole. In terms of geological practice, it means that for mineral exploration the measurement of $\Sigma_{a,R}$ is not possible when the boreholes are dry. Otherwise the S(...) function in Equation 6 becomes much higher than 1. When the borehole is highly absorbing the determination of the grade p of the absorbing element according to Equation 8 is possible, or the localization of the mineralized zone may be made cf. Kashkay et al. (1976) for mercury deposits.

During the same die-away time zone, the thermal neutrons being captured by nuclei give the radiative capture gamma rays. When one observes a given energy line or energy range, the presence and the grade of a particular element can be determined. The problem, however, is rather complicated because of the complexity of the physical phenomenon. Due to the instant emission of the capture gamma lines there is no time shift between the pulsed neutron-neutron and neutron-gamma logs.

The pulsed thermal neutron flux in the infinite medium is:

$$\phi_{th}(r,t) \frac{Q}{[4\pi(L_s^2 + Dt)]^{3/2}} \cdot \exp\left[-t \cdot \nu \cdot \Sigma_a + \frac{r^2}{4(L_s^2 + Dt)} \right] \quad (15)$$

where D is the diffusion coefficient of thermal neutrons in the rock medium, and $\nu = 2200$ m/s is the velocity of thermal neutrons, Q is the fast neutron output. By similar reasoning as for the neutron-gamma method, the time distribution of the neutron capture gamma radiation is with the scattered gamma-ray background subtracted given by

$$N_\gamma(r,t) \approx \nu \cdot \Sigma_{n\gamma} \cdot e^{-t \cdot \nu \cdot \Sigma_a} \cdot F_3(r, L_s, L, D, \mu, t) \quad (16)$$

where again the function $F_3(...)$ depends upon the source-detector distance, rock lithology, its moisture content or porosity and bulk density. $\Sigma_{n\gamma}$ is the macroscopic cross-section for the emission of a given gamma-ray energy line due to the radiative capture of thermal neutrons and it is directly related to the grade of the element being investigated. For strongly absorbing elements, when $\Sigma_{n\gamma} \approx \Sigma_a$, a special delay should be applied to the measurement time window to obtain unequivocal correspondence between the pulsed neutron capture gamma-ray flux and ore grade. This kind of measurement has been reported, as yet, only for mercury detection (Nikulin et al., 1976).

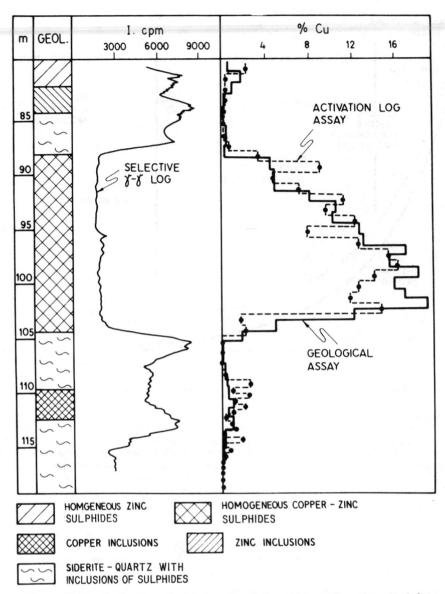

Figure 11.10. Activation Cu-64 log from the Tash – Tau deposit (after Bakhterev et al., 1975).

The last time zone of the pulsed neutron log cycle (Fig. 11.9) belongs to gamma radiation from the radioactive decay of nuclei activated by fast and/or thermal neutrons during the earlier time zones and during the previous cycles. The decay periods $T_{1/2}$ of activated nuclei, being much longer than the time between two consecutive neutron bursts, provides the possibility of observing this activation effect as a build-up of consecutive cycles which would increase the sensitivity of the method. Such a mode of detection introduced first by Givens et al. (1968) in the United States is called the cyclic activation log. This is usually used for oxygen and silicon determination for lithology purposes by fast neutron activation and its application has also been considered for manganese (Muravev et al., 1974). Other applications are also possible, as for example for fluorine by the (n, α) reaction. Cyclic activation logging problems, being at first considered separately from activation logs obtained with steady state sources, can be now, due to the work of Barenbaum and Yakubson (1974), be considered as two versions of the same problem.

Activation logging

The radioactivity of rocks and ores induced by irradiation with fast and thermal neutrons can be observed from a measurement of gamma-ray emission characteristic according to a given decay scheme. When irradiation is performed step by step in the borehole followed by the detection of induced radiation after a given delay time, such stationary logging has similar advantages and disadvantages to that provided by the usual activation method used in the laboratory for the analysis of the rock and/or ore samples. The difference however is that the

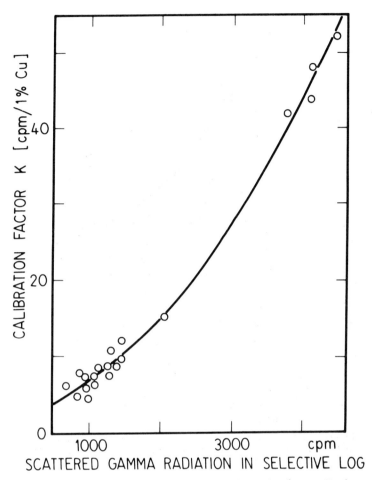

Figure 11.11. Relation between the calibration factor K of the Cu-64 activation log and selective gamma-gamma intensity using a Cs-137 source (after Bakhterev et al., 1975).

calibration curve follows Equation 13, where σ^r is the activation cross-section. Such kinds of logs are used for the long-living isotopes: $^{64}Cu, ^{56}Mn, ^{24}Na$, etc., and because of its incremental nature it is rather time consuming. In complex ores, the calibration factors derived from Equation 13 can be variable due to the variable elemental composition of the rock matrix (variable $(\sigma/A)_R$ and μ). An example of such a measurement for ^{64}Cu is given in Figure 11.10 (Bakhterev et al., 1975a), where the copper grade obtained from an activation log and from a chemical assay are compared. Here the calibration factor varied by a factor of 10 and its values were derived from the selective gamma-gamma log. The correlation of the calibration factor (in cpm/1% Cu) with the selective log readings is given in Figure 11.11.

Some theoretical problems concerning the interpretation of stationary activation logs of layered media were investigated by Vozzhenikov and Zaramenskikh (1975). Also Vozzhenikov and Davydov (1977) have considered, both theoretically and experimentally, the influence of the water-filled borehole diameter, rock porosity and its bulk density on the activation signal from ^{64}Cu and ^{28}Al isotopes. Alsayed and Dumensil (1977) considered the new approach to the semitheoretical calibration of the logging tool for the stationary activation method.

A much more economic version is the continuous activation logger. Here a steady state source is moved down-hole with a constant velocity followed by the detector at a distance large enough not to be influenced by the radiative capture gamma rays. When the neutron source is pulse operated, it is sufficient to perform the measurement in the time window situated at the activation time zone (Fig. 11.9), and a large source-detector distance is not necessary. When the width of this time window is Δt and the time separation between the consecutive neutron bursts is T (T > Δt), the continuous log signal $N_{cont}(d,\nu)$ for a given source-detector spacing d and logging velocity ν is related to the cyclic activation log signal $N_{cycl}(d,\nu)$ by

$$N_{cycl}(d,\nu) = (1 - \Delta t/T) \cdot N_{cont}(d,\nu) \qquad (17)$$

which was first derived by Barenbaum and Yakubson (1974).

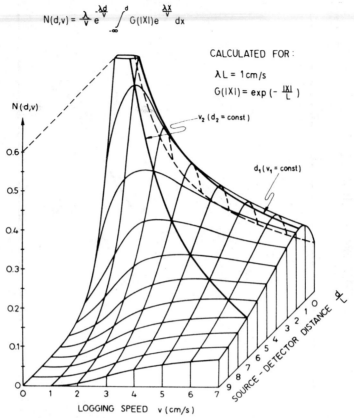

Figure 11.12. Dependence of the activation logging signal $N(d,v)$ on logging speed and source-detector distance in uniform rock medium. (Czubek and Loskiewicz, 1976).

If an infinite medium with a constant grade of activating element is considered, the detector response function, i.e. $N_{cont}(d,v)$, has the behaviour shown in Figure 11.12 (Czubek and Loskiewicz, 1976). The source-detector spacing d in this figure is given in units of the neutron migration length L. It is easy to show that for $d = d_2 = $ const and $d \gg L$, the maximum signal amplitude occurs at the logging speed $v = v_2$:

$$v_2 \approx \lambda \cdot d_2 \qquad (18)$$

where λ is the radioactive decay constant. Equation 18 is valid for all borehole conditions and is marked in Figure 11.12 by the heavy line. When the gamma-ray background is taken into account, Barenbaum and Yakubson (1976) have found:

$$v_2 \approx 0.6 \cdot \lambda \cdot d_2 \qquad (18a)$$

For a given logging speed $v = v_1 = $ const, the maximum activation signal is obtained for some source-detector spacing $d = d_1$ which is shorter than the distance d_2 and usually

$$d_1 \approx L \qquad (19)$$

Such a short source-detector distance is not feasible for steady state neutron sources (due to the radiative capture background around the source), but is quite possible using pulse operated neutron sources with cyclic detection (Fig. 11.9). Unfortunately this optimum condition is very sensitive to borehole conditions and to the rock neutron properties i.e. mostly on its porosity or hydrogen content and bulk density.

Activation logging is used for the detection of many principal or accessory elements in ores. The difficulty is that the detected radioactive isotope can be obtained in different ways from the different primary nuclei, especially when the neutron flux is mixed i.e. is fast and thermal. For example, the following nuclear reactions are possible with the Fluorine-19 isotope:

		Nuclear reaction:			
		$^{19}F(n,\gamma)^{20}F$	$^{19}F(n,2n)^{18}F$	$^{19}F(n,p)^{19}O$	$^{19}F(n,\alpha)^{16}N$
$T_{1/2}$		11.36 s	109.7 min	29.1 s	7.3 s
E_γ MeV		1.631	0.51	0.197; 1.36	6.134; 7.112
% /decay		100	194	97; 59	69; 5
σ (b)	thermal	0.009	-	-	-
	resonance	0.24	-	-	-
	fission	-	7.2×10^{-6}	0.0005	0.0045
	14 MeV	0.01	0.043	0.02	0.05

On the other hand, Nitrogen-16 which is usually used for the detection of fluorine can also be produced by other reactions:

		$^{15}N(n,\gamma)^{16}N$	$^{16}O(n,p)^{16}N$
σ (b)	thermal	2.4×10^{-5}	-
	fission	-	1.95×10^{-5}
	14 MeV	-	0.042

The contribution of different elements in the build-up of ^{16}N has to be considered using the knowledge of the physical and chemical parameters of a given deposit.

The influence of borehole diameter on the activation log readings for the ^{16}N isotope was investigated by Potopakhin et al. (1975) for borehole diameters in the range 90 to 220 mm and for tool diameters 89 and 51 mm for both stationary and continuous logs. Some other methodological aspects of the activation logs obtained using steady state and pulsed neutron sources have been studied by Muravev et al. (1974a, 1974b).

Neutron activation surveys of phosphate ores were used by Matyukhin et al. (1976) and Koshelev et al. (1975d, 1976) by detecting ^{16}N and ^{28}Al. The same isotopes were used by Startsev et al. (1975a, 1975b) and Koshelev et al. (1975a) for the determination of the ore grade of apatite deposits. Fluorspars have also been investigated by the same method by Voynova et al. (1974, 1976) in Uzbekistan who obtained a linear relation between fluorine content and the borehole activation effect. Gorbachev and Petrova (1975a, 1975b) have studied some influencing effects (density, porosity) on the activation results obtained for fluorine. For fluorspars, activation logging with Po-Be and ^{252}Cf sources (for ^{16}N and ^{28}Al) have been carried out by Koshelev et al. (1975a, 1975b), and some methodological properties of the activation log have been studied for these ores. Copper activation by fast neutrons using the borehole neutron generator was also reported by Wylie et al. (1976).

The problems connected with bauxite ore grade determination by activation methods were studied by Blumentsev et al. (1974a, 1974b) and by Shishakin et al. (1974). Aluminum in potassium ores has been determined with this method by Startsev et al. (1975c).

To localize gold-bearing veins, Kuchurin et al.(1976a) have used the stationary activation log of ^{24}Na (80 to 105 minutes per point) to estimate the specific elemental composition of gold-quartz-tourmaline-sulphide deposits.

The presence of sodium in the majority of rocks permitted Bakhterev and Kharus (1975a) to use the activation borehole measurement of rock bulk density. They found that the spectral ratio for the two energy lines of ^{24}Na (1.38 and 2.76 MeV) isotope activated by a Po-Be source with an output of about 10^7 n/s is well correlated with rock bulk density. An example of their results is given in Figure 11.13. The accuracy of density measurements by this method was about 0.05 g/cc.

GEOSTATISTICS OF NUCLEAR LOGGING METHODS

Each nuclear logging method has its own range of penetration which can also be considered in terms of the rock volume, say v_2, sampled by the method. The same can be considered with regard to the chemical assay; the volume, say v_1, of the particular sample is always defined for each type of chemical assay. Let both assays, nuclear and chemical, give an estimate of some geological parameter P e.g. ore grade, density, etc. For a given volume v of the sample from a given volume of the orebody V (V>v), the parameter P has its statistical distribution f (P,v). When the orebody is not homogeneous, the distribution function f (P,v) will be different for different

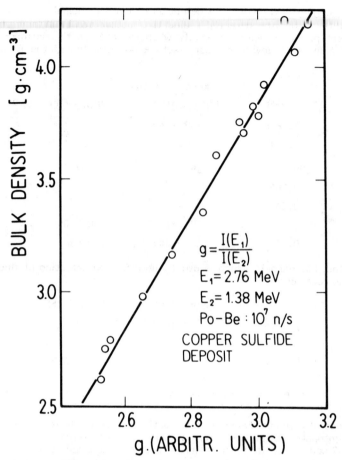

Figure 11.13. Relation between the spectral ratio g of Na-24 activation log and rock bulk density (after Bakhterev and Kharus, 1975).

values of v. For example, when v = V, f (P,V) is the Dirac delta function at P = P, where P is the average value of the ore grade within V. This situation is schematically presented in Figure 11.14. Depending upon the formation heterogeneity, the chemical assay distribution f (P,v_1) will be different from the geophysical assay distribution f (P,v_2), unless v_1 = v_2. Usually, when $v_1 \neq v_2$ and the formation heterogeneity is not negligible, a comparison between the logging data and the core assay is rather problematical.

The problems presented above can be solved by means of the mathematical apparatus of geostatistics. Czubek (1976) has treated them using the geostatistical approach of the French School of G. Matheron. The practical application of the reasoning presented in his paper has been reported by Czubek et al. (1977) in order to obtain the proper calibration curves for nuclear logging tools.

FUTURE DEVELOPMENT OF NUCLEAR LOGGING METHODS

Any forecast for the future development of nuclear well logging methods should be done for each element and for each method separately. This, however, could be valid for a given country only, or even for a given type of deposit. With regard to future progress in general, one can predict to some extent the further development of methods with the neutron and photon generators, or both in one tool, as proposed by Bessarabskiy et al. (1975), or even with generators of high energy protons. On the other hand the need for rather lightweight and small diameter equipment in the mineral industry does not encourage such development. For this reason more careful attention should be given to possibilities other than the application of pulsed particle generators for the time analysis.

The time analysis of induced radiation can be done either with the steady-state neutron or gamma-ray source. This solution although it cannot introduce any new nuclear reaction into logging practice can meet, the requirement for a small tool diameter and makes the equipment more portable. The idea is to treat nuclear logging methods as a stationary time stochastic process at a given borehole depth. By registering the time moments X(t) of the neutron emissions from the source and the time moments Y(t) at which the resulting photons reach the detector, the cross-covariance or even the auto-covariance functions of these two stochastic processes can

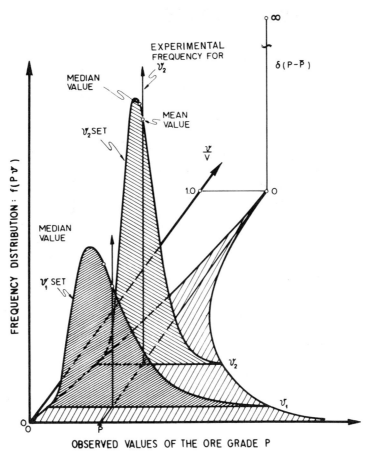

Figure 11.14. Influence of sample size v on the ore grade P distribution f(P,v) within an orebody of volume V (after Czubek, 1975).

be measured. Because neutron emission from the source has a Poisson distribution, detector events are described by the shot noise process with the transmission function h(t) of the system. This function is just the pulsed neutron generator response function (Fig. 11.8). A sketch illustrating this reasoning is presented in Figure 11.15. This kind of measurement seems to be very promising. Some experimental results have already been obtained by Blankov and Kormiltsev (1972, 1974). An example of their results is given in Figure 11.16 which proves the high similarity of the pulsed and cross-correlation experiments.

The stochastic approach can also be applied to the measurement of the amplitude of the decay curve at t=0. This amplitude is directly related to the grade of mineral present. Uranium detection by the fission neutrons should also be possible using this method.

Another possibility, not yet well explored, is the time analysis of radiative capture photons, or photons from inelastically scattered fast neutrons. As an example the gamma-ray lines emitted in cascades i.e. coincident in time because of the very short lifetime of the intermediate energy level, are presented in Figure 11.17 for some iron deposits (Czubek, 1975). By the measurement of time coincidences between the photons of the two energies E_1 and E_2, the presence of a given element can be detected without any influence of the other rock components, which have their own coincident photon pairs of energies E_1 and E_2.

NUCLEAR ASSAYING OF THE IRON – BAUXITE GROUP

In this group iron, manganese and bauxite have been assayed by nuclear methods.

Iron ores

Several laboratories have used different nuclear methods for the assay of iron deposits. Szymborski (1975) and Christell et al. (1976) have used the usual density logging technique, whereas Senko and Zorin (1975) have utilized the selective gamma-gamma method on skarn iron deposits. Ochkur et al. (1976) have applied XRF and neutron-gamma spectrometry for the same purpose. Butyugin (1976) has proposed using the Mössbauer effect with a ^{57}Co source for the assay of iron. The most extensive work has been carried out by the Australian group on hematite

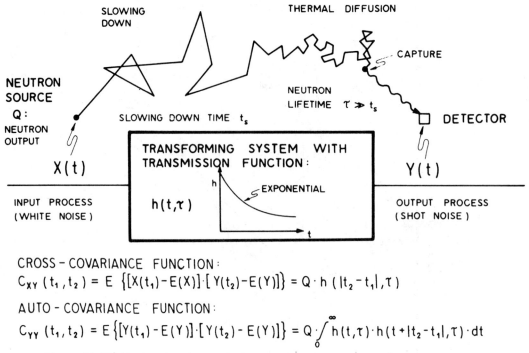

Figure 11.15. *Nuclear logging methods as stochastic processes (after Czubek, 1975).*

deposits (Aylmer et al., 1976, 1978; Charbucinski et al., 1977; Eisler et al., 1976, 1977). The correlation with natural radioactivity, as well as the spectrometry of radiative capture of thermal neutrons in the combined neutron method have been used successfully by this group. The scattered gamma-ray techniques, however, have been the most accurate. They have used the combination of density, P_z and S factor techniques fitting the iron content to the multivariable formula as shown in Figure 11.18. The energy limits for each component of the formula are also given in this figure. The accuracy of iron determination in dry percussion holes of rather large and variable diameter was better than 1 per cent of iron. For water-filled boreholes, neutron-capture gamma-ray spectrometry is a preferable technique for iron assay.

Nuclear assaying of manganese ores

There have not been many papers published during the period 1974-1977 on nuclear logging methods for manganese. Apart from the laboratory research work of Muravev et al. (1974) on the activation method only Ochkur et al. (1976) reported the application of activation and neutron-neutron (thermal) logs on manganese deposits. The reason for this is, perhaps, the recent worldwide interest in marine manganese nodules instead of new minable deposits.

Nuclear assaying of bauxite ores

The main element detected by nuclear methods is aluminum by activation with thermal neutrons. Blumentsev et al. (1974a, 1974b) and Shishakin et al. (1974) have used a ^{28}Al activation logger having a Po-Be source (of activity of about 10^7 n/s) with spectrometric recording above 1.1 MeV. The source-detector spacing was 1.5 m with a 30 to 40 m/h logging speed. To determine the other ore parameters, a selective gamma-gamma logger with a ^{170}Tm source (7 cm source-detector spacing) and a spectrometric measurement capability over the range 20 to 110 keV was used together with a density logger (^{137}Cs source, 32 cm source-detector spacing, recording above 150 keV).

NUCLEAR ASSAYING OF THE BASE METAL GROUP

Deposits of copper, zinc, lead, mercury and barium are discussed in this Section.

Copper ores

Copper ores are usually investigated by the activation logging techniques described earlier and by the XRF method. Activation techniques have been used by Bakhterev et al. (1975a), Wylie et al. (1976) and Christell et al. (1976). The XRF logger and rock face assaying is carried

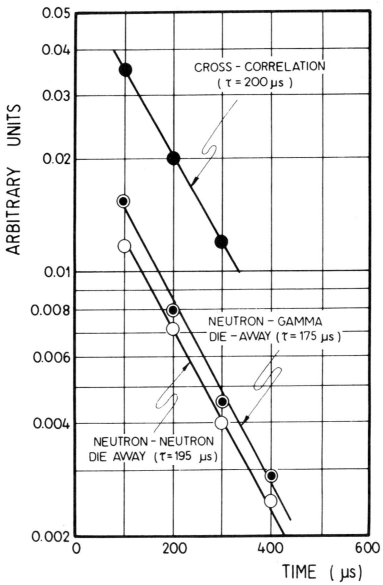

Figure 11.16. Cross-correlation and die-away curves in paraffin wax, (after Blankov and Kormiltsev, 1972).

out by measuring the spectral ratio of the K_α copper line to the scattered photons of energy 14.5 keV (from a ^{109}Cd source), even in multi-element ores with Pb (Ochkur et al., 1974; Grigoryan et al., 1974; Tamrazyan and Popov, 1975; Bolotova and Leman, 1976). Kozlov et al. (1975a, 1975c) have used the selective gamma-gamma method to localize and to determine the quality of the sulphide-copper ores. For the same purpose Grigoryan (1975) has used the epithermal neutron logger.

Zinc, lead and barium ores

Lead and zinc or lead and barium very often occur together in sulphide-type deposits usually in carbonate rocks. Here the XRF logger is the most popular tool. The K_α line of zinc and the L line of lead are used for the simultaneous determination of both elements (Kozlov et al., 1975b; Krasnoperov et al., 1976; Krasnoperov and Zvuykovskiy, 1976; Zgardovskiy et al., 1974) employing ^{119m}Sn, ^{109}Cd or H^3/Zr sources. Proportional counters are usually used and the spectral ratio is recorded. For water-filled boreholes, the close collimation of the source and detector is utilized. Sometimes an additional density logger for lead employing a ^{137}Cs source is used and Cu-Ni filters are utilized to distinguish zinc from iron in the XRF log (Shmonin et al., 1976b). XRF logging speeds vary from 120 up to 350 m/h. When lead only is to be detected, a ^{75}Se source is utilized (Kobelev et al., 1974). When a high accuracy for barium and lead determination is required, the measurement often takes a longer time – up to 4 minutes (Landström, 1976), for the "natural XRF log". In order to localize lead seams with high precision,

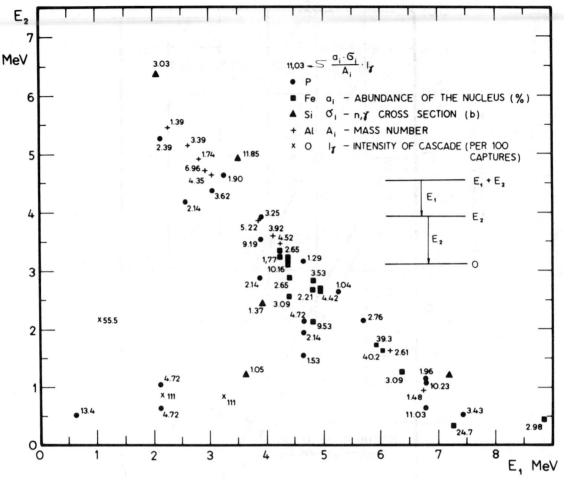

Figure 11.17. *Gamma-ray lines E_1 and E_2 emitted in cascade due to radiative capture of neutron. Example for an iron ore deposit. (Czubek, 1975).*

one can use a density logger with very good results (Yasinovenko, 1975). For Ba-Pb determinations, short gamma-gamma probes (6 to 8 cm) are also used for spectrometric measurements over the range 130 to 150 keV (Shmonin et al., 1976a).

Mercury ores

Mercury is a very convenient element to detect by nuclear methods. Its high atomic number and neutron absorption cross-section permit the application of XRF and neutron methods for mercury assay. The common mercury mineral is cinnabar and the payable grades are from 0.01 per cent Hg up to 5 or 8 per cent in very rich deposits. Sometimes an increased uranium and thorium concentration is correlatable with mercury which provides an opportunity of using the gamma spectrometric survey technique (Antipov, 1975). XRF loggers were used by Mitov et al. (1975) and Balakshin and Kravchenko (1976) employing the L series lines of Hg with quite good agreement with chemical assays (±20% relative). Neutron-neutron logs (simultaneous thermal and epithermal) have been used by Fatkhutdinov (1974), Fatkhutdinov et al. (1974a, 1974b) to estimate the mercury reserves of Hg-Sb deposits. Radiative capture gamma-ray spectrometry in the region around 4 and 6 MeV using a Po-Be source has been used by Balakshin and Kravchenko (1976) and Boyarkin and Kaipov (1974). The agreement with chemical assays was varied from 0.004 up to 0.24 per cent Hg. One of their results is reproduced in Figure 11.19. The "mercury" line I_1 around 4 MeV with the background I_{1b} subtracted and the Ca-Fe line I_2 around 6 MeV with the background I_{2b} subtracted were recalculated to the pure "mercury" intensity I_{13}. A method of interpretation of such I_{13} logs in the layered media was also established.

Egorov et al. (1975) have performed some theoretical and model experiments for mercury determination using the spectral ratio 4-5 MeV/6.1-8 MeV to avoid the influence of iron. Afanasev et al. (1974) have used this method in the Kirgiz Republic and obtained a sensitivity of about 0.06 to 0.08 per cent Hg. In an attempt to avoid the influence of iron an improvement of this method was carried out by Sokolov et al. (1975).

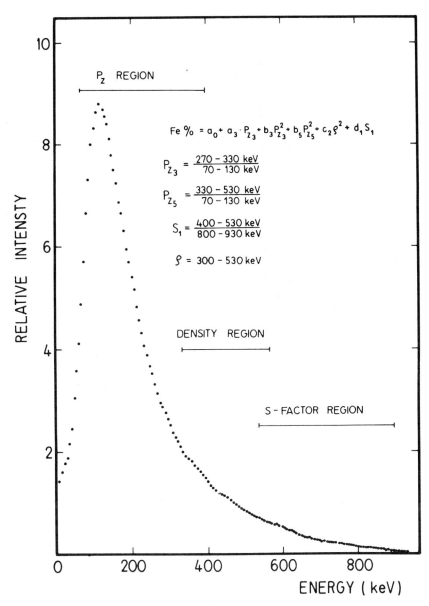

Figure 11.18. An example of the Co-60 scattered gamma-ray spectrum in a dry blast hole used by Charbucinski et al. (1977) to assay iron content.

Pulsed borehole neutron generators have been used to measure the Σ_a in mercury deposits (Putkaradze et al., 1973; Kashkay et al., 1976). The values of Σ_a for mercury ores of different lithologies obtained by Putkaradze et al. (1973) are presented in Figure 11.20.

Another new approach was proposed by Nikulin et al. (1976) applying a spectrometric technique to the radiative capture gamma rays in the pulsed neutron method. Some results of their laboratory experiments are shown in Figures 11.21 and 11.22. The method is based on the detection of the spectral ratio

$$\eta = \frac{I(3-4 \text{ MeV})}{I(> 6 \text{ MeV})}$$

which permits the utilization of an optimal delay time t_d and width of the detection window Δt the measurement of mercury grade in dry boreholes with a negligible influence from iron content.

NUCLEAR ASSAYING OF THE TIN – RARE METAL GROUP

In this group deposits of Sn, W, Mo, Nb, Hf, Ta, Be, Li, Rb-Cs, Cs, Bi, mica, feldspar, Ti, rare earths, Zr are included.

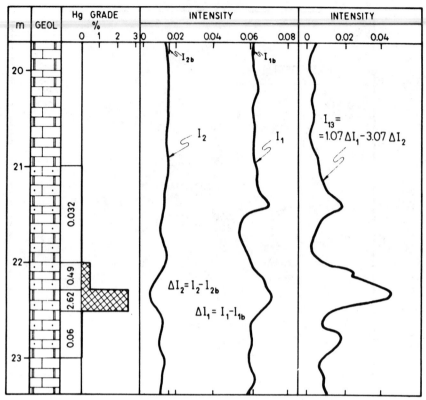

Figure 11.19. Spectral neutron-gamma log from a mercury deposit (after Boyarkin and Kaipov, 1974).

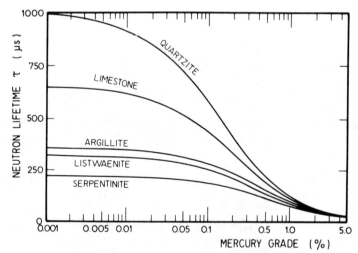

Figure 11.20. Calculated thermal neutron lifetime τ in mercury ores (after Putkaradze et al., 1973).

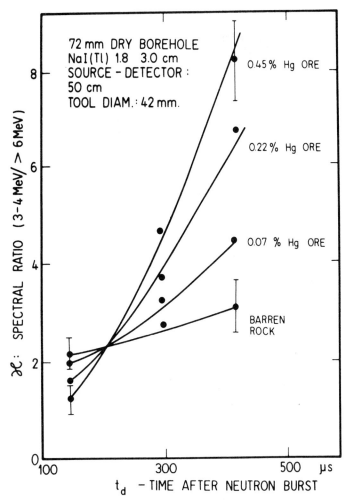

Figure 11.21. Spectral ratios for different delay times t_d in the pulsed neutron-gamma method obtained using mercury ore models (after Nikulin et al., 1976).

Tin deposits

The most commonly-used logging methods are the XRF and Mössbauer effect techniques which give intercomparable results and have the same accuracy as chemical assays. The advantage of the XRF method is the possibility for the simultaneous determination of other elements which occur with tin. Sometimes the activation method is used (Gorbachev et al., 1974; 1975) when the grade of fluorine is correlatable with tin.

XRF logging or in situ ore face assaying for tin have been reported by Grigorkin and Neustroev (1974), Sachuk and Balashov (1974), Afanasev et al. (1974), Bolotova and Leman (1976), Meyer et al. (1976), Ochkur (1976), Balakshin and Kravchenko (1976), Christell and Ljunggren (1976), Ratnikov et al. (1976) Nuclear Geophys. Assay ...(1976), and many others. The usual method was to employ the spectral ratio 25 keV/35 keV to determine tin grade which resulted in detection limits of the order of 0.005 to 0.15 per cent Sn depending upon the characteristics of the deposit. ^{147}Pm and ^{241}Am sources were utilized for this purpose. Some boreholes were water filled. In association with tin, the other elements detected by XRF logging were V, Cu, As, W, Pb, Zn, and Fe.

Cassiterite ores were assayed by the Mössbauer effect either in situ on the rock face or by step by step logging. This work was reported by Goldanskiy et al. (1974), Ochkur (1976), Nuclear Geophys. Assay ... (1976), and others.

Tungsten ores

Tungsten was usually detected by the XRF method (already described) simultaneously with other heavy elements. Kuchurin et al. (1976) reported the application of a selective gamma-gamma logger to detect tungsten grades above 0.2 per cent W. For XRF logging, they have used the spectral ratio 55-65 keV/80-90 keV employing a ^{75}Se source. Manganese was also correlatable with tungsten in this deposit, thus by the activation method grades above 0.4 per cent W have been also determined Ochkur et al. (1974) gave some details of the logging apparatus.

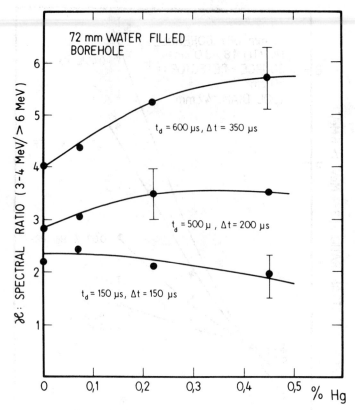

Figure 11.22. Spectral ratios obtained for the different mercury content of ores by the pulsed neutron-gamma method (after Nikulin et al., 1976).

Beryllium and lithium ores

Beryllium ores are always detected by the photo-neutron method, both in surface surveys and in borehole logging. Suvorov and Molochnova (1975) discussed the influence of soil moisture and density, its absorption cross-section and the natural neutron background on the results of carborne surveys. The other measurement parameters for this survey technique were discussed by Brem et al. (1975), whereas Gorev et al. (1975b) and Gorev (1975) have presented the details of the apparatus design. The self-absorption of neutrons in this method was discussed by Kirichenko (1975). For logging methods, the sensitivity of the beryllium assay once of the order of 0.01 per cent BeO (Afanasev et al., 1974) has been improved to the 0.001 per cent level (Grigoreva et al., 1975) and is even more for a more sophisticated logging tool with a 5 mCi ^{124}Sb source (Shestakov, 1975). Some other logging applications are presented by Ochkur (1976) and Krapivskiy et al. (1976).

Krapivskiy (1976) has investigated the feasibility of lithium determination in pegmatite rocks by the combined gamma-neutron and neutron-neutron methods.

Molybdenum ores

Molybdenum has been determined by the XRF method usually together with other elements in boreholes and in surface surveys by Grigoryan et al. (1974) and Bolotova and Leman (1976a) with an accuracy of 0.01 per cent Mo.

NUCLEAR ASSAYING OF THE ULTRABASIC GROUP — CHROMIUM AND NICKEL ORES

Chromium ores have been detected by a combination of spectrometric neutron-capture gamma-ray logging to localize the ore zones with neutron-neutron logging to verify the homogeneity of ores. The spectrometry of the scattered gamma radiation has been used to determine Cr_2O_3 grade. Density logging has also been used in this combined measurement (Feldman et al., 1974; Ochkur et al., 1976). Voznesenskiy (1976) has used the XRF method for chromite exploration.

For nickel ores, Bakhterev and Senko-Bulatnyy (1975) have used spectrometric neutron-capture logging in which the calibration factor was variable because of the varying iron content. This iron content was also determined by the other energy line in the same neutron-capture gamma-ray spectrum.

NUCLEAR ASSAYING OF THE GOLD GROUP

In this discussion gold, antimony and uranium ore deposits are included.

Gold and antimony ores

Gold usually occurs with antimony. Other types of deposits also contain antimony, mercury or tungsten.

Gera (1974) has used selective gamma-gamma logging with a ^{170}Tm source to localize quartz veins. The gold itself has been detected by spectrometric combined neutron-neutron, capture neutron-gamma, activation, selective gamma-gamma and spectrometric gamma logging (Kuchurin et al., 1975a).

Antimony ores have usually been detected by XRF logging with a ^{170}Tm source using the spectral ratio 25 keV/85 keV (Grigorkin et al., 1976; Petrukhin et al., 1976; Ivanyukovich et al., 1976; Ochkur et al., 1974; Afanasev et al., 1974). The agreement of the XRF logging results with chemical assays were always within ±10 to 20 per cent relative. Fatkhutdinov (1974a) and Fatkhutdinov et al. (1974) have used thermal and epithermal neutron logs to determine antimony in Sb-Hg deposits.

Uranium ores

Uranium ores are usually investigated by gamma-ray logging previously discussed. When gamma-ray spectrometry is used, the windows 1.05 to 1.65 MeV and 2.05 to 2.65 MeV are utilized to distinguish the uranium and thorium series. Some experimental aspects of this method have been discussed by Sinitsyn et al. (1974), Gabitov et al. (1974) and Kozynda et al. (1974, 1976). Khaykovich and Yakovlev (1976) have contributed to the problem of the accuracy of the computer interpretation of gamma-ray logs in layered media. Novikov and Ozerkov (1974) have investigated the influence of radon emanation into boreholes on the results of the gamma-ray logs. Some other measurement problems have been discussed by Novikov et al. (1974, 1977).

The frequent radioactive disequilibrium of uranium ores and/or the increasing need for additional remote sensing techniques for uranium exploration has stimulated the development of techniques other than the gamma-ray methods of uranium determination. Czubek's first paper (1972) on the pulsed neutron method gave the theoretical and experimental principles for the prompt and delayed fission neutron detection in uranium ores using a pulsed neutron generator. Next, this idea has been taken up in the United States (see Exploration for Uranium Deposits, 1976; Thibideau, 1977; Renken, 1977; Givens et al., 1976). European laboratories have published some theoretical papers on natural fission (from the spontaneous fission/neutron distribution) for different borehole-layer configurations (Davydov, 1975b) and for a point neutron source in the borehole, but without any discussion of time problems (Davydov, 1975c, 1975d). The theoretical paper of Czubek and Loskiewicz (1976) can be useful in continuous delayed fission neutron logging using the "jerk" source method (Californium Progress 1976, No. 20). It was also found there that the optimum condition for continuous delayed fission neutron logging (cf. Equation 18) is $v = 0.241$ cm/s. Some other theoretical considerations for pulsed delayed fission neutron logging using the Monte-Carlo technique have been presented by Wormald and Clayton (1976).

Another possibility has been proposed by Kartashov and Davydov (1975) for detecting uranium in rock. A photo-neutron source with an output of 10^7 n/s can be used to irradiate uranium-bearing rock in order to generate fission neutrons which can be detected at the fast stage by threshold detectors. The stochastic process approach by the cross-covariance or auto-covariance measurement mentioned in this paper should also be taken into account for uranium measurement by the detection of fission neutrons.

NUCLEAR ASSAYING OF SEDIMENTS, EVAPORITES AND OTHER TYPES OF DEPOSITS

Sulphur deposits

Feldman et al. (1974) have used a combination of neutron and density logging together with the spectrometry of thermal neutron capture gamma rays to determine sulphur content. Neutron gamma spectrometry has been carefully investigated for carbonate-type sulphur deposits by Blinova et al. (1974). The same method has been applied by Niewodniczanski and Palka (1976) and by Chrusciel et al. (1977) on the Polish sulphur deposits. The gamma-ray lines for sulphur (around 4.4 MeV), calcium (around 5.4 MeV) and hydrogen (around 2.2 MeV) using a Po-Be source have been measured. One of their calibration curves is given in Figure 11.23.

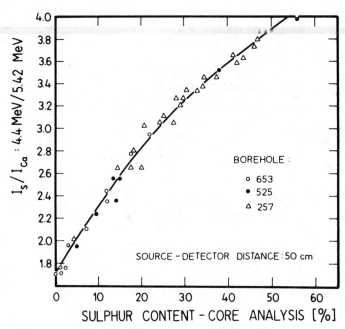

Figure 11.23. Calibration curve for a 50 cm long probe used to determine sulphur content by the neutron gamma spectrometry (after Chrusciel et al., 1977).

Potassium ores

Spectrometric gamma-ray logging was used by Mishin (1976) and Mishin and Gavrilova (1976) to determine the potassium grade and the nonsoluble parts of ore. Other nuclear methods used in the potassium industry are presented in the short monograph by Saturin (1975).

Boron deposits

The boron series has been investigated by Vakhtin et al. (1972, 1973, 1975) by means of the resonance neutron-neutron method using a special resonance detector and some results are shown in Figure 11.8.

Phosphorite, apatite, fluorite and alunite deposits

These deposits are usually investigated by activation logging (Koshelev, 1975; Koshelev et al., 1975a, 1975b, 1975c, 1975d, 1975e, 1976; Startsev et al., 1975a, 1975b; Voynova et al., 1976; Matyukhin et al., 1976). Sometimes the correlation between the P_2O_5 content and ore density gives additional information about the quality of the apatite ores, or a correlation with uranium (Rudyk et al., 1974) is also observed. For alunites Muravev and Yakubson (1975) have observed an increased gamma-ray activity.

FINAL REMARKS

It was not possible to present in this review all applications of the nuclear logging methods to the exploration of minerals during the period 1974 to 1977. Only the most important have been presented here; most theoretical papers have been omitted, especially those concerned with neutron distribution in rock media.

The general trend observed in this branch of applied science is the growing importance of nuclear methods among the range of techniques used in exploration and mining. The most important development, however, is that the accuracy of assays performed using nuclear methods is now comparable with the so-called classical chemical methods being at the same time much cheaper and less time consuming.

On the other hand the application of nuclear methods requires a good understanding of the physical phenomena involved and of the sophisticated field equipment used to perform the detailed energy and time analysis of the recorded radiations. Without this deep knowledge of the physics of these methods, any practical application is very often unsuccessful and a waste of money.

Acknowledgments

It was not possible for me to prepare this review paper without the very kind help of my colleagues and friends from other nuclear geophysical laboratories, namely R.L. Caldwell, Yu.B. Davydov, E.M. Filippov, I.A. Kozachok, D.A. Kozhevnikov, O. Landström, A.P. Ochkur, A.W. Wylie and K.I. Yakubson who have sent me their papers as well as the papers of their colleagues. This has permitted me to be up-to-date with the development of the subjects reviewed in this article. It is my very great pleasure to have the opportunity here to acknowledge this help, a vital contribution to this paper.

References

Afanasyev, A.V., Erkhov, V.A., Kopytov, Yu.Ya., Makarov, Yu.I., Tikhonov, A.I., Ushan, V.S., Shvartsman, Yu.G., and Yudakhin, F.N.
  1974: Application of nuclear geophysics methods in exploration of ore deposits in Kirgiz Republic; in Trudy Upravlenya Geologii Kirg. SSR, No. 3, p. 93-101 (in Russian).

Alsayed, N. and Dumesnil, P.
  1977: Evaluation des possibilités d'analyse élémentaire en forage par activation neutronique; in Nuclear Techniques and Mineral Resources, IAEA, Vienna, p. 265-272.

Antipov, V.S.
  1975: On the possibility of application of the gamma-spectrometric method in regions with mercury mineralization; in Trudy VNIIYaGG, No. 25, Moscow, p. 93-99 (in Russian).

Arakcheev, N.T. and Bondar, V.V.
  1975: Investigation of uranium, thorium and potassium distribution in rocks on the global gamma-ray intensity; in Problems of Geology, Geochemistry and Geophysics of Earth's Crust in White Russia, Minsk, p. 57-62 (in Russian).

Artsybashev, V.A. and Ivanyukovich, G.A.
  1974: On the equivalent atomic number of rocks and ores; Trans. Leningrad Mining Inst., v. 64 (2), p. 127-132 (in Russian).

Aylmer, J.A., Eisler, P.L., Mathew, P.J., and Wylie, A.W.
  1976: The use of natural gamma radiation for estimating the iron content of sedimentary iron formations containing shale bands; in Nuclear Techniques in Geochemistry and Geophysics, IAEA, Vienna, p. 53-74.

Aylmer, J.A., Mathew, P.J., and Wylie, A.E.
  1978: Bulk density of stratified iron ores and its relationship to grade and porosity; Proc. Australasian Inst. Min. Met. (265), p. 9-17.

Bakhterev, V.V. and Kharus, R.L.
  1975: Rock density determination by the activation method; in Nuclear Geophysical Investigations, Acad. Sci. USSR, Ural Branch, Sverdlovsk, p. 34-36 (in Russian).

Bakhterev, V.V. and Senko-Bulatnyy, I.N.
  1975: Some problems of the neutron-gamma ray spectrometric log on the nickel deposit of the silicate type; in Geophysical Methods of Survey and Exploration, No. 1, Sverdlovsk, p. 91-94 (in Russian).

Bakhterev, V.V., Senko-Bulatnyy, I.N., and Akhmetshin, B.Kh.
  1975a: Results of gamma-spectrometric assay of activation method for quantitative copper determination in the Ural deposits; in Nuclear Geophysical Investigations, Acad. Sci. USSR, Ural Branch, Sverdlovsk, p. 37-42 (in Russian).

Bakhterev, V.V., Bausov, A.V., Zyryanov, L.A., Senko-Bulatnyy, I.N., and Shindelman, A.V.
  1975b: Some characteristics of the stabilization system for spectrometers with a light emitting diode as the source of the mark signal; in Nuclear Geophysical Investigations, Acad. Sci. USSR, Ural Branch, Sverdlovsk, p. 58-63 (in Russian).

Balakshin, G.D. and Kravachenko, G.A.
  1976: Development of nuclear physics methods in the Yakutsk territorial geological organization; in Nuclear Geophysics..., Yakutsk, p. 6-10 (in Russian).

Baldin, S.A., Gubin, S.F., Egiazarov, B.G., and Kholomov, M.D.
  1975: Application of X-ray solid state detectors in well logging; Atomnaya Energia, v. 39 (4), p. 436-438 (in Russian).

Baldin, S.A., Dolenko, A.V., Egiazarov, B.G., Krylov, L.N., and Seldyakov, Yu.P.
  1976: Apparatus for geophysical methods; in Build-up of Apparatus Systems for Nuclear Equipment Design, Moscow, p. 108-111 (in Russian).

Barenbaum, A.A. and Yakubson, K.I.
  1974: Theoretical analysis of some peculiarities of continuous activation logging with pulsed neutron source; in Trans. Moscow Inst. Oil Chem. and Gas Industry, No. 111, Publ. House "Nedra", Moscow, p. 147-160 (in Russian).

Barenbaum, A.A. and Yakubson, K.I. (cont'd)
1976: Optimization of continuous activation logging systems; in Razvedochnaya Geofizika, No. 70, Publ. House Nedra, Moscow, p. 161-171 (in Russian).

Belkin, N.V., Grumbkov, A.P., Kolesov, V.I., Guseva, N.S., Tsukerman, V.A., and Tsygankov, V.A.
1975: Measurement of rock densities in boreholes by means of the photon generator; Bull. Acad. Sci. USSR, v. 224 (3), p. 569-572 (in Russian).

Belkin, N.V. and Kolesov, V.I.
1976: Borehole generator of roentgen radiation pulses of 0.5 MV voltage; in Nuclear Geophysical Applications Trudy VNIIYaGG, No. 26, p. 12-16 (in Russian).

Berzin, A.K., Gryaznov, A.L., Rinznik, Ya.M.E., and Sulin, V.V.
1975: Investigation on the layer model of the space-time distribution of thermal neutrons from gamma-neutron reactions; in Geophysics and Astronomy, Information Bull., No. 14, p. 164-170 (in Russian).

Bessarabskiy, Yu.G., Demidov, P.F.P., and Ovsyannikov, S.B.
1975: Pulsed neutron-roentgen tube; Pribory i Tekhnika Eksperimenta (USSR), (1), p. 211-213 (in Russian).

Blankov, E.B. and Kormiltsev, Yu.V.
1972: Possibility of application of correlation functions for the determination of thermal neutron lifetime in rocks using a steady state neutron source; in Nuclear Geophysical Methods, Publ. House Nauka, Novosybirsk, p. 249-253 (in Russian).

1974: Method of non-stationary logging on thermal neutrons with the steady state source; Patent USSR, class G 01 V 5/00, E 21 b 47/00, No. 407260, appl. 15.01.65, iss. 10.04.74.

Blinova, N.M., Muravev, V.V., and Yakubson, K.I.
1974: Sulphur grade determination in alumosilicate rocks using radiative capture gamma ray spectrometry; in Trans. Moscow Inst. Oil Chem. and Gas Industry, No. 111. Publ. House Nedra, Moscow, p. 197-203 (in Russian).

Blumentsev, A.M., Leykin, A.V., and Feldman, I.I.
1974a: Investigations of the choice of the rational combination of nuclear geophysics methods of borehole logging on bauxite deposits; in Exploration Geophysics in USSR in the Beginning of 70's, Publ. House Nedra, Moscow, p. 335-339 (in Russian).

Blumentsev, A.M., Ishchenko, V.I., Leykin, A.V., and Feldman, I.I.
1974b: Method of bauxite determination in conditions of natural occurrences; Patent USSR, class G 01 V 5/00, No. 374567, appl. 9.08.71, iss. 3.06.74.

Bolotova, N.G. and Leman, E.P.
1976a: XRF assay of the multielemental sulphide-cassiterite ores for tungsten with the RRShA-1 and "Mineral-4" equipment; in Geofizicheskaya Apparatura, No. 59, Leningrad, Publ. House Nedra, p. 107-113 (in Russian).

1976b: Methodological peculiarities of XRF assay of multielemental ores on outcrops; in Nuclear Geophysics in Survey and Exploration of Solid Minerals, Yakutsk, 1976, p. 60-61 (in Russian).

Boyarkin, A.P. and Kaipov, R.L.
1974: Problem of the theory and interpretation of results of nuclear physics elemental analysis; in Nuclear Physics Methods of Analysis and Control of Technological Processes, Publ. House Fan, Tashkent, p. 3-50 (in Russian).

Brem, A.A., Gorev, A.B., and Suvorov, A.D.
1975: Carborne beryllium survey; in Trudy VNII Geofiz. Metodov Razvedki, No. 25, Leningrad, p. 55-60 (in Russian).

Butyugin, M.A.
1976: Feasibility of utilization of Mössbauer effect for iron ore assay in natural occurrences; in Nuclear Geophysics..., Yakutsk, p. 33-34 (in Russian).

Charbucinski, J. and Umiastowski, K.
1977: Some factors affecting accuracy in determination of heavy element concentrations in the selective gamma-gamma method; in Nuclear Techniques and Mineral Resources; IAEA, Vienna, p. 281-300.

Charbucinski, J., Eisler, P.L., Mathew, P.J., and Wylie, A.W.
1977: Use of backscattered gamma radiation for determining grade of iron ores in blast holes and development drill holes; Proc. Australasian Inst. Min. Met. (262), June 1977, p. 29-38.

Chernyavskaya, N.A., Filippova, E.I., Meyer, V.A., Filippov, M.M., and Ilina, L.P.
1975: Technology of preparation of powdered roentgen filters on the polyethylene binding agent; in Geofizicheskaya Apparatura, No. 57, Publ. House Nedra, Leningrad, p. 102-107 (in Russian).

Christell, R., Ljunggren, K., and Landström, O.
: 1976: Brief review of developments in nuclear geophysics in Sweden; in Nuclear Techniques in Geochemistry and Geophysics IAEA, Vienna, 1976, p. 21-45.

Chrusciel, E.
: 1976: Borehole spectrometry with the solid-state detector; in Report of the Institute of Nuclear Techniques AGH, No. 95, Krakow. Poland, p. 9-22 (in Polish).

Chrusciel, E., Niewodniczanski, J., Palka, K.W., and Roman, S.
: 1977: Determination of sulphur content in boreholes by neutron capture; in Nuclear Techniques and Mineral Resources, IAEA, Vienna, p. 301-311.

Czubek, J.A.
: 1966: Physical possibilities of gamma-gamma logging; in Radioisotope Instruments in Industry and Geophysics, v. 2, IAEA, Vienna, p. 249-275.
: 1971: Recent Russian and European developments in nuclear geophysics applied to mineral exploration and mining; The Log Analyst, v. 12 (6), p. 20-34.
: 1972: Pulsed neutron method for uranium well logging; Geophysics, v. 37, (1), p. 160-173.
: 1975: Lecture notes on selected problems of nuclear geophysics, CSIRO, Port Melbourne, Victoria, Australia (unpublished).
: 1976: Comparison of nuclear well logging data with the results of core analysis; in Nuclear Techniques in Geochemistry and Geophysics, IAEA, Vienna, 1976, p. 93-106

Czubek, J.A. and Loskiewicz, J.
: 1976: Optimum conditions for uranium detection in delayed neutron well logging; in Exploration for Uranium Deposits, IAEA, Vienna, 1976, p. 471-486.

Czubek, J.A. and Zorski, T.
: 1976: Recent advances in gamma ray log interpretation; IAEA, Advisory Group Meeting on Evaluation of Uranium Resources, Rome, Italy, Nov. 29–Dec. 3, 1976, Paper AG-64/5.

Czubek, J.A., Gyurcsak, J., Lenda, A., Loskiewicz, J., Umiastowski, K., and Zorski, T.
: 1977: Geostatistical method of interpretation of nuclear well logs; in Nuclear Techniques and Mineral Resources; IAEA, Vienna, p. 313-332.

Davydov, Yu.B.
: 1975a: Space distribution of radiative capture gamma rays in boreholes; in Trans. Sverdlovsk Min. Inst. no. 128, Sverdlovsk, p. 120-126 (in Russian).
: 1975b: On the problem of the influence of near borehole zone on the distribution of natural neutron field in the borehole; in Geophysical Methods of Survey and Exploration, No. 1, Sverdlovsk, p. 95-100 (in Russian).
: 1975c: Estimation of the influence of physical and geometrical factors on the delayed fission neutron distribution in boreholes; Atomnaya Energia, v. 39 (1), dep. paper No. 797/7871, p. 49-50 (in Russian).
: 1975d: Space distribution of fission neutrons in multiplying medium crossed by the borehole; Atomnaya Energia, v. 39 (1), dep. paper No. 798/7870, p. 50 (in Russian).

Davydov, Yu.B., Khaov, S.N., and Bakaev, V.P.
: 1976: Investigations of the borehole influence on the results of neutron logging; in Nuclear Geophysics..., Yakutsk, p. 32-33 (in Russian).

Drozhzhinov, Yu.I., Kozhevnikov, D.A., and Moskalev, O.B.
: 1975: Dependence of the response of the slowed down neutron detector upon the hydrogen content of the medium; Bull. Acad. Sci. USSR. Earth's Phys. (4), p. 97-101 (in Russian).

Egorov, E.V., Sokolov, E.A., Vysotskiy, I.B., Makarov, Yu.I., Yaroslavets, V.F., and Volfstein, P.M.
: 1974: Application of spectrometric nuclear geophysical methods of logging for mercury determination in exploration wells; in Exploration Geophysics in USSR in the Beginning of 70's, Publ. House Nedra, Moscow, p. 343-351 (in Russian).

Egorov, E.V., Sokolov, E.A., and Pushchanskiy, V.G.
: 1975: Decrease of the influence of variable rock moisture on the results of neutron-neutron log; in Trudy VNII Geofiz. Metodov Razvedki, no. 25, Leningrad, p. 23-28 (in Russian).

Eisler, P.L., Huppert, P., Mathew, P.J., and Wylie, A.W.
: 1976: A gamma-gamma borehole logging probe for the simultaneous measurement of density, P_z and S-factor; 25th Int. Geol. Congr., Sydney, v. 2, p. 387 (abs.).

Eisler, P.L., Huppert, P., Mathew, P.J., Wylie, A.W., and Youl, S.F.
: 1977: Use of neutron capture gamma radiation for determining of iron ore in blast holes and exploration holes; in Nuclear Techniques and Mineral Resources; IAEA, Vienna, p. 215-228.

Exploration for Uranium Deposits
1976: Proc. Symp. IAEA, IAEA, Vienna, 810 p.

Fatkhutdinov, Kh.N.
1974a: Basic principles of completion of nuclear geophysics methods; in Trans. Central Asia Inst. Colour Metals, no. 9, Tashkent, p. 7-10 (in Russian).

1974b: Nuclear geophysical assay of boreholes and open pits of the Nikitovsk mercury deposit; in Trudy Central Asia Inst. Colour Metals, no. 9, Tashkent, p. 10-15 (in Russian).

Fatkhutdinov, Kh.N. and Urmanov, R.L.
1975: Influence of the rock and ore porosity on the epithermal and thermal neutron distribution; in Trudy VNII Geofiz. Metodov Razvedki, no. 25, Leningrad, p. 20-22 (in Russian).

Fatkhutdinov, Kh.N., Russkin, M.M., and Yuldashev, A.A.
1974: Utilization of the nuclear geophysical data of assays for the reserve estimation of mercury and antimony ores; in Trans. Central Asia Inst. Colour Metals, no. 9, Tashkent, p. 15-18 (in Russian).

Fatkhutdinov, Kh.N., Russkin, M.M., and Urmanov, R.L.
1976: Method of antimony determination in underground holes in complex ores. Patent USSR, class G 01 V 5/00, No. 495626, appl. 14.06.71, No. 1664311, iss. 8.07.76.

Feldman, I.I., Blumentsev, A.M., Karanikolo, V.F., and Zheltikov, A.N.
1974: Nuclear geophysics logging methods for solid minerals deposits; in Exploration Geophysics in USSR in the Beginning of 70's, Publ. House Nedra, Moscow, p. 301-305 (in Russian)

Gabitov, R.M., Novikov, G.F., and Sinitsyn, A.Ya.
1974: Ore gamma spectrometer; Trans. Leningrad Min. Inst., v. 64 (2), p. 105-109 (in Russian).

Galimbekov, D.K.
1975: Application of the similitude principle to the determination of the field of scattered gamma-radiation in the borehole geometry; in Problems of the Theory and Mathematical Methods of Solution, Ufa, p. 45-50 (in Russian).

Galimbekov, D.K. and Soboleva, L.A.
1976: Investigation of scattered gamma ray field for selective gamma-gamma logging in boreholes; in Problems of Physics and Hydro-Dynamics of Oil and Gas, Ufa, p. 89-96 (in Russian).

Galimbekov, D.J., Karanikolo, V.F. and Lukhminskiy, B.E.
1976: Modeling of selective gamma-gamma problems; in Statistical Modeling in Mathematical Physics, Novosybirsk, p. 45-47 (in Russian).

Gamma-Spectrometric Methods in Survey and Exploration of Minerals
1975: Trudy VNIIaGG/Trans. All-Union Sci. Res. Inst. Nucl. Geophys. Geochem., no. 25, Moscow, 186 p. (in Russian).

Gera, D.F.
1974: Application of gamma-gamma log for determination of quartz veins in boreholes; in Exploration Geohysics in USSR at the Beginning of 70's, Publ. House Nedra, Moscow, p. 312-316 (in Russian).

Givens, W.W., Mills, W.R., Dennis, C.L., and Caldwell, R.L.
1976: Geophysics, v. 41 (3), p. 468-490

Goldanskiy, V.I., Dolenko, A.B., Egiazarov, B.G., and Zaporozhets, V.M.
1974: Gamma resonance methods and equipment for the phase analysis of minerals; Atomizdat, Moscow, 1974, 144 p. (in Russian).

Gorbachev, A.N. and Petrova, A.P.
1975a: Investigation of the representative volume of the field fluorometric neutron-activation observations and of the character of their relation to fluor grade; in Trudy VNII Geofiz. Metodov Razvedki, no. 25, Leningrad, p. 47-51 (in Russian).

1975b: Investigations of the factors influencing the results of the field fluorometric neutron-activation determinations; in Trudy VNNI Geofiz. Metodov Razvedki, no. 25, Leningrad, p. 33-37 (in Russian).

Gorbachev, A.N., Karpunin, A.M., and Matukanis, L.F.
1974: Predictive and/or survey consequence of the field neutron-activation fluorometry on some sulphide-cassiterite deposits of the Far East; in Trudy VNII Geofiz. Metodov Razvedki, no. 24, Leningrad, p. 76-79 (in Russian).

1975: On the estimation possibilities of the field neutron-activation fluorometry in exploration for tin deposits; in Trudy VNII Geofiz. Metodov Razvedki, no. 25, Leningrad, p. 51-55 (in Russian).

Gorev, A.V.
    1975: Experience in application of the photon-neutron assay of beryllium ores; in Trudy VNII Geofiz. Metodov Razvedki, no. 25, Leningrad, p. 60-64 (in Russian).

Gorev, A.V., Morozov, A.A., and Khusamutdinov, A.I.
    1975a: Numerical investigation of the relationship between stationary photon-neutron distribution and rock parameters; Geologia i Geofizika, (5), p. 128-132 (in Russian).

Gorev, A.V., Zgardovskiy, B.I., Nazarov, S.S., and Suvorov, A.D.
    1975b: Survey beryllometer Berill-4; in Trudy VNII Geofiz. Metodov Razvedki, no. 25, Leningrad, p. 73-76 (in Russian).

Grigoreva, A.Z., Saltsevich, V.B., and Krapivskiy, A.Z.
    1975: Application of gamma-neutron logging for the survey of orebodies covered by overburden; in Trudy VNII Geofiz. Metodov Razvedki, no. 25, Leningrad, p. 65-73 (in Russian).

Grigorkin, B.S. and Neustroev, A.P.
    1974: Results of work for tin ore grade determination by nuclear geophysical methods on tin ore deposits of the Central Yansk region; in Exploration Geophysics in USSR in the Beginning of 70's, Publ. House Nedra, Moscow, p. 332-335.

Grigorkin, B.S., Ivanyukovich, G.A., Meyer, V.A., Frolova, L.K., Neustroev, A.P., Petrukhin, V.M., and Kalikov, V.D.
    1976: Results of XRF logging on a gold-antimony deposit; in Nuclear Geophysics..., Yakutsk, p. 25-26 (in Russian).

Grigoryan, G.M., Medvedev, Yu.S., and Orlov, V.N.
    1974: Application of the copper-sulphide ore assay in Armenian deposits; Bull. Acad. Sci. Armen. SSR, Earth's Sci., v. 27 (2), p. 62-68 (in Russian).

Grigoryan, R.S.
    1975: Experience with the application of neutron-neutron log on the copper sulphide deposits of South Ural; in Geophysical Methods of Survey and Exploration, no. 2, Sverdlovsk, p. 87-89 (in Russian).

Grumbkov, A.P., Gryaznov, A.L., Ivanov, Yu.M., Marin-Fedorov, S.F., and Tsygankov, V.A.
    1975: Borehole tool GGK-P with generator of gamma photons; in Trudy VNIIYaGG, no. 23, Moscow, p. 26-32 (in Russian).

Grumbkov, A.P., Gryaznov, A.L., Guseva, N.S., Rybochenko, G.V., and Tsygankov, V.A.
    1976: Testing results for GGK-P apparatus with gamma-photon generator; in Nuclear Geophysical Applications Trudy VNIIYaGG, no. 26, p. 30-33 (in Russian).

Gulin, Yu.A.
    1975: Gamma-gamma method of investigations of oil wells; Publ. House Nedra, Moscow, 1975, 160 p. (in Russian).

Ivanyukovich, G.A., Meyer, V.A., and Frolova, L.K.
    1976: On the method of XRF logging for antimony; in Nuclear Geophysics..., Yakutsk, p. 26-27 (in Russian).

Kartashov, N.P. and Davydov, Yu.B.
    1975: On the problem of fission neutron logging with threshold detectors; in Nuclear Geophysics Investigations, Acad. Sci. USSR, Ural Branch, Sverdlovsk, p. 25-28 (in Russian).

Kashkay, M.A., Selekhli, T.M., Sultanov, L.A., and Magribi, A.A.
    1976: Role of physical rock parameters in localization of mercury mineralization on the Agiatagsk and Agkainsk mercury deposits (Low Caucasus); Azerbaijan Branch of VNII Geofiziki, Baku, 11 p. (Manuscript arrived at AzNIINTI Jan. 23, 1976, no. 8) (in Russian).

Khaykovich, I.M. and Yakovlev, V.N.
    1976: Computer interpretation of gamma-ray log and assay results; in Methods of Ore Geophysics, no. 11, VNII Geofiz. Metodov Razvedki, Leningrad, p. 87-93 (in Russian).

Kirichenko, N.M.
    1975: On the problem of the beryllium grade determination in rocks in the field investigations with photon-neutron method; in Geophysics and Astronomy, Information Bulletin, no. 14, p. 158-163 (in Russian).

Kobelev, L.N., Abidov, D.K., and Kaipov, R.L.
    1974: Improvement in the efficiency of geological exploration by means of XRF logging; in Nuclear Physical Methods of Analysis and Control of Technological Processes, Publ. House "Fan", Tashkent, p. 79-86 (in Russian).

Koshelev, I.P.
    1975: Nuclear Physics Methods of the Borehole assays on phosphorite, apatite and feldspar deposits; Published by VNII Razved. Geofiz. (VIRG), Alma-Ata, 212 p. (in Russian).

Koshelev, I.P., Belenko, R.D., Strakhov, G.V., Beyzot, M.Yu., Kovalskiy, V.S., Tebenkov, A.A., and Lahiarova, L.I.
    1975: Application of nuclear geophysics methods for borehole investigations on phosphorite deposits of Krakatau; in Nuclear Physics..., Koshelev, I.P. (ed.), Alma-Ata, p. 175-203 (in Russian).

Koshelev, I.P., Shishakin, O.V., Krasnoperov, V.A., Kamyshev, B.S., Zvuykovskiy, Z.R., and Budnikov, F.G.
    1975b: Application of nuclear physics borehole assays for survey and exploration of the fluorspar deposits; in Nuclear Physics Methods of Borehole Assay in the Deposits of Phosphorites, Apatites and Fluorspars, Alma-Ata, p. 69-94 (in Russian).

Koshelev, I.P., Startsev, Yu.S., and Shvartsman, M.M.
    1975c: Feasibility of application of nuclear geophysics methods for solution of geological assay problems on the apatite deposits of Khibin; in Nuclear Physics Methods of Borehole Assay in the Deposits of Phosphorites, Apatites and Fluorspars, Alma-Ata, p. 95-136 (in Russian).

Koshelev, I.P., Shishakin, O.V., Shvartsman, M.M., Startsev, Yu.S., and Krasnoperov, V.A.
    1975d: Neutron activation log for aluminium and silicon determination; in Nuclear physics methods of borehole assay in the deposits of phosphorites, apatites and fluorspars, Alma-Ata, p. 49-69 (in Russian).

Koshelev, I.P., Krasnoperov, V.A., Shishakin, O.V., Kamyshev, B.S., Startsev, Yu.S., and Shvartsman, M.M.
    1975e: Neutron activation log on fluor (NAK-N^{16}); in Nuclear Physics Methods of Borehole Assay on Phosphorite, Apatite and Feldspar Deposits, Alma-Ata, p. 3-43 (in Russian).

Koshelev, I.P., Startsev, Yu.S., Belenko, R.D., Shvartsman, M.M., and Taushkanov, A.P.
    1976: Application of nuclear logging to characterize the quality of phosphate ores; in Nuclear Geophysics..., Yakutsk 1976, p. 28-30 (in Russian).

Kozachok, I.A.
    1975: Macroscopic equation for neutron slowing-down in geological media; Doklady Akad. Nauk Ukrain. SSR, B (12), p. 1071-1076 (in Russian).

Kozachok, I.A. and Riznik, Ya.M.E.
    1977: Neutron slowing-down characteristics of rocks and collectors at deep levels; Publ. House Naukova Dumka, Kiev, 148 p. (in Russian).

Kozlov, G.G., Kormilitsyn, G.A., Etinger, V.R., and Evsees, V.S.
    1975a: Estimation of copper sulphide ores by gamma-gamma and XRF methods; Razvedka i Okhrana Nedr. (2), p. 50-53 (in Russian).

Kozlov, G.G., Kiryanov, E.A., Berezkin, V.V., and Shchekin, K.I.
    1975b: Lead and zinc determination in multimetallic ores by the XRF method using incoherent scattered radiation; in Trudy VNII Geofiz. Metodov Razvedki, no. 25, Leningrad, p. 107-111 (in Russian).

Kozlov, G.G., Evseev, V.S., Kormilitsyn, G.A., Etinger, V.R., Kostyakov, V.S., and Nikolskiy, V.A.
    1975c: Efficient estimation of the ore quality for the copper sulphide deposits in Ural by means of XRF and gamma-gamma methods; in Trudy VNII Geofiz. Metodov Razvedki, no. 25, Leningrad, p. 11-117.

Kozynda, Yu.O., Novikov, G.F., and Sinitsyn, A.Ya.
    1974: Continuous recording of equivalent uranium and thorium grades in spectral gamma-ray logging; Trans. Leningrad Min. Inst., v. 64 (2), p. 113-116, (in Russian).

    1976: Correction factors for the transient zone in spectral gamma-ray log; in Methods of Ore Geophysics, no. 11, VNII Geofiz. Metodov Razvedki, Leningrad, p. 94-97 (in Russian).

Krapivskiy, E.I. and Brem, A.A.
    1975a: Experimental determination of boundary effects in neutron-neutron log; in Trudy VNII Geofiz. Metodov Razvedki, no. 25, Leningrad, p. 43-46 (in Russian).

    1975b: On the modeling of neutron-neutron logging for absorbing elements; in Trudy VNII Geofiz. Metodov Razvedki, no. 25, Leningrad, p. 7-15 (in Russian).

Krapivskiy, E.I. and Saltsevich, V.B.
    1975: On the applicability of the theory of gamma-neutron logging; in Trudy VNII Geofiz. Metodov Razvedki, no. 25, Leningrad, p. 3-7 (in Russian).

Krapivskiy, E.I., Brem, A.A., Vashestov, Yu.V., and Tsenunin, V.F.
    1976: Application of nuclear logging for exploration of rare metals in pegmatites; in Nuclear Geophysics..., Yakutsk, p. 18-21 (in Russian).

Krapivskiy, E.I.
    1976: Lithium determination in pegmatites by combined gamma-neutron and neutron-neutron logging; in Methods of Exploration Geophysics, no. 21, Leningrad Univ., Leningrad, p. 66-67 (in Russian).

Krasnoperov, V.A. and Zvuykovskiy, Z.P.
    1976: Results of XRF logging in a multimetal deposit in South Kazakhstan; in Geophysical Investigations in Survey and Exploration of Ore Deposits, Alma-Ata, p. 80-84 (in Russian).

Krasnoperov, V.A., Zvuykovskiy, Z.P., Golovin, G.I., and Sheleshko, R.P.
    1976: Increase of the assay accuracy for Zn and Pb by XRF methods in exploration boreholes; in Nuclear Geophysics..., Yakutsk, p. 40-42 (in Russian).

Kuchurin, E.S., Mashkin, A.I., and Lebenzon, L.M.
    1976a: Experience with the application of nuclear geophysics methods of logging for the determination of the quartz-wolframite ore quality in the East-Zabaykal deposits; in Nuclear Geophysics..., Yakutsk, p. 44-49 (in Russian).

Kuchurin, E.S., Zaramenskikh, N.M., and Lebenzon, L.M.
    1976b: Geological efficiency of application of nuclear geophysics logging methods for borehole investigations of the principal types of gold deposits of East Zabaykal; in Nuclear Geophysics..., Yakutsk, p. 49-53 (in Russian).

Landström, O.
    1976: Analysis of elements in boreholes by means of naturally occurring X-ray fluorescence radiation; in Nuclear Techniques in Geochemistry and Geophysics, IAEA, Vienna, p. 47-52.

Leman, E.P., Mitov, V.N., Ochkur, A.P., and Yanshevskiy, Yu.P.
    1975a: Method of XRF logging; Soviet Patent no. 434837, iss. 21.II.75, class G 01 V5/00.

Leman, E.P., Orlov, V.N., and Medvedev, Yu.S.
    1975b: Peculiarities in analysis of heterogeneous media by the scattered radiation in transmission geometry; in Trudy VNII Geofiz. Metodov Razvedki, no. 25, Leningrad, p. 83-100 (in Russian).

Lukhminskiy, B.E.
    1975: Radiation transport in statistically heterogeneous rocks; in Function Theory and Application of Monte Carlo Methods, Ufa, p. 118-129 (in Russian).

Lukhminskiy, B.E. and Galimbekov, D.K.
    1975: Calculation of parameters for the selective gamma-gamma tool; Atomnaya Energia, v. 39 (5), p. 365-366 (in Russian).

Matyukhin, N.B., Borisov, A.P., and Gavrilov, O.I.
    1976: Efficiency of application of nuclear physics methods for survey of rare metals, phosphorites and fluorspars; in Nuclear Geophysics...., Yakutsk, p. 121-124 (in Russian).

Medvedev, Yu.S., Ochkur, A.P., and Leman, E.P.
    1973: Stabilization of the background level in XRF assay of the complex ores; in Geofizicheskaya Apparatura, no. 53, Leningrad, Publ. House Nedra, p. 112-116 (in Russian).

Meyer, V.A. and Filippov, M.M.
    1974: On the application of differential filters in XRF logs; Vestnik Leningrad Univ., (18), p. 133-136, (in Russian).

Meyer, V.A. and Rozuvanov, A.P.
    1974: Automatization of the nuclear geophysical ore assay; Vestnik Leningrad Univ., (24), p. 57-63 (in Russian).

Meyer, V.A., Nakhabtsev, V.S., Ivanyukovich, G.A., and Krotokov, M.I.
    1976: XRF tools for tin determination in boreholes and on the rock face; in Nuclear Geophysics..., Yakutsk, p. 131-132 (in Russian).

Mishin, G.T.
    1976: Determination of the potassium ore quality by the combined measurements of gamma activity of rocks in underground boreholes; in Methods in Exploration Geophysics, no. 21, Leningrad Univ., Leningrad, p. 90-95 (in Russian).

Mishin, G.T. and Gavrilova, L.I.
    1976: Determination of water non-soluble residue in potassium salts by means of gamma-ray logging; in Razvedochnaya Geofizika, no. 72, Publ. House Nedra, Moscow, p. 160-167 (in Russian).

Mitov, V.N., Leman, E.P., Sovtsov, M.I., and Shlakhtich, A.P.
    1975: XRF assay of the ore faces of mercury deposits; in Trudy VNII Geofiz. Metodov Razvedki, no. 25, Leningrad, p. 117-120 (in Russian).

Muravev, V.V., Pokrovskiy, V.A., Strelchenko, V.V., and Yakubson, K.I.
    1974a: Experimental investigation of basic relations of borehole activation analysis with isotopic neutron sources and with a generator of fast neutrons; in Trans. Moscow Institute of Oil-Chemistry and Gas Industry, no. 111, Publ. House Nedra, Moscow, p. 136-147 (in Russian).

Muravev, V.V., Strelchenko, V.V., and Yakubson, K.I.
    1974: Analysis of the efficiency of different neutron-activation methods of Al and Si determination in rocks penetrated by boreholes; in Trans. Moscow Institute of Oil-Chemistry and Gas Industry, no. 111, Publ. House Nedra, Moscow, p. 160-178 (in Russian).

Muravev, V.V. and Yakubson, K.I.
    1975: Spectrometer of natural gamma radiation for combined nuclear geophysical investigations of alunite ores; Trans. VNIIaGG, (25), p. 143-155 (in Russian).

Nakhabtsev, V.S.
    1974: Nomograms for the determination of coherent and non-coherent scattering cross-sections of gamma photons; in Scientific Works of Leningrad University, no. 382, Leningrad, p. 170-180 (in Russian).

    1975: More accurate approximation for differential scattering cross-section of photons on the bond electrons; in Scientific Works of Leningrad University, no. 369, Leningrad, p. 112-116 (in Russian).

    1977: Determination of optimum velocity in XRF logging; in Geofizicheskaya Apparatura, no. 60, Publ. House Nedra, Leningrad, p. 124-127 (in Russian).

Niewodniczanski, J. and Palka, K.
    1976: Some application of gamma spectrometry in borehole logging; in Report of the Institute of Nuclear Technology AGH, no. 95, Krakow, Poland, p. 23-24 (in Polish).

Nikulin, B.A., Gurov, P.N., and Gordeev, Yu.I.
    1976: Experimental investigation of media containing mercury with the method of pulsed neutron-gamma spectrometry; in Razvedochnaya Geofizika, no. 70, Publ. House Nedra, Moscow, p. 171-176 (in Russian).

Novikov, G.F. and Ozerkov, E.L.
    1974: Influence of emanation of uranium ores into boreholes on the results of gamma-ray logging; Trans. Leningrad Min. Inst., v. 64 (2), p. 87-90 (in Russian).

Novikov, G.F., Kozynda, Yu.O., Sinitsyn, A.Ya., and Kalinin, B.V.
    1974: How to take into account the borehole conditions in the interpretation of the spectral gamma-ray log; Trans. Leningrad Min. Inst., v. 64, (2), p. 139-142 (in Russian).

Novikov, G.F., Sinitsyn, A.Ya., and Kozynda, Yu.O.
    1976: Effective attenuation coefficients of gamma radiations for radioactive ores; Atomnaya Energia, v. 40, p. 178-180 (in Russian).

    1977: Specific sensitivity of NaI(Tl) scintillators in the detection of gamma radiation of radioactive ores; Atomnaya Energia, v. 42 (6), p. 495-496 (in Russian).

Nuclear Geophysical Investigations
    1975: Ural Science Center of Academy of Sciences USSR, Ural Branch, Sverdlovsk, 132 p. (in Russian).

Nuclear Geophysics Apparatus
    1976: Trudy VNIIaGG, Trans. All-Union Scientific Research Institute Nuclear Geophysics Geochemistry, no. 26, Moscow, 109 p. (in Russian).

Nuclear Geophysics Assay of Tin Ores in Natural Occurrences
    1976: Trudy VNIIaGG, Trans. All-Union Scientific Research Institute Nuclear Geophysics Geochemistry, no. 28, Moscow, 87 p. (in Russian).

Nuclear Geophysics for Survey and Exploration of Solid Minerals
    1976: Abstracts of papers. Ministry of Geology of RSFSR, Yakutsk, 1976, 148 p. (in Russian).

Nuclear Geophysics Methods in Geology
    1975: Trans. Inst. Geol. Geophys., Siberian Branch, Acad. Sci. USSR, Novosybirsk, 174 p. (in Russian).

Nuclear Geophysics Methods of Survey and Exploration for Bauxite Deposits
    1976: Trudy VNIIYaGG, Trans. All-Union Sci. Res. Inst. Nucl. Geophys. Geochem., no. 29, Moscow, 80 p. (in Russian).

Nuclear Techniques in Geochemistry and Geophysics; Proc.
    1976: IAEA, Vienna, 271 p.

Ochkur, A.P., Leman, E.P., Yanshevskiy, Yu.P., Medvedev, Yu.S., Orlov, V.N., and Volfstein, P.M.
    1974: XRF method of ore assay for colour and rare metals in boreholes and mines; in Exploration Geophysics in USSR in the Beginning of 70's, Publ. House Nedra, Moscow, p. 316-324 (in Russian).

Ochkur, A.P. (editor)
    1976: Gamma methods in ore geology; Leningrad, Publ. House Nedra, p. 408 (in Russian).

Ochkur, A.P., Voznesenskiy, L.I., and Fedorov, S.F.
    1976: Possibilities of nuclear physics methods in the ferrous metal ores assay; in Nuclear Geophysics for Survey and Exploration of Solid Minerals, Yakutsk, 1976, p. 57-59 (in Russian).

Osmonbetov, K.O., Erkhov, V.A., and Kopytov, Yu.Ya.
    1976: Efficiency of application of nuclear-physics methods of assay; Razvedka i Okhrana Nedr, (8), p. 17-19 (in Russian).

Petrukhin, V.M., Grigorkin, B.S., and Neustroev, A.P.
    1976: On the representativity of XRF log on Au-Sb deposits; in Nuclear Geophysics..., Yakutsk, p. 53-55 (in Russian).

Popov, E.P. and Vishnyakov, E.Kh.
    1974: Angular distribution of scattered gamma radiation of the Cs-137 source in the density gamma-gamma method; Trans. Leningrad Min. Inst., v. 64 (2), p. 137-138 (in Russian).

Popov, E.P., Ivanyukovich, G.A., Miroshnichenko, V.M., and Vishnyakov, E.Kh.
    1974: Optimum conditions of measurement in layer differentiation with density gamma-gamma logging; Trans. Leningrad Min. Inst., v. 64 (2), p. 133-136 (in Russian).

Postelnikov, A.F., Shifrin, I.E., and Chekanov, S.S.
    1975: Results of calculating of slow neutron spectra by the Monte-Carlo method; in Trudy VNII Geofiz. Met. Razvedki, no. 25, Leningrad, p. 15-20 (in Russian).

Potopakhin, A.S., Zvuykovskiy, Z.P., and Shishakin, O.V.
    1975: On the influence of variable borehole diameter and tool position on the results of neutron activation of N-16 measurements; in Nuclear Physics Methods of Borehole Assays on Phosphorite, Apatite and Feldspar Deposits, Alma-Ata, p. 44-48 (in Russian).

Pshenichnyy, G.A. and Meyer, V.A.
    1974: On the calculation of scattering and absorption cross-sections of roentgen radiation on atomic electrons; in Scientific Works of Leningrad Univ., no. 382, Leningrad, p. 159-169 (in Russian).

Pshenichnyy, G.A. and Meyer, V.A.
    1975: Estimate of contribution of the coherent and multiple scattered radiation in gamma and XRF methods; in Scientific Works of Leningrad Univ., no. 369, Leningrad, p. 117-124 (in Russian).

Putkaradze, L.A., Kasumov, K.A., Nechaev, Yu.V., Khilov, L.N., and Dolgin, M.G.
    1973: Determination of productive horizons with geophysical methods in old oilfields and in other mineral deposits in Azerbaijan; in Nuclear geophysics and geo-acoustic investigations of cased wells in exploration of old oil and gas fields, Trans. VNIIYaGG, no. 14, Moscow, p. 181-193 (in Russian).

Ratnikov, V.M., Ryabkin, V.K., Timofeeva, M.A., and Shubenok, S.A.
    1976: Nuclear geophysics assay of complex Sn-W ores; in Nuclear Geophysics Assay..., Trudy VNIIYaGG, no. 28, Moscow, p. 16-26 (in Russian).

Renken, J.H.
    1977: Nucl. Sci. Eng., v. 63 (3), p. 330-335.

Rudyk, Yu.M., Volkh, V.A., Mitryushchin, A.N., Sokolov, G.V., and Leypunskaya, D.I.
    1974: Radiometric assay of phosphorites on P_2O_5 in boreholes; in Exploration Geophysics in USSR at the Beginning of 70's, Moscow, Publ. House Nedra, p. 309-312 (in Russian).

Sachuk, V.A. and Balashov, V.N.
    1974: Application of XRF logging for tin assay in exploration holes; in Exploration Geophysics in USSR in the Beginning of 70's, Publ. House Nedra, Moscow, p. 324-328 (in Russian).

Saturin, A.A.
    1975: Nuclear physics methods of potassium determination; Publ. House Nedra, Moscow, 96 p. (in Russian).

Senko, A.K. and Zorin, G.K.
    1975: Experience with the application of selective gamma-gamma logging for iron determination in the skarn deposits of Priangar; in Trudy VNII Geofiz. Metodov Razvedki, no. 25, Leningrad, p. 120-124 (in Russian).

Shestakov, V.V.
    1975: Borehole tool for the photon-neutron assay of blast holes and underground boreholes; in Nuclear Geophysics Investigations Acad. Sci. USSR, Ural Branch, Sverdlovsk, p. 64-66 (in Russian).

Shimelevich, Yu.S., Kantor, S.A., Shkolnikov, A.S., Popov, N.V., Ivankin, V.P., Kedrov, A.I., Miller, V.V., and Polachenko, A.L.
    1976: Physical principles of pulsed neutron methods of borehole investigations; Publ. House Nedra, Moscow, 160 p. (in Russian).

Shishakin, O.V., Taushkanov, A.P., Koshelev, I.P., Ageev, V.V., and Trubov, E.A.
    1974: Rational combination of nuclear geophysical investigations of boreholes on the bauxite deposits of gibbsite type; in Exploration Geophysics in USSR in the Beginning of 70's, Publ. House Nedra, Moscow, p. 339-343 (in Russian).

Shmonin, L.I., Enker, M.B., Kolesov, G.E., Bochek, Yu.V., and Sinelnikov, M.Yu.
    1976a: Nuclear geophysics methods for the assay of lead and barium ores in natural occurrences; in Nuclear Geophysics..., Yakutsk, p. 34-36 (in Russian).

Shmonin, L.I., Mager, E.V., and Gorbachev, A.I.
    1976b: Application of radioisotope methods for lead-zinc ore assaying in pyrite deposits; in Nuclear Geophysics..., Yakutsk, p. 61-64 (in Russian).

Sinitsyn, A.Ya., Kozynda, Yu.O., and Gabitov, R.M.
    1974: Energy stabilization of gamma-ray spectrometers in radioactive ore assaying; Trans. Leningrad Min. Inst., v. 64 (2), p. 123-126 (in Russian).

Sokolov, E.A., Egorov, E.V., and Vysotskiy, I.B.
    1975: Method of elimination of iron influence on the neutron-gamma spectrometric logging results; in Trudy VNII Geofiz. Metodov Razvedki, no. 25, Leningrad, p. 28-33 (in Russian).

Sowerby, B.D.
    1974: Applications of nuclear resonance fluorescence of gamma rays to elemental analysis; Australian Patent no. 454745 class 07.8/G21C-G01 N/, appl. 27.04.71, iss. 21 21.10.74.

Sowerby, B.D. and Ellis, W.K.
    1974: Borehole analysis for copper and nickel using gamma-ray resonance scattering; Nucl. Instruments and Methods, v. 115, p. 511-523.

Sowerby, B.D., Ellis, W.K., and Greenwood-Smith, R.
    1977: Bulk analysis for copper and nickel in ores using gamma-ray resonance scattering; in Nuclear Techniques and Mineral Resources, IAEA, Vienna, p. 499-521.

Startsev, Yu.S., Koshelev, I.P., and Shvartsman, M.M.
    1975a: Detection of apatite ores and the determination of oxidation zones by nuclear well logging methods in the apatite deposits of Khibin; in Nuclear physics methods for borehole assaying in deposits of phosphorites, apatites and fluorspar, Alma-Ata 1975, p. 143-148 (in Russian).

    1975b: Determination of basic payable components of apatite ores of Khibin by means of nuclear logs; in Nuclear physics methods for borehole assaying in deposits of phosphorites, apatites and fluorspar, Koshelev, I.P. (editor), Alma-Ata, p. 149-174 (in Russian).

    1975c: Potassium and acid soluble aluminium determination; in Nuclear physics methods of borehole assaying in deposits of phosphorites, apatites and fluorspars, Alma-Ata, p. 137-142 (in Russian).

Stevens, J.G. and Stevens, V.E.
    1975: Mössbauer Effect Data Index; IFI, Plenum Press, New York.

Suvorov, A.D. and Molochnova, V.A.
    1975: The significance of the influencing factors in carborne photon-neutron surveys; in Trudy VNII Geofiz. Metodov Razvedki, no. 25, Leningrad, p. 38-43 (in Russian).

Szymborski, A.
    1975: The expectancy of application of borehole logging methods for exploration and estimation of iron ores; Kwartalnik Geologiczny (19), p. 413-429 (in Polish).

Tamrazyan, A.A. and Popov, E.S.
    1975: On the possibility of XRF method applied in the conditions of Armenia; in Geophysical and Seismological Investigations of the Earth Crust in Armenia, Publ. Acad. Sci. Armen. SSR, Erevan, p. 138-143 (in Russian).

Thibideau, F.D.
    1977: A pulsed neutron generator for logging; Techn. Inform. Series GEPP-249, Gen. Electr. Co., Neutr. Devices Dep., St. Petersburg, Florida, USA, 14 p.

Tikhonov, A. and Arsenine, V.
    1976: Méthodes de résolution de problèmes mal posés; Edit. Mir, Moscou.

Trudy VNII Geofiz. Metodov Razvedki:
    Methods of exploration geophysics, nuclear geophysics in ore geology; no. 25, Leningrad 1975, 139 p. (in Russian).

Trans. Leningrad Mining Inst. 1974, v. 64(2):
    Geochemical and radioactive methods of survey and exploration of mineral deposits; Leningrad, 175 p. (in Russian).

Umiastowski, K., Buniak, M., Gyurcsak, J., Turkowa, B., and Maloszewski, P.
    1976: Influence of sample granulation on the results of radiometric measurements; Rep. Inst. Nucl. Technol. AGH, no. 94, Krakow (in Polish).

Umiastowski, K., and Buniak, M.
    1977: The influence of rock heterogeneity on the results of gamma-gamma logging; in Nuclear Techniques in Exploration and Mineral Resources, IAEA, Vienna, p. 273-280.

Utkin, V.I., Starikov, V.N., Ermakov, V.I., and Yaushev, K.K.
    1974: Application of the Monte-Carlo method to the problems of selective gamma-gamma method; Bull. Acad. Sci. USSR, Earth's Physics (9), p. 92-96 (in Russian).

Utkin, V.I.
    1975: Method of radioactive logging. Patent USSR, class G 01 v 5/00, no. 379905, appl. 4.03.71, no. 1629285, iss. 30.07.75.

Utkin, V.I. and Burdin, Yu.B.
    1975: On the accuracy of analog recording in gamma-gamma logging; in Nuclear Geophysical Investigations, Acad. Sci. USSR, Ural Branch, Sverdlovsk, p. 52-57 (in Russian).

Utkin, V.I. and Ermakov, V.I.
    1975: Gamma-gamma micrologging; Bull. Acad. Sci. USSR, Earth's Phys. (7), p. 57-65 (in Russian).

Utkin, V.I., Starikov, V.N., Ermakov, V.I., and Yaushev, K.K.
    1975: Solution of the problems of selective gamma-gamma method by the Monte-Carlo technique; in Function theory and application of the Monte Carlo methods, Ufa, p. 130-139 (in Russian).

Utkin, V.I., Gera, D.F., Burdin, Yu.B., and Ermakov, V.I.
    1976: Spectral-angular selection of scattered gamma photons in the gamma-gamma method; Bull. Acad. Sci. Earth's Physics (2), p. 113-118 (in Russian).

Vakhtin, B.S., Ivanov, V.S., Novoselov, Filippov, E.M.
    1972: Radiative-resonance neutron detector for geophysical investigations; Atomnaya Energia, v. 33 (5), p. 928-929 (in Russian).

Vakhtin, B.S., Ivanov, V.S., and Semenov, V.I.
    1973: Intensity and spectrometric investigations of boreholes in boron-bearing rocks; in Application of Nuclear Radiation in Geophysical Investigations, Siberian Branch, Acad. Sci. USSR, Inst. Geol. Geophys., Novosibirsk, p. 45-56 (in Russian).

Vakhtin, B.S., Ivanov, V.S., Sokolov, D.I., and Chernyshev, A.V.
    1975: Experience in application of neutron resonance log for qualitative boron determination in blast holes; in Nuclear geophysical methods in geology, Siberian Branch, Acad. Sci. USSR, Inst. Geol. Geophys., Novosibirsk, p. 16-21 (in Russian).

Varga, J.
    1975: Accuracy and stability in estimation of a differential evaluation method of gamma logging data; Magyar Geofizika, v. 16 (2), p. 41-53 (in Hungarian).

Vladimirov, O.K., Evdokimov, Yu.D., Kalvarskaya, V.P., Karkhu, A.I., and Ochkur, A.P.
    1974: Actual state and future development of ore logging; in Exploration Geophysics in USSR in the Beginning of 70's, Publ. House Nedra, Moscow, p. 368-370 (in Russian).

Voskoboynikov, G.M., Utkin, V.I., Burdin, Yu.B., and Zubaev, G.D.
    1975a: Apparatus for Investigation of Small Diameter Boreholes, Patent USSR, class G 01 v 5/00, no. 370569, app. 20.01.69, no. 1299674, iss. 30.07.75.

    1975b: Measurement Probe for Gamma-gamma Logging; Patent USSR, class G 01 v 5/00, no 333517, app. 20.01.69, no. 1299715, iss. 30.07.75

Voynova, K.P., Savinets, E.A., and Shayakubov, T.Sh.
    1974: Application of geophysical logging methods in survey and exploration of the fluorspar deposits of Uzbekistan; in Geophysical investigations on the Ustuyurt Plateau and its Environs, Publ. House Karakalpakstan, Nukus, Uzbek SSR, p. 135-141 (in Russian).

Voynova, K.P., Savinets, E.A., Khvalovskiy, A.G., and Shayakubov, T.Sh.
    1976: Application of activation logging for survey and exploration of fluorspar (on the example of the Agata-Chibargatinsk deposit); in Fluorite, Publ. House Nauka, Moscow, p. 255-266 (in Russian).

Voznesenskiy, L.I.
    1976: XRF analysis of chromite ores during mining; in Methods of Exploration Geophys., no. 21, Leningrad University, Leningrad, p. 82-87 (in Russian).

Vozzhenikov, G.S. and Zaramenskikh, N.M.
    1975: On the theory of neutron activation; in Trans. Sverdlovsk Min. Inst., no. 107, Sverdlovsk, p. 59-62 (in Russian).

Vozzhenikov, G.S. and Davydov, Yu.B.
    1977: On the theory of neutron activation measurements in boreholes; Atomnaya Energia, v. 42 (3), p. 205, deposited paper no. 897/8830 (in Russian).

Wormald, M.R. and Clayton, C.G.
    1976: Some factors affecting accuracy in the direct determination of uranium by delayed neutron borehole logging; in Exploration for Uranium Deposits, IAEA, Vienna, p. 427-470.

Wylie, A.W., Eisler, P.L., and Huppert, P.
    1976: Method and Apparatus for Detection of Copper (CSIRO); Australian Patent, class 85.2-2, /G 01 V 5/00, G 01N 23/22/, no. 468970, appl. 21.09.71, no. 33747/71, iss. 12.01.76

Yanshevskiy, Yu.P., Medvedev, Yu.S., Ochkur, A.P., Petrukhin, V.M., and Kashintsev, A.A.
    1976a: Application of the spectral intensity method in XRF logging in Nuclear Geophysics for survey and exploration of solid minerals, Yakutsk, 1976, p. 21-22 (in Russian).

Yanshevskiy, Yu.P., Ochkur, A.P., Volfstein, P.M., Medvedev, Yu.S., and Petrukhin, V.M.
    1976b: XRF logging; Leningrad, Publ. House Nedra, p. 140 (in Russian).

Yasinovenko, A.P.
    1975: Decrease of ore depletion using gamma-gamma logging; in Problems of efficiency improvement in East Siberian Mining, Yakutsk, p. 218-223 (in Russian).

Zgardovskiy, V.I., Leman, E.P., Kozlov, G.G., Medvedev, Yu.S., Kotelnikov, V.V., and Orlov, V.N.
    1974: Borehole tool SRPD for the RRShA-1 apparatus; in Geofizicheskaya Apparatura, no. 56, Leningrad, Publ. House Nauka 1974, p. 80-83 (in Russian).

BOREHOLE LOGGING TECHNIQUES APPLIED TO BASE METAL ORE DEPOSITS

W.E. Glenn
UURI-Earth Science Laboratory, Salt Lake City, Utah

P.H. Nelson
Lawrence Berkeley Laboratory, Berkeley, California

Glenn, W.E. and Nelson, P.H., Borehole logging techniques applied to base metal ore deposits; in Geophysics and Geochemistry in the Search for Metallic Ores; Peter J. Hood, editor, Geological Survey of Canada, Economic Geology Report 31, p. 273-294, 1979.

Abstract

Interest in the application of borehole logging to metallic mineral exploration and deposit evaluation has grown substantially in recent years. However, borehole logging tools and techniques were developed primarily by the petroleum industry and neither the tools nor their application are directly suited to the needs of the metallic mineral industry. Lack of generally available commercial logging services for small diameter holes and for the measurement of quantities such as magnetic susceptibility, induced polarization, and ore grade has inhibited the growth of borehole logging in metallic mineral mining.

The need to study new mining technology and to search at greater depths for new ore deposits has created a demand for a slim-hole logging capability. The requirements range from elemental and mineralogical analyses, through properties used in geophysical exploration to bulk rock properties for either conventional or novel mining techniques. This paper reviews the presently available capabilities for downhole analyses for geological information, for geophysical properties and for fluid flow. Much of the review is based upon direct experience with a facility operated in-house by a major metals mining company.

Résumé

Il y a eu au cours des dernières années un intérêt croissant pour l'application de la diagraphie par trou de sonde à l'exploration des minéraux métalliques et à l'évaluation des gisements. Toutefois, les outils et les techniques de diagraphie par trou de sonde ont été mis au point principalement par l'industrie pétrolière; ni les outils ni leur utilisation n'étaient directement adaptés aux besoins de l'industrie des minéraux métalliques. Le manque de services commerciaux de diagraphie accessibles en ce qui concerne les trous de petit diamètre et la mesure de quantités comme la susceptibilité magnétique, la polarisation provoquée et la teneur en minerai a empêché le développement de la diagraphie par trou de sonde dans le secteur des minéraux métalliques.

Le besoin d'étudier de nouvelles techniques d'exploitation et de creuser à des profondeurs plus grandes afin de trouver de nouveaux gisements de minerai a fait naître une demande pour la diagraphie en diamètre réduit. Les besoins vont des analyses élémentaires et minéralogiques jusqu'aux propriétés utilisées en exploration géophysique et aux propriétés de la roche prise dans son ensemble, en ce qui a trait aux techniques d'exploitation classiques ou nouvelles. Le présent document étudie les méthodes actuelles d'analyse, au fond du trou, de l'information géologique, des propriétés géophysiques et de l'écoulement des fluides. La majeure partie de l'étude est basée sur des expériences pratiques à l'aide d'une installation exploitée par une importante société d'exploitation de minéraux métalliques.

INTRODUCTION

Research on the application of well logs to porphyry copper exploration and development began at Kennecott Copper Corp. in 1971. Logging research was justified for many reasons, among which was the need to develop additional techniques for locating and evaluating deeper mineral deposits. Due to a lack of a logging service oriented to metallic mineral applications both in terms of tools and data analyses, Kennecott decided to develop its own logging system in addition to carrying out research on well log applications. This approach allowed a great deal of flexibility for studying what tool specifications were needed, what field techniques were best and what kinds of data were important.

In general, the use of logging in the base minerals industry lags behind that of almost all other sectors of the exploration and geotechnical disciplines. We will not attempt to diagnose reasons for this deficiency, but it means that if the mining industry finds the incentives to apply borehole measurements on a widespread basis tomorrow, it faces reduced start-up costs in terms of tool development. This is not to say that the development and deployment still required will be inexpensive, but with the exception of techniques for mineralogical and elemental analysis, many useful tools have already been developed and are available to base metal explorationists on an "off-the-shelf" basis. Moreover the development of small diameter tools has been accelerating at a rapid pace in the last few years. The stimulation comes not from the base metal mining industry but from other sectors, including,

- production logging for oil and gas,
- uranium and coal exploration and exploitation,
- geothermal exploration and exploitation,
- geological engineering studies, such as power plant siting, tunnelling, etc.

We hope we have made it clear that the true "state-of-the-art" in mineral logging comes not so much from the mineral industry as from the whole spectrum of users of downhole measurements. Since a true overview of all sources and users of slimhole borehole technology is not possible, and would quickly become out-dated, we will simply list some recent and on-going developments in logging technology applicable to the mining industry and devote the remainder of the paper to our own experience in applying different techniques to the exploration and engineering problems encountered within the base metal mining industry.

Recent Developments

1. The development of a borehole assaying tool for nickel, with very encouraging indications of its applicability to copper (Seigel and Nargolwalla, 1975).
2. Recent improvements in the stability of a magnetic susceptibility logging instrument: less than 10 μcgs maximum drift in an operating environmental temperature of 70°C, by the U.S. Geological Survey (J. Scott, pers. comm.).
3. The availability of "high resolution" focussed density tools in slim-hole versions, applicable to the logging of thin seams, veins, and fractures (Fishel, 1976).
4. The continued efforts at quantifying the electrical resistivity and induced polarization response in terms of the sulphide and clay content of the rock (Clavier, et al., 1976; Snyder et al., 1977; Patchett, 1975) and the current on-going development of several commercial and industrial groups of tools for the measurements of polarization and dielectric properties in boreholes.
5. Continuing refinement of precision temperature logging and the analysis of continuous logs (for example, Conaway, 1977) to correlate rock type using thermal conductivity.
6. A calibration facility oriented to base mineral environments under development at the U.S. Bureau of Mines.

The above listing is almost a random sampling of recent developments which might easily be adapted for specific purposes in base mineral exploration and engineering. Another approach is to consider physical and chemical properties which might be measured in situ and applications for which the data will be most useful. Table 12.1 represents one such compilation; it intentionally includes engineering applications as well as exploration. This extension emphasizes our belief that downhole logging will eventually have as much importance for the mining engineer as for the explorationist.

Application of Techniques

In the remainder of this paper we describe an in-house downhole logging facility used for a variety of research and developmental purposes and describe some specific applications and results. Our discussion is keyed to the ten properties listed in Table 12.1; the underlined entries indicate the categories for which examples are given in this paper.

It should be stated that in order to limit the paper we have excluded exploration techniques which combine surface and borehole methods (see Hohmann et al., 1977, for electromagnetic prospecting examples) and cross-hole techniques for investigating large volumes of rock between boreholes (see Lytle et al., 1976, for electromagnetic probing example; R.L. Aamodt, 1976, for examples using acoustic energy; and Daniels, 1977, for IP examples). That is, we confine our presentation to probes in a single borehole with a radius of investigation determined by the rock/fluid properties and the probe geometry.

Keys and MacCary (1971) reviewed the application of borehole geophysics to water resources investigations and much of the material they covered is appropriate to metallic mineral mining applications. From the geotechnical vantage point Van Schalkwyk (1976) reviewed a wide range of equipment which is useful for rock mechanics and engineering studies. Other authors have reviewed the use of borehole instrumentation for various slim-hole applications. For example, Dyck et al. (1975) gave an excellent overview of borehole geophysics applied to metallic mineral prospecting. Each of these three review articles is interesting in terms of its scope and perspective and each contains a useful list of references.

Table 12.1

Physical and chemical in-situ properties obtained by borehole logging with regard to mining applications

	Exploration	Porphyry Copper Characterization	Dump Leaching	Mine Development
Elemental and mineralogical analysis	X̲	X	X	X
Rock type identification and correlation	X̲	X	X	X
Rock quality and fractures	X	X̲		X
Density	X			X
Porosity		X̲	X	
Electrical and magnetic properties	X̲	X		
Temperature and thermal properties		X	X	X̲
Fuid flow		X	X̲	X
Water sampling	X	X	X	X

Underlined entries indicate the categories for which examples are given in this paper.

KENNECOTT COPPER CORPORATION BOREHOLE LOGGING SYSTEM

Kennecott Copper Corp. currently (1977) has three logging trucks, of which only one is designed to measure a broad range of rock properties. The other two are devoted to hydrologic and environmental logging. A brief description of the Kennecott Copper Corp. logging truck (Fig. 12.1 and 12.2) is given here. It is a four-wheel drive Ford F600 truck chassis with a 14 feet (4.3 m) long, 6 feet (1.8 m) wide, 6 feet (1.8 m) high custom-built van on the bed. The van portion is divided into a forward operator/recorder cubicle and a rear mechanical/storage area. A boom is located over the rear centre of the vehicle. The truck is equipped with 1830 m of four-conductor 4.8 mm, steel armour cable with a Gearhart-Owen four-conductor cable head and an extensive set of electronic and mechanical equipment. The electrical power unit is a 6.5 kW gasoline generator. The truck typically carries fourteen different logging tools. The various tools and their primary function(s) are shown in Table 12.2. All tools are 2.25 inches (5.7 cm) or less in diameter.

Many of the tools require only one conductor and the armour for operation, but others such as the IP tool require all four plus the armour. Except for the resistivity and single-point resistance tools, all tools accomplish some electronic signal processing and enhancement downhole before the signal is transmitted to the surface. Uphole modules are mounted in removable nuclear instrumentation modules and do various amounts of signal processing and output the analog signals to one or more channels of a four-channel Texas Instruments chart recorder. A plug-in patch panel (Fig. 12.2) and programmed patch boards (Fig. 12.3) allow the operator to connect tool functions, module functions and auxiliary electronic equipment such as an oscilloscope in any fashion desired or to operate the system in some predetermined fashion by simply plugging in a board. The flexibility of the patch panel is essential for a research and development logging system in that the electronic equipment can be quickly interconnected in a modified or new configuration and a new tool can quickly become operational. A block diagram of the system just described is given in Figure 12.4.

ELEMENTAL AND MINERALOGICAL ANALYSES IN BOREHOLES

One of the most exciting and promising frontiers in mineral logging is the development of tools to analyze for specific elements in the borehole. In the specific case the logging would be a borehole assay for particular ore minerals. In the general case the type of logging would provide an in situ geochemical description of the mineral deposit.

Czubeck (1979) describes the current activity in borehole assaying, most of which is based upon radioactive and fluorescent techniques. A neutron activation method for the borehole assay of copper and nickel has been described by Nargolwalla et al. (1974) and Seigel and Nargolwalla (1975). Their system, called Metalog, available as a routine service, was tested in Anaconda's porphyry copper mine near Yerington, Nevada (Staff, Scintrex Limited, 1976). The test results are very envouraging and are presented in Figure 12.5. Because of limited experience with this technology we will not attempt to describe progress in the field of nuclear analysis, but will discuss a few examples obtained with more conventional methods. It is likely that the results from these other methods will supplement and constrain the nuclear analytical methods when they become available.

The most readily available analytical technique in boreholes is the measurement of the natural gamma radiation. In the mining environment the total count can often be directly correlated with the potassium content from the laboratory analysis of core material as shown in Figure 12.6. In fact, based upon this kind of direct correlation, our logs are routinely scaled in K_2O content in most environments. Where the uranium or thorium content is significant this straightforward procedure does not apply, and spectral information is required to distinguish the three elements. Indeed, spectral logging is applied to uranium exploration and is discussed elsewhere in these proceedings (Killeen, 1979).

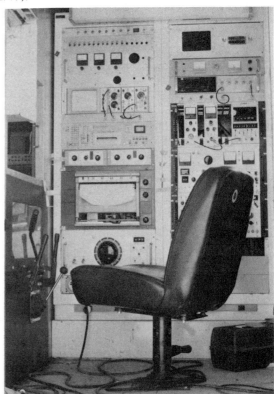

Figure 12.1. Kennecott Copper Company's logging truck.

Figure 12.2. Interior view of operator section of Kennecott Copper Company's logging truck.

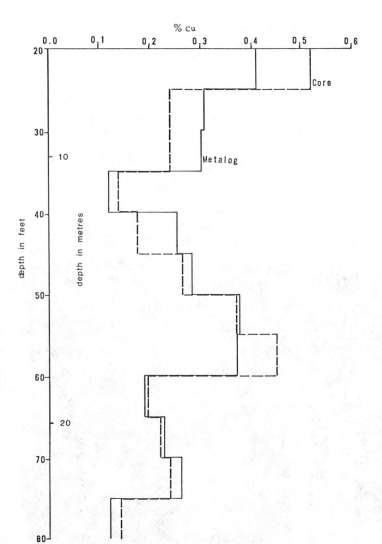

Figure 12.3. Program patch board.

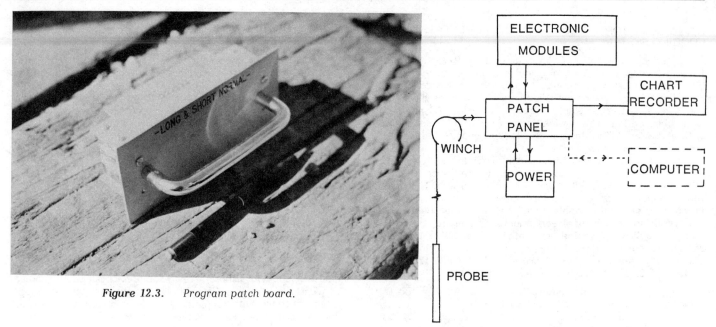

Figure 12.4. Block diagram of logging system

Figure 12.5. Borehole assaying.

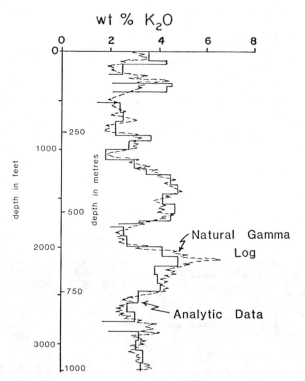

Figure 12.6. Comparison of analytic and well log analysis of K_2O.

Examples of the dependence of the neutron log on both pore water and water bound in hydrous minerals have been presented before (Savre and Burke, 1971; Snyder, 1973; Nelson and Glenn, 1975) and the example of Nelson and Glenn is reproduced in Figure 12.7. The chemically bound water can be so high that it can be directly measured by the neutron tool and hence mask the pore water contributions to the neutron log. We will show in the section on porosity that when the two contributions are mixed, it is still possible to remove the bound water contribution. In any case the bound water can be quantitatively assessed, giving a measure of the hydrated mineral content of the rock, usually translatable as the altered, clay mineral fraction.

Table 12.2

Response of drillhole logging tools to rock properties

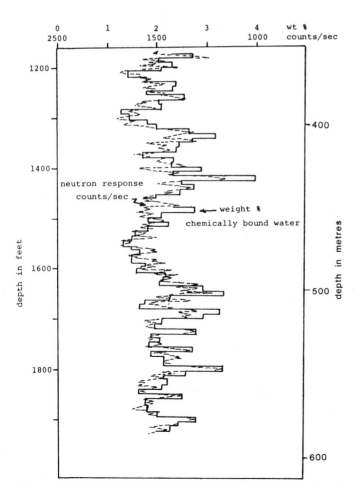

LOGGING TOOLS	Temperature	Hole Diameter	Porosity	Bound Water	Fluid Loss	Potassium	Density	Fracturing, RQD	Sulphides	Magnetite
Temperature	1				4					
Caliper		1			4			4		
Neutron			2	2	3			4		
Natural Gamma						2		4		
Gamma-Gamma (density)			2	4	3		1		2	
Velocity (ΔT)			2		3			2		
Resistivity			3	4	4			3	2	
IP				4					2	
Magnetic Susceptibility										1
Fluid Flow					1					

1 Quantitative measure — high success
2 Quantitative measure — moderate success
3 Always responds qualitatively
4 Often responds qualitatively

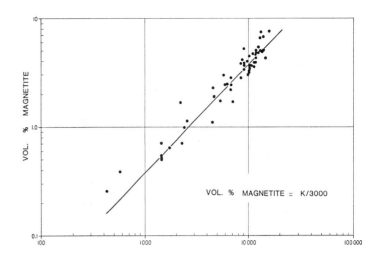

Figure 12.8. Plot of magnetic susceptibility from logs versus volume per cent magnetite.

A third fairly direct measurement is the magnetite content estimated from the magnetic susceptibility tool. Figure 12.8, taken from Snyder (1973), demonstrates the correlation. Here again, as in the case of potassium, is one of the rare cases where a single element or mineral type possesses a distinctive physical property which can be readily measured and is unique to that mineral type in many cases of interest.

An obvious need in mining applications is a direct measure of the sulphide content. Our attempts at this measurement are detailed in subsequent sections on density and electrical properties.

ROCK TYPE IDENTIFICATION AND CORRELATION

Identification of rock types is facilitated where there is a direct correlation between an easily identified mineral species, such as magnetite, and the rock unit. A good example of this is evident at one of Kennecott's porphyry deposits. The copper mineralization is in an andesite pile crosscut by numerous latite dykes. The dykes and andesites show contrasting expressions on both the natural gamma (potassium) and magnetic susceptibility (magnetite) logs. Figure 12.9 illustrates this contrast. The latite between 163 and 186 m shows a high natural gamma response and a near-zero magnetic susceptibility, whereas the andesite exhibits a lower natural gamma response and magnetic susceptibility averaging around 3000 μcgs. Note that the density of the latite is also lower than the density of the andesite. A short interval between 151 and 159 m has a very similar response on all logs as does the latite except that the magnetic susceptibility is not zero. We interpret this interval to be a near miss of the same dyke intersected at 163 m. The result suggests that the drillhole is parallel to a dyke over this interval.

We have noted a striking correlation between resistivity variation and sulphide zoning at one of Kennecott's porphyry copper deposits. The highly variable, low to high, resistivity pattern evident in the upper part of the long and short normal log* for each hole shown on a cross-section in Figure 12.10 is coincident with the pyrite halo in this deposit. The lower cut-off of this variable resistivity is closely correlated with a particular copper grade. Some detailed features of the resistivity logs can be correlated among the drillholes. These correlated features have been associated with both a variation in sulphide veining and the degree of sulphide

Figure 12.7. Comparison of the neutron log with bound water analysis on pulps.

---

* Industry standard: 40 and 16 inches (1.22 and 0.488 m) respectively.

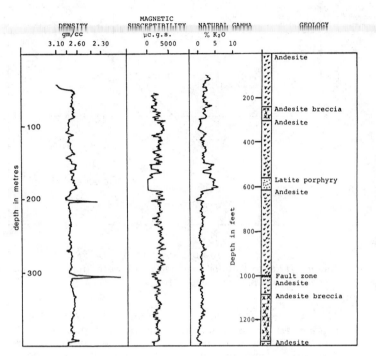

Figure 12.9. Latite dyke correlation on natural gamma and magnetic susceptibility logs.

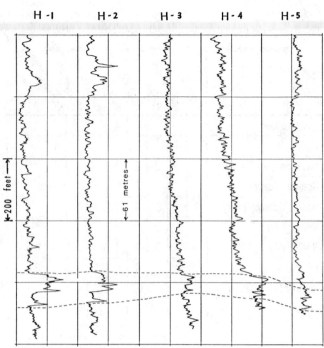

Figure 12.11. Natural gamma logs.

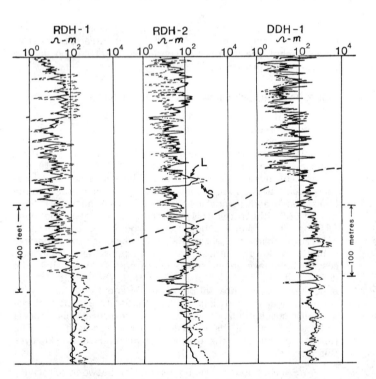

Figure 12.10. Long and short normal resistivity logs.

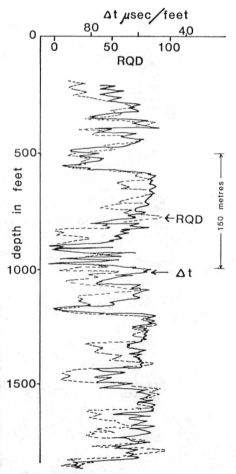

Figure 12.12. Rock quality designation (RQD) from core versus compressional wave transit time measured in a drillhole.

oxidation. Also, at the same deposit, the natural gamma log shows a substantial increase in potassium as the causative intrusive is approached. This feature is depicted in Figure 12.11 where logs from five different drillholes are shown. An interval of high potassium can be correlated on all logs as indicated by the dashed lines in Figure 12.11. The increased potassium is seen in the core as an increase in sericite and orthoclase mineralization.

We have numerous examples of rock-type identification and correlation that we have accomplished at several of Kennecott's porphyry copper deposits and prospects. However, we feel the ones given here are sufficient to illustrate the ability to use well logs both to differentiate rock types and to correlate rock units across a deposit. This information is useful for structural geology studies and for the understanding of the porphyry system geometry.

ROCK QUALITY AND FRACTURE LOCATION

With logging techniques it is possible to determine the mechanical properties of rocks and to study the structural geology. In this section logging relevant to both topics will be described.

Rock Quality Designation (RQD) as used by mine geologists is the measure obtained by summing the length of core pieces greater than 10 cm and dividing the total length of sample, typically 1.5 m to 3 m of core. If all the pieces of core in a 1.5 m section, for example, were greater than 10 cm, the RQD would be 100%, if all pieces were smaller than 10 cm the RQD would be 0%. RQD is an empirical measure of rock competence. We have found that the velocity (reciprocal of the compressional wave transit time) log gives a fairly good measure of RQD where the rock is moderately to well fractured and one example is shown in Figure 12.12. In this case, the velocity log responds primarily to the cracks or fractures in the rock.

Fractures or fractured intervals of rock commonly give a characteristic response on several logs: caliper, velocity, neutron and density. An example is shown in Figure 12.13. For illustration note the response of each log at 308 m where the geologist has noted a fault zone in the core. The caliper shows an increased hole diameter which indicates that material has fallen from the fracture or was plucked out by the bit during drilling. The velocity log shows a low velocity, the neutron shows a high porosity and the density log shows a low density at this depth. Several fractures have been picked from these logs and are indicated by an f beside the caliper log at the appropriate places.

The preceding analysis gives a location only for fractures which appear open in the hole and does not provide strike or dip information.

We have used the seisviewer log offered as a service by Birdwell to determine fracture orientation in drillholes. This tool was first described by Zamanek et al. (1970). An example of this log along with the caliper and velocity logs is shown in Figure 12.14. The interpreted fractures and their orientation are noted beside the seisviewer log. The seisviewer tool sends acoustic pulses toward the borehole wall. The tool rotates and sends a signal to the surface when it passes through magnetic north. The acoustic energy reflected from the borehole wall is monitored and sent to the surface where it is presented on an oscilloscope and photographed. The oscilloscope triggers on the north signal. A smooth borehole wall will generate a strong reflection, the bright areas on the log, and a rough, caved borehole wall will generate a weak reflection, the dark areas on the log. Hence fractures are indicated by dark traces on the log. However, if the tool is not centred in the hole or the hole deviates very much from a cylindrical cross-section, the log will exhibit a

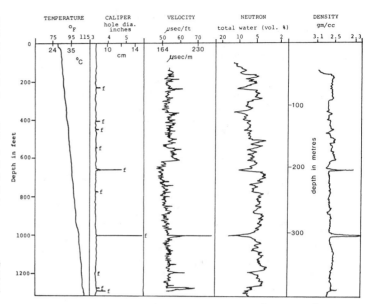

Figure 12.13. Suite of logs showing rock fracture signature on several well logs.

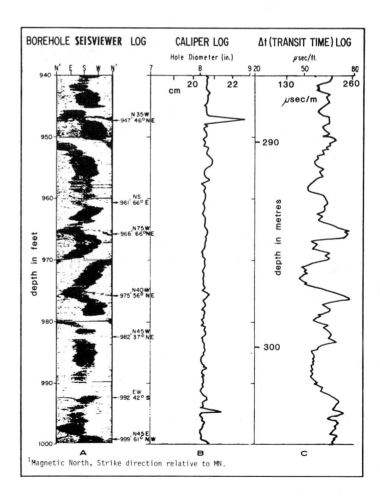

Figure 12.14. Fracture location and orientation obtained from an acoustic seisviewer log.

vertical zebra pattern as seen in Figure 12.14. Often this pattern will obscure the fractures. The tool is also continuously pulled vertically and the resultant log is a planar representation of a cylindrical borehole with north at the left and right of the record and depth along the side. A plane intersecting the drillhole will apear as a sinusoid on the flat surface. The dip magnitude and direction are obtained from the excursion and location of the maximum and the minimum of the sinusoid. The data from the seisviewer log can be studied in a conventional way as illustrated in Figure 12.15. This figure shows a contoured equal-area plot of poles to the fractures and Rose diagrams of strike and dip directions. A plot (not shown) was also made for fractures measured on the surface around the drill site and the data agree extremely well with one exception; understandably, the surface study showed very few near-horizontal fractures.

We believe logging for structure can benefit both the exploration and the mine geologist. Both could use the information to develop a better understanding of the three-dimensional structure of a mineral deposit.

DENSITY LOGGING

Routine Density Estimates

We routinely employ a single-detector, gamma-gamma density probe in NX-size boreholes. The maximum tool diameter is 5.5 cm. In operation the entire tool is held against the borehole wall by a decentralizing arm which extends from the tool. Because only a single detector is used, the measurement is uncompensated, that is, no correction is made for borehole rugosity. In cored holes in hard rock this does not present as great a problem as in soft rock because zones of severe rugosity are relatively infrequent and not much data is lost, although recently it has become possible to apply compensation algorithms to slim-hole density logs (Scott, 1977). We have found that it is a relatively straightforward process to acquire reliable density data in NX boreholes, using core to establish the calibration and referencing the tools with calibration blocks during routine logging operations. Merkel and Snyder (1977) discuss some of the problems of calibrating a slim-hole density tool.

Sulphide Estimates from Density Logs

Because the sulphide minerals are almost twice as dense as the silicates, it should be possible to estimate the sulphide content using the density log, provided that the silicate grain density is constant and the porosity can be determined from another log.

The bulk density of a rock containing three components — sulphides, water-filled pore space, and nonsulphide matrix — is given by

$$\rho_b = S\rho_s + \Phi\rho_f + (1 - \Phi - S)\rho_g \quad (D-1)$$

where ρ_s, ρ_f, ρ_g are the sulphide, fluid and nonsulphide grain densities, and S, Φ are the sulphide content and pore space as volume fractions.

Rewriting and setting the fluid density $\rho_f = 1$ gives

$$\rho_b = \rho_g + S(\rho_s - \rho_g) + \Phi(1 - \rho_g) \quad (D-2)$$

Expressing the sulphides as weight per cent (S') and the porosity as a volume per cent gives

$$\rho_b = \rho_g + 0.01 \frac{\rho_b}{\rho_s} S'(\rho_s - \rho_g) - 0.01\Phi(\rho_g - 1) \quad (D-3)$$

Solving for S' we get

$$S' = 100 \frac{\rho_b - \rho_g + 0.01\Phi(\rho_g - 1)}{\rho_b(1 - \rho_g/\rho_s)} \quad (D-4)$$

Equation (D-4) is used to compute the sulphide logs and is presented graphically in Figure 12.16. Figure 12.16 allows quantitative inspection of the dependence of S' on Φ, ρ_b and ρ_g. In addition, a sensitivity analysis for S' on the four variables ρ_b, ρ_g, ρ_s and Φ was carried out by evaluating the derivatives with respect to each of the four variables. Selecting base values of $\rho_b = 2.70$, $\rho_g = 2.65$, $\rho_s = 5.0$ and $\Phi = 4\%$ (equivalent to S' = 9.1 wt. %), the following errors in the sulphide estimate result:

Perturbations:	$\Delta\rho_b = 0.01$	$\Delta\rho_g = 0.01$	$\Delta\rho_s = 0.01$	$\Delta\Phi = 1\%$
Error in S' (wt.%)	+0.75	-0.81	-0.02	+1.30

Or, equivalently, the following perturbations are required to produce an error in S' of 1.0 weight per cent:

$$\Delta\rho_b = 0.013 \quad \Delta\rho_g = -0.012 \quad \Delta\rho_s = -0.50 \quad \Delta\Phi = 0.77\%$$

Of the four unknowns, changes in the nonsulphide grain density will probably produce the largest errors because there is at present no way to assess grain density in the borehole. The bulk density measurement itself also can be a problem if the count rate is insufficient.

The third variable, the sulphide grain density, is less critical because the sulphide estimate is less sensitive to its fluctuations. The influence can be further reduced if an estimate of the pyrite-to-chalcopyrite ratio is available and no other sulphides are present. Since the pyrite grain density is 5.02 g/cc and chalcopyrite about 4.2 g/cc, we have

$$\rho_s = 4.2c + 5.02p \quad (D-5)$$

where c and p denote the chalcopyrite and pyrite fractions and c + p = 1.

Calling p/c = R gives

$$c = 1/(1 + R)$$
$$p = 1 - c = R/(1 + R)$$

and

$$\rho_s = \frac{1}{(1 + R)}(4.2 + 5.02R) \quad (D-6)$$

Figure 12.17 is a graph of this function.

Finally, the porosity variations are most easily assessed by using the neutron log to compute a neutron porosity for use in the expression which defines the sulphide estimate. Since the neutron porosity incorporates bound water, the porosity estimate will usually be somewhat high and consequently the sulphide estimate will be high unless bound water is taken into account.

Field Examples of Sulphide Estimation from Density Logging

Figure 12.18 presents an empirical test of the method of sulphide estimation. The data were obtained in a sedimentary sequence penetrated by a quartz latite porphyry. The hole is unusually high in pyrite as shown by the heavy liquids analyses as well as by a visual estimate which checked reasonably well with the laboratory analysis. Magnetite occurs only between 381 to 401 m. The density data were reduced using reasonable values in equation D-4, but unfortunately the new density tool utilized had not been properly calibrated at the time the log was obtained so no

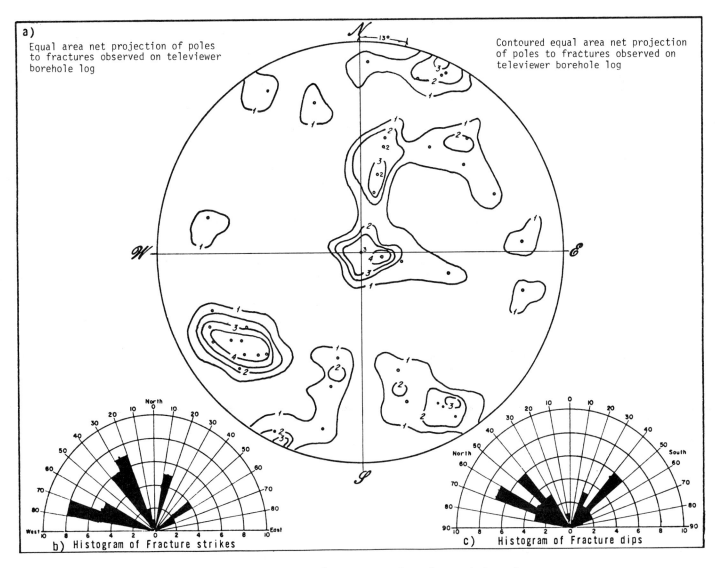

Figure 12.15.   Example of structure analyses from seisviewer log.

scale could be attached to the resulting sulphide estimate. The density-derived sulphide estimate tracks the laboratory values well, and it is reasonable to expect that the agreement can be improved with a proper calibration and judicious use of equation D-4. At least for high sulphide contents (3 to 8 weight per cent), useful estimates of sulphide content should result and the combination of electrically-derived (see next section) and density-derived estimates should be useful.

Figure 12.19 is a second field example which qualitatively illustrates both the sulphides and porosity effects on the density log. The caliper log shows fractures in several intervals but the reader should examine just two, one between 348 to 360 m and the other between 245 to 280 m. The neutron log shows a high porosity for both intervals. The density log shows a lower density for the upper interval which reflects the increased porosity whereas only a part of the lower interval has a lower density. In fact, between 259 m and 265 m the density is relatively high despite the higher porosity indicated by the neutron log. The higher density reflects the galena and sphalerite veining over this interval. A second interval of sulphide veining where no high porosity exists is evident in the density log between 101 m and 131 m. We found that this mode of qualitative log inspection and interpretation is enhanced greatly when several log types are available. Given the present state-of-the-art, we recommend that the suite of logging tools employed in a given hole be as complete as possible.

Other applications of the density tool include 1) control data for the interpretation of gravity surveys, 2) to establish the bulk modulus in combination with acoustic velocity data, and 3) to obtain bulk tonnage estimates.

POROSITY LOGGING

Porosity is commonly estimated from one or more of the following four logs: neutron, gamma-gamma, resistivity and velocity logs. The methods used are well documented in the literature (e.g. Pirson, 1963; Pickett, 1960; Savre and Burke, 1971). However, each log responds to several rock characteristics including porosity. There is an extensive literature describing the response of these logs in various sedimentary rock sequences but little information exists on the response of these tools in igneous or metamorphic rocks (c.f. Ritch, 1975; Nelson and Glenn, 1975; Merkel and Snyder, 1977; Snyder, 1973). One problem is the calibration of tools. No standard facility such as that available for sedimentary rock environments with the API pits at the University of Houston is available for the igneous and metamorphic rock

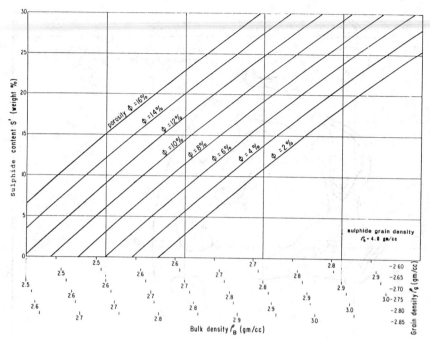

Figure 12.16. Sulphide content versus bulk density, with porosity and grain density varying parameters. Graph is accurate for ρ_g = 2.70, most accurate for other ρ_g at Φ = 8%. Worst S' error is 1.3% for ρ_g = 2.85, Φ = 16%.

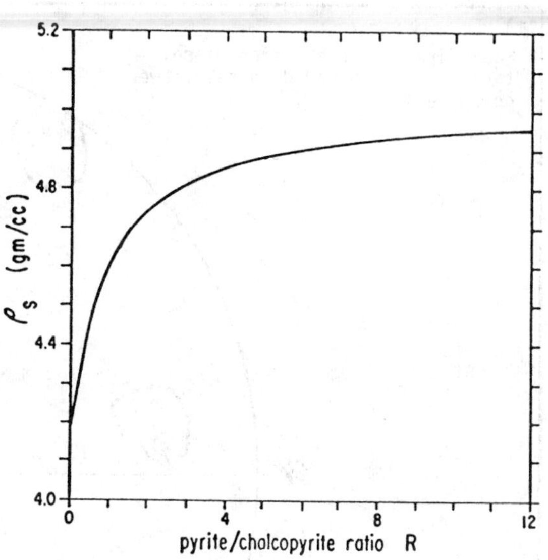

Figure 12.17. Sulphide grain density dependence on pyrite/chalcopyrite ratio R.

environment with its typically very low porosity. The API calibration pits are still used to calibrate tools for the mining industry but it becomes necessary to further calibrate the tools with core analyses. We will discuss tool calibration in more detail in a later section. A second problem is that the response of certain logging tools is significantly influenced by rock properties usually ignored in nonmineral applications.

To be specific, we have found for the low porosity of igneous rocks, typically less than 8 per cent, the water in hydrated minerals accounts for a substantial part and sometimes the total response of the neutron tool (refer to Fig. 12.7). The gamma-gamma density and resistivity log determination of porosity in mineralized rock is significantly affected by the variation in base mineral concentration (Fig. 12.17, 12.18 and 12.19). A porosity determined from either log would be in substantial error. The velocity log is strongly affected by fractures and in low porosity igneous rocks the fractures are as important as "pore" porosity on velocity response (Fig. 12.12). Each of these effects on the four logs has been discussed in previous sections and reference has been made to the appropriate figures in those sections.

Figure 12.20 shows a porosity determination from the velocity and neutron logs. The porosity estimate from the velocity tool is based upon the time-average equation (Wylie et al., 1956). On the basis of substantial laboratory analyses of core for bound water at this mining property we have confidence in uniformly subtracting a 1.35 per cent bound water contribution to the neutron log. The only significant deviation of the two logs occurs over the top 30 m of hole which suggests that the bound water contribution here is less than the 1.35 per cent used to correct the log. The open spike in the porosity determination from the velocity log at 72 m is due to a fractured interval of rock between 70 m and 73 m.

Generally it is difficult to anticipate rock conditions well enough to know which of the four tools will be best suited to estimating porosity and which will be so dominated by another effect that it will be useless. If a porosity estimate is desired, the recommended approach is to utilize all four of the basic tools.

ELECTRICAL PROPERTY LOGGING

Induced polarization (IP) measurements are the most widely applied geophysical exploration technique used in the search for disseminated sulphide mineralization. For this reason alone it is useful to continue the measurement to the subsurface as drilling progresses in order to check the testing of a surface anomaly. Other purposes include the use of electrical properties as a correlation tool (see previous example in Fig. 12.10) and as a direct indicator of the total sulphide content in the borehole. In spite of the obvious benefits, IP logging has not yet been widely applied in exploration programs, partly because of the relative difficulty in carrying out the measurement. Snyder et al. (1977) presented an overview of the IP method in boreholes and also address the difficulties of implementing a reliable system.

The Continuous Resistivity and Induced Polarization Logging Tool of Kennecott Copper Corporation

Kennecott has developed a resistivity-IP logging tool which operates over 1500 m of armoured four-conductor cable and is compatible in every respect with the overall system outlined in Figure 12.4. To overcome the problem of transmitting phase information over long lengths of cable, the signal and reference measurements are made in the downhole sonde, frequency modulated in the sonde, transmitted through the cable, and then demodulated and detected by the surface electronics equipment.

The system provides continuous amplitude and phase information at a nominal logging speed of 9 m per minute. Although the system design does not constrain the operating frequency, continuous sinusoidal transmission and detection at a 10 Hz frequency are used. The phase measurement is

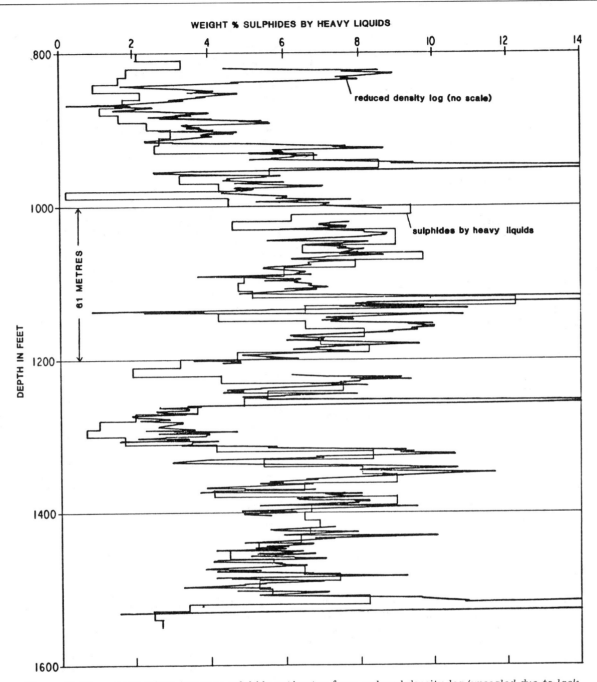

Figure 12.18. Comparison between sulphide estimates from reduced density log (unscaled due to lack of calibration) and total sulphide.

accurate to ±0.1 degree, or about ±2 milliradians, at temperatures up to 77°C. Resistivity is recorded on a logarithmic scale. A unique feature is the ability of the operator to change the downhole amplifier gain from the surface. This is necessary because the phase accuracy is limited to an amplitude range of 35 decibels. Operationally this becomes a nuisance only if the resistivity varies by more than two orders of magnitude within a short vertical distance. With the gain change capability, the dynamic range of the system is 100 decibels at a fixed transmitted current. The ability to control the current level gives another 30 decibels of flexibility. We have made continuous logs in various rock types with resistivities ranging from one ohm-metre to greater than 10 000 ohm-metre (40db). The system can accommodate any electrode array which uses one current electrode on the surface, one current electrode downhole, and two potential electrodes downhole. Most of Kennecott's measurements are made with a one-metre pole-pole (normal) array.

Correction Curves for Borehole IP Measurements

Electrical data from boreholes must be corrected to eliminate the contribution of the borehole fluid. Such borehole correction charts (often called departure curves) for a variety of electrode arrays were computed years ago for resistivity logging (e.g. Schlumberger, 1955). The corresponding chart for induced polarization corrections, which amounts to a derivative of the resistivity chart, was published by Brant et al. (1966). We have recomputed the resistivity

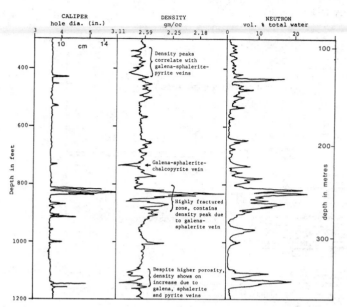

Figure 12.19. Correlation of density log with sulphide veining.

and IP departure curves for the pole-pole array in a borehole of resistivity R_m surrounded by an infinite medium of resistivity R_t (Fig. 12.21 and 12.22). The method of computing the resistivity was taken from Gianzero and Rau (1977). The factor B_2 on Figure 12.21 is

$$B_2 = \frac{R_t}{R_a} \frac{\partial R_a}{\partial R_t}$$

similarly,

$$B_1 = \frac{R_m}{R_a} \frac{\partial R_a}{\partial R_m}$$

where R_a is the apparent resistivity.

A check for numerical accuracy found that $B_1 + B_2$ summed to unity to within three decimal places. The resulting resistivity departure curves checked relatively well with the Schlumberger curves. The IP departure curves checked fairly well with the results of Brant et al. (1966) except for small values of the ratio of electrode spacing to hole diameter, where we believe the curves of Brant et al. to be in error.

Borehole Corrections for Resistivity Measurements

The IP logs can be corrected to give the true or formation value of resistivity if the curves of Figures 12.21 and 12.22 are converted to functions R_t/R_m for the pertinent ratio of electrode spacing to hole diameter, AM/d. One such construction is shown in Figure 12.23 for the case of a one-metre array in an NX borehole. NX boreholes range from 7.6 to 8.3 cm in diameter; we used AM/d = 12.6 for this case.

Conceptually, Figure 12.23 would be used by entering on the right-hand ordinate with the value of $\log(R_a/R_m)$, moving horizontally to the R_a/R_m curve, then downwards to get the value of $\log(R_t/R_m)$. From this point move vertically to intersect the B_2 curve then horizontally to the left to obtain B_2. The phase shift of the rock is then

$$\Phi_t = \Phi_a / B_2$$

In practice, a fifth-order polynomial was fitted to the curves and the logs were corrected to R_t and Φ_t on a digital computer. Note that the correction to the measured phase can be anywhere in the range of 0 to 30 per cent depending upon the resistivity contrast between the rock and the borehole fluid.

This correction procedure requires that the borehole fluid resistivity, R_m, be known. We obtained water samples from the borehole with a water sampler and measured R_m with a fluid conductivity apparatus.

Induced Polarization Logs

Figures 12.24 and 12.25 display IP and resistivity data from two holes containing sulphide mineralization in the southwestern United States. Both logs were obtained in 8.3 cm diameter, water-filled boreholes using the one-metre normal electrode array. DDH-SF1 penetrates andesite and andesite breccia. Sulphide mineralization is pyrite and chalcopyrite, with the pyrite to chalcopyrite ratio ranging from less than one to as high as ten. The borehole fluid resistivity was 14 ohm-metres. Sulphides occur predominantly along veins and also as disseminations.

In Figure 12.24, the erratic nature of the resistivity and phase logs is attributed to the veined character of the sulphides. Estimates of the sulphides content is based upon laboratory analysis (heavy liquid separation) on 3 m composites of the cored samples. These analyses are plotted in the left-hand column on Figure 12.24 and also used to establish the abscissa in Figure 12.26. The data points in Figure 12.26 are 3 m averages of amplitude and phase over the corresponding intervals. The phase values do not display any particular trend as the sulphide content increases. As a result the ratio ϕ/R (also in Fig. 12.26) increases with the sulphide content, roughly following the trendline indicated by the square root of $10\phi/R$. This correspondence is emphasized by overlaying the computed log of $\sqrt{10\phi/R}$ onto the sulphide content in the left-hand column of Figure 12.24, establishing a rough but adequate estimator of the sulphide content based upon the electrical measurements.

The estimator $\sqrt{10\phi/R}$ was first suggested by G.D. Van Voorhis following an exhaustive in-house study at Kennecott Exploration, Inc. of data obtained from measurements on samples from several porphyry copper deposits. The estimator normalizes the polarizability, ϕ, with respect to the resistivity which in turn can vary tremendously depending upon the pore structure of the rock and the nature of the mineralization. The idea is not new, having been previously introduced as the so-called metal factor in frequency-domain IP terminology (see Madden and Cantwell, 1967). What is suggested here, however is that the measure or some variation of it, may be quantitatively useful, providing that the statistical variations can be established and that the exceptions to the rule can be understood. The next example is one such exception.

Figures 12.25 and 12.27 display the results of phase and amplitude measurements in a lithic tuff in drillhole SF2. The sulphides are mostly pyrite occurring as fine disseminations and as blebs, with occasional veins. Magnetite is present only in trace amounts. The borehole fluid is water with a resistivity of 7.1 ohm-metres.

The electrical properties in DDH-SF2 are quite different from those of DDH-SF1. The resistivity is high, ranging between 550 and 1050 ohm-metres and as a result the $\sqrt{10\phi/R}$ estimator is much too low, greatly underestimating the actual sulphide content (see both Fig. 12.25 and 12.27). However the phase in Figure 12.25 is plotted to demonstrate the good correlation between phase and sulphide

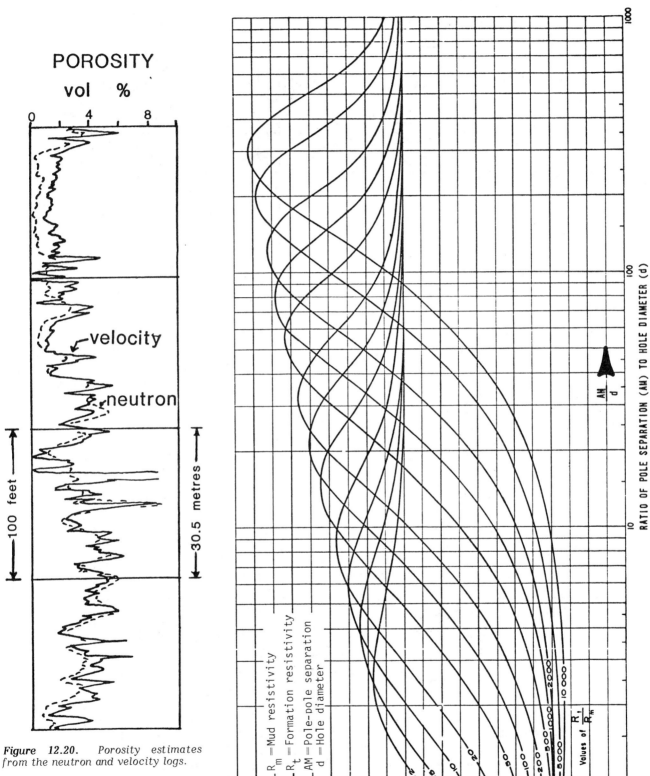

Figure 12.20. Porosity estimates from the neutron and velocity logs.

Figure 12.21. IP departure curves. (Beds of infinite thickness.) Centred pole-pole array.

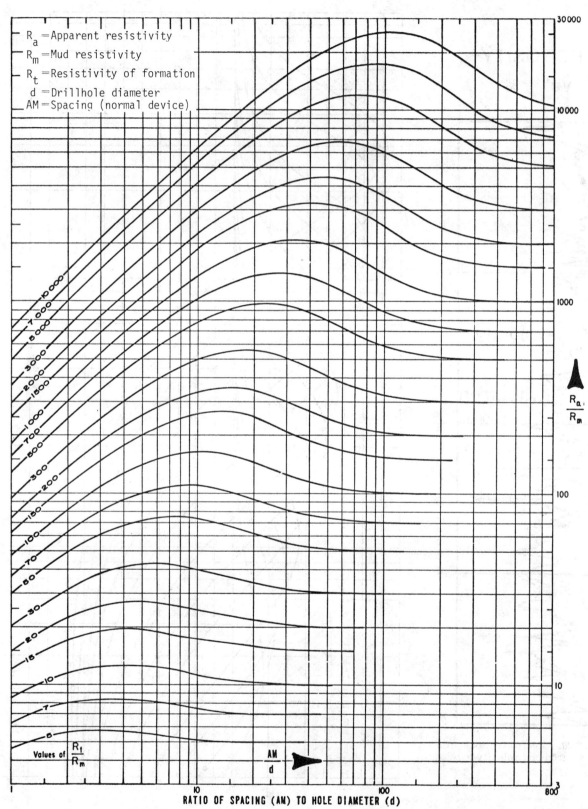

Figure 12.22. Resistivity departure curves. (Beds of infinite thickness.)

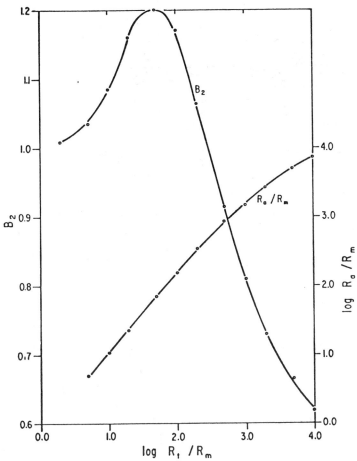

Figure 12.23. Borehole correction factors for the pole-pole array in a borehole penetrating a uniform medium.

content, a correlation which did not occur in DDH-SF1. The same data, replotted on a linear graph show a fairly linear correlation between the phase and sulphide content (Fig. 12.28) with about 15 milliradians of phase shift produced by 1.0 weight per cent sulphides. To emphasize the independence of resistivity from sulphide content, Figure 12.29 displays the porosity computations based upon the neutron log, Φ_N, and upon the resistivity log, Φ_r. The Φ_R trace is based upon Archie's Law using a "cementation exponent" of 2.0 and a pore fluid resistivity of 1.75 ohm-metres. The 2.0 value of the cementation exponent is in accord with the values found by Brace and Orange (1968) for laboratory samples of igneous rocks. The 1.75 value is one-fourth the 7.1 ohm-metres value of the borehole fluid, a fact which can be attributed at least in part to the effects of surface conduction in the rock, as also observed by Brace and Orange. The main point, however, is that the resistivity-determined porosity values are reasonably close to the neutron-porosity, which is not the case when conduction is controlled by the sulphide content.

Dependence of IP Parameters upon Sulphide Content

Although the data base is limited (only three other rock types were examined in the same detail as the two cases discussed above), we propose that the functional dependence of IP parameters upon sulphide content divides into two classes governed by the geometrical distribution of sulphides in rock:

A. Mineralization Predominantly Disseminated

Electrical conduction is controlled by the pore fluids so the resistivity is independent of the sulphide content. Hence the resistivity obeys Archie's Law after the fluid resistivity is modified to account for surface conduction. The phase shift is proportional to the sulphide surface area exposed to the pore fluid, usually manifested by a linear dependence of phase upon sulphide content. Such a dependence of phase upon sulphide content is substantiated by our experience with laboratory measurements on artificial samples and also by prediction from simple mathematical models.

B. Mineralization Predominantly Along Veinlets

In this case, the sulphide mineralization is distributed along intersecting planar features and is sufficiently continuous that electrical conduction is controlled by the sulphide content. Empirically, the resistivity decreases inversely with the square of the sulphide content. The phase is relatively independent of sulphide content, because the abundance of sulphide-electrolyte interfaces does not change much as the sulphide content increases. Empirically, the factor $\sqrt{10\Phi/R}$ or some modification of it provides an estimate of the sulphide content.

These two categories are rather general and are probably not distinct; that is, there will be many instances where the mineralization is equally disseminated and controlled by veining. In fact, using a prototype version of the equipment described herein, Snyder (1973) observed a cube root dependence of the Φ/R factor upon sulphide content, but suggested that the dependence will be a function of rock type and sulphide distribution. The eventual form of a quantitative relation and the determination of its usefulness must await further collection and correlation of empirical data. Because of sampling problems this must be done in the field rather than in the laboratory. Although it is possible to perform surface studies in mines, we believe that borehole methods provide the most expedient means of pursuing this goal.

MAGNETIC SUSCEPTIBILITY LOGGING

The magnetic susceptibility tool, like the induced polarization tool, is more useful in mining applications than in other geotechnical applications. A qualitative use, for rock type identification and correlation, has been demonstrated above (Fig. 12.9). The system used by Kennecott for logging is similar to that described by Broding et al. (1952) and Zablocki (1966).

The magnetic susceptibility tool is also used for quantitative studies in exploration. Figure 12.30 is one example where the magnetic susceptibility measurement in several drillholes yielded the thickness and magnetic susceptibility of the Gila conglomerate (southwestern United States). The volume susceptibility is relatively uniform around 1000 μcgs. These data enabled the exploration geophysicist to improve his aeromagnetic interpretation for this prospect.

A second quantitative use, the direct measurement of the magnetite content has been demonstrated by Snyder (1973) and also in Figure 12.8, but we have not pursued this application any further.

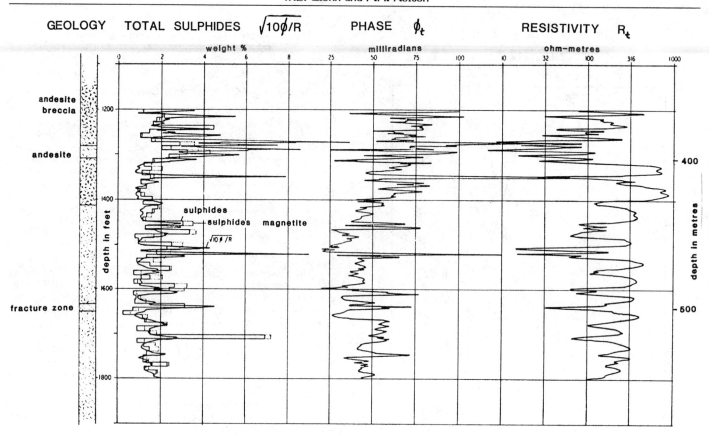

Figure 12.24. *Sulphide data and IP log results corrected for borehole effects in DDH-SF1.*

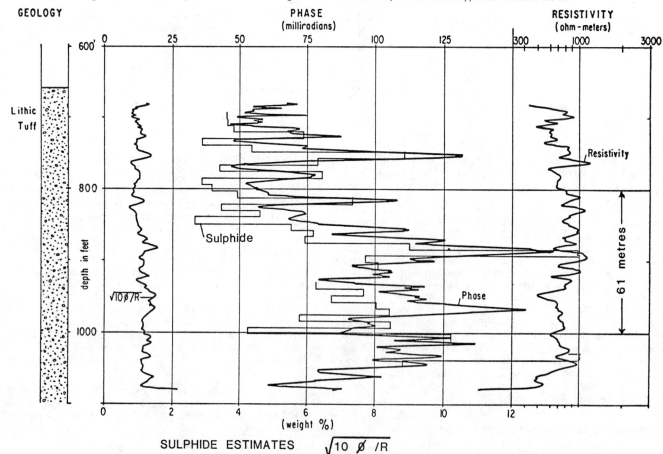

Figure 12.25. *IP logs and sulphide estimates in DDH-SF2.*

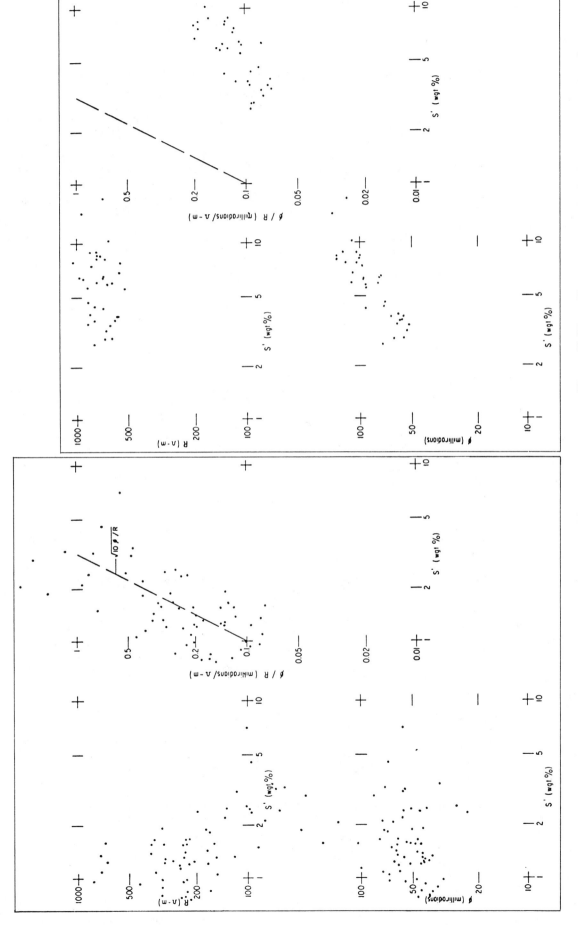

Figure 12.26. IP parameters and sulphide content, DDH-SF1.

Figure 12.27. IP parameters and sulphide content, DDH-SF2.

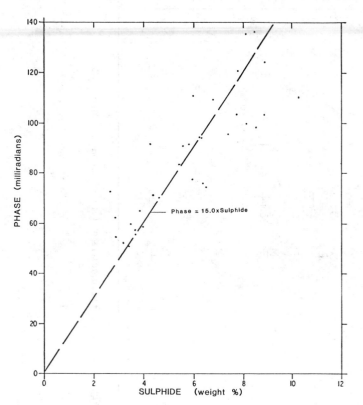

Figure 12.28. Phase values averaged over 10 foot interval versus sulphide content, DDH-SF2.

The tool has been calibrated using magnetic susceptibility measurements on core samples and the calibration is routinely checked with spot core measurements. Figure 12.31 shows an example of this ckeck in an exploration hole in southern Arizona.

TEMPERATURE LOGGING

A temperature log is a continuous record of borehole fluid temperature with depth which will be the geothermal gradient only if the fluid is in equilibrium with the adjacent rocks and if there is no vertical circulation of fluid in the drillhole.

Temperature logs are necessary for the correct interpretation of resistivity logs because of temperature effect on electrical resistance (Alger, 1966; Schlumberger, 1972). The fluid resistivity needs to be corrected for temperature before it is used for porosity analyses or for an estimate of total dissolved solids.

Temperature logs can also be used to examine fluid flow in drillholes (Keys and MacCary, 1971; Wilterholt and Tixier, 1972). One method is to inject cooler surface water into a drillhole for some period of time and obtain temperature logs during injection and at periodic intervals postinjection. These logs are then compared to the preinjection log. The injected fluid will cool the borehole fluid and the zones of fluid exit from the hole. The temperature in the drillhole will recover toward the preinjection temperature profile with zones of fluid exit recovering most slowly. Figure 12.32 illustrates this logging method.

A preinjection and two postinjection temperature profiles are shown in Figure 12.32a and the difference between the post- and pre-injection logs are shown in Figure 12.32b. Also shown are fracture locations picked from the caliper log. The rocks are volcanics intruded by dacite and a number of the contacts show the greatest cooling effects on the temperature logs. The direction of any groundwater flow in these rocks should be strongly influenced by intrusive and volcanic rock contacts.

FLUID FLOW DETERMINATION

In the previous section on temperature logging we noted how the temperature log could indicate points of fluid flow or fluid exit in a drillhole. Some attempt can be made to quantify this flow from the temperature anomalies but the method is very imprecise. The rocks will often show temperature variation beyond the zone of flow and mathematical modelling is not simple (Smith and Steffenson, 1970).

A more direct quantitative measure can be made with a spinner log (c.f. Schlumberger Production Log Interpretation, 1974). However, a spinner tool at best measures flow accurately only above 0.75 m/min. Figure 12.33 shows two flow profiles made in holes used for dump leaching at one of Kennecott's operating properties. The logs show that fluid exit is nonuniform and is exiting primarily out the top portion of the completed hole interval.

CALIBRATION OF LOGGING EQUIPMENT

The American Petroleum Institute has established certain recommended standards for calibration and format of borehole logs; for example, nuclear logs (API, 1974) and electrical logs (API, 1967). API has also established a calibration facility on the University of Houston campus, available to anyone for a set fee. The calibration facility is designed for oil field environments and for nuclear logs. The United States Bureau of Mines has established a calibration facility for the mining and engineering logging applications. This facility provides standards for density, magnetic susceptibility, sonic velocity, resistivity and caliper logs (Snodgrass, 1976) and the facility is available to industry.

Kennecott has calibrated nuclear tools at the API pits and the rest of their tools with core analyses. An example of core calibration of the magnetic susceptibility tool was given in Figure 12.31. Since it is neither practical to calibrate the various tools frequently at the calibration facilities in either Houston or Denver nor to continuously check the calibration with core analyses, Kennecott spot check with core analyses and also verify the tool calibration before and after each log with some fixed reference. For example, the density tool calibration is checked with two blocks of material of known densities before and after each log is made. A geometric array of ferrite rods is used to check the calibration of the magnetic susceptibility tool. The calibration of slim hole tools is not an easy matter and Merkel and Snyder (1977) have discussed this problem in some detail. Waller et al. (1975) examined some basic errors inherent in tool calibration.

CONCLUSIONS

1. Borehole logging methods will gradually become more common in the base metals mining industry for exploration, property assessment and engineering studies. The demand will be a direct function of the ability to obtain in-situ data which is not readily available from core, and the ability to reduce the need for core recovery. In other words, the ability to quantify the elements and minerals which have prime importance to the mining geologist is the key factor in the widespread use of borehole techniques.

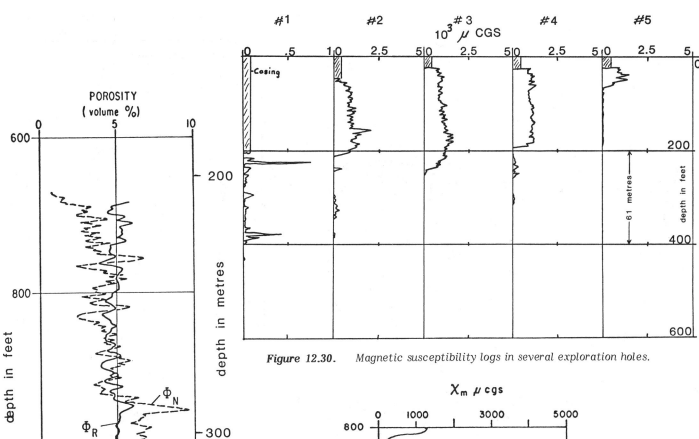

Figure 12.29. Porosity logs in DDH-SF2. Calculated from resistivity using Archie's Law with a 2.0 cementation exponent and a 1.75 ohm-metre mud resistivity. The Φ_N log is computed from the neutron log.

Figure 12.30. Magnetic susceptibility logs in several exploration holes.

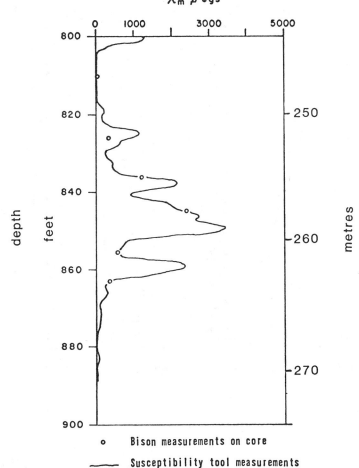

Figure 12.31. Comparison of magnetic susceptibility measurements made in drillhole BB-1 and on BB-1 core samples from southern Arizona.

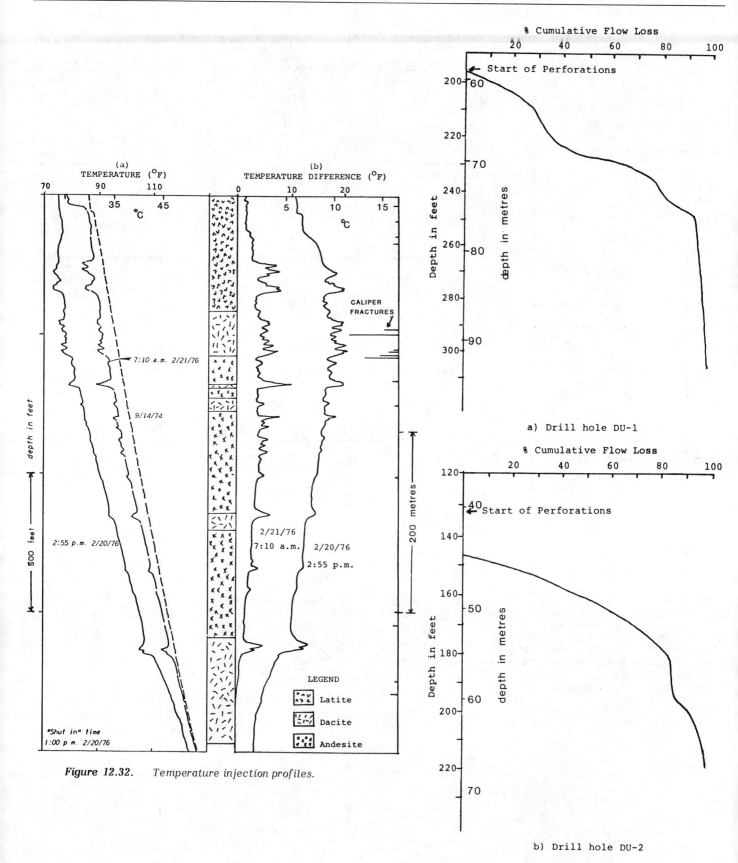

Figure 12.32. Temperature injection profiles.

Figure 12.33. Cumulative per cent fluid loss with depth in two mine dump holes.

2. In general, the added information which can be gleaned from a combination of logs instead of just one or two is well worth the added cost of acquisition. This is true whether the application is a qualitative one such as rock type and correlation, or a quantitative goal such as estimation of sulphide content.

3. With present technology, it is now possible in the base metal environment to make quantitative estimates of porosity, fracture intensity, bulk density, magnetite content, potassium content, and mineralogically bound water. We feel that with some additional effort and care it will soon be possible to add sulphide content to this list.

4. As pointed out by Snyder (1973) these measurements have particular application in the porphyry copper environment where drillholes are often deep and costly. We emphasize that the logs have enhanced value when they are applied on a deposit-wide basis and are integrated into geological and geochemical studies.

5. We agree with an observation made by Dyck et al. (1975) that there is a need for more case history documentation and published empirical relationships for the mining environment such as exists for a wide range of logging applications in the oil and gas industry.

ACKNOWLEDGMENTS

D.D. Snyder initiated the Kennecott logging program in 1972 and formulated many of the practices and designs, including the initial design of the continuous IP logger. D. Purvance performed the numerical computations for the IP departure curves. E.P. Baumgartner acquired and interpreted several of the logs presented in this paper. Dale Green, Roy Lenk, Jim Simmons and Dave Ewing contributed significantly to the development and maintenance of the logging systems. Jonathan Jackson assisted in the recording and analysis of the spinner logs. We thank Kennecott Copper Corporation for permission to publish the material in this paper.

REFERENCES

Aamodt, R.L.
1976: in LASL Hot Dry Rock Geothermal Project July 1, 1975 to June 30, 1976, compiled by A.G. Blair, Jr., W. Tester and J.J. Mortensen; Los Alomas pub. LA-6525-PR.

Alger, R.P.
1966: Interpretation of electric logs in fresh water wells in unconsolidated formations; SPWLA 7th Ann. Logging Symp., Tulsa, Oklahoma, May.

American Petroleum Institute
1967: Recommended practice and standard form for electric logs; API RP 31, 3rd edition.
1974: Recommended practice for standard calibration and form for nuclear logs; 3rd edition.

Brace, W.F. and Orange, A.F.
1968: Further studies of the effects of pressure on electrical resistivity of rocks; J. Geophys. Res., v. 74 (16).

Brant, A.T. and the Newmont Exploration Staff
1966: Examples of induced polarization field results in the time domain; Mining Geophysics, v. I, Soc. Explor. Geophys., Tulsa, Oklahoma.

Broding, R.A., Zimmerman, C.W., Somers, E.V., Wilkelm, E.S., and Stripling, A.A.
1952: Magnetic well logging; Geophysics, v. 17 (1), p. 1-26.

Clavier, C., Heim, A., and Scala, C.
1976: Effect of pyrite on resistivity and other logging measurements; SPWLA 17th Annual Logging Symposium, June.

Conaway, J.G.
1977: Deconvolution of temperature gradient logs; Geophysics, v. 42 (4), p. 823-837.

Czubek, J.A.
1979: Modern trends in mining geophysics and nuclear borehole-logging methods for mineral exploration; in Geophysics and Geochemistry in the Search for metallic ores; Peter J. Hood, ed., Geol. Surv. Can., Econ. Geol. Rep. 31, Paper 11.

Daniels, J.J.
1977: Interpretation of buried electrode resistivity data using a layered earth model; submitted to Geophysics for publication.

Dyck, A.V., Hood, P.J., Hunter, J.A., Killeen, P.G., Overton, A., Jessop, A.M., and Judge, A.S.
1975: Borehole geophysics applied to metallic mineral prospecting; A review; Geol. Surv. Can., Paper 75-31.

Fishel, K.
1976: Applications of electric logs in Appalachian coal exploration; paper presented at 1976 annual meeting of Geol. Soc. Am., Denver, Colorado.

Gianzero, S.C. and Rau, R.
1977: The effect of sonde position in the hole on responses of resistivity logging tools; Geophysics, v. 42 (3), p. 642-654.

Hohmann, G.W., Nelson, P.H., and Van Voorhis, G.D.
1977: Field applications of a vector EM system; presented at 47th annual international meeting and exposition of Society of Exploration Geophysicists, Calgary.

Keyes, W.S. and MacCary, L.M.
1971: Application of borehole geophysics to water resources investigations; Techniques of Water-Resources Investigations of the United States Geological Survey, Book 2, chapter E1.

Killeen, P.G.
1979: Gamma-ray spectrometric methods in uranium exploration: application and interpretation; in Geophysics and geochemistry in the Search for metallic ores; Peter J. Hood, ed., Geol. Surv. Can., Econ. Geol. Rep. 31, Paper 10C.

Lytle, R.J., Lager, D.L., Laine, E.F., and Salisbury, J.D.
1976: Monitoring fluid flow by using high frequency electromagnetic probing; Lawrence Livermore Laboratory pub. UCRL-51979.

Madden, T.R. and Cantwell, T.
1967: Induced polarization, a review; Mining Geophysics, v. II, Society of Exploration Geophysicists, Tulsa, Oklahoma.

Merkle, R.H. and Snyder, D.D.
1977: Application of calibrated slim hole logging tools to quantitative formation evaluation; SPWLA 18th Annual Logging Symposium, June.

Nargolwalla, S.S., Rehman, A., St. John-Smith, B., and Legrady, O.
1974: In situ borehole logging for lateritic nickel deposits by neutron capture-prompt gamma ray measurements; Scintrex Limited, Concord, Ontario, Canada.

Nelson, P.H. and Glenn, W.E.
    1975: Influence of bound water on the neutron log in mineralized igneous rocks; SPWLA 16th Annual Logging Symposium, June.

Patchett, J.
    1975: An investigation of shale conductivity; SPWLA 16th Annual Logging Symposium, June.

Pickett, G.R.
    1960: The use of acoustic logs in the evaluation of sandstone reservoirs; Geophysics, v. 25 (1), p. 250-274.

Pirson, S.J.
    1963: Handbook of well log analysis; Prentice-Hall, Inc., Englewood Cliffs, New Jersey.

Ritch, H.J.
    1975: An open hole logging evaluation in metamorphic rocks; SPWLA 16th Annual Logging Symposium, June.

Savre, W.C. and Burke, J.A.
    1971: Determination of true porosity and mineral composition in complex lithologies with the use of sonic, neutron and density surveys; SPE Reprint Series No. 1, p. 306-341.

Schlumberger
    1955: Resistivity Departure Curves, Document Number 7, Schlumberger Ltd., New York.

    1972: Log interpretation; v. 1 – Principles; Schlumberger Ltd., New York.

    1974: Production log interpretation; Schlumberger Ltd., New York.

Scott, J.H.
    1977: Borehole compensation algorithms for a small-diameter, dual-detector density well-logging probe; SPWLA Trans. 18th Annual Logging Symposium, June.

Seigel, H.O. and Nargolwalla, S.S.
    1975: Nuclear logging systems obtains bulk samples from small boreholes; Eng. Mining J., v. 176 (8), p. 101-103.

Smith, R.C. and Steffensen, R.J.
    1970: Computer study of factors affecting temperature profiles in water injection wells; J. Petrol. Tech., November.

Snodgrass, J.J.
    1976: Calibration models for geophysical borehole logging; U.S. Bureau of Mines, Report of Investigations 8148, p. 21.

Snyder, D.D.
    1973: Characterization of porphyry copper deposits using well-logs; Geophysics, v. 38 (6), p. 1221-1222.

Snyder, D.D., Merkel, R.H., and Williams, J.T.
    1977: Complex formation resistivity, the forgotten half of the resistivity log; SPWLA 18th Annual Logging Symposium, June.

Staff, Scintrex Limited
    1976: Application of the Metalog System for the grade determination of copper in a porphyry deposit; Scintrex Ltd., Concord, Ontario, Canada, p. 36.

Van Schalkwyk, A.M.
    1976: Rock engineering testing in exploratory boreholes; v. 1 of Exploration for Rock Engineering, edited by Z.T. Bieniawski, A.A. Balkema.

Waller, W.C., Cram, M.E., and Hall, J.E.
    1975: Mechanics of log calibration; SPWLA 16th Annual Logging Symposium, June.

Wilterholt, E.J. and Tixier, M.P.
    1972: Temperature logging in injection wells; AIME, SPE-4022, p. 1-11.

Wylie, M.R.J., Gregory, A.R., and Gardner, L.W.
    1956: Elastic wave velocities in heterogeneous and porous media; Geophysics, v. 21 (1), p. 41-70.

Zablocki, C.J.
    1966: Electrical properties of some iron formations and adjacent rocks in the Lake Superior region; Mining Geophysics, v. I, Society of Exploration Geophysicists, Tulsa, Oklahoma.

Zamenek, J., Glenn, E.E., Norton, L.J., and Caldwell, R.L.
    1970: Formation evaluation by inspection with borehole televiewer; Geophysics, v. 35 (2), p. 254-269.

FOCUS ON THE USE OF SOILS FOR GEOCHEMICAL EXPLORATION IN GLACIATED TERRANE

B. Bølviken
Geological Survey of Norway, Trondheim, Norway

C.F. Gleeson
C.F. Gleeson and Associates Ltd., Ottawa, Canada

Bølviken, B. and Gleeson, C.F., Focus on the use of soils for geochemical exploration in glaciated terrane; in Geophysics and Geochemistry in the Search for Metallic Ores; Peter J. Hood, editor; Geological Survey of Canada, Economic Geology Report 31, p. 295-326, 1979.

Abstract

Large portions of the Earth's surface have been glaciated several times during the last two million years. The overburden in these areas is made up of glacial drift, which has been laid down by the action of glaciers and their meltwaters, and thereafter subjected to postglacial processes. In glacial terrane, therefore, geochemical dispersion can be divided into two main classes, (1) syngenetic dispersion, i.e. principally mechanical or particulate dispersal which took place during glaciation and (2) epigenetic dispersion, i.e. chemical and mechanical dispersion which has taken place after glaciation. In combination these processes may result in intricate geochemical dispersion patterns and anomalies that are difficult to interpret. The sampling and analytical methods used should, therefore, be those which will disclose anomalies that are genetically not too complex.

Interpretation of syngenetic patterns presupposes a thorough knowledge of the glacial history. To obtain meaningful results it is frequently necessary to sample tills in section to the bedrock surface. This often requires heavy equipment, and sampling costs may be relatively high. The analytical methods used should employ rigorous chemical digestion as well as mineralogical determination of resistant minerals.

Epigenetic dispersion patterns in glacial overburden can be produced downslope due to metal dispersion in groundwater, or immediately over the bedrock source due to capillary forces, biological activity or gaseous movement of volatile compounds. Mineral deposits in contact with groundwater may act as natural galvanic cells which may result in electrochemical dispersion of metal into the overlying glacial drift. Epigenetic dispersion patterns may be detected in near-surface soils at relatively low sampling costs and by weak chemical extraction.

Empirical evidence supporting these principles is provided by published and unpublished data. This paper reviews those data that have appeared in the western literature during the last decade, the intention being to outline the present state of art in utilizing analysis of soil samples as an exploration tool in a glacial terrane.

If we are to advance the applicability of soil geochemistry in glaciated environments, more research into dispersion mechanisms is required.

Résumé

Au cours des deux derniers millions d'années, les glaces ont recouvert plusieurs fois de grandes parties de la terre. Dans ces régions, les mort-terrains sont constitués d'alluvions glaciaires qui se sont accumulées sous l'effet du déplacement des glaciers et de l'écoulement de leurs eaux de fonte, pour ensuite être soumises aux phénomènes postglaciaires. Dans les terrains glaciaires, on peut donc diviser la dispersion géochimique en deux grandes catégories: 1) la dispersion syngénétique c.-à-d. surtout mécanique ou dispersion des particules survenue au cours de la glaciation et 2) la dispersion épigénétique, c.-à-d. la dispersion chimique et mécanique survenue après la glaciation. Ces phénomènes combinés peuvent donner des modes compliqués de dispersion géochimique et il est difficile d'interpréter les anomalies. Il faut donc avoir recours aux méthodes d'analyse et d'échantillonnage qui permettent de déceler les anomalies dont la complexité n'est pas trop grande sur le plan génétique.

L'interprétation des configurations syngénétiques présuppose une bonne connaissance de l'histoire de la glaciation. Pour obtenir d'excellents résultats, il est souvent nécessaire de prélever des échantillons de till par section jusqu'à la roche en place. Il faut souvent du matériel lourd pour exécuter les travaux, et les coûts de l'échantillonnage peuvent être relativement élevés. Les méthodes analytiques utilisées doivent comporter une digestion chimique rigoureuse et une détermination minéralogique du minerai résistant.

La dispersion épigénétique dans les morts-terrains des glaciers peut se produire vers le bas en raison de la dispersion des métaux dans les eaux souterraines ou immédiatement au-dessus de la roche en place à cause des forces capillaires, de l'activité biologique et des déplacements gazeux de composés volatiles. Les gisements de minerai en contact avec les eaux souterraines peuvent constituer des piles électriques naturelles qui provoquent éventuellement la dispersion électro-chimique des métaux dans les alluvions glaciaires sus-jacentes. La dispersion épigénétique peut être détectée dans le sol près de la surface, au moyen d'un échantillonnage relativement peu coûteux et d'une faible extraction chimique.

Ces principes sont justifiés par les preuves empiriques que constituent d'innombrables données tant publiées qu'inédites. La présente étude a pour objet d'examiner les données fournies dans les revues scientifiques publiées en Occident au cours de la dernière décennie. Il s'agit d'exposer la situation en utilisant l'analyse des échantillons de sol comme instrument d'exploration dans les terrains glaciaires.

Si nous voulons, dans l'avenir, faire progresser l'application de la géochimie des sols dans les milieux glaciaires, il faudra intensifier la recherche sur les mécanismes de dispersion.

INTRODUCTION

In geochemical exploration the word "soil" is commonly used as a synonym for overburden, although in a stricter sense it refers to the upper layered part of the regolith. In this paper the term will be applied to all or part of the earthy unconsolidated material overlying bedrock. Soil sampling will include any type of overburden sampling except sampling of recent stream sediments and other drainage channel material.

Soils can be classified into residual and glacial. Residual soils accumulate in place as a result of weathering of the underlying rock (e.g. tropical soils). Glacial soils or glacial drift, however, are the result of the action of glaciers and their meltwaters on fresh or weathered bedrock and preglacial soils. The use of soils in geochemical exploration in glaciated areas, therefore, creates special problems distinct from those in the tropics and other areas covered by residual soils.

Large parts of the continents have been glaciated several times during the last two million years of the Earth's history. The glaciated regions include nearly all Canada, large parts of the United States north of 40°N (New York), most of Europe north of 50°N (London) and extensive regions of Asia, in particular north of 60°N (Leningrad) (Fig. 13.1). These glaciated regions include some of those parts of the world that are being most actively explored for mineral deposits. This fact is reflected in the large number of publications describing the use of soils for prospecting in glaciated areas that have been published in recent years.

This paper summarizes trends from these works. Our main sources have been Exploration Geochemistry Bibliographies published by The Association of Exploration Geochemists (Hawkes, 1972, 1976), textbooks (Hawkes and Webb, 1962; Levinson, 1974), monographs and symposia proceedings (Cameron, 1967; Kvalheim, 1967; Canney, 1969; Boyle and McGerrigle, 1971; Bradshaw et al., 1972; Jones, 1973a, 1973b, 1975, 1977; Bradshaw, 1975a; Elliott and Fletcher, 1975; Govett, 1976b; Kauranne, 1976a; Legget, 1976), and articles in scientific journals in particular the Journal of Geochemical Exploration. Much of the information on glacial deposits and processes have been taken from Flint (1971); Goldthwait (1971a); Nickol and Björklund (1973) and Dreimanis (1976).

In glaciated areas, geochemical dispersion from a bedrock source to the soils, can be divided into two separate classes, (1) syngenetic dispersion, i.e. principally mechanical or particulate dispersal which took place during the glaciation, and (2) epigenetic dispersion, i.e. chemical and mechanical dispersions which have been taking place during the postglacial period. Any geochemical dispersion pattern presently found in glacial soils will, in principle, be a result of these two types of processes. To interpret the geochemical results of soil surveys in glaciated terrane, it is important to have a thorough understanding of both the syngenetic and epigenetic dispersion processes. A brief account of these two types of processes are given in the following two sections.

SYNGENETIC DISPERSION

Syngenetic geochemical dispersion patterns in glacial soils are formed contemporaneously with the glacial deposits, that is, they are a result of the process of glaciation.

Process of Glaciation

The climate has changed several times during the Quaternary, and in cooler periods there have been extensive glaciations (Flint, 1971). From one or more central areas, the glaciers flowed outwards and towards the sea, the general movement being more or less independent of small variations in the local topography. In most cases only the very high peaks escaped glaciation. On the lowlands, the ice moved as a sheet or spread out as piedmont glaciers. The rate of movement varied considerably over short distances, especially in the marginal zones where subglacial valleys caused locally high velocities. Through abrasion and quarrying, the glaciers eroded and dispersed bedrock and proglacial sediments in fragments ranging from fine powder to large boulders. As the climate gradually became more temperate, a point was reached where the lowland ablation of the glaciers (melting, evaporation and calving) occurred at a faster rate than the accumulation of ice and snow. The glacier front then started to retreat leaving behind a great variety of glacial deposits, collectively called glacial drift.

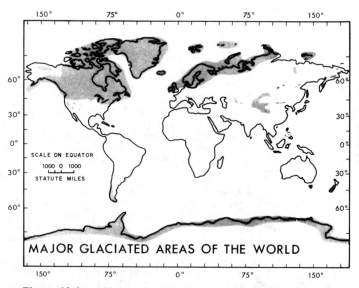

Figure 13.1. Major glaciated areas of the world. After Flint (1971).

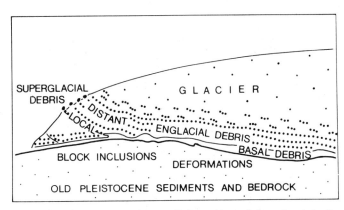

Figure 13.2. Schematical radial section of ice sheet in its terminal zone. After Dreimanis (1976).

According to Dreimanis (1976), glacial drift may be transported by the glacier as (1) basal, (2) englacial and (3) superglacial debris (Fig. 13.2).

1. The basal 1-3 m of a glacier contains most of the rock debris carried by a glacier. This is the main zone of comminution and by crushing and abrasive action the basally transported debris becomes comminuted to its constituent minerals with less than 0.1 per cent of any lithology of clasts surviving beyond 35 km (Goldthwait, 1971b).

2. Compressive flow may move debris upward from basal into englacial positions. Englacial debris and fragments can be observed up to 30 m above the base of a glacier. If englacial debris is transported more or less parallel to the base of a glacier it may survive transport for hundreds of kilometres.

3. Superglacial drift is derived in part, from rock debris falling onto the surface of the glacier and from the transport of englacial debris into the superglacial position at the glacial terminus. Rock may withstand superglacial transport for hundreds of kilometres.

The material carried by the glacier may be reworked and redistributed by meltwater adjacent to the terminus of the glacier.

During the last stages of glaciation, removal of glacier ice resulted in release of impounded proglacial lake waters and in isostatic raising of areas of marine submergence, exposing formerly hidden glacial deposits at the surface.

Classification of Glacial Drift

Based on its sedimentological character glacial drift can be divided into two main groups (1) till and (2) stratified drift (Dreimanis, 1976).

Till is that part of the drift which is nonstratified, poorly sorted and closely packed with a variety of rock and mineral fragments of different sizes.

Stratified drift are sorted and layered sediments showing signs of being deposited in or redistributed by water.

For geochemical exploration purposes, classification of drift based on genesis is of greatest interest, because such classification is the key to tracing syngenetic geochemical anomalies in the drift back to the bedrock source.

Till can be classified genetically into two main groups as shown in Figure 13.3 taken from Dreimanis (1976).

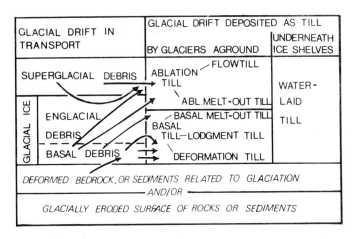

Figure 13.3. Classification of tills. After Dreimanis (1976).

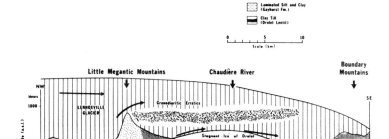

Figure 13.4. Diagrammatic cross-section of glacier showing how englacial debris can form ablation mantle to till. After Shilts (1973a).

Basal till (Lodgment till) is chiefly made up of basal debris (Fig. 13.2). The transportation length could be from a few centimetres up to several kilometres.

Ablation till originates mainly from englacial and superglacial debris (Fig. 13.2) and may have been transported long or short distances depending on circumstances (Shilts, 1976).

If a glacier advances over relatively flat terrain, particles apparently work their way up through the glacier. In this case the ablation till will normally be transported much longer than the basal till.

In mountainous areas where the glacier is flowing through valleys or over peaks, debris may slide or roll on to the ice from adjacent slopes, or be supplied from peaks throughout the thickness of the ice (see Fig. 13.4). In such cases ablation till may be locally derived and contrast sharply in composition with the basal till, which in valleys may be distal. If both basal and ablation till are derived locally, the ablation facies may reflect the resistant rock types that form hills, whereas the basal facies usually reflect softer rocks that form valleys. In all cases the density of sediment carried in the englacial or superglacial mode is less than that carried basally. Ablation till, therefore, has undergone less severe comminution than basal till. Ablation till may also have been reworked by meltwater to a greater extent than basal till. Consequently, ablation till is normally coarser than basal till.

Stratified drift can be divided into (1) ice-contact stratified drift and (2) proglacial sediment (Flint, 1971).

Ice-contact stratified drift deposits are derived from till by partial reworking and redistribution by water adjacent to the terminus of the glacier; esker materials belong to this type.

Proglacial sediments have been transported by meltwater from glaciers and have been deposited at varying distances beyond the ice margin. Glaciolacustrine and glaciomarine sediments belong to this type.

Glacial Epochs

In most areas, the climatic variations have resulted in several advances and retreats of the glaciers. Even though most of the present day landform and glacial deposits were formed during the last glaciation, preglacial or interglacial sediments would, to a varying degree, be intermixed or interbedded with these deposits. Indeed, in many cases older sediments may repeatedly have survived younger glaciations. Glacial deposits formed during the last glaciation, may therefore be underlain by older glacial sediments or preglacial weathering products (Dreimanis, 1976).

Shilts (1975a) has shown that concentrations of ultrabasic components in a till section may change abruptly at certain levels, indicating till sheets of differing provenance (Fig. 13.5). Other descriptions of glacial-event stratigraphy in Canada with relevance to geochemical exploration are given by Alley and Slatt (1976) and Shilts (1976).

In parts of Finland, glacial transport and till stratigraphy have been thoroughly investigated by Kujansuu (1967, 1976), Korpela (1969), Hirvas et al. (1976), Kokkola (1975), Kokkola and Korkalo (1976) and Hirvas (1977).

As reported by Hirvas (1977), observations in 1288 test pits in northern Finland indicated that there have been at least five different episodes of glacial transport that differ in age and direction of ice flow (Fig. 13.6). Various till beds occur that correspond to each of these five stages (Fig. 13.7). The distance travelled by various components in the different till sheets varies, and no formula could be given for the transport distances applicable to the whole of northern Finland.

These findings concerning overburden stratigraphy have important implications when applied to the use of till and other types of glacial drift as sample media in geochemical surveys. To obtain interpretable syngenetic dispersion patterns it is essential that the samples be collected consistently from the same strata.

Glacial Transport

Shilts (1976) stated; "Glaciers appear to disperse material in the form of a negative exponential curve, with the concentration of elements, minerals or rocks reaching a peak in till at or close to the source, followed by an exponential decline in the direction of transport. The parameters of the curve appear to be determined by the physical characteristics of the components and the mode of transport". Shilts then introduced the terms (1) head and (2) tail of the dispersal curve signifying (1) rapid decrease in metal concentration immediately down-ice of the source and (2) a gradual decrease at greater distances (see Fig. 13.8). The tail would normally be the target in reconnaissance sampling, and as such it is often difficult to detect due to low metal contrast in the till. The head, however, shows much better contrast, but needs denser sampling to be detected.

The size of the indicator train is more or less proportional to the area and orientation of the source exposed to glacial erosion (Holmes, 1952; Dreimanis and Vagners, 1969 and Shilts, 1976). Shilts also pointed out that there are two additional factors that determine whether an indicator train can be detected: (1) how easily and accurately the component can be determined with the analytical methods available, and (2) how distinct the component is, i.e. how common the component is in barren rocks of the dispersal area.

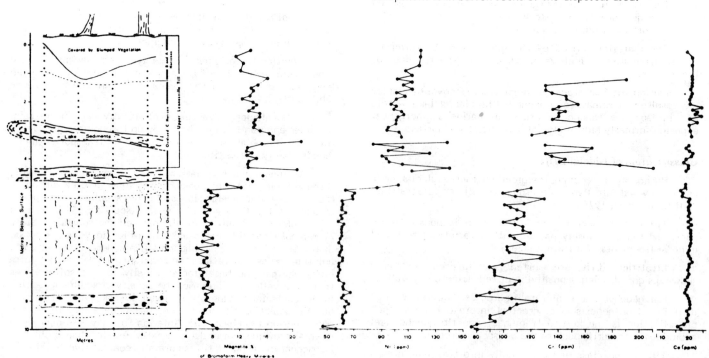

Figure 13.5. Ultrabasic components in multi-till stratigraphic section showing how concentrations can change abruptly at certain levels. After Shilts (1975a).

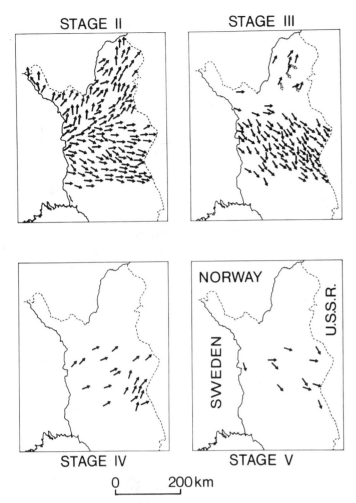

Figure 13.6. Four of the five stages of glacial transport found in northern Finland. After Hirvas (1977).

Pollock et al. (1960), Geoffroy and Koulomzine (1960), Lee (1963, 1965), Wennerwirta (1967 and 1968), Karup-Møller and Brummer (1970), Muntanen (1971), Shilts (1973a, b, 1975a, 1976), Freeman and Ferguson (1973), Nichol and Bjørklund (1973), Szabo et al. (1975) and DiLabio (1976).

It must be remembered that these dispersal trains are three-dimensional; they may surface close to the subcrop of the source or they may come to surface some distance down-glacier from the bedrock origin. Hence to define this third dimension it is necessary to sample the trains in section. The picture may become more complicated if the till is subsequently covered by later glacial deposits such as tills,

Information on transport direction may be obtained from a variety of glacial features including: (1) striations, crescentic marks and ice flow forms; (2) indicator trains and (3) fabrics.

Striations and crescentic marks are grouped by Flint (1971) as small-scale features of glacial abrasion. They are abraded surfaces formed by material at the base of a glacier grinding to the bedrock surface. Most striations and grooves indicate the orientation but not the sense of glacial movements. However, "rat tail" striations, that is tails of soft material in the lee of small, hard inclusions in the bedrock, do indicate the sense as well as azimuth of movement. Crescentic marks gouged out of bedrock are formed at right angles to the direction of glacial movement.

Large ice-flow forms include stoss and lee topography which may consist of knobs or small hills of bedrock each with a gentle abraded surface on the upglacier (stoss) surface and a steeper, rougher, quarried slope on the lee side. Streamlined molded forms such as drumlins have their long axes oriented approximately parallel to the main flow direction of the ice movement.

Typical indicator trains are finger- or ribbon-shaped and sometimes fanlike. They have been used extensively as a reliable method of determining the direction of glacial movement. Flint (1971) presented several examples from North America and Scandinavia, and the results of other studies can be found in articles by: Sauramo (1924), Grip (1953), Okko and Peltola (1958), Dreimanis (1958 and 1960),

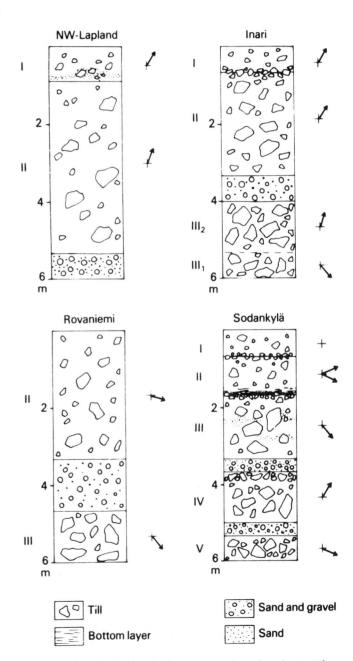

Figure 13.7. Examples of till stratigraphy in northern Finland. Roman numerals to the left of the sections correspond to stages of glacial transport in Figure 13.6. Arrows indicate average direction of glacial flow (north at the top). After Hirvas (1977).

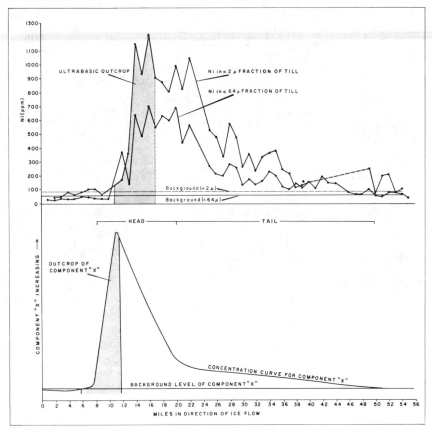

Figure 13.8. Example of actually found (top) and idealized (bottom) glacial dispersal curves showing relationship of the head and tail of the negative exponential curve. After Shilts (1976).

glaciolacustrine, fluvial and bog deposits. In such cases there may be no chemical or physical surface expression of the presence of the concealed metal zone, and the only practical geochemical way to search for the indicator train is to section sample by drilling the lodgment till under the superposed overburden.

Fabric studies may be used in determining the direction of glacial flow. The fabric of till results from the arrangement of its component rock particles; these are often organized so that the majority of clasts lie with their long axis parallel to the direction of flow, while a smaller number lie transverse to this direction (Flint, 1971). Elongated pebbles may dip up-ice; recording of their orientation in till sheets may, therefore, indicate both the sense and azimuth of the movement of the glacier that deposited them. This way of determining the direction of glacial flow was systematized by Krumbein (1939) and Holmes (1941), and has later been used by many workers; the results reported by Toverud (1977) are reproduced in Figure 13.9. Puranen (1977) compared data from such till fabric studies (which is rather laborious) with results obtained by simple measurements of the anisotropy of magnetic susceptibility in oriented till samples. He showed that the estimated magnetite grain orientation in most cases agreed with grain orientation determined by conventional orientation analysis. Puranen also found that measurements of the magnetite content or bulk susceptibility of till could be used to trace magnetite dispersion trains from a bedrock source, and thus may prove to be a valuable method for other aspects of glacial transport studies.

Syngenetic Geochemical Dispersion Patterns in Glacial drift

During glaciation chemical elements were (a) released from their preglacial locations; (b) transported by various mechanisms a short or long distance depending on the circumstances; and (c) deposited at a postglacial position in glacial drift. The resulting geochemical dispersion patterns can be classified according to the type of glacial drift in which they occur. In the following pages various types of syngenetic geochemical dispersion patterns are grouped into (1) anomalies in till, and (2) anomalies in stratified drift.

Syngenetic Anomalies in Till

Till is the most common glacial sediment used for drift prospecting because according to Shilts (1975b):

1. Till is the most widespread type of glacial drift;
2. till sheets can easily be related to specific directions of ice movement;
3. unweathered till usually represents groundup, fresh bedrock;
4. till is easily recognizable; and
5. dispersal trains are larger than the source of the trains.

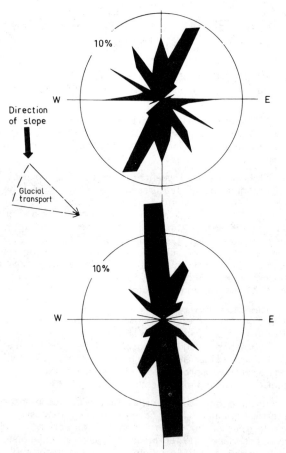

Figure 13.9. Examples from northern Sweden of diagrams showing particle orientation in till. After Toverud (1977).

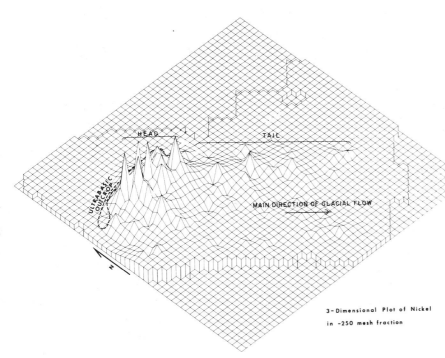

Figure 13.10. Perspective plot of nickel values in the <64μm fraction of surface till in the Thetford Mines ultrabasic indicator train, showing graphically the head and tail regions of dispersal. The tail is traceable for about 60 km in the main direction of glacial flow. After Shilts (1976).

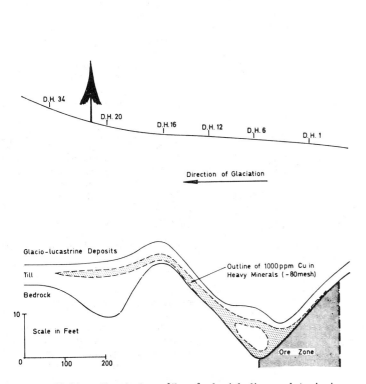

Figure 13.11. Vertical profile of glacial dispersal train in till under glaciolacustrine deposits at Louvem, Val d'Or, Quebec. After Garrett (1971).

Traditionally basal till has been most extensively used as a sampling medium but, as pointed out by Shilts (1976), ablation till could in some cases also be used provided sampling of the same type of material is possible throughout the survey area.

It is important at an early stage in a mineral exploration program to recognize if one is dealing with a glacial fan. Misinterpretation of geochemical results has occurred all too often in the past because of the inability of geologists to recognize such dispersion patterns. DiLabio (1976) has cited one example of such an error which led to the Icon copper deposit north of Chibougamau, Quebec being missed. The ore zone lay undetected for nine years after the initial geochemical soil survey and drilling program was done on the down-ice extension of a geochemical copper indicator fan (i.e. the tail). Two wrong assumptions were made in the interpretation of the geochemical results: (1) It was assumed that the copper soil anomaly was in place when in fact it formed part of a dispersal train; and (2) It was assumed that the source of the anomaly had a steep dip when in fact it was a shallow dipping orebody. In time Icon Sullivan Joint Venture mined some 47 400 tons of copper from this deposit.

In the Thetford Mines area, Quebec, Shilts (1975a, 1976) has shown that a complete syngenetic geochemical indicator train exists. Nickel and chromium values in the minus 62 μm fraction of surface tills from the down-ice area of ultrabasic outcrops show clearly defined anomalies with developed heads and tails, the tail for nickel being traceable for about 60 km in the main direction of glacial flow (Fig. 13.10).

Garrett (1971), Gleeson and Cormier (1971), Skinner (1972) and Shilts (1976) have described situations in the Abitibi clay belt of Ontario, where glaciolacustrine sediments overlie lodgment till. Garrett (1971) sampled lodgment till under glaciolacustrine deposits at the Louvem deposit, Val d'Or, Quebec. The heavy mineral concentrates from the 80-230 mesh (64-180 μm) fraction of the till were analyzed for copper and zinc. Both in horizontal plan and in a vertical section (Fig. 13.11), an anomalous dispersal train occurs in the direction of the regional ice movement. Close to the subcrop of the ore, the anomalous levels for copper and zinc are found at the base of the till, further down-ice the anomaly rises to higher levels in the till.

In Sweden, Brundin and Bergstrøm (1977) collected C-horizon till samples from roadcuts at 1-3 km intervals (sample density about 1 per 10 km^2). Heavy minerals from the till samples were preconcentrated in the field, the concentrates were later investigated mineralogically and analyzed for 27 metals. The distribution pattern of the tungsten in the concentrates (Fig. 13.12) was similar to the results obtained from scheelite grain counting; both analyses disclosed "tail" anomalies greater than 100 km^2. Follow-up work, using the same field techniques and a sample density of 3-5 samples per km^2, delineated a strong "head" geochemical tungsten anomaly within which some 200 scheelite-bearing surface boulders were found. Further exploration work outlined scheelite occurrences in the bedrock. These are now being evaluated.

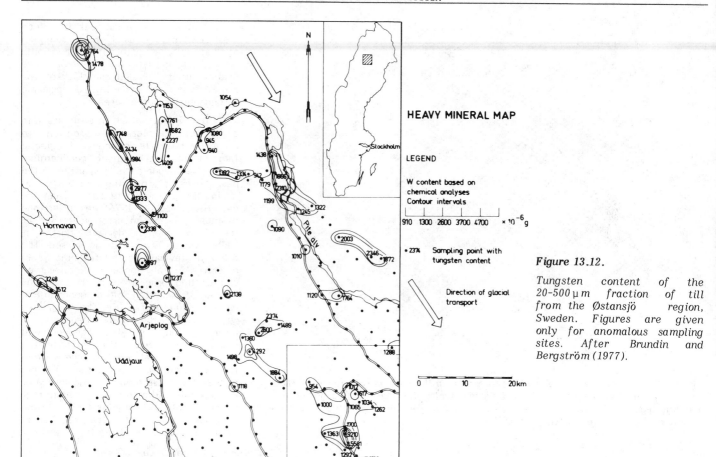

Figure 13.12.

Tungsten content of the 20-500 µm fraction of till from the Østansjö region, Sweden. Figures are given only for anomalous sampling sites. After Brundin and Bergström (1977).

From Finland, Lindmark (1977) has reported results for scheelite in till which agree well with those from Sweden. Till samples were collected along profiles more or less at right angles to the ice direction and from depths that varied between 0.5 and 2.0 m. Scheelite grains from heavy mineral concentrates were identified using an ultraviolet lamp and grain counts were made. The resulting map, showing the distribution of scheelite in the till (Fig. 13.13), outlined a scheelite anomaly approximately 2 km long and 300 m wide, which led to the discovery of a scheelite source near the head of the anomaly.

Syngenetic Anomalies in Stratified Drift

Glacial debris found associated with stratified drift, which includes ice-contact stratified drift and proglacial sediments, generally does not reflect the composition of local bedrock. Therefore such types of glacial sediments should normally not be used to define syngenitic dispersion fans at the detailed level. However, sometimes ice-contact stratified drift is derived from reworking of till of fairly local origin so that it may reflect the local geology. Holmes (1952) showed, for example, that kame terraces in central New York State are composed of material derived only a few kilometres up-ice. Nichol and Bjørklund (1973) cite an example from south-central British Columbia adjacent to the Valley Copper deposit where the bulk of the material in kame terraces along the valley sides consists of local bedrock found upslope. One of these ridges has a copper content locally exceeding 1 per cent in the minus 80 mesh (180 µm) fraction and greater than 0.1 per cent over a considerable area. The source of the copper anomaly is not known.

Eskers are an important type of ice-contact stratified drift which has been used in reconnaissance and semireconnaissance geochemical surveys. In Finland, Virkala (1958) showed that about 60 per cent of the esker material has been transported less than 5 km. The nature of esker sedimentation and the use of Canadian eskers as a sampling medium in geochemical exploration have been studied by Lee (1965, 1968), McDonald (1971), Shilts (1973b), Shilts and McDonald (1975), and Banerjee and McDonald (1976). On the Munro esker in the Kirkland Lake – Larder Lake gold belt of Ontario, Lee (1965, 1968) found that anomaly peaks for counts of diagnostic mineral grains were displaced 5-13 km downstream from the source.

In the Abitibi Clay Belt of Quebec, Cachau-Herreillat and La Salle (1969) sampled 165 km of the Matagami esker at 0.8 km intervals. They found that metals generally were leached from the minus 80 mesh (180 µm) fraction of the weathered near-surface material and that higher trace metal values could be obtained from unoxidized samples at depth (3-4 m). In the latter, higher base metal values, especially for copper, correlated with the presence of volcanic belts crossed by the esker. At one location lead, zinc, silver and gold anomalies were found near known gold-silver occurrences.

Shilts and McDonald (1975) demonstrated that a dispersal train in till can serve as a source for a "fluvial" dispersal train in an esker. Studying the distribution of clasts and trace elements in the Windsor esker, Quebec, which intersects a dispersal train of ultrabasic rocks, they found that the downstream fluvial transport of material may be in the order of 3-4 km. They concluded that hydraulic sorting as well as lateral and vertical variation of depositional environments greatly complicate the study of dispersal of trace elements and clasts in an esker. Shilts (1976) considered that where till (the first derivative of bedrock) is present over wide areas, it is a superior sampling medium to eskers (which may be second derivatives) that traverse only narrow portions of the terrain.

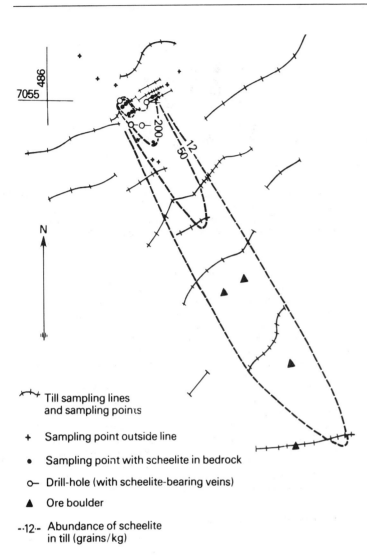

Figure 13.13. Scheelite grains in boulders and till at Tastula, Finland. After Lindmark (1977).

However, where eskers are abundant and especially in areas covered by extensive glaciolacustrine deposits (e.g. Abitibi Clay Belt), eskers provide a unique medium for sampling where they protrude through these sediments. Investigations to date show that eskers may give a good indication of the lithology and geochemistry of the bedrock on a reconnaissance scale.

The materials which make up other types of stratified drift, like glaciolacustrine and glaciomarine deposits, are distal in origin and cannot be expected to produce syngenetic anomalies. Low base metal values in clay were noted by Gleeson (1960) over Mattagami Lake Mines, Quebec where the Zn-Cu-Pb deposit was covered by 6-10 m of glaciolacustrine clays and silts. Similarly, a uniformly low base metal concentration has been noted in varved clays in the Cobalt area, Ontario even directly over the Ag-Ni-Co-As veins (Boyle, 1967). Similar findings have been reported from the Timmins (Fortescue and Hornbrook, 1967) and Uchi Lake (Davenport, 1972) areas of Ontario. In Finland, Kauranne (1976b) has reported little variation in metal concentration between glacial clays over mineralization and clays over barren bedrock.

EPIGENETIC DISPERSION

Epigenetic dispersion processes can be grouped into two classes: (1) Processes involving postglacial weathering of the bedrock as well as transportation and deposition of the products of this weathering. In this connection glacial drift could be considered as a more or less indifferent sheet of overburden responding to the transportation and deposition processes; (2) Postglacial processes leading to modification or reorganization of syngenetic dispersion patterns that already exist in the glacial drift.

In principle, processes (1) and (2) have been occurring simultaneously since the formation of the glacial drift, the result being combined syngenetic and epigenetic dispersion patterns in the drift. However, sometimes the epigenetic dispersion processes may have been negligible: such cases have been treated in the preceding section. Occasionally, the effect of the epigenetic processes may dominate over that of the syngenetic processes. Such cases are discussed in this section. Occurrence of complex syngenetic and epigenetic patterns, probably the most common situation, will be treated in the following section.

Dispersion of the products of postglacial weathering of the bedrock can be grouped into (1) hydromorphic dispersion, (2) electrochemical dispersion and (3) gaseous dispersion.

Hydromorphic Dispersion

Sulphide and other mineral deposits in the bedrock will undergo postglacial weathering some products of which are dissolved in groundwaters. These soluble compounds may later be precipitated, complexed or absorbed in the soil, and, in time, there will be a build-up of metal in the regolith. It has been shown that such processes may be significant for the concentration and distribution of metals in soil even in very cold climates. In subarctic Canada for example, formation of gossans over sulphide deposits is a common feature. Two Scandinavian examples of hydromorphic dispersion are referred to below.

1. The Tverfjellet massive sulphide deposit, Hjerkinn, Norway, Vokes (1976) is located at an elevation of 1000 m above sea level in an area with a mean temperature of 0°C and an annual precipitation of 300 mm. Despite the cold and dry climate (discontinuous permafrost), the deposit has suffered intense postglacial chemical weathering as indicated by strong subsidence phenomena in the subcrop area, very low pH values and high metal contents in the drainage, as well as extensive precipitation of secondary iron oxides downslope in the tills. (Bølviken, 1967; Mehrtens and Tooms, 1973; Mehrtens et al., 1973; Bølviken and Låg, 1977.)

2. At Snertingdal near Gjøvik, Norway there is a lead deposit of the Laisvall-Vassbo type (galena disseminated in quartzitic sandstones) (Bjørlykke et al., 1973). The mineralization which subcrops under 1-2 m of till, is about 2 km long and contains up to 1 per cent Pb over a width of 15-20 m. The lead content of the soil samples shows an anomaly displaced downhill from the lead mineralization (Fig. 13.14). Ice direction is towards the southeast (Sveian, 1979) i.e. uphill, and the anomaly must, therefore, be hydromorphic. This is also demonstrated in Figure 13.15, which shows that directly over the upper (southern) zone of mineralization there is more Pb in the deepest part of the soil than in the topsoil, while farther down the hillside the Pb is enriched in the topsoil.

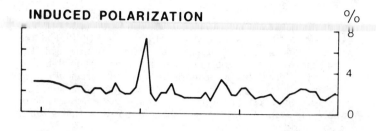
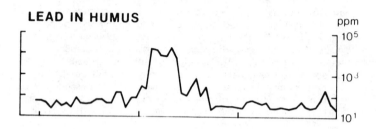
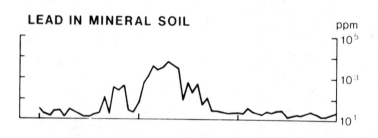
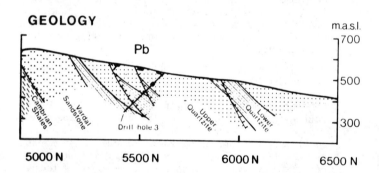

Figure 13.14. Geology, lead in mineral soil (B_2), lead in humus, and induced polarization results, profile 4900 W over the Snertingdal lead deposit, Norway. After Bjørlykke et al. (1973).

Humus plays an important role in the postglacial enrichment of heavy metals. If humus containing soils receive a steady supply of heavy metals, they may gradually become metalliferous and restrictive for natural plant growth, in some cases to the extent of being completely toxic, so that atypical vegetation or even patches of barren soil appear (Fig. 13.16) (Låg et al., 1970; Låg and Bølviken, 1974). Such patches, which may contain as much as 10 per cent of Pb and 5 per cent Cu in the top soil seem to be rather common in the vicinity of sulphide deposits in the hilly parts of Scandinavia, and can be used as a prospecting tool (Bølviken and Låg, 1977). In fact, a 30 million ton copper deposit was recently discovered at Kiruna, Sweden by drilling uphill from areas of natural metal poisoning (Lisbeth Godin, pers. comm.).

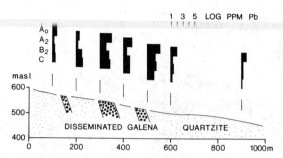

Figure 13.15. Lead values in different soil horizons at the Snertingdal lead deposit Norway, showing hydromorphic metal dispersion. General direction of glacial flow was towards the left (Sveian, 1979). Figure redrawn from Bølviken (1976).

In Figure 13.17, all the observed patches of natural lead poisoning at the Snertingdal lead deposit are plotted. Such a plot is equivalent to a geochemical map of the most intense hydromorphic anomalies in the root zone or humus containing part of the soil profile. It is based upon continuous visual observations, and is therefore fairly complete compared with ordinary geochemical map, showing results from samples taken at discrete sampling sites in a grid. A striking feature of Figure 13.17 is that the geochemical patterns of natural poisoning are very irregular and patchy, the poisoned areas being of varying sizes, elongated, fanshaped etc. This indicates that in searching for hydromorphic anomalies through soil sampling in glaciated terrain, instead of sampling, over a regular grid it may be more fruitful to sample sites where metals are likely to concentrate i.e. (1) sites selected on the basis of vegetation characteristics; (2) at places where the slope changes (3) in seepage areas; (4) in depressions, and (5) at the edges of bogs.

Electrochemical Dispersion

Based upon the existence of self-potential (SP) anomalies associated with ore deposits (see for example Sato and Mooney, 1960; Logn and Bølviken, 1974), an electrochemical model for element dispersion has been suggested (Bølviken and Logn, 1975). The redox potential (Eh) of groundwaters is high near the daylight surface and supposedly low at depth (Bølviken, 1978), therefore vertical Eh gradients must exist in the upper lithosphere. A good electrical conductor, such as an ore deposit penetrating such Eh gradients, will take on the character of a dipole electrode becoming an anode at depths and a cathode near the surface. The system ore deposit/country rock/groundwater consequently can be considered as a galvanic cell where natural electric currents flow carried by electrons in the orebody and by ions in the electrolyte formed by the groundwater. Positive current direction will be downwards in the orebody and upwards in the surroundings. Since overburden generally has better electrical conductivity than bedrock, the ionic current will flow more or less vertically in the country rock and horizontally in the overburden, the current density being highest in the overburden just above the subcrop of the hanging wall of the deposits (Fig. 13.18). Ions will move along the current paths and if during their migration they meet retaining agents like fine grained overburden, Fe-Mn hydroxides, or humus, they may be absorbed or complexed and interchanged for more mobile ions which in turn are released to the electrolyte.

Figure 13.16.
An area of naturally heavy-metal poisoned soil at the Mosbergvik lead zinc deposit, Balsfjord, Norway. Photo Arne Bjørlykke. GSC 203492-L

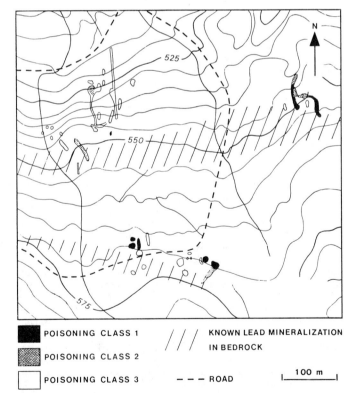

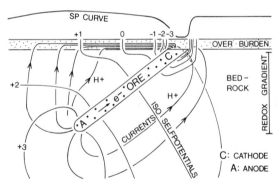

Figure 13.18. *Principles of electrochemical dispersion. Due to vertical redox potential gradients in the upper lithosphere an electrically conducting orebody and its surroundings can be considered as a galvanic cell, with current flow carried by electrons in the orebody and ions in the groundwater.*

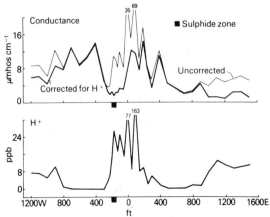

Figure 13.17. *Patterns of naturally lead-poisoned soil and vegetation at the Snertingdal lead deposit, Norway. Increasing degrees of poisoning from class 3 to 1 corresponds to lead concentrations in the upper 5 cm of the soil of approximately 1-3 per cent. After Låg and Bølviken (1974).*

Figure 13.19. *Distribution of H^+ and conductance in slurries of B-horizon soils in water (1 per cent strength) from a profile over the Armstrong A deposit, New Brunswick. After R.E. Uthe, quoted by Govett and Chorck (1977).*

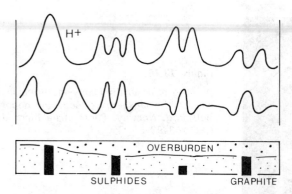

Figure 13.20. Schematic illustrations of H^+ and conductance in surface soils on glacial drift overlying sulphide and graphite conductors at various depths. After Govett (1976a).

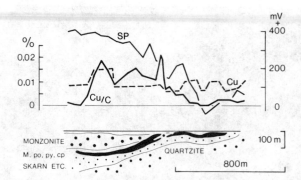

Figure 13.21. Values for copper and ratio copper-to-organic carbon in humus samples and self potentials from a profile over the Laurinoja copper-bearing iron deposit, Finland. After Nuutilainen and Peuraniemi (1977).

Field and laboratory data which seem to corroborate this model have been provided by a number of authors (Bølviken and Logn, 1975; Govett, 1973, 1974, 1975, 1976a; Govett and Chork, 1977; Govett and Whitehead, 1974; Govett et al., 1976; Juve, 1977; Kokkola, 1977; Nuutilainen and Peuraniemi, 1977; Sveshnikov, 1967; Shvartzev, 1976; Thornber 1975a, 1975b). In Canada Govett and Chork (1977), measured conductivity and H^+ in soils over several sulphide deposits and found that soil slurry conductivity seems to correlate with the content of organic carbon in the soils as well as with the localization of mineralization. Govett (1976a) suggested that deeply buried or blind sulphide deposits may be found by (1) a "rabbit ear" H^+ anomaly of two lateral peaks on the sides of a central trough (deep deposit), or an H^+ anomaly of three peaks separated by troughs (moderately deep deposit), (2) conductivity anomalies which are the inverse of the H^+ anomalies and (3) distribution patterns of metals generally similar to those of conductivity (moderately deep deposits) or similar to those of H^+ (very deep deposits) (Fig. 13.19, 13.20). Govett stated that wherever there is a detectable SP anomaly at the surface there will be an anomalous distribution pattern of H^+ in the soils. Nuutilainen and Peuraniemi (1977) also indicated that the content of organic carbon in the soil may possibly influence the development of electrochemical dispersion patterns. They collected humus samples from the A_o soil horizon over known mineralization and analyzed the samples for sodium pyrophosphate soluble base metals and organic carbon following a method described by Antropova (1975). In a profile over the Laurinoja occurrence which is a skarn zone containing an average of 42 per cent Fe, 2.8 per cent S and 0.4 per cent Cu, the measured copper values in the humus show no anomalies over the mineralization. However, when the ratio copper/organic carbon is plotted, a distinct anomaly occurs over the deposit (Fig. 13.21). This anomaly has three outstanding features: (1) The anomaly occurs over the entire inclined part of the deposit, including its deepest extension which is covered by approximately 120 m of barren monzonite and 20 m of till; (2) there is no anomaly above the flat lying and Cu-poor part of the deposit; (3) a pronounced SP anomaly coinciding with the geochemical anomaly shows that the direction of the positive natural current dominantly is from the projection of the deepest part of the deposit towards the outcrop area. Following an electrochemical interpretation (see Fig. 13.18), copper ions are suggested to move from the anodic area more or less vertically in the bedrock and into the overburden where the ions are trapped when they meet humus in the upper and presumably best conducting part of the soil. The flat lying part of the deposit does not produce an anomaly because (1) its copper content is low, and (2) it does not penetrate significant redox gradients.

Another observation indicating that electrochemical dispersion patterns develop in glacial drift above a sulphide deposit, is made by Juve (1977). In 1975, during the preparation for the open-pit mining of the Stekenjokk massive sulphide copper deposit in Sweden, the overburden was stripped off. In the till just above the subcrop of the ore, several tons of native copper were found occurring as a postglacial cement precipitated between the rock fragments and mineral grains of the till. One may wonder why native metals such as Cu, which is unstable under normal atmospheric conditions, can form in glacial drift above a sulphide deposit in an environment normally expected to be acidic and oxidizing. The primary mineralization at Stekenjokk consists of banded, massive to disseminated iron, zinc and copper sulphides lying conformably at the transition between acid pyroclastic rocks and black shales (Juve, 1974). Both ores and shales are highly conductive. The mineralization reaches great depths (700 m below surface), thus penetrating relatively large gradients of the earth's redox potential field. A well-defined surface SP anomaly was found above the deposit prior to the mining activity (Ø. Logn, pers. comm). Probably the cathodic potential of this long sulphide-carbon electrode was low enough to discharge copper ions from groundwater percolating in the zone of the subcrop of the ore during postglacial time. More recently native copper has also been discovered in till over the subcrop of the nearby Joma pyrite deposit in Norway (L.B. Løvaas, pers. comm.). It appears that occurrences of native copper may be fairly common in glacial drift above the subcrops of certain massive sulphide copper deposits.

It should be noticed that the empirical data obtained up to now provide indications only and no proof of the existence of electrochemical dispersion patterns in the overburden above mineral deposits. This subject, geoelectrochemistry,

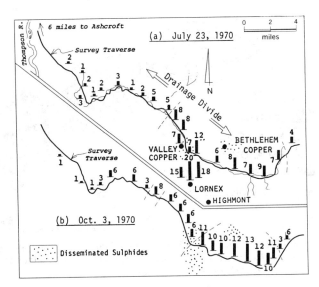

Figure 13.22. Results of SO₂ measurements (in ppb) in soil gas at two different dates from a traverse over the Lornex porphyry copper orebody, Highland valley, British Columbia. (Rouse and Stevens, written comm.).

warrants further research, especially in glaciated areas where conditions might have been such (low temperature, existence of clay layers etc.) as to inhibit the formation of significant geochemical anomalies in the soil by glaciation, oxidation or other commonly accepted processes Under these circumstances glacial drift could be considered as a blanket of more or less neutral background material covering the bedrock source. Possible electrochemical patterns in such a neutral blanket would probably be easier to detect than analogous patterns in residual soil, where the effects of oxidation nearly always dominate over most other modes of dispersion.

Gaseous Dispersion

Release of vapours from metallic deposits does occur in nature largely due to: (1) some of the weathering products of sulphide minerals e.g. H_2S and SO_2 are gases; (2) mercury and several of its compounds have high vapour pressure at low temperatures and (3) two of the radioactive disintegration products of uranium — radon and helium — are gases. Vapour surveys are a relatively recent development in geochemical exploration. In glaciated areas they hold much promise in being able to define epigenetic geochemical anomalies over metal deposits covered by thick drift. They can also be used to trace glacial dispersion trains. The subject has been reviewed by Kravstov and Fridman (1965), Bristow and Jonasson (1972), McCarthy (1972), Ovchinnikov et al. (1973) and Dyck (1976). Most soil-gas surveys carried out to date in glaciated areas have involved analyses for sulphur compounds, mercury, and radon as well as the use of sniffing dogs.

Sulphur Compounds

Rouse and Stevens written comm. reported on the use of sulphur dioxide gas in the detection of sulphide deposits. In a case history study of the Guichon Batholith in the Highland Valley area of British Columbia, they noted that their highest readings from soil gas were obtained over the Lornex porphyry copper orebody which in places is covered by more than 60 m of permeable till. Twelve months after Rouse and Stevens did their survey, Bethlehem Copper announced the discovery by diamond drilling of the J-H orebody just south of Bethlehem Copper's open pit. The new deposit is covered by up to 120 m of till. Background for SO_2 was about 2 ppb, and values over the copper deposits ranged from 7 to 20 ppb (Fig. 13.22).

Measurements of SO_2 in soil gas were made by Meyer and Peters (1973) over several sulphide zones covered by thin till in Newfoundland. They found that in this cool temperate climate there appears to be small but detectable soil-gas anomalies (1.9-7.4 ppb SO_2) over these deposits. The best responses were obtained in warm dry weather, the responses were considerably subdued when measurements were made on cool days.

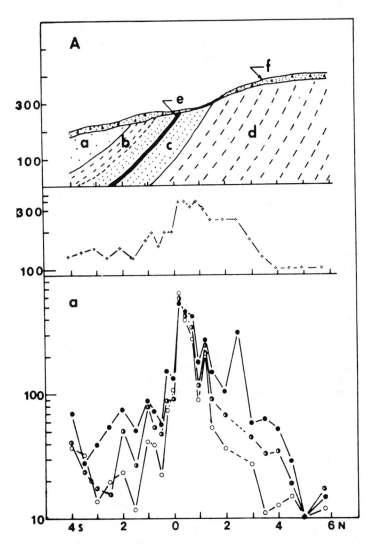

Figure 13.23. Geology (top), gamma-ray activity at the surface (middle) and alpha activity in soil gas (bottom), from a traverse across a subcrop of conglomerate ore at the Rio Algom Quirke mine, Elliot Lake, Ontario. Letter d denotes main conglomerate ore. Alpha activity in soil gas is given for three successive one minute counting intervals represented by dots, half-filled circles, and plain circles, respectively. After Dyck (1969).

Mercury

There are few published reports on soil-gas surveys for mercury in glaciated terrain. Most successful applications of this technique have been confined to arid areas (McNerney and Buseck, 1973). However, Boyle (1967) determined mercury in soil horizons over known mineralization at Cobalt, Ontario and Azzaria and Webber (1969) reported results from various soil traverses in British Columbia and Quebec. Preliminary work by Jonasson (1972) and Bristow (1972) over a Cu-Ag-Hg prospect near Clyde Forks, Ontario proved promising. Soil-gas mercury showed anomalous levels along the strike of mineralized veins.

Meyer and Evans (1973) completed exhaustive studies on mercury in glacial drift over a zinc-lead-cadmium deposit at Keel, Eire. They concluded that mercury levels in soil gas reflected the general pattern found for mercury in soil. The content of mercury vapour in soil gas was enhanced in warm, dry weather. Maximum mercury values in soil gas were found in 7 m thick well drained tills over mineralized faults and in 3 m thick poorly drained soil, over a preglacially decomposed zone of subeconomic sulphides.

Radon

Radon soil-gas surveys have been used successfully by several workers to help locate uranium mineralization under glacial cover. Dyck (1972) reviewed the use of radon methods in Canada. Wennerwirta and Kauranen (1960) reported on their work in Finland where radon soil-gas surveys were able to detect uranium mineralization covered by up to 1.5 m of till. Michie et al. (1973) found that alpha activity in soil-air was useful in the assessment of radioactive anomalies in northern Scotland, where bedrock is covered by several metres of glacial drift and 1 to 8 metres of peat. Dyck (1968, 1969) obtained positive results in his test work using soil-gas radon in sandy moraine over known uranium deposits in the Bancroft and Elliot Lake areas, Ontario (Fig. 13.23). Later in the Bancroft region Dyke et al. (1976) showed that meaningful results for radon could be obtained from snow cover over frozen ground. Morse (1976) worked in the same region and found that soil-gas radon is an effective exploration tool for uranium exploration in areas covered by relatively thin, permeable drift. Beck and Gingrich (1976) completed an alpha-track orientation survey over the "N" zone of the Cluff Lake uranium deposit in Saskatchewan; their results indicated that the survey was capable of outlining buried ore zones where till thickness ranged up to 20 m. Recently soil-gas radon surveys have been successful in delineating uranium mineralization covered by 5-15 m of permeable till in southwestern Quebec. (Pudifin, pers. comm.).

Dogs in Prospecting

The use of dogs to smell out sulphides was first tried by Kahma (1965) and his coworkers at the Geological Survey of Finland. The vapours detected by the dogs are sulphur dioxide, hydrogen sulphide and carbonylsulphide (Kahma et al., 1975). Brock (1972) reported the results of experiments with dogs in Canada and Orlov et al. (1969) have mentioned their use in the USSR. Alsatian dogs are most commonly used, and when trained properly they can detect sulphide-bearing boulders in till more than one metre below surface.

Nilsson (1971) described the training and use of prospecting dogs at the Geological Survey of Sweden. The dogs are bred and trained in the Kennel School of the Swedish Army. Male dogs are best fitted, but the individual qualities of different dogs vary greatly. Training begins at 1-2 years of age. Four months of basic training on snow in the winter is followed by about two months of practical field training on bare ground, at the end of which period the dog is normally working satisfactorily. Full effectiveness is generally achieved in the second summer of field work. A healthy well trained dog can work for about 8 hours a day with occasional rests.

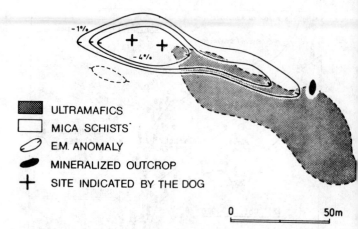

Figure 13.24. Sites where dog-indicated sulphide-mineralization occurs in bedrock under 2m deep overburden, Pielavesi, Finland. After Ekdahl (1976).

Ekdahl (1976) has reported on a detailed investigation using a dog in an area where an outcrop of weakly sulphide-mineralized basic rock was known. On the southern side of this outcrop the dog displayed heightened interest in two places. Geophysical measurements revealed an electromagnetic anomaly over the same area (Fig. 13.24). Three trenches were dug to determine the cause. Disseminated pyrrhotite and chalcopyrite was found in a highly weathered horizon 1-2 m thick. The dog had discovered the mineralization through till 1 m thick overlain by sand and clay also 1 m thick.

COMBINED SYNGENETIC AND EPIGENETIC DISPERSION

Postglacial weathering of glacial drift, modifies original syngenetic dispersion patterns. The distribution of chemical elements presently found in the soil is, therefore, the result of a combination of syngenetic and epigenetic dispersion processes. Interpretation of such combined patterns requires an understanding of both types of processes leading to them.

Weathering of Glacial Drift

The most notable weathering process in the drift is soil formation, by which the upper part of the overburden is transformed into more or less horizontal layers that differ from each other in their properties and composition.

Five major factors govern soil formation (Jenny, 1941): (1) parent material; (2) relief; (3) climate; (4) biological activity, and (5) time. In glaciated areas the time available for the soil-forming processes has been short, and the climate generally cool. The chemical composition of the weathered part of the overburden of these areas is, therefore, often relatively closely related to the composition of the parent material. Depending on local conditions, various types of soil profiles may develop, of which podzols and tundra soils are the most important in glaciated areas (Robinson, 1949).

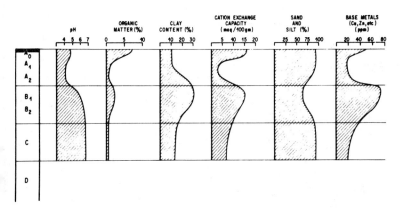

Figure 13.25. Variations in selected physical and chemical properties of different horizons of a generalized podzol profile. After Levinson (1974).

Characteristic features of the podzol profile are: (1) an upper zone rich in organic matter, (A_0-A_1); (2) a bleached layer leached of humus and black minerals, and (3) darker accumulation zones of sesquioxides (B_1-B_2). Variations in selected chemical and physical properties of a generalized podzol profile taken from Levinson (1974) are shown in Figure 13.25. Examples of heavy metal distribution are also given in Figure 13.15.

Shilts (1975a, b; 1976) stated that under Canadian conditions weathering has affected the upper 2-5 m of glacial drift in temperate areas and the active layer in areas of permafrost. In weathered drift, mineral grains derived from sulphide-type mineralization have largely been destroyed by oxidation. The soluble products of the oxidation are scavenged by other weathering products such as finely divided or precipitating hydrous oxides of Fe and Mn as well as clay minerals. If the composition and proportions of these scavenging phases do not change radically over an area, they should scavenge metal in proportion to the metal that was in the unweathered till in the form of sulphides. As a consequence, analysis of the clay fraction could be advantageous in delineating syngenetic geochemical dispersion patterns in till. Although combined syngenetic-epigenetic dispersion patterns are in fact obtained, they can, for interpretation purposes, be regarded as dominantly syngenetic. Many investigations indicate that combined syngenetic-epigenetic patterns are common in northern Europe (Kvalheim, 1967; Wennervirta, 1968; Jones, 1973, 1975, 1977; Kauranne, 1976a). Mehrtens et al. (1973) and Rice and Sharp (1976) reported on successful interpretations of combined syngenetic and epigenetic geochemical patterns in soils in Norway and north Wales. Nurmi (1976) used nitric and citric acid as extracting agents for heavy metals in anomalous and background till samples from Talluskanava Ni-Cu deposit, Tervo, central Finland, and calculated the percentage of syngenetic (glaciomorphic) metals using the formula:

$$\% \text{ Syngenetic Me} = \frac{\text{Me}_{(HNO_3)} - \text{Me}_{(Citr. ac.)}}{\text{Me}_{(HNO_3)}}$$

He found that the Ni and Cu dispersions were mainly syngenetic, the hydromorphic component for both background and anomalous samples being of the order of 10 per cent, while for Co the syngenetic component was approximately 45 per cent and 70 per cent in anomalous and background samples respectively.

A detailed study of the stability of various sulphide minerals in glacial drift at Fäboliden, northern Sweden has been carried out by Toverud (1977). Zn, Pb and to a lesser extent Cu were fairly easily extracted from the till (Table 13.1). Analysis of heavy and light fractions of the till revealed no general or marked differences between the metal content of the two fractions (Table 13.2). Pyrite and chalcopyrite were the only primary sulphide minerals preserved in the till, Zn, Pb and to some extent Cu occurred adsorbed to primary mineral grains or absorbed in secondary minerals. Toverud concluded that the Cu anomalies in the area were dominantly clastic (syngenetic), while the Pb and Zn anomalies were mainly hydromorphically displaced i.e. combined epigenetic/syngenetic.

It should be remembered that the conclusions of Nurmi (1976) and Toverud (1977) are valid only for the particular environments in which the studies were conducted. These Finnish and Swedish investigations are from areas of dominantly intermediate and acid Precambrian rocks.

In the till over Carboniferous carbonate rocks in Ireland, base metals seem to form mainly syngenetic patterns, although some postglacial dissolution of sulphides apparently has occurred. Donovan and James (1967) studied the geochemical dispersion patterns in till overlying the Tynagh Pb-Zn-Ag-Cu-deposit in County Galway, Ireland. The distribution of high and low metal values in the till was found to be erratic and independent of depth and oxidation state of the till thus indicating a clastic (syngenetic) origin.

However, about 10-20 per cent of the total metals in the overburden was readily soluble, and each of the metals analyzed, (i.e. Cu, Zn, Pb, and Hg) was found to concentrate in the clay fraction, which also indicates a hydromorphic (epigenetic) origin for the metals. Postglacial oxidation processes were likewise indicated by the heavy metal contents of the solum, which were somewhat higher than those in the parent till material, the degree of variation within the solum generally being less than twofold. Under certain conditions, however, Cu, Zn, Pb and Hg were highly concentrated in peat near anomalous metal sources, concentrations being up to 0.19 per cent Cu, 7.8 per cent Zn, 0.42 per cent Pb and 0.42 per cent Hg. The areal distribution patterns of Pb, Zn and Cu in soils at a depth of 20-30 cm over the Tynagh deposit are narrow, elongated and stretch for a considerable distance from the orebody in the direction of the last ice movement. From Figure 13.26 it seems evident that the dominant component of the dispersion patterns of lead in soils at Tynagh is syngenetic. This is also the conclusion of Morrissey and Romer (1973), who have described the successful use of soil sampling in mineral exploration in glaciated regions of Ireland. Reconnaissance soil sampling is usually done at approximately 150 m centres on rhombic or square grids. Initial sampling depth generally varies between 23-46 cm. The minus 180 µm (80 mesh) fraction is normally analyzed for hot acid-soluble heavy metals. Any geochemical anomalies found by this procedure are followed up by ground investigation and by resampling on a tighter grid and/or at the maximum depth attainable manually. In promising areas, power augers are used for sampling overburden at depth.

Table 13.1

Cold extractable Cu, Pb and Zn in till fractions as
percentages of total content from section 2 at N. Fäbodliden. After Toverud (1977).

Sample No.	Depth m	Grain size mm	Total content, ppm Cu	Total content, ppm Pb	Total content, ppm Zn	8M HNO₃ Cu	8M HNO₃ Pb	8M HNO₃ Zn	4M HNO₃ Cu	4M HNO₃ Pb	4M HNO₃ Zn	1M HNO₃ Cu	1M HNO₃ Pb	1M HNO₃ Zn	0.50M Citr.acid Cu	0.50M Citr.acid Pb	0.50M Citr.acid Zn	0.25M Citr.acid Cu	0.25M Citr.acid Pb	0.25M Citr.acid Zn	0.05M Citr.acid Cu	0.05M Citr.acid Pb	0.05M Citr.acid Zn	4M Am.acetate Cu	4M Am.acetate Pb	4M Am.acetate Zn	1M Am.acetate Cu	1M Am.acetate Pb	1M Am.acetate Zn
103	0.5	0.06-0.20	120	830	1555	76	100	100	64	100	77	32	92	22	17	60	4	–	na	na	–	31	4	–	51	3	–	29	2
		<0.06	140	825	1830	56	100	71	51	100	60	na	na	na	21	70	4	–	25	5	15	39	4	–	52	3	na	na	na
102	1.0	0.06-0.20	45	340	530	100	100	100	84	92	79	53	97	32	–	71	7	–	65	9	–	44	7	–	44	5	–	19	4
		<0.06	70	560	655	97	100	93	70	100	73	50	100	32	33	71	10	32	68	10	–	45	9	–	41	5	–	27	4
101	1.5	0.06-0.20	75	300	930	100	97	100	84	93	100	61	67	63	–	53	15	–	47	11	–	37	7	na	na	na	–	20	4
		<0.06	100	650	1150	100	100	100	95	100	100	67	100	79	–	60	17	–	55	17	–	40	11	–	26	4	–	17	2
100	2.0	0.06-0.20	40	250	545	na	na	na	na	na	na	na	na	na	–	64	33	–	56	28	–	44	18	–	29	4	na	na	na
		<0.06	125	740	1400	80	100	100	71	100	100	59	93	86	26	57	26	23	57	24	18	38	17	–	27	2	–	16	3
099	2.5	0.06-0.20	85	525	770	61	95	87	56	100	87	56	95	48	–	59	14	–	55	12	–	34	9	–	23	–	–	18	–
		<0.06	130	1315	1085	77	100	100	77	100	100	60	100	79	21	57	24	22	52	23	–	30	16	–	20	3	–	14	–
098	3.0	0.06-0.20	115	805	1065	65	100	94	65	100	94	67	89	83	40	67	45	31	60	40	20	39	29	–	39	4	–	29	3
		<0.06	195	1315	1550	67	100	97	67	100	97	59	100	84	39	65	43	26	62	40	24	–	29	–	37	3	–	25	1
097	3.5	0.06-0.20	195	615	1505	62	98	100	56	98	93	49	89	54	23	60	21	21	57	19	22	44	15	–	39	2	–	31	–
		<0.06	205	880	1655	68	100	100	63	100	100	49	95	57	21	50	18	20	52	18	18	41	15	–	31	2	–	23	2

Notes: na = not analyzed
– = below the detection limit

Results obtained from a zinc anomaly originally defined by manual sampling, were commented upon as follows:

"In most parts of the anomaly, including the area of greatest zinc concentration, power augering showed that zinc values in overburden diminished to background in the upper 1-2 m, but in a relatively restricted area near its centre and extending outside the geochemical anomaly delimited by shallow sampling, they were found to persist or increase with depth. At one point a value exceeding 50 000 ppm was recorded below 2 m of non-anomalous till, and a drillhole collared near this point intersected well-mineralized "suboutcrop".

Uranium seems to offer characteristic examples of combined syngenetic and epigenetic dispersion patterns in glacial drift. In Canada, Klassen and Shilts (1977) analyzed uranium in the clay fraction of till from perennially frozen terrain in the tundra. The till was sampled at a density of 0.5 samples per km^2. The resulting distribution patterns of uranium are of the order of hundreds of metres to hundreds of kilometres. A number of large bedrock source areas of uranium were detected, but the sample density was too small to disclose dispersion trains from some smaller zones of potentially economic-grade mineralization. A more than 10 000 km^2 area of high uranium values appear north of Baker Lake (Fig. 13.27), parts of which are interpreted as being related to high background U values in the underlying volcanic rocks.

Björklund (1976a, b) found that the uranium content in till and stratified drift can be successfully used as a tool in the follow-up of U anomalies obtained by sampling lake sediments in Finland. His results suggest that the U anomalies in till are in part mechanical in origin and that the uranium in near surface samples has been partly leached out. Some observations leading to this conclusion are: (1) at the Paukkajanvaara area, the uranium content of the minus 0.06 mm fraction of the till did not coincide with the known fan of U-mineralized boulders (Fig. 13.28), (2) Strong U anomalies are present in the drift close to bedrock, but are generally lacking in the upper two metres of the till (Fig. 13.29), (3) Anomalous U values were found in stratified drift over U-mineralized bedrock (Fig. 13.29), (4) Good correlation exists between the U contents of the minus 0.06 mm fraction and that of the 0.06-0.25 mm fraction of drift sampled in a semiregional program.

Table 13.2

Contents of copper lead and zinc at various grain sites in light (<2.96 g/cm^3) and heavy (>2.96 g/cm^3) fractions of till. Section 2, N. Fäbodliden, Sweden. After Toverud (1977).

Sample No.	Depth (m)	Grain size (mm)	Sample weight (g) Total	Light fraction	Heavy fraction	*Light fraction (ppm) Cu	Pb	Zn	**Heavy fraction (ppm) Cu	Pb	Zn
103	0.5	0.20-0.60	22.00	21.05	0.95	120	875	1370	100	1800	1600
		0.10-0.20	16.72	14.85	1.87	85	520	1140	100	1500	1300
		0.06-0.10	10.70	9.76	0.94	185	755	2030	70	1000	800
		<0.06	18.75	18.23	0.52	190	770	2025	80	1400	300
102	1.0	0.20-0.60	28.26	26.98	1.28	50	330	455	50	850	300
		0.10-0.20	13.68	11.93	1.75	30	200	285	40	1000	300
		0.06-0.10	5.78	4.90	0.88	25	170	215	40	800	300
		<0.06	2.54	2.22	0.32	30	205	275	50	850	300
101	1.5	0.20-0.60	26.94	25.93	1.01	35	225	400	120	1200	1200
		0.10-0.20	25.95	23.78	2.17	40	195	480	40	1100	300
		0.06-0.10	12.79	10.97	1.87	55	215	630	40	1100	300
		<0.06	11.84	11.31	0.53	110	385	1155	50	1400	300
100	2.0	0.20-0.60	33.45	32.00	1.45	25	205	315	60	900	790
		0.10-0.20	22.11	19.59	2.52	30	170	305	50	1100	300
		0.06-0.10	3.40	2.65	0.75	35	170	390	40	650	300
		<0.06	2.41	2.08	0.33	45	245	475	60	1100	300
099	2.5	0.20-0.60	28.74	27.68	1.06	40	425	430	250	1900	300
		0.10-0.20	28.43	25.88	2.55	40	240	370	60	1300	300
		0.06-0.10	17.19	15.85	1.34	65	625	550	50	2000	300
		<0.06	8.33	8.02	0.31	55	660	465	60	2000	300
098	3.0	0.20-0.60	25.17	23.79	1.38	95	735	1040	50	1000	490
		0.10-0.20	7.42	6.38	1.04	70	570	755	50	900	450
		0.06-0.10	2.47	1.76	0.71	65	515	650	50	1000	480
		<0.06	5.86	5.73	0.13	90	600	740	50	1300	300
097	3.5	0.20-0.60	15.06	12.27	2.79	160	535	1510	50	400	460
		0.10-0.20	4.76	3.72	1.04	120	405	1205	70	500	570
		0.06-0.10	2.25	1.76	0.49	105	425	920	70	700	520
		<0.06	0.67	0.60	0.07	140	950	1200	not	analyzed	

Data obtained by optical emission spectrography,
 *direct reading spectrograph equipped with a tape machine,
**registration on photo plate (Danielson 1968).

Humus as a Sampling Medium

Humus is dead organic matter, or more precisely is a decomposition product of organic origin that is fairly resistant to further bacterial decay. In glaciated regions, humus is a typical and almost ubiqitous constituent of the upper soil horizons. Where the climate is appropriate, organic deposits often accumulate in low lying or poorly drained areas forming bogs, swamps and muskegs due to a local low rate of plant decay in relation to plant production. Humus is capable of complexing practically all of the elements of the periodic table; in particular it forms strong compounds with heavy metals (Szalay and Szilagyi, 1968; Szalay, 1974; Chowdhury and Bose, 1971 and Boyle, 1977). Therefore it is a geochemical sampling medium of great interest. In some cases the geochemical distribution patterns in humus are more or less purely epigenetic in relation to glaciation. Experience shows, however, that glacially dispersed minerals often are the source of those chemical elements presently found in humus, and geochemical dispersion patterns in humus can in many cases be classified as combined syngenetic/epigenetic.

The humus layer of soil profiles has been used as a sampling medium by several research workers (e.g. Boyle and Dass, 1967; Kauranne, 1967a, 1976b; Smith and Gallagher, 1975; Bølviken, 1967, 1976; Kokkola, 1977 and Nuutilainen and Peuraniemi, 1977). Results from soil surveys carried out by Boyle and Dass (1967) in the Cobalt area, Ontario show a greater concentration of Pb, Zn, Cu, As, Sb, Mo, Ag, Co, Mn and Hg in the humus horizons than in the B horizons. Anomalies in the A horizons over known veins gave much better contrast than those obtained from the B horizons. In certain cases, surveys over silver veins covered by 6 m or more of till gave little response when the B horizons were sampled, whereas anomalies in the humus clearly marked the positions of the veins. Similar results for Au were obtained by one of the authors (C.F.G.) in the Duparquet area, Quebec and by Lakin et al. (1974) in the Empire district, Colorado.

Kauranne (1967a) sampled humus from an area containing Ni-bearing boulders and analyzed the samples for cold extractable heavy metals (CxHM). He found a weak anomaly which coincided with occurrences of the nickel boulders and the direction of ice movement. No continuous hydromorphic anomaly in the till was found under the anomaly in humus (Kauranne, 1976b). The apex of the humus anomaly terminated near a diabase dyke containing pentlandite-chalcopyrite mineralization.

It has long been known that some peat bogs are metalliferous, and may contain as much as several per cent of metal in the dry matter. Townsend (1824) remarked on recovery of copper from peat bogs in western Ireland, and Henwood (1857); Ramsay (1881) and Mehrtens et al. (1973) reported on a copper bog in north Wales, with which a new porphyry copper discovery (Rice and Sharp, 1976) was associated. Copper-bearing bogs have also been reported in Montana and Colorado, (Lovering, 1928; Forrester, 1942 and Eckel, 1949). A zinc-rich bog in New York State, has been described by Cannon (1955). In Canada, copper-rich bogs occur near Sackville, New Brunswick (Fraser, 1961a, 1961b and Boyle, 1977) near Bathurst, New Brunswick (Hawkes and Salmon, 1960) and at Highmont, British Columbia (Hornsnail, 1975). A bog high in zinc is found near Daniels Harbour,

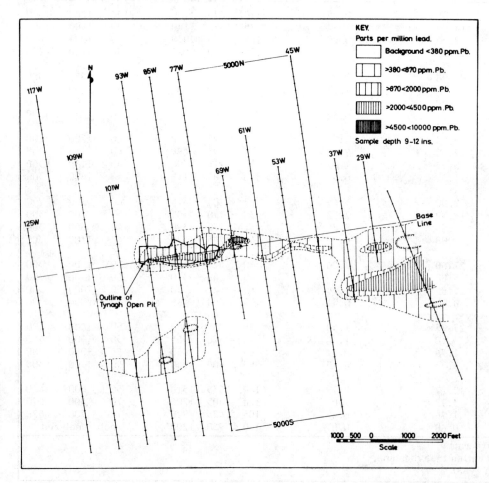

Figure 13.26.

Distribution of lead in minus 180 μm fraction of soil sampled at a depth of 24-30 cm over the Tynagh zinc lead deposit, Galway, Ireland. After Donovan and James (1967).

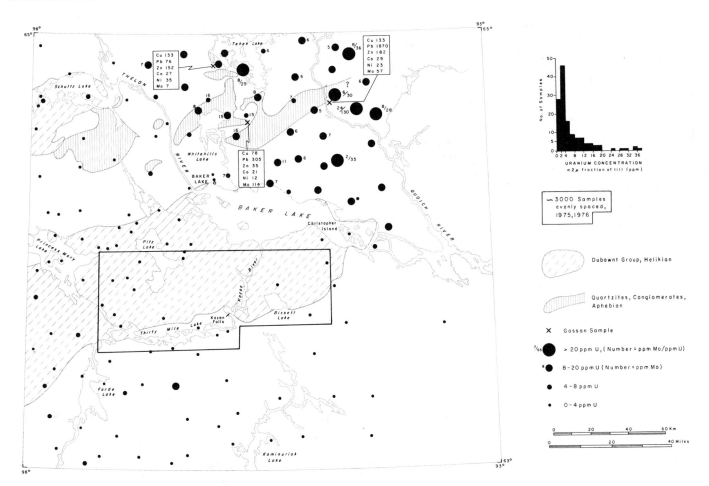

Figure 13.27. *Uranium content in <0.002 mm fraction of till from Baker lake region, Canada. After Klassen and Shilts (1977).*

Newfoundland (Gleeson and Coope, 1967). Donovan and James (1967) mentioned extreme zinc values in a bog near the Tynagh deposit, Co. Galway, Ireland. In Scandinavia, Armands (1967) reported on a uraniferous bog at Masugnsbyn, and Larsson (1976) on a cupriferous bog in the Pajalu district, both in northern Sweden. A related phenomenon of natural heavy-metal enrichment in humus was described in Norway (Låg et al., 1970; Låg and Bølviken, 1974 and Bølviken and Låg, 1977). In the USSR, metalliferous bogs have been reported by Moiseenko (1959), Manskaya and Drozdova (1964), Kochenov et al. (1966) and Albov and Kostariev (1968). Systematic sampling of peat as a prospecting method was pioneered in the 1950s by Salmi (1967), who found anomalous values for heavy metals over a variety of Finnish sulphide and iron deposits. An early review of the literature on geochemical prospecting in peat land has been given by Usik (1969).

Gleeson (1960), studied the distribution of metals in bogs of the Kenora area, Ontario and the Matagami area, Quebec, Canada. The zinc, copper and nickel content of the peat increased with depth and reached a maximum at the peat-clay interface. The bogs studied were underlain by 6-20 m of glacial lake sand, silt and clay. Test work over the Matagami Lake base metal mine proved negative; the dense,

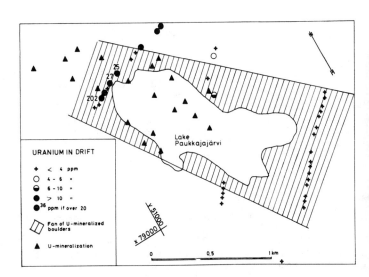

Figure 13.28. *Uranium content in <0.06 mm fraction of glacial drift, Paukkajanvaare, Karelia, Finland. After Björklund (1976).*

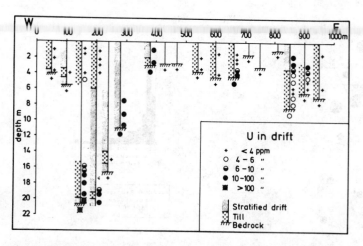

Figure 13.29. Uranium content in <0.06 mm fraction of glacial drift in a vertical section at Paukkajanvaara, Karelia, Finland. The section is located at the northernmost sampling line in Figure 13.28. After Björklund (1976).

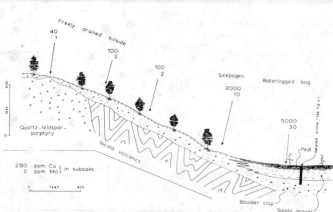

Figure 13.31. Molybdenum and copper in subsoils from a traverse in a study area in British Columbia. After Horsenail and Elliott (1971).

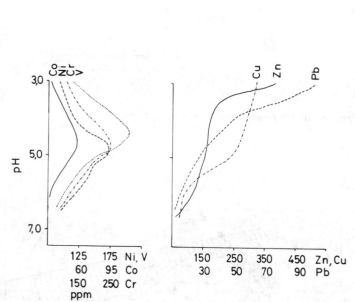

Figure 13.30. Generalized relationships between pH and metal content of 103 peat profiles from various parts of Lapland, Finland. After Tanskanen (1976a).

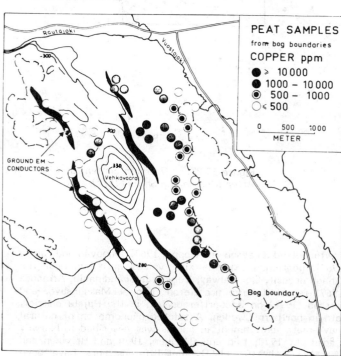

Figure 13.32. Copper content of peat samples in the Vehkavaara area, Pajala district, Sweden. After Larsson (1976).

impermeable glacial lake clays overlying the orebody effectively blocked movement downwards of oxygenated waters as well as the free movement of metal ions from bedrock to surface. When the subcrop of the ore zone was uncovered under about 6 m of clay, the surfaces of the sulphides were fresh and unoxidized.

More recent Canadian investigations including sampling of peat have been published by Fortescue and Hornbrook (1967, 1969). Hornsnail and Elliott (1971), Gunton and Nichol (1974), Bradshaw (1975b), Fortescue (1975) and Hornsnail (1975). Work in northern Europe has been reported by Erämetsä et al. (1969), Hvatum (1971), Erikson (1973, 1976b), Smith and Gallagher (1975), Eriksson and Eriksson (1976a, b), Kokkola (1977), Nieminen and Yliruokanen (1976) and Tanskanen (1976a,b). A description of Russian work has been given by Borovitskii (1976). Tanskanen (1976a) investigated the variation of metal content versus such factors as pH, humification, ash content, and depth in 609 peat samples collected from 103 peat profiles in various parts of central Lapland, Finland. He found that each of the factors had different effects on the various elements: pairs of similarly-behaving elements were Pb-Zn, Cr-V, and Co-Ni. The relationship between pH and heavy metals is shown in Figure 13.30.

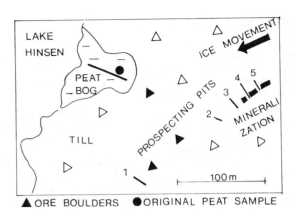

Figure 13.33. *Cu-Zn anomalous peat bog, and prospecting trenches in the Lake Hinsen area, Sweden. After Eriksson and Eriksson (1976a).*

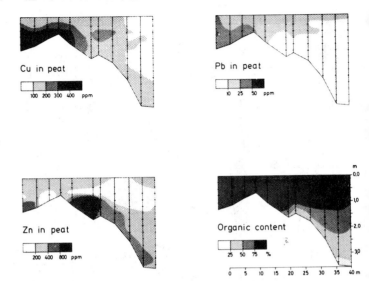

Figure 13.34. *Distribution of Cu, Pb, Zn and organic material in a vertical section of a peat bog near Lake Hinsen, Sweden. For location see Figure 13.33. After Eriksson and Eriksson (1976a).*

The results of Hornsnail and Elliott (1971) (Fig. 13.31) are typical for many of those reported in both Canada and Scandinavia. Metals like Cu, Pb, Zn and Mo move hydromorphically down freely drained slopes, and are enriched in peat if the metal-rich groundwater reaches poorly drained bogs in the lowland. This phenomenon is utilized in a new exploration method "sampling of peat bog margins" which has become a routine prospecting technique in Sweden (Eriksson, 1976b and Larsson, 1976). The Geological Survey of Sweden samples peat bog margins at a depth of less than 0.5 m and at 50-100 m intervals; the method is used to follow-up results from reconnaissance geochemical drainage surveys, (see Fig. 13.32 after Larsson, 1976).

Eriksson (1976b) and Eriksson and Eriksson (1976b) have used peat as a sampling medium in reconnaissance surveys in central Sweden. The samples are taken at a density of approximately 7 per km^2 as close to the border of the bog as possible, and at a depth of about 2 m in the bottom of the organic layer close to the organic/inorganic interphase. Anomalous Zn and Cu values (1130 and 470 ppm Zn and Cu respectively) were found in a bog near Lake Hinsen. Boulder tracing in the vicinity of the anomalous bog led to the discovery of some rich base metal boulders at surface. After trenching and till sampling, Cu/Zn mineralization was found in the bedrock (Fig. 13.33). Later, detailed sampling of the bog revealed distribution patterns (Fig. 13.34) which show that the peat anomaly must be caused by percolating metal-rich groundwater, while the till anomalies most probably are of glacial origin. These interpretations indicate that both mineralized bedrock and till can be sources of high metal contents in bogs.

In another case history the same authors (Eriksson and Eriksson, 1976a) show that a train of boulders containing As, Cu, Pb and Zn was the only source of metals in groundwater creating metal anomalies in a peat bog. This case is an even clearer example of combined syngenetic-epigenetic dispersion. The anomalies, correctly interpreted, can be traced from the peat, through the till and back to the bedrock source. The source was subeconomic skarn-type mineralization containing arsenopyrite, pyrrhotite and chalcopyrite some 1.5 km north of the boulders.

Interpretation of Complex Patterns

In concluding this section two recently published case histories, one Canadian (Cameron, 1977) and one Finnish (Kokkola, 1977) are reported. They both demonstrate that syngenetic and epigenetic dispersion patterns can exist in the same area, and provide examples of how such patterns can be interpreted.

Cameron (1977) reported that in following up lake sediment anomalies in the Agricola Lake area in the eastern part of the District of Mackenzie, massive sulphide mineralization was found in metavolcanic rocks of Archean age. The sulphide zone occurs in a topographic depression (Fig. 13.35) in perennially frozen terrain. Soil samples were collected in a 30 by 30 m grid at a depth of 15-20 cm, and the minus 180 μm (80 mesh) fraction was analyzed for 11 metals. Factor analysis was carried out to identify interelement relationships.

Factor 1 (Table 13.3) represents relatively immobile elements illustrated by the distribution of Au. A syngenetic dispersion train from the sulphide body upslope in the direction of the glacial movement was obtained (Fig. 13.36).

Factor 3 (Table 13.3) is related to elements of intermediate mobility exemplified by Cu. These are to some extent dispersed epigenetically from the mineralized area, and subsequently precipitated down the drainage channel, (Fig. 13.37).

Factor 5 is related to highly mobile elements like Zn, which are dissolved during postglacial time and carried away (Fig. 13.38). An illustration of the historical sequence of dispersion processes taken from Cameron (1977) is reproduced in Figure 13.39.

Kokkola (1977) sampled both till at the overburden bedrock interface and humus in a profile along the ice direction over the Pahtavuoma Cu-Zn deposit in northern Finland. The terrain is characterized by till and bogs, the thickness of the overburden varies from 0 to 6 m. Cu values in the till samples produced an intense, predominantly syngenetic anomaly approximately 2 km long in the down-ice direction from the deposit. In contrast, the Cu content in humus showed an epigenetic anomaly in the immediate vicinity of the mineralization, without any signs of being influenced by glacial movement (Fig. 13.40). Thus this latter dispersion pattern, which could possibly be hydromorphic and/or electrochemical, can be detected by surface sampling of humus, which is cheap compared to deep sampling of lodgment till.

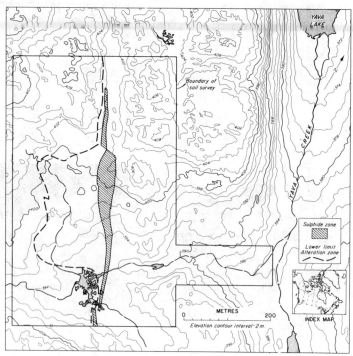

Figure 13.35. *Location of a soil survey area in permafrost environments in the eastern part of the Slave Province, Canada. After Cameron (1977).*

Table 13.3

Varimax factor matrix of logarithmically transformed soil data.
Matrix accounts for 90.2 percent of variance of 284 samples.
Only loadings of 0.25 or greater shown
After Cameron (1977).

	Factor						
	1	2	3	4	5	6	7
Sum of squares	3.6	1.2	1.4	1.2	1.2	0.9	1.3
Mg		0.96					
Ca	-0.25		-0.77	0.40			
Mn		0.47	-0.30	0.46		-0.31	0.50
Fe	0.35			0.83		0.27	
Ni		0.26					0.92
Cu	0.48		0.63	0.29	0.36		
Zn					0.92		
As	0.45			0.29		0.77	
Ag	0.85		0.30	0.25			
Hg	0.83					0.33	
Pb	0.83		0.37				
Au	0.90						

SAMPLING METHODS AND COSTS

Geochemical soil-sampling methods at surface are well known and need no elaboration. However, methods of obtaining overburden samples at depth have been under development for some 20 years and a brief review of these developments will be presented.

The earliest geochemical sampling of till at depth was done in Scandinavia and Canada from hand-dug pits and by the use of hand-operated soil augers. Later Ermengen (1957) in his geochemical studies of the Chibougamau area, Quebec used a "pass-through" type of sampler for sampling till to depths of about 5 m. The sampling tool was designed so that the soil filling the head was continously evacuated under the pressure of newly introduced material as the instrument was hand driven down with a sledge hammer. The sample recovered when the device was pulled out represented the bottom few centimetres of the traversed till. More recently a similar sampling device (Holman-type or Barymin Sampler) driven by small percussion drills has been used in Finland (Kauranne, 1975). Ermengen's studies were followed by those of Gleeson (1960), who used an hydraulic-powered drill mounted on a muskeg tractor and conveyor flight augers to profile to depths of 30 m. Van Tassel (1969) sampled tills in the Keno Hill area, Yukon Territory using an Atlas Copco overburden drill. The tract-mounted drill used a rotary action to penetrate the permanently frozen glacial overburden and compressed air forced the cuttings to surface. This program was successful in indicating a new high-grade silver deposit (Husky orebody) which is presently being mined.

In 1969, Gleeson and Cormier (1971) developed a light-weight percussion sampling system which proved effective in discovering the Louvem zinc deposit which has been exploited near Val d'Or, Quebec. This equipment was effective in sampling up to 30 m of till under glacial lake clays and silts which varied in thickness from 1 to 70 m.

As a result of extensive experimentation, Texasgulf Inc. and Bradley Brothers Diamond Drilling Limited developed a dual-tube rotary drill system which was mounted on a Nodwell tractor and first used in the Timmins area, Ontario. The system uses a dual-tube attached to a tri-cone bit (Skinner, 1972); by rotary action the bit breaks the material which is washed up to surface through the inside tube. This drill has been effective in sampling overburden up to depths of hundreds of metres. In Finland, percussion drills mounted on farm tractors have been used to sample tills at depths of 3 m or more. (Wennerwirta, 1973; Kokkola, 1976)

The costs of overburden sampling vary considerably depending on the type of equipment used, depth and type of overburden and density of the holes; Kokkola (1976) reported that the average cost per sample in Finland in 1974 was $40(US). In Canada, costs (in 1977) using the small percussion drills vary from $3-$8 per metre, for tractor-mounted pneumatic drills they vary from $4 to $15 per metre and for the rotary dual-tube system costs may vary from $10 to $90 per metre. Hence to collect overburden samples at depth may cost from $40 to $300 each in contrast to surface soil samples which may cost from $0.50 to $20 each.

CHEMICAL ANALYSIS OF SOILS

A presentation of various methods for chemical analysis of glacial soils in geochemical exploration is outside the scope of this paper. It should be emphasized, however, that a

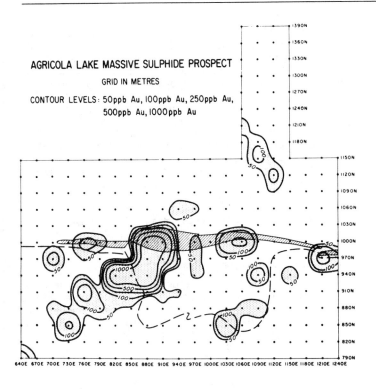

Figure 13.36. Distribution of gold in the minus 0.18 mm fraction of soils taken at a depth of 15-20 cm in a survey area in the eastern part of the Slave Province, Canada (see Fig. 13.35). After Cameron (1977).

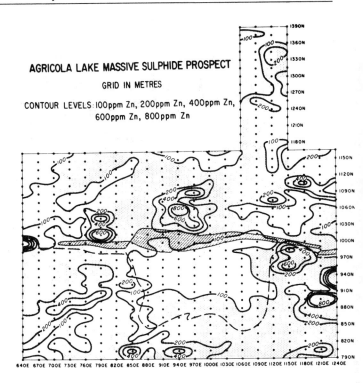

Figure 13.38. Distribution of zinc in the minus 0.18 mm fraction of soils taken at a depth of 15-20 cm in a survey area in the eastern part of the Slave Province, Canada (see Fig. 13.35). After Cameron (1977).

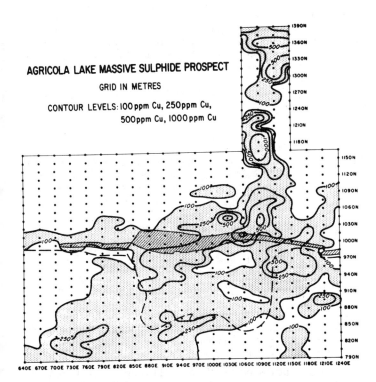

Figure 13.37. Distribution of copper in the minus 0.18 mm fraction of soils taken at a depth of 15-20 cm in a survey area in the eastern part of the Slave Province, Canada (see Fig. 13.35). After Cameron (1977).

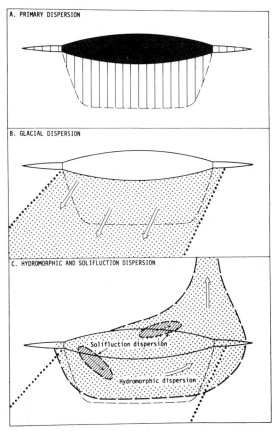

Figure 13.39. Generalized historical sequence of dispersion processes based on data from a study area in the eastern part of the Slave Province, Canada (see Fig. 13.35 to 13.38). After Cameron (1977).

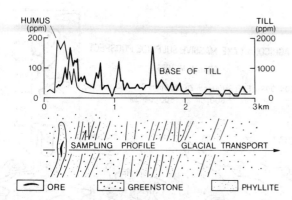

Figure 13.40. *Copper in humus and copper at the base of till from a traverse along the direction of glacial transport across the Pahtavuoma copper deposit, Finland. Redrawn and somewhat generalized from Kokkola (1977).*

meaningful interpretation of analytical data for soils requires knowledge about the mode in which the elements occur in the soil. A discussion of this problem leads to two practical questions (1) which chemical attack should be used on the soil samples, and (2) which grain sizes should be analyzed.

Mode of Occurrence of Trace Elements

Rose (1975) divided the modes of occurrence of trace elements in soils into four important groups: (1) as a major element in trace minerals; (2) as a trace constituent in a mineral of the parent material; (3) as a trace constituent in the lattice of a mineral formed during weathering; and (4) as a trace constituent adsorbed as a counter ion on the surface of an iron or manganese oxide, colloidal clay or organic particle, or in the exchange layer of a clay mineral.

In areas of glaciated terrane purely syngenetic geochemical dispersion patterns would normally comprise modes of occurrence of groups (1) and (2), while purely epigenetic patterns would mainly comprise groups (3) and (4). In complex syngenetic/epigenetic patterns the elements will occur as any possible combination of groups (1), (2), (3), and (4).

Chemical Digestion of Soil Samples

The disclosure of syngenetic dispersion patterns, requires a chemical digestion which is rigorous enough to attack primary minerals. Ideally, this would often require solid phase methods like X-ray fluorescence, and optical spectrography, or use of a perchloric-hydrofluoric acid digestion (Foster, 1971). However, in some cases, depending on the target, less rigorous acid attacks (Ellis et al., 1967; Lynch, 1971; Foster, 1971, 1973; Olade and Fletcher, 1974 and Peachey, 1976) or the simple potassium bisulphate fusion (Harden and Tooms, 1964) will suffice.

The disclosure of epigenetic dispersion patterns would normally not require a very strong chemical attack on the samples. In fact, the results of partial extraction techniques would in some cases give better contrast between anomaly and background than the totals. This was the case in a study of zinc in peats from several bogs in eastern Canada (Gleeson and Coope, 1967). However, Maynard and Fletcher (1973) arrived at the opposite conclusion for the analysis of copper in peat, stating that the greatest contrast was obtained using nitric acid-perchloric digestion. Bradshaw et al. (1974), nevertheless, found that the percentage of cold-extractable metals in mineral soil is increased where the anomalies are a result of hydromorphic accumulation, and increases very dramatically in organic soils under the same conditions. These different results obtained by various workers clearly indicate that partial extractions of metals from soils to help define epigenetic dispersion patterns is a field which warrants further research. An ideal sample attack would be one that selectively dissolves trace elements of a certain mode of occurrence or a certain genesis. Interesting approaches to this problem are provided by: Meyer and Leen (1973) (evolvement of metal vapours); Nurmi (1976), Toverud (1977) (results of partial and total extractions); Aleksandrova (1960), Antropova (1975) (organically bonded metals); Govett (1974, 1975), Nuutilainen and Peuraniemi (1977), Kokkola (1977) (extractants which disclose possible electrochemical dispersion patterns); Chao (1972), Chao and Sanzalone (1973) (dissolution of secondary iron and manganese oxides); Chao and Theobald (1976), Gatehouse et al. (1976) (sequential soil analysis).

Grain-size Fractions for Analysis

Shilts (1975a, b) discussed the mineralogical composition of various size fractions in till. Particles with a diameter of more than 0.25 mm consist mainly of mineral aggregates (rock fragments). Particles from about 0.060 mm to 0.25 mm consist mostly of quartz-feldspar mineral grains with varying amounts of carbonates, heavy minerals and rock fragments. From about 0.002 mm to 0.060 mm, the particles are mostly mineral grains consisting generally of over 90 per cent quartz, feldspars and carbonates, with minor percentages of heavy and phyllosilicate minerals. Particles finer than 0.002 mm consist predominantly of phyllosilicates (micas, clay minerals, chlorite etc.) and of easily crushed secondary oxides (limonite, hematite etc.), with varying accessory amounts of quartz, feldspar and carbonates. The proportions of these textural classes can vary over a limited area; apparently any analysis of till samples that combines two or more of these textural classes may be difficult to compare. In a successful study Shilts (1976) used the minus 0.002 mm fraction for analysis aiming at finding syngenetic dispersion patterns, utilizing the fact that some of the cations released locally from primary minerals in the coarser till fractions would be scavenged by a fairly constant proportion of scavengers in the fine fraction.

From a study of heavy metals in different till fractions in Sweden, Eriksson (1976a) concluded that the greatest contrast between anomalous and background values was obtained from the finest fractions when using HCl/HNO_3 extraction, and from the coarser fraction when analyzing the total metal content.

CONCLUSIONS AND FUTURE TRENDS

The processes and methods when carrying out geochemical soil studies in glaciated terrane are many and varied. An understanding of glacial as well as postglacial processes is paramount if proper interpretation of geochemical soil data is to be accomplished.

Bradshaw et al. (1974) introduced the use of conceptual models in exploration geochemistry. Application of such models based on Canadian case histories was done by Bradshaw (1975a) and later in Europe by Kauranne (1976).

Conceptual models (Fig. 13.41) represent a systematic way of visualizing the various factors that influence metal dispersion and provide a framework into which further data can be fitted. They are of particular importance in glaciated areas where the dispersion processes can be very complex. By expanding this type of presentation to include more and different situations for various elements, a clearer understanding of geochemical processes in relation to the physical and chemical nature of the landscape can be attained. Such clarification will lead to a better interpretation of geochemical data as well as pointing out gaps in our knowledge.

An extensive program of systematic regional till sampling has been established in Finland (Kauranne, 1975; Stigzelius, 1976); large-scale regional till sampling is also being carried out in Canada (Klassen and Shilts, 1977). Such programs have already proved fruitful in evaluating the mineral potential of various areas and are particularly valuable where conventional reconnaissance stream and lake sediment surveys might not be applicable. One can expect that the vast amount of information obtained through this type of survey will greatly improve our knowledge about the use of soils for geochemical prospecting in glaciated areas. However, more studies into Quaternary geology, in particular into the three-dimensional configuration of indicator trains associated with mineral deposits are required, especially in areas covered by glaciolacustrine sediments; these basic studies are needed to help interpret regional till data.

A better understanding of postglacial geochemical dispersion processes is also required if we are to advance the applicability of soil sampling in glaciated environments. Utilization of such processes may be beniticial even in areas where significant glacial dispersal trains are not detectable.

More research is needed in the use and understanding of gaseous dispersion in glaciated terrane. Airborne instruments for detection of gaseous or particulate anomalies caused by mineral deposits are today a real possibility (Barringer, see this volume, Paper 16 (abstract only)).

Studies into electrochemical processes and their role in the movement of ions through bedrock and glacial overburden will continue. The process is probably most important in areas of low relief, some data indicate that it might be operative even where sulphide deposits are covered by thick glaciolacustrine deposits (Govett and Chork, 1977).

A greater use of organic material as a sampling medium can be expected. Utilization of the enrichment of heavy metals by humus has great potential in geochemical prospecting, especially in areas with moderate or high relief where the anomalies formed in humus material can be so strong that the effects of most other dispersion mechanisms are overshadowed (Boyle, 1977) and where vegetation is strongly affected (Bølviken and Låg, 1977), sometimes to the extent that the naturally poisoned areas developed may be detectable by remote sensing techniques (Bølviken et al., 1977).

Very sensitive, rapid and reliable methods for determining chemical constituents of soils are now available to the geochemist. However, we lack knowledge concerning the most applicable extraction techniques that would best differentiate various types of syngenetic and epigenetic anomalies. Further research into this problem provides one of the promising avenues to a more efficient utilization of soils for geochemical prospecting in glaciated terrane.

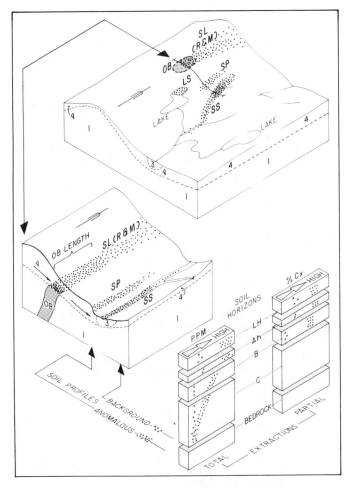

SL(R&M)	=	Residual (R) soil and anomaly
SS	=	stream sediment anomaly
LS	=	lake sediment anomaly
SP	=	seepage anomaly
OB	=	orebody

Figure 13.41. Idealized model for geochemical dispersion of mobile elements in well-drained and poorly-drained ground-till overburden. After Bradshaw (1975).

REFERENCES

Albov, M.N. and Kostariev, I.I.
    1968: Copper in peat bog, formations of the Middle Urals; Soviet Geol., no. 2, p. 132-139.

Aleksandrova, L.N.
    1960: The use of sodium pyrophospate for isolating free humuc substances and their organic-mineral compounds from soil; Soviet Soil Sci., no. 2, p. 190-197.

Alley, D.W. and Slatt, R.M.
    1976: Drift prospecting and glacial geology in the Sheffield Lake – Indian Pond Area, North-central Newfoundland; in Glacial till: an interdisciplinary study R.F. Legget (Ed.); Royal Soc. Can., Spec. Publ., No. 12, p. 249-266.

Antropova, L.V.
    1975: Forms of occurrence of elements in dispersion haloes of ore deposit (In Russian); Nedra, Leningrad, 144 p.

Armands, G.
    1967: Geochemical prospecting of a uraniferous bog deposit at Masugnsbyn, northern Sweden; in Geochemical prospecting in Fennoscandia. A. Kvalheim (Ed.); Interscience Publishers, New York, p. 127-154.

Azzaria, L.M. and Webber, G.R.
    1969: Mercury analysis in geochemical exploration; Can. Inst. Min. Met. Bull., v. 62 (685), p. 521-530.

Banerjee, I. and McDonald, B.C.
  1976: Nature of esker sedimentation; in Glaciofluvial and lacustrine sedimentation A.V. Jopling and B.C. McDonald (Eds.); Soc. Econ. Paleontol. Mineral., Spec. Publ. 23, p. 132-154.

Beck, L.S. and Gingrich, J.E.
  1976: Track etch orientation survey in the Cluff Lake Area, northern Saskatchewan; Can. Inst. Min. Met. Bull., v. 69 (769), p. 104-109.

Björklund, A.
  1976a: The use of till in geochemical uranium exploration; in Exploration of uranium ore deposits; Int. Atomic. Energy Agency, p. 283-295.

  1976b: Pauktajanvaara: Uranium in till above mineralization; in Conceptual models in exploration geochemistry. L.K. Kauranne (Ed.); J. Geochem. Explor., v. 5 (3), p. 287-297.

Bjørlykke, A., Bølviken, B., Eidsvig, P., and Svinndal, S.
  1973: Exploration for disseminated lead in southern Norway; in Prospecting in areas of glacial terrain. M.J. Jones (Ed.); Inst. Min. Metall., p. 111-126.

Bølviken, B.
  1967: Recent geochemical prospecting in Norway; in Geochemical prospecting in Fennoscandia. A. Kvalheim (Ed.); Interscience Publishers, New York, p. 225-253.

  1976: Snertingdal: A lead occurrence found by systematic prospecting; in Conceptual models in exploration geochemistry. L.K. Kauranne (Ed.); J. Geochem. Explor., v. 5 (3), p. 324-331.

  1978: The redox potential field of the earth; in Origin and distribution of the elements. L.H. Ahrens (Ed.); Proceedings of the Second Symposium, Paris — UNESCO, May 1977. Pergamon Press, p. 649-665.

Bølviken, B. and Lag, J.
  1977: Natural heavy-metal poisoning of soils and vegetation: an exploration tool in glaciated terrain; Trans. Inst. Min. Metall. (Section B), v. 86, p. B173-B180.

Bølviken, B. and Logn, Ø.
  1975: An electrochemical model for element distribution around sulphide bodies; in Geochemical Exploration 1974. I.L. Elliott and W.K. Fletcher (Eds.); Elsevier Publ. Co., p. 631-648.

Bølviken, B., Honey, F., Levine, S., Lyon, R., and Prelat, A.
  1977: Detection of naturally heavy-metal poisoned areas by LANDSAT-1 digital data; in Geochemical exploration 1976. C.R.M. Butt and I.G.P. Wilding (Eds.); J. Geochem. Explor., v. 8 (1/2), p. 457-471.

Borovitskii, V.P.
  1976: The application of bog-sampling in prospecting for ore deposits in perennial frost regions; J. Geochem. Explor., v. 5 (1), p. 67-70.

Boyle, R.W.
  1967: Geochemical prospecting research in 1966, Cobalt area, Ontario; Geol. Surv. Can., Paper 66-46, 15 p.

  1977: Cupriferous bogs in the Sackville area, New Brunswick, Canada; J. Geochem. Explor., v. 8 (3), p. 495-527.

Boyle, R.W. and Dass, A.S.
  1967: Geochemical prospecting — use of the A horizon in soil surveys; Econ. Geol., v. 62 (2), p. 274-276.

Boyle, R.W. and McGerrigle, J.I. (Eds.)
  1971: Geochemical exploration; Can. Inst. Min. Met., Spec. Vol. 11, 593 p.

Bradshaw, P.M.D. (Ed.)
  1975a: Conceptual models in exploration geochemistry; J. Geochem. Explor., v. 4 (1), 213 p.

  1975b: Valley copper deposit, British Columbia; in Conceptual models in exploration geochemistry. P.M.D. Bradshaw (Ed.); J. Geochem. Explor., v. 4 (1), p. 100-101.

Bradshaw, P.M.D., Clews, D.R., and Walker, J.L.
  1972: Exploration geochemistry: A series of seven articles reprinted from Can. Min. J., Barringer Research Ltd., 49 p.

Bradshaw, P.M.D., Thomson, I., Smee, B.W., and Larsson, J.D.
  1974: The application of different analytical extractions and soil profile sampling in exploration geochemistry; J. Geochem. Explor., v. 3 (3), p. 209-225.

Bristow, Q.
  1972: An evaluation of the quartz crystal microbalance as a mercury vapour sensor for soil gases; J. Geochem. Explor., v. 1 (1), p. 55-76.

Bristow, Q. and Jonasson, I.R.
  1972: Vapour sensing for mineral exploration; Can. Min. J., v. 93 (5), p. 39-47.

Brock, J.S.
  1972: The use of dogs as an aid to exploration for sulphides; Western Miner. v. 45 (12), p. 28-32.

Brundin, N.H. and Bergstrøm, J.
  1977: Regional prospecting for ores based on heavy minerals in glacial till; J. Geochem. Explor., v. 7 (1), p. 1-19.

Cachau Herreillat, F. and La Salle, P.
  1969: Essai de mise au point de méthods de prospection geochemique utilisant des formations superficielles anciennes: les esker. Ministere de Richesses Naturelles, Quebec, Open File Report.

Cameron, E.M. (Ed.)
  1967: Proceedings, Symposium on geochemical prospecting, Ottawa, April 1966; Geol. Surv. Can., Paper 66-54, 288 p.

  1977: Geochemical dispersion in mineralized soil of a permafrost environment; J. Geochem. Explor., v. 7 (3), p. 301-326.

Canney, F.C. (Ed.)
  1969: International geochemical exploration symposium; Colo. Sch. Mines Q., v. 64 (1), 520 p.

Cannon, H.L.
  1955: Geochemical relations of zinc bearing peat to the Lockport Dolomite; Orleans Co. N.Y.; U.S. Geol. Surv., Bull. 1000D.

Chao, T.T.
  1972: Selective dissolution of manganese oxides from soils and sediments; Soil Sci. Soc. Am. Proc., v. 36, p. 764-768.

Chao, T.T. and Sanzalone, R.F.
1973: Atomic absorption spectrometric determination of microgram levels of Co, Ni, Cu, Pb, and Zn in soil and sediment extracts containing large amounts of Mn and Fe; U.S. Geol. Surv., J. Res., v. 1 (6), p. 681-685.

Chao, T.T. and Theobald, P.K. Jr.,
1976: The significance of secondary iron and manganese oxides in geochemical exploration; Econ. Geol., v. 71 (8), p. 1560-1569.

Chowdhury, A.N. and Bose, B.B.
1971: Role of "humus matter" in the formation of geochemical anomalies; in Geochemical Exploration. R.W. Boyle and J.I. McGerrigle (Eds.); Can. Inst. Min. Met., Spec. Vol. 11, p. 410-413.

Danielson, A.
1968: Spectro chemical analysis for geochemical purposes; in XIII Colloquium Spectroscopicum Internationale; Adam Hilger, London, p. 311-323.

Davenport, P.H.
1972: The application of geochemistry to base metal exploration in Birch-Uchi Lakes Volcano-sedimentary belt, Northwestern Ontario; Unpub. Ph.D. thesis, Queen's University, Kingston, Ontario, 411 p.

DiLabio, R.N.W.
1976: Glacial dispersion of rocks and minerals in the Lac Mistassimi-Lac Waconichi Area, Quebec with special reference to the Icon dispersal train; unpubl. Ph.D. thesis, Univ. Western Ontario, London, Ontario.

Donovan, P.R. and James, C.H.
1967: Geochemical dispersion in glacial overburden over the Tynagh (Northgate) base metal deposit, Westcentral Eire; in Proc. Symposium on geochemical prospecting, Ottawa, April 1966. E.M. Cameron (Ed.); Geol. Surv. Can., Paper 66-54, p. 89-110.

Dreimanis, A.
1958: Tracing ore boulders as a prospecting method in Canada; Can. Inst. Min. Met., Bull., v. 51 (550), p. 73-80.

1960: Geochemical prospecting for Cu, Pb, and Zn in glaciated areas, eastern Canada; 21st Int. Geol. Cong., Norden, Pt. II, p. 7-19.

1976: Tills: their origin and properties; in Glacial till. R.F. Legget (Ed.); Roy. Soc. Can., Spec. Publ. No. 12, p. 11-49.

Dreimanis, A. and Vagners, U.J.
1969: Lithologic relation of till to bedrock; in Quaternary Geology and climate. H.W. Wright (Ed.); Nat. Acad. Sci. Publ. 1701, p. 93-98.

Dyck, W.
1968: Radon-222 emanations from a uranium deposit; Econ. Geol., v. 63 (3), p. 288-289.

1969: Development of uranium exploration methods using radon; Geol. Surv. Can., Paper 69-46, 26 p.

1972: Radon methods of prospecting in Canada; in Uranium prospecting handbook. S.H.J. Bowie, M. Davis and D. Ostle (Eds.); Inst. Min. Metall., p. 212-243.

1976: The use of helium in mineral exploration; J. Geochem. Explor., v. 5 (1), p. 3-20.

Dyck, W., Jonasson, I.R., and Liard, R.F.
1976: Uranium prospecting with ^{222}Rn in frozen terrain; J. Geochem. Explor., v. 5 (2), p. 115-128.

Eckel, E.G.
1949: Geology and ore deposits of the La Plata District, Colorado; U.S. Geol. Surv., Prof. Paper 219, 55 p.

Ekdahl, E.
1976: Pielavesi: the use of dogs in prospecting; in Conseptual models in exploration geochemistry. L.K. Kauranne (Ed.); J. Geochem. Explor., v. 5 (3), p. 296-298.

Elliott, I.L. and Fletcher, W.K. (Eds.)
1975: Geochemical exploration 1974; Elsevier Publ. Co. Amsterdam, 720 p.

Ellis, A.J., Tooms, J.S., Webb, J.S., and Bicknell, J.A.
1967: Application of solution experiments in geochemical prospecting; Inst. Min. Metall., Trans., v. 76, p. B25-B39.

Erämetsä, O., Lounama, K.J., and Haukka, M.
1969: The vertical distribution of uranium in Finnish peat bogs; Soum. Kemistil. B, v. 42 (4), p. 363-370.

Eriksson, K.
1973: Prospecting in an area of central Sweden; in Prospecting in areas of glacial terrain. M.J. Jones (Ed.); Inst. Min. Metall., p. 83-86.

1976a: Heavy metals in different till fractions; in Conceptual models in exploration geochemistry. L.K. Kauranne (Ed.); J. Geochem. Explor., v. 5 (3), p. 383-387.

1976b: Regional prospecting by the use of peat sampling; in Conceptual models in exploration geochemistry. L.K. Kauranne (Ed.); J. Geochem. Explor., v. 5 (3), p. 387-388.

Eriksson, K. and Eriksson, G.
1976a: Strømbo: heavy metals in peat close to a boulder train; in Conceptual models in exploration geochemistry. L.K. Kauranne (Ed.); J. Geochem. Explor., v. 5 (3), p. 342-344.

1976b: Hinsen: heavy metals in peat and till; in Conceptual models in exploration geochemistry. L.K. Kauranne (Ed.); J. Geochem. Explor., v. 5 (3), p. 232-235.

Ermengen, S.V.
1957: A report on glacial geology and geochemical dispersion in the Chibougamau area, Quebec; Unpubl. rep., Que. Dep. Nat. Resour.

Flint, R.F.
1971: Glacial and Quaternary Geology; Wiley, New York, 892 p.

Forrester, J.D.
1942: A native copper deposit near Jefferson City, Montana; Econ. Geol., v. 37 (2), p. 126-135.

Fortescue, J.A.C.
1975: Dorset area, Ontario; in Conceptual models in exploration geochemistry. P.M.D. Bradshaw (Ed.); J. Geochem. Explor. v. 4 (1), p. 159-162.

Fortescue, J.A.C. and Hornbrook, E.H.W.
1967: Progress report on biogeochemical research at the Geological Survey of Canada 1963-1966; Geol. Surv. Can., Paper 67-23, Part I, 143 p.

1969: Progress report on biogeochemical research at the Geological Survey of Canada 1963-1966; Geol. Surv. Can., Paper 67-23, Part II, 101 p.

Foster, J.R.
 1971: The reduction of matrix effects in atomic absorption analysis and the efficiency of selected extraction in rock forming minerals; in Geochemical Exploration. R.W. Boyle and J.I. McGerrigle (Eds.), Can. Inst. Min. Met., Spec. Vol. 11, p. 554-560.

 1973: The efficiency of various digestion procedures on the extraction of metals from rocks and rock forming minerals; Can. Inst. Min. Met., Bull., v. 66 (728), p. 85-92.

Fraser, D.C.
 1961a: A syngenetic copper deposit of recent age; Econ. Geol., v. 56 (5), p. 951-962.

 1961b: Organic sequestration of copper; Econ. Geol., v. 56 (6), p. 1063-1078.

Freeman, E.B. and Ferguson, S.A.
 1973: Tracing float and mineral fragments; Ont. Div. Mines, Misc. Paper 55, p. 43-60.

Garrett, R.G.
 1971: The dispersion of copper and zinc in glacial overburden at the Louvem deposit, Val d'Or, Quebec; in Geochemical Exploration. R.W. Boyle and J.I. McGerrigle (Eds.); Can. Inst. Min. Met., Spec., Vol. 11, p. 157-158.

Gatehouse, S., Russell, D.W., and van Moort, J.C.
 1977: Sequential soil analysis in exploration geochemistry; in Geochemical Exploration 1976. C.R.M. Butt and I.G.P. Wilding (Eds.); J. Geochem. Explor., v. 8 (1/2), p. 483-494.

Geoffroy, P.R. and Koulomzine, T.
 1960: Mogador sulphide deposit; Can. Inst. Min. Met., Bull., v. 53 (578), p. 268-274.

Gleeson, C.F.
 1960: Studies on the distribution of metals in bogs and glaciolacustrine deposits; unpubl. Ph.D. thesis, McGill Univ., Montreal, 221 p.

Gleeson, C.F. and Coope, J.A.
 1967: Some observations on the distribution of metals in swamps in Eastern Canada; Geol. Surv. Can., Paper 66-54, p. 145-165.

Gleeson, C.F. and Cormier, R.
 1971: Evaluation by geochemistry of geophysical anomalies and geological targets using overburden sampling at depth; in Geochemical Exploration. R.W. Boyle and J.I. McGerrigle (Eds.); Can. Inst. Min. Met., Spec. vol. 11, p. 159-165.

Goldthwait, R.P. (Ed.)
 1971a: Till: A symposium; Ohio State Univ. Press, Columbus, Ohio, 402 p.

 1971b: Introduction to Till, Today; in Till: A Symposium. R.P. Goldthwait (Ed.); Ohio State Univ. Press, Columbus, p. 3-26.

Govett, G.J.S.
 1973: Differential secondary dispersion in transported soils and post-mineralization rocks: an electrochemical interpretation; in Geochemical Exploration 1972. M.J. Jones (Ed.); Inst. Min. Metall., p. 81-91.

 1974: Soil conductivity measurements: a technique for the detection of buried sulphide deposits; Inst. Min. Metall. Trans., v. 83, p. B29-B30.

Govett, G.J.S. (cont.)
 1975: Soil conductivities: assessment of an electrochemical exploration technique; in Geochemical Exploration 1974. I.L. Elliott and W.K. Fletcher (Eds.); Elsevier Publ. Co., Amsterdam, p. 101-118.

 1976a: Detection of deeply buried and blind sulphide deposits by measurement of H^+ and conductivity of closely spaced surface soil samples; J. Geochem. Explor., v. 6 (3), p. 359-382.

 1976b: Exploration geochemistry in the Appalachians; J. Geochem. Explor., v. 6 (1/2), 298 p.

Govett, G.J.S. and Chork, C.Y.
 1977: Detection of deeply-buried sulphide deposits by measurement of organic carbon, hydrogen ion and conductance of surface soils; in Prospecting in areas of glaciated terrain 1977. G.R. Davis (Ed.); Inst. Min. Metall., p. 49-55.

Govett, G.J.S., Goodfellow, W.D., and Whitehead, R.E.S.
 1976: Experimental aqueous dispersion of elements around sulphides; Econ. Geol., v. 71 (5), p. 925-940.

Govett, G.J.S. and Whitehead, R.E.S.
 1974: Origin of metal zoning in stratiform sulphides: a hypothesis; Econ.Geol., v. 69 (4), p. 551-556.

Grip, E.
 1953: Tracing of glacial boulders as an aid to ore prospecting in Sweden; Econ. Geol., v. 48 (8), p. 715-725.

Guton, J.E. and Nichol, I.
 1974: The delineation and interpretation of metal dispersion patterns related to mineralization in the Whipsaw Creek area, near Princeton, B.C.; Can. Inst. Min. Met., Bull., v. 67 (741), p. 66-74.

Harden, G. and Tooms, J.S.
 1964: Efficiency of the potassium bisulphate fusion in geochemical analysis; Inst. Min. Metall. Trans., v. 74, p. 129-141.

Hawkes, H.E.
 1972: Exploration geochemistry bibliography. January 1965 to December 1971; Assoc. Explor. Geochem., Spec. Vol. 1, 118 p.

 1976: Exploration geochemistry bibliography. January 1972 to December 1975; Assoc. Explor. Geochem., Spec. Vol. 5, 195 p.

Hawkes, H.E. and Salmon, M.L.
 1960: Trace elements in organic soil as a guide to copper ore; 21st Int. Geol. Cong., Norden, Part 2, p. 38-43.

Hawkes, H.E. and Webb, J.S.
 1962: Geochemistry in mineral exploration; Harper and Row, New York, 415 p.

Henwood, W.J.
 1857: Notes on the copper turf of Merioneth; Edinburgh New Philos. J., v. 5, p. 61-64.

Hirvas, H.
 1977: Glacial transport in Finnish Lapland; in Prospecting in areas of glaciated terrain 1977. G.R. Davis (Ed.); Inst. Min. Metall., p. 128-137.

Hirvas, H., Kujansuu, R., and Tynni, R.
 1976: On till stratigraphy in Northern Finland, Quaternary glaciating in the Northern Hemisphere; Rep. No. 3, Project 73/1/24, p. 256-273.

Holmes, C.D.
  1941: Till fabric; Geol. Soc. Am., Bull., v. 52 (9), p. 1299-1354.

  1952: Drift dispersion in westcentral New York; Geol. Soc. Am., Bull., v. 63 (7), p. 993-1010.

Hornsnail, R.F.
  1975: Highmont Cu-Mo deposits, British Columbia; in Conceptual models in exploration geochemistry. P.M.D. Bradshaw (Ed.); J. Geochem. Explor., v. 4 (1), p. 67-72.

Hornsnail, R.F. and Elliott, I.L.
  1971: Some environmental influences on the secondary dispersion of molybdenum and copper in western Canada; in Geochemical Exploration. R.W. Boyle and J.I. McGerrigle (Eds.), Can. Inst. Min. Met., Spec. vol. 11, p. 166-175.

Hvatum, O.Ø.
  1971: Strong enrichment of lead in surface layers of peat (In Norwegian); Tekn. Ukeblad, No. 27, p. 40.

Jenny, H.
  1941: Factors of soil formation; McGraw-Hill Book Co., New York, 281 p.

Jonasson, I.R.
  1972: Mercury in soil gas applied to exploration for sulphide ores; in Report of Activities, Part A, Geol. Surv. Can., Paper 72-1A, p. 74.

Jones, M.J. (Ed.)
  1973a: Prospecting in areas of glacial terrain; Inst. Min. Metall., 138 p.

  1973b: Geochemical exploration 1972; Inst. Min. Metall., 458 p.

  1975: Prospecting in areas of glaciated terrain; Inst. Min. Metall., 154 p.

  1977: Prospecting in areas of glaciated terrain; Inst. Min. Metall., 140 p.

Juve, G.
  1974: Ore mineralogy and ore types of the Stekenjokk deposit, central Scandinavian Caledonides, Sweden; Sver. Geol. Unders., C. 706, 162 p.

  1977: Formation of native copper in glacial overburden above the Stekenjokk ore body; in Proc. Malmgeologisk Symposium, Trondheim, 23-25 Nov. 1977; B.V.L.I. Bergavdelingen, Techn. Univ. Norway, Trondheim.

Kahma, A.
  1965: Trained dog as a tracer of sulphide bearing glacial boulders; Sedimentology, v. 5 (4), p. 57.

Kahma, A., Nurmi, A., and Mattsson, P.
  1975: On the composition of the gases generated by sulphide-bearing boulders during weathering and on the ability of prospecting dogs to detect samples treated with these gases in the terrain; Geol. Surv. Finland, Rep. Inv. No. 6, 6 p.

Karup-Moeller and Brummer, J.J.
  1970: The George Lake zinc deposit, Wollaston Lake area, northwestern Saskatchewan; Econ. Geol., v. 65 (7), p. 862-874.

Kauranne, L.K.
  1967a: Aspects of geochemical humus investigation in glaciated terrain; in Geochemical prospecting in Fennoscandia. A. Kvalheim (Ed.); Interscience Publishers, New York, p. 261-272.

Kauranne, L.K. (cont.)
  1967b: Facts to be noticed in pedogeochemical prospecting by sampling and analysis of glacial till; in Geochemical prospecting in Fennoscandia. A. Kvalheim (Ed.); Interscience Publishers, New York, p. 273-278.

  1975: Regional geochemical mapping in Finland; in Prospecting in areas of glaciated terrain. M.J. Jones (Ed.); Inst. Min. Metall., p. 71-81.

  1976a: Conceptual models in exploration geochemistry, Norden 1975; J. Geochem. Explor., v. 5 (3), 420 p.

  1976b: Petolahti: Copper and nickel in till and humus; in Conceptual models in exploration geochemistry. L.K. Kauranne (Ed.); J. Geochem. Explor., v. 5 (3), p. 292-296.

Klassen, R.A. and Shilts, W.W.
  1977: Glacial dispersal of uranium in the district of Keewatin, Canada; in Prospecting in areas of glaciated terrain. G.R. Davis (Ed.); Inst. Min. Metall., p. 80-88.

Kochenov, A.V., Zinevyev, V.V., and Lovaleva, S.S.
  1966: Some features of the accumulation of uranium in peat bogs; Int. Geochem., v. 2 (1), p. 65-70.

Kokkola, M.
  1975: Stratigraphy of till at Hitura open-pit Nivala Western Finland, and its bearing on geochemical prospecting; in Prospecting in areas of glaciated terrain 1975. M.J. Jones (Ed.); Inst. Min. Metall., p. 149-154.

  1976: Geochemical sampling of deep moraine and the bedrock surface underlying overburden; in Conceptual models in exploration geochemistry. L.K. Kauranne (Ed.); J. Geochem. Explor., v. 5 (3), p. 395-400.

  1977: Application of humus to exploration; in Prospecting in areas of glaciated terrain 1977. G.R. Davis (Ed.); Inst. Min. Metall., p. 104-110.

Kokkola, M. and Korkalo, T.
  1976: Pahtavuoma: copper in stream sediment and till; in Conceptual models in exploration geochemistry. L.K. Kauranne (Ed.); J. Geochem. Explor., v. 5 (3), p. 280-287.

Korpela, K.
  1969: Die Weichsel-Eiszeit und ihr Interstadial in Peräpohjola (nördliches Nordfinland) im Licht von submoränen Sedimenten; Annis Acad. Scient. Fennicae, Ser. AIII, No. 99, 108 p.

Kravstov, A.I. and Fridman, A.I.
  1965: Natural gases of ore deposits; Doklady Akad. Nauk, SSSR, v. 165 (5), p. 168-175.

Krumbein, W.C.
  1939: Preferred orientation of pebbles in sedimentary deposits; J. Geology, v. 47, p. 673-706.

Kujansuu, R.
  1967: On the deglaciation of western Finnish Lapland; Finland. Comm. Geol. Bull., No. 232, 98 p.

Kujansuu, R.
  1976: Glaciogeological surveys for ore-prospecting purposes in Northern Finland; in Glacial till. An interdisciplinary study R.F. Legget (Ed.); Roy. Soc. Can., Spec. Publ. No. 12, p. 225-239.

Kvalheim, A. (Ed.)
  1967: Geochemical prospecting in Fennoscandia; Interscience Publishers, New York, 350 p.

Låg, J. and Bølviken, B.
1974: Some naturally heavy metal poisoned areas of interest in prospecting, soil chemistry and geomedicine; Norg. Geol. Unders., No. 304, p. 73-96.

Låg, J., Hvatum, O.Ø., and Bølviken, B.
1970: An occurrence of naturally lead-poisoned soil at Kastad near Gjøvik, Norway; Norg. Geol. Unders., No. 266, p. 141-159.

Lakin, H.W., Curtin, G.C., and Hubert, A.F.
1974: Geochemistry of gold in the weathering cycle; U.S. Geol. Surv., Bull. 1330, 80 p.

Larsson, J.O.
1976: Organic stream sediments in regional geochemical prospecting, Precambrian Pajala District, Sweden; in Exploration geochemistry in the Appalachians. G.J.S. Govett (Ed.); J. Geochem. Explor., v. 6 (1/2), p. 233-249.

Lee, H.A.
1963: Glacial fans in till from the Kirkland Lake fault; Geol. Surv. Can., Paper 63-45, 36 p.

1965: Investigations of eskers for mineral exploration; Geol. Surv. Can., Paper 65-14, 20 p.

1968: Glaciofocus and the Munro esker of northern Ontario; in Report of Activities, Part A, Geol. Surv. Can., Paper 68-1A, p. 173.

Legget, R.F. (Ed.)
1976: Glacial till — an interdisciplinary study; Roy. Soc. Can., Spec. Publ., No. 12, 412 p.

Levinson, A.A.
1974: Introduction to exploration geochemistry; Applied Publishing Ltd., Calgary, 612 p.

Lindmark, B.
1977: Till-sampling methods used in exploration for scheelite in Kaustinen, Finland; in Prospecting in areas of glaciated terrain 1977. G.R. Davis (Ed.); Inst. Min. Metall., p. 45-48.

Logn, Ø. and Bølviken, B.
1974: Self potential at the Joma pyrite deposit, Norway; Geoexploration, v. 12 (1), p. 11-28.

Lovering, T.S.
1928: Organic precipitation of metallic copper; U.S. Geol.Surv., Bull. 795-C, p. 45-52.

Lynch, J.J.
1971: The determination of copper, nickel and cobalt in rocks by atomic absorption spectrometry using a cold leach; in Geochemical Exploration. R.W. Boyle and J.I. McGerrigle (Eds.); Can. Inst. Min. Met., Spec. Vol. 11, p. 313-314.

Manskaya, S.M. and Drozdova, T.V.
1964: Geochemistry of organic substances; Nauka Press, Moscow, 315 p. (Russian).

Maynard, D.E. and Fletcher, W.K.
1973: Comparison of total and partial extractable copper in anomalous and background peat samples; J. Geochem. Explor., v. 2 (1), p. 19-24.

McCarthy, J.H.
1972: Mercury vapor and other volatile components in the air as guides to ore deposits; J. Geochem. Explor., v. 1 (2), p. 143-162.

McDonald, B.C.
1971: Sedimentology and pebble transport in the Windsor esker, Quebec; in Report of Activities, Part A, Geol. Surv. Can., Paper 71-1A, p. 190-191

McNerney, J.J. and Buseck, P.R.
1973: Geochemical exploration using mercury vapors; Econ. Geol., v. 68 (8), p. 1313-1320.

Mehrtens, M.B. and Tooms, J.S.
1973: Geochemical drainage dispersion from sulphide mineralization in glaciated terrain, central Norway; in Prospecting in areas of glacial terrain. M.J. Jones (Ed.); Inst. Min. Metall., p. 1-10.

Mehrtens, M.B., Tooms, J.S., and Troup, A.G.
1973: Some aspects of geochemical dispersion from base-metal mineralization in Norway, North Wales and British Columbia, Canada; in Geochemical Exploration 1972. M.J. Jones (Ed.); Inst. Min. Metall., p. 105-115.

Meyer, W.T. and Evans, D.S.
1973: Dispersion of mercury and associated elements in glacial drift environment et Keel, Eire; in Prospecting in areas of glacial terrain. M.J. Jones (Ed.); Inst. Min. Metall., p. 127-138.

Meyer, W.T. and Lam Shang Leen, Y.C.Y.
1973: Microwave-induced argon plasma emission system for geochemical trace analysis; in Geochemical Exploration 1972. M.J. Jones (Ed.); Inst. Min. Metall., p. 325-335.

Meyer, W.T. and Peters, R.G.
1973: Evaluation of sulphur as a guide to buried sulphide deposits in the Notre Dame Bay area Newfoundland; in Prospecting in areas of glacial terrain. M.J. Jones (Ed.); Inst. Min. Metall., p. 55-66.

Michie, U.McL., Gallagher, M.J., and Simpson, A.
1973: Detection of concealed mineralization in northern Scotland; in Geochemical Exploration 1972. M.J. Jones (Ed.); Inst. Min. Metall., p. 117-130.

Moiseenko, V.I.
1959: Biogeochemical surveys in prospecting for uranium in marshy areas; Geochemistry, No. 1, p. 117-121.

Morrissey, C.J. and Romer, D.M.
1973: Mineral exploration in glaciated regions of Ireland; in Prospecting in areas of glacial terrain. M.J. Jones (Ed.); Inst. Min. Metall., p. 45-53.

Morse, R.H.
1976: Radon counters in uranium exploration; in Exploration for Uranium Ore Deposits; Int. Atomic. Energy Agency, Vienna, p. 229-239.

Muntanen, T.
1971: An example of the use of boulder counting in lithological mapping; Geol. Soc. Finl., Bull., v. 43 (2), p. 131-140.

Nichol, I. and Björklund, A.
1973: Glacial geology as a key to geochemical exploration in areas of glacial overburden with particular reference to Canada; J. Geochem. Explor., v. 2 (2), p. 133-170.

Nieminen, K. and Yliruokanen, I.
1976: Kitee: the copper-nickel anomalies of Lietsonsuo peat bog; in Conceptual models in exploration geochemistry. L.K. Kauranne (Ed.); J. Geochem. Explor., v. 5 (3), p. 248-253.

Nilsson, G.
1971: The use of dogs in prospecting for sulphide ores; Geol. Fören. Förh., v. 93 (4), p. 725-728.

Nurmi, A.
1976: Geochemistry of the till blanket at the Talluskanava Ni-Cu ore deposit, Tervo Central Finland; Geol. Surv. Finland, Rep. Inv., No. 15, 84 p.

Nuutilainen, J. and Peuraniemi, V.
1977: Application of humus analysis to geochemical prospecting: some case histories; in Prospecting in areas of glaciated terrain. G.R. Davis (Ed.); Inst. Min. Metall., p. 1-5.

Okko, V. and Peltola, E.
1958: On the Outokumpa boulder train; Geol. Soc. Finl. Comp. Rend., No. 30, p. 113-134.

Olade, M. and Fletcher, W.K.
1974: Potassium chlorate-hydrochloric acid: a sulphide selective leach for bedrock geochemistry; J. Geochem. Explor., v. 3 (4), p. 337-344.

Orlov, A.P., Robonen, V.I., and Kirilenko, G.M.
1969: Geologicheskie poiski a rudorozysknumi sokakami; Nedra, Moskva, 47 p.

Ovchinnikov, L.N., Sokolov, V.A., Fridman, A.I., and Yanitski, L.N.
1973: Gaseous geochemical methods in structural mapping and prospecting for ore deposits; in Geochemical Exploration 1972. M.J. Jones (Ed.); Inst. Min. Metall., p. 177-182.

Peachy, D.
1976: Extraction of copper from ignited soil samples; J. Geochem. Explor., v. 5 (2), p. 129-134.

Pollock, J.P., Schillinger, A.W., and Bur, T.
1960: A geochemical anomaly associated with a glacially transported boulder train Mt. Bahemia, Keewatin County, Michigan; 21st Int. Geol. Cong., Norden: Part II, p. 20-27.

Puranen, R.
1977: Magnetic suceptibility and its anisotropy in the study of glacial transport in northern Finland; in Prospecting in areas of glaciated terrain. G.R. Davis (Ed.); Inst. Min. Metall., p. 111-119.

Ramsay, Sir Andrew
1881: The geology of North Wales; Mem. Geol. Surv. Gt. Br., v. 3, 611 p.

Rice, R. and Sharp, G.J.
1976: Copper mineralization in the forest of Coed-y-Brenin, North Wales; Inst. Min. Metall., Trans., v. 85, B1-B13.

Robinson, G.W.
1949: Soils, their origin, constitution and classification; Thomas Murty, London, 573 p.

Rose, A.W.
1975: The mode of occurrence of trace elements in soils and stream sediments applied to geochemical exploration; in Geochemical Exploration 1974. I.L. Elliott and W.K. Fletcher (Eds.); Elsevier, p. 691-705.

Salmi, M.
1967: Peat in prospecting: application in Finland; in Geochemical prospecting in Fennoscandia. A. Kvalheim (Ed.); Interscience Publishers, New York, p. 113-126.

Sato, M. and Mooney, H.M.
1960: The electrochemical mechanism of sulfide self-potentials; Geophysics, v. 25 (1), p. 226-249.

Sauramo, M.
1924: Tracing of glacial boulders and its application in prospecting; Comm. Geol. Finl., Bull., v. 11 (67).

Shilts, W.W.
1973a: Glacial dispersal of rocks, minerals and trace elements in Wisconsinan till, southeastern Quebec, Canada; Geol. Soc. Am., Mem. 136, p. 189-219.

1973b: Drift prospecting: geochemistry of eskers and till in permanently frozen terrain: District of Keewatin, Northwest Territories; Geol. Surv. Can., Paper 72-45, 34 p.

1975a: Principles of geochemical exploration for sulphide deposits using shallow samples of glacial drift; Can. Min. Met., Bull., v. 68 (757), p. 73-80.

1975b: Common glacial sediments of the Shield, their properties, distribution and possible uses as geochemical sampling media; J. Geochem. Explor., v. 4 (1), p. 189-199.

1976: Glacial till and mineral exploration; in Glacial till. An interdisciplinary study. R.F. Legget (Ed.); Roy. Soc. Can., Spec. Pub. No. 12, p. 205-224.

Shilts, W.W. and McDonald, B.C.
1975: Dispersal of clasts and trace elements in the Windsor esker, southern Quebec; in Report of Activities, Part A, Geol. Surv. Can., Paper 75-1A, p. 495-499.

Shvartsev, S.L.
1976: Electrochemical dissolution of sulphide ores; J. Geochem. Explor., v. 5 (1), p. 71-72.

Skinner, R.G.
1972: Drift prospecting in the Abitibi Clay Belt; Geol. Surv. Can., Open File, No. 116, 27 p.

Smith, R.T. and Gallagher, M.J.
1975: Geochemical dispersion through till and peat from metalliferous mineralization in Sutherland, Scotland; in Prospecting in areas of glaciated terrain. M.J. Jones (Ed.); Inst. Min. Metall., p. 134-148.

Stigzelius, H.
1977: Recognition of mineralized areas by a regional geochemical survey of the till-blanket in northern Finland; in Geochemical Exploration 1976. C.R.M. Butt and I.G.P. Wilding (Eds.); J. Geochem. Explor., v. 8 (1/2), p. 473-481.

Sveian, H.
1979: Gjøvik. Description of the Quaternary geology map 1816 I, Gjøvik (English summary); Norges Geol. Unders., No. 345, 62 p.

Sveshnikov, G.B.
1967: Electrochemical processes in sulphide deposits; Leningrad Univ. (in Russian), 158 p.

Szabo, N.L., Govett, G.J.S., and Lajtai, E.Z.
1975: Dispersion trends of elements and indicator pebbles in glacial till around Mt. Pleasant, New Brunswick, Canada; Can. J. Earth Sci., v. 12 (9), p. 1534-1556.

Szalay, A.
1974: Accumulation of uranium and other micrometals in coal and organic shales and the role of humic acid in these geochemical enrichments; Ark. Min. Geol., v. 5 (1), p. 23-36.

Szalay, A. and Szilagyi, M.
1968: Laboratory experiments on the retention of micronutrients by peat humic acid; Plant. Soil, v. 29 (2), p. 219-224.

Tanskanen, H.
   1976a: Factors affecting the metal contents in peat profiles; in Conceptual models in exploration geochemistry L.K. Kauranne (Ed.); J. Geochem. Explor., v. 5 (3), p. 412-414.

   1976b: Rookkijärvi: The nickel/copper ratio in peat in a gabbro-peridotic area; in Conceptual models in exploration geohemistry. L.K. Kauranne (Ed.); J. Geochem. Explor., v. 5 (3), p. 309-310.

Thornber, M.R.
   1975a: Supergene alteration of sulphides, I. A chemical model based on massive nickel sulphide deposits at Kambalda, Western Australia; Chem. Geol., v. 15 (1), p. 1-14.

   1975b: Supergene alteration of sulphides, II. A chemical study of the Kambalda nickel deposit; Chem. Geol., v. 15 (2), p. 117-144.

Toverud, Ø.
   1977: Chemical and mineralogical aspects of some geochemical anomalies in glacial drift and peat in northern Sweden; Sver. Geol. Unders. CNr. 729, 37 p.

Townsend, J.
   1824: Geological and mineralogical researches during a period of more than fifty years in England, Scotland, Ireland, Switzerland, Holland, France, Flanders and Spain; Gye and Son, Bath,

Usik, L.
   1969: Review of geochemical and geobotanical prospecting methods in peatland; Geol. Surv. Can., Paper 68-66, 43 p.

van Tassel, R.E.
   1969: Exploration by overburden drilling at Keno Hill Mines Limited; Colo. Sch. Mines, Q., v. 64, No. 1, p. 457-478.

Virkala, K.
   1958: Stone counts in the esker of Hämeenlimna, southern Finland; Comm. Geol. Finl., Bull., v. 29 No. 180 p. 87-104.

Vokes, F.M.
   1976: Caledonian massive sulphide deposits in Scandinavia: a comparative review; in Handbook of stratabound and stratiform ore deposits K.H. Wolf (Ed.); Elsevier Publ. Co., v. 6, p. 79-127.

Wennervirta, H.
   1967: Geochemical methods in uranium prospecting in Finland; in Geochemical prospecting in Fennoscandia. A. Kvalheim (Ed.); Interscience Publishers, New York, p. 155-169.

   1968: Applications of geochemical methods to regional prospecting in Finland; Bull. Comm. Geol. Finland, No. 234, 91 p.

   1973: Sampling of the bedrock-till interface in geochemical exploration; in Prospecting in areas of glacial terrain. M.J. Jones (Ed.); Inst. Min. Metall., p. 67-71.

Wennervirta, H. and Kauranen, P.
   1960: Radon measurement in uranium prospecting; Bull. Comm. Geol. Finland, No. 188, p. 23-40.

THE APPLICATION OF SOIL SAMPLING TO GEOCHEMICAL EXPLORATION IN NONGLACIATED REGIONS OF THE WORLD

P.M.D. Bradshaw and I. Thomson
Barringer Research Limited
Rexdale, Ontario

Bradshaw, P.M.D. and Thomson, I., The application of soil sampling to geochemical exploration in nonglaciated regions of the world; in Geophysics and Geochemistry in the Search for Metallic Ores; Peter J. Hood, editor; Geological Survey of Canada, Economic Geology Report 31, p. 327-338, 1979.

Abstract

Early work in the application of geochemistry to mineral exploration revealed evidence of a simple and direct relationship between geochemical patterns in residual soils and those in the underlying bedrock. Many geochemical features are common to soils developed under a wide range of physical and climatic conditions and thus permit the widespread use of geochemistry in exploration. A simple model, based on the ideal pattern, can be applied extensively in the unglaciated regions of the world, provided attention is given to the well understood and documented distortions caused by mechanical movement and hydromorphic dispersion. Significant differences are, however, noted in various places and the prime objective of the exploration geochemist is to correctly recognize these situations and to select techniques of sampling, analysis and interpretation to accommodate them. Under extreme circumstances geochemistry is not effective and its use should be avoided.

In this paper a review is presented of the application of geochemistry in nonglaciated soil environments. The basic geochemical characteristics of these soils are illustrated by the use of idealized models and case histories from around the world. Particular attention is given to effects which create conditions that differ markedly from the idealized situation, including leaching, seepage anomalies, calcrete development, duricrusts and the effects of transported overburden such as alluvium, blown sand, landslides and volcanic ash. In establishing methods, careful consideration must be given to variables such as optimum grain size, soil horizon to be sampled, analytical extraction employed, choice of elements to be analyzed and the interpretation of the various parameters measured and observed.

Résumé

Les premiers travaux sur l'application de la géochimie à la recherche de minéraux démontrent qu'il existe une relation directe entre les compositions géochimiques dans les sols résiduels et celles de la roche en place sous-jacente. Beaucoup de particularités géochimiques sont communes aux sols qui se sont formés par suite d'une grande variation des conditions climatiques et physiques, ce qui a permis l'utilisation considérable de la géochimie dans les travaux d'exploration. Un modèle simple, basé sur la composition idéale, peut être appliqué, à grande échelle, dans les régions de la terre qui ont échappé aux glaciations, à condition de tenir compte des déformations causées par le mouvement mécanique et par la dispersion hydromorphique, qui ont déjà été constatées et appuyées par de nombreux documents. Toutefois, les auteurs remarquent des différences importantes dans divers emplacements et l'objectif premier du géochimiste est de savoir reconnaître parfaitement ces phénomènes et de choisir des techniques d'échantillonnage, d'analyse et d'interprétation qui conviennent. Dans des circonstances exceptionnelles, la géochimie n'est pas efficace et il faut éviter de l'utiliser.

Les auteurs présentent, dans ce rapport, un aperçu de l'application de la géochimie à des milieux qui n'ont pas subi l'action des glaciations. Les particularités géochimiques de base de ces sols sont définies par l'utilisation de modèles optima et de dossiers provenant de toutes les parties du monde. On accorde une attention toute spéciale à des effets générateurs de conditions qui diffèrent passablement de la situation optimale, et qui comprennent le lessivage, les anomalies d'infiltration, la croissance de croûtes calcaires, les croûtes concrétionnées et les effets de la surcharge transportée comme l'alluvion, le sable éolien, les glissements et la cendre volcanique. Lors de l'élaboration des méthodes, il faut tenir compte tout particulièrement de données comme le grain de dimension optimale, l'horizon à échantillonner, le type d'extraction analytique utilisée, le choix des éléments à analyser et l'interprétation des différents paramètres mesurés et observés.

INTRODUCTION

At an early stage in the development of exploration geochemistry it was established that systematic sampling and analysis of soil, where the soil is residual and mineralization is not covered by younger rocks or transported overburden, represent a straightforward and generally reliable method of locating sub-outcropping mineralization. Furthermore, although geochemical anomalies developed in soils under these conditions may be distorted by downslope creep, seepage, and leaching, geochemical techniques for the reliable interpretation of these problems exist. This relatively simple condition probably prevails over as much as 70 per cent of the unglaciated areas of the earth's surface and undoubtedly there are still many places where straightforward collection and analysis of soils will lead to the discovery of new mineralization. However, as exploration continues in any area, sooner or later the geologist is faced with the problem of exploring where the soils are of transported origin (e.g., alluvium, windblown sand, volcanic ash, and landslides) and/or the mineralization is blind under a younger rock cover. (In this paper the term buried deposits refers to those under transported overburden of remote origin while blind deposits are those covered by younger rock.) The correct identification of the presence of such conditions is the first and by no means always simple task of the exploration geochemist. Geochemical methods must then be adapted to meet the challenge of these areas.

This paper reviews the use of geochemistry in areas of residual soil and also examines the role of geochemistry in the more complex environments of transported overburden and blind mineralization. Extensive use of geochemistry over the last ten years has provided sufficient evidence for the confident application of techniques under a wide range of well defined environments. The basic concepts formulated in the early 1960's (Hawkes and Webb, 1962) have been largely proven correct. Advances in the science have refined the application of geochemistry in residual soil environments to the point at which specific procedures may be recommended for particular exploration problems.

RESIDUAL SOILS

Geochemical Anomalies in Residual Soils

During the normal process of weathering and soil formation, trace elements present in the bedrock become incorporated into overlying residual soils. Similar processes prevail over mineralization where the high trace element content of an orebody gives rise to the presence of anomalously high values in overlying residual soils.

As stated in the introduction, the location and identification of these anomalies in residual soil environments represents the most straightforward and direct geochemical method of locating mineralization. The success of this approach is evidenced by an extensive literature of case histories from around the world including Africa – Tooms and Webb (1961); Cornwall (1970); Ellis and McGregor (1967); Philpot (1975); Reedman (1974); Australia – Mazzuchelli (1972); Cox (1975); Mazzuchelli and Robbins (1973); the Southwest Pacific – Coope (1973); Govett and Hale (1967); Asia – Dasgupta (1963); Yellur (1963); Chakrabarti and Solomon (1970); South America – Lewis (1965); Lewis et al. (1971); Montgomery (1971); Thomson and Brim (1976) and North America – Worthington et al. (1976). Indeed, local workers have found indications that, in strictly residual soil and over-limited geographic areas, the strength of a soil geochemical anomaly may be directly related to the grade of mineralization in underlying fresh rock (Ong and Sevillano, 1975; Saigusa, 1975; Tooms and Webb, 1961; Nicholls et al., 1965).

The typical distribution of trace elements in soils overlying base metal sulphide mineralization is shown in Figure 14.1. The normal incorporation of metals in the soils results in the "fan-shaped" distribution depicted. The near-surface part of this fan is generally considerably wider than the soil anomaly near the rock contact. In flat areas the fan is typically symmetrical, while on slopes the fan spreads downslope (by mechanical means) as depicted in Figure 14.1.

The usefulness and importance of soil profile sampling to detect the exact location of the source of the anomaly is frequently overlooked in the application of exploration geochemistry. In a simple fan-shaped anomaly a soil profile taken directly over mineralization shows a geochemical response which remains about the same, or increases with depth, whereas to one side of the anomaly the values decrease with depth. This not only provides much more precise definition of the exact source of the anomaly, but can also give an indication of the orientation of the source. For example, Figure 14.2 from Granier (1973) shows the distribution of copper and gold in a deep tropical soil profile over vein-type mineralization in the Ivory Coast. This example illustrates the manner in which a very broad anomaly at surface, distorted by downslope creep and local chemical redistribution, can be reduced to a narrow anomaly at depth closely related to the location of the mineralization. In addition, a measure of the dip of the vein can be obtained from the soil profile data. The extent to which information obtained from soil profiles can clarify the geochemical nature of an anomaly, its origin and geological relationship is further illustrated by Koksoy and Bradshaw (1969); Montgomery (1971); Granier (1973); Bradshaw et al. (1974) and is well explained by Hawkes and Webb (1962).

As indicated by the above example from the Ivory Coast, provided the soil is residual, a considerable depth of weathering does not necessarily destroy the surface soil response.

In all the examples quoted above, significant and readily identifiable anomalies were located directly over sub-outcropping mineralization. It is important to note in the interpretation of geochemical data that, except where a soil anomaly is modified by some of the processes indicated in later sections of this paper, the anomaly is related to the sub-outcrop of any mineralization and not necessarily the greatest thickness, highest grade, maximum sulphide concentration, etc. of mineralization. This is well illustrated by the work of Scott (1975) at Otjahase, Southwest Africa, where a strong soil anomaly was located over the sub-outcrop of copper-zinc mineralization. The best grades and highest content of sulphide mineralization are, however, between 100 and 150 m downdip resulting in an apparent separation of up to 200 m between the soil geochemical anomaly and the surface projection of the main body of mineralization.

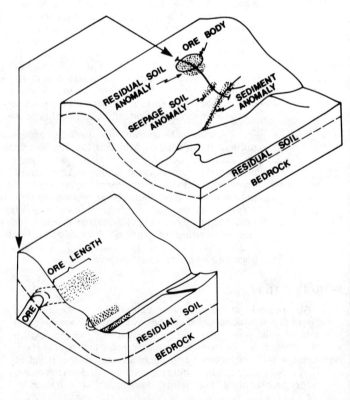

Figure 14.1. Idealized diagram – geochemical anomalies in residual soil over mineralization.

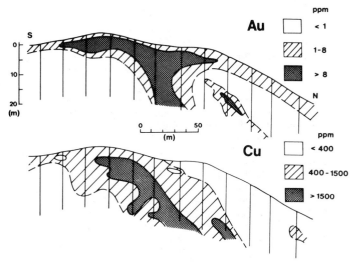

Figure 14.2. Section showing the lateral distribution of gold and copper in residual lateritic soil over a gold-copper vein, Ivory Coast (after Granier, 1973).

Similar relationships often lead to a spatial displacement of geophysical and geochemical anomalies, a fact that can be confusing at first sight. Electromagnetic anomalies, for example, have the strongest response over the greatest conductivity-thickness of a sulphide body (although the strength of the EM anomaly is also modified by the depth of the source). It is evident that, where mineralization has a shallow dip and oxidation has taken place to some depth, there can be a considerable relative separation of the geochemical and geophysical anomalies.

When mineralization sub-outcrops on a slope, the resulting soil anomaly is invariably distorted or displaced downslope by mechanical movement. The degree of displacement may not always be appreciable, but can in some cases have a significant effect on interpretation. However, as shown in Figure 14.2, the distance of downslope displacement and hence the location of the source of the anomaly can usually be accurately determined by soil profile sampling.

Seepage Anomalies

In addition to the incorporation of metals into soil by in situ weathering, a portion of the metals goes into solution in the groundwater and is dispersed in the direction of groundwater flow. (This is normally termed hydromorphic movement.) Around oxidizing sulphide mineralization the groundwaters are typically acidic (as a result of decomposition of the sulphides) and relatively reducing, two conditions which promote solubility of most metals. These metals stay in solution until a change in geochemical environment is encountered. Such a change occurs where the groundwaters come out in streams or lakes. In this case much of the metal is precipitated from solution into the lake or stream sediment resulting in anomalies in these media.

Groundwater can also reach the surface in a variety of other locations where, in response to the change in geochemical environment, metals come out of solution and are precipitated. On occasions, very large concentrations of metals may accumulate in the soil (Fig. 14.1). The principal location of such seepages are at breaks of slope and in low-lying areas, marshes, bogs, etc., although recognition of seepage zones in the field can sometimes be extremely difficult. This process results in the formation of hydromorphically transported anomalies which, although related to mineralization, occur spatially separated from it and often create interpretation problems. These anomalies are frequently very much more intense than the residual anomaly directly over the source. For example, Webb and Tooms (1959, also quoted in Hawkes and Webb, 1962, p. 251) show seepage anomalies 1000 m downslope from known mineralization approximately 10 times stronger than the residual anomaly directly over the mineralization. Seepage anomalies are frequently encountered in humid environments and display a wide range of form and occurrence (Woolf et al., 1966; Cole et al., 1968) with their topographic position often an indication that they are hydromorphic features.

Soil profile sampling is of great assistance in the recognition and discrimination of seepage anomalies. Invariably seepage anomalies are of restricted extent and limited to the near-surface part of soil profiles having no "root" in mineralization. Furthermore, systematic profile sampling of a seepage area may reveal an asymmetry of metal distribution patterns indicating from which direction the metal entered the seepage zone.

An alternative approach to distinguishing residual and seepage anomalies is by the use of different analytical extractions (Hawkes and Webb, 1962). The hydromorphically transported metals tend to be present in secondary minerals with weak bonding and adsorbed or absorbed on the soil material. This metal is easily removed by a weak analytical extraction (such as EDTA) and contrasts with residual anomalies in which metals are typically firmly bonded (Bradshaw et al., 1974). Seepage anomalies are thus often characterized by high absolute values and/or a high percentage of weakly extractable metal.

Soil Differentiation

Weathering processes, both mechanical and chemical, may lead to modification of geochemical patterns between horizons within the soil. Soils typically display both a physical and chemical layering which may be very weakly developed and indistinct in some areas, with no significant effects on the geochemical patterns, or is visibly well developed in other areas, with a strong control on the distribution of metals. To minimize potential problems systematic sampling should be restricted to a single horizon. The so-called B horizon has become the preferred sample medium in routine soil surveys throughout the world. Generally speaking this is a sample collected at a depth of 10-30 cm below any zone of organic accumulation, in a constant horizon (rather than depth) recognized within any one area by distinctive colour or texture. Although the scheme may be modified to suit local requirements, normally only certain extreme conditions require a radical change in soil-sampling procedures. For example, in immature regolithic soils there may be no differentiation at all and thus no preferred sample depth. In other areas surface leaching may be such that a reliable geochemical signature can only be obtained by taking a sample at considerable depth. Also, in the presence of a thin layer of transported material overlying residual soil (alluvium, colluvium, blown sand, etc.) samples may be taken from deeper levels below the exotic overburden. Within the nonglaciated parts of the world a wide range of soil types are encountered. Fortunately, it is not necessary to give a detailed account of the geochemical characteristics of all these soils since the general principles of anomaly formation presented in this paper hold true in the majority of soil types. To illustrate the application of exploration geochemistry under changing residual soil conditions further consideration is given to leached soils and the effects of calcrete development.

Leached Soils

Under certain climatic and topographic conditions, trace elements become leached from surface soils. This condition has been the subject of intensive study over the last ten years and, with understanding, methods have been developed for exploration in these areas. Extensive work, particularly in Western Australia, has shown that despite strong leaching some contrast frequently persists between the metal content of residual soils developed over mineralization and unmineralized rock. As a result, although absolute trace element values may be very different from unleached soils, discrete anomalies related to mineralization may still be discerned, typically with reduced contrast (Scott, 1975; Lord, 1973; Butt, 1976). On occasion leaching can be so intense that the trace element content of residual soils is depleted to the point that no recognizable anomaly exists. This is certainly true for the more mobile elements such as copper and zinc. The metals may be redeposited lower down, forming a zone of secondary enrichment, or be completely removed from the area.

An example of the effects of severe surface leaching is shown by the work of Learned and Boisen (1973) who studied the geochemical response in soils over several porphyry copper prospects in Puerto Rico. In one instance, copper was so leached from residual soils developed on porphyry-copper mineralization that no recognizable copper anomaly could be found. However, it was found that gold, associated with the copper mineralization, formed a significant soil anomaly, undoubtedly because its low chemical mobility prevented its removal from the soils.

Further examples are provided by Nickel et al. (1977) and Butt and Sheppy (1975) from studies of nickel deposits in the Agnew area in Western Australia. In these cases, leaching has removed nickel from the near-surface soil and deposited at least a portion in a zone of secondary enrichment at depths of from 20 to 50 m. In southeastern United States, Worthington et al. (1976) reported that soil geochemistry usually faithfully reflects metal concentrations in the bedrock despite the soils being leached. Anomalies are, however, weaker in the soil than in bedrock.

It is evident that all degrees of surface leaching can occur from insignificant to complete. When severe leaching is expected, routine geochemical practice is to analyze for associated or pathfinder elements such as gold, lead, molybdenum, arsenic, selenium, tin, tungsten and platinum which are less mobile or immobile under the conditions encountered and thus unlikely to be completely removed from the soil.

Calcrete

Calcrete or caliche is a commonly encountered soil development in arid areas, but has generally been ignored or avoided as media for geochemical sampling and its use is virtually unreported in the literature. Calcrete is composed of a cementing and/or replacing carbonate, usually calcite, which can form in almost any type of pre-existing soil by deposition from the soil water (Netterberg, 1971). Physically it can take many forms including powdery disseminations, nodules, pellets, boulders or massive layers. Furthermore, the relative proportions of calcite and host soil may vary considerably over short distances.

Figure 14.3 shows the results of sampling residual soils containing calcrete developed over copper-lead-zinc mineralization at Areachap in the desert environment of the Northern Cape Province of South Africa (Danchin, 1972). The nature of the calcrete varies from a dense massive variety containing occasional quartz and jasper pellets, through powdery to lumpy. There is also a well developed calcretized mixed zone above the bedrock containing relict bedrock fragments in various stages of decomposition with oxidized copper sulphides and copper staining. This sequence was carefully sampled and analyzed for total copper and zinc. Sampling was also carried out through similar calcrete profiles over nickel-copper mineralization in ultrabasic rocks at Jacomyns Pan approximately 120 km from Areachap.

Profiles from both areas show similar results with concentrations of the ore elements decreasing steadily upwards from bedrock to the surface through the calcrete. However, the surface of the calcrete is still anomalous over the mineralization and systematic grid sampling of the uppermost layer of calcrete in these areas delineated the underlying mineralization. The situation is further complicated at Areachap where the top 15 cm of soil above the calcrete is largely composed of windblown material of remote origin which masks any geochemical response from the mineralization. (The complications of windblown sand are discussed in more detail in a later section.) Cox (1975) examined residual soil profiles containing calcrete over nickel-copper mineralization at Pioneer near Norseman, Western Australia. He found that systematic sampling and analysis of the upper surface of the calcrete reflected both the underlying mineralization and lithology. However, both here and in South Africa, the contrast between anomalies over mineralization and background is much lower in the surface calcrete samples than in samples of bedrock. In this case the calcrete horizon is 0.5 - 1.5 m thick.

Worldwide experience in the sampling and the analysis of calcrete in geochemical exploration is limited. However, it appears that certain generalizations can be made. The carbonate material is a secondary precipitate and acts as a dilutant to the original trace element content of the pre-existing soil. Thus in residual soils, as in the examples quoted, the presence of calcrete depresses absolute trace element values and leads to a reduced anomaly contrast. In areas of thin calcrete or where the carbonate represents only a minor portion of the total soil, anomalies related to mineralization and rock types are preserved in the near-surface soil. However, it is probable that where calcrete development is very thick or dominates in the soil, geochemical anomalies may be so depressed that they cannot be discerned using conventional techniques.

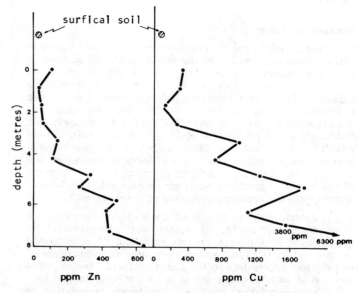

Figure 14.3. Distribution of copper and zinc in soil profile through calcrete over copper-zinc mineralization at Areachap, South Africa (with permission of the Anglo-American Corp.).

The problem can be further aggravated by the presence of several layers of calcrete in the soil. Usually it is only in the lowest calcrete horizon that a reliable geochemical response related to the underlying mineralization may be obtained.

In soils of transported origin (alluvium, colluvium, etc.) trace element distribution patterns will be determined by the composition of the parent material and modified by the presence of the carbonate material. Such soils will not normally show any geochemical patterns related to underlying mineralization.

Duricrust

The terms duricrust, ferricrete and canga are used to describe a hard indurated crust, cap, layer of pellets or nodules formed by dehydration of the upper part of a lateritic soil. The loss of water causes the collapse of the phyllosilicate (clay) lattice and results in a restructuring of the soil constituents into compact, hardened sesquioxides and iron-rich concretionary material. Duricrust forms preferentially on well-drained hills or at the margin of larger hills and plateau surfaces. In the latter situation, the duricrust commonly is thickest close to scarps at the margin of the upland surfaces and is progressively thinner away from the edge. The formation of duricrust is apparently an irreversible process and should drainage or climatic conditions change the duricrust will not revert to a lateritic or other new soil profile.

Duricrust commonly inherits the general geochemical characteristics of its parent material, usually with some modification in the absolute levels of trace elements. For example, Tooms et al. (1965) working in Sierra Leone found that the projection of mineralized veins could be traced in the duricrust horizon of lateritic soils as molybdenum maxima of similar magnitude to those in weathered bedrock. In nearby background areas, however, molybdenum levels are higher in duricrust than in underlying weathered bedrock and contrast between mineralization and background is much lower in the duricrust. In direct contrast the indurated portion of a lateritic profile may become leached and create the problems for geochemical exploration described earlier.

Duricrust resists weathering processes and tends to persist on hills and plateau surfaces as less competent unindurated soil is eroded. This can lead to the formation of fossil duricrusts perched, capping or surrounding younger soils as found over large areas of Western Australia (Butt and Sheppy, 1975). Prolonged erosion and scarp retreat result in the formation of transported detrital duricrusts covering pediment and plateau surfaces, notably in parts of Western Australia (Butt and Sheppy, 1975), central Brazil, and central Africa. Under such conditions geochemical continuity between the duricrust and adjacent residual soils is poor or nonexistent.

Direct comparison of duricrust and residual latosols and other unindurated soils is often difficult and may be impossible. For most routine surveys it is thus recommended that data obtained from duricrust soils be interpreted separately from other soils data.

LITHOLOGICAL ANOMALIES OR FALSE ANOMALIES

The trace metal content of unmineralized rocks is usually uniformly low with respect to mineralization. However, this is not always the case. For example, the high content of nickel, copper and chromium in ultrabasic rocks, zinc and molybdenum in black shale and uranium in many granites, frequently gives rise to anomalies in residual soils which can have the same magnitude as anomalies related to potentially economic sulphide or oxide mineralization. Although basic intrusives and shale are the most frequently encountered sources of false anomalies related to distinctive lithologies, rather than mineralization, they are by no means restricted to these rock types.

An example of soil anomalies related to both rock type and sulphides is provided by work from Brazil (modified from Thomson, 1976) in which soil sampling was carried out over part of the Niquelandia ultrabasic complex and a small altered gabbro body, both of which are intruded into Precambrian schists. Two small bodies of nickel-copper sulphide mineralization are known in the eastern gabbro at locations shown in Figure 14.4; no significant sulphide occurrences have been found in the ultrabasic rocks. The results for total nickel and copper in soils, also shown in Figure 14.4, reveal a very strong and areally large response over the ultrabasic rocks and much smaller and generally weaker anomalies related to sulphide occurrences. On the basis of individual metal distribution patterns alone, it is not possible to distinguish the barren ultrabasic rocks characterized by high concentrations of copper and nickel held, most probably, in the lattice of the silicate minerals from the nickel-copper sulphide mineralization. In this case, anomalies related to mineralization may be differentiated from those related to rock type by the use of metal ratios. In the Niquelandia area, the sulphides are high in copper with respect to nickel contrasting with the ultrabasic rocks in which the nickel is high with respect to copper. Consequently, a change in the copper-nickel ratios (Fig. 14.4) clearly identifies the sulphide mineralization with a further minor anomaly to the west over the ultrabasic rocks. (In the present case, the copper-nickel ratio anomaly in the ultrabasic rocks has not been followed up and its cause is not yet known. Nevertheless, the area of interest requiring follow-up has been significantly reduced by the use of metal ratios.) The fact that the soils in this area are residual is well demonstrated by Figure 14.5 which shows profile samples taken over barren schist, unmineralized altered gabbro, and sulphide mineralization. The profiles not only illustrate the difference in metal concentrations between contrasting rock types, but also demonstrate that there is no significant variation with depth.

Metal ratios as a means of distinguishing between anomalies related to mineralization and lithology have been used elsewhere. For example, copper-nickel ratios have been found useful in exploring areas of ultrabasic rocks over much of southern Africa and in parts of Western Australia (Cox, 1975; Wilmshurst, 1975).

Further assistance in interpretation can frequently be gained by analyzing for elements uniquely associated with either potentially economic mineralization or the unmineralized metal-rich lithology. For example, in the case of nickel-copper mineralization in ultrabasic rocks, cobalt is frequently concentrated in the sulphides in contrast to chromium which occurs either with silicate minerals or as a separate oxide phase (chromite). Consequently, a nickel-copper anomaly high in cobalt, low in chromium, can be up-graded and a similar nickel-copper anomaly, low in cobalt and high in chromium could be downgraded. Wilmshurst (1975) working in Western Australia found platinum, palladium, arsenic and zinc useful pathfinders related to sulphide mineralization and unrelated to barren ultrabasic rocks. Clema and Stevens-Hoare (1973) also working in Western Australia, demonstrated multi-element relationships capable of differentiating leached cappings over nickel sulphides from those developed over unmineralized iron-rich rock types. Experience has shown that once metal associations have been established in one part of a mineral belt, the same associations apply throughout the whole belt and can be used with some reliability (Moeskops, 1977). However, while

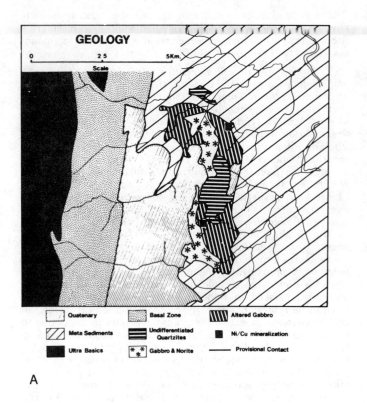
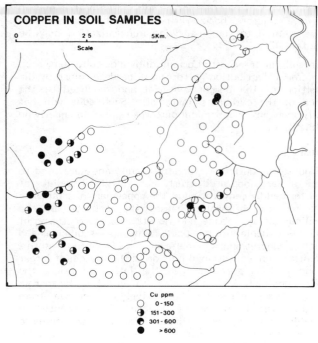
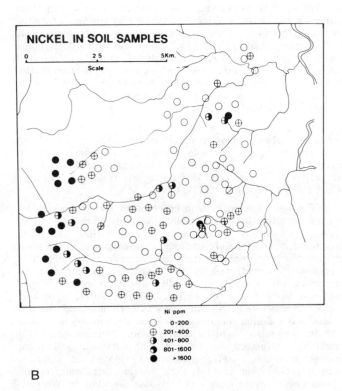
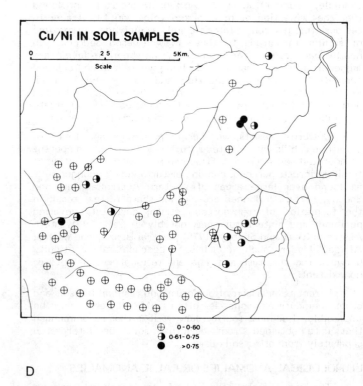

a. Geology
b. Distribution of nickel in near-surface soil samples
c. Distribution of copper in near-surface soil samples
d. Copper/nickel ratio in near-surface soil samples (after Thomson, 1976)

Figure 14.4. The Niquelandia area, Goias, Brazil.

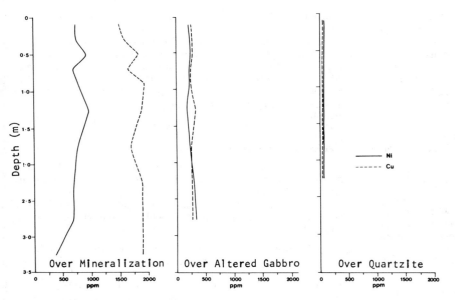

Figure 15.5. Vertical distribution of copper and nickel in profiles through residual soils over copper-nickel sulphide mineralization, altered gabbro, and quartzite, Niquelandia area, Goias, Brazil (after Thomson, 1976).

experience with one mineral belt can assist in determining which associated metals are likely to be of greater assistance in a new belt, these associations must always be confirmed before they are relied upon too heavily.

More recently, workers in Australia have found that the mode of occurrence of nickel and copper in lateritic soils derived from sulphide mineralization is different from similar soils developed over barren rock types. Smith (1977) reports that the association formed by nickel and copper with iron oxides in the lower part of lateritic soil profiles developed over sulphide mineralization is distinct from that found over barren rocks. The use of appropriate selective analytical extraction techniques permitted good discrimination between a sulphide and silicate origin for the raised concentrations of these metals in soil. Studies of this type will no doubt lead to better methods of discriminating anomalies in other geological and climatic environments.

TRANSPORTED SOILS AND BLIND MINERALIZATION

Geochemical Anomalies in Transported Overburden

Any material of remote origin, be it a cover of postmineralization rocks or transported detrital overburden, will usually mask all conventional geochemical expression of the mineralization in overlying soils. (The word conventional is stressed since techniques are under investigation which, at least in limited applications, have provided surface geochemical anomalies through significant thicknesses of bedrock and transported overburden.) The condition is illustrated in Figure 14.6, an idealized model of geochemical dispersion processes from mineralization covered by transported overburden.

There are very few published accounts of "negative" case histories in which soil geochemistry failed because of the presence of transported overburden or postmineralization cover although many certainly exist in company files. Ullmer (1978), however, reports that he found no geochemical response in surface soils over the Sacaton porphyry copper deposit in Arizona where the mineralization is covered by up to 20 m of transported sand and gravel.

The masking affects of an exotic covering thus require modifications in the application of exploration geochemistry techniques (Horsnail and Lovestrom, 1974 and Lovering and McCarthy, 1978) and soil sampling may have no role at all. It is of primary importance, however, that the presence of any cover be recognized otherwise totally erroneous interpretations will be made of soil geochemical data.

The presence of an exotic cover need not necessarily eliminate the application of soil geochemistry. The cover may be discontinuous permitting the positive application of soil sampling in the intervening areas of residual soil. Circulating groundwaters may reach the surface in seepage zones producing hydromorphically transported anomalies related to concealed mineralization. Alternatively, local dispersion processes may permit the upward migration of metal with soil water into the overburden to give anomalies in transported soils directly over mineralization such as those reported by Brown (1970). It is also possible that plants, penetrating the exotic cover, will bring metals up into their foliage which then accumulates in the topsoil when the plant dies giving rise to geochemical anomalies in surface soils over the mineralization.

Alluvium and Colluvium

Alluvium and colluvium are the most readily recognized and most frequently encountered forms of transported overburden. Both materials effectively mask any response from the bedrock and it is essential that the distribution of

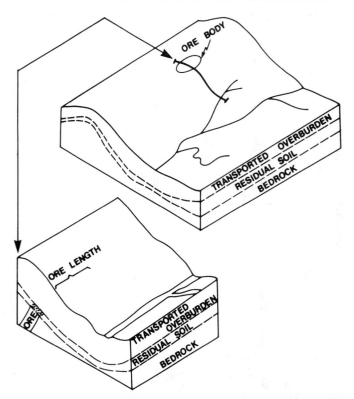

Figure 14.6. Idealized diagram – geochemical anomalies in transported overburden over mineralization.

any such deposits be determined before interpreting soil geochemical data. Chaffee and Hessin (1971) working in Arizona found that soil sampling and analysis for total copper and zinc faithfully reflected the distribution of metals about the Veikol Hill porphyry copper deposit where the soils are residual. However, as soon as 0.5 - 1.5 m or more of alluvial sand and gravel was encountered all response from the underlying mineralization was effectively masked.

Colluvium can be very much more variable than alluvium and range from a minor component of soil, diluting local geochemical patterns, to a complete cover masking all bedrock response. Cox (1975) noted that colluvial material, introduced by sheet wash, is present in some soils at Pioneer, Western Australia. This material dilutes metal values and reduces the contrast of anomalies over copper-nickel sulphides. At the other extreme, Leggo (1977) observed that landslip deposits of stabilized mud avalanche material below steep mountain slopes in Fiji completely mask part of an area of porphyry copper mineralization except where there are windows on ridges and beside streams that have cut through the cover.

Alluvial areas may coalesce to produce extensive valley gravel deposits such as those found in the southwestern United States or merge into lake beds. Playa lakes are found in arid environments throughout the world and are often characterized by Solonchak soils or the development of calcrete. Regardless of the degree to which secondary precipitates may cement or replace these soils it is important to realize that trace element patterns are primarily determined by the initial chemistry of the alluvial sediments. While this material may reflect the geochemistry of the catchmnet area it is unlikely to be able to directly delineate underlying mineralization and is not a useful medium for soil sampling.

Windblown Sand and Loess

Soils over large areas of the world contain at least a portion of windblown material. It must be stressed that, although the geochemical problems in areas of dune sand are obvious, soils in many semi-desert and open plain areas contain a high proportion of windblown material which can cause difficulties.

Aeolian or windblown material tends to be chemically uniform and is usually geochemically featureless. The main effect of the presence of this material in residual soils is to dilute the local geochemical patterns, (Bugrov, 1974; Cox, 1975; Scott, 1975). Bedrock sampling may be used in areas where this dilution is extreme or the overburden is composed entirely of aeolian detritus. However, soil sampling can continue to be of value if the effects of the aeolian material can be removed. Normally only the finer particles of soil (less than 200 μm or approximately 80 mesh) can be moved great distances by the wind. Thus, under most circumstances the effects of dilution by windblown sand can be removed by sieving out the fine fraction of soils.

The character of the problem is well illustrated by the work of Theobald and Allcott (1973) in Saudi Arabia. Figure 14.7 shows the results of analysis of different size fractions of surface samples collected along a traverse over the Uyajah Ring structure. The area is desert with little evidence of a true soil development and most surficial material is sandy-stony regolithic debris with local dune sand accumulations. The data shown in Figure 14.7 reveal consistent distribution patterns which are typical of all elements in the area. There is a conspicuous lack of variation in the finest size fraction which the authors attribute to the homogenizing effect of aeolian material which is predominant in the size range. In progressively coarser fractions there is an increase in the variability of the analytical data. This is interpreted as reflecting a steady increase in the proportion of material derived from the local geology. Indeed, subsequent regional sampling using the -10 +30 mesh fraction was found to reliably define the geology of the ring complex and remove the effects of dilution by windblown material. However, Theobald and Allcott (1973) found that anomalies related to molybdenum-tungsten mineralization could be further enhanced by analysis of nonmagnetic heavy-mineral concentrates recovered from the -10 +30 mesh size fraction. Bugrov (1974) and Bugrov and Shalaby (1975) report similar experience in the deserts of eastern Egypt using the -1 mm + 0.25 mm fraction (-20+70 mesh).

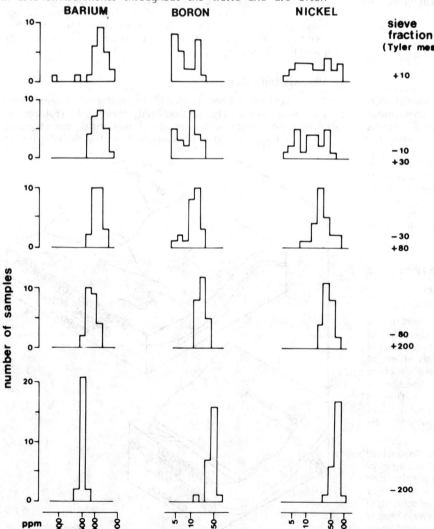

Figure 14.7. Variation of barium, boron and nickel in surface debris as a function of size fraction, Uyaijah Ring Company, Saudi Arabia (from Theobald and Allcott, 1973).

These results suggest that for soil sampling the effects of aeolian dilution can be removed by sieving out the fine fraction and anomalies enhanced by isolating the heavy mineral fraction. Experience has shown, however, that geochemical techniques may have to be adapted to suit local conditions. Bugrov (1974) draws attention to the problem of false anomalies produced by the secondary concentration of resistate heavy minerals in depressions in the desert surface. In direct contrast, Brown (1970) found that in areas with a cover of blown sand on the margin of the Kalahari Desert, copper anomalies were developed in the silt fraction of transported soils overlying copper sulphide mineralization. From an examination of the form and distribution of anomalous copper in the aeolian cover, Brown attributed the anomaly development to the seasonal upward movement of metal-bearing moisture through the overburden. Bugrov (1974) also noted an association of copper with clay-sized material in desert soils over mineralization in eastern Egypt.

In general, however, it appears that when aeolian detritus represents a minor portion of the total soil, the diluting effects may be removed by sieving out the fine fraction. Where windblown material is dominant at the surface alternative procedures may have to be applied and consideration given to sampling residual soil material below the aeolian cover or sampling the bedrock itself. The preferred sampling scheme is best defined by an orientation survey.

Volcanic Ash

A covering of volcanic ash is not commonly encountered in exploration programs since it is restricted to areas of active volcanism. It was, however, the principal source of confusion during early evaluation of the Panguna porphyry copper deposit on Bougainville Island (Mackenzie, 1973).

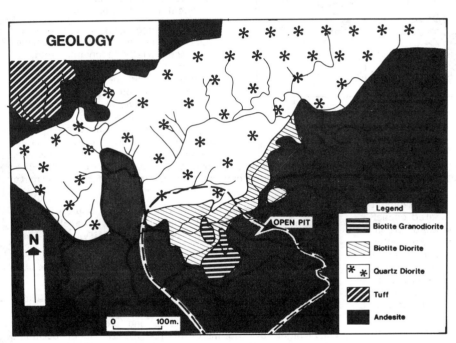

Figure 14.8

Geology of the Panguna area, Bougainville Island (after Baumer and Fraser, 1975).

Figure 14.9

Distribution of copper in stream sediments, Panguna area, Bougainville Island (with permission of RioFinEx).

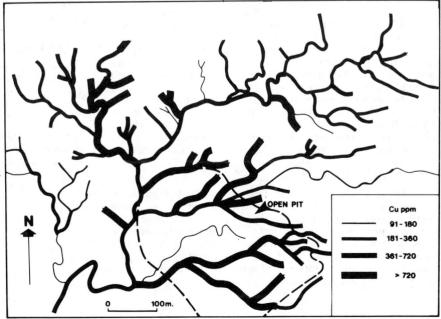

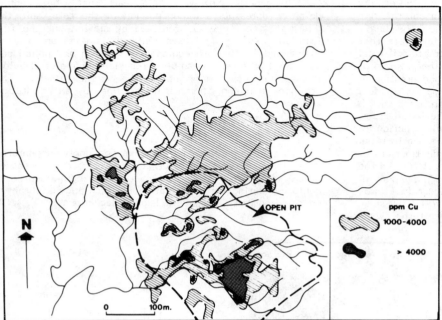

Figure 14.10. Distribution of copper in soils, Panguna area, Bougainville Island (with permission of RioFinEx).

The Panguna deposit is located in youthful mountain terrain covered with dense tropical rainforest. The general geology of the deposit is shown in Figure 14.8 together with the outline of the open pit which contains the principal concentration of mineralization. Weak disseminated copper sulphides extend over a much larger area. Detailed stream sediment sampling in the upper part of the Kawerong Valley reveals an extensive area of high copper values in the general area of the Panguna deposit (Fig. 14.9) which relates to the widespread low-grade disseminated copper mineralization. Soil sampling, however, gave a broken picture with several discrete anomalies (Fig. 14.10), most noticeably in the area of the present open pit separated by areas with low levels of copper. Initially the low copper values found on ridges within the area of anomalous steam sediments were attributed to surface leaching (Mackenzie, 1973). Subsequent investigations, deep augering and pitting, revealed the presence of the volcanic ash blanketing nearly all ridges and spurs to depths varying from a few centimetres to 24 m. The ash is of recent age and is a stratified brown clayey deposit which in a few places carries exotic boulders of agglomerate and breccia. During the early phase of exploration, soil samples were collected by auger at a depth of 0.5 -2 m and field personnel were unable to see any visible difference between the clayey ash and clayey weathered bedrock. Later, it was found that the ash is thickest on slopes and in places extends onto alluvial boulder terraces beside the rivers so that the area of residual soils over the main deposit, as represented by the outline of the open pit, is relatively restricted (Mackenzie, 1973).

Once the problem had been recognized exploration methods were modified to cope with the ash cover. Deep augering was employed to sample weathered rock below the ash, and prior to drilling, efforts made to extrapolate the geology into the blind areas. Systematic drilling ultimately proved the presence of extensive mineralization below the ash cover and alluvial terraces of sufficient grade to sustain a mine (Baumer and Fraser, 1975).

FUTURE DEVELOPMENTS

A detailed discussion of new methods under investigation and future developments is outside the scope of this paper. However, it is probable that effort will be concentrated in certain directions including:

1. The further development of geochemical techniques for locating mineralization beneath a cover of transported overburden or post-mineralization rocks. This includes the identification of seepage of metals from depth, the use of gases which are able to migrate more freely through overburden, continued study of the role of vegetation in the development of anomalies and possible electrochemical dispersion processes.

2. The continuing study of the form and mode of occurrence of metals derived from potentially economic mineralization. The use of selective analytical techniques to isolate metal derived from particular primary or secondary mineral forms shows promise as a way of increasing specificity to mineralization in areas of high background, identifying false anomalies due to rock types and enhancing anomaly contrast where the metal related to mineralization is only a minor component of the total metal in the soil.

The challenge is to devise methods appropriate for problems encountered in the field. Success in developing any technique is dependent on correctly identifying the field conditions.

ACKNOWLEDGMENTS

The authors wish to thank the Anglo American Corporation, Johannesburg, for permission to use data from exploration studies at Areachap and Jacomyns Pan, and C.L. Knight of RioFinEx for his assistance in obtaining information on the geochemistry of Bougainville. The authors are most grateful to Barringer Research Limited for support in the preparation of this paper.

REFERENCES

Baumer, A. and Fraser, R.B.
  1975: Pangua porphyry copper deposit, Bougainville; in The Economic Geology of Australia and Papua New Guinea; C.L. Knight (ed.), Aust. Inst. Min. Met., p. 855-866.

Bradshaw, P.M.D., Thomson, I., Smee, B.W., and Larsson, J.W.
  1974: The application of different analytical extractions and soil profile sampling in exploration geochemistry; J. Geochem. Explor., v. 3, p. 209-225.

Brown, A.
  1970: Dispersion of copper and associated trace elements in a Kalahari sand environment, northwest Zambia; Ph.D. thesis, Univ. of London. (Unpublished)

Bugrov, V.
  1974: Geochemical sampling techniques in the eastern desert of Egypt; J. Geochem. Explor., v. 3, p. 67-75.

Bugrov, V. and Shalaby, I.M.
   1975: Geochemical prospecting in the eastern desert of Egypt; in Geochemical Exploration 1974, I.L. Elliott and W.K. Fletcher (ed.), Elsevier Publ. Co., Amsterdam, p. 523-530.

Butt, C.R.M.
   1976: Mount Keith; in Excursion Guide No. 41C, 25th. Int. Geol. Cong., Canberra, R.E. Smith, C.R.M. Butt and E. Bettanay (ed.), p. 35-39.

Butt, C.R.M. and Sheppy, N.R.
   1975: Geochemical exploration problems in Western Australia exemplified by the Mt. Keith area; in Geochemical Exploration 1974, I.L. Elliott and W.K. Fletcher (ed.), Elsevier Publ. Co., Amsterdam, p. 391-416.

Chaffee, M.E. and Hessin, T.D.
   1971: An evaluation of geochemical sampling in the search for concealed "porphyry" copper-molybdenum deposits on pediments in southern Arizona; in Geochemical Exploration, Can. Inst. Min. Met., Spec. Vol. 11, p. 401-409.

Chakrabarti, A.K. and Solomon, P.J.
   1970: A geochemical case history of the Rajburi antimony prospect, Thailand; Econ. Geol., v. 65, p. 1006-1007.

Clema, J.M. and Stevens-Hoare, N.P.
   1973: A method for distinguishing nickel gossans from other ironstones in the Yilgarn Shield, Western Australia; J. Geochem. Explor., v. 2, p. 393-402.

Cole, M.M., Provan, D.M.J., and Tooms, J.S.
   1968: Geobotany, biogeochemistry and geochemistry in mineral exploration in the Bulman-Waimuna Springs area, Northern Territory, Australia; Inst. Min. Met. Trans., London, v. 77-B, p. 81-104.

Coope, J.A.
   1973: Geochemical prospecting for porphyry copper-type mineralization — a review; J. Geochem. Explor., v. 2, p. 81-102.

Cornwall, F.W.D.
   1970: Discovery and exploration of the Fitula copper deposit, Nchanga area, Zambia; 9th Comm. Min. Met. Cong., Publ. Proc., v. 2, p. 535-560.

Cox, R.
   1975: Geochemical soil surveys in exploration for nickel-copper sulphides at Pioneer, near Norseman, Western Australia; in Geochemical Exploration 1974, I.L. Elliott and W.K. Fletcher (ed.), Elsevier Publ. Co., Amsterdam, p. 437-460.

Danchin, R.
   1972: Aspects of the geochemistry of calcretes and soils from various localities in the Northwest Cape; Anglo American Research Laboratories, Unpublished Report.

Dasgupta, S.P.
   1963: Geochemical prospecting in the Khetri Copper Belt, Rajasthan, India; Proc. Seminar on Geochem. Methods, Min. Res. Devel. Series, No. 21, United Nations, New York, p. 111-116.

Ellis, A.J. and McGregor, J.A.
   1967: The Kalengwa copper deposit in northwestern Zambia; Econ. Geol., v. 62, p. 781-797.

Govett, G.J.S. and Hale, W.E.
   1967: Geochemical orientation and exploration near a disseminated copper deposit, Luzon, Phillipines; Inst. Min. Met. Trans., London, v. 76-B, p. 190-201.

Granier, C.
   1973: Introduction to geochemical prospecting for mineral deposits (in French); Masson et Ci., Paris, France, 143 p.

Horsnail, R.F. and Lovestrom, K.A.
   1974: A geochemical exploration strategy for porphyry copper deposits in a desert pediment environment (abstract); Mining Eng., v. 26, p. 78.

Hawkes, H.E. and Webb, J.S.
   1962: Geochemistry in mineral exploration; Harper and Row, New York, 415 p.

Koksoy, M. and Bradshaw, P.M.D.
   1969: Secondary dispersion of mercury from cinnabar and stibnite deposits, West Turkey; Col. Sch. Mines Quart., v. 64, p. 333-356.

Learned, R.E. and Boisen, R.
   1973: Gold — a useful pathfinder element in the search for porphyry copper deposits in Puerto Rico; in Geochemical Exploration 1972, M.E. Jones (ed.), Inst. Min. Met., London, p. 93-103.

Leggo, M.D.
   1977: Contrasting geochemical expressions of mineralization at Namosi, Fiji; J. Geochem. Explor., v. 8, p. 431-456.

Lewis, R.W., Jr.
   1965: A geochemical investigation of the Caraiba deposit, Bahia, Brazil; U.S. Geol. Surv., Prof. Paper 550-C, p. 190-196.

Lewis, R.W., Jr., Diniz Goncalves, G.N., and
Araujo Mello, V.N.
   1971: Status of geochemical prospecting in Brazil; in Geochemical Exploration, Can. Inst. Min. Met., Spec. Vol. 11, p. 28-31.

Lord, J.E.
   1973: Surface and subsurface geochemistry of the Native Bee-Jasper-Biotite copper prospects, northwest Queensland; J. Geochem. Explor., v. 2, p. 349-365.

Lovering, T.G. and McCarthy, J.H., Jr.
   1978: Conceptual models in exploration geochemistry — The Basin and Range Province of the Western United States and Northern Mexico; J. Geochem. Explor., v. 9, p. 113-279.

Mackenzie, D.H.
   1973: Bougainville — a geochemical case history; Paper delivered at Workshop course arranged by the Australian Mineral Foundation, Inc., Adelaide.

Mazzuchelli, R.H.
   1972: Secondary geochemical dispersion patterns associated with the nickel sulphide deposits at Kambalda, Western Australia; J. Geochem. Explor., v. 1, p. 103-116.

Mazzuchelli, R.H. and Robbins, T.W.
   1973: Geochemical exploration for base and precious metal sulphides associated with the Jimberlana dyke, Western Australia; J. Geochem. Explor., v. 2, p. 383-392.

Montgomery, R.
   1971: Secondary dispersion of molybdenum and associated elements in tropical rainforest environment of Guyana; Ph.D. thesis, Univ. of London (Unpublished).

Moeskops, P.G.
   1977: Yilgarn nickel gossan geochemistry — a review, with new data; J. Geochem. Explor., v. 8, p. 247-258.

Netterberg, F.
　1971: Calcrete in road construction; C.S.I.R. Research Rept. 286, NIRR Bull. 10, 73 p. Pretoria, S. Africa.

Nicholls, O.W., Provan, D.M.J., Cole, M.M., and Tooms, J.S.
　1965: Geobotany and geochemistry in mineral exploration in the Dugald River area, Cloncurry district, Australia; Inst. Min. Met. Trans., London, v. 74-B, p. 695-799.

Nickel, E.H., Allchurch, P.D., Mason, M.G., and Wilmshurst, J.R.
　1977: Supergene alteration at the Perseverence nickel deposit, Agnew, Western Australia; Econ. Geol., v. 72, p. 184-203.

Ong, P.M. and Sevillano, A.C.
　1975: Geochemistry in the exploration of nickeliferous laterite; in Geochemical Exploration 1974, I.L. Elliott and W.K. Fletcher (ed.), Elsevier Publ. Co., Amsterdam, p. 461-478.

Philpot, D.E.
　1975: Shangani — a geochemical discovery of nickel-copper sulphide deposit; in Geochemical Exploration 1974, I.L. Elliott and W.K. Fletcher (ed.), Elsevier Publ. Co., Amsterdam, p. 503-510.

Reedman, J.H.
　1974: Residual soil geochemistry in the discovery and evaluation of the Butiriku carbonatite, southeast Uganda; Inst. Min. Met. Trans., London, v. 83-B, p. 1-12.

Saigusa, M.
　1975: Relation between copper content in soils and copper grade of some porphyry copper deposits in humid tropical regions; in Geochemical Exploration 1974, I.L. Elliott and W.K. Fletcher (ed.), Elsevier Publ. Co., Amsterdam, p. 511-522.

Scott, M.J.
　1975: Case histories from a geochemical exploration program, Windhoek district, Southwest Africa; in Geochemical Exploration 1974, I.L. Elliott and W.K. Fletcher (ed.), Elsevier Publ. Co., Amsterdam, p. 481-492.

Smith, B.H.
　1977: Some aspects of the use of geochemistry in the search for nickel sulphides in lateritic terrain in Western Australia; J. Geochem. Explor., v. 8, p. 259-282.

Theobald, P.K. and Allcott, G.M.
　1973: Tungsten anomalies in the Uyaijah Ring Structure, Kushaymiyah igneous complex, Kingdom of Saudi Arabia; Ministry of Petroleum and Natural Resources, Jeddah, Saudi Arabia.

Thomson, I.
　1976: Geochemical studies in central-west Brazil; D.N.P.M., Ministerio das Minas e Energia, Brasil, 258 p. (Billingual edition)

Thomson, I. and Brim, R.J.P.
　1976: A geoquimica no Projeto Geofisico Brasil-Canada: uma avaliacao preliminar dos resultados; Anais do I Sem. Bras. sob. Tec. Explor. em Geol., Pocos de Caldas, p. 103-104.

Tooms, J.S., Elliott, I., and Mather, A.L.
　1965: Secondary dispersion of molybdenum from mineralization, Sierra Leone; Econ. Geol., v. 60, p. 1478-1496.

Tooms, J.S. and Webb, J.S.
　1961: Geochemical prospecting investigations in the Northern Rhodesian copperbelt; Econ. Geol., v. 56, p. 815-846.

Ullmer, E.
　1978: Sacaton mine area, Pinal County, Arizona; J. Geochem. Explor., v. 9, p. 235-236.

Webb, J.S. and Tooms, J.S.
　1959: Geochemical drainage reconnaissance for copper in Northern Rhodesia; Inst. Min. Met. Trans., London, v. 68-B, p. 125-144.

Wilmshurst, J.R.
　1975: The weathering products of nickeliferous sulphides and their associated rocks in Western Australia; in Geochemical Exploration 1974, I.L. Elliott and W.K. Fletcher (ed.), Elsevier Publ. Co., Amsterdam, p. 417-436.

Woolf, D.L., Tooms, J.S., and Kirk, H.J.C.
　1966: Geochemical surveys in the Labuk Valley, Sabah 1965, Borneo; Malyasia Geol. Survey, Ann. Rept. 1965, p. 212-226.

Worthington, J.E., Jones, E.M., and Kiff, I.T.
　1976: Techniques of geochemical exploration in the southeast piedmont of the United States; J. Geochem. Explor., v. 6, p. 279-295.

Yellur, D.D.
　1963: Geochemical prospecting for lead at Khandia, Baroda district, India — a case history; Proc. Seminar on Geochem. Methods, Min. Res. Dev. Series No. 21, United Nations, New York, p. 131-134.

LITHOGEOCHEMISTRY IN MINERAL EXPLORATION

G.J.S. Govett
University of New Brunswick,
Fredericton, New Brunswick

Ian Nichol
Queen's University, Kingston, Ontario

Govett, G.J.S. and Nichol, Ian, Lithogeochemistry in mineral exploration; in Geophysics and Geochemistry in the Search of Metallic Ores; Peter J. Hood, editor; Geological Survey of Canada, Economic Geology Report 31, p. 339-362, 1979.

Abstract

Lithogeochemistry, as used in this presentation, is defined as the determination of the chemical composition of bedrock material with the objective of detecting distribution patterns of elements that are spatially related to mineralization.

Mineralogical alteration zones in host rocks around mineral deposits have long been recognized and used as indicators of ore. Such alteration zones are the visible manifestations of physical and chemical changes in the host rocks resulting from either primary reactions associated with ore formation or subsequent secondary reactions between the ore and the host rocks. Chemical alteration halos may be more intense, and therefore detectable, over greater distances than mineralogical halos, since the lattice substitution of elements may be detected chemically without having any mineralogical representation. The scale and intensity of changes in the chemistry of the host rock is a function of the genesis of the ore, the chemistry of the host rock, and the nature of the secondary processes. Appreciation of these factors is fundamental to the successful application of lithogeochemistry to mineral exploration.

Lithogeochemistry has application at three levels of exploration: identification of geochemical provinces, favourable ore horizons, plutons or volcanic horizons on a regional reconnaissance scale; recognition of local halos related to individual deposits on a local reconnaissance or follow-up scale; and wall-rock anomalies related to particular ore-shoots on a mine scale.

Lithogeochemistry has been applied on a regional scale to the identification of mineralized areas in granitic and gneissic terrane (Sn, W, Mo, Cu, and U); in areas of basic intrusions (Ni, Cr, and Pt) and basic volcanic rocks (Cu); and in areas of sedimentary rocks (Cu, Pb-Zn, and Au-U). In other cases favourable geological environments have been recognized on the basis of some diagnostic geochemical parameter of lithological units spatially, and probably genetically, associated with mineralization. For example, productive greenstone belts can be distinguished from nonproductive greenstone belts, and cycles of volcanism containing significant massive sulphide deposits can be distinguished from equivalent nonproductive cycles on the basis of enhanced background contents of certain elements; areas of black shale horizons enriched in chalcophile elements may be readily identified, and these horizons may be spatially associated with more favourable horizons – such as quartzite and limestone – for the emplacement of mineralization.

On a more local scale, geochemical dispersion halos of both major and minor elements have been shown to be associated with porphyry copper deposits. Similarly, depletion of Na and Ca, and enhancement of Fe, Mg, and K – together with characteristic distribution patterns of a wide range of trace elements – appear to be a general local-scale feature associated with volcanogenic massive sulphide deposits. Wall-rock dispersion halos have proved useful in detailed exploration, although their nature is very much dependent on local geological conditions.

Various techniques have been employed to detect bedrock anomalies related to mineralization; these include whole rock analysis, measurement of water-soluble elements (especially halogens), and analysis of mineral separates. In some cases multivariate statistical techniques have been used to identify otherwise unrecognized, subtle features of the data or as a means of enhancing very weak anomalies.

The results cited are examples where lithogeochemistry has been or could have been used successfully for the location of particular types of mineralization. They are also indications of the diversified role lithogeochemistry could have in mineral exploration.

Exploration requirements in the future will demand techniques to find mineral deposits in situations where current methods have failed. In particular, presently available routine geochemical methods are unsuitable, or have limited application, for the location of deeply-buried or blind deposits – these are the types of deposits that probably will become common targets for mineral exploration as the more easily detected near-surface targets are exhausted. The subtle lithogeochemical anomalies and element zoning found in the cap-rocks of some types of blind deposits offer scope for their detection, and may also be useful in drill-hole control.

The application of lithogeochemistry is largely a function of the availability of adequate samples either from suitable bedrock exposure or drill core; obviously, it must have limited potential in areas of deep weathering and areas of extensive overburden or soil cover. However, within these constraints, recent developments offer convincing evidence of the potential power of lithogeochemistry as an exploration method.

Résumé

Dans ce rapport, on définit la lithogéochimie comme la détermination de la composition chimique de la roche en place; par cette méthode, on peut déceler le mode de distribution des éléments dans l'espace, dans la mesure où la distribution est liée à la minéralisation.

On a reconnu il y a longtemps l'existence de zones d'altération minéralogique dans les roches favorables qui renferment les gîtes minéraux; on les utilise comme "indicateurs" du minerai. Ces zones d'altération sont les manifestations visibles des modifications physiques et chimiques que subissent les roches favorables par suite des réactions secondaires ultérieures qui ont eu lieu entre le minerai et les roches favorables. Les auréoles d'altération chimique sont parfois plus intenses, par conséquent plus facilement décelables, et cela, sur de plus grandes distances que les auréoles minéralogiques, puisque toute substitution d'éléments apparaît à l'analyse chimique, sans qu'il soit nécessaire d'effectuer des observations minéralogiques. L'étendue et l'intensité des variations de la composition chimique de la roche favorable dépendent du mode de genèse du minerai, de la chimie de la roche favorable et de la nature des processus secondaires. Il est essentiel d'évaluer chacun de ces facteurs, pour pouvoir appliquer avec succès la lithogéochimie à la recherche minière.

La lithogéochimie s'applique à troix niveaux d'exploration: l'identification des provinces géochimiques, des horizons minéralisés favorables, des plutons ou des horizons volcaniques, à une échelle de reconnaissance régionale; l'identification des auréoles locales liées à des gîtes particuliers, à l'échelle de la reconnaissance locale ou des travaux d'exploration détaillée; et l'étude des anomalies qui caractérisent la roche encaissante et sont liées à l'existence de colonnes minéralisées, cela à l'échelle de l'exploitation minière.

On a appliqué, à l'échelle régionale, la lithogéochimie à l'identification de zones minéralisées dans des terrains granitiques et gneissiques (Sn, W, Mo, Cu et U); dans des zones d'intrusions basiques (Ni, Cr et Pt) et de roches volcaniques basiques (Cu); et enfin dans des zones de roches sédimentaires (Cu, Pb-Zn, et Au-U). Dans d'autres cas, on a identifié des milieux géologiques favorables, en fonction de certains paramètres géochimiques diagnostiques caractérisant des unités lithologiques spatialement et sans doute génétiquement associées à une minéralisation. Par exemple, on peut distinguer les zones de roches vertes productives des zones équivalentes non productives, et distinguer les cycles volcaniques pendant lesquels se sont constitués d'importants gîtes sulfureux massifs, des cycles équivalents non productifs, après accentuation des valeurs de fond de certains éléments; on peut facilement reconnaître les secteurs à horizons d'argile litée noire enrichis en éléments chalcophiles, lesquels peuvent être spatialement associés à des horizons plus favorables – tels que quartzites et calcaires – pour la mise en place de minéralisations.

A une échelle plus locale, les auréoles de dispersion géochimique des éléments importants et secondaires sont parfois, comme on l'a démontré, associées à des gîtes porphyriques de cuivre. De même, l'appauvrissement en Na et Ca, et l'enrichissement en Fe, Mg et K – en même temps que le mode de distribution typique d'une vaste gamme d'éléments-traces – semblent constituer un caractère local dû à la présence de gîtes sulfureux massifs, de types volcanogénique. Les auréoles de dispersion dans la roche encaissante se sont avérées utiles pour l'exploration détaillée, bien que leur nature dépende beaucoup plus des conditions géologiques locales.

On a employé diverses techniques pour déceler les anomalies que présente la roche en place du point de vue de la minéralisation; parmi ces techniques, citons l'analyse de la roche entière, la mesure des éléments solubles dans l'eau (en particulier les halogènes), et l'analyse des diverses fractions minérales. Dans certains cas, on a employé des méthodes utilisant plusieurs variables statistiques, pour mettre en relief des résultats moins évidents, jusque là ignorés, ou bien des anomalies très faibles.

Les résultats cités représentent des cas où la lithogéochimie a ou aurait pu donner des résultats satisfaisants pour localiser certains types de minéralisation. Ils indiquent aussi le rôle multiple que pourrait jouer la lithogéochimie dans l'exploration minière.

A l'avenir, il sera nécessaire d'employer de nouvelles techniques d'exploration là où les méthodes habituellement ont échoué. En particulier, les méthodes géochimiques couramment employées actuellement sont inadéquates, ou n'ont que des applications restreintes, pour la localisation des gîtes profondément enfouis ou tout simplement dissimulés – c'est-à-dire les sortes de gîtes qui probablement constitueront les principaux objectifs d'exploration, à mesure que s'épuiseront les minéralisations proches de la surface, plus facilement détectables. Les anomalies lithogéochimiques de faible intensité, et la zonalité des éléments qui caractérisent le chapeau (cap-rock) de certains types de gîtes dissimulés favorisent la détection de ces gîtes, et peuvent sans doute permettre un meilleur choix de l'emplacement des trous de forage.

L'usage de la lithogéochimie dépend largement de la mesure dans laquelle on peut obtenir des échantillons adéquats, provenant soit d'affleurements favorables de la roche en place, soit de carottes de forage; il est évident que la lithogéochimie offre moins de possibilités, dans les zones d'altération profondes, et celles cachées par de vastes terrains de couverture ou une importante couverture de sol. Cependant, malgré ces limitations, les développements récents démontrent de façon convaincante le potentiel de la lithogéochimie en tant que méthode d'exploration.

INTRODUCTION

Lithogeochemistry, or rock geochemistry, as used in this paper, is defined as the study of the chemical composition of bedrock with particular reference to the search for ore deposits. Geochemical patterns in bedrock that can be related to mineralization are referred to as primary patterns or primary dispersion, irrespective of whether the reactions causing the element distribution patterns are syngenetic or epigenetic (James, 1967).

Mineralogical alteration zones in host rocks around mineral deposits have long been recognized and used as indicators of ore. Such alteration zones are the visible manifestations of physical and chemical changes in rocks associated with mineralization. Chemical alteration halos may be more extensive and therefore detectable over greater distances than mineralogical halos, since lattice substitution of elements may occur without any mineralogical change. Recognition of this possibility and the need to develop techniques capable of detecting deeply buried and blind deposits and to improve exploration capabilities have led to the recent widespread interest in lithogeochemistry.

The improved understanding of ore genesis during the last decade is also a significant factor in the increased interest in lithogeochemical techniques. The general features of porphyry ore deposits have been characterized in terms of pre-ore host rock, igneous host rocks, the orebody, hypogene alteration, mineralization and occurrence of sulphides (Lowell and Guilbert, 1970). Whereas a considerable degree of similarity exists among the deposits, aspects in the detailed geological history of a particular deposit (e.g., nature of fluids, degree of fracturing and postmineralization history) might be expected to give rise to significant differences between geochemical responses and mineralization. Understanding of the genesis of volcanic-sedimentary massive sulphide deposits has also vastly improved over the past decade; they are generally regarded as having originated from metal-bearing fumarolic exhalations on the sea floor that were coeval with the associated volcanic or sedimentary activity (Sangster, 1972). Notwithstanding a common genesis, detailed variations in the character of the host rock, tectonic, and postmineralization history of different deposits could be expected to result in variations between the geochemical responses in the wall rock and mineralization.

The chemical composition of a sample of rock reflects the entire geological history of the rock. For example, the composition of a sample of a particular granite may reflect the composition of the primary magma (possibly modified by assimilation of country rock), metasomatic and postmagmatic processes, metamorphism, and, in the case of surface samples, the effects of weathering. The extent to which the effect of a mineralizing event is recognizable in the composition of the host rock depends upon the actual processes that affected the rock. It is thus of critical importance in the evaluation of any lithogeochemical data to consider the nature and effects of all processes that may have affected the sample.

In view of the importance of geological environment on the nature of geochemical dispersion associated with mineralization the main body of this review is divided into two parts: (1) geochemical response in intrusive rocks; and (2) geochemical response in extrusive and sedimentary rocks. Within each geological group, geochemical response is considered in terms of exploration scale:

Regional — large-scale, geochemical responses that are capable of discriminating between productive and barren terrain.

Local and Mine Scale — geochemical responses around individual deposits that can be detected up to 1 to 2 km from a deposit, and the geochemical responses that are limited to the immediate wall rock of a deposit.

This classification is necessarily somewhat arbitrary, and there is some overlap between the two scales of investigation.

The objective of the paper is to present a state-of-the-art review of lithogeochemistry; the source of much of the data, therefore, is published papers, augmented by our personal experiences. The works that we cite were carried out in many different ways — in terms of sampling, analysis, data presentation — and, indeed, with many different objectives. Our lack of first-hand knowledge of many of the areas makes comparisons and evaluation difficult; to minimize some of the problems and to achieve more meaningful comparisons, some of the data have been replotted from the published work and, in so doing, we trust that neither the character of the original data nor the intent of the authors have not been misrepresented.

A number of investigations have been concerned with the distribution of stable isotopes (sulphur, oxygen, hydrogen, and lead) associated with various types of deposits and mineral provinces. The results of these investigations are not considered here, although stable isotope studies may provide a better understanding of the genesis of mineral deposits — information that would be useful in selecting areas for exploration. In a limited number of cases, isotope zoning around mining districts has been noted.

Lithogeochemistry has received considerable attention within the Soviet Union. Although reports indicate notable success and some innovative approaches, the work, in general, is difficult to evaluate and discussion is limited to a consideration of some of the principles of the Soviet work.

The technical viability of lithogeochemistry in mineral exploration is dependent on the existence of some geochemical signature in the rocks associated with mineralization — which normally requires a genetic relationship between the geochemical patterns and the ore-forming processes. A major factor affecting the applicability of lithogeochemistry in exploration is the extent to which features are of general occurrence. The economic viability is related to the cost effectiveness of the procedure in relation to other techniques, and is dependent on the nature of the appropriate sampling, analytical and interpretational procedures.

GEOCHEMICAL RESPONSE TO MINERAL DEPOSITS IN INTRUSIVE ROCKS

Regional Scale

Prior to the present upsurge of interest in lithogeochemistry, geologists and geochemists have sought criteria to discriminate between productive and barren intrusions on a regional scale. Tin seems to be the only element that generally — but not universally — shows enrichment in intrusions associated with tin mineralization in many parts of the world. Quartz-cassiterite-bearing granite bodies have Sn variations in the range 20-30 ppm, compared to 3-5 ppm in barren intrusions (Barsukov, 1967). Granite associated with tin deposits in Transbaikaliya (U.S.S.R.) has 16-32 ppm Sn, compared to barren granite that has less than 5 ppm Sn (Ivanova, 1963).

Tin-bearing granite of northeast Queensland has been shown to have significantly higher Sn contents than nonstanniferous granite (Sheraton and Black, 1973). A granite with a high proportion of samples with Sn contents above background values appears a criterion of potential tin mineralization. The variation in tin content of granite with mineralization is ascribed to a primary uneven distribution of tin in the crust or mantle that was the magma source. Although the areal extent of the granite bodies is not given the Sn content of individual intrusions was characterized on the basis of an average of some 25 samples per intrusive but ranging from 6 to 73 samples for individual bodies. None of the mineralized granite is uniformally high in tin and the characteristic feature is a very variable tin content amongst samples necessitating that sampling must be reasonably extensive if the tin-bearing potential is to be realistically assessed. On the basis of a wider study of granitoids from the Tasman geosyncline in Eastern Australia, granitic rocks associated with cassiterite mineralization have a higher mean Sn content (26 ppm) relative to similar but barren granitoids (3.4 ppm) (Hesp and Rigby, 1975). The samples were collected from outcrops some distance from locations where cassiterite mineralization was observed to avoid the effect of the latter on Sn values. However, the extent to which the higher Sn values are characteristic of the mineralized granite as a whole is not known as the sampling was biased to some extent towards areas of Sn mineralization rather than being representative of the granitoid rocks as a whole (Hesp and Rigby, 1975). Although there are some exceptions to the generalization, particularly in the low range of Sn concentration, all the granitoid samples that have more than

Table 15.1

Variations in composition of mineralized and unmineralized intrusives – Allen et al., 1976

		Mineralized	Unmineralized
K_2O	$\bar{x}$	2.31	1.47
	σ	0.80	1.12
	R	1.12 - 2.77	0.28 - 5.36
Cu	$\bar{x}$	1450	57
	σ	1785	73
	R	17 - 7400	0 - 350
CaO	$\bar{x}$	3.58	4.85
	σ	1.17	1.80
	R	2.0 - 3.62	0.50 - 9.28
Na_2O	$\bar{x}$	2.27	3.67
	σ	1.40	1.82
	R	0.50 - 4.52	0.22 - 6.48
MnO	$\bar{x}$	0.12	0.18
	σ	0.04	0.09
	R	0.07 - 0.21	0.06 - 0.42
Zn	$\bar{x}$	36	92
	σ	14	67
	R	17 - 70	30 - 270
n		20	27

20 ppm have associated tin mineralization (Fig. 15.1); this finding is consistent with most of the work carried out in the U.S.S.R.

Many workers have observed that, aside from tin, there is no noticeable correlation between the abundance of elements in granitic rocks and the presence or absence of mineralization; the lack of correlation is documented for lead and zinc (Tauson, 1967), and for lead, zinc, molybdenum, and tungsten (Barsukov, 1967). In the investigation of mineralized granitic rocks of northeast Queensland referred to earlier, no relation between the lead, zinc, and copper contents of the granite and lead/zinc and copper mineralization was noted, (Sheraton and Black, 1973). This feature was ascribed to the mineralization resulting from processes unrelated to the differentiation of the intrusive.

On the other hand, in situations where there has been hydrothermal alteration, significant differences in mean element content have been observed between mineralized and barren intrusions. For example, the intrusion that hosts the Coed-y-Brenin copper porphyry-type deposit in North Wales (U.K.) is enriched in K_2O, Cu, and Mo, and is depleted in MnO, Na_2O, CaO, and Zn relative to similar but barren intrusions in the same region on the basis of a comparison of mean element content (Allen et al., 1976). With the exception of Cu and Zn, the differences in mean content are slight, and there is considerable overlap in the ranges of the mineralized and unmineralized populations (see Table 15.1). The authors drew attention to the fact that the distribution of the data within the data sets is not normal and thus the statistical data should be treated with caution. The enhancement and depletion effects are interpreted as a reflection of the effects of mineralization and associated hydrothermal alteration. The K_2O, Na_2O, and CaO variations are associated with phyllic alteration, whereas the Zn depletion possibly reflects Zn migration to the pyrite and propylitic zone.

In certain cases, the form of the frequency distribution of an element in a region or a particular rock unit is more diagnostic of mineralization than the mean value of the element (Beus, 1969; Bolotnikov and Kravchenko, 1970;

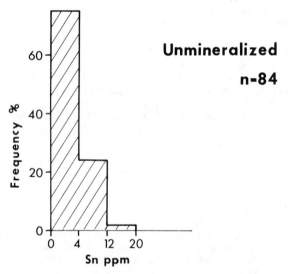

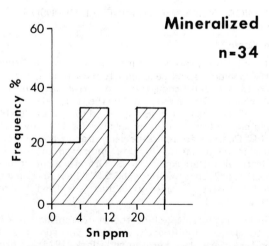

Figure 15.1. Frequency distribution of Sn in mineralized and barren granitic intrusions, Tasman Geosyncline, Eastern Australia (data from Hesp and Rigby, 1975).

Tauson and Kozlov, 1973). The form of frequency distributions for W has been shown to differ between barren and mineralized acidic plutons in the Yukon and Northwest Territories in Canada (Garrett, 1971). The distribution of W in barren plutons (Fig. 15.2) is only slightly skewed, whereas the distribution in mineralized plutons has a more marked positive skew; this is interpreted as reflecting the occurrence of a tungsten mineralizing event due to magmatic processes superimposed upon the primary tungsten distribution. Plutons that are host to porphyry-type disseminated tungsten mineralization, and therefore presumably were closely allied to primary magmatic processes, are more clearly distinguishable from barren plutons than those associated with more local skarn-type mineralization. Garrett concluded that an "average" size pluton could be geochemically characterized by collecting duplicate samples from 15 samples sites in an unzoned pluton.

Attempts have also been made to use variations in element content of a particular mineral phase in a rock to discriminate between barren and productive plutons. Most of the work has been done on biotite (although trace element distribution in feldspar and magnetite have also been investigated). The rationale behind the investigations is the fact that the structure of biotite can readily accommodate minor elements in different states within the lattice and that biotite in productive stocks might contain characteristic minor element contents as a result of hydrothermal alteration associated with the mineralization. Some early work on the distribution of Cu and Zn in biotite from acidic intrusions in the Basin and Range Province of the U.S.A. (an area of major base and precious metal production) by Parry and Nackowski (1963) showed an apparent relation between the contents of Cu and Zn in biotite and the presence of base metal mineralization within a particular stock. The high copper values in biotite from the Basin and Range intrusions were ascribed to the deposition of sulphides in biotite by hydrothermal solutions. Subsequent investigations on biotite from the Basin and Range Province, however, have indicated that there is no clear relation between minor element contents of biotite in intrusions hosting porphyry copper deposits and unmineralized stocks (Jacobs and Parry, 1976); this is attributed to the existence of three genetic types of biotite (magmatic, replacement, and hydrothermal), the composition of which reflects the physical and chemical conditions of crystallization and the postmagmatic history of a stock. Systematic variations in biotite composition of the three types of biotite may be characteristic of individual porphyry copper deposits and thus may serve to distinguish intrusions genetically related to porphyry copper deposits from barren intrusions.

The halogen distributions in biotite within individual plutons of the Basin and Range area are distinctive, but there is no systematic distinction between barren and mineralized intrusions (Parry, 1972; Parry and Jacobs, 1975). The mean Cl content of biotite from intrusions in western North America and the Caribbean was shown to be similar in barren and mineralized plutons, and there is a greater range of values in the former than in the latter (Fig. 15.3) (Kesler et al., 1975a). Similarly neither water-soluble nor total Cl and F content in whole rock is consistently different in barren and mineralized intrusions (Kesler et al., 1973, 1975b). The lack of a clear relation between halogen content and mineralization is attributed to the halogen content of the biotites also being related to magmatic processes.

Some differences have been noted in major and trace element contents in feldspar, biotite, and muscovite from five granite bodies associated with tin, copper, zinc, and lead mineralization in Devon and Cornwall (southwest England) compared to nonmineralized granite elsewhere in England and Scotland (Bradshaw, 1967). An average of eleven samples per stock was used to characterize specific stocks. In the case of biotite, higher mean Rb and Sn contents and lower Sr contents are characteristic of mineralized granite relative to barren granite. In feldspars, mean contents of Rb, Zn, and, to a lesser extent, K, Mn, Pb and Sn are higher and Ca and Sr are lower in mineralized than in nonmineralized stocks. The mean contents of Pb and Zn in feldspar from the granite show a clear distinction in Zn content between barren and mineralized intrusions (Fig. 15.4). The mineralized and barren intrusions, however, are not strictly comparable, since the mineralized intrusions are more highly fractionated than the barren intrusions; only the Weardale granite (identified on Fig. 15.4) is petrochemically similar to the mineralized granite – but has an even lower Zn content than other barren granite. The study was based on relatively few samples (a total of 55 samples from the five mineralized granite bodies

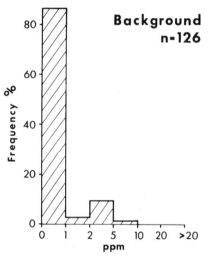
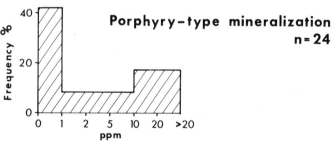
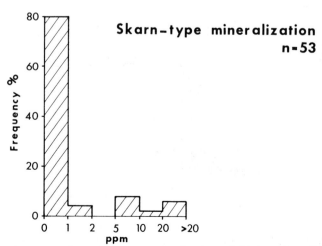

Figure 15.2. Frequency distribution of W in mineralized and barren granitic intrusions, Northern Canadian Cordillera (data from Garrett, 1971).

and a total of 94 from the eight barren granite bodies); although there are clear differences in the mean element contents, there is also considerable overlap between the two groups of granite in terms of element content of individual samples. The differences in mean contents of the ore metals Sn, Pb and Zn between mineralized and unmineralized stocks are attributed to effects of mineralization whereas variations in the mean contents of K, Rb, Sr and Ca are related to varying degrees of fractionation of the granite.

A rather different approach, based on determining the metal content of sulphide minerals, has been adopted to discriminate significantly mineralized from unmineralized ultramafic bodies of the Canadian Shield (Cameron et al., 1971). A total of 1079 samples from 61 locations were investigated and each location classified according to importance of mineralization: an ORE group contains ore grade deposits with more than 5000 tons nickel-copper, a MINORE group contains deposits of ore grade with less than 5000 tons nickel-copper, and a BARREN group contains only minor amounts of sulphide. The samples collected varied according to restraints on sampling imposed by sample availability but attention was focused on obtaining a suite of samples as representative as possible of the individual areas. A sulphide-selective digestion was used to determine the Cu, Ni, and Co contents present in sulphide minerals, based on the assumption that for an ultramafic magma to give rise to significant copper-nickel mineralization, the magma must be enriched in sulphur to such an extent that the solubility products of the metal sulphides will be exceeded, causing metal sulphides to separate. The content of sulphur or metals held as sulphides should be indicative of this condition and hence of possible mineralization. The frequency distributions of sulphide-held Cu, Co, and Ni, and of total sulphur, are all positively skewed; the results are summarized in Table 15.2. Ore-bearing localities (ORE group) are enriched in sulphide-held Ni and Cu (and, to a lesser extent, Co) and total S relative to mineralized localities (MINORE group) and to barren localities (BARREN group). The MINORE and BARREN localities are only distinguishable from one another on the basis of higher sulphide-held Ni content of the MINORE group.

The importance of recognizing the mineralogical site of trace elements has been demonstrated by variations in the mineralogical form of uranium in granite bodies in the U.K. (Simpson et al., 1977); delayed neutron analysis and Lexan plastic fission track studies were carried out to determine the content and form of the uranium. The geometric mean of U content of the Caledonian granite is 3.9 ppm U (118 samples, range 0.5-15.2 ppm), compared to Hercynian granite of southwest England that has a geometric mean of 10.8 ppm U (66 samples, range 3.2-35.5 ppm). The difference between the two granite suites is significant at the 99 per cent confidence level (based on the two-tailed Kolmogorov-Smirnov test). In the Caledonian granite, U occurs in postmagmatic secondary minerals, such as hematite and chlorite, in alteration zones associated with major faults. In southwest England, uranium occurs as mainly primary accessory minerals such as zircon, apatite, and sphene in vein-type deposits closely associated with the granite. It was concluded that the U vein-type mineralization of the Hercynian granite is comagmatic with the uranium-enriched granite and is not attributable to later kaolinization or weathering. Analyses of granite can serve to identify primary uranium provinces with associated vein-type mineralization. In the Caledonides, where uranium is principally associated with later faults and molasse facies sediments, granite sampling can help to identify the enrichment processes and thus aid in the definition of economic targets.

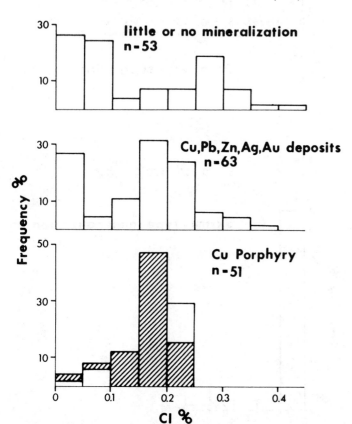

Figure 15.3. Frequency distribution of Cl in biotite from mineralized and barren granitic intrusions, Basin and Range Structural Province, U.S.A. (redrawn from Parry and Jacobs, 1975).

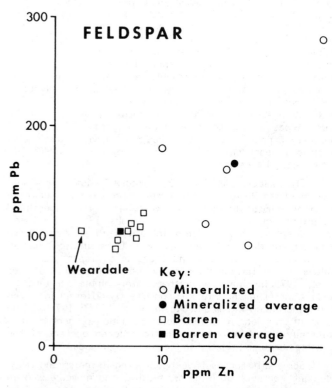

Figure 15.4. Mean contents of Pb and Zn in biotite from mineralized granitic intrusions of southwest England and barren granitic intrusions from northern England and Scotland (data from Bradshaw, 1967).

Table 15.2

Geometric mean content of sulphur and sulphide-held Cu, Ni, and Co in ore-bearing mineralized, and barren ultramafic intrusions in the Canadian Shield
(from Cameron et al., 1971)

	Number of bodies	Number of Samples	Cu, ppm	Ni, ppm	Co, ppm	S, %
ORE groups: (deposits with >5000 tons Ni-Cu)	16	372	67.8	715	57.4	0.166
MINORE group: (deposits with <5000 tons Ni-Cu)	5	91	6.8	560	25.2	0.036
BARREN group:	40	616	6.9	354	31.3	0.031
All deposits:	61	1079	15.2	469	37.9	0.056

Local and Mine Scale

There have been a variety of investigations of geochemical patterns associated with specific mineral deposits within, or closely associated with, intrusive rocks. Porphyry-type ore deposits constitute an important ore type where in the exploration stage it is frequently of significance to distinguish areas of mineralization within intrusions (Lowell and Guilbert, 1970). An evaluation of the nature of geochemical dispersion associated with porphyry copper deposits as a whole is difficult due to variations in sampling procedures and in the elements that have been analyzed by various workers. In view of the importance of porphyry copper deposits as sources of Cu and Mo, it is somewhat surprising that relatively few lithogeochemical studies have been reported. Available data for the distribution of major and trace elements in rocks around copper porphyry deposits are given in Table 15.3. Perhaps the most comprehensive investigations are those carried out on the Highland Valley deposits of British Columbia (Brabec and White, 1971; Olade and Fletcher, 1975, 1976a, 1976b; and Olade, 1977). A total of 1860 samples were used in the study, comprising 1800 from the mineralized areas and 60 from background areas of the Guichon Creek. Surface outcrop samples consisted of 4 kg composites of several fist-sized chip samples and drill core samples comprising composites of 5 cm lengths of drill core taken over 3 m. Analysis was carried out for 30 elements or forms of elements (e.g., total and water extractable Cl). In the Highland Valley, geochemical dispersion in proximity to porphyry copper deposits is related to primary lithology, hydrothermal alteration, and mineralization (Fig. 15.5) Olade and Fletcher (1976a). Anomalous halos of Cu, S, and B variously extend beyond the alteration zones and thus constitute larger exploration targets than the visible alteration halo (Table 15.4), (Fig. 15.5). Rb, Sr, and Ba distributions are closely related to alteration and thus provide useful guides of porphyry zoning. Element ratios were found to be more consistent indicators of mineralization than single element patterns, which show erratic trends due to mineralogical variations.

In a preliminary investigation of geochemical dispersion associated with the Kalamazoo deposit (southwestern United States) whole rock samples of core and cuttings from 3 m intervals, spaced 15 m apart from two drill holes were analyzed for 60 elements (Chaffee, 1976). Trace element distributions, for the most part, are not related to lithology and are thus more significant in terms of mineral exploration than major element distributions. Increases in Cu, Co, B, S, or Se and decreases in Mn, Tl, Rb, and Zn occur with proximity to mineralization within the alteration zone and may be useful in exploration for blind deposits.

The Copper Canyon porphyry copper deposit in Nevada occurs in fractured and altered sedimentary rocks associated with a granodiorite intrusion (Theodore and Nash, 1973). One kilogram composite rock chip samples from areas of $10\ m^2$ were collected over a $16\ km^2$ zone and analyzed for 20 elements. Highest copper concentrations occur over the nonproductive granodiorite rather than associated with the orebodies in the adjacent metasediments. The copper distribution within the sediment has also been affected by premineralization structures and supergene enrichment. The abundance of metal dispersed through rocks at Copper Canyon rather than occurring as fracture fillings of coatings is a more specific indicator of ore deposits. The distribution of high salinity fluid inclusions outlines the orebodies in the metasedimentary rocks more precisely than the pyrite halo. The distribution of the high salinity fluids is considered to reflect the circulation limits of the fluids. These data draw attention to the limited effectiveness of exploration based solely on a consideration of geochemical data and the need for the additional consideration of the petrography of the fluids and alteration.

In the Copper Mountain area of British Columbia where porphyry-type mineralization in volcanic rocks surrounds the intrusive, patterns exist on various scales (Gunton and Nichol, 1975). An orientation survey was carried out to establish the nature of geochemical zoning associated with the deposits and identify appropriate geochemical exploration parameters. On the basis of these results 5 lb composite chip samples were collected as outcrop permitted from 500 foot grid cells and the samples were analyzed for 18 elements or forms of elements. High contents of Na, K, P, Rb, Sr, total Cu and sulphide-held copper are characteristic of the volcanics surrounding the stock relative to background areas thus providing a broader exploration target than the discrete deposits. In significantly mineralized areas surrounding the stock area the host rocks are characterized by 8-fold higher Cu and 1.5-fold higher Sr contents over three-fold more extensive areas than the economic mineralization. The volcanic host rocks associated with the individual Ingerbelle and Copper Mountain deposits are characterized by high soda and potash contents reflecting differences in the nature of metasomatic alteration associated with the two deposits. On the basis of these results it was concluded that in reconnaissance level exploration samples should be taken at a density of ten samples per square mile and in detailed level exploration at 80 samples per square mile with the samples being analyzed for P, Sr, Na, K, and Cu. During sampling attention should be given to recording dominant alteration types, degree of fracturing, and sulphide content.

In a study into the nature of primary dispersion associated with an early Precambrian porphyry-type molybdenum-copper mineralization in northwestern Ontario, fifty 500 g to 1 kg composite samples of rock chips were collected 5-15 m apart from the $5.7\ km^2$ area of the porphyritic granodiorite-quartz monzonite pluton (Wolfe, 1974). The samples were analyzed for Cu, Zn, Pb, Mn, and Mo. Strongly contrasting Cu and Mo dispersion within the stock partly

Table 15.3

Halos of selected elements around some copper porphyry and porphyry-type deposits

Deposit	Na	K	Ca	Mg	Fe	FeS	S	B	Cu	CuS	Zn	Mn	Mo	Hg	Au	Ag	Rb	Sr
Bethlehem JA[1,2,3,4]	-	+	-	-	-	+	+	+	+	+	-	-	+	+			+	-
Valley Copper[1,2,3,4]	-	+	-	-	-	+	+	-	+	+	m	-	+	0			+	-
Lornex[2]	na	na	na	na	na	na	na	+	+	na	m	0	+	na			na	-
Highmount[2]	na	na	na	na	na	na	na	+	+	na	+	-	+	na			na	-
Ingerbelle[5]	+	+	0	0	0	0	na	na	+	+	na	0	na	na	na	na	+	+
Granisle[6]	na	na	na	na	na	na	+	na	+	na	m	na	+	+	+	0	0	0
Bell Copper	na	na	na	na	na	na	+	na	+	na	m	na	+	+	+	0	0	0
Setting Net Lake[7]	na	na	na	na	na	na	na	+	na	0	0	+	nd	na	nd	na	na	
Kalamazoo[8]	-	+?	0	0	+	na	+	+?	+	na	m?	-	+	0	+	+	-	0
Copper Canyon[9]	na	na	na	na	na	na	0	+	na	m	0	+	+	+	+	na	+	
Bio Blanco[10]	na	0	0	na	na	na	na	na	na	na	na	na	na	na	na	na	+	-
El Teniente[10]	na	0	0	na	na	na	na	na	na	na	na	na	na	na	na	na	+	0

na = not analyzed; nd = not detected; + = positive halo; - = negative halo; m = positive halo at margin of deposit. Interpreted in part by the writers from the following sources: Olade and Fletcher, 1975[1], 1976a[2], 1976b[3]; Olade, 1977[4]; Gunton and Nichol, 1975[5]; Jambor, 1974[6]; Wolfe, 1974[7]; Chaffee, 1976[8]; Theodore and Nash, 1973[9]; Oyarzun, 1975[10].

coincides with the known mineralization and also gives some indication of the presence of hitherto unknown mineralization.

Mineralized andesite associated with granodiorite-dacite complex of the Rio Blanco (Chile) deposit have a 3-fold Rb enrichment and an 8-fold Sr depletion relative to fresh andesite (Oyarzun, 1975); at El Teniente a 2-fold to 4-fold enhancement of Rb is characteristic of the alteration zone relative to unaltered andesite (Oyarzun, 1971). No significant variation in Sr content was noted, probably due to the abundant anhydrite present in the potassic zone. At the El Abra porphyry deposit in Chile, no significant anomalous rubidium dispersion is associated with mineralization probably due to lack of chemical contrast between intrusive and intruded rocks (Page and Conn, 1973). At Chuquicamata highly anomalous rubidium contents have been reported (Oyarzun, 1975).

Vein deposits associated with intrusive rocks represent the smallest target for exploration; geochemical halos are also correspondingly small, although narrow wall-rock halos are commonly well defined. Au and Ag anomalies occur in whole rock and ferromagnesian minerals in the Marysville (Montana, U.S.A.) quartz diorite-granodiorite up to 300 m from gold and silver vein deposits within and adjacent to the stock (Mantei and Brownlow, 1967; Mantei et al., 1970). The anomalous contents of Au and Ag in the mineralized areas of the stock were considered to have resulted from the penetration of the wall rock by mineralizing solutions. The presence of Au and Ag and lack of base metal anomalies is possibly related to a higher mobility of Au and Ag or higher contrast in content of Au and Ag of the mineralizing fluids and the adjacent rock than in the case of base metals.

An even more extensive anomalous halo of Mn in biotite in the granodiorite-quartz monzonite Philipsburg Batholith (Montana, U.S.A.) has been described by Mohsen and Brownlow (1971); the Mn content of biotite increases two-fold over 1500 m towards the margins of the intrusion where manganese deposits occur in adjacent sedimentary rocks.

The general feature of geochemical halos around vein-type deposits is illustrated in Figure 15.6 based on the

Table 15.4

Comparison of anomaly extent and contrast for selected elements (from Olade and Fletcher, 1976a)

Element	Bethlehem JA		Valley Copper		Lornex		Highmount	
	E[1]	C[2]	E	C	E	C	E	C
Cu	***	5	***	5	***	5	***	5
S	***	5	***	5	nd	nd	nd	nd
Mo	*	2-5	*	2-5	*	5	*	5
B	*	1-2	0		***	2-5	***	2-5
Hg	**	5	0		nd	nd	nd	nd
Rb	**	1-2	**	1-2	nd	nd	nd	nd
Sr[3]	**	1-2	**	1-2	**	1-2	**	1-2

nd: Not determined

[1]Extent: 0: No anomaly
 *: Anomaly confined to ore zone
 **: Anomaly within alteration envelope
 ***: Anomaly beyond alteration envelope

[2]Contrast: Approximate anomaly to background ratio

[3]Negative anomaly

collection and analyses of chip samples, which shows the distribution of Cu, Pb, Zn, Ag, and Mn in sandstone adjacent to lead-zinc-silver and copper lode deposits in the Park City District of Utah (Bailey and McCormick, 1974). The decay pattern in metal values in wall rock away from the veins is approximately logarithmic — a common pattern that has led most observers to regard this type of dispersion as diffusion-controlled migration of mineralizing fluids from the vein. The anomalous halo has a restricted extent; anomalous Cu and Mn contents are detectable for about 10 m, and anomalous Pb, Zn, and Ag contents are detectable for about 20 m from the vein.

Lithogeochemistry

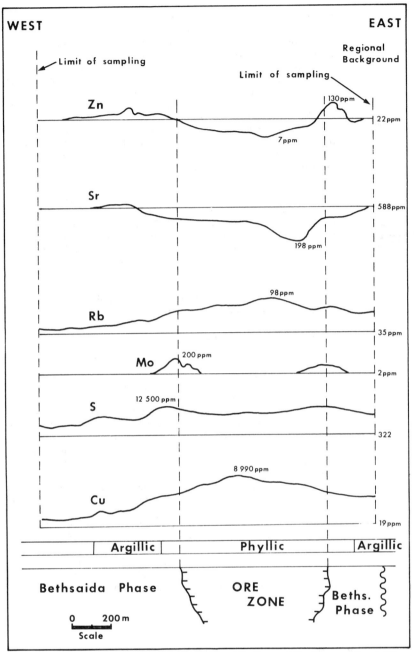

Figure 15.5. Distribution of Cu, S, Mo, Rb, Sr, and Zn in granitic rocks along a traverse over the Valley Copper porphyry copper deposit, British Columbia, Canada (redrawn from Olade and Fletcher, 1976a).

GEOCHEMICAL RESPONSE TO MINERAL DEPOSITS IN EXTRUSIVE AND SEDIMENTARY ROCKS

In this section, the lithogeochemical characteristics of the economically important stratabound sulphide occurrences — variously referred to as "massive", "volcanogenic", or "volcanic-sedimentary" — are considered. Included in this category are deposits that appear to have formed in a subaqueous environment, probably from ascending hot, metalliferous brines, and that lie conformably within mafic volcanic sequences (e.g., Cyprus), felsic volcanic sequences (e.g., Bathurst, New Brunswick, Canada), or in predominantly sedimentary sequences (e.g. McArthur River, Australia). Stratabound deposits of an apparently sedimentary origin (e.g., the Zambian Copperbelt, Kupferschiefer) and vein-type deposits of volcanic association (e.g. Kirkland Lake, Kalgoorlie, Australia) are specifically excluded.

Regional Scale

Attention has been focused by Coope (1977) on the composition of exhalative horizons as a regional indicator of mineralization. Exhalative horizons (i.e., essentially chemical precipitates) commonly occur over a wide area at the same broad stratigraphic horizon as volcanic-sedimentary sulphide deposits. The metalliferous sediments being deposited in the Red Sea represent a sedimentary hydrothermal deposit in the process of formation and show anomalous contents of Hg, Cu, Zn, and Mn in the exhalite sediments extending up to 10 km from the Atlantis II Deep with 6-fold contrasts between mean metal contents in samples adjacent to the deep relative to mean background contents in samples remote from the deep. Anomalous contents of Cu, Zn, and Mn also extend 9 km from the Nereus Deep (Bignell et al., 1976). In both cases the anomalous elements show a gradient of increasing concentration towards the exhalative centres located in the Deeps.

A number of anomalous dispersion patterns in sedimentary rocks around massive sulphides in predominantly sedimentary environments have also been reported. In the limestone overlying the Tynagh base metal deposit in Ireland there is a Mn anomaly in limestone, identifiable in surface and core samples, extending at least 7 km, with up to a 20-fold contrast (Fig. 15.7) (Russell, 1974). Up to a 20-fold range in Mn content exists in samples near mineralization requiring that an adequate number of samples are taken for representative information to be obtained. It is considered that the anomalous concentrations result from deposition of metal from "spent" mineralizing solutions following their escape into the sea subsequent to the major phase of deposition of ore metals in the underlying rocks. High Ba, Pb, Zn contents also occur in the iron formation associated with the mineralization (Russell, 1975). At the Meggen lead-zinc-barium deposit in Germany there is a Mn anomaly in limestone with up to 7-fold contents extending at least 5 km from the deposit (Gwosdz and Krebs, 1977) (Fig. 15.8). The observations based on the analyses of 342 samples again display a 3 to 4-fold range in manganese concentration (particularly near the mineralization). Peak Mn contents lie lateral to, and not immediately over the deposit which is consistent with the model for Mn dispersion in an aqueous environment around an active brine source as predicted by Whitehead (1973).

At the McArthur River (Australia) Zn, Pb, Ag deposit, based on the analyses of samples from six drill holes, anomalous contents of Zn, Pb, As, and Hg extend at least 20 km from the deposit in shale which is the lateral equivalent of the unit hosting the deposit (Lambert and Scott, 1973). Anomalous zinc contents in metasediments in the region of the Kidd Creek deposit have also been recorded (Cameron, 1975). These types of extensive anomalies (Tynagh, Meggen and McArthur River) are interpreted as syngenetic with the deposits and to have resulted from deposition from solutions in euxinic basins in nearshore regions of shallow seas.

The cupriferous pyrite massive sulphides of Cyprus occur in a sequence dominated by theoleiitic basaltic pillow lavas with only minor local sedimentary rocks associated with the deposits. From 20 traverses across the strike of the Troodos volcanic belt 2000 samples were collected and were

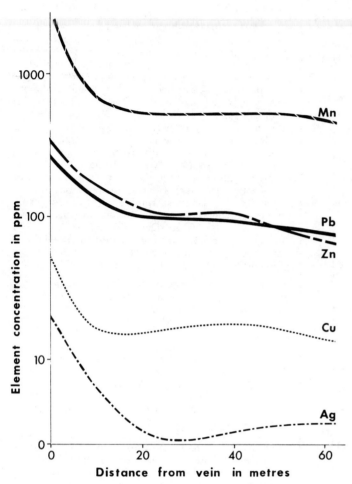

Figure 15.6. Smoothed average distribution of Ag, Cu, Pb, An, and Mn in sandstone adjacent to mercury-copper-lead-zinc vein mineralization, Park City District, Utah, U.S.A. (redrawn from Bailey and McCormick, 1974).

analyzed for up to eleven elements (Govett and Pantazis, 1971). As virtually no fresh surface rocks occur in Cyprus, all the samples have undergone weathering to various degrees, although attention was given to obtaining samples as unweathered as possible. The data indicated that the contents of Cu, Zn, Ni, and Co vary as a function of proximity to mineralization, stratigraphic position, petrology, rock type, and secondary processes. Mineralization occurs characteristically in areas of low Cu and high Zn and Co. No sulphide deposit occurs in a region with a Cu/Zn ratio in basalt which exceeds 1.2; all but two of the deposits occur in zones where the Cu/Zn ratio is less than 1.0, as shown in Figure 15.9 (Govett, 1976).

The salient features of Precambrian volcanogenic massive sulphide deposits in Canada have been described by Sangster (1972). Regional-scale lithogeochemical studies of massive sulphide deposits in the Canadian Shield indicate that they occur in calc-alkaline rather than theoleiitic mafic to felsic volcanic sequences, and that they generally occur in felsic rather than mafic differentiation sequences (Descarreaux, 1973; Cameron, 1975; Wolfe, 1975). The massive sulphide deposits in the Bathurst district of New Brunswick (Canada) similarly occur in calc-alkaline felsic volcanic rocks (Pwa, 1977). Distinctive geochemical features of productive and nonproductive cycles of volcanism have been noted in various areas of the Canadian Shield (Davenport and Nichol, 1973; Cameron, 1975; Nichol et al., 1975; Wolfe, 1975). Productive cycles of volcanism are generally characterized by slightly higher contents of Fe, Mg, and Zn and by lower Na_2O and CaO, contents as illustrated in histograms of Fe_2O_3 and Zn contents from the productive and nonproductive cycles at Uchi Lake (Fig. 15.10) (Davenport and Nichol, 1973). These data are based on the analyses of 1 kg composite chip samples collected at a frequency of 4 samples per square mile over an area of 72 square miles. There is an almost complete overlap in the two populations for Fe_2O_3 and Zn, but the form of the frequency distributions is quite different and the mean contents for Fe_2O_3 and Zn are higher in the productive cycle. The differences between the two cycles are more clearly seen if the degree of fractionation, as represented by the SiO_2 content is taken into account (Fig. 15.11, 15.12) (Nichol, 1975). The Fe_2O_3 and Zn contents of rocks from the productive cycle are clearly higher than those from the nonproductive cycle for equivalent SiO_2 contents. In the latter case these data relate to 1 kg samples of individual

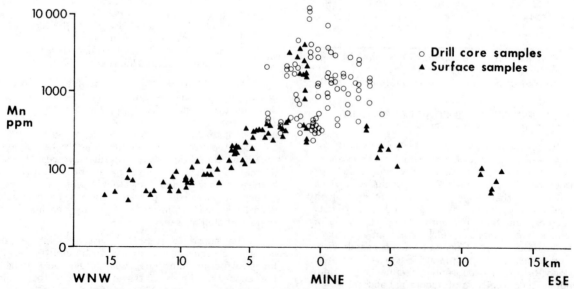

Figure 15.7. Distribution of 0.2 m acetic acid-soluble Mn in limestone that is host to the Tynagh deposit (Pb-Zn-Cu-Ag-$BaSO_4$) deposit, Ireland (redrawn from Russell, 1974).

Lithogeochemistry 349

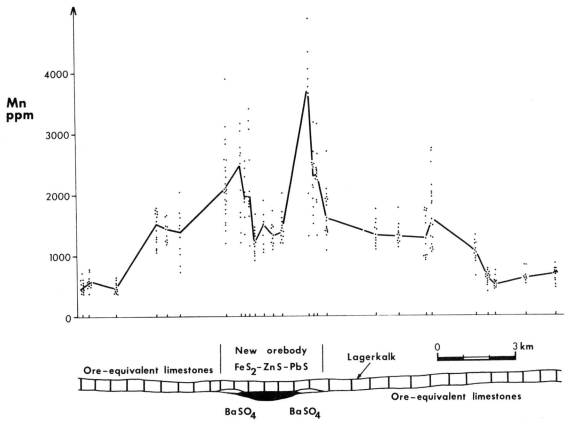

Figure 15.8. Distribution of total Mn in limestone that is host to the Meggan lead-zinc-barium deposit, Germany (redrawn from Gwosdz and Krebs, 1977).

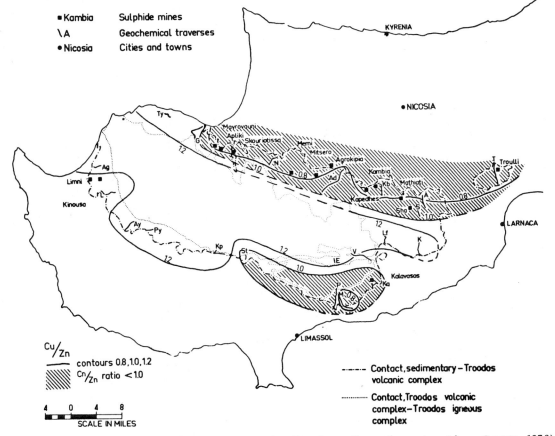

Figure 15.9. Distribution of Cu/Zn ratio in basaltic pillow lava, Cyprus (reproduced from Govett, 1976).

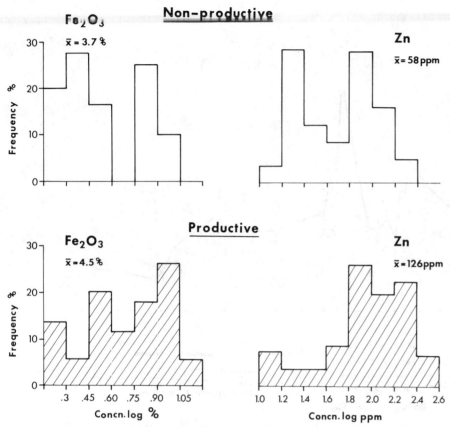

Figure 15.10. *Frequency distributions of Zn and Fe_2O_3 contents in productive and nonproductive cycles of volcanism, Uchi Lake (Davenport and Nichol, 1973).*

textural types rather than composite chip samples. Largely similar differences have been shown to exist between productive and nonproductive cycles of volcanism in a number of mining areas within the Superior Province of the Canadian Shield (Lavin, 1976; Nichol et al., 1975, 1977; and Sopuck, 1977). However, the geochemical distinction of productive cycles is not apparent in all textural types. For example, at Noranda a relatively high Zn content is apparent only in spheroidal flows of the productive cycle and does not appear to exist in the presumably less permeable massive flows (Fig. 15.13) (Sopuck, 1977). Similarly variations in chemical composition with textural type give rise to significant variations in composition. In pyroclastic units of the nonproductive Kakagi Lake area the fragments are more siliceous and soda-rich than the matrix whereas the matrix is enriched in MgO, Fe_2O_3 (3-fold), and Zn (5-fold) (Fig. 15.14) (Sopuck, 1977). Since these are some of the most useful elements indicative of mineralization the nature of the sample needs to be considered during interpretation.

Local and Mine Scale

Investigations of geochemical variations in rocks associated with individual massive sulphide deposits are far more numerous than regional studies (Tables 15.5 and 15.6). Again trace element anomalies are typically quite subtle and are generally most readily detected as population differences. Comparison of the distribution of Zn and Co in background basaltic pillow lavas with that in similar rocks within about 1 km of the Skouriotissa (Cyprus) cupriferous pyrite deposit shows an almost complete overlap in the range of distribution, although the mean value and also the modal value of the anomalous group is higher, as seen in Figure 15.15 (Govett and Pantazis, 1971; Govett, 1972). This type of overlap in element content between background and anomalous populations of rock samples has led to the use of computer-based statistical techniques of data interpretation in an attempt to clarify any differences between populations. The application of discriminant analysis to this type of problem is illustrated with reference to data from a traverse across the Mathiati Mine in Cyprus (Fig. 15.16) (Govett, 1972; Pantazis and Govett, 1973; Govett and Goodfellow, 1975). The difference between a background and an anomalous population has been maximized by calculating the discriminant score that expresses the combined effect of increasing Zn and Co content and decreasing Cu content in pillow lava with proximity to mineralization. Anomalous discriminant scores extend about 2 km from the deposit.

Major element patterns in volcanic rocks around massive sulphides appear to have some common characteristics, regardless of age, economic metals, or the geological environment of the deposits (Table 15.5). Proximity to mineralization is generally indicated by an enrichment in Fe and Mg and a depletion in Na and Ca. The behaviour of K is variable; it is generally enriched relative to Na close to sulphide mineralization but in some cases it is depleted; there are also some cases where Na shows an absolute enrichment. These patterns are best developed in footwall rocks, but, in some situations, they also occur in the hanging wall. Mg enrichment and Na depletion occur adjacent to the Jay Copper Zone (Abitibi, Canada); the very distinct Na anomaly extends up to 500 m from the mineralized zone, but the Na content beyond these limits is still anomalously low (Fig. 17) (Descarreaux, 1973).

The elements that display anomalous dispersion patterns in the wall rock associated with massive sulphide deposits show some variation amongst deposits and the composition of the wall rock also reflects the effects of fractionation. In order to identify a more generally diagnostic indicator of mineralization, a function based on the Fe, Mg, Ca, and Na contents related to mineralization was determined for wall rock associated with a number of massive sulphide deposits in the Canadian Shield (McConnell, 1976). The component of the Fe, Mg, Ca, and Na content due to fractionation was estimated on the basis of the SiO_2 content and subtracted from the observed concentrations, the difference or residual content then possibly being related to mineralization. By combining the residuals of Mg, Fe, Ca, and Na in the form of a standardized net residual a more consistent reflection of proximity of mineralization was obtained. In general the strength of this factor increases towards mineralization and towards the presumable conduit of mineralizing solutions. The distribution of the "standardized net residual" at the South Bay Deposit extends beyond the mapped alteration zone (Fig. 15.18) but at the East Waite deposit it is only as extensive as the alteration zone. The variation in Mg and Ca in footwall felsic volcanic rocks with proximity to the Brunswick No. 12 deposit is shown in Figure 15.19; the Mg/Ca ratio is still three times greater than mean background 175 m from the ore zone (Govett and Goodfellow, 1975).

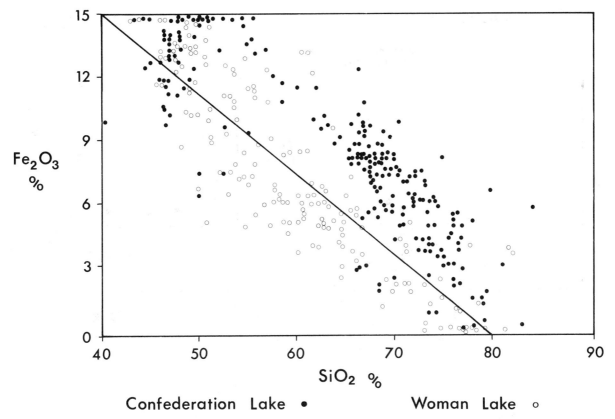

Figure 15.11. Relation between Fe_2O_3 and SiO_2 contents of productive and nonproductive cycles, Uchi Lake (Nichol, 1975).

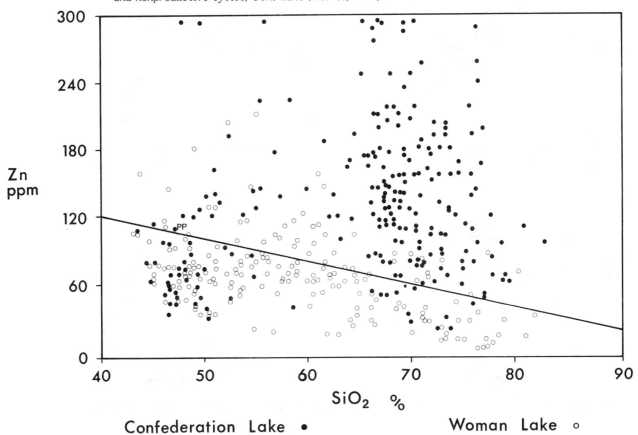

Figure 15.12. Relation between Zn and SiO_2 contents of productive and nonproductive cycles, Uchi Lake (Nichol, 1975).

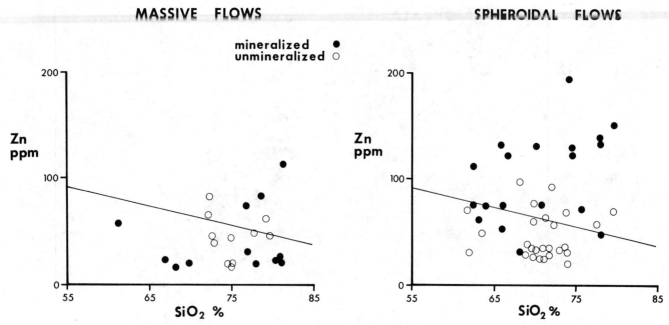

Figure 15.13. Relation between the composition and texture of productive and nonproductive cycles, Noranda (Nichol, 1975).

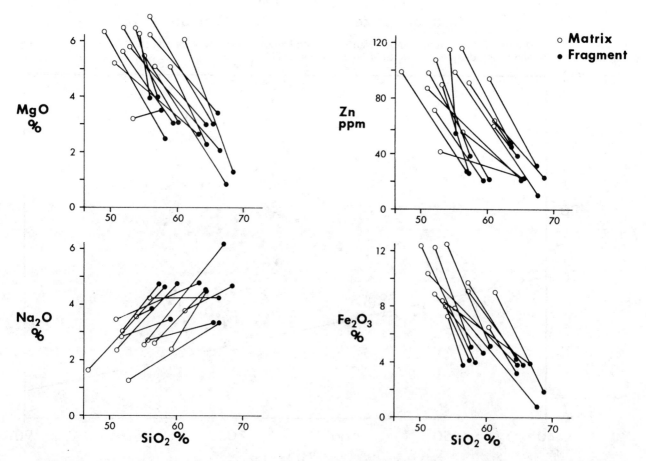

Figure 15.14. Relation in composition of fragment and matrix of pyroclastics at Kakagi Lake (Sopuck, 1977).

Table 15.5

Summary of major element dispersions in relation to
volcanogenic massive sulphide deposits

Deposit	Alteration mineralogy	Elements enriched	Elements depleted	Elements unchanged	Age
Kuroko, Japan Lambert & Sato (1974)	Mon, Ser, Chl, Kaol	K, Mg, Fe, Si	Ca		Cenozoic
Kuroko, Japan Tatsumi & Clark (1972)	Ser, Qtz, Cal	Mg, K	Na, Ca, Fe	Al	Cenozoic
Hitachi, Japan Kuroda (1961)	Cord, Anthoph.	Mg, Fe, Ba	Na, Ca, Sr		Cenozoic
Buchans, Canada Thurlow et al. (1975)	Chl, Ser, Qtz	Mg, Fe, Si	Na, Ca, K		Paleozoic
Heath Steele, Canada Wahl et al. (1975)	Chl, Ser	Mg	Na, Ca		Paleozoic
Brunswick No. 12 Canada Goodfellow (1975)	Chl, Ser, Qtz	Mg, Fe, (Mn), (K)	Na, Ca (Mn), (K)	Al	Paleozoic
Killingdal, Norway Rui (1973)	Chl, Bio, Qtz	Mg, K, Mn	Na, Ca, Si	Al, Ti, Fe (total)	Paleozoic
Skorovass, Norway Gjelsvik (1968)	Chl, Ser	Mg	Na, Ca		Paleozoic
Boliden, Sweden Nilsson (1968)	Chl, Ser, Qtz, Andal	Mg, K, Al	Na, Ca		Proterozoic
Mattabi, Canada Franklin et al. (1975)	Qtz, Carb, Ser, Chld, Chl, Andal, Gar, Kyan, Bio	Fe, Mg	Na, Ca		Archean
Millenbach, Canada Simmons et al. (1973)	Chl, Ser, Anthoph, Cord.	Mg, Fe	Na, Ca, Si		Archean
Mines de Poirier Canada Descarreaux (1973)	Chl, Ser,	Mg, K	Na, Ca	Si	Archean
Lac Dufault, Canada Sakrison (1966)	Chl, Ser	Mg, Fe, Mn	Na, Ca	Al, Ti K, Si	Archean
East Waite, Mobrun, Joutel, Poirier, Agnico-Eagle, Mattabi, Sturgeon Lake, South Bay McConnell (1976)	Qtz, Ser, Chl, Carb, Sauss, Epidote	Mg, Fe	Na, Ca	Si, Al	Archean

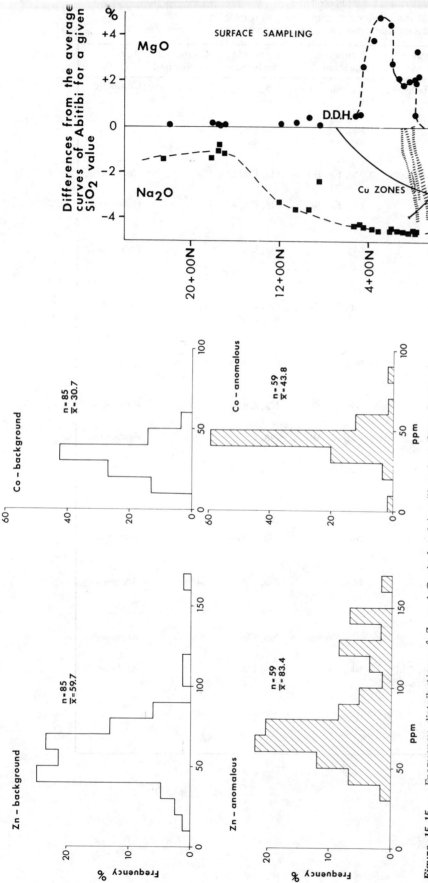

Figure 15.15. Frequency distributions of Zn and Co in basaltic pillow lava from background and anomalous areas within 1 km of the Skouriotissa cupriferous pyrite deposit, Cyprus (data from Govett and Pantazis, 1971; Govett, 1972).

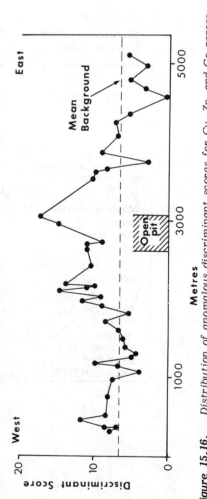

Figure 15.16. Distribution of anomalous discriminant scores for Cu, Zn, and Co across the Mathiati cupriferous pyrite deposit, Cyprus (Govett and Goodfellow, 1975).

Figure 15.17. Distribution of "residual" MgO and Na_2O in felsic volcanic rocks across the Jay copper deposit, Abitibi District, Canada. "Residual" represents the difference between analyzed content and average value in the Abitibi District for corresponding SiO_2 content (redrawn from Descarreaux, 1973).

Table 15.6

Summary of trace-element dispersion in relation to volcanogenic massive sulphide deposits

Deposit	Ore mineralogy	Metal dispersion	Age
Kuroko type, Japan Lambert & Sato (1974)	py, cpy, sph, ga, bar	Zn, Cu, Pb	Cenozoic
Buchans, Canada Thurlow et al. (1975)	sph, ga, cpy, (py, tet, bo, cov)	Zn, Pb, Ba, (Ag) (Cu)	Paleozoic
Brunswick No. 12, Canada Goodfellow (1975)	py, sph, ga, po, cpy, tet, bo	Zn, Pb	Paleozoic
Heath Steele, Canada Whitehead and Govett (1974)	py, sph, ga, cpy	Pb, Zn	Paleozoic
Killingdal, Norway Rui (1973)	py, sph, cpy	Cu, Zn	Paleozoic
Mattabi, Canada Franklin et al. (1975)	cpy, py, sph	Cu, Zn, S	Archean
Millenbach, Canada Simmons et al. (1973)	cpy, py, sph	Zn	Archean
East Waite Mobrun, Joutel, Poirier, Agnico-Eagle, Mattabi, Sturgeon Lake, South Bay Mines, McConnell (1976)	ph, sph, cpy (po)	Zn, Cu, S	Archean

Variations in the nature of footwall and hanging wall anomalies have been noted according to the spatial relation of the deposit to the vent (Wahl, 1978). Extensive Ca and Mg halos occur in both the footwall and hanging wall of deposits located proximal to the vent (A) (Fig. 15.20) whereas a deposit located distal to the vent (B) has an associated footwall anomaly but only a very restricted hanging wall anomaly. The extensive dispersion associated with the proximal deposit is attributed to deposition from acidic fumarolic brines whereas the more restricted dispersion associated with the distal deposit is related to deposition from metal-rich brines that have flowed down the slope from the vent.

The same elements show the same types of trends on both a regional and a local and mine scale in an Archean massive sulphide environment. At the Joutel-Poirier area in the Canadian Shield, relative enrichment of Fe, Mg, and Zn with depletion of Ca, Na, and K, occurs around the deposit on a scale of hundreds of metres; these element distribution patterns also occur regionally on a kilometre scale (Fig. 15.21, Sopuck, 1977). The same situation is evident in the Bathurst district of New Brunswick, where enrichment of Zn and Mg and depletion of Ca and Ma occur both regionally and in the immediate vicinity of individual deposits (Whitehead and Govett, 1974; Goodfellow, 1975; Pwa, 1977; Wahl, 1978). It is reasonable to suppose that similar processes were operative at both the local and regional scale.

SOVIET LITHOGEOCHEMISTRY

It may be surmised from translated articles and other Soviet contributions over the past decade that geochemists in the U.S.S.R. have developed techniques that enable them to successfully detect blind orebodies at depths of hundreds of metres; moreover, they appear to be able to assess whether halos reflect probable economic mineralization or merely the root zone of deposits that have had the economic portion eroded away. The basis of the technique is the establishment of zoning patterns of different elements in halos (Fig. 15.22). Investigations of hydrothermal orebodies have shown that, despite a marked variation in composition and in local geological conditions, the halos conform essentially to a general zoning pattern (Grigoryan, 1974); for steeply dipping bodies the sequence from top to bottom (supra-ore elements to sub-ore elements) is:

Ba-(Sb,As,Hg)-Cd-Ag-Pb-Zn-Au-Cu-Bi-Ni-Co-Mo-U-Sn-Be-W.

Ovchinnikov and Baranov (1972) reported a general zonal sequence in halos around a subvolcanic hydrothermal copper-pyrite deposit of Ba, Ag, Pb, and Zn in the upper zones, and Cu, Mo, Co, and Bi in the lower zones; this is consistent with Grigoryan's sequence.

Soviet geochemists generally use multi-element ratios to reduce the effects of local reversals in zoning sequences, and also to reduce analytical variations (most of their analyses are semi-quantitative spectrographic). Additive halos (addition of different element values standardized to respective background), multiplicative halos (multiplication of element values), and ratios of supra-ore to sub-ore elements are favoured rather than some of the techniques such as regression and discriminant analyses that have found wider use in the Western world. Western geochemists are more concerned with interpreting the absolute content of an element in a halo to the exclusion of its other defining parameter — dimension. Soviet geochemists, on the other hand, use the dimension of halos as an effective interpretive aid by calculating the linear productivity of a halo; this is simply the product of the average content of an element and the width of an anomaly (areal productivities are also used). It has been stated (Grigoryan, 1974) that not only do different deposits have similar zoning patterns, but that their extent, as defined by the change in linear productivity with depth, is also similar.

The application of the procedures of multiplicative and linear productivities is illustrated schematically in Figure 15.22. Elements (a) and (b) give supra-ore halos; elements (c) and (d) give sub-ore halos. The enhancement of anomalies — and the clear distinction between above-ore and below-ore halos — using multiplicative ratios of supra-ore to sub-ore halos is obvious when compared to profiles of elements. Changes in linear productivities with depth for supra-ore element (a) and sub-ore element (d) are also illustrated. The trend of linear productivities of ratios with depth is particularly striking (this schematic trend is typical of those illustrated in the Soviet literature). The practical importance of this type of approach is the apparent possibility of determining the location of blind deposits from surface samples, of predicting the likelihood of the deposits being economic or not, and of accurately locating mineralization from limited drillhole data. The size of the halos (according to Beus and Grigorian (1977), are 200-850 m) are comparable to those detected by Western geochemists, but interpretative procedures seem to be rather more advanced.

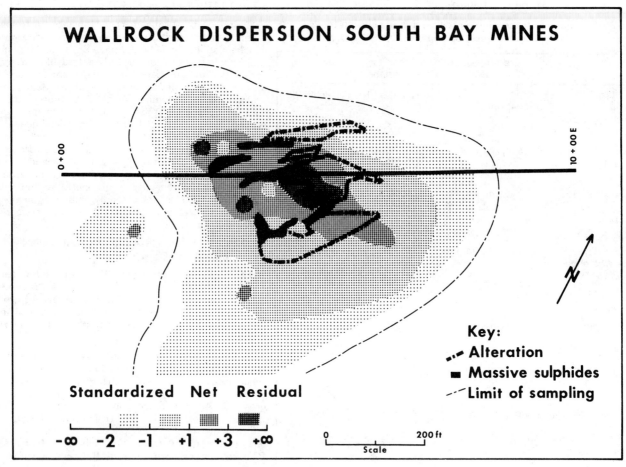

Figure 15.18. Distribution of "standardized net residual" at South Bay Mine (McConnell, 1976).

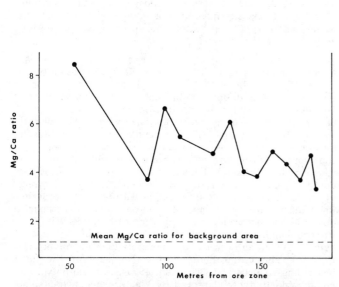

Figure 15.19. Distribution of Mg/Ca ratio in footwall felsic volcanic rocks, Brunswick No. 12 deposit, New Brunswick (reproduced from Govett and Goodfellow, 1975).

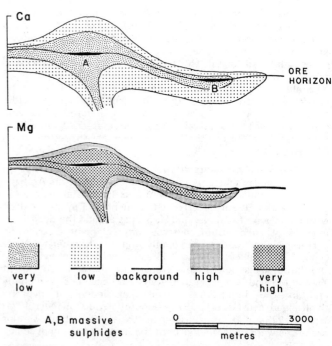

Figure 15.20. Schematic distribution of Ca and Mg in felsic volcanic rocks around proximal (A) and distal (B) massive sulphide deposits, New Brunswick (data from Wahl, 1978).

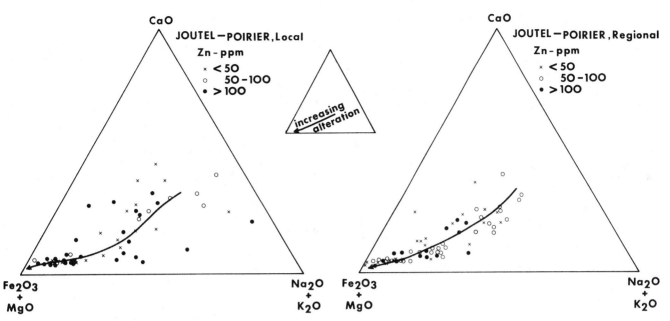

Figure 15.21. Relation between geochemistry and the degree of alteration on the local and regional scale at Joutel-Poirier (Sopuck, 1977).

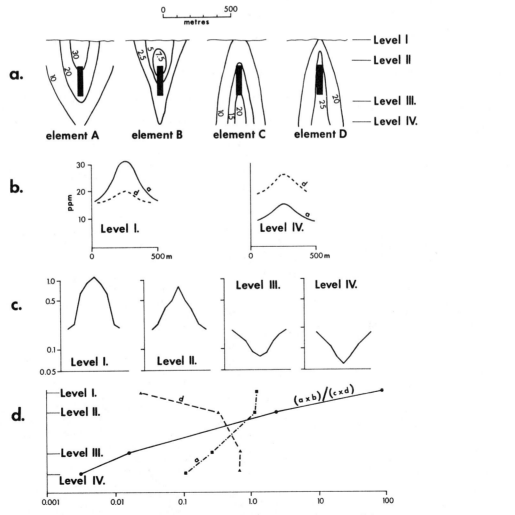

Figure 15.22. Schematic illustration of supra-ore and sub-ore lithogeochemical halos, and anomaly enhancement through the use of multiplicative halos and linear productivities (after Govett, 1977).

Evaluation of State-of-the-Art and Conclusions

The existence of diagnostic lithogeochemical signatures have been recognized on a number of scales associated with different types of deposits. These have a variety of applications in exploration.

a. anomalous metal contents of mineralized relative to unmineralized intrusions e.g., (1) Sn (Australia), W (Canada), Sn (United Kingdom) and Ni, Cu (Canada). These features are attributed to the high metal content of the parent magma that gave rise to the intrusives and the associated deposits. (2) Cu, Mo (United Kingdom) resulting from post-magma hydrothermal activity in the host intrusions.

b. anomalous metal contents of mineralized zones of intrusions relative to unmineralized zones e.g., porphyry copper deposits in general resulting from the permeation of hydrothermal fluids into the intrusion from the centre of mineralization.

c. localized anomalous metal contents in the wall rock adjacent to vein-type mineralization e.g., Pb, Zn, Ag mineralization in Utah as a result of mineralizing fluids diffusing into the wall rock.

d. anomalous metal contents in favourable areas of sedimentary basins for mineralization e.g., McArthur River, Australia; Tynagh, Ireland; Meggen, Germany.

These patterns are related to the exhalation of metal-rich solutions into the sea and subsequent deposition with and decay of certain metal contents over considerable distances from the deposits.

e. anomalous metal contents in productive relative to nonproductive cycles, and on more local scales associated in general with volcanogenic massive sulphide deposits. These features apparently result from hydrothermal activity associated with the phase of mineralization.

These diagnostic geochemical parameters have variously been identified on the basis of whole rock composition, analysis of separate mineral phases or the distinction of subtle variations in the form of populations.

Equally important in terms of drawing attention to the limitations of lithogeochemistry in exploration has been the demonstration that the composition of the whole rock or mineral phases may not be diagnostic of proximity to mineralization. Since the composition of rocks is the result of the overall geological history of the sample the composition represents the effects of a number of processes that have affected the sample. Only in cases where the mineralizing episode is the over-riding process will effects due to mineralization be readily identifiable. In other cases effects on composition due to other causes must be taken into account before any response due to mineralization can be recognized. This feature is illustrated by the need to take into account the effects of fractionation in the interpretation of data relating to igneous rocks if a variety of lithologic types are sampled in order to identify responses due to mineralization. It is thus vital at the interpretation stage to take into account those aspects of the geological history that have a bearing on the geochemistry in order to identify meaningful responses due to mineralization. The lack of general acceptance of lithogeochemistry as an aid in exploration is undoubtedly due to a lack of recognition of diagnostic patterns related to in part petrogenesis, tectonic, and overall geological history.

In many of the examples cited in the review the investigations have been carried out over areas of known mineralization where the geological knowledge is relatively high. Whilst these locations are logical focal points for establishing the nature of dispersion associated with mineralization it may be that in view of the subtlety of many of the responses, they have only been recognized as anomalous responses because of the relatively good geological data base.

The diversity of geochemical responses related to mineralization in terms of anomalous elements, lateral extent, and contrast from different examples of the same type of deposit clearly is an adverse factor affecting the applicability of lithogeochemistry in exploration. This feature is illustrated by the diversity in elements showing anomalous responses, extent, and magnitude of the anomalous dispersions associated with four porphyry-type deposits in the Highland Valley, British Columbia.

In exploration over relatively unexplored terrain the necessary level of geological information to aid interpretation is not available or can only be obtained at considerable expense which must constitute a further constraint on the applicability of the technique. It is probable that a reconnaissance survey carried out over relatively unexplored ground will generate a multitude of subtle anomalies similar to those associated with known mineralization but in the absence of adequate geological information it will be impossible to identify those anomalies reflecting mineralization from those related to other causes.

Notwithstanding these limitations, within the past ten years in the Western World lithogeochemistry has advanced from an essentially academic involvement to a technique used sparingly by the mining industry in mineral exploration. However, we are not aware of any mineral discovery to date that can be attributed to lithogeochemistry.

The use of lithogeochemistry on pre-drilling exploration is dependent upon adequate surface outcrop or expensive surface drilling through overburden to bedrock. Sampling, therefore, is generally more costly than in stream sediment or soil geochemical surveys; crushing and grinding of samples preparatory to analysis is at least five times more expensive than sieving soil or stream sediments.

Nevertheless, there are a number of exploration situations where lithogeochemistry is useful and may be expected to become increasingly important in the future. These are:

1. Regional scale exploration to identify potentially productive plutons and volcanic or sedimentary horizons (anomalous geochemical patterns are detectable over many kilometres).

2. Local scale exploration to locate deeply-buried and blind deposits (geochemical halos have dimensions of more than 100 m and can range up to 1 km).

3. Exploration drilling to assist in drill hole location.

4. Mine scale exploration and underground mapping.

As indicated previously, the nature of geochemical responses in bedrock associated with mineralization varies considerably in terms of anomalous element associations, extent and magnitude of anomalies amongst different examples of the same type of deposit. This situation related in part to conditions in the geological environments prevents the definition of any standardized procedures. Perhaps more than in any other branch of geochemical exploration, it is essential to carry out an orientation survey over known mineralization to establish the nature of geochemical dispersion associated with known mineralization and appropriate sampling, analytical and interpretational procedures. These operational procedures may only be applicable in a precisely similar geological environment.

In terms of sampling the nature of the optimum sample should be established i.e., grab or composite chip sample, and sample interval or density to ensure adequate representivity in terms of the target population. Sample preparation procedure involving preparation of whole rock or mineral separates should be considered. With regard to analytical procedures attention should be given to identifying elements that display diagnostic distributions and ways of analyzing these in terms of obtaining the necessary sensitivity, accuracy, and precision and in some cases the metal held in particular mineral phases e.g., sulphide-held metal. In order to identify the necessary interpretational procedures it will be essential to identify those features of the geological environment that have a bearing on geochemical dispersion.

At the end of 1977 there is abundant evidence that geological events – including those that give rise to mineral deposits – are detectable by lithogeochemistry, given appropriate geological interpretation. Future development and success will depend upon improved understanding of geological processes and the mechanisms of element migration and fixation. The fact that lithogeochemical surveys are more costly to conduct than stream sediment or soil surveys means that the major application of rock geochemistry in the near future is likely to be restricted to areas that have already been well-prospected by more traditional techniques. Moreover, since multi-element data are generally required and interpretation is correspondingly difficult, lithogeochemical techniques are likely to be limited by the availability of advanced analytical technology and highly-trained geochemists. To the extent that the mining industry is prepared to invest in renewed exploration in the well-prospected areas of North America and Europe, lithogeochemistry provides a promising new technique.

REFERENCES

Allen, P.M., Cooper, D.C., Fuge, R., and Rea, W.J.
  1976: Geochemistry and relationships to mineralization of some igneous rocks from Harlech Dome, Wales; Trans. Inst. Min. Met., London, v. 85, p. B100-B108.

Bailey, G.B. and McCormick, G.R.
  1974: Chemical halos as guides to lode deposit ore in the Park City District, Utah; Econ. Geol., v. 69, no. 3, p. 377-382.

Barsukov, V.L.
  1967: Metallogenic specialization of granitoid intrusions; in Vinogradov, A.P., ed., Chemistry of the Earth's Crust, Vol. II (Jerusalem: Israel Program for Scientific Translation), p. 211-231.

Beus, A.A.
  1969: Geochemical criteria for assessment of the mineral potential of the igneous rock series during reconnaissance exploration; Colo. Sch. Mines Quart., v. 64, no. 1, p. 67-74.

Beus, A.A. and Grigorian, S.V.
  1977: Geochemical Exploration Methods for Mineral Deposits; Applied Publishing Co., Wilimette, Illinois, 287 p.

Bignell, R.D., Cronan, D.S., and Tooms, J.S.
  1976: Metal dispersion in the Red Sea as an aid to marine geochemical exploration; Trans. Inst. Min. Met., London, v. 85, p. B274-B278.

Bolotnikov, A.F. and Kravchenko, N.S.
  1970: Criteria for recognition of tin-bearing granites; Akad. Sci. USSR Dokl. Earth Sci. Sect., v. 191, p. 186-187.

Brabec, D. and White, W.H.
  1971: Distribution of copper and zinc in rocks of the Guichon Creek Batholith, British Columbia; in Boyle, R.W. and McGerrigle J.I., ed., Geochemical Exploration, Can. Inst. Min. Met., Spec. Vol. 11, p. 291-297.

Bradshaw, P.M.D.
  1967: Distribution of selected elements in feldspar, biotite, and muscovite from British granites in relation to mineralization; Trans. Inst. Min. Met., London, v. 76, p. B117-B148.

Cameron, E.M.
  1975: Geochemical methods of exploration for massive sulphide mineralization in the Canadian Shield; in Elliott, I.L. and Fletcher, W.K., ed., Geochemical Exploration 1974, Elsevier Publ. Co., Amsterdam, p. 21-49.

Cameron, E.M., Siddeley, G., and Durham, C.C.
  1971: Distribution of ore elements in rocks for evaluating ore potential: nickel, copper, cobalt, and sulphur in ultramafic rocks of the Canadian Shield; in Boyle, R.W. and McGerrigle, S.I., ed., Geochemical Exploration, Can. Inst. Min. Met., Spec. Vol. 11, p. 298-313.

Chaffee, M.A.
  1976: The zonal distribution of selected elements above the Kalamazoo porphyry copper deposit, San Manuel District, Pinal County; J. Geochem. Explor., v. 5, no. 2, p. 145-165.

Coope, J.A.
  1977: Potential of lithogeochemistry in mineral exploration; Paper presented at AIME Annual General Meeting, 7 March 1977, Atlanta, Georgia.

Davenport, P.H. and Nichol, I.
  1973: Bedrock geochemistry as a guide to areas of base-metal potential in volcano-sedimentary belts of the Canadian Shield; in Geochemical Exploration 1972, Jones, M.J. (Editor), Inst. Min. Met., London, p. 45–57.

Descarreaux, J.
  1973: A petrochemical study of the Abitibi volcanic belt and its bearing on the occurrences of massive sulphide ores; Can. Inst. Min. Met., Bull., v. 66, no. 730, p. 61-69.

Franklin, J.M., Kasarda, J., and Poulsen, K.H.
  1975: Petrology and chemistry of the alteration zone of the Mattabi massive sulphide deposit; Econ. Geol., v. 70, no. 1, p. 63-79.

Garrett, R.G.
  1971: Molybdenum, tungsten, and uranium in acid plutonic rocks as a guide to regional exploration, S.E. Yukon; Can. Min. J., v. 92, p. 37-40.

Gjelsvik, T.
  1968: Distribution of major elements in the wall rocks and the silicate fraction of the Skorovass pyrite deposit, Grong area, Norway; Econ. Geol., v. 63, p. 217-231.

Goodfellow, W.D.
  1975: Major and minor element halos in volcanic rocks at Brunswick No. 12 sulphide deposit, N.B., Canada; in Geochemical Exploration 1974, Elliott, I.L. and Fletcher, W.K., ed., Elsevier Publ. Co., Amsterdam, p. 279-295.

Govett, G.J.S.
　　1972: Interpretation of a rock geochemical exploration survey in Cyprus – statistical and graphical techniques; J. Geochem. Explor., v. 1, p. 77-102.

　　1976: The development of geochemical exploration methods and techniques; in Govett, G.J.S. and Govett, M.H., ed., World Mineral Supplies – Assessment and Perspective, Elsevier Publ. Co., Amsterdam, p. 343-376.

　　1977: Presidential Address to the Annual General Meeting of the Association of Exploration Geochemists, Vancouver, B.C., April 1977; J. Geochem. Explor., v. 8, p. 591-599.

Govett, G.J.S. and Goodfellow, W.D.
　　1975: Development of rock geochemical techniques for detecting buried sulphide deposits – a discussion; Trans. Inst. Min. Met., London, v. 84, p. B134-B140.

Govett, G.J.S. and Pantazis, Th.M.
　　1971: Distribution of Cu, Zn, Ni, and Co in the Troodos Pillow Lava Series, Cyprus; Trans. Inst. Min. Met., London, v. 80, p. B27-B46.

Grigoryan, S.V.
　　1974: Primary geochemical halos in prospecting and exploration of hydrothermal deposits; Int. Geol. Rev., v. 16, no. 1, p. 12-25.

Gunton, J.E. and Nichol, I.
　　1975: Chemical zoning associated with the Ingerbelle-Copper Mountain mineralization, Princeton, British Columbia; in Elliott, I.L. and Fletcher, W.K., ed., Geochemical Exploration 1974, Elsevier Publ. Co., Amsterdam, p. 297-312.

Gwosdz, W. and Krebs, W.
　　1977: Manganese halo surrounding Meggan ore deposit, Germany; Trans. Inst. Min. Met., London, v. 86, p. B73-B77.

Hesp, W.R. and Rigby, D.
　　1975: Aspects of tin metallogenesis in the Tasman geosyncline, Eastern Australia, as reflected by cluster and factor analysis; J. Geochem. Explor., v. 4, no. 3, p. 331-347.

Ivanova, G.F.
　　1963: The content of tin, tungsten, and molybdenum in granite enclosing tin and tungsten deposits; Geochemistry, v. 5, p. 492-500.

Jacobs, D.C. and Parry, W.T.
　　1976: A comparison of the geochemistry of biotite from some Basin and Range Stocks; Econ. Geol., v. 71, p. 1029-1035.

Jambor, J.L.
　　1974: Trace element variations in porphyry copper deposits, Babine Lake Area, B.C.; Geol. Surv. Can., Paper 74-9, 30 p.

James, C.H.
　　1967: The use of the terms "primary" and "secondary" dispersion in geochemical prospecting; Econ. Geol., v. 62, p. 997-999.

Kesler, S.E., Issigonis, M.J., Brownlow, A.H., Damon, P.E., Moore, W.J., Northcote, K.E., and Preto, V.A.
　　1975a: Geochemistry of biotites from mineralized and barren intrusive systems; Econ. Geol., v. 70, p. 559-567.

Kesler, S.E., Issigonis, M.J., and Van Loon, J.C.
　　1975b: An evaluation of the use of halogens and water abundances in efforts to distinguish mineralized and barren intrusive rocks; J. Geochem. Explor., v. 4, p. 235-245.

Kesler, S.E., Van Loon, J.C., and Moore, C.M.
　　1973: Evaluation of ore potential of granodioritic rocks using water-extractable chloride and fluoride; Can. Inst. Min. Met. Bull., v. 66, no. 730, p. 56-60.

Kuroda, Y.
　　1961: Minor elements in metasomatic zone related to a copper-bearing pyrite deposit; Econ. Geol., v. 56, p. 847-854.

Lambert, I.B. and Sato, T.
　　1974: The Kuroko and associated ore deposits of Japan: a review of their features and metallogenesis; Econ. Geol., v. 69, p. 1215-1236.

Lambert, I.B. and Scott, K.M.
　　1973: Implications of geochemical investigations of sedimentary rocks within and around the McArthur zinc-lead-silver deposit, Northern Territory; J. Geochem. Explor., v. 2, p. 307-330.

Lavin, O.P.
　　1976: Lithogeochemical discrimination between mineralized and unmineralized cycles of volcanism in the Sturgeon Lake and Ben Nevis areas of the Canadian Shield; M.Sc. thesis, Queen's University, Kingston, Ontario, Canada, 249 p.

Lowell, J.D. and Guilbert, J.M.
　　1970: Lateral and vertical alteration-mineralization zoning in porphyry ore deposits; Econ. Geol., v. 65, p. 373-408.

Mantei, E. and Brownlow, A.
　　1967: Variation in gold content of minerals of the Marysville quartz diorite stock, Montana; Geochim. Cosmochim. Acta, v. 31, p. 225-236.

Mantei, E., Bolter, E., and Al Shaieb, Z.
　　1970: Distribution of gold, silver, copper, lead, and zinc in the productive Marysville stock, Montana; Mineral. Deposita, v. 5, p. 184-190.

McConnell, J.W.
　　1976: Geochemical dispersion in wallrocks of Archean massive sulphide deposits; M.Sc. thesis, Queen's University, Kingston, Ontario, 230 p.

Mohsen, L.A. and Brownlow, A.H.
　　1971: Abundance and distribution of manganese in the western part of the Philipsburg Batholith, Montana; Econ. Geol., v. 66, p. 611-617.

Nichol, I.
　　1975: Bedrock composition as a guide to areas of base metal potential in the greenstone belts of the Canadian Shield; Queen's University, Kingston, Ontario, Unpubl. Rept., 74 p.

Nichol, I., Bogle, E.W., Lavin, O.P., McConnell, J.W., and Sopuck, V.J.
　　1977: Lithogeochemistry as an aid in massive sulphide exploration; in Prospecting in areas of glaciated terrain, Jones, M.J., ed., Inst. Min. Met., London, p. 63–71.

Nichol, I., Lavin, O.P., McConnell, J.W., Hodgson, C.J., and Sopuck, V.J.
　　1975: Bedrock composition as an indicator of Archean massive sulphide environments (abstract); Can. Inst. Min. Met., v. 68, no. 755, p. 48.

Nilsson, C.A.
  1968: Wall rock alteration at the Boliden deposit, Sweden; Econ. Geol., v. 63, p. 472-494.

Olade, M.A.
  1977: Major element halos in granitic wall rocks of porphyry copper deposits, Guichon Creek Batholith, British Columbia; J. Geochem. Explor., v. 7, no. 1, p. 59-71.

Olade, M.A. and Fletcher, W.K.
  1975: Primary dispersion of rubidium and strontium around porphyry copper deposits, Highland Valley, British Columbia; Econ. Geol., v. 70, p. 15-21.

  1976a: Trace element geochemistry of the Highland Valley and Guichon Creek Batholith in relation to porphyry copper mineralization; Econ. Geol., v. 71, p. 733-748.

  1976b: Distribution of sulphur, and sulphide-iron and copper in bedrock associated with porphyry copper deposits, Highland Valley, British Columbia; J. Geochem. Explor., v. 5, no. 1, p. 21-30.

Ovchinnikov, L.N. and Baranov, E.N.
  1972: Endogenic geochemical halos of pyritic ore deposits; Int. Geol. Rev., v. 14, no. 5, p. 419-429.

Oyarzun, M.J.
  1971: Contribution à l'étude des roches volcaniques et phutomiques du Chile; Thèse, Univ. de Paris, 195 p.

  1975: Rubidium and strontium as guides to copper mineralization emplaced in some Chilean andesitic rocks; in Geochemical Exploration 1974, Elliott, I.L. and Fletcher, W.K., ed., Elsevier Publ. Co., Amsterdam, p. 333-338.

Page, B.G.N. and Conn, H.
  1973: Investigacion sobre metodos de prospeccion geoquimica en el yacimiento tipo cobre porfidico, El Abra, provincia de Antofagasta-Chile Instituto Investigaciones Geologicas, Santiago, 40 p.

Pantazis, Th.M. and Govett, G.J.S.
  1973: Interpretation of a detailed rock geochemical survey around Mathiati Mine, Cyprus; J. Geochem. Explor., v. 2, no. 1, p. 25-36.

Parry, W.T.
  1972: Chlorine in biotite from Basin and Range Plutons; Econ. Geol., v. 67, p. 972-975.

Parry, W.T. and Jacobs, D.C.
  1975: Fluorine and chlorine in biotite from Basin and Range Plutons; Econ. Geol., v. 70, p. 554-558.

Parry, W.T. and Nackowski, M.P.
  1963: Copper, lead and zinc in biotites from Basin and Range quartz monzonites; Econ. Geol., v. 58, p. 1126-1144.

Pwa, U. Aung
  1977: Regional rock geochemical exploration, Bathurst District, N.B.; M.Sc. thesis, University of New Brunswick, Fredericton.

Rui, I.J.
  1973: Structural control and wall rock alteration at Killingdal mine, central Norwegian Caledonides; Econ. Geol., v. 68, p. 859-883.

Russell, M.J.
  1974: Manganese halo surrounding the Tynagh ore deposit, Ireland: a preliminary note; Trans. Inst. Min. Met., London, v. 83, p. B65-B66.

  1975: Lithogeochemical environment of the Tynagh base-metal deposit, Ireland, and its bearing on ore deposition; IMM Sect. B, v. 84, p. 128.

Sakrison, H.C.
  1966: Chemical studies of the host rocks of the Lake Dufault Mines, Quebec; Ph.D. thesis, McGill University, Montreal.

Sangster, D.F.
  1972: Precambrian volcanogenic massive sulphide deposits in Canada: A Review; Geol. Surv. Can., Paper 72-22, 44 p.

Sheraton, J.W. and Black, L.P.
  1973: Geochemistry of mineralized granitic rock of northeast Queensland; J. Geochem. Explor., v. 2, p. 331-348.

Simmons, B.D. and the Geological Staff, Lake Dufault Div., Falconbridge Copper Ltd.
  1973: Geology of the Millenbach massive sulphide deposit, Noranda, Quebec; Can. Inst. Min. Met., v. 66, no. 739, p. 67-78.

Simpson, P.R., Plant, J., and Cope, M.J.
  1977: Uranium abundance and distribution in some granites from northern Scotland and southwest England as indicators of uranium provinces; Int. Symp. Inst. Min. Met., London, p. 126-139.

Sopuck, V.J.
  1977: A lithogeochemical approach in the search for areas of felsic volcanic rocks associated with mineralization in the Canadian Shield; Ph.D. thesis, Queen's University, Kingston, Ontario, 400 p.

Tasumi, T. and Clark, L.A.
  1972: Chemical composition of acid volcanic rocks genetically related to formation of the Kuroko deposits; Geol. Soc. Japan J., v. 78, p. 191-201.

Tauson, L.V.
  1967: Geochemistry of rare elements in igneous rocks and metallogenic specialization of magmas; in Chemistry of the Earth's Crust, Vinogradov, A.P., ed., Vol. II (Jerusalem: Israel Program for Scientific Translation), p. 248-259.

Tauson, L.V. and Kozlov, V.D.
  1973: Distribution functions and ratios of trace-element concentrations as estimators of the ore-bearing potential of granites; in Geochemical Exploration 1972, Jones, M.J., ed., Inst. Min. Met., London, p. 37-44.

Theodore, T.G. and Nash, J.T.
  1973: Geochemical and fluid zonation at Copper Canyon, Landor County, Nevada; Econ. Geol., v. 68, p. 565-570.

Thurlow, J.G., Swanson, E.A., and Strong, D.F.
  1975: Geology and lithogeochemistry of the Buchans polymetallic sulfide deposits, Newfoundland; Econ. Geol., v. 70, p. 130-144.

Wahl, J.L.
    1978: Rock geochemical exploration and ore genesis at Heath Steele and Key Anacon deposits, New Brunswick; Ph.D. thesis, University of New Brunswick, Fredericton.

Wahl, J.L., Govett, G.J.S., and Goodfellow, W.D.
    1975: Anomalous element distribution in volcanic rocks around Key Anacon, Heath Steele, B-zone and Brunswick No. 12 sulphide deposits (abstract); Can. Inst. Min. Met., Bull., v. 68, no. 755, p. 49.

Whitehead, R.E.S.
    1973: Environment of stratiform sulphide deposition: variation in Mn:Fe ratio in host rocks at Heath Steele Mine, New Brunswick, Canada; Mineral. Deposita, v. 8, p. 148-160.

Whitehead, R.E. and Govett, G.J.S.
    1974: Exploration rock geochemistry - detection of trace element haloes at Heath Steele Mines (N.B., Canada) by discriminant analysis; J. Geochem. Explor., v. 3, p. 371-396.

Wolfe, W.J.
    1974: Geochemical and biogeochemical exploration research near early Precambrian porphyry-type molybdenum-copper mineralization, Northwestern Ontario, Canada; J. Geochem. Explor., v. 3, p. 25-41.

    1975: Zinc abundance in Early Precambrian volcanic rocks: its relationship to exploitable levels of zinc in sulphide deposits of volcanic-exhalative origin; in Geochemical Exploration 1974, Elliott, I.L. and Fletcher, W.K., ed., Elsevier Publ. Co., Amsterdam, p. 261-278.

THE APPLICATION OF ATMOSPHERIC PARTICULATE GEOCHEMISTRY IN MINERAL EXPLORATION

A.R. Barringer
Barringer Research Ltd., Rexdale, Ontario

Barringer, A.R., The application of atmospheric particulate geochemistry in mineral exploration; in Geophysics and Geochemistry in the Search for Metallic Ores; Peter J. Hood, editor; Geological Survey of Canada, Economic Geology Report 31, p. 363-364, 1979.

Abstract

The atmospheric layer close to the earth's surface carries extensive geochemical information relating to the composition of the underlying terrain. During active mixing, there is an upward flux of both gaseous and particulate material from the surface. Gaseous forms of atmospheric geochemical interest include mercury vapour, halogen vapours, sulphur compounds, and radon. These gases diffuse rapidly and need to be measured very close to the surface to be of value. In general, experiments on the atmospheric geochemical measurement of trace gases have been mainly confined to mercury and radon, and have not led to the development of techniques that have been widely applied. This is due to the inherent problems relating to sensitivity requirements and the effects of rapid dilution.

In the case of atmospheric particulates, comparatively large fluxes of material rise into the atmosphere when mixing conditions are good and with appropriate instrumentation it has been shown feasible to carry out atmospheric geochemical surveys using this material. If it is desired, however, to collect particulate material that relates closely to the underlying terrain, it is important to separate material carried in parcels of rapidly rising air from particulates associated with neutral or sinking conditions. It is also advantageous to use coarse particulate fractions in size ranges larger than 30 microns to minimize the effects of lateral migration and the re-entrainment of particles that have been previously translocated by wind.

Atmospheric particles can be of both inorganic and organic orgin, much of the latter arising from vegetation. It has been established both in the laboratory and in the field that there can be movement of elements through vegetation to leaf surfaces followed by dispersion of these elements into the atmosphere as particulate material. Particulates derived from vegetation are, therefore, related geochemically to the composition of the underlying soils. In practice, material arising from both vegetation and residual soil surfaces can be utilized for geochemical exploration purposes.

Several types of airborne equipment have been developed for carrying out systematic atmospheric geochemical surveys. The spatial resolution of these systems varies according to design between 100 metres and several kilometres. Analytical methods employed have included conventional emission spectroscopy, laser vapourization coupled with emission spectroscopy, x-ray fluorescence spectroscopy, and fission trace-etch counting. The spectroscopic measurements provide analyses for 20 or more elements, including all of the base metals while fission track-etch methods give exceptionally high sensivity and specificity for uranium alone, and are insensitive to radon and bismuth 214.

Optical monitoring and fluidic switching devices have been used to provide for separation of upwelling particulates from material in stagnant and sinking air.

Atmospheric airborne geochemical prospecting appears to offer important potential as a complementary tool to a variety of airborne geophysical methods. In the case of uranium exploration, it can provide information that is unaffected by surface disequilibrium effects and the fine size aerosol interference that can considerably modify airborne gamma-ray spectrometer results.

Résumé

La couche atmosphérique proche de la surface de la terre contient des renseignements géochimiques importants en ce qui a trait à la composition du terrain sous-jacent. Au cours d'une phase active de mélange, il se produit un flux ascendant de matières gazeuses et de particules à partir de la surface. Les formes gazeuses qui sont d'un intérêt particulier pour la géochimie de l'atmosphère comprennent la vapeur de mercure, les vapeurs d'halogène, des composés de soufre et le radon. Ces gaz se diffusent rapidement et l'on doit effectuer leur mesure très près de la surface pour que les données aient une certaine valeur. En général, les expériences sur la mesure par géochimie atmosphérique des gaz à l'état de trace ont été surtout limitées au mercure et au radon; elles n'ont pas conduit à la mise au point de techniques qui ont été utilisées sur une grande échelle. Cette situation est due aux problèmes particuliers concernant les exigences de sensibilité et les effets d'une dilution rapide.

En ce qui concerne les particules atmosphériques, des flux relativement importants de matière s'élèvent dans l'atmosphère lorsque les conditions de mélange sont bonnes et, à l'aide des instruments appropriés, il s'est avéré possible d'effectuer des levés géochimiques de l'atmosphère en se servant de cette matière. Toutefois, si l'on désire recueillir des particules qui ont une relation étroite avec le terrain sous-jacent, il est important de séparer la matière transportée par vagues d'air à ascension rapide des particules en condition neutre ou en retombée. Il est également préférable d'utiliser des fractions granulométriques de particules grossières supérieures à 30 microns afin de minimiser les effets de migration latérale et le rechargage de particules qui ont été préalablement déplacées par le vent.

Les particules atmosphériques peuvent être d'origine inorganique et organique; la majorité de celles de la dernière catégorie proviennent de la végétation. Il a été établi tant en laboratoire que sur place qu'il peut y avoir un mouvement d'éléments passant par la végétation jusqu'à la surface des feuilles, suivi par la dispersion de ces mêmes éléments dans l'atmosphère sous forme de particules. Des particules provenant de la végétation sont, par conséquent, reliées géochimiquement à la composition des sols sous-jacents. Sur le plan pratique, des matières provenant de la végétation aussi bien que des surfaces résiduelles du sol peuvent être utilisées à des fins d'exploration géochimique.

Plusieurs types de dispositifs aéroportés ont été mis au point pour effectuer des levés géochimiques atmosphériques systématiques. La résolution spatiale de ces systèmes varie selon leur conception entre 100 mètres et plusieurs kilomètres. Les méthodes analytiques utilisées comprenaient: la spectroscopie d'émission classique, la vaporisation au laser alliée à la spectroscopie d'émission, la spectroscopie par fluorescence de rayons X, et le calcul par des techniques apparentées à la méthode des traces. Les mesures spectroscopiques fournissent des analyses de 20 éléments ou plus, y compris tous les métaux non précieux, tandis que les techniques apparentées à la méthode des traces donnent une sensibilité et une spécificité exceptionnellement élevées seulement pour l'uranium et sont insensibles au radon et au bismuth 214.

Des dispositifs de contrôle optique et de commutation fluidique ont été utilisées pour permettre la séparation des particules ascendantes de la matière obtenue dans l'air stagnant et descendant.

La prospection géochimique aéroportée des particules atmosphériques semble offrir un potentiel important comme outil complémentaire à une grande variété de méthodes géophysiques aéroportées. En ce qui concerne la recherche de l'uranium, ce genre de prospection peut fournir des renseignements qui ne sont pas touchés par les effets de déséquilibre de surface et l'interférence des aérosols stratosphériques à particules fines qui peuvent modifier considérablement les résultats obtenus à l'aide d'un spectromètre aéroporté à rayons gammas.

ANALYTICAL METHODOLOGY IN THE SEARCH FOR METALLIC ORES

F.N. Ward
U.S. Geological Survey, Denver, Colorado

W.F. Bondar
Bondar-Clegg & Company, Ltd., Ottawa, Ontario

Ward, F.N. and Bondar, W.F., Analytical methodology in the search for metallic ores; in Geophysics and Geochemistry in the Search for Metallic Ores; Peter J. Hood, editor; Geological Survey of Canada, Economic Geology Report 31, p. 365-383, 1979.

Abstract

Atomic absorption and emission spectrography are the methods of analysis most widely used in geochemical exploration. Development of nonflame atomizers, particularly electrothermal devices and reduction cells for atom and metal-hydride generation, has expanded the application of atomic absorption spectrometry by pushing detection limits of many elements well into the parts per billion range and by reducing detection limits for others, such as As, Se, Te, and Sn, to levels useful in lithogeochemical surveys. The recent promotion of inductively-coupled plasma sources for excitation, as well as other variations, such as use of echelle gratings, has increased the number of available spectrographic methods for multielement surveys and has simplified the application of partial extraction techniques in emission spectrography.

Other methods that require mass spectrometers and gas chromatographs are being used to measure volatile indicator elements and compounds such as helium and sulphur gases. Analytical techniques, including those based on voltammetry, ion-selective electrodes, and the use of partial or selective extractions, are finding increased application as analytical tools and as aids in determining metal speciation better to understand geochemical processes of dispersion and concentration.

Current interest in uranium exploration has sparked a major effort to develop new analytical methods or improve existing ones for the determination of uranium and related radionuclides. Exploration geologists may now choose conventional fluorimetry, delayed neutron counting, X-ray fluorescence, laser-induced fluorescence, and nuclear-fission track techniques for the determination of uranium. The choice will depend on sensitivity required, sample media being analyzed, chemical species of the uranium to be determined, turnaround time required, and cost considerations. Two of the methods described, conventional fluorimetry and laser-induced fluorescence, can be adapted for use in the field.

While recent developments of new techniques and apparatus have greatly expanded the number of useful analytical techniques in exploration geochemistry, each has its own problems and limitations as well as its applications. A panacea for analytical problems does not yet exist, except perhaps in the person of the skilled analyst, whose ingenuity in developing and applying new methods augments diligent application of tried and true procedures.

Résumé

Pour l'exploration géochimique, les méthodes d'analyse les plus fréquemment employées sont la spectrophotométrie d'absorption atomique et la spectrographie d'émission. La mise au point de méthodes spectrophotométriques sans flamme, en particulier d'appareils électrothermiques et de cellules réductrices permettant d'obtenir des atomes et des hydrures métalliques, a élargi les applications de la spectrophotométrie d'absorption atomique, en poussant les limites de détection de nombreux éléments jusqu'à la gamme des parties par milliard, et en réduisant les limites de détection d'autres éléments, comme AS, Se, Te et Sn jusqu'à des niveaux utiles pour les levés lithogéochimiques. Récemment, la production de sources de plasma par couplage inductif comme sources d'excitation, ainsi que d'autres techniques telles que l'emploi de réseaux à échelettes, ont permis d'augmenter le nombre de méthodes spectrographiques possibles pour l'analyse d'éléments multiples, et de simplifier l'application des techniques d'extraction partielle en spectrographie d'émission.

D'autres méthodes, qui exigent l'emploi de spectromètres de masse et de chromatographes en phase gazeuse sont employées pour la mesure d'éléments et de composés indicateurs volatiles, comme l'hélium et les gaz soufrés. Les techniques analytiques, en particulier celles basées sur la voltamétrie, les électrodes sélectives, et les méthodes d'extraction partielle ou sélective sont de plus en plus fréquemment appliquées comme outils d'analyse et comme moyen de mieux déterminer les espèces métalliques; ceci permet de mieux comprendre les processus géochimiques de dispersion et de concentration.

L'intérêt actuel pour l'exploration des gîtes uranifères nous a fortement incités à mettre au point de nouvelles méthodes d'analyse, ou à améliorer les méthodes analytiques existantes, pour doser l'uranium et les radionuclides apparentés. Les géologues chargés de l'exploration peuvent maintenant choisir entre les méthodes courantes de fluorimétrie, d'activation neutronique retardée, de fluorescence X, de fluorescence induite par laser, ainsi que les techniques d'observation des trajectoires des particules en chambre de Wilson, pour doser l'uranium. Le choix dépendra du degré de sensibilité requis, du type d'échantillon à analyser, de l'espèce chimique constituée par l'uranium que l'on veut doser, du temps de récupération requis, et des considérations de coût. Deux des méthodes décrites, la fluorimétrie courante et la fluorescence induite par laser peuvent aussi être utilisées sur le terrain moyennant certaines modifications.

Bien que le récent développement de techniques et d'un appareillage tout nouveau ait grandement accru le nombre de techniques d'analyse que l'on peut utiliser en géochimie pour l'exploration des gîtes minéraux, chaque système présente ses propres inconvénients et ses limitations aussi bien que ses applications particulières. Il n'existe pas encore de solution universelle aux problèmes de nature analytique — la solution réside sans doute dans la façon dont procède l'analyste expérimenté, qui, s'il fait preuve d'imagination pour mettre au point et appliquer de nouvelles méthodes, accroît l'efficacité des modes opératoires éprouvés et applicables.

INTRODUCTION

All analytical methodology may be useful in the search for metallic ores, but often economics, short-term needs, and facilities are factors in choosing the procedures and techniques for identifying and quantifying constituents that provide clues to the presence of such ores. Analytical methodology based on chemical methods such as gravimetry and titrimetry, on instrumental methods including the various kinds of spectroscopy, and on other kinds of methodology such as pattern recognition techniques, may be useful.

As generally understood, however, analytical methodology in the search for metallic constituents has to do more with trace methods of analysis; and as for the term "trace", a comment by Hillebrand (1919, p. 32) concerning rock analysis is appropriate: "It may be said with regard to the use of the word 'trace' that the amount of a constituent thus indicated is supposed to be below the limit of quantitative determination in the amount of the sample taken for analysis. It should in general for analyses laying claim to completeness and accuracy, be supposed to indicate less than 0.02 or even 0.01 per cent." Commenting on trace analysis, Sandell (1959, p. 5) said, "The essential feature of a trace analysis is not the determination of a minute quantity of a substance, but the determination of such a quantity in the presence of an overwhelming quantity of other substances which may seriously affect the reaction of the trace constituent." With today's technology both of these statements require modification along the line of amounts determined. Many trace analytical methods are capable of detecting 10^{-12} g or less, e.g., spark source mass spectroscopy and an essential feature of several trace analytical methods is their ability to detect such small amounts with small samples — for example, fission track determination can be made of 1 µg uranium per litre (1 ppb) in a 0.1 to 1 mL water sample (Reimer, 1975).

With the possible exception of titrimetry, all methods of chemical analysis require some sort of instrumentation, and titrimetry is not an exception if one detects the end point of a chemical reaction with a colour or electrical potential change. Methods of chemical analysis for trace amounts of substances require some kind of instrumentation, even including those methods based on colorimetry wherein an instrument is used to measure the colour. Thus in day-to-day jargon the term "instrumental methods of chemical analysis" (Willard et al., 1974) is synonymous with trace methods of analysis.

Karasek (1975) has provided a list of instrumental methods of chemical analysis, and the methods included in Table 17.1 are taken from his work along with others listed by Sandell (1959). The detection limits are compared on the basis of grams of a substance studied and were obtained under optimum conditions. Actual limits often vary by a factor of 10 or more.

Detection limits may be measured not only in terms of mass but also in terms of the signal produced by a detector responding to the mass. Thus the detection limit may be a detectable signal twice that of the noise level. Ordinarily with time-based abscissas the analog signal traces out an area under the curve, but in the case of flameless atomic absorption spectroscopy or gas chromatography with large amplitude signals of short duration, peak heights are more easily measured than areas under the curve. Thus, in flameless atomic absorption spectroscopy the detection limit is conveniently defined as twice the standard deviation calculated from 10 or more replicate peak heights under a given set of conditions.

Detection limits are factors in the choice of methodology in the search for metallic ores, but they are not necessarily decisive factors, especially because they differ remarkably from metal to metal, and also because no one instrumental method of trace analysis is universally applicable, except perhaps for spark source mass spectrometry. The detection limit of zinc determined by atomic absorption spectroscopy is good; that determined by optical emission spectroscopy is poor, whereas the limit of silver by the latter is excellent.

In a bibliography of exploration geochemistry covering the period from 1965 through 1971, Hawkes (1972) listed eight different analytical techniques used to acquire the compositional data required in geochemical exploration as usually practiced in the United States and Canada. Of the eight techniques, procedures based on atomic absorption spectrometry and optical emission spectrography dominate and only scattered instances of the use of other techniques, such as neutron activation, occur in the literature. To be sure, colorimetric procedures were common; indeed, they provided

Table 17.1

Trace methods and detection limits

Method	Detection limits (g)
Electron impact mass spectrometry	10^{-12}
Spark source mass spectrometry	10^{-13}
Ion probe mass spectrometry	10^{-15}
Chemical ionization mass spectrometry	10^{-10}
Neutron activation analysis	10^{-12}
Isotopic dilution	10^{-9}-10^{-16}
Atomic emission spectroscopy	10^{-9}
Flame atomic absorption spectroscopy	10^{-9}
Flameless atomic absorption spectroscopy	10^{-12}
Molecular absorption	10^{-6}
X-ray fluorescence	10^{-7}
Anodic stripping voltammetry	10^{-8}
D.C. polarography	10^{-8}
Pulsed polarography	10^{-10}
Ion selective electrode	10^{-15}
Electron spectroscopy	10^{-10}
Auger spectroscopy	10^{-10}

Table 17.2

Detection limits, by flameless atomic absorption, in picograms
(n.d. indicates no data)

Element	Carbon rod analyzer	L'vov (1961) furnace 1% absolute	Massman (1968) furnace
Zn	1	0.03	0.04
Cd	2	.08	.25
Cu	20	.6	10.
Pb	20	2.0	10.
Ag	1	.1	n.d.
As	1000	n.d.	600

the basis for trace element measurements that triggered a mushroom-like development of geochemical techniques in mineral exploration. Colorimetric procedures are also responsible for geochemical exploration becoming a recognized tool in exploration. However, the widespread development and application of colorimetric procedures based on molecular absorption, in contrast to atomic absorption, occurred in the 1950s, and by the early 1960s the position of such procedures was being undermined by the rapidly developing atomic absorption procedures. Because of their innate sensitivity, apparent simplicity, and the availability of commercial instrumentation, the number of published procedures based on atomic absorption spectrophotometry grew very rapidly. It is not surprising that Hawkes (1976) in his bibliography of exploration geochemistry covering the period from 1972 through 1975 listed 12 different instrumental techniques including over 40 references to atomic absorption determinations.

During the 1960s the application of optical emission spectrographic procedures to exploration geochemistry did not exhibit the flamboyant growth of atomic absorption procedures but experienced a slow and steady growth as workers began to consider that single element analysis, good as it was, did not provide the volume of chemical data attainable with optical emission spectrography. This understanding, coupled with the common practice of assembling field spectrographic laboratories (as established by the U.S. Geological Survey in 1955 and promoted heavily during the 1960s), helped to ensure the extensive use of mobile spectrographic laboratories for attaining the volume of compositional data needed in purely reconnaissance surveys.

The promoters of mobile spectrographic laboratories envisioned the use of such laboratories in the orientation phase of a project to establish diagnostic elements which could then be measured by cheaper analytical methodology; however, several developments occurred to make such data more attractive. The cost of collecting samples increased in greater proportion than the cost of acquiring spectrographic facilities, and geologists and chemists alike learned that personnel could be trained fairly rapidly to make spectrographic analyses.

The idea of movable laboratories is not original with the U.S. Geological Survey. Possibly before World War II, Russian and Scandinavian scientists carted spectrographic equipment into the field to perform such analyses near their study areas. One may say that they had portable laboratories, but the idea of mobile laboratories, specially designed and dedicated vehicles for spectrographic and other kinds of laboratories, is a North American contribution. With the proliferation of time-shared computers and readily available terminals and the faster transport of samples from areas of study to centrally based laboratories, the need for such mobile laboratories is hardly justifiable, especially from the economic side; but many geologists remain reluctant to part with a useful arrangement for private laboratory facilities during part of a field season.

One hardly needs to document the statement that to date most of the analytical methodology used in geochemical exploration has been based first on molecular absorption and then on atomic absorption phenomena, along with methods based on optical emission spectrography. Exceptions to such a statement are also evident, however, especially in the case of naturally occurring radioactive elements such as uranium, thorium, and potassium; but because of the widespread and predominant use of atomic absorption and spectrographic methods in geochemical exploration, we shall limit this discussion to certain innovations in the application of these techniques and the advantages realized therein. A casual glance at the number of presently available analytical methods shows levels of sophistication varying from the simple cold extractible copper test to neutron activation procedures (which require activation facilities and computer treatment of data), to spark-source mass spectrometric methods (which require not only a highly trained staff but rather elaborate instrumentation and data handling facilities). This very proliferation should warn exploration geochemists not to become infatuated by fads in methodology in the search for metallic ores.

The analytical methodology needs to be geared to the problem. If the problem is simply that of locating relatively large targets or favourable areas, then the cold extractible copper procedures (Canney and Hawkins, 1958; Holman, 1956) or the simple spot test for molybdenum minerals recently described by Griffitts et al. (1976) may suffice. The advantage of such methodology is that one can acquire the information on the spot at the sample location and change the sampling as needed. On the other hand, if the problem is one of lithogeochemistry then more sensitive methodology is in order. For example, small amounts of copper dispersed in some manner upward from a buried porphyry require sensitive methods for detection. Similarly, one may need elaborate instrumentation to detect small amounts of uranium or daughter products, for example, in uranium detection by fission track, laser-induced fluorometry, and radon measurements in groundwaters.

In the application of atomic absorption methods in the search for ore deposits, the flame absorption methods may be adequate for elements like copper, zinc etc. in soils and rocks; flameless absorption procedures may be essential when using natural waters as the sampling medium.

We shall here discuss several instrumental trace methods that have been found useful, some that are potentially useful, and others that appear to have limited usefulness in the search for metallic ores. Finally we shall include a discussion of several methods for determining uranium in geochemical exploration with supporting and illustrative data.

ANALYTICAL METHODOLOGY
Innovations in Atomic Absorption Spectroscopy

As a part of the more comprehensive technique of flameless atomic absorption spectroscopy, the various electrothermal devices for atomizing the sample and the sample introduction of the hydrides of elements like arsenic, mercury, and antimony conveyed by an inert carrier gas into a hydrogen flame, are innovations that deserve attention.

Graphite tubes or rods are the most commonly used electrothermal atomization devices, and their use is well established. Graphite furnaces were used by King (1908, 1932) to volatilize elements in spectral studies and later modified by L'vov (1961) for atomic absorption studies in which he achieved sensitivities of 10^{-8} to 10^{-11} g. In the system used by L'vov, a carbon rod with a dried sample was inserted into the hole of a heated atomization tube. The rod was heated and the dried sample volatilized into the confined space of the tube, where it remained in the path of a light beam during diffusion out of the ends of the tube. Working independently, Woodriff and Ramelow (1968) developed a similar furnace, with the advantage that it was used to atomize and maintain the atoms in a free state. In Woodriff and Ramelow's furnace the sample is completely enclosed except for the open ends of the tube, and the resulting uniform temperature helps to eliminate matrix effects. Later Massman (1968) devised a furnace having a hole in the tube wall through which the sample was added, after which the tube was heated to atomize the sample into the path of a line source, as in a conventional atomic absorption instrument.

The graphite rod (better known as the carbon rod) was another type of electrothermal atomizer developed, and West and Williams (1969) used it in atomic absorption and atomic fluorescence analysis. Amos et al. (1971) made a comparison of sensitivities of 11 different elements achieved with the carbon rod atomizer with the L'vov furnace and the Massman furnace. The comparison of six of these elements is shown in Table 17.2 and except as noted the values are in picograms.

The tantalum ribbon developed by Hwang et al. (1971) following the work of Donega and Burgess (1970) is a less commonly used electrothermal atomizer. However, it has several advantages, one of which is the small size of the ribbon and the resulting rapid heat dissipation. Air cooling is adequate, though time consuming. Sensitivities achieved with the ribbon are in the range of 10^{-9} to 10^{-12} g.

All of the electrothermal atomizing devices suffer from spectral interferences, chemical interferences, and background radiation. With certain elements, for example, lead, the graphite tube shows less spectral interference than the carbon rod (Amos et al., 1971), but one cannot generalize. Power requirements of the carbon tube are greater than those of the carbon rod, and hence cooling times between samples are longer. Long cooling times with carbon tubes and rods as well as the tantalum ribbon result in fewer determinations per day.

Maximum temperatures obtained by these devices differ somewhat, the tantalum ribbon being limited by the melting point ~2996°-3000°C. The carbon rod can be heated to about the same temperature, but the life of the rod is limited to 20-40 determinations, and light scattering due to carbon particles is appreciable, causing increases in background absorption especially in visible region of the spectrum. Graphite furnace temperatures of 2800°C are also common.

Hydride generation of volatile elements like arsenic, selenium, antimony, bismuth, germanium, tin, tellurium, and lead (Pollack and West, 1973; Thompson and Thomerson, 1974) by reduction with sodium borohydride followed by atomization and combustion in a heated tube or in a hydrogen-argon or hydrogen-nitrogen flame is another innovation in atomic absorption spectroscopy that merits attention. The generated hydrides are swept along with a carrier gas such as argon or nitrogen, into a heated tube or hydrogen flame positioned in the light path of a hollow cathode or electrodeless discharge lamp. The chemical conversion of arsenic into arsine and its introduction into an argon-hydrogen flame resulted in improved detection limits (Holak, 1969). Chu et al. (1972) eliminated the flame and swept the arsine into a heated absorption tube to achieve better sensitivity, which is possibly due to elimination of flame background and to longer residence times.

Goulden and Brooksbank (1974) automated the procedure for determining antimony, arsenic, and selenium but found that they needed to isolate the hydrides by means of a heated column from products of side reactions prior to combustion in an open-ended heated tube. Combustion of the isolated hydrides produced increases in sensitivity by two orders of magnitude over combustion of the hydrides in a conventional hydrogen-argon entrained air flames. Pierce et al. (1976) combined the heated column with the furnace of Chu et al. (1972) to develop an automated procedure for determining selenium and arsenic in surface waters that have detection limits respectively of 0.019 and 0.011 µg/L.

Hydride generation procedures appear to be relatively free from interferences, although one would expect that any element reducible by sodium borohydride and that forms a volatile hydride would interfere if present in large amounts. In a study of the determination of antimony and arsenic in geological materials, Aslin (1976) concluded that the hydride generation procedures were apparently free from interferences. Iron, cobalt, and copper did not interfere at the four levels studied. Nickel at the 1000 ppb level did interfere with the determination of arsenic, and both nickel and silver interfered with the determination of antimony.

Pierce et al. (1976) noted that copper concentrations in excess of 5 mg/L did compete with selenium and arsenic compounds during reduction, but that the competition could be inhibited by dilution. Also we have noted relatively high blanks in arsenic determinations caused by impurities in the sodium borohydride reagent. Improvements in the manufacture of this reagent are in progress, and despite a few drawbacks, hydride generation procedures provide reasonably reproducible methods for measuring elements like arsenic, selenium, and bismuth in geochemical exploration as well as in monitoring air and water quality in environmental studies.

Innovations in Optical Emission Spectrography

Optical emission spectroscopy has been a principal means of data acquisition in geochemical exploration from the beginning, when Russian scientists transported a spectrograph into the field to support field parties (Ratsbaum, 1939; Fersman, 1952) and Palmqvist and Brundin (1939) set up a stationary spectrographic laboratory to make about 400 determinations per day on ashed plant material. In North America the development of a truck-mounted spectrograph laboratory (Canney et al., 1957) triggered a widespread interest, and although the above authors suggested that spectrographic laboratories be used only in the reconnaissance part of an exploration program because of the cost, truck-mounted spectrographic laboratories soon became an essential part of the data acquisition process, at least in the United States. Data attainable by optical emission spectrography revealed some hitherto unknown mineral assemblages (Curtin et al., 1968).

Currently the optical emission spectrographic system used in geochemical exploration commonly uses d.c. arc excitation and original or replica gratings with photographic readout. The gratings are ruled with 15 000 lines per inch and provide a wavelength coverage from about 2050 to 4850Å in the second order. Ahrens (1950) and others found that d.c. arc excitation was best suited to geological materials especially when the terms semiquantitative and quantitative are carefully defined. Factors in favour of d.c. arc excitation are "simplicity, high concentrational sensitivity, adaptability

Table 17.3

Detection limits of elements

Element	Spectrographic d.c. arc (ppm)[1]	Flame AAS (ng/mL)[2]	Flameless AAS (ng/mL)[2]	ICP Excit (ng/mL)[3]
Zn	3(100)	2	0.006	2
Cd	10	1	.001	2
Cu	0.5	2	.01	1
Bi	20	40	.1	50
Pb	5	20	.06	8
Ag	0.5	2	.0025	4
Co	10	10	.04	3
Ni	5	10	.1	6
U	100	30 000	([4])	30
W	20	3000	([4])	2
Sb	20	100	.2	200
As	100	100	.1	40

[1] U.S. Geological Survey laboratories, written comm.
[2] Slavin et al. (1972).
[3] Fassel and Kniseley (1974a).
[4] No data.

and low cost" (Canney et al., 1957). Recently Timperley (1974) interfaced a dedicated minicomputer with a direct-reading spectrograph for use in data collection in rapid geochemical surveys.

Although a vast amount of semiquantitative spectrographic data has been acquired in geochemical exploration programs using the systems described, changes are inevitable and two such innovations are worth considering. These are plasma sources for excitation and echelle gratings, which have the property of concentrating the energy in the desired order. With regard to plasma sources for excitation most of our discussion will be about inductively coupled plasmas (ICP) which seem to have caught on almost as readily as atomic absorption spectroscopy. A recent quotation is as follows:

"This new technique will probably have a similar impact over the next decade as AAS did over the last" (Barringer Research 1975, written comm.).

Neither innovation is really new: plasmas have been around for several years (Greenfield et al., 1964; Wendt and Fassel, 1965); echelle grating construction was first described by Harrison (1949).

Inductively coupled plasmas are maintained by a high frequency, axial magnetic field in a laminar flow of argon at atmospheric pressures. The discharge does not contact any electrodes in contrast to capacitively coupled plasmas such as high-frequency torch or radio frequency discharges or d.c. plasma jets (Wendt and Fassel, 1965).

Spectrographic systems in common use require a finely powdered sample, often mixed with other materials, to achieve better presentation of sample to the d.c. arc or to simulate matrices whose burn characteristics are well defined. For example, iron oxide and silica may be added to plant ash so that the behaviour of the mixture in an arc resembles that of a granite or silicate rock. In contrast, inductive coupled plasmas take sample solutions with attendant simplification of matrix. Obviously the solution has to be introduced into the plasma, and the process, involving atomization and nebulization, has been accomplished by ultrasonic generation of aerosols (Wendt and Fassel, 1965; Dickinson and Fassel, 1969) and later by pneumatic nebulization (Scott et al., 1974).

Fassel and associates have contributed much to aid in understanding the events taking place in the plasma excitation by developing a model for the plasma that enabled them to effectively optimize a number of experimental factors, such as:

1. desolvation of nebulized solutions prior to entry into plasma.
2. introduction of sample aerosol into the plasma.
3. sensitively tuned rf generator for optimum coupling of the output to the plasma.

When such factors were optimized, they were able to measure concentrations ranging from nanograms to fractional micrograms per mL which effectively extended the range in concentrations measured upward by two to four orders of magnitude. Thus the inductively coupled plasma excitation system has a wide dynamic range with obvious advantages.

The maximum temperature of an argon-supported plasma is of the order of 9000 to 10 000°K, and Fassel and Kniseley (1974b, p. 1158a), stated that "according to our preliminary measurements, the gas temperatures in the axial channel of the eddy current flow region is about 7000°K." They went on to say that this temperature is twice that achieved in the hottest combustion flame. They noted also that residence times of the sample in the plasma before reaching the observing point is about 2.5 ms. Fassel and Kniseley also provided a satisfactory explanation for the improved sensitivities (as compared with flame excitation) for many elements, especially those that form stable monoxide molecules with dissociation energies greater than about 7 electron volts.

A comparison of detection limits for 12 elements is given in Table 17.3.

In selecting these data we have leaned towards the conservative side; for example we choose a value of 6×10^{-12} g for the detection limit of zinc by flameless atomic absorption spectroscopy as compared with 2×10^{-14} g given by Dulka and Risby (1976). We conclude with Fassel and Kniseley (1974b) as follows:

(1) "Inductively coupled plasma excitation for some 30 elements is considerably more sensitive than d.c. arc excitation.
(2) Flame atomic absorption methods have greater sensitivity than d.c. arc excited spectrographic methods.
(3) Flameless atomic absorption methods are considerably more sensitive than flame methods.
(4) Flameless atomic absorption methods are more sensitive than spectrographic methods using either d.c. arc-excitation or inductively coupled plasma excitation."

Echelle gratings have been around for 25 or more years, but until recently attempts to take advantage of the echelle gratings whereby radiation of a given wavelength could be largely concentrated in one order were limited (Richardson, 1953) and of little success.

The echelle is a special kind of diffraction grating ruled with high precision. Its broad, flat grooves are ruled so that the width of each step is several times the height and the spacing between the steps is several times greater than the wavelength of the incident energy. The large number of

Table 17.4

Comparison of detection limits (ng/mL)

(n.d. indicates no data)

Element	Material	ASV diff. pulse	ASV linear scan	AA nonflame
zinc	seawater	0.04	0.04	0.008
cadmium	seawater	.005	.01	.01
lead	seawater	.01	.02	.5
tin	geological material	∼1200[1]	n.d.	n.d.
silver	natural water rain and snow	n.d.	∼400[1]	.0025[2]
mercury	natural water	n.d.	800[3]	n.d.

[1]Bond et al. (1970).
[2]Slavin et al. (1972).
[3]Perone and Kretlow (1965).

orders of the grating, as many as 90, may be separated by an order sorter which results in radiation of a given wavelength being concentrated in one order. A prism with dispersion at right angles to the grating is one kind of order sorter. Successful use of echelle gratings require that the orders be separated, and Danielsson and Lindblom (1972) have developed a spectrograph with a CaF_2 Littrow prism order sorter and an image tube with high sensitivity and resolution. Application of the image tube for spectral analysis required focusing a wide wavelength coverage on the relatively small photocathode, 20 mm in diameter, and easy electronic readoff. The echelle grating with an order sorter met their requirements, and they accordingly developed a stigmatic, coma-compensated echelle spectrographic system with high resolution and considerable dispersion (Danielsson et al., 1974).

In a commercially available d.c. plasma-echelle spectrometer system the temperature of the plasma reaches 6000-8000°K. For elements such as calcium, magnesium, boron, and copper, analytical results are similar to those obtained by atomic absorption spectrometry. Stray light is a problem in determining aluminum. The quartz chimney above the plasma causes interference, and substitution of Teflon for the quartz does not help. Molecular bands interfere with phosphorus determinations, but in general 30-40 samples can be analyzed for 18 elements in about one hour with a precision of slightly more than ±10 per cent.

Echelle gratings are not restricted to monochromators designed specifically for emission. They have also been used in monochromators designed for atomic absorption, and Keliher and Wohlers (1974) made a direct comparison between a line source (hollow cathode lamp) and a continuum source (150 W xenon lamp) using a high resolution echelle spectrometer. They found that the sensitivities obtained with the hollow cathode lamps were slightly superior to those obtained with the xenon lamps using the echelle spectrometer, and although the spectral bandwidth using the later continuum source and the echelle grating is wider than that of the absorbing line, the results were reasonably good.

The relatively poor detection limits of previous continuum systems were attributed to poor signal-to-noise ratios. Wavelength scanning is a viable technique for improving such ratios as long as the wavelength of the absorbing line is within the range of wavelengths scanned.

Veillon and Merchant (1973) described a piezoelectric scanning Fabry-Perot interferometer with a conventional grating monochromator to obtain an overall spectral bandwidth of 0.013Å. This was small enough to suggest the possibility of effectively scanning a wavelength interval of 0.01Å over the 0.1Å monochromator output and measuring the radiation within any 0.01Å interval. As long as the instrumental width of the interferometer is less than that of the absorbing line, the absorbance could be measured with maximum sensitivity and such measurement resulted in improved signal-to-noise ratio. Copper and silver sensitivities were similar to those obtained with line sources.

Thus echelle grating monochromators may find use in atomic absorption as well as in atomic emission methods, and provide the advantages of a single continuum source in place of the large number of line sources heretofore required in atomic absorption spectrometry. With such a development the cost of atomic absorption determinations can obviously be reduced.

Anodic Stripping Voltammetry

Like other instrumental techniques, anodic stripping voltammetry is not new, but recent developments in instrumentation have suggested applications of this method to the solution of several difficult analytical problems involving trace determination of elements like lead, bismuth, cadmium, thallium, tin, and silver.

The fundamental process involved in anodic stripping voltammetry (ASV) have been discussed by several authors, as illustrated by Copeland and Skogerboe (1974). Briefly two steps comprise an ASV measurement: First the analyte species is reduced and concomitantly plated out on an electrode, usually mercury or solid electrodes such as platinum, gold, or silver. Second, the reduced species is oxidized, stripped back into the electrolyte solution by systematically changing the potential in the direction to cause oxidation. Hanging-drop mercury electrodes are also used, but the spherical surface limits the amount of analyte that can be plated out. Alternatively thin-filmed mercury electrodes wherein the mercury is mounted as a film on a substrate such as graphite have been used; they have the advantage of a large surface-area-to-volume ratio, and they can be rotated or stirred during the plating and stripping. At the oxidation potential of each plated analyte species, the Faraday current produced by the oxidation is measured. The stripping current produced by the oxidation of one or more analyte species is proportional to the concentration of the respective species plated out on the electrode and ultimately to the concentration in the solution.

Different waveforms are used to strip the deposited analyte species from the electrode. The common choice is a linear ramp of the potential wherein the latter is scanned at a constant rate over a range covering the potential at which oxidation of the analyte species occurs. The scan rate is most important for achieving maximum stripping current. Infrequently a small sine wave potential is superimposed on the linear ramp, resulting in phase differences between stripping current and the ramping voltage, so that phase-sensitive detection can be used to separate the Faraday current from non-Faraday currents, all at the expense of more sophisticated instrumentation but with significant improvements in signal-to-background ratios. Relatively large amplitude pulses superimposed on the linear potential ramp for short periods are the basis of another stripping technique that provides a choice in measurement periods during the stripping process to reduce non-Faraday currents. Moreover, as suggested by Osteryoung and Christie (1974), the pulsed stripping shows greater sensitivity than linear scan stripping because of better discrimination between Faraday current and the charging current. Pulsed stripping voltammetry also provides considerable signal enhancement due to the replating and reoxidation that takes place near the end of the pulse and during the next pulse, so that the same analyte species is seen repetitively as compared to a single run in linear scan stripping. The net result is a longer residence time using pulsed anodic stripping.

Among the advantages of anodic stripping voltammetric methods are the small size of sample needed due to innate sensitivity, the relatively inexpensive (although highly specialized) instrumentation, and the multielement possibilities, along with its essentially nondestructive aspects. The sensitivities given by Copeland and Skogerboe (1974) are shown in Table 17.4 except as noted. These sensitivities compare favourably with those obtained by flameless atomic absorption spectroscopy, and thus anodic stripping voltammetry may in some cases offer a viable alternative to atomic and molecular absorption techniques.

Selective Ion Electrodes

Selective ion electrodes have been in use since it was found that thin glass membranes could be used to seal off an insulating glass tube containing a dilute solution of hydrogen ion and that a potential would develop across the membrane which could be measured by reference to another solution of fixed concentration. Hence, the glass electrode quickly replaced the hydrogen electrode as well as other kinds such as the quinhydrone electrode to measure the pH of soils and other materials. Such measurements became standard practice in soil and agronomic studies.

In addition to hydrogen ion, the glass membrane electrodes were sensitive to sodium and more than 40 years ago Lengyel and Blum (1934) obtained a Nernstian response with sodium ion — a straight line plot of voltage against logarithm of sodium ion concentration. They predicted the development of glass electrodes that would be sensitive to different metals, but little progress was made until Schwabe and Dahms (1960), using tracer studies, established the fact that although the theory had been accepted for many years, the hydrogen ions did not really pass through the glass membrane to give an electrode potential, but rather, the charges are transported by an ion carrier wherein each charged carrier needs to move only a few atomic diameters before giving up its charge to another carrier. The behaviour of all electrodes was not readily explained by this simple mechanism, but the explanation was adequate to stimulate renewed interest in the development of electrode and associated measurement equipment.

In the meantime a new technology had developed to fabricate glasses of different composition that selectively responded to different cations. By 1958 the Beckman Company[1] was marketing "specific ion electrodes" for elements like sodium, potassium, and silver. The electrodes were far from being specific, hence the more acceptable term selective ion electrodes. The technology of developing selectively responsive glasses is covered by Rechnitz (1967).

As early as 1965 one of us proposed an analytical method for use in geochemical exploration, using the silver content of geological materials as the measurable parameter. Measurement was easily done in the field with simple equipment consisting of the Beckman sodium electrode and a small portable pH meter. The sensitivity of the electrode to silver ions increased with pH and at a pH of 11 to 12 (readily achieved with an organic base such as ethanolamine) as little as 20 µg of silver in 10 mL of solution could be measured. At that time most of our silver measurements were made by means of optical emission spectrography, and the procedure offered a viable alternative at considerable savings in time and costs.

In a theoretical treatment of membrane potentials Eisenman (1969) divided electrodes into three general classes: (a) solid ion exchangers (glass electrodes); (b) liquid ion exchangers; (c) neutral sequestering agents which act as molecular carriers of ions. And in a review paper Pungor and Troth (1970) summarized the electrical behaviour of various ion-selective membranes.

Of the general types, solid ion and liquid ion exchangers comprise most of the present-day practical electrodes, especially in the useful methodology in the search for metallic ores. More than 20 electrodes are available including metals like lead, copper, calcium, all halogens, and different anionic species such as nitrate, cyanide, thiocyanate, perchlorate, and even gases like ammonia and sulphur dioxide. And without doubt the fluoride electrode is not only the best available, but has experienced the widest application. It may be that its superiority accounts for the extensive use.

The fluoride electrode consists of a crystal of lanthanum fluoride doped with europium +2 and cemented in the end of a glass or polyvinyl chloride plastic tube containing a mixture of 0.1M sodium fluoride and 0.1M sodium chloride solutions connected to an outer lead through a silver-silver chloride electrode whose potential is fixed by the chloride ion. The fluoride ion governs the potential at the inner surface of the lanthanum fluoride crystal, and when the electrode is immersed in a fluoride solution, a potential difference occurs across the membrane. The magnitude of such potential is dependent on the ratio of the fluoride ion activities of the inner solution and the outer solution.

The fluoride electrode has been used to measure fluorine in the U.S.G.S. reference rocks (Ficklin, 1970; Ingram, 1970) and to measure fluorine in rocks associated with tin mineralization (Kesler et al., 1973). Farrell (1974) used Ficklin's method to measure fluorine in soils after partial and total extraction in the fluorine province of Derbyshire, England.

The full impact of the successful development of a practical fluoride electrode for measuring small amounts of fluorine can only be appreciated by those older practitioners of analytical chemistry who have struggled to observe the endpoint when a titrating fluoride solution with a thorium nitrate solution in presence of alizarin as indicator.

Friedrich et al. (1973) used the cupric electrode to measure copper ion concentrations in stream water samples from the Ramsback area of Germany. He concurred with Durst (1969) that the electrode was useful for copper concentrations down to 0.6 ppb and was "virtually interference free with respect to the usual divalent cations."

Selective ion electrodes sense activity, which may be a drawback except in dilute solution where activity and concentration are nearly equal. In concentrated solutions where activity and concentration differ markedly, the measurement requires some means of estimating the ionic strength so as to relate activity and concentration. From an analytical perspective one has to repeat ionic strength levels, and this can be achieved either by dilution or by the addition

---

[1] The inclusion of brand or manufacturers' names in this report is for illustrative purposes only and does not imply endorsement by the U.S. Geological Survey.

of a high ionic strength solution to the analyte. The addition of a high ionic strength solution is more practical than dilution because of the innate sensitivity of the electrode. The former, called swamping, tends to nullify small variations due to different kinds of samples.

In a brochure on analytical methods Orion Research (1973) provided information on suitable ionic strength pH adjustor solutions for the different electrodes.

In summary the commercial availability of some 20 different electrodes, the relative cost of a suitable voltmeter, and the ease of making objective analytical determinations under rough conditions are plus factors in the use of ion electrodes in the search for metallic ores.

Miscellaneous Techniques

The following techniques are mentioned primarily to inform the reader of state-of-the-art methodology:

High-pressure liquid chromatography
Photoacoustic spectroscopy
Atomic fluorescence
Zeeman polarization effects.

Liquid chromatography is used to effect separations mostly in biological, clinical, and organic studies; and high pressure is simply a variation resulting in separations not easily accomplished otherwise. High-pressure liquid chromatography may have application in separation of metallic chelates which can then be detected and measured thus providing new methods of trace analysis for metallic constituents. At present, however, such applications are relatively rare and the number of metals measured by such a technique is probably less than a dozen.

Photoacoustic spectroscopy is an old phenomenon which was recently reviewed by Rosencwaig (1975). The solid photoacoustic absorption spectrum qualitatively resembles the solution absorption spectrum of certain materials. In practice, the sample is placed in a sealed, gas-filled cell containing a sensitive microphone for detection. When irradiated by high-intensity chopped monochromatic light, the light absorbed by the sample is changed to heat which raises the temperature of the boundary layer which then expands and contracts at the chopping rate. The microphone detects the alternate expansion and contraction and yields an electrical signal which can be treated in a conventional manner.

Atomic fluorescence is a form of flame spectroscopy wherein a solution of the sample is sprayed into a flame, and the ground state atoms are excited by radiation of the proper frequency from a continuum source in contrast to d.c. arcs, flames, plasmas, and so forth. The excited atoms are deactivated by emission of radiation of the same or lesser frequency. The emitted radiation is proportional to the concentration of excited species. Winefordner and Staab (1964) and Winefordner and Vickers (1964) have exploited this form of spectroscopy; they are responsible for demonstrating the remarkable sensitivities of different elements as zinc, cadmium, and mercury. The state of development of this technique is illustrated by the work of Johnson et al. (1975), who described a procedure for determining 18 elements using a single source and a separated air acetylene flame, and 5 elements with a separated N_2O-acetylene flame in jet engine lubricating oils. They achieved detection limits comparable to those obtained by flame atomic absorption spectrometry and single element hollow cathode lamps. Although the general consensus is that atomic fluorescence determinations are somewhat difficult and costly, the possibilities of multielement determinations of metals in geological materials are attractive.

Although commercial instrumentation utilizing the polarization characteristics of Zeeman split lines to correct for background and other extraneous noise in flameless atomic absorption is available, the advances in the area are rapid enough to warrant delays in major expansions. For example, Hadeishi and McLaughlin (1975) reporting on use of the Zeeman effect in atomic absorption determination of mercury applied a magnetic field to the light source in the direction of propagation of the light beam and used the components of the Zeeman emission lines for absorbing and reference light respectively. During the same year Koizumi and Yasuda (1975) applied the magnetic field to the light source in a direction perpendicular to the propagation of the light beam and were able to determine elements other than mercury, such as lead, cadmium, and zinc (Koizumi and Yasuda, 1976). Later Koizumi et al. (1977) applied the magnetic field to the sample vapour perpendicular to the light beam with improved background correction and steadier baselines. They reported that this particular system was capable of measuring with high sensitivity practically all of the elements determined by conventional atomic absorption spectrometry.

Advantages of the proposed systems are that no sample preparation is necessary, both liquid and solids can be analyzed, and greater extremes in the conditions can be tolerated than in ordinary flameless atomic absorption spectroscopy.

GEOCHEMICAL METHODS OF ANALYSIS FOR URANIUM

Exploration efforts directed towards discovery of new uranium deposits have never been greater than they are now. A measure of this level of effort is evidenced by the fact that in Canada over 60 per cent of all geochemical samples collected in 1976 and 1977 were or are being analyzed for uranium as the element of primary interest. Of the many different methods for measuring uranium, those most frequently used today in exploration geochemistry are listed in Table 17.5 in what is probably the order corresponding to their degree of usage globally in geochemical analyses for uranium.

Conventional Fluorometric Method

The conventional fluorometric method of uranium analysis is based on measurement of the fluorescence produced when uranium is fused into sodium fluoride or other alkali fluoride materials. The resulting fused bead will produce a brilliant fluorescence when it is illuminated with an ultraviolet light at 3550Å. This fluorescence output increases proportionally with increasing uranium concentration and can be measured by means of a suitable instrument such as the Jarrell Ash reflectance-type fluorometer. In practice, a measured aliquot of a sample solution containing uranium, either as a leachate of a solid sample or as a natural water, is placed in a platinum dish and evaporated to dryness. A suitable flux containing a fluoride

Table 17.5

Most commonly used methods for geochemical analysis of uranium

1)— Conventional Fluorometric Method
2)— Delayed Neutron Activation Analysis
3)— X-ray Fluorescence Method
4)— Fission Track Method
5)— Laser-Induced Fluorescence Method

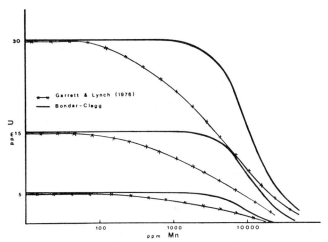

Figure 17.1. Observed fluorometric uranium versus manganese.

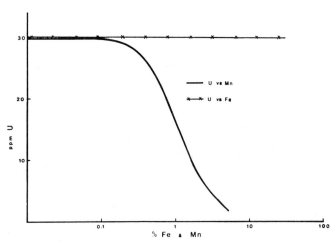

Figure 17.3. Observed fluorometric uranium versus iron and manganese.

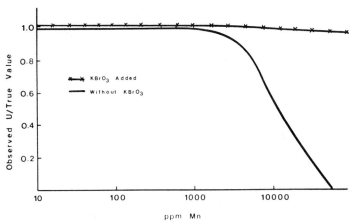

Figure 17.2. Effect of potassium bromate on fluorometric uranium versus manganese.

salt is then added to the dish and the sample is fused. After cooling, the uranium in the fused disc is measured fluorometrically.

A basically simple technique, the fluorometric method offers many advantages. It is relatively rapid and two analysts working together, can produce 300 analyses per work shift. The detection limits of 0.1 ppm for solid samples and 0.05 ppb for 25 mL samples of natural waters routinely attainable by the fluorometric method are adequate for almost all geochemical exploration requirements. Compared to other methods of uranium analysis it is inexpensive both on a cost per sample basis and in terms of the capital investment required to equip a uranium analysis facility. It can be adapted for use in the field. And finally, while capable of providing analytical data for total uranium, it also lends itself most readily to use in the partial or selective extraction procedures often used by geochemists to increase contrast between anomalous and background samples, or to obtain additional data on uranium fractionation in geochemical samples as an interpretive aid. Disadvantages of the method, simply stated, are that the routine fluorometric method of uranium analysis is not as precise at low uranium concentrations (0.1-2 ppm U) as are the X-ray fluorescence and delayed neutron activation methods when analyzing solid samples, and it is not quite as sensitive as the fission track method for the analysis of waters.

One often repeated criticism of the conventional fluorometric method is that both iron and manganese cause "quenching" or suppression of uranium fluorescence. Garrett and Lynch (1976) have shown in a series of control experiments, using manganese alone, that this problem can be severe in geochemical exploration. However it is not necessarily as serious a problem as is commonly believed and furthermore, it can be very easily eliminated.

Figure 17.1 shows observed fluorometric uranium values versus manganese as reported by Garrett and Lynch (1976) who used a high-carbonate flux; and as observed in a separate control experiment by Bondar-Clegg (unpub. data) who used a noncarbonate flux and a slightly different fluorometric procedure. Only the data for 5, 15 and 30 ppm uranium are shown. From these data, it can be seen that depending on the actual fluorometric procedure used, the suppression of uranium fluorescence may not become severe until concentrations approaching 5000 ppm Mn are encountered. Most stream or lake bottom sediments and soils contain less than this amount of manganese.

Ingles (1958) stated that in analyzing samples for uranium, those containing manganese as the principal interfering element can be treated by adding sodium chlorate or similar strong oxidant to precipitate out the interfering manganese.

Figure 17.2 shows the ratio of observed uranium concentration to the true value versus manganese added, with and without the addition of 0.15 gm potassium bromate. Only the data for 30 ppm uranium are shown, although the same results were obtained for other concentrations. The results show clearly that the quenching effect of manganese in concentrations up to 100 000 ppm Mn (10%) can be eliminated by simply adding 0.15 g potassium bromate to the test solution to precipitate the manganese as manganese dioxide. The supernatant liquid is then analyzed in the normal manner. The procedure is simple and does not increase the cost of analysis.

Separate tests, again using a noncarbonate flux, to test the effects of quenching by iron at concentrations up to 30 per cent Fe were also carried out.

The "quenching" effect of iron compared to manganese for increasing amounts of both elements is shown in Figure 17.3. Clearly, the "quenching" effect of iron is considerably less than for manganese and is not likely to be a problem in most geochemical samples.

The fluorometric method is simple and straightforward in theory, but delicate in practice. As such, it is very dependent upon the care of the individual analyst in performing the analysis if acceptable detection limits and precision levels are to be attained. Nonetheless it is very sensitive and can produce highly accurate results.

Delayed Neutron Activation Analysis

The delayed neutron activation method of analysis is based on the detection of delayed neutrons emitted in the decay of the fission products of uranium or other fissile material which is induced to fission by exposing samples to an intense neutron flux in a nuclear reactor. The number of delayed neutrons emitted following irradiation is proportional to the amount of fissile material present in the sample. Neutrons can be detected and selectively counted and the technique is highly specific for fissile material because, except for a number of short-lived light nuclides which have been shown to have negligible interfering effects, fission is the only nuclear reaction which produces nuclides that emit delayed neutrons.

The only naturally occurring fissile nuclides are ^{235}U, ^{238}U and ^{232}Th. ^{235}U is fissioned by slow or thermal neutrons whereas ^{238}U and ^{232}Th are fissioned by fast neutrons. Because the fast neutron flux component in a reactor is usually smaller than the thermal neutron flux component and because the fast neutron cross-section for ^{232}Th and ^{238}U is much lower than the thermal neutron cross-section for ^{235}U, slow thermal fission of ^{235}U greatly predominates. As a result, thorium interference is usually negligible for samples containing thorium in amounts equal to or less than the amount of uranium present. Except for samples which may have extremely high Th/U ratios the delayed neutron activation method can be considered specific for the determination of ^{235}U or uranium, assuming the normal isotopic abundance of ^{235}U in natural uranium. For samples with high Th/U ratios, a correction can be made for thorium interference by making two measurements: once by irradiating the sample in the normal manner with a mixed neutron flux to measure ^{235}U plus ^{238}U and ^{232}Th, and once with the thermal neutrons screened out so that only ^{238}U and ^{232}Th are measured. The ^{235}U content is then derived by difference.

The method as used by Atomic Energy Canada Ltd., Commercial Products Division (Boulanger et al., 1976) is as follows: samples are stacked in an automatic loader and transferred pneumatically to the Slowpoke reactor for irradiation up to a neutron flux of $1 \times 10^{12} \text{ n cm}^{-2} \text{s}^{-1}$. After irradiation, the samples are cooled for 10 s and then transferred to a counting facility consisting of six BF_3 detector tubes embedded in paraffin. By comparing the delayed neutron count to that obtained from standards, the uranium content of the unknown samples is determined. After counting, the samples are ejected into a shielded storage container and held until safe to handle. AECL has determined that by using a 30/10/30 second irradiation/cool/count sequence for geochemical analysis, the system is capable of providing a detection limit of 0.1 ppm with a precision of ±15 per cent at the 1 ppm level, ±10 per cent at 10 ppm U and ±2 per cent at the 100 ppm U level. Sensitivity for thorium is approximately 1 per cent of the sensitivity for uranium. Productivity is approximately 27 complete determinations per man/hour.

For the analysis of solid samples for total uranium, the delayed neutron activation method can be an accurate method which is fairly rapid, more sensitive than X-ray fluorescence, and more precise than fluorometry. Shortcomings of the method are, as a general statement, all cost related. The most obvious disadvantage of the delayed neutron counting method as compared to conventional fluorometry, XRF or laser-induced fluorometry is that one must have access to a nuclear reactor. Another most important shortcoming is that the method does not readily lend itself to measuring partial extractable and selectively extractable uranium in solid samples or to measuring uranium in natural waters. Both can be done but not very easily and not without preliminary separations and/or pre-concentration steps that in most cases are sufficiently complex and time-consuming to make the method slow, expensive, and therefore impractical for routine use in exploration geochemistry.

X-ray Fluorescence Spectrometry

Measurement of uranium by X-ray fluorescence spectrometry is based on measurement of the characteristic secondary X-ray spectra produced when a specimen is irradiated or bombarded with X-radiation of sufficient energy to produce electron transitions within uranium atoms in the sample. The intensity of this secondary radiation of characteristic wavelength is a measure of the amount of uranium present in the sample. Very sensitive detectors are used to measure this characteristic radiation and to distinguish these characteristic X-ray pulses from background radiation. By comparing the intensity of these secondary characteristic X-rays to those obtained from standards, the uranium content of unknown samples is determined.

Because variations in the major constituents of samples being analyzed may produce enhancement or absorption matrix effects which might lead to significant analytical error, a ratio method of calibration is often used. The analyte peak counts are ratioed to the background counts compensating for matrix variations and the ratios compared to calibration curves prepared from artificially prepared or certified natural uranium standard samples. The pressed pellet method commonly used for geochemical samples offers a detection limit of 1 ppm with a precision of ±10-15 per cent at the 10 ppm level. Productivity is approximately 27 determinations per man/hour. Only Rb and Sr, at concentrations in excess of several thousand ppm each, are likely to interfere. Since most rock types, soils, and sediments contain less than these amounts, these elements usually pose no difficulty.

The advantages and disadvantages of the X-ray fluorescence method are similar to those of the delayed neutron activation method. For the analysis of solid samples for total uranium, it can be an accurate method which is fairly rapid and precise. While not as sensitive as either the delayed neutron activation method or the fluorometric method it does nonetheless offer a detection limit of 1 ppm which is adequate for many geochemical applications and has a precision approaching that of the delayed neutron activation method. The disadvantages are that this method, like delayed neutron activation, does not readily lend itself to measuring partial and/or selectively extractable uranium in solid samples or to measuring uranium in natural waters at the low concentrations normally encountered in using these techniques. Again, it can be and has been done, but the time and expense associated with the procedures required to separate and/or concentrate the uranium sufficiently to provide adequate detection limits are simply not justified when other available methods are better suited for such purposes and less expensive on a cost/sample basis. Compared to other methods of uranium analysis, the capital investment required, if the cost of nuclear reactors is included, is higher than for any of the other methods under discussion. If reactor costs are excluded, it then becomes a comparatively inexpensive analytical system for high-precision and high-accuracy uranium measurement with the added bonus that it offers one of the best methods available for the determination of thorium and can of course be used for measuring most elements with atomic number greater than fluorine.

Fission Track Method

The use of fission tracks in uranium analysis is now well known but not yet widely applied. When an atom such as ^{235}U is induced to fission by bombardment with thermal neutrons in a reactor, the atom splits into two more or less equally massive smaller atoms or "fission fragments." These move apart at high velocities producing "tracks" of radiation damage in surrounding matter. The material of the track differs from that of the unaltered solid around it in various ways being, for example, more soluble. This leads to the phenomenon of track etching: if a surface through which fission tracks pass is exposed to an appropriate solvent, the tracks dissolve out as pits to sizes visible in the optical microscope. In practice a sample containing uranium with a normal abundance of ^{235}U is placed in contact with a plastic sheet — usually Lexan — and irradiated. The sheet is then treated to render the fission tracks visible. The number of fission tracks appearing on the surface of the Lexan depends only on the concentration of uranium in the samples and the number of neutrons, or neutron dose, which passed through the samples. By comparing the number of tracks produced by unknown samples against those produced by standards of known uranium concentration during the same irradiation, the concentration of uranium can be calculated by simple proportionality. To date, by far the greatest utilization of the fission track method is in the measurement of low-level uranium concentration in natural water samples.

The procedure for analyzing waters as employed by Bondar-Clegg & Company Ltd. (unpub. data) is as follows: water samples are filtered to separate any suspended matter or sediment present and then acidified to 0.2 M HNO_3. Using a micropipette, 5 μL of sample are placed on Lexan discs and dried, and stacked in a Plexiglas capsule liner along with standards prepared in the same way. When the liner is full, it is placed in a polyethylene irradiation capsule and irradiated in the Slowpoke reactor of AECL to give a neutron dose of 1.6×10^{16} neutrons cm^{-2}. Up to 2400 samples can be irradiated in one capsule. After irradiation, the discs are etched in 6 M NaOH and the tracks for both standards and unknowns are visually counted under a microscope. Calibration curves are constructed from the track counts for the standards irradiated along with the unknowns, and from these, the uranium content of the unknowns is determined.

The method gives a detection limit of 0.01 ppb. By increasing the neutron dose to 2×10^{17} neutrons cm^{-2}, the lowest measurable concentration of uranium is 0.003 ppb. At 10 times the detection limit (0.1 ppb) the theoretical precision of the method based on track counting statistics is ±15 per cent. The actual precision of the method under operating conditions has been studied by running many duplicate analyses. Figure 17.4 shows the results from two separate irradiations of duplicate analyses by the fission track method for a typical precision study.

Total and acid-extractable uranium in rocks and soils have been determined by the fission track method by Bondar-Clegg & Company Ltd. laboratories and good agreement with fluorometric data found. However, at this stage, because of its relative slowness compared to other methods, the fission track method does not offer any significant advantages to recommend its use for such analyses.

At its present state of development, the advantages offered by the fission track method can be summarized as follows: for the measurement of uranium in waters it is the most sensitive method available, with a precision comparable to that of the fluorometric method. It is unaffected by interfering elements. Assuming a reactor is available, the only significant cost involved is labour, as the capital equipment required is minimal. Additionally it requires samples of only a few millilitres, a significant advantage in reducing shipping charges when collecting many samples in remote regions. The principal disadvantage of the method is that with visual counting of tracks, the fission track method is slower and therefore more expensive on a cost/sample basis than conventional or laser-induced fluorescence methods. The reasons for this become obvious as indicated by Figure 17.5, which shows typical fission tracks as seen through a microscope from a sample containing 0.2 ppm U.

Various methods of instrumentally counting the etched fission tracks to speed the process have been investigated by a number of workers. One method of instrumentally counting the etched fission tracks is provided by the type of discharge counters utilized by Cross and Tommasino (1970) and is the approach taken by McCorkell and Yuan (1977). A prototype instrument, which is semiautomatic, has been built and operates as follows: the detector material bearing the tracks to be counted is a disc of Lexan whose thickness is less than the length of a fission track. Tracks therefore etch as holes through the detector. Detectors are pressed between two strips of aluminum-coated mylar, and a high voltage applied between these strips. Discharges take place through the etched tracks and these are counted on a scaler like those used for counting discharges in radiation detectors. The aluminum coating is removed from the mylar at the point of a discharge; therefore only one discharge takes place through each track and the number of discharges counted equals the number of tracks on the detector.

Early analyses of waters by the fission track method were checked quite closely and frequently by fluorometry and it was soon discovered that in some cases, puzzling discrepancies were found between uranium concentrations measured by fluorometry and those measured by fission track method. Investigation of these discrepancies has led to procedures for avoiding them but not to a complete explanation of the phenomenon. Two such cases involved analyses of river and lake waters collected from the Rabbit Lake area of northern Saskatchewan and the James Bay area

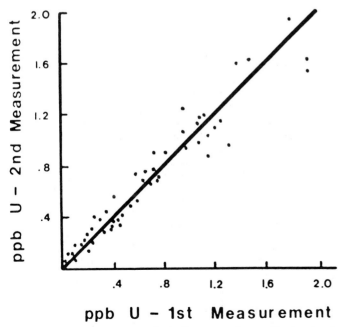

Figure 17.4. Precision of duplicate analyses, fission track method.

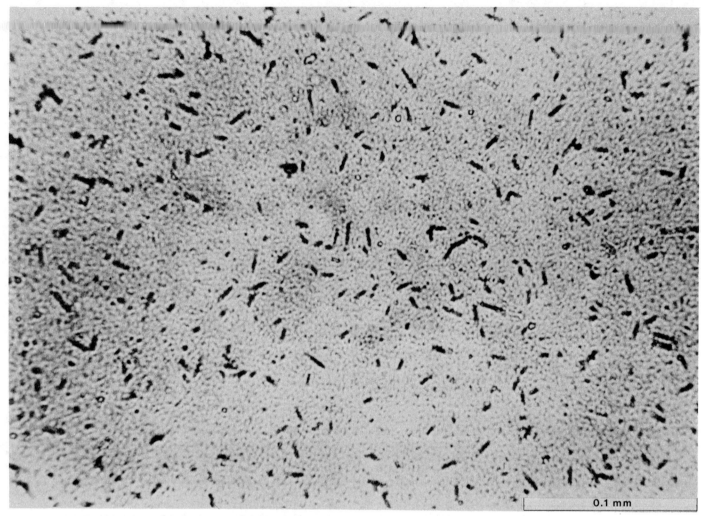

Figure 17.5. Fission tracks. (GSC 203492-A).

Table 17.6

Effect of filtration and acidification on some natural waters
fission track versus fluorometric method

Sample Source and Number		U — ppb			
		Before Acid + Filter		After Acid + Filter	
		Fluorometric	Fission Track	Fluorometric	Fission Track
Rabbit Lake Saskatchewan	1	0.08	ND	0.06	0.07
	2	0.14	0.02	0.15	0.11
	3	0.16	0.07	0.80	0.86
	4	0.27	0.07	0.28	0.23
	5	0.44	0.10	0.56	0.52
	6	4.8	6.7	3.5	3.9
James Bay Quebec	1	0.02	0.18	0.19	0.12
	2	0.07	ND	0.21	0.11
	3	0.35	0.09	0.45	0.34
	4	0.58	0.07	0.62	0.41
	5	0.70	0.03	0.19	0.20
	6	1.2	0.10	1.5	0.93

of northern Quebec. In both cases, analyses of the water samples as received in the laboratory by the fission track method gave results significantly lower than did fluorometric results. It was even more puzzling because the factor by which the fission track and fluorometric results differed varied so widely and because replicate fission track analyses agreed very well except, sometimes, when sediment was present in the samples.

As shown in Table 17.6, when these natural samples were filtered and acidified to 0.2 M HNO_3, the fission track method gave results which were in agreement with the fluorometric values. Moreover when known amounts of uranium were added to these samples, this uranium was correctly measured by the fission track method. Many sediment-free and suspended matter-free natural samples have shown good agreement between fission track and fluorometric results when not filtered and acidified, but no way to distinguish such samples before analysis has been found.

Another puzzling phenomenon observed was that some discrepancies were still found between fission track and fluorometric results for analyses of filtered and acidified artificially prepared standard solutions submitted along with the Rabbit Lake waters. The fission track results were low whereas the fluorometric results were in agreement with the intended concentrations. Known amounts of uranium added to the filtered and acidified standard solutions, however, gave the expected increase in uranium concentration by the fission track method when analyzed immediately.

A possible explanation for these phenomena is that in the unfiltered, unacidified samples, the uranium exists partly in the form of a suspended precipitate (probably a hydrated oxide with similar oxides of other elements and gelatinous organic matter). This may aggregate to varying degrees but being about the same density as the solutions and having almost no strength, it settles little and passes through filters. The 5 µL aliquot taken for fission track analysis may include none of these aggregates whereas the larger 25-50 mL samples taken for fluorometry may contain a representative portion of the aggregates. By acidifying after filtration, these aggregates are dissolved, thereby uniformly distributing uranium throughout the sample, making it representatively available in both the smaller 5 µL fission track aliquot and the larger aliquot used for fluorometry. Incomplete redissolving of the precipitates or aggregates may explain those cases where discrepancies still exist even after filtration and acidification as in the case of the artificially prepared standards just noted.

That some uranium in solution takes this postulated form was found by Reimer (1975). Pond water made 1 ppb in uranium and allowed to stand, produced track clusters on Lexan detectors that had been placed in the water and irradiated. These clusters extended over distances of 25-30 µm and did not appear to be radiating from a particle capable of passing through a 0.45 µm filter. However, the particle could not be removed by such a filter. It is probable that the particles are loose aggregates or gelatinous, and pass such filters by breaking up and reforming.

The explanation offered here to explain the discrepancies sometimes found between fission track and fluorometric methods of analysis is exceedingly *ad hoc* and needs further investigation. In such a comparison, when dealing at such low levels of concentration, one may not be comparing results by two different analytical methods so much as measuring uranium in different forms. The most important conclusion that can be drawn however, is that having now analyzed over 40 000 water samples for uranium by the fission track method, we can say that the fission track technique is an extremely sensitive method of analysis which gives results in agreement with those determined fluorometrically *if* the water samples are filtered and acidified.

Laser-induced Fluorescence

One of the most recent developments in geochemical analysis of uranium is the laser-induced fluorescence method of analysis developed by Scintrex Ltd. It is used primarily for the determination of uranium in natural waters. Like the conventional fluorometric method of uranium analysis, it is based on measurement of the green radiation emitted by uranyl salts under ultraviolet excitation by a suitable photodetector. Basically, the laser-induced fluorescence method of uranium analysis differs from the conventional fluorometric method in two ways. First, instead of measuring uranyl fluorescence in a solid fused disc, it is measured directly in an aqueous sample to which a proprietary reagent trade-named "FLURAN" is added. This reagent increases the sensitivity of direct fluorescence measurements in dilute uranyl solutions to the point where direct analysis of <1 ppb uranium in solution is possible. "FLURAN" also serves to mask the effects of fluorescence quenching agents. The second difference is that a pulsed nitrogen laser is used as the ultraviolet light source. Organic matter normally found in natural waters produces a brilliant but short-lived fluorescence. By applying a short, intense pulse of ultraviolet radiation from a laser source, the fluorescence contributed by uranium is isolated from that produced by organic species in the sample by measuring only those photodetector signals produced after fluorescence of the organic species has decayed to zero.

The principal advantages of the laser-induced fluorescence method of uranium analysis are its operational simplicity and small size which permits its use in the field, thereby providing all those benefits resulting from immediate availability of analytical data to field crews while surveys are in progress. Requiring only small samples, it allows direct, in-field analyses of uranium in natural waters with a sensitivity similar to that of conventional fluorometry and

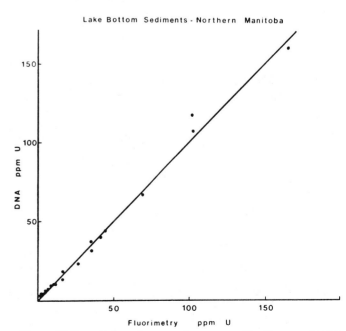

Figure 17.6. Delayed neutron activation versus fluorometry-lake bottom sediments. Fluorometric analyses by Bondar-Clegg; delayed neutron activation analyses by AECL.

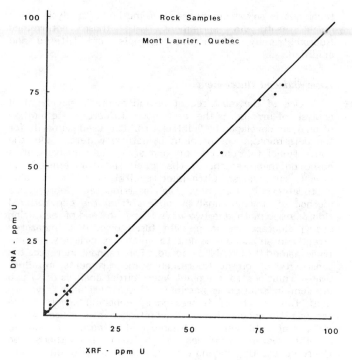

Figure 17.7. *X-ray fluorescence versus delayed neutron activation. XRF analyses by Bondar-Clegg; delayed neutron activation analyses by AECL.*

approaching that of the fission track method. With a moderate amount of additional laboratory apparatus, it can be used in the field to analyze solid samples for uranium as well. As with any newly introduced analytical method of instrumentation, the problems, limitations, or disadvantages do not become apparent until after the method or instrument has been available and in use for some time. Apart from some operational difficulties encountered in earlier designs of this instrument, now said to be corrected, the laser-induced fluorescence method of uranium analysis is still too new to enable any valid critical review of the method's deficiencies, if any.

Comparison of Various Methods of Uranium Analysis

How do the various methods of uranium analysis compare with one another in real situations? Figures 17.6 to 17.10 are scatter diagrams showing uranium concentrations in various types of samples as determined by different combinations of two analytical methods. The fluorometric, fission track and X-ray fluorescence analyses were performed by Bondar-Clegg and the delayed neutron activation analyses were performed by Atomic Energy Canada Ltd., Commercial Products Division. Laser-induced fluorescence data is from Scintrex Ltd. (Robbins, 1977).

Figure 17.6 shows HNO_3 extractable uranium by fluorometry versus total uranium by delayed neutron activation for a series of lake bottom sediments collected in northern Manitoba. Excellent correlation is shown over a concentration range of 1 to 160 ppm U with no evidence of quenching in the fluorometric method. At the lowest concentrations measured one can see that all data points fall very slightly on the DNA side of the line. This most certainly represents uranium tied up in silicates or some other resistate form that is not extracted by HNO_3. This component appears to be constant at about 0.5 ppm over that which is measured by fluorometry and as such, is not distinguishable at higher concentrations.

Figure 17.7 shows randomly selected delayed neutron activation check analyses of earlier X-ray fluorescence data for rock samples from the Mont Laurier area of northwestern Quebec. Total uranium is measured by both methods and good correlation is shown over a concentration range of 1 to 111 ppm U.

In Figure 17.8, from Robbins (1977), laser-induced fluorescence versus fluorometric results for measurements of HNO_3 extractable uranium in rocks are shown. Good correlation over a concentration range of 1 to 3000 ppm is shown.

Figure 17.9 shows fission track results versus fluorometric results for water samples over a concentration range of 0.01 to 1.3 ppb U. The correlation shown is relatively good for this low concentration range.

Results for fluorometric versus laser-induced fluorescence measurements of uranium in waters over a concentration range of approximately 0.2 ppb to about 80 ppb are shown in Figure 17.10. Except for three samples below 1 ppb good agreement between the two methods is shown over a concentration range of 1-100 ppb.

Having made these comparisons, can we draw any conclusions? Except where the uranium concentration is less than 1 ppb, results of these comparison studies appear to suggest that any one of the analytical methods described here could be selected and would provide the exploration geochemist with suitable analytical data of adequate accuracy and precision. This is not always the case! Consider the following true case: one is looking for uranium deposits formed by secondary concentration of uranium minerals precipitated in a reducing environment from groundwater which has accumulated uranium from a basement source.

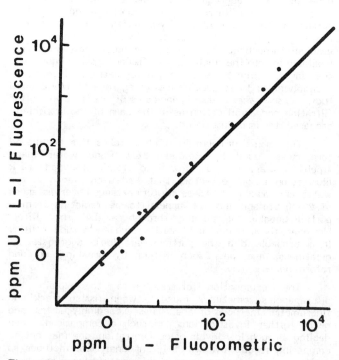

Figure 17.8. *Laser-induced fluorescence versus fluorometric rock and ore samples.*

Geochemical Analysis

Bedrock geochemistry and ground radiometric surveys are being used to test favourable sedimentary environments. Two anomalous areas are found. Both show anomalous radiometric eU values and both are geochemically anomalous in total uranium, as determined by X-ray fluorescence analysis of composited bedrock chip samples. If one anomaly conforms to the model and the other does not, how are they distinguished? In this case it was done by analytical differentiation.

Table 17.7 shows typical results of partial extractions carried out on bedrock samples previously analyzed for total U by X-ray fluorescence, in an attempt to determine those anomalies which did not conform. Comparative results from over a known occurrence, which is ore grade in outcrop and which corresponds to the model used, are shown for comparative purposes. Obviously Anomaly I supports the concept while Anomaly II does not. In this case the fluorometric method using a 2 per cent $Na_2CO_3/5\%$ H_2O_2 extraction, which can be considered almost selective in extracting uraninite/pitchblende from bedrock samples, was most effective in detecting those anomalies of interest. Used in conjunction with a total U measurement the method was also effective distinguishing anomalies of interest from those that are not of interest. In this example, had only HNO_3 extractable uranium or Na_2CO_3/H_2O_2 extractable uranium been measured by fluorometry, the important anomaly would still have been detected. With only a total U measurement, the important anomaly would have been detected, but so were a number of others which subsequently proved of no interest. Proving that an anomaly is of no interest is often more costly than finding one that is. Considering the original concept, selection of an analytical method readily adaptable to measuring HNO_3 extractable, or in this case preferably Na_2CO_3/H_2O_2 extractable uranium in the first place, would have better achieved the objectives of the geochemical survey at lower cost. Note use of the word "readily" because all of the methods described here can be used to measure total uranium and all can be used to measure partially extractable uranium. Some methods just happen to do one or the other a lot easier than some other methods and at lower cost.

Table 17.7

Evaluation of anomalies using total uranium versus extractable uranium

	Radiometric Results	Geochemical Results — ppm U		
		Total U — XRF	Fluorometric HNO_3 — Extractable	Fluorometric Na_2CO_3/H_2O_2 Extractable
Anomaly I	High eU	32	26	16
Anomaly II	High eU	32	4	0.2
Known Occurrence	Very High eU	144	94	45

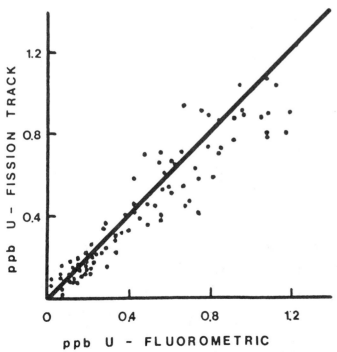

Figure 17.9. Fluorometric versus fission track; water samples from Gatineau Park and James Bay areas, Quebec (McCorkell and Yuan, 1977).

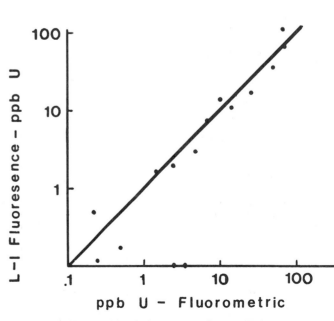

Figure 17.10. Laser-induced fluorescence versus fluorometric method; natural waters from eastern Canada (from Robbins, 1976).

How does one choose the method of geochemical analysis for uranium that best satisfies his needs? Any method selected must satisfy the following four requirements:

1. Must be capable of measuring the desired uranium fraction in a sample at the background concentrations encountered in samples collected.
2. Precision and accuracy must be adequate to recognize significant uranium concentration patterns and changes in these patterns.
3. Turnaround time for reporting results must be reasonable.
4. Cost must be acceptable.

All five methods described in this report can be summarized on the basis of criteria which should influence any such choice intended to satisfy the preceding four requirements. This summary, shown in Table 17.8, provides comparisons for nine criteria upon which a decision might be made to select or to establish an analytical method for geochemical uranium analysis.

Excluded from Table 17.8 are those methods described in the literature (low ppb measurement of U in waters by DNA or XRF for example) which depend on preconcentration of uranium by chelation/solvent extraction, ion-exchange, coprecipitation, or evaporation of large volumes of water, and so forth, to achieve useful detection limits. Such procedures of variable complexity are slow, and with each extra step, additional chances of error are introduced, as are unavoidable losses. Given a choice, direct methods are almost always to be preferred if appropriate sensitivity can be achieved. All detection limits given in Table 17.8 are, with one exception, detection limits for direct measurement of uranium by the methods shown. The one exception is for uranium measurements in water by fluorometry. Since the specified detection limit of 0.05 ppb is readily attained by evaporative concentration of only 20 mL of water and because this can be done simply and rapidly with hundreds of samples at the same time, this one exception was made.

Table 17.8 is self-explanatory and the significant points can be summarized as follows:

(i) Delayed neutron activation provides the best combination of precision and sensitivity for low levels of total uranium measurement, followed by X-ray fluorescence. The conventional fluorometric method is the method best suited for low-level U analyses requiring partial and/or selective uranium extraction procedures for solid samples.

(ii) There is little to choose between fluorometry, fission track and laser-induced fluorescence methods for uranium in natural waters unless extremely low detection limits are required, in which case the fission track method is preferred. The fluorometric method is generally faster than either fission track or laser-induced fluorescence analyses in waters.

Table 17.8

Comparison of five methods of uranium analysis

	Fluorometric	DNA	XRF	Fission Track	Laser-Induced Fluorescence
Detection Limit					
Solids (ppm)	0.1	0.1	1.0	—	—
Waters (ppb)	0.05	—	—	0.01	0.05
Precision (@ 10 x det. limit)	±25%	±10-15%	±15%	±20-25%	±15%*
Productivity analyses/man/hour	20	27	27	13	13-25
Sample size req'd.					
Solids (gm)	0.25	1	5	—	0.05
Waters (mls)	20	—	—	0.005	7
Dependence on analyst's care and skill	High	Low	Low	Moderate	Low
Simple differentiation of total/partial uranium	Yes	No	No	Yes	Yes
Adaptable for field use	Possible	No	No	No	Yes
Approx. capital cost of complete system	$5000	$40 000 + reactor	$140 000.	$1500 + reactor	$15 000
Analytical cost/sample**					
Solids	$2.50-$3.50	$3.00-$15.00***	$2.25-$3.00	—	Depends on users operating cost
Waters	$2.00-$4.00	—	—	$3.75	

*Detection limit for laser-induced fluorescence stated @ twenty x detection limit
**Approximate range of charges by North American commercial laboratories
***DNA cost/sample highly variable because of great variation in charges by various reactor facilities.

(iii) The conventional fluorometric technique is the most flexible analytical method for uranium in solid samples in that it can be easily used to analyze solid samples for total as well as partially extractable uranium unlike DNA or XRF. This capability has not yet been as well established for either the fission track method or the laser-induced fluorescence methods which are comparatively recent developments and whose limitations are therefore not yet fully understood.

(iv) For in-the-field measurements of uranium in natural waters, the laser-induced fluorescence method should be preferred because it has been designed as a truly portable analytical system for field use. Also, it is not as dependent on the operator's skill as is the fluorometric method which is highly dependent upon the care and skill of the analyst, a factor which may partially offset the laser-induced fluorescence system's capital cost. If the primary function of the field laboratory is to analyze soils, rocks or sediments, the fluorometric method is preferred.

(v) Capital cost required to establish an analytical system is usually, but not always, a direct function of the accuracy and precision of that analytical system. As seen in this report the most precise method is the neutron activation delayed counting method but this also requires the greatest capital expenditure to establish.

(vi) Cost of analyses per sample for any given method can be quite variable as shown in this comparison. To paraphrase G.E.F. Lundell, who originally made the following still valid points in 1933, at the 85th Meeting of the American Chemical Society, "Purchasers of analytical services usually get what they pay for and as a rule, are unwilling to pay very much. Since the buyer of analytical services is buying an intangible commodity, in fact an opinion, why pay five dollars when someone else is quoting fifty cents? A very good reason is that the correctness of the opinion cannot be checked except by buying other opinions, which is an expensive as well as a poor way to settle an argument."

This comparison is a summary of the general characteristics and capabilities of the five methods of uranium analyses described. One should not expect that all laboratories employing any of the methods described will, for example, attain the detection limits, precision or productivity figures shown here. Conversely, some laboratories may exceed these figures. This comparison does however, show what is attainable using any of the methods described and what an exploration geochemist should generally expect, as state-of-the-art for routine uranium analysis of geochemical samples.

Often, as we have shown by comparing one analytical method against another in a number of real situations, no obvious clear-cut preference emerges for one analytical method over another. Usually the quality or usefulness of analytical data will depend more on the analyst who performs the analysis than on the analytical method used, or as Carpenter (1972) so succinctly stated in describing the results of a trace element intercalibration study involving several different laboratories and several different analytical methods, "Equally disastrous results were achieved with a few thousand dollars worth of equipment as with devices costing millions of dollars."

SELECTED REFERENCES

Ahrens, L.H.
1950: Spectrochemical Analysis; Addison-Wesley Press Inc., Cambridge, 267 p.

Amos, M.D., Bennett, P.A., Brodie, K.G., Lung, P.W.K., and Matousek, J.P.
1971: Carbon rod atomizer in atomic absorption and fluorescence spectrometry and its clinical application; Anal. Chem., v. 43, p. 211-215.

Aslin, G.E.M.
1976: The determination of arsenic and antimony in Geological Materials by flameless atomic absorption spectrometry; J. Geochem. Explor., v. 6, p. 321-330.

Bond, A.M.
1970: Direct current, alternating current, rapid and inverse polarographic methods for determination of tin (IV); Anal. Chem., v. 42, p. 1165-1168.

Boulanger, A., Evans, D.J.R., and Raby, B.F.
1976: Uranium analysis by neutron activation delayed neutron counting; a paper presented at Canadian Mineral Analysts Conference, Thunder Bay, 1975.

Canney, F.C. and Hawkins, D.B.
1958: Cold acid extraction of copper from soils and sediments — a proposed field method; Econ. Geol., v. 53, p. 877-886.

Canney, F.C., Myers, A.T., and Ward, F.N.
1957: A truck-mounted spectrographic laboratory for use in geochemical exploration; Econ. Geol., v. 52, p. 289-306.

Carpenter, J.H.
1972: Problems in applications of analytical chemistry to oceanography; in Analytical Chemistry: Key to Progress on National Problems, National Bureau of Standards, Spec. Pub. 351, p. 393-419.

Chu, R.C., Barron, G.P., and Baumgarner, P.A.W.
1972: Arsenic determination at sub-microgram levels by arsine evolution and flameless atomic absorption spectrophotometric technique; Anal. Chem., v. 44, p. 1476-1479.

Copeland, T.R. and Skogerboe, R.K.
1974: Anodic stripping voltammetry; Anal. Chem., v. 46, p. 1257A-1268A.

Cross, W.G. and Tommasino, L.
1970: A rapid reading technique for nuclear particle damage tracks in thin foils; Rad. Eff., v. 5, p. 85-89.

Curtin, G.C., Lakin, H.W., Neuerburg, G.J., and Hubert, A.E.
1968: Utilization of humus-rich forest soil (mull) in geochemical exploration for gold; U.S. Geol. Surv., Circ. 562, 11 p.

Danielsson, A. and Lindblom, P.
1972: An echelle spectrograph for image tubes; Physica Scripta, v. 5, p. 227-231.

Danielsson, A., Lindblom, P., and Södermann, E.
1974: Image dissector echelle spectrometer system for spectrochemical analysis; Chemica Scripta, v. 6, p. 5-9.

Dickinson, G.W. and Fassel, V.A.
: 1969: Emission spectrometric detection of the elements at the nanogram per milliliter level using induction-coupled plasma excitation; Anal. Chem., v. 41, p. 1021-1024.

Donega, H.M. and Burgess, T.E.
: 1970: Atomic absorption analysis by flameless atomization in a controlled atmosphere; Anal. Chem., v. 42, p. 1521-1524.

Dulka, J.J. and Risby, T.H.
: 1976: Ultratrace metals in some environmental and biological systems; Anal. Chem., v. 48, p. 640A-653A.

Durst, R.A.
: 1969: Analytical technique and applications of ion-selective electrodes in Ion-Selective Electrodes; Durst, R.A., ed.: National Bureau of Standard's, Spec. Publ. 314, 452 p.

Eisenman, G.
: 1969: Theory of membrane electrode potentials: An examination of the parameters determining the selectivity of solid and liquid ion exchangers and of neutral ion-sequestering molecules; in Ion-Selective Electrodes, National Bureau of Standards, Spec. Publ. 314, p. 1-56.

Farrell, B.L.
: 1974: Fluorine, a direct indicator of fluorite mineralization in local and regional soil geochemical surveys; J. Geochem. Explor., v. 3, p. 227-244.

Fassel, V.A. and Kniseley, R.N.
: 1974a: Inductively-coupled plasma-optical emission spectroscopy; Anal. Chem., v. 46, p. 1110A-1120A.
: 1974b: Inductively-coupled plasmas; Anal. Chem., v. 46, p. 1155A-1164A.

Fersman, A.E.
: 1952: Geochemical and mineralogical methods of prospecting for mineral deposits; Chap. IV, Special Methods of Prospecting: U.S. Geol. Surv. Circ. 127, 37 p. (Transl. from Russian.) Originally published in 1939 as Geokhimicheskie i mineralogicheskie metody poiskov poleznykh isokopaemykh: Izd. Akad. Nauk, SSSR.

Ficklin, W.H.
: 1970: A rapid method for the determination of fluoride in rocks and soils, using an ion-selective electrode; U.S. Geol. Surv. Prof. Paper 700-C, p. C186-C188.

Garrett, R.G. and Lynch, J.J.
: 1976: A comparison of neutron activation delayed neutron counting versus fluorometric analysis in large-scale geochemical exploration for uranium; in Exploration for Uranium Deposits, IAEA, Vienna, p. 321-334.

Goulden, P.D. and Brooksbank, P.
: 1974: Automated atomic absorption determination of arsenic, antimony, and selenium in natural waters; Anal. Chem., v. 46, p. 1431-1436.

Greenfield, S., Jones, I.L., and Berry, C.T.
: 1964: High-pressure plasmas as spectroscopic emission sources; Analyst, v. 89, p. 713-720.

Griffitts, W.R., Ward, F.N., and Alminas, H.V.
: 1976: A simple spot test for molybdenum minerals; Econ. Geol., v. 71, p. 1595.

Hadeishi, T. and McLaughlin, R.D.
: 1975: Isotope Zeeman atomic absorption — a new approach to chemical analysis; Am. Lab., v. 7, p. 57-61.

Harrison, G.R.
: 1949: The production of diffraction gratings, II, The design of echelle gratings and spectrographs; J. Opt. Soc. America, v. 39, July 1949.

Hawkes, H.E.
: 1972: Exploration geochemistry bibliography, Jan. 1965-Dec. 1971; Assoc. Explor. Geochem., Spec. Vol. No. 1, 118 p.
: 1976: Exploration geochemistry bibliography, Jan. 1972-Dec. 1975; Assoc. Explor. Geochem., Spec. Vol. No. 5, 195 p.

Hillebrand, W.F.
: 1919: The analysis of silicate and carbonate rocks; U.S. Geol. Surv., Bull. 700, 285 p.

Holak, W.
: 1969: Gas-sampling technique for arsenic determination by atomic absorption spectrophotometry; Anal. Chem., v. 41, p. 1712-1713.

Holman, R.H.C.
: 1956: A method for determining readily-soluble copper in soil and alluvium — introducing white spirit as a solvent for dithizone; Inst. Min. Metall. Trans., v. 66, p. 7-16.

Hwang, J.Y., Ullucci, P.A., and Smith, S.B., Jr.
: 1971: A simple flameless atomizer; Am. Lab., August, p. 41-43.

Ingles, J.C.
: 1958: A manual of analytical methods for the uranium laboratory, method U.1; in Mines Br., Mono. Ser. 866, Ottawa.

Ingram, B.L.
: 1970: Determination of fluoride in silicate rocks without separation of aluminum using a specific ion electrode; Anal. Chem., v. 42, p. 1825-1827.

Johnson, D.J., Plankey, F.W., and Winefordner, J.D.
: 1975: Multielement analysis via computer-controlled rapid-scan atomic fluorescence spectrometer with a continuum source; Anal. Chem., v. 47, p. 1739-1743.

Karasek, F.W.
: 1975: Detection limits in instrument analysis; Research and Development, July, p. 20-24.

Keliher, P.N. and Wohlers, C.C.
: 1974: High resolution atomic absorption spectrometry using an echelle grating monochromator; Anal. Chem., v. 46, p. 682-687.

Kesler, S.E., Van Loon, J.C., and Bateson, J.H.
: 1973: Analysis of fluoride in rocks and an application to exploration; J. Geochem. Explor., v. 2, p. 11-17.

King, A.S.
: 1908: The production of spectra by an electrical resistance furnace in hydrogen atmosphere; Astrophysical J., v. 27, p. 353-362.
: 1932: Lines of tungsten and rhenium appearing in the spectrum of the electric furnace; Astrophysical J., v. 75, p. 379-385.

Koizumi, H. and Yasuda, K.
: 1975: New Zeeman method for atomic absorption spectrophotometry; Anal. Chem., v. 47, p. 1679-1682.

Koizumi, H. and Yasuda, K. (cont'd.)
  1976: Determination of lead, cadmium and zinc using the Zeeman effect in atomic absorption spectrometry; Anal. Chem., v. 48, p. 1178-1182.

Koizumi, H., Yasuda, K., and Katagama, M.
  1977: Atomic absorption spectrophotometry based on the polarization characteristics of the Zeeman effect; Anal. Chem., v. 49, p. 1106-1112.

Lengyel, B.A. and Blum, E.
  1934: The behavior of the glass electrode in connection with its chemical composition; Trans. Faraday Soc., v. 30, p. 461-471.

L'vov, B.V.
  1961: The analytical use of atomic absorption spectra; Spectrochim. Acta, v. 17, p. 761-770.

Massman, H.
  1968: Comparison of atomic absorption and atomic fluorescence in graphite cuvettes; Spectrochim. Acta, v. 23B, p. 215-226.

McCorkell, R.H. and Yuan, D.Y.H.
  1977: Discharge counter for the determination of uranium in water by the fission track method; Rev. Sci. Instrum., v. 48, no. 8, p. 1005-1009.

Orion Research Inc.
  1973: Analytical methods guide; 5th Ed., 32 p.

Osteryoung, R.A. and Christie, J.H.
  1974: Theoretical treatment of pulsed voltammetric stripping at the thin film mercury electrode; Anal. Chem., v. 46, p. 351-355.

Palmqvist, S. and Brundin, N.H.
  1939: Svenska prospekterings aktiebologet, P.M., angaenda var. geokemiska prospecterings method, Lund. See Brundin, Nils, 1939, Method of locating metals and minerals in the ground, U.S. Pat. 2158980.

Perone, S.P. and Kretlow, W.J.
  1965: Anodic stripping voltammetry of mercury (II) at the graphite electrode; Anal. Chem., v. 37, p. 968-970.

Pierce, F.D., Lamoreaux, T.C., Brown, H.R., and Fraser, R.S.
  1976: An automated technique for the sub-microgram determination of selenium and arsenic in surface waters by atomic absorption spectroscopy; Appl. Spectroscopy, v. 30, p. 38-42.

Pollock, E.N. and West, S.J.
  1973: The generation and determination of covalent hydrides by atomic absorption; Atomic Absorption Newsletter, v. 12, p. 6-8.

Pungor, E. and Troth, K.
  1970: Ion-selective membrane electrodes; Analyst, v. 95, p. 625-648.

Ratsbaum, E.A.
  1939: Field spectroanalytical laboratory for supporting prospecting parties; Razvedka Nedr., No. 1, p. 38-41.

Rechnitz, G.A.
  1967: Ion-selective electrodes; Chem. Eng. News, p. 146-158.

Reimer, G.M.
  1975: Uranium determination in natural water by the fission-track technique; J. Geochem. Explor., v. 4, p. 425-431.

Richardson, D.
  1953: The use of echelles in spectroscopy; Spectrochim. Acta, v. 6, p. 61-65.

Robbins, J.
  1977: Direct analysis of uranium in natural waters; Scintrex Ltd., Toronto, Application Brief, 26 p.

Rosencwaig, A.
  1975: Photoacoustic spectroscopy, A new tool for investigations of solids; Anal. Chem., v. 47, p. 592A-604A.

Sandell, E.B.
  1959: Colorimetric determination of traces of metals; 3rd ed., Interscience Publ., New York, 1054 p.

Schwabe, K. and Dahms, H.
  1960: Investigation of the ion-exchange between a glass electrode and an alkaline solution through the use of radioisotopes; Naturwissenschaften, v. 47, p. 351-352.

Scott, R.H., Fassel, V.A., Kniseley, R.N., and Nixon, D.E.
  1974: Inductively-coupled plasma-optical emission analytical spectrometry; Anal. Chem., v. 46, p. 75-80.

Slavin, S., Barnett, W.B., and Kahn, H.L.
  1972: The determination of atomic absorption detection limits by direct measurements; Atomic Absorption Newsletter, v. 11, p. 37-41.

Thompson, K.C. and Thomerson, D.R.
  1974: Atomic-absorption studies on the determination of antimony, arsenic, bismuth, germanium, lead, selenium, tellurium, and tin by utilizing the generation of covalent hydrides; Analyst, v. 99, p. 595-601.

Timperley, M.H.
  1974: Direct-reading d.c. arc spectrometry for rapid geochemical surveys; Spectrochim. Acta, v. 29B, p. 95-110.

Veillon, C. and Merchant, P., Jr.
  1973: High resolution atomic absorption spectrometry with a scanning Fabry-Perot interferometer; Appl. Spectroscopy, v. 27, p. 361-365.

Wendt, R.H. and Fassel, V.A.
  1965: Induction-coupled plasma excitation source; Anal. Chem., v. 37, p. 920-922.

West, T.S. and Williams, X.K.
  1969: Atomic absorption and fluorescence spectroscopy with a carbon filament atom reservoir, I, construction and operation of reservoir; Anal. Chem. Acta, v. 45, p. 27-41.

Willard, H.H., Merritt, L.L., Jr., and Dean, J.A.
  1974: Instrumental methods of analysis; 5th ed., D. Van Nostrand Co., New York, 860 p.

Winefordner, J.D. and Staab, R.A.
  1964: Determination of zinc, cadmium, and mercury by atomic fluorescence flame spectrometry; Anal. Chem., v. 36, p. 165-168.

Winefordner, J.D. and Vickers, T.J.
  1964: Atomic fluorescence spectrometry as a means of chemical analysis; Anal. Chem., v. 36, p. 161-165.

Woodriff, R. and Ramelow, G.
  1968: Atomic absorption spectroscopy with a high-temperature furnace; Spectrochim. Acta, v. 23B, p. 665-671.

ADVANCES IN BOTANICAL METHODS OF PROSPECTING FOR MINERALS
PART I – ADVANCES IN GEOBOTANICAL METHODS

Helen L. Cannon[1]

U.S. Geological Survey, Denver, Colorado

Cannon, Helen L., Advances in botanical methods of prospecting for minerals. Part I – Advances in geobotanical methods; in Geophysics and Geochemistry in the Search for Metallic Ores; Peter J. Hood, editor; Geological Survey of Canada, Economic Geology Report 31, p. 385-395, 1979.

Abstract

The presence or absence of particular species or varieties of plants in mineralized areas, and the effects of metals on plant growth have been observed and used in the search for concealed ore bodies since the 8th century. In the last ten years, studies of sparsely vegetated areas in wooded country have led to the discovery of lead in Norway and copper in the United States. Botanists have recently observed the actual evolution under stress conditions of new subspecies in mineralized or metal-contaminated ground, and there is now a growing understanding of metal-tolerance mechanisms in various plants. The development of new, highly tolerant races of plants in metal-poisoned ground is much more rapid than was previously supposed. Plants also may rid themselves of metal by dying to the ground each year or by concentrating metal in root cell walls and subsequently growing new adventitious roots. In widely separated areas of Europe, distinctive plant communities have been found to characterize terrain that has an anomalously high content of specific metals. New nickel accumulators have been identified in many countries, and a study by R.R. Brooks of herbarium specimens of previously reported indicator plants has shown many to be true accumulators of specific metals. These and other advances indicate the continuing usefulness of geobotanical methods of prospecting.

Résumé

Dès le VIIIe siècle, on a constaté la présence ou l'absence de certaines espèces ou variétés de plantes dans certaines zones minéralisées, et observé l'effet des éléments métalliques sur la croissance des végétaux, et enfin, utilisé ces observations pour déceler des corps minéralisés enfouis. Au cours des dix dernières années, l'étude de zones de végétation maigre dans une région boisée ont abouti à la découverte de gisements de plomb en Norvège, et de gisements de cuivre aux #tats-Unis. De fait, récemment, des botanistes ont observé, dans un sol minéralisé ou contaminé par des éléments métalliques, l'apparition dans des conditions de stress de nouvelles sous-espèces, et l'on commence à mieux comprendre les mécanismes de tolérance vis-à-vis des métaux. L'apparition de nouvelles races de plantes caractérisées par un niveau élevé de tolérance dans un sol empoisonné par des métaux est beaucoup plus rapide qu'on ne le supposait auparavant. Il peut aussi y avoir élimination progressive du métal, si les plantes meurent chaque année; elles peuvent aussi concentrer les éléments métalliques dans les tissus radiculaires, avant d'acquérir de nouvelles racines adventives. On a constaté, que dans des régions d'Europe très distantes les unes des autres, des communautés végétales distinctives s'étaient établies sur des terrains caractérisés par une teneur anormalement élevée en certains métaux. Dans de nombreux pays, on a identifié de nouvelles plantes concentratrices de nickel, et R.R. Brooks, qui a étudié des spécimens botaniques de plantes indicatrices déjà signalées a démontré que plusieurs d'entre elles concentraient réellement certains métaux. Ces découvertes, parmi d'autres, mettent l'accent sur l'utilité à long terme des méthodes géobotaniques de prospection.

La publication de cet article a été approuvée par le directeur de l'U.S. Geological Survey.

INTRODUCTION

Geobotanical prospecting has not been popular in the geological world because geologists have the mistaken idea that plant relationships in mineralized areas can be seen only by a botanist and that it is necessary to know the whys and wherefores of the distribution of species to be able to use them. Neither of these assumptions is correct. One need not know the exact name of a plant or why it grows where it grows to be able to use it. It is necessary, however, to be able to recognize the species of plants that appear to be most commonly restricted to metal-rich soils wherever they grow, and to observe in detail their distribution in relation to others of the local plant society in relation to the rocks. A plant's distribution may be useful even though it is controlled by a pathfinder element rather than by the most economically valuable element of the suite. The use of geobotany in prospecting through 1971 has been reviewed by Brooks (1972) and by me through 1965 in a previous paper (Cannon, 1971). In the last ten years new indicator plants have been reported, and new uses have been found for old ones. The association of bare areas with mineralized ground has been used for prospecting for lead in Norway and for copper in the northern United States. During this time, great strides have been made in understanding the factors that affect metal tolerance in various plant groups and the actual evolution, under stress conditions, of new varieties or subspecies in mineralized or metal-contaminated ground. We owe thanks for these studies to botanists such as Wilfried Ernst of Germany, H. Wild, G.H. Wilshire and Clive Howard-Williams of Rhodesia, Paul Duvigneaud of Belgium, T. Jaffré of New Caledonia, and Arthur Kruckeberg of the United States.

HISTORY

A short look at the early history of botanical methods of prospecting may help in setting the stage for a discussion of present research. The Chinese are reported to have observed the association of certain plant species with mineral deposits at least as early as the 8th or 9th century and to have been aware of metal uptake. Agricola in 1556 published observations concerning physiological effects of metals on vegetation (Boyle, 1967). It is reported that Thalius noted the

---

[1] Approved for publication by the Director, U.S. Geological Survey

association of **Minuartia verna** with metalliferous soils as early as 1588 (Ernst, 1965). The use of plants in prospecting was described in some detail by Barba (1640) in his book on "Methods of Prospecting, Mining and Metallurgy," which was completed in Potosi, Mexico in 1637 and published in Spain in 1640. Barba states, "Certain trees, marsh plants, and herbs are sometimes indicators of veins. Of these are plants of one type, which appear to be planted on a line; they repeat on the surface the course of the underground vein. Plants growing on top of a metallic vein are smaller and do not show their usual lively colors. This is due to the exhalations of the metals. They injure the plants, which seem emaciated." Lomonosov also noticed the depauperating effects of mineralized soils on plants in 1763 (Malyuga, 1964).

In the 19th century, seven indicator plants were reported in the literature. Of these, three have been confirmed as indicators of mineralized ground: **Polycarpea spirostylis**, discovered by Bailey (1889) on copper soils in Australia; **Viola lutea**, reported by Raymond (1887) as an indicator of zinc deposits in Aachen, Germany; and **Eriogonum ovalifolium**, reported as a silver indicator in Montana by Lidgey (1897). Their distribution and accumulative powers are being studied in detail in these areas today. From 1900 through 1965, more than 100 species indicative of one or more of 24 elements were reported, and in many cases, analyses showing them to be accumulators of the element were also given. Since 1965 many new indicator-accumulator plants for nickel have been reported, and a few for other metals. Much research has been reported on the mechanisms of metal tolerance and the evolution of metal tolerant species.

PLANT TOLERANCE FOR METALLIFEROUS SOILS

Many plants are unable to grow in strongly mineralized soils; such soils commonly have a reduced flora or may, in extreme cases, be entirely bare. In these relatively open areas, there may be an abrupt change in the degree of sunlight, soil moisture, soil temperature, or drainage. The lack of competition may permit the continued growth of relict stands of species that formerly had a wider range, or seeds of tolerant species or varieties from elsewhere may be able to germinate and grow in these areas that are so highly toxic to average plants.

Two types of plants have been shown to be tolerant of highly metalliferous soils. The first type, which includes many indicator plants, is capable of accumulating large amounts of metal in the foliage without excessive harm to the plant. These plants, as reported by Wild (1968) in Rhodesia, concentrate the metal in the leaves, but generally have heavy perennial rootstalks, and die to the ground in dry or cold periods of the year, thus sloughing off a considerable portion of the absorbed metal. The second type, represented most commonly by grasses, tolerates mineralized soils by preventing the toxic element from concentrating in the aerial parts of the plant, either by retention of metals in the root or by a true exclusion of metals at the root surface, possibly owing to a low root cation exchange capacity that permits the entry of monovalent cations but rejects divalent cations (Duvigneaud and Denaeyer-De Smet, 1973). The aerial parts of the second type of plants may contain the same amount of the toxic element regardless of whether it is growing on mineralized or barren ground. The number of such plants may be increased in mineralized areas owing to a lack of competition and their high tolerance for metals.

In metalliferous soils there are several factors that affect the absorption and uptake of metals. First, the absorption depends upon the available metal rather than the total metal in the soil. Varying amounts of metal will be water soluble, exchangeable, or organically bound. Factors that control the availability of metals in mineralized soil are largely unknown. David Crimon (pers. commun., 1975) found that certain Montana soils in areas that support restricted and unusual plant communities contain more EDTA-extractable copper (a measure of available copper) than soils with normal vegetation. Second, an increase in the availability of major plant nutrients such as phosphate or calcium in mineralized soils of low pH commonly affects the absorption of metals. Third, ore-associated elements, such as iron, sulphur, arsenic, cadmium, or selenium, may be deterrents or stimulants to plant growth or may interact with the more abundant metals to decrease or increase their toxic effects. Fourth, there are physiologic differences between plant species, races of the species, and perhaps individual plants that affect metal tolerance and uptake. For instance, the chelated form in which a metal is transported is unique to each species. Finally, the characteristic pH and total ion concentration of cell sap may determine whether a plant can live in a given soil environment and how much metal it can absorb. Plants adapted to a soil with a high cation content generally have a higher total cytoplasmic ion concentration than do plants that are intolerant of mineralized soils.

The location of metals in the cell is also closely associated with metal tolerance. The cell sap (vacuole system) of leaves has been shown to contain copper, zinc, iron, nickel, manganese, and a small amount of lead, but no cobalt or chromium. Bradshaw (1970b) has demonstrated that the greater percentages of copper and zinc occur in the cell wall of tolerant and accumulator plants, and in the mitochondrial fraction in nontolerant plants. Ernst (1972) proposed that the cell walls of metal tolerant plants have a high exchange capacity for heavy metals, and that the older

Table 18.1

Species with metal tolerant races
(taken from Antonovics et al., 1971)

Species	Metal(s) tolerated
GRASSES:	
Agrostis tenuis	Cu, Pb, Ni, Zn
Agrostis stolonifera	Pb, Zn
Agrostis canina	Pb, Zn
Anthoxanthum odoratum	Zn
Festuca ovina	Pb, Zn
Festuca rubra	Zn
Holcus lanatus	Zn
FORBS:	
Alsine (Minuartia) verna	Zn
Armeria maritima	Zn
Campanula rotundifolia	Zn
Linum catharticum	Zn
Melandrium silvestre	Cu
Plantago lanceolata	Zn
Rumex acetosa	Cu, Zn
Silene vulgaris (inflata, cucubalus)	Zn
Taraxacum officinale	Cu
Thlaspi alpestre	Zn
Tussilago farfara	Cu
Viola lutea	Zn

leaves die off as the sites in the cell wall are used up, thus continuously ridding the plant of toxic metals. In the root, most metals are tightly bound in the cortex, but are water soluble in the xylem, where they are available for transport to the upper parts of the plant. Chromium, however, is tightly bound in both the cortex and xylem and remains in the root. Lead is mostly bound in the cortex, but a part is water soluble and available for transport Ernst (1972).

Peterson (1969), using radioactive ^{65}Zn, determined that more than 80 per cent of the Zn absorbed by zinc-tolerant plants of **Agrostis tenuis** separates in the pectate extract. The cation exchange capacity of pectin, a cell wall component, would enable the pectin to remove zinc from the cytoplasm by binding it to the cell wall. Work by Miller et al. (1975) has shown that high lead uptake by soybeans is inversely proportional to the pH and to the soil cation exchange capacity. Timed sequence studies with the electron microscope showed that insoluble amorphous masses of a lead complex are first formed in dictyosome vesicles which then move to and are incorporated in the cell wall of corn roots. Crystals of lead phosphate were also observed to form on the outside of the root (Malone et al., 1974). Normally, if the root becomes clogged with metal the plant dies. However, in monocotyledons, roots are adventitious and can be replaced continuously. This phenomenon probably explains the dominance of grasses and sedges on metalliferous soils. There also is evidence that the growth of tolerant plants is stimulated by small amounts of minor metals; this suggests that the minimum requirements for these elements are greater in tolerant than in nontolerant plants. Antonovics et al. (1971) suggested the alternate possibility that these plants more efficiently inactivate a number of elements and hence have a greater daily requirement for minor metals.

Early work by Bradshaw (1952) showed that metal tolerant populations of **Agrostis tenuis** could grow on mine soils, but that plants transplanted from normal pastures could not survive. Research by Peterson (1969), Bradshaw (1970a), and Smith and Bradshaw (1970) with populations of tolerant species of **Agrostis tenuis, Agrostis stolonifera,** and **Festuca rubra** demonstrated that these plants became established more easily, and produced greater root and shoot growth when they were transplanted to mine waste in Wales than did control plants of **Lolium perenne.** Root growth in solution culture can be studied using a method developed by Wilkins (1957) in which tillers and cuttings are taken from mineralized soils. By this method, the development of tolerant races by a large number of species has been established; these include both grasses and forbs. A few of these reported by Antonovics et al. (1971) are shown in Table 18A.1. The transplantation of plant populations known to have high tolerance for metals would appear to be a powerful tool in stabilizing and reclaiming mine wastes. The rapid natural evolution of populations of high tolerance is accomplished by selection, which is possible through a survival screening process of the few tolerant individuals that normally occur in every generation of an average nontolerant population. There may be only three to four survivor seedlings in many thousand, but these are sufficient to start a new population of highly tolerant plants (Smith and Bradshaw, 1970). The tolerance of evolved races of plants is highly specific for individual heavy metals. Generally, the specialists or indicator plants for one metal are not tolerant of another metal. **Indiofera dyeri** populations, for instance, are 10 times more tolerant of zinc than of copper; **Indiofera setiflora** populations are very tolerant of nickel but of no other metal (Ernst, 1972).

INDICATOR PLANTS

Plant indicators of mineral deposits are species or varieties of plants that give a clue to the chemistry of the rock substrate by their presence or absence on mineralized soils. Most species that have been reported are local indicators. Research during the past 10 years suggests that there are few, if any, universal indicators which do not occur somewhere in the world on nonmetalliferous soils where conditions are particularly favourable for their growth. For this reason, researchers have become more cautious in declaring plants to be indicators, and describe them instead as super tolerant, accumulators, or specialists. In this paper, plants have been given indicator status only where so described by the authors quoted.

Since 1965, relatively few new indicator species have been recognized, but much research has been carried out on the relationships that exist between plant species or varieties and mineralized ground (phytogeoecology). True endemic indicator plant species of any particular metal are rare, and in many cases they survive only in localized refugia because of biotype depletion. That is to say, changing climatic conditions and competition from other closely related species have severely restricted their distribution in general; but, being highly tolerant of metal-rich ground of low pH, they have continued to exist in open, treeless areas of mineralized ground. Given a favourable environment with no competition, they may also be found in isolated areas of unmineralized ground, as on a sunny scree slope.

Recent work by Bradshaw (1959), Duvigneaud and Denaeyer-De Smet (1963), and Denaeyer-De Smet (1970) has shown that physiological ecotypes, which develop in a relatively short time on metalliferous soils, have a much higher tolerance for a particular metal than their counterparts growing on normal soils. These ecotypes may show no unusual morphological differences and can only be identified by root growth studies in solution culture (McNeilly and Bradshaw, 1968; Wilkins, 1957). Although they are true indicators and can grow well only on metaliferous soils, they may be indistinguishable in the field from plants of the same species growing nearby on normal soils.

PLANT COMMUNITIES

In many areas the recognition or delineation of metaliferous soils is aided by observing the patterns of distribution of <u>all</u> species rather than that of a single indicator plant.

The observation of plant communities as an aid in geological mapping was proposed by Karpinsky (1841), and the method has been developed to a fine art in the Soviet Union (Malyuga, 1964). Excellent reviews of this work have been provided by Chikishev (1965) and Viktorov et al. (1964).

"Plant communities" or even the larger unit of "plant associations" may be definitive in outlining particular rock units such as limestone, sandstone, halite, or ultrabasic rocks. Quantitative information concerning the total number of species, their density, and vitality can be obtained on the ground by observations in quadrats or along transects.

The control of plant associations by surface features (physiognomy) and geological substrate was studied by Nicolls et al. (1965) in the Dugald River area of Australia. In areas of base metal mineralization, they found that normal vegetation was absent, and was replaced by specialized plant communities dominated by **Polycarpea glabra** and **Tephrosia** sp. nov. They concluded that geobotany was a useful and inexpensive method that complemented stream-sediment sampling, and that the two methods should be used together when personnel with some botanical training were available.

The tree cover in forested areas or the shrub cover in unforested areas can also be observed and mapped from the air, as is done routinely in Russia, or by remote sensing, as has been used by Cole (1971a, b) in South Africa.

Braun-Blanquet (1951) described the study of plant communities on metal-contaminated soils as a technique necessary to phytosociology, and he devised a system of classifying plant associations and their smaller components. The method provides a means of recording the relative abundance of different but associated species on metalliferous soils. In recent years, the plant communities characteristic of copper, lead, and zinc soils have been examined in Germany by Schwickerath (1931), Ernst (1965, 1968a), and Baumeister (1967); in France by Ernst (1966); and in Great Britain by Shimwell and Laurie (1972) and Ernst (1968b). Ernst found that the associations studied in many of these areas had several species in common — **Viola calaminaria, Thlaspi alpestre, Minuartia verna, Silene vulgaris, Armeria** sp., and **Festuca ovina** — although occasionally one or two were absent owing to climatic differences. He therefore created one phytosociological "order" called **Violetea calaminariae**, which he subdivided into three "families". His work showed that typical communities exist in widely separated mineralized areas, and that their distribution is not controlled solely by habitat, climate, or geography. Such disjunct areas of plants are considered by Stebbins (1942) to contain two types of species: (1) "paleo-endemic" species that formerly had a wide distribution but are now confined to isolated areas and (2) "neo-endemic" species that have developed in response to environmental stress. Such relict communities of formerly widespread species might be caused by glaciation or by widespread volcanic ash fall which has since eroded away. The latter may account for the isolated areas extending from Utah to Texas of **Astragalus pattersoni**, a plant that requires and absorbs large amounts of selenium for survival and at the same time only occurs in areas of measurable radioactivity (Cannon, 1962).

ANOMALOUS GROWTH CHARACTERISTICS

Visible effects of high concentrations of metal on plant growth habits can be observed by on-the-ground studies. The surface area of highest metal content may appear as a "bare" area in a forest. The apparently bare area may actually support small annual plants or perennials that die to the ground each year, thus ridding themselves of a year's accumulation of metal. The trees nearest to a copper or zinc area may exhibit interveinal iron chlorosis because the excess metals interfere with the production of chlorophyll. Trees under such environmental stress have also been observed to turn colour earlier in the fall than those farther from the mineralized area. It is possible that the latter phenomenon, if widespread, may be useful in remote sensing.

Stunted, bushy forms of **Tephrosia longipes** and plants of **Combretum zeyheri** with enlarged fruits have been described by Wild (1968) as true morphological copper ecotypes in Rhodesia. Whether the plants observed in the field have reached a stage of ecotonal development or not is unimportant to the prospector as long as the growth differences can be observed. Stunted trees that tend to be spreading, and of a uniform height, surround the grassy zone of a typical Rhodesian metal anomaly (Wild, 1970). Progressive stunting and chlorosis with increased metal content were observed in rows of vegetables which had been planted across high-zinc areas in drained mucks at Manning, New York (Cannon, 1955).

Measurements made by Jacobsen (1967) of several species in a Rhodesian sampling program showed a progressive decrease in the height of nonspecialized species with increasing soil copper. Studies by Howard-Williams (1971) of the seeds and flowers of **Becium homblei** showed differences in seed weight and corolla shape between populations growing in widely different soil types; the greatest reduction in size occurred in plants growing in heavy-metal soils. Antonovics et al. (1971) have shown that tolerant races of plants are generally dwarfed or prostrate and require less calcium and phosphate.

Geobotanical studies I recently conducted at the Pine-Nut molybdenum deposit in Nevada demonstrated anomalous growth phenomena in several species. The length of the internodes in **Ephedra nevadensis** was greatly extended, producing a wandlike appearance in the plant; and flowers of **Peraphyllum ramossissima** (squaw apple) were white instead of their normal pink colour.

Antonovics et al. (1971) suggested that the morphological changes reported by Malyuga (1964) may well be genetic. The following experience demonstrates that morphological changes are not necessarily genetic, but may be caused by elemental imbalance.

Twenty some years ago I conducted an experimental plot study of mineral uptake by native plants over a period of three years. The plots were treated as shown in Figure 18A.1. Many diverse growth habits were observed. As splits of the same seed collection were sown in each plot, all differences were attributable to the imbalance of nutrients or trace metals rather than to genetic variations. **Euphorbia fendleri** (Fig. 18A.2) grew with an upright habit in the gypsum plot but was completely prostrate and developed nodules on the stems at the crown of the plant in the lime plot. California poppies in the lime plot developed light yellow edges on the petals and some completely yellow flowers, but had dark orange flowers in plots to which phosphate had been added.

GYPSUM & CARNOTITE	GYPSUM	GYPSUM & VANADIUM
SELENIUM & CARNOTITE	SELENIUM	SELENIUM & VANADIUM
CARNOTITE	SELENIUM & THORIUM	VANADIUM
PHOSPHATE & URANIUM	PHOSPHATE	PHOSPHATE & VANADIUM
LIME & CARNOTITE	LIME	LIME & VANADIUM

Carnotite = $K(UO_2)(VO_4) \cdot nH_2O$

Thorium = $SrAl_3 \text{(rare earths)} (PO_4)(SO_4)(OH_6)$

Vanadium = $Na_2H_2V_6O_{17}$

Selenium = Na_2SeO_3

Figure 18A.1. Plan of experimental plot study conducted in 1956-58.

Morphological changes in several species growing in irradiated soil plots were observed. One of the most interesting results was the discovery that **Astragalus pattersoni**, the most useful uranium indicator plant (formerly believed to be entirely dependent on selenium) grew to maturity in the selenium plot, but also in the carnotite plot to which no selenium had been added. **Stanleya** was strongly affected by radiation. Where radioactive ores were added to the soil where the plant spikes were already in flower, the new flowers ceased to produce petals or stamens within a few days, and eventually the pistil produced a new plant asexually.

RECENT USES OF GEOBOTANY FOR SPECIFIC METALS

Copper

Many lower forms of plants are useful in prospecting or at least have been shown to be tolerant of metalliferous soils. The use of indicator "copper mosses" and several liverworts in prospecting was described in the early literature. The resistance to copper of species from several other moss genera has also been described by Ernst (1965), but the plants have not been given indicator status. Certain fungi, bacteria, and algae have also been observed to be associated with copper soils, and the black crust formed by blue-green algae has been used in prospecting (Wild, 1968).

a. *Upright in gypsum plot (GSC 203492-E)*

b. *Prostrate with nodules in lime plot (GSC 203492-H)*

Figure 18A.2. Euphorbia fendleri in plots treated with gypsum and lime.

A small herb, **Eriogonum ovalifolium**, was reported by Lidgey (1897) to be an indicator of silver in Montana (Fig. 18A.3). This plant was observed by Grimes and Earhart (1975) to be the dominant ground cover in a number of bare areas of otherwise forested country in Montana. The areas have since proved to be mineralized with Cu, Pb, Ag, and Zn. Plant samples contained as much as 500 ppm Cu in the dry weight of the leaves and more than 1000 ppm in the roots. As much as 15 ppm Ag in the leaves was detected at another locality. The distribution of several varieties of the species is being studied in different geochemical environments.

In similar clearings in Zambia, Reilly (1967) reported that the indicators **Becium homblei, Vernonia glaberrima, Triumfetta welwitschii,** and **Cryptosepalum maraviense** accumulate copper in their leaves. Grasses growing in the mineralized soil did not accumulate copper. Negative indicators, or cuprifuge plants, were intolerant of soils containing more than 20-40 ppm Cu. Although **Becium homblei**, a mint (Fig. 18A.4), was formerly believed to grow only on soils containing more than 100 ppm Cu, Howard-Williams (1970) reported that the plant has been found in barren ground in both Zambia and Rhodesia, but is confined to certain well defined climatic boundaries. It is also able to grow on high-nickel, lead, and arsenic soils. Ecologic studies show that **Becium homblei** is the dominant forb in a typical grass association but is restricted to soils of high copper content where it is in competition with **Becium obovatum**, which is not tolerant of low-Ca, low-pH, and high-metal soils. Thus, the isolated occurrences of **Becium homblei** are believed to be relict stands. The plant has large underground rootstalks from which leafy shoots arise during each wet season and die off each dry season, thereby ridding itself of the copper. Because of this growth habit, **Becium homblei** is also able to survive periodic man-made bush fires (Howard-Williams, 1972).

Other potential indicator plants were studied by Jacobsen (1968) in the Mangula mining district of Rhodesia. In order to appraise the usefulness of various species, he calculated the specific indicator value (I_{cu}) for a range of copper soils by dividing the plant's tolerance span (Cu maximum minus Cu minimum) by its average soil-copper value (Cu_o). Plants with the lowest I_{cu} values (<12.) are the best indicators for their respective range of soil Cu values. Some of these local indicators for high-copper soils are listed in Table 18A.2. Wild (1968) has also made a study of geobotanical anomalies in Rhodesia and states categorically that "the only taxon definitely distinct at the specific level that is confined to copper soils in Rhodesia is **Becium homblei**, although this species occurs on non-copper soils in Katanga." This suggested to Wild that the plant is in an intermediate stage of development as an endemic. Wild has recognized several subspecies morphological ecotypes as being at an even earlier stage of development: a glabrous, narrow-leafed form of **Justicia elegantula**, S. Moore; an olive-brown **Pogonarthria squarrosa** Pilg.; a stunted bush, **Tephrosia lurida** Sond.; and a form of **Combretum zeyheri** Sond. with unusually large fruits and wavy-winged bracts. These ectoypes appear to be distinct taxa at some infraspecific level and are useful indicators. A species of **Combretum** has also proved useful in indicating copper-ore deposits in South Africa (Cole, 1971a).

Figure 18A.3. **Eriogonum ovalifolium** *silver plant, growing on mineralized soil in Montana. (GSC 203492-G)*

Figure 18A.4. Becium homblei, a copper indicator plant in Rhodesia. (GSC 203492-J)

Table 18A.2

Local indicator plants for high copper soils in Rhodesia (Jacobsen, 1968)

Species	Copper indicator value (I_{Cu})
>5000 ppm Cu in soil	
Bulbostylis contexta	2.07
Eragrostis racemosa	4.07
Monocymbium ceressiiforme	2.82
Trachypogon spicatus	3.91
1900-5000 ppm	
Albizia antunesiana	3.66
Burkea africa	5.34
Ficus burkei	4.09
Heeria reticulata	3.72

Brooks (1977) has analyzed 48 herbarium specimens of 19 species of **Haumaniastrum** (formerly **Acrocephalus**) and reports high-copper and also high-cobalt contents in **H. robertii**, a well known copper indicator plant in Africa, and also in **H. katangense** and **H. homblei**. He suggests that the role of the latter two species as indicator plants should be studied further.

Lewis et al. (1971) have studied the distribution of 25 species in soils of known copper contents in the Monte Alto copper district in Brazil, and reported that 3 plants, **Croton mortbensis, Psidium aracá**, and **Eugenia** sp. grew only in soils containing 175 ppm or more of copper and may be useful indicator plants.

Nicolls et al. (1965) made a thorough study of the geobotanical relationships in the Dugald River area of Australia, making an assessment of all factors thought to govern plant distribution in the area. The characteristic plant associations are replaced over the lodes by a treeless plant assemblage consisting of **Bulbostylis barbata** and **Fimbristylis** sp. nov., which reflect high Cu values in the surface soil; **Polycarpea glabra** and **Tephrosia** sp. nov., most closely related to copper and zinc toxicities; and **Tephrosia** sp. nov., which is tolerant of high lead as well as copper and zinc. The authors demonstrate that the indicator plants are able to exclude copper at relatively low levels in the soil but that large amounts of Cu are absorbed from soils of high Cu content. Analyses of the soils for major nutrients suggest that phosphorus, which increases over the lodes, may have an important influence on plant distribution.

Cobalt

The extraordinary cobalt uptake by **Nyssa sylvatica** var. **biflora**, reported by Beeson et al. (1955), has been confirmed by Brooks et al. (1977) in herbarium specimens from the United States and Southeast Asia. Brooks has also shown that the accumulation of cobalt by other species of **Nyssa** is extraordinarily high, and although the plants also accumulate nickel, the Co/Ni ratio in the plants is 2 or more. The latter ratio is rare in the plant kingdom. A second genus of the Nyssaceae, **Camptotheca**, also accumulates cobalt, but not to the extent of **Nyssa**. As the plants are able to absorb large amounts of cobalt from normal soils, they should be useful in assessing the cobalt status of agricultural soils, but are not indicative of economic cobalt deposits. On the other hand, Brooks' (1977) discovery of unprecendented levels of cobalt in **Haumaniastium robertii**, known previously as a copper indicator plant, raises the question of whether this plant is actually an indicator of cobalt rather than copper. The values range from 1368 to 10 222 ppm cobalt (on a dry weight basis), with a mean of 4304 ppm in six leaf specimens; this is an order of magnitude greater than any values previously reported for cobaltophytes! Brooks' noteworthy method of analyzing small pieces of herbarium specimens sent to him from many countries is providing much new information on the distribution and identification of metal tolerant plants, and should stimulate further research by botanists all over the world.

Zinc and Lead

Denaeyer-De Smet (1970) has found the accumulation of zinc in zinc indicator plants to be generally high but to vary greatly according to species. She studied the accumulation of zinc by nine species growing in high-zinc areas (Fig. 18A.5). The zinc obligate, **Thlaspi sylvestre** ssp **calaminare**, contained 15 700 ppm zinc in the dry weight of the leaves – ten times that of the other species. The leaves and the roots of **Silene cucubalus** var. **humilis** accumulated roughly equal amounts of zinc (1860 ppm). The zinc-vague trees, **Salix caprea, Betula verrucosa** and **Populus tremula** accumulated 1285-1439 ppm zinc in the leaves and became chlorotic. **Armeria halleri** concentrated the zinc strongly in the roots; when the **A. halleri** plants were transplanted from high-zinc soils to normal soils, the zinc rapidly decreased in both the leaves and the roots. Lefebvre (1968), studying several species of **Armeria**, reported the occurrence of ecotype populations having a hereditary tolerance of heavy metals. These ecotypes are able to produce new roots in the presence of high concentrations of zinc, whereas their counterparts obtained from normal soils could not.

Five areas of naturally occurring lead-poisoned soil downslope from known deposits have been found in Norway by Låg and Bølviken (1974). In less advanced stages of poisoning, **Vaccinium** ssp (blueberry) is replaced by **Deschampsia flexuosa** (hair grass). In advanced stages of poisoning, the vegetation is stunted, chlorotic, lacks fruit, or disappears

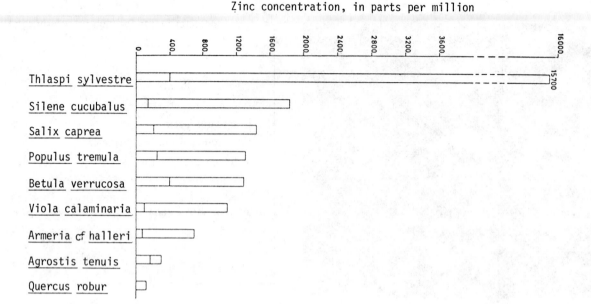

Figure 18A.5. *Graph showing amounts of zinc in the leaves of different species collected from the same zinc-rich biotype at Plombiéres, Belgium (taken from Denaeyer-De Smet, 1970).*

entirely. The largest such bare area is 100 m². The soil here lacks a bleached horizon, is stony, has no covering of sphagnum moss, and averages 24 500 ppm lead. Among the plants that were analyzed, a fern, **Dryopteris lianaeana**, contained more lead (253 ppm) than **Deschampsia flexuosa** (99 ppm). The vegetation of these areas includes no indicator plants, but the authors suggest that the recognition of atypical plant communities and bare areas in otherwise forested country may prove to be an effective prospecting method.

Work by Nicolls et al. (1965) in Australia showed the indicator plant **Eriachne mucronata** to be far more tolerant of high-lead areas than other plants.

Serpentine: Nickel and Chromium

Unusual floras of narrowly endemic species occur on serpentine rocks throughout the world. The areas are commonly characterized by dwarf pine, dwarf shrubs, mosses, lichens, ferns, and certain genera of the laurel, pink, chickweed, and borage families. The endemic floras of soils derived from ultramafic rock in northwest Washington, British Columbia, and Oregon were described in detail by Kruckeberg (1969). He believed that the restriction of these plant species to serpentine rocks is largely determined by their ability to extract enough calcium from acid clay soil, as the Mg/Ca ratio is very high. Although he describes the vegetation in the nickeliferous area of Grants County, Oregon, he does not consider soil metal content as a possible control, nor does he give analyses for Ni or Cr.

Lee et al. (1975) undertook a statistical approach to the problem of controls for the serpentine flora in New Zealand. Soils were collected at random sample sites near two endemic species, **Myosotis monroi** Cheesem. and **Pimelea suteri** Kirk (Fig. 18A.6) and also near three nonendemic species. The five groups of samples were analyzed for Ca, Co, Cr, Cu, K, Mg, Mn, Ni, and Zn, and the data were treated statistically by discriminant analysis. The greatest difference between endemic and nonendemic plants was their magnesium content, and the second greatest difference was nickel. The endemic plants appear to be characterized by an ability to survive in soils having high Mg/Ca ratios and elevated levels of nickel.

The serpentine flora of the Great Dyke in Rhodesia has been studied in detail by Wild (1965). He found that nickel-bearing serpentine produced a depauperate and stunted flora whereas serpentine of low nickel content did not. He attributed the sterility observed in several species to nickel poisoning. The percentage of nickel and chromium in serpentines is often very high, and nickel is accumulated by certain species of plants in very large amounts. Severne (1974) reported a significant negative correlation between nickel and calcium in a nickel accumulator, **Hybanthus floribundus**, in Western Australia. This suggests that Ni can substitute for Ca, a thought which is supported by the recent finding of nickel in pectinates. Possibly nickel is essential to these plants.

Malyuga (1964) listed three indicator plants for nickel: **Alyssum bertolonii**, found in Italy; **Alyssum murale** in Georgia (U.S.S.R.); and **Asplenium adulterium** in Norway. Many nickel accumulator plants have been investigated since that time, and a few of them including 11 additional species of **Alyssum**, have been demonstrated to be indicators of high-nickel ground (R.R. Brooks and O. Verghano Gambi, pers. comm., 1977). **Sebertia acuminata** of the Sapotaceae is endemic to New Caledonia, occurs only on soils derived from ultrabasic rocks, and is found most often on high-nickel soils (Jaffré et al., 1976). A study by Lee and others (1972) of two known hyperaccumulators of nickel from New Caledonia demonstrated that the nickel content of **Hybanthus austrocaledonicus** could be correlated with the total soil nickel content but that the nickel content of **Homalium kanaliense** could not; the latter, therefore, could not be used in prospecting. The ecotype **Hybanthus floribundus** ssp **curvifolius** is restricted to high-nickel soils in Western Australia (Severne, 1974). Lee and others (1977b) have recently found high chromium, in an epiphytic moss that grows on **Homalium guillainii**, a hyperaccumulator of nickel. The moss, **Aerobryopsis longissima** (Doz. and Molk.) Fleisch, had a mean chromium content of 5000 ppm, nearly twenty times that of its host.

Jaffré and Schmid (1974) reported that a tree of the Rubiaceae, **Psychotia douarrei** (C. Brown) Däniker, from New Caledonia, accumulates higher levels of nickel than had ever been reported in a plant. The nickel content ranged from 1.8-4.7 per cent in the dry weight of

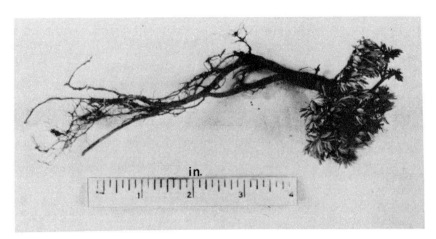

Figure 18A.6. *Pimelea suteri*, endemic serpentine plant in New Zealand. (GSC 203492-B)

the leaves to as much as 9.2 per cent in the roots. They suggested that the nickel uptake is not passive but is actively associated with the plant's physiologic processes.

Brooks et al. (1976) have recently analyzed leaves of 2000 herbarium specimens which included 232 species of **Homalium** and **Hybanthus** from throughout the world. Five additional species of hyperaccumulators (> 1000 ppm Ni, dry weight) were found, all from New Caledonia. As nickel contents greater than 1000 ppm dry weight only occur in plants growing over ultrabasic rocks, the analyses of the 2000 samples have pinpointed a number of Ni-bearing ultrabasic areas. As shown by the work of Lee et al. (1977a), however, the nickel content of a particular hyperaccumulator may or may not correlate with the total content of the soil. Soil analyses for the newly discovered species are therefore, necessary before their true status as indicators of nickel mineralization can be evaluated.

Wild (1971) has described a new species, **Dicoma niccolifera** Wild, a composite, as being almost entirely confined to nickel-bearing soils. It occurs in Zambia and Rhodesia. He believed the species to be an older relict species isolated by biotype depletion rather than adapted modification of another more common species. The plant has been used as a reliable indicator of nickel in Rhodesia. Wild (1970) also found several other species that could be used locally as nickel indicators, although elsewhere they were not confined to nickel soils. **Albizia amara** was a striking indicator of 11 nickel anomalies; it is not normally associated with serpentine. **Turraea nicotica** is of value as an indicator of both serpentine and amphibolite nickel anomalies. **Combretum molle** occurred in all nickel anomalies, and commonly as stunted, pure stands. **Combretum molle** also is common on copper soils, as are the metal-resistant grasses **Loudetia simplex**, **L. flavida**, **Aristida leucophaea**, **Andropogon gayanus,** and others.

SUMMARY

Recognition of sparsely populated or depauperate vegetated areas associated with highly mineralized soil has advanced in the last ten years, and much progress has been made in determining the occurrence and controls of stands of obligate, relict, and evolved ecotypes that may be useful in prospecting.

Plants that tolerate highly metalliferous soils are commonly either deciduous plants that have a high uptake of metals but die to the ground each year, or plants that concentrate the metal in the walls of the root cells but are able to avoid blockage by the continued growth of adventitious roots. The rapid evolution of races of highly tolerant plants in mineralized areas from the few hardy individuals that occur in normal populations has been established. Many of the previously known indicator plants have been shown to be specialized races or ecotypes characterized by an unusual tolerance for a particular metal.

Fewer obligate indicator species have been recognized in the past ten years than during the previous periods, but greater emphasis has been placed on the observation of plant communities as an aid in prospecting. Typical plant communities related to copper, lead, and zinc mineralization with several species in common have been found in widely separated areas of Europe.

Stunting, chlorosis, and morphological effects are common in mineralized areas. Some of these aberrant plants have been shown to be true morphological ecotypes; others are known to result from an imbalance of nutrients. The latter effects can be reversed. Both are useful in prospecting.

Several new nickel accumulators and indicators have been reported from New Caledonia, Western Australia, Zambia, Rhodesia, Europe, and Indonesia.

Analyses by R.R. Brooks of several thousand samples of herbarium specimens of well known indicator plants collected throughout the world have shown many to be true accumulators of specific metals. His analyses have pinpointed two or three previously unknown ultrabasic areas in Indonesia.

The geologist who is prospecting a new area would be advised to look for open areas where the vegetative cover consists of fewer than the normal number of species. The plants may be stunted or prostrate, and the plant community may consist of species unusual for the general area. If such plants are deep-rooted, their presence may offer a means of metal detection not possible from surface soils alone.

REFERENCES

Antonovics, Janis, Bradshaw, A.D., and Turner, R.G.
    1971: Heavy metal tolerance in plants; Adv. Ecol. Res., v. 7, no. 1, p. 1-85.

Bailey, F.M.
    1889: Queensland Flora; Melbourne, Australia.

Barba, A.A.
    1640: Arte de los Metales; Madrid (English translation may have been published in 1923).

Baumeister, W.
    1967: Schwermetall-Pflanzengesellschaften und Zinc-resistenz einiger Schwermetallpflanzen; Angew. Bot., v. 40, p. 185-204.

Beeson, K.C., Lazar, V.A., and Boyce, S.G.
    1955: Some plant indicators of the micronutrient elements; Ecology, v. 36, no. 1, p. 155-156.

Boyle, R.W.
    1967: Geochemical prospecting — Retrospect and prospect; Can. Inst. Min. Met. Bull. 657, p. 44-49.

Braun-Blanquet, Josias
    1951: Pflanzensoziologie; 2nd ed.: Springer, Wien. 2 Aufl. XI 631 p. (240-241) 631 p.

Bradshaw, A.D.
   1952. Populations of **Agrostis tenuis** resistant to lead and zinc poisoning; Nature, v. 169, p. 1098.

   1959: Population differentiation in **Agrostis tenuis** Sibth., part I – Morphological differences; New Phytol., v. 58, p. 208-227.

   1970a: Plants and industrial waste; Trans. Bot. Soc. Edinburgh, v. 41, p. 71-84.

   1970b: Reclamation of toxic metalliferous wastes using tolerant populations of grass; Nature, v. 227, no. 5256, p. 376-377.

Brooks, R.R.
   1972: Geobotany and Biogeochemistry in Mineral Exploration; Harper & Row, New York, 292 p.

   1977: Cobalt and copper uptake by the genus **Haumaniastrum**; Plant and Soil, v. 48, p. 541-545.

Brooks, R.R., Lee, Julian, Reeves, R.D., and Jaffre, Tanguy
   1976: Detection of nickeliferous rocks by analysis of herbarium specimens of indicator plants; J. Geochem. Explor., v. 7, p. 49-57.

Brooks, R.R., McCleave, J.A., and Schofield, E.K.
   1977: Cobalt and nickel uptake by the Nyassaceae; Taxon, v. 26, p. 197-201.

Cannon, H.L.
   1955: Geochemical relations of zinc-bearing peat to the Lockport dolomite, Orleans County, N.Y.; U.S. Geol. Surv. Bull. 1000-D, p. 119-185.

   1962: The development of botanical prospecting for uranium on the Colorado Plateau; U.S. Geol. Surv. Bull. 1085, p. 1-50.

   1971: The use of plant indicators in ground water surveys, geologic mapping, and mineral prospecting; Taxon, v. 20, no. 2-3, p. 227-256.

Chikishev, A.G.
   1965: Plant indicators of soils, rocks, and subsurface water; Consultants' Bureau, New York, 210 p.

Cole, M.M.
   1971a: Biogeographical/geobotanical and biogeochemical investigations connected with exploration for nickel-copper ores in hot, wet summer/dry winter savanna woodland environment; South African Inst. Min. Met., v. 71, no. 10, p. 199.

   1971b: The importance of environment in biogeographical/geobotanical and biogeochemical investigations; in Geochemical Exploration, Can. Inst. Min. Met., Spec. Vol. 11, p. 414-425.

Denaeyer-De Smet, S.
   1970: Considerations sur l'accumulation du zinc par les plantes poussant sur sols calaminaires; Inst. Royal Sci. Nat. Belgique Bull., v. 46, no. 11, 13 p.

Duvigneaud, Paul and Denaeyer-De Smet, S.
   1963: Cuivre et végétation au Katanga; Bull. Soc. Bot. Belgique, v. 96, p. 93-231.

   1973: Considerations sur l'ecologie de la nutrition minérale des tapis vegetaux naturels; Ecol. Plant., v. 8, no. 3, p. 219-246.

Ernst, Wilfried
   1965: Ökologische-Soziologische Untersuchungen der Schwermetall-Pflanzengesellschaften Mitteleuropas unter Einschluss der Alpen; Abh. Landesmus. Naturk. Münster, v. 27, no. 1, 54 p.

   1966: Ökologisch-sociologische Untersuchungen auf Schwermetall-Pflanzengesellschaften Süd Frankreiches und des östlichen Harzvorlandes; Flora, Jena, B., v. 156, p. 301-318.

   1968a: Das Violetum calaminariae westfalicum, eine Schwermetall-Pflanzengesellschaft bei Blankenrode in Westfalen; Mitt. flor.-soz. Arbgemein., v. 13, p. 263-268.

   1968b: Zur Kenntnis der Soziologie und Okologie der Schwermetall-vegetation Grossbritanniens; Ber. dt. bot. Ges., v. 81, p. 116-124.

   1972: Ecophysiological studies on heavy metal plants in South Central Africa; Kirkia, v. 8, pt. II, p. 125-145.

Grimes, D.J. and Earhart, R.L.
   1975: Geochemical soil studies in the Cotter Basin area, Lewis and Clark County, Montana; U.S. Geol. Surv. Open-File Rept. 75-72, 25 p.

Howard-Williams, C.
   1970: The ecology of **Becium homblei** in Central Africa with special reference to metalliferous soils; J. Ecol. (Great Britain), v. 58, p. 745-763.

   1971: Morphological variation between isolated populations of **Becium homblei** (de Wild.) Duvign. and Plancke growing on heavy soils; Vegetatio, v. 23, 3-4, p. 141-151.

   1972: Factors affecting copper tolerance in **Becium homblei**; Nature, v. 237, no. 5351, p. 171.

Jacobsen, W.B.G.
   1967: The influence of the copper content of the soil on trees and shrubs of Molly South Hill, Mangula; Kirkia, v. 6, pt. 1, p. 63-84.

   1968: The influence of the copper content of the soil on the vegetation at Silverside North, Mangula area; Kirkia, v. 6, pt. 2, p. 259-277.

Jaffré, Tanguy, Brooks, R.R., Lee, J., and Reeves, R.D.
   1976: **Sebertia accuminata** – a hyperaccumulator of nickel from New Caledonia; Science, v. 193, p. 579-580.

Jaffré, Tanguy and Schmid, Maurice
   1974: Accumulation du nickel par une Rubiacée de Nouvelle-Caledonie, **Psychotia douarrei** (G. Beau.) Daniker; Comptes Rendus des Séances d L'Academie des Sciences, Paris, v. 278, series D, p. 1727-1730.

Karpinsky, A.M.
   1841: Can living plants be indicators of rocks and formations on which they grow?; Zhur. Sadovodstva, No. 3 and 4.

Kruckeberg, A.R.
   1969: Plant life on serpentinite and other ferromagnesian rocks in northwestern North America; Syesis, v. 2, p. 15-114.

Låg J. and Bølviken, B.
 1974: Some naturally occurring heavy-metal poisoned areas of interest in prospecting, soil chemistry, and geomedicine; Norges Geologiske Undersøkelse, offprint NGU 304, p. 73-96.

Lee, Julian, Brooks, R.R., Reeves, R.D., and Boswell, C.R.
 1975: Soil factors controlling a New Zealand serpentine flora; Plant and Soil, v. 42, p. 153-160.

Lee, Julian, Brooks, R.R., Reeves, R.D., Boswell, C.R., and Jaffré, Tanguy
 1977a: Plant-soil relationships in a New Caledonian serpentine flora; Plant and Soil.

Lee, Julian, Brooks, R.R., Reeves, R.D., and Jaffré, Tanguy
 1977b: A chromium-accumulating bryophyte from New Caledonia; Bryologist.

Lefebvre, Claude
 1968: Note sur un indice de tolérance au zinc chez des populations d'**Armeria maritima** (Mill.) Wiled.; Bull. de la Société royale de Botanique de Belgique, v. 102, p. 5-11.

Lewis, R.W., Jr., Sampaio de Almeida, A.L., Pinto, A.G.G., Ferreira, C.R., Tavora, F.J., and Duarte, F.B.
 1971: Geochemical Exploration of the Monte Alto copper deposit, Bahia, Brazil; U.S. Geol. Surv. Prof. Paper 750-C, p. C141-145.

Lidgey, E.
 1897: Some indications of ore deposits; Austr. Inst. Min. Eng. Proc., v. 4, p. 110-122.

McNeilly, T. and Bradshaw, A.D.
 1968: Evolutionary processes in populations of copper tolerant **Agrostis tenuis** Sibth.; Evolution, v. 22, p. 108-118.

Malone, C., Koeppe, D.E., and Miller, J.E.
 1974: Localization of lead accumulated by corn plants; Plant Physiology, v. 53, p. 388-394.

Malyuga, D.P.
 1964: Biogeochemical methods of prospecting: A translation from Russian 1959 publication by New York Consultants Bureau, 205 p.

Miller, J.E., Hassett, J.J., and Koeppe, D.E.
 1975: The effect of soil properties and extractable lead levels on lead uptake by soybeans; Soil Science and Plant Analysis, v. 6, no. 4, p. 339-347.

Nicolls, O.W., Provan, D.M.J., Cole, M.M., and Tooms, J.S.
 1965: Geobotany and geochemistry in mineral exploration in the Duguld River Area, Cloncurry district, Australia; Inst. Min. Met. Trans., London, v. 74, pt. II, p. 695-799.

Peterson, P.J.
 1969: The distribution of zinc-65 in **Agrostis tenuis** Sibth, and **A. stolonifera** L. tissues; J. Exp. Botany, v. 20, no. 65, p. 863-875.

Raymond, R.W.
 1887: Indicative plants; Am. Inst. Met. Eng. Trans., v. 15, p. 644-660.

Reilly, C.
 1967: Accumulation of copper by some Zambian plants; Nature, v. 215, p. 667-668.

Schwickerath, M.
 1931: Das Violetum calaminariae der Zinkböden in der Umgebung Aachens; Beitr. zur Naturdenkmalpflege, v. 14, p. 463-503.

Severne, B.C.
 1974: Nickel accumulation by **Hybanthus floribundus**; Nature, v. 248, p. 807-808.

Shimwell, D.W. and Laurie, A.E.
 1972: Lead and zinc contamination of vegetation in the southern Pennines; Environ. Pollution, v. 3, no. 4, p. 291-301.

Smith, R.A. and Bradshaw, A.D.
 1970: Reclamation of toxic metalliferous wastes using tolerant populations of grass; Nature, v. 227, no. 5256, p. 376-377.

Stebbins, G.L.
 1942: The genetic approach to problems of rare and endemic species; Madroño, v. 6, p. 241-272.

Viktorov, S.V., Vostokova, Ye. A., Vyshivkin, D.D.
 1964: Short guide to geobotanical surveying; Oxford, Pergamon Press, 158 p.

Wild, H.
 1965: The flora of the Great Dyke of Southern Rhodesia with special reference to the serpentine soils; Kirkia, v. 5, pt. 1, p. 49-86.

 1968: Geobotanical anomalies in Rhodesia, pt. 1 – The vegetation of copper-bearing soils; Kirkia, v. 7, pt. 1, p. 1-71.

 1970: Geobotanical anomalies in Rhodesia, pt. 3 – The vegetation of nickel-bearing soils; Kirkia, suppl. to v. 7, 62 p.

 1971: The taxonomy, ecology and possible method of evalution of a new metalliferous species of **Dicoma** Cass, (Compositae); Mitt. Bot. Staatssamml. München, v. 10, p. 266-274.

Wilkins, D.A.
 1957: A technique for the measurement of lead tolerance in plants; Nature, v. 180, p. 37-38.

ADVANCES IN BOTANICAL METHODS OF PROSPECTING FOR MINERALS
PART II–ADVANCES IN BIOGEOCHEMICAL METHODS OF PROSPECTING

Robert R. Brooks
Massey University, Palmerston North, New Zealand

Brooks, Robert R., Advances in botanical methods of prospecting for minerals. Part II–Advances in biogeochemical methods of prospecting; in Geophysics and Geochemistry in the Search for Metallic Ores; Peter J. Hood, editor; Geological Survey of Canada, Economic Geology Report 31, p. 397-410, 1979.

Abstract

*Biogeochemical methods involving chemical analysis of vegetation have been in use since the Second World War. In Australia some work has been carried out on **Hybanthus floribundus**, a hyperaccumulator of nickel which can indicate nickel-rich rocks. Discriminant analysis of biogeochemical data on bark samples of **Eucalyptus lesouefii** at Spargoville was used to delineate the nature of the bedrock. A large amount of biogeochemical work has been carried out in Canada including work on silver at Cobalt, Ontario. Work in Central Africa has been centred mainly around the copper plant **Becium homblei** and the nickel plant **Pearsonia metallifera**. A number of recent developments in Fennoscandia are described, particularly on bog plants in Finland. A large number of nickel plants have been found over nickeliferous rocks in New Caledonia and many of these have a nickel content which correlates well with the nickel content of the soil. By far the greatest volume of biogeochemical work has been carried out in the Soviet Union. This work is summarized in tabular form. An interesting development in this work is the recognition by Alexander Kovalevsky that plant species and their organs may be classified according to their resistance to metal uptake. "Barrier-free" organs of selected species are clearly the best for biogeochemical prospecting. Most of the biogeochemical work in the United States has been centred around the Denver area and has involved a search for gold, zinc, copper and molybdenum.*

Important recent developments include the use of plant exudates for prospecting. Air sampling of these exudates has been suggested. Chemical analysis of herbarium specimens is another procedure which has been developed and which shows some promise for identification of plant species which may indicate mineral deposits or rocks potentially favourable for mineralization.

Résumé

*Des méthodes biogéochimiques qui s'appuient sur une analyse chimique de la végétation, sont utilisées depuis la seconde guerre mondiale. En Australie, on a effectué des travaux sur l'espèce **Hybanthus floribundus**, plante hyperaccumulatrice de nickel, qui indique la présence de roches nickelifères. On a fait une analyse approfondie des données biogéochimiques sur des échantillons d'écorce d'**Eucalyptus lesouefii** à Spargoville pour définir la nature du socle. Au Canada, il y a eu de nombreuses recherches biogéochimiques en particulier sur l'argent à Cobalt en Ontario. En Afrique Centrale, on s'est plutôt concentré sur la plante concentratrice de cuivre **Becium homblei** et la plante concentratrice de nickel **Pearsonia metallifera**. On a décrit un certain nombre de découvertes récentes en Fenno-Scandie, concernant surtout les plantes de tourbières de Finlande. On a signalé la présence d'un grand nombre de plantes concentratrices à nickel sur des roches nickelifères de Nouvelle-Calédonie; leur teneur en nickel concorde bien avec celle du sol. C'est un URSS qu'a été effectué le plus grand nombre de travaux de biogéochimie et ceux-ci sont présentés sous forme de tableau. Un point intéressant de cette recherche, est la constatation faite par Alexander Kovalevski, que les espèces végétales et leurs organes peuvent être classifiés selon leur résistance à l'absorption de métal. Il est clair que les organes de certaines espèces, où rien ne s'oppose à l'absorption des éléments métalliques conviennent le mieux à la prospection biogéochimique. En général, aux États-Unis, la prospection biogéochimique était concentrée dans la région de Denver et portait sur l'or, le zinc, le cuivre et la molybdène.*

Comme innovation récente d'importance, citons l'analyse des sécrétions végétales parmi les méthodes biogéochimiques. On a suggéré de prélever des échantillons d'air pouvant contenir ces sécrétions. On a mis au point un autre procédé consistant à faire l'analyse chimique de spécimens botaniques, et qui pourrait aider à identifier des espèces végétales indicatrices de gîtes minéraux ou de roches favorables à la minéralisation.

INTRODUCTION

In contrast to geobotanical methods of prospecting which involve visual observations, biogeochemical methods involve chemical analysis of vegetation. The methods were first used just before World War II when Tkalich (1938) found that vegetation could be used to delineate a Siberian iron orebody. Since then, much biogeochemical work has been carried out in the Soviet Union, Canada, the United States, Australasia and Scandinavia.

Analysis of the accumulation of elements in vegetation and the upper humic layer of soils is the basis of biogeochemical prospecting. The mechanisms whereby plants accumulate trace elements are extremely complicated but in essence involve uptake via the root system, the passage of the elements through the aerial parts of the plants into organs such as the leaves and flowers, and finally a return of these elements to the upper layers of the soil when the leaves or flowers wither and fall. The elements are then leached through the various soil horizons and are reaccumulated by vegetation in a series of steps known as the biogeochemical cycle.

Trees and shrubs with long root systems can effectively sample beneath unconsolidated overburden and under favourable conditions can indicate the existence of minerals at depth. This possibility is the greatest advantage of the biogeochemical method, but this advantage is worthless unless accumulation of trace elements from depth is achieved in a reproducible manner and to a degree which is proportional to the concentration of the element or elements which are sought. Although there are a large number of factors which can affect reliability of the biogeochemical method, it is still possible to apply the technique successfully, and the purpose of this review is to discuss approaches adopted in various countries under varying field conditions and to highlight the course of probable future developments. As far as possible, only fairly recent references (mainly after 1970) will be given. The reader is referred to Malyuga (1964) and Brooks (1972) for fuller listing of earlier references.

BIOGEOCHEMISTRY IN SELECTED REGIONS
Australia

The earliest biogeochemical work carried out in Australia was that of Debnam (1955) who examined vegetation in Northern Territory for uranium indicators, but found no species as reliable as soil in indicating ore deposits.

A great deal of work has been carried out throughout Australia by Cole and her co-workers (Cole, 1965; Nicolls et al., 1965; Cole et al., 1968; Elkington, 1969; Cole, 1971). Most of this work was geobotanical in nature and is described elsewhere (see report 18A, this publication).

Other biogeochemical investigations have been reported by Nielsen (1972), Severne (1972), Severne and Brooks (1972) and Cole (1973), indicated the unusually high nickel-accumulating ability of **Hybanthus floribundus** which can contain up to 23% nickel in the ash (i.e. over 1% on a dry-mass basis). The plant chemistry of **H. floribundus** has been studied by Severne (1972), Kelly et al. (1975) and Farago et al. (1975). Although it is now well established that **H. floribundus** is an accumulator of nickel, its role in mineral exploration is less clear. Severne (1972) concluded that **H. floribundus** appeared to be useful as an indicator plant for soils containing more than 0.04% nickel and Cole (1973) deduced that although this species indicates a nickeliferous environment it does not necessarily delineate a nickel sulphide orebody. The same author concluded that the nickel content of this species is related to the concentration of this element in soils and that the plant has some significance for biogeochemical prospecting.

Studies by Nielsen (1972) and Nielsen et al. (1973) have shown the possibility of using discriminant analysis of multielement data to deduce the nature of bedrock from chemical analysis of bark samples of **Eucalyptus lesouefii** from the Spargoville area of Western Australia. Samples were collected from 63 sites of known geology and analyzed for calcium, chromium, cobalt, copper, lead, magnesium, manganese, nickel and zinc. The computer was used to formulate a regression equation of the form:

$$Y = a_1 (Ca) + a_2 (Cr) + a_3 (Co) + a_4 (Cu) + a_5 (Pb) + a_6 (Mg) + a_7 (Mn) + a_8 (Ni) + a_9 (Zn) + C$$

The coefficients a_1-a_9 were chosen to maximize the difference in scores for samples derived from two geological units (ultrabasics and amphibolites). Using the regression equation, it was possible to predict the nature of the bedrock with a certainty of 71%, from the analysis of any one sample. The method has obvious potential for areas where the bedrock is not easily available for analysis.

Other biogeochemical work in Australia includes the studies of Groves et al. (1972) at Herberton, North Queensland who showed that analysis of copper, lead, tin and zinc in **Scleria brownei** and **Coelospermum reticulatum** could be used to delineate an orebody.

Hall et al. (1973) investigated the use of **Melaleuca sheathiana** to delineate nickel mineralization at Norsewood, Western Australia and concluded that analysis of leaf material did indeed indicate mineralization in the bedrock.

To summarize, a number of biogeochemical investigations have been carried out by researchers in Australia during the past 20 years. Most large exploration companies have their own geobotanical/biogeochemical programs, though most of the published work has been done by outside experts working in collaboration with exploration companies.

Canada

A survey of biogeochemistry in Canada cannot fail to give credit to H.V. Warren and his associates from the University of British Columbia, for pioneering work on the method in that country. In the period 1947-1975 nearly 30 scientific papers on this subject were published. Much of this work has been summarized by Warren (1972) but because many of this author's publications predate 1970, and because of the emphasis on later references in this review, the reader is referred to Brooks (1972) for more complete bibliography. Warren's latest work has been concerned with analysis of vegetation for gold (Warren and Hajek, 1973) and with the use of barium/strontium ratios in geochemical prospecting (Warren et al., 1974). Warren must also be given credit for having trained a number of biogeochemists who have carried on this work elsewhere in Canada.

Apart from Warren's group at British Columbia, an appreciable amount of biogeochemical work has been carried out during the past decade by workers at the Geological Survey of Canada. The earliest of these publications are due to Fortescue and Hornbrook (1967a, 1967b, 1969) who reviewed the progress of biogeochemical research at the Geological Survey during the period 1963-1966. Perhaps one of the most interesting developments of this period was the establishment of a mobile biogeochemical laboratory (Fortescue and Hornbrook, 1967a) which was used for the routine analysis of plant material using emission spectrography. These authors have also been responsible for a number of orientation surveys in which the efficacy of the biogeochemical method was tested at various locations in Canada such as: Timmins, Ont., and Gaspé Park, Quebec (Fortescue and Hornbrook, 1969); west-central British Columbia (Hornbrook, 1969a, 1970a); Chalk River, Ont. (Hornbrook, 1970b); Cobalt, Ont. (Hornbrook, 1971, Hornbrook and Hobson, 1972); Coppermine River, N.W.T. (Hornbrook and Allan, 1970).

Elements investigated, included cobalt, copper, lead, manganese, molybdenum, nickel, silver and zinc. The same authors have also been responsible for two other reviews of biogeochemical methods (Hornbrook, 1969a, b; Fortescue, 1970).

It is impossible to review in detail all the work carried out by the above authors, but as a case history, we may briefly consider the work of Hornbrook (1971). Geochemical and biogeochemical exploration methods for silver were compared at Cobalt, Ontario. Plant samples and soils were collected from 452 stations at 25 foot intervals over six traverse lines 100 feet apart and orientated perpendicular to the strike of the silver veins. The most suitable plant organs were spurs of white birch (**Betula papyrifera**) followed by twigs of trembling aspen (**Populus tremuloides**). Anomaly maps for silver, cobalt, copper, nickel, lead, zinc and manganese were compared for birch spurs and soils from the $A_0 + A_1$ horizons.

The principal findings of the survey were that:

1. Silver in soils was the most effective in delineating the silver veins. Individual leakage halos from sub-outcrop fractures were detected and could be related to the projected path of principal ore veins at depth.
2. Other effective anomaly maps were: lead, manganese and zinc in soils and birch spurs, and cobalt and nickel in soils.
3. Copper anomaly maps were not effective.
4. The cobalt anomaly map for birch spurs showed preferential enrichment of cobalt in the western portion of the region. This was delineated by the soil map. The enrichment of cobalt in the western part has been established by drilling and is due to zoning of the silver veins.
5. On many anomaly maps, and most obviously on the map of silver in birch spurs, the most interesting anomalies were associated with surface contamination and not the principal ore veins.

It is clear from this survey that biogeochemistry is not an end in itself, but if combined with other methods can furnish additional useful information.

Other workers in the field of biogeochemistry in Canada include Lily Usik who has reviewed geochemical and biogeochemical prospecting methods in peatlands (Usik, 1969). This excellent review (178 references) gives a good coverage of the field, and is of particular importance because the location of ore deposits beneath organic swamp soil is one of the principal prospecting problems in Canada since water-saturated organic terrain occupies at least 1 250 000 square km (one sixth) of this country.

Further recent biogeochemical investigations are due to staff of the University of British Columbia. Doyle et al. (1973a) studied plant-soil relationships for molybdenum, copper, zinc and manganese in the Yukon. The molybdenum content of several species correlated well with the content of this element in soils. Further investigations on the molybdenum content of plants, soils and bedrock in the same area are reported by Doyle et al. (1973b). Fletcher et al. (1973) investigated the selenium content of Yukon plants and showed a correlation with selenium in soils.

Wolfe (1971, 1974) carried out biogeochemical investigations in eastern Canada. A study of molybdenum mineralization at Setting Net Lake in northwestern Ontario showed that only molybdenum distribution patterns in black spruce needles (**Picea mariana**) showed a reasonable correlation with levels of this element in soils. Copper anomalies in vegetation were extremely shallow in comparison.

Despite the very considerable amount of biogeochemical work carried out in Canada (probably greater than in any other country outside the Soviet Union), there is little evidence that the art has left the hands of the experts and has been taken over by exploration companies on anything approaching a widespread basis.

Central and Southern Africa

By far the greatest proportion of botanical prospecting in central and southern Africa has involved geobotany rather than biogeochemistry (see report 18A, this publication). The pioneering work of Duvigneaud (1958) in Katanga (Zaïre) was later followed by other work in central Africa, mainly centred around the copper indicator **Becium homblei** (Reilly, 1967; Howard-Williams, 1970). Other geobotanical work was carried out by Cole (1971) in South-West Africa and Botswana.

Certainly the most active centre of this type of research is at the University of Rhodesia where Professor Wild and his associates have carried out extensive work, mainly of a geobotanical nature (see report 18a, this publication).

The chromium, nickel, copper and cobalt content of several indigenous species has been investigated by Wild (1974) who found (all on an ash mass basis), 48 000 µg/g chromium in **Sutera fodina**, and 153 000 µg/g nickel and 3300 µg/g cobalt in **Pearsonia metallifera**. The same author Wild (1970) studied the relationship between the nickel content of eight species of plants and the nickel content of the substrate. Some of the species such as **Becium obovatum** (which is also an indicator of copper) and **Securidaca longepedunculata** showed some potential for biogeochemical prospecting.

As far as known, none of the biogeochemical work in central and southern Africa has progressed beyond the stage of looking at elemental levels in vegetation. This is in sharp contrast with geobotany which is very highly developed in central Africa.

Czechoslovakia

The first biogeochemical work in Czechoslovakia was a study by Nemec et al. (1936) on the alleged gold accumulation by **Equisetum arvense** (see also Cannon et al., 1968). Recently, a paper by Matula (1973) has shown that pine needles could be used in the Spissko-Gemersky region to prospect for chalcophile elements over sulphide deposits. The deposit was delineated by an increase of copper levels from 32 µg/g to 340 µg/g (ash mass basis). Similar results for tin and tungsten were obtained over a greisen zone.

Fennoscandia

Finland, Norway and Sweden tend to be grouped together for prospecting methods because of a similarity of terrain and geological conditions. The main problems are associated with glaciation, where the original soil cover has been removed and where the glacial till does not represent bedrock. Under such conditions, vegetation, with its ability to penetrate to bedrock, would seem to be a natural material for sampling. It is not surprising therefore that biogeochemical prospecting began in Scandinavia at about the same time as in the Soviet Union. The earliest work was that of Brundin (1939) who took out a patent for the method. This work was then followed by pioneering investigations by Vogt (1939) in the same year. One of the most useful works on biogeochemical prospecting in Fennoscandia is by Kvalheim (1967) who summarized work carried out in Norway, Sweden and Finland.

Most of the biogeochemical work in Norway has been centred around Trondheim, both at the University and at the Norwegian Geological Survey. Recent publications include a study of vegetation in lead-rich areas (Låg et al., 1969) where lead-tolerance of many species was investigated. In a later paper (Låg and Bølviken, 1974) this work was extended to encompass implications in the fields of soil chemistry and epidemiology.

Though Sweden was the home of much of the original biogeochemical work in the late 1930s, most of the subsequent biogeochemical work has been concerned with pollution studies from heavy metals and in particular with the use of mosses for measuring industrial pollution.

Nevertheless, during the past decade there has been an upsurge of interest in biogeochemical prospecting. For example, Fredriksson and Lindgren (1967) used leaves of **Betula nana** and **Salix polaris** to investigate a sulphide orebody at Västerbotten in northern Sweden. Plants growing over the Levi orebody gave a definite anomaly (3 to 5 times background). Elsewhere the results were unreliable and uptake of copper appeared to be influenced by the degree of drainage.

In investigations over a uraniferous bog at Masugnsbyn, northern Sweden, Armands (1967) measured the alpha activity of leaves and twigs of birch (**Betula nana** and **B. alba**), willow (**Salix**) and alder (**Alnus**). Willow twigs contained up to 860 µg/g uranium in the ash. The fruit and leaves contained up to 450 µg/g. In general there was good agreement between uranium levels in plants and in the peat substrate.

The relationship between heavy metal uptake of spruce (**Picea abies**) needles and variable edaphic factors of the soil was studied by Nilsson (1972) in southern Sweden. Though not specifically oriented towards biogeochemical prospecting, the work nevertheless has important implication for this field. A similar project (Tyler, 1970) involved studying the distribution of lead in a coniferous woodland ecosystem.

Perhaps one of the most extensive Swedish biogeochemical studies in recent years was carried out by Ek (1974) over a sulphide deposit in northern Sweden. Various organs of common trees (birch, pine and spruce) were analyzed for Zn, Cu, Pb, Ni, Co, Cr, Fe, Mg, Ba and Ca. The only material giving anomalously-high values of Cu, Pb and Zn was birch bark. Most of the orebody could be delineated by levels of these three elements in the plant material.

After a long absence from biogeochemical work, Brundin reinvestigated this field by studying the use of organic material (stream peat) in stream sediments as an alternative to inorganic matter (Brundin and Nairis, 1972). This method was found to be superior to conventional methods for regions covered with glacial till. This work was later extended still further by Brundin (1975) using living roots of trees growing on stream banks. He considered that **Carex** roots were a suitable alternative to organic material in streams.

There can be no doubt that most of the biogeochemical work in Fennoscandia has been carried out in Finland in the vicinity of mineral bogs which are so common in that country. Salmi (1956) studied vegetation and peat in the vicinity of the Vihanti mining camp and showed that vegetation could be used to pinpoint anomalous parts of the bog. In a later review (Salmi, 1967), the same author summarized the "state of the art" as regards mineral bogs. Erämetsä et al. (1969) have studied the vertical distribution of uranium and other metals in peat bogs and have shown that distribution is controlled by solid humic acids and results in a concentration of elements at about one quarter of the distance from the bottom of the bog.

Lounamaa (1956) published a monumental paper on elemental abundances in various plants growing on different substrates in Finland. This work which was later extended (Lounamaa, 1967), has provided useful information on expected elemental abundances in various plants growing over specific substrates in Finland.

Kontas has studied molybdenum in till and birch leaves at Sarvisoavi (Kontas, 1976a) and Lahnanen (Kontas, 1976b) but found little response to anomalies present in the till or bedrock.

Yliruokanen (1975a, 1975b) has been responsible for determining uranium, thorium, lead, yttrium and rare earths in Finnish plants.

In the course of analyzing 172 plants (Yliruokanen, 1975a), the author determined normal background levels of these elements. Anomalously high values were only obtained for specimens growing directly over mineralization. Nevertheless (Yliruokanen, 1976b) the rare earth content of the plants appeared to reflect the rare earth content of the associated soils and rocks.

The selenium content of Finnish plants, peats, soils and rocks has been studied by Koljonen (1974, 1975) who reported up to 2.3 µg/g selenium in forest humus overlying sulphide-rich rocks. An investigation of the selenium content of 25 plant species (Koljonen, 1974) gave values ranging from 0.010 µg/g in **Picea excelsa** to 0.420 µg/g in **Alnus incana**. Higher uptakes were unusual because of the low mobility of selenium in Finnish soils.

A thorough investigation of the biogeochemical method was carried out by Bjørklund (1971) who sampled over 4000 birch twigs and soils over the Korsnäs lead deposit in western Finland. Statistical methods were used to improve the biogeochemical data and to give a better plant-soil correlation for Zn, Pb, Co and Cu. This was achieved by standardization of such variables as vegetation type, height of sampled trees and length of sampled shoots.

Because of the nature of the vegetation cover in Finland, particularly in northern latitudes, it is not surprising that a good deal of work has been carried out on the elemental content of mosses and lichens (Lounamaa, 1956). Several other surveys have been carried out and include recent work by Erämetsä and Yliruokanen (1971a, 1971b). Up to 4900 µg/g uranium was found in samples from an abandoned uranium mine, and extensive data are recorded for background (normal) levels of the rare earths and other elements in 90 lichens and 142 mosses from all over Finland.

The overall position of biogeochemistry in prospecting in Fennoscandia is dictated by the overriding problem of exploration in these glaciated areas as summarized by Brotzen et al. (1967). Because of the ubiquitous presence of glacial till, many geochemical methods (such as soil sampling) tend to be unreliable. In such cases, other methods such as biogeochemistry can play a useful role in reinforcing geochemical techniques in the search for elusive anomalies in bedrock.

Germany

A large amount of biogeochemical work has been carried out by W. Ernst in Germany. A useful reference is his 1967 summary of work on plant communities growing over heavy metal deposits (Ernst, 1967). This bibliography comprises references numbering 82 from Europe, 2 from Asia, 6 from Africa, and 6 from North America. Professor Ernst's recent work has been oriented towards plant tolerances to heavy metals, with obvious significance for the fields of geobotany and biogeochemistry in mineral exploration. He is author of a useful book on the subject (Ernst, 1974).

India

Some biogeochemical work has been carried out in India. Chowdhury and Bose (1971) examined humic complexes of several metals and discussed mechanisms whereby these could be removed from the soil. Gandhi and Aswathanarayana (1975) studied a possible base-metal indicator in south India. The plant, **Waltheria indica** appeared to be confined to mineralized ground and contained anomalously high values of Cu, K, Mn, Na, Rb, Sr, and Zn.

Italy

Biogeochemical investigations in Italy have been confined almost entirely to the work of Vergnano Gambi who was instrumental in discovering the first hyperaccumulator of nickel (**Alyssum bertolonii**) in 1947 (Minguzzi and Vergnano, 1948). This was followed by several other papers on the ecology and plant chemistry of this and other serpentine plants (see Brooks, 1972 for a fuller list of papers up to 1970). The same author (Vergnano Gambi et al., 1971) studied manganese uptake by various plants of the Appennines. **Vaccinium myrtillus** showed a remarkable accumulation (nearly 300 µg/g on a dry mass basis). The metabolism of this and other accumulators of manganese seemed to be characterized by a lower uptake of iron and consequently high Mn/Fe ratio.

Middle East

Because of arid conditions and the presence of many deep-rooted plant species, the Middle East would seem to be a good locality for successful use of the biogeochemical method. In Egypt, El Shazly et al. (1971) used **Acacia raddiana** and **A. ehrenbergiana** for prospecting for chalcophile elements such as Co, Cu, Ni, Pb and Zn. Vegetation was superior to alluvia or water in delineating anomalies in bedrock. An investigation of uptake of B, Be and Li by **A. raddiana** showed that the biogeochemical method was superior to alluvium sampling for Li but inferior for the other two elements.

Biogeochemical work has also been carried out by Allcott (1970) in Saudi Arabia.

New Caledonia

Some biogeochemical work and an appreciable amount of geobotanical work has been carried out in New Caledonia by French scientists at O.R.S.T.O.M. (Organisation de la Recherche Scientifique et Technique Outre-Mer) and by New Zealand scientists working in collaboration with them. The geobotanical work is described in Paper 18A of this review. New Caledonia is unusual in having one of the largest ultrabasic areas in the world with a serpentine flora containing a large proportion of endemic species. This flora is noteworthy in that it contains a high proportion of hyperaccumulators of nickel. So far about 15 of these hyperaccumulators (containing >1000 µg/g Ni on a dry-mass basis) have been discovered (Jaffré et al., 1971; Brooks et al., 1974; Jaffré and Schmid, 1974; Brooks et al., 1974; Jaffré et al., 1976). These hyperaccumulating plant species are always found over nickeliferous ultrabasic substrates though they do not necessarily indicate the presence of economic mineralization in the bedrock. Plant-soil relationships have been studied by Lee et al. (1977) for the hyperaccumulators **Homalium kanaliense** and **Hybanthus austrocaledonicus**. They concluded that although the nickel content of **H. austrocaledonicus** was correlated with the nickel content of the soil (i.e. this species was suitable for biogeochemical prospecting), the paucity of statistically-significant plant-soil relationships for nickel and other metals, indicated that organic constituents may have a role in controlling nickel levels in the plant.

New Zealand

Biogeochemical work in New Zealand began in 1965 when Brooks and Lyon (1966) prospected for molybdenum using **Olearia rani**. A molybdenum anomaly was delineated in combination with a soil sampling study. This work was followed by numerous investigations during the succeeding decade involving work on nickel (Timperley et al., 1970a), uranium (Whitehead and Brooks, 1969), tungsten (Quin et al., 1974), copper (Yates et al., 1974a), and zinc and lead (Nicolas and Brooks, 1969).

The theoretical basis of biogeochemical prospecting has been investigated by Brooks (1973) and Timperley et al. (1970b) who showed that the biogeochemical method is less effective for elements which are essential in plant nutrition. This is particularly true for copper and zinc. Unessential elements such as nickel and uranium tend to give much more satisfactory results.

The New Zealand workers have been instrumental in attempting to quantify biogeochemical prospecting by statistical procedures. Timperley et al. (1972a) used multiple regression analysis to reduce the variance of plant/soil ratios for copper and nickel by correcting for a number of chemical and physical variables in plants, soils and the environment. The same workers (Timperley et al., 1972b) applied trend surface analysis to biogeochemical data. Factor analysis of biogeochemical data was used by Yates et al. (1974b) in order to delineate copper anomalies at Coppermine Island, New Zealand.

The New Zealand workers have also investigated edaphic factors controlling a serpentine flora in Nelson Province. Lee et al. (1975) showed that an endemic nickel-accumulating species (**Pimelea suteri**) had a distribution controlled mainly by excess magnesium levels in the soil. Much of the New Zealand work has been reviewed by Brooks (1972).

Soviet Union

The volume of biogeochemical exploration work carried out in the Soviet Union far exceeds that of any other country. The volume is so great, that it is hardly feasible to detail it within the confines of the present review. There are several standard textbooks on biogeochemistry in mineral exploration (i.e. as distinct from geobotany). The most important of these is a work by Malyuga (1964). This book was the first to appear on the subject and has a useful bibliography of nearly 400 references. This was followed by a book by Talipov (1966) which lists 98 Russian references. Other standard texts include Nesvetaylova (1970), Tkalich (1970), Safronov (1971) and more recently, Kovalevsky (1974a). The latter work has 104 references. Kovalevsky is perhaps the most prolific of the Russian biogeochemists. He has also edited (Kovalevsky and Perel'man, 1969) a series of essays on biogeochemical prospecting. This work lists several hundred references. Kovalevsky (1975) is also author of another more specialized text (350 references) on biogeochemical aureoles. Another useful general work (mainly epidemiological but including biogeochemical prospecting) listing over 600 references is a collection of essays by Koval'sky (1974).

Study of the literature shows that a large part of the biogeochemical prospecting research is centred at the Vernadsky Institute (Moscow), Institute of Geology and Geophysics (Tashkent) and at the Buriat Interscience Research Institute (Ulan Ude). Leading biogeochemists include I.K. Khamrabaev and R.M. Talipov (Tashkent), V.V. Koval'sky (Moscow) and A.L. Kovalevsky (Ulan Ude). Because of the difficulty of covering in detail the large volume of Soviet biogeochemical literature, a selection of later references (mainly after 1970) is given in Table 18B.1.

The reader is referred to Brooks (1972) and Malyuga (1964) for Russian references before 1970.

United Kingdom

Although the volume of biogeochemical work carried out by researchers in the United Kingdom is not great, this

Table 18B.1

A summary of recent Soviet literature on Biogeochemistry in Mineral Exploration

Author	Date	Topic or elements investigated
Alekseyeva — Popova	1970	Various
Do Van Ai	1972	B, Ba, Cu, Mn, Ni, Sr
Dvornikov and Ovsyannikova	1972	Chalcophile elements
Grabovskaya and Kuzmina	1971	Be, Li, Mo, Nb, Pb, Sn
Gruzdev and Rubtsov	1972	Ra, Th, V
Ivashov and Bardyuk	1971	Zr
Kovalevsky	1969	Zn
Kovalevsky	1971	Theoretical review
Kovalevsky	1972	V
Kovalevsky	1974a	Standard text
Kovalevsky	1974b	Be
Kovalevsky	1974c	Various
Kovalevsky	1975	Standard text
Kovalevsky and Perel'man (eds.)	1969	Series of essays
Koval'sky	1974	Series of essays
Koval'sky et al.	1973	B
Letova	1970	Various
Malyuga	1964	Standard text
Malyuga and Aivazyan	1970	Various
Melikyan	1972	B
Mitskevich	1971	Various
Molchanova and Kulikov	1972	Radioactive isotopes
Nesvetaylova	1970	Standard text
Ovchinikov and Baranov	1970	Chalcophile elements
Panin and Panina	1971a	Co
Panin and Panina	1971b	Zn
Panin and Schetinina	1974	B
Prozumenshchikova	1972	Mo
Safronov	1971	Standard text
Skarlina-Ufimtseva and Berezkina	1971	Chalcophile elements
Shchulzhenko et al.	1970	Chalcophile elements
Sudnitskaya	1971	Be
Talipov	1966	Standard text
Talipov and Glushchenko	1974	Au
Talipov, Glushchenko et al.	1973	Au
Talipov, Glushchenko et al.	1974	Au and related elements
Talipov, Glushchenko et al.	1975	Au
Talipov, Glushchenko et al.	1976	Au and Sb
Talipov, Karabaev and Akhunkhodzhaeva	1971	Various
Talipov and Khatamov	1973	Various
Talipov and Khatamov	1974	Various
Talipov, Musin et al.	1974	Various
Talipov, Tverskaya et al.	1976	U and Au
Talipov, Yussupov and Khatamov	1970	Au
Yussupov, Talipov, Khatamov et al.	1970	Lichens in prospecting
Yussupov, Talipov, Yussupova et al.	1970	Various
Yussupova et al.	1970	Various

work is nevertheless of considerable historical interest because it was in England that the earliest work was carried out (Brundin, 1939). This work was continued just after World War II by J.S. Webb and his collaborators at Imperial College (e.g. Millman, 1957), again in southwest England. The same workers based at Imperial College have also carried out several other biogeochemical studies overseas (e.g. Nicolls et al., 1965).

A promising development in biogeochemical exploration is the potential of neutron activation analysis when applied to this work. Such studies are now underway at Westfield College, London under the direction of P.J. Peterson. Elements investigated include gold and arsenic (Minski et al., 1977).

United States

Despite the tremendous impetus given to biogeochemical prospecting by the well-known work of the U.S. Geological Survey in the 1950s and 1960s (e.g. Cannon, 1960; Cannon, 1964), the amount of direct research in this field has declined in the past decade. This is partly due to a shift of emphasis into the related fields of pollution, epidemiology and general environmental chemistry. Insofar as such studies involve a study of natural levels of trace elements in vegetation, they may legitimately be classified as investigations of use for biogeochemical prospecting and are therefore included in this review. References are in the main confined to publications since 1970. The reader is referred to Brooks (1972) for a listing of earlier work.

Virtually all biogeochemical prospecting work in the United States has been carried out at the U.S. Geological Survey where pioneering work was carried out by H.T. Shacklette, H.L. Cannon and many others. The investigations of each worker will be considered separately.

Natural levels of eighteen elements in vegetation of Georgia were determined by Shacklette, Sauer et al. (1970) as part of an epidemiological study. Vegetation sampled, included vegetables and eight species of common trees. The absorption of gold by plants was studied by Shacklette, Lakin et al. (1970), Lakin et al. (1974), and has been reviewed by Jones (1970). Laboratory experiments showed that gold uptake was largely a function of the complexing agent used in the test solutions. Natural levels of gold in lodgepole pines and aspens near a gold-bearing vein in Colorado were up to 1.96 µg/g (ash mass) in pine wood and up to 1.0 µg/g (ash mass) in aspen wood. Shacklette (1970) reported mercury levels in 196 native trees and shrubs of Missouri. All specimens contained <0.50 µg/g mercury. However species growing over a cinnabar deposit in the lower Yukon contained up to 3.5 µg/g (**Ledum palustre**) and there was a minimum of 0.5 µg/g in **Betula papyrifera**. The biogeochemical method has been reviewed along with other techniques by Dorr et al. (1971). The same review proposed priorities for future research in tropical regions. A comprehensive survey of trace element levels in vegetation of Missouri has been published in six open-file reports (e.g. Shacklette, 1972a) and by Erdman and Shacklette (1973). The work involved analysis of 19 elements in a large number of smooth sumac (**Rhus glabra**) stems taken from six different vegetational areas of the state. The cadmium content of plants has also been investigated (Shacklette, 1972b). Numerous values are presented for many plant species obtained from different environments in the United States some of the highest values were up to 40 µg/g in the ash of willow (**Salix** sp.) from mineralized areas in Colorado. Much of Shacklette's work on elemental levels in natural vegetation has been summarized in a recent paper (Connor and Shacklette, 1975). Values are given for 48 elements in several hundred specimens of 18 native plant species taken from 147 landscape units in the United States. Perhaps one of the most significant advances in biogeochemical exploration is the use of this technique for on-site inspections of suspected underground nuclear explosions (Shacklette et al., 1970).

Most of H.L. Cannon's publications are listed in Paper 18A of this review because of the preponderance of geobotanical papers in her bibliography. However, mention should be made of an extensive survey carried out by Cannon et al. (1968) on 21 trace elements in the horsetail (**Equisetum**). This investigation showed that the earlier reputation of this species as a gold-accumulator (Nemec et al., 1938) was completely unfounded and that the earlier high values were probably due to analytical error. In recent years Cannon and her co-workers have been increasingly involved in studies of elemental concentration in vegetation as a part of epidemiological work (Cannon, 1970, 1974). The geochemist's involvement with the pollution problem has been discussed by Cannon and Anderson (1971) and by Cannon and Hopps (1971, 1972). The same authors have published a useful paper on problems of sampling and analysis in trace element investigation of vegetation (Cannon et al., 1972).

Another active worker in the field of biogeochemical prospecting is M.A. Chaffee who has analyzed specimens of **Olneya tesota** (ironwood), **Cercidium microphyllum** (foothill palo-verde) and **Larrea tridentata** (creosote bush) for copper, manganese, molybdenum and zinc in the vicinity of porphyry copper deposits in Arizona (Chaffee and Hessin, 1971). Zinc, copper and molybdenum plots in the ash of all three species gave anomalies above a concealed deposit whereas no anomalies were evident in the soil. It was concluded that biogeochemical prospecting would be a useful tool for this particular area. In a review of geochemical techniques applicable in the search for copper deposits, Chaffee (1975) concluded that use of deep-rooted phreatophytes such as **Prosopis juliflora** (mesquite) renders biogeochemical prospecting an effective tool in the search for copper in arid environments.

The mesquite has also been used by Huff (1970) and Brown (1970) for biogeochemical prospecting. In the Pima district of Arizona, Huff (1970) detected anomalous molybdenum concentrations in the ash of mesquite stems collected 13 km away from a mineralized area. Brown (1970) discovered anomalous concentrations of copper in the stem ash of samples of mesquite collected over the Kalamazoo deposit at San Manuel, Arizona.

A brief mention will also be made of the work of Curtin et al. (1970, 1971), who studied the mobility of gold in forest humus and used it for prospecting in Colorado. The humus derived from pine and aspen proved to be a more reliable indication of gold in bedrock, than the soil, pebbles and cobbles beneath the humus. It was concluded that the technique shows promise for regions where bedrock is covered by a transported overburden.

FUTURE DEVELOPMENTS

It is clear that radically new biogeochemical techniques will need to be applied in the future if the method is to prosper. In the past, far too many biogeochemical procedures have merely involved measuring background elemental concentrations in vegetation or trying out the method over previously discovered ore deposits or anomalies. There are two fields in which the method may possibly show significant advances in the future. The first of these involves the use of multielement analysis of vegetation to detect subtle anomalies in the substrate. The data can be processed statistically, by some form of discriminant analysis (e.g. Nielsen et al., 1972) which compensates for the natural variability of the plant/substrate ratio for each element in each species. This sort of procedure will be particularly useful in environments with an overburden not representative of bedrock and particularly where phreatophytes with long root systems can be sampled (e.g. Huff, 1970; Chaffee, 1975).

A new development in biogeochemical prospecting involves the analysis of herbarium specimens. A geobotanical use of herbaria has already been made on a limited scale in the past. For example Persson (1956) noted the collection localities of Swedish herbarium specimens of the copper moss **Mielichhoferia mielichhoferi** and discovered three localities in Sweden with anomalous copper levels in the substrate (one turned out to be an existing copper mine). Later Cole (1971), carried out plant mapping of a mineralized area in southern Africa, identified characteristic plants of the region, and referred back to a herbarium for other collection localities of these species.

Until recently however, chemical analysis of herbarium material has not been feasible, because the size of sample needed for classical methods of analysis had been so great (5-10 g of plant material) that herbarium curators would hardly have tolerated such a disturbance of their collections. With the advent of atomic absorption spectrophotometry and particularly with the development of the ancilliary carbon-rod atomizer, it is now possible to analyze for several elements in tiny leaf samples less than 1 sq. cm in area (i.e. about 0.02 g). Brooks, Lee et al. (1977) used herbarium material to analyze over half of all species (nearly 2000 specimens) of the genera **Hybanthus** and **Homalium** and identified several new hyperaccumulators of nickel (>1000 µg/g in dried leaves) all of which were associated with nickeliferous substrates. The results showed that it was possible to identify most of the world's major ultrabasic areas within the tropical and warm-temperate zones by means of elevated nickel levels in the plants. This work has now been extended to a biogeochemical survey of eastern Indonesia (Celebes and Moluccas) using specimens supplied from several major herbaria. Several specimens of a species from a nickeliferous area in the Celebes were analyzed and resulted in the identification of a new hyperaccumulator of nickel (**Rinorea bengalensis**). Further herbarium specimens of this species collected from all over southeast Asia were then analyzed and resulted in the location of a previously unknown nickeliferous ultrabasic area in West Irian (Brooks and Wither, 1977). This latter specimen had been collected in 1940 by two Japanese botanists, but its significance had remained unknown for over 25 years.

Other herbarium work on species from central Africa (Brooks, 1977), has resulted in the identification of a new hyperaccumulator of cobalt (**Haumaniastrum robertii**) which is only the second known hyperaccumulator of this element (i.e. in addition to the previously known **Crotalaria cobalticola** discovered by Duvigneaud, 1958). It has also been shown that cobalt accumulation is a universal characteristic of the genus **Nyssa** (Brooks, McCleave and Schofield, 1977) and is not confined to **Nyssa sylvatica** var. **biflora** (Beeson et al., 1955).

There is however an inherent paradox in herbarium work of this nature. The more successful it becomes, the greater the demands that will be placed upon herbaria, and the greater will be the resistance of curators to furnishing further material. This reluctance to disturb specimens can be answered to some extent by use of small samples (e.g. 2-3 mg) but then the problem arises as to whether such a small sample is representative or not of the whole leaf.

An exciting new biogeochemical development is the use of plant exudates for mineral prospecting. Curtin et al. (1974) showed that condensed exudates from common conifers such as **Pinus contorta, Picea engelmannii** and **Pseudotsuga menziesii** showed the presence of a large number of trace elements transported from the substrate. The author suggested air sampling of these exudates as a tool in mineral exploration. Further work in this field was carried out by Beauford et al. (1975) who also suggested the possibilities of airborne sampling programs.

Because new ore deposits are becoming progressively more difficult to find, and because such deposits are likely to be found in the more inaccessible (often well vegetated) regions of the earth, it is clear that the biogeochemical method will continue to have a place in future prospecting operations. It is not likely that it will be used by itself, but if used judiciously in combination with other methods, should continue to be a useful component of the exploration geochemist's armoury of techniques.

REFERENCES

Alekseyeva-Popova, N.V.
1970: Elemental chemical composition of plants growing on different rocks of the polar Urals (in Russ.); Bot. Zhur., v. 55, p. 1304-1315.

Allcott, G.H.
1970: Preliminary results from biogeochemical prospecting in Saudi Arabia; in Mineral Resources Research, 1968-1969, Saudi Arab. Direct. Gen. Min. Resour., Riyad, p. 122-124.

Armands, G.
1967: Geochemical prospecting of a uraniferous bog deposit at Masugnbyn, northern Sweden; in Geochemical Prospecting in Fennoscandia (ed. A. Kvalheim). Interscience, New York, p. 127-154.

Beauford, W., Barber, J., and Barringer, A.R.
1975: Heavy metal release from plants into the atmosphere; Nature, v. 256, p. 35-37.

Beeson, K.C., Lazar, V.A., and Boyce, S.G.
1955: Some plant accumulators of the micronutrient elements; Ecology, v. 36, p. 155-156.

Bjørklund, A.
1971: Sources and reduction of metal content variation in biogeochemical prospecting; Geol. Surv. Finl. Bull., 251, 42 p.

Brooks, R.R.
1972: Geobotany and biogeochemistry in Mineral Exploration; Harper Row, New York, 292 p.

1973: Biogeochemical parameters and their significance for biogeochemical prospecting; J. Appl. Ecol., v. 10, p. 825-836.

1977: Cobalt and copper uptake by the genus **Haumaniastrum**; Pl. Soil, v. 48, p. 541-545.

Brooks, R.R., Lee, J., and Jaffré, T.
1974: Some New Zealand and New Caledonian plant accumulators of nickel; J. Ecol., v. 62, p. 493-499.

Brooks, R.R., Lee, J., Reeves, R.D., and Jaffré, T.
1977: Detection of nickeliferous rocks by analysis of herbarium specimens of indicator plants; J. Geochem. Explor., v. 7, p. 49-57.

Brooks, R.R. and Lyon, G.L.
1966: Biogeochemical prospecting for molybdenum in New Zealand; N.Z. J. Sci., v. 9, p. 706-718.

Brooks, R.R., McCleave, J.A., and Schofield, E.K.
1977: Cobalt and nickel uptake by the Nyssaceae; Taxon, v. 26, p. 197-201.

Brooks, R.R. and Wither, E.D.
1977: Nickel accumulation by **Rinorea bengalensis** Wall, O.K.; J. Geochem. Explor., v. 7, p. 295-300.

Brotzen, O., Kvalheim, A., and Marmo, V.
1967: Development, status and possibilities of geochemical prospecting in Fennoscandia; in Geochemical Prospecting in Fennoscandia (ed. A. Kvalheim). Interscience, New York, p. 99-111.

Brown, R.G.
1970: Geochemical Survey of the vicinity of Oracle, Arizona; M.Sc. Thesis, Arizona State University, 56 p.

Brundin, N.H.
1939: Method of locating metals and minerals in the ground; U.S. Pat. 2158980.

1975: Possibilities to use plant roots in biogeochemical prospecting; Unpublished report, 20 p.

Brundin, N.H. and Nairis, B.
1972: Alternative plant samples in regional geochemical prospecting; J. Geochem. Explor., v. 1, p. 7-46.

Cannon, H.L.
1960: The development of botanical methods of prospecting for uranium on the Colorado Plateau; U.S. Geol. Surv. Bull., 1085-A, 50 p.

1964: Geochemistry of rocks and related soils and vegetation in the Yellow Cat area, Grand County, Utah; U.S. Geol. Surv. Bull., 1176, 127 p.

1970: Trace element excesses and deficiencies in some geochemical provinces of the United States; in Trace Substances in Environmental Health III (ed. D.D. Hemphill); Univ. of Missouri, p. 21-44.

Cannon, H.L. (cont'd.)
  1974: Introduction and one of several authors of chapters on lithium and chromium; in The Relation of Selected Trace Elements to Health and Disease, Geochemistry and the Environment, National Academy of Sciences, Washington, 113 p.

Cannon, H.L. and Anderson, B.M.
  1971: The geochemist's involvement with the pollution problem; in Environmental Geochemistry in Health and Disease (eds. H.L. Cannon and H.C. Hopps); Geol. Soc. Am. Mem., v. 123, p. 155-177.

Cannon, H.L. and Hopps, H.C. (eds.)
  1971: Environmental Geochemistry in Health and Disease; Geol. Soc. Am. Mem., v. 123, 230 p.
  1972: Geochemical environment in relation to health and disease; Geol. Soc. Am. Spec. Pap. 140, 77 p.

Cannon, H.L., Papp, C.S.E., and Anderson, B.M.
  1972: Problems of sampling and analysis in trace element investigations of vegetation; Annls. N.Y. Acad. Sci., v. 199, p. 124-136.

Cannon, H.L., Shacklette, H.T., and Bastron, H.
  1968: Metal absorption by Equisetum (horsetail); U.S. Geol. Surv. Bull., 1278-A, 21 p.

Chaffee, M.A.
  1975: Geochemical exploration techniques applicable in the search for copper deposits; U.S. Geol. Surv. Prof. Pap. 907-B, 26 p.

Chaffee, M.A. and Hassin, T.D.
  1971: An evaluation of geochemical sampling in the search for concealed "porphyry" copper-molybdenum deposits in southern Arizona; in Proc. 3rd Int. Geochem. Explor. Symp., Toronto, p. 401-409.

Chowdhury, A.N. and Bose, B.B.
  1971: Role of humus matter in the formation of geochemical anomalies; Can. Inst. Min. Metall. Spec. Vol. No. 11, p. 410-413.

Cole, M.M.
  1965: The use of vegetation in mineral exploration in Australia; 8th Commonw. Min. Metall. Conf., v. 6, p. 1429-1458.
  1971: The importance of environment in biogeographical/geobotanical and biogeochemical investigations; Can. Inst. Min. Metall. Spec. Vol. No. 11, p. 414-425.
  1973: Geobotanical and biogeochemical investigations in the sclerophyllous woodland and shrub associations of the Eastern Goldfields area of Western Australia, with particular reference to the role of Hybanthus floribundus (Lindl.) F. Muell. as a nickel indicator and accumulator plant; J. Appl. Ecol., v. 10, p. 269-320.

Cole, M.M., Provan, D.M.J., and Tooms, J.S.
  1968: Geobotany, biogeochemistry and geochemistry in the Bulman-Waimuna Springs area, Northern Territory, Australia; Inst. Min. Met. Trans., London, Sec. B, v. 77, p. 81-104.

Connor, J.J. and Shacklette, H.T.
  1975: Background geochemistry of some rocks, soils, plants and vegetables in the conterminous United States; U.S. Geol. Surv. Prof. Pap. 574-F, 168 p.

Curtin, G.C., Lakin, H.W., and Hubert, A.E.
  1970: The mobility of gold in mull (forest humus layer); U.S. Geol. Surv. Prof. Pap. 700-C, p. C127-C129.

Curtin, G.C., Lakin, H.W., Hubert, A.E., Mosier, E.L., and Watts, K.C.
  1971: Utilization of mull (forest humus layer) in geochemical exploration in the Empire district, Clear Creek County, Colorado; U.S. Geol. Surv. Bull., 1278-B, 39 p.

Curtin, G.C., King, H.D., and Mosier, E.L.
  1974: Movement of elements from coniferous trees in sub-alpine forests of Colorado; J. Geochem. Explor., v. 3, p. 245-263.

Debnam, A.H.
  1955: Biogeochemical prospecting investigations in the Northern Territory 1954; Aust. Bur. Mineral Resour. Geol. Geophys. Records, 1955-43, p. 1-24.

Dorr, J. van N., Hoover, D.R., Offield, T.W., and Shacklette, H.T.
  1971: The application of geochemical, botanical, geophysical and remote-sensing mineral prospecting techniques to tropical areas: State of the art and needed research; U.S. Geol. Surv. Open File Rep. DC-20, 98 p.

Do Van Ai, Borovik-Romanova, T.F., Koval'sky, V.V., and Makhova, N.N.
  1972: Boron, copper, manganese, nickel, strontium and barium in soils and plants of the Urst-Urst Plateau (in Russ.); Dokl. Vses. Akad. Sel'skokhoz. Nauk, No. 11, p. 4-6.

Doyle, P., Fletcher, W.K., and Brink, V.C.
  1973a: Trace element content of soils and plants from the Selwyn Mountains, Yukon and Northwest Territories; Can. J. Bot., v. 51, p. 421-427.
  1973b: Regional geochemical reconnaissance and the molybdenum content of bedrock, soils and vegetation from the Eastern Yukon; in Trace Substances in Environmental Health, V.VI (D.D. Hemphill ed.), Univ. of Missouri, Columbia, p. 369-375.

Duvigneaud, P.
  1958: The vegetation of Katanga and its metalliferous soils (in Fr.); Bull. Soc. Roy. Bot. Belg., v. 90, p. 127-286.

Dvornikov, A.G. and Ovsyannikova, L.B.
  1972: Biogeochemical dispersion halo of chalcophile elements in the Esaul lead-zinc deposit, Donets Basin (in Russ.); Geokhimiya, No. 7, p. 873-879.

Ek, J.
  1974: Trace elements in till, vegetation and water over a sulphide ore in Västerbotten County, Northern Sweden; Sverig. Geol. Undersök. Avhand., 698, p. 1-50.

Elkington, J.E.
  1969: Vegetation Studies in the Eastern Goldfields of Western Australia with Particular Reference to their Role in Geological Reconnaissance and Mineral Exploration; Ph.D. Thesis, Univ. of London.

El Shazly, E.M., Barakat, N., Eissa, E.A., Emara, H.H., Ali, I.S., Shaltout, S., and Sharaf, F.S.
  1971: The use of acacia trees in biogeochemical prospecting; Can. Inst. Min. Met., Spec. Vol. No. 11, p. 426-434.

Erämetsä, O., Lounamaa, K.J., and Haukka, M.
  1969: The vertical distribution of uranium in Finnish peat bogs; Suomen Kemistilehti, v. B42, p. 363-370.

Erämetsä, O. and Yliruokanen, I.
    1971a: The rare earths in lichens and mosses; Suomen Kemistilehti, v. B44, p. 121-128.

    1971b: Niobium, molybdenum, hafnium, tungsten, thorium and uranium in lichens and mosses; Suomen Kemistilehti, v. B44, p. 372-374.

Erdman, J.A. and Shacklette, H.T.
    1973: Concentrations of elements in native plants and associated soils of Missouri; in Geol. Soc. Am. 7th Ann. Mtg. Boulder. Abs., p. 313.

Ernst, W.
    1967: Bibliography of work on plant communities on soils containing heavy metals, excepting serpentine soils (in Ger.); Excerpta Bot. Sec. B, v. 8, p. 50-61.

    1974: Heavy-Metal Vegetation of the Earth (in Ger.); Fischer Verlag, Stuttgart, 194 p.

Farago, M.E., Clarke, A.J., and Pitt, M.J.
    1975: The chemistry of plants which accumulate metals; Co-ord. Chem. Rev., v. 16, p. 1-8.

Fletcher, K.W., Doyle, P., and Brink, V.C.
    1973: Seleniferous vegetation and soils in the Eastern Yukon; Can. J. Plant Sci., v. 53, p. 701-703.

Fortescue, J.A.C.
    1970: Research approach to the use of vegetation for the location of mineral deposits in Canada; Taxon, v. 19, p. 695-704.

Fortescue, J.A.C. and Hornbrook, E.H.W.
    1967a: A brief survey of progress made in biogeochemical prospecting research at the Geological Survey of Canada, 1963-1965; Geol. Surv. Can., Paper 66-54, p. 111-113.

    1967b: Progress report on biogeochemical research at the Geological Survey of Canada, 1963-1966; Geol. Surv. Can., Paper 67-23, Part I, 143 p.

    1969: Progress report on biogeochemical research at the Geological Survey of Canada, 1963-1966; Geol. Surv. Can., Paper 67-23, Part II, 101 p.

Fredriksson, K. and Lindgren, I.
    1967: Anomalous copper content in glacial drift and plants in a copper-mineralized area of The Caledonides; in Geochemical Prospecting in Fennoscandia (ed. A. Kvalheim), Interscience, New York.

Gandhi, S.M. and Aswathanarayana, U.
    1975: A possible base-metal indicator plant from Mamandur, South India; J. Geochem. Explor., v. 4, p. 247-250.

Grabovskaya, L.I. and Kuzmina, G.A.
    1971: Application of the biogeochemical method during the exploration of rare metal deposits in a region developing continuous permafrost (in Russ.); Vop. Priklad. Geokhim., No. 2, p. 98-108.

Groves, R.W., Steveson, B.G., Steveson, E.A., and Taylor, R.G.
    1972: Geochemical and geobotanical studies of the Emuford district of the Herberton tin field, north Queensland, Australia; Inst. Min. Met. Trans., London, Sec. B, v. 81, p. 127-137.

Gruzdev, B.I. and Rubtsov, D.M.
    1972: Accumulation of thorium, uranium and radium by plants and organic horizons of soils (in Russ.); in Radioecological Investigations in the Biosphere (ed. I.N. Verkhovskaya), Nauka Press, Moscow, p. 112-123.

Hall, J.S., Both, R.A., and Smith, F.A.
    1973: Comparative study of rock, soil and plant chemistry in relation to nickel mineralization in the Pioneer area, Western Australia; Proc. Australas. Inst. Min. Metall., No. 247, p. 11-22.

Hornbrook, E.H.W.
    1969a: Biogeochemical prospecting for molybdenum in west-central British Columbia; Geol. Surv. Can., Paper 68-56, 41 p.

    1969b: The development and use of biogeochemical prospecting methods for metallic mineral deposits; Can. Min. J., v. 90, p. 108-109.

    1970a: Biogeochemical prospecting for copper in west-central British Columbia; Geol. Surv. Can., Paper 69-49, 39 p.

    1970b: Biogeochemical investigations in the Perch Lake area, Chalk River, Ontario; Geol. Surv. Can., Paper 70-43, 22 p.

    1971: Effectiveness of geochemical and biogeochemical exploration methods in the Cobalt area, Ontario; Can. Inst. Min. Met. Spec. Vol. No. 11, p. 435-443.

Hornbrook, E.H.W. and Allan, R.J.
    1970: Geochemical exploration feasibility study within the zone of continuous permafrost; Coppermine River region, Northwest Territories; Geol. Surv. Can., Paper 70-36, 35 p.

Hornbrook, E.H.W. and Hobson, G.D.
    1972: Geochemical and biogeochemical exploration methods research in the Cobalt area, Ontario; Geol. Surv. Can., Paper 71-32, 45 p.

Howard-Williams, C.
    1970: The ecology of **Becium homblei** in Central Africa with special reference to metalliferous soils; J. Ecol., v. 58, p. 745-764.

Huff, L.C.
    1970: A geochemical study of alluvium-covered copper deposits in Pima County, Arizona; U.S. Geol. Surv. Bull., 1312-C, 31 p.

Ivashov, P.V. and Bardyuk, V.V.
    1971: Accumulation of zirconium in plants and soils in a rare metal deposit in the southern Far East region, studied to develop lithogeochemical and biogeochemical prospecting methods (in Russ.); in The Biogeochemistry of the Zone of Hypergenesis (ed. A.S. Khomentovsky). Nauka Press, Moscow, p. 67-78.

Jaffré, T., Brooks, R.R., Lee, J., and Reeves, R.D.
    1976: **Sebertia acuminata:** a nickel accumulating plant from New Caledonia; Science, 193, p. 579-580.

Jaffré, T., Latham, M., and Quantin, P.
    1971: Soils of mining massifs on New Caledonia and their relation to vegetation (in Fr.); Spec. Rep. O.R.S.T.O.M., Nouméa, 26 p.

Jaffré, T. and Schmid, M.
    1974: Accumulation of nickel by a Rubiacea of New Caledonia **Pschotria douarrei** (G. Beauvisage) Däniker (in Fr.); Compt. Rend. Acad. Sci. Paris Sér. D, v. 278, p. 1727-1730.

Jones, R.S.
    1970: Gold content of water, plants and animals; U.S. Geol. Surv., Circ. 625, 15 p.

Kelly, P.C., Brooks, R.R., Dilli, S., and Jaffré, T.
    1975: Preliminary observations on the ecology and plant chemistry of some nickel-accumulating plants from New Caledonia; Proc. Roy. Soc. Sec. B, v. 189, p. 69-80.

Koljonen, T.
    1974: Selenium uptake by plants in Finland; Oikos, v. 25, p. 353-355.

    1975: The behaviour of selenium in Finnish soils; Annls. Agric. Fenn., v. 14, p. 240-247.

Kontas, E.
    1976a: Sarvisoavi: molybdenum in till and birch leaves; J. Geochem. Explor., v. 5, p. 311-312.

    1976b: Lahnanen: molybdenum in till and pine needles; J. Geochem. Explor., v. 5, p. 261-263.

Koval'sky, V.V.
    1974: Geochemical Ecology (in Russ.); Nauka Press, Moscow, 300 p.

Koval'sky, V.V., Letunova, S.V., Altynbay, R.D., and Pelova, Y.A.
    1973: Importance of soil microflora in biogenic migration of boron in biogeochemical provinces with different contents of this element; Sov. Soil Res., v. 5, p. 329.

Kovalevsky, A.L.
    1969: Zinc in plants as a universal biochemical indicator of some types of ore deposits (in Russ.); in Biogeochemical Methods of Exploration for Ore Deposits (eds. A.L. Kovalevsky and A.I. Perel'man), Akad. Nauk, Ulan Ude, p. 187-203.

    1971: Concerning physiological barriers to the absorption of elements by plants; in Microelements in the Biosphere and their Application to Agriculture and Medicine in Siberia and the Far East (in Russ.); Buriat Fil. Akad. Nauk., Ulan Ude, p. 134-144.

    1972: Biogeochemical prospecting for uranium deposits (in Russ.); Atomn. Energ., v. 33, p. 557-562.

    1974a: Biogeochemical Exploration for Ore Deposits (in Russ.); Nedra Press, Moscow, 143 p.

    1974b: Reliability of biogeochemical searches for beryllium according to difficult species and organs of plants (in Russ.); Geokhimiya, No. 10, p. 1575.

    1974c: Conditions for successful use of the biogeochemical method of prospecting for ore deposits; Geochemistry, No. 218, p. 183-186.

    1975: Peculiarities of the Formation of Biogeochemical Aureoles of Ore Deposits (in Russ.); Nauka Press, Novosibirsk, 114 p.

Kovalevsky, A.L. and Perel'man, A.I. (eds.)
    1969: Biogeochemical Exploration for Ore Deposits (in Russ.); Akad. Nauk, Ulan Ude, 289 p.

Kvalheim, A. (ed.)
    1967: Geochemical Prospecting in Fennoscandia; Interscience, New York, 350 p.

Låg, J. and Bølviken, B.
    1974: Some naturally heavy-metal poisoned areas of interest in prospecting, soil chemistry and geomedicine; Norg. Geol. Undersøk. Offprint, No. 304, p. 73-96.

Låg, J., Hvatum, O.Ø., and Bølviken, B.
    1969: An occurrence of naturally lead poisoned soil at Kostad near Gjørvik, Norway; Norg. Geol. Undersøk. Offprint, No. 266, p. 141-159.

Lakin, H.W., Curtin, G.C., and Hubert, A.E.
    1974: Geochemistry of gold in the weathering cycle; U.S. Geol. Surv., Bull. 1330, 80 p.

Lee, J., Brooks, R.R., Reeves, R.D., and Boswell, C.R.
    1975: Soil factors controlling a New Zealand serpentine flora; Pl. Soil, v. 42, p. 153-160.

Lee, J., Brooks, R.R., Reeves, R.D., Boswell, C.R., and Jaffré, T.
    1977: Plant soil relationships in a New Caledonian Serpentine flora; Pl. Soil, v. 46, p. 675-680.

Letova, A.N.
    1970: Experimental Investigations on the Role of Plants in the Migration of the Chemical Elements (in Russ.). Degree Thesis (see Kovalevsky, 1974e), 22 p.

Lounamaa, J.
    1956: Trace elements in plants growing wild on different rocks in Finland; Ann. Bot. Soc. Zoo. Bot. Fenn. Vanamo, v. 9, Suppl. 170, p. 1-196.

    1967: Trace elements in trees and shrubs growing on different rocks in Finland; in Geochemical Prospecting in Fennoscandia (ed. A. Kvalheim), Interscience, New York, p. 287-317.

Malyuga, D.P.
    1964: Biogeochemical Methods of Prospecting; Consultants Bureau, New York, 205 p.

Malyuga, D.P. and Aivazyan, A.D.
    1970: Biogeochemical studies in the Rudnyi Altai (in Russ.); Geokhimiya, No. 3, p. 364-371.

Matula, I.
    1973: Use of biogeochemical methods for prospecting for deposits in the Spissko Gemersky region (in Slovak); Mineralia Slovaca, v. 5, p. 21-27.

Melikyan, M.M.
    1972: Boron biogeochemical provinces of natural grasslands of the Armenian S.S.R. (in Arm.); Izv. Sel'skokhoz. Nauk, v. 15, p. 49-54.

Millman, A.P.
    1957: Biogeochemical investigations in areas of copper-tin mineralization in Southwest England; Geochim. et Cosmochim. Acta, v. 12, p. 85-93.

Minguzzi, C. and Vergnano, O.
    1948: The nickel content of the ash of **Alyssum bertolonii** Desv. Botanical and geochemical considerations (in Ital.); Atti Soc. Tosc. Sci. Nat., v. 55, p. 49-77.

Minski, M.J., Girling, C.A., and Peterson, P.J.
    1977: Determination of gold and arsenic in plant material by neutron activation analysis; Radiochem. Radioanal. Lett., v. 30, p. 179-186.

Mitskevich, B.F.
    1971: Geochemical Landscape Units of the Ukrainian Shield (in Ukr.); Naukova Dumka Press, Kiev, 173 p.

Molchanova, I.V. and Kulikov, N.V.
    1972: Radioactive isotopes in the plant-soil system (in Russ.); Atom. Energ. Comm. Pub., 86 p.

Nemec, B., Babička, J., and Oborsky, A.
1936: On the accumulation of gold in horsetails (in Ger.); Bull. Int. Acad. Sci. Boheme, 1-7, 13 p.

Nesvetailova, N.G.
1970: Prospecting for Ores by Plants (in Russ.); Nedra Press, Moscow, 97 p.

Nicolas, D.J. and Brooks, R.R.
1969: Biogeochemical prospecting for zinc and lead in the Te Aroha region of New Zealand; Proc. Aust. Inst. Min. Metall., No. 231, p. 59-66.

Nicolls, O.W., Provan, D.M.J., Cole, M.M., and Tooms, J.S.
1965: Geobotany and geochemistry in mineral exploration in the Dugald River area, Cloncurry district, Australia; Trans. Inst. Min. Metall., v. 74, p. 695-799.

Nielsen, J.S.
1972: The Feasibility of Biogeochemical and Geobotanical Prospecting at Spargoville, Western Australia; M.Sc. Thesis, Massey Univ., New Zealand.

Nielsen, J.S., Brooks, R.R., Boswell, C.R., and Marshall, N.J.
1973: Statistical evaluation of geobotanical and biogeochemical data by discriminant analysis; J. Appl. Ecol., v. 10, p. 251-258.

Nilsson, I.
1972: Accumulation of metals in spruce needles and needle litter; Oikos, v. 23, p. 132-136.

Ovchinikov, L.N. and Baranov, E.N.
1970: Indigenous geochemical aureoles of pyrite deposits (in Russ.); Geol. Rud., No. 2, p. 10-24.

Panin, M.S. and Panina, R.T.
1971a: Levels of cobalt in plants of the Semipalatinsk region of the Kazak SSR (in Russ.); Biol. Nauki, v. 14, p. 69-75.

1971b: Level of zinc in plants of the Semipalatinsk district of the Kazak SSR (in Russ.); Agrokhimiya, No. 11, p. 122-127.

Panin, M.S. and Shchetinina, V.I.
1974: Boron content of plants of the Semipalatinsk Region of the Kazak SSR (in Russ.); Agrokhimiya, No. 1, p. 106-112.

Persson, H.
1956: Studies of the so-called "copper mosses"; J. Hattori Bot. Lab., v. 17, p. 1-18.

Prozumenshchikova, L.T. and Skripchenko, A.F.
1972: Molybdenum level in plants of the Far East (in Russ.); Uch. Zap. Dal'nevost. Gos. Univ., v. 57, p. 34-48.

Quin, B.F., Brooks, R.R., Boswell, C.R., and Painter, J.A.C.
1974: Biogeochemical exploration for tungsten at Barrytown, New Zealand; J. Geochem. Explor., v. 3, p. 43-51.

Reilly, C.
1967: Accumulation of copper by some Zambian plants; Nature, v. 215, p. 667-668.

Safronov, N.I.
1971: Fundamental Geochemical Methods of Prospecting for Ore Deposits (in Russ.); Nedra Press, Moscow, 216 p.

Salmi, M.
1956: Peat and bog plants as indicators of ore minerals in Vihanti ore field in Western Finland; Bull. Comm. Geol. Finl., v. 175, 22 p.

Salmi, M. (cont'd.)
1967: Peat in prospecting: Applications in Finland; in Geochemical Prospecting in Fennoscandia (ed. A. Kvalheim), Interscience, New York, p. 113-126.

Severne, B.C.
1972: Botanical Methods for Mineral Exploration in Western Australia; Ph.D. Thesis, Massey Univ., New Zealand.

Severne, B.C. and Brooks, R.R.
1972: A nickel-accumulating plant from Western Australia; Planta, v. 103, p. 91-94.

Shacklette, H.T.
1970: Mercury content of plants; in U.S. Geol. Surv., Prof. Paper 713, p. 35-36.

1972a: Geochemical survey of vegetation; in Geochemical Survey of Missouri — Plans and Progress for sixth six-month period (Jan.-June, 1972); U.S. Geol. Surv. Open File Rep., p. 58-79.

1972b: Cadmium in plants; U.S. Geol. Surv., Bull. 1314-G, 28 p.

Shacklette, H.T., Erdman, J.A., and Keith, J.R.
1970: Botanical techniques for on-site inspections of suspected underground nuclear explosions; U.S. Geol. Surv., Tech. Lett., June, 1970, 194 p.

Shacklette, H.T., Lakin, H.W., Hubert, A.E., and Curtin, G.C.
1970: Absorption of gold by plants; U.S. Geol. Surv., Bull. 1314-B, 23 p.

Shacklette, H.T., Sauer, H.I., and Miesch, A.T.
1970: Geochemical environments and cardiovascular Mortality Rates in Georgia; U.S. Geol. Surv. Prof. Pap. 574-C, 39 p.

Shchulzhenko, V.N., Ivanov, M.V., Popov, V.S., Talipov, R.M., and Yussupov, R.G.
1970: Biogeochemical investigations of sulphide deposits and criteria for prospecting for them (for example in several regions of Central Asia); in The Biological Role of Microelements and their Application to Agriculture and Medicine (in Russ.), Nauka Press, Leningrad.

Skarlygina-Ufimtseva, M.D. and Berezkina, G.A.
1971: Importance of biogeochemical dispersion haloes during exploration of sulfide deposits (in Russ.); in The Theory of Questions of Phytoindication (ed. A.A. Korchagin), Nauka Press, Leningrad, p. 188-193.

Sudnitsyna, I.G.
1971: Results of biogeochemical studies on beryllium (in Russ.); in Material from a Biogeochemical Inventory of the Flora of Kirgistan (ed. E.M. Tokobaev), Illum. Press, Frunze, p. 17-26.

Talipov, R.M.
1966: Biogeochemical Exploration for Polymetallic and Copper Deposits in Uzbekistan (in Russ.); Fan Press, Tashkent, 105 p.

Talipov, R.M. and Glushchenko, V.M.
1974: Gold in soils, waters and plant organs in goldfields of the Kuraminsky area (eastern Uzbekistan) (in Russ.); Proc. Symp. Mineralogy and Geochemistry of Gold in Uzbekistan, Tashkent.

Talipov, R.M., Glushchenko, V.M., Lezhneva, N.D., and Nishanov, P.Kh.
1975: The correlation between the gold content of plants and waters in several ore fields of the Kuraminsky Mountains (in Russ.); Uzbek. Geol. Zhur., No. 4, p. 21-26.

Talipov, R.M., Glushchenko, V.M., Nishanov, P., Lunin, A.G., Smigdjanova, M., and Aripova, Kh.
1973: Comparative data for determining micro amounts of gold in plant ash and waters by different methods (in Russ.); Uzbek. Geol. Zhur., No. 4.

Talipov, R.M., Glushchenko, V.M., Nishanov, P., and Smigdjanova, M.
1974: Some regularities of the distribution of gold and accompanying microelements in plants of the Almalisky region (in Russ.); Dokl. Akad. Nauk. Uzbek. SSR, No. 1, p. 53-54.

Talipov, R.M., Glushchenko, V.M., Tverskaya, K.L., and Nishanov, P.
1976: Some peculiarities of the distribution of gold and antimony in plants over ore outcrops of the Chatkalo-Kurmansky region (in Russ.); Uzbek. Geol. Zhur., No. 3.

Talipov, R.M., Karabaev, K.K., and Akhunkhodzhaeva, N.
1971: Some results on biogeochemical investigations in northern Tamditai (western Uzbekistan); in Geology and Mining in Uzbekistan (in Russ.), v. 24, Fan Press, Tashkent.

Talipov, R.M. and Khatamov, Sh.
1973: Biogeochemical studies in the northern part of the central Kyzyl-Kum (in Russ.); Uzb. Geol. Zhur., v. 17, p. 26-31.

1974: The distribution of microelements in plants of the Tamdinsky Mountains (central Kirgistan) (in Russ.); Uzbek. Geol. Zhur., No. 1, p. 23-27.

Talipov, R.M., Musin, R.A., Glushchenko, V.M., Nishanov, P., Machanov, D., and Smigdjanova, M.
1974: Results of biogeochemical investigations in the Almalisky region; in The Metallogeny and Geochemistry of Uzbekistan (in Russ.); Fan Press, Tashkent, p. 109-113.

Talipov, R.M., Tverskaya, K.L., Glushchenko, V.M., and Magdiev, R.A.
1976: Some peculiarities of the distribution of uranium and gold in plants of the Chatkalo-Kuraminsky region (in Russ.); Zap. Uzbek. Otd. VMO, p. 29.

Talipov, R.M., Yussupov, R.G., and Khatamov, Sh.
1970: Biogeochemical and hydrogeochemical prospecting for gold deposits in Uzbekistan; in The Biological Role of Microelements and their Application to Agriculture and Medicine (in Russ.); V.L., Nauka Press, Leningrad.

Timperley, M.H., Brooks, R.R., and Peterson, P.J.
1970a: Prospecting for copper and nickel in New Zealand by statistical analysis of biogeochemical data; Econ. Geol., v. 65, p. 505-510.

1970b: The significance of essential and non-essential elements in plants in relation to biogeochemical prospecting; J. Appl. Ecol., v. 7, p. 429-439.

1972a: Improved detection of geochemical soil anomalies by multiple regression analysis of biogeochemical data; Proc. Aust. Inst. Min. Metall., No. 242, p. 25-36.

1972b: Trend analysis as an aid to the comparison and interpretation of biogeochemical and geochemical data; Ecol. Geol., v. 67, p. 669-676.

Tkalich, S.M.
1938: Experience in the investigation of plants as indicators in geological exploration and prospecting (in Russ.); Vest. Dal'nevost. Fil. Akad. Nauk. SSR, v. 37, p. 3-25.

Tkalich, S.M. (cont'd.)
1970: The Phytogeochemical Method of Prospecting for Ores of Useful Minerals (in Russ.); Nedra Press, Moscow, 175 p.

Tyler, G.
1970: Distribution of lead in a coniferous woodland ecosystem in south Sweden (in Swed.); Grundförbättning, v. 23, p. 45-49.

Usik, L.
1969: Review of geochemical and geobotanical prospecting methods in peatland; Geol. Surv. Can., Pap. 68-66, 42 p.

Vadkovskaya, I.K.
1971: Trace elements in the flora of the Berezina and Sozh River basin (in Russ.); Doklad. Akad. Nauk. Beloruss. SSR, v. 15, p. 751-754.

Vasilev, N.E. and Skripchenko, A.F.
1972: Concentration of tin, lead, copper, titanium, manganese, nickel and barium trace nutrients by forest vegetation on a tin ore deposit of western Sikhote-Alin (in Russ.); Uch. Zap. Dal'nevost. Gos. Univ., v. 57, p. 66-73.

Vergnano Gambi, O., Gabbrielli, R., Lotti, L., and Polidori, V.
1971: Biogeochemical aspects of manganese in the Tosco-Emiliano region of the Appennines (in Ital.); Webbia, v. 25, p. 353-382.

Vogt, T.
1939: Chemical and botanical prospecting at Røros (in Norweg.); K. Norsk. Vidensk. Selsk. Forh., v. 12, p. 82-83.

Voigtkevich, G.V. and Alekseyenko, V.A.
1970: An experiment on the use of the biogeochemical method for the discovery of several ore elements in the Djungar Altai (in Russ.); Izv. Vuzor. Geol. Razved., No. 2, p. 64-69.

Warren, H.V.
1972: Biogeochemistry in Canada; Endeavour, v. 31, p. 46-49.

Warren, H.V., Church, B.N., and Northcote, K.E.
1974: Barium-strontium relationships, possible geochemical tool in search for ore bodies; Western Miner, April.

Warren, H.V. and Hajek, J.H.
1973: An attempt to discover a Carlin-type of gold deposit in British Columbia; Western Miner, Oct.

Whitehead, N.E. and Brooks, R.R.
1969: Radioecological observations in plants of the Lower Buller Gorge Region of New Zealand and their significance for biogeochemical prospecting; J. Appl. Ecol., v. 6, p. 301-310.

Wild, H.
1970: Geobotanical anomalies in Rhodesia. 3, The vegetation of nickel-bearing soils; Kirkia, Suppl. v. 9, p. 1-62.

1974: Indigenous plants and chromium in Rhodesia; Kirkia, v. 9, p. 233-242.

Wolfe, W.J.
1971: Biogeochemical prospecting in glaciated terrain of the Canadian Precambrian Shield; Bull. Can. Inst. Min. Metall., v. 64, p. 72-80.

1974: Geochemical and biogeochemical exploration research near early Precambrian porphyry-type molybdenum-copper mineralization, northwestern Ontario, Canada; J. Geochem. Explor., v. 3, p. 25-41.

Yates, T.E., Brooks, R.R., and Boswell, C.R.
   1974a: Biogeochemical exploration at Coppermine Island; N.Z. J. Sci., v. 17, p. 151-159.

   1974b: Factor analysis in botanical methods of exploration; J. Appl. Ecol., v. 11, p. 563-574.

Yliruokanen, I.
   1975a: Uranium, thorium, lead, lanthanoids and yttrium in some plants growing on granitic and radioactive rocks; Bull. Geol. Soc. Finland, v. 47, p. 471-478.

   1975b: A chemical study of the occurrence of rare earths in plants; Annls. Acad. Sci. Fenn. Ser. A, II Chemica, v. 176, p. 1-28.

Yussupov, R.G., Talipov, R.M., Khatamov, Sh., Yussupova, L.N., and Shchulzhenko, V.N.
   1970: Lichenous flora of ore deposits and their use in prospecting (for example in the Akhangaransky Region of Uzbekistan); in The Biological Role of Microelements and their Application to Agriculture and Medicine (in Russ.); V.I., Nauka Press, Leningrad.

Yussupov, R.G., Talipov, R.M., Yussupova, L.N., Khatamov, Sh., and Shchulzhenko, V.N.
   1970: Biogeochemical investigations on broadleaf forests and their use in prospecting; in The Biological Role of Microelements and their Application to Agriculture and Medicine (in Russ.); V.I., Nauka Press, Leningrad.

Yussupova, L.N., Talipov, R.M., Yussupov, R.G., and Shchulzhenko, V.N.
   1970: Biogeochemical investigations on producers of di- and tri-carboxylic acids and some peculiarities of the behaviour of trace elements in the biosphere; in The Biological Role of Microelements and their Application to Agriculture and Medicine (in Russ.); V.I., Nauka Press, Leningrad.

STREAM SEDIMENT GEOCHEMISTRY

W.T. Meyer
Cities Service Company, Tulsa, Oklahoma

P.K. Theobald, Jr.
U.S. Geological Survey, Denver, Colorado

H. Bloom
Colorado School of Mines, Golden, Colorado

Meyer, W.T., Theobald, P.K., Jr., Bloom, H., Stream Sediment Geochemistry; in Geophysics and Geochemistry in the Search for Metallic Ores; Peter J. Hood, editor; Geological Survey of Canada, Economic Geology Report 31, p. 411-434, 1979.

Abstract

Stream sediments are characterized more by their variability than by their uniformity in composition, grain size, sorting and colour. This variability is a function of the geology, terrain, and climate of the catchment areas sampled by the stream, and it provides the essential ingredient for viable stream sediment surveys. Prospectors have always made use of this fact, initially through observation of mineralized boulders and panning of heavy minerals from the stream bed, and more recently by chemical analysis of the finer fractions of stream sediment. Over the past 25 years, routine application of stream sediment geochemistry has become accepted by government and intergovernmental agencies and the mining industry as the principal method of low-cost reconnaissance exploration in those areas, favoured by the combination of adequate relief and precipitation, where a suitable, integrated drainage system has developed.

At present, surveys that make use of trace analysis of the fine fraction (normally minus-80-mesh) of active sediments, although still the most common, are increasingly being questioned. In areas of extreme climate or where information from the fine sediment is no longer adequate, enhancement of anomalies may require sampling of organic material, of specific coarse fractions of inorganic sediment, of selected mineral groups within the sediment that may be isolated mechanically or chemically, or of a combination of these materials. As a result, variations in sampling and analytical methods have been developed to accommodate regional and climatic differences and differing survey aims (i.e., multi-purpose or strictly prospecting, single or multiple target).

This review outlines the present art with selected examples of surveys by both government and industry. The largest single expense is the sample collection cost, particularly the transportation cost to or between sample sites, and to optimize this investment the tendency is to collect multiple samples at each site and for multi-element analyses. Mathematical and graphical interrelation of all of these components with the geology, geophysics, and physiography of the area under study, usually requiring computer technology, provides a far greater insight into the exploration targets sought today than the single-sample, single-element approach that was so successful in the past.

Résumé

Les sédiment fluviatiles manifestent fréquemment un certain degré de variabilité de composition, granulométrie, triage et couleur. Cette variabilité dépend de la géologie, de la nature du terrain et du climat du bassin-versant qui alimente le cours d'eau, et constitue la base d'une prospection sérieuse des sédiments fluviatiles. Les prospecteurs ont toujours tenu compte de ce fait: autrefois, ils examinaient les blocs minéralisés et recueillaient par lavage les minéraux lourds se trouvant dans le lit du cours d'eau; de nos jours, ils étudient la composition chimique des plus fines fractions granulométriques des sédiments fluviatiles. Depuis 25 ans, les organismes gouvernementaux et intergouvernementaux et l'industrie minière considèrent la prospection géochimique des sédiments fluviatiles comme la meilleure manière, et la moins coûteuse, d'explorer les régions dont la topographie, le régime des précipitations et le réseau hydrographique créent des conditions favorables à ce type de prospection.

Actuellement, on doute de plus en plus de la valeur des relevés qui s'appuient sur l'analyses de la fraction granulométrique fine (généralement maille inférieure à 80) des sédiments actifs, bien que ce soit encore la méthode la plus fréquemment utilisée. Dans les zones où le climat est très rigoureux, ou bien dans lesquelles l'analyses de la fraction sédimentaire fine ne suffit plus, il faut parfois, pour mieux déceler les anomalies géochimiques, recueillir dans le sédiment les débris organiques, certaines fractions plus grandes du sédiment inorganique ou certains groupes de minéraux que l'on peut isoler mécaniquement ou chimiquement, ou bien plusieurs de ces matériaux à la fois. Les méthodes d'échantillonnage et d'analyse se sont donc suffisamment diversifiées au cours des années, pour que l'on puisse tenir compte des différences régionales et climatiques, ainsi que des divers objectifs des levés (par exemple, prospection monominérale ou pluriminérale, objectif simple ou multiple).

Le présent article illustre les progrès de la prospection géochimique par des exemples de levés commandés par le gouvernement et l'industrie. L'étape la plus coûteuse des levés est la collecte des échantillons, surtout en raison des frais de transport vers ou entre les sites d'échantillonnage; on s'efforce actuellement de faire plusieurs prélèvements sur un même site, et d'effectuer le dosage de plusieurs éléments à la fois. On s'applique à établir par des méthodes mathématiques et graphiques une relation entre chacun des éléments et les caractères géologiques, géophysiques et physiographiques de la région étudiée; ces méthodes, qui exigent généralement l'emploi d'ordinateurs, nous renseignent bien mieux sur la valeur des objectifs explorés que les anciennes méthodes, qui ont autrefois fait leurs preuves, de collecte d'échantillons isolés et de dosage individuel des éléments.

INTRODUCTION

Stream sediment geochemistry, although an obvious offshoot of the mineralogical study of alluvium, is a relative newcomer amongst the techniques available to the geologist in search of ore. In a recent review of the historical development of modern geochemical prospecting, Hawkes (1976a) traced the first application of stream sediment sampling in North America to Lovering et al. (1950), who described the results of field investigations, conducted in 1947 and 1948, of the downstream dispersion of copper, lead and zinc from the San Manuel deposit in Arizona. Only 30 years have passed since this initial study, but due to the method's obvious advantages in reconnaissance exploration and to the technological improvements in trace analysis and data processing that have occurred during this period, stream sediment surveys have become accepted as one of the most important low-cost reconnaissance tools available to the present day explorationist.

A variety of sampling and analytical methods have been developed to accommodate regional geologic and climatic variations, and these are further modified to suit the aim of the survey, whether it be multi-purpose or strictly for prospecting. This paper reviews the current state-of-the-art in terms of regional development of techniques and provides some examples of recent applications of stream sediment geochemistry.

GENERAL PRINCIPLES

Stream sediments are characterized more by their variability in composition, grain size, sorting and colour than by uniformity in any of these features. This variability is a function of the geology, terrain, and climate of the catchment area sampled by the stream, and it provides the essential ingredient for viable stream sediment surveys. However, the purpose of geochemical prospecting is to isolate that portion of the variability which reflects the presence of mineralization in the catchment area and, to do this, one must understand how sediment composition is affected by lithology, climate, topography, human activity, vegetation and water chemistry. The following provides a brief review of the physical and chemical controls on stream sediment composition that may be of concern to the exploration geochemist.

Mode of Occurrence

A general discussion of the mode of occurrence of trace elements in stream sediments has been given by Rose (1975). Trace elements can be found in stream sediments as major elements in trace minerals, as trace constituents of primary rock-forming minerals or minerals formed during weathering, as ions adsorbed on colloidal particles or in the lattices of clays, and in combination with organic matter. A study of the transportation of transition metals in the Amazon and Yukon rivers (Gibbs, 1977) has shown that most of the metal in suspended sediment is in the form of crystalline particles and hydroxide coatings. To give examples, the element manganese (Fig. 19.1a) is largely present as a coating, and copper (Fig. 19.1b) is predominantly found in the crystalline or mineral phase.

1. Mineral Phase

Trace elements contained within the lattice of rock-forming minerals and within minerals formed during weathering probably make up the greatest proportion of the background variation in most stream sediments. For the purpose of mineral exploration, contrast between anomalous and background concentrations of metal can be maximized through the use of partial chemical extractions which do not remove lattice-held trace metals from the host mineral. However, study of the efficiency of extraction of metal from minerals using different strengths and combinations of acids (Foster, 1971) has shown that even 5 per cent HNO_3 is capable of extracting appreciable amounts of zinc from pyroxene, limonite, feldspar, biotite, and amphibole (Table 19.1). As the bulk of analytical work on stream sediments is conducted by atomic absorption on hot acid extractions, or by total methods such as emission spectrography and X-ray fluorescence, much of the recorded trace-element variation will be derived from changes in the mineralogical composition of the sediment. This characteristic has been used with some effect in regional reconnaissance to define previously undetected geological units, but in general, geochemical mapping as an aid to geology has been restricted to specific problems in poorly exposed areas (Webb, 1970). It should be noted, however, that trace elements in the lattice of rock-forming minerals (i.e., magnetite and biotite) can provide clues to mineralization, and hence can be anomalous; and that many of the products of weathering are the specific targets (gossans, alteration zones) of the geochemist. Total analysis is often the most effective approach in such cases.

An alternative to partial extraction as a method of suppressing background effects in stream sediment surveys can be provided by multi-element analysis. Rose and Suhr (1971) examined the partition of trace elements among mineral phases of stream sediments in Pennsylvania and found that the most important metal-bearing phases were vermiculite clay (Cu, Zn), iron oxides (Cu, Co, Ni, Zn), and mafic minerals in the sand and silt fraction. Quantitative effects on trace element values resulting from the variation in these major phases were effectively reduced by the application of regression techniques that used the major elements Fe, Mn, Ca, Al, and Mg as predictors of the trace elements. Regression of the values of individual trace elements against R-mode factor scores representing geological and environmental metal associations has also proved useful for background correction of data from varied terrains in Canada (Closs and Nichol, 1975) and Brazil (Meyer, 1977).

Heavy mineral concentrates are being used on a regional scale to mechanically eliminate the effects of dilution by rock-forming minerals. This sample medium greatly enhances anomalies for major metals in trace minerals when the trace minerals have a specific gravity of about 3 gm/cc or more and are resistant to chemical and mechanical destruction in the environment under study. The large initial sample size, often 5 kg or mroe, reduces sampling error, and the great concentration factor, usually three orders of magnitude or more before analysis, often brings the level of some of the rarer or analytically more difficult elements within easy reach of the analyst. Used in this way, heavy mineral concentrates have perhaps the longest recorded history in prospecting of all stream sediment techniques. Mercury prospecting with heavy mineral concentrates was reported by Theophrastus in about 300 B.C. (Hill, 1746). Application of modern, trace element analytical techniques to heavy mineral concentrates in search of trace minerals dates from at least the early 1950s (Overstreet, 1962), when concentrates from the southeastern United States were routinely screened by semiquantitative spectrographic methods. It is probable that these techniques were in use well before those used in the U.S.S.R. (Sigov, 1939).

Heavy mineral concentrates have been used more recently to upgrade analyses for trace constituents of rock-forming minerals. Magnetite (Theobald and Thompson, 1959a) and biotite (Theobald and Havens, 1960) were the first of these minerals investigated and are probably the most extensively known (see, for example, Parry and Nackowski, 1963, and Jacobs and Parry, 1976). The trace metal content of rock-forming minerals appears to define regional metallographic provinces rather than specific exploration targets,

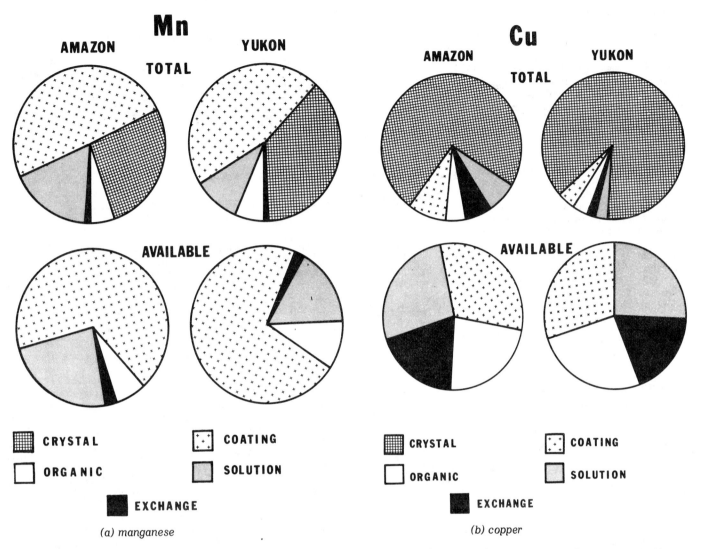

Figure 19.1. Forms in which transition metals are transported by the Amazon and Yukon rivers (after Gibbs, 1977).

hence the technique has been used more by academic or governmental organizations than by the exploration industry. In large virgin areas, multi-element analyses of magnetite concentrates from stream sediment have been used to identify areas most promising for exploration, thus eliminating from the first effort vast areas less likely to contain exploration targets (Theobald, P.K., and Thompson, C.E., unpubl. data).

Separation of the heavy mineral concentrate into several subsamples based on magnetic properties of the minerals leads to far greater enhancement of geochemical patterns than does the analysis of the bulk concentrate. Furthermore, this refinement of the heavy mineral technique often allows the separation of two or more geochemical patterns for a single trace element either in the primary or secondary dispersion of the elements. As shown in the early work on Clear Creek in Colorado, for example, the major part of the tungsten in the bulk concentrate is derived from sporadic scheelite occurrences associated with calc-silicate rocks (Theobald and Thompson, 1959b; Tweto, 1960). None of these are presently known to be economic. The smaller amount of tungsten in the concentrate derived from huebnerite is from the upper zones of the major molybdenum deposits of the Urad-Henderson porphyry system. The two tungsten sources may be easily separated magnetically.

A more dramatic example is found in the work of Alminas et al. (see, for example, Alminas et al., 1972a, b, c) in southwestern New Mexico where the concentrates are routinely split into three fractions using a Frantz Isodynamic Separator[1]: (1) minerals magnetic at 0.2 amp., (2) minerals magnetic between 0.2 and 1 amp., and (3) minerals not magnetic at 1 amp. The third fraction contains most of the common sulphide minerals of the area and direct pseudo-morphic alteration products of these. The second fraction contains iron and manganese oxides precipitated form supergene solutions beyond the immediate environs of the deposits, either in host rocks or in post-ore cover. The anomalies are somewhat more subdued in the second than in the third fraction, but the targets are larger in areas where the host rocks are exposed, and frequently anomalies can be identified in areas of thick post-ore cover.

---

[1] Use of brand names in this report is for descriptive purposes only and does not constitute endorsement by the authors or their institutions.

2. Sorbed Material

Hydromorphic dispersion of metal ions derived from the oxidation of sulphide deposits or from the leaching of oxide ores can result in detectable enrichment of the fine fraction of stream sediments through adsorption of ions on colloidal particles, scavenging by secondary Fe and Mn oxides, ion-exchange reactions with clays, and adsorption and chelation with organic matter (Hawkes and Webb, 1962). Recent work by Hem (1976) has demonstrated the importance of the cation exchange capacity of the fine fraction of stream sediments in controlling the concentration of lead in stream waters. Depending on the cation exchange capacity of the sediment, it was shown that the major portion of ionic lead in river systems falling within the common pH span of 5.7 to 8.5 would be adsorbed on cation exchange sites in the sediment (Fig. 19.2). The high capacity of the fine fraction of stream sediments for metal ion adsorption has been the most important factor contributing to the widespread adoption of silt- and clay-size fractions as the optimum sample material in the search for hydromorphic anomalies. Cold-extractable analytical methods have been particularly useful in isolating the adsorbed metal phase from the fine fraction of sediments.

3. Hydroxide Coatings

In recent years, the ability of iron and manganese oxides to selectively scavenge certain metal ions from surface water has received considerable attention as a means of enhancing contrast between hydromorphic anomalies and background. Scavenging of heavy metals by secondary oxides can take place by any one or a combination of the following mechanisms: coprecipitation, adsorption, surface complex formation; ion exchange, and penetration of the crystal lattice (Chao and Theobald, 1976). Concentration of metals in stream sediments in this manner was demonstrated by Horsnail et al. (1969) in an area where the Eh and pH of groundwaters were substantially lower than those in the stream waters, resulting in precipitation of Mn and Fe oxides in the drainage channel. Canney (1966) had previously noted the relationship of enhanced metal values to ferromanganese accumulation, and had cautioned against the possible identification of spurious anomalies if the Fe and Mn scavenging effect was not taken into account during interpretation.

Whitney (1975) has shown that the prevalence and thickness of oxide coatings increases with increasing particle size, and this was also found to be true for suspended material in the Amazon and Yukon rivers (Gibbs, 1977). The presence of a thicker layer of hydroxides on large grains was made use of by Carpenter et al. (1975), who selectively analyzed the black coating of oxides found on stream boulders and pebbles downstream from base metal mineralization in Tennessee and Georgia. This study demonstrated that the anomaly-to-background ratio for base metals was considerably higher in the oxide coatings than in the minus-80-mesh sediment at the same site (Fig. 19.3a, b) for those base metals (Cu, Zn) concentrated in coatings, while minus-80-mesh sediments gave better results for Pb, which was not scavenged by Fe-Mn oxides. The selective scavenging of certain elements by Fe-Mn oxides has been documented by Nowlan (1976) in a multi-element correlation study of stream sediments and oxide coatings from Maine (Table 19.2).

Table 19.1

Extraction of zinc (ppm) from selected minerals using various acids (after Foster, 1971).

Mineral	$HClO_4$	HCl	HCl-HNO_3	$HClO_4$-HNO_3	HNO_3	25% HNO_3	5% HNO_3	HF-$HClO_4$-HNO_3
Pyroxene	28	23	31	24	23	28	24	105
Limonite	290	290	285	218	84	58	25	285
Feldspar	8	10	10	8	9	11	11	9
Biotite	265	215	216	246	173	170	84	265
Amphibole	255	156	185	157	109	123	79	260

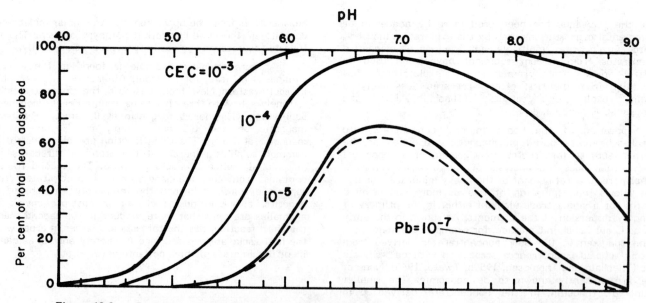

Figure 19.2. Percentage lead absorbed in water-sediment system as a function of pH and cation exchange capacity of sediment. Cation exchange capacity (CEC) indicated in moles/l., $[Na^+] = 10^{-4}$, $[Ca^+] = 10^{-4}$, total lead $= 10^{-6}$ (after Hem, 1976).

Table 19.2

Results of investigation of selective scavenging of elements by Mn-Fe oxides in streams of Maine
(after Nowlan, 1976)

Elements not scavenged by oxides	Elements probably not scavenged by oxides	Elements scavenged weakly by oxides	Elements scavenged strongly by Mn oxides	Elements scavenged strongly by Fe oxides
B	Ag	Cu	Ba	As
Cr	Be	Mo	Cd	In
K	Ca	Pb	Co	-
Mg	Ga	Sr	Ni	-
Rb	La	-	Tl	-
Sc	Sb	-	Zn	-
Ti	Y	-	-	-
V	-	-	-	-
Zr	-	-	-	-

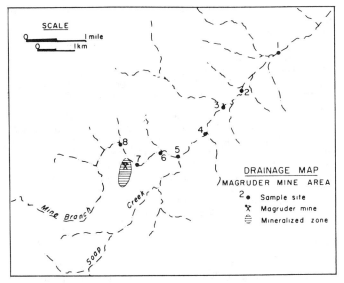

Figure 19.3a. Sample location map for Magruder Mine area, Georgia (after Carpenter et al., 1975).

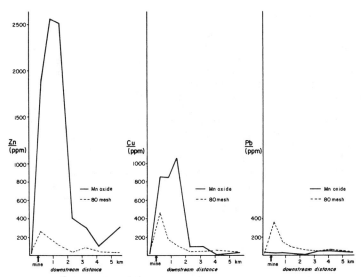

Figure 19.3b. Downstream dispersion for zinc, copper, and lead in minus-80-mesh stream sediments and oxide coatings, Magruder Mine area (after Carpenter et al., 1975).

Selective analysis of Fe and Mn oxides within the fine fraction of stream sediments has also been demonstrated to be efficient in base metal exploration. In a mineral resource assessment of the Tanacross quadrangle, Alaska, Foster et al. (1976) showed that an oxide-specific oxalic acid leach of minus-80-mesh sediment identified base metal occurrences more effectively than total analysis of minus-80-mesh sediment, analysis of heavy minerals, streambank sod, or aquatic bryophytes. Regardless of the size fraction considered, most studies have shown that anomalies are better defined by the ratios of the target metal to Mn or Fe, or to Mn and Fe together. This finding emphasizes the need for analysis of elements not directly related to the desired target in order to interpret the geochemical dispersion of the trace elements under study.

4. Organic Material

Until recently, organic stream sediments have been carefully avoided in exploration sampling because of the difficulty of analysis and high variability of trace element content that result from the mixing of autocthonous and allocthonous organic matter in the stream bed. However, in cold regions of low relief, such as northern Scandinavia, inorganic stream sediments are not universally present, but organic matter tends to accumulate owing to the low rate of decomposition of vegetation. As a result, sampling and analysis of organic stream sediments has become the accepted mode of regional geochemical prospecting conducted by the Geological Survey of Sweden (Larsson, 1976); the method was adopted after extensive orientation studies were conducted by Brundin and Nairis (1972). Their interpretation of the multi-element analysis of the organic sediments employed multiple regression techniques to eliminate the effects of the environmental variables: content of organic matter, Fe, and Mn.

Table 19.3
Summary of causes of spurious stream sediment anomalies, southern Ontario
(after Alther, 1975). Presence of elements are indicated by a cross

Element	Road	Fertilizers	Pesticides	Hogfood	Excreta	Automobile
Pb			X		X	X
Co			X	X	X	
Cu			X	X	X	
Ni				X	X	
Fe		X		X	X	
Mn		X			X	X
Zn			X	X	X	
Mg				X	X	
Ca	X				X	X
K		X		X	X	
Na	X			X	X	X

Seasonal Variations

In 1962, Hawkes and Webb indicated that only two studies on seasonal variation of stream sediment composition had been made: firstly, Govett (1958) showed a reduction in the cold extractable copper content of an anomalous stream during the wet season in Zambia; and secondly, a study by Gower and Barr in British Columbia indicated no significant variation in the cold extractable or total copper content within a four-month period in the summer. Most stream sediment surveys are conducted on the assumption that seasonal variation is minimal, but with the current interest in selective analysis of the Fe-Mn oxides this approach may require reconsideration. In a study conducted in Wales, Fanta (1972) determined that seasonal effects were noticeable in those elements subject to coprecipitation by Fe and Mn oxides, indicating that the precipitate phase of the stream sediment was in a state of delicate balance with the chemistry of the stream and the groundwater.

Recent studies in Goias, Central Brazil (Thomson, 1976) have shown no significant variation in the aqua regia and EDTA soluble Cu and Ni content of minus-80-mesh stream sediments collected downstream from the Americano do Brazil Cu-Ni sulphide prospect at the beginning, middle and end of the rainy season. In the contrasting glaciated environment of northwest Maine, Chork (1977) has shown that nitric acid soluble Cu, Pb, Zn, Co, Ni, Mn, and Fe vary only slightly with the seasons, and should not be of concern in routine stream sediment surveys.

Contamination

By definition, the environment will have been polluted to a greater or lesser degree wherever man has been (Webb, 1975). Contamination is a serious problem in geochemical surveys of heavily industrialized countries or areas with a long mining history. This problem has been brought to the fore by the results of a multi-element regional stream sediment survey of England and Wales carried out by the Applied Geochemistry Research Group at Imperial College, London. As an example, this study showed that the high concentration of arsenic in stream sediments from southwest England (Webb, 1975) was derived from the mining and processing of the arsenic-rich tin and base metal ores of the region, where active lode mining has been carried on since the Middle Ages. As yet, no simple method of applying stream sediment sampling in polluted areas has been developed, therefore, effective exploration in these areas involves time consuming detailed documentation of source areas of potential pollution, often requiring examination of historical mining and industrial records.

Hosking (1971) has provided an excellent review of sources of contamination of drainage due to mining, agricultural, industrial and domestic sources in southwest England. He reports, for instance, that minus-80-mesh stream sediments downstream from the town of Truro contain 400 ppm copper and 600 ppm zinc in an unmineralized area; and that values of 150 to 3000 ppm lead occur in stream sediments for 5 miles below the abandoned mill of the Pengenna mine in northeast Cornwall. While contamination of this degree can be readily recognized, many less spectacular forms of pollution exist in the vicinity of populated and agricultural areas, and the systematic documentation of these forms would facilitate identification of spurious stream sediment anomalies. One case of agricultural pollution has been documented by Alther (1975). This study of recent contamination due to pig farming in southern Ontario identified enhancement of metal values in stream sediments due to a variety of source materials (Table 19.3).

Spurious anomalies are not always due to human activity. A high content of tin in minus-80-mesh stream sediments (as much as 10 000 ppm, see Fig. 19.4a, 19.4b) from an area in southwest England not known to be mineralized was later traced to redistribution of cassiterite from centres of tin mineralization along shorelines of late Miocene-Pliocene age, presently preserved at an elevation of 400-500 feet (Dunlop and Meyer, 1973).

SAMPLING

Stream sediment geochemistry takes a number of different forms depending on the aim of the survey, and the regional or local constraints of physiography and climate which together make up the landscape. The conventional use of stream sediment surveys to detect anomalous dispersion trains downstream from individual mineral deposits is still the prime mode of application undertaken by most in the mineral exploration industry, but the effectiveness of stream sediment as a sampling medium has led to a broadening of the aims of geochemical reconnaissance to include agricultural, geological, and pollution studies (Webb et al., 1968). The design of stream sediment sampling programs has been influenced by this tendency towards generalization, which is most pronounced in government survey programs that require broad-scale resource assessment, and this factor will be considered in the following sections along with specific programs that have single mineral deposits as the principal target.

Types of Survey
Low Density Surveys

Widely spaced stream sediment sampling for the detection of metallogenic provinces in relatively unexplored regions has been demonstrated to be an effective method of delineating broad regions of mineral potential (Garrett and Nichol, 1967; Armour-Brown and Nichol, 1970). The aim of

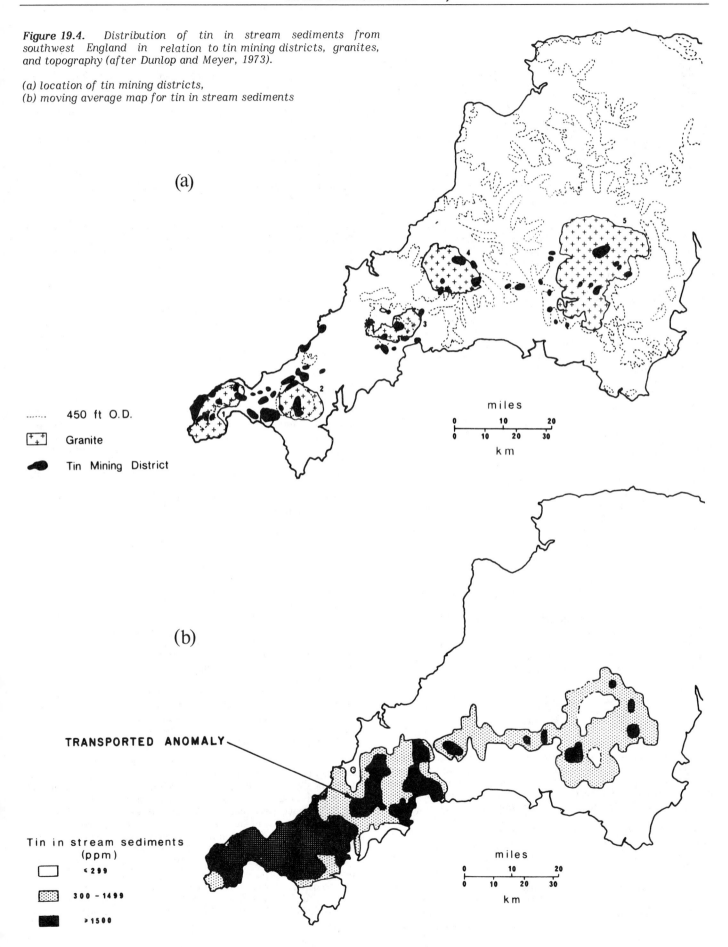

Figure 19.4. Distribution of tin in stream sediments from southwest England in relation to tin mining districts, granites, and topography (after Dunlop and Meyer, 1973).

(a) location of tin mining districts,
(b) moving average map for tin in stream sediments

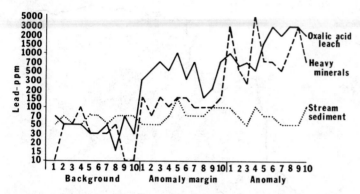

Figure 19.5. Comparison of the lead content of minus-80-mesh stream sediment, the iron-oxide-rich fraction of heavy mineral concentrates, and oxide residue of oxalic acid leachates from minus-80-mesh stream sediment over a suspected buried mineral deposit in the southern San Mateo Mountains, New Mexico. Data contributed by H.V. Alminas and K.C. Watts, U.S. Geological Survey.

this type of survey is not to identify discrete areas of mineralization, but to outline major belts or regions which may warrant further exploration by reconnaissance methods. Sampling density is generally within the range of one sample per 100 to 250 km^2, over areas of 30 000 to 50 000 km^2 or greater, although one survey reached continental proportions when a mining company sampled the total area of Australia amenable to drainage sampling (Duff, 1975). Multi-element analysis is the rule in low density surveys because it maximizes the usefulness of samples obtained at high cost per sample and covers the range of mineralization which can be expected within the area covered. Changes in climate, topography, and broad scale geology, which are inevitable in surveys of this size, can be recognized geochemically by use of statistical methods such as R-mode (Armour-Brown and Nichol, 1970) and Q-mode factor analysis. Both of these techniques were applied by Galbraith (1975) in a research study of an area of 36 000 km^2 in Bahia, Brazil, which delineated geochemical belts or districts on the basis of Cu, Zn, Co, Cr, Ni, Mn, and Ag analysis of samples collected at a density of one sample per 80 to 100 km^2. He concluded that Q-mode was superior to R-mode factor analysis in the classification of sediment samples related to chromite mineralization in this area. Also in Brazil, a 10-element, 400 000 km^2, survey over parts of the states of Goias and Minas Gerais has been carried out at a density of approximately one sample per 115 km^2 by Citco International Minerals Company (W.T. Meyer, unpubl. data).

Regional Reconnaissance

Stream sediment geochemistry excels as an exploration tool in low cost reconnaissance prospecting of large areas for indications of individual mineral deposits, groups of occurrences, or favourable geological environments. Sampling density ranges from one sample per 1 km^2 to one per 25 km^2, depending on type of target and drainage characteristics, and inherent in the reconnaissance concept is the need for more detailed sampling to determine the significance of regional anomalies. Much of the exploration and resource assessment carried out by governmental and intergovernmental agencies falls into this category, as does mining company exploration of large concessions and areas being considered for acquisition.

An example of this type of survey currently underway in the United States is the National Uranium Resource Evaluation (NURE) Program, which aims to systematically sample water and sediment collected from streams, lakes, and wells in order to identify all favourable uranium exploration areas in the United States by 1980 (Dahlem, 1976). Data generated by a program of this scale (a 1977 budget of $6.1 million was reported by Larson, 1977) could also have a significant impact on exploration for other targets, as the multi-element analytical data from samples collected at an average density of one sample per 10 km^2 will be made available to industry. One interesting conclusion from orientation surveys carried out in the southern Appalachians prior to NURE sampling was that the minus-100-mesh stream-bottom sediment gave the same or better information than water or suspended solids, was cheaper to collect and analyze, and presented fewer analytical problems (Ferguson and Price, 1976). The conclusion that stream sediments were preferable to water for reconnaissance drainage surveys for uranium was also reached by Rose and Keith (1976) for eastern Pennsylvania. Also within the United States, reconnaissance geochemistry at a density of one sample per 20 km^2 formed an important part of a survey of the Tanacross quadrangle, Alaska, conducted within the Alaskan Mineral Resource Assessment Program (Foster et al., 1976). This study did not rely on sampling and analysis of the fine fraction of stream sediment alone, but also examined the distribution of trace and major elements in heavy mineral concentrates, stream-bank-sod, aquatic bryophytes, and the soluble oxide fraction of the minus-80-mesh sediment. Results of this survey demonstrated that no one sampling medium or analytical method was effective for all elements in both maturely dissected terrain and mountainous parts of the quadrangle. In the maturely dissected area where hydromorphic dispersion is pronounced, an oxide-selective leach of minus-80-mesh sediment was the most effective medium for locating base metal occurrences, while heavy mineral concentrates were best for detecting tin and tungsten mineralization. Conventional minus-80-mesh stream sediments and heavy mineral concentrates were found to be the most suitable sample media for exploration in the Alaska range, where mechanical weathering is the prevalent destructive process.

A comparison of three of the commonly employed stream sediment techniques is presented in Figure 19.5. In this particular example, a buried, mineralized area in the San Mateo Mountains of New Mexico, the traditional minus-80-mesh stream sediment shows practically no contrast. Over a background of 30 to 50 ppm lead, the maximum value is only 100 ppm. The intermediate-magnetic fraction of the heavy mineral concentrate has a background of 10 to 70 ppm lead, but provides a well defined peripheral zone at 70 to 100 ppm and a distinct, large anomaly of 150 to 5000 ppm lead. This anomaly is attributed to transported, detrital, supergene iron and manganese oxides. The fraction of the sediment soluble in oxalic acid, dominated by iron and manganese oxides but including both the detrital oxides and those precipitated on the sand grains after hydromorphic transport, produces a much broader anomaly. Over a background of 10 to 50 ppm, the intermediate zone is 100 to 700 ppm, and the anomalous zone is 300 to 3000 ppm.

In Canada, more than 90 per cent of the current stream sediment sampling activity of the Geological Survey of Canada is related to the Federal-Provincial Uranium Reconnaissance Program, which commenced in 1975 (R.G. Garrett, pers. comm.). While lake sediments and waters have been adopted as the preferred sampling media in the Canadian Shield (Cameron, 1976), stream sediment sampling was conducted over an area of 75 000 km^2 in southern British Columbia and the Yukon in 1976. The British Columbia portion of the program covered 46 800 km^2 at a mean density of one sample per 13 km^2 (Smee and Ballantyne, 1976), and in the Yukon 2200 stations were sampled over an area of 28 490 km^2, giving the same degree of coverage. In each case, the stream sediments were being submitted for multi-element analysis in addition to the uranium determinations required to fulfill the primary goal of the surveys. Gleeson and Brummer (1976) have recently described a regional stream sediment survey carried out in 1970 and 1971 by Occidental Minerals Corporation of Canada over an area of 64 750 km^2 in the southwestern part of the Yukon Territory. This survey was designed to detect porphyry copper-molybdenum deposits by means of copper, zinc, and molybdenum analysis on the minus-80-mesh fraction of stream sediments collected at a mean density of approximately one sample per 5 km^2. Moving average smoothing proved to be useful in correlating geochemical patterns with the regional lithology, and automatically plotted residual metal maps were used to define specific follow-up targets.

The first countrywide multi-element regional stream sediment program to be undertaken in the United Kingdom was the survey of Northern Ireland carried out by the Applied Geochemistry Research Group (AGRG) at Imperial College, London, in 1967 (Webb et al., 1973). Nearly 5000 samples, collected at a density of one sample per square mile (one per 2.6 km^2) were analyzed for this study. AGRG then sampled England and Wales at the same density in 1969 (Webb, 1975), and collected more than 50 000 samples which have provided the basic data for graduate research on mineral exploration applications (Urquidi-Barrau, 1973; Cruzat, 1973; and Dunlop, 1973), regional geochemistry (Holmes, 1975), data processing, and agricultural and environmental pollution. Regional geochemical sampling by the Institute of Geological Sciences (IGS) commenced in Scotland in 1968. Then, in 1972, under a more detailed mineral reconnaissance program sponsored by the Department of Trade and Industry, stream sediment samples were collected at a density of one per 1.5 km^2 from an area of 1600 km^2 in central Wales (Ball and Nutt, 1974). In Scotland, sediments are taken 0.3 m below the stream surface in order to minimize the effects of Fe and Mn oxides. Samples are wet-sieved to <150 microns (minus-100-mesh) and a split is sent for U analysis by delayed neutron activation, while the remainder of the sample is analyzed by atomic absorption and direct-reading emission spectrography. A major long-term project is the compilation of regional geochemical maps made up of raw data plotted at the sample points using a radiating system of vectors, whose azimuth identifies the element and whose length indicates the concentration. Mineral industry regional surveys in the United Kingdom and Ireland naturally cover less area than comparable programs in other part of the world; a typical regional exploration program in Ireland might cover an area of 1300 km^2 (Horsnail, 1975), with stream sediment samples collected at a density of one sample per 0.85 km^2.

Regional geochemical surveys of the Precambrian Shield areas of northern Sweden are based on the collection of organic stream sediment samples, supplemented by the sampling of heavy minerals in till. In 1975, a total of 22 000 organic stream sediments were collected. Larsson (1976) has described the results of a survey int he Pajola district, in which more than 10 000 organic sediments were collected from an area of 8000 km^2. Multi-element analysis of the ashed samples by X-ray fluorescence and emission spectrography was successful in outlining all previously known base metal occurrences in addition to 40 prospective target areas. In southern Norway, regional stream sediment sampling at road-stream intersections has been described by Bjorlykke et al. (1973) of the Geological Survey of Norway. Wet sieving of stream sediments to minus-80-mesh has been adopted by the Norwegian Survey. While this practice results in loss of some fine material in suspension, it eliminates collecting and shipping large quantities of sample (Bolviken et al., 1976). Both organic and inorganic sediments are collected at an average density of one sample per 1.5 km^2 during regional surveys by the Geological Survey of Finland.

Analysis is by atomic absorption for Co, Cu, Ni, Pb, Zn, Mn, and Cr (Kauranne, 1975). From 1971 to 1974, a total of 23 179 inorganic and 34 942 organic stream sediments were collected. In 1975 alone, the Finnish Geological Survey collected 9500 organic and 7300 mineral stream sediments, and 1500 organic lake sediments – a grand total of 18 400 samples which together cover an area of 23 000 km^2 (L.K. Kauranne, pers. comm.).

Geochemical surveys in France are conducted by Bureau de Recherches Géologiques et Minières (BRGM) for base metals, and by Commissariat`a l'Energies Atomique for uranium. More emphasis is placed on water sampling than stream sediments in the search for uranium, a practice that is also followed by the I.G.S. (Ostle et al., 1972) in the United Kingdom.

Regional stream sediment reconnaissance is actively carried out by most government agencies and mining companies in Central and South America. As an example, a description of geochemical reconnaissance studies carried out in Goias, Brazil, during the pilot phase of the Projecto Geofisico Brasil-Canada has recently been published (Thomson, 1976). The results of these investigations determined the survey parameters for a regional reconnaissance of the State of Goias, which is being undertaken at a density of one sample per 10 km^2 with all samples collected from tributary drainages with catchment areas of less than 25 km^2. Eight elements are being determined by atomic absorption analysis of the portion of the minus-80-mesh fraction that can be extracted using aqua regia. Exploration companies such as Docegeo also use regional geochemistry extensively, as do State mining companies such as Metago (Goias) and CPBM (Bahia). Multi-element regional reconnaissance (one sample per 8-10 km^2) surveys are carried out by Brazilian subsidiaries of international mining companies (Meyer, 1977). In central Guyana, geochemical reconnaissance was undertaken by the Guyana Geological Survey as part of the Potaro-Mazaruni Project, in which 3000 samples were collected from an area of 6000 km^2 (Gibbs, 1974). Semiquantitative optical emission spectrography was used as the analytical method in this study

In Africa, regional and provincial scale geochemical mapping is employed by geologists of the Geological Survey of Zambia engaged in systematic geological mapping. Sample density depends on the scale of mapping, and analyses are for 20 elements by semiquantitative X-ray fluorescence techniques. In Rhodesia, the Institute of Mining Research at the University of Rhodesia has been active in regional geochemical research projects since it was founded in 1969. Recent studies by the Institute include a regional stream sediment reconnaissance of 1350 km^2 of the Sabi Tribal Trust Land at a density of one sample per km^2 (Topping, 1976), and a survey of 1664 km^2 near West Nicholson, Rhodesia, at the same density (Mayfield, 1976). Multi-element analysis was applied in both cases. Most mining companies in southern Africa devote a greater effort to regional soil sampling than to drainage reconnaissance, and Buhlmann et al. (1975) estimated that more than 95 per cent of the samples collected by the major companies in 1973 were taken from soil grids.

A recent survey of stream sediment sampling programs by government agencies in Australia (C.R.M. Butt, pers. comm.) has shown a low level of activity. Only 4250 stream sediments were collected in 1976 and the number is expected to decline in 1977. Government agencies in Australia do not engage in direct exploration, although they occasionally conduct regional or detailed local orientation studies. However, most mineral exploration companies in Australia make regular use of regional stream sediment sampling in the more favourable humid zones; stream sediments are of

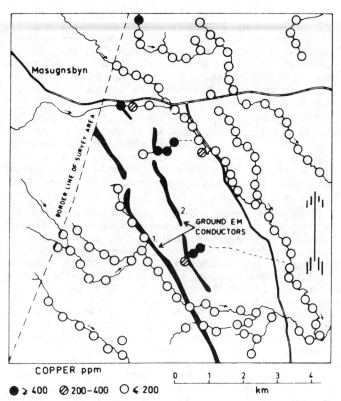

Figure 19.6. Copper content of organic stream sediments in relation to ground electromagnetic conductors, Pajala district, northern Sweden (after Larsson, 1976).

notable exploration successes attributed to UN stream sediment surveys are the discoveries of the Mamut deposit in Malaysia, Cerro Petaquilla in Panama, and Chau-Cha in Equador – all three are porphyry-copper type deposits.

Detailed Surveys

Detailed stream sediment surveys form the logical sequel to reconnaissance sampling after delineation of regional anomalies; they may be the first phase of exploration in established mining districts; and they are a tool for investigation of targets outlined by geological, geophysical or remote sensing studies. The techniques applied in detailed minus-80-mesh stream sediment sampling are essentially the same as those reviewed in the section on regional reconnaissance. The sampling interval is different, of course, and the detailed surveys can make use of seepage samples, bank sediments, and spring precipitates wherever local drainage conditions permit.

In the shield environment of northern Sweden, where organic sediments are collected in the reconnaissance stage of exploration, Larsson (1976) has described a follow-up technique that makes use of bog-margin samples. An organic stream sediment anomaly for copper (Fig. 19.6) was found to be associated with one of a pair of adjacent electromagnetic conductors, originally considered to represent graphitic horizons within greenstones in the Pajala deposit. Follow-up investigations took the form of sampling along the shallow margins of local bogs (usually less than 0.5 m in depth) in order to detect the enhanced metal content derived from adsorption of metals from groundwater in contact with mineralization. The results of the bog-margin survey confirmed the association of copper with the eastern band of conductors (Fig. 19.7), as copper content of the organic material exceeded 1 per cent close to the geophysical target. This follow-up procedure has the advantage in far northern latitudes of being suitable for winter sampling, and could find wider application in similar terrain.

limited use in the more arid areas due to low density of drainage and dilution by wind-blown material. One approach in arid areas has been to use two size fractions, minus-120-mesh and minus-4-plus-16-mesh. The coarse fraction in such studies generally contains gossan fragments and multiple grains cemented by metal-rich iron hydroxides.

Stream sediment surveys have been particularly successful in the Pacific islands of New Guinea, Fiji, Indonesia, and the Philippines in the search for porphyry copper deposits. In Fiji, Leggo (1976) reported the discovery of two porphyry systems on the basis of stream sediment sampling at a density of 2.4 samples per km^2, with the analysis for Cu, Pb, Zn, and citrate-extractable Cu. The dispersion train from these deposits would have provided two or three anomalous samples if a sampling interval of 1 km had been used. The success of stream sediment surveys in the tropical islands of the Pacific can also be demonstrated by the experience of the Australian exploration company, Conzinc Riotinto of Australia Exploration (CRAE). From 1964 and into the 1970s, CRAE conducted stream-sediment surveys in New Guinea and the Solomon Island chain which resulted in drilling of nine porphyry targets, one of which developed into the highly profitable Panguna mine (MacKenzie, 1977). At Panguna, a weak anomaly could still be recognized 28 km downstream. In almost all cases exploration in this environment has been based on analysis of the minus-80-mesh fraction.

The United Nations Development Program (UNDP) often uses regional stream sediment surveys in its mineral exploration programs in a wide variety of environments. According to Brand (1972), a review of all reconnaissance methods employed by the UNDP indicated that geochemical surveys had been more rewarding than airborne geophysics as prospecting aids to basic geological programs. Among the

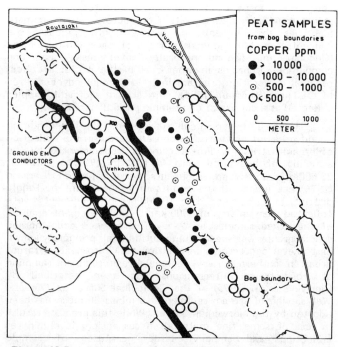

Figure 19.7. Copper in follow-up peat samples from bog margins, Pajala district, northern Sweden (after Larsson, 1976).

A good example of conventional detailed follow-up of a stream sediment anomaly can be found in the investigations of a lead anomaly identified during regional sampling at one sample per 2 km^2 in southern Norway (Bolviken, 1976). The regional anomaly near Snertingdal was detected in samples containing 160 and 130 ppm Pb (Fig. 19.8); subsequent detailed sampling of the drainage system at 100-m intervals outlined an area of interest of about 5 km^2, with peak values in excess of 300 ppm. Systematic soil sampling of the favourable area then served to delineate the target area which was found to be related to a 2 km-long mineralized zone of subeconomic disseminated sulphides (Fig. 19.9).

The benefits of sampling seepage areas in the course of detailed sampling are well demonstrated by the dispersion of copper, molybdenum, and zinc in drainage sediments from the Huckleberry Cu-Mo deposit, British Columbia (Sutherland-Brown, 1975). Significantly higher metal values were generally found in seepages when compared with adjacent stream sediments, and copper seepage values exceeded 1500 ppm in close proximity to the mineralization (Fig. 19.10).

Orientation Surveys

Stream sediment orientation surveys are an essential but often neglected operation used in mineral exploration to establish the field and laboratory parameters that maximize the difference between anomaly and background, and provide the data on which to determine a safe sampling interval for detection of the anomalous dispersion train. The most common type of orientation study takes the form of detailed sampling downstream from a known deposit, preferably one which has not been disturbed by mining operations. It is seldom possible to find a suitable undisturbed deposit within the same physiographic and geologic province as the reconnaissance area and, as a result, it is often necessary to accept the compromise of conducting the orientation on a geologically favourable target which has been disturbed to some extent.

An example of this type of orientation survey can be found in a comparative sampling study conducted on the Caridad porphyry copper deposit, in the semi-arid northwest part of the state of Sonora, Mexico (Chaffee et al., 1976). The Caridad deposit was discovered as a result of a

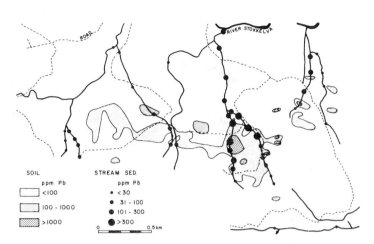

Figure 19.9. Results of detailed follow-up stream sediment and soil sampling of the Snertingdal lead anomaly, southern Norway (after Bolviken, 1976).

reconnaissance stream sediment survey which detected an anomaly 19 km downstream from the mineralized area, but the abandoned small workings in the area may have influenced the size of the anomaly. As part of this pilot study, minus-0.25 mm (minus-60-mesh) stream sediment, heavy mineral concentrates, and trees from arroyos near to and downstream from the deposit were sampled and then analyzed by spectrographic and wet chemical methods for as many as 37 elements. The aim of the study was to determine: (1) which elements other than copper might produce downstream anomalies; (2) the maximum distance at which the deposit could be detected by any sample type; (3) the sample types that would be most effective in revealing the presence of the deposit; and (4) which of the two types of copper analyses, partial or total extraction, would be most effective. Results showed that the porphyry copper deposit at Caridad produced anomalies in stream sediments, heavy mineral concentrates, and two plant species, but that stream sediments gave more useful information than the other sample types investigated. Significant anomalies were detected in molybdenum, zinc, silver, and tungsten in addition to copper. Molybdenum in stream sediments was found to give the longest detectable dispersion train, which extended at least 32 km (Fig. 19.11). Little difference was found between the use of total or cold-extractable copper as a method of locating the Caridad deposit.

Orientation sampling that precedes a regional multi-element survey must be designed to accommodate the aims of the intended survey, i.e. mineral exploration, for a wide variety of targets or multi-purpose geochemical mapping. A good example of an orientation study for a regional geochemical survey is provided by Plant (1971), who studied an area in northern Scotland. She found that grinding of minus-100-mesh stream sediment and collection of samples from second- or third-order streams improved the overall precision, which was a mapping requirement in this area of low geochemical background. Geochemical "noise" produced by coprecipitation of metals with hydrous iron and manganese oxides was reduced by: (1) removal of the fine fraction prior to analysis; (2) grinding of samples; (3) collection of samples beneath the sediment-water interface; and (4) collection of samples from streams greater than first order. This study demonstrates the conflicting requirements of single-target mineral exploration surveys and regional geochemical mapping: an exploration survey for Fe-Mn oxide minerals would seek to amplify the very "noise" that Plant's study tried to minimize, and therefore sampling methods would be completely different.

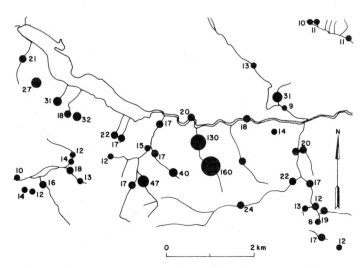

Figure 19.8. Lead content (in ppm) of stream sediments collected at road-stream intersections, Snertingdal area, southern Norway. Area of solid circle is proportional to lead content at each locality (after Bolviken, 1976).

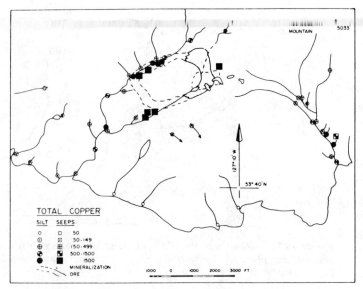

Figure 19.10. Copper in stream sediments and seepages, Huckleberry Cu-Mo deposit, British Columbia (after Sutherland-Brown, 1975).

Large-scale pilot studies are conducted in advance of country wide regional surveys, or where comparisons between new and established methods are required prior to acceptance of modifications of sampling or analytical techniques. The extensive pilot studies recently conducted as part of the Projeto Geofisico Brasil-Canada (PGBC) provided data which facilitated the selection of optimum techniques for both reconnaissance and detailed surveys. This information was obtained through specific orientation studies around known mineral deposits, and the reconnaissance sampling of five relatively large pilot areas (approximately 2500 km² each) at a density of 1 sample per 7-10 km². The study established the validity of stream-sediment sampling in a tropical environment, provided training of survey personnel, and established procedures for systematic sampling of a much larger area (Thomson, 1976). Another pilot study in northern Sweden (Brundin and Nairis, 1972) sampled both organic and inorganic sediments over an area of 400 km² to test the usefulness of organic material in stream sediment surveys.

Field Methods

As alternate sample types and multi-purpose surveys have developed, a greater need has evolved for optimizing sampling design. Miesch (1976) has thoroughly reviewed the statistical basis for optimization of geochemical sampling in general. For practical purposes current stream sediment sampling procedures seldom conform to the unbiased minimum sampling error configuration of the ideal sampling design. Nevertheless, a number of studies of sampling error and survey design (Bolviken and Sinding-Larson, 1973; Howarth and Lowenstein, 1971; Duff, 1975; Sharp and Jones, 1975; and Hawkes, 1976b) have demonstrated a trend towards a more systematic approach to the implementation of stream sediment surveys based on a quantitative assessment of various sampling criteria. In the context of field methods, the present state of the art encompasses the extremes of haphazard sampling without prior planning, and carefully executed systematic surveys designed to achieve a well conceived goal. Some of the current survey techniques and planning consideration are reviewed in the following sections.

Selection of Sample Sites

While orientation surveys often establish the sampling interval required to obtain at least one or two samples from an anomalous dispersion train, other factors help determine the overall sample spacing for regional surveys. Also considered are the completeness of coverage required, logistics, and the total cost allowed for the survey. Most mining company exploration programs tend to err on the side of caution as far as sample density is concerned, based on the valid assumption that a greater sampling density reduces the probability of missing a mineral deposit. Logistics play an important role in setting sample spacing, particularly if the survey takes place in an area of poor access where labour and transportation costs are high. In this case, the prudent policy requires collection of samples with a closer spacing than that estimated from an orientation survey, even if some samples are not analyzed initially. Evidence of an anomaly can then be verified and more closely defined by analysis fo the additional samples without undergoing the expensive alternative of a return to the field area, a practice which can prove particularly important in areas having short periods of access governed by local weather conditions. On the other hand, where access is simplified by the presence of an adequate network of roads crossing the drainage system, the reduction of sampling cost that results through sampling road-stream intersections can provide the overriding factor in determining sample spacing. This was the case in the regional

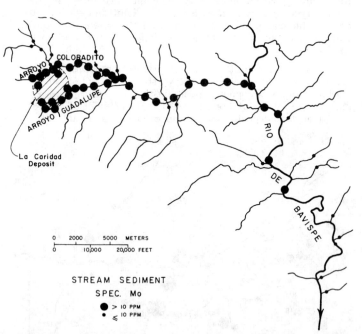

Figure 19.11. Distribution of molybdenum (emission spectrographic analysis) in stream sediments draining the Caridad porphyry copper deposit, Sonora, Mexico (after Chaffee et al., 1976).

sampling undertaken for the Geochemical Atlas of England and Wales (Webb, 1975) and in a reconnaissance survey of 3000 km² in southern Norway (Bolviken and Sindig-Larson, 1973).

In order to quantify the effects of dilution of an anomalous dispersion train and thereby provide a sound basis for establishing sample spacing, and as an aid in interpretation, Hawkes (1976b, c) has developed an empirical formula to relate the composition of sediment samples to the size of the catchment area and the size and grade of the surface expression of the mineralization causing the anomaly. The formula is expressed as:

$$Me_m A_m = A_a (Me_a - Me_b) + A_m Me_b$$

Where Me_m = metal content of mineralized area, Me_a = metal content of anomalous stream sediment, Me_b = metal content of background samples, A_m = surface dimensions of mineralization, and A_a = size of catchment area above anomalous site (see Fig. 19.12 for explanation). For the purpose of establishing a suitable sample spacing, the formula is used to predict the maximum size of drainage basin below which sample collection will result in the detection of an anomaly, assuming the minimum size and grade of an economically acceptable target and the smallest anomaly contrast that can be identified above background levels. For example, if the target is a porphyry copper with an assumed surface area (A_m) of 1 km² and grade (Me_m) of 4000 ppm, a limiting value of 30 ppm would be required if an anomaly contrast of twice background (15 ppm) is adopted.

Substituting these values in a rearranged equation, the area above the anomalous sample site can be calculated as follows:

$$A_a = \frac{A_m (Me_m - Me_b)}{Me_a - Me_b} = \frac{1(4000-15)}{30-15} = 266 \text{ km}^2$$

This area would relate to a linear stream distance of about 15 to 20 km, depending on the shape of the drainage basin, and this figure could be used as the limiting distance between samples to assure the location of a porphyry target under the conditions specified.

At best, this procedure will only provide an approximation of the maximum permissible sample spacing owing to the many assumptions that have to be made: namely, those of constant erosion rate, uniform background, minimal sampling and analytical error, stable sediment water chemistry, and absence of contamination. An important factor which is not dealt with in this formula is the relationship of surface size and grade of mineralization to the economic value of a deposit. Depending on erosion level, a large surface anomaly could result from deep weathering to the base of a mineralized system, whereas a relatively "low-grade" anomaly may signify that the land surface lies above a deposit and erosion has not removed a large mass of the mineralized material (e.g. see Beus and Grigoryan, 1977, p. 186).

Sampling Errors

Discussion of error in a sampling process for exploration must be tempered by a consideration of economics. The object is always to get the desired information as cheaply as possible. Hence the need for orientation studies designed to evaluate the relationship of valued information to error, as the process of trying to minimize or eliminate error can be expensive. Such a study was undertaken in five areas peripheral to the Selway-Bitterroot Wilderness Area in Idaho and Montana. Three size fractions of the stream sediment were compared (coarse = -20+30 mesh medium = -30+80 mesh, fine = -80 mesh). Partial solution techniques were applied to two of these (the medium to fine fractions were treated by the oxalic acid technique described by Alminas and Mosier, 1976), and four magnetic splits were made of the heavy mineral concentrate (sequentially those magnetic at 0.1, 0.5, and 1 amp. on the Frantz Isodynamic Separator and the non-magnetic fraction at 1 amp.). All of the analyses were by optical emission spectroscopy in which 31 elements were sought. The sample design was to pick five areas bordering the Wilderness, one on the west, one on the north, one on the east, and two adjoining each other on the south. These were chosen to reflect a variety of geologic and physiographic environments. Though no significant mineralization is presently known in any of these, clear evidence of low-level mineralization of several types can be seen in the data. Five sites were selected within each area to provide as nearly as possible five adjacent drainage basins of 1-2 km² each. Each site was sampled twice. This plan allows comparison of within-site variance (noise or error) with local variance (between sites, at a scale of a few kilometres) with regional variance (between areas, at a scale of tens of kilometres). The results of the three-level analysis of

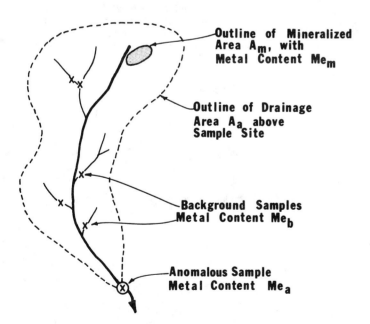

Figure 19.12. Explanation of terms in stream sediment anomaly-dilution formula (after Hawkes, 1976c).

Table 19.4

Summary of the results of a nested analysis of variance of analyses of duplicate samples from 25 sites in five areas bordering on the Selway-Bitterroot Wilderness Area in Idaho and Montana. Each sample was analyzed for 31 elements by optical emission spectroscopy.

Sample medium		Number of elements detected	Number of elements showing significant variation		
			Only within site	Between sites	Between areas
Stream sediment	-20 + 30 mesh	11	2	16	14
	-30 + 80 mesh	9	2	17	15
	-80 mesh	9	2	18	12
Oxalic acid leach of stream sediment	-30 + 80 mesh	9	1	7	21
	-80 mesh	9	1	11	19
Heavy-mineral concentrate	Non-magnetic	7	3	20	12
	-0.5 + 1.0 amp.	9	2	17	16
	-0.1 + 0.5 amp.	9	1	18	18
	Magnetic	8	1	14	19

variance are summarized in Table 19.4. In each sample type, about nine elements could not be detected. For one to three elements most of the variance was within site; i.e. noise or error. For all sample types, a significant proportion of the variance for 7 to 21 elements was attributable to local and regional factors that should be explainable in geologic terms. It should be noted further that two of the three elements discarded because of large within-site variance in the nonmagnetic fraction of the heavy mineral concentrates are tin (cassiterite) and gold. Both are reported in relatively large quantities in only one sample. The apparent local variance results from tin and gold being major components of a sparse mineral, and the instances where these elements are present constitute legitimate "hits". In geochemical exploration, it is often better to blindly ignore the minor statistical nuisance of sampling and analytical error.

In a similar study covering a wide cross-section of climate and geology in Australia, Duff (1975) studies six field areas (see Fig. 19.13 for location) to determine the sampling technique that would be most effective in low-density regional geochemical mapping. Duplicate samples were collected from a number of localities within each pilot area, and the minus-20, -80, and -140 mesh fractions were analyzed for total Co, Cu, Ni, Pb, Zn, Fe, and Mn by atomic absorption. Analysis of variance methods were used to determine: (1) the minimum number of samples required per locality for regional variation to be significant; (2) the best size fraction for each element, and (3) the major source of error, whether sampling or analytical. The summarized conclusions (Table 19.5) indicating the optimum sampling conditions in each case show some broad trends, such as a tendency towards preferring the coarse grain size in more arid areas, but the best choice of mesh size of sediment and number of samples at each site was found to vary from element to element within areas as well as between areas. This study points out the difficulty in selecting a single size fraction and sampling plan to minimize the sampling error for all elements in multi-element surveys, but it should be noted that it is only in those areas where the regional variability for an element is very low that multiple samples from a single site or multiple analyses on a single sample are required to establish valid geochemical patterns. In Area 2, which is underlain by extensive and fairly uniform Tertiary basalts, low regional variation of manganese, rather than high sampling and analytical error, explains the need for multiple samples at a site. In most mineral exploration programs, the variability associated with the desired target would probably be large in relation to the sampling error, and the collection of duplicate samples is only routinely adopted as security against loss in shipping, to provide material for later check analyses in cases of contamination or anomalous metal concentrations, and to establish a value for sampling error. In those situations where sampling and analytical error tends to

Figure 19.13. Location of six field areas used in study of the effect of error on low-density regional geochemical mapping in Australia (after Duff, 1975). See Table 19.5 for summary of results.

be high, use of mapping techniques such as moving-average smoothing produces more reliable regional maps of broad scale features (Howarth and Lowenstein, 1971).

The major component of error in working with either heavy mineral concentrates or stream sediments arises from elements that are major components of trace minerals. As noted earlier, this problem is less severe in heavy mineral concentrates because a larger initial sample is taken, thus (1) increasing the probability that the trace mineral will be found in the sample or, more commonly, (2) increasing the number of particles of the trace mineral present in the sample. The first effect is most important for the precious metals, where a single particle of native metal is highly significant. The sampling problem has been clearly defined from the theoretical standpoint by Clifton et al. (1969). From the practical standpoint, a single example from the Hahns Peak district of northwest Colorado may suffice. Millimetre-size gold particles are common in placer deposits surrounding the mining district. During systematic geochemical sampling in and around the district, stream sediment and heavy mineral concentrates from stream sediment were routinely collected. None of the raw stream sediment samples contained chemically detectable gold.

Table 19.5

Summary of optimum particle sizes and number of samples per locality for six areas. Atomic absorption analysis after a hot mixed acid attack. After Duff (1975)

Element	Area 1	Area 2	Area 3[1]	Area 4	Area 5[2]	Area 6
Co	C 1 A	F 1 A	M 2 A	F 1 A C 2 A	F 1 S	C 1 A F 1 A
Cu	C 1 S	F 1 A	M 1 S	F 1 S C 1 S	F 1 S	F 1 S C 1 S
Ni	C 1 A	F 1 A	M 2 A	C 1 S F 2 A	F 1 S	F 1 S C 1 S
Pb	C 1 A	F 4 A	M 2 A	C 1 A F 2 A	M 2 A F 5 A	M 1 A C 2 A F 5 A
Zn	C 1 S	F 1 A	M 1 S	F 1 S C 1 S	F 1 S	C 2 S F 2 S
Fe	M 1 S C 1 S	F 1 A	M 1 S	C 1 S F 3 S	F 1 S	C 1 S F 3 S
Mn	C 1 S	C 7 S F - S	M 6 S	F 1 (S= (A C 1 (S= (A	F 1 S	F 2 S C 2 A

Key

First symbol: Optimum mesh size,
C = -20 mesh,
M = -80 mesh,
F = -140 mesh

Second symbol: Minimum recommended number of samples per locality.

Third symbol: S or A indicating whether sampling or analytical error is major component of total error.

Notes:
[1] In Area 3 limited minus-20-mesh data were available.
[2] In Area 5 no minus-20-mesh specimens were collected.

Visible gold was noted in the field during panning of the heavy mineral concentrates, and the distribution of this gold helped substantially in understanding the geologic history and metal zoning around the district. Only 50 per cent of the concentrates containing visible gold contained chemically detectable gold when a one-half split of the concentrate was analyzed in toto.

More commonly many particles of the trace mineral are present in the large sample taken for heavy mineral separation. Two examples illustrate the extremes of variability encountered in this situation, both involving discrete tungsten minerals. The first example is from north-central Colorado, where a train of fine grained huebnerite extends down Clear Creek for at least 45 miles from the upper zones of the Urad-Henderson porphyry system (Theobald and Thompson, 1959b). At the lower end of this train, the first measured tungsten value on the heavy mineral concentrate was 300 ppm over a background of 40 or less. On two subsequent years this site was reoccupied by students who had one half hour's experience with a gold pan and one day's instruction in the colorimetric analytical procedure (Ward, 1951). The tungsten contents (ppm) of six separate samples collected, prepared, and analyzed by the students are 200, 400, 600, 200, 200, 400. Tungsten is not detectable in the raw stream sediment at this site.

The second example is from the central part of the Precambrian Shield of Saudi Arabia, where stream sediments and heavy mineral concentrates were collected from dry stream beds on a 1-km grid (Theobald and Allcott, 1975). For control, an untrained sampler was asked to collect 25 samples at a single site, and left to accomplish the task. The hope was that this would produce at least a haphazard if not a random sample. The result was a suite of samples collected systematically from the top of the alluvial fill to bedrock, and the last sample was chipped from the bedrock surface. The mineral sought was scheelite, which forms coarse crystals as much as 2 cm long. Nevertheless, the samples were sieved to minus-10-mesh before concentrating. Twenty-two of the 25 samples were run at random among 500 other samples. Though possible sources of error are maximized in this example both by the coarse grain size of the scheelite and by the stratified nature of the samples, the total range of values obtained was 150 (an extreme value at the low end) to 5000 ppm of tungsten. The mean was 2000 ppm, and the coefficient of variation was 80 per cent. Background tungsten from the other 500 samples was less than 100 ppm.

Field Observations and Analysis

With the advent of computers, it has become common practice amongst government surveys and the exploration industry to record field observations in coded form on data sheets such as the familiar Geological Survey of Canada field card (Garrett, 1974). Apart from the essentials of sample number and location (co-ordinates, map, or airphoto reference), data recorded at the field location can be of two types, quantitative values relating to stream dimension, stream order, measured pH, etc., and qualitative information denoting the opinion of the sampler as to the presence of organic material, precipitates, contamination, etc. The first set of numerical data can be readily combined with analytical values for statistical examination of the data. The second set of observations is less amenable to processing, particularly when the data set is made up of information collected by a number of different survey teams, often recruited at the lowest possible level in order to minimize the cost of the survey. During preparation of the regional survey of England and Wales by the Applied Geochemistry Research Group (RSM), it was found that computer plotting of coded field observations made by well trained sampling teams resulted in patterns which coincided with the field areas covered by different sampling crews. This type of outcome is not uncommon and, as a result, it is probably true that the bulk of the field observations noted in large scale regional surveys are never actually processed. Serious consideration should be given to reducing the number of observations to those items which have proven significance, or which can be reliably quantified. As a general rule, laboratory determination of variables such as content of organic matter and manganese and iron hydroxides provides a more useful measure of sample-site characteristics.

There has been little progress in the development of on-site analytical techniques for stream sediment surveys in recent years. Several field kits are available commercially which enable the sampler to perform colorimetric analysis for Cu, Pb, Zn, or combined heavy metals, largely based on well established methods that use dithizone as the reagent (Bloom, 1955; Holman, 1963; Stanton, 1966) or 2.2' biquinoline for copper (Ward et al, 1963). This type of analysis is particularly useful in the follow-up investigation of base metal anomalies, and can save considerable time in determining the point at which metal enters the stream system, provided the metal occurs in a cold-extractable form. Portable isotope-fluorescence instruments hold promise for an on-site analysis of sediments (Gallagher, 1967; Kunznedorf, 1973), but the poor sensitivity of this method means that field analysis must be restricted to elements which can be concentrated by on-site panning, such as tin in the form of cassiterite.

Laboratory Methods

As analytical methods are reviewed in greater detail elsewhere in this volume, only those procedures related to the preliminary handling of stream sediment samples will be discussed in this section. Included are the methods of sample preparation and the partial extraction techniques most commonly used to evaluate the metal content of selected fractions of the sample.

Sample Preparation

One of the distinct advantages of collecting inorganic stream sediments is the small amount of sample preparation required prior to chemical analysis. In almost all cases, dry sieving through a selected mesh size using noncontaminating screening material (usually nylon or stainless steel mesh) is all that is needed for this type of sample. Sieving can take place at the sample site if the sediments are dry, or if wet-sieving is used (Bolviken et al., 1976), thus facilitating handling by reducing sample weight. The practice of wet-sieving is generally avoided, as fine grains that may contain substantial amounts of adsorbed trace metal tend to be lost during agitation of the sieve. Except in the case of analysis of a selected coarse fraction, grinding of sediment is seldom carried out before chemical decomposition of the sample. However, Plant (1971) found that grinding to minus-200-mesh improved the analytical precision and reduced preferential volatization of mineral coatings relative to detrital material in the case of optical emission spectrography.

Preparation of organic samples is generally more complicated. Samples collected by the Geological Survey of Sweden are dried at 110°C, weighted and ashed in an oxidizing atmosphere at 450°C prior to analysis (Larsson, 1976). After reweighing, the organic content is estimated as the loss on ignition. In Finland, organic samples are ashed at 600°C (Kauranne, 1975). For the Tanacross survey in Alaska, stream-bank sod samples were air dried, sieved through a 2-mm stainless steel screen and roasted until ashing was complete. A split of this sample was saved for gold analysis, and the remainder was sieved through 80 mesh and the fine fraction was retained. Moss samples were dried, the sand and silt was removed, and the samples were pulverized in a Waring blender before ashing in a muffle furnace at a maximum temperature of 500°C (Foster et al., 1976).

Heavy mineral concentrates are among the most difficult to prepare, hence are avoided where a simpler technique is applicable. The accompanying flow sheet (Fig. 19.14) outlines the scheme being used in southwestern New Mexico (H.V. Alminas, written comm., 1975). The similarity of this flow sheet to those presented by Brundin and Bergstrom (1977, p. 4 and 6) for Sweden indicates the general similarity of procedures, and their complexity, now being used in widely separated parts of the world. In general, the procedures include (1) mechanical preconcentration, (2) magnetite separation, (3) laboratory gravity separation, (4) further magnetic or electrostatic separation, and (5) analysis. The goldpan or one of its numerous, regional variants is most commonly used for the initial concentration, though sluices (as in the Swedish scheme), jigs, tables, or combinations of these have been applied. For some problems, this step is sufficient and the rough concentrate may be analyzed directly for the metals or minerals of interest.

Variation in the quantity of magnetite in concentrates collected on a regional basis is often sufficient to control much of the apparent variation in metal content of the rough concentrates (see, for example, Theobald et al., 1967, plate 2F). Further, the high iron content of the magnetite often causes analytical problems. For these reasons, magnetite is usually removed from the concentrate either with a simple hand magnet or a belt or drum magnetic separator. The magnetic concentrate then produced may be analyzed separately if desired. If only the magnetite separate is desired, this may be obtained by direct separation from the stream sediment (Callahan, 1975).

The concentrate remaining after separation of the magnetite may be suitable for direct analysis (Theobald and Thompson, 1959b). Where further separation is desired, the concentrate should be cleaned of light minerals remaining from the rough concentration stage. This is accomplished with a heavy liquid separation usually involving bromoform (specific gravity 2.89) or a similar liquid or sometimes one of the heavier liquids such as methylene iodide (specific gravity 3.3). Heavy liquid separation at this stage serves two purposes: (2) it produces a cleaner final product, minimizing variation due to human or machine variability, and (2) it allows a somewhat rougher and considerably faster initial concentration stage in the field.

Further splitting of the samples is usually accomplished by electromagnetic separation using a Frantz Isodynamic Separator, though strong horseshoe magnets and electrostatic separators have also been used. The number of combinations of properties that may be used to produce subsamples is very large, so some choice of the samples to be analyzed must be made based on the nature of the problem, the physiographic and geologic characteristics of the terrain, and orientation surveys. In the example from New Mexico described earlier, two magnetic splits were analyzed, that not magnetic at 0.1 amp. but magnetic at 1 amp. and that not magnetic at 1 amp. (Alminas et al., 1972a). The magnetite concentrate was isolated but not analyzed. In the Yukon-Tanana upland of Alaska, the electromagnetic separations were made at 0.2 amp. and 0.6 amp. (Tripp et al., 1976). In the Swedish example cited earlier (Brundin and Bergstrom, 1977), the initial sampling plan called for the analysis of six magnetic splits of the concentrate separated by handmagnet and at 0.5, 1, 1.5, and 2 amp. on the Frantz. This was reduced to five splits by using a combination of two heavy liquids, and Frantz separations at 0.7 and 1.3 amp. for different gravity functions.

Selective Extractions

Partial analysis has always been one of the most reliable methods of enhancing the contrast between the metal content of stream sediments related to weathering sulphides and that due to less soluble rock forming minerals. Cold extractions using various buffers (Stanton, 1966) or weak acid attacks (dilute hydrochloric, acetic, EDTA) have been the techniques most frequently used in base metal exploration, but the recent concern over the more precise separation and identification of the form in which metals are present in sediments has led to the use of a far wider range of methods.

Selectivity and effectiveness of the msot common extraction techniques have been reviewed by Rose (1975) and Chao and Theobald (1976). The latter authors proposed the following extraction sequence in order to partition the elements into specific fractions, ignoring the small percentage of adsorbed and exchangeable metal:

1. Mn oxides: 0.1 $\underline{M}$ hydroxylamine hydrochloride in 0.01 $\underline{M}$ HNO_3 at room temperature, 30 minutes;
2. Fe oxides (amorphous): 0.25 $\underline{M}$ hydroxylamine in 0.25 $\underline{M}$ HCl at 70°C, 30 minutes;
3. Fe oxides (crystalline): Sodium dithionate in acetate buffer at pH 4.75 at 50°, 30 minutes;
4. Sulphide minerals: $KClO_3$ + HCl – 4 $\underline{N}$ HNO_3, 20 minute boil;
5. Silicate matrix: $HF-HNO_3$ digestion.

In a similar study, Gibbs (1977) determined the adsorbed metal content of sediments by extraction with 1 $\underline{N}$ $MgCl_2$, which was successful in freeing adsorbed ions within a few minutes without affecting the Fe-Mn oxide coatings or organic material in the sample. The selective extraction of organic material is generally accomplished through the use of the oxidizing agents hydrogen peroxide (Rose and Suhr, 1971) or sodium hypochlorite (Gibbs, 1973), but neither is totally specific if sulphides are present.

For many routine exploration purposes less specific extractions can be adopted, and Bradshaw (1975) has outlined a short list of acceptable extractions to be used in orientation surveys, as follows: hot $HClO_4$ or $HClO_4/HNO_3$ at reflux temperature for at least 4 hours to give "total" metal; 0.5 $\underline{N}$ HCl boiling for 20 minutes to give "hot-extractable" metal; and 0.25 per cent EDTA shaken cold for 2 minutes to give "cold-extractable" metal. Other partial extractions in use for Fe-Mn oxides are the oxalic acid digestion (Alminas and Mosier, 1976), and the citrate/hydroxylamine leach (Whitney, 1975).

INTERPRETATION

Interpretation of geochemical stream sediment surveys begins with the presentation of the sample locations and analytical results in map form, followed by examination of the data for anomalous situations which may be due to the presence of mineral deposits. In the case of single-element maps covering small areas, empirical methods are often all that are required to identify abnormal metal concentrations.

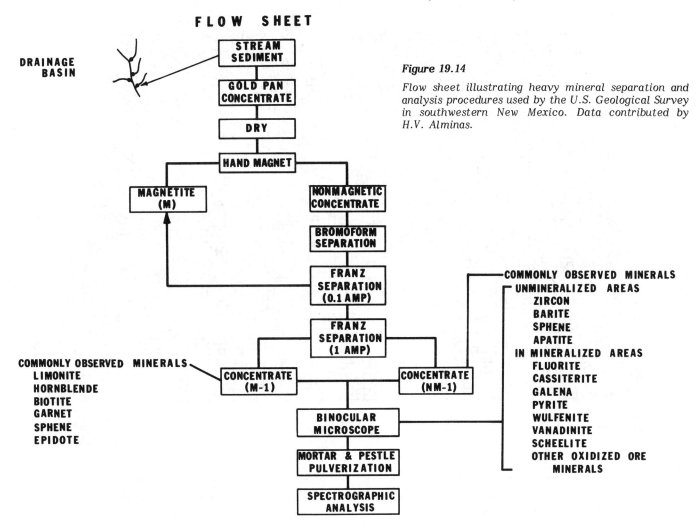

Figure 19.14

Flow sheet illustrating heavy mineral separation and analysis procedures used by the U.S. Geological Survey in southwestern New Mexico. Data contributed by H.V. Alminas.

However, multi-element data are not easy to interpret in two dimensions, even using overlays, unless the values are statistically combined using one of the many techniques available for simplifying data. With the advent of extensive regional surveys and multi-element analytical techniques, computer processing and machine plotting have become important parts of the interpretation procedure. While no attempt will be made to describe the various statistical techniques in detail, some examples of both empirical and statistical interpretation procedures will be outlined in the following paragraphs.

Geochemical Maps

As a first step in making single-element maps, point localities are customarily plotted, whether by hand or machine, giving the analytical value at the map co-ordinates of the sample site, either by recording the numerical concentration or by use of a symbol representing a range of values. Machine plotting of single-element maps can be accomplished symbolically by means of line printer overprinting methods such as Howarth's grey-level mapping program PLTLPI (Howarth, 1971) or the SYMAP program (Dudnick, 1971) favoured by Chapman (1975). Drum or flat-bed plotters are also commonly used for mapping point information, and for contouring metal values. Contouring of stream sediment data from detailed surveys is generally of limited value due to the displacement of the stream sediment sample downstream from the source area, but the validity of contour maps can be improved by relocating the sample points upstream from the true localities (J. Galbraith, pers. comm.). Regional smoothed data can, however, be effectively represented by machine contouring, and the contoured maps, can, in turn, be converted into three dimensional perspective plots (Fig. 19.15).

In the case of multi-element maps, point-source information presents a more difficult display problem. In an experiment to determine the optimum machine plotting technique for illustrating multi-element stream sediment data as point symbols on a single map, Rhind et al. (1973) investigated symbols such as bar graphs, graduated circles, pie diagrams, and wind roses by considering the amount of information that could be displayed in terms of space and cost. As a result of this study, the wind-rose symbol was selected and has since been adopted for machine plotting of the IGS geochemical maps. Three concentration variables can also be effectively combined using the subtractive colour-mixing technique developed for the laser plotter by Lowenstein and Howarth (1973) and later adapted for line-printer mapping (Howarth and Lowenstein, 1976). This particular method is generally more suited to smoothed regional data having few gaps in the survey coverage.

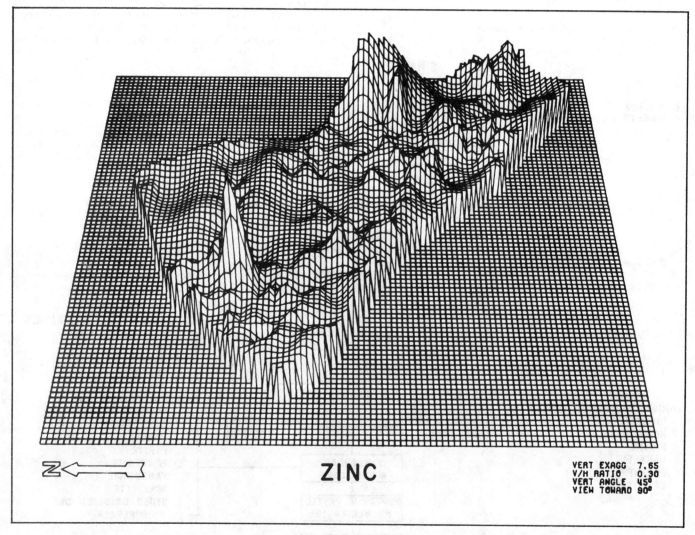

Figure 19.15. Three-dimensional perspective view of zinc distribution in stream sediments from an area of 1100 km² in Central America.

Visual evaluation of mapped stream sediment data, when viewed in combination with geology and topography, can go a long way towards identifying the changes in background resulting from differences in lithology and environment. Grouping of metal values for mapping can be based on log probability plots, percentiles, arithmetic or geometric classes, or multiples of an established background (e.g. 50th percentile). Spatial coincidence of enhanced concentrations of more than one element are often diagnostic of particular types of mineralization (e.g. Cu-Mo with porphyries) and can also be recognized through the examination of overlays or combined element maps, as outlined by Chaffee (1977).

Empirical Methods of Interpretation

Over the years, the experience gained from case studies has led to the development of empirical methods of data analysis for certain exploration situations. An example of this approach can be found in the dilution formula of Hawkes (1976b, c) mentioned earlier in the discussion of sample spacing, and in the method for field interpretation of stream sediment surveys developed by Conzinc Riotinto of Australia Exploration (CRAE) for porphyry copper exploration in tropical terrain (MacKenzie, 1977). This latter method is based on many case histories which indicate that the drainage value is a direct reflection of the average soil value of the catchment. One example that demonstrates this relationship is the Panguna copper deposit on Bougainville Island (Fig. 19.16), where the 2-km² soil anomaly averaging 3100 ppm combined with an 18 km² background of 400 ppm results in an overall average of 600 ppm for the 20 km² catchment area. The recorded stream sediment value of 670 ppm at the outlet closely corresponds to the calculated average soil content of 660 ppm. The value of this approach lies in the fact that once the size and tenor of a target has been defined, the required excess copper concentration over background in stream sediments and be calculated, provided that mechanical transportation of soil to stream takes place under tropical weathering conditions similar to those in the pilot areas.

Study of the geochemical distribution of elemental ratios is also of value in a variety of exploration situations. In the case where sampling precision is relatively poor, mapping of metal ratios can be more effective than single-element maps in outlining different lithologies and sources of mineralization. This was found to be true for X-ray fluorescence analysis of heavy mineral concentrates from north Sunderland and the Cheviot Hills (Leake and Aucott, 1973). In any selective analysis procedure making use of secondary oxides of iron and manganese, interpretation is customarily based on the ratios of metal/Mn, metal/Fe, and metal/Mn + Fe in order to reduce the effect of the amount of oxide in each sample (Chao and Theobald, 1976).

Where relatively small numbers of samples and small survey areas are involved, log probability plots can be used to estimate the threshold values which provide the best separation of a population into background and anomalous groups (Tennant and White, 1959; Lepeltier, 1969). This procedure has recently been described in detail, with suitable examples, by Sinclair (1976). However, too much reliance should not be placed on log probability plots as a means of recognizing anomalies in stream sediment data, as this type of frequency distribution study is unrelated to the spatial distribution of the data. As a result, no area should be eliminated from further consideration simply because it shows no indications of an anomalous distribution on a log probability plot; the data should be mapped and examined in detail. Most geochemists have observed the phenomenon described by Chaffee (1977) whereby the log probability plots of large data sets tend to be unimodal, regardless of the presence or absence of anomalies related to ore deposits in the survey area.

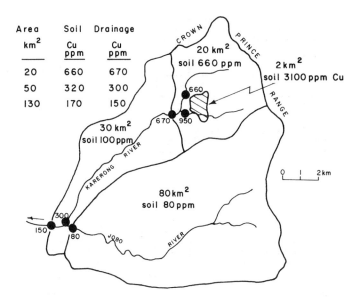

Figure 19.16. Copper in stream sediments and soils in the vicinity of the Panguna deposit, Bougainville Island (after MacKenzie, 1977).

Statistical Methods

Statistical methods of data analysis have become an essential part of regional geochemical reconnaissance using the multi-element approach, and most mining and exploration companies now have access to computing and automatic plotting facilities suitable for handling large data sets. As an example, in the mineral exploration department of Cities Service Company, regional multi-element data are routinely processed via remote-terminal time-sharing operations, which control statistical procedures, and are plotted by line-printer, pen, or electrostatic plotters (Fig. 19.17). Manipulation of contour plots is also carried out by use of an interactive video display system. Statistical programs available include those accessed through standard statistical packages, and log probability plots, trend-surface, empirical discriminant, and Q- and R-mode factor analysis for large sets of data. With these facilities at hand, no difficulty was experienced in processing a recent project involving 40 000 samples with as many as nine analyses per sample.

Historically, statistical methods for interpretation of regional stream sediment data developed from the use of trend-surface, moving-average, and factor analysis, particularly R-mode (Nichol and Webb, 1967). All these methods, together with cluster analysis (Obial and James, 1973), are based on statistical manipulation of the data without prior input of parameters such as known geological or environmental controls. This type of analysis is particularly suited to areas in which the geology is poorly known, as in most "grass-roots" exploration projects. For those regions where the geology is mapped in greater detail, techniques such as discriminant analysis, which requires a prior assignment of training areas, provide an alternate method of interpretation. A recent application of this technique to regional stream sediment data from the U.K. (Castillo-Munoz and Howarth, 1976) gave about a 60 per cent success rate in classifying reconnaissance samples on the basis of small training areas selected as typical of known lithology or mineralization, despite the presence of extensive glacial overburden. The method also served to outline anomalous sample sites related to past smelting or mining activity, and localized areas of manganese scavenging.

Multiple regression has been used with some success to allow for variations in stream sediment matrix due to changes in secondary environment or bedrock composition in the

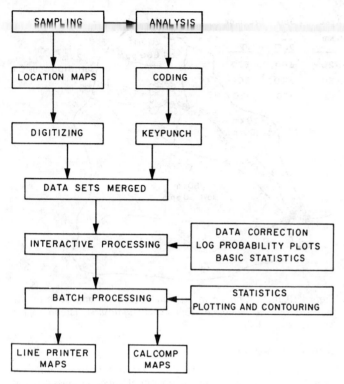

Figure 19.17. Data handling scheme for typical stream sediment survey conducted by Citco International Minerals.

catchment area (Dahlberg, 1968; Rose et al., 1970; Chatupa and Fletcher, 1972), and factor analysis prior to regression has also been used for this purpose (Closs and Nichol, 1975). Attempts have also been made to use multiple regression techniques to predict mineral resources (De Geoffrey and Wignall, 1970), and a variation of this method of combining geological criteria, historical mining records, and regional stream sediment data has been used to estimate the base metal resources of northwest England (Cruzat and Meyer, 1974). Combination of stream sediment results with geophysical and geological variables has also been accomplished by means of discriminant analysis in a study designed to evaluate the potential of the Triassic province in Pennsylvania for magnetite deposits (Rose, 1972).

Computer modelling techniques for interpretation of stream sediment data have recently been described by Culbert (1976). This approach makes use of a terrain and transport model to describe the drainage system, with input in the form of: (1) co-ordinates of sample sites and turning points on streams, (2) catchment areas of creeks, (3) lithology of grid squares, and (4) elevations of grid squares. The program then stimulates the flow pattern and dilution between sequential samples and produces an interpretation similar to the one that would result from the previously mentioned empirical methods involving dilution equations.

DISCUSSION

This review of stream sediment geochemistry in 1977 has emphasized recent changes in routine methods that have been brought about by the application of the technique to environments previously considered to be unfavourable for drainage surveys, and by the wider use of stream sediments for multi-purpose regional mapping. It is likely that these methods will continue to change, not only because new environments are being explored but also because areas formerly prospected by more conventional surveys are being subjected to re-examination using more refined methods.

First-pass regional exploration is rapidly reducing the number of unexplored areas, and future stream sediment surveys will increasingly be devoted to second- or third-pass sampling programs using different sampling, analysis, or interpretation procedures.

While data processing and analytical costs per unit have tended to remain stable or decline in recent years due to technological improvements in instrumentation and transportation, labour costs have increased in line with trends in worldwide inflation. As a result, efficient stream sediment surveys require optimization of sampling and analytical procedures to increase the amount of useful information that can be extracted from each sample. In many situations, this goal is best achieved through the use of multi-element analysis and computer-aided interpretation.

ACKNOWLEDGMENTS

The authors would like to acknowledge the co-operation of Cities Service Company in supplying technical assistance in preparing this review. In particular, the efforts of Travis Brady in the preparation of slides and figures are acknowledged.

REFERENCES

Alminas, H.V. and Mosier, E.M.
1976: Oxalic acid leaching of rock, soil, and stream-sediment samples as an anomaly-accentuation technique; U.S. Geol. Surv., Open File Rept. 76-275, 25 p.

Alminas, H.V., Watts, K.C., and Siems, D.L.
1972a: Maps showing lead distribution in the Winston and Chise quadrangles and in the west part of the Priest Tank quadrangle, Sierra County, New Mexico; U.S. Geol. Surv. Misc. Field Studies Map MF 398.

1972b: Maps showing molybdenum distribution in the Winston and Chise quadrangles and in the west part of the Priest Tank quadrangle, Sierra County, New Mexico; U.S. Geol. Surv., Misc. Field Studies Map MF 399.

1972c: Maps showing silver and gold distribution in the Winston and Chise quadrangles and in the west part of the Priest Tank quadrangle, Sierra County, New Mexico; U.S. Geol. Surv., Misc. Field Studies Map MF 400.

Alther, George R.
1975: Geochemical analysis of stream sediments as a tool for environmental monitoring: A pigyard case study; Geol. Soc. Am. Bull., v. 86, p. 174-176.

Armour-Brown, A. and Nichol, Ian
1970: Regional geochemical reconnaissance and the location of metallogenic provinces; Econ. Geol., v. 65, p. 312-330.

Ball, T.K. and Nutt, M.J.C.
1974: Preliminary reconnaissance of central Wales; Inst. Min. Met. Trans., London, Sect. B, v. 83, p. 66-67.

Beus, A.A. and Grigoryan, S.V.
1977: Geochemical exploration methods for mineral deposits; Applied Publishing Ltd., Illinois, 287 p.

Bjorlykke, A., Bolviken, B., Eidsvig, P., and Svinndal, S.
1973: Exploration for disseminated lead in southern Norway; in Prospecting in areas of Glacial Terrain, Inst. Min. Met., London, p. 111-126.

Bloom, H.
1955: A field method for the determination of ammonium citrate-soluble heavy metals in soils and alluvium; Econ. Geol., v. 50, p. 533-541.

Bolviken, B.
1976: Snertingdal: A lead occurrence found by systematic prospecting; J. Geochem. Explor., v. 5, p. 324-331.

Bolviken, B., Krog, J.R., and Naess, G.
1976: Sampling techniques for stream sediments; J. Geochem. Explor., v. 5, p. 382-383.

Bolviken, B. and Sindig-Larsen, R.
1973: Total error and other criteria in the interpretation of stream sediment data; in Geochemical Exploration 1972, Inst. Min. Met., London, p. 285-295.

Bradshaw, P.M.D. (ed.)
1975: Conceptual models in exploration geochemistry; J. Geochem. Explor., v. 4, p. 1-213.

Brand, H.
1972: United Nations mineral survey programs; 24th Int. Geol. Congr., Montreal, Abstr., p. 498.

Brundin, N.H. and Bergstrom, J.
1977: Regional prospecting for ores based on heavy minerals in glacial till; J. Geochem. Explor., v. 7, no. 1, p. 1-19.

Brundin, N.H. and Nairis, B.
1972: Alternative sample types in regional geochemical prospecting; J. Geochem. Explor., v. 1, p. 7-46.

Buhlmann, E., Philpott, D.E., Scott, M.J., and Sanders, R.N.
1975: The status of exploration geochemistry in southern Africa, in Geochemical Exploration 1974, Elsevier Publ. Co., Amsterdam, p. 51-64.

Callahan, J.E.
1975: A rapid field method for extracting the magnetic fraction from stream sediments; J. Geochem. Explor., v. 4, no. 2, p. 265-267.

Cameron, E.M.
1976: Geochemical reconnaissance for uranium in Canada: Notes on methodology and interpretation of data; in Report of Activities, Part. C, Geol. Surv. Can., Paper 76-1C, p. 229-236.

Canney, F.C.
1966: Hydrous manganese-iron oxide scavenging: Its effect on stream sediment surveys, abstr.; Geol. Surv. Can., Paper 66-54, p. 267.

Carpenter, R.H., Pope, T.A., and Smith, R.L.
1975: Fe-Mn oxide coatings in stream sediment geochemical surveys; J. Geochem. Explor. v. 4, p. 349-363.

Castillo-Munoz, R. and Howarth, R.J.
1976: Application of the empirical discriminant function to regional geochemical data from the United Kingdom; Geol. Soc. Am. Bull., v. 87, p. 1567-1581.

Chaffee, M.A.
1977: Some thoughts on the selection of threshold values as practiced in the Branch of Exploration Research of the U.S. Geological Survey; Assoc. Explor. Geochem., Newsletter 21, p. 14-16.

Chaffee, M.A., Lee-Moreno, J.L., Caire, L.F., Mosier, E.L., and Frisken, J.G.
1976: Results of geochemical investigations comparing samples of stream sediment, panned concentrate, and vegetation in the vicinity of the Caridad porphyry copper deposit, northern Sonora, Mexico; U.S. Geol. Surv., Open File Rept. 76-559, 34 p.

Chao, T.T. and Theobald, P.K.
1976: The significance of secondary iron and manganese oxides in geochemical exploration; Econ. Geol. v. 71, p. 1560-1569.

Chapman, R.P.
1975: Data processing requirements and visual representation for stream sediment exploration geochemical surveys; J. Geochem. Explor., v. 4, p. 409-423.

Chatupa, J. and Fletcher, K.W.
1972: Application of regression analysis to the study of background variations in trace metal content of stream sediments; Econ. Geol., v. 67, p. 978-980.

Chork, C.Y.
1977: Seasonal, sampling and analytical variations in stream sediment surveys; J. Geochem. Explor., v. 7, p. 31-48.

Clifton, H.E., Hunter, R.E., Swanson, F.J., and Phillips, R.L.
1969: Sample size and meaningful gold analysis; U.S. Geol. Surv., Prof. Paper 625-C, 17 p.

Closs, L.G. and Nichol, I.
1975: The role of factor and regression analysis in the interpretation of geochemical reconnaissance data; Can. J. Earth. Sci., v. 12, p. 1316-1330

Cruzat, A.C.E.
1973: Application of regional stream sediment geochemistry in forecasting base metal production, northern England; unpubl. Ph.D. thesis, Univ. London.

Cruzat, A.C.E. and Meyer, W.T.
1974: Predicted base metal resources of northwest England; Inst. Min. Met. Trans., London, v. 83, Sect. B, p. 131-134.

Culbert, R.R.
1976: A multivariate approach to mineral exploration; Can. Min. Metall. Bull., v. 69, no. 766, p. 39-52.

Dahlberg, E.C.
1968: Application of a selective simulation and sampling technique to the interpretation of stream sediment copper anomalies near South Mountain, Penn.; Econ. Geol., v. 63, p. 409-417.

Dahlem, D.H.
1976: A national hydrogeochemical sampling program for uranium; Soc. Min. Eng./Am. Inst. Min. Met. Eng., Preprint 76-L-38, 22 p.

De Geoffrey, J. and Wignall, T.K.
1970: Statistical decision in regional exploration: application of regression and Bayesian classification analysis in the southwest Wisconsin zinc area; Econ. Geol., v. 65, p. 769-777.

Dudnick, E.E.
1971: SYMAP: Users reference manual for synagraphic computer mapping; Rept. 71-1, Dept. Architecture, Univ. Illinois at Chicago Circle, Ill., 114 p.

Duff, J.R.V.
  1975: Sources of error in some geochemical data from Australian stream sediments; unpubl. Ph.D. thesis, Univ. London, 389 p.

Dunlop, A.C.
  1973: Geochemical dispersion of tin in stream sediments and soils in southwest England; unpubl. Ph.D. thesis, Univ. London.

Dunlop, A.C. and Meyer, W.T.
  1973: Influence of late Miocene-Pliocene submergence on regional distribution of tin in stream sediments, southwest England; Inst. Min. Met. Trans., London, Sect. B, v. 82, p. 62-64.

Fanta, P.
  1972: Effects of seasonal variations on iron, manganese and associated metal contents of stream sediments and soils; unpubl. M. Phil. thesis, Univ. London.

Ferguson, R.B. and Price, V.
  1976: National Uranium Resource Evaluation (NURE) program — hydrogeochemical and stream sediment reconnaissance in the eastern United States; J. Geochem. Explor., v. 6, p. 103-118.

Foster, H.L., Albert, N.R.D., Barnes, D.F., Curtin, G.C., Griscom, A., Singer, D.A., and Smith, J.G.
  1976: The Alaskan Mineral Resource Assessment Program: Background information to accompany folio of geologic and mineral resource maps of the Tanacross quadrangle, Alaska; U.S. Geol. Surv., Circ. 734, 23 p.

Foster, J.R.
  1971: The reduction of matrix effects in atomic absorption analysis and the efficiency of selected extractions on rock forming minerals; Can. Inst. Min. Met., Spec. Vol. 11, p. 554-560.

Galbraith, J.
  1975: Regional stream sediment geochemistry and data analysis in northeast Bahia, Brazil; Ph.D. thesis, Univ. Idaho, 277 p.

Gallagher, M.J.
  1967: Determination of molybdenum, iron, and titanium in ores and rocks by portable radioisotope X-ray fluorescence analyzer; Inst. Min. Met. Trans., London, Sect. B, v. 76, p. 155-164.

Garrett, R.G.
  1974: Field data acquisition methods for applied geochemical surveys at the Geological Survey of Canada; Geol. Surv. Can., Paper 74-52, 36 p.

Garrett, R.G. and Nichol, I.
  1967: Regional geochemical reconnaissance in eastern Sierra Leone; Inst. Min. Met. Trans., London, Sect. B, v. 76, p. 97-112.

Gibbs, A.K.
  1974: Regional geochemistry in Guyana; unpubl. M.Sc. thesis, Univ. London, 168 p.

Gibbs, R.J.
  1973: Mechanisms of trace metal transport in rivers; Science, v. 156, p. 1734-1737.
  1977: Transport phases of transition metals in the Amazon and Yukon Rivers; Geol. Soc. Am. Bull., v. 88, p. 829-843.

Gleeson, C.F. and Brummer, J.J.
  1976: Reconnaissance stream-sediment geochemistry applied to exploration for porphyry Cu-Mo deposits in southwestern Yukon Territory; Can. Min. Metall. Bull., v. 69, no. 769, p. 91-103.

Govett, G.J.S.
  1958: Geochemical prospecting for copper in Northern Rhodesia; unpubl. Ph.D. thesis, Univ. London.

Hawkes, H.E.
  1976a: The early days of exploration geochemistry; J. Geochem. Explor., v. 6, p. 1-12.
  1976b: The downstream dilution of stream sediment anomalies; J. Geochem. Explor., v. 6, p. 345-358.
  1976c: Selection of sample sites in stream sediment reconnaissance; Preprint 76-L-1, Am. Inst. Min. Met. Eng.; Annual Meeting, February 1976.

Hawkes, H.E. and Webb, J.S.
  1962: Geochemistry in mineral exploration; Harper and Row, New York.

Hem, J.D.
  1976: Geochemical controls on lead concentrations in stream water and sediments, Geochim. Cosmochim. Acta, v. 40, p. 599-609.

Hill, John (translator)
  1746: History of stones by Theophrastus, 300 B.C., (?)London.

Holman, R.H.C
  1963: A method for determining readily-soluble copper in soil and alluvium; Geol. Surv. Can., Paper 63-7.

Holmes, R.
  1975: The regional distribution of cadmium in England and Wales; unpubl. Ph.D. thesis, Univ. London.

Horsnail, R.F.
  1975: Strategic and tactical geochemical exploration in glaciated terrain: illustrations from Northern Ireland; in Prospecting in areas of glaciated terrain, Inst. Min. Met., London, p. 16-31.

Horsnail, R.F., Nichol, I., and Webb, J.S.
  1969: Influence of variations in secondary environment on the metal content of drainage sediments; Colo. Sch. Mines Q., v. 64, p. 307-322.

Hosking, K.F.G.
  1971: Problems associated with the application of geochemical methods of exploration in Cornwall, England; Can. Inst. Min. Met., Spec. Vol. 11, p. 176-189.

Howarth, R.J.
  1971: FORTRAN IV Program for grey-level mapping of spatial data; Mathematical Geol., v. 3, no. 2, p. 95-121.

Howarth, R.J. and Lowenstein, P.L.
  1971: Sampling variability of stream sediments in broadscale regional geochemical reconnaissance; Inst. Min. Met. Trans., London, Sect. B, v. 80, p. 363-372.
  1976: Three-component colour maps from lineprinted output; Inst. Min. Met. Trans., London, v. 85, Sect. B, p. 234-237.

Jacobs, D.C. and Parry, W.T.
  1976: A comparison of the geochemistry of biotite from some Basin and Range stocks; Econ. Geol., v. 71, no. 6, p. 1029-1035.

Kauranne, L.K.
1975: Regional geochemical mapping in Finland; in Prospecting in areas of glaciated terrain, Inst. Min. Met., London, p. 71-81.

Kunzendorf, H.
1973: Non-destructive determination of metals in rocks by radioisotope X-ray fluorescence instrumentation; in Geochemical Exploration 1972, Inst. Min. Met., London, p. 401-414.

Larson, L.T.
1977: Geochemistry – 1976 Annual Review; Min. Eng., v. 29, no. 2, p. 57-63.

Larsson, J.O.
1976: Organic stream sediments in regional geochemical prospecting, Precambrian Pajala district, Sweden; J. Geochem. Explor., v. 6, p. 233-250.

Leake, R.C. and Aucott, J.W.
1973: Geochemical mapping and prospecting by use of rapid automatic X-ray fluorescence analysis of panned concentrates; in Geochemical Exploration 1972, Inst. Min. Met., London, p. 389-400.

Leggo, M.D.
1976: Contrasting geochemical expressions of copper mineralization at Namosi, Fiji; 25th Int. Geol. Congr., Sydney, Abstr., p. 448-449.

Lepeltier, C.
1969: A simplified statistical treatment of geochemical data by graphical representation; Econ. Geol., v. 64, p. 538-550.

Lovering, T.S., Huff, L.C., and Almond, H.
1950: Dispersion of copper from the San Manuel copper deposit, Pinal County, Arizona; Econ. Geol., v. 45, p. 493-514.

Lowenstein, P.L. and Howarth, R.J.
1973: Automated colour-mapping of three-component systems and its application to regional geochemical reconnaissance; in Geochemical Exploration 1972, Inst. Min. Met., London, p. 297-304.

MacKenzie, D.H.
1977: Empirical assessment of anomalies in tropical terrain; Assoc. Explor. Geochem., Newsletter 21, p. 6-10.

Mayfield, I.
1976: Regional geochemical drainage reconnaissance near west Nicholson; in Viewing, K.A., 7th Annual Report, Inst. Mining Res., Univ. Rhodesia, p. 51-53.

Meyer, W.T.
1977: Regional stream sediment interpretation procedures in a tropical environment, Central Bahia, Brazil (abs.); Min. Eng., v. 29, no. 1, p. 75.

Miesch, A.T.
1976: Sampling designs for geochemical surveys – syllabus for a short course; U.S. Geol. Surv., Open File Rept. 76-772, 128 p.

Nichol, I. and Webb, J.S.
1967: The application of computerized mathematical and statistical procedures to the interpretation of geochemical data; Proc. Geol. Soc. London, no. 1642, p. 186-198.

Nowlan, G.A.
1976: Concretionary manganese-iron oxides in streams and their usefulness as a sample medium for geochemical prospecting; J. Geochem. Explor., v. 6, p. 193-210.

Obial, R.C. and James, C.H.
1973: Use of cluster analysis in geochemical prospecting with particular reference to southern Derbyshire, England; in Geochemical Exploration 1972, Inst. Min. Met., London, p. 237-257.

Ostle, D., Coleman, R.F., and Ball, T.K.
1972: Neutron activation analysis as an aid to geochemical prospecting for uranium; in Uranium Prospecting Handbook, Inst. Min. Met., London, p. 95-107.

Overstreet, W.C.
1962: A review of regional heavy-mineral reconnaissance and its application in the southeastern Piedmont; Southeast. Geol., v. 3, no. 3, p. 133-173.

Parry, W.T. and Nackowski, M.P.
1963: Copper, lead, and zinc in biotites from Basin and Range quartz monzonites; Econ. Geol., v. 58, no. 7, p. 1126-1144.

Plant, J.
1971: Orientation studies on stream sediment sampling for a regional geochemical survey in northern Scotland; Inst. Min. Met. Trans., London, Sect. B, v. 80, p. 324-345.

Rhind, D.W., Shaw, M.A., and Howarth, R.J.
1973: Experimental geochemical maps; J. Brit. Cartographic Soc., Dec. 1973, p. 112-118.

Rose, A.W.
1972: Favorability for Cornwall-type mangetite deposits in Pennsylvania using geological, geochemical and geophysical data in a discriminant function; J. Geochem. Explor., v. 1, p. 181-194.

1975: The mode of occurrence of trace elemtns in soils and stream sediments applied to geochemical exploration; in Geochemical Exploration 1974, Elsevier Publ. Co., Amsterdam, p. 691-705.

Rose, A.W., Dahlberg, E.C., and Keith, M.L.
1970: A multiple regression technique for adjusting background values in stream sediment geochemistry; Econ. Geol., v. 65, p. 156-165.

Rose, A.W. and Keith, M.L.
1976: Reconnaissance geochemical techniques for detecting uranium deposits in sandstones of northeastern Pennsylvania; J. Geochem. Explor., v. 6, p. 119-138.

Rose, A.W. and Suhr, N.H.
1971: Major element content as a means of allowing for background variation in stream-sediment geochemical exploration; Can. Inst. Min. Met., Spec. Vol. 11, p. 587-593.

Sharp, W.E. and Jones, T.L.
1975: A topologically optimum prospecting plan for streams; in Geochemical Exploration 1975, Elsevier Publ. Co., Amsterdam, p. 227-235.

Sigov, A.P.
1939: Shlikovye Izyskaniia (Prospecting by heavy mineral studies); Transactions of the Ural Scientific Research Institute of Geology, Prospecting and Economic Mineralogy and Sverdlovsk Mining Institute, no. 4, 64 p. (Abstracted by H.E. Hawkes, in U.S. Geol. Surv., Open File Rept. issued 1953(?)).

Sinclair, A.J.
1976: Applications of probability graphs in mineral exploration; Assoc. Explor. Geochem., Spec. Vol. 4, 95 p.

Smee, B.W. and Ballantyne, S.B.
1976: Examination of some Cordilleran uranium occurrences; in Report of Activities, Part. C, Geol. Surv. Can., Paper 76-1C, p. 255-258.

Stanton, R.E.
1966: Rapid methods of trace analysis for geochemical application; Edw. Arnold, London.

Sutherland-Brown, A.
1975: Huckleberry Cu-Mo deposit, British Columbia; in Conceptual Models in Exploration Geochemistry, J. Geochem. Explor., v. 4, no. 1, p. 72-75.

Tennant, C.B. and White, M.L.
1959: Study of the distribution of some geochemical data; Econ. Geol., v. 54, p. 1281-1290.

Theobald, P.K. and Allcott, G.H.
1975: Tungsten anomalies in the Uyaijah ring structure, Kushaymiyah igneous complex, Kingdom of Saudi Arabia, section A – geology and geochemistry of the Uyaijah ring structure; Saudi Arabian Project Report 160, U.S. Geol. Surv., Open File Rept. 75-657, 86 p.

Theobald, P.K., Jr. and Havens, R.G.
1960: Base metals in biotite, magnetite and their alteration products in a hydrothermally altered quartz monzonite porphyry sill, Summit County, Colorado (abs.); Geol. Soc. Am. Bull.,v. 71, no. 12, pt. 2, p. 1991.

Theobald, P.K., Jr., Overstreet, W.C., and Thompson, C.E.
1967: Minor elements in alluvial magnetite from the inner Piedmont belt, North and South Carolina; U.S. Geol. Surv., Prof. Paper 554-A, 34 p.

Theobald, P.K., Jr. and Thompson, C.E.
1959a: Reconnaissance exploration by analysis of heavy mineral concentrates (abs.); Min. Eng., v. 11, no. 1, p. 40.

1959b: Geochemical prospecting with heavy-mineral concentrates used to locate a tungsten deposit; U.S. Geol. Surv., Circ. 411, p. 1-13.

Thomson, Ian
1976: Geochemical studies in central west Brazil; Final report of the pilot phase of the Projeto Geofisico Brasil-Canada, DNPM, Brazil, 258 p.

Topping, N.J.
1976: Regional geochemical drainage reconnaissance in the tribal trust lands; in Viewing, K.A., 7th Annual Rept., Inst. Min. Res. Univ. Rhodesia, p. 49-51.

Tripp, R.B., Curtin, G.C., Day, G.W., Karlson, R.C., and Marsh, S.P.
1976: Maps showing mineralogical and geochemical data for heavy-mineral concentrates in the Tanacross quadrangle, Alaska; U.S. Geol. Surv., Misc. Field Studies Map MF-767.

Tweto, Odgen
1960: Scheelite in the Precambrian gneisses of Colorado; Econ. Geol., v. 55, no. 7, p. 1406-1428.

Urquidi-Barrau, F.
1973: Regional geochemical variations related to base metal mineralization in Wales; unpubl. Ph.D. thesis, Univ. London.

Ward, F.N.
1951: A field method for the determination of tungsten in soils; U.S. Geol. Surv., Circ. 119, 4 p.

Ward, F.N., Lakin, H.W., Canney, F.W. et al.
1963: Analytical methods used in geochemical exploration by the U.S. Geological Survey; U.S. Geol. Surv., Bull. 1152, p. 25.

Webb, J.S.
1970: Some geological applications of regional geochemical reconnaissance; Proc. Geol. Assoc., v. 81, p. 585-594.

1975: Environmental problems and the exploration geochemist; in Geochemical Exploration 1974, Elsevier Publ. Co., Amsterdam, p. 5-17.

Webb, J.S., Lowenstein, P.L., Howarth, R.J., Nichol, I., and Foster, R.
1973: Sampling, analytical and data processing techniques (for provisional geochemical atlas of Northern Ireland); Applied Geochem. Research Group, Imperial College, London, Tech. Comm. no. 61.

Webb, J.S., Nichol, I., and Thornton, I.
1968: The broadening scope of regional geochemical reconnaissance; 23rd. Int. Geol. Cong., Prague, v. 6, p. 131-147.

Whitney, P.R.
1975: Relationship of manganese-iron oxides and associated heavy metals to grain size in stream sediments; J. Geochem. Explor., v. 4, p. 251-263.

LAKE SEDIMENT GEOCHEMISTRY APPLIED TO MINERAL EXPLORATION

W.B. Coker, E.H.W. Hornbrook, and E.M. Cameron
Geological Survey of Canada

Coker, W.B., Hornbrook, E.H.W., and Cameron, E.M., Lake sediment geochemistry applied to mineral exploration; in Geophysics and Geochemistry in the search for Metallic Ores; Peter J. Hood, editor; Geological Survey of Canada, Economic Geology Report 31, p. 435-478, 1979.

Abstract

Lakes comprise a complex system of interplay between various physical, chemical and biological factors. The distribution of heat and suspended and dissolved substances, including gases, as well as compositional variation within the sediment, are factors which can affect the dispersion of trace elements in the waters, and the accumulation or depletion of trace elements in the sediments of lakes.

A knowledge of the processes by which a metal is mobilized, transported, precipitated, and possibly remobilized, is of prime concern in order to comprehend possible controls on the metal. The principal mechanisms affecting base metal transport, accumulation and fixation into bottom deposits are thought to involve: scavenging of metals by organic matter; sorption and coprecipitation by hydrous oxides; sorption by clay minerals; chemical processes involving hydrolytic reactions; and variations in the gross physical-chemical nature of the sediments.

The range of possible physicochemical-limnological conditions within the lacustrine environment emphasizes the complexity of this regime. Variations in these conditions can affect the nature of metal response in lake bottom materials, with respect to mineralization, in different geographic — climatic and geological environments.

In recent years considerable attention has been focused on investigating the role of lake sediment geochemistry as a guide to mineralization, mostly within the Shield regions of the northern hemisphere. These regions can contain extensive glacial overburden and, being of low relief, are characterized by indefinite and disorganized drainage systems. Investigations have therefore been initiated because conventional geochemical exploration techniques have found limited application in many of these areas of the Shield.

In Canada studies of the relationships between lake sediment geochemistry and various types of mineralization have been performed by government agencies, the mining industry and university groups. Features which have contributed to the effectiveness of lake sediment geochemistry for reconnaissance mineral exploration within Shield areas include: the great abundance of lakes; centre-lake bottom sediments constituting a homogeneous sample medium for trace metal accumulation; the amenability of this procedure to rapid helicopter sampling techniques thus allowing large areas to be covered rapidly and relatively cheaply.

The effectiveness of lake sediment geochemistry for regional reconnaissance has been established in the northern permafrost regions of the Canadian Shield. However, the problems appear to be somewhat more complex in regions of the southern Shield, south of the zone of discontinuous permafrost, where, although some success has been achieved, only limited work has been done. The problems here include increased biologic activity; widely varying limnological environments; and an apparently less intense rate of weathering compared to northern Shield regions. Over the Shield as a whole it is clear that the effectiveness of the method varies with the element and type of deposit sought and with local terrain and overburden conditions. Because of all these considerations much research is required on a number of fundamental aspects, for example: on the rates of weathering of various types of mineralization in different regions of the Shield, on the transport and accumulation of metals within organic-rich environments, and on the physical and chemical processes which operate during and after glacial activity.

The techniques of lake sediment geochemistry are currently at a stage where they are being refined and evaluated and are being tested in different regions. The main application of lake sediment geochemistry for mineral exploration has been within the Canadian Shield and adjoining areas. The problems that originally led to its application in these regions are present in other regions. In North America these include parts of the Cordillera and Appalachians, and abroad the Scandinavian Shield.

Résumé

Les lacs sont le siège de tout un système complexe de réactions entre différents facteurs physiques, chimiques et biologiques. La distribution de la chaleur, les substances dissoutes et en suspension, y compris les gaz, ainsi que les différences de composition entre divers types de sédiments, sont autant de facteurs qui peuvent affecter la dispersion des éléments en trace que peut contenir l'eau et l'accumulation ou l'épuisement de ces éléments dans les sédiments lacustres.

Pour bien comprendre comment on peut contrôler la teneur d'une eau en un métal donné, il est essentiel de bien connaître les processus par lesquels ce métal est mobilisé, transporté, précipité et peut-être remobilisé. On pense que les principaux mécanismes affectant le transport, l'accumulation et la fixation d'un métal de base dans les sédiments de fond font intervenir: l'extraction des métaux par la matière organique; l'adsorption et la coprécipitation par les oxydes hydratés; l'adsorption par les minéraux argileux; des processus chimiques mettant en jeu l'hydrolyse; les variations dans la nature physico-chimique des sédiments.

La gamme des conditions limniques et physico-chimiques possibles en milieu lacustre augmente la complexité de ce régime. Les variations de ces conditions peuvent affecter le comportement du métal se trouvant dans les sédiments de fond, en ce qui concerne la minéralisation, selon le contexte géographique-climatique et le contexte géologique.

Ces dernières années, l'attention des chercheurs s'est concentrée sur le rôle que la géochimie des sédiments lacustres peut jouer dans la découverte de minéralisation, en particulier dans les boucliers de l'hémisphère nord. Ces régions peuvent présenter de grandes étendues de couverture glaciaire, et avoir un relief peu accusé; elles sont caractérisées par des réseaux hydrographiques peu structurés, aux frontières mal définies. Des recherches ont donc été entreprises car les techniques classiques de la prospection géochimique s'appliquent mal dans de nombreuses régions du Bouclier.

Au Canada, plusieurs organismes du gouvernement, l'industrie minière, et certaines universités ont fait des études sur les relations existant entre la géochimie des sédiments lacustres et les différents types de minéralisation. Plusieurs facteurs contribuent à faire de la géochimie des sédiments lacustres appliquée à la prospection minérale de reconnaissance dans les régions du Bouclier un outil efficace, entre autres: la grande abondance des lacs; le fait que les sédiments au centre d'un lac constituent un exemple de milieu homogène pour l'accumulation de métaux en traces; cette méthode se prête aux techniques d'échantillonnage rapide par hélicoptère qui permettent de couvrir rapidement et économiquement de grandes étendues.

L'efficacité de la géochimie des sédiments lacustres appliquée à la reconnaissance régionale des zones de pergélisol de la partie nord du Bouclier canadien a été démontrée. Mais, les problèmes semblent être un peu plus complexes dans les régions sud du Bouclier, au sud de la zone de pergélisol discontinu, où, malgré quelques résultats intéressants, très peu de travaux ont été effectués. Dans ces régions, on rencontre en effet des difficultés: l'activité biologique est plus intense, le contexte limnologique varie beaucoup plus et l'érosion est moins active que dans le nord du Bouclier. Il est bien évident que l'efficacité de cette méthode appliquée au Bouclier varie avec l'élément et le type de gisements cherchés, avec le type de terrain rencontré et avec les conditions de la couverture. Compte tenu de toutes ces considérations, on doit alors faire beaucoup de recherches sur un grand nombre d'aspects fondamentaux, comme par exemple: le taux d'altération des différents types de minéralisation dans les différentes régions du Bouclier, le transport et l'accumulation des métaux dans les milieux riches en matière organique et les processus physiques et chimiques entrant en jeu pendant et après l'activité glaciaire.

Les techniques de la géochimie des sédiments lacustres en sont actuellement au stade de la mise au point et de la réévaluation; elles sont aussi mises à l'essai dans différentes régions. La principale application de la géochimie des sédiments lacustres pour l'exploration minière se fait dans le Bouclier canadien et les régions avoisinantes. On trouve ailleurs les problèmes qui nous ont amenés à appliquer ces Techniques dans ces régions. En Amérique du Nord, il s'agit d'une partie de la Cordillère et des Appalaches; à l'étranger c'est le cas du Bouclier scandinave.

INTRODUCTION

The adoption of lake sediment and water geochemistry for mineral exploration by geologists and geochemists, is still in a relatively juvenile stage compared to the intensive study of the physical, chemical and biologic nature of lakes by limnologists. Over the past 100 years limnologists have accumulated a vast amount of information on lakes. However, relatively little work has been done on trace element cycles in the lacustrine environment. Only recently, because of environmental concern over pollution, and, to a certain extent, because of the application of lake sediment geochemistry to mineral exploration, has any concentrated attention been focused on the study of trace metals in the lacustrine environment.

It has only been since the late 1960s or early 1970s that any concerted effort has been made in investigating the role of lake sediment geochemistry as an indicator of mineralization. The majority of studies have taken place within the Precambrian Shield regions of the northern hemisphere. This is primarily because more conventional geochemical exploration techniques have found limited application in Shield areas with extensive glacial overburden and low relief commonly characterized by indefinite and disorganized drainage systems.

Many geologists and geochemists, while they may utilize lake sediment geochemistry as part of an exploration program, are largely unaware of the characteristics and physicochemistry of the lacustrine environment. An attempt has therefore been made to review and summarize the relevant limnologic and chemical literature in order to outline the physical, chemical and biological processes operative on metal distributions in the lacustrine environment, and their importance to the interpretation of lake sediment chemical data. The final part of the paper reviews the application of lake sediment geochemistry to mineral exploration including a discussion of sampling equipment, logistics, costs and selected case histories.

THE LAKE WATER — LAKE SEDIMENT ENVIRONMENT

Since lake sediment and water compositions are affected by the physicochemical processes active in the lake regime, these processes will be reviewed briefly. This will allow an understanding of the factors to be considered when interpreting lake sediment and water geochemical data for use in mineral exploration. It is virtually impossible to discuss the geochemistry of lake sediments as a subject completely divorced from lake waters. Therefore, lake waters will be discussed, as required, to elucidate various aspects of the geochemistry of lake sediments. More detailed information on limnology, the scientific study of lakes, is available in texts by Hutchinson (1957, 1967), Ruttner (1963), Frey (1963), Kuznetsov (1970), and with particular reference to the application of lake water and sediment geochemistry to mineral exploration, by Levinson (1974).

Origin and Morphometry of Lakes

Inland lakes cover a relatively small portion of the earth's land surface — only about 1.8 per cent or 2.5 million km^2. The number of lakes in several countries such as Sweden, central Finland, Canada and the central northeastern United States of America, particularly within the recently glaciated Precambrian Shield regions of these countries, is very large and hence the dimensions of most lakes are relatively small. In the Canadian Shield most lakes are small, less than 10 hectares (1 hectare = 2.471 acres = 10 000 m^2), and relatively shallow, usually less than 9 m deep (Cleugh and Hauser, 1971).

Lakes may be formed by a variety of processes. Almost all lakes within the Precambrian Shield regions of Sweden, Finland and Canada occupy depressions that are of glacial and/or tectonic origin. Lake basins formed by glacial action predominate in these Shield regions, as well as in the Canadian Cordillera and Appalachian regions. Lake basins formed by glacial action can be placed into four major categories: (1) those formed in cirques and mountain valleys as a result of mountain glaciation; (2) those occupying glacial rock basins formed by scour action on peneplains or in shallow valleys, due to continental glaciation. The position of the lakes in (1) and (2) is a function of variation in lithology and/or the existence of joints, faults and shatter belts; (3) those developed in glacial or postglacial sediments such as in kettles, subglacial channels, or irregularities in ground moraine; (4) those dammed by surficial materials including ice, moraine, etc. Of these four, most lakes in Shield regions can be placed in category (2).

Lake waters originate as any one or combination of stream waters, groundwaters, surface runoff, rain and snow. Lake basins are ephemeral, geologically, and grow ever shallower with time, through erosion of the enclosing rim and through sedimentation: thus the lifespan of a lake is limited.

The Physicochemical Nature of Lakes

Lakes comprise a complex system involving interplay between various physical, chemical and biological factors. The complex natural processes active in the lake environment are presented in graphical form in Figure 20.1. The distribution of heat and suspended and dissolved substances, including gases absorbed from the atmosphere, as well as compositional variation within the sediment, are factors which can affect the dispersion of trace elements in the waters, and the accumulation or depletion of trace elements in the sediments of lakes. Limnological environments may be classified in terms of water column temperature, oxygen, and dissolved and suspended matter profiles as illustrated in Figure 20.2.

Thermal Properties of Lakes

Deeper lakes in temperate regions are generally subject to gradual heating of the waters in the spring and summer resulting in a characteristic form of thermal stratification (Fig. 20.2) consisting of: (1) the **epilimnion,** an upper region of generally uniformly warm, circulating and fairly turbulent water; (2) the **hypolimnion,** a bottom layer of cold and relatively undisturbed water separated from the epilimnion by (3) the **metalimnion,** an intermediate region in which the temperature gradient, the **thermocline,** is steepest. The distribution of heat in lakes depends largely on the mixing effect of wind and, to a lesser extent, on convection currents. The metalimnion, which acts as a barrier between the epilimnion and hypolimnion, indicates the limit of mixing from the surface.

Not all lakes display this classic type of thermal stratification and there are many different forms of stratification brought about by the form, size, depth, and location of the lake basin, volume of through-flow, and effects of climate.

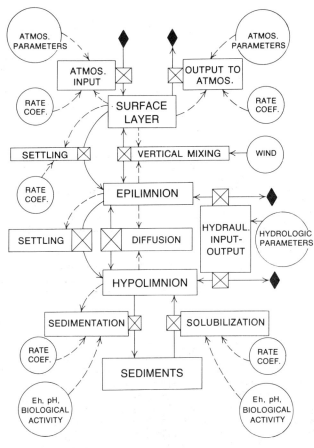

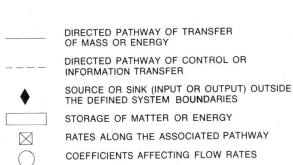

Figure 20.1. Graphical representation of the complex natural processes operative in the lacustrine environment (from Sain and Neufeld, 1975).

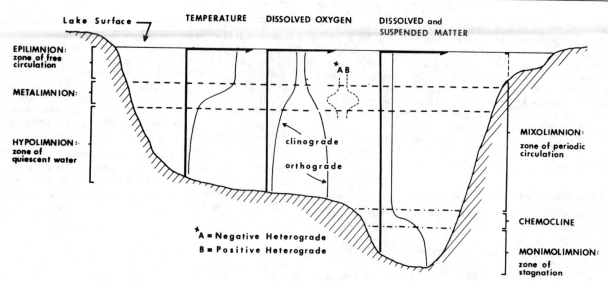

Figure 20.2.   Stratification within lake environments (from Nichol et al., 1975).

Lakes may be further classified into thermal types on the basis of the number of periods of circulation per year (mixing of surface to bottom waters due to temperature and as a result, density uniformity). The types of interest are: (1) **monomictic**, water circulation once a year; (2) **dimictic**, freely circulating twice a year, spring and fall, inversely stratified in winter, directly in summer. A type of dimictic lake, termed **meromictic**, is one in which water circulation does not extend to the base of the water column. The water mass that does not take part in the mixing is termed the **monimolimnion** and the expression **chemocline** has been applied to the boundary zone that separates the bottom waters stabilized by dissolved or suspended substances, from the overlying less dense waters; (3) **polymictic**, as applied to shallow northern temperate lakes (Schindler, 1971), display stratification by heating during the day with complete mixing occurring during the cooler evenings; (4) **unstratified**, display uniform temperature distribution and can go through freely circulating periods at which time the circulation is complete to the bottom.

Mainly dimictic, although some polymictic and unstratified lakes, were reported in the Canadian Shield of northwestern Ontario (Conroy, 1971; Schindler, 1971; and Coker and Nichol, 1975). While deep lakes examined within the northern region of the Canadian Shield displayed well marked temperature stratification and were classified primarily as dimictic, most shallow lakes were unstratified (Brewer, 1958; Jackson and Nichol, 1975; Klassen et al., 1975; and Shilts et al., 1976). In northern Saskatchewan (Arnold, 1970) and in southern Ontario (Jonasson, 1976; Coker and Jonasson, 1977a, b) the lakes studied displayed temperature profiles having approximately constant temperature or decreasing temperature with depth, generally characteristic of unstratified and dimictic lakes respectively. Variations in the thermal stratification and the resulting nature of the circulation of lake waters can play a definite role in determining the distribution of suspended and dissolved substances, including gases, within the lake water column. The distribution of heat, more than any other single parameter, governs the chemical nature of the water column with the thermocline, when present, restricting the exchange of materials between sediment/hypolimnion and the epilimnion, and hence influencing whether trace metals remain in solution and disperse in the waters or precipitate and accumulate in the sediments of the lakes.

Dissolved Gases

Of the gases present dissolved in lake waters, oxygen, carbon dioxide and hydrogen sulphide, along with their various gaseous compounds and dissociated ionic forms, are perhaps most important in determining the movement and fixation of trace elements in lakes.

At times of lake water circulation, the gases in the lake waters can become entirely in equilibrium with those in the atmosphere. This equilibrium will, however, depend on the atmospheric pressure (and therefore on the altitude), lake water temperature, the completeness of mixing, the length of the circulation period, and complexed species formed. In most cases an equilibrium between the gases in the water and air is reached and the whole lake is then replenished with atmospheric gases from top to bottom. These gases then can become enriched or depleted within the lake water column after the period of circulation.

Lakes can be classified on the basis of biological productivity related to the dissolved oxygen content and nutrient supply. Lakes with a rich biota can supply more oxidizable organic debris to the lake bottom and thereby deplete oxygen concentration in bottom waters. In temperate regions, lakes are classified into two groups: (1) **eutrophic**, mature lakes with considerable organic matter and nutrients which can result in a high plankton population. The oxygen distribution in such lakes is generally **clinograde**, displaying decreases in oxygen content in the metalimnion and hypolimnion in a manner almost parallel to the temperature curve (Fig. 20.2); (2) **oligotrophic**, youthful lakes with high oxygen and low nutrient content. These lakes display an **orthograde**, oxygen distribution, one in which the oxygen content remains almost uniform in the whole water column (Fig. 20.2).

Oxygen stratification may be found in all gradations between the extremes of oligotrophic and eutrophic lakes. In some lakes, very striking maxima in oxygen concentrations can develop in the metalimnion, during stratification, as a result of photosynthesis; this type of distribution is termed **positive heterograde** (B, Fig. 20.2). Many cases are also known of marked minima, **negative heterograde**, (A, Fig. 20.2) caused by accumulation of oxidizable materials.

Studies of relatively deep lakes, greater than 10 m, in the Canadian Shield of northern Saskatchewan indicated the presence of both oligotrophic and eutrophic lakes (Arnold, 1970). In the Shield lakes of northwestern Ontario shallow (less than 4 m), unstratified or polymictic lakes are oligotrophic being nearly saturated with oxygen all summer and having uniform oxygen content with depth, (Schindler, 1971). Lakes from 5 to 12 m had bottom waters that became anaerobic, or nearly so, by late summer. Lakes greater than 12 m showed a definite oxygen depletion in their bottom waters and good oxygen stratification displaying decreasing oxygen content with depth. These deep lakes were classified as eutrophic. Similar observations were made in other Shield lakes of northwestern Ontario (Conroy, 1971; Coker and Nichol, 1975). In the northern regions of the Canadian Shield, near Yellowknife, lakes were classified as eutrophic or oligotrophic and in certain cases oxygen maxima occur in the metalimnion (Jackson and Nichol, 1975); in the Kaminak Lake area the lakes are oligotrophic with the highest oxygen levels occurring either in the metalimnion or hypolimnion (Klassen et al., 1975; Shilts et al., 1976). In the Nechako Plateau of the Canadian Cordillera, Hoffman and Fletcher (1976) identified two contrasting limnological environments: (1) large (> 3 km in length) oligotrophic lakes and; (2) smaller lakes, associated with swampy areas of low relief, classified as eutrophic or **dystrophic** (brown-water lakes with very low lime content and a very high humus content, often characterized by a severe poverty of nutrients (Ruttner, 1963).

The carbon dioxide profile, including bicarbonate, is often roughly the inverse of the oxygen distribution during the summer period of stratification (Hutchinson, 1957). This is due to the close association of the carbon cycle with the dissolved oxygen content of the lake waters. In the water column high biological activity can result in decreases in the carbon dioxide content and increases in the oxygen content, even to supersaturation levels, due to photosynthesis in the upper **trophogenic** layer (region of photosynthetic production) (Ruttner, 1963). Respiration and oxidation processes, primarily the bacterial oxidation of organic matter, in the lower **tropholytic** region (region of breakdown), can cause enrichment in carbon dioxide (or its salts) and depletion in oxygen. Also, one of the main products of the oxidation of organic matter is the sulphate ion, which under anaerobic conditions is generally reduced to sulphide species, H_2S, HS^-, and S^{2-} (Hutchinson, 1957; Garrels, 1960; Kuznetsov, 1970 and 1975).

The presence of oxygen, carbon dioxide and/or hydrogen sulphide, along with their various gaseous compounds and dissociated ionic forms, determine whether aerobic oxidizing or anaerobic reducing conditions exist through the water column. The nature of the electrochemical conditions of the water column determines whether dissolved trace elements remain in solution, or are precipitated and accumulated in the underlying sediment, and whether selected trace elements are retained within the surface sediments, or released through dissolution or desorption back into the overlying water.

pH

The usual pH range for open lakes is between 4 and 9, with lakes in regions of acid rocks displaying pH values below 7 and lakes over calcareous rocks exhibiting values well over 8 (Hutchinson, 1957; Baas Becking et al., 1960). In most lakes which have near neutral pH, the pH is regulated by the carbon dioxide — bicarbonate — carbonate system (Hutchinson, 1957; Garrels and Christ, 1965). In general, oligotrophic waters tend to have slightly acid pH, whereas eutrophic waters tend to be more alkaline. In most cases the pH of the waters tends to be about 7 and oligotrophic cannot be distinguished from eutrophic on the basis of pH. The form of the pH profile tends to follow that of temperature, with the surface waters having higher pH values than the deeper waters. The lower pH in the deeper waters can be attributed to the release of carbon dioxide from respiration reactions and/or the bacterial oxidation of organic matter and the subsequent hydrolysis to H_2CO_3. Also, the pH of the surface waters will be increased as carbon dioxide is consumed during photosynthesis. In a number of geochemical studies of Canadian Shield lakes, pH values were recorded ranging from 3.0 to 8.7 with extremely acid pH values being associated with lakes adjacent to oxidizing sulphide mineralization and alkaline pH values being associated with lakes in areas of carbonate lithologies (Arnold, 1970; Conroy, 1971; Schindler, 1971; Allan et al., 1973a; Closs, 1975; Coker and Nichol, 1975; Jackson and Nichol, 1975; Klassen et al., 1975; Jonasson, 1976; Meineke et al., 1976; Shilts et al., 1976; Cameron, 1977; Cameron and Ballantyne, 1977; Coker and Jonasson, 1977a, b; Maurice, 1977a, b).

Dissolved and Suspended Solids

In normal fresh water the total content of dissolved solids consists of only a few salts: the carbonates (including bicarbonates), sulphates and chlorides of the alkali (Na, K) and alkaline-earth (Ca, Mg) elements, silicic acid, and small amounts of nitrogen and phosphorus compounds (Hutchinson, 1957; Ruttner, 1963; Kuznetsov, 1970). Compounds of iron and manganese can also reach significant concentrations in waters under suitable anaerobic conditions (Juday et al., 1938; Mortimer 1941, 1942, and 1971; Hutchinson, 1957; Livingston, 1963; Ruttner, 1963; Mackereth, 1965; Arnold, 1970, and Kuznetsov, 1970). In addition, there are minute concentrations of trace elements in solution, generally at the parts per billion (ppb) level, together with various mineral colloids and suspensions (Riley, 1939; Turekian and Kleinkopf, 1956; Hutchinson, 1957; Ruttner, 1963; Gorham and Swaine, 1965; Allan and Hornbrook 1970; Arnold, 1970; Kuznetsov, 1970; Dyck et al., 1971; Hornbrook and Jonasson, 1971; Allan et al., 1973a,b; Cameron, 1977; Coker and Jonasson, 1977a,b; Maurice, 1977a,b).

Water conductivity may be regarded as a measure of the ionic material present in the water. There is generally a direct relationship between conductivity, bicarbonate alkalinity and the pH of lake waters (Coker, 1974) and consequently, as has been demonstrated by many surveys in the Canadian Shield, surface lake waters with high conductivities will occur in carbonate-enriched terrain, whereas lower conductivities will occur in granitic terrains (Armstrong and Schindler, 1971; Conroy, 1971; Closs, 1975; Coker and Nichol, 1975; Jackson and Nichol, 1975; Klassen et al., 1975; Semkin, 1975; Dean and Gorham, 1976; Shilts et al., 1976; Cameron and Ballantyne, 1977; Coker and Jonasson, 1977a,b). Cameron (1977) noted a trend of increasing conductivity and acidity and decreasing alkalinity and pH in the surface waters of lakes in positions varying from remote to adjacent to actively oxidizing sulphide mineralization (see Fig. 20.6).

Organic substances occur in solution in both suspended and dissolved forms within natural waters. Concentrations of dissolved organic matter in natural waters, expressed as the amount of carbon per unit volume of water, are commonly in the range 0.1 to 10 mg/L (Stumm and Morgan, 1970). Organic substances that occur in solution frequently exceed by several fold those in particulate form (Birge and Juday, 1934; Kuznetsov, 1970). Suspended organic matter consists mainly of: (1) living, and the remains of dead and decayed plankton, bacteria and algae; (2) vegetation and animal-derived detritus from the littoral zone of the lake, from around the edge of the lake, and that brought in by wind, drainage and precipitation. Dissolved organic matter, including colloidal organic substances, is composed primarily of: (1) intermediate decomposition products such as amino acids, fatty acids, alcohols, hydrocarbons, proteins, etc.; (2) substances resistant to further degradation — a group broadly named "aquatic humus" (Swain, 1958; Kuznetsov, 1970). Humic matter is

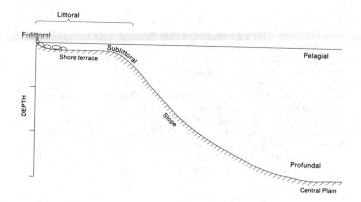

Figure 20.3. Lake habitat subdivisions (after Ruttner, 1963).

derived, in part, by leaching and/or eroding of soils and swamps (**allochthonous** origin – material formed outside the lake) and, in part, from the cellular constituents and exudates of indigenous aquatic organisms (**autochthonous** origin – material formed within the lake) (Hutchinson, 1957; Bordovsky, 1965; Flaig, 1971; Nissenbaum and Kaplan, 1972). It is concentrations of humic substances which give the brown to yellow colour characteristic of the waters of many lakes in Shield areas. The structure of humus, as it is known, is very complex and still in dispute. The following classification of humic substances is now generally accepted: (1) humins, (2) fulvic acid, (3) humic acid (Flaig, 1971).

Lake Sediments — Major Composition

The composition of lake bottom deposits (sediments) depends on: lake morphology, biota types and abundances, drainage, glacial history of the area including type of glacial deposit, and geographical (climatic) and geological location. According to composition these deposits are partly inorganic, partly organic; according to origin they are either **autochthonous**, having been formed in the lake itself by life processes or physical-chemical processes separating them from the water, or **allochthonous**, having been introduced from outside the lake by inflowing water, falling dust, precipitation, etc. (Hutchinson, 1957; Ruttner, 1963).

The inorganic allochthonous particles introduced into a lake undergo sorting according to their size and density. In the vicinity of effluents, and within the littoral zones (Fig. 20.3) of many lakes (due to sorting by wave action and periglacial processes), materials derived from the glacial and postglacial sediments of the drainage basin are deposited and reworked into zones of coarser and finer materials. The finer portions are distributed more uniformly throughout the lake, settling out over the entire lake bottom and contributing substantially to the inorganic component of sediments in the deepest and/or central (profundal – see Fig. 20.3) portion of the lake basin. Where present (in quantity, generally south of the treeline (see Fig. 20.11)) the organic component of allochthonous sediments contains a considerable amount of humic matter, derived from the leaching of the organic layer of soils and swamps, coarser organic detritus from vegetation around the edge of the lake, and to a lesser extent animal materials, such as hair, exudates, chitinous parts of insects, etc. The contribution of airborne dust can be considerable, particularly in the spring when plants are in bloom and lake surface waters are commonly covered with a yellow film of pollen and spores, which eventually are incorporated into the sediment.

Autochthonous sediments consist primarily of two types: (1) precipitates that form external to living processes, as a result of physical-chemical changes (i.e.: the precipitation of compounds of Fe, Mn, Ca, Si etc.); (2) plant and animal remains from the lake community along with their inorganic and organic integuments and supporting materials. After being worked over by bottom animals and bacteria, in an environment at least periodically supplied with oxygen, the organic and inorganic materials in the lake centre (profundal) basin forms a very characteristic, finely divided sediment of grey, greyish brown to brown-black colour and at times of elastic consistency, called **gyttja** (Ruttner, 1963).

Lake sediments, particularly from areas within the Canadian Shield, have been grouped into three distinct classes each of which has broad chemical, physical and mineralogical characteristics that are relatively uniform (Jonasson, 1976): (1) **organic gels** (gyttja) are materials commonly found in the deeper waters (profundal basin) of organic-rich lakes and are abundant in most parts of the Canadian Shield where there is a deciduous-coniferous forest cover. Almost completely organic in composition, they have a strong odour (H_2S), indicative of reducing conditions, and are thixotropic. As their coherence often is very low the gels may disperse freely into the lake waters if the sediment-water interface is disturbed. Their origin probably lies in the sedimentation of fine, dispersed particles of vegetation debris (pollen, spores etc.), and of coagulating colloids of dissolved organic matter (organic acids such as humates, fulvates, etc.). Relatively little coarse organic debris or mull is present in these samples or organic gels which occur in the central and/or deepest parts of a lake. These gels dry to a hard, dark, lustreless, homogeneous cake which is difficult to break and does so conchoidally; (2) **organic sediments** are widespread throughout most lakes of the Shield especially in shallow waters either near shores or near inflows. They are also the most abundant material in swamps and marshes, and may be regarded as mixtures of organic gels, organic debris, and inorganic sediments (mainly silts and clays but also some sands, gravels and boulders may be present); (3) **inorganic sediments** implies various combinations of boulders, gravel, sand, silt, marl and clay with inorganic oxides and hydroxides (precipitates and colloids) with virtually no organic matter, and with little or no regard paid to mode of deposition or derivation of materials. Inorganic sediments commonly occur throughout lake basins in areas of the Shield which have few or no trees and in shallow waters in most lakes, especially at the shores of a lake, and near inflows and outflows. They may originate from the winnowing action of waves on shoreline sediments and soils, or may occur as silty deposits in deeper waters.

Although the classification of Jonasson (1976) probably applies to lake sediments from most areas of the southern Canadian and Fennoscandian Precambrian Shields, as well as to the Canadian Appalachia, a limited number of detailed compositional studies of lake sediments from these areas, as summarized in Table 20.1, have produced additional data on the nature of the major composition of lake bottom deposits.

Studies carried out on the compositional nature of lake sediments from lakes in the northern Canadian Shield, characterized by open woodland and/or tundra landscape, illustrate some distinct differences, and yet similarities, to the composition of sediments from lakes in the south of the Shield below the treeline. In the tundra of the Coppermine River area, District of Mackenzie, nearly all nearshore lake bottom samples were composed of inorganic clastic material (boulders to silts) with only a few being organic (sludges or algal gels) or clayey (Allan, 1971).

Near Yellowknife, District of Mackenzie, an area of predominantly open woodland, nearshore sediment composition varied from sand to silt to clay to decomposed and relatively undecomposed organic material (Jackson, 1975). Towards lake centres the sediment became more homogeneous in composition consisting either of clay, silt-sized particles or organic-rich ooze. In nineteen lakes examined in detail the organic content of sediments, as determined by L.O.I. (450°C), varied from <1% to 85%.

Table 20.1

Compositional nature of lake sediments from the southern part of the Canadian and the Fennoscandian Precambrian Shields

Area	Organic content	Composition
Lac La Ronge – Flin Flon area Saskatchewan, Canada. (Arnold, 1970)	L.O.I. (700°C): 2 to 58%	Silicate fraction: quartz, garnet, minor Ferromagnesian minerals. Feldspars in nearshore sediments. Clay mineral: illite. Heavy mineral separate: hematite, magnetite and pyrite.
East-central Saskatchewan, Canada. (Lehto et al., 1977)		Sand, silt and clay sized mineral matter Organic fraction: algae, plant fibre, diatoms and the remains of fish and small crustaceans.
Northwestern Ontario, Canada. (Brunskill et al., 1971)	L.O.I. (900°C): 18 to 62%. Organic carbon: 8 to 34%. L.O.I.: C ratio is 2.7 (low organics) to 2.3 (high organics)	Major minerals: quartz, plagioclase, potassium feldspar, illite, chlorite and kaolinite. Organic matter, water, Si and Al are main components of sediments
Red Lake – Uchi Lake area, northwestern Ontario, Canada. (Timperley and Allan, 1974)	L.O.I. (450 to 550°C): 5 to 50%.	Sediments in the deep part of large lakes: maximum L.O.I. of 40%, grey-green ooze; organics are humic colloids and algal remains; composed of quartz feldspar, minor clays and micas coated with organics, free-floating organics, skeletal remains of plankton, pollen and spores. Sediments in shallow lakes and areas of impeded drainage: L.O.I. could exceed 60%; organic matter derived from shoreline and aquatic vegetation. Ferromanganese nodules noted.
Northwestern Ontario, Canada. (Coker and Nichol, 1975)	L.O.I. (450°C): 14 to 61%. L.O.I. : C ratio is 2.4	Lakeshore sediments: silt, fine sand and coarser materials. Lake-centre sediments: silt, clay and organics. Silt, clay and organics are most abundant and homogeneous in lake centre (profundal) sediments. Primary minerals: quartz, plagioclase, potash feldspars, amphiboles, micas and rock fragments. Secondary minerals: illite, kaolinite, chlorite with trace vermiculite, layered clays, and montmorillonite.
Minnesota, U.S.A. (Dean and Gorham, 1976)	L.O.I. (550°C) of profundal sediments: 6 to 60%. Mean L.O.I.: C ratio is 2.1.	Mean carbonate content (L.O.I. (1000°C-550°C)) for profundal sediments is 16% with lakes in carbonate lithologies averaging 26%. Carbonate minerals are calcite, dolomite and aragonite. Clastic component of profundal sediments contain little sand-sized material consisting mainly of approximately equal amounts of silt and clay-sized material. Primary minerals: quartz, plagioclose and orthoclase. Secondary minerals: illite, kaolinite and chlorite.
Minnesota, U.S.A. (Meinke et al., 1976).	L.O.I. (800°C): 16 to 85%	Sediments of small lakes: exhibit H_2S odour, contain fibric-organic, fine-grained organic and clastic material. Large lakes: nearshore sediments composed of sand and/or gravel with depth. Medium lakes: organic-rich gelatinous sediment.
Southwestern Finland. (Koljonen and Carlson, 1975)	L.O.I. (550°C): 6 to 80%	Quartz, plagioclase and potassium feldspar identified in all samples.
Eastern Finland. (Bjorklund et al., 1976)	L.O.I. (550°C): 0 to 98%	Organic gels.

(L.O.I. = loss-on-ignition)

Numerous lakes occupy depressions on the perennially frozen glaciated tundra in the Kaminak Lake area, District of Keewatin. The shallow margins of these lakes are characterized by periglacial features such as polygonal patterns, frost-heaved boulders, and mudboils (Shilts and Dean, 1975). Digitate, cobble-covered ribs and boulder-filled troughs, composed of till and thought to be the subaqueous equivalents of mudboils, commonly form a crenulate pattern in shallow shelves adjacent to till-covered shores. Sediments of zones away from the lake margins are generally soft, loosely consolidated grey to olive-grey silts and, locally, yellowish brown silty clays which sometimes grade into an underlying firm, grey, coarse sand-silt (Klassen et al., 1975). These sediments are composed predominantly of quartz and feldspar fragments. Organic matter in the sediments, as determined by L.O.I. (450°C) varies from <1% to 27% averaging 6%. The upper silt material varies from structurally uniform to finely laminated. A stiff to watery, gel-like sediment that flocculates rather than dispersing in the lake water is thought to represent true modern lake sediment in the area (Shilts et al., 1976). This gel is either absent or very thin, rarely exceeding 1m, over large parts of many lake basins. The surface of the sediment is generally a reddish orange colour, and black or orange (manganese/iron precipitates) horizontal bands are commonly observed through the sediment (Klassen et al., 1975; Shilts et al., 1976). Also, in some lakes a significant thickness of massive to laminated, grey to pink silty clay, interpreted as marine sediment, was found to underlie the gel and overlie a unit interpreted as till (Shilts et al., 1976).

Two distinct types of sediment, nearshore and centre-lake, were found in lakes in the tundra landscape of the Agricola Lake area, District of Mackenzie (Cameron, 1977). Nearshore sediments show patterned features and are heterogeneous mixtures of material from cobble to clay size. The second type, centre-lake, was described by Williams (1975) as consisting of two varieties: (1) having a reddish uppermost layer overlying soupy brownish material; and (2) consisting of bright reddish surface material overlying brownish material over bluish grey sediment.

In general the stratigraphy of lake bottom materials from northern and southern Canadian Shield lakes is very similar. The upper strata of freshwater lake sediments comprises relatively modern organic-bearing materials. This sediment is thickest and contains the greatest amount of organic material in lakes of the southern Shield and is relatively thin and areally restricted, where present at all, in lakes of the northern Shield. This is largely because lakes of the south have been sites of accumulation through a longer period of time and have also had higher organic productivity both within the lake itself and within the associated drainage basin. Inorganic components of these modern organic sediments, which are generally minor, are derived from glacial and/or postglacial sediments within the associated drainage basin. In nearshore areas, where the modern organic sediments may be largely absent, the inorganic materials are washed glacial or postglacial sediments. Underlying the modern lake sediments can be deposits of glacial-lacustrine and/or marine sediments in areas that have been covered by glacial lakes and/or subjected to marine invasion. In areas that have not been postglacially submerged, modern organic material generally directly overlies till, the next lowest stratigraphic unit. In some cases the modern organic material may lie directly over bedrock and/or paleosol material or these materials may constitute the lake bottom itself. The thickness and types of bottom deposits in any individual lake can vary both horizontally and vertically and any of the stratigraphic units can form the lake bottom depending on the sedimentation history of the lake. Mineral exploration surveys which collect material from lake centres in the southern parts of the Canadian Shield and, in limited cases in the northern Shield, are most likely collecting the organic lake sediment facies. In contrast, surveys collecting material from the mineral sediment found around the margins of Shield lakes are not necessarily collecting modern lake sediment but are generally obtaining glacial, glacial-lacustrine or marine sediments, or slumped soils which have been subjected to some reworking, including wave action and, in addition, in the north of the Shield, to various periglacial processes.

FACTORS AFFECTING TRACE METAL DISPERSION AND ACCUMULATION IN THE LAKE ENVIRONMENT

A knowledge of the processes by which a metal is mobilized, transported, precipitated, and possibly remobilized, is of prime concern in order to comprehend possible controls on that metal's dispersion, accumulation, and fixation into lake bottom materials. Certain of these processes, operative within the lacustrine environment, and the nature of lake bottom deposits, have been previously described.

The principal mechanisms affecting trace metal transport, accumulation, and fixation into bottom deposits are thought to involve: (1) scavenging of metals by algal and plankton blooms and other organic matter, both particulate and dissolved; (2) sorption and coprecipitation by hydrous iron and manganese oxides; (3) sorption by clay minerals; (4) chemical processes involving hydrolytic reactions and both complexed and dissolved ions, for example sulphide, carbonate, hydroxide; and (5) variations in the gross physical-chemical nature of the sediments. Interactions between these various mechanisms commonly occur. One model for metal transport and accumulation in lakes within the forested portion of the Canadian Shield is presented in Figure 20.4 (Timperley and Allan, 1974).

Aquatic Biota

Trace metal scavenging by algal and plankton blooms and other suspended and dissolved organic particles followed by deposition and incorporation of the metal-bearing materials into the lake bottom deposits has been documented (Kuznetsov, 1970; Morris, 1971; Andelman, 1973; Gibbs, 1973; Knauer and Martin, 1973; Leland et al., 1973; de Groot and Allersma, 1975; Trollope and Evans, 1976). Excretory products of plankton are also a source of many trace elements (Boothe and Knauer, 1972). After death or moulting of plankton settling occurs. The efficiency of metal transport to the sediment depends in part on the rate of decay during settling of the dead organism, test, moulted exoseletion, or excretory product.

Bacterial activity in the surficial sediments causes further decay at the sediment-water interface and, after burial, may result in increases in trace element concentrations in the sediments (Leland et al., 1973). Trace metals may accumulate in the upper 5 to 20 cm of lake sediment by biological and geochemical mechanisms as well as recent cultural loading (Mortimer, 1942 and 1971; Gorham and Swaine, 1965; Mackereth, 1966; Cline and Upchurch, 1973). Upward migration of heavy metals may occur because of dewatering due to compaction and unidirectional ion migration, but, to a much greater extent, migration appears to be due to a bacterial mechanism (Cline and Upchurch, 1973). Bacteria can also cause a drop in Eh and pH; which will cause release of the complexed metals. The heavy metals may then be transported upward either on bubble interfaces, in a gaseous complex or as soluble organic complexes. When the metal reaches the biologically active surficial sediment, which is generally oxidizing in nature (Mortimer, 1942 and 1971; Gorham, 1958), it is immobilized as a new organic complex or an inorganic precipitate. Therefore, heavy metals have a tendency to remain at the sediment-water interface.

If the surficial sediments are reducing in nature, and in fact the overlying waters are also, the heavy metals may in fact be mobilized into the waters.

Aquatic flora and fauna can concentrate many trace metals to levels much higher than those existing in the waters (Leland et al., 1973; Trollope and Evans, 1976). Concentration factors depend upon the physicochemical interactions of each metal with other environmental parameters and organisms, as well as the nutritional requirements of individual species. In lakes having aquatic flora and fauna as a significant source of organic matter, such as many lakes of the southern Canadian Shield as opposed to those of the north, or in which waters are highly productive and sedimentation is rapid, the influence of organisms on trace element distribution may be significant.

Organic Matter

Organic matter probably plays an important role in the complexing of heavy metals in lake waters and sediments. Within surficial waters metal-organic interaction is generally accepted to involve chelation of trace elements with humic matter (Bowen, 1966; Saxby, 1969). Of the organic species involved, humic acids, fulvic acids and humins are most abundant. These substances are probably also important in pore waters and interstitial waters of sediments representing as they do end members of humic matter degradation. Micro-organisms are, in part, responsible for the decomposition of the higher weight organic acids to produce smaller, more soluble fragments (Kuznetsov, 1975). Micro-organisms are also important in mobilizing certain metals from the sediments by means of very specific biochemical interactions (Wood, 1974 and 1975; Wong et al., 1975; Chau et al., 1976).

Humic and fulvic acids behave as negatively charged species in solution. Neutralization of this charge by interacting metal ions, metal oxide colloids, or adsorption onto clay particles can lead to flocculation of the colloids and subsequently, further coprecipitation of metals. Mechanisms by which metallic ions from natural waters are absorbed or complexed by such organic matter, have been extensively discussed (Krauskopf, 1955; Curtis, 1966; Schnitzer and Khan, 1972). The ability of solid humic matter to physically and chemically adsorb metals from aqueous solution has been documented (Rashid, 1974). Jackson and Jonasson (1977) suggested a probable order of binding strength for a number of metal ions onto humic or fulvic acids; $UO_2^{++} > Hg^{++} > Cu^{++} > Pb^{++} = Ca^{++} > Zn^{++} > Ni^{++} > Co^{++}$. The partitioning of trace elements in the organic phases of lake sediments from the Elliot Lake and Sturgeon Lake areas, Ontario was examined by Schaef (1975). Results indicated that Zn is preferentially concentrated in the fulvic component while, U, Cu and Pb are concentrated in the humic component.

Another function of organic matter in water-sediment interactions concerns the ability of soluble organic acids to chemically leach and even dissolve minerals, extracting a variety of elements including trace metals (Baker, 1973; Guy and Chakrabarti, 1976). Interactions between metal ions, minerals, organic sediments and water seem to be mutually destructive of all solid species. The ultimate result of the breakdown of a mineral or metal-sediment complex by organic acids must be to promote the remobilization of that metal ion until the adsorption, flocculation, polymerization, precipitation cycle takes it out of solution again.

Organic matter is also capable of acting as a reducing agent. In this way selected metal ions can be stabilized by complexation with certain organic acids or humic acid compounds (Rashid and Leonard, 1973; Theis and Singer, 1973).

From the studies that have been carried out to date it appears that metal-humic matter interactions are potentially important in the following geochemical processes: (1) leaching of metals from solid mineral phases; (2) dispersion within drainage basins as soluble or colloidal metal-organic species; and (3) concentration and fractionation of many elements within organic-rich sediments.

In the flat-lying, tree-covered terrain characteristic of the southern Canadian Shield and Fennoscandian Shield, and in the terrain of the Canadian Appalachians, the incidence of organic matter is high and metal-organic interactions are predominant. Streams and other waters which supply lakes are commonly fed by waters from swamps, marshes and muskeg. The dissolution residence time of waters within organic trash ensures considerable quantities of humic materials are present. Silts and clays are often coated with organic matter. By contrast, lakes in Shield areas above the treeline and from the alpine Cordilleran regions are fed by waters derived mainly from snow melt which contains very little dissolved organic material. Absorption of metals directly into clays, rock flour, and hydrous metal oxides and dissolution of mineral particles are the predominant water-sediment interactions.

Hydrous Metal Oxides

The geochemical cycles of iron and manganese in the exogenic environment have been extensively reviewed in the literature, notably by Mortimer (1941, 1942 and 1971), Hutchinson (1957), Krauskopf (1957, 1967), Hem (1964), Gorham and Swain (1965), Mackereth (1966), Jenne (1968), and Stumm and Morgan (1970). There are also a large number of publications that describe the occurrence, composition and origins of freshwater ferromanganese deposits in European and North American lakes (viz.; Kindle 1932, 1935 and 1936; Twenhofel and McKelvey, 1941; Lundgeer, 1953; Beals, 1966; Delfino and Lee, 1968; Rossman and Callender, 1968, 1969; Harriss and Troup, 1969; Troup, 1969; Arnold, 1970; Cronan and Thomas, 1970, 1972; Dean, 1970; Edgington and Callender, 1970; Schoettle and Friedman, 1971; Damiani et al., 1973; Sozanski, 1974; Timperley and Allan, 1974; Coker and Nichol, 1975; Cook and Felix, 1975; Jackson and Nichol, 1975; Klassen et al., 1975; Koljonen and Carlson, 1975; Robbins and Callender, 1975).

Under oxidizing conditions, hydrous oxides of iron and manganese are excellent scavengers of trace elements; however, under reducing conditions they are made soluble and may result in increases in concentrations of cations and anions in overlying waters (Mortimer, 1941, 1942 and 1971; Hutchinson, 1957). Upon their release from rocks and overburden as divalent ions, iron and manganese respond similarly to changing Eh-pH conditions, but iron is oxidized and precipitated at lower Eh-pH fields than manganese, activities being equal. Much of the iron and manganese may be transported colloidally in water as oxide hydrosols, likely stabilized by interactions with organic matter (Hem, 1971). Incoming runoff can bring iron and manganese into the lake as: (1) oxide coatings on mineral grains; (2) in solution and organic complexes; (3) adsorbed on solids; and (4) incorporated in organic solids. The proportion of the metals arriving in solution and in colloidal form, as complexes and in organic solids, will be high in the swampy creeks (rich in humic content and generally low in pH) that characterize much of the carbonate-poor forested Shield regions of the northern hemisphere. In the better aerated, sometimes rapid influents of the barren northern Canadian Shield and alpine Canadian Cordilleran regions oxide-coated particles and hydrous oxide complexes can attain predominance. On entering the lake most of the streamborne particles will

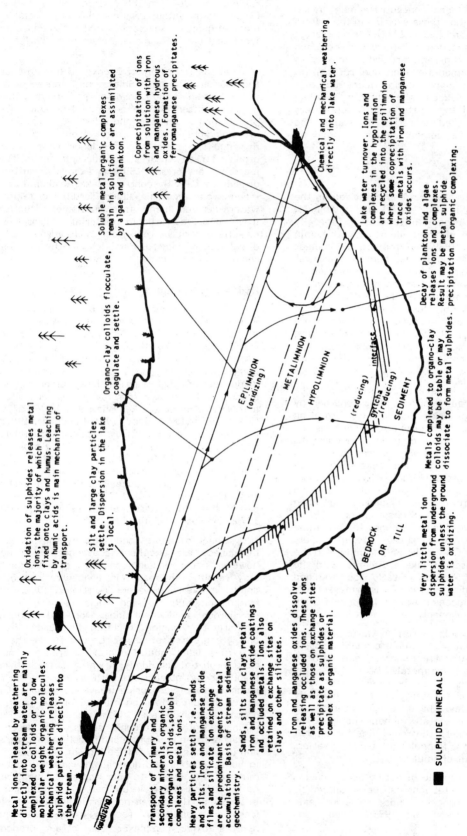

Figure 20.4. One model for mineral weathering, metal transport and metal accumulation within lakes of the forested portion of the Canadian Shield (from Timperley and Allan, 1974).

settle onto the bottom. When manganese and iron are brought into a lake basin which has oxidizing conditions, they will be oxidized and eventually precipitated, along with any trace elements they have absorbed, and incorporated into the sediment. In the forested areas of the Shield, the lake centre (profundal) sediments are generally characterized by a microzone of oxidizing sediment, at the sediment-water interface, overlying highly reducing organic-rich material (gyttja) (Mortimer, 1942 and 1971; Gorham, 1958). Littoral zone material is generally in an oxidizing environment. In the barren northern parts of the Shield the littoral and profundal sediments are generally oxidizing in nature, although under winter ice cover reducing conditions could result and are perhaps one explanation for the cyclic oxide banding (Klassen et al., 1975) noted in some sediments. If the sediments are reducing in nature, as is the general case in most profundal sediments in lakes from forested areas of the Shield, chemical dissolution of the hydrous oxides of maganese and iron can occur via two mechanisms: (1) the formation of iron or manganese complexes with organic acids (Baker, 1973); and (2) reduction of the hydrous oxides to more soluble species (Hem, 1960; Theis and Singer, 1973). Once this occurs the metals can then migrate laterally and up through interpore spaces in the sediments, along potential gradients, until oxidizing conditions are encountered and they are reprecipitated as crusts or nodular to discoid concretions in or atop the oxidized surface profundal or littoral zone sediments (Rossman and Callender, 1968). Other proposed models of concretion formation usually combine diagenetic and hydrogenous sources of iron and manganese (Troup, 1969; Terasmae, 1971; Cronan and Thomas, 1972). The effect of biota as direct oxidizers of iron and manganese has been widely asserted (Kindle, 1935; Troup, 1969; Kuznetsov, 1975), although some writers are skeptical about the role of microorganisms in the overall precipitation of the oxides (Krauskopf, 1957; Stumm and Morgan, 1970). If during the process of summer oxygen stratification, anaerobic conditions develop in the hypolimnion and the surface profundal sediments become reducing, manganic oxides and complexes in the sediments could be reduced and mobilized into the overlying lake waters. If oxygen becomes sufficiently depleted, the bottom water conditions will eventually lead to the reduction of ferric iron which would facilitate its migration from the sediments into the overlying waters as well. This phenomenon also affects the trace elements that are coprecipitated with the iron and manganese. However, as the conditions in the bottom waters become reducing in nature, the formation of carbon dioxide and hydrogen sulphide (or their salts) in appreciable quantities is facilitated. These can result in further solution of the trace elements as bicarbonates or polysulphides or reprecipitation of the trace elements as carbonates or sulphides. When a lake reaches the eutrophic stage all its sediments and most of its waters are reducing, thereby preventing the formation of higher oxides.

The scavenging effect of both iron and manganese hydroxide precipitates on trace metals in the lacustrine environment and the resulting false anomalies (Hawkes and Webb, 1962; Levinson, 1974) in lake sediments has been noted in the course of several exploration-oriented lake sediment studies (e.g. Arnold, 1970; Davenport et al., 1975a,b; Timperley and Allan, 1974; Coker and Nichol, 1975 and 1976; Jackson and Nichol, 1975; Hoffman and Fletcher, 1976; Hornbrook and Garrett, 1976; Cameron, 1977). Horizontal bands of precipitated iron and manganese are distinctive features of profundal sediments from lakes in the Kaminak Lake area, District of Keewatin although the trace metal content of the surface zones of the precipitate do not differ from those of the underlying sediment (Klassen et al., 1975). Various statistical methods have been employed to compensate for such false trace metal (mainly Zn, Co and Ni) anomalies, caused as a result of coprecipitation with iron and manganese. These include ratioing (Coker and Nichol, 1975 and 1976; Jackson and Nichol, 1975) although this method was objected to by Clarke (1976) who like Spilsbury and Fletcher (1974), Davenport et al. (1975b) and Hornbrook and Garrett (1976) preferred to employ regression analysis. In general, the percentage of lakes in which iron and manganese concentrations are high enough to result in falsely anomalous trace metal occurrences is low. Such lakes can be readily identified by the abnormally high levels of iron and manganese present in their profundal sediments. Also, as has been suggested by Coker and Nichol (1975) and others, sampling below the surficial oxidizing layer of sediment into the reducing sediment, which is generally achieved by most sampling devices in use today, will avoid many of the problems of dealing with the more chemically-active surficial sediment.

Iron and manganese oxides are certainly important species in organic systems but their role as direct absorbers of metal ions is overshadowed by competition from the more reactive humic materials and organo-clays or obscured by coatings of organic matter. Moreover, these oxides are unstable in certain organic rich sediments, particularly if reducing. In lakes found in the barren tundra of Canada and in the Canadian Cordilleran regions, the role of organic matter in water — sediment interactions can be relatively unimportant compared with the influences of hydrous iron and manganese oxides.

Clay Minerals

The relative importance of clay minerals in the transport and accumulation of trace metals within the lacustrine environment is poorly understood. The charge characteristics of clays, which impart their ion-exchange properties, originate from isomorphous replacements and the broken edges of crystal surfaces (Leland et al., 1973). The basic principles governing the selectivity characteristics of clays for different cations are valence, hydrated ionic radius, electronegativity, and the free energy of formation. Ionic potential (charge/radius) is a useful parameter for predicting the affinities of clays for different cations. The order of difficulty in displacement of cations is approximately (Mitchell, 1964): $Cu^{++} > Pb^{++} > Ni^{++} > Co^{++} > Zn^{++} > Ba^{++} > Rb^{+} > Sr^{++} > Ca^{++} > Mg^{++} > Na^{+} > Li^{+}$ but relative positions in such a series vary with concentration in solution and the nature of the substrate. Complexing agents seem to alter the affinity of clay minerals for different cations. Several complexing agents are present in lacustrine environments but how they may affect the exchange properties of clay minerals or retention of heavy metals is not well known.

Examination of the inorganic clay-sized fraction of some profundal sediments from lakes in the southern Canadian Shield revealed that the true clay minerals present, mainly illite, chlorite and kaolinite, generally constitute less than half of the clay-size fraction. Quartz, plagioclase and potassium feldspar were the main minerals present. In the overall composition of certain profundal sediments true clay minerals were a very minor component (Arnold, 1970; Brunskill et al., 1971; Timperley and Allan, 1974; Coker and Nichol, 1975; and Dean and Gorham, 1976). The cation exchange capacity of illite, chlorite and kaolinite is relatively low for clays and very low relative to humic matter which undoubtedly coats the clay minerals in many instances.

Although clay minerals can be important in retention and transport of trace elements, most prominently in the Canadian Cordillera, and barren tundra portions of the Canadian Shield, their relative role must be evaluated with caution. Most recent literature emphasizes the importance of organic matter and hydrous oxides in the transport of heavy metals; particularly in the forested Shield regions of the northern hemisphere.

Table 20.2

The main forms of trace metal ions found in aquatic systems for pH's from 5 to 9.5 and various main constituent concentrations (from Morel et al., 1973).

	Species Accounting for more than 90%	Species Accounting for a few per cent
Fe	$Fe(OH)_2^+$, $FePO_4(s)$, $Fe(OH)_3(s)$, $FeCO_3(s)$, $FeS(s)$, $FeSiO_3(s)$	
Mn	Mn^{++}, $MnCO_3(s)$, $MnO_2(s)$, $MnS(s)$	$MnHCO_3^+$, $MnSO_4$, $MnCl^+$
Cu	Cu^{++}, $Cu_2CO_3(OH)_2$, $CuCO_3$, $Cu(OH)_2(s)$, $CuS(s)$	$CuSO_4$
Ba	Ba^{++}, $BaSO_4(s)$	
Cd	Cd^{++}, $CdCO_3(s)$, $Cd(OH)_2(s)$, $CdS(s)$	$CdSO_4$, $CdCl^+$
Zn	Zn^{++}, $ZnCO_3(s)$, $ZnSiO_3(s)$, $ZnS(s)$	$ZnSO_4$, $ZnCl^+$
Ni	Ni^{++}, $Ni(OH)_2(s)$, $NiS(s)$	$NiSO_4$
Hg	$HgCl_2$, $Hg(OH)_2(s)$, $HgS(s)$, $Hg(liq)$, HgS_2^{2-}, $Hg(SH)_2$	
Pb	Pb^{++}, $PbCO_3(s)$, $PbO_2(s)$, $PbS(s)$	$PbSO_4$, $PbCl^+$
Co	Co^{++}, $CoCO_3(s)$, $Co(OH)_3(s)$, $CoS(s)$	$CoSO_4$, $CoCl^+$
Ag	Ag^+, $AgCl$, $Ag_2S(s)$	
Al	$Al_2Si_2O_7(s)$, $Al(OH)_3(s)$	AlF^{+2}, AlF_2^+

The species of each metal are listed in the order they are found with increasing pH. ((s) solid precipitate)

Solubility Chemistry

The ability of a trace metal ion to remain in solution in an aqueous system is limited by the stability of the compounds it forms by reactions with the other components in solution (i.e: carbonate, chloride, sulphate and hydroxide, etc.). The Eh-pH conditions of the water are the main controlling factors as has been discussed at length by Baas Becking et al. (1960), Garrels and Christ (1965), Krauskopf (1967) and others.

Morel et al. (1973) listed the main forms of trace metal ions found in aquatic systems for pH ranging from 5 to 9.5 and various main constituent concentrations (Table 20.2). The species are listed in the order in which they are found with increasing pH. The general pattern is well known: the free ions tend to predominate at low pH, the carbonate and then the oxide, hydroxide or even silicate solids precipitate at higher pH. Because the complexing or precipitating ligands are usually in large excess of the trace metals, they mediate few important interactions (competition) among those metals.

As has been previously discussed, organic ligands can play a major role in natural waters by complexing trace metal ions and also by mediating interactions among trace metal ions. However, the relative stability of soluble metal humates or fulvates in competition with other complexing ions such as carbonate, sulphide and hydroxide is not well known. The adsorption influences induced by the presence of solid substrates are also little known. In one experimental laboratory study the presence of humic acid strongly inhibited trace metal complexation and precipitation by the inorganic ligand (carbonate, sulphide or hydroxide) (Rashid and Leonard, 1973). Hydroxide appeared most effective in causing trace metal precipitation, in competition with soluble metal humate formation, followed by sulphide and then carbonate. However, at the lower final pH of the hydroxide reactions, H^+ ions can interfere in the precipitation reactions.

One of the main products of the oxidation of organic matter is the sulphate ion which in reducing environments is reduced chemically or bacterially to sulphide species, H_2S, HS and S^{2-} (Hutchinson, 1957; Garrels, 1960; Kuznetsov, 1970 and 1975). Within the bottom reducing muds of lakes, metals may be partitioned to varying degrees between inorganic sulphides and organic complexing agents. Timperley and Allan (1974) attempted to determine the forms of binding of Cu, Fe and Zn in organic-rich reducing muds. One of the main conclusions drawn was that Cu was held mainly by sulphide precipitation, whereas organic complexing was more important for Zn and Fe. In another study carried out by Jackson (1977) almost the reverse conclusions were drawn. The degree of affinity of trace metals for sulphide, relative to degree of affinity for organic matter, correlated strongly with the standard entropy of the metal sulphides, decreasing in the order Hg > Cd > Cu > Fe > Zn (Jackson, 1977). Thus, the stability of the metal sulphides tends to control the partitioning of metals between sulphide and organic matter. Sulphide appears much more effective than organic matter in preventing remobilization of Zn, Cd and Fe from the mud. The behaviour of Cu suggested strong affinity for organic chelating agents.

The role of the carbon dioxide-bicarbonate-carbonate cycle in the dispersion and accumulation of trace metals in the lacustrine environment can be significant. Both iron and manganese and most trace metals, including Cu, Zn, Ni, Co and Pb are immobile under oxidizing alkaline (high pH) conditions and mobile under reducing acid (low pH) conditions (Baas Becking et al., 1960; Hawkes and Webb, 1962; Garrels and Christ, 1965; Krauskopf, 1967; Andrews-Jones, 1968). In the Sturgeon Lake area of northwestern Ontario the dispersion of Zn from a massive sulphide source into the centre of adjacent Lyon Lake was restricted when the relatively neutral (pH ∿ 7) waters of the inflowing stream encountered the relatively alkaline (pH ∿ 8) lake waters (Fig. 20.5) (Coker and Nichol, 1975). The effect of increasing alkalinity on restricting trace metal dispersion in the Agricola Lake area of the barren northern Shield was noted by Cameron and Ballantyne (1975) and Cameron (1977) (Fig. 20.6). The influence of bicarbonate ion on uranium solubility has long been recognized (Bowie et al., 1971) and was postulated as a potential influence effecting the lack of correspondence between anomalous levels of U in water and sediment from the same lake (Cameron and Hornbrook, 1976; Coker and Jonasson, 1977a,b; Maurice, 1977a,b). Maurice (1977a) stated that at low alkalinities the lake sediments are in most cases more useful than the lake waters in indicating uranium dispersion in the Baffin Island study area and the converse applies for high alkalinities. Jackson (1975) designed a laboratory study over pH ranges from 4 to 9 in order to simulate interactions in the system dissolved metal ion-clay-dissolved natural organic acids. He found the behaviour of Ni was strongly controlled by interference reactions with carbonate and hydroxide and also that Pb and Fe favoured organic complexing, Cu dispersion was by sorption to organic-rich sediments, and that his experimental scheme was inadequate for Zn.

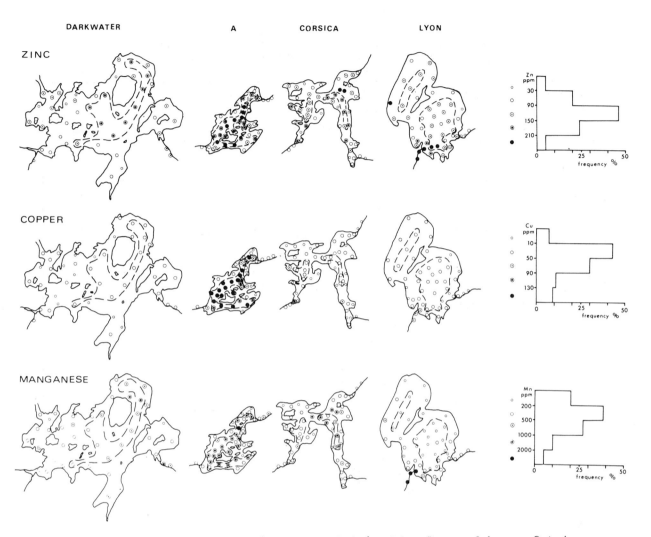

Figure 20.5. The distribution of Zn, Cu and Mn in four lakes, Sturgeon Lake area, Ontario (after Coker and Nichol, 1975).

Lake Sediment Physicochemistry

The physicochemical characteristics of lake bottom deposits play a major role in determining the distribution of trace metals within an individual lake. One of the most influential variables is the organic carbon content of the bottom deposits. Deep-water, generally centre-lake (profundal) sediments from lakes in the southern Canadian Shield generally have very homogeneous organic contents as shown in Figure 20.7 for lakes in northwestern Ontario (Coker and Nichol, 1975). Jonasson (1976) found this same feature in lakes examined in detail in the Grenville geological province of Ontario. A similar, although not so clear cut relationship, due to factors discussed previously was found in lakes of the northern Canadian Shield (Jackson and Nichol, 1975; Klassen et al., 1975). Most trace metals tend to be enriched in the organic sediments, a factor which is most probably due to the nature and strength of metal-organic binding and perhaps increased ion-exchange capacity of organic sediments over inorganic types. As a result, the highest and most uniform concentrations of trace metals generally occur in the deep central areas (profundal basins) of each lake where the sediments generally have the highest and most homogeneous organic contents (Coker and Nichol, 1975; Jonasson, 1976). This feature is illustrated diagrammatically (Figs. 20.5 and 20.8) and arithmetically (Tables 20.3 and 20.4), as indicated by the generally smaller coefficients of variation for trace metals in lake-centre sediments compared to lakeshore materials (Coker and Nichol, 1975; Jonasson, 1976). The presence of organic carbon in relatively highly variable quantities in shoreline (littoral) materials strongly influences the levels of trace metals found.

However, each lake is an entity as its sediments have individual characteristic amounts of organics which can vary considerably from lake to lake within a large survey area. This particular phenomenon has been examined by Garrett and Hornbrook (1976), for Zn, from some 3850 lake-centre sediment samples collected from northern Saskatchewan (Fig. 20.9). The Zn content of the centre-lake bottom sediments was found to rise linearly with increasing organic content, as measured by L.O.I. at 500°C, only at low levels (0 to 12% L.O.I.). At these low levels the Zn distribution patterns will therefore be partly controlled by the amount of organic material present and may not truly reflect the chemistry of the drainage basin. Where there is an excess adsorption capacity in the sediments (L.O.I. > 12%), the Zn distribution patterns should reflect the chemistry of the drainage basin. The observed decrease in Zn values with very high organic contents (> 50% L.O.I.) may be reflecting a dilution factor introduced by the ever increasing load of organic material to a system which is not receiving more Zn. Preliminary studies of the relationship of Zn and U with both Fe and Mn and of U with L.O.I. revealed a similar general

Table 20.3

Arithmetic mean, range and coefficient of variation of Zn, Cu and Mn in sediments from four lakes, Sturgeon Lake area, Ontario (from Coker and Nichol, 1975)

Lake Population		Lyon C	Lyon S	A C	A S	Corsica C	Corsica S	Darkwater C	Darkwater S
Zinc	$\bar{x}$	79	58	115	51	84	29	106	44
	R	53-435	12-92	68-220	17-111	65-110	16-41	90-137	21-120
	V	114	31	35	49	11	24	12	60
Copper	$\bar{x}$	14	11	85	21	26	8	23	10
	R	10-17	<4-18	56-121	<4-60	24-28	4-12	21-25	4-37
	V	12	38	20	77	7	27	6	91
Manganese	$\bar{x}$	94	151	712	283	492	216	410	326
	R	70-203	66-292	125-1465	115-877	190-1199	127-426	221-675	187-432
	V	38	40	45	64	74	35	38	26
	n	18	25	24	16	10	17	10	13

Population: C = Lake-centre sediments. (profundal)
S = Lake-shore sediments. (littoral)
$\bar{x}$ = Arithmetic mean (quoted in ppm).
R = Highest and lowest value (quoted in ppm).
V = Coefficient of variation ((Standard deviation/$\bar{x}$) × 100).
n = Number of samples.

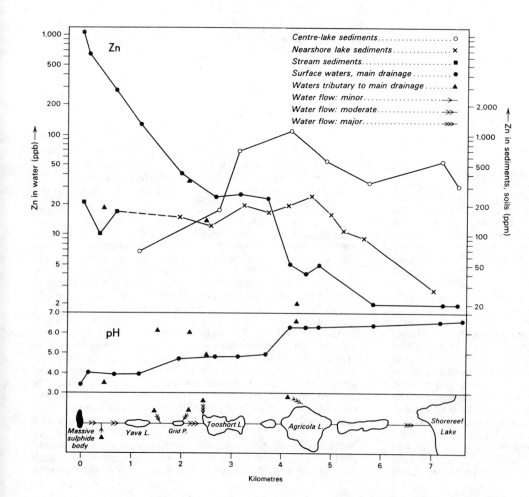

Figure 20.6.

Distribution of Zn in lake and stream sediments and waters, Agricola Lake system, District of Mackenzie (after Cameron, 1977).

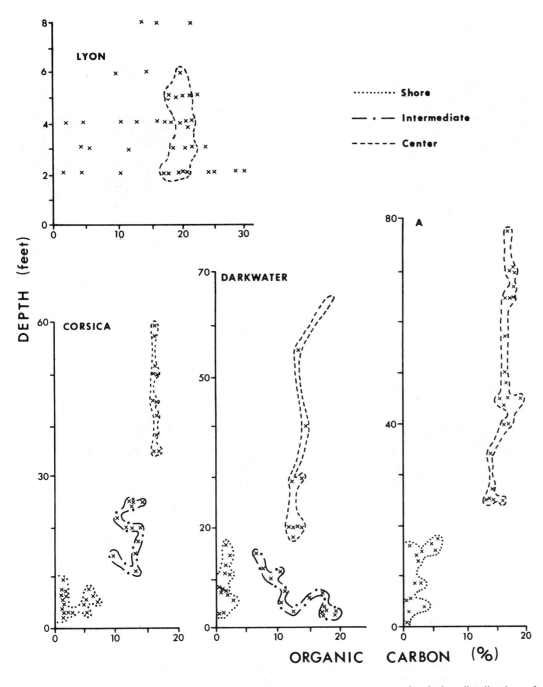

Figure 20.7.
Relation between organic carbon content of sediments and water depth within four lakes, Sturgeon Lake area, Ontario (from Coker and Nichol, 1975).

behaviour to that of Zn with L.O.I. for the lake-centre sediments samples from northern Saskatchewan (Garrett and Hornbrook, 1976).

The relationship between the Zn and organic contents of 275 reconnaissance lake-centre sediments from the Sturgeon Lake area in northwestern Ontario was similar to that found in sediments from northern Saskatchewan (Coker and Nichol, 1976). However, the use of ignited Zn data (analysis based on a fixed weight of ashed residue) was thought to constitute a more homogeneous data set, which is relatively independent of the amount of organic matter, over a larger range of data than for unignited Zn data (analysis of a fixed weight of dried sample).

Bjorklund et al. (1976) also used analysis of ignited organic-rich lake sediments to examine the distribution of U in lake sediments from Karelia, Finland. In addition they examined the distribution of U in material lost-on-ignition from organic-rich sediments (i.e. approximately the organic part) and also the U contents expressed as relative deviation from a polynomial function, approximated through the medians of uranium contents constructed from a graph of U in ignited sediment versus L.O.I. The significance of the various interpretational methods has not been thoroughly studied to date.

The relationship between the U and organic contents of sediments from several individual lakes in northern Saskatchewan was examined by Lehto et al. (1977). A roughly linear relationship exists for each lake and the slope of the U versus organic curve was unique for any given lake and was felt to be a function of the local geological-geochemical-limnological environment. Similar relationships for Cu and Zn versus organic matter, iron and manganese where shown for

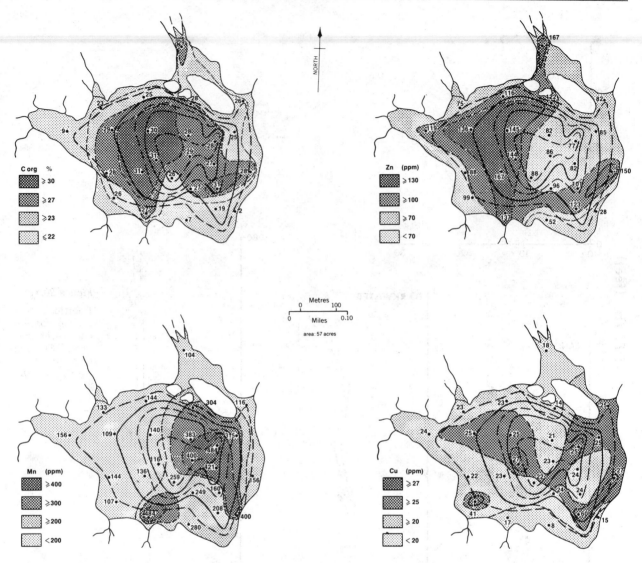

Figure 20.8. *Organic carbon, Zn, Mn and Cu distribution in organic sediments of Perch Lake, Ontario (after Jonasson, 1976).*

Table 20.4

Geochemical comparisons of silty shoreline (littoral) and organic deep water (profundal) sediment samples from Perch Lake, Ontario (after Jonasson, 1976).

Element	Shoreline Silts		Organic gels		Organic enrichment factor
	X_s	$C_v\%$	X_o	$C_v\%$	$\dfrac{X_o}{X_s} = R$
Zn (ppm)	44	52	106	33	2.41
Cu (ppm)	5.5	83	24	40	4.36
Mn (ppm)	103	60	236	54	2.29
C org (%)	8.4	132	24.7	31	2.94

X_s = arithmetic mean for shoreline silts. (littoral)
X_o = arithmetic mean for organic gels. (profundal)
$C_v\%$ = coefficient of variation.

Lake Sediment Geochemistry

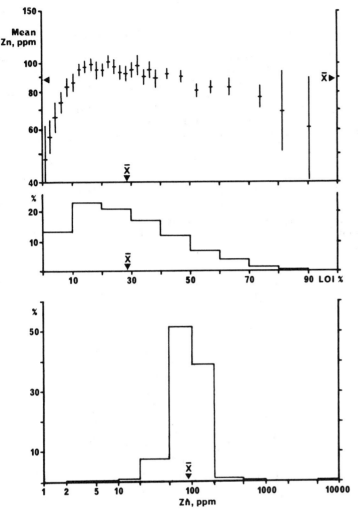

Figure 20.9. Graphic display of the relationship between Zn and L.O.I., and histograms for Zn and L.O.I. from 3844 centre-lake bottom sediments, east-central Saskatchewan (from Garrett and Hornbrook, 1976).

selected single lakes from the Kaminak Lake area, District of Keewatin, Yellowknife area, District of Mackenzie, and Sturgeon Lake area, Ontario (Nichol et al., 1975). Lehto et al. (1977) stated that by sampling organic-rich sediments from lake centres the chemical data obtained would tend to plot on the outer ends of the metal versus organic matter curve for each individual lake which, when examined together would give the curved distribution as found by Garrett and Hornbrook (1976) in Figure 20.9. They further suggest that since the trace metals appear to be associated with the organic fraction of lake sediments the metal values should be normalized for varied dilution by recalculating to 100% organic matter (essentially similar to calculating metal values on a loss-on-ignition basis).

Another method of determining lake sediment data, as suggested by Parslow (1977), involves employing a constant volume of sample rather than a constant weight.

In general, regardless of the method of analyzing the sediments (although use of ignited or constant volume data have not received much application to date), it appears that a homogeneous and ubiquitous sample medium which does not exhibit severe matrix problems can be found in organic-rich lake-centre sediments with L.O.I.'s generally ranging between a lower limit of 10 to 20% and an upper limit of 50 to 70%.

LAKE SEDIMENTS AND MINERAL EXPLORATION

The range of physicochemical-limnological conditions present within the lacustrine environment, as previously outlined, should emphasize the complexity of this regime. Variations in these conditions can effect the nature of metal transport, and accumulation in lake bottom materials, with respect to mineralization in different geographic — climatic and geological environments.

Within the last decade universities, government agencies, and the mineral exploration industry have carried out research studies and orientation, regional reconnaissance and follow-up lake sediment surveys in the search for various types of mineralization. The majority of these surveys have taken place within the Canadian Precambrian Shield and to a lesser extent within the Cordilleran and Appalachian regions of Canada. In addition, some lake sediment surveys have taken place in the Precambrian Shield, Cordilleran and Appalachian regions of the United States of America and the Precambrian Shield of Fennoscandia.

Sampling Equipment, Logistics and Cost

A wide variety of sampling devices have been used by geochemists to collect lake sediments. One type of lake sediment sampler currently in wide use in North America, by both government and industry, is the 1976 model Geological Survey of Canada sampler (Fig. 20.10), or some device similar to it. This sampler has successively evolved, with extensive modifications, from a sampler developed by the Geological Survey of Canada and the Newfoundland Department of Mines and Energy in the early 1970s (Hornbrook et al., 1975a; Hornbrook and Garrett, 1976). The sampler now features a nose design similar to a previously developed Finnish lake sediment sampler (Bjorklund et al., 1976). The nose section of the 1976 model sampler (Fig. 20.10) has a stainless-steel butterfly valve inset, which prevents loss of sample material by automatically closing, below the collected sample, upon retrieval of the sampler. This sampler was not designed to collect lake bottom materials with a large coarse clastic component or coarse organic debris component, as these types of sample compositions are not desired. It was designed to achieve rapid sampling of organic sediments from profundal lake basins.

Many other types of lake bottom samplers have been, and still are used. The bucket-type Ekman Birge-bottom dredge, which closes by means of a messenger sent down from surface, works well and collects large quantities of material in organic muds and gels, but is inefficient in coarse clastic materials or coarse organic debris (Closs, 1975; Coker and Nichol, 1975). The clam-type Petite Ponar grab sampler (Klassen et al., 1975), and the Kel Scientific Instruments mud snapper (Hoffman and Fletcher, 1976), both of which are triggered shut on impact, are efficient in clastic materials but not in organic muds or gels. These clam- and bucket-type samplers are generally not as amenable to rapid sample collection from the float of a helicopter as are the pipe-like devices similar to the 1976 model Geological Survey of Canada sampler. The Phleger corer has been used to collect lake sediment cores (Closs, 1975) as have other specially designed sediment corers (Coker, 1974; Meineke et al., 1976), all of which employ plastic core barrel liners. Nearshore lake bottom materials have been collected by hand or with extension posthole augers (Allan et al., 1973b) and with a home-made tube device with a one-way valve (Dyck, 1974). A combination of home-made extension augers, telephone-spoon shovel, and sample tubes, driven by a portable overburden drill, were used by Hornbrook and Gleeson (1972) to collect lake bottom materials through winter ice cover. Collection of lake sediment grab samples and core sections has also been carried out by scuba divers (Jonasson, 1976; R.A. Klassen, pers. comm.).

Table 20.5

A summary of lake sediment surveys carried out in the Canadian Cordillera, Shield and Appalachians.

Location Number and Location	Geology	Mineralization	Glacial Sediments	Density, Area, Site Location	Size Fraction, Elements	Survey Type	Reference Number
1 Central, British Columbia Nechako Plateau	volcanics and intrusives	Mo	ground moraine and fluvio-glacial deposits	>1/24 km² 16,000 km² centre-lake	-80 mesh Cu, Mo, Pb, Zn, Ni, Mn, L.O.I., Sr, Ba, Cr, Co, Ga, Ag, V, Ti, Bi, Zn, Sb	A-1	35
2 Central Interior British Columbia	volcanics and intrusives	Mo	ground moraine	1/10 km² centre-lake	Mo	B-1	50
3 South-central British Columbia	volcanics and intrusives	Cu	ground moraine and fluvio-glacial deposits	various nearshore	-80 mesh Cu, Fe, Mn, Zn, K	B-1	34
CANADIAN SHIELD, NORTHERN (north of latitude 60°00'N)							
4 Coppermine River Area NWT	Proterozoic basalts and sediments	Cu	ground moraine minor glacio-marine sediments	1/26 km² 3400 km² nearshore	-80 mesh Cu, Zn	A-2	2
5 Bear Province NWT	Archean and Proterozoic volcanics, sediments and granites	U	ground moraine	1/26 km² nearshore	-250 mesh U, Si, Al, Fe, Mg, Ca, Ti, Mn, Ba, Na, K, Zn, Cu, Pb, Ni, Co, Ag, As, Hg	A-1	12
6 High Lake area, N. Slave Province NWT	volcanics	Cu, Zn	ground moraine	1/26 km² 700 km² nearshore	-250 mesh Zn, Cu, Pb, Ni, Co, Ag, Si, Al, Fe, Mg, Ca, Ti, Mn, Ba	B-1	3
7 Hackett River, E. Slave Province NWT	volcanics	Zn, Pb, Ag, Cu	ground moraine	1/26 km² 700 km² nearshore	-250 mesh Zn, Cu, Pb, Ni, Co, Ag, Si, Al, Fe, Mg, Cu, Ti, Mn, Ba	B-1	3
8 Agricola Lake, E. Slave Province NWT	metavolcanics metasediments	Zn, Cu, Pb, Ag, Au, As, Cd, Hg	ground moraine	various nearshore centre-lake	-250 mesh Zn, Cu, Pb, Ni, Co, Ag, Fe, Mn, Hg, As	B-1	11
9 Bear Slave Provinces NWT	Archean and Proterozoic volcanics and sediments	Cu, Zn, Au, U	ground moraine	1/26 km² 93 000 km² nearshore	-250 mesh Zn, Ag, Mn, Li, U, As, Sb, Cu, Pb, Sn, V, Mo, Cr, Co, Ni, Be, La, Y, Zr, Sr, Ba, Ti, Ca, Mg, Fe, K	A-1	4
10 Yellowknife NWT	Archean metavolcanics, metasediments and granites	Cu, Pb, Zn, Ag, Au	ground moraine glaciolacustrine deposits	* centre-lake	-80 mesh Cu, Zn, Pb, As, Ag, Mn, L.O.I., Co, Ni, Fe	B-1	42
11 Yellowknife NWT	Archean volcanics, sediments	Cu, Pb, Zn, Ag, Au	ground moraine and glacio-lacustrine deposits	1/26 km² 2600 km² nearshore	-250 mesh Cu, Pb, Zn, Ag, Co, Ni, Fe, Mn	A-1	52
12 East Arm, Great Slave Lake, NWT	Archean volcanics, sediments and granites	Cu, Pb, Zn, Ag	ground moraine glaciolacustrine sediments	1/2.6 km² 260 km² centre-lake	-80 mesh Cu, Zn, Pb, As, Ag, Mn, L.O.I., Co, Ni, Fe	A-1	42
13 Nonacho Lake, NWT	sediments volcanics granites	U, Cu	ground moraine glaciolacustrine	1/1 km² 1400 km² centre-lake	-80 mesh U, Zn, Cu, Pb, Ni, Co, Ag, Mn, As, Mo, Fe, Hg, L.O.I.	B-2	46 48

Table 20.5 (cont'd)

Location Number and Location	Geology	Mineralization	Glacial Sediments	Density, Area, Site Location	Size Fraction, Elements	Survey Type	Reference Number
14 Kaminak Lake, NWT	Archean metavolcanics, metasediments and intrusives	Cu, Zn, Ni	ground moraine glaciomarine	* centre-lake	-80 mesh Cu, Zn, Pb, Ag, Fe, Mn, Co, Ni	B-1	44
15 Baffin Island, NWT	Proterozoic granite, migmatite and quartzfeldspar gneiss	U	ground moraine and glaciomarine permafrost	1/3 km^2 323 km^2 1/13 km^2 3800 km^2 centre-lake	-200 mesh U, Fe, Mn, Cu, Pb, Zn, Ni, Mo, Ag, V, Cr, Be, La, Y, Sr, Ba, Co, Ti	B-1	47
16 Cape Smith – Wakeham Bay, N. Quebec	serpentinized ultramafic sills	Ni, Cu	ground moraine permafrost	65 km^2 nearshore	-80 mesh Ni, Cu	B-1	1
CANADIAN SHIELD, SOUTHERN (south of latitude 60°00'N)							
17 Beaverlodge, Saskatchewan	metasediments granites metavolcanics and sediments	U, Cu, Co, Ni, Pb, Zn, V, Hg, Pt, Au	ground moraine glaciolacustrine sediments	1/0.6 km^2 77 km^2 nearshore	-60 mesh Ra, U, Zn, Cu, Pb, Ni, Fe, Mn, organic content	B-1	29
18 Rabbit Lake, N. Saskatchewan	Proterozoic Athabasca sandstone metasedimentary rocks	U	ground moraine	>1/13 km^2 1300 km^2 centre-lake	-80 mesh U, Zn, Cu, Pb, Ni, Co, Ag, Mn, Fe, Mo, L.O.I.	B-1	13
19 N. Saskatchewan	Athabasca sandstone	U	ground moraine glaciolacustrine sediments	1/6 km^2 6000 km^2 centre-lake	U, Cu, Pb, Zn, Ni, Co, Fe, Mn	B-1	53
20 Key Lake, N. Saskatchewan	Athabasca sandstone Archean metasediments	U, Ni	ground moraine	>1/1 km^2	U, Ni, Cu, Pb, Zn	B-1	53 55
21 Mudjatik Lake, Saskatchewan	mafic granulites, felsic gneiss	U	ground moraine	3600 km^2 nearshore centre-lake	-60 mesh Hg, Co, Ni, Cu, Pb, Zn, U, Mo	B-1	33
22 Stanley area, Saskatchewan	volcanics sediments intrusives	Cu, Au	ground moraine glaciolacustrine sediments	440 km^2 nearshore	Cu, Zn, Ni, Pb, Mo, U, Co, Fe, Mg, Mn, Na, Hg	B-1	32
23 Southeast Saskatchewan	Archean volcanics sediments and intrusives	Cu, Ni, Au, U	ground moraine glaciolacustrine sediments	>1/1 km^2	U, Cu, As, Zn	B-2	45
24 La Ronge, Saskatchewan	volcanics sediments granites	Cu, Zn, Au, Ag	ground moraine glaciolacustrine deposits	*	Cu, Zn, Ni, Fe, Mn, Co	B-1	7

*one or more lakes sampled in detail from a variety of site locations for various lake bottom materials.

Survey Type   A Regional 1 Reconnaissance survey
                          2 Detail survey
              B Research 1 Orientation survey
                          2 Follow-up survey

Location Number refers to site number on location map, Figure 20.11
Reference Number refers to source in the bibliography

Table 20.5 (cont'd)

Location Number and Location	Geology	Mineralization	Glacial Sediments	Density, Area, Site Location	Size Fraction, Elements	Survey Type	Reference Number
25 East-central Saskatchewan	Archean volcanics sediments intrusives	Cu, Ni, Au	ground moraine glaciolacustrine sediments	1/13 km² 52 000 km² centre-lake	-80 mesh Zn, Cu, Pb, Co, Ni, Ag, Fe, Mn, Hg, As, Mo, U, L.O.I.	A-1	39 40 41
26 N.W. Manitoba	Archean granites Aphedian sediments intrusives	U	ground moraine glacio-fluvial deposits	1/1 km² 900 km² centre-lake	-80 mesh Zn, Cu, Pb, Ni, Co, Ag, Mn, As, Mo, Fe, Hg, L.O.I., U, F	B-2	16
27 Fox Lake Manitoba	Archean volcanics sediments and intrusives	Cu, Zn	ground moraine glaciolacustrine sediments	*	Cu	B-1	9 14
28 Saskatchewan west of Flin Flon, Manitoba	volcanics sediments granites	Cu, Zn, Au, Ag	ground moraine glaciolacustrine	*	Cu, Zn, Ni, Fe, Mn, Co	B-1	7
29 Wintering Lake, Manitoba	Archean volcanics sediments intrusives	Cu, Ni	glaciolacustrine deposits	*	Cu, Ni	B-1	8
30 Red Lake-Uchi Lake, Ontario	volcanics sediments	Au	ground moraine glaciolacustrine deposits and outwash	*	Cu, Zn, Mn, Fe, H₂S, HS⁻, S²⁻, L.O.I.	B-1	56
31 Northwestern Ontario	Archean volcanics sediments and intrusives		ground moraine	*		B-1	10
32 Upper Manitou Lake, N.W. Ontario	volcanic sedimentary and intrusive rocks	Cu, Zn, Ag, Ni	ground moraine glaciolacustrine deposits	* centre-lake	-80 mesh Zn, Ni, Mn, Cu, Pb, Ag, Fe, Co, L.O.I.	B-1	17 18
33 Sturgeon Lake, N.W. Ontario	volcanic sedimentary and intrusive rocks	Cu, Zn, Ag, Ni	ground moraine glaciolacustrine deposits	* centre-lake	-80 mesh Zn, Ni, Mn, Cu, Pb, Ag, Fe, Co, L.O.I.	B-1	17 18
34 Shebandowan Lake, N.W. Ontario	volcanic, sedimentary and intrusive rocks	Cu, Zn, Ag, Ni	ground moraine glaciolacustrine deposits	* centre-lake	-80 mesh Zn, Ni, Mn, Cu, Pb, Ag, Fe, Co, L.O.I.	B-1	17 18
35 Manitouwadge Lake, N.W. Ontario	volcanic sedimentary and intrusive rocks	Cu, Zn, Ag, Ni	ground moraine glaciolacustrine deposits	* centre-lake	-80 mesh Zn, Ni, Mn, Cu, Pb, Ag, Fe, Co, L.O.I.	B-1	17 18
36 Elliot Lake, Ontario	Aphebian sediments	U	ground moraine glaciolacustrine deposits	* centre-lake	-80 mesh U, Sc, Y, Pb, Fe, Mn, As, Ag, Cu, Ni, Zn, Zr, L.O.I.	B-1	15
37 Sudbury, Ontario	ultramafic irruptive metasediments and volcanics	Ni, Cu, Pb, Zn	ground moraine glaciolacustrine deposits	* nearshore	-200 mesh Ni, Cu, Zn, Pb, Ag, Cd, Fe, Mn, As, Hg, Sb, Mo, L.O.I.	B-1	6
38 Timmins-Val d'Or area, Ontario and Quebec	volcanics sediments intrusives	Cu, Zn, Au	ground moraine glaciolacustrine sediments	1/8 km² 34 000 km² various	-230 mesh Cu, Pb, Zn, Ni, Mn, As, Ag, Mo	A-1	30 31 36 37

Table 20.5 (cont'd)

Location Number and Location	Geology	Mineralization	Glacial Sediments	Density, Area, Site Location	Size Fraction, Elements	Survey Type	Reference Number
39 Renfrew area, Ontario	Proterozoic granites gneisses and carbonate rocks	U, Mo, Pb, Zn	ground moraine	1/4.6 km² 1150 km² centre-lake	-80 mesh Cu, Zn, Fe, Mn, Co, Ni, Mo, U, L.O.I.	B-1	19 20
40 Lanark area, Ontario	Grenville terrain, marbles, etc.	Hg, Ag	ground moraine	* various	Zn, Pb, Cu, Fe, Mn, Hg, As, Ni, Mo, U, C, CO_3, S, L.O.I.	B-1	5 43
41 Chibougamou, Quebec	volcanics basic intrusives granites	Cu, Au, Ag	ground moraine glaciolacustrine sediments	*	-80 mesh cxHM	B-1	6 54
42 Kaipokok area, Labrador	Proterozoic volcanics and sediments	U, Cu	ground moraine	* nearshore	-80 mesh U, Cu, C	B-1	51
APPALACHIA							
43 Bathurst, New Brunswick	Paleozoic sediments and volcanics	Pb, Zn, Cu	ground moraine	*	-80 mesh cxHM	B-1	54
44 Daniel's Harbour, Newfoundland	Lower Paleozoic carbonates	Zn, Pb	ground moraine glaciomarine deposits	1/5.2 km² 900 km² centre-lake	-80 mesh Zn, Cu, Pb, Co, Ni, Ag, Mn, Fe, L.O.I.	B-1	38
45 Western Newfoundland	Lower Paleozoic carbonates	Zn, Pb	ground moraine glaciomarine deposits	1/3.1 km² 7800 km² centre-lake	-80 mesh Pb, Zn, Mn, Fe U, L.O.I.	A-1	21 24
46 Burlington Peninsula, Newfoundland	volcanics sediments granite gneiss	Cu, Au, Ag	ground moraine	1/2.7 km² 1300 km² centre-lake	-80 mesh Cu, Pb, Zn, Co, Ni, Ag, Mn, Fe, L.O.I., As, Mo, F, U	A-2	22 25 27
47 New Bay Pond area, Newfoundland	volcanics	Cu, Zn, As	ground moraine	1/2.3 km² 250 km² centre-lake	Cu, Pb, Zn, Co, Ni, Ag, Mn, Fe, Hg, L.O.I.	B-1	38
48 Southwestern Newfoundland	Carboniferous sediments		ground moraine glaciomarine deposits	1/5.3 km² 1800 km² centre-lake	Cu, Pb, Zn, Ba Sr, U, Mn, Fe L.O.I.	A-2	49
49 Burrin Peninsula, Newfoundland	volcanics sediments granites	F	ground moraine	1/3 km² 3900 km² centre-lake	U, F, L.O.I.	A-2	26
50 Avalon Peninsula, Newfoundland	sediments volcanics granites		ground moraine	1/3.8 km² 4800 km² centre-lake	Cu, Pb, Zn, Co, Ni, Mo, Ag, As, F, Mn, Fe, U, L.O.I.	A-2	23 25

* one or more lakes sampled in detail from a variety of site locations for various lake bottom materials.

Survey Type A Regional 1 Reconnaissance survey
                         2 Detail survey

             B Research 1 Orientation survey
                         2 Follow-up survey

Location Number refers to site number on location map, Figure 20.11
Reference Number refers to source in the bibliography

It is important, as Coker and Nichol (1975) and others have pointed out, that the top 5 cm, to as much as 10 cm, of the sediment at the defined sediment-water interface be discarded because it is subject to complex redox reactions as described by Mortimer (1942 and 1971). If the discard is not automatic as in the case of the 1976 model sampler then it should be done prior to bagging the sample.

Water sampling is often an integral part of a lake sediment survey. Water samples have been collected by Closs (1975) at various depths using, for example, a Hydro Products Van Dorn water sampler. Water sampling is amenable to automation in helicopter supported surveys. Cameron and Durham (1975) and Durham and Cameron (1975) have described a helicopter-mounted automated water-collecting system with pumps, tubes, reservoirs and instruments for measuring pH, conductivity and temperature with an onboard digital readout system. To avoid surface contaminants surface water samples should be collected below (i.e.: arms length at least) the water-air interface. Water samples collected below the surface epilimnion in the hypolimnion can be subject to widely varied physicochemical (i.e.: oxygen, temperature, pH, Eh) conditions from one lake to another as previously discussed.

Lake sediment sample-site densities have ranged from detailed sampling of all available bodies of water (i.e.: Coker and Jonasson, 1977a,b) some lakes more than once or in detail (i.e.: Jonasson, 1976), up to wide interval reconnaissance survey densities of one sample per 13 km^2 (i.e.: Hornbrook and Garrett, 1976) and as great as one sample per 90 km^2 as used by Barringer Research Limited in an area of northern Ontario – Manitoba in the late 1960s (Bradshaw, pers. comm.) (see Table 20.5).

The sample density employed should be a function of the type and objectives of the survey modified by a knowledge of the mobility and characteristic geochemical dispersion of the elements related to the intended target (see Table 20.5). Distributions of mobile elements are defined by wide-interval reconnaissance surveys and of relatively immobile elements by detailed surveys. The routine Geological Survey of Canada Uranium Reconnaissance Program (U.R.P.) sampling density of one site per 13 km^2 satisfactorily defines the regional distribution of a mobile element like uranium. The effectiveness of this density in the Great Bear Lake area, District of Mackenzie, was tested and confirmed for uranium distribution by an extensive and independent resampling program in 1975 (Hornbrook, 1977). However, in Newfoundland, where base metal targets had priority, reconnaissance surveys for the less mobile elements (Cu, Pb, Zn, Co, Ni) had to be carried out at a greater sample density (one site per 2.7 km^2 to one site per 5.3 km^2) to satisfactorily define element distribution (Davenport and Butler, 1975; McArthur et al., 1975). Most detailed or follow-up lake sediment surveys are carried out at densities equal to or greater than one site per 1 km^2.

Theoretically, in a given geochemical survey, the density chosen would be the ideal one to satisfactorily define the least mobile element desired. In practice, the design of most multi-element surveys incorporates some compromise where the density is better than required for highly mobile elements but frequently not entirely sufficient to satisfactorily define the least mobile element. This primarily occurs because of the increasing costs of detailed high-density sampling over a given area relative to wide-interval sampling coverage. If the sampling density used in a given area is not sufficient for the key elements, as should be determined by orientation studies, then such a survey is a wasted effort because the area cannot be confidently excluded from further exploration for these elements.

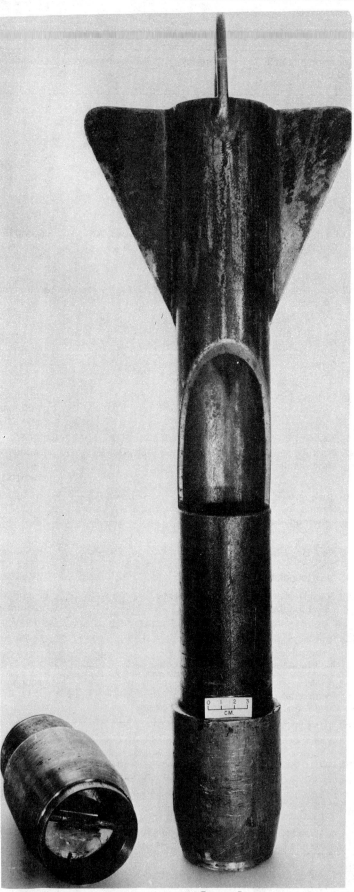

Figure 20.10. The 1976 model Geological Survey of Canada lake sediment sampler.

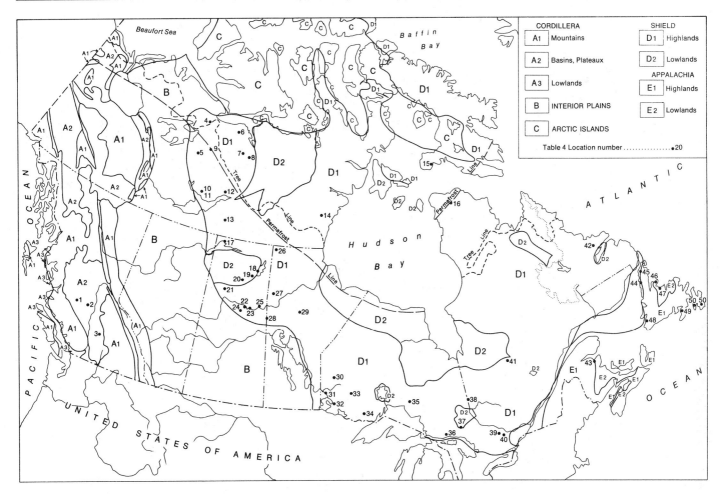

Figure 20.11. Lake Sediment Investigations in the Canadian Cordillera, Shield, and Appalachia (Numbers on map refer to Table 20.5).

Most lake sediment and/or water surveys are helicopter-supported usually by a Hughes 500C or a Bell Jet Ranger 206B type of turbine-powered rotary-wing aircraft because of their speed and manoeuverability. Frequently, as in the case of all U.R.P. geochemical surveys, a winching device and external working platform are installed on the helicopter fuselage and one of the floats respectively, to facilitate and increase the efficiency of sediment sampling operations.

Normally, in Geological Survey of Canada-U.R.P. operations, a helicopter-sampling team is composed of three people: the pilot; a navigator-notetaker who records data and also collects the water sample; and a person in the rear compartment who collects the sediment sample. It must be emphasized that the capability of the pilot plays a significant role in sampling operations particularly in terms of co-operation in traverse planning, navigation, site selection and overall flight path utilization to minimize flight time not productively used for sampling.

On reconnaissance surveys utilizing one helicopter and two crews, 500-1000 sites may be visited each week and samples collected at almost all of these. It is not always possible to collect a satisfactory sediment sample at every lake that is visited. Routine U.R.P. lake sediment and water surveys are carried out at a rate in excess of 15 sample sites per hour, while on traverse, and 12 sites per hour, calculated on a basis of total flying time. Total time includes nonproductive ferry flights and time spent at sites where no samples were collected. When the density is increased to one sample site per 6 km^2, the sampling rate increases to 18 or 20 sites per hour. Maurice (1976) achieved 21 sites per hour on Baffin Island at a follow-up study density of one site per 1.2 km^2.

On day-long helicopter traverses, to manage in an organized fashion several hundred collected samples, it is desirable to use trays or partitioned boxes to contain the bags and bottles. The trays also permit order to be maintained in the numbering sequences of bags and bottles. Although the bags are usually constructed of high wet-strength paper with water-resistant glue they will not withstand rough treatment. Plastic bags are sometimes used but require more handling for subsequent drying. Most traverse data are recorded on some form of field computer compatable cards by the navigator-notetaker. Cards of this nature have been described by Garrett (1974). Obviously, all operational procedures in the helicopter are designed to avoid errors and unnecessary time loss while carrying out satisfactory collection of sediment and water samples. Heated drying tents are required in the field for sorting and checking of sample numbers as well as drying sediment samples prior to shipping.

There is a marked absence of logistical and cost data and descriptions of equipment and procedures in most published accounts of lake sediment and water surveys. Such information on the 1974 Saskatchewan survey may be found in Hornbrook and Garrett (1976). Average costs can be estimated despite the great number of variable factors which contribute to the overall survey costs. Although the collection costs are usually the highest cost component;

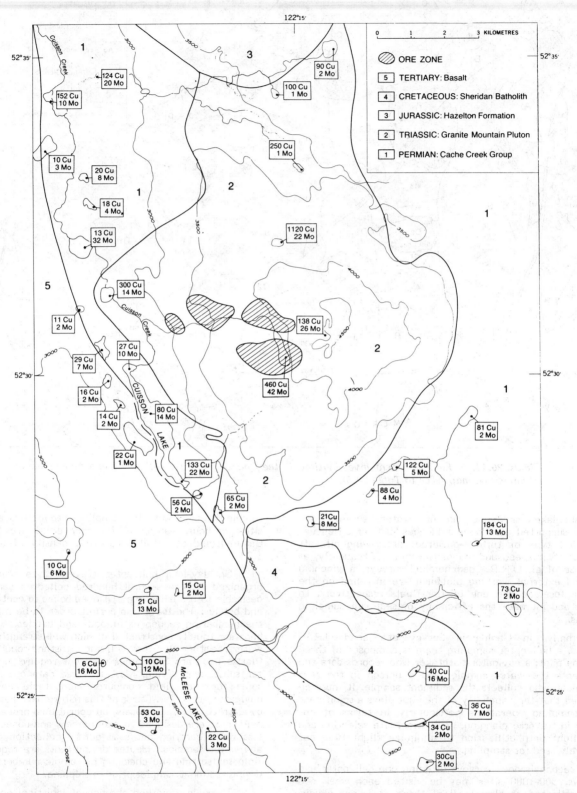

Figure 20.12. Distribution of Cu and Mo in lake sediments in the vicinity of the Gibraltar copper-molybdenum porphyry deposit, B.C. (Cu and Mo values in ppm). (Data supplied by Rio Tinto Canadian Exploration Co. Ltd.)

sample preparation, analyses, data compilation, map production and interpretation costs can be significant in multi-element surveys. Approximate 1977 collection costs for U.R.P. lake sediment and water surveys at a density of one sample per 13 km^2 range from $40.00 a site in southern developed areas to $50.00 a site in northern or remote areas. An average cost of collection would be $3.50 per km^2 for large scale reconnaissance geochemical surveys. Sample preparation and analytical costs will vary tremendously. In U.R.P. surveys, where lake sediments are disaggregated, ball milled and sieved prior to analysis for 12 or 13 elements and their organic content, and lake waters are analyzed for 2 elements and pH, these costs are significant. These costs together with data processing, base map compilation, data compilation and production of element maps and data listings may constitute an additional cost equal to up to 50 per cent of sample collection costs. Thus, it is obvious that there is no fixed quotable cost per sample site or km^2 unless all relevant cost factors are considered.

Lake Sediment Geochemistry Applied to Mineral Exploration in Canada

The association of elevated levels of trace metals in lake sediments adjacent to mineralization was observed as early as the mid 1950s in New Brunswick and Quebec by Schmidt (1956). However, for all practical purposes, the rapid increase in the utilization of lake sediments and waters as sample media for geochemical exploration did not occur until the early 1970s.

By the mid 1970s research and development was very active in the Shield and several lake sediment reconnaissance surveys had been carried out. The application of lake sediment geochemistry in exploration is now widespread in the Canadian Shield. Some users of the method are experiencing difficulties, but this may often be attributed to inappropriate application of the method and frequently an inability to properly interpret the complex data obtained.

The most extensive use of lake sediment geochemistry in the search for mineralization has been within Canada, primarily in the Shield but also in the Cordilleran and Appalachian regions, as indicated in Figure 20.11 and summarized in Table 20.5.

Cordillera

The use of lake sediments for geochemical reconnaissance in the Cordillera was initiated by M.B. Mehrtens and A.G. Troup of Rio Tinto Canadian Exploration Co. Ltd. In 1970, 8000 km^2 in the Quesnel area of south-central British Columbia were sampled using a float-equipped helicopter. This work demonstrated that lake sediment samples could be collected more economically than stream sediments in this heavily forested region. In total, Rio Tinto sampled 54 000 km^2 in British Columbia and the Yukon with the largest area covered being 30 000 km^2 between Quesnel and the United States–Canada border. Sampling of centre-lake sediments by means of an Ekman-Birge dredge was at an average density of one per 10 km^2.

The concentrations of copper and molybdenum in lake sediments collected by Rio Tinto, from lakes around the Gibraltar copper-molybdenum porphyry deposit (326 megatonnes of 0.37% Cu and 0.01% Mo) are shown in Figure 20.12. Sampling was post-discovery, but prior to the commencement of mining and any possible contamination of the associated drainage system. These data show strong anomalies for both Cu and Mo in close proximity to the deposit, with downstream dispersion of these metals being clearly evident.

In 1970, Hoffman and Fletcher (1972) collected a few nearshore lake-bottom samples as part of a multi-media geochemical investigation in south-central British Columbia. A Cu anomaly was found associated with a known mineralized zone in syenites. These authors subsequently carried out a survey of 16 000 km^2 in the Nechako Plateau, British Columbia. Organic-rich centre-lake sediments were taken from 500 lakes (Hoffman and Fletcher, 1976). The survey revealed regional geochemical variations in the concentrations of a number of elements in the lake sediments directly related to variations in underlying bedrock geology. For instance, lake sediments with high levels of Ni and Cr were found up to 10 km and 20 km respectively down-ice from ultramafic intrusives. Anomalies for Cu, Mo, Pb and Zn outlined mineralization, or lithologies favourable for mineralization. While considerable variation in the trace metal content of samples collected from single lakes was found, this variation did not obscure multi-lake anomalies related to known mineralization.

In general, the number of lakes per unit area is much less in the Cordillera than in the Canadian Shield. Within the Cordillera lakes are more abundant in the interior plateaus than in the moutain ranges (Fig. 20.11). It is primarily within the plateau regions that there are sufficient lakes to allow sampling at reconnaissance densities of one site per 10 to 20 km^2.

In Shield areas where low relief generally prevails, the widespread dispersion of indicator elements along drainages must largely depend on their movement in solution. It should be recognized that in parts of the Cordillera and in similar mountainous regions, mechanical transport may play a more significant role in metal movement.

Northern Shield

The northern Shield is here considered to be that portion of the Canadian Shield north of latitude 60°N. Most of this region is north of the treeline and is underlain by permafrost (Fig. 20.11).

The first reported study of the use of lakeshore materials for mineral exploration in the northern Shield was by Allan (1971). In 1970 he sampled lakeshore materials from lakes in a 4000 km^2 area, underlain mainly by basaltic rocks, near Coppermine on the Arctic coast. The Cu content of samples collected at a density of 1 site per 26 km^2 indicated the widely disseminated copper mineralization that occurs within the basalts. Allan and Hornbrook (1970) and Allan (1971) found chemical weathering active in this area of continuous permafrost. A good regional correlation exists between Cu in the nearshore sediments and Cu in the associated lake waters. This discovery provided a basis for further investigations of nearshore materials for regional geochemical reconnaissance, since for application at the reconnaissance-level wide dispersion of indicator elements in the surface environment is essential, and in typical Shield terrain of low relief such dispersion must be primarily hydromorphic.

A more extensive study was carried out in 1971 by Allan, Cameron and Durham, involving the sampling of rocks, lake waters and lakeshore materials from seven areas, including some mineralized and some barren, within the Bear and Slave geological provinces of the northwestern Shield. Results from this work (Allan et al., 1973a) formed the basis for the first large scale reconnaissance survey, the Bear-Slave operation, carried out in 1972 over an area of 93 000 km^2. Helicopter supported sampling of nearshore lake bottom materials at a site density of 1 per 26 km^2 was completed in six weeks (Allan et al., 1973b). Geochemical maps for the survey area, released in 1973 (Allan and Cameron, 1973), contained data for U, Zn, Pb, Cu, Ni, K, Fe, Mn and

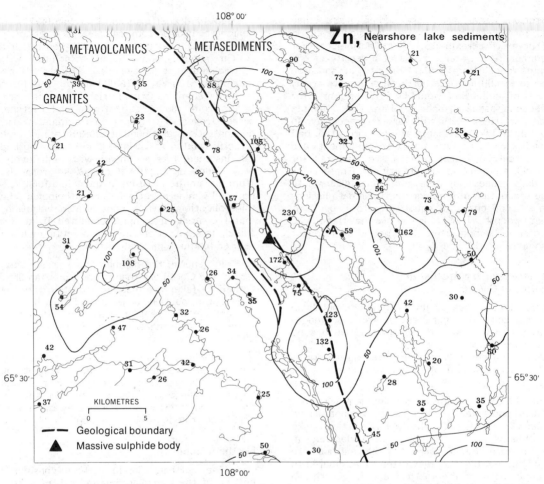

Figure 20.13. Distribution of Zn in nearshore lake bottom materials, Agricola Lake area, N.W.T. (after Cameron and Durham, 1974b).

organic content. Follow-up work was carried out by the Geological Survey of Canada in the Bear Province (Cameron and Allan, 1973) and the east part of the Slave Province (Cameron and Durham, 1974a,b) and throughout the survey area by the mining industry. The activities in the east part of the survey area resulted in the first mineral discovery in the Shield attributable to reconnaissance level nearshore lake bottom sampling — the Agricola Lake massive sulphide body.

While this work was being carried out by the Geological Survey of Canada, students of I. Nichol at Queens University were carrying out more detailed geochemical investigations of the lake environment in the Yellowknife and Kaminak Lake regions. The former area represents a transitional environment between the northern and southern Shield, since it is forested but within the zone of discontinuous permafrost. After an initial reconnaissance survey (Nickerson, 1972) nineteen mineralized and barren areas were studied in detail (Jackson and Nichol, 1975). They found centre-lake sediments to be more useful for geochemical reconnaissance than nearshore materials because they are more homogeneous within lakes and they better reflect the presence of mineralization. Jackson and Nichol (1975) found that Fe and Mn oxide precipitates in the sediments could be an important control on the distribution of many base metals under suitable physicochemical conditions.

The Kaminak Lake area is within the zone of continuous permafrost, north of the treeline. Work commenced in 1973 and is summarized by Klassen et al. (1975). For reasons that are not presently understood no relationship was found between lake sediment metal levels and known mineralization. Work in this region has been continued by Klassen, Shilts and co-workers at the Geological Survey of Canada emphasizing detailed physical and chemical studies of lake sediment and water regimes (Shilts et al., 1976).

As has been discussed, both nearshore materials and centre-lake sediments have been sampled for mineral exploration purposes in the northern Shield. In general, the former have **not** been produced by normal processes of lacustrine sedimentation. Instead, they are subaqueous equivalents of glacial and postglacial sediments on the margins of lakes and show similar patterned features, such as mudboils and rib and trough structures (Shilts and Dean, 1975). Like glacial and postglacial sediments, shoreline materials are heterogeneous mixtures of material from cobble to clay in size. The central, fine grained portions of both the surface (i.e. Shilts, 1971) and subaqueous (i.e. Allan, 1971) mudboils have been sampled for mineral exploration purposes.

The Canadian mining industry has made much use of various sample media from lakes for geochemical exploration in the northern Shield. The principal targets have been massive sulphides and uranium mineralization. Virtually all of this work is unpublished. In 1976, approximately $400 000 was spent by the industry on lake sediment and lake water surveys in the Northwest Territories (W.A. Padgham, pers. comm.). This comprised two thirds of their total geochemical expenditures in the Northwest Territories.

Figure 20.13 shows a strong anomaly for Zn in lakeshore materials in the eastern part of the Slave Geological Province (Cameron and Durham, 1974b). For the region, the geometric mean Zn content of these shore deposits is 32 ppm. These data were obtained in 1972 and 1973 and the Agricola Lake Zn-Cu-Pb-Ag-Au bearing massive sulphide, near the centre of the anomaly, was discovered in 1974. It occurs in steeply-dipping metavolcanic rocks near their contact with metasedimentary strata. The extensive nature of the anomaly is caused by widely scattered mineralization, presumably related to the massive sulphide body, and by widespread dispersion of mobile base metals in the present drainage system.

More detailed data for secondary dispersion from the Agricola Lake massive sulphide body are shown in Figure 20.6. In the proximal portion of the eastward-draining lake-stream system, the waters are acid and Zn migrates in solution (Cameron, 1977). Down-drainage the pH increases as the mineralized waters mix with those of other streams. This results in precipitation of Zn in both nearshore materials and centre-lake sediments. Down-drainage from Agricola Lake there is a rapid decline in the Zn content of the nearshore materials, roughly paralleling the decline of this element in the waters, but Zn in centre-lake sediments continues to be strongly anomalous. Cameron (1977) suggested that this difference is possibly caused by the nearshore material having sorbed Zn from the waters in situ while the centre-lake sediments contain fine grained particulates that have been transported down-drainage after having sorbed Zn. The evidence of more widespread dispersion in centre-lake sediments, together with their greater homogeneity compared to nearshore materials, indicates they are more suitable than the latter for regional reconnaissance sampling. In addition, with modern sampling equipment (see Fig. 20.10), centre-lake sediments can be sampled more rapidly than nearshore materials.

Until the late-nineteen sixties it was widely believed, in North America, that chemical weathering was minimal in permafrost environments and consequently that many geochemical approaches to mineral exploration would be unsuitable in these regions. Experience in the northern Shield, the Yukon, Alaska, Scandinavia and the U.S.S.R. has shown that this is not the case. Indeed, the mobility of many elements appears to be greater in the northern Shield than in the south, one possible reason perhaps being because of a lesser amount of fixing by organic material. The studies referenced earlier in this section indicate that U, F, Zn, Cd, Cu, Ni, Co and Mo are relatively mobile in the surface environment of the northern Shield and therefore should be given prime consideration as indicator elements for reconnaissance lake sediment surveys. Less mobile elements include Pb, Ag, Au, Hg and As. Once target areas have been defined by lake sediment reconnaissance, water geochemistry using mobile indicator elements can be of considerable assistance in more precisely defining targets (Cameron and Ballantyne, 1975).

Southern Shield

The southern Shield represents that portion of the Canadian Shield, generally south of latitude 60°N, within Alberta, Saskatchewan, Manitoba, Ontario, Labrador and Quebec (Fig. 20.11).

This region is characterized by a tremendous variation in environmental factors that influence weathering, transport and, eventually, precipitation and sedimentation of trace elements in lake sediments. Although most of the region is south of the limit of continuous permafrost more than one half lies within the zone of discontinuous permafrost and also, is south of the treeline (see Fig. 20.11).

In northern Saskatchewan lake sediment sampling began in the early 1970s and has been employed extensively in the last few years. Early work was conducted by the Saskatchewan Research Council (Arnold, 1970; Haughton et al., 1973), Saskatchewan Geological Survey, (Sibbald, 1977) and by a few exploration companies (Dunn, pers. comm.). In 1974, a 51 800 km^2 region of east-central Saskatchewan was covered by an organic lake-centre sediment-sampling reconnaissance program (Hornbrook et al., 1975b, 1977; Hornbrook and Garrett, 1976). The program was jointly undertaken by the Geological Survey of Canada and the Province of Saskatchewan.

Lake sediment chemical data from this program defined single and multi-element regional trends and local highs (i.e. for U and Cu-Zn-Pb) in the survey area that frequently coincided with known areas of mineralization. Several anomalous areas of unknown mineral potential were also defined. Correlation and regression studies showed that Fe, Mn and organic content do not appear to play important roles as scavengers of trace metals and therefore do not cause the occurrence of significant false anomalies. Statistical studies demonstrated that the surficial environment is not adversely affecting the raw data for interpretive purposes, and the relationship and interaction of the elements are primarily a reflection of bedrock, and geological and chemical processes. Extensive statistical treatment, beyond separation of the data on a bedrock catchment basin basis, was not required as it did not substantially improve interpretation of the data. This survey has been followed-up by Lehto et al. (1977) of the Saskatchewan Research Council and by the mineral exploration industry.

In 1975, the Saskatchewan Geological Survey undertook reconnaissance lake sediment studies along the western portion of the unconformity between Precambrian and Phanerozoic rocks. At the same time they started a four year study of lakes peripheral to the Athabasca Sandstone. Reconnaissance lake-sediment geochemistry was further studied in 1976 by the Saskatchewan Geological Survey (Rameakers and Dunn, 1977) and by the Saskatchewan Research Council (Lehto et al., 1977). The method was actively utilized by the Saskatchewan Mining and Development Corporation and many other exploration companies amounting to more than $250 000 worth of exploration in 1976 (Dunn, pers. comm.). In 1977, under the joint Federal-Provincial Uranium Reconnaissance Program (Darnley et al., 1975), a reconnaissance lake sediment survey was carried out over a 12 000 km^2 area covering the north-eastern extension of the Wollaston Trend, up to the Manitoba border. These data (Geological Survey of Canada, 1978c) were released in June, 1978.

Both the Key Lake – Highrock Lake and Rabbit Lake areas provide examples of the effectiveness of reconnaissance lake sediment sampling. Tan (1977) has described the geochemical exploration program conducted during 1973 and 1974 that was part of a much larger uranium exploration program, carried out by Uranerz Exploration and Mining Limited and partners, which led to the discovery of the Key Lake uranium-nickel deposits in 1975-76. It was also pointed out that organic-rich lake sediments were more effective as a sample medium than swamp or soil material, at both the reconnaissance and detail level, and in generating distinct anomalies related to the two orebodies.

The geology and geochemistry of the eastern margin of the Athabasca basin, including the Key Lake area, which unconformably overlies crystalline basement rocks, have been described by Ramaekers and Dunn (1977). Figure 20.14 shows as yet unpublished data form lake sediment samples collected in the vicinity of Key Lake in 1975 by C.E. Dunn, Geological Survey of Saskatchewan. Sample density was one per 7 km^2. Lake sediments from lakes 2-12 km down the glacial trend to the southwest of the Key Lake uranium-nickel deposits contain up to 2000 ppm uranium. The total dispersion halo extends down-ice and drainage at least 30 km at which distance 10 ppm uranium (2.5 times background) is commonplace. Smaller associated halos of nickel, cobalt, zinc and arsenic are also present but not shown in Figure 20.15. Uranium and nickel distributions in organic-rich lake centre sediments have clearly defined the location of the Key Lake deposits and give evidence of the potential of the method in similar environments.

The Rabbit Lake uranium deposit has given rise to markedly different geochemical patterns (Ramaekers and Dunn, 1977). Uranium is present in excess of 1000 ppm in the sediments of Rabbit Lake itself, and several hundred ppm in the neighbouring lakes to the east, but is only marginally anomalous 5 km from Rabbit Lake where its waters eventually drain into Wollaston Lake. In another lake sediment survey carried out by Cameron and Ballantyne (1977) in the Rabbit Lake area a large regional U anomaly was outlined (an area of 216 km^2 as enclosed by the 5 ppm contour in Fig. 20.15). The anomalous lakes trend down-ice from the deposit and appear to be an example of the glacial dispersion of uranium-bearing detritus, followed by leaching of the U from this detritus into the lake system. In addition, it is possible that satellite deposits of U could also be influencing the distribution patterns of U in the lake sediments. Lake waters within the area of anomalous sediments are also anomalous in U and F. In Saskatchewan, centre-lake bottom sediment sampling has been extensively and successfully employed over the last few years, more so than in other areas of the Canadian Shield.

Extensive glaciolacustrine sediments formed by Lake Agassiz have rendered much of the Province of Manitoba, except for the northern part, unsuitable for routine reconnaissance lake sediment surveys. The diverse influence of glaciolacustrine sediments in inhibiting geochemical response in the Abitibi Clay Belt of Ontario and Quebec has been described by Gleeson and Hornbrook (1975a). Lake sediment surveys have been used for mineral exploration in Manitoba but there are very few published accounts of such work other than that by Clews (1975) at Fox Lake and by Bradshaw (1975) at Wintering Lake.

In Manitoba, utilization of lake sediment geochemistry for mineral exploration by the mining industry received a major impetus as a result of the lake sediment surveys carried out under the joint Federal-Provincial Uranium Reconnaissance Program in 1975 and 1976.

In these reconnaissance surveys centre-lake organic-rich sediments from the profundal basins of suitable lakes, and surface lake waters, were collected at an average density of 1 sample per 13 km^2. The 1975 survey (Hornbrook et al., 1976a, b, c and d) and the 1976 survey (Geological Survey of Canada, 1977b) covered 87 300 km^2 of northern Manitoba north of latitude 58°N and west of longitude 95°W. The most interesting resultant dispersion patterns were those exhibited by uranium and associated elements in the northwest corner of the survey area (NTS 64N) along the extension of the Wollaston Trend. This particular area has undergone very active exploration and follow-up activities.

To evaluate the 1975 reconnaissance data, follow-up surveys were conducted in selected areas by the Geological Survey of Canada (Coker, 1976). Follow-up survey methods included: high density (1 site per km^2) lake sediment and water sampling; airborne gamma-ray spectrometry surveys by the Geological Survey Skyvan (1 km line spacing) and a helicopter-mounted McPhar Spectral four-channel spectrometer (0.25 km line spacing); and ground investigations where overburden and bedrock were sampled. An association was found to exist between uranium and

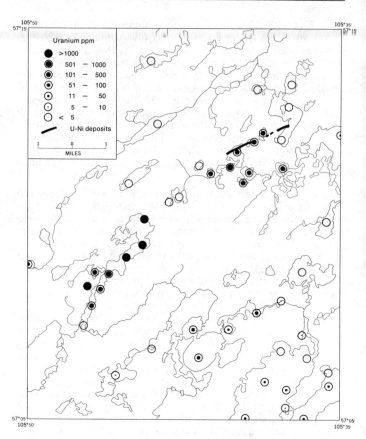

Figure 20.14. Distribution of U in lake sediments in the vicinity of the Key Lake U-Ni deposit, Saskatchewan. (Data supplied by C.E. Dunn, Geological Survey of Saskatchewan.)

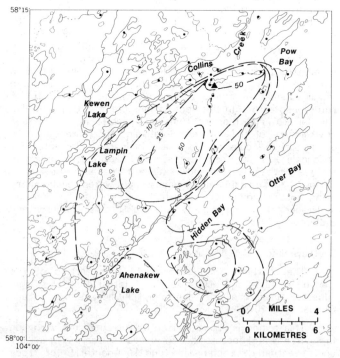

Figure 20.15. Uranium (ppm) in lake sediments near the Rabbit Lake uranium deposit, Saskatchewan (after Cameron and Ballantyne, 1977). Location of deposit shown by solid triangle.

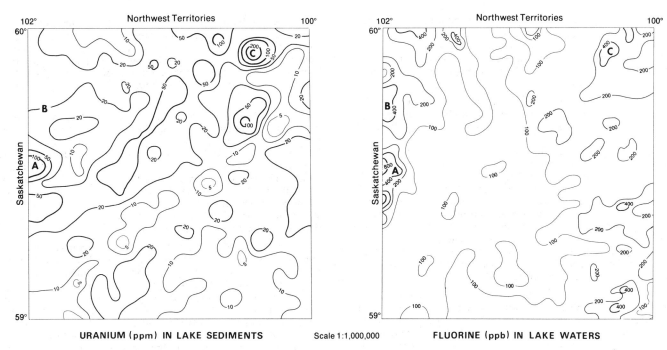

Figure 20.16. *Distribution of U in lake sediments and F in lake waters (sample density 1 site per 13 km^2) northwestern Manitoba (64N).*

fluorine in lakes within the Hudsonian granitoid bodies in the area (see A, B and C in Fig. 20.16). At a detailed sampling scale of one sample per km^2, uranium and fluorine results were found to define precisely the extent of the granitoid bodies, even indicating zoning within them. The association of other elements with uranium was felt to be useful for discriminating between high level regional uranium anomalies related to elevated uranium levels in bedrock (U association with F) and other relatively lower levels of uranium of possible economic significance (U alone or associated with Ni and As) within the geological environment of northwestern Manitoba.

Active follow-up studies by the mineral exploration industry has continued into 1977 with positive results. Detailed lake sediment and water sampling has been used in conjunction with other geological, geochemical and geophysical follow-up methods.

In recent years a great deal of lake sediment work has been carried out in Ontario. Among the earliest regional surveys was that of Hornbrook and Gleeson (1972, 1973) and Gleeson and Hornbrook (1975a,b) where 34 000 km^2 were sampled at a density of one sample per 8 km^2 in the Abitibi Clay Belt of Ontario and Quebec. This survey demonstrated the limitations of lake sediment geochemical response in a clay terrain. Jonasson (1976) began intensive research into the hydrogeochemistry of two small lakes in the Lanark area of southeastern Ontario in 1970 and eventually isolated and summarized many of the complex problems in the application and interpretation of lake sediment geochemistry. Contemporaneously, lake sediment and water investigations were being carried out in Ontario by Coker (1974), Coker and Nichol (1975, 1976), and by Timperley and Allan (1974). In northwestern Ontario, Coker investigated the nature of and factors affecting trace element accumulation in lake sediments in order to evaluate the feasibility of using lake sediments for reconnaissance mineral exploration. Investigations were carried out in the following areas: Sturgeon Lake, Upper Manitou Lake, Shebandowan Lake and Manitouwadge Lake. This work revealed that the deepest central region of a lake basin provides the most representative and homogeneous source of sample media.

Interpretation of the regional lake sediment chemical data demonstrated that the distribution of individual metals generally failed to identify adjacent mineralization due to the complex physicochemistry of the lake environment. However, element compositions, when ratioed with manganese (Zn/Mn, Ni/Mn), or when regression analysis was employed, could identify lakes adjacent to massive sulphide or Ni-Cu mineralization (Coker, 1974). Such data manipulation may often be required to remove the effects on trace element distributions in lake sediments caused by coprecipitation or varying pH (see previous discussion). The work of Timperley and Allan (1974) in the Red Lake – Uchi Lake area was concerned with investigating the use of gyttja as a prospecting medium and with providing some idea of the overall physicochemical system involved in the lake regime. Brunskill et al. (1971) investigated the relationship of the chemistry of the surface lake sediments to that of the overburden and bedrock in the Experimental Lakes Area in northwestern Ontario. Closs (1975) carried out an orientation survey to examine the geochemistry of lake sediments in the Elliot Lake region. His work showed that contamination from mining activity can be a problem and that correlation analysis and regression analysis are an approach to data evaluation that should not be overlooked in order to recover a maximum of information from the data. The work of Allan and Timperley (1975) was specifically concerned with centres of intensive mining activity (i.e. Sudbury) and the resultant contaminating effect from such activity on the chemistry of lake sediments. They found widespread evidence of heavy metal contamination in the upper 5 to 10 cm of the lake sediment column. In their conclusions they emphasized that the use of dredging devices should be avoided for sampling and that samples should be collected by coring-type devices, at least 10 cm below the lake sediment surface, to prevent the inclusion of contaminated sample media.

In the Grenville geological province west of Ottawa near Renfrew, Ontario an orientation survey was carried out to permit testing of geochemical methods to ascertain their responses to typical Grenville geological and environmental influences. This survey, by Coker and Jonasson (1977a,b) provided the basis for the 22 300 km² reconnaissance centre-lake bottom sediment and surface lake water survey completed in 1976. The reconnaissance survey was a joint undertaking with the Ontario Government under the Federal-Provincial Uranium Reconnaissance Program (Geological Survey of Canada, 1977a).

In the orientation survey, 246 lake sediment and 276 lake water samples were collected from all bodies of water in the area including lakes of all sizes, ponds, beaver dammed ponds, true swamps and flooded marshes. Data from the 1150 km² area (NTS 31F 07) are presented in Figure 20.17. This detailed level of sampling proved efficient in outlining favourable geology and perhaps certain structures with possible mineral potential. Exact correspondence of sediment and water uranium anomalies is not achieved. However, examined together, they reinforce each other and do outline geological features; for example, the Hurd Lake granite, where there is an annulus of elevated uranium values in both water and sediments. Field inspection with scintillometers confirmed the presence of radioactive mineralization in pegmatites and skarns peripheral to the Hurd Lake granite.

A definite value was found in interpreting the hydrogeochemical dispersion patterns in terms of elemental association (a simplistic cluster analysis, grouping values greater than the mean plus one standard deviation) based on a knowledge of the trace and minor-element chemistry of known mineral assemblages in the area. The same scale of sampling also seemed to be of value in seeking lead-zinc prospects in Grenville marble and skarns and also for locating new molybdenum occurrences in metamorphosed sediments. The broad extent of the anomalies outlined indicates that reconnaissance scale lake sediment geochemical sampling at 1 site per 13 km², using lakes, the larger ponds and true swamps, would be successful in locating zones of interest for detailed follow-up surveys.

Other than the work of Schmidt (1956), Hornbrook and Gleeson (1972, 1973), Gleeson and Hornbrook (1975a) and Allan and Timperley (1975) there is very little published on lake sediment geochemistry in Quebec. The early investigations of Schmidt (1956) in the Chibougamau area, Quebec revealed that the anomalous distribution of metals in lake sediments was related to adjacent mineralization. Hornbrook and Gleeson's work has been previously described. Allan and Timperley's (1975) Chibougamau area studies were concerned primarily with industrial heavy metal contamination and their approach and conclusions are similar to that described for their Sudbury, Ontario investigations. However, lake sediment sampling has and is being, used by the exploration industry in Quebec but their data remain unpublished to date. For example, the James Bay Development Corporation used centre-lake bottom sediments to explore over 100 000 km² during 1973-75 and this work is continuing. Minatidis and Slatt (1976) have demonstrated the usefulness of nearshore sediments as a follow-up technique after reconnaissance surveying. Their work in the Kaipokok region of Labrador defined U and Cu enriched zones, some of which were known and others which represent new discoveries. Thus, systematic nearshore sampling of materials around the periphery of several lakes in a mineralized area may provide a rapid method of delineating local mineral-rich areas.

Appalachia

Schmidt (1956) carried out geochemical investigations near Bathurst, New Brunswick where, similar to his work at Chibougamau, Quebec, the presence of anomalous metal contents in lake sediments related to adjacent mineralization, was observed. In 1972, on insular Newfoundland, lake sediment orientation studies were carried out in the New Bay Pond area by Hornbrook and in the Daniel's Harbour area by Hornbrook and Davenport (Hornbrook et al., 1975a). In 1973, following the orientation studies, a reconnaissance centre-lake sediment survey was carried out over 7800 km² on Lower Paleozoic carbonate rocks in western Newfoundland, including the Daniel's Harbour area (Davenport et al., 1974, 1975).

The Daniel's Harbour orientation study successfully determined optimum procedures for reconnaissance geochemical exploration for zinc mineralization in the St. George and Table Head groups of carbonate rocks. Figure 20.18 shows that the distribution of zinc in organic centre-lake bottom sediments, collected over 900 km² at an average density of one sample per 5.2 km², has delineated the Daniel's Harbour locality containing known zinc deposits. This study was carried out prior to the development of the zinc deposits by Newfoundland Zinc Mines Ltd.

The frequency distribution of the zinc data is lognormal and the contour intervals in Figure 20.18 were arbitrarily chosen at 0.5, 1.5, 2.5, 3.5 standard deviations above the mean (150, 400, 1000, 2700 ppm Zn respectively). The major zinc deposits and related showings are revealed by a multi-station anomaly where the lake sediments contain zinc concentrations ranging from 6250 to 14 500 ppm. Only one of the sampled lakes may have been contaminated by trenching on an adjacent showing. Other weaker zinc anomalies are present along the eastern margin of the study area where zinc showings are known to occur. Zinc content of samples is weakly correlated with the iron and organic content. It is not correlated with the manganese content. However, similar zinc distribution patterns can be produced by plotting either untreated zinc data or residual zinc data after regression with iron and/or organic content.

In the 1973 reconnaissance area, known zinc and lead mineralization typically occurs as clusters of pods or veins which may occupy an area of several square kilometres. In the Daniel's Harbour area, a sufficient number of zinc sulphide bodies fortuitously suboutcrop and supply zinc in detectable amounts to adjacent lakes, but in other more remote areas of the reconnaissance survey this may not be the case. To maximize the possibility of detecting anomalies due to mineralization, where only a small portion of a deposit intersects the bedrock surface, the reconnaissance survey sample density was increased to an average of one sample per 2.6 km².

The distribution of zinc in the reconnaissance survey detects the zinc deposits and related showings of Newfoundland Zinc Mines Ltd. and duplicates the results of the orientation study data from the previous year. Elsewhere in the total reconnaissance survey area numerous other zinc anomalies were found and are described in Davenport et al. (1974).

From 1974 to 1976 the Newfoundland Department of Mines and Energy have carried out four more lake sediment surveys totalling 11 800 km². These are: Burlington Peninsula, 1300 km², Davenport and Butler (1975, 1976) and Davenport et al. (1976b); southwestern Newfoundland, 1800 km², McArthur et al. (1975); Burrin Peninsula, Davenport et al. (1976a); Avalon Peninsula, 4800 km², Davenport and Butler (1976), Davenport et al. (1975, 1976). These surveys were directed toward base metal exploration and were therefore carried out at sample densities ranging from one sample per 2.7 km² up to 5.3 km². The centre-lake bottom sediment and surface lake water surveys, begun in 1977, are to cover 24 000 km² of insular Newfoundland and 144 000 km² of Labrador and are to be completed in 1978. Coverage in 1977 was 73 000 km² in Labrador and 17 400 km² on insular Newfoundland. The 1977-78 surveys are carried out under the Federal-Provincial Uranium Reconnaissance

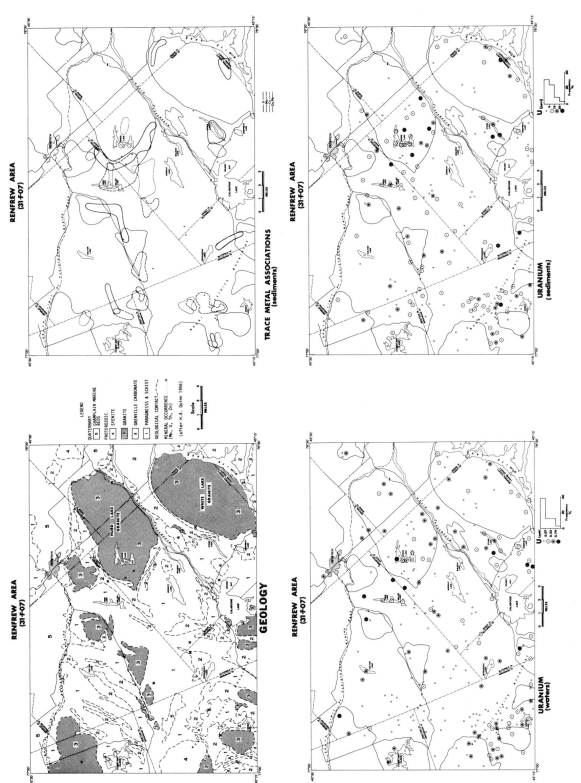

Figure 20.17. General geology, distribution of U in waters and in sediments and the association of trace metals in lake sediments, Renfrew area (31F 07), Ontario (from Coker and Jonasson, 1977a).

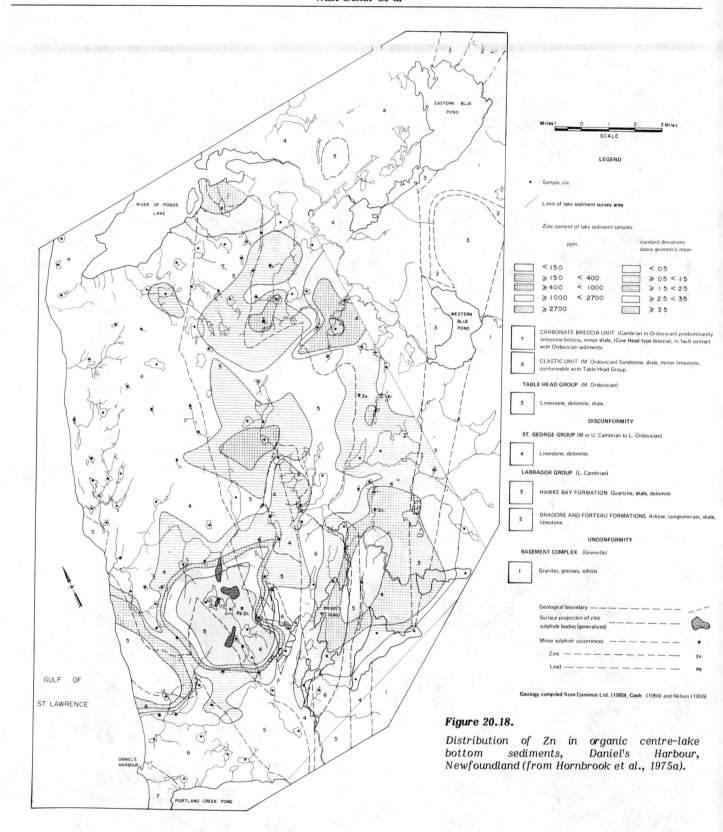

Figure 20.18.

Distribution of Zn in organic centre-lake bottom sediments, Daniel's Harbour, Newfoundland (from Hornbrook et al., 1975a).

Program. Follow-up surveys will be carried out to assess the data by the Geological Survey of Canada and the Newfoundland Department of Mines and Energy.

National Geochemical Reconnaissance

It is only relatively recently that geochemical surveys that are national in scope, and complimentary to national geological and geophysical surveys, have been undertaken. In Canada the first National Geochemical Reconnaissance (N.G.R.) surveys commenced in 1975, the immediate stimulus being the Federal-Provincial Uranium Reconnaissance Program (U.R.P.). In the Shield, and in similar terrain, such as insular Newfoundland, the primary sampling media are lake sediments and waters. The lake sediments are analyzed for U, Zn, Cu, Pb, Ni, Co, Mo, Ag, As, Hg, Mn, Fe and loss-on-ignition (L.O.I.). Waters are measured for U, F and pH. Areas sampled under the auspices of the N.G.R.-U.R.P. programs for lake sediments and waters, to the end of 1977 are shown on Figure 20.19 (Hornbrook et al., 1976a, b, c, d, e, f, g, h, i; Geological Survey of Canada 1977a, b, c, d and 1978a, b, c, d, e).

The sampling density for this reconnaissance work is an important consideration as it must provide an adequate level of information, but at the same time have wide enough spacing that costs and speed of coverage are reasonable. The chosen compromise was one sample per 13 km^2. It is not the aim of the program to identify individual ore deposits, but rather to delineate regional trends where mineralization is likely to occur. Despite this, the sample spacing has proven sufficient to outline anomalous lake sediments associated with a number of uranium and base metals deposits.

In addition to the short-term objectives of the Uranium Reconnaissance Program, the National Geochemical Reconnaissance data will provide a long-term data base for a variety of geoscientific and environmental purposes. In order that the data be consistent from year to year, standardized sampling and analytical techniques are used and quality control of the data is emphasized. Thus each batch of 20 analyses contains the following:

16 Routine Reconnaissance samples

1 Cell Duplicate Sample. This is collected from the same 13 km^2 grid cell as one of the reconnaissance samples, but form a different lake. The two samples allow the measurement of within-cell sampling variance.

1 Lake Duplicate Sample. Collected from the same lake as the cell duplicate sample. Allows estimation of within-lake sampling variance.

1 Analytical Duplicate Sample. A split from one of the routine reconnaissance samples. Allows estimation of analytical variance.

1 Control Reference Samples. A standard sample inserted to measure the analytical accuracy of the batch.

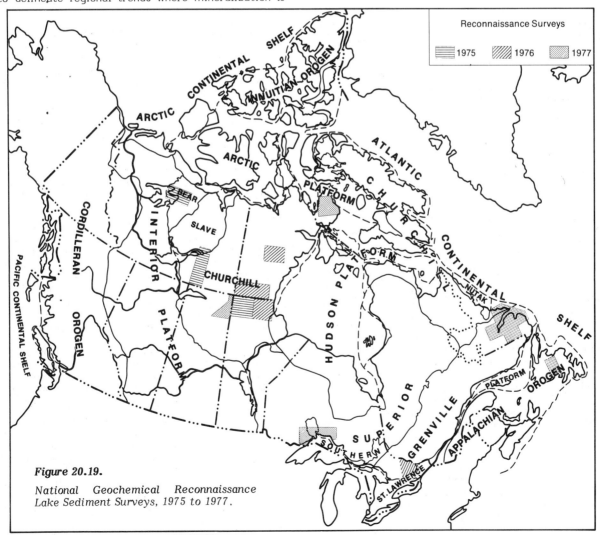

Figure 20.19.

National Geochemical Reconnaissance Lake Sediment Surveys, 1975 to 1977.

In 1977, N.G.R.-U.R.P. surveys were carried out over 230 000 km^2 of Canada and of this 142 00 km^2 were by lake sediment and water surveys. Sampling is carried out during the summer months and data released by the Geological Survey of Canada, and appropriate provincial or territorial governments, during the spring or early summer of the following year.

Lake Sediment Geochemistry Applied to Mineral Exploration in the United States of America and Fennoscandia

There is very little published information on the use of lake sediment geochemistry in mineral exploration outside that in Canada.

United States of America

About the only published account of the application of lake-sediment geochemistry for mineral exploration in the United States of America is the work of Meineke et al. (1976) carried out in 1974-75 over a portion of the Precambrian Shield in Minnesota. The survey involved the collection of some 275 lake sediment samples from 75 lakes over an area of 520 km^2 in the eastern Lake Vermilion-Ely area, St. Louis and Lake counties, Minnesota. Several significant anomalies were located by the survey. Anomalous Cu was found in a lake near an interesting copper prospect. Copper, Pb, Ti and Zn appear to reflect bedrock composition; Cr, Mg, and Ni reflect both bedrock composition and glacial dispersion.

In addition, the United States Energy Research and Development Administration (ERDA) commenced reconnaissance lake sediment surveys in 1976 in Alaska in support of the National Uranium Resource Evaluation Program (NURE) (Sharp and Aamodt, 1976). By the end of the 1979 field season most of Alaska will have been sampled. Both centre-lake sediments and waters are being collected at a density of one site per 23 km^2.

The mineral exploration industry has carried out reconnaissance lake sediment surveys in the Cordillera, Precambrian Shield and Appalachia of the United States of America, although to date there are no published accounts of these surveys.

Fennoscandia

There are portions of the Fennoscandian Precambrian Shield very similar in nature to the Canadian Precambrian Shield and consequently amenable to the application of lake sediment geochemistry for mineral exploration.

The Geological Survey of Finland has been particularly active in studying the application of lake sediment geochemical methods (Bjorklund et al., 1976; Bjorklund and Tenhola, 1976; and Tenhola, 1976). Organic-rich sediments were collected by the G.S.F. over a 6000 km^2 area in Karelia, eastern Finland. In addition to several small anomalies, the investigations indicated an extended zone of anomalously high uranium in lake sediments along the contact between Karelian schists and Pre-Karelian rocks. It was also found that central lake sediment samples may sometimes fail to indicate small isolated mineral occurrences in adjacent bedrock. A comparison between lake and stream sediments indicated that stream sediments may be used to complement lake sediment grids in areas of low lake density in the area studied.

In addition to the active use of the lake sediment method in Finland, similar work is being carried out on a modest scale in Norway and Sweden. In the northwestern U.S.S.R. helicopter supported lake sediment surveys are reportedly being carried out (L.K. Kauranne, pers. comm.).

SUMMARY

The application of lake sediment and water geochemistry to mineral exploration did not begin until the late 1960s and early 1970s. Since that time there has been a rapid increase in the application of the technique, particularly within the Canadian Precambrian Shield, but also within the Cordilleran and Appalachian regions of North America and the Precambrian Shield of Fennoscandia.

The success of this exploration technique may be attributed to the demonstrated ability of lake sediments to reflect the presence of nearby mineralization. In addition, centre-lake sediments are usually homogeneous and may be sampled relatively easily and economically.

It is evident that while the technique is viable for reconnaissance-level mineral exploration in several regions, there is clearly still much work to be done to understand the processes operative on trace metals within the lacustrine environment itself, and within lakes located in different physiographic, climatic-geographic, and geological environments. A knowledge of the processes by which a metal is mobilized, transported, precipitated, and possibly remobilized, is of prime concern in order to comprehend possible controls on that metal's dispersion, accumulation and fixation into lake bottom materials.

In lakes having aquatic flora and fauna as a significant source of organic matter, or in which waters are highly productive and sedimentation is rapid, the influence of organisms on trace element distribution may be significant. In the flat-lying, tree-covered terrain characteristic of the southern Canadian Shield and Fennoscandian Shield, and in the terrain of the North American Appalachia, the incidence of organic matter is high and metal-organic interactions are predominant. The presence of organic matter can enhance trace element mobility, by forming mobile-soluble organic complexes or retard it, by direct precipitation of insoluble organic complexes or sulphides. The occurrence of abundant swamps or marshes around or in close proximity to a lake may restrict trace element movement into the lake itself. By contrast, lakes from Shield areas above the treeline and from the alpine Cordilleran regions are fed by waters derived mainly from snowmelt and containing very little dissolved organic material. Here, absorption of metals directly onto clays, rock flour, and hydrous metal oxides and dissolution of mineral particles are the predominant water-sediment interactions.

Under oxidizing conditions, hydrous oxides of iron and manganese are excellent scavengers of trace elements; however, under reducing conditions they are solubilized and may result in increases in concentrations of cations and anions in overlying waters. The scavenging effect of both iron and manganese hydroxide precipitates on trace metals in the lacustrine environment and the resulting false anomalies in lake sediments has been noted. Iron and manganese oxides are certainly important species in organic systems but their role as direct absorbers of metal ions is generally overshadowed by competition from the more reactive humic materials and organo-clays or obscured by coatings of organic matter. Moreover, these oxides are unstable in reducing organic-rich sediments.

In general, the stratigraphy of lake-bottom materials from northern and southern Canadian Shield lakes is very similar. There is an upper strata of gel-like sediment, containing a variable quantity of organic material, which has formed since the recession of the glaciers. This modern organic sediment is thickest and contains the greatest amount of organic material in lakes of the southern Shield and is relatively thin, areally restricted, and sometimes absent in lakes of the northern Shield. It is invariably present in lake centres in the southern Shield and Appalachians and occurs in the centres of most lakes in the northern Shield and Cordillera. In contrast to centre-lake sites, surveys collecting material from the mineral sediment found around the margins of lakes are not necessarily collecting modern organic lake sediments, but are most often collecting glacial, glacial-lacustrine or marine sediments, or soils which have been subjected to some reworking, including wave action and, in addition, in the north of the Shield, to various periglacial processes.

Most trace metals tend to be enriched in the modern organic sediments, a factor which is most probably due to the nature of the metal-organic binding strength and perhaps increased ion-exchange capacity of organic sediments over inorganic types. As a result, the highest and most uniform concentrations of trace metals generally occur in the modern-organic sediment found in the deep central areas (profundal basins) of most lakes. In general, it appears that a profundal sediment is a homogeneous sample medium which does not exhibit severe matrix problems. It has also been shown that in certain situations (i.e. insufficient density of suitability sized lakes, relatively locally derived till) systematic nearshore sampling of materials around the periphery of several lakes in a mineralized area may provide a rapid method of delineating local mineral-rich areas.

The range of physicochemical-limnological conditions present within the lacustrine environment should emphasize the complexity of this regime. Variations in these conditions in different geographic — climatic and geological environments, with respect to mineralization, can affect the nature of metal transport, and accumulation in lake bottom materials.

Water sampling is often an integral part of a lake sediment survey as a knowledge of the distribution of many elements in the sediments often needs to be supplemented by information on their distribution in the overlying waters. This additional information can often provide some insight into the effects of variations in certain physicochemical factors (pH, Eh, alkalinity, Mn, Fe and organics, etc.) which might inhibit or prolong the dispersion of a given trace element in solution in the lacustrine environment. Therefore, an improved data base for interpretation of lake sediment data can often be obtained by the collection and analyses of both lake waters and sediments. Surface lake waters can generally be collected at a much faster, and hence cheaper, rate than lake sediments. However, the analytical methods currently available enable only selected elements (i.e. U, Zn, Cu etc.), generally present at very low levels (ppb or less), to be determined with relatively low precision.

The sample density employed should be a function of the type and objectives of the survey modified by a knowledge of the mobility of the different elements in the surface environment and their distribution in rocks around the intended target. If mobile elements are used as indicators, wide interval reconnaissance surveys (1 sample per 5 to 20 km^2) are often adequate. But use of immobile elements as the principal indicators will require more detailed sampling even for reconnaissance.

At the reconnaissance level of sampling, gross bedrock differences can be discerned and regional trends are clearly outlined frequently depicting appropriate element associations. As a follow-up to reconnaissance-level sampling, detailed sampling at 1 sample per km^2 or several per individual lake, at inflows and around the margins where groundwater is thought to play a role, have proved effective in outlining potential economic mineralization.

---

*Numerals refer to Table 20.5.

Sophisticated computer processing of lake water and sediment data is not always necessary although suitable computer programs for data compilation, sorting and simple statistical determinations can often simplify and clarify the data for interpretation. In many instances the interelement relationships have been found to be of more interpretative value than the absolute magnitude of a single element in a given sediment or water sample. However, in the more complex limnological environments it may be necessary to adopt involved interpretational techniques such as metal ratios or regression analyses to screen out non-significant features of the data and focus attention on components of the data related to mineralization.

ACKNOWLEDGMENTS

A great many people from both the federal and provincial governments and from the mining community have contributed to this review paper, far too many to individually acknowledge. However, there are some who do require special mention. Dr. C.E. Dunn, Saskatchewan Geological Survey, and A.C. Troup and C.D. Spence, Rio Tinto Canadian Exploration Co. Ltd., who respectively supplied data from lake sediment surveys carried out around the Key Lake deposit, Saskatchewan and the Gibraltar deposit, British Columbia. Dr. P. Davenport, Newfoundland Dept. of Mines, and Dr. L.G. Closs, Ontario Geological Survey, who supplied figures, information and logistical data for the paper. Dr. L.K. Kauranne, Geological Survey of Finland, provided information on the application of lake sediment geochemistry to mineral exploration in Fennoscandia and the USSR. The paper has greatly benefited from the, as yet, unpublished manuscript. "The nature of metals – sediment – water interactions in natural waterbodies, with emphasis on the role of organic matter", by Dr. K.S. Jackson and Dr. G.B. Skippen of Carleton University, and Dr. I.R. Jonasson of the Geological Survey of Canada. Dr. I.R. Jonasson and R.A. Klassen, Geological Survey of Canada, kindly provided critical comment on the first sections of the manuscript. Preparation of figures was carried out by D. Kurfurst.

REFERENCES

Allan, R.J.
1971: Lake sediment: a medium for regional geochemical exploration of the Canadian Shield; Can. Min. Metall. Bull., v. 64 (715), p. 45-99.

*1. 1973: Surficial dispersion of trace metals in Arctic Canada: A nickel deposit, Raglan area, Cape Smith – Wakeham Bay belt, Ungava; in Report of Activities, Part B, Geol. Surv. Can., Paper 73-1B, p. 9-19.

Allan, R.J. and Hornbrook, E.H.W.
1970: Development of geochemical techniques in permafrost, Coppermine River Region; Can. Min. J., v. 91 (4), p. 45-49.

2. Allan, R.J., Lynch, J.J., and Lund, N.G.
1972: Regional geochemical exploration in the Coppermine River area, District of Mackenzie: A feasibility study in permafrost terrain; Geol. Surv. Can., Paper 71-33, 53 p.

Allan, R.J. and Cameron, E.M.
1973: Zinc content of lake sediments, Bear-Slave Operation, District of Mackenzie; Geol. Surv. Can., Map 10-1972 (3 sheets).

3. Allan, R.J., Cameron, E.M., and Durham, C.C.
　　1973a: Lake geochemistry – a low sample density technique for reconnaissance geochemical exploration and mapping of the Canadian Shield; In Geochemical Exploration, 1972 (M.J. Jones, Editor), Inst. Min. Metall., p. 131-160.

4. 　　1973b: Reconnaissance geochemistry using lake sediments of a 36,000 square mile area of northwestern Canadian Shield; Geol. Surv. Can., Paper 75-50, 70 p.

5. Allan, R.J., Cameron, E.M., and Jonasson, I.R.
　　1974: Mercury and arsenic levels in lake sediments from the Canadian Shield; In Proc. First Int. Mercury Congress, 1974, v. 2, p. 93-119.

6. Allan, R.J. and Timperley, M.H.
　　1975: Prospecting using lake sediments in areas of industrial heavy metal contamination; In Prospecting in areas of glaciated terrain, 1975 (M.J. Jones, Editor) Inst. Min. Metall., p. 87-111.

Andelman, J.B.
　　1973: Incidence, variability, and controlling factors for trace elements in natural, fresh waters; In Trace metals and metal-organic interactions in natural waters (P.C. Singer, Editor), Ann Arbor Science Pub., Ann Arbor, Michigan, p. 57-87.

Andrews-Jones, D.A.
　　1968: The application of geochemical techniques to mineral exploration; Colorado Sch. Mines, Min. Ind. Bull., v. 11 (6), 31 p.

Armstrong, F.A.J. and Schindler, D.W.
　　1971: Preliminary chemical characterization of waters in the Experimental Lakes Area, northwestern Ontario; J. Fish. Res. Bd. Can., v. 28 (2), p. 171-187.

7. Arnold, R.G.
　　1970: The concentrations of metals in lake waters and sediments of some Precambrian lakes in the Flin Flon and La Ronge areas; Sask. Res. Council, Geol. Div., Circ. 4, 30 p.

Baas Becking, L.G.M., Kaplan, I.R., and Moore, D.
　　1960: Limits of the natural environment in terms of pH and oxidation-reduction potentials; J. Geol., v. 68, p. 243-284.

Baker, W.E.
　　1973: The role of humic acids from Tasmanian podzolic soils in mineral degradation and metal mobilization; Geochim. Cosmochim. Acta., v. 37, p. 269-281.

Beals, H.L.
　　1966: Manganese-iron concentrations in Nova Scotia lakes; Marit. Sediments, v. 2, p. 70-72.

Birge, E.A. and Juday, C.
　　1934: Particulate and dissolved organic matter in inland lakes; Ecol. Monogr., v. 4, p. 440-474.

Bjorklund, A. and Tenhola, M.
　　1976: Karelia: Regional prospecting for uranium; J. Geochem. Explor., v. 5 (3), p. 244-246.

Bjorklund, A.J., Tenhola, M. and Rosenberg, R.
　　1976: Regional geochemical uranium prospecting in Finland; In Exploration for Uranium Ore Deposits, I.A.E.A., Vienna, p. 283-295.

Boothe, P.N. and Knauer, G.A.
　　1972: The possible importance of fecal material in the biological amplification of trace and heavy metals; Limnol. Oceanogr., v. 17, p. 270.

Bordovsky, O.K.
　　1965: Accumulation and transformation of organic substances in marine sediment; Marine Geol., v. 3, p. 1-114.

Bowen, H.J.M.
　　1966: Trace Elements in Biochemistry; Academic Press, London.

Bowie, S.H.V., Ostle, D., and Ball, T.K.
　　1971: Geochemical methods in the detection of hidden uranium deposits; In Geochemical Exploration (R.W. Boyle and J.I. McGerrigle, Editors), Can. Inst. Min. Metall., Special Vol. 11, p. 103-111.

8. Bradshaw, P.M.D.
　　1975: Wintering Lake Cu-Ni prospect, Manitoba; J. Geochem. Explor., v. 4(1), p. 188-189.

9. Bradshaw, P.M.D., Clews, D.R., and Walker, J.L.
　　1973: Exploration geochemistry; Barringer Research Ltd., Rexdale, Ontario, 50 p.

Brewer, M.C.
　　1958: The thermal regime of an Arctic lake; Am. Geophys. U. Trans., v. 39 (2), p. 278-284.

10. Brunskill, G.J., Povoledo, D., Graham, B.W., and Stainton, W.P.
　　1971: Chemistry of surface sediments of sixteen lakes in the Experimental Lake Area, northwestern Ontario; J. Fish. Res. Bd. Can., v. 28 (2), p. 277-294.

11. Cameron, E.M.
　　1977: Geochemical dispersion in lake waters and sediments from massive sulphide mineralization, Agricola Lake area, Northwest Territories; J. Geochem Explor., v. 7 (3), p. 327-348.

12. Cameron, E.M. and Allan, R.J.
　　1973: Distribution of uranium in the crust of the northwestern Canadian Shield as shown by lake sediment analysis; J. Geochem. Explor., v. 2 (2), p. 237-250.

Cameron, E.M. and Durham, C.C.
　　1974a: Follow-up investigations on the Bear-Slave geochemical operation; in Report of Activities, Part A, Geol. Surv. Can., Paper 74-1A, p. 53-60.

　　1974b: Geochemical studies on the eastern part of the Slave structural province, 1973; Geol. Surv. Can., Paper 74-27, 22 p.

Cameron, E.M. and Ballantyne, S.B.
　　1975: Experimental hydrogeochemical surveys of the High Lake and Hackett River areas, N.W.T.; Geol. Surv. Can., Paper 75-29, 19 p.

Cameron, E.M. and Durham, C.C.
   1975: Further studies of hydrogeochemistry applied to mineral exploration in the northern Canadian Shield; in Report of Activities, Part C, Geol. Surv. Can., Paper 75-1C, p. 233-238.

Cameron, E.M. and Hornbrook, E.H.W.
   1976: Current approaches to geochemical reconnaissance for uranium in the Canadian Shield; In Exploration for Uranium Ore Deposits, I.A.E.A., Vienna, p. 241-266.

13. Cameron, E.M. and Ballantyne, S.S.
   1977: Reconnaissance-level geochemical and radiometric exploration data from the vicinity of the Rabbit Lake uranium deposit; Can. Min. Metall. Bull., v. 70 (781) p. 76-85.

Chau, Y.K., Wong, P.T.S., Silverberg, B.A., Luxon, P.O., and Bengert, G.A.
   1976: Methylation of selenium in the aquatic environment; Science, v. 192, p. 1130-1131.

Clarke, D.E.
   1976: The relation of lake sediment geochemistry to mineralization in the northwest Ontario region of the Canadian Shield — A discussion; Econ. Geol., v. 71 (5), p. 952-955.

Cleugh, T.R. and Hauser, B.W.
   1971: Results of the initial survey of the Experimental Lakes Area, northwestern Ontario; Jour. Fish. Res. Bd. Can., v. 28(2), p. 129-137.

14. Clews, D.R.
   1975: Fox Lake Cu-Zn deposit, Manitoba; J. Geochem. Explor., v. 4 (1), p. 181-182.

Cline, J.T. and Upchurch, S.B.
   1973: Mode of heavy metal migration in the upper strata of lake sediment; In Proc. 16th Conf. Great Lakes Res., 1973, Int. Assoc. Great Lakes Res., p. 349-356.

15. Closs, L.G.
   1975: Geochemistry of lake sediments in the Elliot Lake Region, District of Algoma: A pilot study; Ont. Div. Mines., OFR 5125, 49 p.

Coker, W.B.
   1974: Lake sediment geochemistry in the Superior Province of the Canadian Shield; Unpublished Ph.D. thesis, Queen's University, Kingston, 297 p.

16. Coker, W.B.
   1976: Geochemical follow-up studies, northwestern Manitoba; in Report of Activities, Part C, Geol. Surv. Can., Paper 76-1C, p. 263-267.

17. Coker, W.B. and Nichol, I.
   1975: The relation of lake sediment geochemistry to mineralization in the northwestern Ontario region of the Canadian Shield; Econ. Geol., v. 70 (1), p. 202-218.

18.   1976: The relation of lake sediment geochemistry to mineralization in the northwest Ontario region of the Canadian Shield — A reply; Econ. Geol., v. 71(5), p. 955-963.

19. Coker, W.B. and Jonasson, I.R.
   1977a: Geochemical orientation survey for uranium in the Grenville province of Ontario; Geol. Surv. Can., O.F. 461, 45 p.

20.   1977b: Geochemical exploration for uranium in the Grenville province of Ontario; Can. Min. Metall. Bull., v. 70(781), p. 67-75.

Conroy, N.
   1971: Classification of Precambrian Shield lakes based on factors controlling geological activity; Unpublished M.Sc. thesis, McMaster Univ., Hamilton.

Cook, D.O. and Felix, D.W.
   1975: Ferromanganese deposits in the Saranac Lake system, New York; J. Great Lakes Res., v. 1(1), p. 10-17.

Cronan, D.S. and Thomas, R.L.
   1970: Ferromanganese concretions in Lake Ontario; Can. J. Earth Sci., v. 7, p. 1346-1349.

   1972: Geochemistry of ferromanganese oxide concretions and associated deposits in Lake Ontario; Geol. Soc. Am. Bull., v. 83, p. 1493-1502.

Curtis, C.D.
   1966: The incorporation of soluble organic matter into sediments and its effect on trace element assemblages; In Advances in Organic Geochemistry (G.D. Hobson and M.C. Louis, Editors), Pergamon Press, Oxford, p. 1-13.

Damiani, V., Morton, T.W., and Thomas, R.L.
   1973: Freshwater ferromanganese nodules form the Big Bay section of the Bay of Quinte, northern Lake Ontario; In Proc. 16th Conf. Great Lakes Res., Int. Assoc. Great Lakes Res., p. 397-403.

Darnley, A.G., Cameron, E.M., and Richardson, K.A.
   1975: The Federal-Provincial Uranium Reconnaissance Program; In Uranium Exploration '75, Geol. Surv. Can., Paper 75-26, p. 49-68.

Davenport, P.H., Hornbrook, E.H.W., and Butler, A.J.
   1974: Geochemical lake sediment survey for Zn and Pb mineralization over the Cambro-Ordovician carbonate rocks of western Newfoundland; Nfld. Dept. Mines Energy, Min. Dev. Div., O.F.R., St. John's, Nfld.

21.   1975: Regional lake sediment geochemical survey for zinc mineralization in western Newfoundland; In Geochemical Exploration 1974 (I.L. Elliott and W.K. Fletcher, Editors), Assoc. Explor. Geochem, Special Pub. No. 2, Elsevier, p. 555-578.

22. Davenport, P.H. and Butler, A.J.
   1975: Geochemical stream and lake sediment surveys of the eastern part of the Burlington Peninsula, Newfoundland; Nfld. Dept. Mines Energy, Min. Dev. Div., O.F. 785, St. John's, Nfld.

23. Davenport, P.H., Butler, A.J., and Howse, A.F.
    1975a: A geochemical lake sediment survey over the Harbour Main Group and the Bull Arm Formation, Avalon Peninsula, Newfoundland; Nfld. Dept. Mines. Energy, Min. Dev. Div., O.F. 879, St. John's, Nfld.

24. Davenport, P.H., Hornbrook, E.H.W., and Butler, A.J.
    1975b: Regional lake sediment geochemical survey for zinc mineralization in western Newfoundland; In Geochemical Exploration, 1974 (I.L. Elliott and W.K. Fletcher, Editors), Assoc. Explor. Geochem, Special Pub. No. 2, Elsevier, p. 555-578.

25. Davenport, P.H. and Butler, A.J.
    1976: Uranium distribution in lake sediment in western Newfoundland and the Avalon and Burlington Peninsulas; Nfld. Dept. Mines Energy, Min. Dev. Div., O.F. 904, St. John's, Nfld.

26. Davenport, P.H., Butler, A.J., and Dibbon, D.
    1976a: Uranium distribution in lake sediment on the Burin Peninsula, Newfoundland; Nfld. Dept. Mines Energy, Min. Dev. Div., O.F. 944, St. John's, Nfld.

27. Davenport, P.H., Butler, A.J., and Howse, A.F.
    1976b: Distributions of arsenic, molybdenum and fluorine in lake sediment on the eastern part of the Burlington Peninsula, Newfoundland; Nfld. Dept. Mines Energy, Min. Dev. Div., O.F. 893, St. John's, Nfld.

28. 1976c: A geochemical lake sediment survey over the Harbour Main Group and Bull Arm Formation, Avalon Peninsula, Newfoundland; Part II, distributions of residual Zn, Co, Ag, As, and F; Nfld. Dept. Mines Energy, Min. Dev. Div., O.F. 894, St. John's, Nfld.

Dean, W.E.
    1970: Fe Mn oxidate crusts in Oneida Lake, N.Y.; In Proc. 13th Conf. on Great Lakes Res., Internat. Assoc. Great Lakes Res., p. 147.

Dean, W.E. and Gorham, E.
    1976: Major chemical and mineral components of profundal surface sediments in Minnesota lakes; Limno. Oceanogr., v. 21(2), p. 259-284.

Delfino, J.J. and Lee, G.F.
    1968: Chemistry of manganese in Lake Mendota, Wisconsin; Envir. Sci. Tech., v. 2, p. 1094-1100.

Durham, C.C. and Cameron, E.M.
    1975: A hydrogeochemical survey for uranium in the northern part of the Bear Province, N.W.T.; in Report of Activities, Part C, Geol. Surv. Can., Paper 75-1C, p. 331-332.

29. Dyck, W.
    1974: Geochemical studies in the surficial environment of the Beaverlodge area, Saskatchewan; Geol. Surv. Can., Paper 74-32, Part B., p. 21-30.

Dyck, W., Dass, A.S., Durham, C.C., Hobbs, J.O., Pelchat, J.C., and Galbraith, J.H.
    1971: Comparison of regional geochemical uranium exploration methods in Beaverlodge area, Saskatchewan; In Geochemical Exploration (R.W. Boyle and J.J. McGerrigle, Editors), Can. Inst. Min. Metall., Spec. Vol. 11, p. 132-150.

Edgington, D.N. and Callender, E.
    1970: Minor element geochemistry of Lake Michigan ferromanganese nodules; Earth Planet. Sci. Lett., v. 8(2), p. 97-100.

Flaig, W.
    1971: Some physical and chemical properties of humic substances as a basis of their characterization; In Advances in Organic Geochemistry (H.A.V. Gaertnes and H. Wehrer, Editors), p. 49-67.

Frey, D.G.
    1963: Limnology in North America; Univ. Wisconsin Press, Madison, Wisconsin, 734 p.

Garrels, R.M.
    1960: Mineral Equilibra at Low Temperature and Pressure; Harper and Brothers, New York, N.Y., 254 p.

Garrels, R.M. and Christ, C.L.
    1965: Solutions, Minerals and Equilibrea; Harper and Row, New York, 450 p.

Garrett, R.G.
    1974: Field data acquisition methods for applied geochemical surveys at the Geological Survey of Canada; Geol. Surv. Can., Paper 74-52, 36 p.

Garrett, R.G. and Hornbrook, E.H.W.
    1976: The relationship between zinc and organic content in centre-lake bottom sediments; J. Geochem. Explor., v. 5(1), p. 31-38.

Geological Survey of Canada
    1977a: Regional lake sediment geochemical reconnaissance data, eastern Ontario; Geol. Surv. Can., O.F. 405 (N.T.S. 31C N/2, Map NGR-1-76) and 406 (N.T.S. 31F, Map NGR-2-76).

    1977b: Regional lake sediment geochemical reconnaissance data, northeastern Manitoba; Geol.Surv. Can., O.F. 407 (N.T.S. 64I, and 54L W/2, Map NGR-3-76) and 408 (N.T.S. 64P and 54M W/2, Map N.G.R.-4-76).

    1977c: Regional lake sediment geochemical reconnaissance data, southern District of Keewatin; Geol. Surv. Can., O.F. 413 (NTS 54A, Map N.G.R.-9-76), 414 (N.T.S. 65B, Map NGR-10-76) and 415 (NTS 65C, Map NGR-11-76).

    1977d: Regional lake sediment geochemical reconnaissance data, Baker Lake, N.W.T.; Geol. Surv. Can., O.F. 416 (NTS 55M, Map NGR-12-76) and 417 (NTS 65P, Map NGR-13-76).

Geological Survey of Canada (cont'd)
  1978e: Regional lake sediment geochemical reconnaissance data, south-central Newfoundland (parts of NTS 1M, 2D, 11P and 12A); Geol. Surv. Can., O.F. (in preparation).

Gibbs, R.J.
  1973: Mechanisms of trace metal transport in rivers; Science, v. 180, p. 71-73.

30. Gleeson, C.F. and Hornbrook, E.H.W.
  1975a: Semi-regional geochemical studies demonstrating the effectiveness of till sampling at depth; In Geochemical Exploration, 1974 (I.L. Elliott and W.K. Fletcher, Editors), Elsevier, Amsterdam, p. 611-630.

31. 1975b: Nighthawk Lake area, Ontario; J. Geochem. Explor., v. 4(1), p. 178-181.

Gorham, E.
  1958: Observations on the formation and breakdown of the oxidized microzone at the mud surface in lakes; Limno. Oceanogr., v. 3, p. 291-298.

Gorham, E. and Swaine, D.
  1965: The influence of oxidizing and reducing conditions upon the distribution of some elements in lake sediments; Limno. Oceanogr., v. 10, p. 268-279.

de Groot, A.J. and Allersma, E.
  1975: Field observations on the transport of heavy metals in sediments; in Heavy metals in the aquatic environment (P.A. Krenkel, Editor), Pergamon Press, New York, p. 85-95.

Guy, R.D. and Chakrabarti, C.L.
  1976: Studies of metal-organic interactions in model systems pertaining to natural waters; Can. J. Chem., v. 54, p. 2600-2611.

Harriss, R.C. and Troup, A.C.
  1969: Freshwater ferromanganese concretions: chemistry and internal structure; Science, v. 166, p. 604-606.

32. Haughton, D.R.
  1975: Geochemistry of bedrock, overburden, lake and stream sediments in the Stanley area (west half), Saskatchewan; Geol. Div., Sask. Res. Council, Rep. 15, 30 p.

33. Haughton, D.R., Arnold, R.G., and Smith, J.W.J.
  1973: Geochemistry of bedrock, overburden, lake and stream sediments in the Mudjatik area, Sakatchewan; Geol. Div., Sask. Res. Council, Rep. 12, 24 p.

Hawkes, H.E. and Webb, J.S.
  1962: Geochemistry in mineral exploration; Harper and Row, New York, 415 p.

Hem, J.D.
  1960: Complexes of ferrous iron with tannic acid; U.S. Geol. Surv. Water Supply Paper 1459-D.

  1964: Decomposition and solution of manganese oxides; U.S. Geol. Surv. Water Supply Paper, 1667-B, p. 14-33.

Hem, J.D. (cont'd)
  1971: Study and interpretation of the chemical characteristics of natural water; U.S. Geol. Surv. Water Supply Paper 1473.

34. Hoffman, S.J. and Fletcher, W.K.
  1972: Distribution of copper at the Dansey-Rayfield River property, south-central British Columbia, Canada; J. Geochem. Explor., v. 1(2), p. 163-180.

35. 1976: Reconnaissance geochemistry of the Nechako Plateau, British Columbia, using lake sediments; J. Geochem. Explor., v. 5(2), p. 101-114.

Hornbrook, E.H.W.
  1977: Geochemical reconnaissance for uranium utilizing lakes of the Canadian Shield; In Uranium in Saskatchewan (C.E. Dunn, Editor), Sask. Geol. Soc., Spec. Pub. No. 3, p. 125-156.

Hornbrook, E.H.W. and Jonasson, I.R.
  1971: Mercury in permafrost regions; occurrence and distribution in the Kaminak Lake area, N.W.T.; Geol. Surv. Can., Paper 71-43, 13 p.

36. Hornbrook, E.H.W. and Gleeson, C.F.
  1972: Regional geochemical lake bottom sediment and till sampling in the Timmins – Val d'Or region of Ontario and Quebec; Geol. Surv. Can., O.F.R. 112.

37. 1973: Regional lake bottom sediment moving average – residual anomaly maps in the Abitibi region of Ontario and Quebec; Geol. Surv. Can., O.F.R. 127.

38. Hornbrook, E.H.W., Davenport P.H., and Grant, D.R.
  1975a: Regional and detailed geochemical exploration studies in glaciated terrain in Newfoundland; Nfld. Dept. Mines Energy, Min. Dev. Div., Report 75-2, St. John's, Nfld.

39. Hornbrook, E.H.W., Garrett, R.G., Lynch, J.J., and Beck, L.S.
  1975b: Regional geochemical lake sediment reconnaissance data, east-central Saskatchewan; Geol. Surv. Can., O.F.R. 266.

40. Hornbrook, E.H.W. and Garrett, R.G.
  1976: Regional geochemical lake sediment survey east-central Saskatchewan; Geol. Surv. Can., Paper 75-41, 20 p.

Hornbrook, E.H.W., Garrett, R.G., and Lynch, J.J.
  1976a: Regional lake sediment reconnaissance data, northwestern Manitoba, N.T.S. 64J; Geol. Surv. Can., O.F. 320.

  1976b: Regional lake sediment reconnaissance data, northwestern Manitoba, N.T.S. 64K; Geol. Surv. Can., O.F. 321.

  1976c: Regional lake sediment reconnaissance data, northwestern Manitoba, N.T.S. 64N; Geol. Surv. Can., O.F. 322.

  1976d: Regional lake sediment reconnaissance data, northwestern Manitoba, N.T.S. 64O; Geol. Surv. Can., O.F. 323.

Hornbrook, E.H.W., Garrett, R.G., and Lynch, J.J. (cont'd)
1976e: Regional lake sediment geochemical reconnaissance data, Nonacho Belt, east of Great Slave Lake, N.W.T., N.T.S. 75C; Geol. Surv. Can., O.F. 324.

1976f: Regional lake sediment geochemical reconnaissance data, Nonacho Belt, east of Great Slave Lake, N.W.T., N.T.S. 75F; Geol. Surv. Can., O.F. 325.

1976g: Regional lake sediment geochemical reconnaissance data, Nonacho Belt, east of Great Slave Lake, N.W.T., N.T.S. 75K; Geol. Surv. Can., O.F. 326.

1976h: Regional lake sediment geochemical reconnaissance data, Great Bear Lake, N.W.T., NTS 86K; Geol. Surv. Can., O.F. 327.

1976i: Regional lake sediment geochemical reconnaissance data, Great Bear Lake, N.W.T., NTS 86L and parts of 96I; Geol. Sur. Can., O.F. 328.

Hornbrook, E.H.W., Garrett, R.G., Lynch, J.J., and Beck, L.S.
1977: Regional lake sediment geochemical reconnaissance data, east-central Saskatchewan; Geol. Surv. Can., O.F.R. 488.

Hutchinson, G.E.
1957: A Treatise on Limnology; v. 1, Geography, Physics, and Chemistry; Wiley and Sons, New York, 1015 p.

1967: A Treatise on Limonology; v. 2, Introduction to Lake Biology and the Limnoplankton; Wiley and Sons, New York.

Jackson, K.S.
1975: Geochemical dispersion of elements via organic complexing; Unpublished Ph.D. thesis, Carleton Univ., Ottawa, Canada, 344 p.

Jackson, K.S. and Jonasson, I.R.
1977: An overview of the problems of metals-sediment-water interactions in fluvial systems; In Proc. Workshop, Fluvial Transport of Sediment – Associated Nutrients and Contaminants, Kitchener, Ont., Oct. 20-22, 1976 (H. Shear and A.E.P. Watson, Editors), Publ. Int. Joint. Comm., Windsor, Ont., p. 255-271.

Jackson, R.G.
1975: The application of lake sediment geochemistry to mineral exploration in the southern Slave Province of the Canadian Shield; Unpublished M.Sc. thesis, Queen's University, Kingston, 306 p.

42. Jackson, R.G. and Nichol, I.
1975: Factors affecting trace element dispersion in lake sediments in the Yellowknife area, N.W.T. Canada; Assoc. Int. Limnol., Proc., v. 19(1), p. 308-316.

Jackson, T.A.
1977: A biogeochemical study of heavy metals in lakes and streams, and a proposed method for limiting heavy metal pollution of natural waters; Paper presented at 20th S.I.L. Congress, Copenhagen, Denmark, August 1977, Environ. Geol., in press.

Jenne, E.A.
1968: Controls on Mn, Fe, Co, Ni, Cu and Zn concentrations in soils and waters: The dominant role of hydrous Mn and Fe oxides; In Trace inorganics in water, Adv. Chem. Ser., No. 73, p. 337-387.

43. Jonasson, I.R.
1976: Detailed hydrogeochemistry of two small lakes in the Grenville geological province; Geol. Surv. Can., Paper 76-13, 37 p.

Juday, C., Birge, E.A., and Meloche, V.M.
1938: Mineral content of the lake waters of northeastern Wisconsin; Trans. Wis. Acad. Sci. Arts. Lett., 23, p. 233-248.

Kindle, E.M.
1932: Lacustrine concentrations of manganese; Am. J. Sci., v. 24, p. 496-504.

1935: Manganese concentrations in Nova Scotia lakes; Roy. Soc. Can. Trans., Sec. IV, v. 29, p. 163-180.

1936: The occurrence of lake bottom manganiferous deposition in Canadian lakes; Econ. Geol., v. 31, p. 755-760.

Klassen, R.A.
1975: Lake sediment geochemistry in the Kaminak Lake area, District of Keewatin, N.W.T.; Unpublished M.Sc. thesis, Queen's University, Kingston, 209 p.

44. Klassen, R.A., Nichol, I., and Shilts, W.W.
1975: Lake geochemistry in the Kaminak Lake area, District of Keewatin, N.W.T.; Assoc. Int. Limnol., Proc., v. 19(1), p. 340-348.

Knauer, G.A. and Martin, J.H.
1973: Seasonal variations of cadmium, copper, manganese, lead and zinc in water and phytoplankton in Monterey Bay, California; Limnol. Oceanogr., v. 18, p. 597-604.

Koljonen, T. and Carlson, L.
1975: Behaviour of the major elements and mineral in sediments of four humic lakes in southwestern Finland; Fennia, 137, p. 47.

Krauskopf, K.B.
1955: Sedimentary deposits of rare metals; Econ. Geol., 50th. Anniversary Volume, p. 411-463.

1957: Separation of manganese from iron in sedimentary processes; Geochim. Cosmochim. Acta., v. 12, p. 61-84.

1967: Introduction to Geochemistry; McGraw-Hill Inc., New York, 721 p.

Kuznetsov, S.I.
1970: The Microflora of Lakes and its Geochemical Activity (Carl H. Oppenheimer, Editor); U. of Texas Press, Austin, Texas, 503 p.

1975: The role of microorganisms in the formation of lake bottom deposits and their diagenesis; Soil Science, v. 119(1), p. 81-87.

45. Lehto, D.A.W., Arnold, R.G., and Smith, J.W.J.
    1977: Lake sediments as a media for exploration in Saskatchewan; In Uranium in Saskatchewan (C.E. Dunn, Editor), Sask. Geol. Soc., Spec. Pub. No. 3, p. 100-124.

Leland, H.V., Shukla, S.S., and Shimp, N.F.
    1973: Factors affecting distribution of lead and other trace elements in sediments of southern Lake Michigan; In Trace Metals and Metal-Organic Interactions in Natural Waters (P.C. Singer, Editor), Ann Arbor Science Publ., Ann Arbor, Michigan, p. 89-129.

Levinson, A.A.
    1974: Introduction to Exploration Geochemistry; Applied Publishing Ltd., Calgary, Alberta, Canada, 612 p.

Livingstone, D.A.
    1963: Chemical composition of rivers and lakes; In Data of Geochemistry (M. Fleischer, Editor), U.S. Geol. Surv., Prof. Paper 440, Chap. G., p. 61.

Lundgreer, P.
    1953: Some data concerning the formation of manganiferous and ferriferous bog ores; Geol. Foren. Stockholm. Forh., v. 75, p. 277-297.

Mackereth, F.J.H.
    1965: Chemical investigation of lake sediments and their interpretation; Proc. Roy. Soc., B161, p. 285-309.

    1966: Some chemical observations on post-glacial lake sediments; Phil. Trans. Roy. Soc. London, Ser. B250, p. 165-213.

46. Maurice, Y.T.
    1976: Detailed geochemical investigations for uranium and base metal exploration in the Nonacho Lake area, District of Mackenzie, N.W.T.; in Report of Activities, Part C, Geol. Surv. Can., Paper 76-1C, p. 259-262.

47.   1977a: Geochemical methods applied to uranium exploration in southwest Baffin Island; Can. Min. Metall. Bull., v. 70(781), p. 96-103.

48.   1977b: Follow-up geochemical activities in the Nonacho Lake area (75F, K), District of Mackenzie; Geol. Surv. Can., O.F.R. 489.

49. McArthur, J.G., Davenport, P.H., and Howse, A.F.
    1975: A geochemical reconnaissance survey of the Carboniferous Codroy Bay – St. George Basin, Newfoundland; Nfld. Dep. Mines Energy, Min. Dev. Div., O.F. 787, St. John's, Nfld.

50. Mehrtens, M.G., Tooms, J.S., and Troup, A.G.
    1973: Some aspects of geochemical dispersion from base-metal mineralization within glaciated terrain in Norway, North Wales and British Columbia; In Geochemical Exploration, 1972 (M.J. Jones, Editor), Inst. Min. Metall., p. 105-115.

Meineke, D.G., Vadis, M.K., and Klaysmat, A.W.
    1976: Gyttja lake sediment exploration geochemical survey of eastern Lake Vermilion-Ely area, St. Louis and Lake Countries, Minnesota; Minn. Dep. Nat. Res., Report 73-3-1, 53 p.

51. Minatidis, D.G. and Slatt, R.M.
    1976: Uranium and copper exploration by nearshore lake sediment geochemistry, Kaipokok region of Labrador; J. Geochem. Explor., v. 5(2), p. 135-144.

Mitchell, R.L.
    1964: Trace elements in soils; In Chemistry of the Soil (F.E. Bear, Editor), Reinhold Publishing Corp., New York, p. 320.

Morel, F., McDuff, R.E., and Morgan, J.J.
    1973: Interactions and chemostasis in aquatic chemical systems: Role of pH, pE, solubility and complexation; In Trace Metals and Metal-Organic Interactions in Natural Waters (P.C. Singer, Editor), Ann Arbor Science Publ., Ann Arbor, Michigan, p. 157-200.

Morris, A.W.
    1971: Trace metal variations in sea water of the Menai Straits caused by a bloom of Phaeocystis; Nature, v. 233, p. 427-428.

Mortimer, C.H.
    1941: The exchange of dissolved substances between mud and water in lakes. Part I and II; J. Ecol., v. 29, p. 280-329.

    1942: The exchange of dissolved substances between mud and water in lakes. Part III and IV; J. Ecol., v. 30, p. 147-207.

    1971: Chemical exchanges between sediments and water in the Great Lakes – speculations on probable regulatory mechanisms; Limno. Oceangr., v. 16(2), p. 387-404.

Nichol, I., Coker, W.B., Jackson, R.G., and Klassen, R.A.
    1975: Relation of lake sediment composition to mineralization in different limnological environments in Canada; In Prospecting in areas of glaciated terrain – 1975 (M.J. Jones, Editor), Inst. Min. Metall., p. 112-125.

52. Nickerson, D.
    1972: An account of a lake sediment geochemical survey conducted over certain volcanic belts within the Slave structural province of the Northwest Territories during 1972; D.I.N.A., O.F.R., 1972, 22 p.

Nissenbaum, A. and Kaplan, I.R.
    1972: Chemical and isotopic evidence for the in situ origin of marine humic substances; Limnol. Oceanogr., v. 17, p. 570-582.

Parslow, G.R.
    1977: A discussion of the relationship between zinc and organic content in centre-lake bottom sediments; J. Geochem. Explor., v. 7(3), p. 383-384.

53. Ramaekers, P.P. and Dunn, C.E.
    1977: Geology and geochemistry of the eastern margin of the Athabasca Basin; In Uranium in Saskatchewan (C.E. Dunn, Editor), Sask. Geol. Soc., Spec. Pub. No. 3, p. 297-322.

Rashid, M.A.
    1974: Absorption of metals on sedimentary and peat humic acids; Chem. Geol., v. 13, p. 115-123.

Rashid, M.A. and Leonard, J.D.
  1973: Modification of the solubility and precipitation behaviour of various metals as a result of their interactions with sedimentary humic acid; Chem. Geol., v. 11, p. 89-97.

Riley, G.A.
  1939: Limnological studies in Connecticut: Part I. General limnological survey; Part II. The copper cycle; Ecol. Monogr., v. 9, p. 66-94.

Robbins, J.A. and E. Callender
  1975: Diagenesis of manganese in Lake Michigan sediments; Am. J. Sci., v. 275, p. 512-533.

Rossman, R. and Callender, E.
  1968: Manganese nodules in Lake Michigan; Science, v. 162, p. 1123-1124.

  1969: Geochemistry of Lake Michigan manganese nodules; In Proc. 12th Conf. on Great Lakes Res., Int. Assoc. Great Lakes Res., p. 306-316.

Ruttner, F.
  1963: Fundamentals of Limnology; 3rd Edition, Univ. of Toronto Press, Toronto, 295 p.

Sain, K.S. and Neufeld, R.D.
  1975: A dynamic model of biogeochemical cycle of heavy and trace metals in natural aquatic systems; Paper presented at 2nd Int. Symposium on Environmental Biogeochemistry, Burlington, Ontario, Canada, 20 p.

Saxby, J.D.
  1969: Metal-organic chemistry of the geochemical cycle; Rev. Pure Appl. Chem., v. 19, p. 131-150.

Schaef, D.G.
  1975: A preliminary investigation into organic extraction procedures and partitioning of elements between organic phases of lake sediments from the Elliot Lake area; Unpublished B.Sc. thesis, Queen's University, Kingston, 54 p.

Schindler, D.W.
  1971: Light, temperature, and oxygen regimes of selected lakes in the Experimental Lakes Area, northwestern Ontario; J. Fish. Res. Bd. Can., v. 28(2), p. 157-169.

54. Schmidt, R.G.
  1956: Adsorption of Cu, Pb and Zn on some common rock forming minerals and its effect on lake sediments; Unpublished Ph.D. thesis, McGill University, Montreal

Schnitzer, M. and Khan, S.U.
  1972: Humic Substances in the Environment; Marcel Dekker Inc., New York, N.Y.

Schoettle, M. and Friedman, G.
  1971: Fresh water iron-manganese nodules in Lake George, N.Y.; Geol. Soc. Am. Bull., v. 82, p. 101-110.

Semkin, R.G.
  1975: A limnogeochemical study of Sudbury area lakes; Unpublished M.Sc. thesis, McMaster University, Hamilton, 248 p.

Sharp, R.R. Jr., and Aamodt, P.L.
  1976: Hydrogeochemical and stream sediment survey of the National Uranium Resource Evaluation Program, October – December 1975; LA – 6346 – PR Progress Report, United States ERDA Contract W-7402-ENG, 36, 86 p.

Shilts, W.W.
  1971: Till studies and their application to regional drift prospecting; Can. Min. J., v. 92(4), p. 45-50.

Shilts, W.W. and Dean, W.D.
  1975: Permafrost features under Arctic lakes, District of Keewatin, Northwest Territories; Can. J. Earth Sci., v. 12(4), p. 649-662.

Shilts, W.W., Dean, W.E., and Klassen, R.A.
  1976: Physical, chemical and stratigraphic aspects of sedimentation in lake basins of the eastern Arctic Shield; in Report of Activities, Part A, Geol. Surv. Can., Paper 76-1A, p. 245-254.

Sibbald, T.I.I.
  1977: Geochemistry of near-shore lake sediments, Sandy Narrows area, Saskatchewan; Sask. Geol. Surv., Rep. No. 170 (in press).

Spilsbury, W. and Fletcher, W.K.
  1974: Application of regression analysis to interpretation of geochemical data from lake sediments in central British Columbia; Can. J. Earth Sci., v. 11, p. 345-348.

Sozanski, A.G.
  1974: Geochemistry of ferromanganese oxide concretions and associated sediments and bottom waters from Shebandowan Lakes, Ontario; Unpublished M.Sc. thesis, University of Ottawa, Ottawa, p. 104.

Stumm, W. and Morgan, J.J.
  1970: Aquatic Chemistry; Wiley-Interscience, New York.

Swain, F.M.
  1958: Geochemistry of Humus; In Organic Geochemistry (I.A. Breger, Editor), Int. Series of Monographs on Earth Sciences, v. 16, p. 87-147.

55. Tan, B.
  1977: Geochemical case history in the Key Lake area; In Uranium in Saskatchewan (C.E. Dunn, Editor), Sask. Geol. Soc., Spec. Publ. No. 3, p. 323-330.

Tenhola, M.
  1976: Ilomantsi: Distribution of uranium in lake sediments; J. Geochem. Explor., v. 5(3), p. 235-239.

Terasmae, J.
1971: Notes on lacustrine manganese – iron concretions; Geol. Surv. Can., Paper 70-69, 13 p.

Theis, T.L. and Singer, P.C.
1973: The stabilization of ferrous iron by organic compounds in natural waters; In Trace metals and metal-organic interactions in natural waters (P.C. Singer, Editor), Ann Arbor Publ. Inc., Michigan, p. 303-320.

56. Timperley, M.H. and Allan, R.J.
1974: The formation and detection of metal dispersion halos in organic lake sediments; J. Geochem. Explor., v. 3, p. 167-190.

Trollope, D.R. and Evans, B.
1976: Concentrations of copper, iron, lead, nickel and zinc in freshwater algal blooms; Environ. Pollution, v. 11, p. 109-116.

Troup, A.C.
1969: Geochemical investigations of ferromanganese concretions from three Canadian lakes; Unpublished M.Sc. thesis, McMaster University, Hamilton, 82 p.

Turekian, K.K. and Kleinkopf, M.D.
1956: Estimates of the average abundance of Cu, Mn, Pb, Ti, Ni and Cr in surface waters of Maine; Geol. Soc. Am. Bull., v. 67, p. 1129-1131.

Twenhofel, W. and McKelvey, V.
1941: Sediments in freshwater lakes; Am. Assoc. Pet. Geol. Bull., v. 25, p. 826-849.

Williams, J.D.H.
1975: Limnological investigations in the Agricola Lake area; in Report of Activities, Part A, Geol. Surv. Can., Paper 75-1A, p. 227-228.

Wood, J.M.
1974: Biological cycles for toxic elements in the environment; Science, v. 183, p. 1049-1051.

1975: Metabolic cycles for toxic elements; Proc. Int. Conf., Heavy Metals in the Environment, Toronto, Abstr. No. A-5.

Wong, P.T.S., Chau, Y.K., and Luxon, P.L.
1975: Methylation of lead in the environment; Nature, v. 253, p. 263.

APPLICATION OF HYDROGEOCHEMISTRY TO THE SEARCH FOR BASE METALS

W.R. Miller
U.S. Geological Survey, Denver, Colorado

Miller, W.R., Application of hydrogeochemistry to the search for base metals; in Geophysics and Geochemistry in the Search for Metallic Ores; Peter J. Hood, editor; Geological Survey of Canada, Economic Geology Report 31, p. 479-487, 1979.

Abstract

As the world's exposed mineral deposits become depleted, new techniques need to be developed to aid in the search for buried deposits. If circulating water comes into contact with mineralization at or below the surface, and particularly within the zone of oxidation, certain elements may concentrate above the natural background and form aqueous dispersion halos. The detection and interpretation of these halos form the basis of hydrogeochemical prospecting.

The behaviour of trace components in water that is in contact with mineralization depends on the type of mineralization, the enclosing rocks, and the chemical and hydrological environments. In addition, the geochemical and physical environment will affect the mobility of trace elements. Factors that determine the formation and shape of aqueous dispersion halos include: (1) the physical and chemical properties of the migration forms of the elements; (2) the composition of the mineralization; (3) climate; (4) topography; (5) composition and permeability of the formation containing or covering the mineralization; (6) direction of groundwater flow; and (7) geologic environment including structure.

Interpretation of hydrochemical data includes both the treatment of single variables and the use of multivariate techniques. The treatment of single variables consists of the plotting of the concentrations of those pathfinder elements that should occur with the expected type of mineralization and that should be mobile for the given chemical and physical environment. The use of multivariate techniques allows more sophisticated interpretation of water analyses, particularly pattern-recognition techniques. Water chemistry used in conjunction with thermodynamic data can also be used to interpret the geochemical data and to gain a better understanding of the geochemical cycles of selected elements in the weathering environment.

Résumé

Etant donné que les gisements connus de minéraux à travers le monde s'épuisent rapidement, il faut mettre au point de nouvelles techniques afin de faciliter la recherche de gisements souterrains. Lorsque l'eau d'érosion vient en contact avec une minéralisation à la surface ou en profondeur, en particulier à l'intérieur de la zone d'oxydation, certains éléments peuvent se concentrer au-dessus du fond naturel et former des auréoles de dispersion aqueuse. La détection et l'interprétation de ces auréoles constitue la base de la prospection hydrogéochimique.

Le comportement des composantes à l'état de trace dans l'eau qui est en contact avec la minéralisation dépend du type de minéralisation, des roches encaissantes et des milieux chimique et hydrologique. De plus, le milieu géochimique et physique influe sur la mobilité des éléments à l'état de trace. Les facteurs qui déterminent la formation et le modelé des auréoles de dispersion aqueuse comprennent: 1) les propriétés physiques et chimiques des formes de migration des éléments; 2) la composition de la minéralisation; 3) le climat; 4) la topographie; 5) la composition et la perméabilité de la formation contenant ou recouvrant la minéralisation; 6) la direction de l'écoulement des eaux souterraines, et 7) le milieu géologique, y compris la structure.

L'interprétation des données hydrogéochimiques comprend à la fois le traitement des variables simples et l'utilisation de techniques multi-variables. Le traitement des variables simples consiste en l'enregistrement graphique des concentrations d'éléments des indicateurs minéraux qui peuvent exister dans le type de minéralisation prévu et qui seraient mobiles pour le milieu chimique et physique donné. L'utilisation de techniques multivariables permet une interprétation plus poussée des analyses de l'eau; c'est le cas des techniques permettant d'en confirmer la composition. La chimie de l'eau, utilisée conjointement avec des données thermodynamiques, peut aussi être utilisée pour l'interprétation de données géochimiques, de façon à permettre une meilleure compréhension des cycles géochimiques des éléments choisis dans le milieu d'actions météoriques.

INTRODUCTION

Water is called the "universal solvent". All natural substances will dissolve in water to some measurable extent, given proper circumstances and time. Therefore the waters of an area will have a certain natural background concentration for various elements. In addition, water is capable of penetrating below the earth's surface to depths as great as 2 km and dissolving natural substances. Some of this solute-containing water will eventually return to the surface and become part of the natural water system of an area. The water chemistry will depend on the source and history of the water. If the water has come into contact with mineralization at or below the surface, and particularly if within the zone of oxidation, certain elements will increase in concentration above the natural background and form geochemical dispersion halos. The detection and interpretation of these aqueous-dispersion halos form the basis for hydrogeochemical prospecting.

Types of samples used for hydrogeochemical prospecting are waters from springs and seeps; groundwater from wells; surface waters from streams, rivers, and lakes; snow; and groundwater from drillholes and mine workings.

The advantages of using water as a sampling medium for geochemical exploration are: 1) water is capable of penetrating below the earth's surface, and providing a third dimension in the search for hidden ore deposits; 2) water is usually a homogeneous and representative sample of the watershed; 3) water requires no sample preparation except for filtering and acidifying; 4) elements in water under oxidizing conditions, generally form large dispersion halos; 5) water is particularly adapted to the detection of large low-grade deposits; and 6) water is involved with most low temperature-pressure geochemical processes and can be useful in the interpretation of other geochemical data.

Unfortunately, because water is usually moving and in a dynamic state, the aqueous dispersion halos associated with mineralization can be affected by changes through time, such as those caused by rainfall or season. Disadvantages of hydrogeochemical prospecting therefore include the following: 1) changes through time occur in the water chemistry at the sampling point, owing to variability in discharge of the water or fluctuations of the chemical contents of water with season and rainfall; 2) analyses of trace components in water can be difficult, time-consuming, and expensive in the low ranges ($\mu g/L$); 3) sources of water may not be present for the needed sampling density; and 4) low concentrations of trace components in water are susceptible to contamination and storage problems.

Because of these disadvantages, hydrogeochemical prospecting for base metals with few exceptions has not been used to any general extent outside of the Soviet Union. However recent advances in both analytical techniques and data interpretation have reduced or eliminated many of these disadvantages.

DISCUSSION

Generally, trace elements in water can be used as pathfinders for mineralization. A trace element can be defined as normally occurring in water in concentrations less than 0.1 mg/L. Sources for trace elements in water are usually the enclosing rocks and soils, but other sources, such as pollution, are possible. The migration of trace elements in water depends on both intrinsic and extrinsic factors of migration. Intrinsic factors depend on the structure of the electron shell of the atom, and an example of this is the complexing capacity of an element. Extrinsic factors depend on the geochemical environment; for example, copper, lead, and zinc are more mobile in an acid environment, whereas molybdenum and arsenic are more mobile in an alkaline environment.

The behaviour of trace components in water that is in contact with mineralization depends on the type of mineralization, the enclosing rocks (the chemical activity of the rocks or their physical nature such as fracturing), and the chemical and hydrological environments. The chemical reactions that control the trace element concentrations in water include solution, precipitation, and adsorption. The mechanisms for migration of elements in water include ionic, complexing, colloidal, sorption, suspension, and electrochemical.

SAMPLING AND ANALYTICAL METHODS

Natural water is a continuously changing medium. The manner in which water is collected is more important than with most other media; collection must be performed carefully and consistently. Clean, acid-rinsed plastic bottles should be used for collection. Before sampling, the bottle should be rinsed several times with the water to be sampled. Wells should be allowed to flow prior to sampling. Turbid water should be filtered using a 0.45 μm membrane filter. To insure a consistent method, it is a good technique to filter all samples that are to be analyzed for trace elements. Filtering removes the large colloids and sediments which upon addition of acid may contribute to contamination (Kennedy et al., 1974).

The sample that is to be analyzed for trace elements should be acidified to pH<2 with metal-free nitric or hydrochloric acid.

Storage of samples before analyses should be in a cool place out of exposure to the sun to minimize growth of algae. If this is not possible, chloroform should be added to retard algal growth.

On-site determinations for reconnaissance exploration purposes are usually kept to a minimum, but may include determinations of pH or conductivity. For detailed exploration or orientation studies, additional field determinations may be needed. Possible determinations include: 1) temperature; 2) flow rate; 3) pH; 4) redox potential; 5) dissolved gases such as O_2, CO_2, and H_2S; 6) conductivity; 7) specific components such as Fe^{+2} and HCO_3^-.

For dilute waters, pH can generally be satisfactorily determined at the end of the day by the use of potentiometer at the campsite or motel. For more accurate pH measurements, a potentiometer can be used to measure pH in the field.

On-site descriptions will differ depending on the purpose of the survey. Possible on-site descriptions include: 1) sample source; 2) width and depth; 3) turbulence or the degree of aeration; 4) presence of suspended or organic material; 5) presence of precipitates, moss or algae; 6) local bedrock geology; 7) dominant vegetation; 8) elevation and geographic location in relation to the watershed; and 9) possible contaminants. In addition to the above, when sampling water wells, the following should be included: 1) water well configuration, tanks, pressure system; 2) depth of well and aquifers; 3) hydrological information; and 4) owner's address.

Recent advances in equipment and techniques allow methods that have sufficient sensitivities for trace-element analyses. Some of the methods that are used for trace-element analysis are shown on Table 21A.1.

An atomic adsorption unit coupled with a resistively heated device is capable of the sensitivities needed for many trace-element analyses. This method is probably the most useful one at the present time. Emission spectrometry using a plasma source and ion chromatography have come into use and will probably become more important in the near future.

METHODS

The geochemical and physical environment will affect the mobility of trace elements. Factors that determine the formation and shape of aqueous dispersion halos include: 1) the physical and chemical properties of the migration forms of the elements; 2) the composition of the mineralization; 3) climate; 4) topography; 5) composition and permeability of the formation containing or covering the mineralization; 6) direction of groundwater flow; and 7) geological environment including structural setting.

During intense oxidation of an ore deposit, an increase of 100 times above background in associated trace components is not unusual. Under reducing conditions, the contrast of most trace components will be less than under oxidizing conditions. Table 21A.2 lists the most frequently

Table 21A.1

Some analytical methods used for analysis
of trace elements in water

Methods	Comments
Atomic absorption spectrometry	Trace metals usually concentrated by solvent extraction techniques.
Atomic absorption spectrometry/ resistively heated devices	Usually no prior concentration, provides good sensitivity and at the present time it is probably the best methods for trace metal analyses.
Emission spectrometry (DC-Arc)	Multielement analyses, sample must be concentrated by evaporation, precipitation, or other means.
Emission spectrometry (Induction coupled plasma)	Multielement analyses, potentially one of the better methods.
Anodic stripping voltammetry	Means for differentiating metal species.
Ion chromatography	Multielement analysis, mostly anions.
Colorimetric or spectrophotometric	Usually slow.
Fluorimetric	U analysis, usually slow.
Ion selective electrodes	Poor sensitivity except for F^-.
Neutron activation analysis	Sensitive method but expensive.

Table 21A.2

The most frequently encountered and maximum contents (in ppb)
around mineralization of a number of microcomponents in
neutral waters of the supergene zone (after Shvartsev et al., 1975)

Element	Most frequently encountered contents (regional background)		Maximum contents found	Contrast ratio
	Waters of oxidizing setting	Waters of gley setting		
Barium	2.0 - 10	2.0 - 2.5	500	400
Beryllium	0.1 - 0.8	0.05 - 0.3	300	3 000
Boron	10 - 50	-	20 000	2 000
Bromine	20 - 200	-	20 000	1 000
Vanadium	0.5 - 2.0	0.5 - 1.5	90	180
Tungsten	0.1 - 0.5	-	200	2 000
Iodine	1 - 10	-	5 000	5 000
Cadmium	0.1 - 0.5	0.05 - 0.5	260	5 200
Cobalt	0.5 - 3	0.2 - 10	68	340
Lithium	5 - 10	-	10 000	2 000
Manganese	10 - 50	20 - 400	15 000	1 500
Copper	2 - 8	1 - 5	1 000	1 000
Molybdenum	1 - 5	0.2 - 3	8 000	40 000
Arsenic	1 - 5	0.1 - 2.0	100	1 000
Nickel	0.8 - 5	0.5 - 2	200	400
Tin	0.1 - 0.5	0.1 - 0.5	60	600
Mercury	0.5 - 3.0	-	20	40
Lead	1 - 8	0.5 - 4	250	500
Selenium	0.5 - 1.0	-	50	100
Silver	0.1 - 0.6	0.05 - 0.5	50	1 000
Strontium	5 - 50	3 - 20	1 000	300
Antimony	1 - 10	0.5 - 5.0	1 300	2 600
Uranium	0.5 - 5	0.3 - 3.0	500	1 700
Fluorine	50 - 1 000	-	11 500	230
Chromiun	5 - 10	0.1 - 5	600	6 000
Zinc	5 - 50	2 - 20	2 500	1 250

encountered trace-element concentrations in water. The physical environment also influences the mobility of trace components in water. Hydrogeochemical prospecting in different physical environments requires different approaches. A summary for hydrogeochemical prospecting in different environments is shown in Table 21A.3.

The types of expected mineralization also need to be considered in the selection of pathfinder elements for hydrogeochemical prospecting. Examples are the selection of copper, lead, sulphate, and zinc to detect polymetallic mineralization, or the selection of molybdenum and copper to detect porphyry-metal-type mineralization. Pathfinder elements associated with different types of mineralization are shown in Table 21A.4.

INTERPRETATION

Interpretation of hydrogeochemical data includes both the treatment of single variables and the use of multivariate techniques. The treatment of single variables consists of the plotting of the concentrations of those pathfinder elements that should occur with the expected type of mineralization and that should be mobile for the given chemical and physical environment. Several pathfinders are usually selected. Ratios such as Cu/Zn or Pb/Zn can be useful (since zinc is more mobile than copper or lead, the ratio will usually increase toward mineralization). Corrections for changes through time due to rainfall or season can sometimes be made by normalizing the pathfinder elements with conductivity or components such as chloride or sodium that are little affected by chemical controls in water. In a similar manner, other effects on the water chemistry can be detected and sometimes corrected. Examples of maps (Miller and Ficklin, 1976) using a single variable of a study conducted in the White River National Forest, Colorado are shown in Figure 21A.1. An intense copper anomaly occurs along Cataract Creek where the concentration of copper in water reaches 480 µg/L. A less intense copper anomaly is also present west of Hunter Peak. A zinc anomaly, which coincides with the copper anomaly, also occurs along Cataract Creek. Concentrations of zinc in water reach 35 µg/L. In addition an intense molybdenum anomaly, where the concentrations of molybdenum reach 450 µg/L, occurs west of Hunter Peak. The results indicate two main areas of interest, the area along Cataract Creek and the area west of Hunter Peak.

Just as recent advances in equipment and techniques have allowed the more effective use of water for geochemical prospecting, recent use of mathematical and statistical methods has allowed more sophisticated interpretation of water analyses.

Because the main purpose of a geochemical survey is to detect anomalous areas, pattern-recognition techniques are advantageous. An example of the use of this type of technique (Miller and Ficklin, 1976) is an extended form of Q-mode factor analysis and is shown in Figure 21A.2; which uses the same data as Figure 21A.1. A five-factor model explains 96 per cent of the total variation. Three factors are interpreted as being related to the mineralization. The first factor is represented by loadings on calcium and sulphate and is interpreted as reflecting pervasive pyrite

Table 21A.3

A summary for the application of hydrogeochemical prospecting in different environments

	Areas of permafrost	Areas of arid conditions	Areas of swamps	Areas of mountainous terrain
Characterization of water	Geochemical movement is partly electrochemical, with water in the active zone above the permafrost gradually becoming enriched in trace components.	High values for pH and salinity and slightly high background values for many trace elements but with low contrast.	Low pH values and flow rates and slightly high background values for many trace elements but with low contrast.	Chemically aggressive waters. The maximum contrast is obtained for moderate relief and rainfall; higher relief or rainfall tends to decrease contrast.
Mobility of elements	Mineralized bodies that emerge at the surface usually have oxidation zones and high contrast (difference between anomalous and background values); dispersion patterns traceable in the active zone by both surface and spring waters.	Fair to moderate mobility of many elements because of flat landscape and lack of chemically aggressive waters. Elements capable of forming negatively-charged ions or soluble compounds are the most mobile such as U, Mo, V, As, and Se.	Movement and precipitation of chemical elements are governed mainly by the concentration, pH, and contents of humic and fulvic acids, and organic material in contact with the water.	Good mobility of many elements because of the aggressive waters, the greater relief, and the more likely chance for oxidizing conditions.
Optimum time for sampling	Either the first part or the last part of the summer.	When the number of springs and ephemeral streams will be the greatest, usually in the spring and fall.	Most times except during runoff or flooded periods.	Usually after spring runoff when the streams have returned to normal flow or during low streamflow, which is usually during the fall and winter.
Abundance of possible sample sites	Good sample density is possible for both spring and surface waters.	Fair to poor sampling density, usually ephemeral streams and some springs	Abundant water, good sampling density.	Usually abundant water with good sampling density for both spring and surface waters. First-order streams and springs are probably the best source for water samples.
General comments	Chemical weathering reduced because of cold temperatures, organic material is abundant.	There is an accumulation of salts and secondary minerals and the lack of organic material.	Dispersion patterns of mineralization may be characterized by a set of elements which may not correspond exactly to their paragenic associations in the mineralization. Dispersion patterns are reduced because of little dissection and less favourable mobility under more reducing conditions and abundant organics.	The dissection of terrain increases the possibilities for the intersection of mineralization and oxidizing conditions.

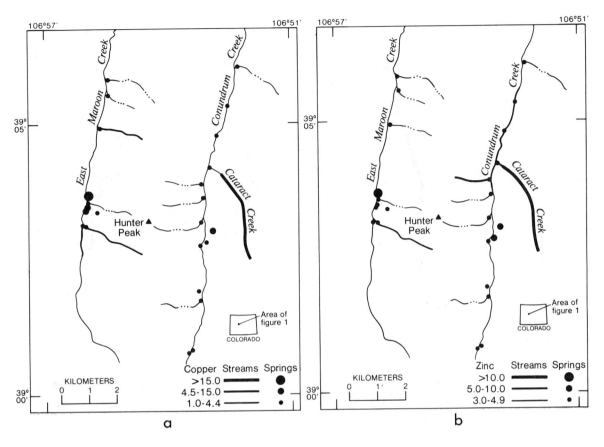
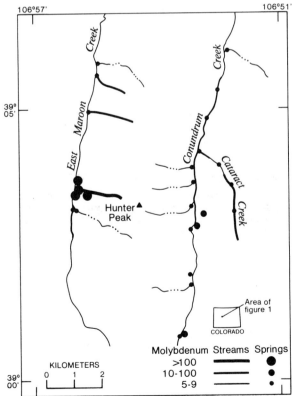

Figure 21A.1

Map showing concentration (μg/L) of (1a) copper, (1b) zinc, and (1c) molybdenum in water collected from streams and springs, White River National Forest, Colorado.

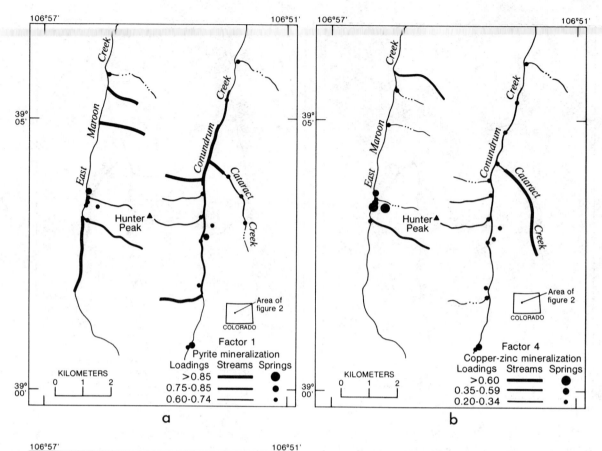
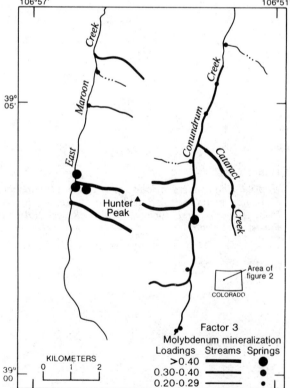

Figure 21A.2

Map showing locations of loading on (2a) factor 1, pyrite mineralization, (2b) factor 4, copper-zinc mineralization, and (2c) factor 3, molybdenum mineralization.

Table 21A.4

Hydrogeochemical indicator elements for different types of deposits (after Beus and Grigorian, 1975).

Type of deposit	Indicator elements of orebodies	
	Strongly oxidizing	Weakly oxidizing
Copper-pyrite	Cu, Zn, Pb, As, Ni, Co, F, Cd, Se, Ge, Au, Ag, Sb	Zn, Pb, Mo, As, Ge, Se, F
Polymetallic	Pb, Zn, Cu, As, Mo, Ni, Ag, Cd, Sb, Se, Ge	Pb, Zn, As, Mo, Ni
Molybdenum	Mo, W, Pb, Cu, Zn, Be, F, Co, Ni, Mn	Mo, Pb, Zn, F, As, Li
Tungsten-beryllium	W, Mo, Zn, Cu, As, F, Li, Be, Rb	W, Mo, F, Li
Mercury-antimony	Hg, Sb, As, Zn, F, B, Se, Cu	Ag, As, Zn, B, F
Gold ore	Au, Ag, Sb, As, Mo, Se, Pb, Cu, Zn, Ni, Co	Ag, Sb, As, Mo, Zn
Tin ore	Sn, Nb, Pb, Cu, Zn, Li, F, Be	Sn, Li, F, Be, Zn
Titaniferous magnetite	Ti, Fe, Ni, Co, Cr	Ni, Fe
Spodumene	Li, Rb, Cs, Mn, Pb, Nb, Sr, F, Ga	Li, Rb, Cs, F
Copper-nickel	Ni, Cu, Zn, Co, Ag, Ba, Sn, Pb, U	Ni, Zn, Ag, Sn, Ba
Beryllium-fluorite	Be, F, Li, Rb, W	Be, F, Li
Baritic-polymetallic	Ba, Sr, Cu, Zn, Pb, As, Mo	Be, Sr, As, Mo

mineralization. The weathering of pyrite releases sulphate and hydrogen ions to the natural waters of the area. The hydrogen ions are then exchanged for calcium ions during the chemical attack of rock minerals, particularly plagioclase and calcite, which are abundant in the area, and this exchange is reflected in the water chemistry. The first factor generally affects waters in the area surrounding Hunter Peak.

Factor 4 (Fig. 21A.2) is interpreted as reflecting copper-zinc mineralization and is represented by loadings on copper, zinc, hydrogen ion, fluoride, and sulphate. This factor shows up strongly in waters along Cataract Creek, in a manner similar to copper and zinc as shown in Figure 21A.1, but factor 4 also shows a general zoning pattern around Hunter Peak, usually within and partly coinciding with the pyrite mineralization factor.

Factor 3 (Fig. 21A.2) is interpreted as reflecting molybdenum mineralization and is represented by loadings of molybdenum, potassium, silica, fluoride, and bicarbonate. The area showing the largest influence by this factor centres around Hunter Peak. Two streams draining the east side of Hunter Peak have only background concentrations of molybdenum (Fig. 21A.1), but loadings for factor 3 for waters of these two streams are anomalous showing the advantage of this technique in detecting patterns. The zoning — with the molybdenum mineralization pattern in the centre, surrounded by the copper-zinc mineralization pattern, and this in turn surrounded by a pyrite mineralization pattern — is interpreted as a possible porphyry molybdenum deposit. In addition a second area of interest is indicated along Cataract Creek and is interpreted as representing hydrothermal vein-type mineralization. A major mineral exploration company is now using this method to explore for molybdenum mineralization in Colorado.

Trend surface analysis, regression analysis, cluster analysis, and weighted sums are a few of many multivariate techniques available for use for interpretation.

The use of thermodynamic data in making equilibrium calculations or the construction of stability diagrams may also prove helpful for interpretation even though organo-metallic reactions are not taken into account. Equilibrium calculations are useful in predicting the forms of migrations of elements in natural waters within given limits for Eh, pH, and activities of aqueous species. The discovery of water with near-equilibrium concentrations of trace metals with respect to corresponding secondary minerals is a good indication of mineralization.

SUMMARY

Conclusions for the application of hydrogeochemical prospecting are: 1) waters collected from first-order streams, springs, and wells are probably the best hydrogeochemical media; 2) when selecting associated and indicator elements for a hydrogeochemical survey, it is important to take into account types of mineralization that may be present and the physical and chemical environment; 3) hydrogeochemical surveys should be carried out within short periods of time to reduce temporal effects; 4) in humid areas low-water level is usually the best time to sample, in arid areas the best time to sample is when the number of springs producing water is at a maximum; 5) false anomalies may be distinguished by using the water chemistry; false anomalies usually have a less-varied suite of trace elements; 6) water is capable of penetrating below the earth's surface and detecting hidden mineralization; and 7) hydrogeochemical prospecting is most effective when combined with other media and methods to obtain maximum information. An example of effective combination of media for geochemical exploration would be the combined use of water and heavy-mineral concentrates as geochemical sampling media. This would be an effective means of obtaining maximum geochemical information because both secondary and primary dispersion patterns may be detected.

Reports on hydrogeochemical prospecting are appearing with more frequency in the literature: an excellent review of hydrogeochemical methods used in the Canadian Shield was given by Boyle et al. (1971). In addition a list of papers (available in English) on hydrogeochemical prospecting for base metals during the last ten years appears in the Selected Bibliography. As those deposits that are easily discovered by conventional methods are gradually exhausted, new methods will become increasingly important. Research needed to make the technique more useful in the future include: 1) improved analytical techniques particularly multielement methods; 2) a better understanding of geochemical cycles in different environments, particularly for lesser-studied elements such as Sn and W; 3) a better understanding of chemical changes in water due to seasonal and rainfall effects; 4) a better understanding of the role of organo-metallic complexes; and 5) the use of interstitial water as a geochemical sampling medium. In addition, a better union between hydrogeochemistry and hydrology is needed to better understand the role of water in the subsurface environment. Although much work remains to be done, hydrogeochemical prospecting will undoubtedly become more important during the next ten years.

SELECTED BIBLIOGRAPHY

Allan, R.J., Cameron, E.M., and Durham, C.C.
  1973: Reconnaissance geochemistry using lake sediments of a 36,000-square mile area of northwestern Canadian Shield (Bear-Slave Operation, 1972); Geol. Surv. Can., Paper 72-50, 70 p.

Allan, R.J. and Hornbrook, E.H.W.
  1971: Exploration geochemistry evaluation study in a region of continuous permafrost, Northwest Territories, Canada; in Geochemical Exploration, Can. Inst. Min. Met., Spec. Vol. 11, p. 53-66.

Barnes, I., Hinkle, M.E., Rapp, J.B., Heropoulos, C., and Vaughn, W.W.
  1973: Chemical composition of naturally occurring fluids in relation to mercury deposits in part of north-central California; U.S. Geol. Surv., Bull. 1382-A, 19 p.

Bell, H. and Siple, G.E.
  1973: Geochemical prospecting using water from small streams in central South Carolina; U.S. Geol. Surv., Bull. 1378, 26 p.

Beus, A.A. and Grigorian, S.V.
  1975: Geochemical exploration methods for mineral deposits; Applied Publishing Ltd., Wilmette, Ill., 287 p.

Boyle, R.W., Wanless, R.K., and Stevens, R.D.
  1970: Sulfur isotope investigation of the lead-zinc-silver-cadmium deposits of the Keno Hill-Galena Hill area, Yukon, Canada; Econ. Geol., v. 65(1), p. 1-10.

Boyle, R.W., Hornbrook, E.H.W., Allan, R.J., Dyck, W., and Smith, A.Y.
  1971: Hydrogeochemical methods – application in the Canadian Shield; Can. Inst. Min. Met. Bull., v. 64(715), p. 60-71.

Brown, E., Skougstad, M.W., and Fishman, M.J.
  1970: Methods for collection and analysis of water samples for dissolved minerals and gases; U.S. Geol. Surv., Techniques Water-Resources Inv., Book 5, Chap. A1, 160 p.

Brundin, N.H. and Nairis, B.
  1972: Alternative sample types in regional geochemical prospecting; J. Geochem. Explor., v. L, p. 7-46.

Cameron, E.M.
  1977: Geochemical dispersion in lake waters and sediments from massive sulphide mineralization, Agricola Lake area, Northwest Territories; J. Geochem. Explor., v. 7, p. 327-348.

Cameron, E.M. and Ballantyne, S.B.
  1975: Experimental hydrogeochemical surveys of the High Lake and Hackett River areas, Northwest Territories; Geol. Surv. Can., Paper 75-29, 19 p.

Collins, B.I.
  1973: Concentration controls of soluble copper in a mine tailings stream; Geochim. et Cosmochim. Acta, v. 37(1), p. 69-75.

Corbett, J.A., Deutscher, R.L., Giblin, A.M., Mann, Q.W., and Swaine, D.J.
  1975: Austr. CSIRO, Min. Res. Lab., Ann. Rept. Vol., p. 17.

Dall'Aglio, M.
  1971: Comparison between hydrogeochemical and stream-sediment methods in prospecting for mercury; in Geochemical Exploration; Can. Inst. Min. Met., Spec. Vol. ll, p. 126-131.

Dall'Aglio, M. and Gigli, C.
  1972: Storage and automatic processing of hydrogeochemical data; 24th Int. Geol. Congr., Montreal, Rept. Sess., v. 16, p. 49-57.

Dall'Aglio, M. and Tonani, F.
  1973: Hydrogeochemical exploration for sulphide deposits – Correlation between sulphate and other constituents; in Geochemical Exploration 1972, Proc. Int. Geochem. Explor. Symp., London, No. 4, p. 305-314.

De Geoffroy, J., Wu, S.M., and Heins, R.W.
  1967: Geochemical coverage by spring sampling method in the southwest Wisconsin zinc area; Econ. Geol., v. 62, p. 679-697.

  1968: Selection of drilling targets from geochemical data in southwest Wisconsin zinc area; Econ. Geol., v. 63, p. 787-795.

Durham, C.C. and Cameron, E.M.
  1975: A hydrogeochemical survey for uranium in the Northwest Territories; in Report of Activities, Part C, Geol. Surv. Can., Paper 75-1C, p. 331-332.

Dyck, Willy
  1974: Geochemical studies in the superficial environment of the Beaverlodge area, Saskatchewan; Geol. Surv. Can., Paper 74-32, 30 p.

Gibbs, R.J.
  1973: Mechanism of trace metal transport in rivers; Science, v. 180(4081), p. 71-73.

Gleeson, C.F. and Boyle, R.W.
  1975: The hydrogeochemistry of the Keno Hill area, Yukon Territory; Geol. Surv. Can., Paper 75-14, 22 p.

Goleva, G.A.
  1968: Hydrogeochemical prospecting of hidden ore deposits; Nedra, translated from Russian by Secretary of State, Canada – copy retained by Geological Survey of Canada Library, 512 p.

Handa, B.K.
  1972: Indirect hydrogeochemical methods for prospecting of copper, lead, and zinc sulphide ores with discussion; in Base Metals, Part II, no. 16, p. 751-761.

Hem, J.D.
  1970: Study and interpretation of the chemical characteristics of natural water; U.S. Geol. Surv., Water-Supply Paper 1473, 2nd ed., 363 p.

Hoag, R.B., Jr. and Webber, G.R.
  1976: Hydrogeochemical exploration and sources of anomalous waters; J. Geochem. Explor., v. 5(1), p. 39-57.

Horsnail, R.F. and Elliott, I.L.
  1971: Some environmental influences on the secondary dispersion of molybdenum and copper in western Canada; in Geochemical Exploration; Can. Inst. Min. Met., Spec. Vol. 11, p. 166-175.

Jonasson, I.R. and Allan, R.J.
  1973: Snow sampling medium in hydrogeochemical prospecting in temperate and permafrost regions; in Geochemical Exploration 1972; Proc. Int. Geochem. Explor. Symp., London, No. 4, p. 161-176.

Kennedy, V.C., Zellweger, G.W., and Jones, B.F.
  1974: Filter pore-size effects on the analysis of Al, Fe, Mn, and Ti in water; Water Resources Res., v. 10, p. 785-790.

Kolotov, B.A. and Rubeykin, V.Z.
  1970: One pattern in composition of natural waters in the upper part of the supergene zone in mountainous folded belts; Dokl. Acad. Sci. USSR, Earth Sci. Sec., v. 191, p. 211-212.

Lalonde, J.
  1973: Dispersion of fluorine in the vicinity of fluorite deposits; in Report of Activities, Part A, Geol. Surv. Can., Paper 73-1A, p. 57.

Levinson, A.A.
  1974: Introduction to exploration geochemistry; Applied Publishing Ltd., Calgary, 612 p.

Miller, W.R. and Ficklin, W.H.
  1976: Molybdenum mineralization in the White River National Forest, Colorado; U.S. Geol. Surv., Open File Rep. 76-711, 29 p.

Perhac, R.M.
  1972: Distribution of Cd, Co, Cu, Fe, Mn, Ni, Pb, and Zn in dissolved and particulate solids from two streams in Tennessee; J. Hydrol., v. 15(3), p. 177-186.

Schwartz, M.O. and Friedrich, G.H.
  1973: Secondary dispersion patterns of fluoride in the Osor area, Province of Gerona, Spain; J. Geochem. Explor., v. 2, p. 103-114.

Shvartsev, S.L., Udodov, P.A., and Rasskazov, N.M.
  1975: Some features of the migration of microcomponents in neutral waters of the supergene zone; J. Geochem. Explor., v. 4, p. 433-439.

Siegel, F.R.
  1974: Applied geochemistry; Wiley-Interscience, New York, 353 p.

Singer, P.C. (ed.)
  1973: Trace metals and metal-organic interactions in natural waters; Ann Arbor Sci. Pub., 380 p.

Sinha, B.P.C.
  1972: Further hydrogeochemical evidence for zinc-copper-lead mineralization in the Gola River catchment, Kumaon Himalayas; in Base Metals, Part II, no. 16, p. 727-730.

Stumm, W. and Morgan, J.J.
  1970: Aquatic chemistry: an introduction emphasizing chemical equilibria in natural waters; Wiley-Interscience, New York, 583 p.

Udodov, P.A., Shvartsev, S.L., Rasstazov, N.M., Matusevich, V.M., and Solodovnikova, R.S.
  1973: A manual of methods used in hydrogeochemical prospecting for ore deposits; Nedra, translated from Russian by Secretary of State, Canada, copy retained by Geological Survey of Canada Library, 270 p.

Vales, V. and Jurak, L.
  1975: Hydrogeochemical prospecting for fluorite and barite in the Zelezne Mountains; SB. Geol. Ved. Loriskova Geol., Mineral. 17, p. 59-94.

APPLICATION OF HYDROGEOCHEMISTRY TO THE SEARCH FOR URANIUM

W. Dyck

Geological Survey of Canada

Dyck, W., Application of Hydrogeochemistry to the search for uranium; in Geophysics and Geochemistry in the Search for Metallic Ores; Peter J. Hood, editor; Geological Survey of Canada, Economic Geology Report 31, p. 489-510, 1979.

Abstract

As the world's surficial mineral deposits become depleted new methods of search for subsurface deposits will have to be developed. The hydrogeochemical technique can be applied to groundwaters but the interpretation of the results is more difficult than for surface waters because of the addition of another dimension.

For the successful application of the hydrogeochemical method of prospecting a thorough knowledge of the chemistry of the elements, the composition of the rocks and the movement of the water are essential. At the surface this information is acquired relatively easily, below the surface with some difficulty.

A literature search reveals that by far the greatest amount of work on hydrogeochemical prospecting is published in Russian journals. While few prospectors can use these publications directly, several translation services have translated numerous major works.

The development of rapid and accurate methods of detection of uranium and its decay products in natural waters at very low concentrations has made the hydrogeochemical method a powerful tool in the search for U ore deposits.

The elements of interest in hydrogeochemical exploration for uranium are uranium and its decay products-radium, radon, and helium. Each element has specific radiochemical and/or geochemical properties which make it a useful tracer for uranium ore deposits. Recent studies indicate that isotopic data of these elements are useful in the interpretation of simple abundance anomalies and the estimation of ore potential.

Uranium is easily oxidized to the hexavalent state in the presence of oxygen in natural waters. Its mobility in surface and near surface waters is greatly enhanced by the complexing action of carbonates and humates in neutral and basic waters, of sulphates in acid waters, and of phosphates and silicates in neutral waters. Organic matter adsorbs uranium strongly and is responsible for decreasing migration of the uranyl ion in surface waters. The great abundance of bicarbonates in groundwaters of sedimentary rocks results in intensive leaching and wide dispersion of uranium in the ground in the zone of oxidation.

The hydrogeochemical techniques employing radium and/or radon are best suited to detailed or semi-detailed investigations of radioactive occurrences. Their ease of detection and short range make them good tracers for pinpointing uranium occurrences or outlining radioactivity too weak for the gamma-ray spectrometer or the fluorimeter.

Helium, because of its inertness and great mobility has the potential of revealing uranium orebodies through much greater thickness of cover than any other geochemical or geophysical technique, but the exact same factors can also cause false anomalies.

Résumé

Au fur et à mesure de l'épuisement des gîtes minéraux superficiels dans le monde entier, il faudra mettre au point de nouvelles méthodes d'exploration des gîtes de subsurface. On peut appliquer les techniques hydrogéochimiques aux eaux souterraines, mais l'interprétation des résultats est plus ardue que dans le cas des eaux de surface, en raison de l'introduction d'une autre dimension.

Pour pouvoir appliquer avec succès la méthode hydrogéochimique de prospection, il est essentiel de connaître parfaitement la chimie des éléments, la composition des roches et le mouvement des eaux. A la surface, il est relativement facile d'obtenir cette information, par contre, en profondeur, la situation est moins facile.

Un examen de la documentation montre que jusque-là, la majorité des travaux relatifs à la prospection hydrogéochimique ont paru dans des revues scientifiques soviétiques. Comme peu de prospecteurs sont capables de lire le texte original, des services de traduction ont été chargés de traduire un grand nombre des principales publications à ce sujet.

La mise au point de méthodes rapides et précises de détection de l'uranium et de ses produits de désintégration dans les eaux naturelles à des concentrations minimes, explique que la méthode hydrogéochimique soit devenue un instrument précieux pour l'exploration des gîtes uranifères.

Les éléments considérés au cours de l'exploration hydrogéochimique de l'uranium sont l'uranium et ses produits de désintégration — le radium, le radon, et l'hélium. Chaque élément a des propriétés radiochimiques et ou géochimiques spécifiques, qui en font un traceur utile pour la recherche des gîtes uranifères. Des études récentes indiquent que les données isotopiques relatives à ces éléments facilitent l'interprétation des anomalies simples, et l'évaluation du potentiel minier.

L'uranium s'oxyde facilement à l'état hexavalent en présence d'oxygène dans les eaux naturelles. Sa mobilité dans les eaux de surface et peu profondes est nettement renforcée par la faculté des carbonates et humates à former des complexes avec l'uranium dans les eaux neutres et basiques, ou des sulfates dans les eaux acides, et la faculté des phosphates et silicates à former des complexes dans les eaux neutres. La matière organique absorbe fortement l'uranium, et réduit la migration de l'ion uranyle dans les eaux de surface. La grande abondance des bicarbonates dans les eaux souterraines qui traversent les roches sédimentaires favorise une dispersion et un lessivage importants et étendus de l'uranium dans le sol, dans la zone d'oxydation.

Les techniques hydrogéochimiques où l'on utilise le radium ou le radon, ou les deux à la fois, sont celles qui conviennent le mieux à une exploration détaillée ou semi-détaillée des venues radioactives. La facilité avec laquelle on les détecte et leur faible domaine d'influence en font de bons traceurs, pour localiser les gîtes uranifères ou délimiter toute radioactivité trop faible pour le spectromètre gamma ou le fluorimètre.

En raison de son inactivité chimique et de sa grande mobilité, l'hélium permet de détecter les corps uranifères à travers une couverture beaucoup plus épaisse, mieux que toute autre technique géochimique ou géophysique, mais ces mêmes caractères peuvent aussi donner lieu à de fausses anomalies.

INTRODUCTION

Hydrogeochemical prospecting is still in its infancy, but mankind has used it since the dawn of history with the aid of human senses, sight, taste and smell (Boyle, 1967; Boyle and Garrett, 1970). Modern ultrasensitive methods of detecting trace elements are partly responsible for the increased use of hydrogeochemistry in exploration. Perhaps equally important in the development of hydrogeochemical methods of exploration is the ever increasing demand for minerals and the need to detect buried deposits. The element uranium, more than any other, has become an important metal and presently receives more exploration attention than gold.

This paper describes the state of the art of hydrogeochemical exploration methods for uranium ore deposits. Because of the great wealth of published information, references with a few exceptions will be limited to the last ten years; and even then the bibliography will be incomplete because of lack of time and access to certain publications. The state of the art of a prospecting method for a certain element reflects current demand for that element, present knowledge of the geochemistry of the element, and technological advancements. The demand for uranium and technological advancements have increased greatly in the last decade. While our knowledge of uranium geochemistry is steadily increasing, certain aspects of its chemical behaviour in the natural environment have been known for sometime (Rankama and Sahama, 1950; Baranov, 1961; Hawkes and Webb, 1962; Vinogradov, 1963; Krauskopf, 1967; Levinson, 1974). The essential features of the geochemistry and methods of search for uranium in the hydrosphere are summarized in the following pages.

PRINCIPAL RADIOCHEMICAL AND GEOCHEMICAL PROPERTIES OF URANIUM

Geological models play a vital role in pointing to regions of high ore potential. But one must be on guard against the tendency to let established principles negate yet undiscovered geological settings. As important as anologous settings are, the actual location of a U deposit is carried out by means of instruments and techniques that respond to the inherent radiochemical and geochemical properties of U and its decay products. It is therefore important that the prospector be familiar with them.

The ultimate origin of U remains a mystery, but isotope work suggests that there were periodic injections of U into the crust from the mantle (Cherdyntsev, 1971). Pure U metal comprises essentially two isotopes ^{238}U (99.3%) and ^{235}U (0.7%). A 3rd isotope, ^{234}U, results from the decay of ^{238}U but its isotopic abundance is only 0.006%. Even so, it plays an important role in studies of the geochemistry of U, particularly in disequilibrium studies (Cherdyntsev, 1971; Osmond and Cowart, 1976; Cowart and Osmond, 1977). These U isotopes have nearly identical physicochemical properties and hence are never found separate in the natural state. But it is actually the ^{235}U that is used in nuclear reactors. Fortunately the isotopic abundance is quite constant except one ore deposit in Gabon, reportedly has only 0.4% ^{235}U in a small section of high grade ore and there is evidence that this orebody was a natural reactor at one time (Nicolli, 1973).

The principal decay products of ^{238}U and main radioactive emissions are shown in Figure 21B.1. The elements of interest to the exploration geochemist are U, Ra, Rn, and He. Several methods are available for the determination of U in sub-ppb amounts in waters, the Ra-Rn couple is relatively easily measured and is quite specific for U, and mass spectrometric techniques are now available that will measure natural levels of He with relative ease. Some of the other elements in the series are either difficult to detect or have too short a half life, or are not specific for U. The various characteristic gamma rays emitted by the decay products and the branching decay modes are not shown in Figure 21B.1. For a classic account on nuclear and radiochemistry, the reader is referred to Friedlander et al. (1964).

Geochemically U is a strongly lithophile and oxyphile element; it has never been observed in the native state in nature. The most important mineral of U is UO^2 in uraninite and pitchblende. Some other uranium minerals found in nature include hydroxides such as bequerelite, uranomicas e.g. $(2UO_2 \cdot 3H_2O) K[UO_2/VO_4] \cdot 1\ 1/2\ H_2O$, arsenates, vanadates, uranates, carbonates, silicates, phosphates, and sulphates. Its close association with Th in crystalline rocks is believed to be due to the fact that the ionic radii and simple chemistry of U^{IV} and Th^{IV} are quite similar (.97 and .95 respectively).

What makes hydrogeochemical prospecting possible are chemical processes which can be presented symbolically by the following interactive steps:

$$UO_{2(s)} + H_2O \underset{\text{red.}}{\overset{\text{ox.}}{\rightleftharpoons}} UO_2^{2+}{}_{(aq)} \underset{H^+}{\overset{OH^-}{\rightleftharpoons}} UO_2^{2+}{}_{(s)} + H_2O$$

Not shown in these processes is the importance of CO_2, HCO_3^- and CO_3^- in the leaching and complexing of U in the natural environment. In fact Tugarinov (1975) claims that the CO_2 regime is the main factor in dissolving, transporting, and depositing U. At CO_2 concentrations of the order of 80 g/L and elevated temperatures U is leached from rocks at great depth and deposited from ascending solutions at CO_2 concentrations of 5 g/L or less; CO_2 is lost by degassing or reactions with enclosing rocks.

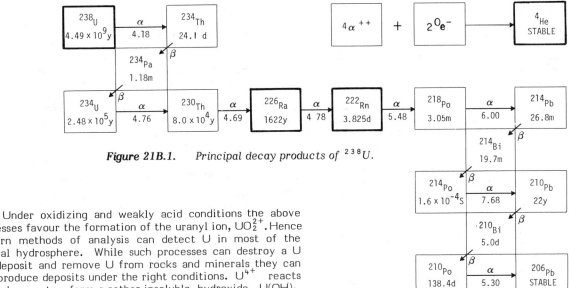

Figure 21B.1. Principal decay products of ^{238}U.

Under oxidizing and weakly acid conditions the above processes favour the formation of the uranyl ion, UO_2^{2+}. Hence modern methods of analysis can detect U in most of the natural hydrosphere. While such processes can destroy a U ore deposit and remove U from rocks and minerals they can also produce deposits under the right conditions. U^{4+} reacts with bases to form a rather insoluble hydroxide $U(OH)_4$ ($K_i=10^{-46}$ at 25°C). This corresponds to a U^{4+} concentration of about 0.02 ppb at a pH of 5. Because the hydroxide is unstable and dehydrates, $U(OH)_4 \to UO_2 + 2H_2O$, actual concentrations of U^{4+} are even smaller than that. Some of the pH-Eh relationships of U have been described in some detail by Hostetler and Garrels (1962) and Langmuir and Applin (1977). Under appropriate conditions of pH and Eh water samples in the midst of a U deposit can be void of detectable amounts of U. Fortunately for the prospector U^{4+} has a strong affinity for O_2. Hence in the zone of oxidation, which can be from several tens to several hundred metres thick, U^{4+} is oxidized to the rather soluble UO_2^{++} ion. This oxygen is supplied mainly by air dissolved in water during the hydrologic cycle, although auto-oxidation of uraninite also takes place. This latter process is believed to be partly responsible for the occurrence of pitchblende, the mineral with a varying composition of UO_2 and UO_3, depending on the age and preservation of the mineral. As oxygen-bearing water finds its way through microfissures, made more abundant in the vicinity of U minerals as a result of radiation damage, it oxidizes and hydrolyses the U^{4+}. This process of solution of U^{4+} is aided even further by the formation of soluble complexes such as bicarbonate, carbonate, sulphate, fluoride, phosphate and silicate. A prominent role in the complexing of U species is played by organic matter particularly humic acids. In basic waters more of the humic acids dissolve and hence keep U in solution. In neutral or lightly acidic waters U is complexed by soluble fulvic acids or absorbs on the solid phase of organic matter and falls to the bottom of lakes and streams. For details of our present state of knowledge on the geochemistry of U the reader is referred to Krauskopf (1967), Vinogradov (1967), Yermolayev et al. (1968), Mann (1974a, 1974b), Roylance (1973) and Langmuir and Applin (1977).

Before describing field and analytical methods and results of the hydrogeochemical technique the problem of radioactive and chemical disequilibrium should be discussed. There are several pitfalls in the use of radiometric techniques in the search for surficial U deposits. These pitfalls result from the weathering of geologic materials which causes fractionation of U from its decay products. In the zone of oxidation the mobility of U is generally greater than that of its decay products. This means that the surface and near-surface waters will contain relatively more U than decay products according to decay chain predictions. Much of this U reacts chemically and partitions between soluble complexes, such as humic acids and carbonates, and the organic matter and clays in the bottom of lakes and streams.

Conversely, the soils and rocks from which this U is leached contain an excess of U-decay products. Under reducing conditions U is not mobile, in fact, it precipitates from solution. The greater apparent mobility of Ra under reducing conditions comes about in an indirect way. Although relatively insoluble, in natural waters, Ra will go into solution particularly in the presence of chlorides. However, actual Ra concentrations are, as a rule, much lower than those calculated from solubilities of its most insoluble salts, sulphates and carbonates. Most likely Ra coprecipitates with its more abundant group members Ba and Ca and freshly forming hydrous oxides of Fe^{3+} and Mn^{4+}. Fe and Mn in their lower oxidation states will remain in solution so that salts of Ra, Ba and Ca in concentrations lower than their solubilities have no host to coprecipitate on. Conversely, under reducing conditions the soils and rocks retain U and become depleted in Fe and Mn and certain decay products of U, particularly Ra.

In order to avoid pitfalls it is therefore essential to measure not only U and Ra or Rn but also such parameters as salinity and alkalinity and oxidation-reduction couples such as $Fe^{3+} - Fe^{2+}$. In prospecting, relatively inexpensive approximations of the above parameters can be obtained by measuring conductivity, dissolved O_2 Eh and pH. The reader who wishes to acquaint himself in more detail with the problem of radioactive disequilibrium may consult Dooley et al. (1966), Starik et al. (1967), Cherdyntsev (1971), Cowart and Osmond (1974, 1976, 1977), Osmond and Cowart (1977). For details on laboratory leaching experiments see Starik et al. (1967), Szalay (1967), Szalay and Samsoni (1970), Gavshin et al. (1973), Roylance (1973).

Recently Cowart and Osmond (1977) measuring radioactivity ratios of ^{234}U and ^{238}U, found a sharp rise in this ratio in groundwater aquifers downdip from sandstone-type uranium deposits, even though the total uranium content in these waters decreased. The reasons for this rise in activity ratio is not quite clear. One possible explanation may be the fact that ^{234}U is bound more loosely than ^{238}U because of recoil during the two decay steps from ^{238}U to ^{234}U and hence is easier oxidized and mobilized during auto-oxidation in U ore zones where radiation damage is more pronounced than in country rock.

Seasonal variations will also affect the size and intensity of an anomaly. When precipitation varies during a season the amount of water in lakes and streams varies correspondingly resulting in inverse relationship between U content and rate of discharge (Ridgley and Wenrich-Verbeek, 1978) or size of lake or stream (Dyck et al., 1970). In lakes and streams ionic species vary less than dissolved gases. During the course of a season Dyck and Smith (1969) found that Rn levels increased considerably in lakes and streams covered with ice whereas U concentrations remained the same in lakes but decreased in streams during the winter months. Lake bottom waters from the Key Lake, Saskatchewan area contained only 25% less U in early summer than in the winter but the average Rn content was lower by a factor of 3 and the average He content by a factor of 8 (Dyck and Tan, 1978). Korner and Rose (1977) and Rose and Keith (1976) found large fluctuations in the U and Rn content of streams with time but were unable to relate them clearly to seasonal variations but rather ascribed them to fluctuations in proportion of groundwater to surface water and to amount of precipitation. Germanov et al. (1958) illustrate clearly the effect atmospheric precipitation has on the U content of spring waters, dropping from 10 ppb to about 1 ppb with a rise in precipitation from 200 mm to 500 mm.

The relative change in concentration of U and Ra (Rn) with depth is illustrated in Figure 21B.2, and observed concentration levels and ratios for various environments are given in Table 21B.1. The variations with depth are rather generalized and exceptions will occur depending on the groundwater flow pattern in a particular region. Smith et al. (1961) found an increase in Rn content with depth of well when areas of several square miles were considered but over large areas, the depth of a well did not influence the Rn content noticeably. In Eastern Maritime Canada U, Rn, and He in well waters increased with depth as indicated by positive, though weak, correlation coefficients (Dyck, Chatterjee et al., 1976). In the Cypress Hills, Saskatchewan area (Dyck et al., 1977) by contrast both U and Rn levels decreased with depth. Obviously to explain such observations requires a great amount of effort and knowledge about the variables responsible for such changes. These factors are: Geography and topography, geology, type of U mineralization and location, atmospheric precipitation, groundwater movement, and rock type.

ADVANCES IN ANALYTICAL TECHNIQUES

The four elements (U, Ra, Rn, and He) of interest to the uranium hydrogeochemist vary greatly in their properties. Hence each requires specific modes of preservation and analysis. Advances in sample handling and analysis are therefore described below for each element separately.

Uranium Analysis

The collection and storage of water samples for U analysis is less of a problem than for other trace elements because it forms stable complexes with the ever present bicarbonates, carbonates and sulphates in natural waters. Even so, when U concentrations drop to 0.1 ppb or lower, loss of U to the walls of the containers in which samples are stored and labware used during analysis becomes appreciable. Hence, it is advisable that such samples be treated with a complexing agent or acidified prior to analysis to a pH of 1.

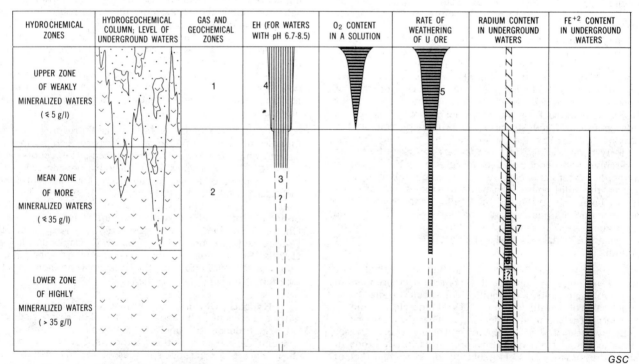

1. Atmospheric O_2; atmospheric and partially soil N_2; atmospheric and soil CO_2.
2. Atmospheric and partially soil N_2; atmospheric and soil CO_2.
3. From 300 (200?) to zero.
4. From 525 to 300 mv, more seldom to 250 (200?).
5. In a solution up to $n \times 10^{-2}$ g/l.
6. In oreless areas.
7. In ore areas.

Figure 21B.2. Hydrogeochemical zonality in rocks void of organic matter (after Germanov et al., 1958).

Table 21B.1
Types, conditions of formation, and chemical and radioactive constituents of natural waters
(After Novikov and Kapkov, 1965)

| Radiological type of water | Genetic type of water | Hydrogeological type of water | Conditions of formation | | | | Gas composition | Chemical composition | $\frac{^{234}U}{^{238}U}$ | Radioactivity | | | | |
			Lithological	Structural	Hydrodynamic					Rn pCi/L	Ra g/L	U g/L	$\frac{Rn}{Ra}$	$\frac{Ra}{U}$
Radon (bearing)	weathering joints	Fracture and ground	massifs of acid magmatic rocks	Open structures (foothills)	Intensive water exchange	O_2, CO_2	HCO_3^-		>10^4	10^{-12}–10^{-13}	10^{-6}–10^{-7}	>>1	≤1
	emanating collectors	Ground and head			Various	CO_2, N_2			≤10^5	10^{-12}	10^{-6}	>>1	1–3
	Tectonic joints in granites	Head-fracture and vein			Intensive water exchange	CO_2 N_2	Ca, Mg t to 100°C		<10^4	10^{-12}	10^{-6}	>>1	2–3
Radium (bearing)	Sedimentary and metamorphic rocks	Stratal and fracture head	Sedim. and metamorphic rocks	Closed structures (basins)	Very restricted exchange	H_2S CH_4	Na, Ca, I, Br, Cl		<10^4	10^{-9}–10^{-10}	10^{-7}–10^{-8}	≤1	>>1
Uranium (bearing)	Surface water reservoirs	Closed reservoirs and streams	Considerable evaporation from reservoirs without outlets			O_2			<10^3	10^{-11}–10^{-12}	10^{-2}–10^{-4}	~1	<1
Uranium Radium (bearing)	Sedimentary and metamorphic rocks rich in disseminated U	Stratal and fracture, ground and head	Sedimentary and metamorphic rocks	Open structures	Intensive exchange	O_2	Various	1.5–2.5	<10^3	10^{-10}–10^{-12}	10^{-3}–10^{-5}	~1	~1
				Semi-open structures	Restricted exchange	CH_4	"	1.5–2.5	<10^3	10^{-9}–10^{-10}	10^{-5}	~1	>1
Uranium Radon (bearing)	Fracture zones of magmatic rocks; oxidation of ore deposits	Fracture and/or stratal	Acid magmatic rocks	Fully open Structures	Intensive exchange	O_2	"	2–5	10^4	10^{-11}–10^{-13}	10^{-4}–10^{-5}	>1	<1
			U ore deposits			CO_2		~1		10^{-10}–10^{-12}	10^{-2}–10^{-5}	>1	<1
Uranium Radon Radium (bearing)	Zone of oxidation of U ore deposits	As above	Ores with rich U inclusions	Open and semi-open structures	As above	O_2 CH_4	"	~1	10^4–10^6	10^{-8}–10^{-10}	10^{-1}–10^{-4}	>1	<1
			Ores with disseminated U			CO_2		1–10	10^4–10^6	10^{-8}–10^{-11}	10^{-1}–10^{-5}	>1	~1
Radon Radium (bearing)	Primary ore and U ore reduction zones	Fracture-stratal-head	As above	As above	Restricted exchange	H_2S CO_2	"		10^4	10^{-8}–10^{-10}	10^{-6}–10^{-7}	>1	>>1

This acidification can lead to false anomalies if suspended matter is present in the water (Wenrich-Verbeek, 1977). In surface waters this is usually a fluffy organic substance which has U adsorbed on it. Filtering such samples prior to acidification removes not only the suspended matter but also some of the dissolved U by adsorption on the filters. There is therefore no real satisfactory way of treating such samples short of analyzing in situ. Recently Parslow (1977) has worked out an attractive method which involves dropping a tea bag filled with ion exchange resin into a freshly collected water sample. NURE uses a similar method. This not only circumvents the acid-suspended matter problems and minimizes the adsorption on labware but also preconcentrates the sample, making it possible to determine very low concentrations of U (0.05 ppb) with good precision.

Surface waters in arid regions and groundwaters in sedimentary basins usually contain sufficient dissolved solids and U (> 0.1 ppb) so that adsorption on walls of bottles is no problem. However, acidification is still advisable, for groundwaters usually contain enough Fe and Mn in the lower valence states to form precipitates on exposure to air which then carry U down by adsorption. Frequently carbonates precipitate upon loss of CO_2.

By far the largest number of U analyses in waters have been carried out by the fluorimetric technique. There are probably as many variations on this method as there are laboratories. In low conductivity surface waters it can detect about 0.2 ppb U using a 5 ml sample with sample pretreatment (Thatcher and Barker, 1957; Samsoni, 1967; Smith and Lynch, 1969. However, in the presence of Fe, Mn, and other dissolved salts the U fluorescence is quenched resulting in poor analytical precision and loss of sensitivity. For surface waters in humid zones this presents no problem but waters from arid regions, or groundwaters may be affected. To get around this, various solvent extraction and ion exchange techniques are employed (Centanni et al., 1957; Danielson et al., 1973; Parslow, 1977).

More recently nuclear reactors are being employed for U analysis. The two methods, delayed neutron counting analysis and fission track counting are providing quantitative U determinations although the DNC method requires preconcentration of U in natural waters, for its detection limit is about 1 microgram (Ostle et al., 1972). The fission track method can detect 0.01 ppb U in a drop of water, (Reimer, 1975; Fleischer and Delany, 1976), and indicate the proportion of U in true solution and in suspended solids. However, fission track counting is still tedious when done manually and rather expensive by image analysis. A novel innovation of the fluorimetric method, employing laser excitation is presently undergoing evaluation as a rapid technique for U in natural waters (Robbins, 1977). This instrument has the potential of analyzing for U in situ at the 0.1 ppb level.

For U concentrations greater than 1 ppb a colorimetric technique can be used (Hunt, 1958; Smith and Chandler, 1958), and for concentrations of > 0.01% even a spot test can be employed to identify U in minerals (Goldstein and Liebergoot, 1973).

Radon-222 and Ra-226 Analysis

Because ^{222}Rn has a rather short half life ($t_{1/2}$ = 3.8 days) it depends greatly on the presence of Ra for its continued existence. Hence, methods suitable for Rn determinations will usually also measure Ra. There are a great many methods (based on the ionization chamber principle, - the ZnS (silver activated) scintillation phosphors, charged electrode collection method, alpha track etching, nuclear diode alpha counting) for measuring Rn in atmospheric air and soil emanations. For Rn in water determinations, the ZnS counter is used mostly, although the ionization chamber technique can also be employed; but the latter is more cumbersome for comparable sensitivity. With care the ZnS method can detect 1 pCi/L Rn and hence 1 pCi/L Ra (Lucas, 1957; Baranov, 1961; Higgins et al., 1961; Rushing, et al. 1964; Novikov and Kapkov, 1965; Sedlet, 1966; Dyck, 1969; Allen, 1976; Naguchi and Wakita, 1977). The classical method of coprecipitation of Ra with $BaSO_4$, while more sensitive than the Rn-Ra method requires very large samples and is therefore too cumbersome for prospecting. Although the ZnS method is relatively simple to use it does require somewhat more skill than that required for a scintillometer. Three factors require special attention for high quality data. 1) Selection of sample site; very turbulent streams and lakes will contain less radon than quiet ones. Samples more than 5 to 10 metres from the bottom of a lake or stream will seldom contain measurable concentrations of Rn. 2) Loss of Rn during collection and storage. When domestic wells are sampled the aerator on kitchen sink taps should be removed. Glass bottles filled completely and closed tightly with screw caps or bottle caps will avoid Rn escape during transport and storage. 3) Stray light can easily introduce false counts into the instrument. Hence regular background checks and two successive sample counts are advised.

Helium Determinations

The predominant technique of measuring He employs a mass spectrometer. It is specific in that it identifies the element and determines the amount simultaneously. Without some form of enrichment most mass spectrometers can detect He in the ppm range only. By condensing the major gas components into activated charcoal cooled with liquid nitrogen, and operating the mass spectrometer in the static mode, fractions of ppm may be measured. A portable battery operated instrument based on the mass spectrometer principle and suitable for the detection of 50 ppm or more He has been described by Eremeev et al. (1973). This instrument is capable of analyzing up to 30 samples per hour but lacks the sensitivity required for near-surface samples. A portable He analyzer capable of measuring 1% or more has been described by Sonnek et al. (1965). It works on the principles of chromatography and thermistor response.

He analyses were carried out before the days of the mass spectrometer with the aid of a manometer, liquid nitrogen and activated charcoal. This method is still being used today by some scientists (Penchev, 1969). More recently He leak detectors have been modified for field work (Goldak, 1974; DeVoto et al., 1976; Reimer, 1976; Dyck and Pelchat, 1977). There is also on the market a chromatographic technique using a He ionization detector with sufficient sensitivity for natural samples (Carlo Erba, 1976).

One of the most difficult steps in the He method is the retention of the He in the sample until it is analyzed. For gases, high-vacuum valves and containers are required which make the sampling of large numbers of samples expensive. Clarke and Kugler (1973) have adapted a method involving annealed copper tubing and special pinch clamps for the collection of water samples. At the Geological Survey of Canada water samples are collected in 300 ml soft drink-type glass bottles. Filled completely and capped properly these samples will retain He for a long time (about 10% loss per month) (Dyke et al., 1976). The determination of the isotope ratio of ^{4}He/^{3}He is much more difficult than total He determination and can only be carried out with a high-sensitivity and high-resolution mass spectrometer. Since there is only about 1 ppm ^{3}He in natural He, a mass spectrometer has to measure precisely 1 part of less in 10^{12} parts of air. To achieve such a low detection limit, special procedures and apparatus such as preconcentration of sample, operation of mass spectrometer in the static mode, high-gain multipliers, etc., are employed. A resolution of 600 or more is required in order to separate the ^{3}He peak from the mass 3 peaks produced by (^{1}H^{2}D) and ^{3}H. Details of two such analytical facilities have been described by Mamyrin et al. (1970) and by Kugler and Clarke (1972). Such complex instruments and procedures make He isotope determination very expensive.

APPLICATION OF PRINCIPLES AND TECHNIQUES TO U PROSPECTING

The action of water on soils, rocks, and minerals is the basis for hydrogeochemical prospecting. The hydrologic cycle provides the continuity for this action. The moment rain touches the earth a series of chemical and physical forces come into play. One of the best summaries of the fate of U and Ra in the hydrologic cycle is given by Germanov et al. (1958). No prospector should venture forth without the knowledge contained in that report. Although 20 or more years old and dealing with eastern European settings mainly, its findings can be applied equally well everywhere to groundwater prospecting for U. Other noteworthy publications on the subject resulting from the U boom in the fifties are by Denson et al. (1956), Fin (1956), UN (1955, 1958), Phoenix (1960). In wilderness areas such as northern Canada surface waters are being used as a complement to lake and stream sediment surveys. To focus in some detail on past experiences the elements U, Ra and Rn, and He will be discussed separately. In a general way U is best suited for prospecting on a regional scale in the zone of oxidation; Ra and Rn are well suited for semidetailed and detailed investigations in either the oxidizing or reducing environments and He, although not yet well understood, seems to be most suitable for detailed subsurface prospecting.

Uranium

Surface lake and stream waters have been used extensively for U exploration. In humid regions U concentrations are generally low and it is desirable to have a method that can detect 0.05 ppb U particularly in granitic terrane and terrane with extensive overburden (Chamberlain, 1964; MacDonald, 1969; Meyer, 1969; Dyck et al., 1970; Durham and Cameron, 1975; Dyck and Cameron, 1975; Cameron and Hornbrook, 1976; Jonasson, 1976; Cameron and Ballantyne, 1977; Coker and Jonasson, 1977). Because of its great mobility U in waters is an excellent regional tracer and sample densities of 1 sample/30 km^2 will outline significant U mineral zones (Dyck, 1975) see also Figure 21B.3. Since organic matter adsorbs U strongly and carries it to the bottom of lakes and streams, sediments are sometimes the preferred medium (Cameron and Hornbrook, 1976; Ferguson and Price, 1976), depending on the cost of sample collection and use of samples. However, the competition for the uranyl ion between soluble complexes and solid organic matter will distort the size and intensity of anomalies (Dyck, 1975; see also Maurice, 1977). It is therefore important that notice be taken of the organic matter in the sediment and the alkalinity of the water in a survey area. A simple approximation of alkalinity is obtained by measuring the conductivity of the water and of organic matter by measuring the volume of unit weight of sediment. Breger and Deul (1955) also recognized the ability of organic material to adsorb uranium from water. However, in alkaline waters U adsorption is weaker than in neutral waters (Doi et al., 1974; Lopatkina, 1970). Similarly there is evidence that in acidic waters U adsorption decreases. Salinity or conductivity were found to be useful in interpreting stream water U anomalies (Dall'Aglio, 1971) and in the searching of calcrete-type uranium deposits in arid regions (Premoli, 1976). A knowledge of the presence of carbonate rocks also helps in the interpretation of U anomalies. Sergeyeva et al. (1972) have shown that UO_2CO_3 is moderately soluble in neutral and alkaline waters.

Haglund et al. (1969) found that U in limestones is easily leached out. The effect of alkalinity and organic matter on the partition of uranium in stream water and sediment is illustrated in Figure 21B.4.

In areas where U mines are in operation U values in waters tend to rise to tens and even hundreds of ppb due to contamination attesting to the high mobility of U but confusing the prospector in terms of assigning significance to an anomaly in nearby virgin territory. A similar rise in U values is observed in streams in arid and semi-arid regions particularly in regions with radioactive coal seams (Boberg and Runnells, 1971). In mountainous terrain with carbonate rocks alkaline stream waters with ppb levels of U are common (Illsley, 1957; Ballantyne, 1976). In areas with known U mineralization alkaline lakes without outlets can attain several thousand ppb U (Culbert and Leighton, 1978). Similar results were obtained by Kyuregyan and Kochargan (1969) from less saline waters in the Caucasian district of Russia. Saline lakes and their deposits contain much less U in the absence of mineralization (Bell, 1955; 1960). Rose and Keith (1976) found large variability in Pennsylvania stream waters and hence opted for sediment as the preferred medium. Although stream and lake sediments cost more to collect they do have the advantage of averaging out seasonal fluctuation of U and other elements observed in waters (Rose and Keith, 1976).

As surface water becomes groundwater it picks up CO_2 from decaying organic matter and carbonates and becomes a more effective leaching agent. Initially Ca and Mg dominate the population of cations but these are gradually replaced by Na and K as the water sinks and moves through clay minerals. Trace elements also go into solution during this leaching process, particularly the oxyphile elements such as U. The concentration of total dissolved solids is also a function of the annual precipitation and relief. Significant U concentration levels are usually greater than 1 ppb in groundwaters from sedimentary environments so that analytical requirements are less stringent, but interferences from quenchers are more pronounced. The effect of rainfall, relief, and total dissolved solids is evident in the groundwater results obtained from two sedimentary environments, Eastern Maritime Canada and Cypress Hills, Saskatchewan. To see the effect of these parameters it is necessary to compare U and Rn patterns and concentrations. These are illustrated in Figures 21B.5, 21B.6, 21B.7, and 21B.8. In the Maritimes where the annual precipitation is 80 cm and relief is gentle the average conductivity, U and Rn concentrations of 1700 well waters from a 25 000 km^2 region were 301 micromhos/cm, 1.0 ppb and 857 pCi/L respectively (Dyck et al., 1976a). There is good spatial and mathematical correlation between these variables. It should be noted here that the positive association of uranium and flourine in alkaline waters observed in this study and by others (Doi et al., 1975; Ballantyne, 1976; Boyle, 1976; Culbert and Leighton, 1978) suggests a common source such as radioactive pegmatites or minerals such as U-rich flourapatite. Thermodynamic calculations by Langmuir and Applin (1977) show that fluorine complexes with U only in fairly acid media (pH <4.5) and the phospate ion forms stable U complexes in weakly acid and neutral waters (pH 4.0 to 7.5). Little is known about these complexes in natural waters and some research is in order. In the Cypress Hills, relief is pronounced and average annual precipitation about 25 cm. There, 900 well water samples from a 15 000 km^2 area gave 1460 micro mhos/cm, 11.5 ppb U, and 355 pCi/L on the average. We see here a significant drop in precipitation and Rn content and an even more significant rise in U and conductivity compared to the Maritime data. Furthermore, in the Cypress Hills area the Rn highs cluster in the high country and the Cypress Hills Formation whereas U is concentrated on and down the slopes of the hills. Where relief becomes gentle, as in the north and

Figure 21B.3.

Uranium in surface lake waters, Beaverlodge, Saskatchewan, 1969; Effect of sample density.

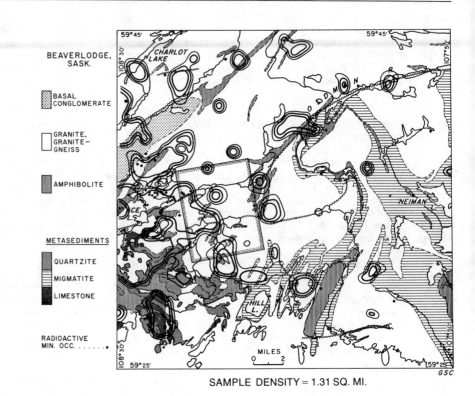

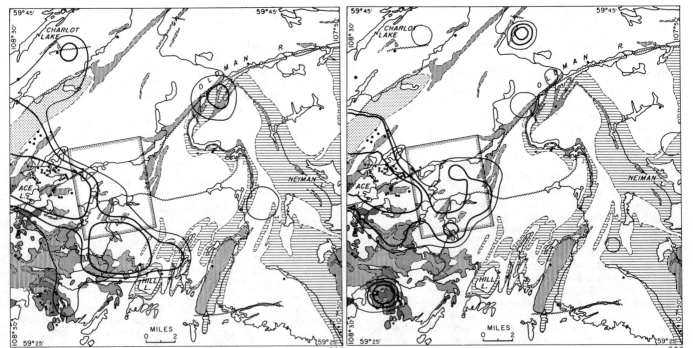

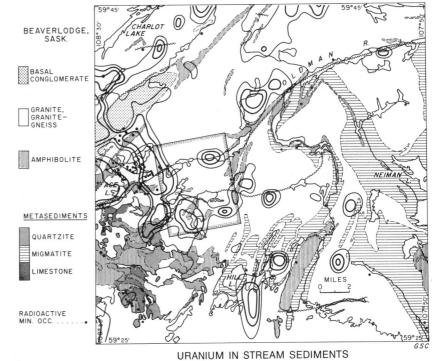

Figure 21B.4.

Effect of alkalinity and organic matter on the partition of U in stream water and sediment, Beaverlodge, Saskatchewan.

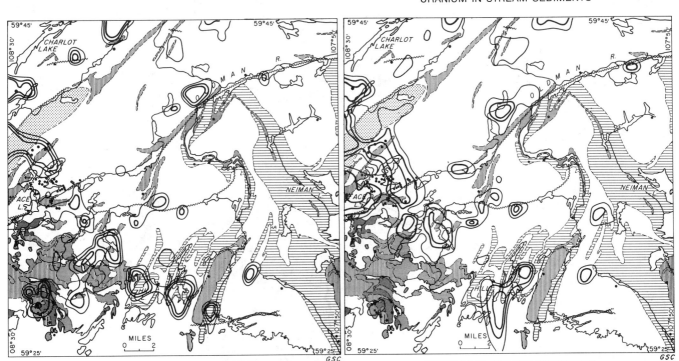

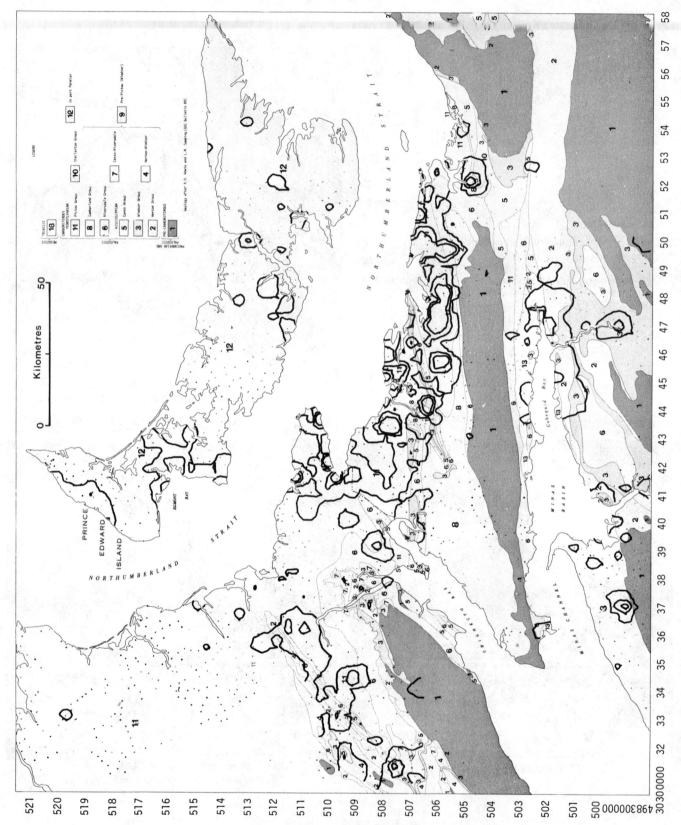

Figure 21B.5. Uranium in well waters in ppb, contours $\bar{X} = 1.0$, $\bar{X} + S = 4.0$, $\bar{X} + 2S = 7.0$, Eastern Maritime Canada.

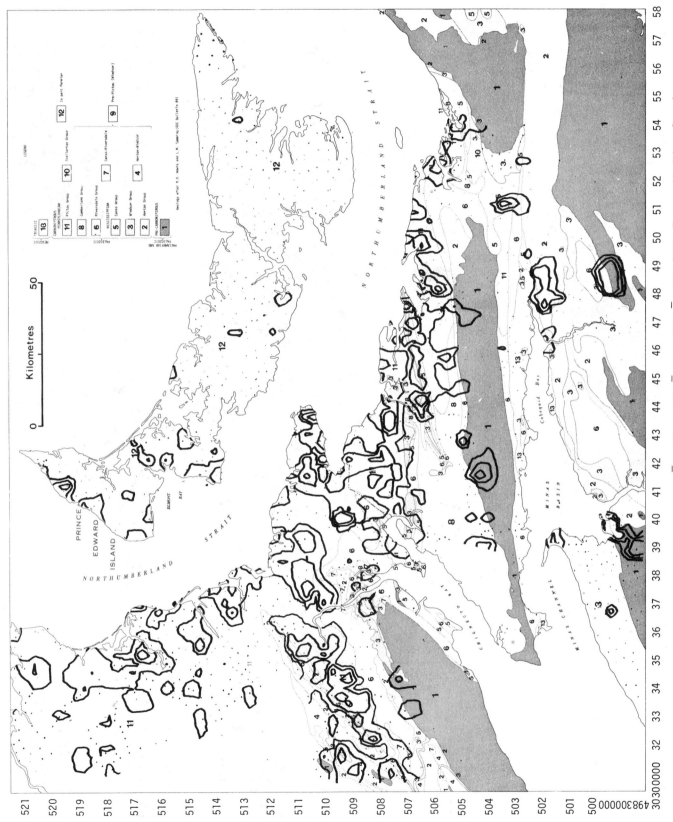

Figure 21B.6. Radon in well waters in pCi/L, contours $\bar{X}+.25S=1060$, $\bar{X}+S=1670$, $\bar{X}+2S=2500$, Eastern Maritime Canada.

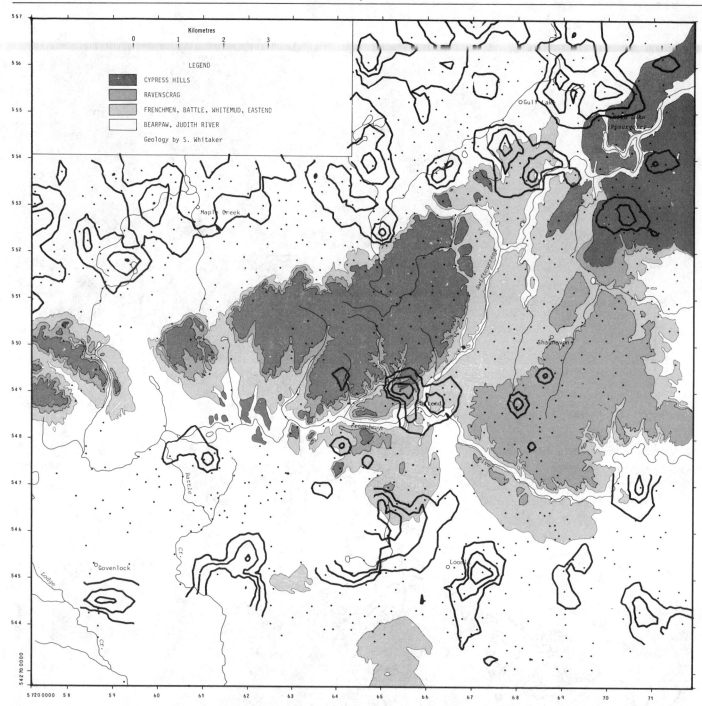

Figure 21B.7. Uranium in well waters in ppb, contours $\bar{X} + .25S = 17$, $\bar{X} + S = 32$, $\bar{X} + 2S = 53$, Cypress Hills, Saskatchwan.

north east of the area, U and Rn highs more or less overlap and coincide with the Cypress Hills Formation. The lower Rn values suggest lower Ra concentrations and hence less U mineralization in the Cypress Hills than in the Maritimes even though the absolute U concentrations suggest the opposite. Similarly Korner and Rose (1977) found that Rn in groundwaters was anomalous near U mineralization but U not. However, rate of water turnover is not the only factor, the porosity of the ground and the type of sediments also influence Rn and Ra release into the waters. Even so, the two elements U and Rn make a powerful team in the search for U mineralization at depth. But it is absolutely essential in the interpretation that the investigator have a good knowledge of the behaviour of these two elements. Even then, in order to know for a fact, drilling is ultimately required.

The present U boom has redirected industry and government funds to the search of U in an unprecedented scale and hydrogeochemistry is playing an important role in this search (King et al., 1976; ERDA, 1977; USGS, 1977; Darnley et al., 1975). Unfortunately a rather useful publication dealing with all aspects of U exploration is not available in English (Novikov and Kapkov, 1965). Hence the author feels obliged to summarize excerpts from this book. Table 21B.1 is a summary of a vast amount of information on the types, conditions of formation and chemical and radioactive constituents of natural waters. The hydrogeochemical method has been used extensively and profitably in Russia. Under favourable conditions this method can detect U deposits at considerable depth. In mountainous terrain deposits buried 300 to 400 m, and in foothill regions 50 to

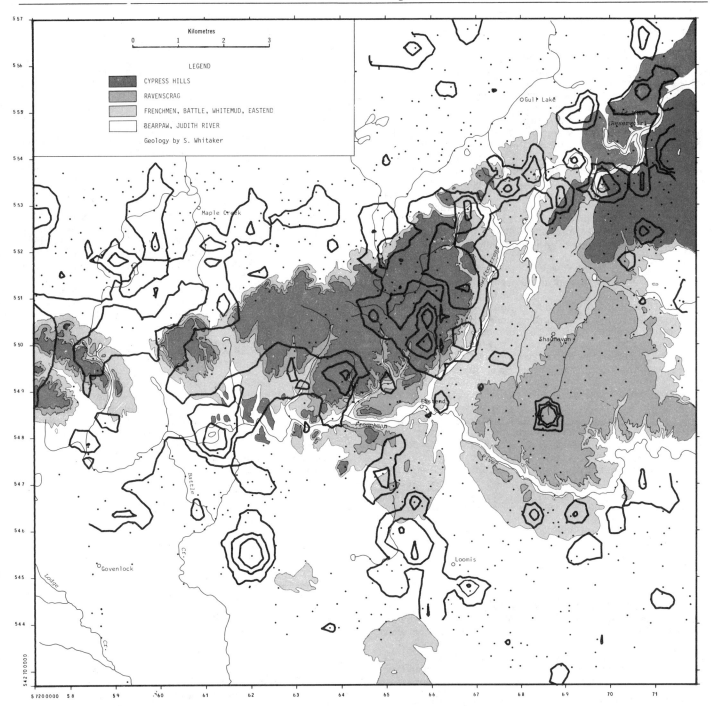

Figure 21B.8. Radon in well waters in pCi/L, contours $\overline{X} + .25S = 439$, $\overline{X} + S = 689$, $\overline{X} + 2S = 1023$, Cypress Hills, Saskatchewan.

70 m have been detected. The interpretation of hydrogeochemical results are rather difficult because they depend on so many environmental factors, including climate, chemistry of the elements, geology, mineralogy, hydrodynamics, etc. of a region. Of outmost importance is the background concentration of an element in an area. In mountainous regions, a U concentration of 10^{-6} g/L may be anomalous, whereas in arid regions evaporation of water can give backgrounds of the order of 10^{-4} g/L. Therefore a rise in the uranium content is of greater interest if it is corrected for total dissolved solids or conductivity in the water.

Radon-222 and Radium-226

To treat U separately from its decay products, as is being attempted in this discussion, is not really the best way to carry out a hydrogeochemical exploration program. Each element in the U decay series is unique in some respects and hence is suited best for certain conditions and a certain phase in the exploration program. Radon-222, because of its short half life can never migrate far away from its immediate parent Ra-226, and since Ra is usually not found in detectable amounts (~1×10^{-12} g/L) in surface waters, the Rn in such waters in larger amounts must come from Ra in solids such as rocks, soils and sediments. Although the solubility product of Ra salts is seldom reached in natural waters, it invariably gets adsorbed as sulphates and carbonates on surfaces of rocks and minerals. In the zone of oxidation it is also coprecipitated by hydrous oxides of iron and manganese. Only in the vicinity of strong sources of very saline waters will the Ra concentration rise to 10^{-10} g/L or even 10^{-8} g/L. The link between Rn in water and Ra in sediments must be firmly implanted in the prospectors mind. For he could easily miss a deposit in the bottom of a large deep lake if he sampled the surface water for Rn only. Rn will not travel much beyond 6 m by true diffusion, although mechanical agitation by wind on lakes or a flowing stream can increase the Ra-Rn separation to 50 to 100 m. A second factor is the emanation efficiency of Rn of the solid through which the water moves. This seldom reaches 20% and is usually only a few percent. Even though Ra is so immobile in the surficial environment, the law of dynamic equilibrium demands that some of it go into solution and since it has a relatively long half life (1600 years), it can migrate considerable distances, perhaps several kilometres, in well established aquifers at concentrations below the detection limit of most analytical techniques, adsorbing and desorbing continually on the walls of the solids through which the water moves. While most of this Ra is adsorbed on surfaces at any one time, the Rn emitted by it enters the water phase easily. Thus it happens that water from taps in the town of Bancroft contains easily detectable amounts of Rn even though the lake water that this water comes from has no detectable Rn levels. Similarly old domestic well casings when logged with a gamma-ray probe can have much higher activities than fresh holes drilled right beside them. Accumulations of Ra are particularly prominent in groundwaters from depth, where reducing conditions prevail, which enter the zone of oxidation. As Fe and Mn oxides precipitate, Ra is coprecipitated. Also deep waters usually contain large amounts of CO_2 which escape when the waters reach atmospheric pressure. Thus CO_2 escape causes Ca and Mg carbonates to precipitate, again carrying Ra down with them. This phenomenon is particularly evident in mineral springs (Cadigan and Felmlee, 1977). Waters from acidic rocks are usually more radioactive than waters circulating in basic rocks. Waters with intensive circulation and intensive flow are weakly radioactive. Groundwaters with a limited circulation tend to become mineralized and may become strongly radioactive in acidic rocks enriched in uranium. In mountainous areas with rugged relief, waters near the peaks are commonly weakly radioactive but at the foot of the mountains one can encounter highly radioactive springs even in the absence of U ore, although usually, some mineralization is necessary to produce highly radioactive waters, such as secondary mineralization in fractures through which the water moves.

Several criteria are given below which will help in deciding on the significance of radioactive anomalies in groundwater:

(i) A threefold or greater increase in the content of Rn or Ra compared to the background of a region.

(ii) Occurrence of anomalous amounts of all four elements (Rn, Ra, U, and He).

(iii) Increased content of indicators such as Mo, Pb, Cu, Zn, As, P, V, Ni, F.

(iv) A sharp rise in the Rn concentration after a rain or thaw period of up to 10 times normal levels in the presence of U ore; not more than a fourfold rise above natural levels in the absence of U deposits.

The use of the radioactivity of waters in prospecting, practiced in Eastern Europe extensively (Baranov, 1961; Novikov and Kapkov, 1965), has only recently found wider acceptance in North America. No doubt the complexity of the method and the large variations in Rn levels (largely due to its gaseous nature and short half life) have contributed to this reluctance (Rogers and Tanner, 1956; Rogers, 1958; Makkaveev, 1960; Smith et al., 1961).

With good equipment and operators, the method can be applied to surface lake and stream waters at sample densities of about one sample per 3 km^2 or higher (Boyle et al., 1971). In sedimentary basins with well established groundwater regimes and gentle terrain this method outlines radioactive areas at sample densities of 1 sample/13 km^2 (Dyck et al., 1976a). The distribution of Rn in well waters in Eastern Maritime Canada is shown in Figure 21B.6 and for the Cypress Hills area in Figure 21B.8. These results have already been discussed briefly in the previous section. Rn tests in a phased multi-method approach to U exploration has provided useful information in delineating drilling targets (King et al., 1976). Korner and Rose (1976) found Rn in wells in Pennsylvania useful in reconnaissance scale exploration, however Rn in stream waters was not considered as useful because of low erratic Rn levels. Wenrich-Verbeek et al. (1976) applied factor analysis to Ra data from radioactive springs and conclude that Ra in spring waters can be an indicator for deeply buried U deposits.

These varied experiences and seemingly conflicting reports on the effectiveness of Rn and Ra in hydrogeochemical prospecting confirm the complexity of their geochemistry, the lack of sufficient research, or careless application of the method. It is only through first hand experience and through understanding of the elements and the environment in which they are tested that a prospector will find ore using the hydrogeochemical technique.

Helium

The fact that He is produced during the radioactive decay of U and Th makes it a potential tracer for U and Th ore deposits. Each time an alpha particle, from the many nuclear transformations in the decay series of U and Th, loses its charge it becomes a He atom. Thus, for this discussion, the decay schemes of the three naturally radioactive series can be represented by the simplified decay schemes:

$$^{238}U = {}^{206}Pb + 8\,{}^{4}He$$
$$^{235}U = {}^{207}Pb + 7\,{}^{4}He$$
$$^{232}Th = {}^{208}Pb + 6\,{}^{4}He$$

Being inert and very small, He does not react chemically but has a great tendency to escape into fissures and thence into the groundwater regime. From there it may eventually be expelled into gas-tight underground pockets or into the atmosphere. Atmospheric air at the present time contains 5.2 ppm He by volume. Air dissolved in surface waters contains 2 ppm He. This surface water He background is quite stable in lakes, streams and shallow wells. Dissolved gases in groundwater can contain over 100 000 ppm He (Dyck, 1976).

There appears to be very little literature on the use of He for U prospecting, however, lately Clarke and Kugler (1973) have shown that the He content in groundwaters near U mineralization does rise. But, He escapes rapidly into the atmosphere or is carried away by circulating groundwaters in permeable soils. In stagnant groundwater or wet clays, a He gradient may be observed over a deposit.

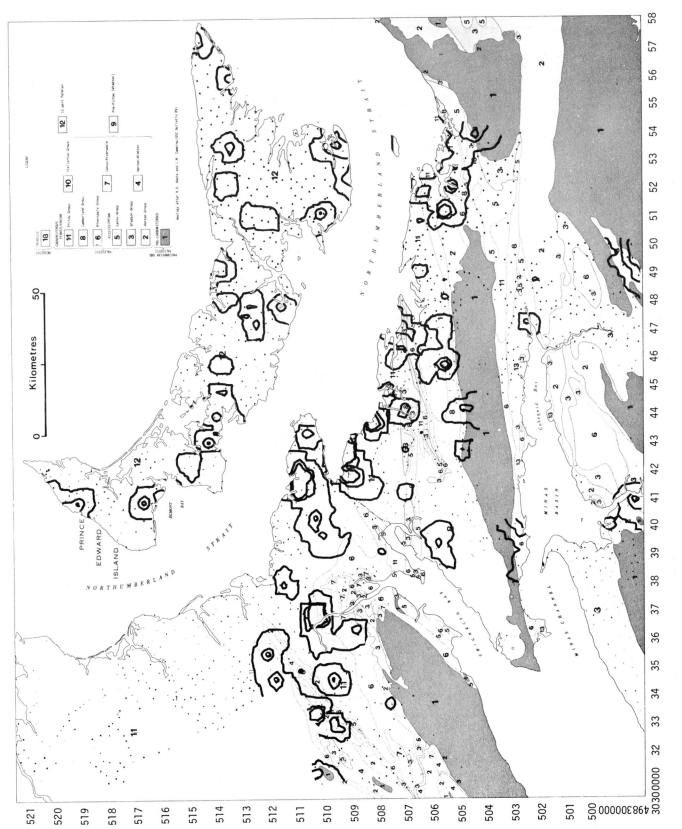

Figure 21B.9. Helium in well waters in microccs/L, contours $\bar{X} = 550$, $\bar{X} + S = 4350$, $\bar{X} + 2S = 8160$, Eastern Maritime Canada.

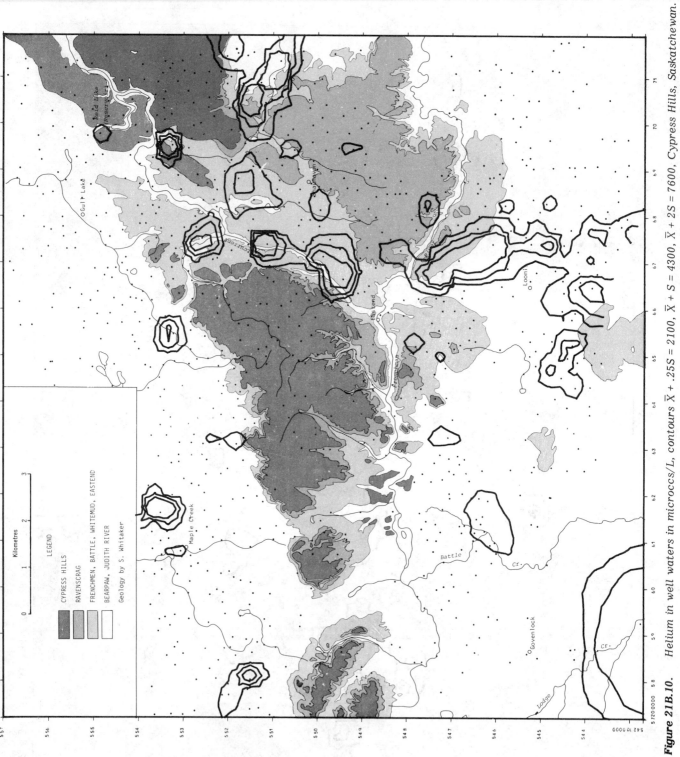

Figure 21B.10. Helium in well waters in microccs/L, contours $\bar{X} + .25S = 2100$, $\bar{X} + S = 4300$, $\bar{X} + 2S = 7600$, Cypress Hills, Saskatchewan.

A summary of He occurrences and its use in mineral exploration is given by Dyck (1976). Most of the He observed in groundwaters and springs has escaped into the water systems from rocks and minerals, particularly the U- and Th-rich basement rocks. Thus most regional He anomalies will reflect fault and fracture zones rather than U ore deposits (Ovchinnikov et al., 1973; Eremeev et al., 1973). However, in areas of high U potential, a He anomaly in groundwater or lake waters can point to a U deposit buried at depth. Recent tests by the author indicate that in the case of lakes, a thermal gradient is necessary to detect a He gradient. Such gradients exist in the winter in northern regions by virtue of an ice cover. While handling of water samples in subzero weather is a problem, particularly for He analysis, because rigid containers are required, the build up of He (and Rn) under the ice in bottom lake waters can be significant in the presence of U mineralization, relative to summer conditions. For example, the average net He content in 87 lake water samples from the vicinity of the Key Lake, Saskatchewan U ore deposits was 8 times higher in the winter than in the summer. Total Rn levels were 3 times higher in winter than in summer. Pogorski et al. (1976) found higher than average He concentrations in waters from the Bancroft area. Clarke et al. (1977) using ^{4}He, ^{3}He and ^{3}H determinations found positive correlation between U mineralization and He concentrations in central Labrador. Although both He isotopes are of radiogenic origin their mode of production differs and theoretical considerations suggest and experimental data confirm higher ^{4}He/^{3}He ratios in gases from U rich rocks and U ore deposits (Gerling et al., 1971; Kamminsky et al., 1971; Tolstihlin et al., 1969). Since ^{3}H decays to ^{3}He, ^{3}H measurements are necessary to correct the ^{3}He measurements. ^{3}H is radioactive and is produced by cosmic rays, nuclear reactors and hydrogen bombs also release large amounts of ^{3}H in the atmosphere continuously. Hence ^{3}He can be used for estimating lake water residence times which permits the evaluation of relative intensities of He anomalies.

Groundwaters, easily obtained in inhabited areas from domestic wells, are not subject to the temperature gradient experienced by surface waters although seasonal fluctuations in precipitation will cause some fluctuation in water tables and flow rates. The He maps of the two Canadian environments discussed earlier under U and Rn are shown in Figures 21B.9 and 21B.10. The Maritime He pattern coincides well with the U and Rn patterns suggesting a common source such as U mineralization or U-rich acid volcanics or volcanic ash. The Saskatchewan He pattern on the other hand, while much stronger, follows the CH_4 pattern very closely on the regional scale suggesting the known oil and gas accumulations at depth as the common source. Just from probability considerations alone, it is obvious that few if any of these He anomalies point to U ore directly but rather that they reveal regional geological and structural features at depth. However, structural mapping with the aid of gases is a powerful tool for prospecting for ore deposits in general (Ovchinnikov et al., 1973). Recently Reimer and Otton (1976) found that groundwaters and soil gases downstream from a roll-type U deposit gave anomalous He. The problem of discerning between He from basement rocks and He from U ore deposits will always be with the prospector, much in the same way as are the U and Rn-Ra anomalies in the hydrosphere. But combined, these elements increase the chances of detecting buried ore.

CONCLUSIONS

A great deal of work on the geochemistry of uranium and decay products has been carried out to date. Much of this work is published in Russian journals but translations of several major papers are available in English. This geochemical knowledge has been applied successfully to the search of U deposits in many parts of the world.

The elements of interest in hydrogeochemical exploration for uranium are uranium and its decay products: radium, radon and helium. Each element has specific radiochemical and geochemical properties which make it a useful tracer for uranium ore deposits.

Uranium is easily oxidized to the hexavalent state. Its mobility in surface and near surface waters is enhanced by the complexing action of corbonates in neutral and basic waters, of sulphates and fluorides in acid waters, of phosphates in neutral and slightly acid waters, and of silicates in neutral waters. Solid organic matter adsorbs uranium strongly and is responsible for limiting migration of the uranyl ion in surface waters. However, dissolved organic matter is also an important complexing agent of uranium and can enhance its dispersion under appropriate conditions. The greater abundance of bicarbonates in groundwaters within certain sedimentary rocks results in intensive leaching and wide dispersion of uranium in the zone of oxidation.

The hydrogeochemical techniques employing radium, and radon or both are best suited to detailed or semi-detailed investigations of radioactive occurrences. Their ease of detection and short range of dispersion from source make them good tracers for pinpointing uranium occurrences or outlining radioactive zones too weak for direct detection by gamma-ray spectrometry or fluorimetry. Conflicting reports on the success of the Rn/Ra method of prospecting point to the need for further research into the behaviour of these elements particularly that of Ra, since it controls its short-lived daughter, Rn.

Helium, because of its chemical inertness and great mobility, has the potential to reveal uranium orebodies through much greater thickness of cover than any other geochemical or geophysical technique, but exactly the same factors can also produce anomalies unrelated to mineralization. Tests of the usefulness of ^{3}He/^{4}He ratios as a means of differentiating between He from U deposits and basement rocks and ^{3}H measurements as a means of determining residence times of water reservoirs are urgently needed to evaluate the He method as a U prospecting tool for buried deposits.

Springs have served man as guides to buried minerals since the dawn of history. In inhabited areas wells can do the same for 20th century man. In uninhabited areas systematic drilling and testing of aquifers will become one of the tools of search in the future.

REFERENCES

Allen, J.
1976: Development of a portable radon detection system; Open File Report GJBX-50(76), Energy Res. and Dev. Admin., Grand Junction, Colo., 47 p.

Ballantyne, S.B.
1976: Canada - British Columbia uranium reconnaissance program. Regional stream sediment and water survey, 1976; Geol. Surv. Can. Open File Rep. 341 and 409.

Baranov, V.I.
1961: Radiometry, Translation, AEC-tr-4432, 381 p.

Bell, K.G.
1955: Uranium in precipitates and evaporites; U.S. Geol. Surv, Prof. Paper 300, p. 381-386; U.N. Int. Conf. Atomic Energy 6, p. 520-524.

1960: Deposition of uranium in salt-pan basins; in Shorter contributions to general geology, 1959; U.S. Geol. Surv., Prof. Paper 354, p. 161-169.

Boberg, W.W. and Rummells, D.D.
1971: Reconnaissance study of uranium in the South Platte River, Colorado; Econ. Geol., v. 66, p. 435-480.

Boyle, D.R.
1976: The geochemistry of fluorine and its applications in mineral exploration; Ph.D. Thesis, Royal School of Mines, London, 386 p.

Boyle, R.W.
1967: Geochemical prospecting — Retrospect and prospect; Can. Inst. Min. Metall. Trans. v.LXX p. 1-6.

Boyle, R.W. and Garrett, R.G.
1970: Geochemical prospecting — A review of its status and future; Earth Sci. Rev., v. 6, p. 51-75.

Boyle, R.W., Hornbrook, E.H.W., Allan, R.J., Dyck, W., and Smith, A.Y.
1971: Hydrogeochemical methods — Application in the Canadian Shield; Can. Min. Metall. Bull. v. 64, p. 60-71.

Breger, I.A. and Deul, M.
1955: The organic geochemistry of uranium; U.S. Geol. Surv., Prof. Paper 300, p. 505-510.

Brundin, N.H. and Nairis, B.
1972: Alternative sample types in regional geochemical prospecting; Geol. Explor., v. 1, p. 7-46.

Butler, A.P.
1969: Ground water as related to the origin and search for uranium deposits in sandstones; Contribution to Geology, Wyoming uranium mine, p. 81-85.

Cadigan, R.A. and Felmlee, J.K.
1977: Radioactive springs geochemical data related to uranium exploration; J. Geochem. Explor., v. 8, p. 381-395.

Cameron, E.M. and Ballantyne, S.B.
1977: Reconnaissance-level geochemical and radiometric exploration data from the vicinity of the Rabbit Lake uranium deposit; Can. Inst. Min. Met. Bull., v. 70 (781), p. 76-85.

Cameron, E.M. and Hornbrook, E.H.W.
1976: Current approaches to geochemical reconnaissance for uranium in the Canadian Shield; in Explorations for uranium ore deposits, Int. Atom. Energy Agency, Rep. IAEA-SM-208/31, p. 241-266.

Carlo, Erba
1976: Gas Chromatograph fracto vap mod. 2700 with helium ionization detector; Carlo Erba Serumentazione STFV 2700-E, 4p.

Centanni, A., Ross, A.M, and DeSesa, A.
1956: Fluorimetric determination of uranium; Anal. Chem., v. 28 (11), p. 1651-1657.

Chamberlain, J.A.
1964: Hydrogeochemistry of uranium in the Bancroft-Haliburton Region, Ontario; Geol. Surv. Can., Bull. 118, 19 p.

Cherdyntsev, V.V.
1971: Uranium-234; Israel Program of Scientific Translations, Jerusalem, 254 p.

Clarke, W.B. and Kugler, G.
1973: Dissolved helium in ground water: a possible method for uranium and thorium prospecting; Econ. Geol., v. 68, p. 243-251.

Clarke, W.B., Top, E., Beavan, A.P. and Gandhi, S.S.
1977: Dissolved helium in lakes: Uranium prospecting in the Precambrian terrain of central Labrador; Econ. Geol., v. 72, p. 233-242.

Coker, W.B. and Jonasson, I.R.
1977: Geochemical exploration for uranium in the Grenville Province of Ontario; Can. Inst. Min. Metall. Bull., v. 70 (781), p. 67-75.

Corner, L.A. and Rose, A.W.
1977: Radon and streams and groundwaters of Pennsylvania as a guide to uranium deposits; The Pennsylvania State University, GJO1659-2, 152 p.

Cowart, J.B. and Osmond, J.K.
1974: ^{234}U and ^{238}U in the Carrizo sandstone aquifer of South Texas; in Symp. on isotopic techniques in groundwater hydrology, Vienna; IAEA-SM-182/35.

1976: Uranium isotopes in groundwater as a method of prospecting for sandstone type deposits; 25th Int. Geol. Cong., Sydney, v. 2, p. 439.

1977: Uranium isotopes in groundwater; their use in prospecting for sandstone-type uranium deposits; J. Geoch. Explor., v. 8, p. 365-379.

Culbert, R.R. and Leighton, D.B.
1978: Uranium in alkaline waters. Okanagan area, B.C.; Can. Inst. Min. Metall. Bull., v. 71 (793), p. 103-116.

Danielson, A., Roennholm, O., Kjellstroem, L.E., Ingman, F.
1973: Fluorimetric method for the determination of uranium in natural waters; Talanta, v. 20 (2), p. 185-192.

Dall'Aglio, M.
1971: A study of the circulation of uranium in the supergene environment in the Italian Alpine range; Geochim. Cosmochim. Acta, v. 35 (1), p. 47-59.

Darnley, A.G., Cameron, E.M., and Richardson, K.A.
1975: The Federal-Provincial uranium reconnaissance program; in Uranium exploration '75, Geol. Surv. Can., Paper 75-26, p. 49-68.

Denson, N.M., Zeller, M.D., and Stephens, J.G.
1956: Water sampling as a guide in the search for uranium deposits and its use in evaluation of widespread volcanic units as potential source beds for uranium; U.S. Geol. Surv., Prof. Paper 300, p. 673-686.

DeVoto, R.G., Mead, R.H., Berquist, L.E., and Martin, J.P.
1976: Helium detection in uranium exploration; paper presented at ERDA Geophysics Conference, Grand Junction, Col.

Doi, Kazumi, Hirono, Shuichiro, and Sakamaki, Yukio
1974: Uranium mineralization by groundwater in sedimentary rocks, Japan; Econ. Geol. v. 70 (4), p. 628-646.

Dooley, J.R., Granger, M.C., and Rosholt, J.H.
1966: Uranium-234 fractionation in the sandstone type uranium deposits of the Ambrosia Lake district, New Mexico; Econ. Geol., v. 61, p. 1362-1382.

Durham, C.C. and Cameron, G.M.
1975: A hydrogeochemical survey for uranium in the northern part of the Bear Province, Northwest Territories; in Report of Activities, Part C, Geol. Surv. Can., Paper 75-1C, p. 331-332.

Dyck, W.
1969: Field and Laboratory Methods Used by the Geological Survey of Canada in Geochemical Surveys. No. 10. Radon Determination Apparatus for Geochemical Prospecting for Uranium; Geol. Surv. Can., Paper 68-21, 30 p.

1974: Geochemical studies in the surficial environment of the Beaverlodge area, Saskatchewan; Geol. Surv. Can., Paper 74-32, 30 p.

1975: Geochemistry applied to uranium exploration; Geol. Surv. Can., Paper 75-26, p. 33-47.

1976: The use of He in mineral exploration; J. Geochem. Explor., v. 5, p. 33-20.

Dyck, W. and Cameron, E.M.
1975: Surface lake water uranium-radon survey of the lineament lake area, District of Mackenzie; in Report of Activities, Part A, Geol. Surv. Can., Paper 71-1A, p. 209-212.

Dyck, W. and Pelchat, J.C.
1977: A semiportable helium analysis facility; in Report of Activities, Part C, Geol. Surv. Can., Paper 77-1C, p. 85-87.

Dyck, W. and Smith, A.Y.
1969: The use of radon-222 in surface waters in geochemical prospecting for uranium; in Proc. Int. Geochem. Explor. Symp., F.C. Canney ed. Q. Color. Sch. Mines, v. 64 (1), p. 223-236.

Dyck, W. and Tan, B.
1978: Seasonal variations of helium, radon and uranium in lake waters near the Key Lake uranium deposit, Saskatchewan; J. Geochem. Explor. (in press)

Dyck, W., Campbell, R.A., and Whitaker, S.H.
1977: Well water uranium reconnaissance, southwestern Saskatchewan; Sask. Geol. Soc. special publication No. 3, ed. by C.E. Dunn, p. 157-168.

Dyck. W., Chatterjee, A.K., Gemmell, D.E. and Murricane, K.
1976: Well water trace element reconnaissance, Eastern Maritime Canada; J. Geochem. Explor., v. 6, (1/2), p. 139-162.

Dyck, W., Dass, A.S., Durham, C.C., Hobbs, J.D., Pelchat, J.C. and Galbraith, J.M.
1970: Comparison of regional geochemical uranium exploration methods of the Beaverlodge area, Saskatchewan; Can. Inst. Min. Metall., Spec. Vol. 11, p. 132-150.

Dyck, W., Pelchat, J.C., and Meilleur, G.A.
1976: Equipment and procedures for the collection and determination of dissolved gases in natural waters; Geol. Surv. Can., Paper 75-34, 12 p.

ERDA
1977: Symposium of hydrogeochemical and stream-sediment reconnaissance for uranium in the United States; U.S. ERDA and Bendix Field Engineering Corp., Grand Junction, Colo., Abstracts of papers; 30 p.

Eremeev, A.N., Sokolov, V.A., Solovov, A.P., and Yanitskii, I.N.
1973: Application of helium surveying in structural mapping and ore deposits forecasting; in Geochemical Exploration 1972; Inst. Min. Met., London, p. 183-193.

Ferguson, R.B. and Price, V.
1976: National uranium resource evaluation (NURE, program-Hydrogeochemical and stream sediment reconnaissance in the Eastern United States); J. Geochem. Explor., v. 6 (1/2), p. 103-117.

Fin, P.F.
1956: Hydrogeochemical exploration for uranium; in Contributions to the geology of uranium and thorium, Page, L.R. (ed.); U.S. Geol. Surv., Prof. Paper 300, p. 667-671.

Fleischer, R.L. and Delany, A.C.
1976: Determination of suspended and dissolved uranium in water; Anal. Chem., v. 48 (4), p. 642-645.

Friedlander, G., Kennedy, J.W., and Miller, J.M.
1964: Nuclear and Radiochemistry; 2nd Edition, John Wiley and Sons Inc., New York, 585 p.

Gavshin, V.M., Bobrov, V.A., Pyalling, A.O. and Reznikov, N.V.
1973: The two types of uranium accumulation in rocks by sorption; Geochem. Int., v. 10 (3), p. 682-690.

Gerling, E.K., Mamgrin, B.A., Tolstihlin, I.N. and Yakovleva, S.S.
1971: Helium isotope composition in some rocks; Trans. from Geokhimiya No. 10, p. 1209-1217.

Germanov, A.I., Batulin, S.G., Volkov, G.A., Lisitsin, A.K. and Serebrennikov, V.S.
1958: Some regularities of uranium distribution in underground waters; in Survey of Raw Material Resources, Proc. 2nd UNIC on Peaceful Uses of Atomic Energy, Geneva, 1958, v. 2, p. 161-177.

Goldak, G.E.
1974: Helium-4 mass spectrometry for uranium exploration; Preprint 74-L-44, Amr. Inst. Min. Eng., 12 p.

Goldstein, D. and Liebergott, E.K.
1973: New spot test for the identification of uranium in minerals; Anal. Chim. Acta, v. 63 (2), p. 491-492.

Haglund, D., Freedman, G., and Miller, D.
1969: The effect of fresh water on the redistribution of uranium in carbonate sediments; J. Sed. Petrol., v. 39 (4), p. 1283-1296.

Hawkes, M.E. and Webb, J.S.
1962: Geochemistry in mineral exploration; Harper and Row, New York, 375 p.

Higgins, Frederick B. Jr., Grune, Werner N., Smith, Benjamin M. and Terrill, James G. Jr.
1961: Methods for Determining Radon-222 and Radium-226; J. Am. Water Works Assoc., v. 53 (1), p. 63-74.

Hostetler, P.B. and Garrels, R.M.
1962: Transportation and precipitation of uranium and vanadium at low temperatures, with special reference to sandstone-type uranium deposits; Econ. Geol., v. 57 (2), p. 137-167.

Hunt, E.C.
1958: The examination of natural waters for uranium in Derbyshire; Dep. Sci. Ind. Res., Sci. Rep. N.C.L./A.E. S38, Middlesex, England.

Illsley, C.T.
1957: Preliminary report on hydrogeochemical exploration in the Mt. Spokane area, Wash.; U.S. Atom Energy. Comm., Open File MFC No. 7233, 19 p.

1961: Hydrogeochemical reconnaissance for uranium in the Stanley area, south-central Idaho; U.S. Atom. En. Comm., Division of Tech. Inf. RM-140, 25 p.

James, G. Jr.,
1961: Methods for Determining Radon-222 and Radium-226; J. Amer. Water Works Assoc., v. 53 (1), p. 63-74.

Jonasson, I.R.
1976: Detailed hydrogeochemistry of two small lakes in the Grenville Geological Province; Geol. Surv. Can., Paper 76-13, 37 p.

Kamenskiy, I.L., Yakutseni, V.P., Mamyrin, B.A., Anufriyev, S.G. and Tolstihlim, I.N.
1971: Helium isotopes in nature; Trans. from Geokhimiya No. 8, p. 914-931.

Kyuregyan, T.N. and Kocharyan, A.G.
1969: Migration forms of uranium in carbonate waters of a Caucasian district; Int. Geol. Rev., v. 11 (10), p. 1087-1089.

King, J., Tauchid, M., Frey, D., and Basset, M.
1976: Exploration for uranium in southwestern Australia: a case History; in Exploration for uranium ore deposits; Int. Atom. Energy Agency, 22 p.

Korner, L.A. and Rose, A.W.
1977: Radon in streams and ground waters of Pennsylvania as a guide to uranium deposits; The Pennsylvania State University, Mineral Conservation Section Report GJO-1659-2, 152 p.

Krauskopf, Konrad B.
1967: Introduction to geochemistry; McGraw-Hill Book Co., New York, p. 526-530.

Kugler, G. and Clarke, W.B.
1972: Mass spectrometric measurements of ^3H, ^3He, and ^4He produced in thermal-neutron Ternary fission of ^{235}U; evidence for short range ^4He; Phys. Rev., ser. C, v. 5 (2), p. 551-560.

Langmuir, D. and Applin, K.
1977: Refinement of the thermodynamic properties of uranium minerals and dissolved species, with application to the chemistry of groundwaters in sandstone-type uranium deposits; in uranium-thorium symposium, 1977; U.S. Geol. Surv., Circ. 753, p. 57-60.

Levinson, A.A.
1974: Introduction to Exploration Geochemistry; Applied Publishing Ltd., Calgary, 612 p.

Lopatkina, A.P., Komarov, V.S., Sergeyev, A.N., and Andreyev, A.G.
1970: On concentration of uranium by living and dead peat-forming plants; Geochem. Int., v. 7 (2), p. 277-282.

Lucas, Henry F.
1957: Improved low-level alpha-scintillation counter for radon; Rev. Sci. Instr., v. 28 (9), p. 680-683.

McDonald, J.A.
1969: An orientation study of the uranium distribution in lake waters, Beaverlodge District, Saskatchewan; in Q. Col. Sch. Min., v. 64 (1), (Int. Geoch. Explo. Symp.) p. 357-376.

Makkaveev, A.A.
1960: Problems and methods of radiohydrogeological exploration-radon from ores and rocks; Trudy Inst. Geol. Nauk, Akad. Nauk Belorus. S.S.R., No. 2, p. 209-215.

Mamyrin, B.A., Anufriyev, G.S., Kamenskiy, I.L., Tolstikhin, I.N.
1970: Isotopic composition of atmospheric helium; Dokl. Akad. Nauk S.S.S.R., v. 195 (1), p. 188-189.

Mann, A.W.
1974a: Calculated solubilities of some uranium and vanadium compounds in pure carbonated waters, as a function of pH; CSIRO Min. Res. Labs., Rept. FP6, p. 1-39.

1974b: Chemical ore genesis models of the precipitation of Carnotite in Calcrete; CSIRO Min. Res. Labs., Rept. FP7, p. 1-18.

Maurice, Y.T.
1977: Geochemical methods applied to uranium exploration in southwest Baffin Island; Can. Inst. Min. Metall. Bull. v. 70 (781), p. 96-103.

Meyer, W.T.
1969: Uranium in lake water from the Kaipokok region, Labrador; in Q. Colo. Sch. Min., v. 64 (1), (Int. Geoch. Explo. Symp.), p. 377-394.

Naguchi, M. and Wakita, H.
1977: A method of continuous measurement of radon in groundwater for earthquake prediction; J. Geophys. Res., v. 82 (2), p. 1353-1357.

Nicolli, M.
1973: Oklo phenomenon; CETAMA Bull. Inform., No. 4, p. 48-51 (in French).

Novikov, G.F. and Kapkov, Yu. N.
1965: Radioactive methods of prospecting; NEDRA, Leningrad, 759 p. (in Russian).

Osmond, J.K. and Cowart, J.B.
1976: The theory and uses of natural uranium isotopic variations in hydrology; Atomic Energy Rev., v. 14 (4), p. 621-679.

Ostle, D., Coleman, R.F., and Ball, T.K.
1972: Neutron activation analysis as an aid to geochemical prospecting for uranium; Uranium Prospecting Handbook, Inst. Min. Met., London, p. 95-109.

Ovchinnikov, L.N., Sokolov, V.A., Fridman, A.E., and Yanitskii, I.N.
1973: Gaseous geochemical methods in structural mapping and prospecting for ore deposits; in Geochemical Exploration 1972, Inst. Min. Met., London, p. 177-182.

Parslow, G.R.
1977: The extraction of uranium ions from water; a rapid and convenient method of sampling using prepackaged ion-exchange resins; J. Geochem. Explor. (in press)

Penchev, N.P.
1969: Contribution to the analysis and the geochemistry of the nobel gases; Bulgarian Acad. Sci., Communications, Dep. Chem., v. 2 (3), p. 603-618.

Phoenix, D.A.
1960: Occurrence and chemical character of groundwater in the Morrison Formation; U.S. Geol. Surv., Prof. Paper 320, p. 55-64.

Pogorski, L.A., Quirt, G.S., and Blascheck, J.A.
  1976: Helium surveys: a new exploration tool for locating uranium deposits; Chemical Projects Ltd. Tech. Paper CPL-6/76.

Premoli, C.
  1976: Formation of and prospecting for uraniferous calcretes; Austr. Min, April, p. 13-16.

Rankama, K. and Sahama, Th. G.
  1950: Geochemistry; Univ. of Chicago Press, Chicago, 911 p.

Reimer, G.M.
  1975: Uranium determination in natural water by the fission-track technique; J. Geochem. Explor., v. 4 (4), p. 425-431.

  1976: Design and assembly of a portable helium detector for evaluation as a uranium exploration instrument; U.S. Geol. Surv., Open File Rep. 76-398, 17 p.

Reimer, G.M. and Otton, J.K.
  1976: Helium and soil gas and well water in the vicinity of a uranium deposit Welch County, Colorado; U.S. Geol. Surv., Open File Rep. 76-699, 10 p.

Ridgley, J.L. and Wenrich-Verbeek, K.J.
  1978: Scatter diagrams and correlations of uranium in surface water versus discharge, conductivity, and pH at various locations throughout the United States; U.S. Geol. Surv., Open File Rep. 78-581, 331 p.

Robbins, J.
  1977: Direct analysis of uranium in natural waters; Ann. Meeting, Can. Inst. Min. Metall., Ottawa, April 18-20.

Rogers, A.S. and Tanner, A.B.
  1956: Physical behaviour of radon; U.S. Atom. Energy Comm., Rep. TEL-620.

Rogers, A.S.
  1958: Physical behaviour and geological control of radon in mountain streams; U.S. Geol. Surv., Bull. 1052E, p. 187-211.

Rose, R.W. and Keith, M.L.
  1976: Reconnaissance geochemical techniques for detecting uranium deposits in sandstones of northeastern Pennsylvania; J. Geochem. Explor., v. 6 (1/2), p. 119-137.

Roylance, J.G. Jr.
  1973: Fundamental aspects of uranium geochemistry applied to the genetic study of sandstone-type uranium concentrations; Presented at the Vanguard symposium on mineral deposit models, 22-26 March 1971, Spokane, Wash., 44 p.

Rushing, D.R., Garcia, W.J., and Clarke, D.A.
  1964: Analysis of effluents and environmental samples; in Symposium on Radiological Health and Safety in Mining and Milling of Nuclear Materials, Int. Atom. En. Agency, Vienna, 26-31 Aug. 1963, p. 187-230.

Samsoni, Z.
  1967: Investigation of factors affecting the accuracy and sensitivity of the fluorimetric determination of uranium; Microchimica Acta., v. 1, 88-97.

Sedlet, Jakob
  1966: Radon and Radium; in Treatise on Analytical Chemistry, Part 2, Interscience Publ., John Wiley and Sons, New York, v. 4, p. 219-366.

Sergeyeva, E.I., Nikitin, A.A., Khodakovsky, I.L., and Naumov, G.B.
  1972: Experimental investigation of equilibria in the system UO_3- CO_2- H_2O in 25-200°C temperature interval; Geochem. Int., p. 900-911.

Smith, A.Y. and Lynch, J.J.
  1969: Field and laboratory methods used by the Geological Survey of Canada in geochemical surveys. No. 11: Uranium in soil stream sediment and water; Geol. Surv. Can., Paper 69-40, 9 p.

Smith, B.M., Grune, W.N., Higgins, F.B., and Terrill, J.G.
  1961: Natural radioactivity in groundwater supplies in Maine and New Hampshire; J. Am. Water Works Assoc., v.53 (1), p. 75-88.

Smith, G.H. and Chandler, T.R.D.
  1958: A field method for the determination of uranium in natural waters; in Proc. 2nd UN Int. Conf. on the Peaceful Uses of Atomic Energy, v. 2, p. 148-152.

Sonnek, R.A., Klingman, G.C., and Seitz, C.L.
  1965: A portable helium analyser; U.S. Bur. Min., Rep. Inv. 6600, 15 p.

Starik, I.E., Kuznetsov, Yu. V., Petryaev, E.P., and Legin, V.K.
  1967: Certain aspects of the geochemistry of radioactive-isotopes; in Chemistry of the earth's crust (A.P. Vinogradov, Ed.), v. 1, p. 396-412 (I P S T, 1967).

Szalay, A.
  1964: Cation exchange properties of humic acids and their importance in the geochemical enrichment of UO_2^{++} and other cations; Geochem. Cosmochemo. Acta, v. 28, p. 1605-1614.

  1967: The role of humic acids in the geochemistry of uranium and their possible role in the geochemistry of other cations; in Chemistry of the earth's crust (A.P. Vinogradov, Ed.), v. 2, p. 456-471.

Szalay, A. and Samsoni, Z.
  1970: Investigations on the leaching of uranium from crushed magmatic rocks; Proc. Sci. Council (Japan-Int. Ass. Geoch. Cosmochem (Tokyo), p. 261-272.

Thatcher, L.L. and Barker, F.B.
  1957: Determination of uranium in natural waters; Anal. Chem., v. 29 (11), p. 1575-1578.

Tolstihlin, I.N., Kamenskiy, I.L., and Mamyrin, B.A.
  1969: An isotope criterion of the origin of helium; Trans. from Geokhimiya No. 2, p. 201-204.

Tugarinov, A.I.
  1975: Origin of uranium deposits; in Recent contributions to geochemistry; (Tugarinov, A.I., Ed.) I P S T translation, Jerusalem, p. 293-302

United Nations
  1955: Proc. 1st UN Int. Conf. on the Peaceful Uses of Atomic Energy; v. 6, Anal. Chem., v. 29 (11), p. 1575-1578.

  1958: Geochemical prospecting; Proc. 2nd UN Inter. Conf. on the Peaceful Uses of Atomic Energy; v. 2, Survey of Raw Material Resources; p. 123-211.

U.S. Geological Survey
  1977: Proceedings of the Uranium-Thorium Symposium; U.S. Geol. Surv., Circ. 753, 75 p.

Vinogradov, A.P. (ed.)
1963: The essential features of the geochemistry of uranium; Akademiya NAUK USSR; (Ottawa), Foreign Languages Division, 1965, 630 p., in 5 parts.

1967: Chemistry of the earth's crust; v. 1 and 2, Proc. of the Geoch. Conf.; Israel Program of Scientific Translations, Jerusalem.

Wenrich-Verbeek, K.J.
1977: Uranium and coexisting element behaviour in surface waters and associated sediments with varied sampling techniques used for uranium exploration; J. Geochem. Explor., v. 8, p. 337-355.

Wenrich-Verbeek, K.J., Cadigan, R.A., Felmlee, J.K., Reimer, G.M., and Spirakis, C.S.
1976: Recent developments and evaluation of selected geochemical techniques applied to uranium exploration; Int. Atom. Energy Agency, Vienna, SM-208/18, 17 p.

Yermolayev, N.P., Zhidikova, A.P. and Zarinskiy, V.A.
1968: Transport of uranium in aqueous solutions in the form of complex silicate ions; Trans. from: Geokhimiya, No. 7, p. 813-826.

REMOTE SENSING IN THE SEARCH FOR METALLIC ORES: A REVIEW OF CURRENT PRACTICE AND FUTURE POTENTIAL

Alan F. Gregory
Gregory Geoscience Ltd., Ottawa, Canada

Gregory, Alan F., Remote Sensing in the search for metallic ores: A review of current practice and future potential; in Geophysics and Geochemistry in the Search for Metallic Ores; Peter J. Hood, editor; Geological Survey of Canada, Economic Geology Report 31, p. 511-526, 1979.

Abstract

Remote sensing denotes the aerospace practices of measuring the ultraviolet, visible, infrared and microwave radiations emitted and reflected from the surface of the Earth and from the atmosphere. As defined here, remote sensing excludes the more conventional methods of geophysics and geochemistry.

Aerial photography, side-looking airborne radar and airborne thermal infrared scanning are remote-sensing techniques that can currently provide the mineral industry with high-resolution data to meet specific requirements at acceptable costs. With respect to mineral exploration, other types of airborne remote sensing are largely experimental.

Several systems for remote sensing from satellites can be utilized for mineral exploration. Undoubtedly, the most universally available and most cost-effective type of remote sensing from the prospector's point of view is multispectral scanning by NASA's series of experimental Landsat satellites. Cloud-cover permitting, these satellites provide repetitive data of moderate resolution at low cost for many parts of the world.

Neither airborne nor orbital remote sensing are "stand alone" exploration techniques. Their effectiveness is optimized by integration, in the usual exploration manner, with other sets of data from geological, geophysical and geochemical surveys. While visual interpretation is the most widely used method of analysis, digital processing can provide improved images, enhancements and spectral discriminations that may meet specific exploration needs. However, digital analysis as a general tool for mineral exploration should still be viewed as largely experimental.

While atmospheric attenuation, heavy vegetational cover and "inadequate" spectral and spatial resolutions may cause problems, the major current limitations of remote sensing for mineral exploration are: (1) a lack of significant penetration below the surface of the ground, and (2) an inability to classify rocks and soils except in a generalized way. Despite these limitations, remote sensing, and in particular the complementary tools of Landsat and aerial photography, can provide much useful information at relatively low cost.

The principal contemporary benefits accruing from remote sensing in the search for metallic ores are: rapid regional reconnaissance, access to remote areas, and mapping of geological structures and formational continuity. While discovery of mineralization through both airborne and orbital remote sensing has been reported, such discovery of an economic mineral deposit has not been documented in the literature.

National remote sensing programs, both in Canada and abroad, have generally failed to provide adequate incentives for essential development of practical applications in contradistinction to strong support for development of technology to generate and process data.

In the future, airborne remote sensing will develop as a specialized source of detailed information to meet recognized goals. In particular, aerial thermography will be more widely used in arid terrains. Visual interpretation of Landsat data is already developing into a widely-used exploration tool for regional reconnaissance. Digital analysis of such data will soon achieve practicality for a few specific applications, particularly in arid regions. However, because of the increasing emphasis on subsurface exploration, relevant development of remote sensing should be focused in a co-ordinated program of demonstration projects. Specific objectives for such projects might include: (1) inexpensive interpretive techniques; (2) case histories related to rock alteration, soil contrasts and biophysical anomalies associated with orebodies; (3) classification of linears and their relation to ore; (4) integration of remote sensing with other exploration data, and (5) demonstration of techniques in an exploration mode.

Résumé

On désigne par télédétection les techniques aérospatiales mises en oeuvre pour mesurer les rayonnements ultraviolet, visible, infrarouge et de micro-ondes, émis par la surface de la Terre et réfléchis par elle, ou émis par l'atmosphère. Ainsi définie, la télédétection exclut les méthodes classiques de la géophysique et de la géochimie.

Les techniques de télédétection peuvent fournir à l'industrie minérale des données très fouillées pour répondre à certaines demandes particulières tout en restant à des prix relativement bas; ces techniques sont, la photographie aérienne, le radar à faisceau latéral et le balayage thermique en infrarouge. En ce qui concerne la prospection minérale, les autres techniques de télédétection aéroportée sont encore à l'état expérimental.

Pour la prospection minérale, on peut utiliser plusieurs systèmes de télédétection à partir de satellites. Le plus répandu et le plus rentable du point de vue du prospecteur, est sans conteste le balayage multispectral des satellites expérimentaux Landsat de la NASA. Lorsque la nébulosité est favorable, ces satellites fournissent, à intervalles répétés pour plusieurs régions du globe, des données exploitables, avec une résolution moyenne et pour un coût peu élevé.

Les techniques d'exploration aéroportée et de télédétection orbitale ne sont jamais utilisées seules. Leur efficacité est optimale si on les intègre, comme on fait toujours en prospection, à d'autres données provenant de levés géologiques, géophysiques et géochimiques. L'interprétation visuelle est la méthode d'analyse la plus largement utilisée, mais le traitement des données numérales peut donner de meilleures images plus contrastées, avec une bonne séparation spectrale qui peuvent répondre à certains besoins spécifiques de la prospection. Cependant, l'analyse numérique prise comme une technique générale d'exploration minérale doit être considérée comme une méthode largement expérimentale.

Alors que l'atténuation atmosphérique, la densité de la végétation et l'insuffisance des pouvoirs de résolution spectrale et spatiale peuvent créer des problèmes, les principales limitations de la télédétection pour l'exploration minérale sont: 1) le manque de pénétration dans le sol, 2) l'incapacité de définir de façon suffisamment spécifique les roches et les sols. Malgré ces limitations, la télédétection, et en particulier les outils complémentaires que sont les satellites Landsat et la photographie aérienne, peuvent donner des renseignements très utiles à des prix relativement bas.

Les principaux avantages actuels de la télédétection dans la recherche des gîtes métallifères sont: reconnaissance régionale rapide, accès aux régions les plus lointaines, possibilité de faire le levé des structures géologiques et de suivre la continuité des formations. On a signalé une découverte de minéralisation par les techniques de télédétection aéroportée et orbitale, mais on n'a pas décrit dans la littérature la découverte d'aucun gisement minéral présentant un intérêt économique.

Les programmes nationaux de télédétection, au Canada et à l'étranger n'ont pas, en général, réussi à encourager la mise au point d'applications pratiques alors qu'ils ont porté, avant tout, sur la mise au point des techniques de production et de traitement des données.

Dans l'avenir, la télédétection deviendra une source spécifique de renseignements détaillés permettant de répondre à des objectifs définis. En particulier, la thermographie aérienne sera plus largement utilisée en terrain aride. L'interprétation visuelle des données Landsat est en voie de devenir une technique d'exploration largement utilisée en reconnaissance régionale. L'analyse numérique des données Landsat deviendra bientôt pratique courante pour certaines applications spécifiques, en particulier dans les régions arides. Cependant, étant donné l'importance croissante que prend la prospection par les méthodes de sub-surface, l'expansion de la télédétection devrait se concentrer sur un programme coordonné d'opérations de démonstration. Les objectifs spécifiques de ces opérations pourraient comprendre: 1) des techniques d'interprétation peu coûteuses, 2) des cas concrets se rapportant à l'altération des roches, aux contrastes des sols et aux anomalies biophysiques associées à des gîtes minéraux; 3) une classification des linéaments et de leur relation avec la présence de minerais; 4) l'intégration de la télédétection aux autres données d'exploration; 5) la démonstration des techniques dans des conditions de prospection.

INTRODUCTION

In conducting their search for metallic ores, explorationists are faced with a severe problem of selecting and acquiring information that can advance their specific work. A great variety of data, or factual knowledge, may feed into an exploration program. The sources of such data are many, their scale of observation is greatly variable and the potential volume of their details is vast. For example, in addition to personal observation in the field, sources of exploration data may include: thin sections for microscopic analysis; samples of rock, soil and water for chemical analysis; geophysical maps for extrapolating beyond the outcrop; aerial photographs for stratigraphic trends; and radioactive ages for chronological relationships. Recently, these sources have been augmented by spectral surveys from airborne and orbital platforms. Each source of data, in its own way, presents a wealth of detail. No one source can stand alone as the fount of geological information and none can serve as a substitute for another in the sense of completely replacing it.

This paper is concerned primarily with remote sensing and the integration of relevant data into information, or instructional knowledge, which may be used to make appropriate decisions in an exploration program. As is well known, such information includes: (1) peripheral factors e.g. terrain conditions, current weather, environmental changes; (2) broad inferences and conditional analogies that lead to the recognition of exploration targets; (3) specific indicators that suggest the presence of mineralization; and (4) definitive analyses that establish the presence of ore. Remote sensing can, and regularly does, provide some information in each of the first three classes. However, it should be immediately obvious that no type of remote sensing, however defined, can provide the specific information required to establish the presence of ore.

WHAT IS REMOTE SENSING?

In its broadest scope, remote sensing can be defined as: the measurement or acquisition of information about some property of an object or phenomenon by means of a recording device that is not in physical or intimate contact with the feature under study (after Reeves et al., 1975, p. 2102). However, such a definition includes, among others, such exploratory practices as geophysics and geochemistry as well as astronomy and medical diagnosis. For some years, a few practitioners have argued that the concept of remote sensing should be narrowed to represent a practical paradigm i.e. an accepted model or body of knowledge which serves, for a time, to guide the collective research and practice of a specialized community of scientists or technologists (after Kuhn, 1970). Thus, the following practical definition is used herein (c.f. Gregory, 1972; Gregory and Moore, 1975):

<u>Remote sensing</u> *denotes the aerospace practices of measuring the ultraviolet, visible, infrared and microwave radiations emitted and reflected from the surface of the Earth and from the atmosphere.*

So defined, remote sensing has little capability for measuring below the surface of the ground, except for the limited penetration of low-frequency microwaves. This restricted meaning includes most work reported as remote sensing (c.f. Gregory, 1972; Anon., 1977) but does not satisfy the purists. On the other hand, there are valid reasons for including gamma radiation in the definition although current practice tends to place it in exploration geophysics. The argument, however, is really pedantic because the paradigm of remote sensing will surely expand and ultimately subdivide. Over the years, geophysics has split into exploration geophysics, solid earth geophysics and atmospheric physics and then, subsequently, exploration geophysics has split into magnetics, gravity, electromagnetics and the other practical paradigms that were represented in this symposium.

FUNCTIONS OF A REMOTE SENSING SYSTEM

A restricted definition of remote sensing does not imply that the technology of remote sensing comprises a relatively few tools. Indeed, there is great variety in the choice of possible components for a remote sensing system. However, for the purposes of this review, the components may be grouped together in terms of functions, of which the following four are principal from the applications point-of-view:

1. Sensing: The potential number of sensors required to cover the remote sensing spectrum is vast; some sensors are as old as the camera while others are as new as the Fraunhofer line discriminator;

2. Coverage: The platforms which carry sensors constrain the velocity, altitude and scale, field of view and repetition of the actual sensing as well as the means of transmitting the data for subsequent operations;

3. Preprocessing: Digital and/or photographic processing may be required to present the data set in a format suitable for interpretation;

4. Interpretation: This is the essential function of the remote sensing system in that useful information is derived from the data by mechanized, digital, optical and/or visual techniques, all supported by human judgment.

Sensing

A great array of sensors is available (Reeves et al., 1975, p. 235-537) to measure the intensity of radiation as functions of wavelength, time, geometry, phase change or polarization. No single instrument can do all these things well, or even satisfactorily, because each sensor emphasizes some parameters at the expense of others. Imaging sensors stress spatial resolution. Maximum resolvable detail is attained in the visible band. In the ultraviolet and thermal infrared, resolution is about one order of magnitude less while radar is almost another order of magnitude lower. Non-imaging sensors relate intensity to factors other than scene geometry. Active sensors, such as radar, provide their own illumination while passive sensors measure radiation that originated elsewhere. Each sensing system records some attribute of a cover class (e.g. vegetation, water, soil, buildings, and, here-and-there, rocks) as measured through the atmosphere.

Thus, the sensor is the definitive component of the remote sensing system in that it specifies the range of radiation parameters that can be measured. The more specific the sensor, the more carefully it must be matched to the spectral characteristics of the target. Such matching is one key to practical remote sensing.

Coverage

Sensors and ancillary equipment are carried on platforms that provide access and mobility relative to the target. Many types of platforms are available (Reeves et al., 1975, p. 538-588). Free and tethered balloons, powered aircraft of all sizes, helicopters, drones, dirigibles, sailplanes and manned and unmanned spacecraft all have been used at one time or another because of particular advantages. Depending upon the specific sensing system, data may be returned to base station by telemetry or by physical return of films and magnetic tapes. In many cases, the user's choice of platform is more limited than the preceding list suggests because of restricted availability or high cost for a particular platform. On the other hand, a wide variety of data collected from restricted platforms, especially at high altitudes, may be available if the user contacts the responsible agencies.

Thus, the type of platform may markedly affect scale, spatial resolution, field of view and format of raw data, as well as periodicity for repeated acquisition and temporal analyses. Hence, choice of platform and related coverage are important considerations in planning for remote sensing.

Preprocessing

As used here, preprocessing refers to operations performed on a set of remotely sensed data in order to compile that data in the format most suitable for subsequent interpretation. Preprocessing includes three different types of operations, which are here referred to as: (1) restoration, which removes effects that are inherently extrinsic to the target; (2) modification, which purposely alters the data-set to match standard conditions or to emphasize selected characteristics in the data; (3) correlation, which interrelates and registers several sets of data on a common geometric base. Typical restoration includes correction for geometric and radiometric distortions in the sensing system and clean up of artifacts and noise. Common modifications include reduction to standard conditions for atmospheric transmission and/or illumination; contrast stretching; density slicing; spectral classification; addition, subtraction and ratioing of bands; selective filtering and edge enhancement (c.f. Reeves et al., 1975, p. 688-710). Often such modification may improve the interpretability of the data although it may destroy their original character. Correlation primarily involves manipulation of data sets to register them on a common base e.g. aeromagnetics or radar with Landsat, or change detection using similar data for different dates (c.f. Fischer et al., 1976). It may also include statistical analyses to define the probabilities of relationships between features in the data sets.

The choice of preprocessing needs to be carefully considered because it can significantly affect the quality and, in some cases, the very success of subsequent interpretation. While digital processing may blur the distinction between preprocessing and interpretation, the functions are inherently and distinctly different.

Interpretation

In general, interpretation is the operation that assigns significance to the preprocessed remotely sensed data and, thereby, defines their utility. Spectral data are seldom sufficient, in themselves, to identify an object. With a few notable exceptions, the concept of unique spectral signatures is not generally valid in any field of remote sensing. Shape, texture, pattern and context are important, and usually essential, clues to the identification of geological features (c.f. Grossling and Johnston, 1977, p. 34-37). Unfortunately, detailed geometric and contextual data are not readily incorporated into digital systems.

While interpretation is not always amenable to rigid statistical expression, it is a probabilistic statement about the target and its environment. The greater the knowledge concerning the capabilities and limitations of the system and its target, the greater will be the probability of correct interpretation. Obviously, if the target is well known, there is little need for interpretation except as calibration or verification of technique. The interpretative process comprises three stages that may appear to be an iterative continuum: (1) analysis, which identifies, selects and calibrates useful observables from the more numerous elements of the data set; (2) synthesis, which puts the selected observables together in the context of the user's need; and (3) explanation, which integrates complementary data and human judgment with remote sensing to derive significant information. Feature extraction, thematic classification, mapping, mensuration and change detection are the principal activities of interpretation. There are many visual, machine-assisted and digital methods for solving the related problems (Reeves et al., 1975, p. 711-787, 869-1076). Automation and digital processing may accelerate these interpretive operations but, as practically foreseen, they will not obviate the need for human judgment especially with respect to geomorphic expression of rocks exposed to different climatic conditions (c.f. Doyle, 1977; Gregory, 1973, p. 89; Grossling and Johnston, 1977, p. 37-39; Reeves et al., 1975, p. 1072).

Experience has shown that not all professional resource scientists are good interpreters (c.f. Reeves et al., 1975, p. 1057). There are striking differences between individual interpreters, especially with respect to consistency, accuracy, colour discrimination and other personal factors. Imagination (which is the synthesis of new concepts), prior experience and accumulated knowledge are also key factors in interpretation. Accordingly, the quality of interpretations, be they digital, machine-assisted or manual, will continue to be influenced by the specific capabilities of the human beings involved in the process.

In the context of mineral exploration, remotely sensed data are surrogates for observables related to geology, among other things, as seen through the atmosphere. Interpretation of remotely sensed data by an experienced geologist is no less factual than the interpretation of visual observations on an outcrop, although the scale of observation and the size and nature of the observable may be different. Few geological observables are seen in their totality on either outcrops or images. Obviously, the closer the geologist is to the rock, the smaller the observable that he can utilize; and, also, the more detail he is exposed to. Hence, he must become more selective. On the other hand, observables of grand scale simply may not be apparent without a broad overview, such as that provided by Landsat. Thus, several sets of remotely sensed data may be required to advance mineral exploration.

In summary, the quality of any interpretation depends greatly on three choices: data, interpretive process and human interpreter. All should be carefully matched to the specific need for information in an exploration program.

Operational Remote Sensing Systems

Over the past few years, numerous potential applications for remote sensing have been suggested. Regardless of how beneficial they may appear, potential applications remain potential until three factors are assessed: the need for the derived information, the timeliness of interpretation and the cost-effectiveness compared to other methods of acquiring comparable information. There is a fundamental difference between <u>operational</u> <u>sensing</u> <u>systems</u> and <u>operational sensing to meet specific exploration objectives</u>. Failure to recognize this difference can cause financial problems for the unwary user. At this relatively early stage in the evolution of remote sensing, many sensors are still being used in experiments to develop practical applications. However, contemporary applications in mineral exploration can be recognized for some sensing and interpretive systems which are operational and which are, or could be, used systematically with reasonable economy.

APPLICATIONS OF REMOTE SENSING IN MINERAL EXPLORATION

General

While complete data do not appear to be available, there is abundant evidence that the mineral industry is a major user, if not the principal user, of remotely sensed data. Of the many sensors available, only aerial photography, side-looking airborne radar, aerial thermography and Landsat have immediate potential for systematic use in mineral exploration. Of these, aerial photography and Landsat are currently receiving the greatest use, presumably because of their relatively low cost (see p. 520). For example, exploration companies and consultants involved in the search for minerals and petroleum are major purchasers of Landsat data in Canada, the United States of America, Australia and, probably, South Africa (Wukelic et al., 1976; Gregory and Morley, 1977; U.K. Remote Sensing Society, 1975; Sabins, 1974). However, the split between use in ore exploration and use in oil exploration is not clear. The latter use may be greater although the former is significant and increasing on a worldwide scale.

Airborne Remote Sensing

Aerial Photography

Undoubtedly, conventional low-altitude aerial photography is the contemporary system of remote sensing that is most widely used in mineral exploration. As is well known, black-and-white photographs with high spatial resolution are systematically acquired and extensively used in the geological mapping of rocks and structures.

Monoscopic and stereoscopic techniques for interpreting black-and-white photographs and their relevant geological applications have been well described elsewhere (Allum, 1966; Miller, 1961; Newton, 1971; Reeves et al., 1975, p. 911-941; Tator, 1960). Accordingly, no attempt is made here to outline either the practice or the benefits for mineral exploration. Instead, attention is directed to a few of the newer techniques that can also be applied to the search for ore.

Colour and colour-infrared (CIR) films have added a new and slightly more costly dimension to aerial photography (Smith and Anson, 1968). While these films have a potential for improving feature identification, it is uncertain whether their capabilities with respect to geology are worth the additional cost unless specific colour contrasts are of principal interest in the target area. Anderson (1963) noted that outcrop areas are more readily discerned on colour photography than on black-and-white at comparable scale. He concluded that colour film would be preferable if the criteria for recognition of rock types were based on colour contrasts that might not be separable on black-and-white films. Gilbertson et al. (1976) reported that both colour and false colour (CIR) were ineffective in identifying rock types (schists and amphibolites) and in distinguishing gossans from soils under arid climatic conditions. On the other hand, Slaney (1975) noted that "many, if not most, gossans known to be present" could be recognized on colour photography for part of the Canadian tundra. In addition, Offield (1976) reported that coloured aerial photographs have long been used in the search for surficial discolorations associated with sedimentary uranium deposits in arid areas.

The practical utility of multiband aerial photography in mineral exploration is equally uncertain. Discrimination between rock types, especially sedimentary rocks, does not appear to be practical (Raines and Lee, 1974). Geobotanical differences, especially in areas of heavy vegetation, are dominant but may sometimes be useful in mapping rocks and structure (Reeves et al., 1975, p. 1231-1238). Gilbertson et al. (1976) were able to enhance differences between outcrop and overburden and to distinguish between gossan and some rock types. They concluded, however, that most differences can also be detected on conventional photography and that, because of the cost, multiband photography would not be an effective exploration tool.

On the other hand, B. Bølviken (Norges Geologiske Undersolkelse, Trondheim, Norway; pers. comm., 1977) reported that multiband photography has detected natural poisoning of birch forests over copper sulphide deposits in Norway. Differences in texture, pattern and tone that are related to severe damage to the vegetation are visible also on black-and-white aerial photographs. However, on such photographs, the area of copper poisoning is difficult to distinguish from common bogs (c.f. Bølviken et al., 1977). The multiband images also defined a surrounding spectral anomaly, possibly stressed vegetation, that appears to coincide with a major geochemical anomaly.

In summary, black-and-white photography continues to be widely used for general exploration. However, colour, colour-infrared and multiband photography can provide more specific data where the metallic mineralization is known to cause geological and/or botanical anomalies that are detectable with specified film/filter combinations.

Side-Looking Airborne Radar (SLAR)

Because of its unique capability for imaging through clouds, haze and darkness, side-looking airborne radar has a contemporary application for obtaining timely and moderately detailed data about the terrain under conditions that are not favourable for photography (Jensen et al., 1977; Reeves et al., 1975, p. 443-475). Resolution, however, is commonly less than for camera systems.

SLAR is an active microwave sensing system that illuminates terrain to the side of the aircraft and records the backscatter or reflected returns on either magnetic tape or photographic film. The SLAR image represents a continuous strip of the terrain but the image is not a "snapshot". It is produced by a line-scanning technique and, hence, may not have fixed two-dimensional geometry. In addition, fore-shortening (or layover) may be a problem in areas of high relief.

There are two basic types of SLAR. Real aperture radar (RAR) is less costly but its spatial resolution (30-100 m) varies with depression angle, range, and altitude. Synthetic aperture radar (SAR) is more expensive because it utilizes elaborate signal-processing techniques. However, it provides a finer resolution (10-30 m) that can be invariant with range and also a multispectral capability.

Most SLAR imagery is at small scales (1:150 000 to 1:500 000) but enlargements of up to 5X may be useful. Radiometric calibration is not usually required except for quantitative measurements e.g. spectral discrimination of materials on the ground.

Geometric corrections can be made, which for synthetic aperture radar, result in positional errors of less than 150 m (Peterson, 1976). Landsat can also be used to control the geometry of SLAR mosaics. Further, SLAR can replace one spectral band in a Landsat image to provide a synergistic combination of radiations from two different parts of the spectrum (Harris and Graham, 1976).

The utility of SLAR depends on the physical nature of the terrain and on the orientation of the sensing system relative to topographic features. Depression angles must be matched to the relief of rugged terrain. Shadowing may cause severe loss of detail, especially at low angles of illumination.

Photointerpretive techniques, similar to those used with airphotos but augmented by experience with radar, comprise the basis of practical application (Reeves et al., 1975, p. 982-1057; Parry, 1977). Pseudo-stereoscopic viewing is preferable, provided appropriate images are available.

Surface roughness is the principal determinant of tone on a SLAR image. Specular, or mirror-like, reflection results from smooth surfaces. A diffuse, or scattered, reflection is returned from rough terrain. Intensity, or brightness, varies with angle of incidence and roughness of the surface. Electrical properties of the surface comprise another major determinant of tone. The complex dielectric constant of a material varies directly with contained water. Penetration is greatest and reflection least for low water content.

As might be expected, texture, shape and pattern are the major discriminants used in interpreting SLAR data. Microtexture, or speckle, is an inherent random noise in the system which is commonly expressed as a fine texture in specular reflectances. Mesotexture in the SLAR data represents oriented, small-scale features in plant communities, as well as minor topographical relief. Macro-textures represent the coarser geomorphic elements of the terrain and are the principal features used in geological interpretation. Slope of surface and relative relief can also be estimated (Reeves et al., 1975, p. 1006-1008).

SLAR is most useful for the study of near-surface geological features because current systems have very limited penetration into the ground or overlying vegetation. Geomorphic patterns and textures are well expressed; hence, a knowledge of local geomorphology is important. SLAR images can assist in distinguishing surficial materials, outcrop boundaries, fractures, foliation and other structure (Reeves et al., 1975, p. 1224-1231, 1271-1289; Dellwig and Moore, 1977). Distinctions between rock types may be apparent in lightly vegetated or arid regions, especially if there is associated microrelief of the order of the radar wavelength (ca 3 cm) or greater, as on alluvial fans or volcanic flows. In the presence of abundant vegetation, such distinctions are difficult to make unless different vegetational communities are also associated with the rock types. However, it is worth noting that many of these features are also displayed on airphotos, especially if the photos are acquired at low solar inclinations (c.f. Reeves et al., 1975, p. 1174-1175; Lyon et al., 1970).

Synthetic aperture radar has been flown over large areas of remote tropical rain forest where persistent cloud cover precludes systematic aerial photography. The final maps are mosaics at a scale of 1:250 000. Such maps are available for all of Brazil and parts of Venezuela, Colombia, Peru and Bolivia. Other SLAR coverage is available for smaller areas in Panama, United States and other countries. Such maps provide valuable geographic and geological information for large areas about which little was previously known.

Where SLAR data are widely and inexpensively available (because of government support as in Brazil and USSR), they have been used extensively in mineral exploration. Kirwan (Hood, 1977, p. 9-10) noted that synergistic analysis of SLAR and Landsat has given major impetus to geological mapping and mineral exploration in Brazil.

The major value of current SLAR systems for mineral exploration is related to the acquisition of regional information about remote, poorly mapped areas obscured by persistent cloud. SLAR data can be acquired very rapidly, regardless of weather or lighting conditions. The systems can be flown at high altitudes to present relatively detailed, multi-interest images of large areas. Under such conditions, the delineation of drainage and the mapping of structure and other geological features may be well worth the additional cost to exploration, especially if the data can be applied subsequently in other disciplines relevant to natural resources and regional development.

Aerial Thermography

Airborne line scanners are complex optical-mechanical instruments that can make single-band or multispectral measurements in parts of the ultraviolet, visible and infrared regions of the electromagnetic spectrum (Reeves et al., 1975, p. 375-397). Radiation intensity is measured by a sensing system that scans a line perpendicular to the track of the aircraft. Hence, the data do not have fixed geometry and must be corrected for a variety of distortions related to movement of the aircraft (Reeves et al., 1975, p. 953-970). The electrical signal from the detector can be recorded on magnetic tape or on strip film. At present, aerial photography is a less expensive alternative for surveying the reflective radiations. UV scanning has not developed practical applications. Hence, only aerial thermography has the imminent potential of providing new data for mineral exploration.

Most thermal infrared (TIR) systems operate in two atmospheric windows: 3.5 to 5.5 micrometres and 8 to 14 micrometres. In those bands, thermometric temperature and emissivity of the surficial materials are the principal determinants of signal strength. The emissivities of rocks and soils depend largely on the roughness of their surfaces and on their chemical composition, especially the moisture content. The thermometric (or true) temperature depends on the heating or cooling history of the material and on its thermal inertia or rate of heat transfer. Water, for example, has high thermal inertia and thus warms or cools very slowly. Consequently, its diurnal temperature range is small. On the other hand, rock has low thermal inertia and a much larger diurnal temperature range. The combination of different emissivities and different thermometric temperatures in rocks and soils can produce large contrasts in radiometric (or apparent) temperature.

Time of day and season of the year are important parameters affecting both data acquisition and subsequent interpretation. Emission of TIR radiation from the surface of materials varies with time because of differential heating and cooling which in turn are related to intensity and angle of incidence of the solar illumination. Thus, two different materials may have contrasting radiometric temperatures at one time but will not be separable at other times when such temperatures are similar i.e. during diurnal or seasonal crossovers (c.f. Spectral Africa, 1977).

For geological purposes, nighttime imagery is superior to daytime imagery. Predawn appears to be the best time for thermography directed to geological objectives (Reeves et al., 1975, p. 964; Spectral Africa, 1977). The predawn conditions minimize solar contribution and emphasize reradiation of energy and, hence, the physical properties of rocks and soils. Maximum thermal contrast is attained early in the evening but effects of air temperature and shadowing may be significant also. Thus, in order to define specific requirements, an understanding of local diurnal and seasonal conditions should be obtained.

The standard TIR product is a black-and-white image which is a spatial record of radiometric temperatures within the field of view. Usually, cold areas are dark grey to black and warm areas are light grey to white. The image may, or may not, be corrected for geometric distortions. The mapping of TIR contrasts with reasonable resolution requires careful planning to minimize and correct geometric distortions so that a mosaic can be prepared. However, such corrections are not usually adequate for a precise planimetric presentation.

Practical interpretation is based on standard photogeologic principles, augmented by experience with TIR. Radiometric temperature contrasts may be enhanced by density slicing and/or colour coding. Calibrated TIR data may be obtained and contoured maps of temperature can be prepared. However, such data have had limited use in the search for metallic ores. Artifacts, unrelated to geology, may appear in the TIR data as a result of electronic noise, level shifts, wind and weather conditions and blooming or exaggeration of hot spots (Sabins, 1973; Quiel, 1974; Slaney, 1971).

Because of the many variables affecting aerial thermography, extensive field work is required to identify subtle radiometric contrasts. In the past, practical geological applications have been limited to major thermal contrasts such as volcanoes, hot springs and fumaroles and major emissivity/thermal contrasts such as moist soil in and over fault zones or thin, dry soil over bedrock (Fig. 22.1). Extensive tests in arid alluvial terrains, however, have recently shown that remarkable detail concerning bedrock structure and lithology can be obtained by aerial thermography during the dry season (Spectral Africa, 1977). Few comparable details are recognizable on relevant conventional aerial photographs. While there is much promise in the technique of discriminating rock types on the basis of their thermal inertia, it has not yet become an operational technique for mineral exploration (Gillespie and Kahle, 1977).

Orbital Remote Sensing

Meteorological and Manned Spacecraft

Systematic remote sensing from space began in 1958 with the meteorological satellites, particularly the Nimbus and current NOAA systems. From the viewpoint of a photogeologist, meteorological sensing systems have improved in resolution (to about 0.5 km) but they still are useful only for very small-scale studies, repetitive observations of large regional features and, of course, weather factors that determine operational conditions (Reeves et al., 1975, p. 565-577, 583-586, 1244; Gregory and Moore, 1976, p. 155). The geological information that can be derived from these telemetered black-and-white images (both visible and thermal infrared) is simply inadequate for most applications in mineral exploration (c.f. Reeves et al., 1975, p. 972-973).

Manned spacecraft have provided more detailed data in several bands of the electromagnetic spectrum for those regions of the Earth over which the spacecraft have passed. None of these systems was designed specifically for remote sensing, although the Mercury, Gemini, Apollo and Skylab programs all returned useful, and sometimes spectacular, images that can be used in mineral exploration. The resolution (of up to 10 m) is adequate for many geological applications (Maffi and Simpson, 1977; Cassinis et al., 1975; Reeves et al., 1975, p. 577-583, 1244-1247). However, the images are limited in geographic location to the lower latitudes, they are rarely repetitive and they are not current. For these reasons, such data have not received much attention in Canada although they may be very useful where relevant coverage is available (c.f. Fischer et al., 1976).

 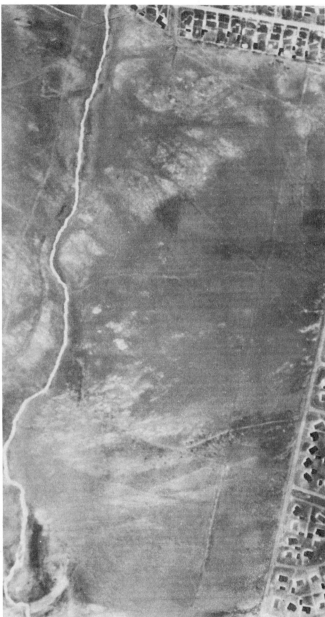

(GSC 203492-F)   (GSC 203492-C)

Figure 22.1. (a) Conventional panchromatic aerial photograph and (b) pre-dawn thermal IR scanner image (8-14 micrometres) acquired within a few weeks of each other, Witpoortjie fault area, West Rand, South Africa. Scale: approximately 1:5000. The fault, which is represented by the abrupt termination of the light grey to white outcrop and sub-outcrop, is one of the most obvious features on the thermal image. Note also the faulted shale horizon (dark grey) in the lower third of the image. For further details, see Spectral Africa, 1977.

Photograph and image courtesy of Spectral Africa (Pty) Ltd., P.O. Box 976, Randfontein, South Africa.

Landsat (ERTS)

On July 23, 1972, NASA launched the experimental Earth Resources Technology Satellite (ERTS-1) to begin an era of systematic and relatively detailed exploration of the world's land areas from space. A second, similar satellite was launched January 22, 1975 and both were renamed as Landsat-1 and Landsat-2. These satellites provided data from a 4-band multispectral scanner (MSS) that recorded in two bands of the visible spectrum and two bands of the reflective infrared (Williams and Carter, 1976; Short et al., 1976). Weather permitting, each Landsat collects MSS data every 18 days over preselected areas in its near-polar, sun-synchronous orbit. Coverage of all land masses at least once is a principal goal of the program. Landsat-1, which operated well beyond its designed lifetime, is no longer operational. When both satellites were operating, data were collected every 9 days. A similar satellite, Landsat-3, was launched on March 3, 1978, to record the same 4 bands plus an additional thermal infrared band as well. These satellites also carry a data collection system that relays data by telemetry from remote ground platforms via Landsat to a central readout terminal.

The remotely sensed data provided by the current Landsat system comprise: four black-and-white images in visible green, visible red and two non-thermal infrared bands. Relevant computer compatible tapes (CCTs) are also available. The CCTs provide 64 levels of radiation intensity with an inherent resolution of 80 m, whereas the photographic products have the same resolution and the equivalent of 16 levels of intensity.

False colour composites are usually prepared for any scene with three appropriate bands (c.f. Pl. 3). Digital processing is used to locate ground control, to perform radiometric and geometric corrections and to compile the image. The standard products [1] (false colour and 4 bands) are near-orthographic prints with common, pre-selected scene geometry and a scale of 1:1 million. Each scene represents about 34 000 km^2 (or about 13 000 sq. miles). Ground resolution is of the order of 80 m for many observable geological features but smaller targets with high contrast are often recorded. Other types of data that may be acquired include: transparencies, enlargements, mosaics, digital enhancements, digital transforms and, for Canada only, microfiche for one infrared band (MSS 6).

Most systematic interpretation of Landsat data for mineral exploration is based on the recognition of tonal contrasts, shapes and patterns using standard photogeologic techniques adapted to the selected wavebands (Gregory and Morley, 1977; Wukelic et al., 1976; Gregory and Moore, 1975). While prints are widely used in visual interpretation, transparencies may be usefully projected at various scales onto airphotos and topographic, geological or geophysical maps. Such projections can be used for systematic interpretation with a "zoom" capability for focusing on detail at scales as large as 1:30 000 provided that the targets have dimensions of 100 m or more, depending upon relative contrast.

The most common type of interpretation (i.e. monocular, single band analysis) should take into account the wavelength being interpreted because the black-and-white images are not directly analogous to ordinary aerial photographs [2]. Stereoscopic viewing of Landsat images is feasible but is inherently limited by the geometry of the sensing system. Such stereoscopy is unidirectional along the scan line rather than omnidirectional as in aerial photography. Stereoscopic capability improves with increasing geographic latitude or occasional orbit overlaps. In these cases, the technique can be exploited to study major geological features such as incised river valleys or mountain belts but it is not suitable for most photogrammetric purposes (c.f. Welsh and Lo, 1977). In addition, pseudostereoscopic viewing of the same scene in two bands (i.e. band-lap stereo) or at two times (time-lap stereo) may also provide information useful for mineral exploration.

A colour composite is prepared from 3 bands which may be variously coded by colour. Some composites approximate natural colour but the most common type simulates aerial colour infrared (CIR) film, a product which has been a standard tool of photointerpreters for many years. On the Landsat equivalent of this latter product, healthy vegetation appears red because chlorophyll is one of the best reflectors of the infrared radiation (which is coded red) while clear, deep water is a strong absorber and thus appears black.

A heavy cover of vegetation may obscure bedrock features. However, Kirwan (Hood, 1977, p. 9) noted that controlled colour composites have improved the interpretability of Landsat images, especially for terrains covered by tropical rainforest. Similar improvements can be achieved by digital processing (c.f. Offield et al., 1977) and by selective filtering in projection systems.

Visual thematic classifications have been used to map mine wastes, geological formations, laterites and placers, as well as environmental and vegetational themes. Such classifications, are based on spectral response, texture and context. Digital classifications of spectral data appear to be on the threshold of practicality for a few specific geological applications in arid areas (see page 521) but, in most cases, further research appears warranted to define costs and applicability to mineral exploration. Density slicing and other forms of optical-mechanical analysis have not found major applications in the search for ore.

Unlike most aerial photography Landsat data are not always acquired under optimal conditions. Further, seasonal aspects of climate, weather and illumination provide different views that may preclude, hinder or assist the interpretation of geology. Therefore, it is preferable to visually select data for each interpretation. Such selection of optimal data should consider the following interrelated factors: coverage, scene quality, seasonal aspects, waveband, format and delivery schedule (Gregory and Moore, 1975).

Repetitive Landsat data have proven very useful in mineral exploration, primarily because various seasonal phenomena enhance geological features. Data acquired in appropriate seasons can provide the following enhancements (Gregory and Moore, 1975):

1. <u>Low inclination of the sun</u>, giving a pseudo-radar effect with shadows enhancing even minor topography;

---

[1] Standard products and ISISFICHE for Canada may be purchased from: Integrated Satellite Information Services Ltd., P.O. Box 1630, Prince Albert, Sask., S6V 5T2. For other parts of the world, standard products may be purchased from EROS Data Centre, U.S. Geological Survey, Sioux Falls, S.D. 57198, U.S.A., or from relevant national agencies.

[2] Images in MSS 5 (visible red) have tones and contrasts that will be most familiar to the photogeologist as they approximate the common panchromatic photograph, or, more closely, the less common orthochromatic photograph.

2. Uniform surface of a thin cover of snow or sand obscures terrain noise and serves to emphasize topography and penetrative structures especially in conjunction with low sun;

3. Residual snow and ice remaining in the solar lee of hills and depressions enhances associated linears;

4. Meltwater swells drainage channels and temporarily fills minor drainage features to enhance linears, particularly through low reflectance in the infrared bands;

5. Soil moisture patterns following rainfall in semiarid regions may enhance soil and related bedrock contrasts;

6. Preferential growth of vegetation may enhance associated rocks and soils because of the high infrared reflectance of vegetation; relative vigour of growth is measurable.

The chief value of Landsat data to the search for ore lies in the regional view and the moderate level of detail that is presented in either a single frame or a mosaic. As can be readily seen through comparison of relevant maps and images at the same scale, many Landsat images portray greater detail in topography and geology than do the comparable maps. With experience, Landsat images can be used at scales of 1:50 000 or larger although they lack the detail of photographs at similar scales. However, Landsat does not provide definitive information about rock composition or the third dimension of depth below the surface.

While neglect of spectral content in the four bands is not recommended, photogeological interpretation alone can provide much geological information that is valuable in the search for ore. As such, these simple techniques may be readily adopted by an experienced field geologist. Information interpreted from Landsat can assist mineral exploration with: mapping of major geological units; discrimination of rock classes (such as alluvium, sedimentary rocks, metamorphic rocks, granitic gneisses, intrusions and volcanic cones and flows); mapping of structures[1], including linears; and detection of alteration zones and other broad surface indications of mineralization. Such geological mapping has already effected significant savings in time, by factors of 3 to 10, relative to conventional mapping at similar scales. Many examples of these benefits have been published recently (Palabekiroglu, 1974; Slaney, 1974; Barthelemy and Dempster, 1975; Gregory and Moore, 1975; Viljoen et al., 1975; Short et al., 1976; Williams and Carter, 1976; Iranpaneh, 1977; Shazly et al., 1977; Woll and Fischer, 1977). Alteration zones in arid regions may be visually delineated (c.f. Schmidt, 1973; Smith et al., in press) but digital modification can provide significant improvements (see page 521).

In addition, Landsat images can be used in appropriate enlargements as relatively detailed base maps for reconnaissance, as control for preparing mosaics of aerial photographs and, under favourable conditions, as a base map for compiling airborne geophysical surveys.

Costs of Remote Sensing Applied to the Search for Ore

The costs of acquiring and interpreting remotely sensed data are estimated in Table 22.1. Few definitive costs have been published except for aerial photography. Hence, the tabulation should be construed as a preliminary approximation. It is based on many bits of information including price lists, published papers, personal communications and experience, all of which were updated to 1977 dollars. Even if actual costs vary by a factor of 2 or 3, the trends remain obvious. The current cost of acquiring Landsat data is much less than the cost of interpretation. This is rarely true for other types of remote sensing except under special but analogous circumstances i.e. when the data have been acquired by government for multiple use and are sold across the counter at nominal prices. Otherwise, the cost of new airborne sensing is greater than the cost of interpretation and much greater than the total cost of interpreted Landsat data. While significant savings can be effected for airborne surveys of very large areas, the cost per unit area will still be greater than current Landsat costs. Cost, however, should be balanced against the desired level of detail which is inherently greater for airborne surveys.

Equally noticeable is the fact that the total cost for small scale photography is only about twice the total cost for Landsat if the airborne data are available across the counter at a nominal price. In Canada, the wide availability of black-and-white aerial photography is comparable to that of Landsat. Also, total costs for interpreted information are low for both types of sensing. Undoubtedly, these two factors (i.e. wide availability of data and low cost of visual analysis) are major reasons for the emphasis on those two types of sensing by the exploration industry in Canada (c.f. page 514).

Although they lack the detail of aerial photographs, Landsat data have a number of unique advantages relative to aerial photography, including:

(1) synoptic scale, which makes it possible to view major features in their entirety;

(2) constant raked illumination, which results in uniform presentation of features over a large area;

(3) spectral sensing, which serves to emphasize contrasts in reflectance and may ultimately assist in classifying materials;

(4) repetitive imaging, which provides seasonal enhancements and facilitates monitoring of the terrain;

(5) moderate resolution, useful for identifying many features with dimensions greater than 80 m and smaller features with high contrast; and,

(6) global coverage, for all land masses between latitudes 81°N and 81°S, without interruption at national boundaries.

However, the principal benefit to mineral exploration from any remote sensing lies in the quality of geological information that can be derived from the data. Except for very special conditions, this paper shows that conventional aerial photography and Landsat can provide more useful information per exploration dollar than other types of remote sensing (as defined herein). Landsat has advantages relative to its generality and spectral/temporal content while conventional aerial photography provides detail. The systems, thus, are complementary rather than competitive sources of data.

CHALLENGES TO THE ADVANCE OF REMOTE SENSING IN THE SEARCH FOR ORE

At the present time, the role of remote sensing in mineral exploration is undergoing quiet assessment in Canada and elsewhere in the world. In part, this reflects the current low level of exploration activity in this country. There is little doubt, though, that simple practical applications of remote sensing have been recognized and are being integrated with other exploration tools.

Further advances of remote sensing with specific application to mineral exploration will require that the mining industry face up to several major challenges and decide which challenge, if any, it wishes to accept. There are two interrelated types of challenge: technical and financial.

---

[1] Note that the restricted range of sun azimuths introduces a bias against structures subparallel to that azimuth.

Table 22.1

Approximate 1977 costs[1] for acquisition and interpretation of remotely sensed data

Data	Format	Acquisition Cost[2] (per 1000 sq. km)		Interpretation Cost[4] (per 1000 sq. km)	Range of Total Costs[5] (per 1000 sq. km)
		over the counter	new mobilization[3]		
Landsat	Colour Transparency (1:1 million)	$0.40[6] ($12)	—	$180-$1000 (visual)	$190-$1000
	Colour & 4 bands, transparencies (1:1 million)	$1.30[6] ($44)	—		
	CCT (1 tape)	$4.50[6] ($150)	—	$2000-$9000+ (digital)	$2200-$9000+
Air photos (stereo coverage; 22.8 cm format)	b & w 1:50 000	$42	$1000-$2300	$700-$1500 (visual)	$740-$3800
	b & w 1:20 000	$250	$2400-$4000	$3000-$9000 (visual)	$3300-$13 000
	colour 1:50 000	$100	$1500-$4500	$700-$1500 (visual)	$800-$6000
	colour 1:20 000	$640	$3900-$5000	$3000-$9000 (visual)	$3600-$14 000
Multiband photography	4 bands, 1:20 000	—	$5000-$20 000	$8000-$30 000 (visual)	$13 000-$50 000
SLAR	strips & mosaics (1:250 000)	$4-$8	$8000-$20 000	$1000-$5000 (visual)	$1000-$25 000
Aerial Thermography (IR)	strips & mosaics (1:50 000)	—	$5000-$10 000	$1000-$2500 (visual)	$6000-$13 000

[1] Costs were estimated from many sources, inflated to 1977 Canadian dollars and rounded to two significant figures.

[2] Single purpose, single coverage.

[3] Including a mobilization cost of $1000.

[4] Single theme mapping; includes professional salaries and overhead, but wide range reflects amount of detail in image, scale, objectives of project and method of interpretation.

[5] Exclusive of field studies.

[6] Subject to minimal cost of data for 1 scene as given in brackets.

This section of the paper is concerned primarily with the former although, without doubt, the greatest challenge is financial.

Technical Challenges

A multitude of specific sensing needs that are related to particular systems can be quickly identified by any experienced technologist e.g. improvements in spectral and spatial resolution of a sensor, refinements to corrections for attenuation or development of a new algorithm to enhance contrasts. Geologists and geophysicists can identify needs related to characteristic spectral radiances for rocks or better understanding of the ubiquitous linears which have well publicized, but poorly defined, relationships to ore (c.f. Carter et al., 1977; Gilluly, 1976). All such problems, however, are details — though important details — which do not alter the fact that the current capabilities of remote sensing can now provide significant assistance in the search for ore.

However, there are two broader technical challenges which, if they are approached in orderly way, can be overcome to advance not just remote sensing but also mineral exploration in general.

These challenges, which are related to the practicality of digital preprocessing (see page 513), are not inherently system specific, although current focus is on Landsat. These challenges are:

1. Correlation of multiple data sets, and

2. Digital modifications to assist geological interpretation.

Correlation of Multiple Data Sets

The principles of synergism, or integration of different sets of data to produce more or better information than the sets can separately provide, are receiving much current attention in remote sensing. While he may not recognize the terminology, the experienced explorationist is familiar with the principles and intuitively practices them in his everyday work. He uses converging lines of evidence and optimizes his interpretation by integrating all relevant geological, geophysical, geochemical and topographical data. Commonly, this is done by mental approximations, by overlaying maps at the same scale or by projecting an image of one set of data onto another data base. The limitations for multiple data sets are obvious. Recently, digital techniques have been developed to register Landsat data with other data, including Landsat data for different times as well as SLAR and geophysical data (c.f. Fischer et al., 1976; Harris and Graham, 1976; Anuta, 1977; Reeves et al., 1975, p. 1100). Perhaps the most significant trend with respect to mineral exploration is the ultimate development of precise location of data relative to ground control or other field data (c.f. Raynolds and Lyon, 1976).

With respect to practical application, these developments are still experimental in that their specific advantages and costs for use in mineral exploration have not been established. Their potential for correlation and probability analyses could be of value in mineral exploration if we can decide what data sets should be combined and what goals should focus the synergism.

The visual correlation of Landsat prints and transparencies with other (e.g. geophysical) data is now a practical simple method of synergistic analysis. Further improvements are warranted, including: better image quality, a broader range of photographic enhancements and augmented optical projection systems with magnifications of the order of 30X to 100X. Further research is required to determine the need for cosmetic improvements, contrast stretching, preselection of density levels, band ratios, etc., and, especially, to define the most useful products for specific conditions.

Digital Modification to Assist Geological Interpretation

Background

Optional cosmetic improvements can now be effected on Landsat images, usually at considerable expense. Noise, striping and dropped lines may be removed (Goetz et al., 1975) and definition improved (Longshaw et al., 1976) to obtain a more aesthetically pleasing image. However, the image must still be interpreted to derive geological information. While there appears to be an improvement in interpretability for some images, it has not yet been established that the improvement is generally worth the additional cost. In fact, it now seems likely that cosmetic improvements and reductions to standard conditions (c.f. Ahern et al., 1977) are not essential for geological interpretation. The human interpreter can handle adequately the relatively small interference and bias, except for subtle geological contrasts that require enhancement in any case.

Other modifications, however, have exciting potential. Enhancements can serve to emphasize subtle spectral contrasts that may have lithologic significance (Goetz et al., 1975). Also, spectral classifications and transformations can define areas with similar spectral characteristics and may help to discriminate rock types (Podwysocki et al., 1977). While exciting, these modifications contain a major pitfall: it is too easily — and too commonly — assumed that spectral classes (often called thematic classes by digital analysts) are the same as user-defined thematic classes for rocks and soils on the surface of the ground. Fortunately, the magnitude of this problem is declining with increasing involvement of geologists in digital processing.

The thematic classification of rocks by remote sensing is hindered by many almost overwhelming factors. First of all, the spectral radiance of rocks in situ is rarely characteristic in any available waveband and usually requires textural information for even generalized classification (c.f. Podwysocki et al., 1977; Rowan et al., 1976). Further, outcrops of rock comprise a minor part of many landscapes although massive exposures are not uncommon. Frequently, the outcrops are obscured by weathered surfaces, lichen, moss, dust, rain or snow or an overhanging canopy of vegetation. And as every geologist knows, each rock type is represented by a range of compositions. Because of these factors, the absolute identification of rock types by remote sensing is a highly improbable task. However, spectral discrimination (or separation of spectral units and subtle contrasts) may, under certain conditions, facilitate future mapping of rock units and alterations.

Arid and semiarid areas

Digital processing techniques have been developed to retain textural information in a modified image. Under arid conditions, these images may be useful in discriminating among rock types (Podwysocki et al., 1977; Rowan et al., 1976; Goetz et al., 1975). Spectral classifications, devoid of texture, do not appear to be useful for mapping similar arid areas (Siegal and Abrams, 1976).

Discrimination of altered zones related to ore deposits has received much attention. Current work is based mainly on the enhancement of reflectance minima resulting from absorption bands representing ferric iron (primarily limonite, jarosite, etc.) and hydroxyls in clays (c.f. Pl. 4). Rowan et al. (1976, 1977) used stretched ratio enhancements to discriminate both limonitic and siliceous hydrothermal alterations in a Tertiary volcanic complex in the Nevadan desert. Shales and siltstones, being similar in composition to the altered rocks, were not separable. In Western Australia, Smith et al. (1978) used digital analysis to improve on their visual thematic classification of hydrothermal silicate alteration in flow tops in a Keeweenawan-type volcanic sequence. Current success seems to be dependent on minimal shadowing from flat to slightly undulating terrain, flat formational dips that provide broad targets, and a weathered profile that accentuates the spectral contrast between altered flow tops and less altered basalts. Discriminatory capabilities were similarly assessed for iron oxides associated with uranium deposits in flat lying, pallid sedimentary rocks of Wyoming (Offield, 1976; Vincent, 1977) and iron oxides associated with mercury in the McDermitt caldera of Nevada and Oregon (Raynolds and Lyon, 1976). Similar modifications to Landsat data have delineated anomalies considered to represent sulphates and alteration zones associated with porphyry coppers in Pakistan (Schmidt, 1976; Schmidt and Berstein, 1977). In the latter case, no significant amounts of ferric iron were present in the alteration. Schmidt (1976) also tested his technique in a quasi-exploration context and selected 23 targets for prospecting in areas adjacent to his training site. Seven of these areas contained hydrothermally altered rock, mostly porphyry (Schmidt, pers. comm., 1977).

All of these feasibility studies used data for arid to semiarid sites. In all cases, known alteration zones were delineated although errors were more numerous than desirable. Errors of commission (misidentifications or false alarms) and errors of omission (nonrecognition of known occurrences) were both experienced. On the other hand, it is debatable whether such errors are any more numerous than those experienced with other exploration techniques, especially at a comparable early stage in the development of those techniques.

Moderate vegetational cover

To date, digital modifications of Landsat data seem to work best in areas with little or no vegetation. Grasses and trees can significantly mask and alter the spectral reflectance of the ground as measured by sensors in aircraft or satellites (c.f. Siegal and Goetz, 1977). A cover of about 30 per cent green vegetation may overwhelm the reflectance of underlying rocks and soils. Indeed, if such materials have low albedo, the reflectance may be altered beyond recognition by only 10 per cent green cover. On the other hand, dead or dry vegetation has minimal effect on reflectance from the ground (Levine, 1975).

The staff of Gregory Geoscience Limited have attempted to adapt some of these digital classifications for use on the tundra where there may be abundant grass but no tree canopy. Spectral classifications based on band ratios were not successful in identifying gossans associated with the Muskox intrusion but they partially discriminated between several rock types (H.D. Moore, pers. comm., 1976). Subsequent work followed up on rock discrimination in an area near the Great Bend of the Coppermine River, N.W.T. where the dominant lithologies are flatly dipping basalt, sandstone and dolomite with sandy alluvium. Various ratios of MSS bands and algebraic combinations of ratios were tried but separations were incomplete (P. Chagarlamudi, pers. comm., 1977). An attempt was then made to strip out the vegetational contribution by developing a complex algorithm that minimized the effect of vegetation. On a pixel-by-pixel basis, this classification provided encouraging results but further testing has not been completed.

Spectral geobotanic anomalies

Vegetation covers over two thirds of the land surface of the world and the root systems sample the solutions in underlying soils and bedrock. Geobotanical anomalies are known to be related to metallic mineralization under certain conditions and relevant surveys are used in mineral exploration. It is not surprising, then, that the concept of measuring related spectral anomalies by remote sensing has been investigated for several years.

Except for relatively short transient changes, natural vegetation is always in equilibrium with the soil, moisture and climate. Contemporary research indicates that vegetation with an intake of excess metal ions or complexes will develop stress symptoms. Relative to unstressed vegetation, such symptoms are manifested by both geometric and spectral differences e.g. extinction of some species, unique plant associations, stunting, sparse foliage, wilting or chlorosis, among others. It is possible, then, that plant communities could serve as surrogates for the chemical state of soils and soil solutions and, hence, for mineralization. Related biophysical anomalies could be expected to have dimensions of a kilometre or more.

Aerial multiband photography and modified Landsat data have been used to test the feasibility of mapping biophysical anomalies related to stress. As noted previously, multiband photography has serious limitations as an exploration technique, despite some seeming successes (c.f. Canney, 1975). Evidence from Landsat is equivocal. Anomalies have been delineated and related to copper poisoning in the boreal birch forests of Norway (Bølviken et al., 1977), and to molybdenum poisoning in the pine and juniper forests of Nevada (Lyon, 1977). The specific causes of spectral anomalies in both areas are uncertain. Low ratios for band 7/band 5 (or band 4) may result from spectral contrasts such as chlorosis, from textural contrasts such as sparse foliage or from some combination of the two. At the Norwegian test site, clearings unrelated to copper poisoning have a normal grass cover. Hence, their high 7/5 ratios serve to separate them from the poisoned areas. No similar anomalies were found in tropical rain forest over a porphyry copper deposit (Lyon, ibid.).

Experience has shown that the link between geobotanical anomalies and ore may be tenuous. The link between a remotely sensed anomaly and ore will be even more so, in part because there are other natural causes of stress. Nevertheless, current results should encourage further work if only to define the most favourable environment in which to continue development. It may be, for example, that the best time to detect metal toxicity in vegetation is during the time when the regional plant community is under uniform stress so that metal poisoning is accelerated in plants adjacent to mineralization. Such uniform stress could result from irregular but persistent droughts in temperate climates or from the annual dry season in more arid areas.

Summary

The successful, though limited, detection of Landsat anomalies related to metallic mineralization and geobotanic anomalies points to an obvious extension: a more detailed study with more specific bands in an airborne multispectral scanner. Above all, however, successful discrimination of alteration zones, rocks or geobotanical anomalies by sensors in aircraft or satellites must be shown to be cost-effective in an exploration context. To date, the costs appear to be too high for general use although they may be acceptable for specialized surveys in areas where specific useful contrasts have been identified.

Financial Challenges

Remote sensing is an array of interdisciplinary technologies that depends upon many fields of science. This requirement for broad knowledge, plus specialization, is a strength in that it builds on converging lines of evidence to reach decisions. Unfortunately, it is also a weakness because the scope cuts across accepted boundaries of science, practice and administration. Nothing shows this better than the lack of a co-ordinated program for developing practical applications of remote sensing in Canada. Over the past seven years, the federal government has invested millions of dollars in the technology and methodology of remote sensing (c.f. annual reports of the Canadian Advisory Committee on Remote Sensing). As a consequence, Canadians are in the forefront as far as acquiring data and creating products of high quality. During that seven years, however, neither government nor industry in Canada has invested even a small fraction of those millions in the development of practical applications. This has resulted in a severe gap in credibility with respect to the use of remote sensing — and not only in the mineral industry.

Five years ago, it seemed obvious (Gregory, 1973, p. 89-91) that the innovators of remote sensing faced the same problems that are experienced by any other innovator of a new product — the problems of identifying and assessing the market. The principal determinants of successful sales (i.e. use of a product) are two interrelated factors: technical quality of the product and acceptability to the user. Both must be high. The evidence to date shows that current remote sensing products are consistently of high quality. What then can be done for acceptability to the user and, more specifically, the user in mineral exploration?

With respect to mineral exploration in particular, the key requirement still seems to be a need for demonstration projects, both visual and digital, with a practical cost-effective focus. This paper, backed up by relevant detail in the references, shows that certain types of remote sensing can now make significant contributions to practical mineral exploration. It is no criticism of those involved to say that these practices are relatively simple. How can they be otherwise in a time when investment by Canadian industry in research and development is rapidly declining and when the government contribution to the development of practical applications of remote sensing, especially for geology and mineral exploration, is almost negligible?

A co-ordinated program could bring together many ideas about the practical application of remote sensing to the search for ore. Such ideas may be technical and related to making sensors or data more specific to the needs of mineral exploration. They may be geological and related to understanding what we have sensed and what we should sense. As isolated projects, such ideas are difficult to sell to either government or industry. But by combining the splendid capabilities of government agencies with the practical experience of the exploration industry, the ideas could be filtered and developed into practical applications.

Of course, the major challenge is funding in the current economic climate. Perhaps the solution is a shared common fund with interdisciplinary support for the development of practical applications. The annual cost need not be great and initially could be less than the annual cost of a small exploration program. Most of the money would be expended on salaries and office expenses. Hardware and technology, already supported by a co-ordinated program, might continue to be shared. A comparable vehicle to support the development of practical applications should achieve comparable advances. Industry and government can work together towards such an end. It is not the lack of an operational satellite that confounds the development of practical applications of remote sensing, but the lack of a co-operative will to do something practical with the available data — even for mineral exploration.

SUMMARY AND FORECAST

In several large countries around the world, conventional aerial photography and Landsat images comprise complementary data that are widely used in mineral exploration. Conventional photography provides detail while Landsat data have advantages related to their overview and spectral/temporal content. Both tools are relatively simple and inexpensive to apply in the search for ore, although neither is specific with respect to the precise location of ore. The practices of photogeology are well established and can be adapted to Landsat with little difficulty. In general, Landsat images are so inexpensive per unit area and their information content is so comprehensive, that no regional exploration program should proceed without preliminary analysis of carefully selected images. Digital analysis of Landsat data has an encouraging potential that can, in some cases, be exploited. However, in general, further development is required to define limiting conditions and costs as applied to mineral exploration.

Side-looking airborne radal (SLAR) and aerial thermography can meet specific needs in the search for ore. Under certain conditions, one or the other may be the only means of providing essential exploration data. In some countries, SLAR is becoming more widely available through government acquisition. In such countries, SLAR imagery is widely used in geological mapping and mineral exploration. Aerial thermography yields significant amounts of geological information that may not be apparent from conventional aerial photography, especially for arid areas with sparse outcrop.

In the future, airborne remote sensing will develop as a specialized source of detailed information. There will be increased emphasis on multispectral scanners, selection of diagnostic wavelengths and digital processing to meet specific needs. Orbital remote sensing will provide higher resolution, perhaps with a zoom capability, although still retaining synopticity. Increased government participation in the acquisition of regional data seems probable because the high costs for large areas can be shared by several users.

The gap between remotely sensed data and information needed for mineral exploration will decline as government and industry recognize the need for co-ordinated effort in order to develop future reserves of ore. Much of that ore lies buried under soil and vegetation. In learning to deal effectively with those constraints, many new tools will be needed, including remote sensors to map new attributes of geology e.g. alteration patterns in residual soils, ghost linears that reflect buried structures and biophysical anomalies related to geobotanical stresses. To further these ends, a co-ordinated program for developing specific applications of remote sensing should be started now.

ACKNOWLEDGMENTS

Numerous contributions in the form of papers, manuscripts and comments were received from correspondents in Australia, Canada, Italy, Norway, South Africa and the United States of America. The author sincerely acknowledges their great assistance, regrets that he cannot recognize them individually and assumes full responsibility for his analysis of their contributions.

REFERENCES*

Ahern, F.J., Goodenough, D.G., Jain, S.C., Rao, V.R., and Rochon, G.
    1977: Atmospheric corrections at CCRS; Proc. 4th Canadian Symposium on Remote Sensing, p. 583-594.

Allum, J.A.E.
    1966: Photogeology and regional mapping; Pergamon Press, Oxford, U.K.

Anderson, D.T.
    1963: Colour photography as an aid to geological interpretation; in Air Photo Interpretation in the Development of Canada, Part IV, p. 2-9, Dep. Energy, Mines and Resources, Ottawa.

Anonymous
    1977: What photogrammetric engineering and remote sensing is; Photogramm. Eng. and Remote Sensing, v. 43 (5), p. 451.

Anuta, P.E.
    1977: Computer assisted analysis techniques for remote sensing data interpretation; Geophysics, v. 43 (3), p. 468-481.

Barthelemy, R. and Dempster, A.
    1975: Geological interpretation of the ERTS-1 satellite imagery of Lesotho, and possible relations between lineaments and kimberlite pipe emplacement; Proc. 10th Internat. Symposium on Remote Sensing of Environment, v. II, p. 915-924.

*Bølviken, B., Honey, F., Levine, S.R., Lyon, R.J.P., and Prelat, A.
    1977: Detection of naturally heavy-metal-poisoned areas by Landsat digital data; J. Geochem. Explor., v. 8, p. 457-471.

Canney, F.C.
    1975: Development and application of remote sensing techniques in the search for deposits of copper and other metals in heavily vegetated areas — Status Report June/75; U.S. Geol. Surv. Project Report (IR) NC-48.

---

* A short list of the more significant references with respect to mineral exploration includes the special issue of Geophysics (v. 43, no. 3, 1977) and the compendia and references marked with an asterisk herein.

Carter, W.D., Lucchitta, B.K., and Schaber, G.G.
  1977: Preliminary lineament map of the conterminous United States; Proc. 11th Internat. Symposium on Remote Sensing of Environment, v. II, p. 1543 (summary).

Cassinis, R., Lechi, G.M., and Tonelli, A.M.
  1975: Application of Skylab imagery to some geological and environmental problems in Italy; Proc. NASA Earth Resources Survey Program, Houston, p. 851-867.

Dellwig, L.F. and Moore, R.K.
  1977: Tradeoff considerations in utilization of SLAR for terrain analysis; in Woll and Fischer, 1977 (which see), p. 293-306.

Doyle, F.J.
  1977: Photogrammetry: the next two hundred years; Photogramm. Eng. and Remote Sensing, v. 43 (5), p. 575-577.

El Shazly, W.M., Abdel Hady, M.A., El Ghawaby, M.A., and Khawasik, S.M.
  1977: Application of Landsat satellite imagery for iron ore prospecting in the western desert of Egypt; Proc. 11th Internat. Symposium on Remote Sensing of Environment, v. II, p. 1355-1364.

Fischer, W.A., Hemphill, W.R., and Kover, A.
  1976: Progress in remote sensing (1972-1976); Photogrammetria, v. 32, p. 33-72.

*Gilbertson, B., Longshaw, T.G., and Viljoen, R.P.
  1976: Multispectral aerial photography as exploration tool IV-V; An application in the Khomas Trough Region, South West Africa; and cost effectiveness analysis and conclusions; Remote Sensing of Environment, v. 5, p. 93-107.

Gillespie, A.R. and Kahle, A.B.
  1977: Construction and interpretation of a digital thermal inertia image; Photogramm. Eng. and Remote Sensing, v. 43 (8), p. 983-1000.

Gilluly, J.
  1976: Lineaments — ineffective guides to ore deposits?; Econ. Geol., v. 71, p. 1507-1514.

Goetz, A.F.H., Billingsley, F.C., Gillespie, A.R., Abrams, M.J., Squires, R.L., Shoemaker, E.M., Lucchitta, I., and Elston, D.P.
  1975: Application of ERTS images and image processing to regional geologic problems and geologic mapping in northern Arizona; Tech. Report 32-1597, Jet Prop. Lab., Cal. Inst. Tech., Pasadena, Cal.

Gregory, A.F.
  1972: What do we mean by remote sensing?; Proc. First Canadian Symposium on Remote Sensing, p. 33-37, Canada Centre for Remote Sensing, Ottawa.

  1973: A possible Canadian role in future global remote sensing; Can. Aero. and Space J., v. 19 (3), p. 85-92.

Gregory, A.F. and Moore, H.D.
  1975: The role of remote sensing in mineral exploration with special reference to ERTS-1; Can. Inst. Min. Met. Bull., v. 68 (757), p. 67-72.

  1976: Recent advances in geologic applications of remote sensing from space; in Astronautical Research, 1973, p. 1-18, Pergamon Press.

Gregory, A.F. and Morley, L.W.
  1977: An overview of Canadian progress in the use of Landsat data in geology; Proc. First Annual W.T. Pecora Memorial Symposium (1975), U.S. Geol. Surv. Prof. Paper 1015, p. 33-42.

Grossling, B.F. and Johnston, J.E.
  1977: Gap between raw remote sensor data and resources and environmental information; Remote-Sensing Application for Mineral Exploration, edited by B.L. Smith; Dowden, Hutchison & Ross, Inc., Stroudsburg, Pa.

Harris, G., Jr. and Graham, L.C.
  1976: Landsat-radar synergism; paper presented to Commission VII, 13th Congress, International Society of Photogrammetry, Helsinki; available from Goodyear Aerospace Corp.

Hood, P.J.
  1977: Mineral exploration trends and developments in 1976; Can. Min. J., v. 98 (1), p. 8-47.

Iranpanah, A.
  1977: Geologic applications of Landsat imagery; Photogramm. Eng. and Remote Sensing, v. 43 (8), p. 1037-1040.

Jensen, H., Graham, L.C., Porcello, L.J., and Leith, E.N.
  1977: Side-looking airborne radar; Sci. Am., v. 237 (4), p. 84-95.

Kuhn, T.S.
  1970: The structure of scientific revolutions; in International Encyclopedia of Unified Science, v. II, no. 2, Univ. Chicago Press.

Levine, S.
  1975: Correlation of ERTS spectra with rock/soil types in Californian grassland areas; Proc. 10th International Symposium on Remote Sensing of Environment, v. II, p. 975-980.

Longshaw, T.G.
  1976: Application of an analytical approach to field spectroscopy in geological remote sensing; Modern Geology, v. 5, p. 201-210.

Longshaw, T.G., Viljoen, R.P., and Hodson, M.C.
  1976: Photographic display of Landsat-1 CCT images for improved geological definition; IEEE Trans. on Geosci. Electr., v. GE-14 (1).

Lyon, R.J.P., Mercado, J., and Campbell, R., Jr.
  1970: Pseudo radar; Photogramm. Eng., v. 36, p. 1257-1261.

Lyon, R.J.P.
  1977: Mineral exploration applications of digitally processed Landsat imagery; in Woll and Fischer (1977), p. 271-292.

Maffi, C. and Simpson, C.J.
  1977: Skylab photography for geological mapping; Aust. Bur. Miner. Resour., Geol. Geophys. J., v. 2, p. 17-19.

Miller, V.C.
  1961: Photogeology; McGraw-Hill, New York.

*Newton, A.R.
  1971: The uses of photogeology: a review; Trans. Geol. Soc. S. Africa, v. LXXIV, part 3 (Sept./Oct.), p. 149-171.

Offield, T.W.
  1976: Remote sensing in uranium exploration; in Exploration for Uranium Ore Deposits, International Atomic Energy Agency, Vienna, p. 731-744.

Offield, T.W., Abbott, E.A., Gillespie, A.R., and Loguercios, S.O.
    1977: Structure mapping on enhanced Landsat images of southern Brazil: tectonic control of mineralization and speculations on metallogeny; Geophysics, v. 42 (3), p. 482-500.

Palabekiroglu, S.
    1974: The value of ERTS-1 imagery for mineral exploration; Proc. Second Canadian Symposium on Remote Sensing, v. 2, p. 464-470.

Parry, J.T.
    1977: Interpretation techniques for X-band SLAR; Proc. Fourth Canadian Symposium on Remote Sensing, p. 376-394.

Peterson, R.K.
    1976: The correction of anamorphic scale errors in holographic radar imagery; paper presented to Working Group on Geometry of Remote Sensing at the 13th Congress, International Society of Photogrammetry, Helsinki; available from Goodyear Aerospace Corp.

Podwysocki, M.H., Gunther, F.J., and Blodget, H.W.
    1977: Discrimination of rocks and soil types by digital analysis of Landsat data; Preprint X-923-77-17, NASA.

Quiel, F.
    1974: Some limitations in the interpretation of thermal IR imagery in geology; Information Note 062874, Lab. Applic. Rem. Sens., Purdue Univ., West Lafayette, U.S.A.

Raines, G.L. and Lee, K.
    1974: An evaluation of multiband photography for rock discrimination; in Proc. 3rd Annual Conference on Remote Sensing of Earth Resources, University of Tennessee Space Institute.

Raynolds, R.G. and Lyon, R.J.P.
    1976: Satellite remote sensing of the McDermitt Caldera, Nevada-Oregon; Stanford Remote Sensing Laboratory Technical Report 76-4, Stanford Univ., Cal. 94305.

*Reeves, R.G., Anson, A., and Landen, D. editors-in-chief
    1975: Manual of remote sensing, 2 vols., 2144 p.; Am. Soc. Photogramm., Falls Church, Va., U.S.A. 22046.

Rowan, L.C.
    1975: Application of satellites to geologic exploration; Am. Sci., v. 63 (4), p. 393-403.

*Rowan, L.C., Wetlaufer, P.H., Goetz, A.F.H., Billingsley, F.C., and Stewart, J.H.
    1976: Discrimination of rock types and detection of hydrothermally altered areas in south-central Nevada by the use of computer-enhanced ERTS images; U.S. Geol. Surv., Prof. Paper 883, 35 p.

Rowan, L.C., Goetz, A.F.H., and Ashley, R.P.
    1977: Discrimination of hydrothermally altered and unaltered rocks in visible and near infrared multispectral images; Geophysics, v. 42 (3), p. 522-535.

Sabins, F.F., Jr.
    1973: Recording and processing thermal IR imagery; Photogramm. Eng., v. 39, p. 839-844.

    1974: Oil exploration needs for digital processing of imagery; Photogramm. Eng., v. 40 (10), p. 1197-1200.

Schmidt, R.G.
    1973: Use of ERTS-1 images in the search for porphyry copper deposits in Pakistani Baluchistan; in Symposium on Significant Results Obtained from ERTS-1, NASA report SP-327, v. 1A, p. 387-394.

    1976: Exploration for porphyry copper deposits in Pakistan using digital processing of Landsat data; U.S. Geol. Surv., J. Res., v. 4 (1), p. 27-34.

Schmidt, R.G. and Bernstein, R.
    1977: Evaluation of improved digital processing techniques of Landsat data for sulfide mineral prospecting; in Woll and Fischer (1977), p. 201-212.

*Short, N.M., Lowman, P.D., Jr., Freden, S.C., and Finch, W.A., Jr.
    1976: Mission to earth: Landsat views the world; NASA report SP-360.

Siegal, B.S. and Abrams, M.J.
    1976: Geological mapping using Landsat data; Photogramm. Eng. and Remote Sensing, v. 42 (3), p. 325-337.

Siegal, B.S. and Goetz, A.F.H.
    1977: Effect of vegetation on rock and soil type discrimination; Photogramm. Eng. and Remote Sensing, v. 43 (2), p. 191-196.

Slaney, V.R.
    1971: An assessment of thermal infrared scanning; in Report of Activities, Part B, Geol. Surv. Can., Paper 71-1B, p. 66-68.

    1974: Satellite imagery applied to earth science in Canada; Proc. Symposium on Remote Sensing and Photo Interpretation, International Society for Photogrammetry, pub. by Canadian Institute of Surveying, Ottawa.

    1975: Colour photography in the Beechey Lake Belt, District of Mackenzie; in Report of Activities, Part A, Geol. Surv. Can., Paper 75-1A, p. 65.

Smith, J.R. and Anson, A. (editors)
    1968: Manual of colour aerial photography; 550 p., Am. Soc. Photogramm., Falls Church, Va., U.S.A. 22046.

*Smith, R.E., Green, A.A., Robinson, G., and Honey F.R.
    1978: Use of Landsat-1 imagery in exploration for Keweenawan-type copper deposits; Remote Sensing of Environment; v. 7, no. 2, p. 129-144.

*Spectral Africa (Pty) Ltd.
    1977: Better mapping through remote sensing; Coal, Gold & Base Minerals of Southern Africa, v. 25, no. 9 (Sept.), 10 p.

Tator, B.A. (editor)
    1960: Photo interpretation in geology; in Manual of Photographic Interpretation, p. 169-342, Am. Soc. Photogramm., Washington, D.C.

Tonelli, A.M.
    1975: Contribution of ERTS-1 and Skylab missions to regional studies in Italy; J. Brit. Interplanetary Soc., v. 28 (9-10), p. 647-652.

U.K. Remote Sensing Society
    1975: Newsletter no. 7 (Nov.), p. 14.

*Viljoen, R.P., Viljoen, M.J., Grootenboer, J., and Longshaw, T.G.
    1975: ERTS-1 imagery: an appraisal of applications in geology and mineral exploration; Min. Sci. Eng., v. 7 (2), p. 132-168.

Vincent, R.K.
   1977: Uranium exploration with computer-processed Landsat data; Geophysics, v. 42 (3), p. 536-541.

Welch, R. and Lo, C.P.
   1977: Height measurements from satellite images; Photogramm. Eng. and Remote Sensing, v. 43, no. 10, p. 1233-1241.

*Williams, R.S., Jr. and Carter, W.D. (editors)
   1976: ERTS-1: a new window on our planet; U.S. Geol. Surv., Prof. Paper 929, 362 p. incl. illust.

*Woll, P.W. and Fischer, W.A. (editors)
   1977: Proc. First Annual William T. Pecora Memorial Symposium (1975); U.S. Geol. Surv. Prof. Paper 1015.

Wukelic, G.E., Stephan, J.G., Smail, H.E., Landis, L., and Ebbert, T.F.
   1976: Final report on survey of users of earth resources remote sensing data; Battelle Columbus Laboratories, Columbus, Ohio, 43201.

Discussion

V.R. Slaney: In the South African example of aerial thermography used to map faults, was the panchromatic image acquired at the same time as the thermal image?

Reply: The thermography and panchromatic photography were acquired separately but within a few weeks of each other. At scales of about 1:10 000, the 8-14 micrometre band consistently provided more geological detail than either the panchromatic photograph of the second TIR channel (3.5 to 5.5 micrometres). The Witpoortjie fault is a prime example of this discrimination. No seasonal changes have been observed in the area since the imagery was obtained (Viljoen, R.P., pers. comm.).

V.R. Slaney: What remote-sensing techniques would you recommend to: (a) a field geologist mapping a quarter degree sheet in the Northwest Territories; and (b) a mining company operating in northern Ontario or Quebec?

Reply: In both cases, interpretation of LANDSAT colour composites in transparency format for the regional analysis plus interpretation of selected black-and-white aerial photographs for clarification. Recommendations for more detailed analysis would depend on specific terrain conditions and needs for information at the particular location.

K.A. Morgan: Can we anticipate that the resolution of new satellites will be better than the 80 m of the current LANDSAT system?

Reply: The next LANDSAT, to be launched early in 1978, will have one sensor with about twice the current resolution. Higher resolution has been requested for a number of purposes. A major constraint on increased resolution results from the greater volume of data and associated problems of telemetry.

K.A. Morgan: In applying numerical data processing techniques to integrate LANDSAT data with other digital data acquired near surface (e.g. low level gamma ray spectrometry), I anticipate the necessity of degrading the resolution of such near-surface data.

Reply: The high resolution data may not require degradation. We have optically integrated LANDSAT spectral data with aerial photographs and aeromagnetic data to obtain synergistic benefits without degrading any data.

T. Findhammer: Could you tell us something more about research into thermal inertia in Canada?

Reply: To the best of my knowledge, no such work is in progress in this country. The U.S.G.S. in Denver and Jet Propulsion Laboratory in Pasadena both have programs related to thermal inertia.

B. Bølviken: I disagree that multispectral photography has restricted use in exploration. The occurrence of natural poisoning of vegetation is a common feature associated with sulphide deposits. Such occurrences can be detected by multispectral aerial photography which, therefore, is an important tool for prospecting.

Reply: The point of dissension here is one of semantics rather than science. Natural poisoning of vegetation by metallic elements in ore deposits has been demonstrated in some terrains (e.g. your studies in Norway). However, other studies mentioned in my review have not revealed analogous biophysical anomalies in other terrains. Hence, at this point in time, multiband photography should be considered as a specialized technique to be used where the occurrence of detectable poisoning has been established. It is, thus, not yet a tool for general exploration. In essence, further research is required to define the range of applications and costs.

COMPUTER COMPILATION AND INTERPRETATION OF GEOPHYSICAL DATA

Allan Spector
Allan Spector and Associates Ltd., Don Mills, Ontario

Wilf Parker
Dataplotting Services Inc., Don Mills, Ontario

Spector, Allan and Parker, Wilf, Computer compilation and interpretation of geophysical data; in Geophysics and Geochemistry in the Search for Metallic Ores; Peter J. Hood, editor; Geological Survey of Canada, Economic Geology Report 31, p. 527-544, 1979.

Abstract

This paper details the progress in the application of the digital computer to the compilation and the interpretation of geophysical data in the period 1967-1977. Probably the most dramatic feature of this period has been the dramatic drop in the cost of computers. Miniaturized electronics has reduced the cost and size of today's computers.

Much more accurate, versatile, and faster computer graphics devices have become available. The most important hard-copy devices are the large flatbed plotters, the fast drum plotters, and the electrostatic plotters. A colour plotter has become available and will be very useful for producing full colour geophysical and geological maps. Digitizers are larger and more accurate. Interactive computer terminals, connected to local or remote computers, allow fast editing and processing of data.

Semi-automatic computer compilation programs, using interactive terminals, permit the fast processing and plotting of large volumes of geophysical data. A good deal of this progress has been spurred by the development of reliable digital acquisition systems for airborne surveys, especially aeromagnetic, gamma ray spectrometer and electromagnetic (EM) surveys.

Not quite as dramatic, but surely significant advances have been made in the area of data analysis and interpretation. Several survey contractors have the software and hardware facilities to perform the following operations:

(a) Aeromagnetic data:

- *computation of synthetic anomalies for arbitrarily complex models and also interactive modelling,*
- *matched filtering, downward and upward continuation, vertical gradient, magnetic pole reduction, pseudo-gravity transformation facilitated by Fourier transform techniques and susceptibility mapping.*

(b) Gravity data: much more correlation and closer interaction with geological data plus iterative modelling for semi-automatic interpretation and various anomaly enhancement operations.

(c) Input electromagnetic data: conductivity-width and depth computation employing computer graphics facilities.

(d) Gamma ray spectrometer data: background definition, stripping, ratio computation and quantitative estimates of radioactive element composition.

Résumé

Le présent document fait état des progrès accomplis dans l'utilisation du calculateur numérique pour le rassemblement et l'interprétation des données géophysiques pour la période de 1967 à 1977. La caractéristique la plus étonnante de cette période a probablement été la chute dramatique du coût des ordinateurs. La miniaturisation des éléments a eu pour effet de réduire le coût et les dimensions des ordinateurs modernes.

On dispose maintenant d'appareils graphiques automatisés beaucoup plus précis, polyvalents et rapides. Les appareils d'impression les plus importants sont les gros traceurs à plat, les traceurs rapides à tambour et les traceurs électrostatiques. Il existe maintenant un traceur couleur: il sera très utile pour la production de cartes géologiques et géophysiques en couleur. Les convertisseurs analogiques/numériques sont plus gros et plus précis. Les terminaux de dialogue, reliés aux ordinateurs satellites et locaux, permettent le traitement et la correction rapides des données.

Les programmes de rassemblement semi-automatiques, utilisant des terminaux de dialogue, permettent le traitement et le traçage rapides d'un grand nombre de données géophysiques. Une bonne partie de cette évolution a été motivé par la mise au point de systèmes efficaces de saisie numérique pour les levés aériens, principalement les levés aéromagnétiques, les levés électromagnétiques et les levés de spectromètre à rayons gamma.

Des progrès moins spectaculaires, mais tout aussi importants, ont été réalisés dans le domaine de l'analyse et de l'interprétation des données. Plusieurs arpenteurs privés possèdent les installations de matériel et de logiciel pouvant effectuer les travaux suivants:

(a) Données aéromagnétiques:

— *calcul des anomalies synthétiques pour les modèls arbitrairement complexes et aussi des modèles de dialogue,*

— *le filtrage assorti, la suite descendante et ascendante, le gradient vertical, la réduction du pôle magnégique, la transformation pseudo-gravité facilité par les techniques de transformation de Fourier et la cartographie de susceptibilité.*

(b) Données gravimétriques: relation et interaction plus serrées entre les données géologiques, en plus des modèles de dialogue pour l'interprétation semi-automatique et divers travaux de mise en valeur des anomalies.

(c) Données électromagnétiques en entrée: le calcul de la largeur et de la profondeur de conductivité au moyen des appareils graphiques d'ordinateur.

(d) Données de spectromètre à rayons gamma: la définition de la zone de fond, le stripage, le calcul des rapports et les évaluations quantitatives de la composition des éléments radioactifs.

INTRODUCTION

This paper describes the current state-of-the-art of computer application to both the compilation and the analysis of geophysical data intended for mining exploration. We have attempted to differentiate what is available on a commercial basis from that which is undergoing research and development. We shall be basically reviewing advances in computer hardware and software since 1967. Thus, this paper may be regarded as a sequel to a similar paper given by West et al. (1970) at the 1967 Niagara Falls conference. The reader is also directed to two other state-of-the-art papers on allied topics by Grant (1972) and by Reford (1976).

The authors would like to introduce themselves as a geophysicist and a geologist, who, over the last 10 years, have been specifically involved in the computer processing of geophysical data — on a contractual basis.

This paper is not intended as an exhaustive study of the subject but rather a bird's eye view dated 1977 over Toronto. Whereas in 1967 mining geophysicists were not really making great use of computers, in 1977 it is difficult to find a geophysicist who is not making use of, or is at least familiar with, data processing techniques. The electronic calculator, the programmable calculator, the minicomputer, and the microcomputer have all come into common usage during the past decade as a result of the miniaturization of electronic circuitry.

Four main areas are covered in this paper:

(1) digital acquisition,

(2) computer processing hardware, including terminals and plotting facilities,

(3) computer software, and

(4) anticipated future developments.

ADVANCES IN COMPUTER HARDWARE

Reliable Digital Acquisition

Probably the most significant improvement in the state-of-the-art since 1967 has been the development of reliable digital acquisition systems for airborne geophysical survey use. All major airborne survey contractors in Canada now have such systems. A minicomputer or microprocessor forms the nucleus of all the state-of-the-art systems. A stored computer program controls the acquisition of both digital and analogue data from the geophysical sensors. It also controls the formatting of the data onto magnetic tape and verifies that each tape record has been properly written on the tape.

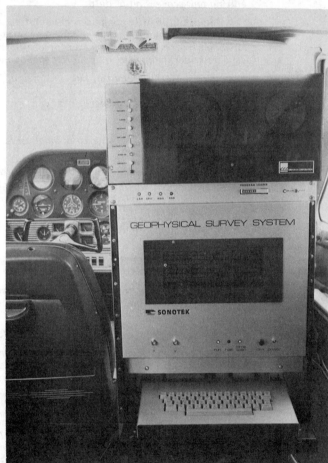

Figure 23.1. Model IGSS Minicomputer-controlled digital acquisition system; Sonotek Ltd., Mississauga, Ontario. (GSC 203492-M)

Each system may also contain a cathode ray tube (CRT) display, a keyboard and in some cases a low speed printer in addition to the ubiquitous analog chart recorder for display of the digital data.

The following North American companies have built state-of-the-art digital acquisition systems:

1) Applied Geophysics, Salt Lake City, Utah

2) Geometrics, Sunnyvale, California, U.S.A.

Figure 23.2. Geac Minicomputer system; Dataplotting Services Inc., Don Mills, Ontario. (GSC 203492-N)

3) Geoterrex, Ottawa, Canada

4) Kenting, Ottawa, Canada

5) McPhar Instruments, Toronto, Canada

6) Sander Geophysics, Kanata, Ontario

7) Sonotek, Mississauga, Ontario, Canada.

This list gives an indication of the number of systems available in North America; other systems have been fabricated in Europe and elsewhere. Figure 23.1 shows a digital acquisition system, manufactured by Sonotek, installed in a single engine light aircraft. Sonotek has stated that the system is so reliable and easy to operate that it has been flown without an instrument operator with the pilot and copilot/navigator alone in the survey aircraft.

The Minicomputer

Today's minicomputer systems are similar in capability to the large 1967 systems, such as the IBM 7094 and the Univac 1108 computers, at about one-tenth of the purchase price. A typical minicomputer system presently costs about $150 000 U.S. Some of the more popular systems are manufactured by Digital Equipment Corp., Data General, and Hewlett Packard. There are many other reputable manufacturers and a user should select the system most suited to his needs and his budget.

Both batch mode and interactive mode are important on any system. In batch mode, a job is submitted to the computer and the user waits for the job to be completed and examines the results. In interactive mode a job is submitted to the computer and intermediate results are usually

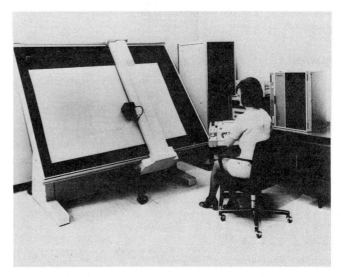

Figure 23.3. Gerber Model 22 Flatbed Plotter; Dataplotting Services Inc., Don Mills, Ontario. (GSC-203492-P)

displayed on a terminal; the computer may request data or instructions from the user in order to continue with the computations.

Figure 23.2 shows a minicomputer system which consists of the following components: a central processing unit (CPU) with 64K (K=1024) bytes of memory, a 9-track magnetic tape drive, a 60 million character disk drive, a 300 line per minute printer, a 300 card per minute card reader, and 3 CRT terminals.

Interactive Graphics Terminals

The most popular interactive graphics terminals are made by Tektronics. These terminals are usually connected to a computer system so that graphics data such as curves, profiles, and contours can be displayed on a CRT. The functions that a user is allowed to perform at the graphics terminal are totally dependent upon the program in the computer system that controls the terminal.

Plotting Devices

In 1967, the drum plotter was the only commonly-available plotting device. By 1977, the drum plotter was still greatly utilized but its speed and quality were vastly improved. Flatbed plotters and electrostatic plotters have now come into general use. A colour plotter has just become commercially available.

The drum plotter remains the most popular plotting device because of its cost and ease of operation. Plots can be created that are up to 36 inches (91.4 cm) in width and up to 120 feet (36.6 m) in length. Plot width and length varies from model to model. From one to four ballpoint or ink pens are used simultaneously to create the plot.

The flatbed plotter is more expensive and more difficult to operate than the drum plotter but it can produce a much higher quality plot. Ink on mylar plots can be of drafting quality. The largest flatbed plotter in general use can produce drawings up to 4 feet by 7 feet (1.2 x 2.1 m). Larger flatbed plotters are used for aircraft and ship design.

The electrostatic plotter, available in widths up to 72 inches (182.9 cm), can produce a plot several times faster than a drum or flatbed plotter but the quality is not as good. The resolution is 0.005 inch (0.013 cm) (200 dots/inch) compared to an increment size of 0.0002 inch (0.0005 cm) for a flatbed plotter. Figure 23.3 shows a Gerber Model 22 flatbed plotter.

ADVANCES IN COMPILATION SOFTWARE

Lower cost computer time, interactive terminals, and better quality plotters have encouraged a great deal of development in programming for the purposes of data compilation and data analysis. The Geological Survey of Canada specifications for airborne gamma ray spectrometer and magnetometer data have also required that survey contractors in Canada become very computer conscious.

Most computer groups or computer departments involved in the processing of geophysical data have developed their own computer programs and subroutines. The methods can vary widely from group to group and are dependent on such things as the type of computer used, the type of data being processed, and the preferences of the individuals involved. The most important advances have been made in gridding and contouring programs (Walters, 1969; Crain, 1970; Crain, 1972; Wren, 1975). In 1967, contour maps were plotted on drum plotters, with a coarse grid size, and were often not considered to be a final product. Today, greatly improved software and better quality plotters have permitted the production of high quality final contour maps.

The cost of computer compilation is generally less than the cost of manual compilation but the great advantage of computer compilation is its speed. Large volumes of data can be processed many times faster by computer than by manual methods.

EXAMPLES OF SOFTWARE APPLICATION

Aeromagnetic Survey Compilation

The key to processing large volumes of digital data is the early detection of errors. The detection process starts with the digital recording system in the aircraft and continues in the data processing centre. The data processing centre must be capable of quickly performing editing operations. This is best accomplished using an interactive terminal to view the data and enter any corrections.

The following operations, involved in the compilation of a typical aeromagnetic survey for mining exploration, demonstrate the extent to which computer graphics are utilized.

(a) Digital Data: The digitally-recorded survey data are copied from magnetic tape to a computer disk file. The data are reformatted into a format which is readily usable by the computer. Any "bad" data are displayed on a CRT terminal and are corrected by compilation personnel.

(b) Flight Path Position Data: The flight line is digitized from the base map and entered into the computer. A computer program calculates the average speed of the survey aircraft between the picked points and plots these speeds in the form of a bar chart so that inconsistencies are easily recognized. This is generally referred to as a "speed check".

(c) Rough Contour Maps: Several preliminary or rough contour maps are drawn to establish whether errors may be detected in the contoured data, e.g. herringbone patterns indicate poor levelling.

This is an iterative process of finding errors, correcting the data and plotting another contour map. From three to six "rough" contour maps may be necessary to complete the editing operation.

(d) Final Contour Map: The final contour map, such as the one shown in Figure 23.4, is plotted with black ink on a stable base material. A flatbed plotter generally gives the best quality and accuracy. Flight lines and fiducial locations may be plotted onto the contour plot or onto a separate overlay. In order to achieve a high quality final map it is necessary to use a small grid size (0.25 cm or smaller) which accurately represents the original data and to use a contouring method which gives an accurate and visually pleasing presentation of the data. Holroyd (1974) has described the computer-oriented aeromagnetic data compilation system developed at the Geological Survey of Canada.

Aeromagnetic Interpretation

Since 1967, a much wider acceptance of the computer and computer graphics (Smith et al., 1972) in the interpretation of geophysical data has occurred. Firstly, the availability of the data in digital form has allowed the interpreter a great deal more freedom to assess the data, particularly in profile form at whatever scale he chooses. To some interpreters, a surprising amount of useful information can be gleaned from profiles e.g., the two-gamma faults discussed by Friedberg (1976). In 1967, computer-plotted profiles of aeromagnetic data were the normal product of high sensitivity surveys for petroleum exploration, because only in these surveys was the higher cost of digital acquisition really accepted.

Figure 23.5 shows the advantages gained by digitally recording the output from the magnetometer in the aircraft with the magnetometer on the ground serving as a diurnal monitor; namely

1) compensation for diurnal variation,

2) altitude correction, and

3) rectification to constant horizontal scale.

Other operations carried out to aid in the analysis of aeromagnetic profiles include various kinds of filtering (especially to remove high frequency noise) and the production of vertical/horizontal derivatives to help differentiate overlapping anomalies.

Model curves are generated by the computer on a routine basis to assist the analyst in the identification of anomalies, the location of contacts, the computation of depth to magnetic basement, and to give various other forms of information. Figure 23.6 shows an example of magnetic anomalies computed over a prism model for various combinations of prism width and strike length. Model curves of this nature help familiarize the analyst with anomaly characteristics at a particular magnetic latitude.

Automatic Aeromagnetic Profile Interpretation

Major development work in the field of automatic anomaly analysis was carried out just prior to 1967 by a research group associated with Aero Service of Philadelphia. The technique that they developed was called Werner Deconvolution (Hartman et al., 1971). In the deconvolution technique, anomalies were first resolved and then both prismatic and laminar models were used to determine depth, horizontal position, dip and magnetization of the causative source. Figure 23.7 shows an example of its application, taken from a 1977 Aero Service brochure. This profile illustrates how the following major geological features are discerned;

(A) major vertical contact,

(B) vertical dyke,

Figure 23.4. *A typical aeromagnetic contour map; Dataplotting Services Inc., Don Mills, Ontario.*

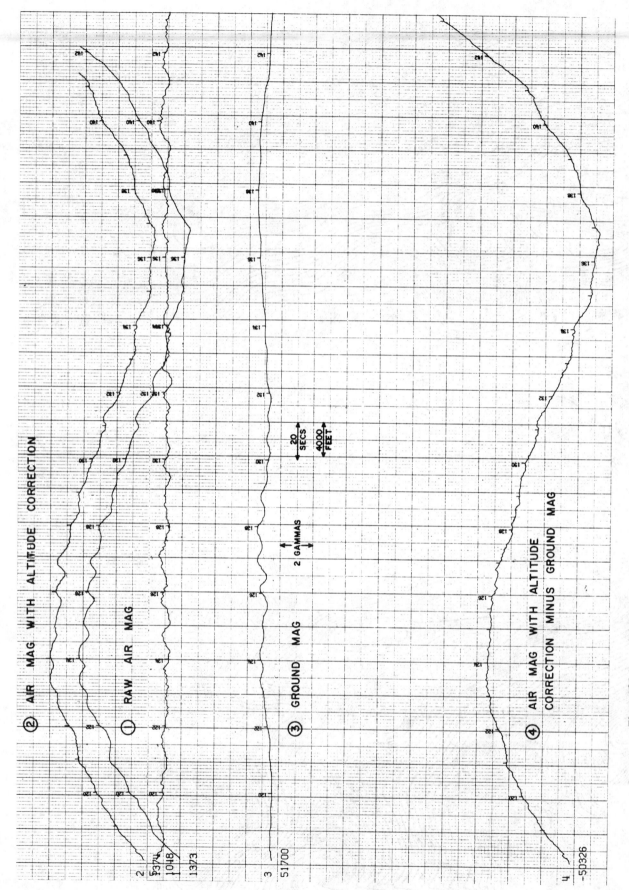

Figure 23.5. Correction of aeromagnetic data for diurnal and altitude variation; Geoterrex Ltd., Ottawa.

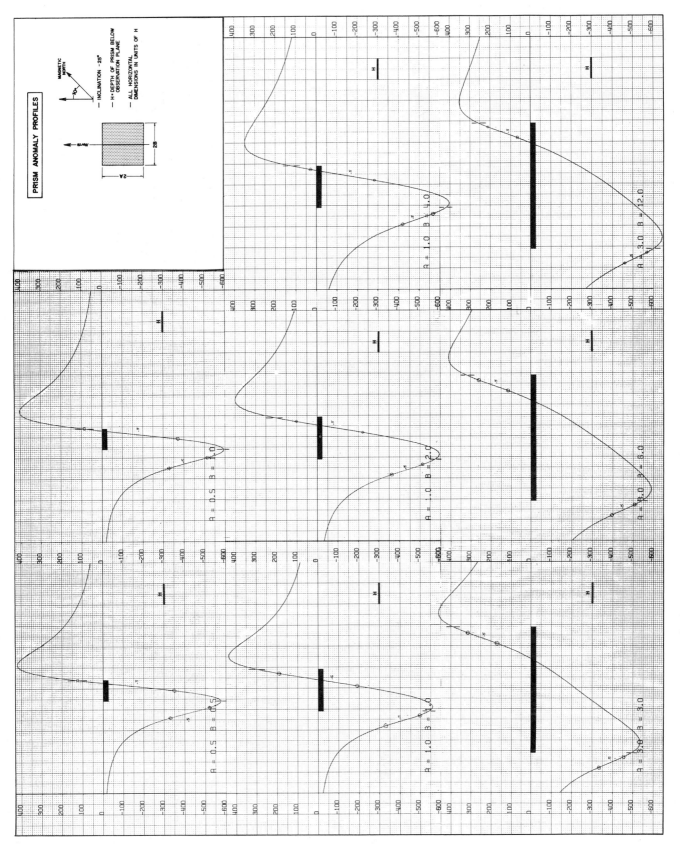

Figure 23.6. Synthetic magnetic profiles over a vertical prism model at latitude 25°S for various geometries.

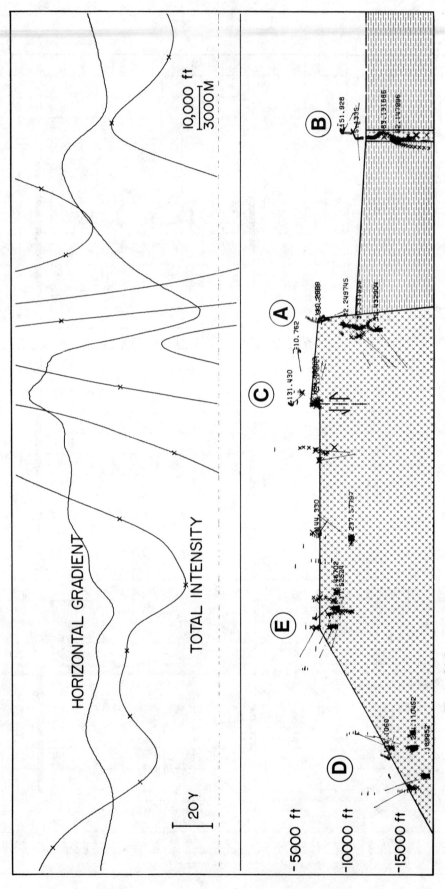

Figure 23.7. Automatic aeromagnetic interpretation by Werner Deconvolution, Aero Service Corp., Houston.

(O'Brien, 1971), which maps basement depth and structure by utilizing Fourier and Hilbert transformations. A semi-automatic method of magnetic curve fitting has been developed by McGrath and Hood (1973) in which a wide range of geological features may be synthesized using combinations of the thin plate model. Using an iterative procedure, the computer program achieves a best least-squares fit, as is shown in Figure 23.8.

Some of the major obstacles in the application of the various methods listed above are described as follows:

(1) the data must be carefully edited in advance;

(2) there is a problem in defining background or regional levels;

(3) two-dimensional causative bodies are often assumed;

(4) a very high degree of manual interaction is often required to synthesize the results into plausible geology; and

(5) the high cost of computer processing versus manual or graphical methods must be considered.

Aeromagnetic Map Analysis

In the last 10 years, there has been a very strong swing toward the computer processing of airborne geophysical data, particularly aeromagnetic data. Today, there are few geophysical contractors that do not have access to packaged programming to do a wide range of analytical operations, including the computation of complex model anomalies. Particular emphasis, however, has been placed on the application of Fourier transformation for purposes of analysis of the data in the frequency domain and especially to perform linear filtering. Because of this heavy emphasis, this paper elaborates on the subject of spectrum analysis, matched filtering and the various other linear filtering operations that are possible through Fourier transformation and are only feasible techniques if the computer is utilized in carrying them out.

Spectrum Analysis and Matched Filtering

Matched filtering of aeromagnetic maps has been found to be particularly useful in areas of volcanic cover e.g., the southwestern U.S.A. to obtain the following information:

(a) identification of buried intrusives, as well as regional structure concealed by the volcanic cover, and

(b) variations in the thickness of the volcanic cover, i.e., where it is excessively thick.

Matched filtering is based on an analysis of the computed energy spectrum of an aeromagnetic map. From spectrum analysis, a picture of the physical make-up of the data is gained. In the spectrum, contributions in the magnetic data from the following sources can be distinguished:

(a) shallow or near-surface features,

(b) regional lithologic and structural features, or

(c) deep-seated features.

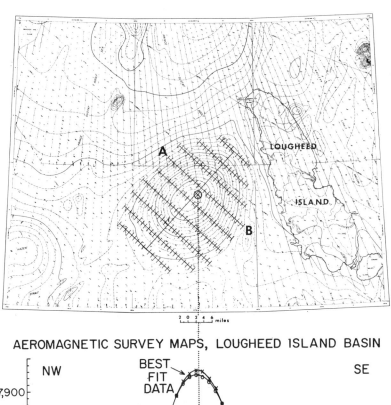

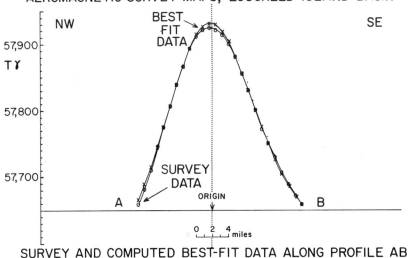

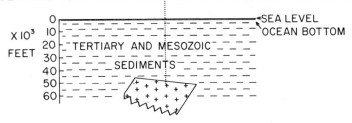

Figure 23.8. Least-squares fit of model to aeromagnetic data (from McGrath and Hood, 1973).

(C) fault of small throw, and

(D) and (E) reversely magnetized zones.

However, a great deal of effort is still required to synthesize all of this output information to interpret basement configuration. A similar automated interpretation technique has been developed recently by Compagnie Generale de Geophysique of Paris and is offered by Geoterrex of Ottawa. Compu-Depth is another computer-oriented interpretation technique offered by Geometrics of California

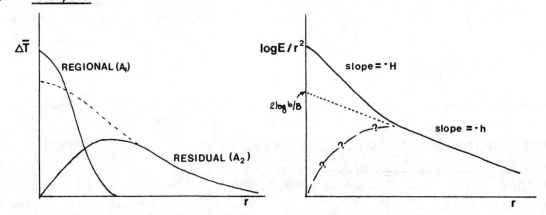

Figure 23.9. *Energy spectrum analysis and design of a matched filter.*

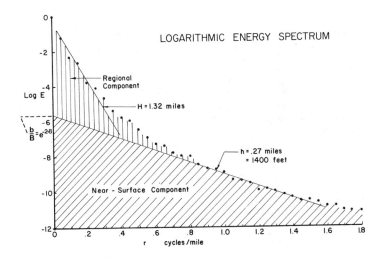

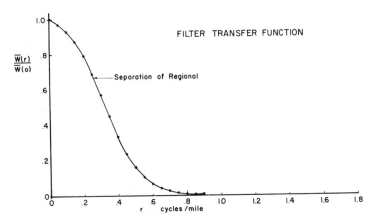

Figure 23.10. *An example of a logarithmic energy spectrum and a corresponding filter transfer function.*

Figure 23.9 describes some of the computational steps involved:

— computing an estimate of the Fourier Transform or "Complex Spectrum" after first multiplying the data by a "Data Window" to avoid distortion in the spectrum caused by abrupt edge effects;

— taking the modulus of the Complex Spectrum to obtain the Energy Spectrum; and

— averaging the Energy Spectrum with respect to azimuth on the frequency plane in order to view the drop-off of the logarithmic spectrum with radial frequency.

In Step 3, spectrum analysis is done. Based on the appearance of the spectrum, i.e., changes in the slope of the spectrum curve, the spectrum is divided into two components:

(a) shallow origin or Near-Surface Component; the slope of high frequency part of the curve gives us the average depth to magnetic sources: h

(b) the Regional Component which dominates the low frequency or the long wavelength part of the spectrum.

The complex spectrum may be approximated as

$$\Delta \overline{T}(r) = A_1(r) + A_2(r)$$

with the following approximations:

$$A_1(r) = Be^{-Hr} \quad \text{(Regional Component)}$$
$$A_2(r) = be^{-hr} \quad \text{(Near-Surface Component)}$$

so that

$$E(r) = (Be^{-Hr}(1 + \frac{b}{B}e^{(H-h)r}))^2$$
$$= (A_1 \cdot (W)^{-1})^2$$
$$= ((\text{Regional Component}) \cdot (\text{Filter Transfer Function})^{-1})^2$$

Matched filtering consists of multiplying the Fourier Transform of an aeromagnetic map by the Transfer Function W in order to separate out the Regional Component. The parameters for definition of W; H, h and b/B, are determined directly from a graphical analysis of the energy spectrum curve, i.e., its logarithmic, radial component.

It is implicit in the preceding discussion that the slopes of each of the two parts of the spectrum curve are due to differences in depth between (a) the Near-Surface Sources and (b) the Regional Sources. Actually, the slope of the spectral curve, to a large extent, is decided by the size or cross-section of the causative sources; the larger the source the more long wavelength spectral composition and therefore, the greater the slope of the spectral curve.

According to Spector (1968), a correction can be applied to the spectral curve to correct for the size effect, if some measure of average body size can be made.

For the limited purposes of the analysis, it is preferable to lump the depth and size effects together, i.e., to treat the slope as indicating an apparent depth, H, which we understand, is in excess of the true average depth to the deeper magnetic basement. Figure 23.10 shows an actual example of a computed energy spectrum and the corresponding filter transfer function.

Figure 23.11 shows a comparison between the original aeromagnetic map and the filtered result, the Regional Component. The major feature in the filtered map is a large granitic intrusive.

Other Forms of Linear Filtering

Figure 23.12 shows how matched filtering is just one of several types of linear filter operators that can be effectively used through Fourier transformation as follows:

1) downward/upward continuation,

2) magnetic pole reduction,

3) pseudo-gravity transformation (see Fig. 23.13).

In addition, susceptibility mapping, a computer processing service offered by Paterson, Grant and Watson Ltd. of Toronto, has been applied rather extensively. An example is shown in Figure 23.14. The upper part of the figure is part of an aeromagnetic map published by the Geological Survey of Canada; the contour interval is 10 gammas. The lower part is the corresponding susceptibility contour map; contour interval is 0.5×10^{-3} e.m.u. A similar process described as "Magnetization Mapping" is offered by Geometrics of California.

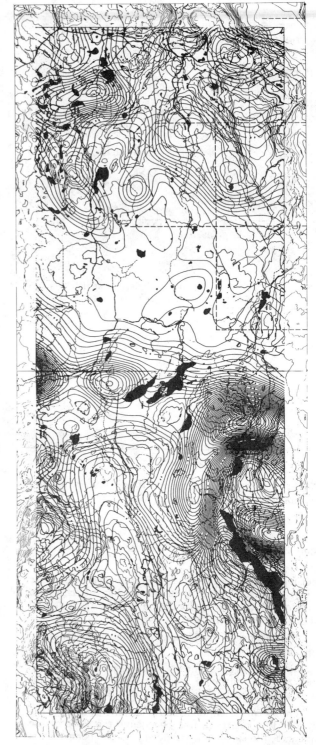

Figure 23.11. An example of matched filtering; suppression of surficial volcanics.

DECONVOLUTION OF AEROMAGNETIC DATA

1. Compute Fourier Transform:

$$\overline{\Delta T}(f_x, f_y) = \int_{-\infty}^{\infty} \int_{-\infty}^{\infty} \Delta T(x,y) \cdot e^{-2\pi i(f_x \cdot x + f_y \cdot y)} dx\, dy$$

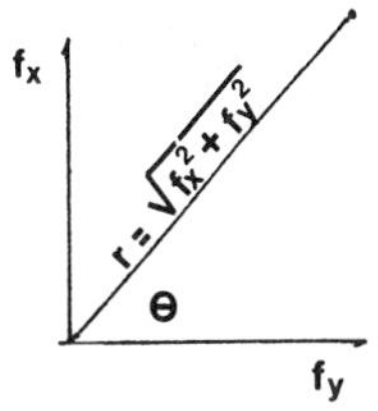

2. Apply Filter Transfer Function (W)

$$\boxed{\overline{\Delta T}_F = \overline{\Delta T} \cdot W}$$

where

$W_{MF} = \{1 + \frac{b}{B} e^{(H-h)r}\}^{-1}$ <u>Matched Filtering</u>

$W_{DC} = e^{Zr}$ <u>Downward Continuation</u>

$W_{MP} = \{\sin I_E + i\cos I_E \cdot \sin(D_E + \theta)\}^{-2}$ <u>Magnetic Pole Reduction</u>

$W_{PG} = \frac{G}{4T_o} \left(\frac{\Delta\rho}{\Delta k}\right) \frac{e^{\Delta h r}}{r} \cdot W_{MP}$ <u>Pseudo Gravity Transformation</u>

where

$\frac{b}{B}$, H, h: matched filter parameters from energy spectrum analysis,

Z: the depth of continuation,

I_E, D_E: inclination and declination of the earth's magnetic field,

G: Universal Gravitational Constant

T_o: Geomagnetic Field Intensity

Δk: average susceptibility contrast

$\Delta\rho$:   average density contrast

Δh: mean separation between aeromagnetic survey altitude and gravity survey elevation.

3. Compute Inverse Fourier Transform:

$$\Delta T_F(x,y) = \int_{-\infty}^{\infty} \int_{-\infty}^{\infty} \overline{\Delta T}(f_x, f_y) \cdot W \cdot e^{2\pi i(f_x \cdot x + f_y \cdot y)} df_x\, df_y$$

Figure 23.12.   *Deconvolution of aeromagnetic data; various types of linear filtering.*

Computer Processing of Airborne Electromagnetic (AEM) Data

Compilation of Input EM Data

The Input EM method was developed by Barringer Research of Toronto and is in extensive use. Two survey contractors are licensed to fly Input: Questor Surveys of Toronto and Geoterrex of Ottawa. The Input data collected by Questor are processed in the following manner: (1) the data are copied from the digitally recorded data tape to a computer disk file, (2) the data are then displayed on a graphics terminal and an operator flags EM anomalies and associated magnetic highs, and (3) a computer program calculates the value of the conductivity-thickness product from the anomaly and plots a map of the anomalies. The use of the computer/plotter combination allows the processing of large volumes of data in a much shorter period of time than by manual methods. The cost is also less for large volumes of data, than by manual methods.

Compilation of Dighem Data

The Dighem EM system was developed by Barringer Research of Toronto and is flown by Dighem Ltd. of Toronto (Fraser, 1972). The data are processed by computer and a stacked profile plot is produced for each line of data. All recorded data as well as several computer calculated profiles are plotted. The calculated profiles show anomaly enhancement and the suppression of surface conductivity. An example of a Dighem profile over the Montcalm orebody in Quebec is shown in Figure 23.15.

The Computer Processing of Gamma ray Spectrometer Data

The computer processing of gamma ray spectrometer data has undergone important developments during the past decade which are likely to continue for many more years. Larger crystal sizes and improved technology has resulted in more reliable data and computer processing techniques have improved the presentation of the data, either in the form of contour maps or stacked profiles. The recording of "full spectrum" data, up to 1024 channels, is one of the latest survey developments to become commercially available.

The computer processing operations for the compilation of gamma ray spectrometer data consist of the following steps:

(a) determination and subtraction of background radiation levels,

(b) correction of the data for variations in survey altitude, and

(c) corrections for Compton scattering.

These operations were described in detail by Grasty (1972).

Due to the presence of a fairly large component noise it is often desirable to filter the data prior to contouring. Total count and uranium contour maps are normally produced, while thorium, potassium, and the ratio maps (U/Th, U/K, Th/K) are produced selectively. Perhaps the most useful presentation is the stacked profile. This is a separate data plot for each flight line and usually nine parameters are displayed: Total Count U, Th, K, U/Th, U/K, Th/K, altimeter, and magnetometer. Figure 23.16 is an example of such a plot.

Processing of Ground Geophysical Data

Gravity Data

There has been a continued shift towards utilization of computer processing of gravity data for two main reasons:

(a) to reduce the time and cost of data reduction, and

(b) to increase the ability to distinguish what is significant in the measurements.

Low cost, portable desktop programmable computers have become an essential requirement to perform preliminary data reduction after each day's survey production. An example of a field minicomputer is the Hewlett-Packard Model 9820A which has a memory capacity of 1477 words. Gravity in measurements can be stored on a cassette tape.

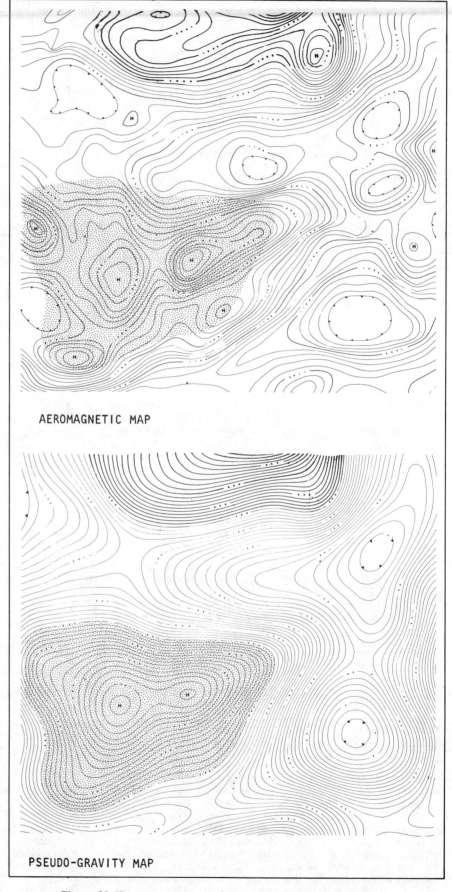

Figure 23.13. *An example of pseudo-gravity transformation.*

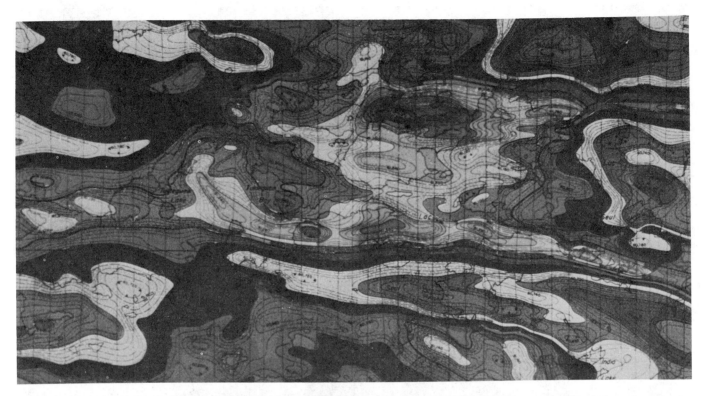

Figure 23.14. *An example of susceptibility mapping; Paterson, Grant and Watson Ltd., Toronto. (GSC 203492-O)*

Terrain correction programs have been developed by a number of contractors and government agencies, e.g. Stacey and Stephens (1970). There are many types of gridding and contouring packages available which are particularly suited for gravity data, e.g., Hessing et al. (1972). Linear filtering through Fourier transformation has been introduced gradually to attack the fundamental problem of regional/residual separation, in place of the older trend-fitting approaches. Grant (1972) has given an excellent discussion of this problem.

Iterative model-fitting interpretation programs are widely used, particularly for problems that can be solved using a single density interface, e.g., the thickness of a buried salt mass, whose upper or lower surface is known in advance.

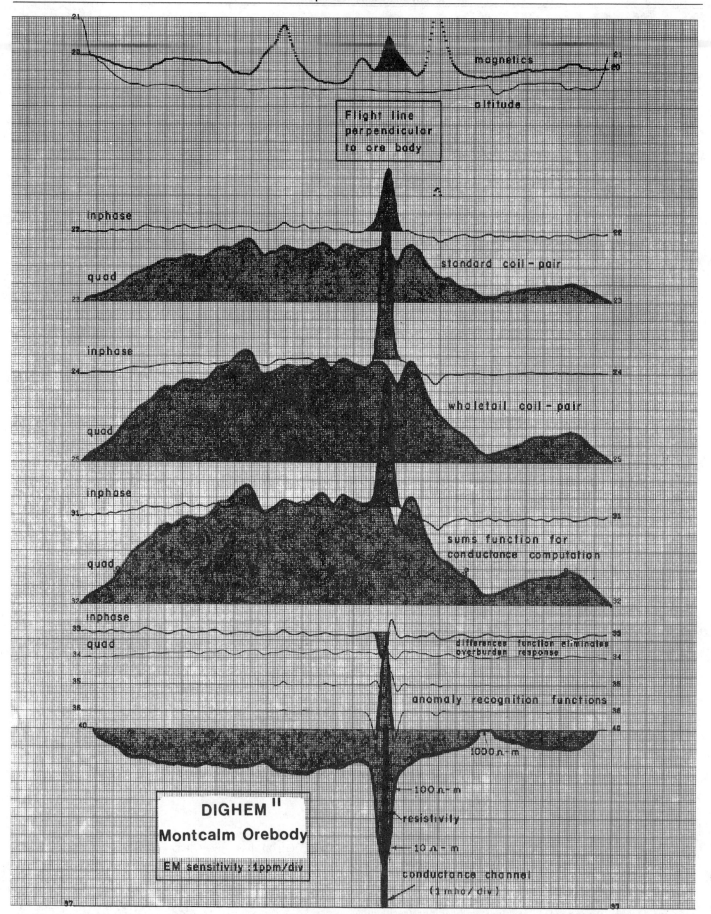

Figure 23.15. Dighem II computer processed data over the Montcalm Deposit; Montcalm Township, Ontario, Dighem Ltd., Toronto.

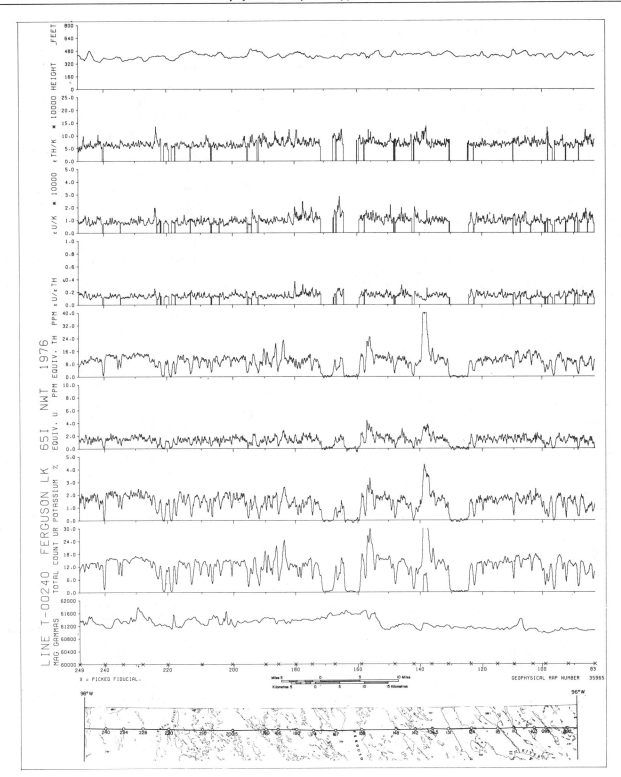

Figure 23.16. Stacked profile representation of spectrometer survey data; Dataplotting Services Inc., Don Mills, Ontario.

Magnetometer and Electromagnetic Data

Computer compilation of ground magnetometer and electromagnetic data is becoming quite common. With regard to VLF electromagnetic surveying, computer processing is useful for the compilation of such data because of the need for filtering (smoothing) and for computing the horizontal derivative of the in-phase component.

FUTURE TRENDS

Increased Computer Utilization

If the present trend to lower prices of computer hardware is maintained, we can expect an even greater percentage of geophysicists to acquire either digital computers or computer services. Specifically, interactive graphics displays will become more and more commonplace as an ideal tool both for the initial screening of survey data and subsequent analysis including semi-automatic anomaly interpretation.

The Man/Computer Interface

Computer languages will continue to become more powerful and easier to use and computer operating systems and user programs will allow a greater usage of interactive terminals by the geophysicist. The computer is a powerful tool for the geophysicist and easier usage will expand its utilization.

Software Development

With multi-sensor surveys involving, in the case of gamma ray spectrometer data, simultaneous measurement of 512 or 1024 channels of data, there is a major demand for more comprehensive data analysis programming, e.g., the anomaly picker devised by Dighem (see Figure 23.15). Pattern recognition algorithms are currently the subject of much research, particularly the identification of soil and rock types from gamma ray spectrometer data.

ACKNOWLEDGMENTS

We wish to thank Michael Reford of Geoterrex, M.T. Holroyd and P.H. McGrath of the Geological Survey of Canada, Arthur Loveless of Barringer Research, Carl Gehring of Aero Service, and Scott Hogg of Northway Consultants Ltd., for supplying us with examples of geophysical data processing used in this paper.

REFERENCES

Crain, E.R.
1972: Review of gravity and magnetic data by processing systems; J. Can. Soc. Explor. Geophys., v. 8 (1), p. 54-76.

Crain, I.K.
1970: Computer interpolation and contouring of two-dimensional data — a review; Geoexploration, v. 8 (2), p. 71-86.

Fraser, D.C.
1972: A new multicoil aerial electromagnetic prospecting system; Geophysics, v. 37 (3), p. 518-537.

Friedberg, J.L.
1976: The two-gamma fault; paper presented at 46th Annual Meeting, Soc. Explor. Geophys., Houston.

Grant, F.S.
1972: Review of data processing and interpretation methods in gravity and magnetics, 1964-71; Geophysics, v. 37 (4), p. 647-661.

Grasty, R.L.
1972: Airborne gamma-ray spectrometer data-processing manual; Geol. Surv. Can., Open File No. 109.

Hartman, R.R., Teskey, D.J., and Friedberg, J.L.
1971: A system for rapid digital aeromagnetic interpretation, Geophysics, v. 36 (5), p. 891-918.

Hessing, R.C., Lee, H.K., Pierce, A., and Powers, E.N.
1972: Automatic contouring using bicubic functions; Geophysics, v. 37 (4), p. 669-674.

Holroyd, M.T.
1974: The aeromagnetic data automatic mapping system (ADAM); in Report of Activities, Part B, Geol. Surv. Can., Paper 74-1B, p. 79-81.

McGrath, P.H. and Hood, P.J.
1973: An automatic least squares multimodel method for magnetic interpretation; Geophysics, v. 38, p. 349-358.

O'Brien, D.P.
1971: An automated method for magnetic anomaly resolution and depth-to-source computation; Proc. Sym. on Treatment and Interpretation of Aeromagnetic Data, Berkeley, California.

Reford, M.S.
1976: State-of-the-art in magnetics; Proc. 46th Annual Meeting, Soc. Explor. Geophys., Houston.

Smith, R.B., Warnock, J.E., Stanley, W.D., and Cole, E.R.
1972: Computer graphics in geophysics; Geophysics, v. 37 (5), p. 825-838.

Spector, A.
1968: Spectral analyses of aeromagnetic data; unpublished Ph.D. Thesis, University of Toronto, 250 p.

Stacey, R.S. and Stephens, L.E.
1970: Procedures for calculating terrain corrections for gravity measurements; Publ. Dom. Obs., Ottawa, v. 39 (10), p. 348-363.

Walters, R.F.
1969: Contouring by machine: A user's guide; Am. Assoc. Pet. Geol., Bull., v. 53 (11), p. 2324-2340.

West, G.F., Grant, F.S., and Martin, L.
1970: Geophysical applications of modern computer systems; in Mining and Groundwater Geophysics 1967 (L.W. Morley, Ed.), Geol. Surv. Can., Econ. Geol. Rep. 26, p. 191-201.

Wren, A.E.
1975: Contouring and the contour map: a new perspective; Geophys. Prosp., v. 23 (1), p. 1-17.

COMPUTER-BASED TECHNIQUES IN THE COMPILATION, MAPPING AND INTERPRETATION OF EXPLORATION GEOCHEMICAL DATA

R.J. Howarth
Applied Geochemistry Research Group, Imperial College of Science and Technology, London, England

L. Martin
Computer Applications and Systems Engineering, Rexdale, Ontario

Howarth, R.J. and Martin, L., Computer-based techniques in the compilation, mapping and interpretation of exploration geochemical data; in Geophysics and Geochemistry in the Search for Metallic Ores; Peter J. Hood, editor; Geological Survey of Canada, Economic Geology Report 31, p. 545-574, 1979.

Abstract

Computer methods used for planning, quality control, presentation, and interpretation of exploration geochemical data may be broadly classified into purely numerical methods and mapping, in which sample location is considered. This paper reviews both current industrial and governmental practice, and research techniques of potential application.

Correct organization of field sampling, and laboratory quality control will minimize bias in the data. The majority of industrial surveys are still conducted on a single-element interpretation basis. National geological survey agencies have a growing awareness of the utility of regional multi-element studies for resource appraisal, and data-base management and retrieval systems greatly assist operation of these larger exploration programs. Data treatment is consequently mainly limited to derivation of the mean, standard deviation, etc. and histograms, although computer-based data transformation and analysis of frequency distributions can improve definition of thresholds and class boundaries for mapping. Geochemical maps usually consist of graded symbols at the sample locations, although contour-like maps are preferable for display of the regional geochemical patterns and are often based on moving average smoothing. Proper attention to cartographic design can greatly improve the geochemist's acquisition of information from maps. Multivariate statistical analysis is mainly used to obtain inter-element correlations, to correct for possible Mn/Fe scavenging, and (less frequently) for determination of element associations by principal components or factor analysis, and discriminant or cluster analysis.

While in some cases the lack of suitable computer hardware has limited the application of these methods, in others the restraining factor is the lack of appreciation of their power and flexibility when used intelligently. Geochemists are now gaining a better understanding of the range of statistical and mapping techniques available to them and powerful, relatively inexpensive, minicomputers are becoming more widely available. It is to be hoped that the standard of quantitative analysis of geochemical data will improve in consequence.

Résumé

Les méthodes de l'informatique utilisées pour la planification, le contrôle de la qualité, la présentation et l'interprétation des données d'exploration géochimique peuvent, d'une manière générale, regrouper les méthodes purement numériques et la cartographie, qui tient compte de la localisation des échantillons. Cette étude examine les pratiques gouvernementales et industrielles courantes et les techniques d'application potentielle de la recherche.

Une organisation convenable de l'échantillonnage sur le terrain et du contrôle de la qualité en laboratoire, permettra de minimiser les erreurs au niveau des données. La majorité des levés industriels sont encore effectués d'après une base d'interprétation qui ne tient compte que d'un seul élément. Les levés géologiques à l'échelle nationale tiennent de plus en plus compte de l'utilité des études régionales portant sur plusieurs éléments, pour l'appréciation des ressources; d'autre part, les systèmes d'extraction et de gestion des fichiers de données aident beaucoup au fonctionnement de ces plus importants programmes d'exploration. En conséquence, le traitement des données vise surtout à l'obtention de moyennes, d'écarts types, etc., de même que d'histogrammes, même si la transformation des données et l'analyse des distributions de fréquences à l'aide d'ordinateurs peut améliorer la définition des seuils et des limites de classes pour fins de cartographie. Sur les cartes géochimiques, les points d'échantillonnage sont indiqués par des symboles tramés; cependant, les cartes de type topographique sont mieux adaptées à la représentation des profils géochimiques régionaux et se fondent souvent sur le lissage des moyennes mobiles. Le fait de consacrer l'attention qui convient à la conception cartographique peut de beaucoup améliorer, pour le géochimiste, l'acquisition de renseignements à partir de cartes. L'analyse à plusieurs variables est surtout utilisée pour l'obtention de corrélations entre éléments, pour compenser la contamination Mn/Fe possible et (moins souvent) pour la détermination d'associations d'éléments à l'aide de leurs principales composantes ou analyse factorielle, et pour l'analyse discriminante ou de dispersion.

Tandis que dans certains cas le manque de matériel informatique approprié a limité l'application de ces méthodes, dans d'autres le facteur restrictif est le manque d'appréciation de leur efficacité et de leur flexibilité lorsqu'utilisées intelligemment. Les géochimistes comprennent actuellement de mieux en mieux la gamme des techniques statistiques et cartographiques disponibles, et des mini-ordinateurs puissants et relativement peu coûteux deviennent de plus en plus accessibles. Il est à espérer que la qualité de l'analyse quantitative des données géochimiques en sera en conséquence améliorée.

INTRODUCTION

We aim in this paper to review current practice in the acquisition of exploration geochemical data and methods for its subsequent treatment by numerical and mapping techniques as an aid to interpretation. Since it is not possible within the scope of this paper to pursue all the topics raised in equal depth, an extensive bibliography has been included. The review has been divided into four main parts: (1) quality control and methodology in both the field and the laboratory; (2) requirements of the geochemical map, its compilation and presentation; and statistical interpretation of the data using (3) univariate, and (4) multivariate methods.

QUALITY CONTROL AND METHODOLOGY

Quality Control of Laboratory Analyses

One of the fundamental aspects of any geochemical exploration program is the adequate supervision of laboratory results. It is surprising how many organizations are content to send samples for analyses (either to within-house or contractor laboratories) with little or no attempt to monitor analytical precision or accuracy. With luck, analytical errors may be reflected by spurious patterns on the map which lead to identification of the erroneous data (e.g. Lockhart, 1976) but this is by no means a certainty. Various authors have drawn attention to the importance of randomizing samples prior to submission to the laboratory in order to convert systematic to random errors, and to facilitate recognition of an abnormal situation (Miesch, 1964, 1967b, 1971; Plant, 1973; Plant et al., 1975; Howarth, 1977). It may be argued that it is not convenient during a rapid reconnaissance to randomize all samples prior to analysis. In a long term program, such as that practised by a national survey, it may be possible, however, to randomize all samples for one map sheet at a time. Randomization of samples (and preferably standards and duplicates also) at the batch level will certainly be more beneficial than not doing so. One should always attempt to define both within- and between-batch error levels, since the latter is often the larger effect. It should be realized that careful choice of suitable standards (Allcott and Lakin, 1975; Hill, 1975a, b; Plant et al., 1975) and their submission at random locations in the analytical stream is an integral part of the control process. It may be argued that randomization could lead to sample renumbering errors when passing through the laboratory. Use of pre-randomized number sequences in the field which are subsequently sorted into sequential order for laboratory submission (Plant, 1973) have been found very satisfactory (Plant et al., 1975) and with adequate supervision of personnel no problems have arisen with more complex laboratory-based schemes (e.g. Howarth, 1977). The unequivocal identification of laboratory-induced error in maps from unrandomized samples (e.g. analyzed in traverse order) is not possible in many cases; even if it is suspected, later subtraction of such bias may be impossible since the contribution of local spatial variation will not be known exactly. Whether the additional time and care involved in sample randomization can be justified in a rapid exploration survey on the grounds of cost is a different question, but from the statistical and interpretational viewpoint it is undeniably highly desirable.

Because of the difficulty of obtaining suitable standards in situations where a multiplicity of rock types or stream sediments may be analyzed together (aspects to be considered are: amount of standard; lithological and mineralogical similarity to samples; adequate concentration range for elements of interest; long-term segregation of bulk standard, etc.), attention has turned increasingly to the use of duplicate analyses as an additional control method (Garrett, 1969, 1973a; James, 1970; Michie, 1973; Plant et al., 1975; Howarth and Thompson, 1976; Miesch, 1976b; Thompson and Howarth, 1976, 1978). Recent simulation studies (Howarth and Thompson, 1976) have shown that many of the standard laboratory practices in recording results (such as rounding up to integer ppm values; quoting negative machine readings, normally resulting from a statistical estimation of a near-zero concentration, as zero; or values below some presumed detection limit as 0.5 of this limit) prior to determining accuracy and precision will lead to an over-optimistic bias in the estimation of these parameters. The practice of quotation of within-batch precision rather than between-batch values also leads to optimistically biased results. However, any indication of analytical precision is better than the more usual situation of none.

A rapid method of control based on duplicate analyses and suitable for long production runs, which leads to a measure of total laboratory error variability (i.e. analytical error plus subsampling error) in terms of precision as a linear function of concentration has been described elsewhere (Thompson and Howarth, 1976, 1978). This model is in contrast with the usual assumptions of either constant absolute error (standard deviation) or constant relative error (coefficient of variation) and appears to conform well with most laboratory situations. Analysis of variance techniques are still in most common use for estimating and partitioning field and laboratory sources of error and are discussed below. Other types of control chart methods based on estimates of analytical precision are also in use (e.g. Hill, 1975b).

In many cases large-scale exploration programs are concerned with routine analysis of less than five elements. However, the increasing availability of rapid multi-channel analytical techniques, such as plasma-source emission spectrography, means that problems of efficient multi-element quality control must be solved for upwards of 20 elements at a time. This is a challenging situation and adequate criteria for batch rejection and re-analysis on a multi-element basis are currently being investigated in a number of laboratories.

Geochemical Sampling

There is now increasing awareness of the effect of sampling strategies in geochemical exploration, and this has placed emphasis on sampling media for which an individual sample acts as an indicator for a large geographical area (although the samples are taken from a single location). For example, stream sediments, lake sediments, stream or lake waters, and tills may all derive from relatively large "catchment" areas, whereas soil and rock samples generally are representative of more localized areas.

While it is difficult to attach actual costs to particular types of survey method, this has been done in some cases. For example, Cameron and Hornbrook (1976) illustrate cost functions in relation to regional lake geochemical sampling. Undoubtedly any sampling medium with an inbuilt vector property will prove advantageous in a reconnaissance situation. Aspects of field sampling methodology have been discussed by Miesch, 1964, 1967b, 1971, 1976a, b; Garrett, 1969, 1973a, 1977; DeGeoffroy and Wu, 1970; Dahlberg, 1971; Howarth and Lowenstein, 1971, 1972; Kayser and Parry, 1971; Plant, 1971; Smith, 1971; Bolviken and Sinding-Larsen, 1973; Hodgson, 1973; Connor et al., 1974; Sharp and Jones, 1975; Sinclair, 1975; Cameron and Hornbrook, 1976; Hawkes, 1976; Chork, 1977; and David, 1977. In many of these cases analysis of variance methods have been used to partition field and laboratory sources of error (e.g. Garrett, 1969, 1973a, b; Howarth and Lowenstein, 1971, 1972; Plant, 1971; Bolviken and Sinding-Larsen, 1973; Michie, 1973; Plant et al., 1975; Cameron and Hornbrook, 1976; Miesch, 1976a, b; Sinding-Larsen, 1977; and Chork, 1977). However, it has been suggested recently (I. Clark, pers. comm., 1977) on the basis of mine sampling data that the basic assumptions on which classical analysis of variance rests may be sufficiently violated in practice for the results of methods of assessing field variability using this approach to be misleading. This appears to be partly related to spatial distribution of the data as well as to nonconstancy of variance with concentration, although it is not yet certain just how important these effects may be in exploration geochemical data.

Choice of Sample Spacing

If prior knowledge exists of the likely target size, shape, and orientation, then exact solutions may be obtained for optimum grid sampling techniques. These are summarized by Sinclair (1975) in a geochemical context. For example, if an elliptical target with major to minor axes of a:b is anticipated, then the grid should be oriented parallel to the axes of the ellipse with spacings of $\sqrt{2a}$ and $\sqrt{2b}$ respectively. Singer (1972) gives a computer program for calculating the exact probability of locating an elliptical or circular target with either a square, rectangular or hexagonal grid. More recently Garrett (1977) has illustrated the use of computer simulation methods to determine the detection probabilities for similar targets with irregularly spaced sampling points, such as those more usually encountered in lake or stream sediment sampling plans. His results reinforce the consideration of appropriate choice of sample spacing. For example, an elliptical target with a 1:2 size ratio and major axis length of 5 miles would have an 0.95 probability of detection with a sample density of 1/5 square miles, this falls to 0.50 at 1/10 square miles. However, for a smaller target of the same shape but major axis length of 4 miles the probabilities are approximately 0.70 and 0.20, and with a major axis length of 2 miles they have fallen to below 0.01 in either case. As Sinclair (1975) pointed out, it is the knowledge of target size (which in most exploration programs will be the size of the secondary dispersion geochemical halo) which is unfortunately so difficult to obtain. Hawkes and Webb (1962) suggested 500 feet to 10 miles for drainage and areal soil anomalies, and 5 to 500 feet for localized soil and biogeochemical anomalies as maximum dimensions.

As a result of this difficulty, a recent development has been the interest in selection of optimum sample spacing based upon the spatial properties of the data. This is a reflection of the inherent correlation between values at adjacent sample locations, which will increase as the inter-sample distance falls, and conversely, at extreme distances the concentrations become uncorrelatable. Hodgson (1973), Dijkstra and Kubik (1975), Sinclair (1975) and Dijkstra (1976), have discussed this from the point of view of auto-correlation.

An alternative approach is the application of the geostatistical theory of Professor G. Matheron, founder of the Centre de Morphologie Mathématique, Fontainebleau[1]. His Theory of Regionalized Variables (1957, 1962, 1963) has been successfully applied in the mining industry to ore reserve calculations and optimization of drillhole locations (see for example David and Dagbert, 1975; Journel, 1975; Guarascio et al., 1976; David, 1977; and Alldredge and Alldredge, 1978). In exploration geochemistry, a variety of investigations are in progress (Sinclair, 1975; Crossant, 1977; S.A.M. Earle and R. Sinding-Larsen, pers. comm., 1977) concerning its utility for the optimization of sample spacing in soil traverses or grids and are showing encouraging results.

Following Matheron's treatment, if z_{x_i}, the observed element concentration at a location x_i, is one possible realization of a certain random function $Z(x_i)$ representing all possible concentration values which could be obtained at x_i, and $z_{x_{i+h}}$ is the observed concentration at a point a distance h away from the first, then the <u>intrinsic hypothesis</u> is that the differences in concentration between all pairs of points separated by distance h (e.g. $z_{x_i} - z_{x_{i+h}}$ and $z_{x_j} - z_{x_{j+h}}$) are considered as different realizations of the same random increment $\Delta(h) = Z(x) - Z(x+h)$, and are independent of the location of x and x+h. This condition of stationarity applies, from the intrinsic hypothesis, to both the mean and variance of $\Delta(h)$. If it is also assumed that the mean is zero, then this behaviour may be studied experimentally by the semi-variogram:

$$\gamma^*(h) = \frac{1}{2n} \sum_{i=1}^{n} \left[z(x_i) - z(x_{i+h}) \right]^2$$

where n is the number of pairs of points a distance h apart. This function is suitable even in the presence of gaps in the data, but (as a rule of thumb) requires preferably at least 50 points in a traverse. Different shapes of semi-variograms may be obtained depending on the nature of the underlying spatial behaviour of the phenomenon (see, for example David, 1977). The most commonly observed pattern with soil data is for $\gamma^*(h)$ to rise fairly rapidly from a small initial value at the shortest sample spacing used (corresponding to all sources of variation at a scale smaller than that of the sampling interval) until it becomes approximately constant for large values of h. Figure 24.1 shows an example taken from a study in the Mendip Hills lead-zinc area of southwest England by the Applied Geochemistry Research Group (AGRG) of Imperial College, London. The rising portion of the curve allows an estimation to be made of the <u>range</u> of influence of the element, that is the distance above which the concentration values will be statistically independent.

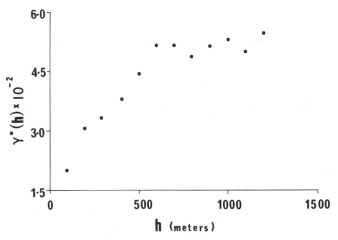

Figure 24.1. Semi-variogram for log (Zn) in B-horizon soil samples over Carboniferous Limestone Series of Mendip Hills, Somerset, England, 70 samples taken at 100 m intervals along traverse. Range of influence is approximately 750 m (Information supplied by S.A.M. Earle, AGRG, 1977).

---

[1] Where 'geostatistical' is used in this paper the term refers specifically to Matheron's theory rather than to the general usage of statistics in the earth sciences.

Semi-variograms could be made in various directions (i.e. along and across strike) to assess the variability of range with direction, since on average the values of the regionalized variables will become independent at distances exceeding the range. Setting the sample spacing in a subsequent investigation as approximately equal to the range is a useful criterion, but the optimum solutions will vary with the type of semi-variogram. If multi-element data are available, the interval may be set with regard to the overall ranges found for the various elements. Sinclair (1975) pointed out that from the prospecting point of view the critical sampling interval would need to be determined from both background and anomalous areas (and there may be a shortage of data for the latter) since the range for a background region will probably not be related to that from an anomalous area, and geological information on likely target size and shape would still be required. Even if the geostatistical approach is not yet a practical evaluation tool in prospecting, widespread application should give a valuable insight into the ranges of different elements in particular environments, thus leading to a better understanding of the fundamentals of the dispersion phenomena with which we are concerned.

Sharp and Jones (1975) have devised a topologically based method for detailed sampling of a drainage basin stream network using the results of sequential (on-site) analyses which may well be useful for more detailed geochemical exploration, say in the follow-up stage of a regional program. Regional stream sediment surveys are generally planned on the basis of the size of the feature which it is required to delineate.

In all cases the decision whether to operate a phased reconnaissance or a single high density sampling program will depend ultimately on time and cost criteria. The latter approach will in many cases lead to gross oversampling to determine potential target areas, whereas a phased program may extend over more than one field season. Miesch (1976a, b) has devised elegant optimization techniques, based on analysis of variance, to distinguish between very broad scale units pre-defined on a geological or geographical basis, but the resulting sampling densities appear to be rather low for exploration geochemical purposes.

The Role of Sampling and Analytical Error

It is often assumed in planning a survey that a small target (e.g. geochemical halo) will be successfully recognized with one sample. Naturally, there will be the possibility of additional geochemical, geological or geobotanical information which may assist, but in many instances initial recognition will depend on the presence of an "anomalous" concentration value. Decisions affecting identification of such values will depend not only on the natural geochemical environment (many known orebodies are not distinguished geochemically at the surface), but also on what transforms (if any) are to be applied to the data (see for example the discussion in Govett et al., 1975), and on the variance of the geochemical determinations.

A general linear model for increasing analytical plus sampling variance with element concentration seems to be applicable in a wide variety of laboratory situations (Thompson and Howarth, 1976). It is therefore quite possible as a consequence of this that the observed concentrations from target locations on occasion could be below the threshold set by the investigator to define "anomalous" samples and vice versa. If the form of the increase of variance (which could include field sampling error if duplicates were taken at this level) with concentration is known, then it is possible to predict the probability that background samples will be wrongly classed as anomalous (false alarm), and anomalous samples wrongly classed as

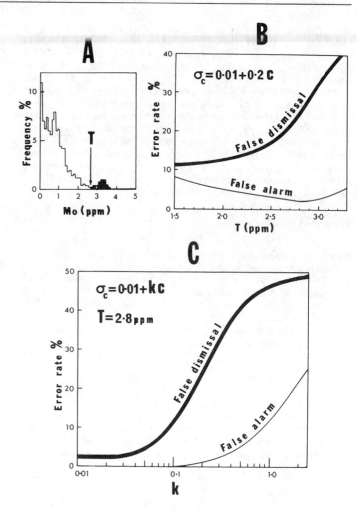

A) A priori frequency distribution for Mo from primary reconnaissance stream sediment survey, showing 'anomalous' population above threshold, T.

B) Expected error rates for misclassification of background samples (false alarm) or anomalous samples (false dismissal) in a subsequent survey, assuming a constant analytical system equation, as a function of T.

C) Expected error rates assuming constant threshold (T) with worsening analytical precision (k in the system equation).

Figure 24.2. Example of effect of analytical precision on expected accuracy of geochemical target recognition. For full explanation see text.

background (false dismissal). The false dismissal probability may be obtained by integration of the expected probability distributions for all observed concentrations corresponding to samples whose true concentrations lie above the defined threshold, and the false alarm probability is obtained in a similar manner from samples below this threshold. A normal distribution of errors is assumed in this model.

Consider an example in which an orientation survey for molybdenum in stream sediments has yielded a bimodal frequency distribution (Fig. 24.2A). Analyses have been determined to 0.1 ppm, and the total error variability of the system (including both analytical and subsampling errors) has in this case been established as being adequately represented by the equation $\sigma_c = 0.01 + 0.2C$, where σ_c is the standard deviation for the normal error distribution about a true concentration C. Let us further assume that in the area under investigation it is probable that the maximum concentration which could contribute significantly to the anomalous population in a subsequent reconnaissance survey is expected to be 5 ppm (i.e. we exclude the rare events with concentrations above this from the integration). Figure 24.2B shows the resulting false dismissal and false alarm error rates to be expected subsequently if the threshold for defining the "anomalous" concentrations is changed from 1.5 to 3.3 ppm. Without the unacceptably high false alarm rate from the background population (necessitating many check analyses or follow-up surveys), a significant proportion (24 per cent) of the anomalies could go undetected even were the threshold to be placed at the optimum boundary of 2.8 ppm (Fig. 24.2A). The false alarm and false dismissal rates for the same data are plotted in Figure 24.2C as the slope (k) in the equation $\sigma_c = 0.01 + kC$ is increased from 0.01 to 1.0 (i.e. the errors increase) given a constant threshold of 2.8 ppm. This figure dramatizes the need for laboratory quality control as a prerequisite, but also demonstrates that only when the magnitude of total field plus laboratory error is known can one gain a reliable knowledge of the likelihood of anomaly detection among the population of all sites sampled in an area. This does not, of course, include loss of targets which went undetected because no sample fell within their range of influence. There will always be a danger that subtle anomalies (e.g. related to deeply buried mineralization) may be missed if they are in the "background" range. This chance is minimized if the data are plotted in map form, but careful attention must naturally be given to the selection of class intervals. Even then, no method will guarantee complete success.

Discussion

It should be apparent that one of the major problems in cost-effective geochemical exploration is the adequate definition of the geochemical target in terms of size, shape, orientation and likelihood of correct recognition on the basis of the chemical analysis of the sample. Long-term acquisition of reliable data from geostatistical studies should assist in definition of the spatial behaviour of the elements of interest, both in background and anomalous areas.

If one wishes to find as many targets of fixed size as possible in a given area from a single field sampling exercise, then there is no recourse but to put in blanket coverage at a suitable density. For a fixed number of samples the density should be made as low and even as possible over the whole area of interest (other factors permitting). However, a phased approach based on an initial lower density regional survey to identify areas more suitable for later detailed examination would be generally preferable. Rapid low-cost analytical techniques are generally considered suitable in the early stages of such work.

A sampling medium which represents a large geographical area from a point-source sample (e.g. lake or stream sediment) will be of assistance in areal reconnaissance of suitable terrain. Where soil or rock sampling has to be undertaken, the price of more localized sample representivity will inevitably be a larger number of samples. Systematic control of field and laboratory practice should help to ensure that a target will stand the best chance of recognition. It should not be overlooked that the increasing availability of multi-element information on a routine basis could well enhance the probability of recognition of a potential target (if only because the joint probability of false dismissals would be lowered).

Data Storage and Retrieval

The computer forms an ideal medium for the storage of geochemical data and related field observations. Several commercial organizations are known to be investing a great deal of effort in building up data storage and retrieval facilities, since the ease with which data base interrogation may be undertaken is felt to justify the installation of such a facility even without a requirement for sophisticated statistical processing of the information.

Gordon and Hutchison (1974), Hutchison (1975) and Merriam (1976, 1977) include papers describing a wide variety of computer-based systems for storage and retrieval of geological data. Major files of exploration geochemical data are held in computer-compatible form by national or state geological surveys, academic and industrial organizations in many countries. It is becoming more common for field data collected in the course of a geochemical survey to be entered directly on forms designed as keypunch documents for subsequent computer entry (e.g. U.S. Geological Survey, 1969; Garrett, 1974a; Plant et al., 1975; Baucom et al., 1977).

The general philosophy involved in such systems may be gleaned from the papers included in Gordon and Hutchison (1974), Hutchison (1975) and Merriam (1976, 1977). A variety of generalized data base management systems may be bought, or leased, from commercial manufacturers thus ensuring long-term maintenance and updating. However, several organizations with sufficient in-house facilities have preferred to develop systems specifically for geological data. Systems such as GAS (Garrett, 1974b), GRASP (Bowen and Botbol, 1975), GEXEC (Plant et al., 1975; Jeffery and Gill, 1976), SIGMI (Kremer et al., 1976), NDMS (Ferguson et al., 1977) and System 2000 (Davidson and Moore, 1978) are typical of those currently being used for geochemical data files. A wide variety also exist tailored to the requirements of the field geologist or igneous petrologist, with which we are not concerned here. There are those who would argue that data base management packages supplied by major software companies or the computer manufacturers are preferable to home developed packages, on the grounds of reliability, standards of documentation and, perhaps of greatest importance, long term maintenance and update of the system. Against this must be set the fact that such packages are seldom tailored for the geochemist's requirement, particularly if a single system embracing data management, statistical treatment and graphics facilities is required.

In the long run, it is the considerable ease with which statistical data and simple graphical displays and maps can be obtained, using integrated data base management facilities to prepare the requisite information for management decision-making, which will justify the cost and effort required to create and maintain this type of facility. For many of the larger national and industrial organizations, this justification has now been amply proven.

GEOCHEMICAL MAP COMPILATION

Location is the key element of exploration data. Because the eye is well adapted to recognition of patterns in spatial data, maps have the capacity to present exploration information with great impact, and the spatial component becomes an integral part of the compilation and interpretational process. Comparison of geochemical data with topographic, geological or geophysical information is rendered a much easier task when all are in map form at the same scale, and the geochemical map has thus become a useful and familiar tool to the explorationist.

Types of Geochemical Map

Map usage can be divided into three broad categories generally related to scale: (a) the regional atlas, at scales upwards of 1:250 000; (b) reconnaissance maps, at scales between 1:25 000 and 1:250 000; and (c) local detail, generally prepared at the follow-up stage at scales between 1:100 and 1:25 000. Within each class variety exists in both the type of data, and the way in which it is presented.

Regional Geochemical Atlases

While not specifically exploration oriented, this type of atlas provides the broadest view of large portions or the totality of a country. Typically, sampling densities will be well below $1/km^2$, over an area in excess of 10 000 km^2, and information for 12 to 30 elements will be included, usually presented as moving-average smoothed maps. To the explorationist such maps can provide useful information by defining geochemical 'provinces' within which more intensive exploration surveys can be framed.

In general, three main lines of approach have emerged: firstly, extrapolation of the results of detailed studies in type localities considered to be representative of large areas; secondly, very low-density multi-stage (hierarchical) sampling of large units considered to be relatively distinct; and thirdly, low-density sampling of the entire survey area, maintaining where possible a uniform coverage.

The first two methods have generally been applied to broad units, often related to soil type, and while of value to environmental geochemistry (Tourtelot and Miesch, 1975), have little application from the mineral exploration standpoint.

The third approach, however, can provide a rapid cost-effective tool for the identification of broad regions within which conventional mineral exploration surveys might be applied, and in this context is analogous and strictly complementary to national geological and soil maps and to satellite imagery. Generally speaking, samples representative of a catchment zone (such as lake sediment or stream sediment) have been preferred for this type of survey.

Garrett and Nichol (1967) and Armour-Brown and Nichol (1970) demonstrated the ability of stream sediment sampling, at densities sparser than one sample per 180 km^2, to identify major mineralized belts in Sierra Leone and Zambia. The regional lake geochemistry sampling program currently being undertaken by the Geological Survey of Canada (Hornbrook and Garrett, 1976; Garrett, 1977) at a mean density of one sample per 13 km^2 has successfully identified a number of major geochemical anomalies, in some cases related to known mineralization, in the 421 000 km^2 covered so far. A regional geochemical atlas of Uganda (Geological Survey of Uganda, 1973) has been completed, based on cuttings from wells bored for water, and stream sediment sampling where suitable drainage exists, at overall mean densities of about one sample per 70 and 260 km^2 respectively. Regional geochemical atlases of Northern Ireland (Webb et al., 1973) and England and Wales (Webb et al., 1978) based on stream sediment sampling at a mean density of one sample per 2.8 km^2 have shown many features of interest from the geological and environmental points of view, in addition to the major mineralized districts.

An extensive hydrogeochemical and stream-sediment reconnaissance of the 7 827 617 km^2 of the conterminous United States and Alaska was initiated in 1975 as part of the U.S. Department of Energy National Uranium Resource Evaluation (NURE) program, with the aim to obtain uranium data for stream and lake water, groundwater and stream and lake sediments. It is hoped to complete the 621 NTMS 1°x2° quadrangles, with average sampling densities of the order 1/13 km^2 and 1/25 km^2 for stream sediment and groundwater samples respectively, by 1983. Currently, some 60 quadrangles have been sampled and are being analyzed for up to 40 elements (Bendix Field Engineering Corp., 1976, 1977, 1978). This project will undoubtedly provide a source of multi-element regional geochemical atlas material of immense interest.

Clearly, the compilation of national multi-element geochemical atlases is a responsibility which lies with government. In the context of resource appraisal (particularly in the Third World), the diversity of requirements for mineral exploration, agriculture, public health, land-use and pollution emphasize the necessity for a multidisciplinary approach aimed at establishing the best methods of acquiring and presenting regional geochemical information, and for research to develop interpretational criteria to meet the differing interests of all the eventual users (Webb and Howarth, in press). In many parts of the world even a preliminary broad-scale coverage will take a considerable time to accomplish. In the meantime, reconnaissance geochemical surveys of very extensive areas have been, and are being, undertaken by national and international agencies and industry.

Reconnaissance Geochemical Maps

Surveys in this class are mainly undertaken by government agencies and industry. The objectives are to identify areas of economic interest and select specific targets for more detailed follow-up investigation. Physical and climatic conditions, and (in some cases) the degree of man-made interference determine the kind of sampling media used. The more common sampling media include stream sediment, lake sediment, and soil, till and rock where appropriate. Lake, stream and spring or well waters are less commonly used. Sampling densities generally range between $2/km^2$ and $1/25$ km^2. Semi-airborne geochemical surveying techniques (such as the Barringer Research Surtrace method based on sampling the surface microlayer from the air) hold considerable promise for the future.

The most valuable and commonly produced geochemical map types show firstly, the sample location and secondly, the element analytical value at the sample site. These are being supplemented with increasing frequency by various types of 'derived' maps which include those in which some kind of geochemical surface is fitted to the data, either to show broad-scale trends or deviations from them, the results of statistical or numerical manipulation of the data, or maps of multi-element combinations.

While many industrial reconnaissance scale surveys are still restricted to, say, 3 to 5 elements, in recent years there has been an increasing tendency for surveys undertaken by government agencies to cover a wider range of elements (e.g. Plant and Rhind, 1974; Wolf, 1974, 1975; Bendix Field Engineering Corp., 1976, 1977, 1978; Larsson, 1976; Cockburn, 1977; Baucom et al., 1977; Nichols et al., 1977).

Local Geochemical Maps

The results of an intensive detailed survey, often with sampling densities of 10 to over $100/km^2$, can be presented in any of the ways described under reconnaissance. Soil samples are most commonly used for this type of survey, and since the area is usually small can be considered relatively homogeneous, so that contouring the element concentrations is often used.

Map-Plotting Equipment

There are now a wide variety of possible media on which maps may be produced. Criteria for their choice include: the purpose of the survey (e.g. government or industrial use); ease of availability; size of output, print quality and ease of reproduction; physical stability of the final product; the duration for which the map will be required; and special characteristics such as colour plotting capability.

Pen plotters are probably the most widely used output medium for maps. The older drum plotters often suffered from paper stretch and inadequate inking of symbols, but have been much improved recently. Flat-bed plotters can attain very high cartographic quality and can cope with large map sheets. They usually have the ability to scribe high quality masters on to specially stable material for subsequent reproduction, and many have the capability of optical projection of symbols on to stable photographic film. Black and white line symbols are used generally, although some plotters have a colour pen capability. Unless optical plotting is used, large areas of tonal symbols may be tedious to produce. The lineprinter is one of the cheapest and most widely available media and is commonly used for posting data values or plotting tonal maps using overprinting. Special printer chains or coloured print ribbons can be used if required. Large maps can be made by sticking parallel strips of output together, but paper stretch may be a problem. Electrostatic or thermal printers produce similar output, although more flexibility in character generation may be achieved (allowing square symbols and a wide range of tonal symbols) but again paper stability may be a problem.

Computer-generated microfilm is becoming increasingly available. The cost per 35 mm frame is generally low, although high quality photographic treatment is required to preserve accuracy in the final enlargement. Most modern microfilm plotters have very good line definition, some with a variable line intensity capability. Colour microfilm plotters are still uncommon, although effective colour-separated masters for coloured maps may be generated on black-and-white film (Schweitzer, 1973; Mancey and Howarth, 1978).

Plotters using a laser beam to produce repeatable variable intensity images (with 16 or more grey levels) on sheets of dimensionally stable photographic film, are now being used extensively for production of seismic cross-sections and remote sensing imagery. They are all capable of producing larger original images than microfilm, and are well suited to the generation of colour separated masters for subsequent colour printing (e.g. Webb et al., 1978), although some devices can now plot directly onto colour film.

A new type of colour plotter using jets of cyan, magenta and yellow ink to draw directly onto paper was originally developed in Scandinavia (Smeds, 1976) and is now coming into general use. Early problems with ink colour and stability have now been overcome, and recent examples of maps produced with this type of plotter are of very good quality (J.G. Knudsen, pers. comm., 1978).

Exploration geochemists concerned with the rapid examination of geochemical map data as an aid to decision-making, rather than elaborate final copy, have at their disposal an enormous range of interactive computer display terminals, with which the ease of changing parameters (such as: class intervals, the number of contours and their levels, user-controlled values determining the type of contouring or moving average calculation to be performed) and redisplaying the result immediately is maximized.

Alphanumeric terminals can display text or a number of special characters (with up to 7 colours on some terminals) and may be used for lineprinter-like output. Graphic terminals, which can display accurate line drawings, or in special cases many grey levels or colour, are available in a wide variety of costs and capabilities. The two basic types are (1) storage tube displays, in which a large quantity of line information is drawn on a special long-persistence phosphor, and to change the image the whole screen must be replotted; and (2) "refresh" displays in which the image is rapidly and continuously redrawn, as with a television picture, which gives a greater flexibility for particular types of graphic operations such as real-time rotation of figures in two or three dimensions. This last facility is particularly useful for the inspection of three-dimensional views of geochemical surfaces (Botbol, 1977).

While ease of map production, time and cost may be the ultimate criteria to the geochemist, it is inescapable that as the complexity of the data increases the visual design aspects of the display play an increasingly important part in helping the geochemist to extract the maximum of information from the map.

Geochemical Basemap Requirements

Before examining the various types of geochemical map we consider the requirements for the basemap on which the geochemical data are to be plotted. The relevant factors are contents, scale and map projection, and reproduction process.

The main function of the basemap is to provide ready and adequate information so that a sample site can be located geographically. Where available, geological and possibly geophysical information may be included since it will be relevant to the interpretation as well as the geochemistry. The use of overlays may be helpful, as an overly complex basemap may be self-defeating when it comes to recognition of the geochemical patterns, particularly at the regional scale. The topographic information in the basemap may be derived from maps published by government agencies (where available, there are still many parts of the world not mapped at scales below 1:75 000), or from air photograph mosaics. It is usually necessary to enhance, by tracing or selective reproduction, the features relevant to the survey.

The map scale will depend on the area of the survey, the density of sampling and the kind of data to be included. A good choice will result in a small number of maps of convenient size on which the relevant information can be shown with clarity. Hawkes and Webb (1962) suggested that for a map in which individual sites are to be shown, a scale such that sample points are usually about on-quarter inch apart results in a pleasingly legible map. A confused overlapping of symbols should be avoided at all costs.

The choice of a particular reproduction process depends on the availability of plotting and reproduction equipment, cost and user requirements, and may range from dyeline reproduction to photo offset-lithography or other methods in government survey work. It should be borne in mind however, that unless the basemap and subsequently reproduced geochemical maps are on dimensionally stable materials significant stretching may occur in a relatively short time. The addition of the computer-generated geochemical data to the basemap can be done by plotting on to it directly, by plotting on a blank medium and then jointly reproducing the two to generate the final map, or by use of transparent overlays.

Sample Location Map

This is produced during the survey by recording the position and corresponding sample identification on a topographic or other working field map, or air photograph.

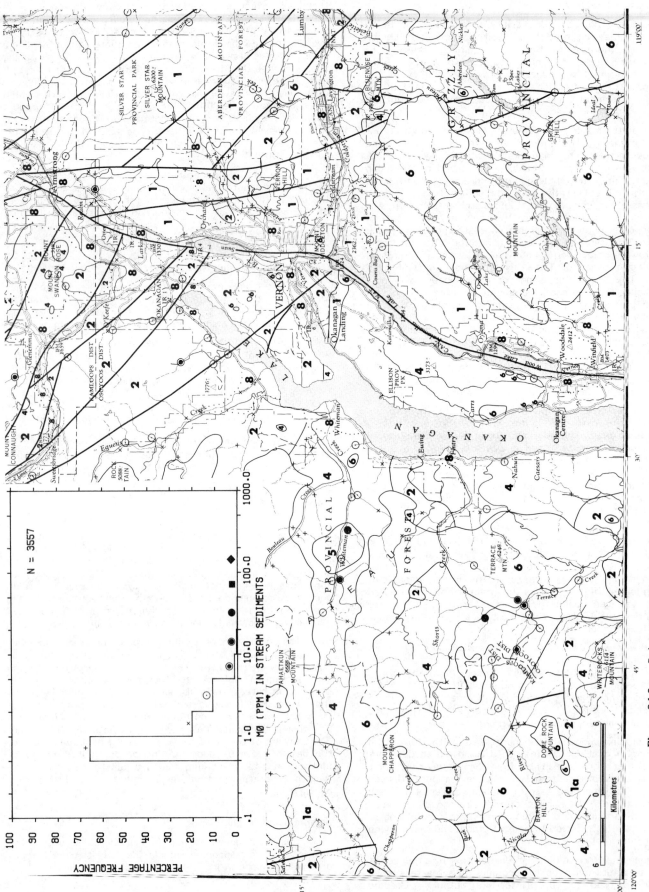

Figure 24.3. Point-source measured value map with graded symbols for molybdenum in stream sediments. Note the geological information on the basemap. (Geological Survey of Canada – British Columbia Ministry of Mines and Petroleum Resources joint Open File 410, NTS sheet 82 L, Vernon, B.C.)

Experience has shown that human recording of location co-ordinates in the field is prone to significant error (e.g. inadvertently reversing digits in the numbers). Automatic recording of site co-ordinates directly from the field document can conveniently be carried out by placing it on an X-Y digitizer (of which several types are now available); the operator moves a cursor from point to point, entering the location co-ordinates automatically. Relevant topographical or geological information may also be encoded at the same time for later computer-aided drafting (U.S. Geological Survey, 1969). The computer can subsequently be used to convert relative site positions in terms of the digitizer table co-ordinates to UTM or other grid co-ordinates and to make any appropriate transformations for the required map projection.

Computer-assisted drafting of the sample location map will often save time and will certainly improve accuracy in the final result.

Geochemical Map Presentation Techniques

In this section we review a variety of different methods used to make geochemical maps, all of which may be implemented by computer plotting, and the computer has widespread applications as a tool to aid the selection of appropriate class-intervals, even if the maps are subsequently drawn by hand. While we illustrate the methods with reference to elemental concentration values, they are equally applicable to the mapping of associated information such as pH, stream characteristics, overburden thickness, etc.

A recent survey of methods of selection for class-intervals (Evans, 1977) found at least 16 have been used in practice. Of these, the use of: 'natural breaks' in the frequency distribution; class width defined as a proportion of the overall standard deviation; arithmetic or geometric progressions of class width; and percentile classes (in which each class contains a known percentage of the data) appear to be the choices most often used by the geochemist. (Class selection is returned to under Univariate Statistics, below.) It should not be overlooked that there is usually an unstated assumption that every map should permit its users to unambiguously identify the class to which each individual symbol belongs. The upper limit to the number of distinct shadings or sizes of symbol which the human eye can distinguish is generally taken as seven or eight, and would only exceptionally exceed ten. If the map is to be used simply as a spatial data bank, then writing the actual values onto the map is clearly desirable. However, if it is to be used to study the overall spatial pattern of the map, to assess contrasts between different places (e.g. look for geochemical 'anomalies') or to compare the patterns of different element distributions then some kind of symbolism is desirable. Reducing the number of classes may achieve simplification at the expense of local contrasts.

Values may be shown on a map by numbers, symbols, symbol-number combinations, and contour-like patterns. We consider first, methods related to showing the value at a discrete site (point-symbol maps), and secondly maps emphasizing the distribution of regional or background patterns of variation, such as contour or trend-surface maps.

Point-Symbol Geochemical Maps

Sample Value Map

The simplest type of map is one in which the element concentration is written at the collection site as a numerical value. This avoids possible bias on the part of the mapmaker by not putting the data into symbolic classes for presentation, but unless the scale is carefully chosen to avoid overwriting of concentration values, and the values themselves are colour coded into broad classes, it may be extremely difficult to make out patterns without subsequent contouring by hand (with all its attendant difficulties, see below).

Symbol Map

A widely used method is the designation of the sampling location by a symbol. By varying in shape, size and internal detail, the symbol can serve the dual role of indicating both site and element value. A convenient set of symbols can thus be used in a map, each symbol denoting a class or a range of element values. The choice of classes can be based on the results of numerical evaluation of the data and the symbols will then indicate some level of significance. This can be a useful aid for anomaly identification.

Symbol-only maps have the advantage of clarity and minimum overprinting (Fig. 24.3). They are most useful where element values have a narrow dynamic range so that a symbol will represent a value within a few units of significance and analytical resolution. A wide variety of black and white symbols may be easily drawn using a pen or light-spot projection flat-bed plotter (e.g. Garrett, 1973b), or microfilm (e.g. Baucom et al., 1977).

Proportional symbols have been used in a variety of ways. A proportional width 'worn diagram' method has been used for manually produced maps of stream sediment surveys, whereby the stream is enclosed by parallel lines whose separation is proportional to the element value along the sampled section of the stream (e.g. Gregory and Tooms, 1969). It is however, very laborious, of limited value and unsuited to automated drafting. Proportional line lengths are being used for reconnaissance maps of Scotland (Plant and Rhind, 1974; Experimental Cartography Unit, 1978). Dickinson (1973) in discussion of this type of symbol noted that while it is easy to estimate visually (either unaided, or with the help of some kind of scale) and is thus useful to determine concentration at a given location, "the linear nature of the symbol makes it very difficult to accommodate a range of quantities of any great extent . . . as its size increases the visual 'weight' of the bar becomes more and more detached from the actual locality it is supposed to symbolize and the message of the map becomes rather vague from a distributional point of view". Experimental tests (Rhind et al., 1973) confirmed the difficulty of assessment of background values with this type of symbol, although it may be useful for giving an idea of local variability (J. Plant, pers. comm., 1978). Provided overlap of adjacent symbols is not disadvantageous, proportional circles or squares can better encompass a wide dynamic range than linear symbols, area being proportional to concentration. Errors in estimation are generally lower using proportional squares (Dickinson, 1973), although circular symbols appear to have been used more frequently in geochemical studies (e.g. Dall'Aglio, 1971; Barringer, 1977).

The use of graded (shaded) symbols seems to have been much more widespread than that of all types of directly proportional symbols (e.g. Webb et al., 1964; Garrett, 1973b; Hornbrook and Garrett, 1976; Baucom et al., 1977; Cockburn, 1977). Colour-coded dots have occasionally been used (e.g. Nichol et al., 1970), each colour being based on a black and white master produced from microfilm (Nichol et al., 1966).

The limitations of the completely symbolic map become apparent with wide dynamic ranges which must be covered either by a very large number of symbols, thereby losing simplicity and convenience of use, or by having symbols represent classes of values too wide for adequate resolution.

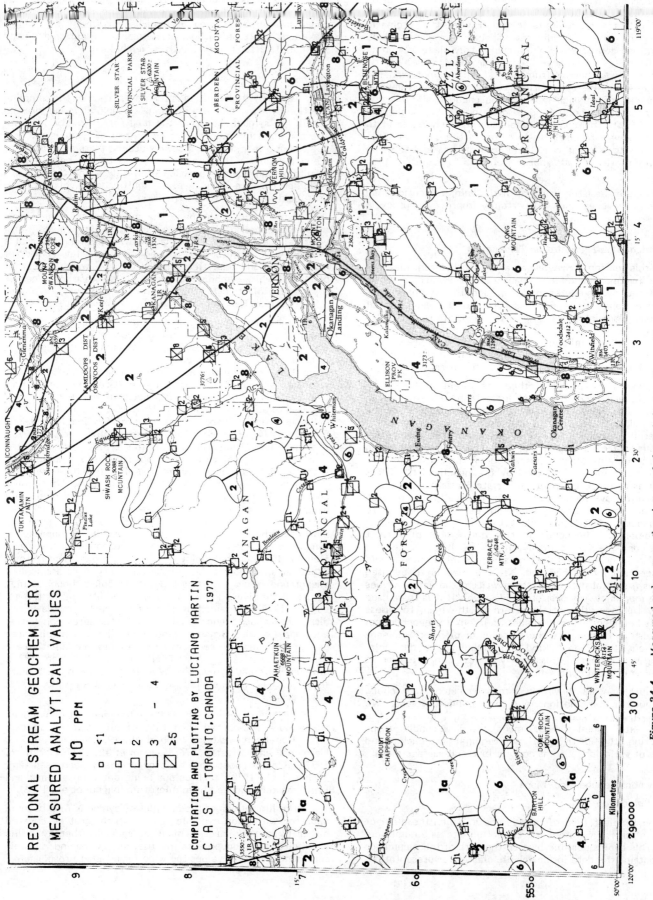

Figure 24.4. Measured value map showing graded symbols together with actual values for molybdenum in stream sediments. Map is portion of GSC Open File 410, NTS sheet 82 L, Vernon, B.C.

Symbol-number Map

By explicitly printing the measured element value next to a proportional or graded location symbol (Fig. 24.4), the geochemist has a convenient graphic representation of the element values, yet is not limited to accepting the interpretation of significance inherent in the classes chosen for the map. While the added information in such maps may create some problems of legibility and overprinting, these can be minimized by choosing a suitable scale, and shifting the position of the numbers where necessary. The advantages of this kind of presentation usually greatly outweigh the minor drawbacks, and it is widely used by many exploration geochemists.

Background-related Symbol Map

While qualitative considerations based on experience, intuition and general knowledge of the area have a significant role, a quantitative evaluation of the geochemical information should be very useful. The geochemist will usually start this evaluation by applying a basic statistical treatment to the total data set. From inspection of the frequency distribution histogram, the cumulative curve, or knowledge of the mean and standard deviation, 'background' and 'threshold' limits may be defined, and the element classes selected accordingly. The samples could now be divided into statistically anomalous and 'background' classes.

While this selection process will satisfy the statistician, we must remember that our targets are spatial geological anomalies, not just statistical. Since the population consisted of all the samples, it is clear that for statistical significance to be geologically relevant the data should have one uniform background population. Reconnaissance surveys usually cover several lithologies, each providing a distinct geochemical background. This creates the possibility that samples which would be anomalous within their local environment will not be statistically anomalous in the total population. Conversely, a value which appears anomalous in the total population might be geologically insignificant if the sample is from an area of high local background.

The background-related symbol map (which Martin terms the 'justified measured value map') is similar to the symbol-number map described above, as it includes the actual element value and a grading symbol. However, whereas in the earlier map the grading was based on overall statistical parameters, the symbol now reflects the significance of a sample within a group with similar background. To prepare this type of map one must first know the background determining factors (such as underlying lithologic units) and use these to divide the data into groups of compatible samples. Statistical parameters are then computed for each group and are used to produce the justified grading by one of the standard methods of class selection. The computer is invaluable for this type of work, making use of integrated data base manipulation and statistical tools.

Maps Emphasizing the Regional Geochemical Distribution

The emphasis so far has been placed on individual sample locations. However, for further interpretation it may be useful to take into account variations in the topography of the regional geochemical surface of element concentration values. Before considering suitable methods to do this, it is essential to realize that for the extrapolation from the data points to a more general surface to be valid, the area under consideration must be considered homogeneous, and the density of the sample points must be sufficiently representative of a continuous surface. In reconnaissance surveys these prerequisites are seldom met and therefore in most cases contouring of untreated concentration values may not be justified; it may however be possible at a local scale, or at a broad regional scale.

Following Rhind (1975) we may divide surface-fitting methods in to two broad classes: firstly, global methods in which a generalized "trend surface" represented by a polynomial equation, is fitted to the entire data set; and secondly, local methods which include contouring in general, with moving average and kriging being regarded as different types of interpolation functions.

Trend Surfaces

Very broad-scale regional variation is approximated by fitting a surface of simple geometric form, normally a polynomial equation in two dimensions involving powers and cross-products of the (x,y) spatial co-ordinates and (z) the element concentration values. The 'degree' of the fitted surface corresponds to the highest power term used. Trend-surfaces are usually 1st (linear), up to 5th (quintic) degree (Fig. 24.5), although higher powers have occasionally been used. So-called Fourier surfaces are similar and involve sine and cosine terms but do not seem to have been used for exploration geochemistry problems. Trend-surface analysis requires a certain amount of care both with programming (e.g. Agterberg and Chung, 1975) and interpretation, as the

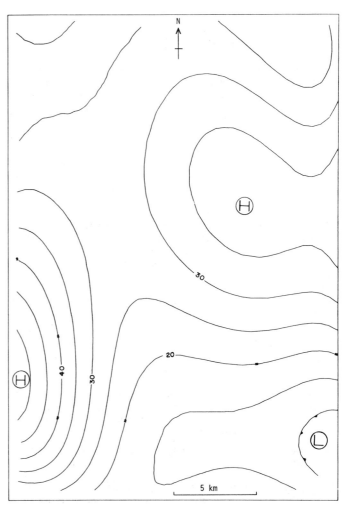

Figure 24.5. Derived regional map: 5th degree polynomial trend surface used to approximate the regional component; 5 ppm contour interval.

orientation of the surface may be sensitive to the number of data points and their disposition (Howarth, 1967; Norcliffe, 1969; Chayes, 1970a; Doveton and Parsley, 1970; Shaw, 1977).

The element value at a location is assumed to be a product of both the geological environment at large (the regional component, represented by the trend-surface) and local variations imposed on this surface (residuals). Trend-surface analysis separates these two components in an arbitrary fashion. The residual value (difference between the measured concentration and the regional surface value) may be used as an indicator of anomalous mineralization, and mapped (e.g. in symbol-number form), class selection being based on statistical analysis of the residual values. Examples of trend analysis in exploration geochemistry include Connor and Miesch, 1964; Nackowski et al., 1967; Nichol and Webb, 1967; Cameron, 1968; Bosman et al., 1971; Brabec and White, 1971; and Lepeltier, 1977.

Garrett (1974b) concluded, based on both his and E.M. Cameron's experience, that "the method has a limited application to either very broad regional surveys, or very detailed local surveys, and in either case the order of the surfaces should not exceed quartic at the maximum. Bearing in mind the heterogeneity of individual rock-units and whole survey areas we are not convinced that a polynomial model is appropriate as we do not have, in the vast majority of cases, a sufficient knowledge of the true pattern of areal variation and the included discontinuities".

Contour and Related Maps

One of the commonest methods of representation of the regional component of the geochemical surface is by contouring. Contour maps may be produced manually or by computer techniques. It is a common fallacy that all humans can produce "good" contour maps, because one has only to give the same point source data to a number of persons to contour to find out how much subjective variation can be introduced.

The steps involved in computer contouring from a given set of data points all involve some subjective choices of methodology but once the parameters for a particular method have been chosen, subsequent contouring is completely objective and reproducible providing the sampling density is adequate. The initial choice is the size of the square mesh on to which the data values are to be interpolated; related to this is the method used to choose the number and disposition of the data points surrounding each grid node (nearest n-neighbours, quadrant or octant search) on which the node value will be based, and the distance weighting assigned to each point (inverse squared distance is commonly used). There are a large number of possible algorithms upon which the interpolation is based, typically between three and ten points per node are commonly used.

Most contour interpolation techniques are exact at the data points (although this is not necessarily true of methods based on polynomial interpolation), but it is quite possible for spurious maxima and minima to be generated in intermediate areas as artifacts of the methods used (although these may be of small magnitude). There are few comparative studies of the results of different contouring algorithms (Walden, 1972) and no general guidelines have yet emerged about techniques best suited to geochemical data.

The geostatistical Theory of Regionalized Variables (mentioned earlier in the context of sampling) provides in kriging or universal kriging a method yielding an optimal exact linear interpolator (the kriging estimators match the true values at the locations of the data points), which is also unbiased, and optimum in that the estimation variance is minimum (see Huijbregts and Matheron, 1971; Olea, 1975; Guarascio et al., 1976; David, 1977). The weights used in the kriging estimator reflect the spatial autocorrelation within the surface and the distribution of the data points. The estimation variance is a superior measure of probable error in the map compared with the more usual variance of residuals from a fitted surface, or variance of the data values within a cell.

This approach has so far been mainly applied to problems in ore reserve analysis and has yet to be explored as a tool for geochemical mapping.

Because of the noisy nature of much geochemical data (field and analytical variability), moving average methods have long been applied to extraction of the local background variations at broad-scale regional or more detailed scales, density of the data points permitting. A square, or circular, window is passed across the map area and the average value of all data points within it is assigned to the grid node (or, more rarely, to the centroid of the data point locations). Variables include window size, degree of overlap (if any) between successive window positions, and the weighting function used. The resulting map is usually contoured (Nichol et al., 1966; DeGeoffroy et al., 1968; Wolf, 1974, 1975; Hornbrook and Garrett, 1976; see Figs. 24.6 and 24.7). An alternative approach has been to present the map in a cell form, using small square or rectangular cells (the latter corresponding to the lineprinter print positions). This avoids the computational overheads of contour-threading, while still presenting the spatial patterns of variation satisfactorily

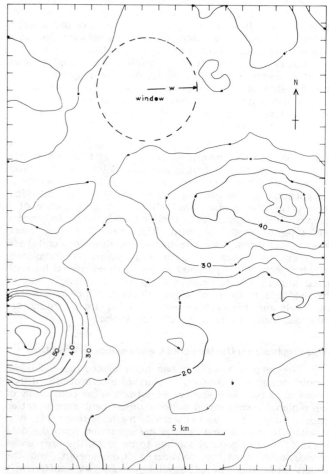

Figure 24.6. Derived regional map: Weighted moving average using 3 km radius window applied to the same original data used to obtain Figure 24.5, 5 ppm contour interval.

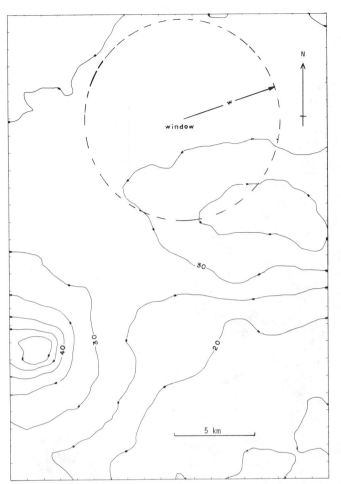

Figure 24.7. Derived regional map: As Figure 24.6 but with 6 km radius window.

(Howarth, 1971a; Howarth and Lowenstein, 1972, 1974; Webb et al., 1973). More recently, the use of recursive line-by-line moving average methods has enabled the processing of data sets in the order of 50 000 samples with minimal use of computer storage (Webb et al., 1968; Fig. 24.8).

For interpretational effectiveness the regional contours and the residual values are often plotted on the same map (see Fig. 24.9). Among the more significant advantages of this method is the dual use of the regional component as a flexible, continuously variable background estimator, and as a direct aid to geologic mapping. The full realization of the potential benefits depends on the choice of a computational technique and parameters for the regional component which optimize the sampling density and best reflect the underlying geological features of the survey area.

While computer-contoured maps can look aesthetically pleasing, too often the geochemist is unaware of the method used to produce them. One should always be prepared to discuss the exact basis of the interpolation and contouring techniques used with the mapmaker. Facilities such as contour annotation, smooth contours, suppression of contours where they bunch too closely together, a variety of line styles, and blanking out of contours in areas of inadequate control are available in some contouring packages, but sophisticated graphics are computationally more expensive to produce.

Alternative methods for enhancing anomalous patterns using mapping techniques have been in experimental use at AGRG, London, for some time and include high-pass, picture frame and Kolmogorov-Smirnov filters (Howarth, 1974). Current work with regional geochemical data from the United Kingdom, Canada and the United States is yielding encouraging results.

Multi-element Geochemical Maps

In many cases the interaction of more than one element may be of relevance to the interpretation of the geochemical map. A simple example would be the mapping of elemental ratios to manganese in stream sediments from areas where manganese coprecipitation is known to take place. Multi-arm 'wind rose' symbols, in which arm length is proportional to concentration and each element has a specific arm orientation, have been used (Rhind et al., 1973; Plant and Rhind, 1974; Experimental Cartography Unit, 1978) for point-symbol maps at a reconnaissance scale. Colour-mixing techniques for three elements at a time (in which each element is represented by either a cyan, yellow or magenta map, with an intensity proportional to concentration; the resulting colour mixtures of the three maps when superimposed show relative compositions) have been successfully used at a regional scale for maps produced by lineprinter (Howarth and Lowenstein, 1976), laser plotter (Lowenstein and Howarth, 1973; Webb et al., 1978), and microfilm (Mancey and Howarth, 1978). A related technique for radiometric data has been described by Lindén and Åkerblom (1976).

These are all essentially cartographic devices which attempt to use graphical, rather than statistical, means to optimize presentation of multi-element data for comprehension by the geochemist. They can be regarded as further aids to interpretation and would generally be thought of as complementary to the original single-element maps.

Statistical methods which have been used in conjunction with multi-element interpretation include multivariate regression and the mapping of factor scores, and are considered later.

UNIVARIATE STATISTICS

It is apparent from information we have obtained from many countries that the majority of geochemical exploration programs encountered in industry tend to be considered on a single-element basis, from the points of view of both data description and interpretation. There are indications that national geological surveys, research institutions and universities in the more advanced countries are applying multivariate techniques, but these have not become widespread in industry. The reasons for this are probably twofold: (1) the lack of routine analysis for, say, 5 to 30 elements means that the problem of interpretation of complex multi-element interrelationships is partly avoided (although the investigator may thereby be missing helpful information); and (2) the lack of adequate computer facilities for geochemists in many countries makes effective use of more than the most elementary statistical techniques impossible. This situation can be expected to change rapidly during the next few years as computers are now becoming very much cheaper, as well as more powerful, in the mini- and micro-computer market. A third problem is the time and personnel involved in preparation of the data for entry to the computer (Lepeltier, 1977) which usually necessitates writing data on coding forms for punching of cards or, more recently, direct key-to-disc or terminal entry systems. Data-logging devices are ideally required for ease of transfer of multi-element data for analytical systems with a throughput of several hundred samples per day.

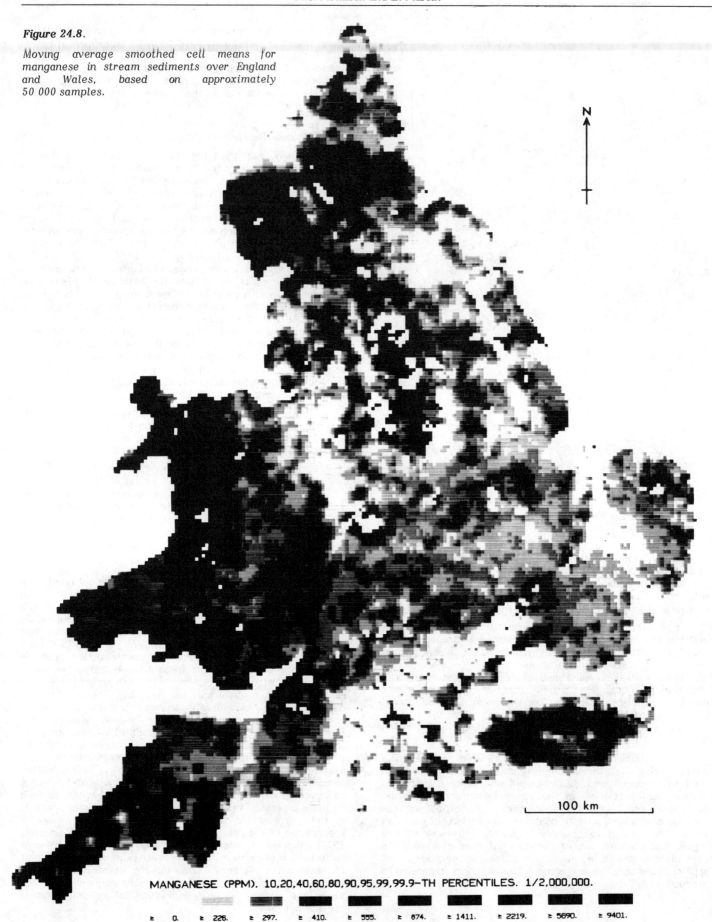

Figure 24.8.
Moving average smoothed cell means for manganese in stream sediments over England and Wales, based on approximately 50 000 samples.

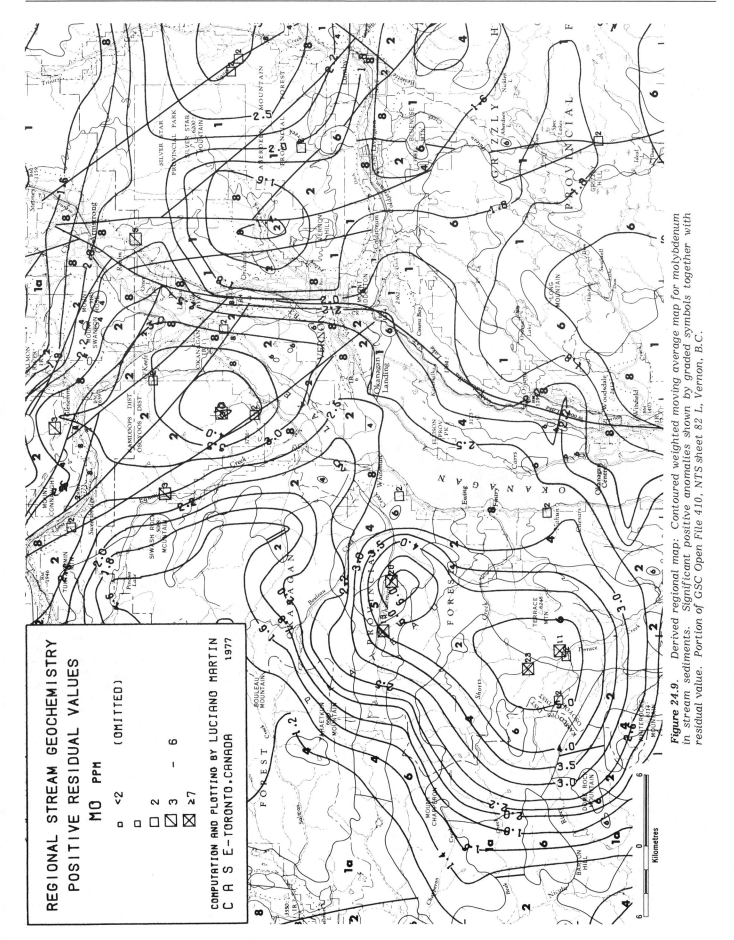

Figure 24.9. Derived regional map: Contoured weighted moving average map for molybdenum in stream sediments. Significant positive anomalies shown by graded symbols together with residual value. Portion of GSC Open File 410, NTS sheet 82 L, Vernon, B.C.

It is arguable that for the interpretation of only a few elements, the computer offers little advantage over a roomful of people collating data manually and plotting it on maps. If one is content with single "one-off" exercises with inherent human error, this is probably a fair statement. However, once the data have been encoded, the computer affords the flexibility to repeat investigations rapidly with changed parameters. This is equally true of the study of a frequency distribution, where choice of class-width can be critical, as of more complex situations such as decomposition of frequency distributions, multivariate statistics, or computer mapping (discussed earlier). The computer however has been used successfully in many projects in exploration geochemistry for the relatively unsophisticated tasks of converting instrumental output to concentrations, and generating histograms prior to the preparation of maps (even if these are plotted by hand). As mentioned earlier, one of the initial time-consuming tasks is the preparation of the data for the computer.

Elementary Statistics

Miesch (1967a, 1967b) reviews the fundamental univariate aspects of geochemical error and estimation of abundance. However, one of the practical problems which can arise in calculation of even elementary descriptive statistics such as the mean and variance, is their accurate calculation without storage of all the values in the computer at one time, and on a machine in which the number of binary "bits" representing the concentration value is relatively small (it can commonly vary from word-lengths of 8 to 60 bits for single precision arithmetic on different computers). Ling (1974) reviewed possible algorithms for the computation of sample means and variances and recommended two pass methods using $M_1 = \Sigma x_i/n$ followed by $M_2 = M_1 + \Sigma(x_i - M_1)/n$ and $S_2 = \Sigma(x_i - M_1)^2 - 1\left[\Sigma(x_i - M_1)\right]^2/n$, where n is the number of samples. The variance is then $S_2/(n - 1)$. For discussion of one-pass algorithms (which have to be chosen to some extent with regard to the data characteristics) the reader is referred to Ling's paper. A similar approach can be used to build up matrices of cross-products in the multivariate case. If equal class widths are used for histogram computation, the frequency counts may be built up without storing all the values of the element in core at one time, and algorithms for cumulative frequency plots on the lineprinter are available (e.g. May, 1972). Burch and Parsons (1976) recently proposed a very simple new test for normality based on use of the cumulative curve.

Threshold Selection

The major interest in univariate statistical analysis has focussed on the problems of anomaly threshold determination and the decomposition of polymodal frequency distributions. Criteria based on a cut-off point of the arithmetic mean plus 2.0 standard deviations (Hawkes and Webb, 1962; Lepeltier, 1969), or less commonly utilizing the "geometric deviation" (calculated as for the standard deviation but differencing with regard to the geometric mean) or empirical criteria, are being abandoned by several institutions in favour of either percentile cut-offs based on the actual frequency distribution (i.e. the concentration values exceeded by known percentages of the data), or the results of frequency distribution splitting (see below). The appropriate percentiles to use will depend not only on the form of the frequency distribution, but also on the size of the data set (e.g. the 95th percentile for 1000 samples would include 50 points, but the equivalent would be 500 points for 10 000 samples, which might be undesirably large in spatial terms in a final map). A percentile-based criterion however does have the advantage of exact comparability from one element to another irrespective of the forms of their frequency distributions. An inconvenience is that the threshold would be different for data from different areas, whereas a purely arbitrary criterion can remain fixed during the lifetime of a continuing project. The penalty of the latter is usually insensitivity to differing regional background values. The interrelationship of threshold and sampling-plus-analytical error has been discussed earlier. More complex geochemical criteria have also been advanced for anomaly recognition (e.g. Dubov, 1973; Cachau-Hereillat, 1975).

Splitting Frequency Distributions

Techniques for the decomposition of mixed frequency distributions are now fairly well known, and a useful survey has been given by Clark (1976). Probably the best known graphical methods in geochemistry are those of Lepeltier (1969) and Sinclair (1974, 1975) which are based on analysis of the cumulative distribution of concentration values. The splitting of the cumulative distribution into straight line segments by Lepeltier (1969) implies the relatively unusual situation of mutually truncated populations whereas the fitting of a smooth-curve, used by Sinclair and others, assumes mutually overlapping populations, which is probably more realistic. Bolviken (1971), Parslow (1974) and Sinclair (1975) illustrate families of smooth cumulative curves which may be encountered with two mixed normal distributions, and McCammon (1977) gives a Fortran subroutine for calculating percentiles of such distributions. The graphical approach is straightforward to apply, but may be somewhat subjective (it can be difficult to pick the inflection point(s) by eye), and may give misleading results for small populations.

Numerical methods are well suited to computer evaluation. Clark and Garnett (1974) described estimation of parameter values by a nonlinear least-squares technique, which minimizes the sum of squared differences between the observed and expected frequencies up to each class end-point, based on the cumulative normal distribution. A Fortran program (Clark, 1977) has been successfully implemented on a mini-computer and is well suited for interactive use; it has given excellent results with minimum chi-square values in comparative goodness-of-fit tests against other methods (Clark, 1976).

The temptation when splitting multi-modal distributions by either graphical or numerical techniques must be to fit a new distribution corresponding to each small irregularity in the cumulative curve. Here one must have regard to the confidence belt around the observed distribution (e.g. fitting bounds on the basis of the Kolmogorov-Smirnov statistic, cf. Stevens, 1974), and it is reasonable to suppose that in the context of a geochemical map the populations might also be geographically distinct (O. Celenk, pers. comm. 1972; I. Clark, pers. comm. 1976-7; Govett et al., 1976). Examples of the decomposition of frequency distributions in exploration geochemistry are cited by Brabec and White, 1971; Lenz, 1972; Bolviken and Sinding-Larsen, 1973; Sinclair, 1974; Montgomery et al., 1975; Sinding-Larsen, 1977; and Lepeltier, 1977, and graphical methods are known to be in wide use among mining companies.

Techniques also exist for dealing with censored frequency distributions (i.e. in which the number of determinations falling below detection limit is known, but their exact values are not), both graphically on the basis of the cumulative curve (Sinclair, 1976) and from the point of view of estimation of mean and standard deviation corrected for the degree of censoring (Cohen, 1959, 1961, also cited in Miesch, 1967b; Selvin, 1976) or the computer method of Hubaux and Smiriga-Snoeck (1964). Such corrections can be of great importance when it is required to compare abundance estimates from two or more data sets in which a significant proportion of censored data occur.

Data Transformation

A problem of continuing interest in the exploration context is how helpful it is to apply logtransformation to "pseudo-lognormal" distributions. Such distributions are commonly positively skewed and have a high kurtosis, thus superficially resembling a lognormal distribution, and are often composed (in exploration geochemical data at least) of aggregates of smaller populations over a large geographical region. Recent papers (Link and Koch, 1975; Govett et al., 1975) and ensuing correspondence (Chapman, 1976b) have suggested that logtransform may not always be appropriate for estimation of the mean, owing to introduction of a (generally small) bias; and that logtransformation will often enhance the contribution of background with deleterious effect on the recognition of high concentration values particularly in a multivariate situation. Miesch (1977) suggests that a small bias in the estimate of the mean may be preferable to a very wrong answer because the method of estimation is inefficient, and also that logtransformation is desirable to reduce dependency of error variance on the magnitudes of the means (a common situation in analytical data, as discussed earlier): this is particularly important for analysis of variance and it enables one to examine proportional differences in the data. It is clear from Chapman's reply (1977) that, for him at least, this controversy hinges on the recognition of anomalous populations and the interpretability of the results gained from logtransformed data. Many would agree with Miesch (1977) that although "there are many situations in geochemistry where a log transformation of the data is inappropriate, ... log transformations are helpful far more often than not".

A more general nonlinear power transform exists for which the lognormal distribution is a special case (Box and Cox, 1964):

$$\left. \begin{array}{l} z = (x^\lambda - 1)/\lambda, \lambda \neq 0 \\ z = \log_e x, \quad \lambda = 0 \end{array} \right\} \quad x > 0$$

The value of λ being chosen to minimize departures of either skewness, or skewness and kurtosis, from that of the normal distribution. Continuing investigations at AGRG, London, on a variety of data sets suggest that this provides a most useful alternative to the usual logtransformation. Initial results are encouraging and a variety of computer methods exist to optimize λ (Howarth and Earle, 1979). This transform is further discussed in the next section.

MULTIVARIATE STATISTICS

In contrast to univariate analysis, in which the computer is often performing functions that could be undertaken manually (although once the geochemical data have been encoded, it is generally much faster and more accurate than the human), multivariate statistical analysis allows us to investigate complex inter-element relationships that would be prohibitively time-consuming to undertake by hand, even if it were humanly possible.

Our recent survey of computer usage among mining companies, national geological surveys, research institutions and universities suggests that multivariate statistics are being little used among exploration geochemists as a whole, but where adequate computer facilities exist regression and correlation analysis are the most common applications; principal components or factor analysis, discriminant analysis, and cluster analysis being relatively little used. Davis (1973) gives a very readable introduction to these techniques.

In most institutions reliance is generally placed on existing statistical packages for general use. Slysz (1973) gave an interesting survey of the contents of various standard packages for multivariate data analysis with comparative time and cost evaluations, and showed some major performance differences among them. Anyone considering which package to use would be well advised to consult his paper, although comparative evaluations of accuracy could have usefully been included. Those which seem to be used most often by exploration geochemists are BMD/BMDP (Dixon, 1975), SPSS (Nie et al., 1975) and, in earlier work, SSP (IBM Corp.). A vast range of special programs are also available (particularly for factor and principal components analysis, discriminant analysis, and cluster analysis).

It is surprising how often, when re-running older "standard" computer programs with the more powerful debugging compilers now available, erroneous code appears to have been released on the unsuspecting! Users are cautioned particularly to ensure that they have reliable algorithms for operations such as matrix inversion and determination of eigenvalues[1] since these can make a significant difference to the result of a calculation. However, programs tailored for the handling of the very large data sets commonly encountered in exploration geochemistry have generally to be developed by the users themselves.

Correlation

Correlation is a measure of the extent to which one element varies either sympathetically or antipathetically with another. The familiar Pearson linear correlation coefficient (Pearson, 1899) is easily calculated (e.g. Davis, 1973) and can be very informative. It was shown in the discussion of univariate statistics how the necessary cross-products of deviations from the mean could be accumulated without having to store a large data-matrix in the computer. However, as was pointed out by Howarth (1973a) and Chapman (1976a), the correlation matrix may be seriously affected by: erroneous or "wild" data values; the presence of more than one geochemical population in the data; very common sample types which may overwhelm the characteristics of less usual samples; scaling of the data, e.g. logtransformation, will affect the correlation coefficients obtained; and testing the significance of the Pearson coefficient requires that the elements are approximately normally distributed. The number of samples needed to reliably estimate the multivariate correlation matrix should preferably be at least 10 times the number of elements involved (Howarth, 1973a; Trochimczyk and Chayes, 1977). Alternatives such as the non-parametric Spearman (1904) and Kendall (1938) rank correlation coefficients are more complex to compute (see, for example, Siegel, 1956) and may be too time-consuming to use in practice (Chapman, 1978).

In many cases a matrix of Pearson correlation coefficients between all element pairs will give much useful information on the structure of inter-element relationships

---

[1] The NAG (NAG Central Office, Oxford OX2 6NN, England) and IMSL (IMSL Inc., Houston, Texas 77036) program libraries are examples of software in which particular attention has been paid to numerical accuracy.

for little computational effort. However, it is usually wise to check abnormally high or low correlations, by making scatter plots of the data, to guard against spurious effects. Chapman (1976a) contains a cautionary example in which the correlation between Cu and Mo in a set of 211 stream-sediment samples was reduced from 0.938 to 0.173 following the removal of one spurious sample, logtransformation of the original data (including the "wild" sample) yielded a correlation value of 0.226. This suggests that comparative evaluation of untransformed and logtransformed data may have its benefits. Rank correlation coefficients of the original data would also have shown lack of correlation.

A problem which still remains unresolved is concerned with determining the significance of correlation coefficients in data which add to a constant sum (e.g. 100 per cent). Such data were termed "closed" by Chayes (1960) who first pointed out that spurious correlations could be introduced into data by the effects of closure (e.g. by the percentaging operation), often showing up as apparent strong negative correlations. Chayes and Kruskal (1966) devised a statistical test to determine whether a given correlation coefficient derived from such data might be nonzero. Miesch (1969) suggested that while their criterion might be used to test the hypothesis that none of the variables in a specific geochemical problem are correlated in excess of the correlation attributed simply to the effects of closure, if this hypothesis is rejected it is still impossible to tell which specific pairs of variables are truly correlated geochemically. Chayes (1970b, 1971) rebutted this, but a recent re-examination of the problem (Kork, 1977) has confirmed that identification of significant correlations is essentially impossible even if the number of variables and sample size are large.

Very often, careful analysis of a correlation matrix will give sufficient information about element associations to form hypotheses, but is not capable of the more sophisticated decomposition possible with principal components or factor analysis techniques.

Regression

Linear regression is generally used, in a predictive sense, to fit an equation of the form:

$$Y = a_0 + a_1 x_1 + a_2 x_2 + \ldots\ldots\ldots a_m x_m$$

where Y is the dependent element and x_1 to x_m are the independent elements (Davis, 1973). For example, it may be required to regress zinc as a function of manganese to assess the effect of possible manganese scavenging. In general, error can only be associated with one variable; usually the X values are assumed error free and the regression line is fitted so that the sum of squared deviations of the Y values from the line is minimized. The situation becomes more complex if both X and Y values are assumed to have independent errors associated with them, weighted regression techniques must be used, or the "reduced major axis" (e.g. Chapman, 1976a) can be fitted. If the errors in X and Y are not independent, correct fitting techniques will be highly complex. For a discussion of the related theory of "functional analysis" the reader is referred to Mark and Church (1977).

Practical applications in exploration geochemistry have almost entirely been concerned with correction of element concentration values for supposed manganese and iron scavenging. In many cases so-called stepwise multiple regression techniques have been used which fit the regression equation only to those elements which contribute significantly (in a statistical sense) to the observed variation in the dependent variable, omitting those which do not from the equation. A useful survey of such methods appears in Berk (1978). Applications illustrating regression on organic content, and/or iron and manganese for stream sediments include: Rose et al., 1970; Rose and Suhr, 1971; Brundin and Nairis, 1972; Austria and Chork, 1976; and Larsson, 1976. It should be noted that the strength of the regression coefficients are not consistent from one study to another and can vary significantly between areas in the same region of similar environment and geology (O. Selinus, pers. comm., 1977). Computation of a global regression equation over a very large area may be unwise for similar reasons to those discussed earlier for correlation coefficients.

Other applications of regression in exploration geochemistry include: DeGeoffroy et al., 1968; Dahlberg, 1969; DeGeoffroy and Wignall, 1970; Rose et al., 1970; Hesp, 1971; Chatupta and Fletcher, 1972; Dall'Aglio and Giggli, 1972; Pelet and de Jekhowsky, 1972; Timperley et al., 1972; Dall'Aglio and Tonani, 1973; Smith and Webber, 1973; Culbert, 1976; and Rose and Keith, 1976. In some cases nonlinear functions may be more appropriate. Cameron and Hornbrook (1976) illustrate a spectacular quadratic relationship between mean U and L.O.I. for lake sediments in the Canadian Shield.

Factor Analysis and Related Techniques

"Factor" analysis was one of the first multivariate analytical methods to become widely used among geologists and geochemists. While it has been helpful in many cases, misrepresentation of its capabilities in the early years probably served more to bring computer applications into disrepute in exploration geochemistry than to promote the power of the computer as an aid to problem solving.

For those who require an in-depth survey of the subject, the recent textbook by Jöreskog et al. (1976) is set in a geological context, and a straightforward introduction will be found in Davis (1973). The intention underlying the use of principal components or factor analysis in exploration geochemistry has been generally to separate the element associations inherent in the structure of the correlation matrix into a number of groups of elements which together account for the greater part of the observed variability of the original data. The intention being to represent the large number of elements in the original data by a smaller number of "factors", each of which is a linear function (transformation) of the element concentrations, thus giving a greater efficiency in terms of information compression over the original data, and hopefully also gaining something in interpretability. Techniques taking the correlation (or covariance) matrix as a starting point are generally referred to as R-mode.

In principal components analysis (PCA), the interelement covariance matrix for m elements is decomposed into an orthogonal (uncorrelated) set of new variables which are linear combinations of the original element values. The eigenvectors and related eigenvalues are extracted in such a way that the first eigenvector (which may be regarded as the coefficients of the first linear transform equation) accounts for as much as possible of the total variability of the data, the second for as much as possible of the residual variance, and so on; m eigenvectors are needed to account for all of the variability of the original data. The eigenvectors will not necessarily be of unit length when based on the covariance matrix. However, since in many cases the measurement scales are different, it is common practice to treat the data as standardized (weighting all the variables equally in terms of measurement scale by subtracting the mean from each observation and dividing by the standard deviation). The covariance matrix calculated from standardized data is identical with the correlation matrix, and in the majority of geological PCA studies the correlation matrix is used as the basis. In order to ensure reliable estimation of the multivariate structure of the data, the correlation matrix should be based on a data set with preferably at least 10 times as many samples as variables (Howarth, 1973a; Glasby et al., 1974; and Trochimczyk and Chayes, 1977).

In factor analysis, the correlations among the m elements are assumed to be accountable by a model in which the variance for any element is distributed between a number of "common" factors (with at least two elements significantly associated with it), and a remaining number of "unique" factors (with only one significantly associated element). The remaining variance being accounted for by an error term. The correlation matrix is the starting point and the resulting (normalized) eigenvectors are of equal unit length. It is hypothesized that the number of common factors (k) is much less than m, and in the psychological studies from which the factor analysis approach originated, k was generally postulated a priori. A subjective element enters factor analysis at this stage. There are no statistically based tests which are universally agreed to define the significant number of common factors. The usual criteria are: a certain cumulative percentage of the overall variance is accounted for by k factors; those factors with corresponding eigenvalues exceeding 1.0 are considered to be "significant"; a plot of eigenvalues versus factor number shows a "significant" break in slope (Cattell's scree test; Cattell, 1966); factors associated "significantly" with more than one element are retained; factors which make sense from the point of view of geochemical interpretability are retained. In order to obtain more interpretable results, the original orthogonal linear transform may be rotated in a geometrical sense (although it is performed algebraically) until the coefficients for the elements are maximized in contrast, that is to say high ones tend to ±1.0 and low ones to zero, in order to obtain a "simple structure". The varimax rotation method (Kaiser, 1959) is most commonly used although others have also been investigated (e.g. Chapman, 1978). Some investigators have also dispensed with representing the data variability in terms of a set of uncorrelated transformed variables, preferring so-called "oblique" solutions in which the transformed variables are themselves correlated (usually obtained by promax rotation, e.g. Chapman, 1978). There is some disagreement among investigators whether this is justifiable in geochemical terms, quite apart from the difference it may make to the interpretability of the data.

The end-product of the transform equations, when element values for the actual samples have been substituted in, are generally known as the "factor scores". Trochimczyk and Chayes (1978) recommend that if the components have been obtained from the covariance matrix, scores should be based on unstandardized data; and if components have been based on the correlation matrix, then the scores should be based on the standardized element values to ensure that the resultant score values are uncorrelated.

Thus we see that PCA gives a clearly based mathematical solution, and in terms of data compression, a set of PCA linear transform equations truncated at the k-th eigenvector is well defined. In contrast, the various factor analytic approaches are more subjective in choice of possible methods, and in difficulties of deciding how many factors should be retained (unless there is, unusually, clear a priori evidence about how many common factors might be expected to operate on the data) since the aim is to determine the best estimate of the number of common factors operating on the observed variates whether it be large or small. It should be understood by the geochemist that the interpretation of factors as the underlying causes giving rise to the observed element behaviour is not implicit in the model. Separation into "background" and "mineralization" element groups may occur in some cases, but factor analysis is not in itself capable of sorting such associations out of the data in all cases. The appearance of such groups merely reflects the structure of the covariance (correlation) matrix which is subject to the same guidelines as those discussed earlier under correlation and regression. Occasionally rank correlation coefficients have been used in the initial matrix (R.P. Chapman, N.J. Marshall, pers. comms., 1975, 1976), although the use of a linear transform based on an essentially nonlinear coefficient poses some theoretical problems. Applications of R-mode PCA or factor analysis in exploration geochemistry include: Garrett and Nichol, 1969; Saager and Esselaar, 1969; Nairis, 1971; Nichol, 1971; Wennervita et al., 1971; Brundin and Nairis, 1972; Rose, 1972; Summerhayes, 1972; Garrett, 1973b; Obial and James, 1973; Smith and Webber, 1973; Glasby et al., 1974; Saager and Sinclair, 1974; Wennervirta and Papunen, 1974; Shikawa et al., 1975; Ek and Elmlid, 1976; Chapman, 1978; and Mancey and Howarth, 1978.

A related technique, known as Q-mode factor analysis (Davis, 1973) attempts to find samples of extreme compositional type, and to represent all other samples in terms of mixtures of these "end-members", beginning with a matrix of intersample (cosine theta) similarity coefficients which is decomposed by factor analytic methods. In general, the results seem to have been more easily interpretable in petrologically oriented cases where the concept of end-member mixtures has genuine significance, than in studies based on stream-sediment data. Cluster analysis methods rather than Q-mode factor analysis, are now tending to be used for problems where it is required to assign the samples to similar groups. Examples of Q-mode factor analysis in exploration geochemistry include: Nichol and Webb, 1967; Garrett and Nichol, 1969; Armour-Brown and Nichol, 1970; Dawson and Sinclair, 1974; Saager and Sinclair, 1974; Chapman, 1975, 1978; and Culbert, 1976.

Two related techniques based upon the R-mode approach, but which allow both samples and element associations to be shown on the same diagram as an aid to interpretation, have recently been applied in exploration geochemistry. These are correspondence analysis (David and Dagbert, 1975; Teil, 1975) and the eigenvector biplot display (Sinding-Larsen, 1975); full details will be found in the references cited. These techniques will certainly find application to exploration geochemistry in the future, in accord with the wider usage of the R-mode techniques.

The effect of data transformation on the structure of the correlation matrix is illustrated by preliminary results from a study of 895 stream sediment samples from a regional geochemical survey of Denbighshire, UK, originally carried out by Nichol et al. (1970). PCA was performed on the basis of standardized results for Co, Cr, Cu, Fe_2O_3, Ga, Mo, Mn, Ni, Pb, Ti, V and Zn (all ppm except Fe_2O_3, per cent) from untransformed data, logtransformed data, and data subjected to the nonlinear Box and Cox (1964) transform (discussed earlier in the section on Univariate Statistics). The data include contributions from various background lithologies as well as minor areas of mineralization and contamination in the vicinity of smelting and mining operations.

It is apparent from Figure 24.10 that the effect of a global logtransformation is to induce negative skewness for all of the elements, although this is not as severe as the original positive skewness; kurtosis is also reduced. For some elements (Ga, Ti and V) which were approximately normally distributed originally, the effect of logtransformation has been particularly unhelpful. The Box-Cox transform has improved the approach to a normal distribution in all cases.

Comparison of the eigenvalues (Table 24.1) and unrotated PCA loadings (Fig. 24.11) for the three solutions indicates that the effect of the logtransform could (in this example) be to induce rather high positive correlations

Table 24.1

Eigenvalues (E.V.) and cumulative variance (Cum. Var.) for 895 stream sediment samples from regional geochemical survey of Denbighshire (U.K.) showing effects of nonlinear data transformation on correlation matrix

	Untransformed		Logtransformed		Box-Cox (1964)*	
	E.V.	Cum. Var. %	E.V.	Cum. Var. %	E.V.	Cum. Var. %
1	4.067	33.9	6.125	51.0	5.374	44.8
2	1.779	48.7	1.940	67.2	2.132	62.5
3	1.589	61.9	0.895	74.7	1.024	71.1
4	1.258	72.4	0.706	80.6	0.748	77.3
5	0.907	80.0	0.586	85.4	0.633	85.6
6	0.591	84.9	0.418	88.9	0.462	86.4
7	0.541	89.4	0.376	92.1	0.443	90.1
8	0.374	92.5	0.301	94.6	0.360	93.1
9	0.292	95.0	0.225	96.4	0.275	95.4
10	0.222	96.8	0.194	98.1	0.224	97.3
11	0.206	98.5	0.124	99.1	0.189	98.9
12	0.174	100.0	0.109	100.0	0.135	100.0

*$(x^\lambda - 1)/\lambda$

between some of the element pairs. Following the Box-Cox transform the eigenvectors and unrotated PCA loadings (Fig. 24.11) show a tendency to have the high coefficients move towards ±1.0, and the low coefficients towards zero compared with the logtransform, although the general structure in terms of element associations remains the same. This suggests that alternatives to the logtransform could be very helpful in increasing the interpretability of multivariate data analysis in exploration geochemistry. A wider investigation is currently in progress at AGRG and is showing interesting results on a number of data sets.

The application of PCA-based factor analysis to a large regional data set, based on 21 432 moving average cell means for 22 elements from the Wolfson Geochemical Atlas of England and Wales (Webb et al., 1978) covering approximately 50 000 mi^2, has shown in preliminary results that although known areas of mineralization form a very small proportion of the whole, relevant mineralization associations can be identified despite expected difficulties associated with the size of the data set and multiplicity of background types associated with regional lithological changes. The initial factor score maps have been colour-combined to further increase their information content (Mancey and Howarth, 1978), and are providing much useful information.

Cluster Analysis

The aim of cluster analysis techniques is to group the samples together without a priori knowledge, on the basis of similarity in terms of their elemental compositions, in order to separate areas of similar background composition, and identify samples of unusual composition. Generally the Euclidean distance measure is used:

$$d_{ij} = \left[\sum_{q=1}^{m} (x_{iq} - x_{jq})^2 \right]^{1/2}$$

where x_{iq} and x_{jq} are the q-th measurements on the i-th and j-th samples respectively, q=1, ...m. To increase computational speed the related metric:

$$d_{ij} = \sum_{q=1}^{m} |x_{iq} - x_{jq}|$$

known as the City block or "Manhattan" distance, may also be used.

PCA is often used as a suitable decorrelation preprocessor to cluster analysis, to transform the original data into a set in which the new variables are genuinely uncorrelated (and hence orthogonal for the calculation of d). It should be remembered that if the last few components are not retained in this case, then information characterizing the less common groups of samples may have been sacrificed, and it is probably safer to retain all components resulting from the PCA transform initially.

A detailed review of the numerous possible methods is not appropriate here (see, for example, Everitt, 1974, or Duda and Hart, 1973). It is sufficient to point out that the applicability of many techniques to very large sets of geochemical data is limited by the necessity to keep track of all possible pairwise distances within the data set, which will soon build up a computationally impossible number of items. In order to overcome this difficulty samples are either: 1) grouped into classes if their composition is within a threshold distance of an existing class centre, represented by its mean composition; or 2) a random subset of the original data is clustered using a preferred method and these clusters are subsequently taken as training information for discriminant analysis (discussed below) which classifies all the remaining samples in to one of these groups. Either method is reasonably fast to implement and can be tailored to existing computer requirements. Techniques of the former type generally aim to obtain compact classes, in the sense that between-group variance exceeds within-group variance and it is possible in some methods (e.g. ISODATA, Ball and Hall, 1966) to split or merge the clusters as the calculation

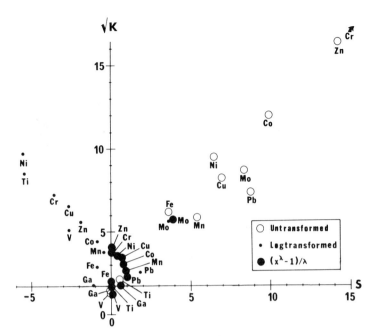

Figure 24.10. Effect of data transformations on symmetry (skewness, S) and peakedness (kurtosis, K) of frequency distributions for 12 elements in 895 stream sediment samples from a regional survey of Denbighshire, England. Origin of the axes placed at expected values for these parameters for a Normal distribution; note that ordinate is scaled as $\sqrt{K}$.

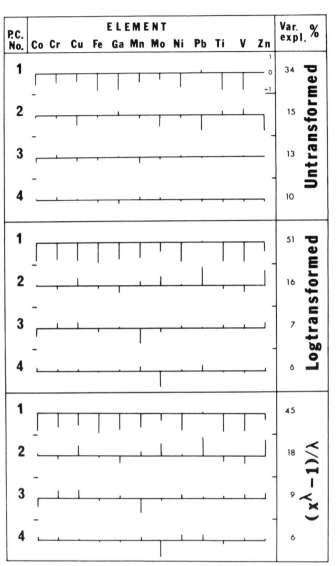

Figure 24.11. Effect of data transformations on the loadings of the first four unrotated principal components for the data of Figure 24.10. Line length indicates strength of loading (between -1 and +1); percentage of overall variance accounted for by each component also shown.

proceeds in order to optimize the clustering criterion. While they may be somewhat artificial as regards "naturalness" of the resultant classification they have the advantage of speed. More complex techniques in current use include the tree (dendrogram) methods (Obial, 1970; Lenthall, 1972; Hesp, 1973; Hesp and Rigby, 1973; Howarth, 1973a; Obial and James, 1973; and Glasby et al., 1974) yielding a hierarchical structure which may be better suited to interpretational requirements, or attempt to obtain a completely natural classification based on optimal two-dimensional representation of the multi-dimensional data structure by methods of the nonlinear mapping type (Garrett, 1973b; Howarth, 1973a, b). Sinding-Larsen (1975) implemented a technique similar to ISODATA following a PCA type of transform, and Duval (1976) used a distance threshold grouping in combination with Q-mode factor analysis.

Detailed studies based on approximately 900 regional stream sediment samples from Denbighshire, UK, on the basis of 12 elements, using hierarchical cluster analysis and nonlinear mapping (Castillo-Muñoz, 1973) and ISODATA (Crisp, 1974), suggest that scaling effects can be important in all three techniques. "Anomalous" samples of unusual composition could be detected by all the methods used and in the untransformed data these were generally clearly identified. In general however, the groups based on logtransformed data were comparable for the three methods and gave more interpretable results. ISODATA was very much the fastest, taking approximately 300 seconds on the CDC 6400 central processor to process the entire data set. Nonlinear mapping in contrast took approximately 1200 seconds for 500 samples.

These results suggest that methods based on group mean compositions are more suitable for processing large sets of exploration data where rapid division in to broad groups is required, although a number of rather critical parameters must be empirically set by the user to control the clustering process. The group mean compositions could always be reclustered by another method if it is required to verify the nature of the groupings. It must be borne in mind that with any clustering method there is no well-defined criterion for the number of clusters present, and the final interpretation is somewhat subjective.

Discriminant Analysis

In contrast to cluster analysis, discriminant analysis aims to classify samples of unknown type into one of a number of pre-defined groups (e.g. mineralized or barren) on the basis of their elemental compositions. In all methods, irrespective of the nature of the decision rule (discriminant function) used, the starting point is to have an established "training set" of samples representing each group. This should be as large as possible for each class, to ensure adequate statistical representivity (the rule of thumb for the number of samples per class of at least 10 times the number of variables also applies here for most reliable performance, although as few as 3 times the number of variables is adequate if training samples are hard to obtain). Ideally, once the decision criteria have been determined, an independent "test set" of samples should be classified on the basis of the established decision rules to find out how well they perform before applying them to known data. A detailed discussion will be found in Howarth (1973a), in a geochemical context, or Duda and Hart (1973).

Although extensively used in other branches of the earth sciences and in igneous petrology, discriminant functions have not been widely used in exploration geochemical problems to date. The reason for this is almost certainly the practical difficulties associated with obtaining adequate data for formulating the training sets. The majority of applications have used Fisher (1936) linear discriminant functions (Wignall, 1969, 1970; Dahlberg, 1970; Cameron et al., 1971; Govett and Pantazis, 1971; Lenthall, 1972; Rose, 1972; Govett, 1973; Pantazis and Govett, 1973; Shiikawa et al., 1974; Whitehead and Govett, 1974; Chapman, 1975, 1978; Govett et al., 1975; and Culbert, 1976) and because of the inherent simplicity of the method (Davis, 1973), and availability of suitable programs in packages such as BMDP (Dixon, 1975), and SPSS (Nie et al., 1975), it continues to be widely used in industry. An alternative method, which does not depend on the multivariate normal assumptions inherent in the Fisher linear discriminant is the "empirical discriminant function" (Howarth, 1973a, c) which is a form of potential function (Duda and Hart, 1973). This has proven very successful in regional geochemical survey classification for multi-class applications (Howarth, 1971b, 1972, 1973a; Castillo-Muñoz and Howarth, 1976) and is being used increasingly for exploration problems (Gustavsson and Bjorklund, 1976; C.Y. Chork, pers. comm., 1976; N. Gustavsson, pers. comm., 1977). Occasionally discrimination problems can be solved graphically by a suitable choice of elements (Govett, 1972; Joyce and Clema, 1974; Bull and Mazzuchelli, 1975).

It is known that there is an increasing interest among mining companies in techniques for selecting the "best" subset of elements from a set of multi-element analyses to distinguish between mineralized and unmineralized bedrock, with the aim of obtaining reliable discriminant functions. A variety of methods are available for this type of problem (Howarth, 1973a, c; Duda and Hart, 1973). Once the discriminant functions have been evaluated subsequent classification is extremely rapid, even when several thousand samples are screened. It is for this reason that combination of discriminant analysis following cluster analysis of a subset of a large data set (the groups of samples obtained becoming the training set for the discriminant function) is a very powerful technique for dealing with a large data set.

A somewhat related technique, known as Characteristic Analysis, has been developed by Botbol et al. (1978) as an extension of a method originally devised by him for the study of mineral assemblages. Geochemical and other information, such as geology or geophysical survey data, is expressed in binary (0/1) form. A value 1 indicates favourability for exploration in a local spatial context, and is defined on the basis that the measured value for a variable in a map cell is higher than its surrounding neighbours. Thus second derivatives of geological, geochemical or geophysical parameters can be treated in a similar manner. Training sets are established for known mineralized areas and comparison of the characteristics of the matrix of joint "favourable" occurrences for the training set with the coded binary data for the unknown cells allows their similarity to the training set to be determined, thus establishing the regional classification. Results from the Coeur d'Alene district, Idaho (Botbol et al., 1978) are very encouraging.

Discussion

The multivariate techniques described in this section are complex and are best performed on a digital computer because of necessity for speed and accuracy, irrespective of the problems associated with the large data sets produced in exploration geochemistry. The majority of the techniques can generally be programmed for storage requirements of the order of $m^2/2$ for m elements, but (as discussed above) cluster analysis methods generally require storage of the order of $n^2/2$ intersample similarity values for n samples and are therefore more difficult to carry out without very large computer facilities, unless methods storing only the group mean compositions are used.

However, taken as a whole, the multivariate approach offers an immensely more powerful means of data analysis than univariate techniques because of the ability to study the interaction of the various elements. It is often extremely difficult to reliably predict the joint behaviour of a number of elements from a study of their univariate characteristics

As instrumental facilities increase the ease with which multi-element analyses become routinely available at low cost, investigators will be forced to turn to multivariate techniques in order to make optimum use of the information at their disposal. The published case histories cited in this survey show that in most instances multivariate analysis provided greater insight into the geochemical behaviour of the multi-element survey. The basic dilemma for the exploration geochemist remains as discussed by Lepeltier (1977), that the application of powerful computer methods to the data resulting from exploration projects often awaits collation of the total data set, and may then come too late to aid the decision-making (the time of which is often dictated by political or economic factors outside the geochemist's control). In the long term, however, a better understanding of the geochemical relationships governing observed geochemical behaviour, the discovery of common element associations, and interrelationship with geophysical data (e.g. Culbert, 1976) etc. must be beneficial to prediction of criteria for the likelihood of occurrence of a geochemical target.

At the present state-of-the-art of geochemical data interpretation it is probably true that what we need is not a widespread development of new methods of analysis, but adaption of existing techniques to the small-computer environment and the modification of methods for the treatment of large data sets. It is also highly desirable that standard data sets be made available on which comparative evaluations of existing methods may be made (the basis for this is currently being established by the Centre de Recherche Pétrographique et Géochimique, Nancy). Release of "old" exploration geochemical data by mining companies for research purposes, once confidentiality restrictions can be lifted, would be very helpful for this type of study. A growing impetus will probably come from national geological surveys, since the establishment of multi-purpose multi-element regional geochemical surveys is now being seen as increasingly desirable in the context of resource evaluation and environmental monitoring. It may be confidently predicted that with the gradual spread of multi-element analysis and availability of powerful microprocessors, together with an increasingly numerate education of geologists and geochemists, the routine application of multivariate analysis in exploration as an aid to general understanding of geochemical behaviour will become established.

Perhaps with the spread of minicomputers and the ability to transmit data from remote sites to a central computer by short-wave radio (Botbol, 1975, 1977), integration of field operations with computer processing for short-term turn-around of results will become easier. It is unfortunate that in many cases the power and flexibility of the computer for mapping or statistical analysis are not being utilized until it is too late.

ACKNOWLEDGMENTS

We would like to thank the many individuals and organizations who responded to our requests for information on current practice in the application of computers for exploration geochemical data processing. Our thanks also go to the referees for their helpful comments on the manuscript, and for encouragement to track down some references from the early history of statistics. Miss J. Nielsen and Mrs. J. O'Donnell are also thanked for typing versions of this manuscript.

SELECTED BIBLIOGRAPHY

Agterberg, F.P. and Chung, C.F.
   1975: A computer program for polynomial trend surface analysis; Geol. Surv. Can., Paper 75-21, 51 p.

Allcott, G.H. and Lakin, H.W.
   1975: The homogeneity of six geochemical exploration reference samples; in Geochemical Exploration 1974, I.L. Elliot and W.K. Fletcher (ed.), Elsevier Publ. Co., Amsterdam, p. 659-681.

Alldredge, J.R. and Alldredge, N.G.
   1978: Geostatistics: a bibliography; Int. Stat. Rev., v. 46, p. 77-88.

Armour-Brown, A. and Nichol, I.
   1970: Regional geochemical reconnaissance and the location of metallogenic provinces; Econ. Geol., v. 65, p. 312-330.

Austria, V. and Chork, C.Y.
   1976: A study of the application of regression analysis for trace element data from stream sediment in New Brunswick; J. Geochem. Explor., v. 6, p. 211-232.

Ball, G.H. and Hall, D.J.
   1966: ISODATA, an interactive method of multivariate data analysis and pattern classification; IEEE Int. Communications Conf., p. 116-117.

Barringer, A.R.
   1977: A multi-disciplinary approach to airborne mineral exploration; Unpub. ms. (presented at Univ. California, Berkeley, 22 April 1977), Barringer Research Ltd., Rexdale, 64 p.

Baucom, E.I., Price, V., and Ferguson, R.B.
   1977: Savannah River Laboratory hydrogeochemical and stream sediment reconnaissance. Preliminary raw data release Winston-Salem 1° x 2° NTMS area. North Carolina, Virginia, Tennessee; E.I. DuPont de Nemours Co., Savannah River Laboratory, Aiken, S. Carolina, DPST-77-146-1/GJBX-66(77), 98 p.

Bendix Field Engineering Corporation
   1976: Annual NURE Report 1976; Bendix Field Engineering Corporation, Grand Junction, Colorado, Rept. GJBX-11(77), 75 p.

   1977: Symposium on hydrogeochemical and stream-sediment reconnaissance for uranium in the United States; Bendix Field Engineering Corporation, Grand Junction, Colorado, Rept. GJBX-77(77), 468 p.

   1978: NURE 1977. National Uranium Resource Evaluation annual activity report, May 1978; Bendix Field Engineering Corporation, Grand Junction, Colorado, Rept. GJBX-11(78)R, 91 p.

Berk, K.N.
   1978: Comparing subset regression procedure; Technometrics, v. 20, p. 1-6.

Bolviken, B.
   1971: A statistical approach to the problem of interpretation in geochemical prospecting; Can. Inst. Min. Met., Spec. Vol. 11, p. 564-567.

Bolviken, B. and Sinding-Larsen, R.
   1973: Total error and other criteria in the interpretation of stream-sediment data; in Geochemical Exploration 1972, M.J. Jones (ed.), Inst. Min. Met., London, p. 285-295.

Bosman, E.R., Eckhart, D., and Kubik, K.
   1971: Application of automatic data processing to problems of exploration geochemistry; Geol. Mijnb., v. 50, p. 768-770.

Botbol, J.M.
   1975: Field site data processing. A high frequency radio communication link between field camp and computer; Am. Inst. Min. Eng., Preprint 75-L-18, Annu. Mtg., New York, Feb. 1975, 16 p.

   1977: The use of the storage cathode ray tube (SCRT) in geochemical exploration; Sci. Terre, Ser. Inform. No. 9, p. 93-103.

Botbol, J.M., Sinding-Larsen, R., McCammon, R.B., and Gott, G.B.
   1978: A regionalized multivariate approach to target selection in geochemical exploration; Econ. Geol., v. 73, p. 534-546.

Bowen, R.W. and Botbol, J.M.
   1975: The Geologic Retrieval and Synopsis Program (GRASP); U.S. Geol. Surv., Prof. Paper 966, 87 p.

Box, G.E.P. and Cox, D.R.
   1964: An analysis of transformations; J. Roy. Statist. Soc., Ser. B, v. 26, p. 211-243.

Brabec, D. and White, W.H.
   1971: Distribution of copper and zinc in rocks of the Guichon Creek batholith, British Columbia; Can. Inst. Min. Met., Spec. Vol. 11, p. 291-297.

Brundin, N.H. and Nairis, B.
   1972: Alternative sample types in regional geochemical prospecting; J. Geochem. Explor., v. 1, p. 7-46.

Bull, A.J. and Mazzucchelli, R.H.
   1975: Application of discriminant analysis to the geochemical evaluation of gossans; in Geochemical Exploration 1974, I.L. Elliott and W.K. Fletcher (ed.), Elsevier Publ. Co., Amsterdam, p. 219-226.

Burch, C.R. and Parsons, I.T.
   1976: "Squeeze" significance tests; Appl. Stat., v. 25, p. 287-291.

Cachau-Hereillat, F.
1975: Towards quantitative utilization of geochemical exploration: The threshold and its estimation in soil surveys; in Geochemical Exploration 1974; I.L. Elliott and W.K. Fletcher (ed.), Elsevier Publ. Co., Amsterdam, p. 183-189.

Cameron, E.M.
1968: A geochemical profile of the Swan Hills Reef; Can. J. Earth Sci., v. 5, p. 287-309.

Cameron, E.M. and Hornbrook, E.H.W.
1976: Current approaches to geochemical reconnaissance for uranium in the Canadian Shield; in Exploration for Uranium Ore Deposits, IAEA, Vienna, p. 241-266.

Cameron, E.M., Siddeley, G., and Durham, C.C.
1971: Distribution of ore elements in rocks for evaluating ore potential: Nickel, copper, cobalt and sulphur in ultramafic rocks of the Canadian Shield; Can. Inst. Min. Met., Spec. Vol. 11, p. 298-314.

Castillo-Muñoz, R.
1973: Application of discriminant and cluster analysis to regional geochemical surveys; Unpubl. Ph.D. thesis, London University, 258 p.

Castillo-Muñoz, R. and Howarth, R.J.
1976: Application of the empirical discriminant function to regional geochemical data from the United Kingdom; Bull. Geol. Soc. Amer., v. 87, p. 1567-1581.

Cattell, R.B.
1966: Handbook of multivariate experimental psychology; Rand McNally, Chicago, p. 174-243.

Chapman, R.P.
1975: Data processing requirements and visual representation for stream sediment exploration geochemical surveys; J. Geochem. Explor., v. 4, p. 409-423.

1976a: Limitations of correlation and regression analysis in geochemical exploration; Inst. Min. Met. Trans., London, Sect. B, v. 85, p. 279-283.

1976b: Some consequences of applying lognormal theory to pseudolognormal distributions; Math. Geol., v. 8, p. 209-214.

1977: Logtransformations in exploration geochemistry; Math. Geol., v. 9, p. 194-198.

1978: Evaluation of some statistical methods of interpreting geochemical drainage data from New Brunswick; Math. Geol., v. 10, p. 195-224.

Chatupa, J. and Fletcher, K.
1972: Application of regression analysis to the study of background variations in trace metal content of stream sediments; Econ. Geol., v. 67, p. 978-980.

Chayes, F.
1960: On correlation between variables of constant sum; J. Geophys. Res., v. 65, p. 4185-4193.

1970a: On deciding whether trend surfaces of progressively higher order are meaningful; Bull. Geol. Soc. Amer., v. 81, p. 1273-1278.

1970b: Effect of a single nonzero open covariance on the simple closure test; in Geostatistics, D.F. Merriam (ed.), Plenum Press, New York, p. 11-22.

1971: Ratio Correlation; Univ. Chicago Press, Chicago, 99 p.

Chayes, F. and Kruskal, W.
1966: An approximate statistical test for correlations between proportions; J. Geol., v. 74, p. 692-702.

Chork, C.Y.
1977: Seasonal, sampling and analytical variations in stream sediment surveys; J. Geochem. Explor., v. 7, p. 31-47.

Clark, I.
1977: ROKE, a computer program for non-linear least-squares decomposition of mixtures of distributions; Comput. Geosci., v. 3, p. 245-256.

Clark, I. and Garnett, R.H.T.
1974: Identification of multiple mineralization phases by statistical methods; Inst. Min. Met. Trans., London, Sec. A, v. 83, p. 43-52.

Clark, M.W.
1976: Some methods for statistical analysis of multimodal distributions and their application to grain-size data; Math. Geol., v. 8, p. 267-282.

Cockburn, G.H.
1977: Atlas Géochimique des Sédiments du Ruisseau La Grande Rivière; Min. Rich. Nat. Quebec, 500 p.

Cohen, A.C.
1959: Simplified estimators for the normal distribution when samples are singly censored or truncated; Technometrics, v. 1, p. 217-237.

1961: Tables for maximum likelihood estimates: Singly truncated and singly censored samples; Technometrics, v. 3, p. 535-541.

Connor, J.J., Tidball, R.R., Erdman, J.A., Ebens, R.J., and Felder, G.L.
1974: Geochemical survey of the Western Coal Regions; U.S. Geol. Surv., Open File Rep. 74-250, 38 p.

Connor, J.J. and Miesch, A.T.
1964: Analysis of geochemical prospecting data from the Rocky Range, Beaver County, Utah; U.S. Geol. Surv., Prof. Paper 475-D, p. 79-83.

Crisp, D.A.
1974: Application of multivariate methods to geochemistry: The evaluation of a new technique; Unpub. M.Sc. thesis, London Univ., 113 p.

Croissant, A.
1977: La géostatistique comme outil dans la prospection géochimique; Sci. Terre, Ser. Inform. No. 9, p. 129-144.

Culbert, R.
1976: A multivariate approach to mineral exploration; Can. Inst. Min. Met. Bull., v. 69, p. 39-52.

Dahlberg, E.C.
1969: Use of model for relating geochemical prospecting data to geologic attributes of a region, South Mountain, Pennsylvania; Quart. J. Colo. Sch. Mines, v. 64, p. 195-216.

1970: Generalized Bayesian classification: K classes; Discussion; Econ. Geol., v. 65, p. 220-222.

1971: Algorithmic development of a geochemical exploration program; Can. Inst. Min. Met., Spec. Vol. 11, p. 577-580.

Dall'Aglio, M.
1971: Comparison between hydrogeochemical and stream-sediment methods in prospecting for mercury; Can. Inst. Min. Met., Spec. Vol. 11, p. 126-131.

Dall'Aglio, M. and Gigqli, C.
1972: Storage and automatic processing of hydrogeochemical data; Proc. 24th Inter. Geol. Congr., Montreal, v. 16, p. 49-57.

Dall'Aglio, M. and Tonani, F.
1973: Hydrogeochemical exploration for sulphide deposits: Correlation between sulphate and other constituents; in Geochemical Exploration 1972, M.J. Jones (ed.), Inst. Min. Met., London, p. 305-314.

David, M.
1977: Geostatistical ore reserve estimation; Elsevier Publ. Co., New York, 364 p.

David, M. and Dagbert, M.
1975: Lakeview revisited: variograms and correspondence analysis — New tools for the understanding of geochemical data; in Geochemical Exploration 1974, I.L. Elliott and W.K. Fletcher (ed.), Elsevier Publ. Co., Amsterdam, p. 163-181.

Davidson, A. and Moore, J.M.
1978: Omo River Project data management system: an appraisal; Comput. Geosci., v. 4, p. 101-113.

Davis, J.C.
1973: Statistics and data analysis in geology; Wiley, New York, 550 p.

Dawson, K.M. and Sinclair, A.J.
1974: Factor analysis of minor element data for pyrites, Endako molybdenum mine, British Columbia, Canada; Econ. Geol., v. 69, p. 404-411.

DeGeoffroy, J. and Wignall, T.K.
1970: Statistical decision in regional exploration: Application of regression and Bayesian classification analysis in the southwest Wisconsin zinc area; Econ. Geol., v. 65, p. 769-777.

DeGeoffroy, J. and Wu, S.M.
1970: Design of a sampling plan for regional geochemical surveys; Econ. Geol., v. 65, p. 340-347.

DeGeoffroy, J., Wu, S.M., and Heins, R.W.
1968: Selection of drilling targets from geochemical data in the southwest Wisconsin zinc area; Econ. Geol., v. 63, p. 787-795.

Dickinson, G.C.
1973: Statistical mapping and the presentation of statistics; (2nd ed.), Edward Arnold, London, 195 p.

Dijkstra, S.
1976: Simple uses of covariograms in geology; Geol. Mijnb., v. 55, p. 105-109.

Dijkstra, S. and Kubik, K.
1975: Autocorrelation studies in the analysis of stream sediment data; in Geochemical Exploration 1974, I.L. Elliott and W.K. Fletcher (ed.), Elsevier Publ. Co., Amsterdam, p. 141-161.

Dixon, W.J.
1975: BMDP Biomedical computer programs; Univ. California Press, Berkeley, 792 p.

Doveton, J.H. and Parsley, A.J.
1970: Experimental evaluation of trend surface distortions induced by inadequate data-point distributions; Inst. Min. Met. Trans., London, Sect. B, v. 79, p. 197-207.

Dubov, R.I.
1973: A statistical approach to the classification of geochemical anomalies, in Geochemical Exploration 1972, M.J. Jones (ed.), Inst. Min. Met., London, p. 275-284.

Duda, R.O. and Hart, P.E.
1973: Pattern classification and scene analysis; Wiley, New York, 482 p.

Duval, J.S.
1976: Statistical interpretation of airborne gamma-ray spectrometric data using factor analysis; in Exploration for Uranium Ore Deposits, IAEA, Vienna, p. 71-80.

Ek, J. and Elmlid, C.G.
1976: Tjarrovare: Chromium in till; J. Geochem. Explor., v. 5, p. 349-364.

Evans, I.S.
1977: The selection of class intervals; Trans. Inst. Br. Geog., New Ser., v. 2, p. 98-124.

Everitt, B.
1974: Cluster analysis; Heinemann, London, 122 p.

Experimental Cartography Unit
1978: Maps from the Shetlands data bank. Geochemistry: nickel and cobalt; Geogr. Mag., v. 50, p. 748-749.

Ferguson, R.B., Maddox, J.H., and Wren, H.F.
1977: Data management and analysis systems for large-scale hydrogeochemical reconnaissance; Comput. Geosci., v. 3, p. 453-458.

Fisher, R.A.
1936: The use of multiple measurements in taxonomic problems; Ann. Eugenics, v. 7, p. 179-188.

Garrett, R.G.
1969: The determination of sampling and analytical errors in exploration geochemistry; Econ. Geol., v. 64, p. 568-574.

1973a: The determination of sampling and analytical errors in exploration geochemistry. Reply to discussion; Econ. Geol., v. 68, p. 282-283.

1973b: Regional geochemical study of Cretaceous acidic rocks in the northern Canadian Cordillera as a tool for broad mineral exploration; in Geochemical Exploration 1972, M.J. Jones (ed.), Inst. Min. Met., London, p. 203-219.

1974a: Field data acquisition methods for applied geochemical surveys at the Geological Survey of Canada; Geol. Surv. Can., Paper 74-52, 36 p.

1974b: Computers in exploration geochemistry; Geol. Surv. Can., Paper 74-60, p. 63-66.

1977: Sample density investigations in lake sediment geochemical surveys of Canada's uranium reconnaissance program; in Symposium on Hydrogeochemical and Stream-sediment Reconnaissance for Uranium in the United States; Bendix Field Engineering Corporation, Grand Junction, Colorado, Rept. GJBX-77(77), p. 173-186.

Garrett, R.G. and Nichol, I.
1967: Regional geochemical reconnaissance in eastern Sierra Leone; Inst. Min. Met. Trans., London, Sect. B, v. 76, p. 97-112.

1969: Factor analysis as an aid in the interpretation of regional geochemical stream sediment data; Quart. J. Colo. Sch. Mines, v. 64, p. 245-264.

Geological Survey of Uganda
   1973:   Geochemical Atlas of Uganda, Entebbe, 43 p.

Glasby, G.P., Tooms, J.S., and Howarth, R.J.
   1974: Geochemistry of manganese concretions from the northwest Indian Ocean; New Zealand J. Sci., v. 17, p. 387-407.

Gordon, T. and Hutchison, W.W. (Ed.)
   1974: Computer use in projects of the Geological Survey of Canada; Geol. Surv. Can., Paper 74-60, 108 p.

Gordon, T. and Martin, G.
   1974: Computer-based data management in the Geological Survey of Canada —a preliminary appraisal; Geol. Surv. Can., Paper 74-60, p. 7-14.

Govett, G.J.S.
   1972: Interpretation of a rock geochemical exploration survey in Cyprus — statistical and graphical techniques; J. Geochem. Explor., v. 1, p. 77-102.

Govett, G.J.S., Goodfellow, W.D., Chapman, R.P., and Chork, C.Y.
   1975: Exploration geochemistry — distribution of elements and recognition of anomalies; Math. Geol., v. 7, p. 415-446.

Govett, G.J.S. and Pantazis, Th.M.
   1971: Distribution of Cu, Zn, Ni and Co in the Troodos Pillow Lava Series; Inst. Min. Met. Trans., London, Sect. B, v. 80, p. 27-46.

Gregory, P. and Tooms, J.S.
   1969: Geochemical prospecting for kimberlites; Quart. J. Colo. Sch. Mines, v. 64, p. 265-305.

Guarascio, M., David, M., and Huijbregts, C.
   1976: Advanced Geostatistics in the Mining Industry; Ridel, Dordrecht, 461 p.

Gustavsson, N. and Bjorklund, A.
   1976: Lithological classification of tills by discriminant analysis; J. Geochem. Explor., v. 5, p. 393-395.

Hawkes, H.E.
   1976: The downstream dilution of stream sediment anomalies; J. Geochem. Explor., v. 6, p. 345-358.

Hawkes, H.E. and Webb, J.S.
   1962: Geochemistry in mineral exploration; Harper and Row, New York, 415 p.

Hesp, W.R.
   1971: Correlations between the tin content of granitic rocks and their chemical and mineralogical composition; Can. Inst. Min. Met., Spec. Vol. 11, p. 341-353.

   1973: Classification of igneous rocks by cluster analysis; Acta Geol. Acad. Sci. Hung., v. 17, p. 339-362.

Hesp, W.R. and Rigby, D.
   1973: Cluster analysis of rocks in the New England igneous complex, New South Wales, Australia; in Geochemical Exploration 1972, M.H. Jones (ed.), Inst. Min. Met., London, p. 221-235.

Hill, W.E.
   1975a: Analytical standards for the mineral industry; unpub. ms. (Presented at AIME Mtg., Broadmoor), 16 p.

   1975b: The use of analytical standards to control assaying projects; in Geochemical Exploration 1974, I.L. Elliott and W.K. Fletcher (ed.), Elsevier Publ. Co., Amsterdam, p. 651-657.

Hodgson, W.A.
   1973: Optimum spacing for soil sample traverses; in Application of Computer Methods in the Mineral Industry, S. Afr. Inst. Min. Met., Johannesburg, p. 75-78.

Hornbrook, E.H.W. and Garrett, R.G.
   1976: Regional geochemical lake sediment survey, east-central Saskatchewan; Geol. Surv. Can., Paper 75-41, 20 p.

Howarth, R.J.
   1967: Trend surface fitting to random data — an experimental test; Amer. J. Sci., v. 265, p. 619-625.

   1971a: Fortran IV program for grey-level mapping of spatial data; Math. Geol., v. 3, p. 95-121.

   1971b: Empirical discriminant classification of regional stream sediment geochemistry in Devon and east Cornwall; Inst. Min. Met. Trans., London, Sect. B, v. 80, p. 142-149.

   1972: Empirical discriminant classification of regional stream sediment geochemistry in Devon and east Cornwall. Reply to discussion; Inst. Min. Met. Trans., London, Sect. B, v. 81, p. 115-119.

   1973a: The pattern recognition problem in applied geochemistry; in Geochemical Exploration 1972, M.J. Jones (ed.), Inst. Min. Met., London, p. 259-273.

   1973b: Preliminary assessment of a non-linear mapping algorithm in a geological context; Math. Geol., v. 5, p. 39-57.

   1973c: Fortran IV programs for empirical discriminant classification of spatial data; Geocom Bull., v. 6, p. 1-31.

   1974: The impact of pattern recognition methodology in geochemistry; Proc. 2nd Intl. Joint Conf. Pattern Recognition, Copenhagen, IEEE Pub. 74CH0885-4C, p. 411-412.

   1977: Automatic generation of randomized sample submittal schemes for laboratory analysis; Comput. Geosci., v. 3, p. 327-334.

Howarth, R.J. and Earle, S.A.M.
   1979: Application of a generalized power transform to geochemical data; Math. Geol., v. 11, p. 45-48.

Howarth, R.J. and Lowenstein, P.L.
   1971: Sampling variability of stream sediments in broad-scale regional geochemical reconnaissance; Inst. Min. Met. Trans., London, Sect. B, v. 80, p. 363-372.

   1972: Sampling variability of stream sediments in broad-scale regional geochemical reconnaissance. Discussion; Inst. Min. Met. Trans., London, Sect. B, v. 81, p. 122-123.

   1974: Data processing for the Provisional Geochemical Atlas of Northern Ireland; Applied Geochemistry Research Group, London, Tech. Comm. 61, 8 p.

   1976: Three-component colour maps from lineprinter output; Inst. Min. Met. Trans., London, Sect. B, v. 85, p. 234-237.

Howarth, R.J. and Thompson, M.
   1976: Duplicate analysis in geochemical practice. Part 2. Examination of proposed method and examples of its use; Analyst, v. 101, p. 699-709.

Hubaux, A. and Smiriga-Snoeck, N.
1964: On the limit of sensitivity and the analytical error; Geochem. Cosmochim. Acta, v. 28, p. 1199-1216.

Huijbregts, C. and Matheron, G.
1971: Universal kriging (an optimal method for estimating and contouring in trend surface analysis); Can. Inst. Min. Met., Spec. Vol. 12, p. 159-169.

Hutchison, W.W. (Ed.)
1975: Computer-based systems for geological field data; Geol. Surv. Can., Paper 74-63, 100 p.

Ingamells, C.O., Engels, J.C., and Switzer, P.
1972: Effect of laboratory sampling error in geochemistry and geochronology; Proc. 24th Inter. Geol. Congr., Montreal, Sect. 10, p. 405-415.

James, C.H.
1970: A rapid method for calculating the statistical precision of geochemical prospecting analysis; Inst. Min. Met. Trans., London, Sect. B, v. 79, p. 88-89.

Jeffery, K.G. and Gill, E.M.
1976: The design philosophy of the G-EXEC system; Comput. Geosci., v. 2, p. 345-346.

Jöreskog, K.G., Klovan, J.E., and Reyment, R.A.
1976: Geological Factor Analysis; Elsevier Publ. Co., Amsterdam, 178 p.

Journel, A.
1975: Geological reconnaissance to exploitation — a decade of applied geostatistics; Can. Inst. Min. Met., June 1975, p. 75-84.

Joyce, A.S. and Clema, J.M.
1974: An application of statistics to the chemical recognition of nickel gossans in the Yilgam Block, Western Australia; Proc. Australs. Inst. Min. Met., No. 252, p. 21-24.

Kaiser, H.F.
1959: Computer program for Varimax rotation in factor analysis; Educ. Psychol. Meas., v. 19, p. 413-420.

Kayser, R.B. and Parry, W.T.
1971: A geochemical exploration experiment on the Texas Canyon Stock, Cochise Country, Arizona; Can. Inst. Min. Met., Spec. Vol. 11, p. 354-356.

Kendall, M.G.
1938: A new measure of rank correlation; Biometrika, v. 30, p. 81-93.

Kork, J.O.
1977: Examination of the Chayes-Kruskal procedure for testing correlations between proportions; Math. Geol., v. 9, p. 543-562.

Kremer, M., Lenci, M., and Lesage, M.T.
1976: SIGMI: A user-oriented file-processing system; Comput. Geosci., v. 1, p. 187-193.

Larsson, J.O.
1976: Organic stream sediments in regional geochemical prospecting, Precambrian Pajala district, Sweden; J. Geochem. Explor., v. 6, p. 233-249.

Lenthall, D.H.
1972: The application of discriminatory and cluster analysis as an aid to the understanding of the acid phase of the Bushveld Complex; Econ. Geol. Res. Unit, University of Witwatersrand, Inform. Circ. No. 72, 70 p.

Lenz, C.J.
1972: Evaluation of geochemical data; in Application of Computer Methods in the Mineral Industry, M.D.G. Solomon and F.N. Lancaster (ed.), S. Afr. Inst. Min. Met., Johannesburg, p. 73-74.

Lepeltier, C.
1969: A simplified statistical treatment of geochemical data by graphical representation; Econ. Geol., v. 64, p. 538-550.

1977: Le role de l'ordinateur en exploration géochimique; Sci. Terre, Ser. Inform. No. 9, p. 15-38.

Lindèn, A.H. and Åkerblom, G.
1976: Method of detecting small or indistinct radioactive sources by airborne gamma-ray spectrometry; in Geology, Mining and Extractive Processing of Uranium, M.J. Jones (ed.), Inst. Min. Met. London, p. 113-120.

Ling, R.F.
1974: Comparison of several algorithms for computing sample means and variances; J. Amer. Stat. Ass., v. 69, p. 859-866.

Link, R.F. and Koch, G.S.
1975: Some consequences of applying lognormal theory to pseudolognormal distribution; Math. Geol., v. 7, p. 117-128.

Lockhart, A.W.
1976: Geochemical prospecting of an Appalachian porphyry copper deposit at Woodstock, New Brunswick; J. Geochem. Explor., v. 6, p. 13-33.

Lowenstein, P.L. and Howarth, R.J.
1973: Automated colour-mapping of three-component systems and its application to regional geochemical reconnaissance; in Geochemical Exploration 1972, M.J. Jones (ed.), Inst. Min. Met., London, p. 297-304.

Mancey, S.J. and Howarth, R.J.
1978: Factor score maps of regional geochemical data from England and Wales; Applied Geochemistry Research Group, Imperial College, London, 2 sheets.

Mark, D.M. and Church, M.
1977: On the misuse of regression in earth science; Math. Geol., v. 9, p. 63-75.

Matheron, G.
1957: Théorie lognormale de l'échantillonage systématique des gisements; Ann. Mines, v. 9, p. 566-584.

1962: Traité de geostatistique appliquée; Editions Technip, Paris, Vol. 1 (1962), 334 p., Vol. 2 (1963), 172 p.

1963: Principles of geostatistics; Econ. Geol., v. 58, p. 1246-1266.

May, J.P.
1972: PROBPLT: A computer subroutine for plotting a cumulative frequency distribution on a probability scale; Geol. Soc. Am. Bull., v. 83, p. 2867-2870.

McCammon, R.B.
1977: BINORM — A Fortran subroutine to calculate the percentiles of a standardized binormal distribution; Comput. Geosci., v. 3, p. 335-339.

Merriam, D.F. (Ed.)
1976: Capture, management and display of geological data; with special emphasis on energy and mineral resources; Comput. Geosci., v. 2, p. 275-376.

1977: Computer software for the geosciences; Comput. Geosci., v. 3, p. 385-545.

Michie, U. McL.
1973: The determination of sampling and analytical errors in exploration geochemistry. Discussion; Econ. Geol., v. 68, p. 281-282.

Miesch, A.T.
1964: Effects of sampling and analytical error in geochemical prospecting; in Computers in the Mineral Industries, G.A. Parks (ed.), Stanford Univ., Stanford, p. 156-170.

1967a: Methods of computation for estimating geochemical abundance; U.S. Geol. Surv., Prof. Paper 574-B, 15 p.

1967b: Theory of error in geochemical data; U.S. Geol. Surv., Prof. Paper 574-A, 17 p.

1969: The constant sum problem in geology; in Computer applications in the earth sciences, D.F. Merriam (ed.), Plenum Press, New York, p. 161-176.

1971: The need for unbiased and independent replicate data in geochemical exploration; Can. Inst. Min. Met., Spec. Vol. 11, p. 582-584.

1976a: Geochemical survey of Missouri: Methods of sampling, laboratory analysis and statistical reduction of data; U.S. Geol. Surv., Prof. Paper 954-A, 39 p.

1976b: Sampling designs for geochemical surveys: Syllabus for a short course; U.S. Geol. Surv., Open File Rep. 76-772, 140 p.

1977: Logtransformations in geochemistry. Discussion; Math. Geol., v. 9, p. 191-194.

Montgomery, J.H., Cochrane, D.R., and Sinclair, A.J.
1975: Discovery and exploitation of Ashnola porphyry copper deposit, near Keremeos, B.C.: A geochemical case study; in Geochemical Exploration 1974, J.L. Elliott and W.K. Fletcher (ed.), Elsevier Publ. Co., Amsterdam, p. 85-100.

Nackowski, M.P., Mardirosian, C.A., and Botbol, J.M.
1967: Trend surface analysis of trace chemical data, Park City district, Utah; Econ. Geol., v. 62, p. 1072-1087.

Nairis, B.
1971: Endogene dispersion aureoles around the Rudtjebacken sulphide ore in the Adak area, northern Sweden; Can. Inst. Min. Met., Spec. Vol. 11, p. 357-374.

Nichol, I.
1971: Future trends of exploration geochemistry in Canada; Can. Inst. Min. Met., Spec. Vol. 11, p. 32-38.

Nichol, I., Garrett, R.G., and Webb, J.S.
1966: Automatic data plotting and mathematical and statistical interpretation of geochemical data; Geol. Surv. Can., Paper 66-54, p. 195-210.

1969: The role of some statistical and mathematical methods in the interpretation of regional geochemical data; Econ. Geol., v. 64, p. 204-220.

Nichol, I., Thornton, I., Webb, J.S., Fletcher, W.K., Hennahil, R.J., Khaleelee, J., and Taylor, D.
1970: Regional geochemical reconnaissance of the Denbighshire area; Rep. Inst. Geol. Sci., No. 70/8, 40 p.

Nichol, I. and Webb, J.S.
1967: The application of computerized mathematical and statistical procedures to the interpretation of geochemical data; Proc. Geol. Soc. Lond., No. 1642, p. 186-199.

Nichols, C.E., Butz, T.R., Cagle, G.W., and Kane, V.E.
1977: Uranium geochemical survey in the Crystal City and Beeville Quadrangles, Texas; Union Carbide Corp., Oak Ridge Gaseous Diffusion Plant, Oak Ridge, Tennessee, Rept. K/UR-5, GJBX-19(77), 324 p.

Nie, N.H., Hull, C.H., Jenkins, J.G., Steinbrenner, K., and Bent, D.H.
1975: Statistical package for the Social Sciences; McGraw-Hill, New York, 675 p.

Norcliffe, G.B.
1969: On the use and limitations of trend surface models; Canad. Geog., v. 13, p. 338-348.

Obial, R.C.
1970: Cluster analysis as an aid in the interpretation of multi-element geochemical data; Inst. Min. Met. Trans., London, Sect. B, v. 79, p. 175-180.

Obial, R.C. and James, C.H.
1973: Use of cluster analysis in geochemical prospecting, with particular reference to southern Derbyshire, England; in Geochemical Exploration 1972, M.J. Jones (ed.), Inst. Min. Met. London, p. 237-257.

Olea, R.A.
1975: Optimum mapping techniques using regionalized variable theory; Kansas Geol. Surv., Series on spatial analysis, No. 2, Univ. Kansas, Lawrence, 137 p.

Pantazis, Th.M. and Govett, G.J.S.
1973: Interpretation of a detailed rock geochemistry survey around Mathiati mine, Cyprus; J. Geochem. Explor., v. 2, p. 25-36.

Parslow, G.R.
1974: Determination of background and threshold in exploration geochemistry; J. Geochem. Explor., v. 3, p. 319-336.

Pearson, K.
1899: Data for the problem of evolution in man. III. On the magnitude of certain coefficients of correlation in man, etc.; Proc. Roy. Soc., London, v. 66, p. 23-32.

Pelet, R. and De Jekhowsky, B.
1972: Etude statistique sur ordinateur de la géochimie de certain elements dans la formations sédimentaires du Bassin de Paris; Proc. 24th Inter. Geol. Congr., Montreal, v. 16, p. 60-75.

Plant, J.
1971: Orientation studies on stream sediment sampling for regional geochemical survey in Scotland; Inst. Min. Met. Trans., London, Sect. B, v. 80, p. 324-344.

1973: A random numbering system for geochemical samples; Inst. Min. Met. Trans., London, Sect. B, v. 82, p. 64-65.

Plant, J., Jeffery, K., Gill, E., and Fage, C.
   1975: The systematic determination of accuracy and precision in geochemical exploration data; J. Geochem. Explor., v. 4, p. 467-486.

Plant, J. and Rhind, D.
   1974: Mapping minerals; Geogr. Mag., Nov. 1974, p. 123-126.

Rhind, D.
   1975: A skeletal overview of spatial interpolation techniques; Computer Appl., Nottingham, v. 2, p. 293-309.

Rhind, D., Shaw, M.A., and Howarth, R.J.
   1973: Experimental geochemical maps; Cartog. J., Dec. 1973, p. 112-118.

Rose, A.W.
   1972: Favorability for Cornwall-type magnetite deposits in Pennsylvania using geological, geochemical, and geophysical data in a discriminant function; J. Geochem. Explor., v. 1, p. 181-194.

Rose, A.W., Dahlberg, E.C., and Keith, M.L.
   1970: A multiple regression technique for adjusting background values in stream-sediment geochemistry; Econ. Geol., v. 65, p. 156-165.

Rose, A.W. and Keith, M.L.
   1976: Reconnaissance geochemical techniques for detecting uranium deposits in sandstones of northeastern Pennsylvania; J. Geochem. Explor., v. 6, p. 119-137.

Rose, A.W. and Suhr, N.H.
   1971: Major element content as a means of allowing for background variation in stream-sediment geochemical exploration; Can. Inst. Min. Met., Spec. Vol. 11, p. 587-593.

Saager, R. and Esselaar, P.A.
   1969: Factor analysis of geochemical data from the Basal Reef, Orange Free State Goldfield, South Africa; Econ. Geol., v. 64, p. 445-451.

Saager, R. and Sinclair, A.J.
   1974: Factor analysis of stream sediment geochemical data from the Mount Nansen area, Yukon Territory, Canada; Miner. Deposita, v. 9, p. 243-252.

Schweitzer, R.H.
   1973: Mapping urban America with automated cartography; U.S. Dept. of Commerce, Bureau of the Census, Washington, 21 p.

Selvin, S.
   1976: A graphical estimate of the population mean from censored normal data; Appl. Stat., v. 25, p. 8-11.

Sharp, W.E. and Jones, T.L.
   1975: A topologically optimum prospecting plan for streams; in Geochemical Exploration 1974, I.L. Elliott and W.K. Fletcher (ed.), Elsevier Publ. Co., Amsterdam, p. 227-235.

Shaw, B.R.
   1977: Evaluation of distortion of residuals in trend surface analysis by clustered data; Math. Geol., v. 9, p. 507-517.

Shiikawa, M., Wakasa, K., and Tono, N.
   1975: Geochemical exploration for kuroko deposits in north-east Honshu, Japan; in Geochemical Exploration 1974, I.L. Elliott and W.K. Fletcher (ed.), Elsevier Publ. Co., Amsterdam, p. 65-76.

Siegel, S.
   1956: Nonparametric statistics for the Behavioral Sciences; McGraw-Hill, New York, 312 p.

Sinclair, A.J.
   1974: Selection of threshold values in geochemical data using probability graphs; J. Geochem. Explor., v. 3, p. 129-149.

   1975: Some considerations regarding grid orientation and sample spacing; in Geochemical Exploration 1974, I.L. Elliott and W.K. Fletcher (ed.), Elsevier, Amsterdam, p. 133-140.

   1976: Applications of Probability Graphs in Mineral Exploration; Ass. Explor. Geochem., Spec. Vol. 4, 95 p.

Sinding-Larsen, R.
   1975: A computer method for dividing a regional geochemical survey area into homogeneous subareas prior to statistical interpretation; in Geochemical Exploration 1974, I.L. Elliott and W.K. Fletcher (ed.), Elsevier Publ. Co., Amsterdam, p. 191-217.

   1977: Comments on the statistical treatment of geochemical exploration data; Sci. Terre, Ser. Inform., No. 9, p. 73-90.

Singer, D.A.
   1972: ELIPGRID, a Fortran IV program for calculating the probability of success in locating elliptical targets with square, rectangular and hexagonal grids; Geocom Programs No. 4, Geosystems, London, 16 p.

Slysz, W.D.
   1973: Software Compatibility in the Social Sciences; in Proc. Educom. Fall Conf., Princeton, p. 235-243.

Smeds, B.
   1976: Ink jet colour plotter for computer graphics; Rep. Lund Inst. Techn., Dep. Elect. Measurements, No. 1/1976.

Smith, F.M.
   1971: Geochemical exploration over complex mountain glacial terrain in the Whitehorse Copperbelt, Yukon Territory; Can. Inst. Min. Met., Spec. Vol. 11, p. 265-275.

Smith, E.C. and Webber, G.R.
   1973: Nature of mercury anomalies at the New Calumet Mines area, Quebec, Canada; in Geochemical Exploration 1972, M.J. Jones (ed.), Inst. Min. Met. London, p. 71-80.

Spearman, C.
   1904: The proof and measurement of association between two things; Amer. J. Psychol., v. 15, p. 72-101.

Stevens, M.A.
   1974: EDF statistics for goodness of fit and some comparisons; J. Amer. Stat. Ass., v. 69, p. 730-737.

Summerhayes, C.P.
   1972: Geochemistry of continental margin sediments from northwest Africa; Chem. Geol., v. 10, p. 137-156.

Teil, H.
   1975: Correspondence factor analysis: an outline of its method; Math. Geol., v. 7, p. 3-12.

Thompson, M. and Howarth, R.J.
1976: Duplicate analysis in geochemical practice. Part 1. Theoretical approach and estimation of analytical reproducibility; Analyst, v. 101,

1978: A new approach to the estimation of analytical precision; J. Geochem. Explor., v. 9, p. 23-30.

Timperley, M.H. et al.
1972: The improved detection of geochemical soil anomalies by multiple regression analysis of biogeochemical data; Proc. Austral. Inst. Min. Met., v. 242, p. 23-36.

Tourtelot, H.A. and Miesch, A.T.
1975: Sampling designs in environmental geochemistry; Geol. Soc. Amer., Spec. Paper 155, p. 107-118.

Trochimczyk, J. and Chayes, F.
1977: Sampling variation of principal components; Math. Geol., v. 9, p. 497-506.

1978: Some properties of principal components; Math. Geol., v. 10, p. 43-52.

U.S. Geological Survey
1969: Proceedings of the Symposium on map and chart digitizing; U.S. Geol. Surv., Comput. Contr. No. 5, 81 p.

Van Trump, G. and Miesch, A.T.
1977: The U.S. Geological Survey RASS-STATPAC system for management and statistical reduction of geochemical data; Comput. Geosci., v. 3, p. 475-488.

Walden, A.R.
1972: Quantitative comparison of automatic contouring algorithms; Kansas Oil Exploration (KOX) Project Rept., Kansas Geol. Surv., 137 p.

Webb, J.S., Fortescue, J., Nichol, I., and Tooms, J.S.
1964: Regional geochemical maps of the Namwala Concession area, Zambia; Geol. Surv., Maps 1-1X.

Webb, J.S. and Howarth, R.J.
Regional geochemical mapping; in Proceedings of the Discussion Meeting on Geochemistry and Health, The Royal Society, London. (in press)

Webb, J.S., Nichol, I., Foster, R., Lowenstein, P.L., and Howarth, R.J.
1973: Provisional Geochemical Atlas of Northern Ireland; Applied Geochemistry Research Group, Imperial College, London, 36 p.

Webb, J.S., Thornton, I., Thompson, M., Howarth, R.J., and Lowenstein, P.L.
1978: The Wolfson Geochemical Atlas of England and Wales; Oxford University Press, Oxford, 74 p.

Wennervirta, H., Bolviken, B., and Nilsson, C.A.
1971: Summary of research and development in Scandinavian countries; Can. Inst. Min. Met., Spec. Vol. 11, p. 11-14.

Wennervirta, H. and Papunen, H.
1974: Heavy metals as lithogeochemical indicators for ore deposits in the Iilinjarvi and Aijala fields, SW-Finland; Bull. Geol. Surv. Finl., No. 269, 22 p.

Whitehead, R.E.S. and Govett, G.J.S.
1974: Exploration rock geochemistry detection of trace element haloes at Heath Steel Mines (N.B., Canada) by discriminant analysis; J. Geochem. Explor., v. 3, p. 371-386.

Wignall, T.K.
1969: Generalized Bayesian classification: K classes; Econ. Geol., v. 64, p. 571-574.

1970: Generalized Bayesian classification: K classes; Reply to discussion; Econ. Geol., v. 65, p. 221-222.

Wolf, W.J.
1974: Geochemical distribution of zinc, nickel and copper in volcanic rocks, Ben Nevis Township and parts of Clifford and Pontiac Townships, District of Timiskaming; Ontario Div. Mines, Prelim. Geochem. Maps P.915, P.916 and P.917.

1975: Zinc abundance in early Precambrian volcanic rocks: its relationship to exploitable levels of zinc in sulphide deposits of volcanic-exhalative origin; in Geochemical Exploration 1974, I.L. Elliott and W.K. Fletcher (ed.), Elsevier Publ. Co., Amsterdam, p. 262-278.

NOTE:

Enquiries for all reports with GJBX numbers should be addressed to:

Bendex Field Engineering Corp.,
Technical Library,
P.O. Box 1569,
Grand Junction,
Colorado, U.S.A.
C081501

SOME ASPECTS OF INTEGRATED EXPLORATION

J.A. Coope
Newmont Exploration of Canada Ltd., Toronto, Ontario

M.J. Davidson
Newmont Exploration Ltd., Tucson, Arizona, U.S.A.

Coope, J.A., and Davidson M.J., Some aspects of integrated exploration; in Geophysics and Geochemistry in the Search for Metallic Ores; Peter J. Hood, editor; Geological Survey of Canada, Economic Geology Report 31, p. 575-592, 1979.

Abstract

For many years, regional techniques of mineral exploration incorporating airborne geophysical methods and other remote-sensing techniques applied to identify broad geological parameters have been used under a variety of geological and geomorphological conditions. Such use of disciplines has served to broadly classify areas based primarily on broad features such as physical properties and prospective mineral potential.

The better understanding of the geological processes leading to the formation of mineral deposits, together with the enterprising development of geophysical, geochemical and other types of exploration tools during recent years, has provided a basic framework for an integrated application of techniques on a more detailed scale and with the potential of improved cost and exploration effectiveness.

Variations in geophysical and geochemical backgrounds can provide critical data which serve to elucidate geological interpretation and the evaluation of anomalous conditions. The variations are commonly related to primary geological features or larger scale, ore-forming processes and can yield critical information when the connection is recognized and properly interpreted. Magnetic, resistivity, electrical, lithogeochemical and other selected techniques are particularly adaptable in this more detailed integrated application.

Different disciplines and data-gathering methods within disciplines can yield the same or equivalent information interpretable in a geological context. However, recognition of the physio-chemical fingerprint of an ore environment requires a multidisciplined, integrated approach when signatures observed from any one discipline are too subtle to generate adequate confidence levels by themselves. In any particular situation there is a sequence of application and areal coverage which will define targets at a minimum cost.

Résumé

Depuis plusieurs années, on utilise des techniques régionales d'exploration minière, intégrant des méthodes géophysiques aéroportées et autres techniques de télédétection pour définir des paramètres géologiques larges dans divers milieux géologiques et géomorphologiques. Ces disciplines ont servi à définir grosso modo des zones, surtout en fonction de leurs caractères physiques et de leur potentiel minier.

Une meilleure compréhension des processus géologiques qui engendrent des gîtes minéraux, et la mise au point ces dernières années de méthodes d'exploration géophysique, géochimique et autres, permettent d'intégrer les diverses techniques à une échelle plus détaillée et sans doute de réduire les coûts et de rendre l'exploration plus efficace.

Les variations du fond géophysique et du fond géochimique peuvent fournir des données essentielles pour l'interprétation géologique et l'évaluation des anomalies. Ces variations sont généralement liées à des éléments géologiques primaires ou, à une échelle plus grande, à des processus de genèse des gîtes minéraux. Elles peuvent aussi apporter des renseignements essentiels lorsque la corrélation est reconnue et proprement interprétée. Les méthodes magnétique, de résistivité, électrique, lithogéochimique et autres techniques appropriées peuvent facilement être intégrées, à une échelle plus détaillée.

Les différentes disciplines et les méthodes de recueil des données dans ces disciplines peuvent donner les mêmes renseignements ou des renseignements équivalents, qui peuvent être interprétés dans un contexte géologique. Cependant, l'identification des caractères physico-chimiques d'un milieu minéral exige que l'on adopte une approche pluridisciplinaire et intégrée, lorsque les observations recueillies au moyen d'une méthode particulière sont trop peu convaincantes. Dans toute situation, on peut adopter une série de procédés, et un mode de couverture aérienne, de manière à définir les objectifs d'exploration au coût le plus bas possible.

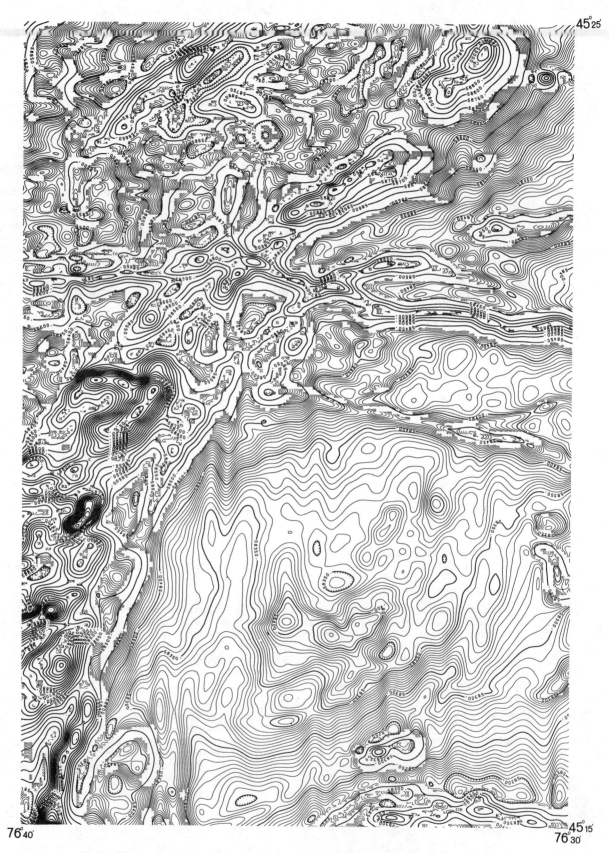

Figure 25.1. Total magnetic field map of part of NTS 31 F/7 southern Ontario. (Reproduced from Hood et al., 1976.)

DESCRIPTIVE NOTES

This total field contour map was compiled from data recorded during aeromagnetic survey operations by a self-orienting rubidium-vapour magnetometer which was installed in the tail stinger of a Beechcraft B80 Queenair aircraft. The data was digitally-recorded with a resolution of 0.02 gammas. A second boom mounted above the tail stinger at a distance of 2.08 metres forms a vertical gradiometer system.

Flight altitude was 500 feet above ground at 1000 feet average flight line spacing and double control lines were flown at an average spacing of 4 miles.

The data was edited, compiled, levelled and gamma values for contouring interpolated on a square grid (0.1" grid spacing at the published map scale) by automatic computer processes.

The automatic levelling process employs the two components of the double control line and the short segments of traverse which connect them where they are not exactly co-incident. This data is used to minimize and distribute non-geological contributions from the total magnetic field profile along the control line. The corrected control lines are used to level the traverse by a method of minimal sum-total adjustment.

The final grid was contoured and plotted using the automatic contouring program and digital plotter facilities of the Department of Energy, Mines and Resources, Computer Science Centre.

Airborne survey was carried out in April 1975 and digital compilation by Resource Geophysics and Geochemistry Division, Geological Survey of Canada. The Queenair aircraft of the Geological Survey of Canada was flown under contract to Kenting Earth Sciences Ltd.

No correction has been made for regional variation.

The photo and map base for this map was compiled by Survey and Mapping Branch, Department of Energy, Mines and Resources.

ISOMAGNETIC LINES (absolute total field)
50-100 gammas
5 gammas
Magnetic depression

Flight altitude: 500 feet above ground level
Contour interval: 5 gammas

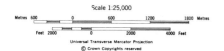

GEOLOGICAL SURVEY OF CANADA   COMMISSION GÉOLOGIQUE DU CANADA
DEPARTMENT OF ENERGY, MINES AND RESOURCES
MINISTÈRE DE L'ÉNERGIE, DES MINES ET DES RESSOURCES

PART OF 31 F/7
ONTARIO
TOTAL MAGNETIC FIELD

Scale 1:25,000

Universal Transverse Mercator Projection
© Crown Copyrights reserved

OPEN FILE
DOSSIER PUBLIC
339
APRIL 1976
GEOLOGICAL SURVEY
COMMISSION GÉOLOGIQUE
OTTAWA

INTRODUCTION

Effective "integrated exploration" requires communication and co-operation between the various mineral exploration disciplines. The concept of integrated utilization of exploration methods and tools can be applied on all scales — from the regional reconnaissance of large geographic areas to very detailed application in the search for extensions of known orebodies. A review of this broad field would require consideration of the techniques that have been reviewed during the Exploration '77 Symposium together with several exploration methods which have not. The latter methods include the practical application of trenching, percussion and diamond drilling, and isotope geology and geochemistry.

This paper makes general reference to some fundamental aspects of exploration integration, and numerous principles are reviewed and illustrated which are consistent with good exploration practice.

THE IMPORTANCE OF GEOLOGY

The predominant preoccupation of the Exploration '77 Symposium with geochemical and geophysical methods should not detract attention from the fundamental foundation of mineral exploration which is a knowledge and understanding of the geology of mineral environments. Significant advances in the geological understanding of mineral environments have been made in the past 20 to 25 years. These advances which reflect a greater understanding of the processes of mineralization and the localization of economic quantities of metals, provide the basic foundation for the development, refinement and sophistication of mineral exploration techniques. All interpretations, whether based on geophysical or geochemical measurements, must relate to geology to be meaningful and, consequently, neither the exploration geochemist nor the exploration geophysicist can function independently of the exploration geologist.

Exploration is progressive. In many parts of the world, the last two decades have been marked by the rapid development and application of many exploration geophysical and exploration geochemical techniques. Many of these developments have been extremely successful and have directly indicated sub-outcropping or near-surface orebodies and extended the search capabilities into areas of extensive cover and difficult accessibility.

This period of rapid development and success nurtured, among many, a reliance on the airborne EM method or the stream sediment method or some other technique to identify "anomalies" for detailed geological scrutiny. In some localities, e.g. the Bathurst Camp, New Brunswick, this approach to exploration led to the establishment of several mining operations in areas where the geological controls on the mineral deposits were poorly known or misunderstood. Follow-up studies in these areas have corrected this imbalance. As time progresses, and, as it becomes apparent that additional ore reserves in these regions will only be found in blind deposits at increasing depths beyond the limits of detection by conventional geochemical and geophysical methods, the importance of geology becomes overwhelming and new exploration tools have to be applied based on the latest geological understanding of the mineral environment of interest.

It has been proven many times that, for some combination of geological or other reasons, an orebody of the type sought will not respond dramatically to a specific exploration technique, and will not produce a prominent anomaly. Total reliance on a geophysical or geochemical tool as the leading technique in an exploration program is therefore predestined to imperfection.

The example provided by the discovery of the South Bay orebody at Confederation Lake in northwestern Ontario where a relatively weak airborne INPUT anomaly became enhanced by the interpretation of the geological relationships in nearby outcrops, illustrates the advantage of the integration of geological mapping and interpretation into the exploration sequence. Knowledge of geology and an understanding of mineral environments serves to classify anomalies detected by the application of geophysical and geochemical techniques much more satisfactorily than does reliance on the magnitude of a physical or chemical concentration.

To many, these comments will appear to be a statement of the obvious. Nevertheless, the fundamental consideration of geology must be emphasized in the context of good exploration practice. This contribution is intended to illustrate how thoughtful integration of geological expertise with the enterprising development and application of geochemical and geophysical

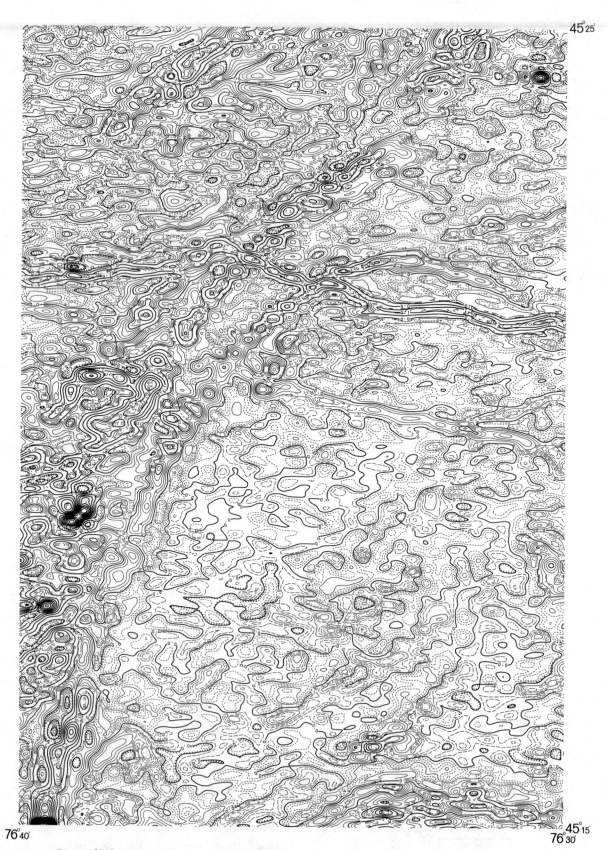

Figure 25.2. Vertical gradient magnetic map of part of NTS 31 F/7 southern Ontario. (Reproduced from Hood et al., 1976.)

Integrated Exploration

DESCRIPTIVE NOTES

This map is based on in-flight digitally recorded high sensitivity aeromagnetic data obtained with two self-orienting Rubidium vapour magnetometers installed in twin tail booms inboard a Beechcraft B80 aircraft. The magnetometer heads are separated by a distance of 2.08 metres with each measuring the total magnetic field to a resolution of 0.02 gammas.

Flight altitude was 500 feet above ground at 1000 feet average flight line spacing and double control lines were flown at an average spacing of 4 miles.

The data was edited, compiled, levelled and gradient values for contouring interpolated on a square grid (0.1" grid spacing at the published map scale) by automatic computer processes.

The vertical gradient data was filtered with a digital operator to remove noise spikes and instrument hash. The vertical gradient data from the tie lines was not used to compile the map, instead each line was individually adjusted as required.

The final grid was contoured and plotted using the automatic contouring program and digital plotter facilities of the Department of Energy, Mines and Resources, Computer Science Centre.

Airborne survey was carried out in April 1975 and digital compilation by Resource Geophysics and Geochemistry Division, Geological Survey of Canada. The Queenair aircraft of the Geological Survey of Canada was flown under contract to Kenting Earth Sciences Ltd.

The photo and map base for this map was compiled by Surveys and Mapping Branch, Department of Energy, Mines and Resources.

EQUIPOTENTIAL LINES (vertical gradient field)

.5 gammas/meter
.1 gammas/meter
.025 gammas/meter (above 0.0)
.025 gammas/meter (below 0.0)
Magnetic depression

Flight altitude: 500 feet above ground level
Contour interval: .1 gammas/meter
Intermediate contour interval between +/-1.0 gammas/meter : 0.025 gammas/meter

GEOLOGICAL SURVEY OF CANADA   COMMISSION GÉOLOGIQUE DU CANADA
DEPARTMENT OF ENERGY, MINES AND RESOURCES
MINISTÈRE DE L'ÉNERGIE, DES MINES ET DES RESSOURCES

PART OF 31 F/7 ONTARIO VERTICAL GRADIENT

Scale 1:25,000

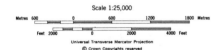

Universal Transverse Mercator Projection
© Crown Copyrights reserved

INDEX MAP

OPEN FILE
DOSSIER PUBLIC
339
APRIL 1976
GEOLOGICAL SURVEY
COMMISSION GÉOLOGIQUE
OTTAWA

---

methods can accommodate the demands of our expanding societies for the discovery of elusive economic mineral deposits in a selection of environments. In addition, examples will highlight how penetrative interpretation of geophysical and geochemical data can extend the understanding of geological relationships at depth or beneath overburden beyond the direct observation of the geologist.

INTEGRATED EXPLORATION

Opinions as to what constitutes "integrated exploration" vary widely. Actually, many who consider themselves to be adherents, are taking only partial advantage of the benefits and cost-effectiveness that full integration offers. True integrated exploration does not begin at some late stage in an exploration program after a geophysical or geochemical (or some other) survey has been completed. The full benefits can only be obtained when integration is effected at the earliest stage of the program. At this time the geologist should raise all the possible geological questions whose answers may be significant to the program, and the geophysicist and the geochemist should consider the various techniques which are capable of providing partial or complete answers to these questions under the conditions of the proposed survey. It is a rare case when a given type of geophysical or geochemical survey cannot provide answers to more than one geological question provided the survey planners are aware of these multiple questions in time to design appropriate field procedures.

Geophysical Aids to Geological Mapping

The geophysicist must convert field data into physical property representations that can be communicated and which are consistent with general geological principles and with whatever is known about the geology. The step from physical property maps to interpreted geology is best performed by the geologist and the geophysicist working jointly. To function efficiently in this fashion the geologist must be aware of the physical properties associated with various geological units, even to the extent of mapping these properties in the field regardless of whether or not they appear to be of economic significance. This would include such features as magnetite content, total sulphide content, the presence of bentonitic clays, graphite, magilmenite exsolution crystals, porosity and apparent resistivity contrasts. During the process of interpretation, the geophysicist is aware of the confidence levels associated with various physical property representations and is best equipped to modify these towards greater consistency with the emerging geological picture. The geologist, on the other hand, is best equipped to assess the confidence levels associated with the geological information and to formulate the overall geological model or models most consistent with all the observations.

Geophysical methods are often classified as "direct" when they are used to sense economic mineralization and "indirect" when they provide more general geological information. It is important to realize that this distinction is not inherent in the method but in the questions asked of the method. It is also important to appreciate that many anomalous responses which appear to mask or obscure the direct detection of mineralization are capable of yielding significant geological information. So called "geological noise" can be converted into a geological signal by merely asking the right question.

For example, consider a case where IP was being used to outline sulphide mineralizaiton in a thick sequence of dolomites. A large formational IP response in the general area of the known mineralization was found to be due to carbonaceous material in a black fetid dolomite. Recognizing that the depositional environment of the dolomite was ideal for the precipitation of metallic sulphides if the metal ions entered the reducing basin, the IP program was expanded with broad-spaced regional lines to locate and map paleochemical environments.

In the search for uranium-bearing roll fronts in sedimentary basins, the first phase is commonly groundwater hydrogeochemistry to establish areas of anomalous uranium and radon. The second phase could be resistivity surveying to map quickly and cheaply the meandering sand channels and establish sand thicknesses. In certain environments where pyrite has been deposited near roll fronts and where bentonitic clays are not ubiquitous, IP can be used to reduce the area where drilling is required.

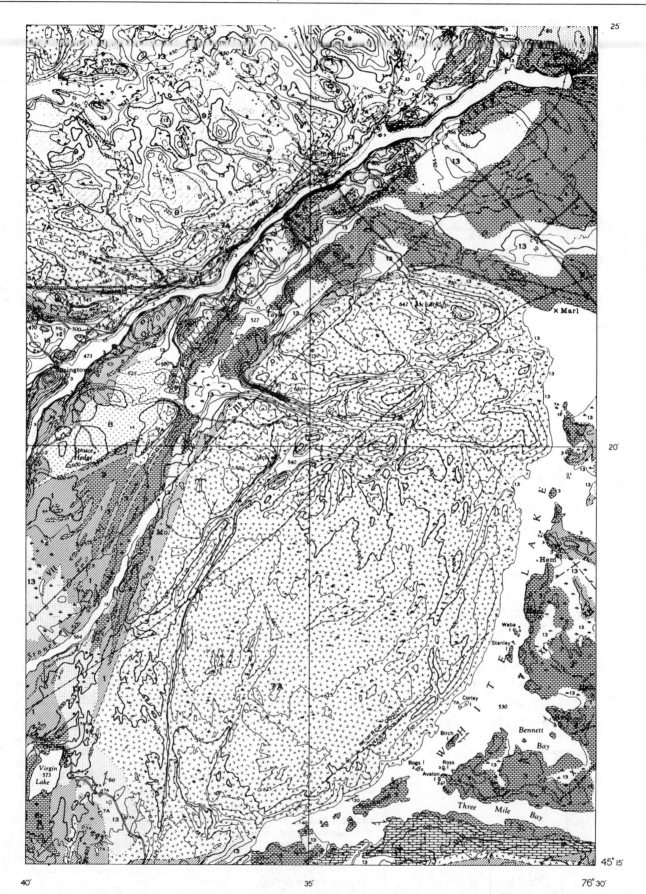

Figure 25.3. Geological map of part of NTS 31 F/7 southern Ontario. (Reproduced from Hood et al., 1976.)

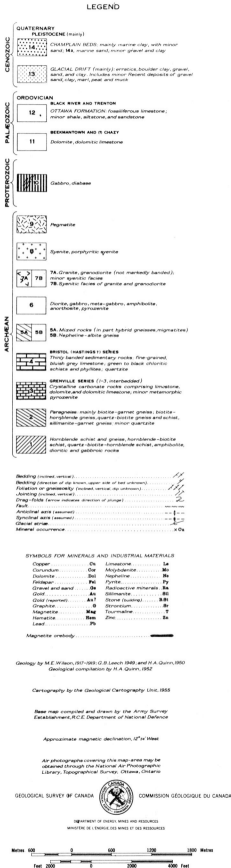

Where there is lithological variation and major rock units can be classified according to resistivity contrasts, it is possible to map these units broadly with resistivity measurements made during the completion of routine IP surveys. Such an approach has a direct application in areas of extensive overburden where the distance between outcrops is such that geological units cannot be mapped with satisfactory accuracy.

Magnetic surveys have, of course, been widely used to aid in the geological mapping of areas where sufficient magnetization contrast exists between rock units. Two papers by Boyd (1969) and Morley (1969) in the 1967 Symposium documented the capability of the magnetic, gravity and seismic methods to provide significant geological information. Papers presented during the 1977 Symposium and published herein described the geological mapping capability of airborne techniques such as γ-ray spectrometry (for mapping alteration in some terranes) and the potential of E Phase, Radiophase and other VLF EM methods, and high resolution aeromagnetic surveys.

The magnetic gradiometer and gradient techniques are a positive development which permit resolution of composite or complex magnetic anomalies into their individual constituents and, on the same basis, automatically remove the regional magnetic gradient to better define shallower features which may be of interest. In practice the method allows a sharper delineation of geological units with contrasting susceptibilities. The illustrations in Figures 25.1 and 25.2 compare gradient magnetic data with total intensity magnetic data for an area in Southern Ontario (Hood et al., 1976) and it is immediately apparent that the gradient map reflects the geological details (Fig. 25.3) much more precisely. Late dykes, striking approximately west-northwest across the granite body, are prominent in the magnetic data but were not observed during the geological survey.

The three examples cited utilizing the IP and resistivity survey responses illustrate the adaptation of other geophysical techniques to similar mapping roles.

In a recent article, Pelton (1977) explained how a careful study of the position and magnitude of peaks in the plot of IP phase angle versus frequency can be used to discriminate between graphite and massive sulphide mineralization, magnetite and nickeliferous pyrrhotite and between barren pyrite halos and disseminated economic mineralization in some porphyry copper systems.

These few examples illustrate, in a general way, how a more objective interpretation and adaptation of conventional techniques can complement the broader geological picture and add information that is of direct and immediate value in the assessment of truly anomalous responses which may be related to ore.

There is an obvious requirement for the development of the "indirect" application of geophysical techniques in order to satisfy the need to carefully examine and geophysically characterize the large scale environment of known deposits. This expertise, when developed to a satisfactory level, can then be adapted to down-hole geophysical surveying as mine-finding exploration is extended below the first hundred metres.

Geochemical Aids to Geological Mapping

It has been noted that, as exploration progresses in areas where shallow mineral deposits have been found and the search expands to explore for blind deposits at increasing depths, knowledge and understanding of the geology of mineral environments becomes the basis for the development and application of sophisticated geochemical, geophysical and other techniques that may be used effectively in these advanced exploration programs.

As an example, one can trace the recent development of lithogeochemical exploration methods. It became apparent early in the history of geochemical prospecting that trace element backgrounds varied according to the lithology and composition of rock types. Nickel, for example, enters into the crystal structures of olivines and copper tends to be higher in certain pyroxenes.

Different background levels in geochemical data in areas uncomplicated by mineralization can be correlated with rock lithologies and, in greater detail, with mineralogical associations. This correlation is obviously more direct in the case of the lithogeochemical method but similar correlations are possible with certain soil, stream sediment and other data sets (Fig. 25.4). Multi-element data and computer programming can help refine these correlations.

The correlation between geochemical data and mineralogy is relatively easy to demonstrate and understand. Partial analysis or "speciation" of lithogeochemical sample material can be used to differentiate qualitatively elements in different mineral phases such as certain sulphides, silicates, and oxides.

For many years, geochemical exploration was concerned primarily with the measurement of trace element quantities. With the advance of analytical expertise, major element analysis has become incorporated into the exploration geochemical spectrum. More recent instrumental adaptation, particularly the XRD method, has allowed the quantitative measurement of minerals, and the work of Hausen and Kerr (1971), and Figures 25.5A to 25.5D and Franklin et al. (1975) has clearly shown the adaptation of this method to map quantitatively the distribution of alteration minerals.

Franklin and his co-authors (1975), utilizing XRD techniques, have identified a pipelike alteration zone characterized by manganiferous siderite extending at least 300 m directly below the Mattabi massive sulphide deposit. Copper and zinc are concentrated in anomalous proportions in the upper central part of the pipe. The distribution of siderite and dolomite is shown in Figures 25.6 and 25.7. Figure 25.8 is a composite illustration of the generalized distribution of alteration and mineralization based on the Franklin et al. studies.

Quantitative XRD geochemistry plus trace and major element data can geochemically describe a geological environment and complement geological observation and understanding by portraying subtle or invisible variations which might otherwise go unnoticed. In the mineralized environment of the volcanogenic massive sulphide deposit, for example, the portrayal would reveal alteration patterns of chloritic, sericitic, feldspathic, carbonate and silica intensity which are, in part, visible and, in part, invisible to the eye of a trained geologist plus major and trace element variations in primary dispersion patterns indigenous to or, superimposed on, geochemical backgrounds related to rhyolitic and andesitic volcanic products.

Such a geochemical overlay on the geological environment is of immense value to the geologist in the investigation of the genesis of volcanogenic mineralization. In turn, when this understanding is developed, these geochemical relationships can be applied in prospecting in other favourable geological situations. In such a prospecting effort, it will be apparent that the exploration geochemist will be dependent on the knowledgeable exploration geologist for guidance in selecting critical lithogeochemical samples.

The integration of the geochemical and geological disciplines has been illustrated by results from the Canadian Shield which indicate that, in certain mineralized districts, the felsic and intermediate formations in volcanic cycles hosting mineralization contain significantly higher background zinc contents than equivalent rocks in unmineralized cycles (Nichol et al., 1975) (Fig. 25.9).

A note of caution should be registered with respect to the geology and related geochemistry of volcanogenic environments. Conspicuous variations in the geology and alteration associated with productive deposits are frequently recognized from mining camp to mining camp. For example, the relatively restricted distribution of chlorite at Mattabi in the Sturgeon Lake Camp, (Franklin et al., 1975) and Figure 25.8, is not a feature of the Noranda Camp where chlorite is more common and is widely distributed in the footwall alteration zones. Consequently a parallel study to the Mattabi study by Franklin et al. on a Noranda Camp deposit would outline a footwall alteration zone and other distinctive features related to the host and enclosing rocks, but the distribution and dominance of chlorite would be strongly contrasting.

Similarly, the Confederation Lake-Woman Lake relationship (Fig. 25.9) may not necessarily be reproduced in other camps although based on a geological understanding of these environments relationships of this or allied types can be anticipated in other volcanogenic areas.

Illustrations of how geological knowledge of a mineral environment can indicate useful exploration techniques can be selected from papers in the literature dealing with volcanogenic environments.

It can be demonstrated that anomalous amounts of metal occur in exhalite horizons contemporaneous in age with volcanogenic massive sulphide deposits. In their study of the dispersion of metals from the submarine exhalative bodies in the Red Sea, Holmes and Tooms (1972) were able to demonstrate that dispersion of metals from the metalliferous brines is taking place through normal seawater. Furthermore, this dispersion is detectable in both a soluble form in the water and in suspended particulate matter (Fig. 25.10 and 25.11). The net effect of the dispersion process is reflected in the upper few centimetres of the surface sediments, (Fig. 25.12), and it is apparent that the distance of anomalous dispersion is measurable in miles.

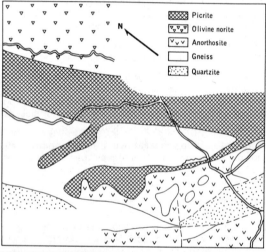

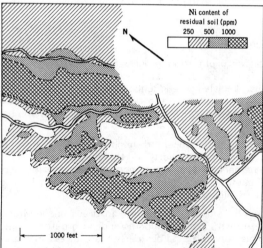

Figure 25.4. Relationship between geology and the pattern of nickel in residual soil, Nguge Region, Tanzania. (Colluvial and alluvial overburden occur flanking main rivers.) After Coope (1958) Courtesy Hawkes, H.E. and Webb, J.S., (1962) 'Geochemistry in Mineral Exploration', Harper and Row, New York, N.Y.

Integrated Exploration

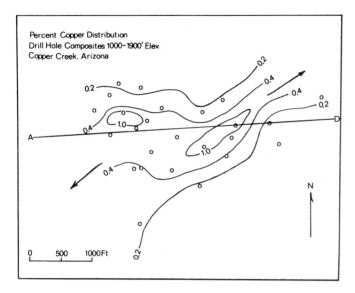

A. Per cent copper distribution at Copper Creek, Arizona – drillhole omposites 1000-1900 ft. elev.

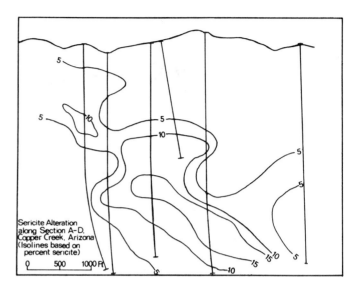

C. Sericite alteration along Section A-D, Copper Creek, Arizona (isolines based on per cent sericite).

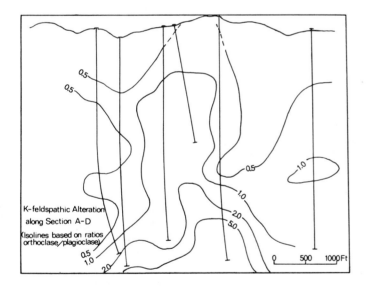

B. K-feldspathic alteration along Section A-D, Copper Creek, Arizona (isolines based on orthoclase/plagioclase ratios).

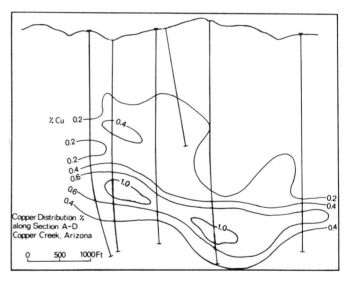

D. Copper distribution along Section A-D, Copper Creek, Arizona.

Figure 25.5. Per cent copper, sericite alteration, K-feldspar alteration, Copper Creek, Arizona. (Reproduced from Hausen and Kerr, 1971.)

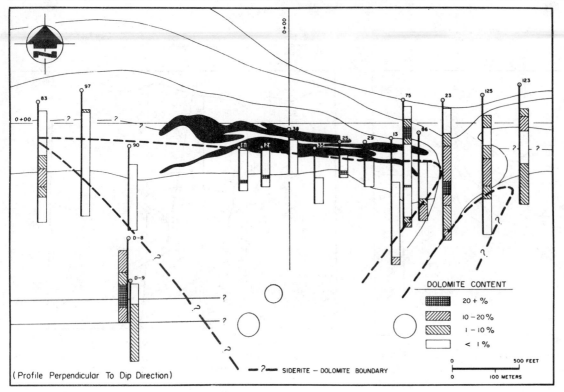

Figure 25.6.
Distribution of dolomite in the Mattabi Mine area. Reproduced from Franklin et al. (1975).

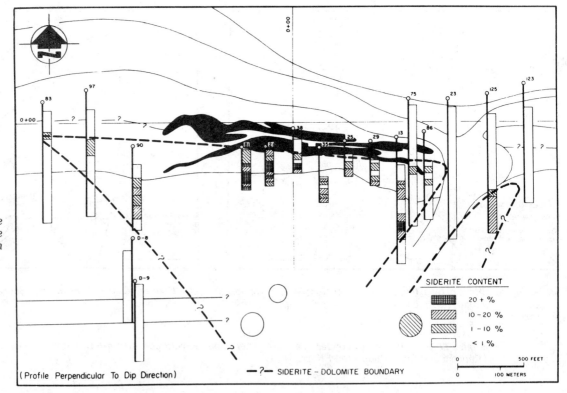

Figure 25.7.
Distribution of siderite in the Mattabi Mine area. Reproduced from Franklin et al. (1975).

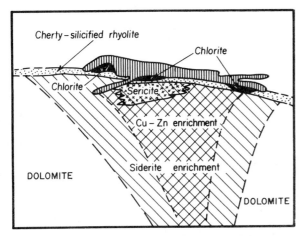

Figure 25.8.
Generalized distribution of alteration types in the footwall rocks of the Mattabi Mine. Reproduced from Franklin et al. (1975).

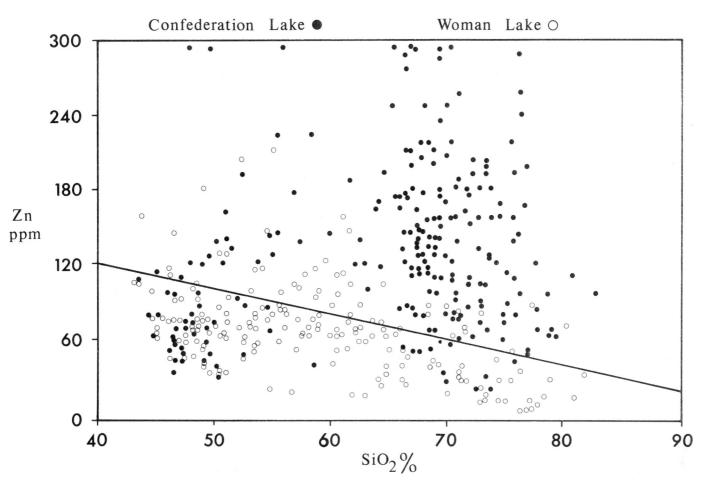

Figure 25.9. *Comparison of zinc with SiO_2 between the mineralized Confederation Lake Volcanic Cycle and the unmineralized Woman Lake Volcanic Cycle, Uchi Lake Area, Ontario. (After Govett and Nichol, 1978.)*

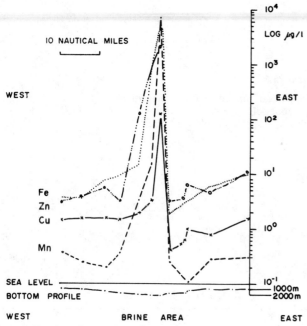

Figure 25.10. Variation of dissolved metal species in near-bottom water, west-east across Atlantis II Deep (no samples from brine). (After Holmes and Tooms, 1972.)

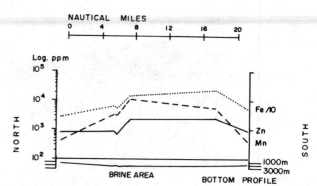

Figure 25.11. Variation in particulate metal contents in near-bottom water, north-south across Nereus Deep. (After Holmes and Tooms, 1972.)

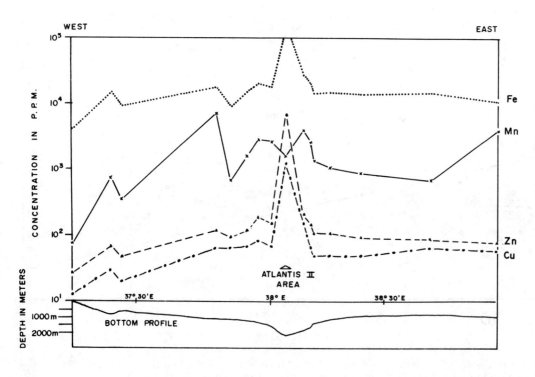

Figure 25.12. Cu, Zn, Fe, Mn contents of surface sediments along traverse across Atlantis II Deep. (After Holmes and Tooms, 1972.)

More recent work by Bignell (1975) and Bignell, Cronan and Tooms (1976), records anomalous geochemical halos of Fe, Mn, Cu, Zn and Hg detectable in surface sediments 3 km to 10 km from the metalliferous deposits in the Atlantic II Deep and the Nereus Deep (Fig. 25.13 and 25.14). Background values in these diagrams have been calculated from data from sediments collected along the axial valley sides within the Red Sea, away from the mineralized deeps.

The evidence from the modern environments of the Red Sea clearly supports the concept of the exhalative process advocated by numerous observers of massive sulphide deposits in ancient terranes and immediately suggests that contemporary sediments deposited at the same time as polymetallic massive sulphide mineralization in ancient exhalative environments could contain anomalous amounts of metals spatially related to economic mineral deposits. Such contemporary sediments include the chemically precipitated exhalites.

Figure 25.15 is a geological map published by Cameron (1977) of a volcanogenic massive sulphide deposit initially indicated in a lake sediment sampling survey in the Northwest Territories (Cameron and Durham, 1974a). The body is known as the Yava Syndicate or the Agricola Lake deposit. The exhalite horizon related to the mineralization, (known locally as the "B" horizon), is highly sericitized and the general geological relationships of the area are shown in Figure 25.16.

Sampling of rock outcrops beyond the limits of Figure 25.15 proved the presence of anomalous levels of Zn, Cu and Pb along the extension of the massive sulphide-bearing stratigraphy (Cameron and Durham, 1974b). Anomalous values for zinc occurring along or near the "B" horizon exhalite and its projection are located approximately 300 m and approximately 900 m beyond the suboutcropping massive sulphides (Fig. 25.17). Such anomalous dispersion most likely occurred during the exhalative period when the massive sulphide deposit was being formed.

There is an indication, therefore, that anomalous dispersion in horizons contemporaneous with massive sulphides is detectable in some Precambrian volcanogenic environments more than 300 m from the massive sulphides. This has significant implications with respect to the potential of lithogeochemistry in mineral exploration in that, in vertically dipping stratigraphy, indications of significant, blind massive sulphide concentrations buried to depths of 300 m or more could be detectable in samples of exhalite horizons from surface outcrops.

Russell (1975) described unusually high zinc values and anomalous manganese values in an exhalative sedimentary iron formation and contemporaneous limestone facies 1 km to 7 km from the Tynagh base metal deposit in Lower Carboniferous rocks in Ireland (Fig. 25.18). Gwosdz and Krebs (1977) described somewhat similar dispersion extending 5 km from the Meggen deposit in Germany (Fig. 25.19).

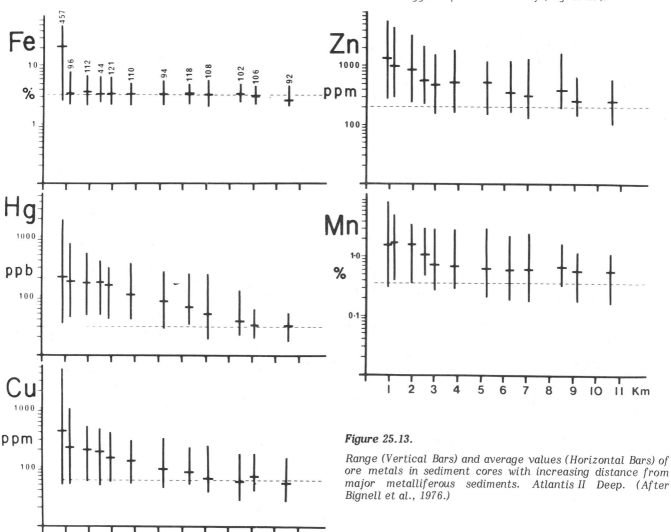

Figure 25.13.

Range (Vertical Bars) and average values (Horizontal Bars) of ore metals in sediment cores with increasing distance from major metalliferous sediments. Atlantis II Deep. (After Bignell et al., 1976.)

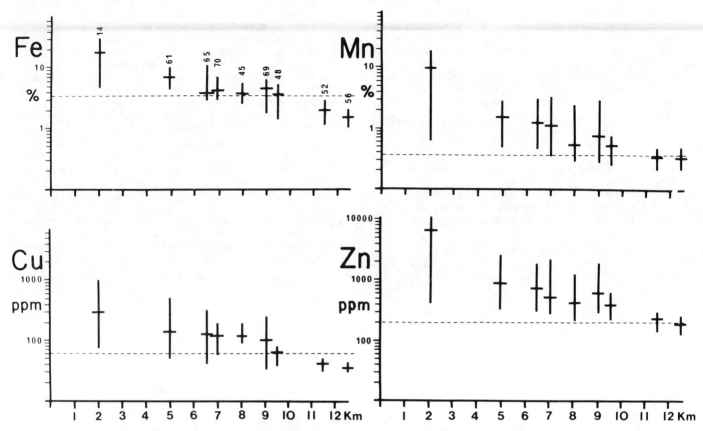

Figure 25.14. Range (Vertical Bars) and average values (Horizontal Bars) of ore metals in sediment cores with increasing distance from major metalliferous sediments. Nereus Deep. (After Bignell et al., 1976.)

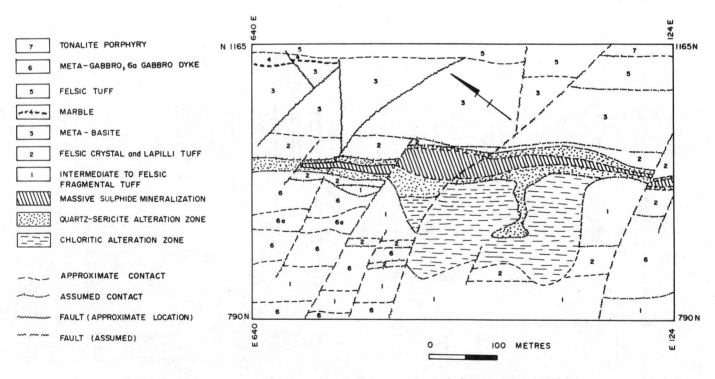

Figure 25.15. Geology of the Agricola Lake deposit. (After Cameron, 1977.)

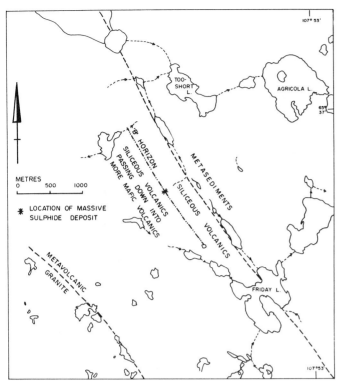

Figure 25.16. General Geology-Agricola Lake area. (After Cameron and Durham, 1974b.)

The dispersion patterns that are described are compatible with the volcanogenic model. Investigations involving careful sampling of exhalite horizons require the input and guidance of the exploration geologist familiar with the model and appreciative of the fact that the significant exhalite horizons can be thin and relatively inconspicuous.

The complete integration of the lithogeochemical technique with geology can be illustrated by noting the potential role of lithogeochemistry in the development and refining of geological concepts leading to a better understanding of the ore environments and ore genesis.

Potentially Useful Research Integrating Geophysics, Geochemistry and Geology

Pertinent information on mineralization genesis has been obtained from a geochemical study of active geological processes in such localities as hotsprings areas where mineral-bearing solutions are circulating, ocean floors where processes related to plate-tectonism are active, and areas such as the Red Sea and Vulcano where volcanogenic processes have deposited sulphide mineralization. As has been noted with reference to Red Sea data, geochemical observations in recent environments have served to confirm or explain conceptual mineralization processes advocated by observers of mineralized environments in ancient terranes. Parallel studies in other recent environments may provide pertinent information leading to a better understanding of the genesis of other types of mineral deposits, which, indirectly, will lead to the design of more effective exploration programs applicable throughout the geological succession.

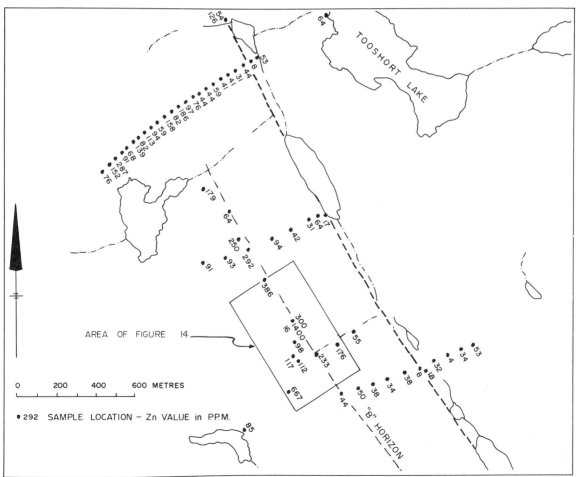

Figure 25.17. Zinc content of rock samples from the Agricola Lake area. (After Cameron and Durham, 1974.)

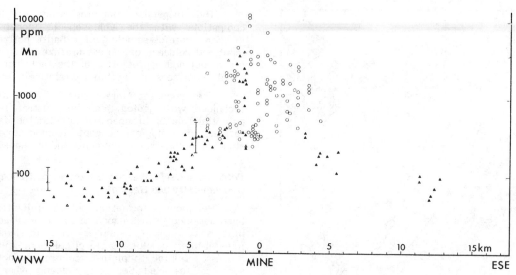

Figure 25.18. Manganese content of Waulsortian limestone horizon reflecting exhalative disperson around the Tynagh deposit (Mine). Circles represent drill core samples, triangles represent outcrop samples. (Reproduced after Russell, 1974.)

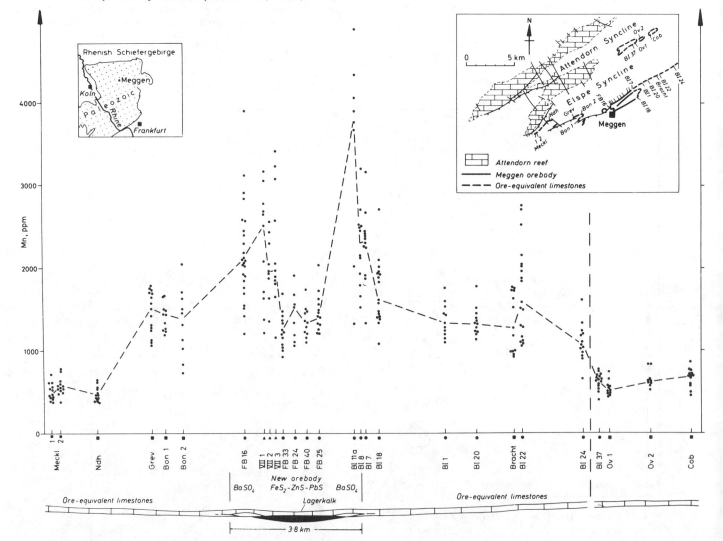

Figure 25.19. Manganese distribution in Lagerkalk and ore-equivalent limestones in area of Meggen ore deposit: southeastern flank of Elspe syncline separated from anticline structure between Elspe and Attendorn syncline by vertical broken line; other broken line links median values (■ quarry, ▲ underground working, ● drillhole). (Reproduced from Gwosdz and Krebs, 1977.)

An area where meaningful geophysical results might be anticipated would be research on the post-mineralization impact of a large sulphide mass on the host environment. In arid climates, the effect of acid weathering on the physical and chemical properties of the surrounding rock is poorly documented although there is some evidence that IP – responsive clay minerals are formed locally. Electrochemical effects driven by the oxidation of a sulphide mass may also produce post-mineral changes in the host rocks which are detectable at a distance. Such a mechanism for dispersion of elements is currently understudy and a considerable amount of interesting data is being accumulated (Govett, 1973; Govett et al., 1976; Bolviken and Logn, 1975). Comprehensive geochemical studies of the weathering zone overlying known mineralized environments similar to those described by Nickel et al. (1977) and those being carried out in other CSIRO programs in Australia, may also reveal important geochemical parameters which would be pertinent in the identification of favourable environments from surface observations.

Another area for fruitful investigation concerns the alteration mineralogy of the iron minerals. Magnetite and the various hematites are very difficult to distinguish by most analytical techniques including XRD, yet these minerals have relatively distinct magnetic properties and appear to be sensitive indicators of chemical environments. Some work of this type has been done by US Geological Survey personnel using high resolution susceptibility measurements to characterize the chemical environment of uranium roll fronts.

COST EFFECTIVENESS

Just as integrated exploration offers significant opportunities to the exploration geologist so the selection and application of the variety of available techniques presents an important challenge.

There are numerous examples on record of areas being systematically covered by one technique after another in the hope that the accumulation of data may reveal all significant geological secrets. This is a bland, unimaginative approach to exploration.

Cost-effectiveness is of prime importance in good exploration practice. Optimum cost-effectiveness is based on considering the incremental cost of obtaining additional information from each survey in an integrated program and on establishing the proper sequence of surveys, mapping and drilling so that each stage builds on the body of knowledge already obtained. Confirmation of geological information which has already been obtained with an adequate confidence level, is both time consuming and expensive.

The old adage "you will never find a mine until you break rock" cannot be denied and it should be the objective of all exploration teams to strive for the geological fact as expeditiously as possible and equate the cost of diamond drilling or other definitive and factual geological data acquisition against the added time and cost of additional discriminatory geophysical and geochemical data gathering.

Optimum cost-effectiveness requires a flexible approach consistent with the level of geological knowledge, and programming in which the various exploration decision points are carefully considered in advance. Within an integrated program, there are more stages at which to place a decision point which can take advantage of appropriate techniques to provide reliable negative or positive information.

The problem of differentiation of EM anomalous zones in Precambrian terranes of the Canadian Shield covered by extensive and complexly transported overburden, has been attacked in several ways. The high incidence of graphite in these regions has made systematic drilling of the EM zones expensive and unjustifiable. Various criteria based on the strength of the anomalies, and their length and position relative to broad, complex, major anomalous trends have also been applied with mixed and often negative results. Overburden drilling and sampling has been adapted to the problem and, by examination of trace element values and heavy mineral concentrates in basal tills, differentiation of graphite, barren sulphides and metal-bearing sulphides can be achieved under certain conditions. Favourable targets thus differentiated are then drilled.

Provided the parameters of size and depth of any desired massive sulphide deposit are properly defined, surface gravity techniques are capable of differentiating less dense graphitic sources from the denser, sulphide-rich sources which are more likely to be orebodies.

Overburden geochemical drilling, under favourable conditions, has the property of discrimination between barren and metal-bearing sulphide accumulations, but the cost of the geochemical drilling, especially in complex glacial overburden deeper than 30 m, becomes significant when anomalies are being investigated. The gravity technique, in concert with seismic and elevation surveys adapted for rapid, inexpensive coverage under Canadian Shield conditions, can quickly and relatively cheaply, eliminate anomalies related to graphite and other low density sources. Cost-savings achieved utilizing the gravity method at this stage of a program have been known to offset the additional diamond drilling costs required to examine remaining targets related to all higher density sulphide concentrations and also offset the extended time plus overburden drilling and related analytical costs that would be required to define the fewer targets by the alternative approach.

The principle of cost-saving and cost-efficiency is also illustrated by a much simpler example from Basin and Range Province areas where considerable exploration has been directed towards evaluation of pediment areas for buried porphyry copper deposits. Following a careful study of available geology, certain pediment areas have been selected for drilling to determine if significant features correlatable with porphyry copper mineralization in the pediment bedrock could be identified. Repeatedly, this approach proved unsuccessful and led to the abandonment of the property because, after acquisition of land, sometimes at a significant cost, holes were drilled thousands of feet without penetrating the valley gravels.

An integrated program has been developed with identical objectives but, instead of the first decision point in the program being the identification of a significant geological feature in pediment bedrock, the first decision point was based on an estimation of the depth of the pediment gravel. Relatively cheap gravity profiles carried out on public roads across the valleys quickly established the location of the Range Front faults bounding the pediments, and also provided adequate estimates of gravel thickness. In this integrated program every drillhole tested bedrock, and many areas were rejected on the basis of excessive depth of bedrock.

CONCLUSIONS

The strength of integrated exploration is the fact that the geophysicist and the geochemist can sense many aspects of a geological environment that the geologist cannot observe. The conversion of geophysical and geochemical data into geological information is usually ambiguous and imprecise without the input of the geologist and the objective of integrated exploration is to convert data into geological information at a satisfactory level of precision.

Integrated exploration does not come into being automatically. Exploration geophysics, exploration geochemistry and exploration geology have, to some extent, evolved separately from different roots, and competition, rather than co-operation, has been the norm. Much of the integrated exploration practiced today has resulted from the insistence of enlightened exploration managers rather than the natural tendencies of individuals. Fortunately, once integrated exploration has taken root under whatever impetus, most individuals have been quick to appreciate the benefits and have accepted the philosophy wholeheartedly.

The contribution of the geologist is important. Integrated exploration is dependent on geologists being able to formulate meaningful questions which relate to the environment of mineral deposits and which, in some way, can be answered by changes in physical and chemical properties.

Geophysical and geochemical techniques of today can be designed to be cost-effective in the detection of the relatively gross physical and chemical property contrasts associated with the direct detection of mineralization and in the provision of critical and useful geological information relevant to the environment of mineral deposits.

Objective application of this capability and expertise in advancing the geological understanding of the genesis of mineral deposits will perpetuate and improve the geological momentum which we have witnessed over the past 20 to 25 years. This will ensure the progressive development of our exploration capabilities in the face of the challenging demand of our various societies.

REFERENCES

Bignell, R.D.
1975: The geochemistry of metalliferous brine precipitates and other sediments from the Red Sea; unpublished Ph.D thesis, Imperial College, London.

Bignell, R.D., Cronan, D.S., and Tooms, J.S.
1976: Metal dispersion in the Red Sea as an aid to marine geochemical exploration; Inst. Min. Met. Trans. (Sect B: Appl. Earth Sci.) v. 85, p. B274-B278.

Bolviken, B. and Logn, O.
1975: An electrochemical model for element distribution around sulphide bodies; in Geochemical Exploration 1974, Elliott, I.L. and Fletcher, W.K. (Eds.), 5th Int. Geochem. Explor. Symp., Elsevier Publ. Co., Amsterdam, p. 631-648.

Boyd, D.
1969: The contribution of airborne magnetic survey to geological mapping; in Mining and Groundwater Geophysics/1967, Geol. Surv. Can., Econ. Geol. Rept. no. 26, p. 213-227.

Cameron, E.M.
1977: Geochemical dispersion in mineralized soils of a permafrost environment; J. Geochem. Explor., v. 7(3), p. 301-326.

Cameron, E.M. and Durham, C.C.
1974a: Geochemical studies in the eastern part of the Slave structural province, 1973; Geol. Surv. Can., Paper 74-27.

1974b: Follow-up investigations on the Bear-Slave geochemical operation; in Report of Activities, Part A, Geol. Surv. Can., Paper 74-1A, p. 53-60.

Coope, J.A.
1958: Studies in geochemical prospecting for nickel in Bechuanaland and Tanganyika; unpublished Ph.D thesis, Imperial College, London.

Franklin, J.M., Kasarda, J., and Poulsen, K.H.
1975: Petrology and chemistry of the alteration zone of the Mattabi massive sulphide deposit; Econ. Geol., v. 70, p. 63-79.

Govett, G.J.S.
1973: Differential secondary dispersion in transported soils and post-mineralization rocks: an electrochemical interpretation; in Geochem. Explor. 1972; Jones, M.J. (Ed.), 4th Int. Geochem. Expl. Symp., Inst. Min. Met., London, p. 81-91.

Govett, G.J.S., Goodfellow, W.D., and Whitehead, R.E.S.
1976: Experimental aqueous dispersion of elements around sulfides; Econ. Geol., v. 71, p. 925-940.

Govett, G.J.S. and Nichol, I.
1978: Lithogeochemistry in mineral exploration; Paper 25 in Geophysics and Geochemistry in the Search for metallic ores, P.J. Hood (Ed.), Geol. Surv. Can., Econ. Geol. Rept. 31.

Gwosdz, W. and Krebs, W.
1977: Manganese halo surrounding Meggen ore deposit, Germany; Inst. Min. Met. Trans., Sect. B, v. 86, p. B73-B77.

Hausen, D.M. and Kerr, P.E.
1971: X-ray diffraction methods of evaluating potassium silicate alteration in porphyry mineralization; Can. Inst. Min. Met. Spec. Vol. 11, p. 334-340.

Hawkes, H.E. and Webb, J.S.
1962: Geochemistry in mineral exploration; Harper and Row, New York, 415 p.

Holmes, R. and Tooms, J.S.
1972: Dispersion from a submarine exhalative orebody; in Geochem. Explor. 1972, Jones M.J. (Ed.), Inst. Min. Met., London, p. 193-202.

Hood, P.J., Sawatzky, P., Kornik, L.J., and McGrath, P.H.
1976: Aeromagnetic gradiometer survey, White Lake, Ontario; Geol. Surv. Can., Open File 339.

Morley, L.W.
1969: Regional geophysical mapping; in Mining and Groundwater Geophysics – 1967, Geol. Surv. Can., Econ. Geol. Rept. no. 26, p. 249-258.

Nichol, I., Lavin, O.P., McConnell, J.W., Hodgson, C.J., and Sopuck, V.J.
1975: Bedrock composition as an indicator of Archean massive sulphide environments; Abstract of Paper presented at the 77th Annual General Meeting of the Canadian Institute of Mining and Metallurgy, Toronto, Can. Inst. Min. Met. Bull, v. 68, no. 755, p. 48.

Nickel, E.H., Allchurch, P.D., Mason, M.G., and Wilmshurst, J.R.
1977: Supergene alteration at the Perseverence nickel deposit Agnew, Western Australia; Econ. Geol., v. 72, p. 184-203.

Pelton, W.H.
1977: New IP method may discriminate between graphite and massive sulphides; N. Miner, v. 63, no. 27, Sept. 15, 1977.

Russell, M.J.
1974: Manganese halo surrounding the Tynagh ore deposit, Ireland; a preliminary note; Inst. Min. Met. Trans., London, Sect. B, v. 83, p. B65-B66.

1975: Lithogeochemical environment of the Tynagh base metal deposit, Ireland and its bearing on ore deposition; Inst. Min. Met. Trans., London, Sect. B, v. 84, p. B128-B133.

EXPLORATION DISCOVERIES, NORANDA DISTRICT, QUEBEC
(Case History of a Mining Camp)

J. Boldy
Gulf Minerals Canada Limited, Toronto, Ontario, Canada

Boldy, J., Exploration discoveries, Noranda district, Quebec (Case History of a Mining Camp); in Geophysics and Geochemistry in the Search for Metallic Ores; Peter J. Hood, editor; Geological Survey of Canada, Economic Geology Report 31, p. 593-603, 1979.

Abstract

This paper is a review of exploration discovery in the Noranda district over a 60-year period and illustrates the varied and interrelated role played by prospecting, geology, geophysics and geochemistry in the discovery of massive sulphide deposits.

Discovery case histories are examined and data compiled, showing the primary and contributory exploration methods which were successfully employed. To date, 19 sulphide deposits have been discovered in the Noranda district. Based on primary exploration methods, 4 resulted from surface prospecting, 8 from geological methods, 6 from geophysical surveys and 1 from the use of pathfinder lithogeochemistry. A valid set of prospecting criteria can be established and applied from these case histories.

Each case history is judged within its own time frame, and is characterized by a certain exploration methodology which cannot be divorced from the exploration philosophy prevailing at the time. For this reason, the discoveries are related to the Exploration Life Cycle of the district. The Cycle is divided into three phases: 1) The Early Discovery Years (1920-1935), dominated by prospecting discoveries, 2) The Middle Discovery Years (1935-1955), balanced between ground geophysical discoveries and those resulting from empirical geological methods, and 3) The Later Discovery Years (1955-1977), dominated by volcanogenic geological discoveries, airborne geophysical discoveries, and the utilization of pathfinder geochemistry as an exploration aid to locate blind ore deposits. Today, exploration in the Noranda district is one of diminishing return using currently available technology. It is proposed that an accelerated discovery rate can be achieved — particularly in the search for deeply buried, blind deposits — by further development and application of pathfinder lithogeochemistry, coupled with an appreciation of the role of volcanism in ore genesis.

Résumé

Ce rapport fait l'inventaire des découvertes dans le district de Noranda sur une période de 60 ans, et fait ressortir le rôle varié et mutuel de la prospection de la géologie, de la géophysique et de la géochimie dans la découverte des gisements de sulfures massifs.

On a étudié quelques cas particuliers relatifs à ces découvertes et compilé des données montrant ainsi les méthodes d'exploration primaire et secondaire qui avaient été employées avec succès. A ce jour, 19 gisements de sulfure ont été découverts dans le district de Noranda. Dans le cas des méthodes d'exploration primaire, 4 gisements ont été découverts au moyen de travaux de surface, 8 par des méthodes géologiques, 6 par des levés géophysiques et 1 à partir d'indicateurs géochimiques. On peut établir et appliquer un ensemble valable de critères de prospection à partir de ces cas particuliers.

Chaque cas est jugé dans le cadre de l'époque; il est caractérisé par une certaine méthodologie de l'exploration, inséparable de la philosophie de l'exploration à l'époque considérée. Pour cette raison, les découvertes font partie du cycle d'exploration du district de Noranda. Ce cycle est divisé en trois phases: 1) les premières années de la découverte (1920-1935), dominées par les découvertes par prospection, 2) les années intermédiaires de la découverte (1935-1955), avec un équilibre entre les découvertes dues à la prospection géophysique au sol et aux méthodes géologiques, et 3) les dernières années de découverte (1955-1977), dominées par les découvertes de gîtes volcaniques, celles faites par suite des levés géophysiques aéroportés et de levés géochimiques, où l'on utilise des indicateurs géochimiques pour localiser les gisements métallifères n'affleurant pas. Aujourd'hui, l'exploration dans le district de Noranda n'est plus rentable avec la technologie actuelle. Pour atteindre un taux de découverte élevé, on propose — particulièrement dans la recherche des gisements profonds et n'affleurant pas — de perfectionner les méthodes géochimiques où l'on utilise des "indicateurs" en tenant compte aussi du rôle du volcanisme dans la genèse du minerai.

INTRODUCTION

The Exploration '77 Symposium was a state-of-the-art review of the various geophysical and geochemical techniques employed in mineral exploration. Without doubt, technological aids have played a vital part in target definition and subsequent discovery, by detecting certain anomalous physical and/or chemical properties associated with metallic ores. However, as important as these disciplines are in modern exploration, the role of geology should not be underestimated or ignored. An appreciation of the geological environment under investigation is equally vital, in order to enhance the probabilities of discovery.

This paper is a personal geohistorical review of the history-of-the-art of massive sulphide exploration and discovery in the Noranda district over the past 60 years or so. It is a case history of a mining camp, viewed in the context of

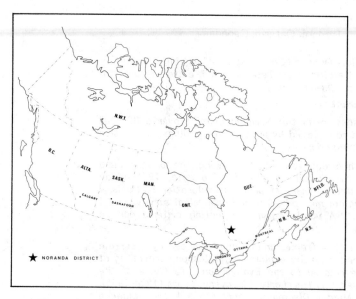

Figure 26.1.   Location of Noranda district, Quebec.

certain significant exploration developments, which aided in establishing one of the world's premier base-metal mining districts located in the Abitibi region of northwestern Quebec (Fig. 26.1). Like all histories, it may be somewhat subjective. In any event, it attempts to bring out the interrelationship of the various exploration disciplines, each of which was successful in the context of its time, and each of which owes a debt to those of other disciplines, and to those who were earlier on the scene. It is the author's belief that exploration effectiveness will be increased by studying the exploration history of a mining district in depth, and in its entirety, particularly the evaluation and use of various prospecting criteria which led to discovery.

NORANDA SULPHIDE DEPOSITS

The geological setting of the Noranda district deposits has been adequately covered in the literature, namely by Wilson (1941), Gilmour (1965), Dugas (1966), Sangster (1972), and Spence and de Rosen-Spence (1975). The bibliography which has been compiled, comprises principally broad concepts and selected references of case history technology. Detailed geological descriptions of the various deposits may be obtained by referring to these publications.

The term 'massive sulphides' refers to mineralization containing greater than 50 per cent sulphides composed of varying proportions of pyrite, pyrrhotite, sphalerite and chalcopyrite, with lesser amounts of galena, magnetite and other metallic minerals, not all of which are necessarily present in any one deposit. The massive sulphide component of a deposit consists of one or more lensoid masses which are essentially conformable with the host stratigraphy. Disseminated sulphides tend to flank the massive sulphides in the stratigraphic footwall, generally a more or less altered felsic volcanic unit. This alteration zone is often crudely pipelike in form. Disseminated sulphides also occur along the plane of ore-hosting rocks, generally a cherty-tuffaceous unit, a product of terminal felsic volcanism.

Although there are certain exceptions to the rule, deposits tend to be associated with the terminal phases of felsic pyroclastic volcanism. Most sulphide masses, with the exception of one or two of the larger deposits in the Noranda district, occur as fairly discrete bodies; the maximum dimensions are generally less than 400 m along the plane of the host rocks. By nature, these deposits have a variable response to geophysical and geochemical search techniques, being dependent in large part on their depth of burial below surface. It is vital therefore to have a proper understanding of the geology and geometry of these deposits, in order to obtain a sharper focus for target definition.

The Noranda district is the most prolific of Canadian Precambrian massive sulphide clusters, aggregating 114 million tons with an average grade of 2.14% copper, 1.37% zinc, 0.59 oz. silver, and 0.12 oz. gold per ton (Boldy, 1977). This tonnage is obtained from 19 deposits, 85 per cent of them located within (16 km) of the major Horne deposit. This tonnage is equivalent to 30 per cent of the (volcanogenic) base metal tonnage of the Abitibi area of northwest Quebec and northeast Ontario, and in total is equivalent to 20 per cent of Canada's Precambrian (volcanogenic) base metal tonnage. Currently, the Millenbach deposit remains in production, the Corbett and Macdonald deposits are under various stages of development, and the New Insco, Magusi River, and Mobrun deposits are considered as reserves for the future. Although primarily a base metal district, the Noranda area was a gold mining district in years past. Sixteen deposits have been mined out, aggregating 17 million tons, grading 0.16 oz. gold per ton.

EVALUATION DISCOVERY TECHNOLOGY

Reference to Table 26.1 and Figures 26.2 and 26.3, illustrate the discovery role played by prospecting, geology, geophysics and geochemistry in the Noranda district. The primary and contributory roles played by these disciplines have been examined and evaluated for all the various Noranda district deposits. Often a difficult decision had to be made when deciding which of several methods was the one primarily responsible for discovery. However, in order to understand the real significance of the various successful methodologies employed, the discoveries should be placed in their time frame, so that one can appreciate the interplay of ideas and their application during the course of an exploration program. Each case is governed by the following points:

(1) the available knowledge/data base for the period;

(2) the availability of applicable technological search tools;

(3) the prevalence and diffusion of new exploration concepts (often considered unorthodox at the time);

(4) the appropriate application of ideas and techniques and their proper execution in the field.

There is no doubt that some discoveries resulted from a complete and correct analysis of the evidence at hand, however, cases do exist where serendipity played an overly large part in discovery! Perhaps if the whole truth were known, most discoveries resulted when adequate preparation met opportunity, and was firmly grasped.

RELATIONSHIP OF DISCOVERY TECHNOLOGY TO EXPLORATION LIFE CYCLE AT NORANDA

Reference to Table 26.2 illustrates the relationship of discovery to the Exploration Life Cycle of the Noranda district. Exploration at Noranda may be conveniently divided into three phases, each roughly separated by a hiatus of about seven barren discovery years.

The Early Discovery Years (1920-1935)

As might be expected, prospecting discoveries dominated this period. Ed Horne staked the sulphide gossans near Osisko Lake in Rouyn Township in 1920, whilst intermittently prospecting the area for gold since 1911. In 1923, a drilling program was initiated which was successful in intersecting a major copper-gold zone beneath the pyritized

Table 26.1

Compilation of discovery technology, volcanogenic sulphide deposits, Noranda district, Quebec

	Deposit	Year of Discovery	Prospecting	Geology - Mineralization	Geology - Alteration	Geology - Lithology-Stratigraphy	Geology - Structure	Geology - Other	Geophysics - Aerial EM	Geophysics - Ground EM	Geophysics - Ground Mag.	Geophysics - I.P.	Geophysics - Gravity	Geochemistry - Soil	Geochemistry - Litho-Geochem	Geochemistry - Mercury Pathfinder	Depth of Deposit from Surface
EARLY YEARS (1920-1935)	HORNE - UPPER 'H'	1923	P	X			X										SURFACE
	AMULET - UPPER 'A'	1925	P	X	X												SURFACE
	AMULET 'C'	1925	P	X		X											SURFACE
	OLD WAITE	1925	P	X	X												SURFACE
	ALDERMAC	1925	X	X	X						P						-30'
	AMULET 'F'	1929		X	X	X		P									-125'
	HORNE - LOWER 'H' *	1931		P		X	X	X									-1300'
MIDDLE YEARS (1935-1955)	AMULET - LOWER 'A' *	1938		X	P	X											-700'
	JOLIET	1940	X	P		X											-25'
	MACDONALD	1944	X	X		X				P							-25'
	QUEMONT	1945				X	X			P							-200'
	D'ELDONA	1947	X	P		X	X										-500'
	EAST WAITE	1949				X		P									-1300'
LATER YEARS (1955-1977)	MOBRUN	1956				X				P					X		-30'
	VAUZE	1957				P	X										-25'
	DUFAULT - NORBEC	1961				P	X										-1100'
	DELBRIDGE	1965		X		X					X					P	-300'
	DUFAULT - MILLENBACH	1966				P	X	X								X	-2300'
	MAGUSI RIVER	1972				X			P								-50'
	NEW INSCO	1973				X			P								-50'
	DUFAULT - CORBETT	1974				P	X									X	-2300'

P = PRIMARY DISCOVERY METHOD
X = CONTRIBUTORY METHOD
* = SUBSEQUENT DISCOVERY

rhyolites. This discovery was to become the famous Horne mine, and with it, the Noranda organization was born (Roberts, 1956). It is interesting that within a year of its discovery, the property was surveyed using the spontaneous polarization (SP) technique, the first such geophysical survey conducted in Canada (Kelly, 1957). Although most of the ore lenses were discovered without reference to the SP results, the results did confirm the presence of a multiple series of sulphide lenses, some of which did not outcrop.

By 1925, other gossans were discovered (8 km) to the northwest of the Horne mine. These were the Amulet Upper A and Amulet C deposits in Dufresnoy Township, and the Old Waite deposit (2.4 km) to the northwest in Duprat Township.

The first geophysical discovery of the period was made in 1925 in a location (16 km) to the west of the Horne mine in Beauchastel Township. There, a simple dip-needle magnetic survey led to the discovery of the Aldermac deposit which

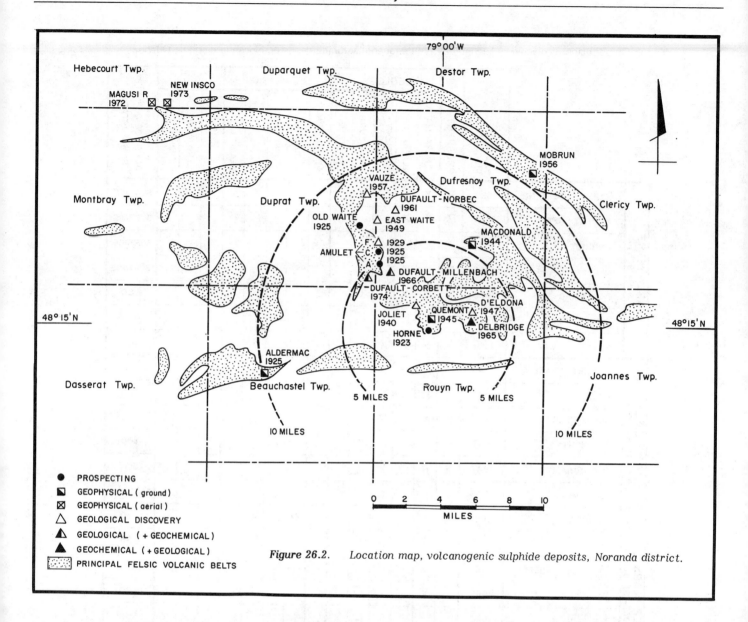

Figure 26.2. Location map, volcanogenic sulphide deposits, Noranda district.

was located under low ground, about 100 m north of mineralized and altered felsic volcanics. The alteration (cordierite-hornfels) was recognized to be similar to that associated with the Amulet Upper A and Amulet C deposits.

In 1929, a drill program based on geological interpretations intersected the Amulet F deposit, which was located 1.6 km to the north of the Amulet C deposit. Mineralized and altered felsic volcanic outcrops existed to the west of this discovery, and the drilling had intersected massive sulphides at a shallow depth, on the downdip extension of the (Amulet) rhyolite-andesite contact.

The period closes in 1931 with the discovery of the large Lower H deposit in the Horne mine during the course of underground development. It was noted at the time that the massive sulphide lenses in the mine area were located in a major fault wedge, within which the felsic volcanics were brecciated, sericitized and silicified. This specific structural setting was considered vital for the location and formation of this major ore deposit (Price, 1948).

The Middle Discovery Years (1935-1955)

This period is characterized by the empirical phase of geological exploration, where certain associations of lithology, mineralization, alteration and structure known to be favourable for ore deposition, were traced and investigated by drilling. Emphasis was placed in locating "favourable" fault or fold-controlled structures, within which mineralizing solutions rising from depth could replace physically or chemically receptive lithologies.

In 1938, after a period of exploration stagnation, the Amulet Lower A deposit was discovered 210 m beneath the surface outcrop of the smaller Amulet Upper A deposit. The two deposits were physically separated but were linked by a common alteration pipe of cordierite-anthophyllite which had pierced the favourable (Amulet) rhyolite-andesite contact. This contact hosted the Amulet C deposits to the north. In 1940, the Joliet deposit (copper-bearing siliceous flux ore) was discovered 1.6 km to the northwest of the Horne mine, where disseminated sulphides had been previously noted in sheared rhyolitic rocks. The deposit subcropped beneath shallow overburden.

Two major geophysical discoveries were made in 1944-1945. The first of these was the Macdonald deposit located in Dufresnoy Township 8 km to the northeast of the Horne mine. A subsurface conductor was defined by ground EM in an area containing scattered base metal bearing sulphide occurrences along a rhyolite-granodiorite contact. It was the first ground EM survey to be successful in locating a massive sulphide deposit in the Noranda district. The deposit occurred under 8 m of overburden.

The major geophysical discovery however, was that of the Quemont deposit, discovered in 1945. This property lies adjacent to the Horne mine, generally north of the Horne Creek fault, which up to then was thought to be the bounding structure for the emplacement of ore in this sector of the mining district. Much surface and underground exploration had been done without success during the previous 20 years. The successful geophysical technique utilized was a ground magnetic survey which outlined several magnetic anomalies. These were systematically drilled by Mining Corporation. Massive sulphides were found to be associated with one of the magnetic anomalies located along the north shore of Osisko Lake. This major discovery was a classic example of the importance of target definition and location.

In 1947, following the excitement generated by the Quemont discovery, empirical geological investigations continued. One of these resulted in the discovery of the D'Eldona deposit, 4 km to the east of the Quemont mine. Here, weakly disseminated pyritic sulphides had been noted on surface close to a northeasterly-trending diabase dyke (a continuation of the dyke that passes near the Horne mine), which cuts through a sequence of porphyritic rhyolites, not unlike those rhyolites that hosted the Quemont deposit 4 km to the west. A subsequent drilling program at this location resulted in the discovery of a pyrite-sphalerite lens 150 m below surface.

The final episode in this period occurred in 1949 with the discovery of the East Waite deposit, in a step-out hole, following the third deepening of an assessment drillhole which had been originally started in 1938! This deposit is located in Dufresnoy Township, 1.2 km to the northeast of the Old Waite deposit which had been discovered in 1925. The new deposit was down-dip from surface and occurred on the same lithological contact as that of the Old Waite deposit. This contact had been recognized in the assessment drillhole. The deposit was located 400 m below surface, and was underlain by a classic alteration pipe; in addition, the deposit appeared to be associated with a domal structure on the upper surface of the (Waite) rhyolite, and its position was masked by 0.4 km thickness of (Amulet) andesite, in part intruded by diorite. This discovery was to have an important influence on geological thinking in the district.

Following the geophysical and geological developments at the close of this period, probably one of the earliest attempts to utilize lithogeochemical methods in prospecting in the Amulet area was carried out by Riddell (1950). The investigation involved the determination of trace amounts of copper and zinc peripheral to the Amulet deposits, some of which outcropped, and were associated with cordierite-anthophyllite alteration pipes. Although it is not known whether or not this technique led to any discovery it did outline a broad, anomalous zone, and could be related to one or two of the deposits.

The Later Discovery Years (1955-1977)

With the continued use of ground EM systems which had been successfully employed in many massive sulphide districts, much time and effort were expended conducting EM surveys in the Noranda district. One such survey (truck-mounted EM) was initiated by Rio Tinto and was conducted

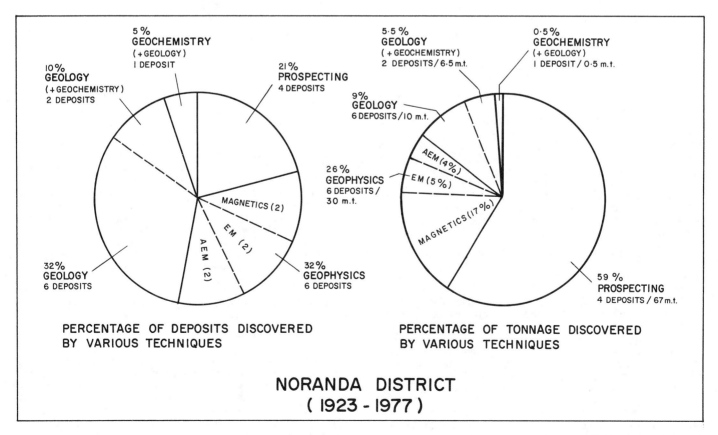

Figure 26.3. Percentage of deposits and tonnage discovered by various prospecting techniques in the Noranda district.

Table 26.2

Relationship of mineral exploration technology to the exploration life cycle of the Noranda district, Quebec

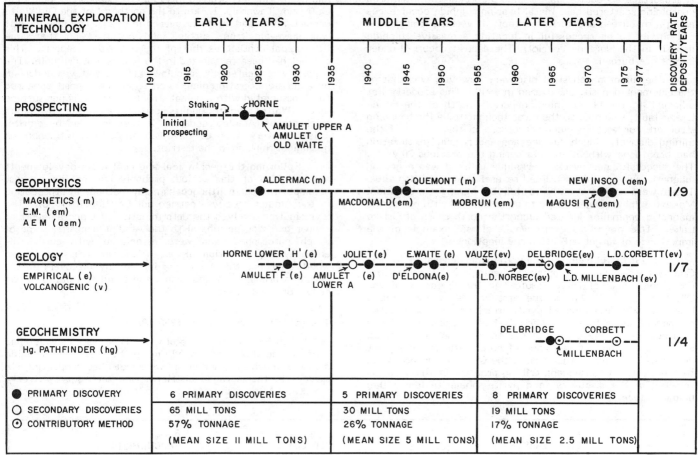

along road networks known to be underlain by felsic volcanics. This resulted in the discovery of the Mobrun deposit in northeast Dufresnoy Township in 1956, located 17.7 km to the northeast of the Horne mine (Siegel et al., 1957). In addition, a gravity survey helped confirm the presence of a massive sulphide deposit prior to drill investigation of the anomaly, and a tonnage estimate was made from the gravity results by Goetz (1958). An airborne EM test survey, carried out over this deposit later, has been described by Paterson (1961). Sixteen years were to elapse between this discovery and others made by geophysical methods.

The period commenced with a consolidation of empirical geological knowledge gained over the previous 30 years from the efforts of earlier workers. However, major advances in geological thinking were made in Australia in 1953 (King, 1965; 1976), and in Europe (Oftedahl, 1958), with the advancement of new concepts of ore genesis. After studies on the Broken Hill deposit in Australia, Haddon King, a Consolidated Zinc Corporation geologist, came to the conclusion that certain sulphide deposits should be considered as an integral part of their lithological environment. His ideas may be briefly paraphrased as follows: "The (sedimentary) concept in mineral exploration.....promises to lift ideas of ore occurrence out of the realm of miracles and put it, for the first time, where it belongs--in the scheme of natural events. In this view, ore occurrences should become not only understandable--perhaps even predictable." A similar viewpoint was developed in Canada during the course of investigations in the Bathurst camp in New Brunswick in the mid 1950s (Stanton, 1959; Cheriton, 1960; Miller, 1960; Sullivan, 1968).

Approximately concurrent with these developments in ideas of ore genesis, certain Noranda-based Consolidated Zinc geologists, namely Edwards, Gilmour and Spence, were also developing ideas that the massive sulphide deposits in Noranda were probably contemporaneous with their volcanic host rocks. Furthermore, they believed that major ore controls could be related to particular time — stratigraphic periods of felsic volcanism — and that localizing (volcanic) structures would be found associated with certain phases of felsic (eruptive) volcanism. In addition, they believed that the hydrothermal mineralizers, a terminal product of volcanism, would precipitate metals during the waning stages of a particular volcanic cycle, not unlike that of hotsprings and fumeroles (Gilmour, 1965; Spence, 1967; Spence and de Rosen-Spence, 1975).

The discovery of the Vauze deposit in 1957 by Consolidated Zinc Corporation, 2.4 km northwest of the East Waite deposit resulted from a revolutionary interpretation of existing data. Although of modest proportions, it was the first to be discovered using empirical methods, coupled to embryonic volcanogenic concepts. This discovery opened up a large subsurface area known to be underlain by the favourable (Waite) rhyolite, but covered by an extensive sequence of (Amulet) andesite.

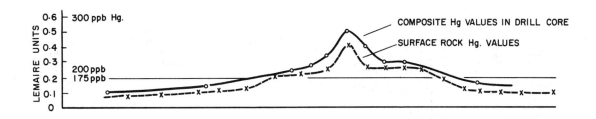

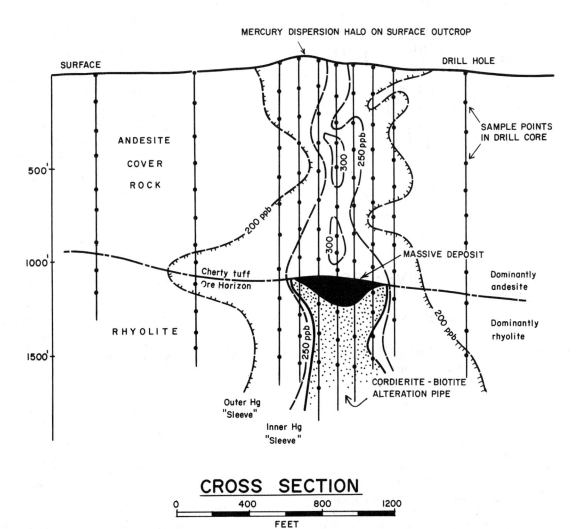

Figure 26.4. Section — Relationship of mercury dispersion halo to a blind massive sulphide deposit, Noranda district.

Much painstaking analysis of data was undertaken by various companies, based on interpreting drilling information, particular emphasis being placed on determining the subsurface configuration of the Waite rhyolite using isopachs and structural contour plotting of this felsic unit. This unit fortuitously dipped at a gentle angle, and combined with repetition of stratigraphy due to faulting, was within reach of surface exploratory drillholes. In addition, it was noted that the deposits in this sector of the district tended to occur along trends of 070° and 350° relative to the location of the East Waite deposit. In 1961, after a long barren period, hole 124 drilled by Lake Dufault Mines discovered the Norbec deposit in Dufresnoy Township, 2 km northeast of the East Waite, at a depth of 335 m. This discovery marked a turning point in the fortunes of Lake Dufault Mines.

Geochemical literature regarding the use of mercury as a pathfinder to discover blind sulphide deposits (Fursov, 1958; Ozerova, 1959; Hawkes and Williston, 1962) came to the attention of Falconbridge Exploration in 1963. In addition, previously published data on geochemical investigations of the East Tintic district, Utah (Lovering et al., 1948), revealed that meaningful information could be obtained by analyzing the trace element content of fractures uprake from blind deposits.

Analysis by Falconbridge of surface and drill core samples taken around the recently discovered Norbec deposit using a simple S-1 Lemaire detector, revealed easily detectable anomalous amounts of mercury in the 150-300 ppb range. A distinct primary dispersion halo was found to exist

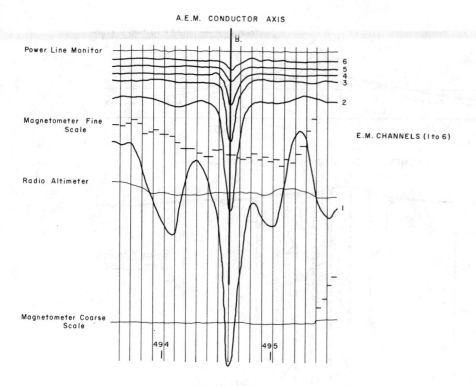

Figure 26.5. AEM Input survey profile, Magusi River district, Noranda district.

in cover rock above the deposit encompassing an area 460 by 210 m. The surface halo covered an area approximately four times the area of the underlying alteration pipe which is characteristic of this type of deposit (Fig. 26.4).

The detailed lithogeochemical investigation of the Norbec deposit and other blind deposits in the district was carried out over the following two years. It was found that the mercury was confined to a minute meshwork of fractures (i.e. grid-type alteration in Noranda terminology), suggestive that the leakage of mercury had indeed occurred from below, leaving an easily detectable three-dimensional imprint in the surrounding lithologies. In addition, the largest and highest amplitude mercury anomalies (300-1200 ppb) tended to be associated with massive sulphide deposits that contained appreciable amounts of zinc, silver or telluride mineralization. The application of this practical tool helped focus attention on certain areas which were considered prospective for deeply buried, blind sulphide deposits (Boldy, 1963).

An appreciation of the significance of mercury as a geochemical pathfinder coupled with a realization of the importance of volcanic stratigraphy in ore localization, resulted in Falconbridge discovering the Delbridge deposit in 1965; this deposit is 4 km east of the Horne – Quemont complex, and 0.4 km south of the old D'Eldona workings. It was the first deposit to be discovered using pathfinder lithogeochemistry in the Noranda area. In addition, frequency-domain IP surveys were utilized in defining weakly-disseminated pyritic mineralization occurring in the stratigraphic footwall of the blind deposit, which was subsequently found to be located 90 m below surface, and extended to a depth of 350 m from surface. IP surveys have been proven to be a useful exploration tool for massive sulphides (Hallof, 1960; Hendrick and Fountain, 1971).

In 1966, and more recently in 1974, the Millenbach and Corbett deposits were discovered by Lake Dufault Mines, in an area approximately 2 km distant from the Amulet A deposit. Both these new discoveries occurred at depths of about 700 m below surface. The discoveries were made using a variety of empirical (lithology, stratigraphy, and structural) information, particularly a deciphering of the structural setting, based on a fairly immense data bank gathered over the years from previous drilling programs. Pathfinder geochemistry (mercury and certain other elements) was a contributory aid at the Millenbach and also at the Corbett deposits. In the latter area, the leakage anomalies were horizontally offset in plan-view, which added complications to ready interpretation at that time.

After a sixteen year lapse since the previous successful geophysical discovery, an airborne Input EM survey conducted in 1972 by Questor Surveys on behalf of the Quebec Department of Natural Resources, (re) discovered a strong, isolated, magnetic conductor in Hebecourt township, 34 km northwest of the Horne mine (Fig. 26.5). This airborne EM system has been described by Lazenby (1973). This anomaly had been previously drilled with inconclusive results. Following a correct analysis of the geophysical data, drilling by the Iso-Copperfields group discovered the Magusi River deposit (Jones, 1973). A case history study was carried out by the McPhar group using various geophysical methods (Fountain, 1972). In 1973, a detailed Dighem AEM survey located the adjoining New Insco deposit (Fraser, 1974) on ground that had been previously acquired with stratigraphic concepts in mind.

It is interesting to note that the use of AEM surveys, which is a standard survey technique today, found two deposits after a period of almost 60 years of active exploration. If this tool had been available years earlier, perhaps up to half of the deposits discovered to date could have been located using this exploration tool.

Reference to Table 26.3 illustrates the tonnage and grade figures of the various volcanogenic sulphide deposits discovered to date in the Noranda district. At current metal prices the gross value of the volcanogenic sulphide deposits is in excess of $6 billion.

CONCLUSIONS

The Noranda district can be considered to have reached a mature stage of exploration development over the past sixty years. This is due in part to Nature's bounty, and in part to a certain tenacity and approach to exploration by those active on the local scene. The diffusion of ideas on theories of ore genesis, international in scope, coupled with the use of various technological aids, have all had an impact on

Table 26.3

Tonnage-grade data, volcanogenic sulphide deposits, Noranda district

Deposit	Status	Size (M/Tons)	% Cu	% Zn	oz. Ag	oz. Au
Horne	Past producer	60.26	2.20	--	0.40	0.17
Quemont	Past producer	16.35	1.20	1.80	0.54	0.12
Amulet A	Past producer	5.30	5.12	5.50	1.40	0.04
Magusi	Prospect	4.11	1.20	3.60	0.90	0.03
Norbec	Past producer	4.00	2.80	4.71	1.40	0.03
Joliet	Past producer	3.70	1.12	--	0.25	0.01
Millenbach	Producer	3.58	3.69	4.73	1.72	0.03
Macdonald	Inactive producer	3.25	0.07	4.77	0.63	0.02
Mobrun	Prospect	3.00	0.62	2.30	0.60	0.05
Corbett	Active development	2.93	2.92	1.98	0.59	0.03
Aldermac	Past producer	2.07	1.48	--	0.20	0.01
E. Waite	Past producer	1.50	4.13	3.30	0.90	0.05
O. Waite	Past producer	1.25	4.70	3.00	0.60	0.03
New Insco	Prospect	1.15	2.11	--	0.50	0.03
Amulet C	Past producer	0.60	2.12	8.50	2.00	0.02
Delbridge	Past producer	0.40	0.55	8.60	2.00	0.07
Vauze	Past producer	0.37	2.94	1.00	0.80	0.02
Amulet F	Past producer	0.28	3.54	3.40	0.10	0.02
D'Eldona	Past producer	0.10	0.30	5.00	0.76	0.12
		114.20	2.14	1.37	0.59	0.12

Note: 1-Grade and tonnage figures include production and reserves.

2-District metal content: Copper   2,455,000 tons
                          Zinc   1,542,000 tons
                          Silver   67,378,000 ounces
                          Gold   13,704,000 ounces

discovery, irrespective of the disciplines employed and the companies involved in discovery. No single person or company can afford to be too smug about their achievements. In return, many ideas and concepts advanced and developed by those who explored the Noranda district over a period of many years, saw these concepts develop into valid prospecting guidelines, and become of use to others engaged in volcanogenic massive sulphide exploration within the Noranda district, and elsewhere in the Precambrian Shield. Today, the use of the term Noranda-type i.e. a volcanogenic massive sulphide deposit, is firmly established in the geological literature (Gilmour, 1976).

The Early Discovery Years (1920-1935) were dominated by prospecting discoveries with geophysics and geology playing a minor role. The Middle Discovery Years (1935-1955) demonstrated the value of ground EM and magnetic surveys, coupled to the development and use of empirical geological methods. The Later Discovery Years (1955-1977) commenced with the development of volcanogenic concepts and coupled to a strong empirical base, resulted in a string of discoveries. During this period, well-designed ground EM and airborne EM systems still continued to discover ore in the peripheral areas of the district. In addition, the advent of the use of mercury geochemistry as a pathfinder to ore, helped focus attention on certain prospective areas and was successfully employed in targeting deeply buried blind deposits.

The review of various case histories of discovery illustrates the depth and scope of successful exploration technology applied over the years in the Noranda district. Currently, blind ore discoveries will probably continue to be made between 300 m and 900 m below surface. Deep ore search however, is difficult and expensive, and requires a long-term, adequate drilling budget. Today, we are faced with the problem of diminishing return using currently available technological search tools. However, an accelerated discovery rate will reward those who have an appreciation of the role of volcanism in ore genesis, coupled to the continued development and use of lithogeochemical pathfinder techniques. The presence of ghost mining districts is a reminder to all explorationists of what may lie ahead if we fail to meet the challenge. In any event, the development of successful exploration technology in the Noranda district can be considered a monument to the spirit of human endeavour, in this most heart-breaking of pursuits — the quest for ore!

ACKNOWLEDGMENTS

I am indebted to innumerable explorationists, especially to those who actively explored the Noranda district over a period of many years. Amongst the many, I would like to acknowledge the aid of several who helped compile the list on Discovery Technology. They are W. Bancroft, Bud Hogg, Peter Price, and Roger Pemberton — Noranda Mines Ltd.; Colin Spence — Rio Tinto Canadian Exploration; Hugh Squair — Selco Mining Corporation; Stan Charteris and Gordon Walker — Falconbridge Nickel; George Archibald and Bert Sakrison — New Insco Mines Ltd.; Mike Knuckey — Lake Dufault Division — Falconbridge Copper; Hugh Jones — Geophysical Engineering Ltd.; and E. Hart — Geological Consultant. In addition, an appreciation is extended to W.G. Robinson and D.H. Brown, and the late A.S. Dadson, with whom I had the pleasure of having many shared experiences in the Noranda district whilst in the employ of Falconbridge between 1962-1968.

The writer acknowledges the permission of the senior management of Gulf Minerals Canada Limited to publish this historical review. However, he assumes responsibility for the text and for any conclusions reached in the article.

SELECTED BIBLIOGRAPHY

Baragar, W.R.A.
 1968: Major element geochemistry of the Noranda volcanic belt, Quebec — Ontario; Can. J. Earth Sci., v. 5, p. 773-790.

Boldy, J.
 1963: Mercury dispersion halos as exploration targets; unpubl. Falconbridge report.
 1977: (Un)Certain exploration facts from figures; Can. Min. Metall. Bull., v. 70 (781), p. 86-95.

Cheriton, C.G.
 1960: Anaconda exploration in the Bathurst district of New Brunswick; Am. Inst. Min. Eng. Trans., v. 217, p. 278-284.

Descarreaux, J.
 1973: A petrochemical study of the Abitibi belt and its bearing on the occurrences of massive sulphide ores; Can. Min. Metall. Bull., v. 66 (730), p. 61-69.

Dreimanis, A.
 1960: Geochemical prospecting for Cu, Pb and Zn in glaciated areas, eastern Canada; 21st Int. Geol. Cong., Copenhagen, pt. 11, p. 7-19.

Dresser, J.A. and Denis, T.C.
 1949: Geology of Quebec; Econ. Geol., v. 3; Quebec Dep. Mines, Geol. Rept. 20, 562 p.

Dugas, J.
 1966: The relationship of mineralization to Precambrian stratigraphy in the Rouyn-Noranda area, Quebec; Precambrian Symposium, Geol. Assoc. Can., Spec. Paper no. 3, p. 43-56.

Fountain, D.K.
 1972: Ground geophysical data over Magusi River sulphide deposit, Noranda; McPhar case history study.

Fraser, D.C.
 1974: Survey experience with the Dighem AEM System; Can. Min. Metall. Bull., v. 67 (744), p. 97-103.

Friedrich, G.H. and Hawkes, H.E.
 1966: Mercury dispersion halos as ore guides for massive sulphide deposits, West Shasta district, California; Mineral Deposita, v. 2, p. 77-88.

Fursov, V.Z.
 1958: Halos of dispersed mercury as prospecting guides at Ashisai lead-zinc deposits; Geokhim No. 3, p. 338-344.

Gilbert, J.E.
 1953: Geology and mineral deposits of Northwestern Quebec; Geol. Soc. Am. and Geol. Assoc. Can. Field guidebook No. 10, p. 1-12.

Gilmour, P.
 1965: The origin of the massive sulphide mineralization in the Noranda district, northwestern Quebec; Geol. Assoc. Can. Proc., v. 16, p. 63-81.
 1976: Some transitional types of mineral deposits in volcanic and sedimentary rocks; in Handbook of stratabound and stratiform ore deposits, Elsevier Sci. Pub. Co., v. 1, p. 111-160.

Gleeson, C.F. and Cormier, R.
 1971: Evaluation by geochemistry of geophysical anomalies and geological targets, using overburden sampling at depth; Can. Inst. Min. Met., Spec. Vol. 11, p. 159-165.

Goetz, J.F.
 1958: A gravity investigation of a sulphide deposit; Geophysics, v. 23 (3), p. 606-623.

Goodwin, A.M.
 1965: Mineralized volcanic complexes in the Porcupine - - Kirkland Lake — Noranda region; Econ. Geol., v. 60, p. 955-971.

Hallof, P.G.
 1960: Uses of induced polarization in mining exploration; Am. Inst. Min. Eng. Trans., v. 217, p. 319-327.

Hawkes, H.E. and Williston, S.H.
 1962: Mercury vapour as a guide to lead-zinc-silver deposits; Min. Cong., v. 48, p. 30-33.

Hendrick, D.M. and Fountain, D.K.
 1971: Induced Polarization as an exploration tool, Noranda area; Can. Min. Metall. Bull., v. 64 (706), p. 31-38.

Hutchinson, R.W., Ridler, R.H., and Suffel, G.G.
 1971: Metallogenic relationships in the Abitibi belt; a model for Archean metallogeny; Can. Min. Metall. Bull., v. 64 (708), p. 48-57.

Jones, H.
 1973: The Copperfields-Iso, Magusi River discovery; 41st Ann. Meet., Prospectors and Developers Association, Toronto.

Kelly, S.F.
 1957: Spontaneous polarization survey on Noranda mines property, Quebec (1924); in Method and case histories in mining geophysics, Can. Inst. Min. Met., Spec. Vol., p. 290-293.

King, H.F.
 1965: The sedimentary concept in mineral exploration; 8th Commonwealth Mining and Metallurgical Congress, Australia: Explor. Min. Geol., v. 2, p. 25-33.

King, H.F. (cont'd)
1976: Development of syngenetic ideas in Australia; in Handbook of stratabound and stratiform ore deposits, Elsevier Sci. Pub. Co., v. 1, p. 161-182.

Lalonde, J.P.
1976: Fluorine — An indicator of mineral deposits; Can. Min. Metall. Bull., v. 69 (769), p. 110-122.

Lazenby, P.G.
1973: New developments in the Input airborne EM system; Can. Min. Metall Bull., v. 66 (732), p. 96-104.

Lovering, T.S., Sokoloff, V.P., and Morris, H.T.
1948: Heavy metals in altered rock over blind ore bodies, East Tintic district, Utah; Econ. Geol., v. 43, p. 384-399.

Miller, L.J.
1960: Massive sulphide deposits in eugeosynclinal belts: Abstract; Econ. Geol., v. 55 (6), p. 1327-1328.

Oftedahl, C.A.
1958: A theory of exhalative — sedimentary ores; Geol. Foeren. Stockh., Foerh. 80, p. 1-19.

Ozerova, N.A.
1959: The use of primary dispersion halos of mercury in the search for lead-zinc deposits; Geokhim, No. 7, p. 793-802.

Paterson, N.R.
1961: Helicopter EM test, Mobrun orebody, Noranda; Can. Min. J., v. 82 (11), p. 53-58.

Price, P.
1948: Horne mine; in Structural geology of Canadian ore deposits; Can. Inst. Min. Met., v. 1, p. 763-772.

Riddell, J.E.
1950: A technique for the determination of traces of epigenetic base metals in rocks; Que. Dep. Mines, Prelim. Rep. No. 239, 23 p.

Roberts, L.
1956: Noranda; Clarke, Irwin Pub. Co. Ltd.

Roscoe, S.M.
1965: Geochemical and isotopic studies, Noranda and Matagami area; Can. Min. Metall. Bull., v. 58 (641), p. 965-971.

Sakrison, H.C.
1971: Rock geochemistry — its current usefulness on the Canadian Shield; Can. Min. Metall. Bull., v. 64 (715), p. 28-31.

Salt, D.J.
1966: Tests of drillhole methods of geophysical prospecting on the property of Lake Dufault mines, Noranda; in Mining Geophysics, Soc. Explor. Geophys., v. 1, p. 206-226.

Sangster, D.F.
1972: Precambrian volcanogenic massive sulphide deposits in Canada: A review; Geol. Surv. Can., Paper 72-22, 44 p.

Sangster, D.F. and Scott, S.D.
1976: Precambrian strata-bound massive Cu-Zn-Pb sulphide ores of North America; in Handbook of stratabound and stratiform ore deposits, Elsevier Sci. Pub. Co., v. 6, p. 129-222.

Sears, W.P.
1971: Mercury in base metal and gold ores of the Province of Quebec; in International Geochemical Exploration, Symposium, 3rd, Toronto, 1970, p. 384-390.

Sharpe, J.I.
1965: Summary of field relations of Mattagami sulphide masses bearing on their distribution in time and space; Can. Min. Metall. Bull., v. 58 (641), p. 951-964.

1967: Metallographic portrait of the Noranda district, Quebec; in Centennial Field Excursion Volume, NW Quebec and NE Ontario; Can. Inst. Min. Met., p. 62-64.

Siegel, H.O., Winkler, H.A., and Boniwell, J.B.
1957: Discovery of the Mobrun Copper Limited sulphide deposit, Noranda mining district, Quebec; in Methods and case histories in mining geophysics, Can. Inst. Min. Met., p. 236-245.

Spence, C.D.
1967: The Noranda area; in Centennial Field Excursion Volume, NW Quebec and NE Ontario, Can. Inst. Min. Met., p. 36-39.

Spence, C.D. and de Rosen-Spence, A.F.
1975: The place of mineralization in the volcanic sequence at Noranda; Econ. Geol., v. 70, p. 90-101.

Stanton, R.L.
1959: Mineralogical features and possible mode of emplacement of the Brunswick Mining and Smelting orebodies, Gloucester County, New Brunswick; Can. Min. Metall. Bull., v. 52 (570), p. 631-643.

Sullivan, C.J.
1968: Geological concepts and the search for ore, 1930-1967; in The Earth Sciences in Canada, A Centennial Appraisal and Forecast, Roy. Soc. Can., Spec. Pub. 11, p. 82-99.

Tatsumi, T. (Editor)
1970: Volcanism and ore genesis; Univ. Tokyo Press, 448 p.

Van De Walle, M.
1972: The Rouyn-Noranda area. Precambrian geology and mineral deposits of the Noranda — Val d'Or — Matagami — Chibougamau greenstone belts, Quebec, 24th Int. Geol. Cong., Guide to excursions A41-C41, p. 41-51.

Wilson, M.E.
1941: Noranda district, Quebec; Geol. Surv. Can., Mem. 229, 169 p.

1948: Structural features of the Noranda-Rouyn area; in Structural geology of Canadian ore deposits, Can. Inst. Min. Met. Spec. Vol. 1, p. 672-683.

EXPLORATION CASE HISTORIES OF THE ISO AND NEW INSCO OREBODIES

W.M. Telford
McGill University, Montreal, Canada

Alex Becker
Ecole Polytechnique, Montreal, Canada

Telford, W.M. and Becker, Alex, Exploration case histories of the Iso and New Insco orebodies; in Geophysics and Geochemistry in the Search for Metallic Ores; Peter J. Hood, editor; Geological Survey of Canada, Economic Geology Report 31, p. 605-629, 1979.

Abstract

The Iso and New Insco base metal deposits, located in Hébécourt Township about 20 miles (32 km) northwest of Rouyn, Quebec, were both discovered by airborne EM in 1972, the former directly as a result of the Input survey of Rouyn-Noranda area made by the Quebec Ministry of Natural Resources, the latter by Dighem survey carried out shortly after. Although these bodies are relatively small, only limited magnetic and EM ground follow-up were necessary to locate drill targets which indicated economic grade mineralization at both sites.

Following the initial drilling, a relatively large amount of geochemical and geophysical work was carried out on both properties. Some of this work was done to establish possible lateral and depth extensions of the mineralization; much of it was done for test purposes and with the aim of preparing case histories. The geochemical work included B-horizon soil and basal-till analyses for various elements. There was a variety of geophysical surveys — magnetic, gravity, various EM, induced polarization, self potential, telluric, magnetotelluric and electrical logging of drill holes.

Geologically these deposits are located on the north edge of the rhyolite, dacite and andesite flows bounding the Dufault and Flavrian granodiorite intrusives. It is suggested that this geological environment is somewhat different from that of the typical Noranda-area orebody, which occurs near an original volcanic vent.

The two deposits differ in detail, Iso being larger, with a higher grade of Zn than Cu and no magnetic signature, while the New Insco mineralization is mainly Cu with a strong magnetic response due to pyrrhotite. Apart from size, they are similar in strike, dip, depth, and the character of host rock and overburden. Both are excellent targets for most geophysical and geochemical survey techniques, although the small dimensions of the New Insco body make quantitative EM interpretation difficult.

Résumé

Les gîtes de métaux de base de Iso et New Insco, situés dans le canton de Hébécourt à environ 32 km (20 milles) au nord-ouest de Rouyn au Québec, ont été tous deux découverts au cours d'un levé EM aéroporté en 1972; dans le premier cas, la découverte du gisement résultait directement du levé INPUT du secteur de Rouyn-Noranda, levé effectué par le ministère des Richesses naturelles du Québec, dans le second cas, la découverte résultait d'un levé DIGHEM effectué peu de temps après. Bien que ces corps minéralisés aient eu une taille relativement faible, quelques levés magnétiques et EM au sol ont ensuite suffi pour localiser des objectifs de forage, qui indiquaient une minéralisation d'intérêt économique sur les deux sites en question.

Du point de vue géologique, ces gisements se situent sur le rebord septentrional des coulées rhyolitiques, dacitiques et andésitiques qui limitent les masses granodioritiques intrusives de Dufault et Flavrien. On suggère que ce milieu géologique diffère quelque peu de celui qui caractérise la masse minéralisée typique du secteur de Noranda, qui se situe près de la cheminée volcanique initiale.

Après le forage initial, on a effectué sur les deux propriétés un nombre relativement élevé de travaux géochimiques et géophysiques. Une partie de ceux-ci ont servi à reconnaître des extensions possibles de la minéralisation, latérales et profondes, pour les forages d'essai, et aussi la préparation des études de cas types. Pour le dosage des divers éléments, l'analyse géochimique concernait l'horizon de sol B et le till de fond. On a effectué toute une variété de levés géophysiques — magnétiques, gravimétriques, divers levés EM, des levés de polarisation induite, de polarisation spontanée, des levés par les méthodes telluriques, magnétotelluriques, ainsi que la diagraphie électrique des trous de sondage.

Les deux gisements diffèrent dans les détails; celui d'Iso est plus vaste, et se distingue par une teneur du minerai plus élevée en Zn qu'en Cu, et par l'absence de signature magnétique, tandis que la minéralisation de New Insco consiste surtout en Cu, et estcaractérisée par une puissante réponse magnétique due à la présence de pyrrhotine. Excepté leurs dimensions, ils présentent des ressemblances du point de vue de la direction, du pendage, de la profondeur, et du caractère de la roche favorable et des terrains de couverture. Tous deux constituent d'excellents objectifs pour l'application de la plupart des techniques de levé, bien que les dimensions réduites du corps minéralisé de New Insco rendent une interprétation EM quantitative difficile.

INTRODUCTION

Location

The Iso and New Insco properties are situated on Lots 48 and 40-43, respectively, Range 1, Hébécourt Township, approximately 20 miles (32 km) northwest of Rouyn, Quebec, at the northwest end of Lac Duparquet. The Iso deposit lies about 1 mile (1.6 km) west of New Insco (Fig. 27.1).

Initial Discovery

Although there had been sporadic exploration with limited drilling in the area over a considerable time, both these deposits were airborne EM discoveries. The Iso anomaly appeared on the original data from the Input EM survey of the Noranda area, made public by the Quebec government in August 1972. New Insco, which has a very limited strike length, was first detected during a Dighem survey flown at a lower altitude in November 1972. Figures 27.2 and 27.3 display, respectively, sections of the original Input EM and Dighem anomaly maps; the cluster of anomalies south of Magusi River, obvious in both figures, marks the Iso zone, while the New Insco anomaly appears only in Figure 27.3.

Survey Sequence

Following the airborne EM surveys, limited ground geophysics, consisting of magnetic, vertical-loop and horizontal-loop EM surveys, were carried out to outline both mineral zones. Preliminary drilling in November 1972 indicated ore grade sulphides at Iso, while the discovery hole at New Insco was drilled in January 1973.

Considerable additional ground geophysical surveys and some geochemical work were done subsequently for two purposes. During 1973 basal-till geochemistry and frequency-domain IP surveys were undertaken primarily to determine possible lateral and depth extensions of the mineral zones on both properties. Later surveys, carried out to test new equipment and mainly to provide further data for case histories, included self potential, various types of EM, telluric, magnetotelluric, gravity, and seismic refraction surveys and electrical logging in diamond-drill holes.

Regional Geology

The rhyolites in the area west of Rouyn-Noranda are shown in Figure 27.4. Assuming that these define a depositional dome which was a volcanic seamount surrounded by lower areas in which pyroclastics, sediments and very fluid flows would be emplaced simultaneously with the volcanism on this 'Mount Noranda', the materials flowing outward in several directions were of widely varying composition, with felsic lavas being more common in the rocks which now underlie Hébécourt Township than in Duparquet Township to the east. The next peripheral area containing such large amounts of felsic material is 20 miles (32 km) to the east in the Clericy syncline.

The Noranda dome is cored by granodiorite batholiths which may represent subvolcanic intrusives associated with the volcanics. Because it was stiffened by thick rhyolite horizons and granodiorite batholiths, the dome remained undeformed whereas the intervolcanic tuff and sedimentary basins were isoclinally folded in the areas between it and other neighbouring volcanic centres of similar type. Both the Iso (Magusi River) and New Insco deposits are located on the north edge of the rhyolite, dacite and andesite flows which bound the Dufault and Flavrian granodiorite intrusives (Fig. 27.4).

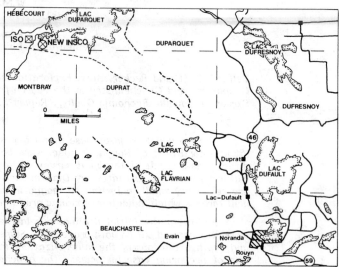

Figure 27.1. Location map, New Insco and Iso base metal deposits, northwestern Quebec.

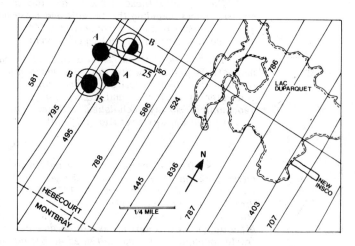

● 6 CHANNELS   ◐ 4 CHANNELS
◓ 5 CHANNELS   ⊙ 4 CHANNELS WITH MAGNETIC CORRELATION

Figure 27.2. Input-EM map, location of anomalies.

Detail on Geophysical Methods

While most of the geophysical surveys — magnetic, gravity, IP, horizontal-loop EM, etc. — carried out on these properties are conventional, some of the less commonly-used techniques and the methods for displaying their data require explanation.

Dighem Survey

This helicopter-borne EM equipment (Fraser, 1972, 1974) is capable of detailed mapping with the flight lines being spaced 300-400 feet (90-120 m) apart and being flown at an altitude around 150 feet (45 m). In addition to the conventional coaxial vertical transmitter and receiver coils mounted fore and aft, the boom contained two additional receiver coils, one horizontal, the other with the plane vertical in the minimum coupling position (so-called fishtail configuration). The latter is called the strike-sensitive coil (see also Figures 27.3 and 27.11).

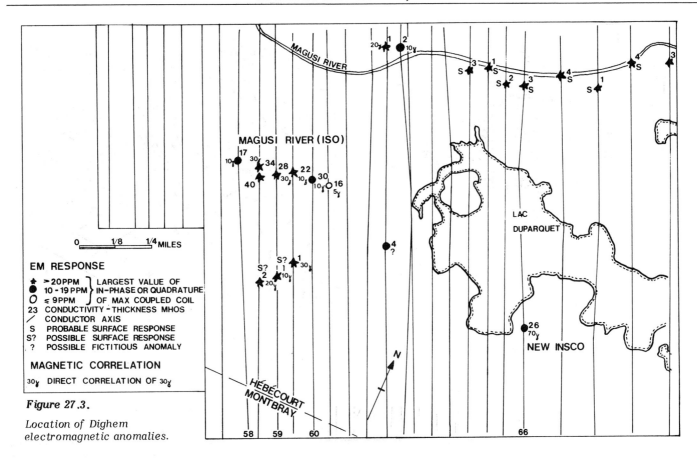

Figure 27.3.

Location of Dighem electromagnetic anomalies.

VLF-EM Survey

The contours in Figure 27.12 are obtained from a conventional VLF-EM survey by averaging the dip angles from several stations. If $\theta_1.....\theta_4$ are dip angles at stations 1, ..4, the contour value is:

$$C_{23} = (\theta_3 + \theta_4) - (\theta_1 + \theta_2)$$

plotted midway between stations 2 and 3 (Fraser, 1969). The sign of C_{23} is positive in the vicinity of proper crossovers, negative for reverse crossovers. Negative values are discarded; in the process there is also a smoothing effect which reduces the geological noise inherent in the relatively high frequency VLF signals.

EM-25 Ground EM Survey

The Geonics EM-25 unit was originally designed for use in areas of highly conductive overburden (Paterson, 1973). It is a two-coil ground EM system operating at 50-70 Hz which may be employed in two survey configurations. With the fixed transmitter arrangement, a large single-turn rectangular transmitting loop is laid out with the long axis oriented more or less perpendicular to survey lines, similar to Turam EM. Readings are obtained with a small receiver loop in a vertical plane normal to the near side of the transmitter, being inclination in degrees and per cent (%) ellipticity of the polarization ellipse. These quantities correspond to the in-phase and quadrature components, respectively, of the secondary magnetic field produced by a conductor. Examples of this type of survey are found in Figures 27.18 to 27.22. In the moving transmitter configuration, the transmitter loop is a 100 foot (30 m) square or circular loop located at various stations along the traverse line. The receiver coil occupies several stations, varying between 400 and 1600 feet (120-480 m) from each transmitter set-up, again recording in-phase and quadrature readings.

Magnetotelluric and Telluric Surveys

Magnetotelluric data (Fig. 27.16) were obtained with a unit designed by Professor V.G. Pham and constructed at Ecole Polytechnique in Montreal. The equipment operates at a range of frequencies between 1 and 2000 Hz; four sharply-tuned channels may be used at one time. The magnetic field detector is a ferrite-core coil, bandwidth 1-2000 Hz, which is generally oriented to pick up the H_y (strike axis) component, while the telluric field E_x is measured with metal electrodes spaced 100 feet (30 m) apart along the profile line. Signals are integrated for 30 seconds (or longer to improve the lowest frequency response) and the apparent resistivity is derived from the relation:

$$\rho_a = |E_x/H_y|^2/5f$$

plotted midway between the electrodes. In some cases, generally to increase survey speed, only the telluric field is measured along the profile and the relative telluric field is plotted for the same station as above.

Drillhole EM Logging

The EM log in Figure 27.27 was obtained by John Betz with his DHEM unit, which has a multi-turn square transmitter coil mounted at surface and a ferromagnetic-core solenoid and receiver downhole. This equipment is used in two modes, known as min-coupled and max-coupled. In the former mode, the transmitter loop is rotated to obtain a minimum signal in the receiver; in the latter mode the Tx

Figure 27.4. Simplified regional geology, Noranda area.

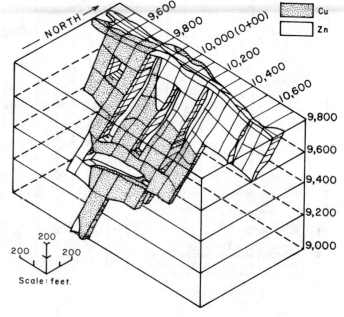

Figure 27.7. Isometric view of Iso Orebody.

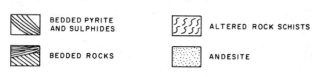

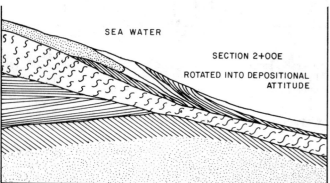

Figure 27.5. Idealized section of Iso-type deposition.

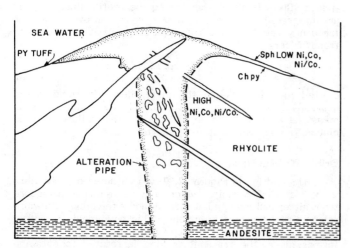

Figure 27.6. Idealized section of Noranda-type deposit (Vauze, Norbec, etc.).

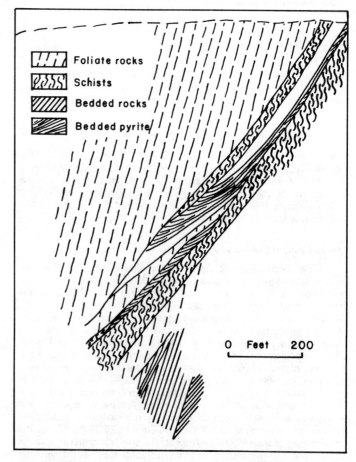

Figure 27.8. Fine layer and structure section, Line 2+00E, Iso Orebody.

loop is rotated for a maximum. System frequency is 735 Hz. In a barren area, the plane of the transmitter coil will contain the drillhole at minimum signal; when the downhole receiver solenoid is in the vicinity of zones of anomalous conductivity, the transmitter must be rotated about a horizontal axis and generally the minimum signal obtained is larger than in barren areas and is called the residual voltage (RV). Transmitter tilt angle and RV are related to the in-phase and quadrature components of the anomalous response respectively. The max-coupled mode is essentially a backup for the other measurement, to ensure that a conductive zone whose attitude provides poor coupling in the min-coupled position may not be missed.

ISO OREBODY

Following the discovery hole, pattern drilling outlined 5.8 million tons of ore averaging 1.13 per cent Cu, 2.72 per cent Zn, 0.82 oz/ton Ag, and 0.022 oz/ton Au contained in a tabular body striking east-west for about 1800 feet (550 m), dipping 50° south and extending more than 1000 feet (300 m) downdip. Clay and gravel overburden about 40 feet (12 m) thick covers the area.

Local Geology

Information on the geology of the Iso area is derived almost entirely from the excellent paper delivered by Hugh Jones at the Prospectors' and Developers' Meeting in Toronto in March 1973.

Rocks mapped by the Quebec Ministry of Natural Resources geologists, in the vicinity of the Magusi River north of the postulated extension of the Rouyn-Noranda district rhyolites, have been designated generally as dacites, although they may be more mafic than dacitic. Rocks from drill cores in the neighbourhood of the Iso deposit are flows and pyroclastic-sedimentary units of andesitic and rhyolitic composition which, compared to the thick flows in the main Rouyn-Noranda volcanic centre, are relatively thin. These flows may represent the periphery of 'Mount Noranda'. The sulphide mineralization is contained between two schistose units which are mainly sheared forms of andesitic lavas.

The sequence of deposition has been studied using samples from the deepest holes (e.g. M-19 on Line 2+00E). There is no evidence that the beds have been overturned, although some tilting may have taken place between the explosive events which provided the flows. An idealized section of deposition is shown in Figure 27.5. The pyroclastic-sedimentary units grade from coarse bottom layers to thin sedimentary caps, many of which contain considerable calcium carbonate. This is compatible with precipitation at some distance from the volcanic centre. Although the pyroclastic flows are thicker in the 'Clericy syncline', they are otherwise similar to the Magusi and in their relation to the rocks of the main Rouyn-Noranda centre, suggesting that the Magusi area is an extension of the 'Clericy syncline' about 20 miles (32 km) to the east.

Analyses for major oxides of numerous samples of the volcanic rocks from DDH M-19 show that there are two definite groups, designated rhyolitic (70% silica) and andesitic (52% silica) and that rocks remote from the ore zone are sodic rhyolite above and tholeiitic basalt below it. It is concluded that the country rocks occur in a syncline north of the main Rouyn-Noranda centre, in an area where a tongue of felsic volcanics may have extended from the original topographic high into a depositional basin. This environment may be contrasted with that suggested for the typical deposits of the Rouyn-Noranda area such as at the Norbec and Vauze mines (Roscoe, 1965; Spence, 1966), where the sulphide bodies appear to lie on what was originally a gentle volcanic dome (Fig. 27.6). The rhyolitic flows may be as thick as 1800 feet (550 m) directly below the deposits, decreasing to a few hundred feet between them; pyroclastics and sediments are rare.

Iso Ore Zone

Overall the Iso deposit has a tabular shape within which the orebody, enriched in Cu and Zn, forms a thick portion of a more extensive sheet of Cu-Zn-bearing pyrite. This is illustrated in the isometric view of Figure 27.7, with some sections cut out to show the layering. The pyrite sheet is continuous, with no apparent gaps caused by faulting, folding or the intrusion of dykes. Thickness varies from a few feet at the edges to a maximum 110 feet (33 m). Abrupt variations in thickness occur along both strike and dip, indicating possible disruption by low angle faults, migration within an originally regular body, or an irregular mode of deposition. Some high grade ore sections coincide with thicker parts of the sheet.

Sulphide Composition

The economic sulphides in this pyritic body are chalcopyrite and sphalerite, with galena as a common accessory. Microprobe studies show that the Cu, Zn, and Pb are confined to their appropriate sulphides (as above) and are not present as solid solutions in pyrite or in one another. Pyrrhotite occurs as very minor disseminations within pyrite grains. Arsenopyrite has also been identified. Both magnetite and specularite are found in minor amounts, the former probably being the source of the weak magnetic anomaly associated with the orebody. Native gold and silver have been observed, associated with coarser and finer grained chalcopyrite respectively. However, it is thought that much of the silver occurs as a sulphide or sulphosalt.

The carbonate-quartz-chlorite gangue is rich in dolomite compared to the siliceous gangue of typical Rouyn-Noranda deposits, perhaps because of deposition in a carbonate-rich basin rather than the high silica environment of the volcanic centre.

Table 27.1

Process	Iso Deposit	Rouyn-Noranda Type Deposit
Deposition	Down flank of volcanic seamount	Nipples on a broad dome
Alteration	Schist envelope	Pipe extending considerably below ore zone, diameter ≃ orebody.
Dyking	Simple; semi-conformable gabbro	Several dyke compositions and orientations, cutting alteration pipe which seems to be locus of activity.
Layering Cu/Zn Zoning	Numerous layers rich in Cu, Zn sulphides	Piled to form equi-dimensional deposit; distinct Cu/Zn zoning, Zn on top, extending out on volcanic surface.

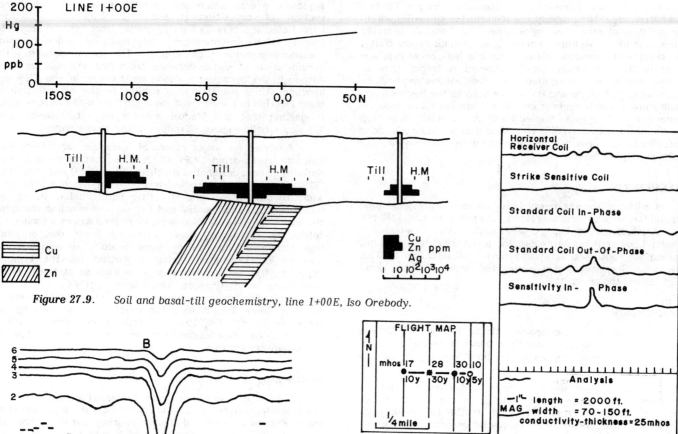

Figure 27.9. Soil and basal-till geochemistry, line 1+00E, Iso Orebody.

Figure 27.11. Dighem data, flight record profile, Iso Orebody.

Figure 27.10. Input data, flight record profile, Iso Orebody.

Zoning and Layering

Chalcopyrite and sphalerite are confined to layers within the deposit, the thickness being in the order of tens of feet. The layers can usually be correlated from one drillhole to another. Copper-bearing layers may occur on either the hanging wall or footwall side and there may be more than one such layer present. Frequently a layer on the hanging wall downdip will cross over to appear on the footwall near surface (e.g. Fig. 27.8). Zn is predominant at both ends of the deposit, while Cu concentrates toward the centre.

On a small scale the major sulphides and gangue minerals are delicately layered and bedded. About 1 per cent of the sulphide mineralization is found in thin cross-cut veinlets or as fracture coatings; up to 10 per cent occurs in coarse interconnecting blebs. Both these fractions were mobile at a later date than the fine grained sulphides which comprise the bulk of the ore. The coarse and fine forms have somewhat different compositions and the coarse sulphides are associated with coarser carbonates in the gangue: the coarse sphalerite is deficient in iron compared to the fine grained variety.

In the fine grained part of the deposit, pyrite is the predominant mineral, with fairly consistent grain size and texture; the ore minerals occur as fine inclusions in the pyrite and as matrix for pyrite crystals and aggregates. There appear to be two distinct sizes, either 20 or 100 µm approximate diameter. Crystal and aggregate forms of pyrite include numerous agglomerations, some nodules and radial textures in which internal fine patterns are overgrown with coarser crystals. Inclusions of the other sulphides and gangue minerals are trapped in these aggregates. The nodules are similar to those found in pyritic black shale and in pyrite sands around recent fumaroles. A few 100 µm crystals in a matrix of spongy pyrite have been observed, resembling the Norbec ores. The existence of both delicate textures and the generally fine grained pyrite suggests little drastic reorganization during metamorphism.

From drill core measurements, the fine layering appears to be conformable with the sulphide body where the deposit is thin, whereas it is at a definite angle to the overall layering in the thicker sections. A schematic for the section on

Line 2+00E in Figure 27.8 illustrates this: the angle at which the fine layers are stacked as they cross the thick layering from hanging wall to footwall is the same as that of the thick layers crossing the sulphide section. This suggests the bedding of detrital material deposited in an active fluid environment.

Alteration

Nearby rocks in both the hanging wall and footwall sections of the Iso orebody have been altered to carbonate, chlorite and sericite schists. Occasionally, identifications of areas of former andesite or gabbro dykes can be made as a result of the presence of relict amygdules or leucoxene disseminations respectively. The altered rock is foliated with micaceous minerals and banded by the carbonates; core angles indicate the schistosity varies between the dip of the foliation (steeply south) and that of the strata (50°S). Within this schist zone, an early pervasive iron-rich chloritization and a later cross-cutting veined magnesian chlorite have been identified. Veinlets are confined to rocks within 100 feet (30 m) of the orebody. Soda and potash are significantly different (respectively leached and enriched) and disseminated sulphides are more common in the schist zone than in the wall rocks, whereas there is no enrichment of either magnesium or iron in the alteration zone. Perhaps the pyrite evolved in situ during creation of the schist.

The schist zone seems conformable with the sulphide body and the general stratigraphy. It is possible, however, that this zone of alteration and shearing could extend farther into the footwall, since numerous drillholes terminate in schist. Beyond this schist zone the rocks are quite fresh, with flow banding, amygdules and volcanic texture generally preserved. Here the foliation is weak and dips are steeply south.

Comparison with Other Deposits

The size of the Iso body is well within the range of other deposits in the Rouyn-Noranda, Joutel-Poirier and Mattagami mining camps. However, there appear to be fundamental differences between the Iso body and the so-called Rouyn-Noranda type. These are summarized in Table 27.1 (see also Figs. 27.5, 27.6).

Geochemical Surveys, Iso Orebody

The results of the basal-till geochemical sampling on Line 1+00E are presented in Figure 27.9. Mercury values from the top surface of the clay overburden are plotted over a distance of 200 feet (60 m) in the vicinity of the orebody. Values range from 85 to 145 ppb Hg, which appears to be rather high; however, there is no apparent relation to the ore zone. Basal-till Hg analysis gave maximum values of 150 ppb directly over the conductor, with a background of 60 ppb on the flanks. These are not particularly anomalous. Three sphalerite ore samples gave 540, 650, and 730 ppb Hg, whereas two samples of chalcopyrite contained 25 and 40 ppb.

Basal-till sampling results for Cu, Zn and Ag are illustrated by the horizontal bars at the bottom of the three holes located at 1+20S, 0+20S, 0+80N. Values to the left of the holes are for the complete till sample, to the right for the heavy mineral fraction. All three metals show high values over the orebody. At 1+20S they are still somewhat anomalous, although the decrease from 0+20S is rather large and is larger in the complete till samples than for the heavy mineral. The direction of glacial movement is north-south.

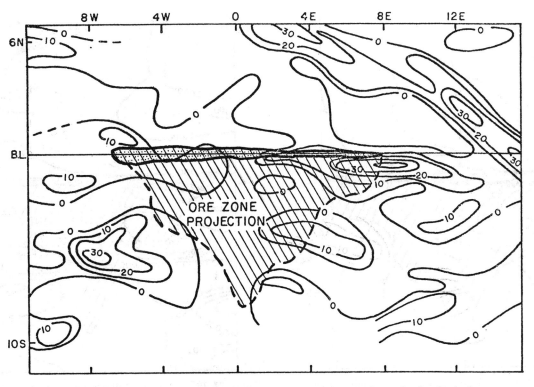

Figure 27.12.   VLF-EM contours; contours in degrees tilt angle, Iso Orebody.

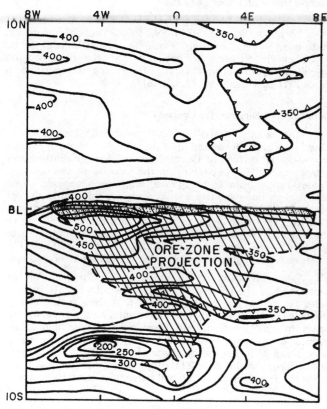

Figure 27.13. Vertical field magnetic contours; contour interval 25 gammas, Iso Orebody.

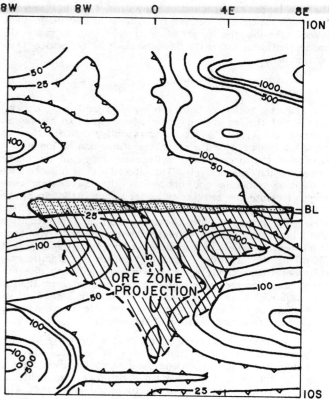

Figure 27.15. EM16R contours; contours of ρ_a Ωm. $f = 18.6$ kHz, Iso Orebody.

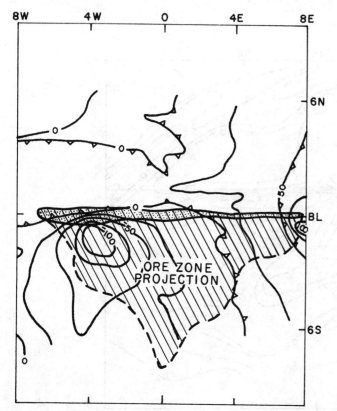

Figure 27.14. Self potential contours; contour interval 25 mV, Iso Orebody.

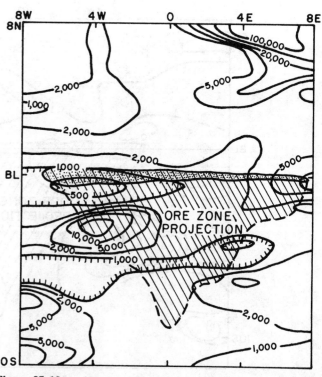

Figure 27.16. Magnetotelluric contours; contours of ρ_a Ωm. $f = 250$ Hz, Iso Orebody.

Geophysical Surveys, Iso Orebody

Input and Dighem (Airborne Electromagnetic) Profiles

Figure 27.10 presents Input EM (Mark 5) profiles of 6 channels from one flight line over the Iso ore zone. The AEM anomaly is clearly seen on all channels. The huge fluctuations in Channel 1 are doubtless a reflection of the varying thickness of the conductive overburden. The conductivity-thickness (σt) product obtained from the Input EM survey was 12 mhos, roughly the same as for New Insco (Lazenby, 1973).

A profile from the Dighem helicopter survey is displayed in Figure 27.11. The insert shows a section of the flight map for the area. Going from top to bottom are the quadrature profiles for the two minimum-coupled receiver-coils, then the in-phase and quadrature profiles from the standard coaxial maximum-coupled receiver coil and at the bottom the same in-phase profile for the coaxial receiver coil (labelled 'sensitivity in-phase') but amplified considerably. Analysis of these data indicated a conductor of about 2000 foot (610 m) strike length, 70-150 feet (20-45 m) thick, with a σt value of 25 mhos. The latter compares with 70 mhos obtained in the Dighem survey over the New Insco deposit.

VLF-EM Survey

Contoured data obtained from the VLF-EM survey carried out using the Crone Radem unit are presented in Figure 27.12. Except around Line 8+00E, where the ore zone is pinching out, there is no apparent correlation with the Iso conductor. The response is particularly weak in the west end of the zone. Anomalies centred at about 8+00W, 5+00S and 3+00E, 7+00N lie in areas of rather high surface resistivity, as determined later by a Geonics EM16R VLF survey (see Fig. 27.15). These anomalies appear to be mainly due to fluctuations in the thickness of the conductive clay overburden.

Ground Magnetic Survey

Contours of the vertical magnetic field, displayed on a 2000 by 1600 foot (610 by 490 m) grid in Figure 27.13, were obtained from the results of a fluxgate magnetometer survey (Fountain and Fraser, 1973). There is very little magnetic relief over the Iso deposit — the maximum variation is about 300 gammas — doubtless because of the lack of pyrrhotite and/or magnetite. Rock types change from rhyolite to diorite to andesite in a south-north sequence (see Fig. 27.17 to 27.22). Although the average magnetic susceptibilities of these rocks increase in the same order, they are not much different and there is no north-south gradient to indicate this. However, the axis of the conductor is quite well marked by mild maximum contours along the baseline, particularly in the west where a south dip is indicated. Magnetic lows in the south and northeast grid areas may be reflections of thicker overburden (see Fig. 27.21), but this is by no means definite.

Self Potential Survey

Data from the SP survey are contoured in Figure 27.14. There is a 100 mv negative anomaly near the baseline at 8+00E whose asymmetry indicates a south dip. The steep negative gradient of 100 mv on Line 4+00W between 1+50S and the baseline is characteristic of a vertical contact between rock types of different electrochemical properties, rather than a sulphide slab. Apart from these two features the SP map is featureless. Neither provides a satisfactory picture of the Iso deposit, which has practically pinched out on Line 8+00E.

EM16R and MT Surveys

Apparent resistivity contours obtained with the Geonics EM16R and MT units are presented in Figures 27.15 and 27.16 respectively. These surveys were carried out simultaneously. The EM16R data were obtained using the Seattle transmitter which operates at 18.6 kHz. Although the MT survey was made at four frequencies (3, 21, 250, and 1200 Hz) only the 250 Hz contours are shown in Figure 27.16, since the results were not sufficiently illuminating to warrant displaying all the data. MT readings showed a high noise level, particularly for the H-field at low frequencies.

Considering Figures 27.15 and 27.16 in conjunction with the VLF-EM contours of Figure 27.12, there are obvious similarities between the first two but very little correlation with the VLF-EM results. Both sets of ρ_a contours outline the Iso conductor along the baseline mainly on the west half of the grid and there is good correlation between the two resistivity zones in the southwest and northeast corners. Presumably the 18.6 kHz response reflects the conductive overburden much more than the 250 Hz MT, which may explain the less predominant east-west trend in the former.

Both EM16R and MT surveys measured only a single E- and H-component (E_x and H_y for MT, E_y and H_x for Seattle using the EM16R). Consequently, these contour plots are valid only for a layered geometry: lateral conductivity variations produce anisotropy. However, the VLF stations at Cutler, Maine and Panama were not available and lack of time prevented completion of the MT survey in two directions.

Detailed Geophysical Survey Profiles

Data from the four preceding ground geophysical surveys have been displayed in contour form, partly to condense the material, mainly because they were not particularly significant. In the following, IP, EM, gravity, and seismic refraction results are shown in profiles with accompanying vertical sections.

Line 8+00W

Vertical-loop and horizontal-loop EM (VLEM, HLEM) and induced polarization (IP) and apparent resistivity (ρ_a) profiles for Line 8+00W are illustrated in Figure 27.17, together with the incomplete vertical section. This line is beyond the end of the zone (see Fig. 27.7), which is predominantly rich in Zn in the western portion. DDH M-51 intersected 36.6 feet (11.2 m) of mineralization as shown. The other holes appeared barren.

The geophysical profiles generally reflect the presence of minor mineralization along Line 8+00W, apart from the vertical-loop EM profile, which has a clear crossover at 0+50S. This is to be expected, since VLEM usually responds for some distance beyond the end of a conductor when the transmitter loop is located on top of it. A mild HLEM anomaly at 0+50S suggests a rather poor conductor (σt = 24 mhos) at a depth of about 60 feet (18 m).

While the IP metal factor is anomalous at the same station, PFE values are insignificant. High resistivity to the south is clearly evident on the ρ_a profile; this correlates with the contours of Figures 27.15 and 27.16. Additional evidence that this resistivity high may be close to surface was seen in Figure 27.12, where the crossover at 5+00S on Line 8+00W suggests thinning of conductive overburden, or perhaps a shallow resistive bed.

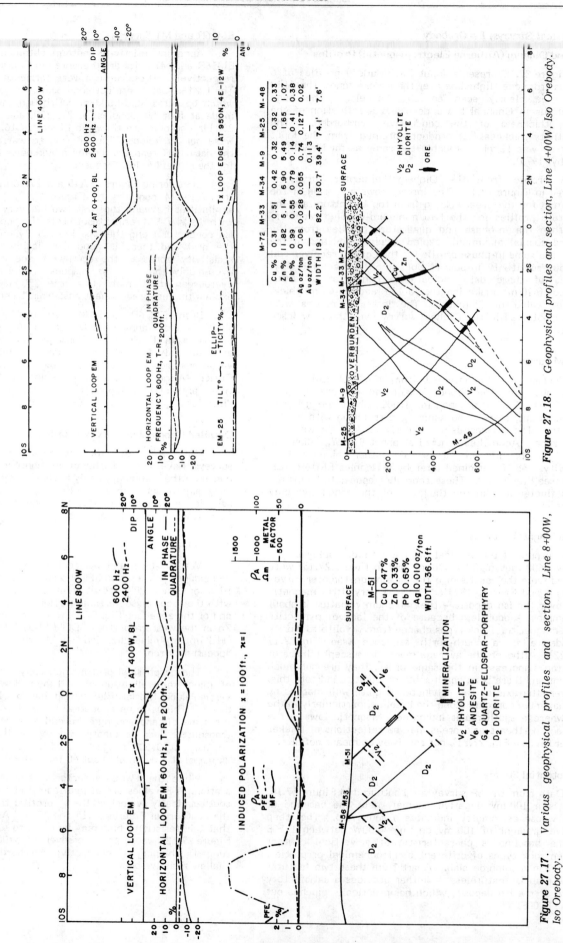

Figure 27.18. Geophysical profiles and section, Line 4+00W, Iso Orebody.

Figure 27.17. Various geophysical profiles and section, Line 8+00W, Iso Orebody.

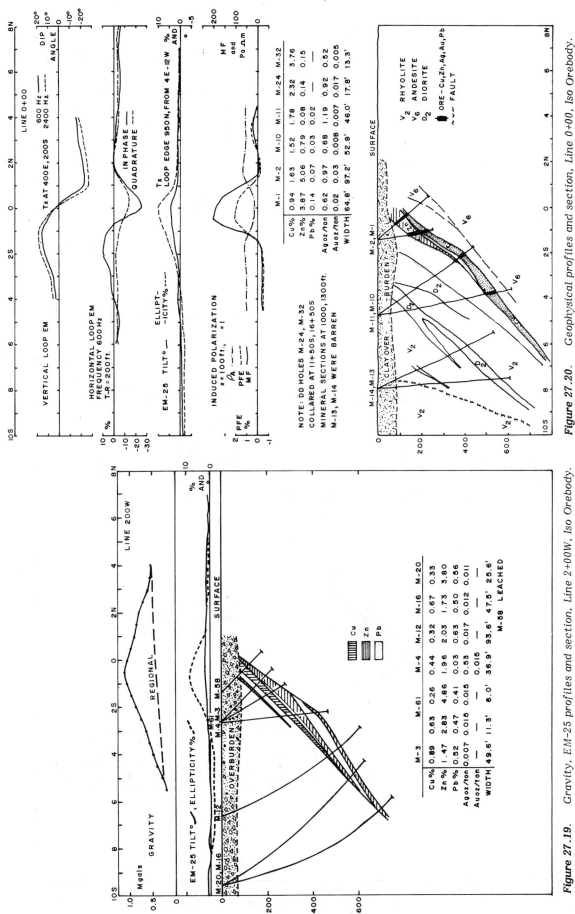

Figure 27.20. Geophysical profiles and section, Line 0+00, Iso Orebody.

Figure 27.19. Gravity, EM-25 profiles and section, Line 2+00W, Iso Orebody.

Table 27.2

Parameters for Iso Conductor from HLEM Thin Sheet Curves

P	f	st	t	z	P	f	st	t	z	P	f	st	t	z
ft	Hz	mhos	ft	ft	ft	Hz	mhos	ft	ft	ft	Hz	mhos	ft	ft
200	222	130	0	55	400	222	175	20	60	600	222	130	50	55
"	444	150	10	55	"	444	105	20	50	"	444	80	50	50
"	888	122	5	60	"	888	60	40	40	"	888	35	60	30
"	1777	55	5	50	"	1777	22	60	40	"	1777	14	85	10
266	600	80	75	42	(from Figure 27.20)									

ℓ = coil spacing
σ = conductivity, mhos/m
t = conductor thickness
f = frequency of operation of HLEM unit
z = depth to top

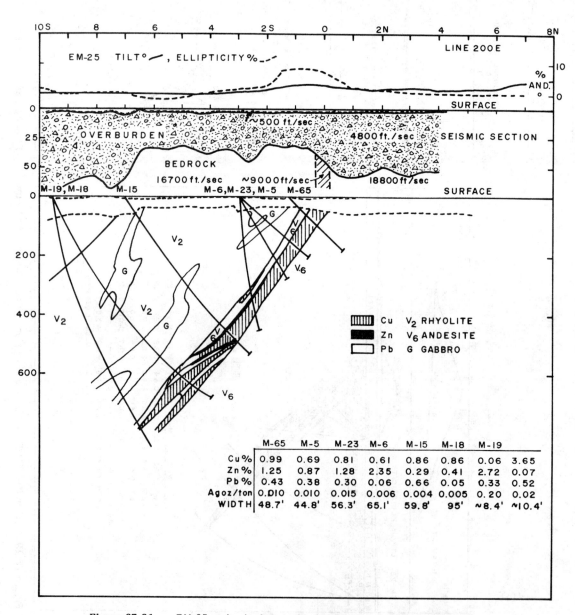

Figure 27.21.   EM-25, seismic data and section, Line 2+00E, Iso Orebody.

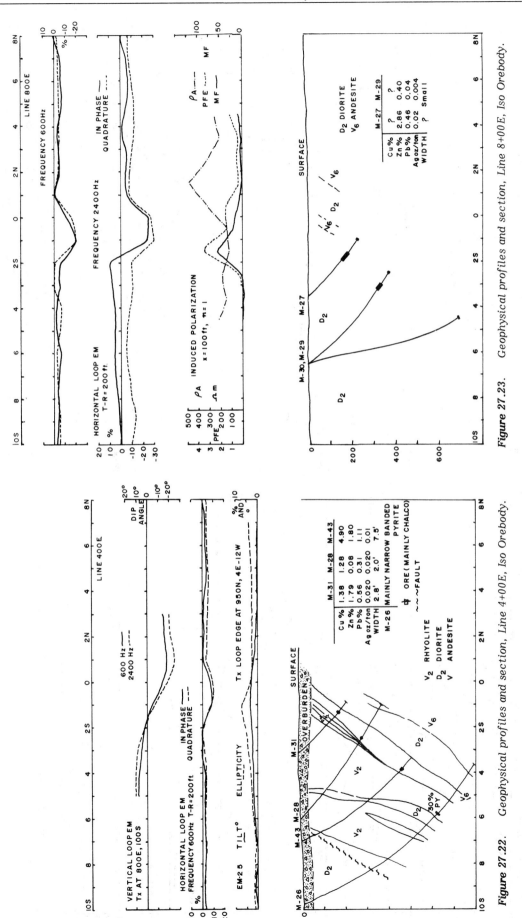

Figure 27.23. Geophysical profiles and section, Line 8+00E, Iso Orebody.

Figure 27.22. Geophysical profiles and section, Line 4+00E, Iso Orebody.

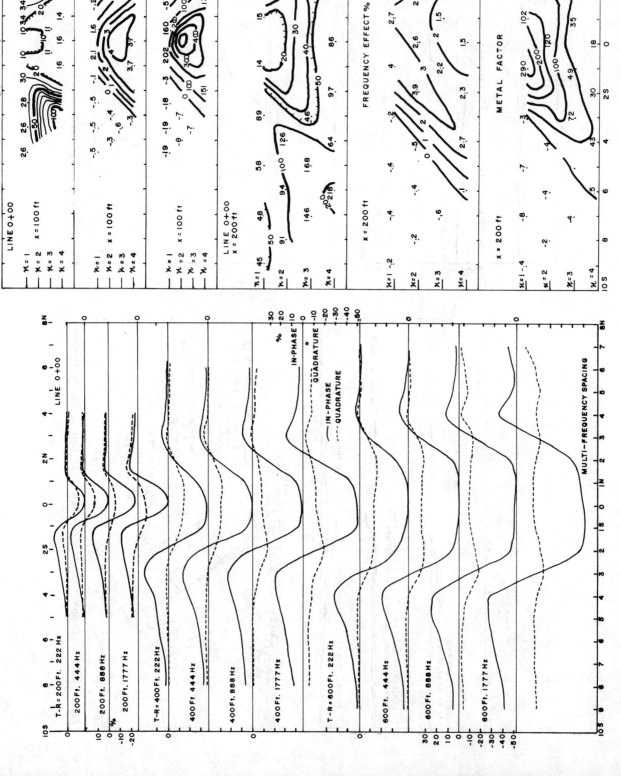

Figure 27.24. Horizontal-loop EM profiles, multifrequency-spacing, Line 0+00, Iso Orebody.

Figure 27.25. IP pseudo-depth plots, Line 0+00, double-dipole spread, Iso Orebody.

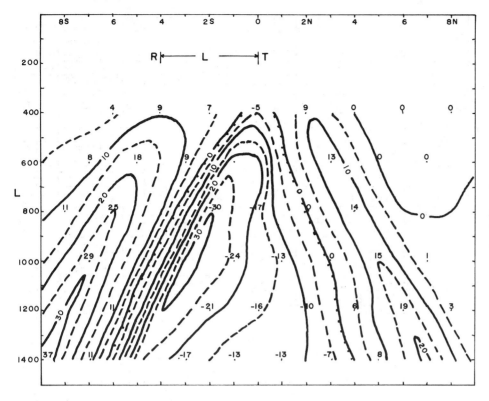

Figure 27.26.

EM-25 quadrature pseudo-section, Line 0+00; contours in per cent ellipticity, Iso Orebody.

Line 4+00W

Vertical-loop EM, horizontal-loop EM, and EM-25 profiles along Line 4+00W are presented in Figure 27.18. The vertical section shows Line 4+00W to be situated directly over the ore zone, which is covered by about 70 feet (20 m) of overburden. From the core logs in Figure 27.18, this is clearly the Zn-rich part of the mineralization. All three profiles are more or less anomalous in the area slightly south of the baseline. The HLEM data indicate a σt value of about 50 mhos and a depth of only 35 feet (11 m), when applied to the thin dipping sheet model characteristic curves.

Line 2+00W

Figure 27.19 shows a gravity profile along with EM-25 results over the vertical section of Line 2+00W, together with the mineralization observed in the drillholes. The gravity anomaly is about 0.7 mgals after removing the marked regional. Tonnage calculations based on this anomaly (Fountain and Fraser, 1973), are given as 3400 tons/linear foot, or 6.8 million tons for a strike length of 2000 feet (610 m). These are not necessarily all ore grade sulphides.

Line 0+00

The profiles and section for Line 0+00 are shown in Figure 27.20. Here the Cu and Zn concentrations have increased appreciably, particularly at depth; Zn decreases with depth and Pb is generally low.

The electrical profiles all show clear anomalies at 0+00 or slightly south, characteristic of a conductor with south dip. From the appropriate thin sheet model curves for the HLEM method, a depth of about 42 feet (13 m) and a σt product of about 80 mhos was calculated.

The IP profiles are quite conventional. Variation of IP and ρ_a response with depth is clarified in Figure 27.25, where it is evident that all three parameters peak in the section between n=2 and n=3 for x=100 feet (30 m), that is, 150-200 feet (45-60 m) below surface.

Line 2+00E

Figure 27.21 has been inserted mainly to illustrate the value of using the shallow seismic refraction technique to map the bedrock terrane. The EM-25 profile and vertical ore section are also shown. Interpretation of the seismic data was done by Hales' (1958) method. Velocities were quite uniform, averaging 4800 ft/s (1465 m/s) in the overburden and 18 000 ft/s (5485 m/s) in bedrock. There is a distinct low velocity zone in the bedrock averaging 11 000 ft/s (3350 m/s), which coincides generally with the ore zone and extends somewhat downdip.

The EM-25 profile is similar to those on adjacent lines, with the ellipticity anomaly predominating. In the 400 foot (120 m) section between Lines 2+00W and 2+00E the average ellipticity/tilt ratio indicates a σt product of 25-35 mhos and the asymmetry suggests a 60° south dip.

Line 4+00E

From the vertical section in Figure 27.22 and from Figure 27.7 it is clear the ore zone has grown thinner and extends only about 500 feet (150 m) downdip. The EM profiles reflect the decreased width of mineralization. The VLEM profiles locate the top of the conductor at 1+50S and indicate the south dip, while the maximum dip angles have decreased compared to Lines 2+00E and 0+00. The HLEM response has also decreased; modelling gives a depth of 55 feet (17 m) and a σt of 17 mhos. The EM-25 response is similarly smaller.

Line 8+00E

Geophysical profiles and vertical section are illustrated in Figure 27.23. Neither the depth of cover nor the detailed mineral zone section is known for this line, although it appears from Figure 27.7 that the latter is somewhat wider near surface and of the same depth extent compared to Line 4+00E. The 2400 Hz HLEM data suggest the south dip

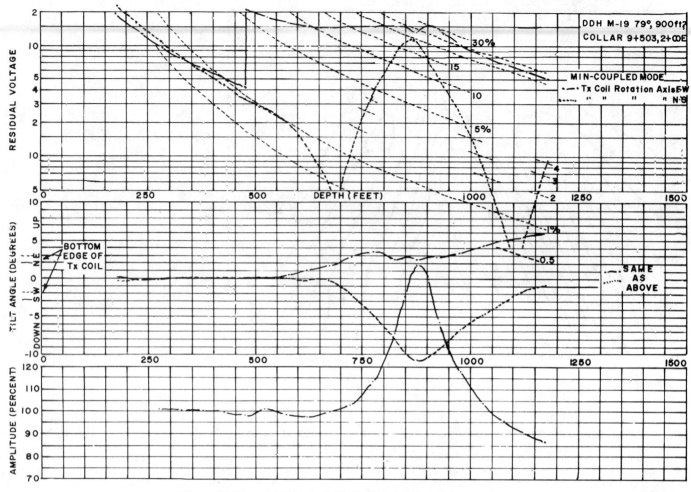

Figure 27.27. EM logs in DDH M-19 (after Betz), Iso Orebody.

more obviously than does the 600 Hz HLEM profile. The depths and σt products obtained from the thin sheet model are 25 feet (8 m) and 21 mhos at 600 Hz, about 10 feet (3 m) and 14 mhos for 2400 Hz.

The ρ_a profile from the IP survey also suggests a south-dipping conductor. At larger dipole spreads (x=200, 300 feet (60, 90 m)), the ρ_a values increase to the north. On this line the PFE anomaly is larger than on Lines 8+00W and 0+00. Furthermore, both PFE and metal factor increase more with expanded spreads than on the other two lines and shift the location of the ore zone somewhat south. Hence the IP data indicate possible disseminated mineralization at depth.

Maxmin Horizontal-Loop EM Profiles

A suite of horizontal-loop EM profiles obtained with the Apex Parametrics Maxmin unit at four frequencies — 222, 444, 888, 1777 Hz — and three coil separations — 200, 400 and 600 feet (60, 120, 180 m), is displayed in Figure 27.24 for Line 0+00. Parameters obtained from thin sheet characteristic curves (60° dip), using pertinent data from these profiles, are tabulated below. Estimates of σt, t and z from the HLEM profiles of Figure 27.20 are presented in Table 27.2; these should be closest to corresponding Maxmin values at ℓ = 200 feet (60 m), f = 444 Hz. The σt and z values are approximations, while the thicknesses (t) are merely crude estimates.

In general the agreement is not particularly good between the results of the two HLEM surveys, nor with the true values of σt, t and z. However, it is obvious that the cross-sectional geometry is not that of a thin sheet.

Pseudo-Depth Sections

IP and EM-25 pseudo-depth sections for Line 0+00 are shown in Figures 27.25 and 27.26 respectively. A shallow conductor of limited depth extent is indicated by the IP contours; the south dip is not apparent.

The EM-25 section in Figure 27.26 clearly has a more attractive appearance than the IP plot, although the depth of the anomaly appears too large. The absence of readings for L < 400 feet (120 m) is due to the minimum Tx-Rx spacing used, being 400 feet (120 m), with the moving transmitter mode of the EM-25. The transmitter coil was north of the receiver stations in all cases, since this configuration provides better coupling to a conductor dipping south. Note that the vertical scale in Figure 27.26 represents the full Tx-Rx spacing; it should be divided in half to be equivalent to Figure 27.25.

EM Logging

Results of the Betz DHEM survey in DDH M-19 are presented in Figure 27.27. The hatched scales accompanying the residual voltage (RV) log are a superimposed grid of per cent vs depth values used to convert the RV readings to the quadrature component of the response. Appropriate location of this grid is determined by calibration of the tilt angle vs receiver signal in a barren section of the hole.

There is a very strong response in the max-coupled (so-called AMP) signal, as well as for RV and tilt (Tx coil N-S), at about 875 feet (265 m). The RV and tilt logs for the E-W transmitter orientations are not so anomalous. Betz reports this as an indication of a large conductive zone within 25 feet (7.5 m) of the hole and probably to the west. DDH M-19 shows two rather thin mineral sections at an approximate depth of 900 feet (275 m).

NEW INSCO OREBODY

The New Insco deposit (see Fig. 27.1) is considerably smaller than Iso, being about 1 million tons averaging 2.5 per cent Cu, 0.25 oz/ton Ag, with very little Zn. The geometry, shown in Figure 27.28, indicates its limited depth extent (about 600 feet (180 m)), strike length (less than 400 feet (120 m)) and south dip. The overburden is similar to that at the Iso orebody. No information was available with regard to the local geology.

Geochemical and Ground Geophysical Surveys, New Insco Orebody

Line 0+00

The geochemical, magnetic, gravity and HLEM profiles along Line 0+00 are shown in Figure 27.29. From the vertical section, there is clearly little mineralization of appreciable width. The drill section shows 1.3 feet (40 cm) of 1.4 per cent Cu at 140 feet (43 m) in DDH H73-1; 6 feet (180 cm) of 1.8 per cent and 3 feet (90 cm) of 6.1 per cent Cu at 490 and 540 feet (150 and 165 m) respectively in H73-2. This mineralization has not been considered to be continuous over the intervening 275 feet (85 m) between the holes.

The magnetic and B-horizon soil geochemical profiles are quite uninteresting. No basal-till geochemical sampling was done on this line. There is a minor EM anomaly centred at 12+25N which, from its shape and from model curves, indicates a thin conductor dipping south from about 50 feet (15 m) below surface and having a σt product of 11 mhos. The broad gravity anomaly of 0.4 mgals, centred at 12+00N and persisting to the north, does not fit this interpretation. Nor would it appear to be due to either a density contrast in rock types – since dacite is generally of lower density than diorite – or an overburden anomaly. It may be due to disseminated mineralization. Thus the profiles on this line indicate that the main ore zone has pinched out slightly to the east.

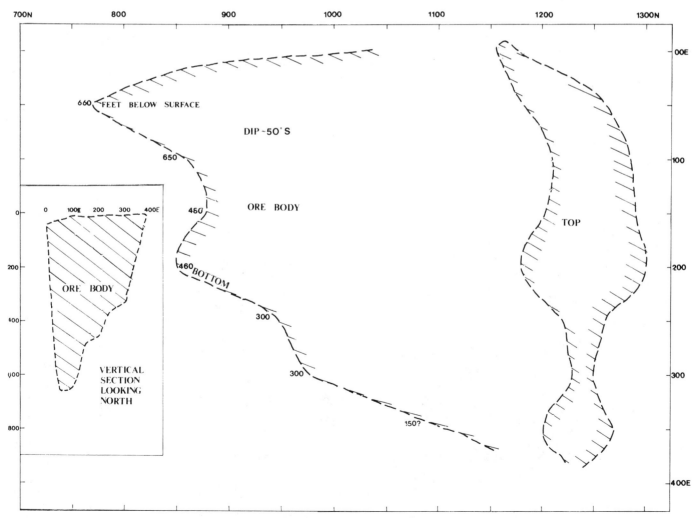

Figure 27.28. Surface projection and vertical section, New Insco Orebody.

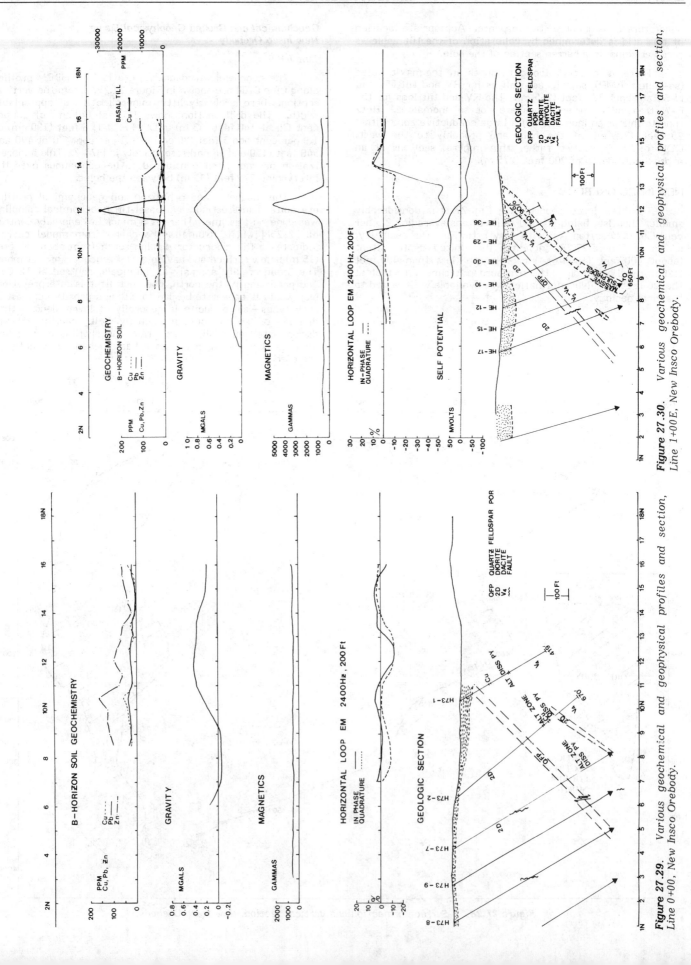

Figure 27.29. *Various geochemical and geophysical profiles and section, Line 0+00E, New Insco Orebody.*

Figure 27.30. *Various geochemical and geophysical profiles and section, Line 1+00E, New Insco Orebody.*

Line 1+00E

Figure 27.30 displays the geochemical, magnetic, gravity, HLEM and SP profiles on Line 1+00E, with the section. Projected to surface, the top of the ore zone has a horizontal width of about 75 feet (23 m) and it bottoms out at 650 feet (200 m). The average grade is about 2.6 per cent Cu and there is a high concentration of pyrrhotite and pyrite throughout.

The SP response is surprisingly small for a massive sulphide at shallow depth. This may be due to the type of overburden. From soil sampling of the B-horizon, there appeared to be solid clay below about 10 inches (25 cm). Clay can have a masking effect on surface SP response, judging from previous surveys in this area and in the Quebec Eastern Townships.

The huge (2.9 per cent) basal-till Cu anomaly at 12+00N is on the south flank of the ore zone rather than directly over it. This is attributed by G.F. Archibald to the fact that the orebody subcrops 5-10 feet (1.5-3 m) above bedrock and may have been stripped off by glaciation. There is no reflection of this Cu anomaly in the surface soil geochemical profile.

The sharp magnetic anomaly peaking at 12+25N indicates a sheet or dipole type of causative body of limited depth extent and steep dip. Presumably the pyrrhotite in the ore zone is the source, although this is a large anomaly for pyrrhotite. The peak is sharper than would be produced by a uniform distribution of magnetic mineral across the ore section; its thickness could hardly be greater than 10 feet (3 m). However, there seems little doubt that the magnetic anomaly is associated with the ore zone, since it has practically disappeared on Line 3+00E and there is no indication of it on Line 0+00. There is no mention of pyrrhotite on the detailed drill section of Line 0+50E and none on 3+50E above a depth of 420 feet (130 m).

The HLEM profile indicates a good conductor dipping south and having a width of about 60 feet (18 m). From characteristic curves the depth to the top is more than 10 feet (3 m) and the σt value is about 55 mhos.

The gravity profile peaks somewhat south of the ore zone centre at bedrock, although if a meter reading had been taken at 12+50N the maximum might have shifted somewhat. In any case, the peak should be slightly downdip. However, the profile suggests a slab dipping north rather than south since the positive gravity persists to the north as on Line 0+00. Presumably this is due to disseminated mineralization.

Line 2+00E

Figure 27.31 presents additional profiles on Line 2+00E — multifrequency telluric, IP and apparent resistivity plots — along with those on the previous figures. On Line 2+00E the ore zone is 120 feet (36 m) wide at the top and has an average grade of 2.9 per cent Cu. There is also considerable massive pyrite and pyrrhotite through the section. DDH HE-42 intersects ore grade Cu at 350, 400 and 425 feet (107, 122, 130 m), the last appearing to be continuous as far as HE-14, that is, over 80 feet (24 m).

On Line 2+00E the basal-till Cu anomaly is on the north flank of the section. The 200 ppm soil anomaly at 11+00N presumably is a reflection of the bedrock anomaly caused by transport of material during north-south glacial movement.

There is a good SP anomaly of −200 mv on Line 2+00E, with the peak at 13+00N corresponding to that for the basal-till Cu anomaly. Asymmetry of the anomaly indicates a south dip.

The IP profiles on Line 2+00E display the double-dipole traverse at x = 100 feet (30 m), n=1, the shallowest spread. There are strong IP anomalies at 12+50N. Although not evident on the scale used, the ρ_a profile has a minimum of about 9 Ωm at the same station and is generally less than 100 Ωm between 7+00N and 14+00N,

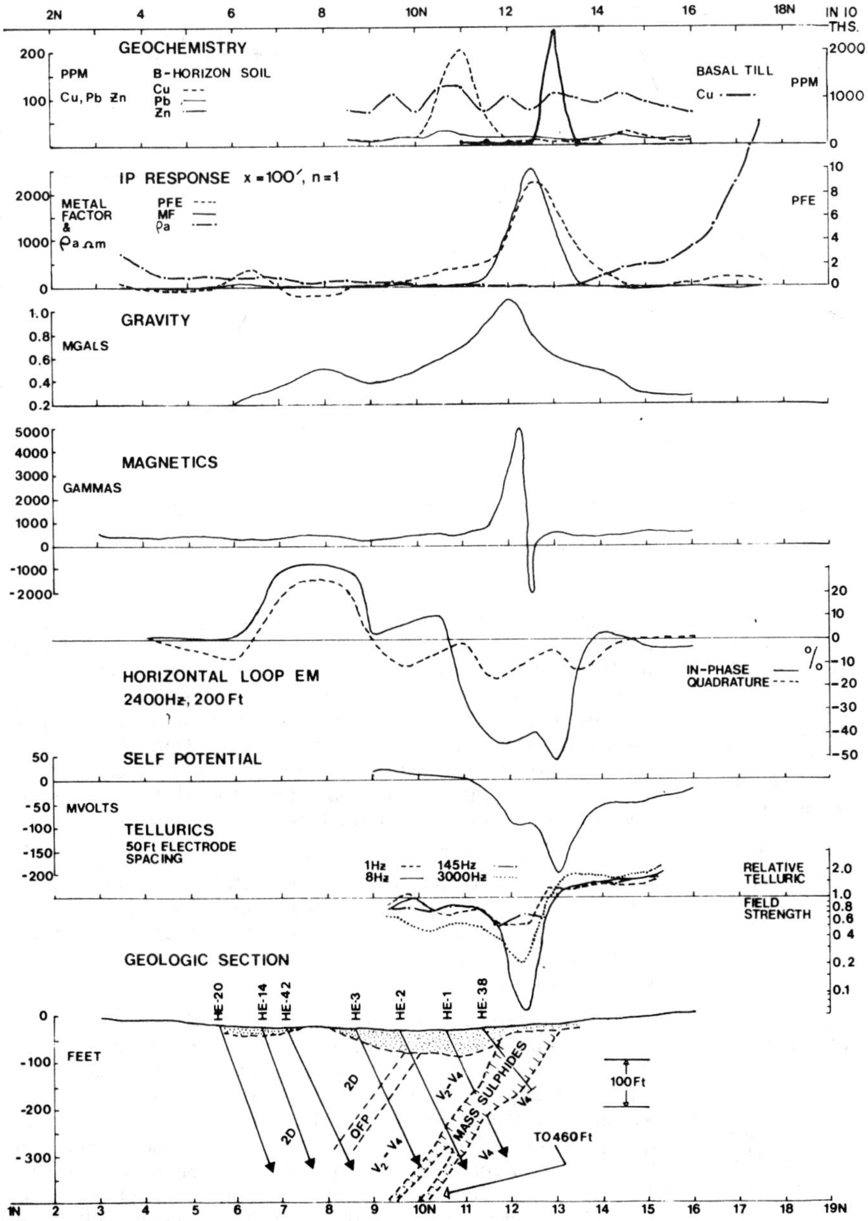

Figure 27.31. *Various geochemical and geophysical profiles and section, Line 2+00E, New Insco Orebody.*

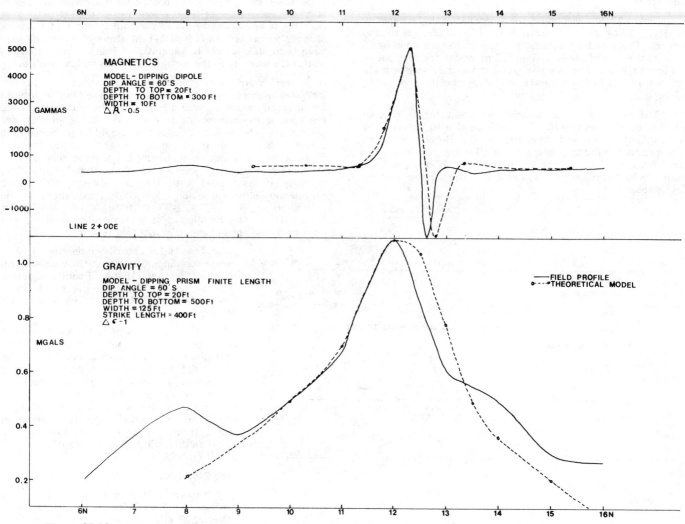

Figure 27.32. Matching of magnetic and gravity profiles by model shapes, Line 2+00E, New Insco Orebody.

reflecting the conductive clay overburden. Resistivity increases at both ends of the line, very sharply to the north. There is also a minor PFE peak at 6+50N which is not matched on either of the other profiles.

The telluric profiles on Line 2+00E also show a sharp anomaly over the top of the ore zone, particularly for 8 and 3000 Hz. Surprisingly the 145 Hz channel is least affected at this station: one would expect that the response to a conductor of limited depth extent such as this would vary directly with the frequency. Background noise, including conductive overburden, may be responsible. The asymmetry of the anomaly indicates a south dip, although the large telluric response north of station 14+00N may be mainly due to higher resistivity in the area.

As on Line 1+00E, there is a single 5000 gamma peak on the magnetic profile at 12+30N which, on this line, is followed immediately to the north by the characteristic negative tail associated with a thin sheet or dipole dipping south.

The HLEM response on Line 2+00E indicates a shallow zone of high conductivity at least 100 feet (30 m) wide. The shape of this anomaly is somewhat confusing, because the maximum negative value and the steep slope of the real component are on the north flank, while the positive over-shoot is to the south. These characteristics conflict, suggesting north and south dips respectively. The rather peculiar anomaly shape may be due to the width and shallow depth of the conductor, that is, some saturation in the negative response. This argument is reinforced by the appearance of double peaks in both real and imaginary profiles, particularly the latter. From thin sheet model curves, a depth of about 12 feet (4 m) and a σt product of 40 mhos was calculated. Obviously the thin sheet is not appropriate as a model for the New Insco orebody.

On Line 2+00E, the gravity anomaly clearly indicates a prism section with south dip. The north flank response is less persistent here than on lines to the west. A minor positive excursion at station 8+00N is somewhat similar to the gravity anomaly on Line 1+00E. There is no obvious explanation for these features.

An attempt was made to match the magnetic and gravity anomalies of Line 2+00E, as shown in Figure 27.32. As mentioned earlier, the magnetic anomaly is so sharp that an extremely small cross-section is required to provide sufficiently steep slopes. Although the match is reasonably good, the strike length is unrealistic and the susceptibility too large. The latter could be improved by reducing the depth to 10 feet (3 m), which is nearer the actual value. On the other hand, if a strike length of 200 feet (60 m) is used, as it should be, the anomaly becomes too broad. In matching the gravity anomaly the problem is to make the model wide enough to

Iso and New Insco Orebodies, Quebec

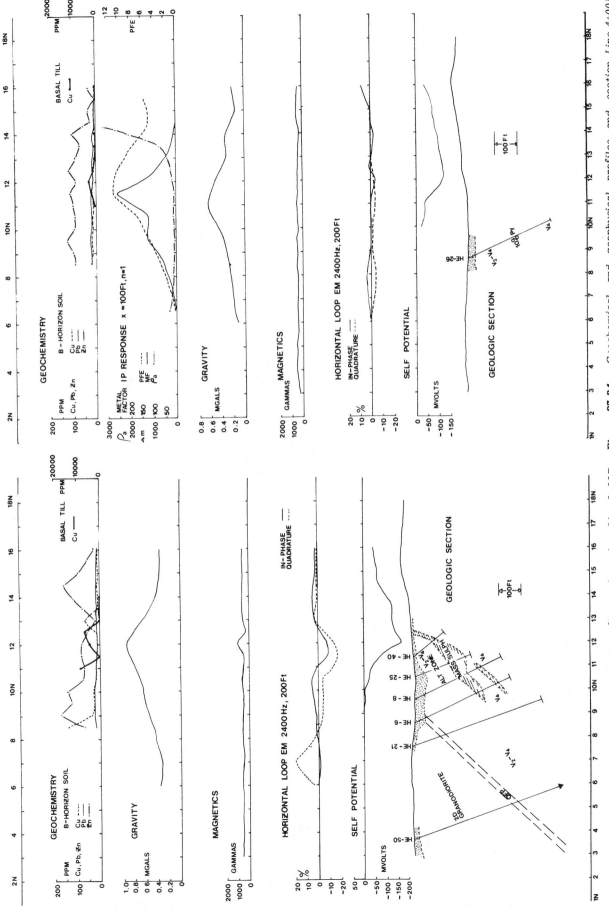

Figure 27.34. Geochemical and geophysical profiles and section, Line 4+00E, New Insco Orebody.

Figure 27.33. Geochemical and geophysical profiles and section, Line 3+00E, New Insco Orebody.

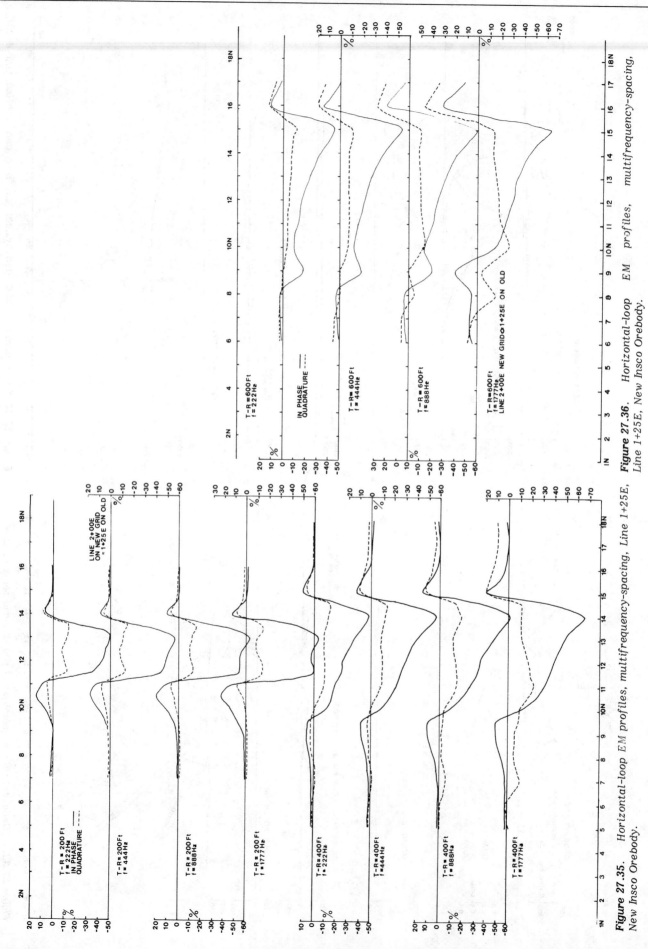

Figure 27.36. Horizontal-loop EM profiles, multifrequency-spacing, Line 1+25E, New Insco Orebody.

Figure 27.35. Horizontal-loop EM profiles, multifrequency-spacing, Line 1+25E, New Insco Orebody.

increase the slopes on the flanks. More heavy material is required than can be contained in the prism section. The rather high density contrast is not abnormal.

Line 3+00E

Figure 27.33 presents the geophysical and geochemical data obtained along Line 3+00E, where the ore cross-sectional area has shrunk and the average grade has decreased to 1.4 per cent Cu. A smaller section below the main zone, intersected by drillholes HE-8 and HE-6, averages 3.3 per cent Cu, but appears of limited extent. Also the amount of pyrrhotite in the sulphides has decreased, particularly in the shallow portion of the zone.

The basal-till Cu anomaly has decreased to 0.6 per cent at 12+50N on the north edge of the bedrock ore exposure. There is a second Cu peak at 11+00N. As on Line 2+00E, the soil geochemistry is anomalous for Cu about 400 feet (120 m) south. Glaciation is the probable cause of both these anomaly displacements.

The SP peak is quite strong, located 50 feet (15 m) downdip from the basal-till Cu maximum. The profile shape indicates a south dip. There is very little magnetic response, indicating the decrease in pyrrhotite.

The HLEM anomaly has also decreased and the profile shape reflects a very thin conductor dipping south. From model curves the depth is about 20 feet (6 m) and the σt product about 3 mhos — a surprisingly poor conductor. Gravity response, on the other hand, remains strong. The south dip is evident, but the gravity anomaly is wider than one would expect.

Line 4+00E

The drill section for Line 4+00E (see Fig. 27.34) shows little mineralization in the single drillhole HE-26. The ore zone appears to have pinched out 50 feet (15 m) west. However, only the barren magnetic and HLEM results support this assumption. The 0.3 per cent Cu basal-till anomaly at 12+00N is supported by an appreciable SP response. The IP profiles show a strong, broad PFE peak (larger than on Line 2+00E) and a metal factor maximum, both at 11+50N, although the latter is only 10 per cent of the maximum on Line 2+00E. This is due to the higher resistivity. A broad gravity anomaly is still present, reduced somewhat in amplitude from previous lines.

Maxmin Horizontal-Loop EM Profiles

Figures 27.35 and 27.36 display Apex Maxmin HLEM profiles from Line 1+25E. A new set of grid lines was offset 75 feet (23 m) west of the original; hence Line 2+00E on the new grid is actually 1+25E on the old and these profiles should resemble those of Line 1+00E.

Figure 27.35 shows eight profiles from Line 1+25E for four frequencies and three transmitter-receiver coil spacings. The fourth profile, 1777 Hz at 200 feet (60 m) spacing, should match most closely with the HLEM profile of Figure 27.30 on Line 1+00E.

The profiles for the 200 foot (60 m) spacing in Figure 27.35 have a conventional appearance at all four frequencies. However, at spacings of 400 and 600 feet (120, 180 m), (the latter appears in Fig. 27.36), they become increasingly distorted. This is due to the short strike length of the New Insco ore zone relative to such coil separations. An increasing fraction of the primary flux passes around the conductor rather than through it. The New Insco orebody is neither thin enough nor long enough to resemble a thin sheet model under these conditions. In Table 27.3, information obtained from HLEM data, when compared to thin sheet characteristic curves, is listed for the twelve profiles of Line 1+25E.

Clearly the thin sheet model becomes unrealistic as the coil separation increases. Estimates of thickness and depth, reasonably good for ℓ = 200 feet (60 m), become too large at 400 and 600 feet (120, 180 m), while the σt product and derived conductivity decrease steadily as ℓ increases. This subject has been discussed in some detail by West (1973) and Betz (1973). To realize a good estimate of σt for a conductor of the dimensions of the New Insco orebody, it would be necessary to use a coil separation of about 50 feet (15 m) and a much lower frequency — around 1 Hz. By extrapolation of the Maxmin survey data to these parameters, Betz has concluded that the true σt value should be about 9000 mhos.

Pseudo-Depth Sections

An assortment of pseudo-depth section plots is shown in Figures 27.37 and 27.38 for Lines 1+25E and 2+00E. Real-component horizontal-loop EM data appear in Figure 27.37. The upper four diagrams correspond to IP sections, both in the plotting arrangement and in general appearance. That is, they become broader at depth, show a steeper gradient on the north, and give no indication of the dip. The lower three

Table 27.3

Parameters for New Insco Conductor from HLEM Thin Sheet Curves

ℓ	f	t	mIP	z	σt	σ	ℓ	f	t	mIP	z	σt	σ
ft	Hz	ft	ohms	ft	mhos	mhos/m	ft	Hz	ft	ohms	ft	mhos	mhos/m
200	222	60	0.107	15	490	27	400	888	105	0.856	20	58	2
"	444	70	0.214	15	300	14	"	1777	100	1.712	20	28	0.9
"	888	70	0.428	10	175	8	600	222	145	0.321	60	170	4
"	1777	70	0.856	10	95	4	"	444	145	0.642	50	130	3
400	222	105	0.214	25	245	8	"	888	150	1.284	50	60	1.3
"	444	100	0.428	20	170	6	"	1777	5	2.568	30	14	9?

ℓ = coil spacing;   t = conductor thickness;   z = depth to top;
μ = permeability = $4\pi \times 10^{-7}$ h/m; $\omega = 2\pi f$; σ = conductivity

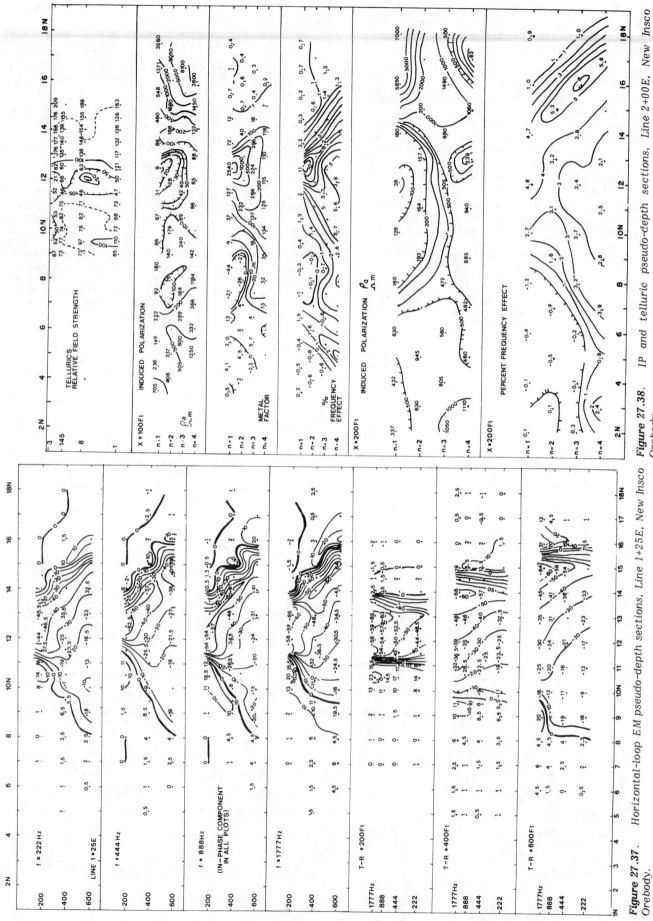

Figure 27.37. Horizontal-loop EM pseudo-depth sections, Line 1+25E, New Insco Orebody.

Figure 27.38. IP and telluric pseudo-depth sections, Line 2+00E, New Insco Orebody.

sections in Figure 27.37 use frequency for the ordinate as in MT plots. They are clearly reflections of the profiles, since the anomalous zone widens with increasing Tx-Rx separation and the dip appears to be north. There is some suggestion that the conductor is shallow and of limited depth extent.

IP and telluric pseudo-sections are shown in Figure 27.38. All the IP sections for x = 100 feet (30 m) locate the top of the zone correctly, but the direction of dip is not evident. There is some indication that the zone is of limited depth extent. In the next two sections (x = 200 feet (60 m)), however, the response is diffused to such an extent that the anomalous PFE zone is displaced about 300 feet (90 m) north of its real position. In the top diagram the telluric contours indicate a narrow conductor which appears to have its highest concentration at considerable depth, due to the 8 Hz low at station 12+00N.

DISCUSSION AND CONCLUSIONS

Detailed geological and geochemical studies of the Iso orebody have shown it to be fundamentally different from the conventional Rouyn-Noranda type deposit, both in its original deposition and subsequent alteration.

The lack of magnetic signature associated with the Iso orebody is due to the absence of pyrrhotite and magnetite, whereas for the New Insco orebody, the magnetic anomaly correlates closely with the richest section of the ore zone. Presumably this is due to the presence of massive pyrrhotite in the orebody.

The insignificant SP response over the Iso ore zone may be due to the conductive clay overburden or, more likely, the high Zn (sphalerite) content of the ore, particularly in the shallow portion. Correlation between SP and basal-till Cu values is excellent at New Insco on Lines 2+00E, 3+00E and 4+00E but not on Line 1+00E: between basal-till and soil geochemistry the correlation is only fair.

The EM-25 moving transmitter field system produced a highly interesting pseudo-depth section at Iso, while the IP survey results were fairly conventional. The MT survey, however, did not provide particularly good data, probably because of noisy signals.

The New Insco orebody is a poor target for IP, unless one were to use an abnormally small electrode spacing. Whether the IP anomalies on Lines 4+00E, 6+00E and 8+00E (the last two have not been included here) reflect appreciable disseminated mineralization is not known, since no drill data were available. The persistence of the broad gravity high to the east reinforces the IP results.

Variable spacing HLEM data are better at Iso than at New Insco because of the larger dimensions (particularly strike length) of the former. The New Insco deposit is something of a problem for interpretation by conventional electrical methods.

In general the wealth of geochemical and geophysical information provided by these studies supports the evidence obtained from pattern drilling of both the Iso and New Insco orebodies with regard to their geometry and to some extent their physical characteristics. That is to say, they are excellent targets for a variety of geophysical and geochemical survey techniques.

ACKNOWLEDGMENTS

We wish to thank all those who contributed survey data and other pertinent information to make this case history possible, including G.F. Archibald, J.E. Betz, Bondar Clegg & Co., D.C. Fraser, Geophysical Engineering Limited, R.D. Hutchinson, McPhar Geophysics, New Insco Mines, Noranda Exploration Ltd., J.E. Riddell, Salamis & Associates, SOQUEM, and Terraquest Surveys Ltd. The Mineral Exploration Research Institute (IREM/MERI) of Montreal financed the compilation through a grant from the Department of Energy, Mines and Resources, Ottawa.

REFERENCES

Betz, J.E.
1973-4: Test program of continuously portable EM systems; Profiles — copyright 1973, text — copyright 1974.

Fountain, D.K. and Fraser, D.C.
1973: Geophysical analysis of the Magusi River sulphide deposit; 43rd Annual Meeting, Soc. Explor. Geophys., Mexico City, October 1973 (Abs.).

Fraser, D.C.
1969: Contouring of VLF EM data; Geophysics, v. 34, no. 6, p. 958-967.

1972: A new multicoil aerial electromagnetic prospecting system; Geophysics, v. 37, no. 3, p. 518-537.

1974: Survey experience with the Dighem airborne EM system; Can. Inst. Min. Met. Bull., v. 67, no. 744, p. 97-103.

Hales, F.W.
1958: An accurate graphical method for interpreting seismic refraction lines; Geophys. Prosp., v. 6, no. 3, p. 285-314.

Johnson, A.E.
1966: Mineralogy and textural relationships in the Lake Dufault Ore, Northwest Quebec; Ph.D. Thesis, Univ. Western Ontario.

Lazenby, P.G.
1973: New development in the Input airborne EM system; Can. Inst. Min. Met. Bull., v. 66, no. 732, p. 96-104.

Paterson, N.R.
1973: Some problems in the application of ELF-EM in high conductivity areas; Symp. on EM Explor. Methods, KEGS-Univ. of Toronto, May 1973, Paper 28.

Roscoe, S.M.
1965: Geochemical and isotopic studies, Noranda and Mattagami areas; Can. Inst. Min. Met. Bull., v. 58, p. 965-971.

Spence, C.J.
1966: Volcanogenetic setting of the Vauze base metal deposit, Noranda District, Quebec; 68th Ann. Meet., Can. Inst. Min. Met., April 1966 (Abs.).

West, G.F.
1973: Some effects to look for in EM interpretation; Symp. on EM Explor. Methods, KEGS-Univ. of Toronto, May 1973, Paper 25.

THE DISCOVERY AND DEFINITION OF THE LESSARD BASE METAL DEPOSIT, QUEBEC

Laurie E. Reed
Selco Mining Corp. Ltd., Toronto, Ontario

Reed, Laurie E., *The discovery and definition of the Lessard Base Metal Deposit, Quebec; in Geophysics and Geochemistry in the Search for Metallic Ores; P.J. Hood, editor; Geological Survey of Canada, Economic Geology Report 31, p. 631-639, 1979.*

Abstract

In 1971, prospector Antoine Lessard, using a ground VLF-EM instrument, identified an electromagnetic conductor during a search for the source of copper-nickel sulphide float found in the lac Frotet area of northern Quebec. Lessard outlined the conductive zone to its apparent limits with VLF-EM and magnetic surveys. Diamond drilling to test the conductor intersected copper-zinc sulphides in a favourable Precambrian volcanic environment. Subsequent drilling outlined a deposit containing 1.46 million tons to a depth of 1700 feet (520 m).

The presence of copper-zinc sulphides in the initial drill core was sufficiently encouraging to carry out more extensive geophysical surveys including airborne Input EM, ground horizontal-loop EM, induced polarization, gravity and mise-à-la-masse. These surveys have provided useful information about the deposit and its environment. Each method has supplied guides to the drilling program by showing some different aspect of the deposit. Discrimination between the sulphides and nearby peridotite bodies became a necessary requirement for the geophysical surveys. Clear discrimination was achieved by the magnetometer, airborne Input EM and ground horizontal-loop EM. Induced Polarization and VLF-EM surveys produced similar responses over the sulphide and peridotite bodies. The gravity survey did not produce an anomaly over the sulphides. Mise-à-la-masse was particularly informative both on surface and down holes. It is apparent however, that the initial VLF-EM survey made the major contribution to the discovery and definition of the near-surface portions of this deposit.

Résumé

En 1971, le prospecteur Antoine Lessard, en utilisant au sol un appareil pour levés EM-VLF (méthode électromagnétique aux très basses fréquences radio) a identifié un conducteur électromagnétique, alors qu'il recherchait la source de débris minéralisés contenant les sulfures de cuivre et de nickel, que l'on avait découverts dans le secteur du lac Frotet, dans le nord du Québec. Lessard a tracé les limites apparentes de la zone conductrice en effectuant des levés magnétiques et EM-VLF. Des forages au diamant que l'on a faits pour explorer le conducteur ont recoupé des sulfures de cuivre et de zinc dans un milieu volcanique précambrien favorable. Par la suite, des forages ont permis de délimiter un gîte contenant 1.46 million de tonnes, à une profondeur de 520 m (1,700 pieds).

La présence de sulfures de cuivre et de zinc dans la carotte de forage initiale a été un élément assez encourageant pour que l'on entreprenne des levés géophysiques plus poussés, en particulier des levés aéroportés EM par la méthode INPUT, des levés EM au sol par la méthode des bobines horizontales et coplanaires, des levés de polarisation induite, gravimétriques, et de mise à la masse. Ceux-ci ont apporté des informations utiles sur le gisement et son environnement. Chaque méthode a contribué à orienter le programme de forage, en révélant un caractère particulier du gisement. Pour faire les levés géophysiques, il a été nécessaire de pouvoir établir une distinction entre les sulfures et les masses de péridotite proches. On a pu clairement établir cette distinction, en effectuant des levés magnétométriques, des levés aéroportés EM par la méthode INPUT et par la méthode des bobines horizontales disposées au sol. Les levés de polarisation induite et EM-VLF ont donné des réponses similaires au-dessus des corps composés de sulfures et de péridotite. Le levé gravimétrique n'a pas indiqué d'anomalie au-dessus des sulfures. La méthode de mise à la masse a apporté des renseignements particulièrement importants, à la fois au sol et dans les trous de forage. Cependant, on se rend compte que c'est grâce au levé initial EM-VLF que l'on a découvert et pu définir les portions de ce gisement proches de la surface.

INTRODUCTION

The Lessard copper-zinc-silver deposit is located in the Frotet-Troilus greenstone belt some 360 miles (580 km) north of Montreal and 58 miles (93 km) north of the town of Chibougamau, Quebec, at approximately 50°30' north and 74°40' west (Fig. 28.1 and 28.2). The overall trend of this small Archean greenstone belt is northeasterly. The trend of the southern half of the belt, in which the Lessard Deposit occurs is east-southeast. The belt is some 50 miles (80 km) long and 25 miles (40 km) wide. It lies west of the Grenville front and north of the Abitibi greenstone belt. The Frotet-Troilus belt consists of volcanic and sedimentary rocks intruded by granite, gabbro and ultramafic bodies. The sulphide deposit is situated at the top of a narrow sequence of felsic volcanic rocks at a contact with overlying mafic volcanic flows.

Drilling to date on the deposit has indicated a reserve of 1.46 million tons of 1.73% copper, 2.96% zinc, 1.1 oz. of silver per ton, and 0.019 oz. of gold per ton, to a depth of 1700 feet (520 m). A dilution factor of 15 per cent was allowed. The deposit does not appear viable under current economic conditions. The zone is open at depth with the deepest hole, at a vertical depth of 1600 feet (490 m), having a grade of 3.8% copper, 3.1% zinc, 3.3 oz. of silver per ton and 0.05 oz. of gold per ton over a true thickness of 20 feet (6 m). Further exploration including the use of drillhole geophysics is contemplated.

INITIAL DISCOVERY OF THE LESSARD DEPOSIT

The discovery of the deposit in 1971 was the result of persistent work by prospector Antoine Lessard who was attracted to the area by copper-nickel float which had been found in 1958 (Murphy, 1962) some 4.5 miles (7.2 km) southwest of the deposit. Prospecting northeast along the trend but in the opposite direction to the latest glacial ice movement, Lessard discovered chalcopyrite in quartz within a gabbro at lac Strip, south of lac Frotet, and staked a number of claims.

A search for buried conductive sulphides was carried out using a Crone Radem VLM electromagnetic (EM) instrument (Crone, 1977). This instrument employs signals from VLF transmitters to detect subsurface conductivity contrasts. Dip angles of the magnetic field component were read. Traverses were made along claim lines (0.25 mile (0.4 km) intervals east-west and north-south). A strong conductor was found southwest of the lake and a detail grid was traversed to the limits of the conductive zone (Fig. 28.3).

Lessard found that the conductor changed strike so that it was necessary to read lines at orthogonal and diagonal directions to the initial east-west lines. It was also necessary to use different VLF transmitters depending on the local strike of the conductor. The station at Cutler, Maine (17.8 kHz) was used for conductors having an east-west and northwest-southeast strike, while the station at Balboa, Panama (24.0 kHz) was used for conductors having a north-south strike.

Lessard also carried out a magnetometer survey on this grid using a pocket magnetometer made by L.A. Levanto Oy of Finland (Hood, 1967). His survey showed the VLF-EM conductor to be magnetic. The results of a more recent magnetometer survey are shown in Figure 28.7.

The VLF-EM data was filtered using Fraser's (1969) technique to move the data by 90° in order to change the cross-overs into peaks and to reduce noise. Contours of the filtered data are shown in Figure 28.3. The strongest portion of the conductor is arcuate. A weaker north-south component appears to the west.

The source of the conductor does not outcrop, although gabbro, peridotite and andesite outcrop near the conductor. Therefore, the identification of the zone by geophysical surveys played the major role in the discovery after the discovery of the copper-nickel sulphide float.

At this stage the property was brought to the attention of Muscocho Explorations Limited, and then in turn to Selco Mining Corporation Limited. Subsequent work on the property has been managed by Selco on behalf of a joint venture between Selco and Muscocho.

Limited confirmation of the conductor was made using a vertical-loop electromagnetic instrument (Ward, 1967). The vertical-loop survey (not shown) located conductors at each of the first four drillholes. Then, the four holes appearing on Figure 28.3 were drilled. Holes L1, L2, and L4 intersected copper-zinc sulphides in felsic volcanics. Hole number L3 identified graphite slips in serpentinized peridotite.

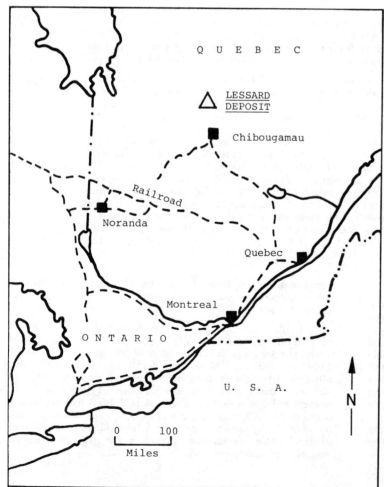

Figure 28.1. Index map

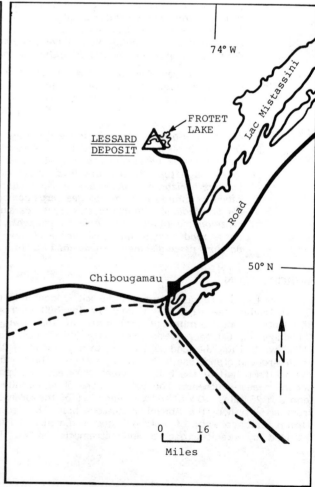

Figure 28.2. Index map

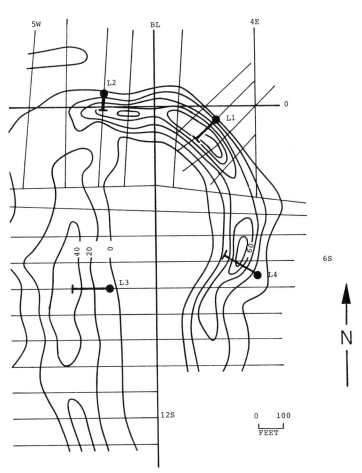

Figure 28.3. VLF-EM survey on the discovery grid. The contour interval is 20 filtered degrees. L1, L2, and L4 are the discovery drillholes.

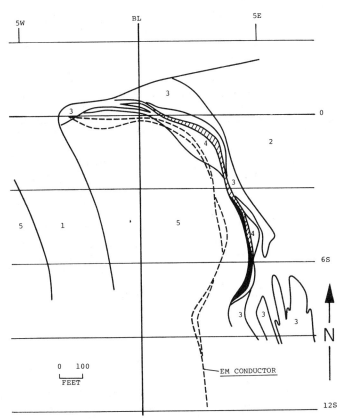

Figure 28.4. Geology of the Lessard Deposit at 400 feet (122 m) below surface (legend on Fig. 28.5).

DETAILED GEOPHYSICAL FOLLOW-UP

After the first four holes were drilled, a new grid was cut using the original grid as a base. Lines were generally cut with a 100 foot (30 m) line spacing. A number of these lines have been left out of figures accompanying this paper. However the instrument data or trends from the data on these lines are presented. Magnetometer, horizontal-loop electromagnetic, induced polarization (IP), gravity, and mise-à-la-masse surveys were carried out during the next two years. A Mark VI Input airborne electromagnetic survey carried out in the region also covered the deposit. These surveys provided definiton of the ore zone and guided the drill program as it progressed.

GEOLOGY OF THE LESSARD DEPOSIT

Most geological knowledge of the sulphide zone and its immediate environment comes from diamond-drill core since outcrop is sparse near the deposit. A plan of the 400 foot (122 m) elevation (Fig. 28.4) and a cross-section at 600S (Fig. 28.5) show the relationship of the sulphide zone to lithology. (The trace of the surface electromagnetic conductor defined in Fig. 28.6 is indicated in Fig. 28.4). The sulphide mineral assemblage, alteration, and volcanic stratigraphy suggest that the deposit is similar to other volcanogenic deposits in the Canadian Shield described by Sangster (1972). A description of the local and regional geology of the deposit, drawn from Selco maps and reports, is presented by Bogle (1977).

The sulphides are confined to a felsic volcanic unit at, or stratigraphically below, a contact with mafic volcanic rocks. Within the felsic unit there are rhyolite flows and intermediate to felsic tuffs. Argillaceous units are occasionally seen within the tuffs, below, and marginal to the sulphides. The rocks have been overturned so that the stratigraphically-lower felsic rocks are above the mafic rocks. Dips are generally to the east and north, although in places, they are nearly vertical. The felsic rocks are truncated by a gabbro sill which bounds the felsic rocks to the east. To the north, the volcanics, including the sulphide zone, are terminated by a serpentinized peridotite intrusion. Serpentinized peridotite also defines the western margin of the mafic volcanics.

Mineralization is in the form of stringer to massive sulphides. The mineralized zone is a few feet to over fifty feet (15 m) wide. The stringer sulphides occur stratigraphically below the massive sulphides. In the massive sulphides, pyrite predominates over pyrrhotite and sphalerite and chalcopyrite are in about equal proportions. In the stringer zone, pyrrhotite and chalcopyrite are the dominant sulphide minerals.

GEOPHYSICAL SURVEYS

Horizontal-loop EM Surveys

The electromagnetic conductor initially identified by VLF-EM (Fig. 28.3) was more completely defined by a horizontal-loop EM survey. Some of the profiles are seen in Figure 28.6. This survey employed a McPhar VHEM instrument using a coil separation of 200 feet (61 m) and a frequency of 600 Hz.

The resulting arcuate anomaly corresponds exactly with the strong VLF-EM conductor identified by Lessard. The strong EM response between 0 and 3S east of the base line indicates the shallowest part of the zone. This was subsequently confirmed by drilling. The response to the west along line zero, indicates that the zone remains shallow. The amplitude of response, however, drops as the sulphides become thinner and terminate between 3W and 4W. The diminishing response east of the base line south of line 6S occurs as the main body of the sulphides plunges toward the south.

Horizontal-loop EM profiles suggest that near-surface dips are very nearly vertical. This was confirmed by drilling (Fig. 28.5). Chalcopyrite, pyrrhotite, and pyrite, in both massive and stringer zones, were identified in the drilling as the cause of the EM conductor.

A complex response west of the base line on line 9S identified the VLF-EM conductor over the peridotite body. Graphitic slips and serpentine seen in drillhole L3 are the likely source of the weak negative quadrature. The positive in-phase appears to be a high magnetic susceptibility response generated by magnetite in the peridotite.

Magnetic Surveys

The vertical-field magnetic surveys were repeated using a McPhar M-700 fluxgate magnetometer and the results are shown in Figure 28.7. The two prominent highs west and north on the grid identify the peridotite bodies which contain magnetite. The high response at the western end of line zero also has its source in peridotite, although it is possible to confuse this with the responses just to the east, which have their origin in pyrrhotite in the sulphide zone. The responses from the pyrrhotite, which are occasionally bipolar, follow the arcuate form of the conductor.

A comparison of VLF-EM and magnetic responses over the two peridotite bodies demonstrates that the westerly body is weakly conductive while the northerly body which has a similar magnetic intensity, is not conductive. The cause of these differences has not been revealed by drilling.

The magnetic surveys have not clearly discriminated between the gabbro and the volcanic rocks. The decrease of magnetic response to the southeast, however, does correlate with an increase in felsic rocks. The gabbro to the east is not distinctively magnetic and has a similar response to the mafic volcanics west of the ore zone.

Airborne EM Survey

A Mark VI Input EM survey was flown by Questor Surveys Ltd. of Toronto. Details of the system are given by Lazenby (1973). The direction of the profile, presented in Figure 28.8 is reversed from normal in order to match the presentation of the ground responses in this figure. The locations of the Input EM flight line over the deposit and the resultant anomalies are shown in Figure 28.6.

The six-channel conductor C identifies the main sulphide zone. The leading anomaly, B, also has its source in this sulphide zone. Anomaly B results from the asymmetry of the Input system which generates a secondary leading anomaly over a conductor which is vertical or dips toward the approaching aircraft (Palacky and West, 1973). The weaker, poorer, conductor D identifies the serpentinized peridotite. Anomaly A on Figure 28.8, which looks much like anomaly D, also has its source in a peridotite body about a mile north of the Lessard Deposit. Uneconomic sulphide stringers were identified as the source of anomaly A.

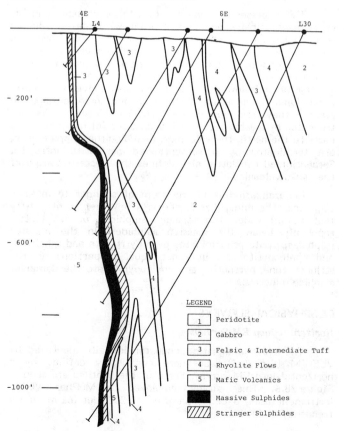

Figure 28.5. Geology Section 6S.

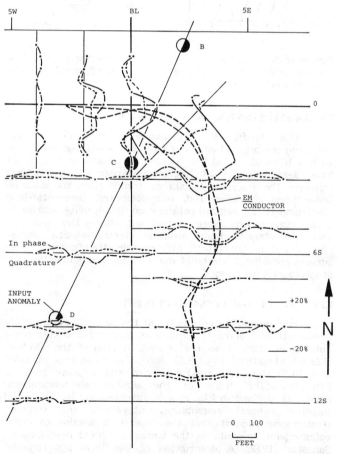

Figure 28.6. Horizontal-loop EM survey (using 200 foot (61 m) coil separation) and airborne Input anomalies.

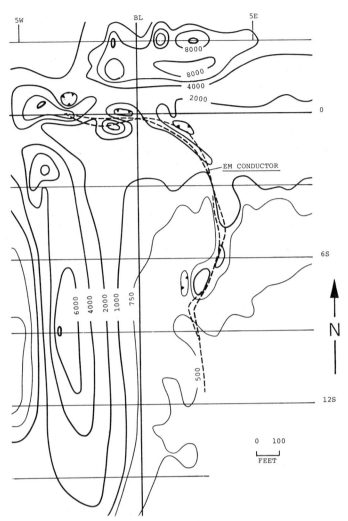

Figure 28.7. Contours of the vertical magnetic field (in gammas). Thicker contours have a 2000 gamma interval except for the 1000 gamma level. Thinner contours have a 250 gamma interval.

Induced Polarization and Resistivity Surveys

An induced polarization survey using a McPhar frequency-domain instrument (Madden, 1967; Hendrick and Fountain, 1971) covered the zone south from line 3S. Contours of per cent frequency effect, shown in Figure 28.9, are for a dipole-dipole array having an "a" spacing of 200 feet (61 m) at n = 1. The frequencies used for the survey were 5.0 and 0.3 Hertz. The location of the EM conductor is plotted for reference. Although it is clear that frequency effect responses occur over the sulphide zone, definition of the zone is masked by the overlapping responses of the peridotite to the west. Similarly, the resistivity component of the survey, shown in Figure 28.10, displays a markedly low resistive response over the peridotite. The resistive low between lines 3S and 6S east of the base line identifies the sulphide zone.

Pseudo-sections of the IP and resistivity response on line 6S (Fig. 28.11) show that while individual anomalies occur over the sulphide zone, large responses from the serpentinized peridotite mask the sulphide response at large electrode separations. The apparent IP effect from the peridotite is slightly higher than from the sulphides, while apparent resistivities of the peridotite are considerably lower than that of the sulphides.

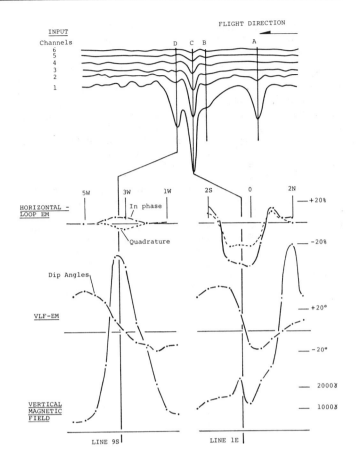

Figure 28.8. Profiles comparing geophysical methods over sulphides (right) and peridotite (left).

Very high resistivities to the east correlate with gabbro. Similar high resistivities to the west indicate that bedrock to the west may be gabbro as well.

Gravity Survey

A gravity survey over the deposit yielded no detectable anomaly from the sulphides. A profile of the Bouguer gravity on line 6S shown in Figure 28.11 is typical of responses in the area. The lack of a gravity response from the sulphides is due to the fact that the main mass of sulphides occurs 400 feet (120 m) below the surface. The geological section on line 6S (Fig. 28.5) shows the thickest sulphides are between 600 and 800 feet (180 to 240 m) from surface. Nearer surface, the sulphides are thinner and in stringer form. These do not provide a significant gravity target.

Mise-à-la-masse Survey

A mise-à-la-masse survey (Parasnis, 1967) employed a current electrode placed in the sulphide zone in a drillhole at a depth of about 550 feet (170 m) from surface. Current was maintained at 1.0 amp. at a frequency of 5.0 Hz. Infinite current and potential electrodes were placed 3500 feet (1066 m) north and south of the survey area respectively. Voltages were measured on surface every 50 feet (15 m) along lines at an interval of 100 feet (30 m) and every 50 feet (15 m) down available holes.

A number of features related to the distribution of sulphides in the zone are indicated by the distribution of voltages on the surface shown in Figure 28.12, and down holes shown in Figures 28.13 and 28.14.

The arcuate shape of the contours follows the shape of the electromagnetic conductor. The highest values (over 1700 millivolts) are found at the strongest EM conductor, east of the base line between lines 0 and 3S. These identify the shallowest part of the sulphide zone. Elsewhere a ridge of high values occurs along the length of the EM conductor. Voltages decrease along the ridge in both directions from the peak, indicating increasing depth to the top of the sulphide zone. The steep gradient off the ridge of the anomaly west along line zero, appears to indicate that the sulphide zone is narrow and limited in depth extent. The voltages flanking the sulphides are reduced however, by resistive lows from peridotite bodies to the north and southwest. The gradual voltage drop south of the southerly end of the EM conductor (south of line 12S), suggests that the sulphide zone plunges to the south. The low gradients off the ridge of the anomaly from lines 3S to 15S indicate the zone extends to greater depth south of 3S than north of it. Contours are more open east of the ridge than west of it suggesting an easterly dip. The low resistivities of the peridotite to the west, combined with the high resistivities of the gabbro to the east however, probably distort the contours so that the dip interpretation is suspect.

The depression south of line 15S is part of a long, linear low response lying nearly east-west across the strike of the sulphides. A fault, producing low resistivities in bedrock, or a bedrock depression is indicated. Drilling has not extended far enough to confirm this.

Voltage measurements down holes in section 6S (Fig. 28.13) identify a peak response of over 1700 millivolts which generally corresponds with the location of the sulphides traced in from Figure 28.5. Apparent discrepancies occur in hole L30 where peak voltages are observed not only in the main sulphide zone (location C), near the bottom of the hole, but also higher up at locations A and B. The contours on Figure 28.13 connect the high values at A and B to the high values in the hole above, while high values at C do not connect. This is not so much a representation of what is really happening but is a condition forced by the limitations of available data. The high values at A and B do not correlate with ore intersections, but do identify 10 to 20 per cent pyrrhotite with minor chalcopyrite in siliceous volcanics. It would seem that an electrical connection (possibly by way of sulphides) exists between sulphides at A and B and the main sulphide zone C.

The mise-à-la-masse survey in holes on section 1W at the north end of the deposit, and shown in Figure 28.14, indicates a different voltage distribution than that of section 6S. Only one hole, L7, has intersected sulphides. The other

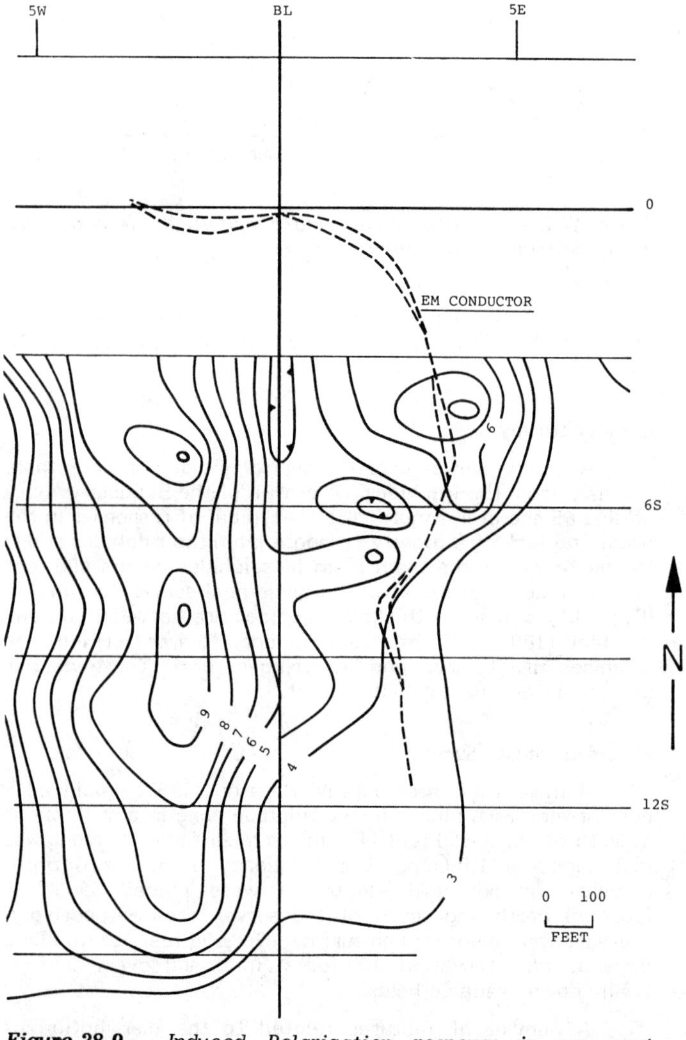

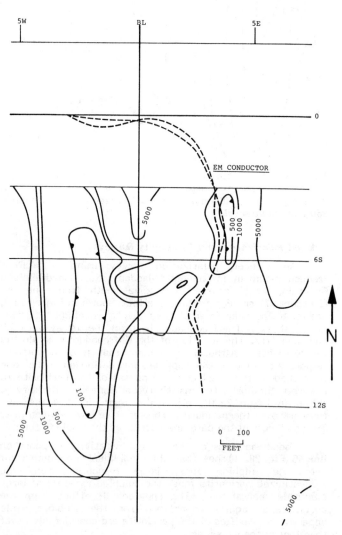

Figure 28.9. Induced Polarization response in per cent frequency effect using a dipole-dipole array with a = 200 feet (61 m) and n = 1.

Figure 28.10. Resistivity response in ohm-metres using the same electrode array as in Figure 28.9.

hole in the section, L13, may not have been drilled far enough to intersect sulphides. However, the apparent dip seen in the trend of the voltages and the low voltages at the bottom of L13 give little encouragement for the possibility of intersecting any. Rocks on the same horizon as those containing the sulphides in hole L7 are intersected near the bottom of hole L13 but contain no significant sulphides.

The small size of the sulphide zone on section 1W, compared with that of section 6S, is apparent from the distribution of the voltages. Sharp gradients appear close to the smaller part of the body on 1W, while more gentle gradients occur around the larger part of the body on line 6S. As noted earlier, however, rocks adjacent to the sulphides influence the voltage pattern. On section 1W, voltages drop rapidly to the north, in part because of the low resistivities in the peridotite. On section 6S, higher resistivities in gabbro, east of the sulphides contribute to the lower voltage gradient to the east.

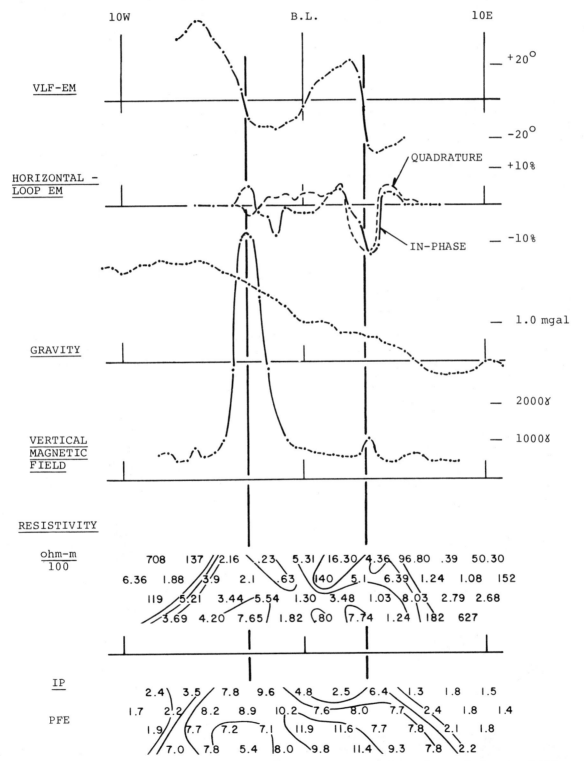

Figure 28.11. *Profiles and pseudo-sections on line 6S comparing geophysical methods over sulphides (right) and peridotite (left). IP and resistivity uses the same electrode array as in Figure 28.7.*

COMPARISON OF GEOPHYSICAL METHODS

The profiles in Figures 28.8 and 28.11 over the sulphide and serpentinized peridotite conductors provide a useful comparison of geophysical methods. The sulphide conductor, anomaly C on Figure 28.8, and the anomaly east of the base line on Figure 28.11 have good Input EM horizontal-loop EM, VLF-EM, IP and resistivity responses. The peridotite conductor, anomaly D on Figure 28.8, and the anomaly west of the base line on Figure 28.11 have a poor Input EM and irregular horizontal-loop EM, fairly good but broad VLF-EM, good IP and good (i.e. low) resistivity responses. Taken together, there is a clear separation of response from the two different sources by these methods. The Input EM and the horizontal-loop EM responses discriminate most effectively between the sulphide conductor and the serpentinized peridotite.

The magnetic responses in Figures 28.8 and 28.11 over the sulphides and the peridotite are quite different. The bipolar 800 gamma anomaly from the sulphides shown in Figure 28.8 looks insignificant beside the 8000 gamma anomalies over the peridotite bodies north on line 1E and on line 9S. An easterly dip to the peridotite is indicated by the asymmetrical shape of the magnetic anomalies on lines 6S (Fig. 28.11) and 9S (Fig. 28.8).

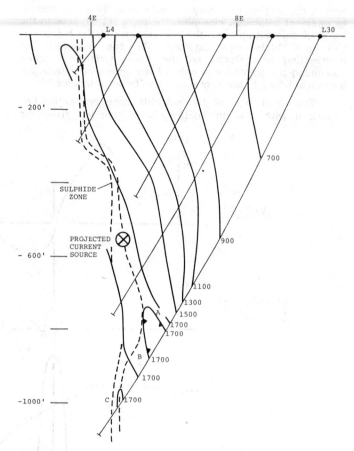

Figure 28.13. Mise-à-la-masse survey on surface and in drillholes on section 6S. Readings are in millivolts. Current electrode is located in sulphides about 50 feet (15 m) south of the section.

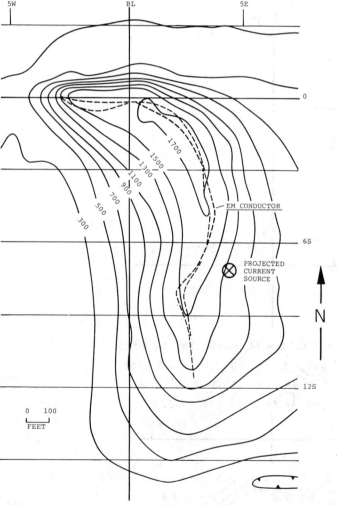

Figure 28.12. Mise-à-la-masse survey. Contours are in millivolts for readings taken on surface. Current electrode is located in sulphides 550 feet (167 m) below surface.

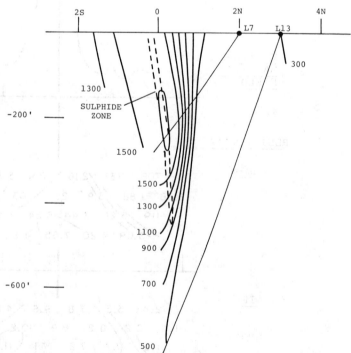

Figure 28.14. Mise-à-la-masse survey on surface and in drillholes on section 1W. Readings are in millivolts. The current electrode near 6S does not appear in this section.

A comparison of Input EM and horizontal-loop EM responses may be made using the apparent conductivity-thickness (σt) products (Grant and West, 1965; Palacky and West, 1973). A vertical half-plane source has been assumed for both ground and airborne responses. The early Input EM channels (numbers 1 to 3; Fig. 28.8) show a σt of 9 mhos while the later channels (number 3 to 6) show a σt of 18 mhos. The σt response from the ground instrument on line 1E just under the flight line is 7 mhos. This compares favourably with the early channel Input EM response. Variability of conductivity – thickness is evident however, as a higher σt response of 26 mhos is observed by the horizontal-loop EM on the diagonal line just east of the flight line (Fig. 28.6).

The duality of the σt of the airborne EM anomaly suggests that the zone has two conductive components. It is not clear if these two airborne EM responses have their origin at a single location, or if two sources occur along strike from each other. The latter case is indicated by the responses on the ground. If the two sources occur together, however, it is suggested that highly conductive sulphides causing the higher σt value are part of a larger but less conductive unit causing the lower σt value. The dual Input EM response, then, may represent the massive and stringer sulphides which are observed together in drill core.

CONCLUSIONS

The Lessard Deposit is probably a nearly ideal electromagnetic target. Highly conductive sulphides found near surface in resistive rocks in a steeply-dipping attitude are easily detected by a number of electromagnetic systems. The presence of conductive and magnetic serpentinized peridotite nearby only marginally interferes with the resolution of the zone by geophysical methods. While IP, resistivity and VLF-EM surveys identified the sulphides, these methods detected similar responses over the peridotite. A clear discrimination between the sulphides and the peridotite was achieved by the Input EM and horizontal-loop electromagnetic methods. The magnetometer survey as well, discriminates between the sulphides and peridotite by identifying magnetic fields of different character over these bodies.

The persistent work of prospector Antoine Lessard using simple geophysical instrumentation which led to the discovery of the deposit, is not downgraded by the fact that more sophisticated instruments also detect the deposit. The use of sophisticated instrumentation beyond the discovery phase is justified by the definition of the sulphide deposit and the discrimination of the deposit from nearby peridotite bodies by these instruments.

ACKNOWLEDGMENTS

The author wishes to thank Selco Mining Corporation Limited and Muscocho Explorations Limited for permission to publish this paper. Data presented in the paper were collected as part of the exploration program funded by these companies.

The geology was obtained from reports and maps by D.A. Hutton and I.F. Downie. Special thanks to J.E. Rackley, S. Christopher and M. Safranek for drafting, A. Melanson for typing, and D.A. Hutton, Dr. H.S. Squair and Mrs. D.J. Reed for reviewing the manuscript. Errors and omissions are the responsibility of the author. Input is a Registered Trademark of Barringer Research Ltd.

REFERENCES

Bogle, E.W.
1977: The primary dispersion associated with the Lac Frotet volcanic cycle, Quebec; unpubl. MSc. thesis, Queen's University, Kingson, Ontario, v. 1, 349 p., v. 2, 368 p.

Crone Geophysics Ltd.
1977: Radem information sheet; Crone Geophysics Limited, Mississauga, Ontario.

Fraser, D.C.
1969: Contouring of VLF-EM data; Geophysics, v. 34 (6), p. 958-967.

Grant, F.S. and West, G.F.
1965: Interpretation theory in applied geophysics; McGraw-Hill Book Co. Inc., New York, 584 p.

Hendrick, D.M. and Fountain, D.K.
1971: Induced polarization as an exploration tool, Noranda Area, Quebec; Can. Inst. Min. Metall. Bull., v. 64 (706), p. 31-38.

Hood, P.
1967: Magnetic surveying instrumentation – a review of recent advances; in Mining and Groundwater Geophysics, Geol. Surv. Can., Econ. Geol. Rept. No. 26, p. 3-31.

Lazenby, P.G.
1973: New development in the Input airborne EM system; Can. Inst. Min. Metall. Bull., v. 66 (732), p. 96-104.

Madden, T.R. and Cantwell, T.
1967: Induced polarization, a review; Mining Geophysics, Soc. Expl. Geophys., v. 2, p. 373-400.

Murphy, D.L.
1962: Preliminary report on Frotet Lake area; p. 476; Que. Dep. Nat. Resourc., 8 p.

Palacky, G.J. and West, G.F.
1973: Quantitative interpretation on Input AEM measurements; Geophysics, v. 38 (6), p. 1145-1158.

Parasnis, D.S.
1967: Some recent geolectrical measurements in the Swedish sulphide ore fields illustrating scope and limitations of the methods concerned; in Mining and Groundwater Geophysics, Geol. Surv. Can., Econ. Geol. Rept. No. 26, p. 298-301.

Sangster, D.F.
1972: Precambrian volcanogenic massive sulphide deposits in Canda: a review; Geol. Surv. Can., Paper 72-22, 44 p.

Ward, S.H.
1967: The electromagnetic method; Mining Geophysics, Soc. Expl. Geophys., v. 2, p. 224-372.

IZOK LAKE DEPOSIT, NORTHWEST TERRITORIES, CANADA: A GEOPHYSICAL CASE HISTORY

George Podolsky
Texasgulf Inc., Golden, Colorado

John Slankis
Texasgulf Inc., Toronto, Ontario

Podolsky, G. and Slankis, J.A., Izok Lake deposit, Northwest Territories, Canada: A geophysical case history; in Geophysics and Geochemistry in the Search for Metallic Ores; Peter J. Hood, editor; Geological Survey of Canada, Economic Geology Report 31, p. 641-652, 1979.

Abstract

An exploration program in the Northwest Territories was begun by Texasgulf Inc. in 1971. The first encouraging results were found a year later in an area about 250 miles (400 km) north of Yellowknife. The field program, consisting of geological reconnaissance, airborne geophysics, intensive ground follow-up and drilling, resulted in the discovery of two small, high-grade massive sulphide deposits in an area just east of the southern end of Takiyuak Lake. Both these deposits had associated airborne EM anomalies.

Attention from this initial success was temporarily diverted by the investigation of gossan zones, often containing high-grade mineralization, which did not respond to electrical methods. By 1974, interest had shifted some 30 miles (48 km) to the south, where a mineralized showing and high-grade sphalerite boulders had been found on the shores of a small lake, subsequently named Izok Lake.

In the spring of 1975, a ground VLF EM survey over the lake ice detected a conductor, not far from the high-grade float. Subsequent drilling through the spring of 1977 established a high-grade, massive Zn-Cu-Pb-Ag sulphide body in excess of 12 million tons, the Izok Lake deposit. In addition to obtaining a very good picture of the deposit from geophysical work, the Izok Lake massive sulphide body has been used as a test case for several field techniques including the newly developed UTEM system.

Texasgulf's exploration program in the Northwest Territories commenced with the standard approach of geologic reconnaissance followed by airborne geophysics and then by ground surveys using conventional geophysical methods. In the six years culminating in 1976, many different geophysical methods were tried and some were found to be particularly suited to conditions prevailing in the Territories. Due to the relatively thin overburden and high resistivity of the country rock, electrical surveys often approximated the idealized case. Magnetic and gravity surveys also gave good results, but in all cases a thorough appreciation of the underlying geology was necessary to avoid an erroneous interpretation.

Résumé

La Texasgulf Inc. a entrepris en 1971, un programme d'exploration dans les territoires du Nord-Ouest. Elle a obtenu les premiers résultats encourageants une année plus tard, dans une zone située à environ 250 milles au nord de Yellowknife. Le programme d'exploration, qui consistait en une reconnaissance géologique supplémentaire, en levés géophysiques aériens, en travaux ultérieurs détaillés au sol, et en forages, a permis de découvrir deux petits gîtes sulfureux massifs, de teneur élevée, dans un secteur situé immédiatement à l'est de l'extrémité sud du lac Takiyuak. Ces deux gîtes ont été caractérisés par des anomalies EM décelées par levé aéroportée.

Lors de l'inspection de zones de chapeau de fer (gossan), souvent caractérisées par une minéralisation de teneur élevée, ou qui ne donnaient pas de réponses lors de l'utilisation de méthodes électriques, on a laissé de côté ce succès initial. En 1974, le centre d'intérêt s'est déplacé vers 30 milles au sud, où des indices de minéralisation et des blocs riches en sphalérite ont été observés sur les rives d'un petit lac.

Au printemps 1975, un levé EM-VLF au sol effectué au-dessus du lac englacé a permis de déceler un conducteur, non loin des débris minéralisés de forte teneur. Par la suite, des forages effectués pendant le printemps 1977 ont révélé la présence d'un corps sulfureux massif de teneur élevée, d'une capacité supérieure à 12 millions de tonnes, le gisement du lac Izok. Non seulement les travaux géophysiques ont donné une image très précise du corps sulfureux massif du lac Izok, mais encore, celui-ci a servi de cas-type, pour la mise à l'épreuve de plusieurs techniques d'exploration sur le terrain, en particulier le nouveau système UTEM!

Le programme d'exploration de la Texasgulf dans les territoires du Nord-Ouest a commencé par la méthode habituelle de reconnaissance géologique, suivie de levés géophysiques aéroportées, puis de levés au sol par des méthodes géophysiques conventionnelles. Pendant les six années précédant 1977, on a essayé diverses méthodes géophysiques, et constaté que certaines d'entre elles étaient particulièrement bien adaptées aux conditions régnant dans les territoires du

Nord-Ouest. En raison de la minceur relative des terrains de couverture, et de la résistivité élevée de la roche encaissante, les résultats qu'ont donnés les levés électriques se sont souvent rapprochés du cas idéal. Des levés magnétiques et gravimétriques ont aussi donné de bons résultats, mais dans tous les cas, il a été nécessaire de faire une évaluation approfondie des conditions géologiques.

INITIAL DISCOVERIES – TAKIYUAK LAKE REGION

A geological reconnaissance of the southeastern area of Takiyuak Lake, some 250 miles (400 km) north of Yellowknife (Fig. 29.1) was initiated by Texasgulf Inc. in 1971. The program concentrated on mapping and sampling gossan or stain zones over an area previously depicted as metasediments on published maps, but reinterpreted as metavolcanics by Texasgulf geologists. An airborne EM survey was flown early in 1973 and produced numerous EM anomalies, some of which even corresponded to a few of the sampled gossans. However, the gossan yielding the best assays resulted in a one line EM anomaly at only twice the background response. Subsequent ground geophysical surveys gave a more complete picture of this anomaly designated as Hood River (H.R.) 10, showing it to be due to a complex, narrow, Y-shaped sulphide body of limited length. The work included VLF EM, magnetic, horizontal-loop EM at three coil spacings, and gravity surveys (Fig. 29.2a-d). Of note is the excellent correlation between all the geophysical data. The magnetic response proved to be characteristic of the stronger sulphide conductors encountered. Estimated tonnage is less than 1 000 000 tons, but the grade is high.

The first truly geophysical success, designated as H.R. 41, proved to be an even smaller sulphide body which, nevertheless, produced an airborne EM response similar to that of H.R. 10. The correlation between the various sets of geophysical (Fig. 29.3a, b) is not as good, probably due to the presence of a granite intrusive.

By the summer of 1974, after having tested a number of other barren conductors, more attention was devoted to mineralized zones which did not respond to airborne EM, ground EM, or even VLF EM methods. One typical example, H.R. 46, is just north of H.R. 10. Here, grab samples from scattered surface gossans assayed at better than 2 per cent combined copper-zinc. Magnetic and horizontal-loop EM surveys showed nothing and a VLF EM survey produced only minor crossovers (i.e. 4° peak to peak). However, gradient IP surveys (Fig. 29.4a), in which surprisingly no problems were encountered in getting sufficient current into the ground, produced chargeability and resistivity anomalies that corresponded to a zone of minor, but interesting mineralization. A telluric survey (Fig. 29.4b) using a Texasgulf-built receiver operating at 8, 145, and 5000 Hz, showed comparable apparent resistivity profiles. Only small differences existed between the 5000 Hz and 8 Hz responses so, for clarity, only the latter is shown in Figure 29.4b. With little or no overburden and high host rock resistivities, it was apparent that resistivity profiling at VLF frequencies would be a practical exploration technique.

Farther to the west, the VLF EM technique (Fig. 29.5a) was utilized to outline a long, apparently narrow northeasterly trending conductor (anomaly H.R. 462) which occurred along a geological contact zone. The horizontal-loop EM responses were not encouraging and were complicated by topography. Magnetic surveys confirmed the trend, but showed no anomaly over the conductor. A drillhole at 65W on line 26S penetrated a short section of high-grade, stringer copper mineralization, but additional holes in the immediate area were disappointing. A subsequent VLF EM resistivity survey (Fig. 29.5b) probably gave the best picture of sulphide distribution along this conductor.

Up to this time, little more than technical success had been obtained. It was obvious that massive sulphides were to be found in the area, but they tended to be small and were generally poor EM targets (except for EM systems capable of high definition). The examples discussed might have merited greater priority had they been within several miles of an operating mine, but the fact they were discovered at all, and followed up with drilling, is due largely to the geological control afforded by the extensive outcrop.

It was considered that the rather low apparent conductivity of the sulphide systems was due to their small size, the high metamorphic grade of the country rock, and its consequent high resistivity. The VLF EM technique proved to be very effective as a reconnaissance tool, but could not be relied upon for consistent data (in a qualitative sense). Also, magnetic anomalies correlated well with the more massive sulphide occurrences, but not with stringer mineralization. Up to the fall of 1974, all geophysical survey work had been carried out in the summer.

IZOK LAKE DEPOSIT, NORTHWEST TERRITORIES

Fortunately, while the geophysical crews were involved in follow-up work, the geologists carried out a reconnaissance mapping, sampling, and staking program. These endevours resulted in interest being focused on an area about 30 miles (48 km) south of Hood River. In the course of staking claims around one of the weakly mineralized showings, a geologist and his assistant located high-grade sphalerite boulders on the shore of what subsequently became known as Izok Lake. The size and distribution of these boulders and their grade of mineralization suggested that the source had to be either beneath, or in the immediate vicinity of the lake. With this indication of the presence of mineralization and other encouraging geological evidence, an airborne EM survey using the Kenting in-phase/out-of-phase vertical coaxial system operating at a frequency of 390 Hz was planned for the spring of 1975 to be complemented by ground follow-up, prior to break-up, of all targets not entirely on land.

Figure 29.1. Map showing location of Takiyuak Lake area and the Izok Lake deposit, Northwest Territories.

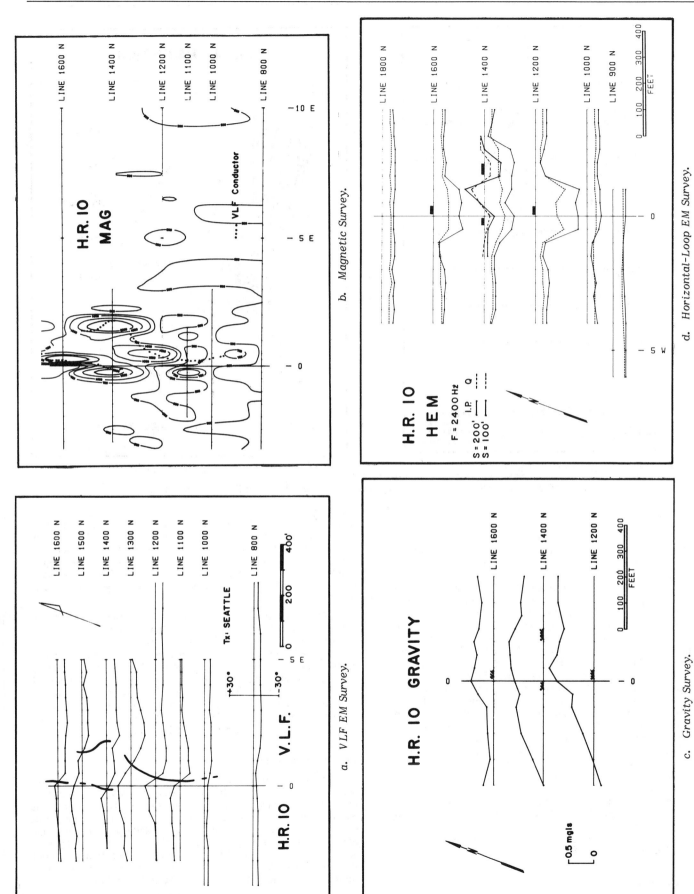

Figure 29.2. Ground geophysical surveys carried out on the Hood River 10 Anomaly.

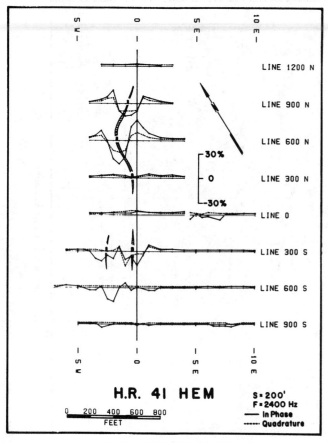
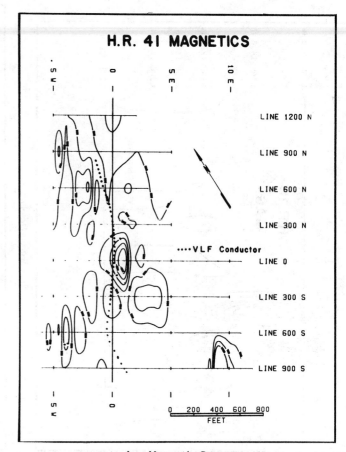

a. Horizontal-Loop EM Survey.   b. Magnetic Survey.

Figure 29.3. Ground geophysical surveys carried out on the Hood River 41 Anomaly.

1975 Geophysical Surveys

The first area to be ground checked in 1975 was the possible eastward extension onto Izok Lake of an onshore zone of weak mineralization opposite the area of the high-grade boulders. This resulted in the detection of a very strong VLF EM conductor towards the middle of the lake, but well beyond any apparent extension of the conductor previously located on shore. A grid was then laid out across the lake ice and surveyed by a number of geophysical methods including VLF EM, the Geonics EM-16R VLF resistivity system, horizontal-loop EM at 100 and 200 foot (30 and 61 m) spacings, and fluxgate magnetometer. The work was curtailed by deteriorating lake ice conditions and the need to cover other areas, but only after enough data had been obtained to outline a drill target. On May 27, 1975 from the first drill location spotted at 6 + 50S, line 22E on the grid, the discovery of the Izok Lake orebody was established with the intersection of about 20 feet (6 m) of 14 per cent zinc. Ironically, the airborne EM program, hampered by bad weather and by snow cover which persisted longer than anticipated, did not cover the area of Izok Lake until the third hole was in progress. Of the major massive sulphide volcanogenic deposits discovered in the Archean of North America since the mid-fifties, Izok Lake can be said to be an exception in that it was not located as a direct result of the follow-up of an airborne EM survey. It missed this classification by two weeks.

VLF EM Survey

The VLF EM survey (Fig. 29.6a), using the Seattle transmitter (NPG 18.6 kHz), detected what appeared to be two major conductors running parallel to one another generally south of the base line. Two other minor conductive zones were also mapped approximately 800 and 1200 feet (240 and 370 m) south of the base line. Vertical field strengths as well as dip angles were recorded. Figure 29.6a shows essentially the interpretation made in the field. The open circles indicate weak conductors, generally from minor inflections.

Of the conductors shown, only the southernmost extends onto the mainland; the others, apart from the portions crossing two small islands, are entirely under water. Note that a weak conductor has been indicated (by open circles) starting at 9 + 00N on line 16E and ending at approximately 16 + 00N on 28E. Apart from the latter and the crossover at 2 + 50S on 8E, all the conductor trends shown on this map are due to sulphide mineralization. These conductors eventually became identified as the Main or Central Zone, North Zone, South Zone, and Northwest Zone (Fig. 29.6b).

VLF Resistivity Survey

This was the first time that the Geonics EM-16R system was utilized by Texasgulf and this might not have occurred under normal circumstances. Fortunately, the survey was carried out by a graduate student who used the technique as part of his thesis work. Although considerable success had been attained in prior years with low frequency telluric surveys, it was felt that the unit utilized had an input impedance too low to get results over the lake ice. The input impedance of the EM-16R is rated at 100 megohms.

Measurements were made of both resistivity and phase-angle over a limited portion of the grid. The results (Fig. 29.7a, b) show a pattern similar in configuration to the

VLF EM survey. The Central and North Zones, particularly the Central Zone, now show appreciable width and continuity. From this data, the Northwest Zone appears to be a separate entity.

The resistivity anomalies are quite sharp and well defined. Average background resistivities appear to be around 2000 ohm-m, but vary throughout from about 1500 to 6000 ohm-m. The lower figure probably represents a water resistivity although this measurement was never actually made.

Anomalous resistivities are as low as 2 ohm-m. The hachured area in Figure 29.7a includes zones having 50 ohm-m resistivity or less. The reader should note that in all cases, apart from the South Zone where resistivities equal 500 ohm-m or greater, the phase angle remains high (i.e. 45°) over the entire span of the 500 ohm contour. This feature assumed increasing significance, though for a time the anomaly was ascribed to the presence of lake bottom sediments. As it happened, this method resulted in the production of one of the better maps of the deeply buried sulphides.

Horizontal-Loop EM Survey

The grid was next surveyed with a horizontal-loop EM unit using a coil spacing of 200 feet (61 m) and frequency of 2400 Hz. The most encouraging aspect of this survey was that it not only indicated very high conductivity for the North and Central Zones, but also showed substantial widths to each zone, especially on lines 22E on the Central Zone and 26E on the North Zone. The Central Zone appears to lack the continuity established by the VLF EM data at both east and west ends whereas the North Zone shows up as a minor conductor of fairly high In-phase Quadrature ratio.

To check the continuity and apparent widths of the strong conductors, a portion of the grid was resurveyed with a 100 foot (30 cm) coil separation. The resultant data provided greater definition, but essentially confirmed the 200 foot (61 m) work (see 1975 compilation-Fig. 29.8a).

On the basis of the data from these surveys, the geophysicist on the project located the first hole at 6 + 50S on line 22E to drill at -60° to the north.

Magnetic Surveys

The 1975 magnetic survey was done with a vertical field fluxgate magnetometer. Previous experience with the small sulphide bodies to the north indicated that appreciable percentages of magnetite and/or pyrrhotite could be expected to occur with ore grade sulphides.

Due to the pressure of time, and deteriorating ice conditions, magnetic survey coverage was quite erratic and generally not too satisfactory. A number of anomalies were recorded, but their full significance was not appreciated until a proper interpretation was made late in the summer. Figure 29.9a is actually a computer contour plot produced in 1976, but it corresponds closely with the final hand drawn version of the 1975 magnetic survey results. When the strong correlation between magnetic anomalies and sulphides became apparent, susceptibility measurements were obtained on the sulphide sections from several of the holes. This not only proved that magnetic susceptibilities, even within the massive sphalerites, were anomalously high throughout the sulphides, but showed a distinct variation in susceptibilities between the sulphide lenses comprising the Central Zone.

Several other anomalous areas were mapped apart from the Central Zone. The reader should note that no distinct magnetic anomaly was recorded for the North Zone although a narrow anomaly was traced out north of the base line. The anomaly to the northwest indicated a target somewhat deeper than the Central Zone and possibly larger in areal extent. The high readings along line 36E were thought to be the result of a mistie until it was noted that this corresponded to a narrow diabase dyke mapped on shore.

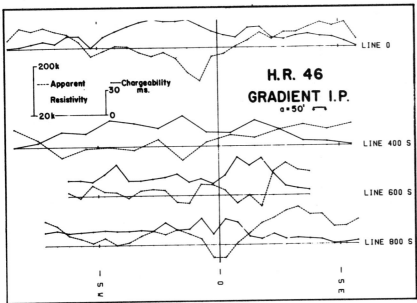

a. Gradient Array IP Survey.

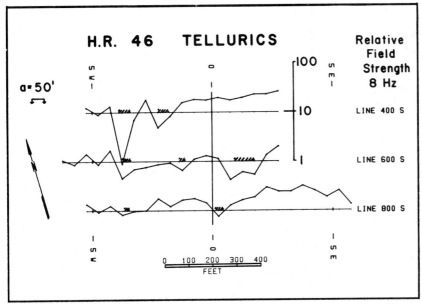

b. Telluric Survey – 8 Hz.

Figure 29.4. Ground geophysical surveys carried out on the Hood River 46 Anomaly.

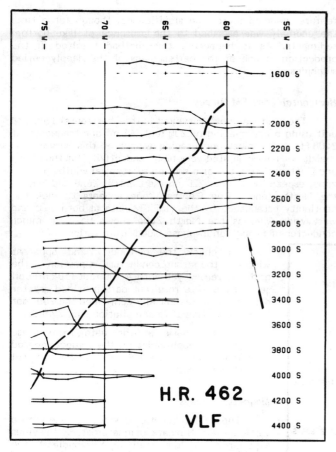
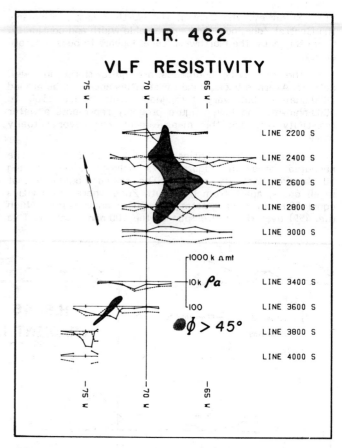

a. *VLF EM Survey.* b. *VLF Resistivity Survey.*

Figure 29.5. *Ground geophysical survey carried out on the Hood River 462 Anomaly.*

Airborne EM Survey

Despite the fact that the airborne EM survey was flown during the drilling program, the AEM results at least gave the assurance that the target had been fairly well delimited by the ground work. From previous experience at Hood River, there was some concern about the magnitude of the AEM anomaly that would be obtained, but the anomaly over the Izok Lake deposit was in excess of ten times background.

The airborne EM survey had originally been laid out to cover this particular area in two directions, an indication of the importance that was attached to this area and the complexity of the geology, and it is interesting to report that the stronger AEM responses were recorded on lines flown approximately parallel to the grid lines.

Although coincident airborne magnetic anomalies were recorded, the overall magnetic picture is dominated by the effect of a series of northwest-trending diabase dykes, one of which cuts the Izok Lake deposit at the east end.

Geology

A paper was published by P.L. Money and J.B. Heslop (1976) of the exploration staff of Texasgulf, which essentially dealt with the geology of the Central Zone. The data presented are applicable to the other zones with only minor changes in zinc-copper ratios.

In summary, the Izok Lake sulphide deposit is a fairly straightforward example of a volcanogenic copper-zinc-lead deposit. The rocks are highly metamorphosed and recrystallized, but what was formerly mapped as amphibolite is now considered to be a meta-andesite, the siliceous metasediments (i.e. gneisses) include felsic metavolcanics, and the hybrid "lit-par-lit" gneisses are a mixed metasedimentary-metavolcanic sequence intruded by numerous granitic bodies.

Drilling Program

By the end of the first week in June 1975, the drill had to be moved off the ice onto the southern island in the lake. Drilling continued into the summer on the Main Zone from the island as well as the South Zone from the south shore. At the conclusion of the summer season, approximately 7 million tons of high-grade copper, zinc, lead, and silver mineralization had been indicated. Most of the North Zone remained untested and the anomaly to the northwest had not been touched.

Despite the extremely encouraging results and the remarkable correlation between some of the geophysical work and the drill data, it was felt that more geophysical survey coverage, particularly ground electromagnetic surveys to obtain greater depth penetration than could be obtained from a 200 foot (61 m) horizontal-loop EM survey, was needed. Magnetic survey coverage was neither uniform nor consistent and with the correlation that had been obtained between massive sulphides and magnetic survey results, it was felt that a higher order of accuracy would provide a greater potential to detect sulphides at depth. Accordingly, a very comprehensive program of EM, magnetic, gravity, and some experimental surveys (mise-à-la-masse) was planned and carried out during the spring of 1976 prior to a second phase of diamond drilling.

Izok Lake Deposit

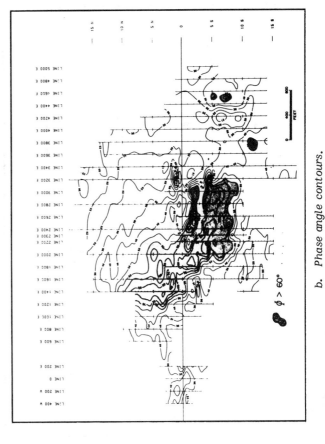

a. Apparent resistivity contours.

b. Phase angle contours.

Figure 29.7. VLF resistivity survey of Izok Lake, NWT.

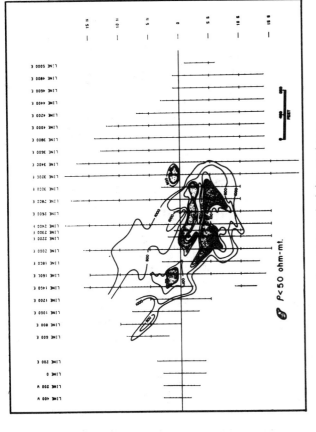

Figure 29.6a. 1975 VLF EM survey of Izok Lake using the Seattle (NPG) transmitter.

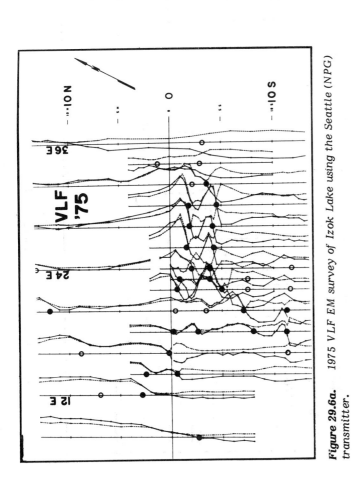

Figure 29.6b. Same VLF EM results as Fig. 6a but showing principal conductive zones. Open and solid circles are poor and good conductors respectively.

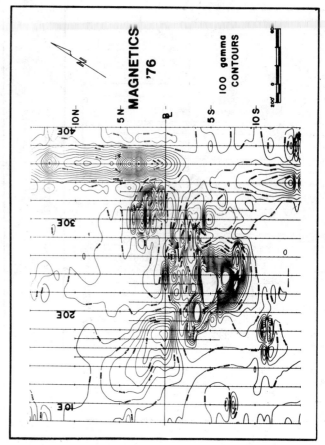

a. 1975 Survey.

b. 1976 Survey.

Figure 29.9. Magnetic surveys of Izok Lake, NWT.

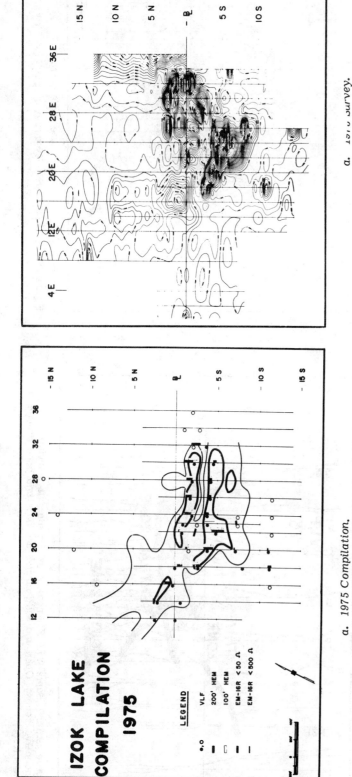

a. 1975 Compilation.

b. 1976 Compilation.

Figure 29.8. Electrical surveys of Izok Lake, NWT.

1976 EXPLORATION PROGRAM

The 1976 exploration program was quite unusual in that there had been time to digest all the results from 1975, to evaluate the geological and geophysical parameters, and to take a second approach to the project. The work, therefore, benefitted from more planning, more people, a greater redundancy of equipment, and more time allotted for the program.

The actual field work began in mid-March 1976 with the re-establishment of the survey grid from the previous year. The field work was scheduled so that the priority geophysical surveys, that is the 400 foot (122 m) 800 foot (244 m) horizontal-loop EM, gravity and magnetic surveys were carried out by mid-April 1976, before the resumption of drilling. By early May, the program had been essentially completed, including the drilling of about 700 holes through the six-foot (1.8 m) thick ice to measure the water depths and to spot electrodes from ice surface to lake bottom at each point.

Horizontal-Loop EM Surveys
(see 1976 compilation Fig. 8b)

In the initial review of the results of the 1976 horizontal-loop EM work, two aspects of the data were considered significant. First of all, although this was admittedly viewed with the clearer vision of hindsight, it was felt that the combination of 200, 400, and 800 foot (61, 122 and 244 m) coverage would have permitted a better understanding of the Central Zone sulphides had this work been carried out in 1975. More importantly, the similarities in the data, allowing for differences in amplitudes, for the Central Zone and Northwest Zone strongly indicated the presence of another sulphide body in the Northwest Zone of even greater areal extent than, and possibly of equal thickness to the Central Zone.

The differences in response between the 800 and 400 foot (244 and 122 m) curves (Fig. 29.10a, b) on lines 16E and 14E could only be accounted for by the occurrence of a relatively flat-lying body at a depth greater than 160 feet (50 m), but probably less than 240 feet (73 m). Furthermore, the response from shallow penetration geophysical methods indicated the presence of a sheet, which subcropped along its south margin, was relatively narrow along this margin, and dipped moderately (say 30-40°) to the north from the subcrop edge.

Considerably more horizontal-loop EM surveying of a detailed nature was done than can be presented in this paper. In general, the 1777 Hz response was stronger than that obtained at 444 Hz, but the two showed an identical pattern. It is debatable whether one may have interpreted this as evidence of a thin sheet response since this observation holds for even the thickest sulphides (e.g., section 24 E). A series of east-west profiles tended to suggest a northeasterly trend which was not borne out by the drilling. Based on the high shoulders for the profiles, a much deeper sulphide sheet had been postulated to the north of the base line, east of line 22E. Limited drilling has not confirmed this interpretation.

It was disappointing that the field work did not extend the Central Zone beyond 34E. One interesting facet to the 800 foot (244 m) horizontal-loop EM coverage is that the South Zone cannot be detected because of the coil spread used, body size, and depth extent. Perhaps the most startling fact brought out by this work was the remarkable correlation between the 800 foot (244 m) HEM and EM 16R data, particularly in reference to the 500 ohm-m resistivity contour. This is apparent from a comparison of the 1975 and 1976 compilations (Fig. 29.8a, b).

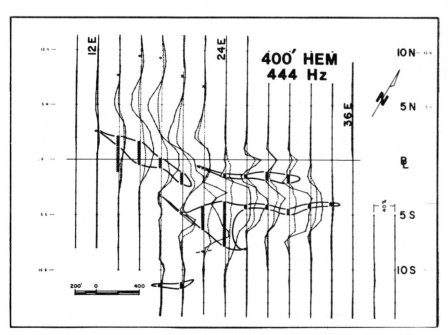

a. 1976 survey using 800 foot (244 m) coil spacing.

b. 1976 survey using 400 foot (122 m) coil spacing.

Figure 29.10. Horizontal-Loop EM survey of Izok Lake, NWT.

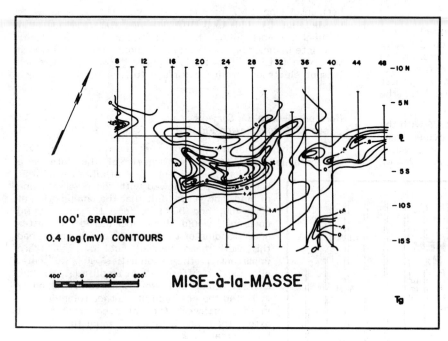

Figure 29.11.   Mise-à-la-masse survey of Izok Lake, NWT.

1976 VLF EM Survey Results

The VLF EM survey was rerun in 1976 in an attempt to achieve more uniform and more extensive coverage. The 1975 and 1976 VLF EM results are almost identical, but additional weak anomalies were recorded off the east end of the Central Zone. The persistence of the anomalous trends appears to be more related to structure than to mineralization, but much remains to be done in correlating this information with the geology.

VLF Resistivity Survey

The 1975 VLF resistivity survey was filled in and expanded during 1976, but the new data did not improve the previous interpretation. It was gratifying to obtain a good correlation with repeated profiles over anomalous areas. The major change appeared to be a higher background resistivity due, probably to different ice conditions.

Gravity Survey

The gravity work enhanced the possibilities of mineralization being associated with the Northwest Zone. The free air data not only showed the two-lobed nature of the Central Zone, but also indicated a gravity high beneath the Northwest Zone. When one considered the greater depth to sulphides and the fact that water depths in this area were generally greater than 30 feet (9 m), it seemed that a total excess mass roughly equal to that of the Central Zone was present. It was subsequently found that the interpretation of the geophysical results was correct, but the geological inferences were presumptuous.

Magnetic Surveys

The 1976 ground magnetic survey (Fig. 29.9b) was designed to cover the entire grid over the lake as uniformly and intensively as possible. The results from the 1975 magnetic survey and susceptibility measurements on drill core were encouraging, and it was felt that magnetic surveys were probably the best and most sensitive tool to use to delineate the extent of known sulphides and detect new ones at depth. A proton magnetometer was used to achieve the high order of accuracy considered necessary.

After the second day of the survey, severe magnetic storms were detected with fluctuations exceeding 40 gammas over a period of two or three minutes and occurring at random intervals during a span of several hours. Thereafter, magnetic activity was monitored with a station magnetometer, and readings taken during magnetic storms were simply discarded.

Several factors are readily apparent in comparing the 1976 results to those obtained in 1975. There is no doubt about the existence of the diabase dyke at the east end of the grid. The apparent "gap" south of the base line was checked with traverses along 4S and 6S between lines 34 and 36E, and it was determined that the dyke had a true width of 25 feet (7.6 m) or less. The northwest magnetic anomaly is more regular in shape, and in fact, gives an excellent representation of the subsurface extent of the sulphides, probably better than the EM results. Again, there is little evidence of deeply buried sulphides to the east of line 34E, although more detailed work away from the influence of the diabase dyke might have shown some trends. The magnetic anomalies in the area of North Island (i.e., lines 28E to 34E, just north of the base line) were found to be due to magnetite stringers within the metavolcanics.

Mise-à-la-masse Survey

A mise-à-la-masse survey (Fig. 29.11) gave the same anomalous pattern as did the other electrical methods over the Central and North Zones, but the Northwest and South Zones were not as well defined. It is suspected that the latter response might have been improved by a better placement of the in-body current electrode, but only the drillholes that were available could be utilized. The operating frequency was 100 Hz and electrical contact to lake bottom was achieved by wires suspended from the surface ice. Measurements were made with General Radio wave analyzers with an input impedance of approximately 500 megohms.

The strong anomaly on line 8E, and the strong trend near the base line from 36E to 48E are probably due to weak, near-surface conductors of structural origin. Drilling in the vicinity of the northernmost anomaly on line 40E did not intersect mineralization. To the south, along the same line, a small disseminated sulphide lens was confirmed by the 1975 drilling program. It was noted during the survey that the induced response, that is, with the 100 foot (30 m) surface dipole not connected to the electrode wires, was about half the measured electrical potential. A similar phenomenon was noted during the VLF resistivity survey.

UTEM Survey

The University of Toronto EM system (UTEM) was tried in the spring of 1977, with excellent results. Figure 29.12a shows the overall interpretation and provides good definition of the Central and Northwest Zones as well as distinct UTEM anomalies for the North and South Zones. Line 12E is reproduced in Figure 29.12b to show the UTEM response over a deeper portion of the Northwest Zone. The interpreted depth of just over 300 feet (91 m) is shown in the small sketch above the profiles.

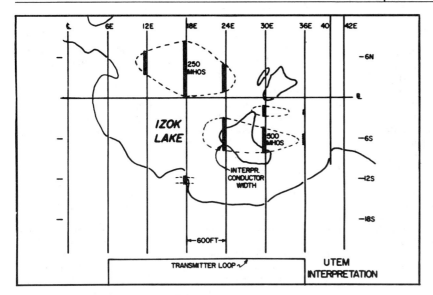

Figure 29.12.

1977 UTEM survey of Izok Lake, NWT.

a. *Interpreted results.*

b. *Profile 12E.*

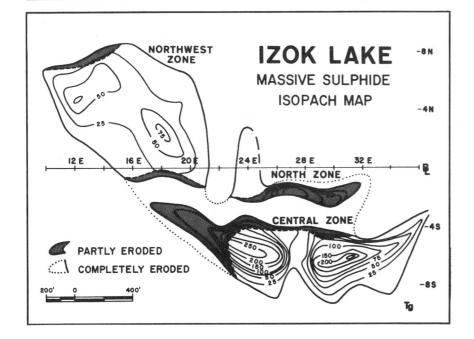

Figure 29.13.

Isopach Map of Izok Lake deposit showing total thickness of massive sulphide lenses.

DRILLING RESULTS

Figure 29.13 is an isopach map of the massive sulphides froming the Izok Lake deposit obtained from the drilling results. The reader should note that the average depth to sulphides within the Northwest Zone is greater than 200 feet (61 m). The summary report on the work that was completed just prior to the start of the 1976 drilling program (i.e. everything but the mise-à-la-masse survey and some magnetic coverage), accurately predicted the ultimate configuration of the Northwest Zone. However, the estimate of the overall thickness of the Northwest Zone was somewhat exaggerated. In drilling the Northwest Zone it was found that, rather than the expected average sulphide thickness of 120 feet (37 m), only a 40 foot (12 m) sheet was intersected that was overlain by up to 200 feet (61 m) of amphibolites whose average specific gravity was about 3.05 g/cm^3 (or 0.33 to 0.35 higher than the country rock). These amphibolites had been mapped around the lake, but had not been encountered in our 1975 drilling, and consequently were an unpleasant surprise. The effect of the amphibolites is illustrated by a gravity profile along line 12E (Fig. 29.14a) wherein a simple two-dimensional model was used. This may be contrasted to a profile along line 24E (Fig. 29.14b) where the gravity anomaly is entirely due to sulphides.

From the drilling completed to the end of 1976, the estimated tonnage figure for the Izok Lake deposit was approximately 12.2 million tons grading 13.73 per cent zinc, 2.83 per cent copper, 1.43 per cent lead, and 2.05 oz./ton silver (with minor values in gold).

CONCLUSIONS

In retrospect, one might conclude that the Izok Lake deposit was subjected to a surfeit of geophysical investigations, but the temptation, on a target such as this, was too strong to resist. Within the rather stringent constraints of time, weather, operating conditions, and the need to provide the geologists with the guidelines for a drilling program subject to the same constraints, it is felt that the work carried out was totally successful. In addition, a wealth of experience was accumulated in applying geophysical methods in the Arctic, and on this particular type of volcanogenic massive sulphide deposit.

Quite obviously, the high frequency VLF EM technique works extremely well in this high resistivity environment, but is still prone to misleading anomalies. The horizontal-loop EM systems are very effective within the limitations imposed by their geometry and such EM surveys should not be limited to a single coil spacing in the ground follow-up of airborne EM anomalies. Magnetic surveys are essential for both the geophysical the geological interpretations. A gravity survey may be very decisive in delineating high density massive sulphides, but can also be misleading if the geology is not fully appreciated and density contracts exist within the country rock.

The mise-à-la-masse method works very well, but is subject to the same shortcomings as the VLF EM technique with the minor near-surface conductors tending to be accentuated. The response provided by IP is only marginally better than with the telluric technique (or VLF Resistivity) except for deep-seated targets. The gradient IP array poses no problems and provides the fastest and most efficient geophysical survey coverage.

One could justifiably feel quite proud of the contribution that geophysics has made to the discovery and evaluation of this very significant massive sulphide deposit, and equally proud of the people, both staff and students, who did the work. The authors wish to thank the management of Texasgulf Inc. for allowing publication of this paper, and for providing the encouragement to prepare it.

REFERENCE

Money, P.L. and Heslop, J.B.
    1976: Geology of the Izok Lake massive sulphide deposit; Can. Min. J., v. 97(5), p. 24-27.

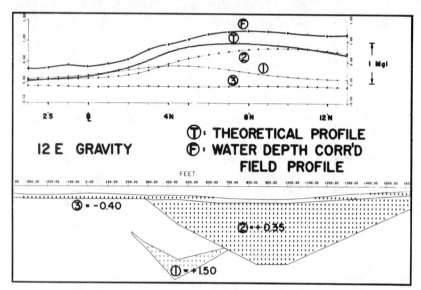

a. Profile 12E with interpretation.

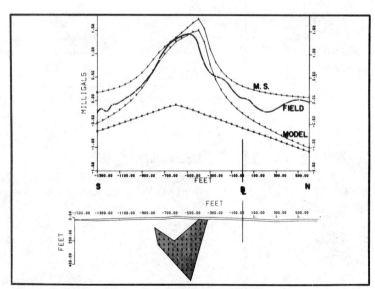

b. Profile 24E with interpretation.

Figure 29.14. *Gravity profiles across Izok Lake, NWT.*

GEOPHYSICAL EXPLORATION AT THE PINE POINT MINES LTD. ZINC-LEAD PROPERTY, NORTHWEST TERRITORIES, CANADA

Jules J. Lajoie and Jan Klein
Cominco Ltd., Vancouver, Canada

Lajoie, Jules J. and Klein, Jan, Geophysical exploration at the Pine Point Mines Ltd., zinc-lead property, Northwest Territories, Canada; in Geophysics and Geochemistry in the Search for Metallic Ores; Peter J. Hood, editor; Geological Survey of Canada, Economic Geology Report 31, p. 653-664, 1979.

Abstract

Pine Point is a zinc-lead carbonate camp with about 40 known orebodies on the south shore of Great Slave Lake, Northwest Territories, Canada. The ore host is a Devonian barrier reef complex. The ore mineralogy consists, in order of relative abundance, of sphalerite, marcasite, galena, pyrite, and occasional pyrrhotite. The gangue consists of dolomite and calcite. Mineralization occurs as open-space filling in tectonically-prepared ground and solution-collapse structures. The area is mostly covered by glacial till and muskeg which averages 15 m in thickness.

In 1963, comparative tests of electromagnetic (EM), self-potential, and induced polarization (IP) methods clearly demonstrated that IP is the most powerful exploration tool for this district. EM is unsuccessful because of poor electrical continuity of the conducting minerals in the orebodies. In subsequent years, extensive induced polarization surveys resulted in the discovery of many of the Pine Point orebodies. The majority produce strong IP anomalies due to subcropping mineralization. There is often little or no coincident resistivity anomaly. The gradient array was commonly used at first, and, subsequently, improvements in instrumentation permitted the use of multiple separation arrays for better target detection and delineation. Exploration case histories and tests are discussed to emphasize particular aspects of exploration in Pine Point with the IP method. Gravity surveys respond to sinkholes and the larger, more massive orebodies. Magnetic surveys produce only weak anomalies from minor pyrrhotite which is sometimes associated with the ore. The seismic reflection method may be helpful in detecting collapse structures but is not a cost effective tool in comparison with the IP method. The background IP geologic noise level is very low but telluric noise is often severe. Careful analysis of the IP data aids in lithologic correlation studies.

Résumé

Pine Point est une région de carbonates minéralisés en plomb et zinc, comprenant environ 40 masses minéralisées situées le long de la côte sud du Grand lac des Esclaves, dans les Territoires du Nord-Ouest au Canada. La roche encaissante est un complexe dévonien de récif barrière. Le minerai se compose, par ordre d'abondance relative, de sphalérite, marcasite, galène, pyrite et parfois pyrrhotine. La gangue consiste en dolomie et calcite. La minéralisation se présente comme un remplissage de fissures dans un terrain fracturé et dans des structures d'effondrement par dissolution. La région est principalement couverte de dépôts glaciaires et de sol de marais, qui totalisent une épaisseur moyenne de 15 m.

En 1963, des essais comparatifs des méthodes électromagnétiques, de polarisation spontanée, et de polarisation provoquée ont montré clairement que la polarisation provoquée est l'outil d'exploration le plus efficace dans cette région. Les méthodes électromagnétiques ne donnent pas de résultats concluants à cause de la faible continuité électrique des minéraux conducteurs dans les corps minéralisés. Dans les années ultérieures, des levés détaillés de polarisation provoquée ont abouti à la découverte de plusieurs masses minéralisées à Pine Point. La plupart d'entre elles ont créé de fortes anomalies de polarisation provoquée, dues aux minéralisations proches de la surface. La désposition des électrodes suivant la méthode du gradient a d'abord été utilisée, puis, avec les perfectionnements de l'appareillage, un mode de disposition des électrodes suivant des lignes multiples a été adoptée, pour mieux déceler et localiser les corps minéralisés. Plusieurs exemples d'exploration et d'essais de prospection sont donner ici pour bien montrer les aspects particuliers de l'exploration pratiquée à Pine Point par la méthode de polarisation provoquée. Les levés gravimétriques enregistrent en particulier la présence de dolines et de corps minéralisés massifs et de grande taille. Les levés magnétiques enregistrent seulement de faibles anomalies dues à la présence de petites quantités de pyrrhotine parfois associée au minerai. La méthode de sismique réflexion peut aider à détecter les structures d'effondrement mais elle n'est pas rentable comparée à la méthode de la polarisation provoquée. Le bruit de fond résultant du milieu géologique est très faible en polarisation provoquée, mais le bruit tellurique est souvent gênant. L'analyse détaillée des données de la polarisation provoquée facilite la corrélation des niveaux lithologiques.

INTRODUCTION

The Pine Point zinc-lead district is located on the south shore of Great Slave Lake in Canada's Northwest Territories, approximately 800 km directly north of Edmonton, Alberta (Fig. 30.1). It is accessible by road and railroad and there is daily air service into the town of Pine Point with a present (1977) population of about 2000. The regional topographic relief is about 50 m and much of the area is swampy.

In 1898, the first few claims were staked by a local fur trader as a result of mining interest generated by prospectors en route to the Klondike gold fields of the Yukon. In 1926, a property examination by the Consolidated Mining and Smelting Company, now Cominco Ltd., revealed a geological similarity to the famous Tri-State zinc-lead district of southeastern Missouri, and an option was secured on some mineral claims. During the late 1940s, a large concession was acquired by Cominco enclosing the known mineralization. In 1951, Pine Point Mines Ltd. was formed to finance exploration on the property. A production decision was finally reached in 1963, after successful negotiations with the federal government for construction of a northern railway.

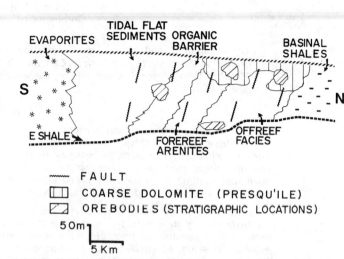

Figure 30.3. *Simplified geological cross-section of barrier complex. Note the exaggerated vertical scale.*

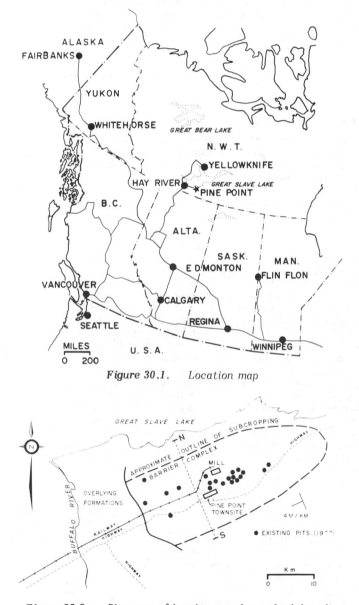

Figure 30.1. Location map

Figure 30.2. *Plan map of barrier complex and mining pits.*

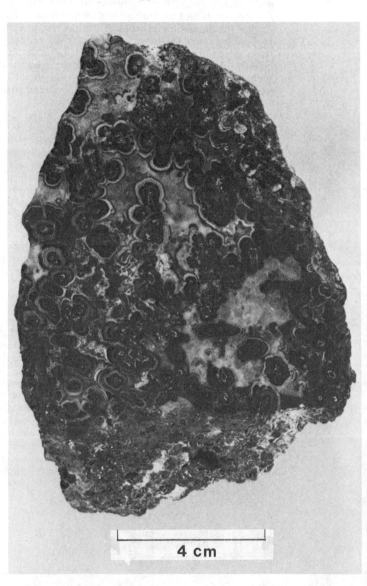

Figure 30.4. *Sample of sphalerite and galena mineralization in a dolomite gangue. (GSC 203492-I)*

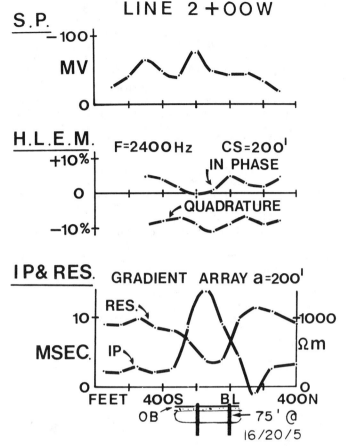

Figure 30.5. Geophysical test profiles over orebody N-42, line 2+00W, 1963 (Huntec MK 1). Ore grade is in per cent Pb, Zn, and Fe.

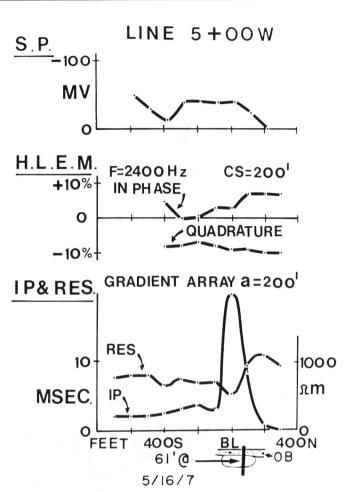

Figure 30.6. Geophysical test profiles over orebody N-42, line 5+00W, 1963 (Huntec MK 1). Ore grade is in per cent Pb, Zn, and Fe.

At this time, the company had 8.8 million tons of ore averaging 2.6% lead and 5.9% zinc in several orebodies which had been found by drilling. The induced polarization (IP) method which was first introduced to Pine Point in 1963, played a very important role in discovering more orebodies, and in significantly extending the ore reserves of the district. The total known ore is now about 74 million tons averaging 2.8% lead and 6.3% zinc in about 40 orebodies. About half of this ore has now (1977) been mined, mostly by open pit methods.

Previous articles have presented geophysical survey results in the Pine Point district, mainly IP, using time domain (Seigel et al., 1968; Paterson, 1972) and frequency domain (Hallof, 1972) methods. This paper describes the geophysical work done at Pine Point from an historical perspective, and discusses some geophysical aspects of particular interest.

GEOLOGY

The host of the zinc-lead mineralization is a Devonian barrier reef complex (Skall, 1975) which subcrops (i.e. outcrops below the overburden) on the south shore of Great Slave Lake, approximately as shown in Figure 30.2. The reef strikes east-northeast and dips about 4 metres per kilometre to the southwest, where it is overlain by more recent Devonian formations. The overburden consists of glacial drift and muskeg whose depth varies from about 5 to 30 m with an average of about 15 m over the area. Figure 30.2 shows the location of the pits which now exist on the property. Sulphide mineralization subcropped in all but one of the orebodies shown here.

In cross-section, the reef exhibits the major characteristics of a clastic reef complex as shown in Figure 30.3. The environmental facies encountered from south to north are evaporites, tidal flat sediments, organic barrier, forereef arenite facies, offreef facies, and basinal marine shales. Note that there is a vertical exaggeration of 100 in Figure 30.3. The development of the barrier reef during the Devonian was accompanied by a greater rate of sedimentation in the evaporite basin to the south. This was at least partly responsible for faulting and fracturing within the reef complex, parallel to the strike of the reef. Later, these fractures served as conduits for magnesium-rich fluids which dolomitized parts of the reef to a coarse, vuggy dolomite, referred to locally as the Presqu'ile facies.

The tectonic activity acted as ground preparation for the deposition of Mississippi Valley-type deposits. There is a vast supply of sulphur in the evaporite basin to the south. Sources of metals are postulated to be the basinal shales to the north, the carbonate pile itself, or some other deeper source. Within the reef complex, the porous carbonate rocks allow easy migration of metal brines and sulphur-bearing solutions.

The prime requirement for the formation of the ores is the availability of open spaces in the rock for precipitation of sulphides. At Pine Point, the main causes of open spaces are: a) karsting and solution breccias which developed during marine regressions; b) faulting and fracturing and especially the intersection of fault zones; and c) dolomitization which produces a rock with high porosity. Traps created by pinch outs and facies changes also play a role in ore deposition. Figure 30.3 shows the original stratigraphic positions of four major, partially-eroded, subcropping orebodies.

Most of the orebodies can be subdivided into two basic shapes: massive orebodies which are roughly equidimensional and tend to have a somewhat higher grade, and tabular orebodies which are restricted vertically and extend horizontally either in one horizontal direction (run type) or two horizontal directions (blanket type).

The ore mineralogy is simple. It consists of crystalline and colloform sphalerite, marcasite, galena, pyrite, and occasional pyrrhotite. The zinc to lead ratio averages about two to one, and varies widely for different orebodies. The gangue consists of calcite and dolomite. A typical sample of ore is shown in Figure 30.4, where galena mineralization is surrounded by colloform sphalerite and the gangue is dolomite. A common physical property of the ore is the lack of electrical continuity of the conducting minerals: galena, pyrite and marcasite. In Figure 30.4, the galena is surrounded by sphalerite which is a nonconductive and nonpolarizable mineral. Generally, however, the conducting paths are interrupted by calcite and dolomite gangue. This explains why the orebodies do not respond to standard electromagnetic exploration methods.

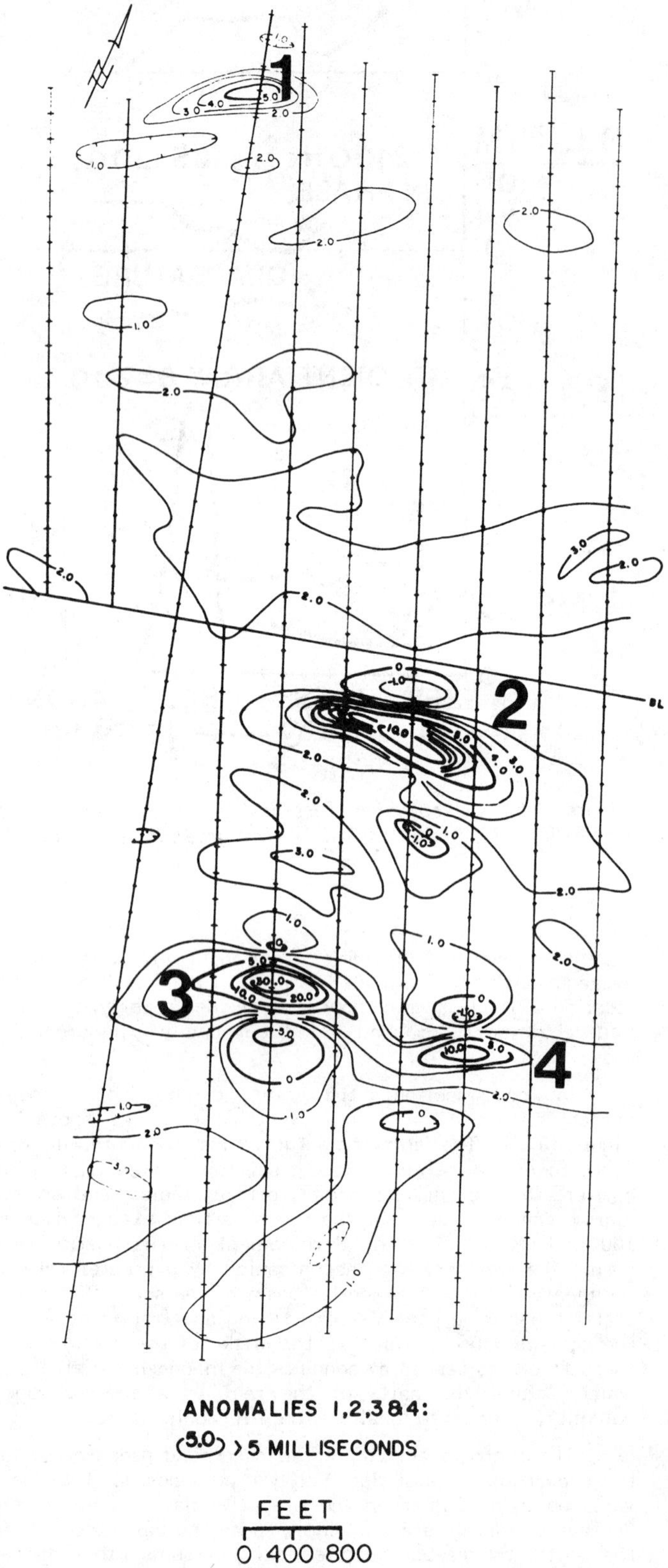

Figure 30.7. Chargeability contour map of 1964 gradient-array IP survey (Huntec MK 1).

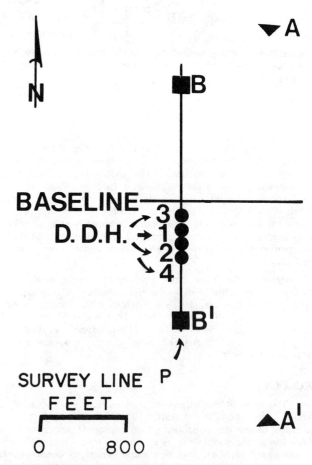

Figure 30.8. Plan of mineralized holes on line P and positions of current electrodes (AA' and BB') for gradient-array survey on line P.

GEOPHYSICS

First Geophysical Tests

Comprehensive geophysical tests were performed during the spring of 1963. Hunting Survey Corporation Limited was contracted to carry out tests of self-potential, electromagnetic, and induced polarization geophysical methods over orebody N-42. The results of these tests on two lines three hundred feet (91 m) apart are shown in Figures 30.5 and 30.6. Representative sections through the orebody are drawn to scale at the bottom of these figures. The station interval for the data points is 100 feet (30 m).

The self-potential results shown at the top of Figure 30.5 show moderate activity on both lines. However, there is no anomaly which could be attributed to mineralization.

For the horizontal-loop electromagnetic (HLEM) data, a coil separation of 200 feet (61 m) and a frequency of 2400 Hz were used. There is possibly a weak quadrature anomaly near the base line on line 2W (Fig. 30.5) where a drillhole has a high grade 75-foot (23 m) intersection of 16% Pb/20% Zn/5% Fe. On line 5W (Fig. 30.6) however, there is no evidence of an EM anomaly over the orebody. Further reconnaissance work with EM produced weak anomalies similar to that on line 2W and these correlated well with swampy ground. Seigel (1968) showed Turam results over the Pyramid No. One orebody which confirm the lack of response to electromagnetic methods.

The IP data in Figures 30.5 and 30.6* were acquired using a gradient array with a potential electrode spacing of 200 feet (61 m). It is evident that strong chargeability anomalies coincide directly with the orebody on both lines. The background chargeability level is between 2-3 ms and is remarkably flat; thus, the anomalies are about seven times background. A distinct resistivity low occurs on line 2W where a drillhole shows a combined lead-iron grade of 21%. On line 5W, where the combined lead-iron grade is only about 12%, a weak resistivity anomaly is only barely detectable over the background variations.

Initial High Success Rate

The geophysical test of 1963 marked the beginning of a new exploration era at Pine Point. The following year, about 250 line miles of time-domain gradient-array IP were surveyed. Figure 30.7 shows an example of IP discoveries made during the extraordinarily high success period experienced during 1964. The lines are about 500 feet (152 m) apart. There are four chargeability anomalies with amplitudes greater than 5 ms. Each is caused by an orebody.

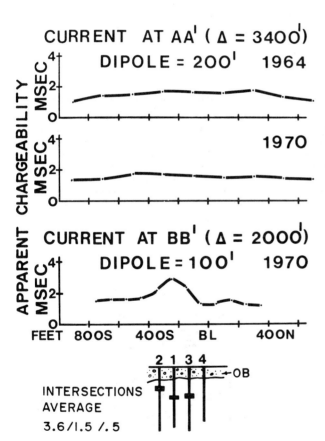

Figure 30.9. Gradient-array IP profiles obtained in 1964 and 1970 on line P (Huntec MK 1). Δ is the current electrode separation. Grade is in per cent Pb, Zn, and Fe.

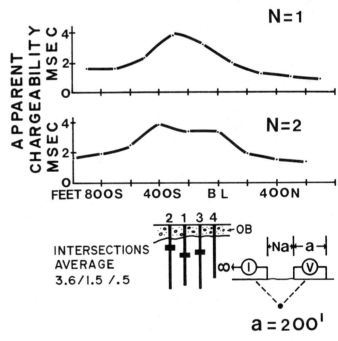

Figure 30.10. Pole-dipole array IP profiles on line P (Huntec MK 1). Grade is in per cent Pb, Zn, and Fe.

---

* In figure captions denoting "Huntec MK 1", the data were acquired with a Huntec MK 1 time-domain receiver which integrates the secondary voltage from 15 ms to 415 ms after cessation of the 2 s ON current pulse. This is then normalized to the received voltage of the current pulse. The chargeability units are therefore mvolts-second per volt or, simply ms. In figure captions denoting "Huntec MK 3", the data were acquired with a Huntec MK 3 time domain receiver which integrates the secondary voltage from 120 ms to 1020 ms after cessation of a 2 s ON current pulse.

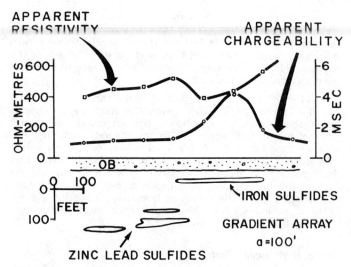

Figure 30.11. Example of an indirect discovery using time-domain IP on line M (Huntec MK-1) showing apparent resistivity and chargeability profiles. OB is overburden.

Note the very low variation in background level of chargeability; this is typical for much of the Pine Point district. The anomaly in the centre of Figure 30.7 is the N42 orebody on which the first IP tests were carried out. It had been previously found by grid drilling. The other three orebodies occurred nearby within areas which had been previously tested with drillholes spaced roughly 1000 feet (305 m) apart. However, none had intersected the orebodies. They were discovered directly as a result of the survey shown in Figure 30.7, actually within a few days of each other.

Advantages of Multi-separation Arrays

The high success rate of the early surveys eventually slowed down with the completion of the drill testing of all obvious anomalies; experimentation then began to determine more diagnostic survey parameters for ore delineation. The following example illustrates a field experiment where the target consisted of short mineralized intersections along line P with grades of about 3.6% Pb/1.5% Zn/0.5% Fe in three holes 100 feet (30 m) apart. They are shown in plan in Figure 30.8 and in section in Figures 30.9 and 30.10. The upper profile in Figure 30.9 is a 1964 gradient-array survey with a potential electrode separation of 200 feet (61 m). The current electrodes were located at A and A', 3400 feet (1034 m) apart and 800 feet (244 m) away from the survey line. There is no indication of an anomaly to hint at the mineralization. As shown in the centre profile, the survey was repeated in 1970, using the identical parameters and the results were essentially the same. For the lower profile in Figure 30.9, the current electrodes were moved to B and B', directly on the survey line, 2000 feet (610 m) apart, and the potential electrode spacing was reduced to 100 feet (30 m). As can be seen on Figure 30.9, this electrode configuration produced a definite anomaly coincident with the mineralization. However, surveying large areas with such small electrode spacings and survey blocks is costly.

Next, a two-separation pole-dipole array was tried with an electrode spacing of 200 feet (61 m). The resultant anomaly shown in Figure 30.10 is stronger and more definite than that obtained with the gradient array. On second separation (N=2), the expected double peaking of the anomaly with the strongest peak on the side of the current electrode can be seen. Such field tests showed that, here at least, multi-separation arrays are better focused for the detection of weak mineralization than the gradient array. Also, they make it easier to interpret the depth and the lateral limits of polarizable sources. The multi-separation arrays became practical with improvements in instrumentation in both time domain and frequency domain methods, which allowed measurements of the lower signal levels at higher separations.

Example of an Indirect Discovery

The examples previously discussed demonstrate the ideal one-to-one correlation between IP anomalies and economic mineralization. Actually, during the initial 1963 and 1964 surveys, such correlation was very common and everyone expected the first hole drilled into an anomaly to intersect an orebody. In some cases, however, IP anomalies led to discoveries indirectly and this section describes an example of this.

In Figure 30.11, the iron sulphides nearly subcrop whereas the zinc-lead sulphides are deeper. The data on line M are from a gradient-array IP survey with a potential electrode spacing of 100 feet (30 m). A weak but definite anomaly coincides directly with the iron sulphide mineralization. The first drillhole confirmed the source of this anomaly. However, due to favourable geological indicators, drilling continued and economic mineralization was eventually found. Careful analysis of the detailed IP data can aid in guiding drilling in the search for hidden economic mineralization.

IP Anomaly Over a Small, Massive, Shallow Orebody

The larger orebodies at Pine Point may be discovered relatively easily with multi-separation IP surveys, especially if they are at or near subcrop. The smaller orebodies, however, offer a little more challenge and this example illustrates the discovery of a small massive-type orebody which is a very difficult drill target. On a property the size of Pine Point, there are many drillholes with interesting intersections of mineralization and it is difficult to decide which ones to follow-up first. Diamond-drill hole (DDH) A

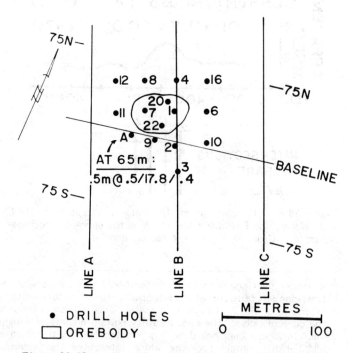

Figure 30.12. Plan of orebody R, drillholes, and IP lines.

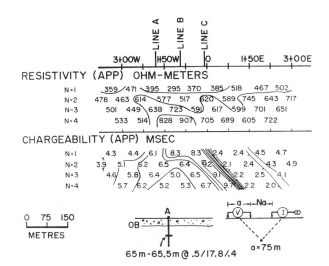

Figure 30.13. IP-resistivity data obtained on baseline of orebody R using pole-dipole array (Huntec MK 3). Grade is in per cent Pb, Zn, and Fe.

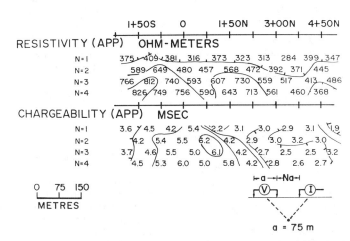

Figure 30.15. IP-resistivity data obtained on line C of orebody R (Huntec MK 3).

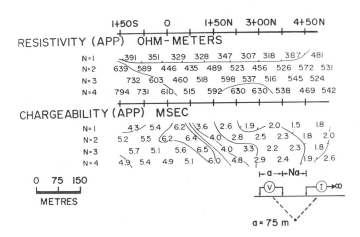

Figure 30.14. IP-resistivity data obtained on line A of orebody R (Huntec MK 3).

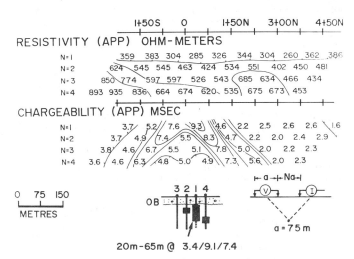

Figure 30.16. IP-resistivity data obtained on line B of orebody R (Huntec MK 3). Grade is in per cent Pb, Zn, and Fe.

located near orebody R as shown in Figure 30.12 is typical. At a depth of 65 m, it intersected 0.5 m with a grade of 0.5% Pb/17.8% Zn/0.4% Fe. The base line was surveyed with a pole-dipole IP array and the results are shown in Figure 30.13 in pseudosection form. The pole (moving current electrode) is to the right and the electrode spacing is 75 m. The top pseudosection shows apparent resistivity data while the bottom shows the apparent chargeability data. DDH A with its intersection is shown at the bottom of Figure 30.13. An IP anomaly is evident and it has a few important characteristics. First, the shape of the anomaly indicates a shallow source. However, the strongest chargeabilities do not occur on first separation (N=1) as might be expected. Instead, the strongest chargeabilities occur at separations of N=2, 3 and 4, on the pole-side "pant-leg" of the anomaly, the highest being 9.7 ms. Second, the mineralized intersection in DDH A does not explain the high chargeabilities of over 9 ms. Third, the centre of the anomaly is displaced from DDH A. An anomaly with these characteristics, that is, a shape indicating a shallow source but with the strongest chargeabilities occurring on higher separations, is recognized as evidence of a subcropping body of polarizable material which is just off the survey line to one side or the other.

Subsequently, lines A, B, and C which are about 100 m apart, were surveyed. Figure 30.14 shows the data obtained on line A. There is an apparent chargeability high of greater than six ms in a background of about two ms and it has a poorly developed anomaly shape. The anomaly is centred just north of the baseline.

Figure 30.15 shows the IP-resistivity data obtained on line C. The results are similar to those obtained on line A, that is, a broad region of moderate chargeability with the highest readings on second and third separations. Figure 30.16 shows the IP-resistivity data obtained on line B, the centre line. This is a typical Pine Point IP anomaly over subcropping mineralization with the pole-dipole array. The anomaly shape indicates a shallow source with the strongest chargeability value occurring on first separation, on the pole side, and with the chargeability amplitudes decaying down each "pant-leg" of the anomaly from 9.3 to 7.3 ms. Note that there is no significant apparent resistivity anomaly coincident with the IP anomaly on this line. It is apparent that line B passes directly over the polarizable source. DDH 1 intersected 45 metres of mineralization grading 3.4% Pb/9.1% Zn/7.4% Fe at subcrop. Subsequent drilling outlined a small but high grade orebody whose horizontal dimension along line B was less than 60 m.

An example such as this shows that IP is a very practical tool for finding orebodies of small lateral dimensions at Pine Point. Geological exploration through drilling is useful to define areas of good potential. However, when it comes to pinpointing the target for drilling, IP is the more cost-effective exploration tool.

Example of Lithologic Correlation

So far, the discussion has been mainly concerned with the location of IP anomalies and their subsequent drilling to discover economic mineralization. Another use for the geophysical data is in lithologic correlation studies. Figure 30.17 shows a pseudosection of apparent resistivity, chargeability, and metal factor* data on line N. A pole-dipole array was used, the pole being to the north. The electrode spacing was 75 m. One can easily pick out two different regions from these data. In the northern half of the section, both the apparent resistivity and chargeability data appear to indicate a two-layer earth, while, in the southern half, a halfspace appears to be a more appropriate model. In the northern region, there is a good positive correlation between the apparent resistivity and chargeability data which would lead one to suspect that the variations observed in these data are due to the same geological units. Because of this positive correlation, the metal factor data acts as a filter of the layer effect.

In the northern region, an average of the apparent resistivities and chargeabilities from A to B, denoted in Figure 30.17, for each separation, yields the values shown in Figure 30.18. Figure 30.18 also shows a simple two-layer resistivity and chargeability model which gives a reasonable fit with the observed data. The thickness of the top layer in the numerical model is 80 m.

A correlation with drillhole information (Fig. 30.19) shows that the high resistivity and chargeability layer may be explained by a facies which could be described as a fine, sucrosic, and argillaceous dolomite with poorly developed bedding. The high resistivity is most likely the result of poorer permeability and the higher chargeability is probably a function of the clay and iron content in this facies.

Limits of Detectability and Usefulness of the Induced Polarization Method in Pine Point

The following two examples demonstrate the limits of detectability of the IP method at Pine Point.

In the first, hole no. 1 (Fig. 30.20) was drilled in an area of favourable geology and intersected 56 feet (17 m) grading 3.5% Pb/6.4% Zn/7.6% Fe at a depth to top of 203 feet (62 m). However, 16 more holes were drilled on a grid of 100 feet (30 m) and no sulphides were intersected. A 4-separation pole-dipole IP survey was carried out on line L and the data presented in Figure 30.21, with the drill intersection shown at the bottom. The background in chargeability is about 1.8 ms and a weak anomaly with a peak of 2.8 ms, only 1 ms above background, occurs on second separation. Note that the contour interval is only 0.25 ms. From the shape and amplitude of this anomaly, one can safely conclude that it is caused by the mineralization found in DDH 1 at a depth of 200 feet (61 m). This is a good field example for demonstrating how small a target can be detected at moderate depths in areas of very low background variations in chargeability.

The second example demonstrates how the effectiveness of the IP method drops off completely when the orebody contains no significant amounts of the conducting minerals — galena, marcasite, and pyrite. Figure 30.22 shows an IP and apparent resistivity pseudosection over a "run-type" orebody which passes under line S and consists mostly of sphalerite which does not produce an IP effect. The electrode interval chosen for this test survey was only 25 m in an attempt to optimize the detection and resolution of this shallow orebody which is at a depth of 30-40 m. There is no anomaly coinciding with the ore zone. This can be explained by the grade of combined conducting sulphides of lead and iron which is only about 1.3%, and not significantly anomalous in this particular area of the property. Thus the only possibility of mapping the extension of the ore zone is by studying its relation to patterns in the goephysical data.

The Gravity Method

After IP, the next logical geophysical tool to use is gravity. Seigel (1968) quoted densities of 2.65 and 3.95 gm/cc for the host limestone and ore in the Pyramid orebody. There has been no attempt to further establish rock densities, however, because of the widely varying relative amounts of sphalerite, galena and pyrite from one orebody to the other. The densities of these minerals are 3.7, 7.5 and 5.0 respectively. Also, the degree of porosity can vary significantly in the host rock in the vicinity of an orebody. The following examples will attempt to show the advantages and limitations of the gravity method.

Figure 30.23 shows the residual Bouguer gravity and IP data over orebody A. There is a chargeability anomaly of about 10 ms above background within a broad region of low apparent resistivities of about 100 ohm-metres. Coincident with the IP high is a residual Bouguer gravity anomaly of 0.5 mgals. There is obviously very little doubt that these anomalies are due to underlying mineralization. One of the most interesting holes drilled on this orebody intersected 72 feet (22 m) of ore grading 8.9% Pb/17.5% Zn/13.4% Fe. Orebody A has 3.5 million tons of ore grading 4.2% Pb and 9.5% Zn. This example illustrates that gravity data form an excellent complement to the IP data. The gravity method responds not only to the conductive ore minerals but also to sphalerite and may therefore indicate concentrations of zinc ore which do not show up on the IP data.

Figure 30.24 shows IP and gravity data over a sinkhole on line U. Sinkholes occur very frequently at Pine Point and are usually filled with sand, gravel, granite boulders, and limestone breccia whose average density might therefore be expected to be less than the host limestones and dolomites. This sinkhole is characterized by a moderate chargeability anomaly of about 4 ms above background and resistivity anomaly of 200 ohm-metres below background. The residual gravity data show a weak negative anomaly of 0.4 mgals. Two holes did not intersect solid bedrock after drilling about 150 feet (46 m) in an area where the overburden thickness is known from nearby drilling to be about 40 feet (12 m).

The last two examples suggest that it is entirely logical to conclude that, given an IP anomaly, a coincident positive gravity anomaly indicates mineralization while a negative gravity anomaly indicates a sinkhole. Figure 30.25 demonstrates the danger of being restricted by any model,

---

* Metal factor is defined herein as apparent chargeability divided by apparent resistivity, multiplied by 1000.

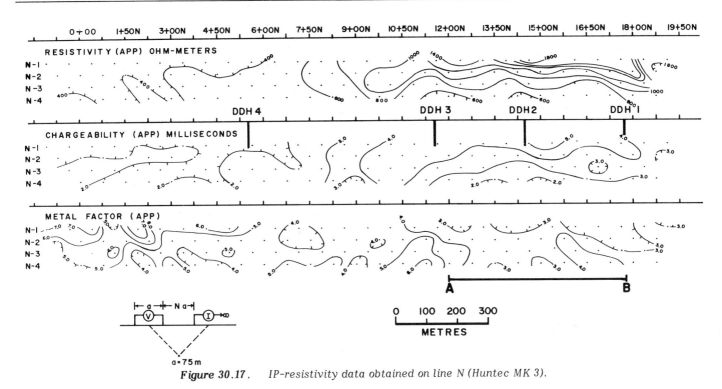

Figure 30.17. *IP-resistivity data obtained on line N (Huntec MK 3).*

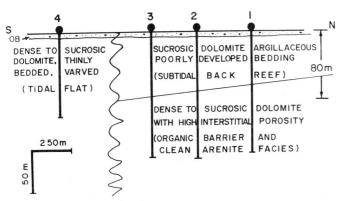

Figure 30.18. *Model fitting of IP-resistivity data obtained on line N.*

Figure 30.19. *Geological cross-section of line N.*

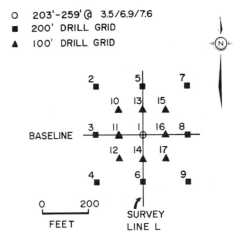

Figure 30.20. *Plan of drillholes in the vicinity of line L.*

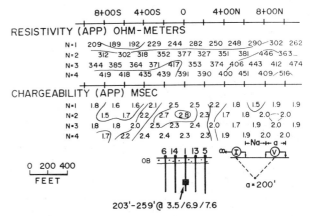

Figure 30.21. *Time-domain IP-resistivity data on line L over deep and localized mineralized intersection (Huntec MK 3). Grade is in per cent Pb, Zn, and Fe.*

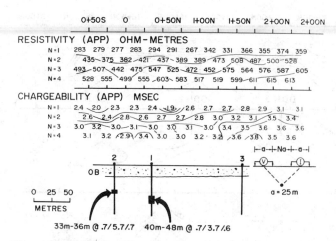

Figure 30.22. Time-domain IP-resistivity data on line S over run-type orebody consisting mostly of sphalerite (Huntec MK 3). Grade is in per cent Pb, Zn, and Fe.

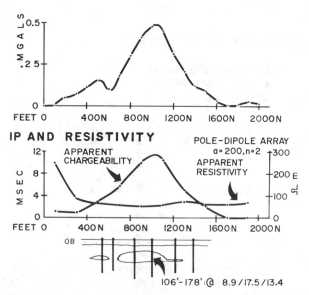

Figure 30.23. Time-domain IP-resistivity and Bouguer gravity data over orebody A (Huntec MK 1). Grade is in per cent Pb, Zn, and Fe.

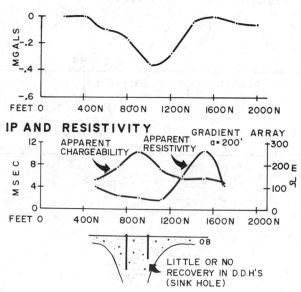

Figure 30.24. Time-domain IP-resistivity and Bouguer gravity data over a sinkhole on line U (Huntec MK 1).

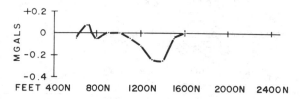

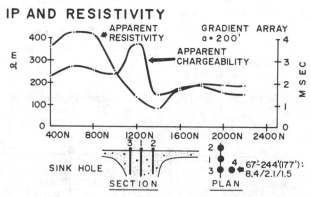

Figure 30.25. Time-domain IP-resistivity and Bouguer gravity data over a sinkhole and nearby orebody W (Huntec MK 1). Grade is in per cent Pb, Zn, and Fe.

even one as simple as that just described. In Figure 30.25, there is a chargeability anomaly of about 2 ms above background, a resistivity anomaly of about 300 ohm-metres below background and a negative residual gravity anomaly of 0.2 mgals. As might be expected, the three drillholes shown in section intersected sinkhole material down to about 150 feet. Drillhole no. 4, however, shown in plan view and only 100 feet (30 m) to the east of the other three, intersected 177 feet grading 8.4% Pb/2.1% Zn/1.5% Fe. Further drilling outlined a small orebody adjacent to the sinkhole. The drilling was guided by that part of the chargeability anomaly which was intermediate in amplitude between background and the peak over the sinkhole in the remainder of this survey area. Actually, in this case, the gravity data were acquired on a test basis after all drilling was completed.

A moderate amount of exploration experience using gravity has shown that most anomalies were due to varying overburden thickness. Therefore, to be truly effective, gravity surveying should be accompanied by refraction seismic surveying to determine the variations in overburden thickness in order to apply terrain corrections to the gravity data. However, this considerably lowers the cost-effectiveness of the gravity method in reconnaissance surveys at Pine Point in comparison with the IP technique and may not be feasible for general use.

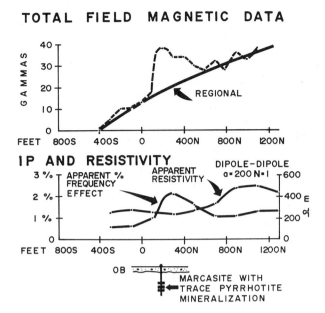

Figure 30.26. Frequency-domain IP-resistivity and magnetic data over iron sulphide deposit on line Y (McPhar P660, 0.3-5 Hz).

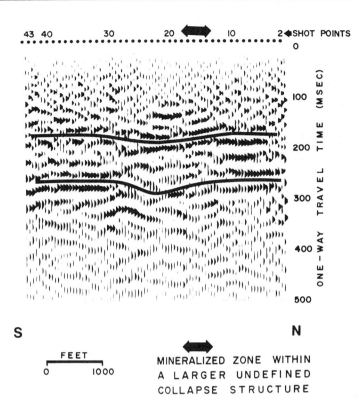

Figure 30.27. Reflection seismic data over orebody A. Shot spacing = 110 feet (34 m). Group offset = 1320 feet (402 m).

The Magnetic Method

Small amounts of pyrrhotite, the magnetic iron sulphide, are sometimes found with pyrite and marcasite mineralization which, in turn, occur with the ore. Figure 30.26 shows frequency-domain, IP-resistivity and magnetic profiles over one extremity of a noneconomic deposit consisting mainly of barren iron sulphides. The frequency-domain IP results were obtained using a McPhar P660 system. Of the holes drilled into this deposit, the drill hole shown had the highest concentration of pyrrhotite. Coincident with the IP anomaly is a magnetic anomaly of about 20 gammas which was confirmed by repeating the survey. However, since Pine Point is in the auroral zone, the area is subjected to severe magnetic storm activity, and since only a 20 gamma anomaly was obtained over a deposit with higher amounts of pyrrhotite than usual, it is unlikely that the magnetic method can be a significant exploration tool at Pine Point in comparison with IP.

The Reflection Seismic Method

Some experimentation has been carried out with the seismic reflection method in order to determine its possible usefulness as a mapping tool, a direct ore finder, and to find disturbances in the bedrock such as faults and collapse structures which may lead to mineralization. Figure 30.27 shows a seismic section on a line which passes over orebody A whose lateral extent is shown by the thick arrow. The orebody is contained within a large area of collapse. The data were processed with a 25-75 Hz bandpass filter. Several good reflections appear at the ends of the line and there is a definite depression from shot points 13 to 27. The shallowest continuous reflection appears at a one-way travel time of about 160 ms and this definitely puts it below the ore zone. It is possible that the depression in the reflected events is a velocity slow down anomaly; that is, reflections from deeper horizons were slowed down in passing through a region of lower velocity in the collapsed area surrounding the mineralization. However, Figure 30.27 shows the best results of the experiment. The remaining seismic data acquired during the test were not as encouraging.

It is possible that very high resolution seismic reflection surveys such as those carried out for nuclear site investigation may be more successful. Again, however, the cost effectiveness of such surveys would not compare with that of the IP method on the Pine Point Mines property.

CONCLUSIONS

Most of the Pine Point Mississippi Valley-type deposits contain sufficiently high concentrations of the conducting minerals galena, marcasite, and pyrite, to be good IP targets. The poor electrical continuity in bulk is a common characteristic of these ores and explains why the orebodies do not respond to standard electromagnetic methods. The chargeability anomalies produced by the ore are usually of the order of 10 ms (although chargeability values in excess of 30 ms were observed in one case). This is not very large when one considers that normal background variations in chargeability in many volcanic terranes are often as high as this. Therefore, the success of the IP method at Pine Point depends largely on the very low and uniform chargeability of the host limestones and dolomites. Weak variations in the chargeability can be measured which can sometimes be correlated to the lithology. Production surveying is however often hindered by high telluric activity since Pine Point is in the auroral zone. The gravity method is useful as a complementary tool especially in anomaly detailing, and responds to sphalerite mineralization which is not polarizable. At the present time, EM, magnetic and seismic methods appear to have little application as direct ore-finding tools in comparison with the IP method. Thus, the IP method is by far the most cost-effective geophysical exploration tool on the Pine Point Mines property.

ACKNOWLEDGMENTS

We wish to thank Cominco's Northern Group based in Yellowknife, N.W.T. and Pine Point Mines Ltd., for permission to publish this paper. The geological data were acquired from the geological staff of Pine Point Mines Ltd. The IP data was taken from surveys performed for Pine Point Mines Ltd. by Canadian geophysical contractors.

REFERENCES

Hallof, P.G.
    1972: The induced polarization method; in Proc. 24th Int. Geol. Cong., Montreal, Sec. 9, p. 64-81.

Paterson, N.R.
    1972: The applications and limitations of the IP method — Pine Point area, N.W.T.; Can. Min. J., v. 93 (8), p. 44-50.

Seigel, H.O., Hill, H.L., and Baird, J.G.
    1968: Discovery case history of the Pyramid ore bodies, Pine Point, Northwest Territories, Canada; Geophysics, v. 33, p. 645-656.

Skall, H.
    1975: The paleoenvironment of the Pine Point lead-zinc district; Econ. Geol., v. 70, p. 22-47.

ON THE APPLICATION OF GEOPHYSICS IN THE INDIRECT EXPLORATION FOR COPPER SULPHIDE ORES IN FINLAND

M. Ketola
Outokumpu Oy, Exploration, Finland.

Ketola, M., On the application of geophysics in the indirect exploration for copper sulphide ores in Finland; in Geophysics and Geochemistry in the Search for Metallic Ores; Peter J. Hood, editor; Geological Survey of Canada, Economic Geology Report 31, p. 665-684, 1979.

Abstract

Exploration for sulphide ores in Finland is becoming increasingly difficult. Most of the easily detectable economic sulphide ores, whose location is indicated either by outcropping mineralization or by measurable geophysical anomalies, have probably been discovered. Hence measurements conducted by conventional geophysical methods are less and less able to pinpoint an anomaly caused by an ore deposit. For the time being, however, the survey data collected by various geophysical results can often be utilized in indirect exploration, in which they serve as a supplement to the geological information available on the survey area. This paper presents some examples to illustrate the principle of combining various geophysical methods and adapting them to the exploration for sulphide ores in Finland.

Résumé

L'exploration des minerais sulfurés en Finlande devient de plus en plus difficile. La plupart des gîtes sulfurés "facilement repérables", ayant une valeur économique, et dont la position est indiquée soit par des affleurements minéralisés, soit par des anomalies géophysiques mesurables, ont probablement déjà été découverts. Ceci explique que les mesures découlant des méthodes géophysiques conventionnelles permettent de plus en plus rarement de localiser des anomalies dues à la présence d'un gisement métallifère. Mais pour l'instant les résultats de divers levés géophysiques, qui apportent un appoint d'information géologique, permettent souvent une exploration indirecte dans la région étudiée. Dans le présent rapport quelques exemples illustrent la méthode, qui consiste à combiner diverses méthodes géophysiques, et à les adapter à l'exploration des minerais sulfurés en Finlande.

INTRODUCTION

In Finland, where the bedrock is intensely metamorphosed and is almost completely of Precambrian age, sulphide ore prospecting concentrates mainly on areas with all but ubiquitous graphite-bearing schists. These rocks, which in Finland are usually called black schists or phyllites, contain variable amounts of pyrrhotite and pyrite. Because the thickness of the overburden is mostly less than 30 m and its resistivity, except in coastal areas, is high, varying according to Puranen (1959) from 50 to 30 000 Ωm, black schists are normally delineated by means of electrical methods. If the black schists contain appreciable amounts of pyrrhotite, as they often do, they can also be detected by magnetic measurements.

Prospecting for sulphide ores in black schist areas is however becoming increasingly more difficult. Most of the easily detectable economic sulphide deposits (i.e. ores) which can be localized either by means of a mineralized exposure or a conspicuous geophysical anomaly, have probably already been found.

It is typical of black schist areas that the number of anomalies resulting from magnetic and electrical surveys is far too high to make it economically feasible to check them all by diamond drilling. Seldom can a geologist establish the cause of an anomaly on an exposure. Pyrrhotite-bearing graphite schists and associated mineralization have usually been eroded more deeply by glacial ice than the surrounding country rock and thus they are mostly covered by overburden.

A fair number of methods are available for classifying geophysical anomalies and selecting explorationally-interesting anomalies in black schist areas, e.g. the simultaneous use of a combination of geophysical and geochemical methods. The contact polarization curve method developed in the Soviet Union and which uses the anodic and cathodic electrochemical reactions produced at the contact of mineralization with country rock, enables anomalies caused by graphite to be discerned from those caused by base metal mineralization. The method has been discussed in several publications, e.g. those by Ryss (1973), Parasnis (1974) and the U.S. Exchange Delegation in Mining Geophysics (1976).

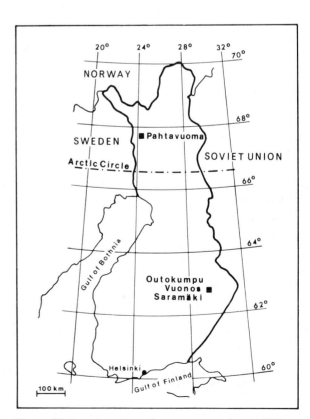

Figure 31.1. Location of copper orebodies in Finland.

Because it is difficult to discriminate between geophysical anomalies caused by economic sulphide mineralization and anomalies caused by black schists and other rocks, geophysical data are also utilized in so-called indirect exploration, in other words, in supplementing geological information. In the black schist areas of Finland, geophysical surveys are mainly conducted by magnetic, electromagnetic and gravity methods. The IP method is used only in special cases, owing to the intense electromagnetic anomalies caused by the black schist zones. Copper ores that occur in the black schist areas pose a very difficult exploration problem. This paper deals with geophysical results and their use in indirectly exploring for copper ores within black schist areas.

The copper deposits of Vuonos and Saramäki in the Outokumpu area in eastern Finland and the Pahtavuoma deposit, associated with the greenstone area of Kittilä, in northern Finland, have been selected as examples. The average copper content at Vuonos is about 2.5 per cent, at Saramäki 0.7 per cent and at Pahtavuoma 1.0 per cent. A map of Finland showing the location of the deposits discussed is presented in Figure 31.1.

The ores of Vuonos and Saramäki are of the Outokumpu-type, which means that they are compact copper ores with some cobalt. The ores occur in a geological formation that is surrounded by mica gneiss and is characterized by serpentinites, dolomites, skarns, quartz rocks and black schists. By means of aerogeophysical maps, this rock complex known as the Outokumpu association, can be traced for more than 240 km.

The Outokumpu deposit, averaging 3.80 per cent Cu and 0.20 per cent Co, was discovered in 1910. The orebody lies in a depression in association with serpentinite-quartzite formations. The total length of the ore deposit is 4 km. The width fluctuates from 200 to 400 m in the central part of the deposit but tapers off to the southwest at a depth of 150 m. At the extreme northeastern end, the orebody outcrops. The ore varies in thickness from a few metres to 40 m, the average being about 10 m. At the northeastern end the deposit dips 30° to 60° SE but is almost subhorizontal in the southwestern end. Some of the publications in which the geology of the Outokumpu and Vuonos ores is discussed, are those by Vähätalo (1953), Isokangas (1975) and Gaál, Koistinen and Mattila (1975). The latter paper includes a slingram (horizontal loop EM) map of the Outokumpu zone covering an area of 150 km².

THE GEOPHYSICS OF THE ORE DEPOSITS
The Vuonos orebody, eastern Finland

The Vuonos orebody, located about 4 km northeast of Outokumpu, was found in 1965. Preliminary drilling demonstrated that the Vuonos area, underlain by thick serpentinite-quartzite formations, was analogous to the Outokumpu area. Lithogeochemical investigations of drill cores, analyzed for sulphide copper, nickel and cobalt, showed that the copper content and the nickel-to-cobalt ratio in the serpentinites at Vuonos were anomalous compared with the serpentinites in the areas free from cobalt mineralization. Häkli (1963) has shown that in areas where serpentinites are not associated with copper-cobalt mineralization their nickel-to-cobalt ratio is similar to that in ultramafic rocks and varies from 20:1 to 25:1. Follow-up drilling led to the discovery of the Vuonos ore, whose average Ni-to-Co ratio is 2:3.

The geological and lithogeochemical investigations of the Outokumpu area have been reported by Huhma and Huhma (1970).

The Vuonos orebody is about 3.5 km long and 50 to 100 m wide. The average thickness is 5 to 6 m but it may attain as much as 20 m. The ore tapers out gradually towards both ends. The southwestern end is at a depth of about 60 m and the northeastern at a depth of about 200 m. Hence the orebody is blind.

The magnetic, slingram and gravity maps of the Vuonos area are presented in Figure 31.2 and show that the subhorizontal copper ore is not detectable by those methods. Figure 31.3 shows the geophysical results along profile y = 194.25 directly across the Vuonos orebody. The petrophysical data is plotted in profile form along the holes drilled in the profile. Resistivity values were obtained by in situ measurements by the single-point method, the smallest measured resistance being 1 Ω. The susceptibility and density determinations were done in the laboratory on drill core samples about 10 cm long, at intervals of 1 m for the whole length of the hole. Considering the geological and petrophysical information, the density of the conducting chrysotile serpentinite overlying the ore is low and the susceptibility higher than that of the surrounding rock types. The negative slingram anomalies are due to the black schists that, in conjunction with serpentinite, also cause magnetic anomalies. The susceptibility of the skarn is low, but the density is higher than that of other rock types, and so these rocks give rise to the positive gravity anomaly measured on the profile. Due to its horizontal position and great depth, the high density conducting orebody cannot be localized by means of geophysical methods because the anomalies produced by the country rocks camouflage the weaker anomaly produced by the ore.

The Saramäki orebody, eastern Finland

The Saramäki copper ore of Outokumpu-type is located in a formation that consists of skarns, serpentinites, quartzites and black schists surrounded by mica gneiss. The ore, which was found by drilling, outcrops. Nontheless it is not possible to localize it on the basis of the geophysical results depicted in Figure 31.4. The geology and resistivity results plotted along the drillholes show that the whole formation is a good conductor. The negative slingram anomalies are mainly caused by the black schists. The susceptibility values suggest that the formation is heterogeneously magnetized and that its upper part produces the strong magnetic anomaly. The gravity anomaly is generated by the whole formation, the density value of which differs clearly from that of the surrounding mica gneiss.

The Pahtavuoma orebody, northern Finland

The four ore lenses, found after the discovery of some copper-bearing outcrops at Pahtavuoma, are situated in a phyllite schist zone in the southern part of the Kittilä greenstone formation. The phyllite zone strikes from east to west. The phyllites are graphite-bearing and cause the strong slingram and VLF-EM anomalies shown in Figure 31.5. The VLF-EM map suggests that the phyllites are notably more continuous along the strike than does the slingram map, on which the anomalies are discontinuous from east to west. This is probably due to the higher frequency and better depth penetration of the VLF-EM method. The outcrops of the two ore lenses (A ore, the largest, and Ulla ore) are depicted on the geophysical maps in Figure 31.5. It is clear from the magnetic and gravity maps that the phyllite zones and ore lenses cannot be located by geophysical techniques. One reason is that the pyrrhotite content of the phyllites is very low. In Figure 31.6, the geophysical and petrophysical data of the profile y = 11.9 from Figure 31.5 are presented together with the geological information. The magnetic curve drawn in the upper part of the figure is almost nonanomalous, and the

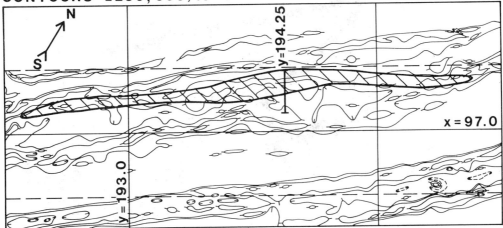

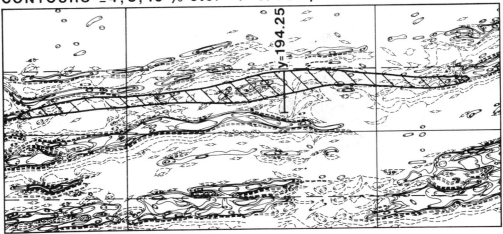

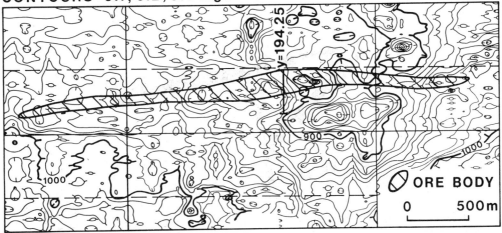

Figure 31.2. Geophysical maps of the Vuonos area.

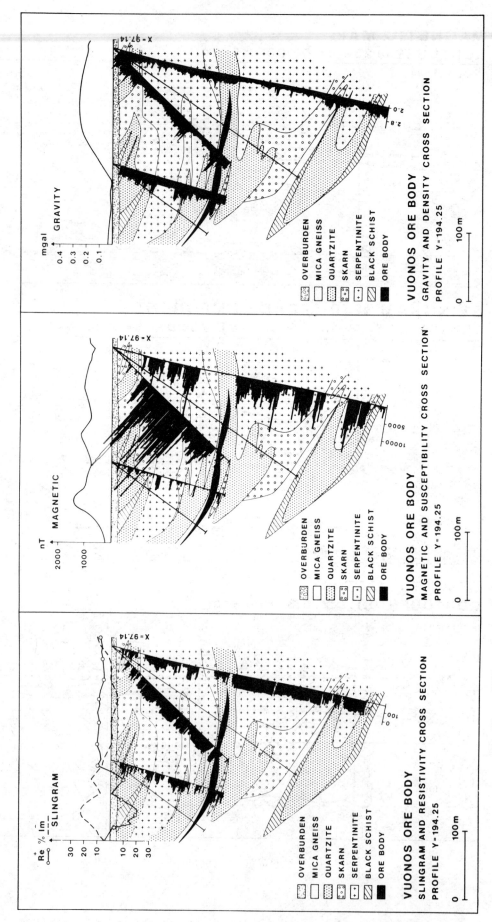

Figure 31.3. Geophysical and petrophysical data from profile $y = 194.25$ in the Vuonos area. The values on the drillhole profiles are respectively in Ω, $\times 10^{-5}$ SI and gm/cc.

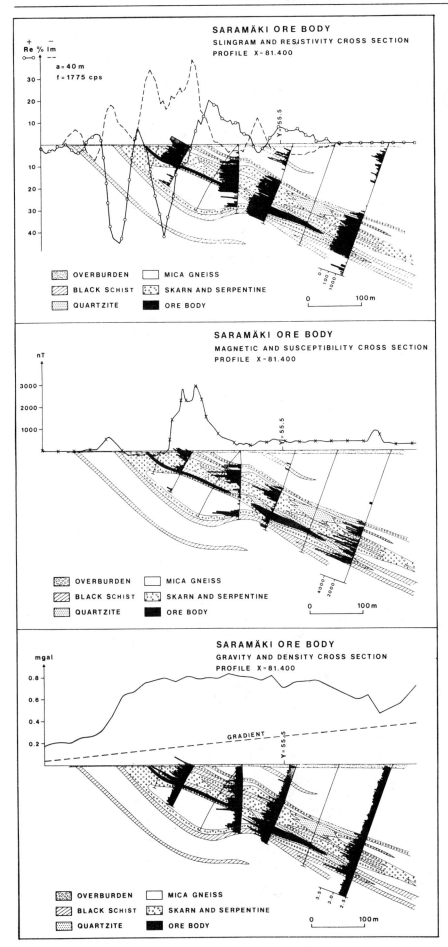

susceptibility determinations made on drill cores gave values that were too low to warrant their inclusion in Figure 31.6. The resistivity measurements conducted in drillholes demonstrate that the slingram anomalies are caused by phyllites and that the ore cannot be detected electromagnetically. In the lower part of Figure 31.6, next to the drillholes, the average density values for different rock types have been drawn. The density of phyllites is lower than that of greenstones. Although the alternation of phyllites and greenstones is not clearly seen from the gravity curve due to the steep gradient, these rocks can be localized on the residual anomaly map.

Kokkola and Korkalo (1976) have described the till- and stream-sediment investigations in the Pahtavuoma area. Soil surveys, whose purpose was to classify geophysical anomalies, turned out to be difficult to interpret owing to the existence of five till beds and layers deposited by ice that moved successively in different directions and because the shape of a geochemical anomaly in till depends on the direction of glacial transport.

GEOPHYSICAL METHODS IN GEOLOGICAL MAPPING

Because it has not been possible by means of geophysical methods to localize directly the copper ores at Vuonos, Saramäki and Pahtavuoma, survey data have and will continue to be used for indirect exploration in these areas.

The following chapter gives some examples of the use of airborne, ground and drillhole methods in the indirect exploration for sulphide ores.

Airborne geophysical surveys

The first phase of exploration of extensive areas entails airborne magnetic and electromagnetic (AEM) surveys. These are presently being conducted at a low flight altitude and with a dense line spacing, because the general aerogeophysical maps made by the Geological Survey of Finland from a flight altitude of 150 m already cover the whole country. The aim of the low-altitude airborne geophysical surveys is to obtain a picture of the magnetized and conductive rock types of the area under investigation in as great a detail as possible. Because the thickness of the overburden in Finland averages only 8 m, rigid airborne electromagnetic systems are eminently suitable for mapping the magnetic and conductive rock types in black schist areas that suboutcrop under the overburden.

Figure 31.7 shows an aeromagnetic and AEM map of the Miihkali area about 10 km north of Saramäki surveyed from an altitude of 40 m with a 125 m linespacing. The AEM

Figure 31.4.

Geophysical and petrophysical data from profile x = 81.4 over the Saramäki orebody.

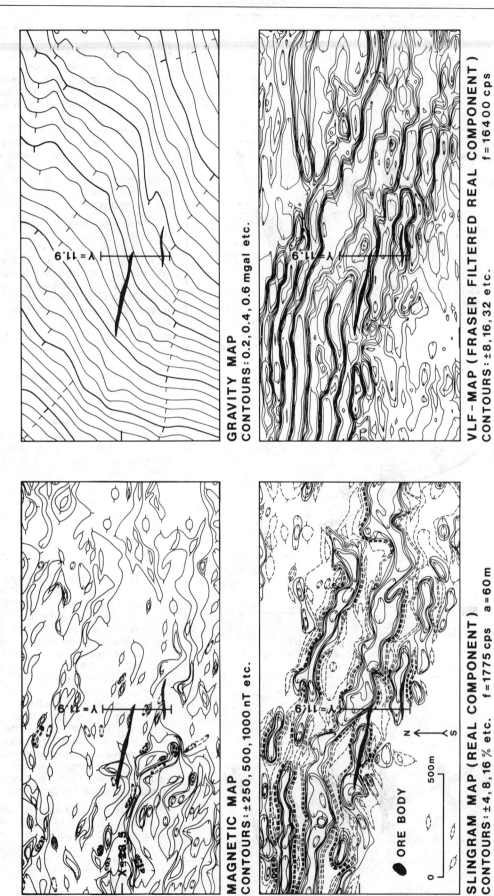

Figure 31.5. Geophysical maps of the Pahtavuoma area. The VLF-EM data have been filtered using the technique of Fraser (1959).

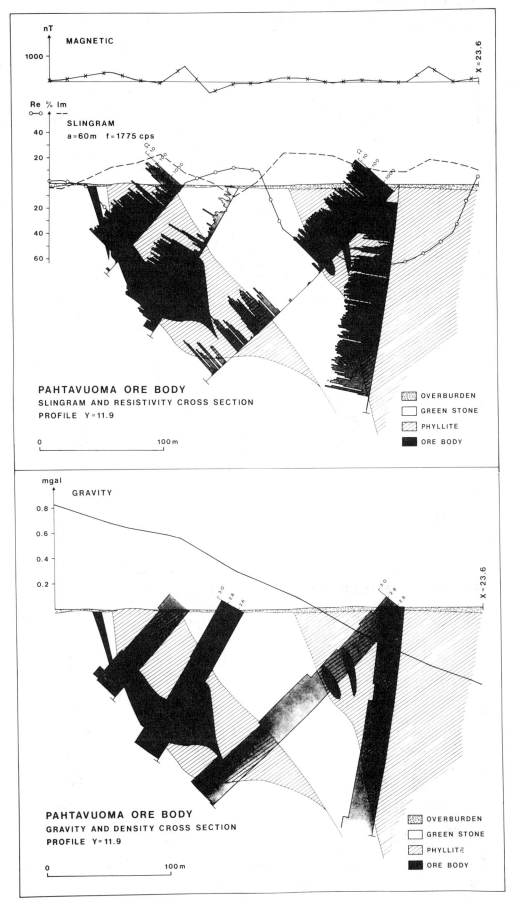

Figure 31.6. *Geophysical and petrophysical data from profile y = 11.9 in the Pahtavuoma area.*

AEROMAGNETIC MAP
CONTOURS: 200, 400, 600 nT etc.   FLIGHT ALTITUDE 40 M

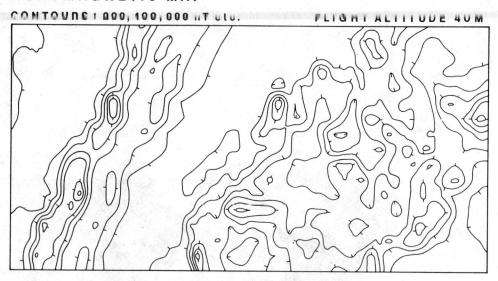

AEROELECTROMAGNETIC MAP (WING TIP, REAL COMPONENT)
CONTOURS: 200, 400, 600 ppm etc.   FLIGHT ALTITUDE 40 M

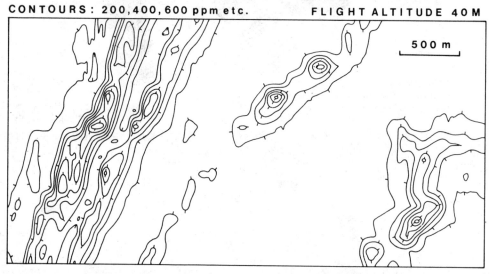

SLINGRAM MAP (REAL COMPONENT)
CONTOURS: −4, +8, ±16, ±32 % etc.   f = 1775 cps   a = 60 M

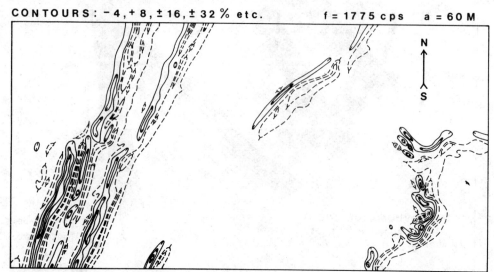

Figure 31.7. Comparison between airborne and ground geophysical data from the Miihakali area.

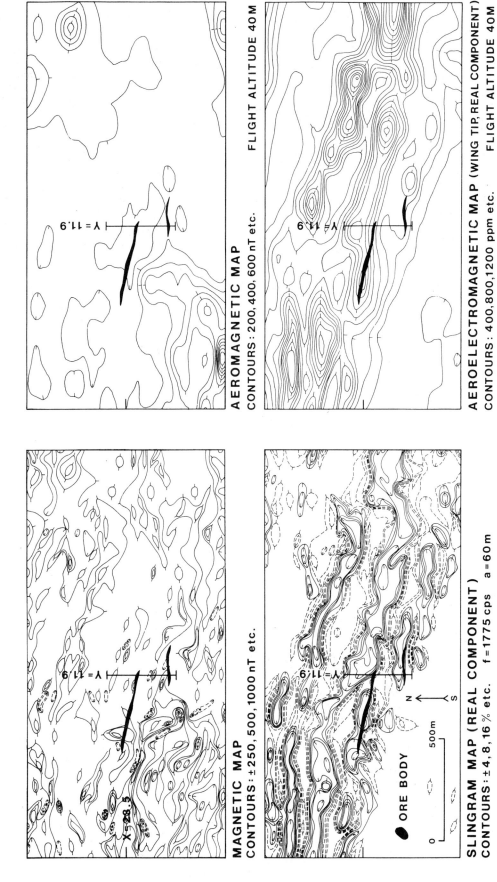

Figure 31.8. Comparison between airborne and ground geophysical data from the Pahtavuoma area.

measurements were obtained using the Coplanar-coil wing-tip system. For the sake of comparison, the slingram real component map of the area investigated by airborne surveys is also shown in Figure 31.7. The correlation between the AEM and ground EM results is excellent bearing in mind the inevitable broadening of the AEM anomalies with height. Experience has shown that if the flight lines are perpendicular to the strike of the conducting zones and the flight-line positioning is sufficiently accurate, the Coplanar-coil wing-tip AEM system gives results that closely correspond to those of the slingram method. If the direction of the flight lines coincides with the strike of the conducting zones, AEM measurements do not produce results as good as those obtained by ground EM surveys, because the latter allow a ready change of measurement direction to meet the requirements of the strike of the geological formations.

The airborne and ground geophysical surveys of the Pahtavuoma area are displayed and compared in Figure 31.8. The east-west striking, phyllite belt is indicated by the real component map of the AEM survey. The anomalies of the AEM map display a good correlation with slingram anomalies. The profiles in Figure 31.9 confirm the good correlation between the wing-tip and slingram results above the phyllite formation that enclose the orebodies in the Pahtavuoma area. The Pahtavuoma case indicates that AEM surveys provide detailed geological information in areas where the magnetite and pyrrhotite contents of the bedrock are so small that only weak anomalies occur. Thus the AEM technique may be used as a geological mapping tool in such areas.

Ground geophysical surveys

The detailed results of low-altitude airborne magnetic and electromagnetic surveys have reduced the need for ground measurements, so these have been concentrated on those areas favourable for the occurrence of ores. In order to augment geological information as much as possible these areas are investigated using several ground geophysical methods. The most commonly used are magnetic, slingram and gravity measurements carried out with a 20 m station spacing and 50-100 m line separation. The geophysical results of aeromagnetic and AEM measurements are sometimes supplemented by a gravity survey conducted over the whole flight area with a grid density of 0.5 to 1 km.

Copper-cobalt ores of Outokumpu type are associated with serpentinite-skarn-quartz rock formations, and so geophysical measurements can be used to localize these explorationally important rock types. Figure 31.10 presents the geophysical results from a profile in the Miihkali area. Several holes have been drilled on this profile to check the anomalies. The petrophysical information from some of the holes drilled in the profile is plotted in the lower part of Figure 31.10. A broad chrysotile serpentinite formation with low resistivity (r in ohms), low density (ρ in gm/cc) and high susceptibility ($k \times 10^{-5}$ SI) and which is represented by intersection 4 in hole D 6 can be localized by the positive magnetic and slingram anomalies and the well developed negative gravity anomaly. The positive gravity anomaly in the middle of the profile is caused by a skarn formation, whose average density (ρ), according to intersection 3, is 2.95 gm/cc. Pyrrhotite-bearing black schists produce strong magnetic and slingram anomalies.

Figure 31.11 shows the magnetic, gravity and slingram maps of the Usinjärvi area. A rounded anomaly is visible in the middle of the area. The negative gravity anomaly and the positive slingram anomaly suggest that the magnetized formation is conductive and that its density is low. It is a chrysotile serpentinite body in the northern part of which there are black schists that cause negative slingram anomalies.

In the localization of ultrabasic formations, especially when prospecting for nickel ores, systematic gravity surveying has turned out to be of much practical importance. Presented in Figure 31.12 A and B are the results of magnetic, slingram and gravity measurements performed in two different black schist areas. In both areas the black schists cause strong slingram anomalies. Figure 31.12 B shows a gravity anomaly caused by a peridotite body, in the northeastern end of which there is nickel mineralization that also causes a weak magnetic and slingram anomaly. A strong positive anomaly is depicted on the gravity map in Figure 31.12 A. This anomaly is produced by a peridotite body that cannot be discovered by means of a magnetic survey as the magnetic results indicate.

Experimental reflection seismic measurements have been conducted in the Outokumpu area to trace the deep-seated continuation of the geological formations. The measurements and interpretation were the work of a team from the Institute of Seismology in Helsinki University under the leadership of Dr. E. Penttilä. Figure 31.13 is a simplified reflection seismic cross-section obtained across the

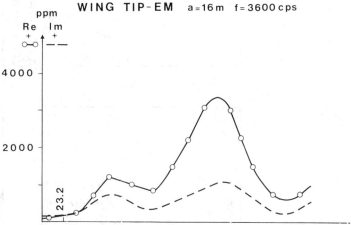

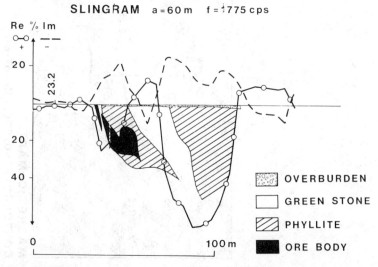

Figure 31.9. Comparison between airborne and ground electromagnetic data from profile $x = 11.9$ in Pahtavuoma area.

Outokumpu area. The location of the Outokumpu-association and the most clearly reflecting horizons have been marked in Figure 31.13 on the basis of the reflector distances. The reflections, of unknown origin, suggest that the Outokumpu formation might continue albeit thinly, down to a depth of about 1.5 km. The dip of the continuations begin to slope more gently at the greater depth. Thus the seismic technique can be of value in delineating the extension of important ore-associated formations at depth.

Figure 31.14 displays the magnetotelluric survey results in the Saramäki area where the continuation of the ore-associated formation was followed down to depth. As a result of this survey the conductive formation can partly be located. Anomalous area A indicates the upper part of the conductive formation. The anomalous area B correlates with the proven deeper parts of the formation and indicates further continuation to depth. The results do not indicate the presence of the ore-associated formation at depths between 300-400 m. The magnetotelluric surveys were carried out with French equipment fabricated by Societé ECA, acquired by the University of Oulu. The output of the apparatus was the apparent resistivity at nine frequencies from 8 to 3700 Hz. In the interpretation of the magnetotelluric results, an interactive computer program was adapted, which used a layered earth model and hyperparabola minimization process presented by Lakanen (1975).

The localization and establishment of deep-seated continuations of geological formations is important when planning deep and expensive drillholes for finding new ore deposits. It is obvious from the foregoing examples that the use of reflection seismic and magnetotelluric investigations is of considerable benefit in indirect exploration.

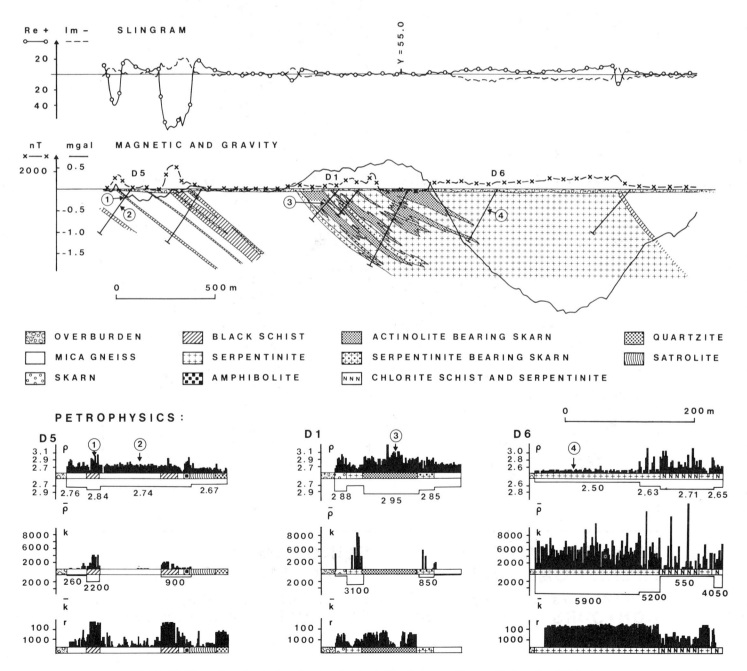

Figure 31.10. Geophysical and petrophysical data from profile x = 91.0 in the Miihkali area.

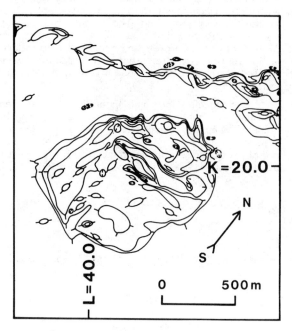
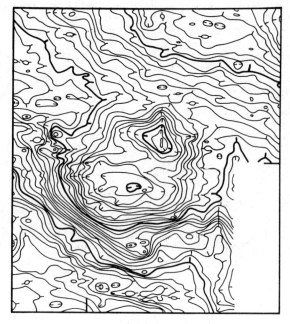

MAGNETIC MAP
CONTOURS: ±250, 500, 1000, 2000 nT etc.

GRAVITY MAP
CONTOURS: 0.2, 0.4, 0.6 mgal etc.

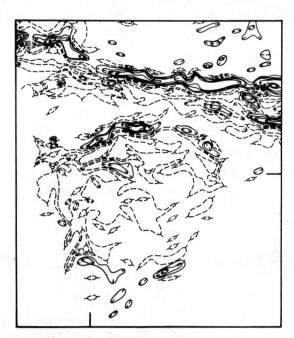
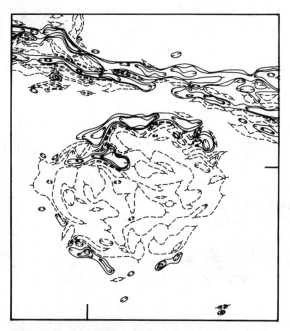

SLINGRAM MAP (REAL COMPONENT)
f = 1775 cps, a = 60 m
CONTOURS: ±4, 8, 16 % etc.

SLINGRAM MAP (IMAGINARY COMPONENT)
f = 1775 cps, a = 60 m
CONTOURS: ±4, 8, 16 % etc.

Figure 31.11. Geophysical results from the Usinjärvi area.

MAGNETIC MAPS
CONTOURS: ±250, 500, 1000 nT etc.

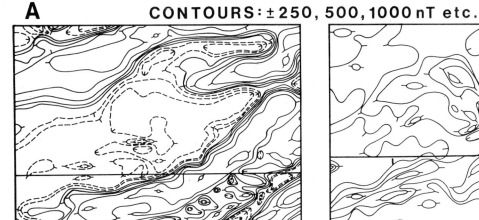
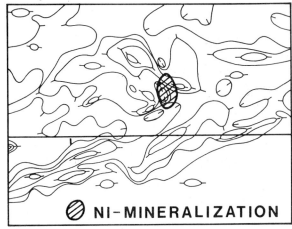

SLINGRAM MAPS (REAL COMPONENT)
CONTOURS: ±4, 8, 16 % etc.   f = 1775 cps   a = 60 m

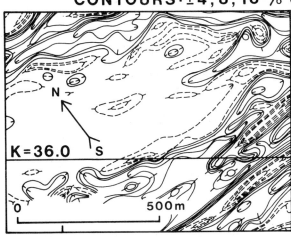
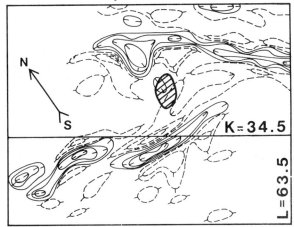

GRAVITY MAPS
CONTOURS: 0.1, 0.2, 0.3 mgal etc.

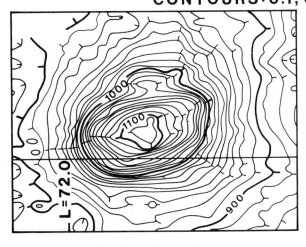
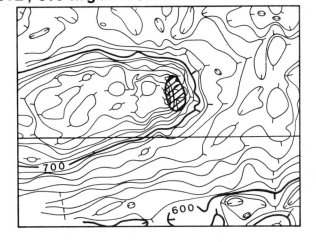

KERIMÄKI AREA   **LAUKUNKANGAS AREA**

Figure 31.12.   Geophysical results from the Kerimäki and Laukunkangas areas.

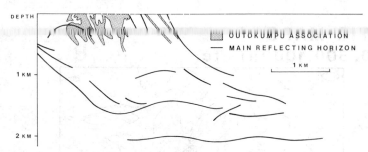

Figure 31.13. Reflection seismic cross-section through the Outokumpu area.

Borehole surveys

The inner structure of formations rich in black schists cannot usually be investigated by electrical ground methods. The conductive subsurface schists shield the underlying rocks, even though the depth penetration of the geophysical methods adopted would otherwise be sufficient. In some black schist areas it is possible to use electrical drillhole logging to investigate geological formations.

Figure 31.15 shows profile y = 12.0 from the Pahtavuoma area, where resistivity and charged-potential (mise-à-la-masse) measurements were obtained in the holes drilled into the orebody. The resistivity measurements demonstrate that there is a clear conductivity contrast between phyllites and greenstones. Although it was not possible to solve directly the connection between the ore intersections by means of charged-potential measurements, they nevertheless present an indirect means, because the host rock of the ore is a highly conductive phyllite.

The results of the charged-potential measurements, drawn in the form of the isoanomaly cross-section, reveal that ore intersections 1 and 2 are not located in the same phyllite formation, there being a marked increase in potential between the intersections. The holes drilled from the adit marked in Figure 31.15 confirm that a narrow weakly-conducting greenstone zone is sandwiched in phyllites.

The upper part of Figure 31.16 shows a slingram profile measured over the Outokumpu formation together with the holes drilled into it. The resistivity results are marked at the drillholes as are the intersections in which the conducting black schists and serpentinites occur in association with skarns. The geological picture given of the formation by drilling is not well established. To clarify the geological information charged-potential measurements were conducted using a number of grounding systems. The results from two of the grounding systems are depicted in Figure 31.16, and demonstrate that the formation is composed of various units. The results of grounding system 1 indicate that the formation surrounded by mica schist is a syncline. The results also demonstrate that there is a good conductivity connection between the intersections of the rock types favourable for the occurrence of ore in the southeastern part of the formation. The results of grounding system 2 are from another part of the formation, where the black schist-serpentinite intersections are at the same potential.

Ketola (1972) has published the results of the charged potential measurements conducted in the Saramäki (earlier called Miihkali) area. The ore formation was also investigated by means of SP measurements obtained on the surface and in drillholes. It can be seen from Figure 31.17 that the formation produces a clear negative SP anomaly, the intensity of which is − 200 mV at the surface. There is only a weak response in the SP anomaly curve above the black schists, which cause negative slingram anomalies.

The surface and drillhole SP data were processed by the method described by Logn and Bølviken (1974). The field anomalies were divided into two parts, a time-dependent anomaly of a separate conductor in the formation ECP (electronic current potential) and an anomaly caused by the whole formation ICP (ionic current potential). Shown in the lower part of Figure 31.17 is the cross-section, compiled on the basis of the ICP data, which demonstrates that the anomaly caused by the formation is reversed along the dip from negative to positive. The positive anomaly may be extensive, as was demonstrated by Semjonov (1975) using some practical examples. Hence, it is not always possible to localize accurately the lower end of the formation.

INTERPRETATION OF THE GEOPHYSICAL ANOMALIES AND THEIR GEOCHEMICAL CLASSIFICATION

Before the start of drilling or geochemical sampling the geophysical data are interpreted in order to evaluate the dimensions, positions and petrophysical properties of anomalous formations. Ketola, Ahokas, Liimatainen and Kaski (1975) and Ketola, Liimatainen and Ahokas (1976) have investigated the feasibility of two- and three-dimensional interpretation methods and petrophysical determinations.

Depicted in Figure 31.18 are the two-dimensional interpretations of a magnetic and gravity anomaly curves measured over the Saramäki orebody. The geophysical and petrophysical data of this profile have been discussed above. The result of the first magnetic interpretation with effective susceptibility values (k), remanence omitted, is seen in the upper part of Figure 31.18. In the other magnetic interpretation, which suggests a more gentle dip and smaller susceptibilities (k_r) for anomalous formations, the remanence has been taken into account. In the interpretation, the average ratio of remanent to induced magnetization was 10:1 and the inclination and declination of the remanence were 45° and 90°, respectively (compared to 75° and 7° respectively for the earth's magnetic field) measured on some oriented samples drilled from black schist exposures. Susceptibility determinations on drill cores demonstrate that the remanence-corrected interpretation gives the better susceptibility and dip estimates to the magnetized upper part of the formation depicted in Figure 31.4. The gravity anomaly is caused by the skarn rocks, which in the light of interpretation and drilling data are located mainly between the magnetized parts of the formation, but partly also within them. Thus the magnetic and gravity data cannot be combined in the interpretation although the dip of the adjacent formations must be compatible i.e. 30°.

The results from profile x = 81.4 given by the magnetic, slingram and gravity interpretations were put to good use when planning geochemical humus sampling with the purpose of classifying geophysical anomalies. Beneath the magnetic and gravity profiles in the upper part of Figure 31.19 there is a petrophysical cross-section constructed by means of the two-dimensional interpretation results on the basis of which the sites of humus samples were selected. The petrophysical interpretation reveals the magnetized, conducting and anomalously high density sections. The copper contents of humus, showing an anomaly above the copper mineralization is presented together with slingram and geological data in the lower part of Figure 31.19. A strong but rather narrow humus anomaly indicates that geochemical sampling, the aim of which in this case was to check the magnetic and conducting sections, should be done with a short enough sample spacing. When planning sampling it is also worth employing the results of geophysical interpretation.

According to Wennervirta (1973), the Saramäki copper mineralization also manifests itself as an intense geochemical till anomaly. The samples were taken by percussion or pneumatic drilling from the depth of maximum penetration immediately above the surface of the bedrock.

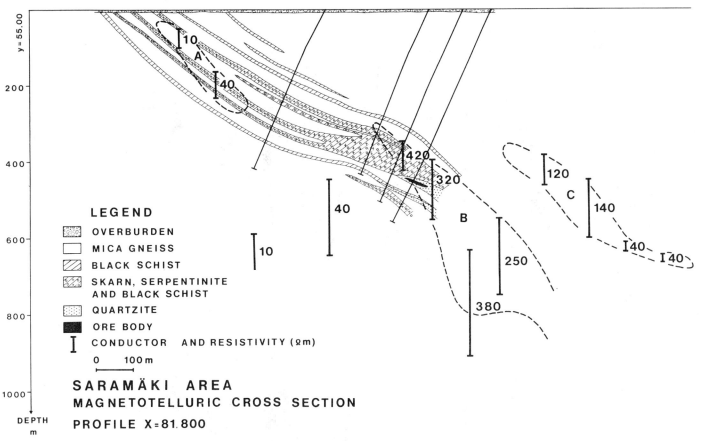

Figure 31.14. Magnetotelluric cross-section in the Saramäki area.

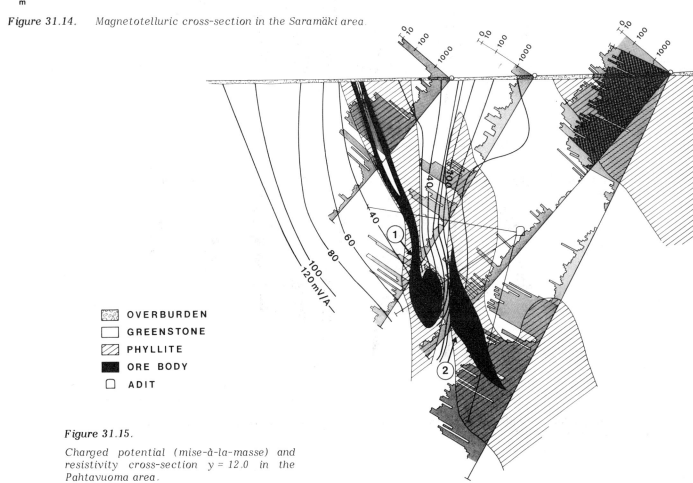

Figure 31.15.

Charged potential (mise-à-la-masse) and resistivity cross-section y = 12.0 in the Pahtavuoma area.

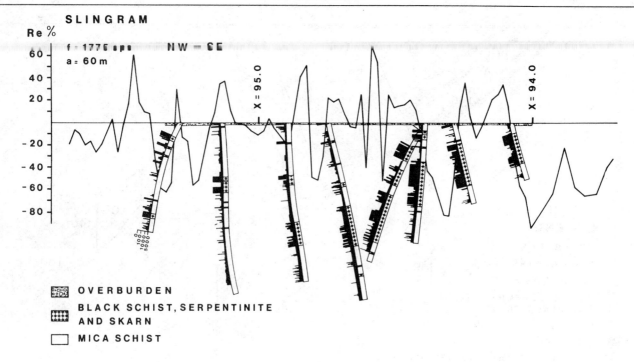
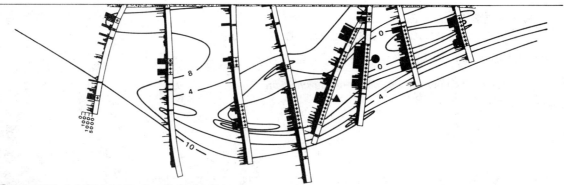
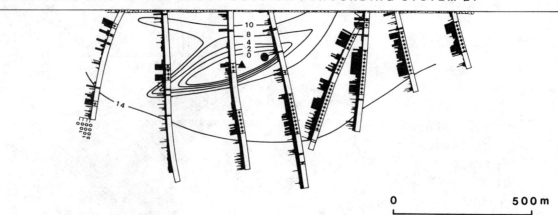

Figure 31.16. Charged potential (mise-à-la-masse) and resistivity cross-section y = 176.0 in the Outokumpu area.

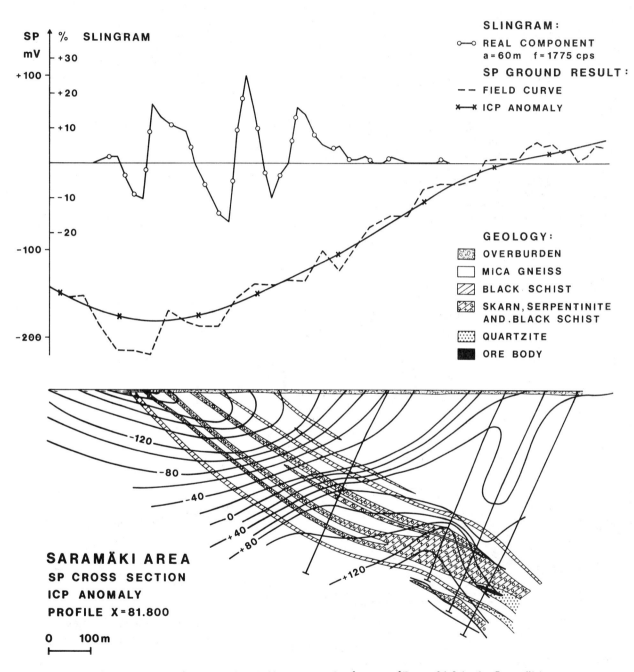

Figure 31.17. Self-potential and slingram results from profile x = 81.8 in the Saramäki area.

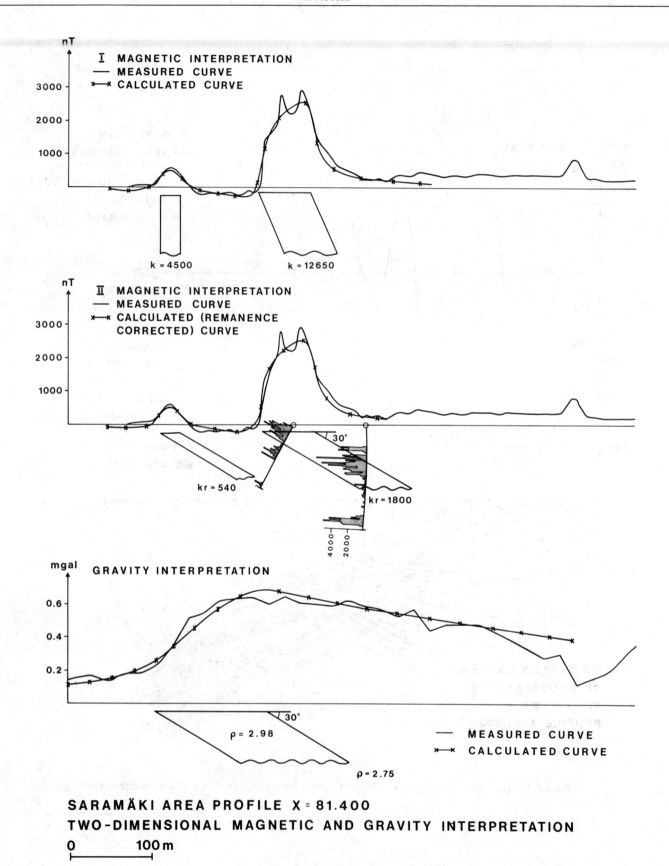

Figure 31.18. Interpretation of magnetic and gravity data from profile x = 81.4 in the Saramäki area.

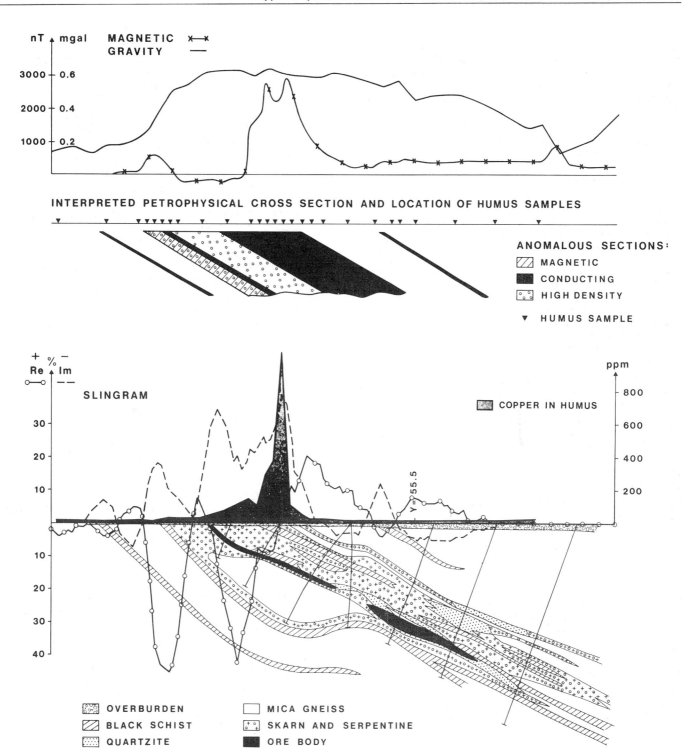

Figure 31.19. Use of geophysical interpretation results in the planning of geochemical sampling.

CONCLUSIONS

In this paper, an attempt has been made to demonstrate the feasibility of geophysical methods in the indirect exploration of sulphide ores in a graphite-bearing schist environment. It is evident that as exploration becomes more difficult, geophysics has increasingly to be applied to indirect exploration.

The simultaneous and effective use of different sciences, such as geology, geophysics and geochemistry, is necessary in order to find new orebodies.

ACKNOWLEDGMENTS

Thanks are due to Outokumpu Oy, Exploration, for permission to publish the data included in this paper. The author is especially grateful to Mr. M. Laurila, the Head of the Geophysical Department, who has led and taken an active part in geophysical investigations in all survey areas and whose positive attitude made it possible to complete this paper. Examples were collected from the extensive archives of Outokumpu Oy, which cover the last 25 years. I am happy to acknowledge my gratitude to Mr. M. Liimatainen, the geophysicist responsible for the charged-potential measurements at Pahtavuoma and the SP investigations in the Saramäki area, Mr. T. Ahokas, the geophysicist who planned the charged-potential measurements in the Outokumpu-zone and Messrs. E. Lakanen, A. Ruotsalainen and P. Kaikkonen, the geophysicists who performed the magnetotelluric measurements in the Saramäki area and interpreted the results. Mr. J. Longi performed the magnetic and gravity interpretation calculations and Mr. Kokkola, geochemist, the humus investigations in the Saramäki area. The manuscript was translated by Mrs. G. Häkli and Mrs. M. Sarkki, whom I thank for their pleasant co-operation.

REFERENCES

Fraser, D.C.
    1959: Contouring of VLF-EM data; Geophysics, v. 34(6), p. 958-967.

Gaál, G., Koistinen, T., and Mattila, E.
    1975: Tectonics and stratigraphy of the vicinity of Outokumpu, North Karelia, Finland; Geol. Surv. Finland, Bull. 271, 67 p.

Häkli, A.
    1963: Distribution of nickel between the silicate and sulphide phases in some basic intrusions in Finland; Geol. Surv. Finland, Bull. 209, p. 1-54.

Huhma, A. and Huhma, M.
    1970: Contribution to the geology and geochemistry of the Outokumpu region; Geol. Soc. Finland Bull., v. 42, p. 57-88.

Isokangas, P.
    1975: The mineral deposits of Finland; Lis. thesis, Univ. of Helsinki, Finland, 128 p.

Ketola, M.
    1972: Some points of view concerning Mise-à-la-masse measurements; Geoexploration, v. 10(1), p. 14-21.

Ketola, M., Ahokas, T., Liimatainen, M., and Kaski, T.
    1975: Some case histories of the interpretation of magnetic and gravimetric anomalies by means of two- and three-dimensional models; Geoexploration, v. 13(3), p. 137-169.

Ketola, M., Liimatainen, M., and Ahokas, T.
    1976: Application of petrophysics to sulfide ore prospecting in Finland; Pageoph, v. 114, p. 215-234.

Kokkola, M. and Korkalo, T.
    1976: Pahtavuoma copper in stream sediment and till; J. Geochem. Explor., v. 5(3), p. 280-287.

Lakanen, E.
    1975: Geofysikaalinan tulkinta graafisen tietokonekäsittelyn avulla; Geofysiikan päivät, Oulu.

Logn, Ø, and Bolviken, B.
    1974: Self potentials at the Joma pyrite deposit, Norway; Geoexploration, v. 12(1), p. 11-28.

Parasnis, D.S.
    1974: Some present-day problems and possibilities in mining geophysics; Geoexploration, v. 12(213), p. 116-118.

Puranen, M.
    1959: Aerosähköisistä mittausmenetelmistä, Espoo.

Ryss, Yu.S.
    1973: The search and exploration for ore bodies by the contact method of polarization curves; Nedra, Leningrad.

Semjonov, M.V.
    1975: Osnovy poiskov i izutsenija koltsedannopolimetallitseskin rudnyh polei geofizitseskimi metodami; Nedra, Leningrad, p. 118-141.

U.S. Exchange Delegation in Mining Geophysics
    1976: A Western view of mining geophysics in the U.S.S.R.; Geophysics, v. 41(3), p. 320-321.

Vähätalo, V.
    1953: On the geology of the Outokumpu ore deposit in Finland; Geol. Surv. Finland, Bull. 164, p. 7-10.

Wennervirta, H.
    1973: Sampling of the bedrock-till interface in geochemical exploration; Papers presented at a symposium in "Prospecting in areas of glacial terrain", organized by the Institution of Mining and Metallurgy, Trondheim, Norway.

GEOPHYSICAL AND GEOCHEMICAL METHODS USED IN THE DISCOVERY OF THE ISLAND COPPER DEPOSIT, VANCOUVER ISLAND, BRITISH COLUMBIA

K.E. Witherly

Utah Mines Ltd., Toronto, Ontario

Witherly, K.E., Geophysical and geochemical methods used in the discovery of the Island Copper Deposit, Vancouver Island, British Columbia; in Geophysics and Geochemistry in the Search for Metallic Ores; Peter J. Hood, editor; Geological Survey of Canada, Economic Geology Report 31, p. 685-696, 1979.

Abstract

The Island Copper Mine of Utah Mines Ltd. is a large, low-grade copper-molybdenum deposit located 354 km northwest of Vancouver, B.C. The deposit lies in the Bonanza Volcanic formation of lower to mid-Jurassic age. The ore zones are in andesitic, pyroclastic rocks in the hanging wall and footwall of a quartz-feldspar porphyry dyke. At the commencement of production, ore reserves totalled 257 million tonnes averaging 0.52 per cent copper and 0.017 per cent molybdenum.

Exploration work on the deposit began in early 1966 when Utah Construction and Mining Co. (now Utah Mines Ltd.) optioned approximately 175 claims located on the north side of Rupert Inlet. Initial work consisted of geological mapping, soil geochemistry, ground magnetic and IP surveying. Considerable dependence was placed on the latter three tools due to the extensive overburden cover on the property. The geochemical and geophysical surveys produced an abundance of anomalies over the property.

During the first year of exploration, testing of geochemical and geophysical anomalies was done with a small company-owned diamond drill which could penetrate depths of up to 75 m under favourable conditions. Although copper and magnetite had a known association in the area, initial direct testing of the best magnetic anomalies located only subeconomic copper mineralization. The IP anomalies tested intersected some low-grade copper with pyrite, but more often just barren pyrite. It was observed from the initial testing that the total sulphide content of the anomalies tested was fairly constant, i.e. when the chalcopyrite picked up, the pyrite content dropped. This made anomaly discrimination based on amplitudes quite difficult.

In view of the poor resolution the geophysical anomalies provided, it was decided to test the copper soil geochemical anomalies before any more geophysical work would be done. The orebody was discovered as a result of testing one such geochemical anomaly with several shallow drillholes. Deeper drilling of the deposit was initiated soon after and the orebody was delineated a little more than two years later in the Spring of 1969. A VLF-EM survey over the deposit did not respond to the ore mineralization itself but did show up a number of significant structural features.

Although geophysics did not assist directly in the discovery of the Island Copper deposit, its contributions to the development of geological thinking in the area have been valuable. A large part of the recent interest in the northern part of Vancouver Island stemmed from a joint British Columbia Department of Mines/Geological Survey of Canada aeromagnetic survey flown in 1962. Examination of the geophysical data made subsequent to the discovery of Island Copper show that the geophysical results support and enhance information on structure, alteration and mineralization both in and around the orebody. Such information is not only of interest in its own right but as well provides a valuable data base to draw upon for exploration in similar geological environments.

Résumé

La mine de cuivre Island de l'Utah Mines Ltd. est formée d'un important gisement pauvre de cuivre-molybdène, 354 km au nord-ouest de Vancouver, en Colombie-Britannique. Le gisement repose dans la formation volcanique Bonanza du Jurassique inférieur et moyen. Les zones de minerai se trouvent dans des roches pyroclastiques à andésite dans la lèvre supérieure et la lèvre inférieure d'un dyke porphyrique de quartz-feldspath. Au début de la production, les réserves de minerai atteignaient 257 millions de tonnes d'une teneur en cuivre d'environ 0,52 pour cent et d'une teneur en molybdène d'environ 0,017 pour cent.

Les travaux d'exploration du gisement ont commencé au début de 1966, lorsque l'Utah Construction and Mining Co. (aujourd'hui l'Utah Mines Ltd.) a choisi environ 175 claims sur la rive nord de l'inlet Rupert. Les travaux préliminaires comprenaient la cartographie géologique, l'étude géochimique des sols, des levés magnétiques et des levés par méthode PP. On avait beaucoup misé sur les trois dernières méthodes, étant donnée la couche importante du terrain de couverture.

Au cours de la première année d'exploration, des vérifications d'anomalies géochimiques et géophysiques ont été effectuées à l'aide d'un petit foret au diamant que possédait la société; cet outil de forage pouvait atteindre des profondeurs de 75 m dans des conditions favorables.

Les levés géochimiques et géophysiques ont révélé un grand nombre d'anomalies sur toute l'étendue du terrain. Bien que l'on ait su que, dans cette région, le cuivre et la magnétite étaient associés, la vérification directe initiale des meilleures anomalies magnétiques n'a indiqué qu'une minéralisation de cuivre non susceptible d'être exploitée de façon rentable.

Les anomalies vérifiées par la méthode PP ont livré un peu de cuivre pauvre contenant de la pyrite, mais surtout de la pyrite stérile. On a remarqué lors de la vérification initiale que le contenu total en sulfure des anomalies vérifiées était à peu près constant, c'est-à-dire que lorsque la chalcopyrite augmentait, la pyrite diminuait. Cette constation a rendu assez difficile la distinction des anomalies en fonction des amplitudes.

Compte tenu du faible pouvoir de résolution qu'avaient fourni les anomalies géophysiques, il fut décidé de vérifier les anomalies géochimiques des sols contenant du cuivre avant que d'autres travaux géophysiques ne soient entrepris. Le gisement en amas fut découvert à la suite de la vérification d'une telle anomalie géochimique au moyen de plusieurs trous de forage peu profonds. Le forage du gisement à plus grande profondeur fut entrepris peu de temps après, et le gisement en amas fut délimité un peu plus de deux ans plus tard, au printemps de 1969.

Un levé électromagnétique à très basse fréquence effectué sur l'ensemble du gisement n'a pas réagi à la minéralisation elle-même, mais a révélé un nombre important de particularités structurales.

Bien que la géophysique n'ait pas contribué directement à la découverte du gisement de cuivre Island, elle a influé grandement sur l'opinion que se font les géologues de la structure de la région. Une grande partie de l'intérêt manifesté dernièrement à ce sujet dans la partie nord de l'île Vancouver fait suite à un levé aéromagnétique effectué en 1962 par le ministère des Mines de la Colombie-Britannique et la Commission géologique du Canada. L'analyse des données géophysiques effectuées à la suite de la découverte de la mine de cuivre Island indique que les résultats d'ordre géophysique corroborent des renseignements sur la structure, l'altération et la minéralisation tant à l'intérieur qu'autour du gisement en amas. De tels renseignements ne sont pas seulement intéressants en eux-mêmes, mais ils fournissent également une base de données de valeur pour décider de travaux d'exploration dans des milieux géologiques semblables.

INTRODUCTION

The Island Copper (ISCU) deposit, controlled and operated by Utah Mines Ltd. (formerly Utah Construction and Mining), is located on the north shore of Rupert Inlet (50°36'N, 127°37'W) approximately 16 km south of Port Hardy on northern Vancouver Island (Fig. 32.1). The town of Port Hardy is 354 km northwest of Vancouver, B.C. and is serviced daily by scheduled flights from Vancouver. Port Hardy and the northern portion of Vancouver Island can also be reached by surface transportation using either public roads or a combination of provincial ferries and roads.

Elevations on the property range from sea level to 150 m. The area around the deposit is heavily timbered with spruce, hemlock and cedar. Glacial deposits with thicknesses of up to 75 m cover most of the area. Average annual precipitation at Port Hardy is 1900 mm, which includes an average of 600 mm of snow. The yearly temperature range is from -7°C minimum to 27°C maximum, with an annual mean of 8°C.

The orebody was discovered after more than four years of exploration activity that stemmed from the release of results of an aeromagnetic survey flown from July to September 1962 over the whole northern portion of Vancouver Island. This survey was a joint effort between the British Columbia Department of Mines and the Geological Survey of Canada. Release of the aeromagnetic survey results in 1963 stimulated considerable exploration interest in the area, with both companies and individuals participating in the follow-up. As part of their programs, a number of major mining companies, including Utah Mines Ltd., reconnoitered the Rupert Inlet area (Fig. 32.2), but found no encouraging evidence of economic mineralization. A small island, called Red Island, located slightly less than 2 km to the southeast of the present pit, was known to have good copper grades associated with abundant magnetite in volcanics. Prospectors at the time however, felt that the occurrence was only a high-grade pod and not part of a larger, mineralized system.

Soon after the release of the aeromagnetic data, Gordon Milbourne, a local prospector, ran a dip-needle survey over a localized magnetic high, situated just north of Rupert Inlet. Two years later, Milbourne returned to the area and located float and outcrops containing native copper and chalcopyrite, just west of Frances Lake (Fig. 32.2). Subsequent trenching revealed high-grade chalcopyrite in volcanic rocks.

Utah Mines Ltd., examined Milbourne's ground in October 1965, and concluded that the property warranted follow-up. After some preliminary exploration work in the Fall of 1965, Utah signed an agreement with Milbourne in January 1966.

Because outcrops on the property were very sparse, due to an extensive cover of glacial overburden, it was decided that an integrated geological, geochemical and geophysical approach should be used in evaluating the property. To facilitate this evaluation, a survey grid totalling 160 line km was established over the property. The lines were spaced 152 m apart with survey stations every 30.5 m. The lines were oriented N22°E, but local magnetic disturbances occasionally caused significant deflections in a surveyed line.

The choice of geochemical and geophysical parameters to be measured was based largely upon the initial test surveys done in 1965. Since only copper had been found in any significant amounts on the property, copper was the only element for which the soil samples were analyzed. Magnetic surveys were considered an important tool, due to the known affinity copper and magnetite showed in the area. Induced polarization was also felt to be useful, since considerable pyrite was found in association with the known copper mineralization.

Shallow drilling, with a company-owned X-ray machine, began early in 1966 to evaluate the mineralized area at the southwest end of Frances Lake. As geochemical and geophysical surveys progressed, the X-ray drill was used to test anomalies located elsewhere on the property. The initial criterion for drilling was that the anomalies be coincident with geochemical, magnetic and IP highs. On this premise, after some initial drilling north of Frances Lake, some interesting observations were made. Firstly, testing of the best magnetic anomalies encountered only low-grade copper mineralization; secondly, pyrite seemed ubiquitous in the area, with its distribution on a local scale being apparently unrelated to economic copper mineralization. Thus, primarily for these reasons, it was decided early in 1966 to give the geochemical anomalies the highest priority in the drilling program. Later that year, in the course of testing one such geochemical high located several kilometres to the southeast of the original prospect area, ore-grade mineralization was encountered in three of four X-ray holes.

Soon after these first shallow intercepts were confirmed with deeper drilling, the exploration tempo gradually built up, and by 1968 there were four large diamond-drill rigs working on the property. When drilling of the deposit was completed in May 1969, a total of 128 holes had been drilled for an aggregate length of over 35 000 m. The results of the drilling program outlined a deposit containing 240 million tonnes of copper and molybdenum ore, grading 0.52 per cent copper and 0.017 per cent molybdenum, extending from surface to a depth of 300 m, along a strike length of about 1700 m. Mill construction began in late 1969 and the first shipment of concentrates was made in December 1971. The current production from the mill is approximately 34 500 tonnes per day, and the current reserves (1976) are about 230 million tonnes.

GEOLOGY OF NORTHERN VANCOUVER ISLAND AND THE ISLAND COPPER DEPOSIT

Regional Geology

The rocks of Vancouver Island, north of Rupert and Holberg Inlets, are primarily volcanic and sedimentary units ranging in age from Upper Triassic to Lower Cretaceous (Muller et al., 1974). Intrusive rocks of early to middle Jurassic age intrude the central and upper portions of the North Island sequence and locally are associated with copper-molybdenum mineralization.

The Karmutsen Formation is the oldest rock unit in the area and constitutes the largest volume of the various units. Rocks of the Karmutsen are predominantly porphyritic and amygdaloidal basalt flows. The next oldest unit is the Quatsino Formation, which is generally a massive limestone with rare interbeds of tuffaceous material. Following the Quatsino Formation is the Parson Bay Formation. This is a transitional unit that separates the Quatsino Formation from the Bonanza Formation. The Parson Bay Formation is comprised of black calcareous siltstones, shales and limestones with shaly interbeds.

The Bonanza Formation is mid to lower Jurassic in age. The volcanic unit consists of bedded and massive tuffs, formational breccias and rare amygdaloidal and porphyritic flows. Porphyritic dykes and sills intrude the lower part of the unit. Unconformably over the Bonanza lie rocks of Lower Cretaceous age, consisting primarily of conglomerates, siltstones, sandstones and greywackes.

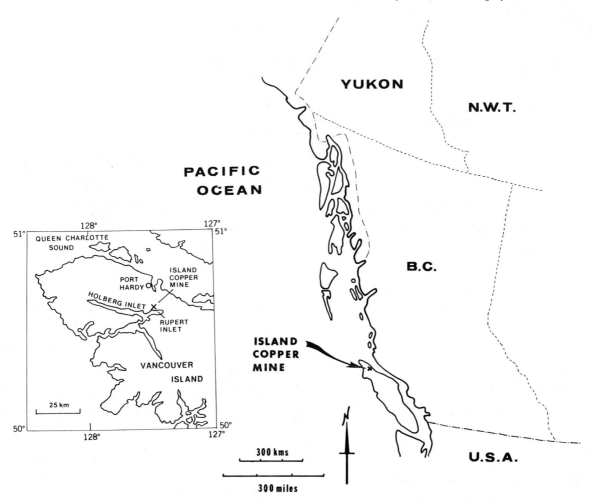

Figure 32.1.   *Location map — Island Copper Mine, northern Vancouver Island, British Columbia.*

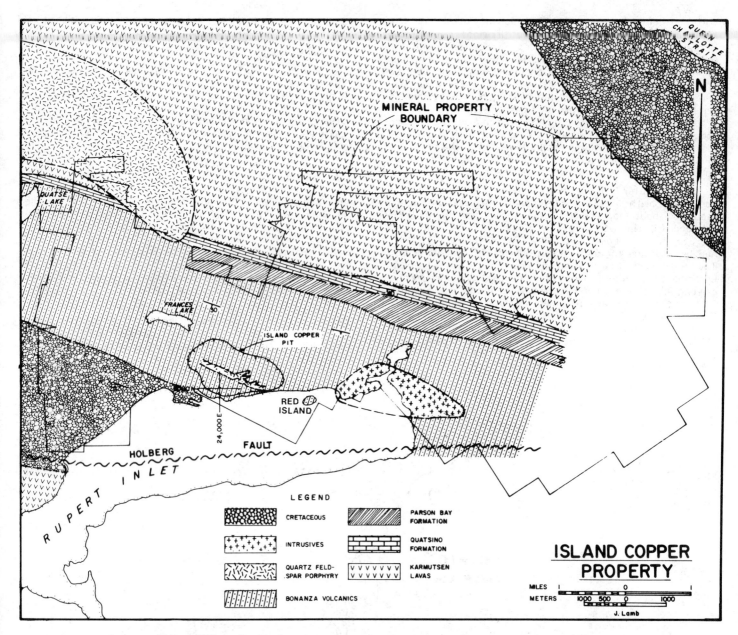

Figure 32.2. Regional geology of Island Copper Mine (after Cargill et al., 1976).

A northwest-trending zone of intrusive stocks extends from the east end of Rupert Inlet to the Queen Charlotte Sound in a zone about 56 km long and up to 6.5 km wide. The stocks range in composition from diorite to quartz monzonite.

Quartz-feldspar porphyry dykes occur along the southern edge of the zone of stocks. These dykes tend to be quite small, usually less than 30 m wide and 150 m long. Felsic dykes and sills occur around the margins of some of the intrusive stocks. They are generally a few metres wide and less than 100 m long and cut through rocks of the Karmutsen, Quatsino and Bonanza formations. Andesite dykes which cut the Karmutsen, Quatsino and Parson Bay formations are apparently feeders to Bonanza-age volcanism.

Geology and Mineralogy of the Island Copper Deposit

The geology on the north side of Rupert Inlet is a representative section of the rocks just discussed in the regional setting. The rocks increase in age going northeasterly from the shoreline and dip at approximately 30° to the south. The regional strike is about N70°W, with several other prominent northwesterly and northeasterly trends superimposed on the regional strike. A small granodiorite stock lies at the east end of Rupert Inlet (Fig. 32.2). The mineralized quartz-feldspar-porphyry dyke at ISCU is 4 km to the northwest from this stock along a N70°W bearing. the ISCU dyke lies in the upper part of the Bonanza Formation and dips to the northeast approximately normal to the bedding of the Bonanza rocks. Several kilometres to the northwest of the ISCU dyke lies another granodiorite body of larger proportions than the one at the east end of Rupert Inlet. This intrusive body outcrops lower in the geological section and lies predominantly in the Karmutsen Formation.

The ISCU deposit lies in the upper portion of the Bonanza Volcanic Formation. The deposit is localized in and adjacent to a quartz-feldspar-porphyry dyke that strikes at N70°W, and dips at about 60° to the northeast (Fig. 32.3). The dyke at the pre-mining bedrock surface has a strike length of slightly over a kilometre and an irregular width that varies from 180 m to less than 20 m. The dyke is granodiorite in composition, the same as the stock at the east end of Rupert Inlet. These two bodies are believed to be genetically of the same origin.

Several periods and directions of faulting are apparent in the vicinity of the deposit. The most prominent structure adjacent to the deposit is the End Creek fault (Fig. 32.3), which dips steeply to the south on the southwest side of the dyke. This fault is believed to be post-mineralization and has an unknown magnitude and direction of offset. The End Creek fault has a width of approximately 60 m and has a strike that has been mapped and inferred for a length of several kilometres. A second major fault zone strikes N65°E, roughly through the centre of the deposit (Fig. 32.3). As with the End Creek fault, this structure has a steep dip and possesses an unknown amount of offset.

Two major stages of metamorphism have been recognized at ISCU (Cargill et al., 1976). The first stage is called the contact metamorphic stage, which occurred when the dyke initially intruded the volcanic pile. The major alteration minerals formed in the host rocks at this time were biotite, chlorite and epidote, in a direction moving away from the dyke. After this initial alteration occurred, the hydrothermal system around the dyke changed as the dyke cooled. As a result of further heating, a second stage of alteration — the wall rock stage, took place. The characteristic minerals formed during this stage were chlorite, sericite, pyrophyllite, dumortierite and a rusty-orange dolomite. This last alteration sequence affected both the quartz-feldspar-porphyry dyke as well as the surrounding volcanic rocks.

There are three associated breccia phases identified with the deposit: (1) pyrophyllite; (2) marginal; and (3) Yellow Dog breccia. The pyrophyllite breccia lies over the top of the dyke in a hood-like fashion at the northwest end of the deposit and extends over 1100 m in a northwesterly direction. The breccia fragments originate both from the dyke and from highly altered volcanic rocks. The marginal breccias lie on both sides of the dyke in a sheet-like fashion and extend down-dip to a considerable depth. Composition varies from predominantly volcanic fragments, away from the dyke, to almost totally quartz-feldspar-porphyry fragments, adjacent to the dyke. The Yellow Dog breccia derives its name from the rusty-brown colour of veinlets of ferro-andolomite. The Yellow Dog breccia cuts through the middle of the deposit and has north and northeast trends. The fragments are primarily highly altered volcanic rocks with quartz and carbonate vein fillings.

The ore minerals at Island Copper are chalcopyrite and molybdenite. The orebody is composed of a hanging wall zone and a footwall zone around the quartz-feldspar porphyry dyke. The ore zone, on both sides of the dyke, is about 60 m to 180 m wide and 1700 m long, extending to over 300 m below the surface. The bulk of the ore is found in the volcanic rocks adjacent to the dyke, although some parts of the dyke are mineralized. Most of the chalcopyrite occurs in veinlets on slip surfaces and occasionally as disseminations. Molybdenite occurs primarily on slip surfaces, but occasionally in quartz veinlets.

Other prominent minerals are pyrite and magnetite. Pyrite is two to three times more abundant than chalcopyrite in the ore zone and is also locally abundant in the alteration halo around the dyke. The mode of occurrence for pyrite is disseminations and fracture fillings. Magnetite occurs in both the volcanic rocks and in the dyke. In the volcanic rocks the magnetite is found as small disseminations and with chlorite pseudomorphs of mafic minerals. There appears to be very little primary magnetite in the Bonanza volcanics themselves; therefore, the bulk of the magnetite is believed either to have been introduced, or more likely, altered from original iron-rich minerals. Recent magnetic susceptibility studies confirm in a qualitative way, the affinity between magnetite and chalcopyrite in the area (J. Flemming, pers. comm., 1977).

GEOPHYSICAL SURVEYS OF THE ISLAND COPPER DEPOSIT

Aeromagnetic Survey

The Federal/Provincial aeromagnetic survey was flown on north-south lines with a nominal 800 m (0.5 mile) line spacing. The total magnetic field was recorded at a mean elevation of 305 m above the ground surface. Results were diurnally corrected and compiled on 1:63 360 scale topographic maps (Anonymous, 1964) with a 10 gamma minimum contour interval. The major feature of the aeromagnetic survey contours north of Holberg and Rupert Inlets (see Fig. 32.4) is a northwesterly-trending magnetic high which correlates closely with the line of Jurassic stocks discussed in the Regional Geology portion of this report. The strongest magnetic responses occur along the edges of the stocks in contact with the Karmutsen rocks. Some skarn highs also appear over the Quatsino Limestone Formation, but these are generally small. The smaller stocks are generally homogeneous, low amplitude highs on the airborne data. Some of the larger stocks however, show considerable magnetic zoning inside their boundaries. At the far southeasterly end of the major axis of the magnetic high and offset slightly to the southwest, lies a small, isolated high. The ISCU deposit underlies the southern part of this magnetic anomaly.

The aeromagnetic contours around the ISCU deposit (Fig. 32.4) show a prominent anomaly centred just northwest of the pit. The anomaly as defined by the 4000 gamma contour is shaped like a lopsided arrow head, with the southern side drooped down over the top of the deposit. The peak values of the zone lie roughly parallel to the north shore of Frances Lake and extend some distance to the west and east. The complete anomaly is approximately 6 km long, striking N65°W, with a width varying from 1 km to 1.5 km. A smaller high dominates the southeast portion of the overall feature and is situated at the edge of Rupert Inlet. Several kilometres north of the ISCU high lies a larger but somewhat less intense anomaly. This correlates with a granodiorite stock that straddles the Karmutsen and Quatsino formations. A third and much weaker high occurs at the east end of Rupert Inlet and correlates with the southeast edge of the granodiorite stock mapped in the area.

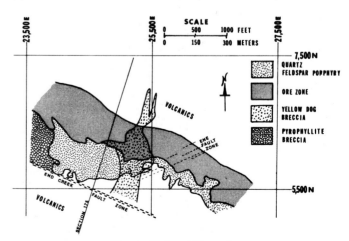

Figure 32.3. Pit geology.

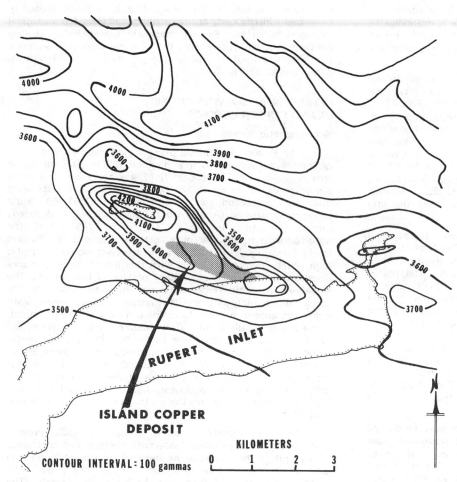

Figure 32.4. Aeromagnetic survey over the Island Copper mine area (from Aeromagnetic Map 1734G, Geol. Surv. Can.).

Ground Magnetic Surveys

The ground magnetic surveys were carried out along the 160 km of survey grid by several two-man crews who took both magnetic readings and obtained soil samples. To extend the survey data, 50 stations were observed over Frances Lake using an aluminum boat, so as to better define the anomaly immediately north of the lake. The instruments used were a Jalander fluxgate and an Askania Gfz torsion magnetometer — both are vertical field instruments with repeatable sensitivities of about ± 20 gammas and ± 5 gammas respectively. The normal diurnal corrections were applied to the readings and the results plotted and contoured on 1:2400 scale base maps. A background level of 1000 gammas was selected to reference the local magnetic variations against.

The ground magnetic survey resolved the aeromagnetic anomaly observed over the ISCU deposit into two distinct zones of positive magnetic anomalies (Fig. 32.5). The largest and most continuous of the two zones lies several kilometres to the northwest of the orebody. This zone is called the Frances Lake Zone, having a strike of N73°W, a length of 2.8 km and an irregular width that averages about 300 m inside the 2500 gamma contour. The magnetic relief along the Frances Lake Zone is locally 4000 gammas above background, with the most intense portion just northwest of the lake. This is the area that Milbourne originally investigated with a dip-needle instrument. The other zone of positive magnetic anomalies is called the Pit Zone. This zone is roughly parallel to the Frances Lake Zone, but is offset about 700 m to the southwest with an overlapping stagger of 1400 m. The Pit Zone strikes at N65°W and has an observed length of 3.3 km. Observed length is a significant qualification since the southeast end of the zone ends at the shoreline. Red Island is located on the same southeast trend at a distance of 1.2 km from the shoreline (Fig. 32.2). On the aeromagnetic map (Fig. 32.4) the extension of the Pit Zone to include Red Island is apparent so the total length of the zone, both on land and underwater, is almost 5 km. The Pit Zone's width as defined by the 2500 gamma contour is variable but is generally slightly less than 700 m (Fig. 32.6). Four major anomalies stand out in the zone, of which the two central highs are located directly over the orebody. A third high lies off the northwest end of the deposit and the fourth high is a partially defined zone lying along 1100 m of shoreline.

Even though the Frances Lake and Pit Zones are spatially quite close to each other, there are several differences in their magnetic expressions. In a semi-quantitative sense, the differences can be listed as:

1. The Frances Lake Zone is quite continuous in appearance; it is not broken-up like the Pit Zone.

2. An examination of the anomaly frequencies over the two zones shows the magnetic sources in the Frances Lake Zone are very near surface, while in the Pit Zone, particularly at its northwest end, the bulk of the causative magnetic body is located at a substantial depth. The anomaly over the northwest end of the pit has a source depth based on the half-width rule (Riddell, 1966) of about 215 m.

With regard to the first point of difference, corroborating information was obtained from a detailed magnetic susceptibility mapping program on core and outcrop in the pit at various bench levels. The instrument used was a small, hand portable magnetic susceptibility meter model PP-2A, manufactured by Elliot Geophysical Company, Tucson. The survey results in general substantiated the interpretation of the surface magnetic data in that the distribution of magnetite around the orebody was found to be quite irregular. Drilling of the Frances Lake and Pit Zones both before and after the exploration phase substantiated the second observation — early testing of the Frances Lake Zone encountered magnetite at a shallow depth, while more recent drilling at the northwest end of the pit found the percentage of magnetite to increase sharply 240 m below the pre-mining surface (vs 215 m based on the half-width rule).

Although there are apparent differences between the Frances Lake and the Pit Zones, there is also evidence from the magnetic data suggesting that the two zones are closely related in origin and may at one time, have been a more or less continuous feature. Support for this hypothesis comes in part from the geometrical relationships between the magnetic highs and lows around the deposit. The lows of particular interest are two magnetic troughs that trend N15°E; one commences from the southeast end of the Frances Lake Zone and strikes to the northeast and the other commences 1400 m to the northwest of the Pit Zone and extends to the southwest (Fig. 32.5). The separation between

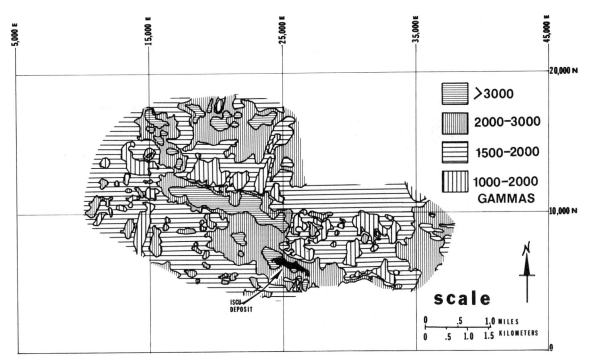

Figure 32.5. Regional ground magnetic survey.

the troughs on a northwest-southeast line is about 1400 m and the displacement on a northeast-southwest line is about 730 m. These figures are almost identical to the displacement and offset values between the Frances Lake Zone and the Pit Zone. The conclusion then is that the two zones of magnetic high might well have originally formed together along the same structural trend but that post-mineralization faulting has displaced the zones in a right lateral sense to the present positions. The origin of the magnetic lows is not certain, but they are probably due to the presence of a major fault that was once active, resulting in the local destruction or remobilization of magnetite. One other prominent N15°E magnetic trend is a line normal to the southeast end of the Frances Lake Zone. If this lineament is carried to the southwest, it lines up with the Yellow Dog breccia in the pit, which, in turn separates the two magnetic highs in the deposit.

Other magnetic features of interest include the End Creek fault which lies on the southern flank of the Pit Zone. This structure appears in profile form on a number of lines as a positive peak with a distinctive negative on the south side. The amplitudes of the positive and negative varies considerably from line to line but the pattern indicates local removal of magnetite along the fault zone with some remobilization to the footwall side. This signature lies on the south side of the larger magnetic highs caused by deeper sources from within the deposit. Another area of interest is around the two highs located in the northern and northwestern portion of the survey area. The anomaly in the northwest part of the grid appears to be an extension of the west end of the Frances Lake Zone. Interestingly, the northwest zone commences with a dog-leg from the Frances Lake Zone at a bearing of N29°W. The northern part of the outlined anomaly lies adjacent to a granodiorite stock mapped off the northwestern corner of the survey grid. The northern anomaly correlates with a mapped diorite stock and the magnetic response seems to be due to the presence of original magnetite in the rock.

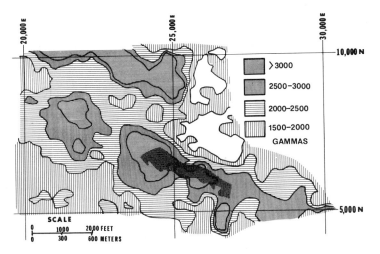

Figure 32.6. Pit area ground magnetic survey.

Induced Polarization Survey

The equipment used for all the induced polarization (IP) exploration work carried out on the ISCU property was a Hewitt 100 time-domain IP system. This unit is a physically-linked transmitter-receiver system with a rated output of 800 watts. The receiver integration periods are similar to the Newmont standard (Dolan, 1967) and in comparison field tests, the conversion factor between the two systems was about 1:1. The Wenner electrode array was used except for the last survey over the orebody when the pole-dipole array was utilized. A regional survey along the logging roads used dipoles of 152 m and 183 m spacing while the pole-dipole work used a potential electrode spacing of 61 m with N=1 and N=2 separations. The extremely damp conditions on the property, especially during the winter months, caused the IP equipment to malfunction a number of times which was partly the reason why the IP method was not utilized more in the initial exploration stage.

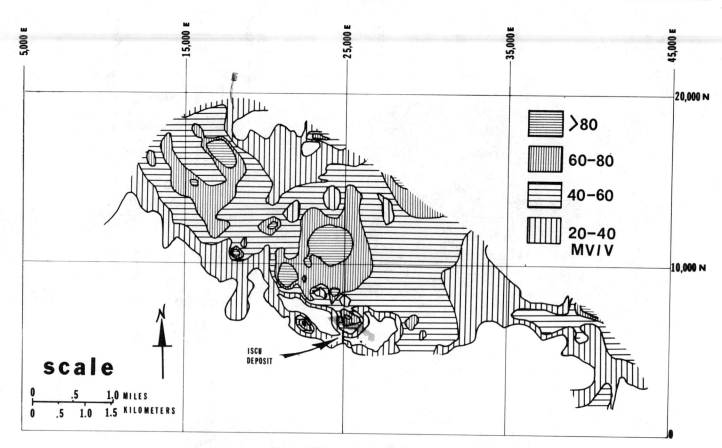

Figure 32.7. Regional IP survey.

The induced polarization technique was used initially to investigate specific areas of geological and geochemical interest around the west end of Frances Lake. Numerous anomalies were obtained and it was found by subsequent drilling that the best IP highs were due mostly to pyrite. Although this information was useful in an indirect manner, in that it could help outline zones of alteration, the magnetic and geochemical survey results provided much less expensive and more diagnostic means by which to locate copper mineralization. Largely for this reason, little IP work was done during 1966 and once the deposit was found, only the immediate area of the deposit was surveyed using IP.

In 1968, as part of a program to evaluate the remaining ground held by Utah Mines Ltd. around the deposit, a regional IP survey was conducted utilizing the extensive system of logging roads on the property. The results of this survey (Fig. 32.7) shows a large zone of anomalous chargeability as outlined by the 40 ms contour trending at N70°W across the property. The zone has an observed length of 6.4 km and a width of about 2 km. The anomaly ends 1500 m past the southeast end of the pit and is open to the northwest. There are several clusters of intense highs (>80 mv/v) within the overall zone. These clusters vary in size, but there appear to be at least four pairs with the members aligned at about N50°E ±10°. Interestingly, these pairs of IP highs all straddle anomalous magnetic zones. The Frances Lake Zone is straddled by two such pairs and the magnetic high in the northwestern corner of the property is almost encircled by an IP high, open only to the northwest.

The resistivity results provide little in the way of unique diagnostic information, either on the structure or mineralization in a regional context. The dominant feature is a northwesterly-trending break running roughly across the property, as indicated by the 500 ohm-metre contour. Northeast of the break, resistivities are generally greater than 500 ohm-metres, while to the southwest of the contour, they tend to be less than 500 ohm-metres. This corresponds roughly to the contact between the Bonanza-Parson's Bay formations to the southwest and the Quatsino-Karmutsen formations to the northwest.

The results shown in Figure 32.8 (chargeability) and Figure 32.9 (resistivity) are the N=1 data from the detailed pole-dipole survey over the deposit. The chargeability results around the deposit show two bands of high straddling the dyke and the pyrophyllite breccia cap on a more or less continuous basis, with the high on the north side of the dyke being much more intense and extensive than the zone on the south side. This difference is felt to be probably due in part to vertical movement along the End Creek Fault which has removed most of the pyritized wall rock on the southwest side of the dyke.

It is apparent from Figure 32.8 that there are several pairs of chargeability highs straddling the orebody and the pyrophyllite cap in a similar fashion to those already noted in the regional results. The pairs of anomalies around the deposit are noticeably smaller in area and somewhat less in amplitude than the pairs around Frances Lake and in the northwest corner of the property, but they show the same northeasterly alignment and similar straddling of magnetic highs.

Besides the major zone of anomalous IP, a second anomalous zone is noted in Figure 32.7 at the far southeastern corner of the survey. This anomaly is adjacent to the granodiorite stock at the east end of the Inlet, and could be considered a continuation of the main chargeability zone, interrupted by Rupert Inlet.

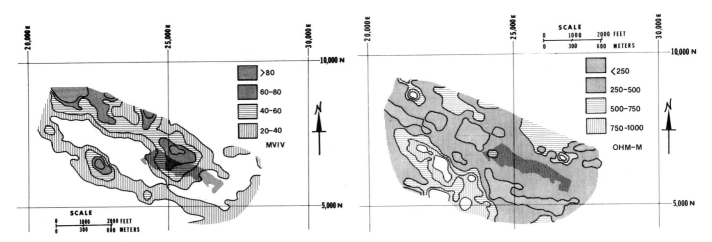

Figure 32.8. Pit area IP survey.

Figure 32.9. Pit area resistivity survey.

In assessing the IP results, consideration must be given to the unknown dilution effects caused by the large variations in overburden thickness around the deposit. The chargeability anomalies on the ISCU property are felt to be due to a combination of sulphide and non-sulphide responses. Pyrite is present both in the ore and in the adjacent rocks as noted earlier. Together with chalcopyrite, sulphides in the pit total about 5 per cent combined, whereas pyrite reaches 15 per cent locally north of the deposit. Alteration clays around the deposit also appear to make a significant contribution to the observed chargeability response. Interestingly, the pyrophyllite breccia contains a small but significant amount of finely disseminated pyrite. On the IP results, however, the breccia cap stands out as a pronounced chargeability low. Some graphitic material is found in sediments north of Frances Lake, however the sediments are spatially quite restricted and are not believed to have contributed significantly to the overall IP picture.

The regional sulphide zoning as indicated by the IP results with respect to the magnetite in and around the deposit, appears to be an interesting variation of the pyrite-shell-IP-donut pattern sometimes suggested for the Southwest United States-type porphyry copper deposits. Geologically, the mineralization at ISCU appears to have been emplaced when an extensive zone of sulphides was first introduced along a northwesterly grain. This was followed by the alteration of some of the sulphide iron to magnetite. At about the same time, faulting (probably) disrupted the original northwesterly zone of weakness and some sulphides were selectively redistributed along northeasterly structures. Later, post-mineral faulting then further dislocated the sulphide distributions to their present position.

Around the pit area, the resistivity results (Fig. 32.9) were a little more diagnostic than those obtained in the regional survey. A portion of the End Creek Fault correlated with a resistivity low on its hanging wall side. Northwest and slightly southwest of this low, a pronounced zone of high resistivity shows up with a low on its northeastern side. The pyrophyllite breccia cap extending from the northwestern end of the dyke is located between the resistivity ridge and trough and where the cap apparently ends, so does the ridge-trough resistivity feature. The character over the deposit itself is quite uniform, with the resistivity values ranging from 200 to 300 ohm-m.

A typical cross-section through the deposit is shown in Figure 32.10. This figure presents the magnetic, geochemical, chargeability and resistivity profiles across the orebody and quartz feldspar porphyry dyke, and illustrates the spatial relationship of the ore with the geophysical and geochemical data.

VLF-EM Survey

In the early drilling around the southwestern corner of Frances Lake, some massive copper mineralization was found in a near-surface vein system. To try to assist in the detection of such mineralization, a limited VLF-EM survey was run over the area. A Geonics EM-16 receiver was employed with the Jim Creek (Seattle), Washington station (48°12'N, 121°55'W) being used as the transmitter. This transmitter provided a good alignment with respect to the regional geological strike on the property. After some initial positive results over mineralization around Frances Lake, it was decided to conduct a survey over the newly found orebody using the EM-16. Although the overburden generally had a fairly high resistivity, there were many poor surface conductors (swamps, beaver ponds, etc.) where the EM-16 could be expected to respond. The VLF-EM results are presented in Figure 32.11 and because of their low amplitude character, little in the way of quantitative information was gained from the results at the time of the survey. Subsequent examination of the data showed that several sets of conductors axes could be correlated with legitimate structural and lithological features. Three of the most distinct sets of crossovers were adjacent to magnetic highs and two of the three conductor axes straddled the orebody. One other major conductor axis picked out the edge of the Bonanza volcanic-Cretaceous sediment contact west of the pit. This contact can be traced for a distance of almost 2 km, based on the EM-16 results. Most of the survey was carried out along grid lines 152 m apart with survey stations every 30.5 m. This type of coverage although likely too coarse for massive sulphide detection, gave satisfactory results in mapping major structures.

GEOCHEMICAL SURVEY OF THE ISLAND COPPER DEPOSIT

The geochemical survey of the ISCU property was carried out in conjunction with the ground magnetic survey during the first half of 1966. Soil samples were taken every 30.5 m along lines that were 152 m apart (the same distance used in the magnetic survey). A total of 4203 samples were collected along 160 km of grid.

The area around the ISCU deposit has a variable cover of glacial till, peat and moss which ranges in thickness from less than one metre to over 75 m. The soil samples were taken with a mattock in areas where the organic cover was thin and with a 1.2 m long auger where the cover was thicker. The B soil horizon was sampled wherever possible. In all, however, 22 per cent of the stations could not be sampled due to excessive organic cover, lakes or lack of soil development over bedrock.

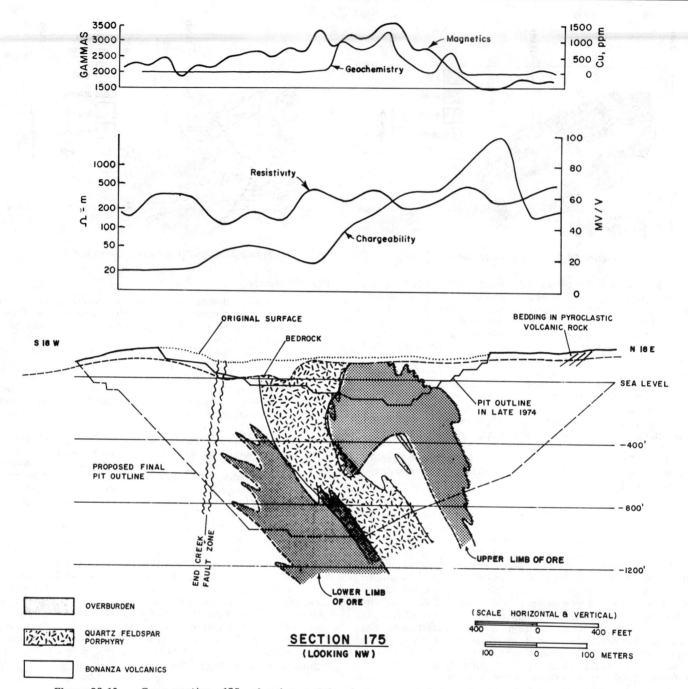

Figure 32.10. Cross-section 175; showing geochemical, ground magnetic, IP and resistivity profiles.

The soil samples were first dried, ground and then sieved through an 80 mesh screen. The residues were then analyzed spectrographically for total copper content in parts per million (ppm). In order to assist in interpreting the geochemical results, some statistical treatment of the data was applied. A semilogarithmic frequency distribution histogram of the geochemical data is presented in Figure 32.12. Based partly on the statistical analysis of the geochemical, but more on empirical results derived from the drilling, an anomalous threshold of 100 ppm was decided upon. Values between 100 to 200 ppm were considered slightly anomalous, between 200 to 500 ppm were anomalous and above 500 ppm were considered significantly anomalous.

Several large anomalous zones were identified inside the survey block (Fig. 32.13), the largest being a N75°W trending zone located roughly in the west central portion of the property (approximately 10 000W to 15 000N, 10 000E, to 25 000E). The zone is composed of a number of discrete anomalies linked together by an envelop of above-background values. The zone terminates quite abruptly to the east but splits into a number of small parallel-trending zones in the west. Several relatively discrete geochemical anomalies with approximately N70°W trends were found to lie south of Frances Lake (Fig. 32.2), one at the southwest end of the lake centred at 9460N, 18 700E and the other several kilometres to the southeast centred at 6720N, 24 450E.

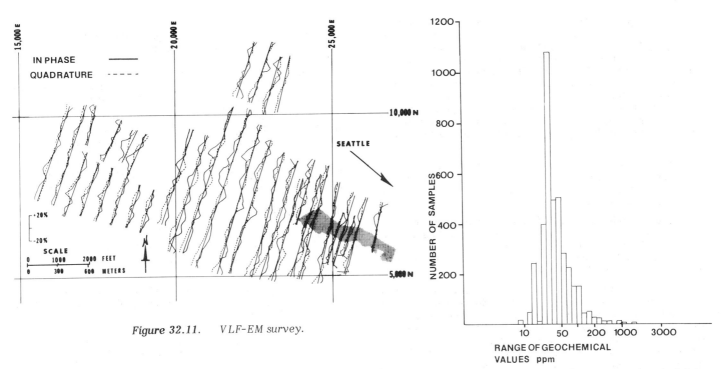

Figure 32.11. VLF-EM survey.

Figure 32.12. Histogram of copper geochemical data.

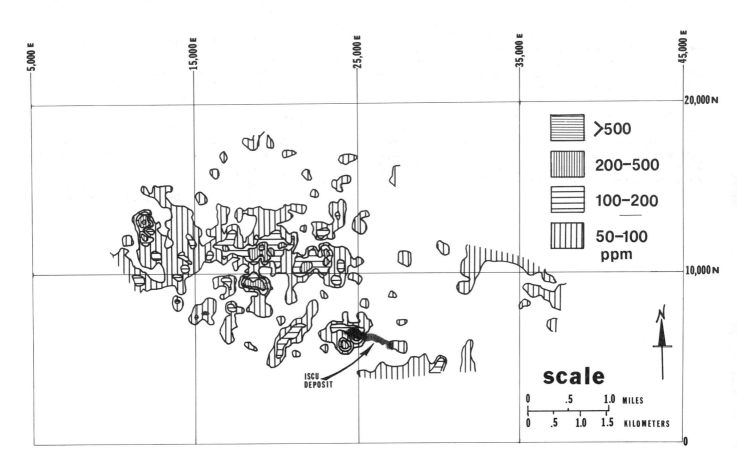

Figure 32.13. Regional geochemical survey.

The major geochemical anomaly, just north of Frances Lake, was tested in a number of places along strike and the underlying bedrock was found to be mostly volcanic rock containing subeconomic copper (0.2% to 0.25% range) with a variable amount of magnetite. The Frances Lake magnetic zone is coincident with a large part of the geochemical zone and based on the observed correlation between copper and magnetite, it was felt that considerable low-grade copper might occur along the length of the magnetic anomaly. A limited amount of drilling has tended to bear this hypothesis out.

The geochemical anomaly southwest of Frances Lake, coincides in part with the more massive vein-type mineralization as was first located by Milbourne and later confirmed from drilling by Utah Mines Ltd.

The anomaly southeast of Frances Lake is centred over a portion of the ISCU deposit. Subsequent drilling of this anomaly indicated the area of the anomaly above the 200 ppm level corresponded fairly well with the part of the orebody overlain by less than 9 m of overburden.

Testing of other geochemical anomalies on the property most often encountered some low-grade copper mineralization in the bedrock. This observation encouraged the reliance upon the geochemical anomalies as being a prime source of drill targets, since, even though much of the glacial till had been transported, a significant amount of the soil at a given site was apparently locally derived. Extensive drilling of the anomaly over the deposit did show, however, that more than 15 m of the overburden could inhibit the surface expression of even subcropping ore-grade mineralization.

CONCLUSIONS

Of all the survey techniques carried out in the discovery of the Island Copper (ISCU) deposit, a soil geochemical survey was the most successful due to its advantages of speed, cost and detection of the specific element of interest. The problem of thick overburden subduing the geochemical expression was apparent on the property but did not present a serious problem.

Magnetic surveys were the most useful of the geophysical tools used at ISCU; an aeromagnetic anomaly initially focused attention on the area around ISCU, and later both during and after the exploration phase, ground magnetic surveys provided invaluable information concerning the structure and mineralization around the deposit.

Induced polarization surveys did not compare with geochemical and magnetic surveys in their effectiveness in directly locating mineralized bedrock at ISCU. However, after a regional perspective using IP was obtained, it allowed more detailed work to be put into an intelligible context, which in turn directed attention to some unique structures and alteration effects in the area of the deposit.

A VLF-EM survey demonstrated its usefulness as an aid to tracing structures, even in the presence of thick overburden and numerous poor near-surface conductors.

The author would like to acknowledge permission of Utah International Inc., to publish and to thank the various people who have assisted in the preparation of this paper.

REFERENCES

Anonymous
  1964: Aeromagnetic Map 1734G, Geol. Surv. Can., Aeromagnetic Series, Scale 1:63 360.

Cargill, D.G.
  1975: Geology of the "Island Copper" Mine, Port Hardy, British Columbia; unpubl. Ph.D. thesis, Univ. British Columbia.

Cargill, D.G., Lamb, J., Young, M.J., and Rugg, E.S.
  1976: Island Copper; in Porphyry Deposits of the Canadian Cordillera; Can. Inst. Min. Metall. Special Vol. 15, p. 206-218.

Dolan, W.
  1967: Considerations concerning measurement standards and design of pulsed IP equipment; in Proceedings of a Symposium on Induced Polarization, Univ. California, Berkeley.

Hewitt, L.
  1965: Hewitt 100 I.P. System Technical Manual.

Muller, J.E. et al.
  1974: Geology and mineral deposits of Alert Bay — Cape Scott map area, Vancouver Island, B.C.; Geol. Surv. Can., Paper 74-8.

Riddell, P.A.
  1966: Magnetic observations at the Dayton iron deposit, Lyon County, Nevada; in Mining Geophysics, v. 1, Case Histories, Soc. Explor. Geophys., p. 418-428.

Young, M.J. and Rugg, E.S.
  1971: Geology and mineralization of the Island Copper deposit; in Western Miner, v. 44 (2), p. 31-38.

GEOPHYSICAL AND GEOCHEMICAL CASE HISTORY OF THE QUE RIVER DEPOSIT TASMANIA, AUSTRALIA

S.S. Webster
Abminco N.L., Wayville, South Australia

E.H. Skey
Abminco N.L., Kalgoorlie, Western Australia

Webster, S.S., Skey, E.H., Geophysical and geochemical case history of the Que River deposit, Tasmania, Australia; in Geophysics and Geochemistry in the Search for Metallic Ores; Peter J. Hood, editor; Geological Survey of Canada, Economic Geology Report 31, p. 697-720, 1979.

Abstract

The Que River deposit in northwestern Tasmania, discovered in 1974, comprises several separate massive sulphide lenses located within an area 800 m by 100 m. The lenses occur within a sequence of pyritic dacites and andesites approximately 300 m wide over a strike length of 4 km. The lenses are vertical with an average width of 9 m. One lens is predominantly pyrite and chalcopyrite, the others being predominantly pyrite, sphalerite and galena. Outcrop of massive sulphides is nonexistent.

The exploration area was selected within a well-mineralized belt of Cambrian calc-alkaline volcanics marking the eastern edge of the Dundas Trough. The initial reconnaissance involving geological traverses and stream sediment sampling covered an area of 60 km^2. Several areas of anomalous geochemistry were located in favourable rock types.

Progress of the reconnaissance program and follow-up investigation was impeded by dense rainforest and rugged terrain. Accordingly an airborne electromagnetic survey was flown. Though this technique had not been an ore-finder in Australia, the geophysical environment in Tasmania was such that application of the method was warranted. A conductor was immediately identified in one area of anomalous stream sediment geochemistry.

The target was subsequently delineated by soil geochemistry and ground electromagnetic techniques. Initial drilling proved the conductor to be a single lens of predominantly copper and iron sulphides. Additional drilling intersected a comparatively major zone of zinc, lead and iron sulphides which was not detected by the electromagnetic surveys, but was expressed by soil geochemistry. An integrated orientation survey showed that the induced polarization technique, combined with soil geochemistry, optimized drill target definition.

Résumé

Le gisement de la Que River dans le nord-ouest de la Tasmanie, découvert en 1974, comprend plusieurs lentilles distinctes de sulfures massifs dans une zone de 800 m sur 100 m. Les lentilles se trouvent dans une série de dacites pyriteuses et d'andésites, mesurant environ 300 m de large sur 4 km le long de la structure. Les lentilles sont verticales et ont une épaisseur moyenne de 9 m. Une lentille contient surtout de la pyrite et de la chalcopyrite, et les autres contiennent surtout de la pyrite, de la sphalérite et de la galène. Les affleurements de sulfures massifs sont inexistants. La zone d'exploration a été choisie à l'intérieur d'une zone bien minéralisée formée de roches volcaniques calco-alcalines du Cambrien, marquant la bordure orientale de la fosse de Dundas. La reconnaissance initiale, comprenant les cheminements géologiques et l'échantillonnage des dépôts fluviatiles, a couvert une surface de 60 km^2. Plusieurs zones d'anomalies géochimiques ont été découvertes dans des roches favorables.

Le programme de reconnaissance et les travaux ultérieurs ont été retardés à cause d'une forêt humide et dense et la topographie accidentée. Un levé électromagnétique aéroporté a alors été effectué. Même si cette méthode n'a jusque là pas permis de découvrir de gisements en Australie, le milieu géophysique en Tasmanie se prête à son application. Un conducteur a été immédiatement identifié dans une zone d'anomalies géochimiques à l'intérieur de sédiments fluviatiles. Le conducteur a été délimité par des techniques de levé géochimique des sols et de levé électromagnétique au sol. Un premier forage a montré que le conducteur était une lentille isolée contenant surtout des sulfures de fer et de cuivre. D'autres forages ont rencontré une zone relativement importante de sulfures de zinc, de plomb et de fer qui n'avait pas été détectée par les levés électromagnétiques, mais qui avait été signalée par la géochimie du sol. Une étude systématique a montré qu'en combinant la méthode de polarisation provoquée à l'exploration géochimique des sols, on pouvait très bien définir legisement.

INTRODUCTION

The Que River deposit, within the main mining district of northwestern Tasmania, is situated 2 km east of the Murchison Highway, approximately 25 km northeast of Rosebery (Fig. 33.1).

An Exploration Licence was acquired in 1970 within the Mount Read volcanics, a Cambrian calc-alkaline suite of pyroclastics, lavas and intrusives, which form an arcuate belt 10 km wide and 240 km long marginal to the Precambrian Tyennan nucleus. To the west is the Dundas Trough composed

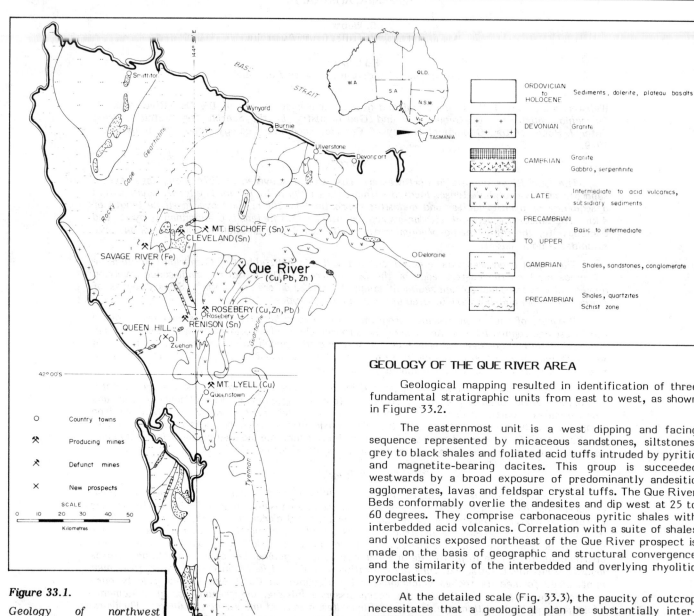

Figure 33.1.
Geology of northwest Tasmania (after Williams, 1976; Gee, 1967).

of late Proterozoic and early Palezoic sediments which are partly a facies equivalent of the volcanics (Corbett et al., 1974).

The Mt. Read volcanics were chosen for exploration as they host the Rosebery Zn-Pb-Cu deposit and the Mt. Lyell Cu-Au deposit. These deposits are important mineral producers in Tasmania and though they have been in production for most of this century, the volcanics have not been well prospected. This is mainly due to the dense vegetation and rugged terrain of the area, however improved access and modern geochemical and geophysical practices, have facilitated exploration of such difficult areas.

GEOLOGY OF THE QUE RIVER AREA

Geological mapping resulted in identification of three fundamental stratigraphic units from east to west, as shown in Figure 33.2.

The easternmost unit is a west dipping and facing sequence represented by micaceous sandstones, siltstones, grey to black shales and foliated acid tuffs intruded by pyritic and magnetite-bearing dacites. This group is succeded westwards by a broad exposure of predominantly andesitic agglomerates, lavas and feldspar crystal tuffs. The Que River Beds conformably overlie the andesites and dip west at 25 to 60 degrees. They comprise carbonaceous pyritic shales with interbedded acid volcanics. Correlation with a suite of shales and volcanics exposed northeast of the Que River prospect is made on the basis of geographic and structural convergence and the similarity of the interbedded and overlying rhyolitic pyroclastics.

At the detailed scale (Fig. 33.3), the paucity of outcrop necessitates that a geological plan be substantially interpreted from the drilling results (Fig. 33.4). The subvertical sequence from east to west consists of, from the bottom, footwall andesitic pyroclastics, unaltered but with traces of sphalerite and galena; a porphyritic dacitic unit containing "stringer" mineralization; heavily pyritized lower dacitic pyroclastics with sericite-carbonate-silica alteration and disseminated to massive base metal sulphides; barren dacitic lavas which form a wedge between the lower sequence and the upper sequence and western ore lenses; several repetitions of barren dacites and mineralized pyroclastics containing the major galena-sphalerite ore lenses; hanging-wall andesitic-pyroclastics, unaltered and virtually devoid of sulphide mineralization. The eastern (S) ore lens consists of bands and veins of coarsely crystalline pyrite, which is also locally framboidal or coloform. Galena, sphalerite and chalcopyrite occur within the pyrite host and associated silica-carbonate gangue. In part this lens is composed of massive pyrite with chalcopyrite only.

The western (P) lenses commonly exhibit bands in the range 1 mm to 1 cm of pyrite, sphalerite and galena with minor chalcopyrite. Framboidal and coloform textures are microscopically visible. Gangue minerals include silica, carbonate, sericite and barite.

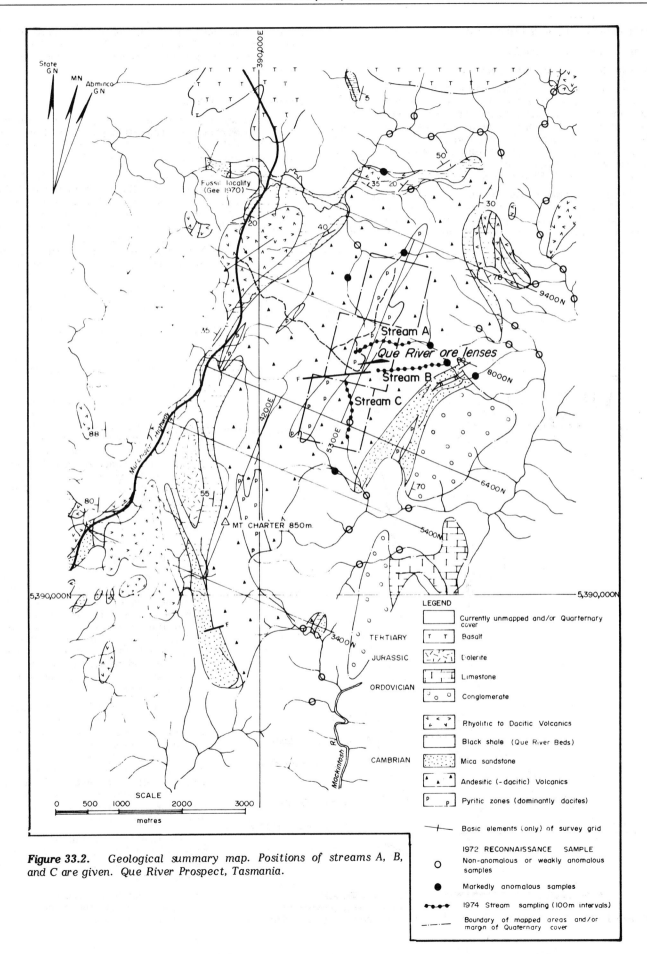

Figure 33.2. Geological summary map. Positions of streams A, B, and C are given. Que River Prospect, Tasmania.

GEOCHEMICAL SURVEY PROGRAM

Stream Sediment Geochemistry

Regional Reconnaissance Program

During 1970-71, 276 stream sediment samples were collected throughout the property with a sample density of approximately 3 to 5 samples per square kilometre. After sieving at minus 20 mesh, the fine grained gravel and silt was pulverized and digested in hot perchloric acid, then analyzed by atomic absorption spectrophotometry (AAS) for copper, lead and zinc. Results from the vicinity of the Que River prospect are shown in Figure 33.5.

Stream sediment values of the order 45 ppm Cu, 300 ppm Pb and 340 ppm Zn occurred in the vicinity of the later identified Que River prospect and were recognizably anomalous in a regional sense. Inspection of the metal values in adjacent samples within the volcanics revealed the local background to be of the order 15-20 ppm Cu, 20-00 ppm Pb, 50-100 ppm Zn, thus further enhancing the character of the anomaly.

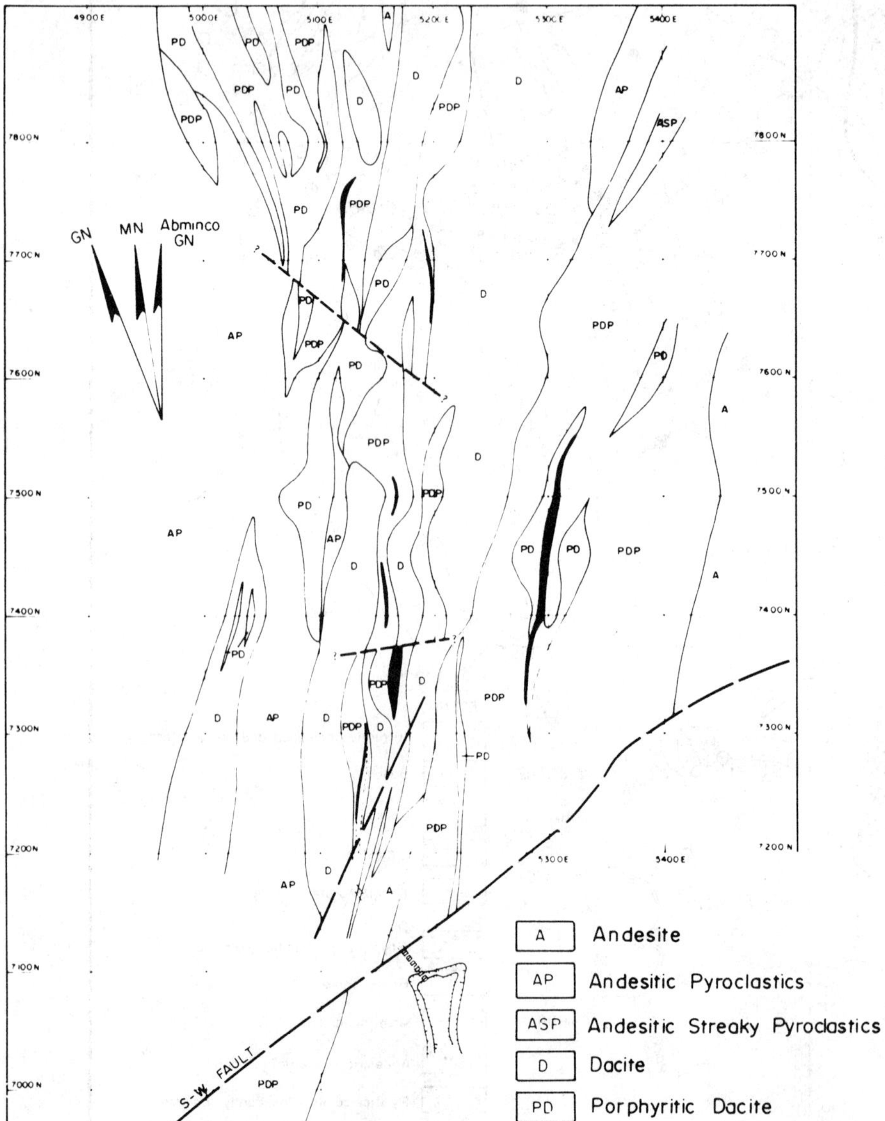

Local Orientation Program

Three streams draining the prospect were sampled in detail for geochemical orientation purposes. This program was conducted concurrently with the grid geophysics and geochemistry in 1974 and also during 1976.

The geographic disposition of streams A, B and C relative to the prospect are illustrated in Figure 33.2. All samples were dried and fractionally sieved. Copper, lead and zinc were determined by AAS after digestion in hot perchloric acid. Iron was determined by stannous chloride titration against standard potassium dichromate following potassium bisulphate fusion and hydrochloric acid dissolution.

The range of results achieved for -200, -100 +200, -40 +100 and -40 (100%) mesh size fractions is shown in Figures 33.6, 33.7 and 33.8. In streams A and B in particular there is a marked tendency for metals to be of greater value in the (progressively) finer fractions. The minus 40 mesh results approximate to an average of the individual fraction analysis.

In general, lead and iron values peak nearer to the source than copper or zinc, reflecting the greater mobility of the latter, though this pattern is confused by probable additions of metals, particularly zinc, from footwall sources. Sampling was not pursued far enough downstream to encompass the complete geochemical dispersion, however broad spaced sampling was confirmed as an acceptable reconnaissance technique.

Soil Geochemistry

During the preliminary program samples were collected at 50 m intervals from the A and C soil horizons using hand screw augers. After drying and sieving to obtain the minus 80 mesh fraction, (for analytical convenience) the samples were digested and analyzed in the same way as the stream sediments. Iron was analyzed as an indicator of pyrite and as a potential lithological marker.

Over sericitized pyroclastics a superficial A horizon soil of black or dark brown humus, 5 to 45 cm thick, is underlain by grey clays which are patchily ironstained and occasionally overlie massive gossan above fresh pyritic rock. This (C) horizon is typically 50 cm to 3 m in thickness and is usually underlain by rotten grey ironstained pyroclastics to a depth of 20 m or more.

Virtually no C horizon soil occurs above silicified zones. Fresh sulphides may be seen by removal of the thin humus-rich surface layer.

On sulphide-poor dacites, pink to fawn clays or feldspathic sands are present beneath the A horizon.

Figure 33.3. Interpretetive surface geology, Que River Prospect, Tasmania.

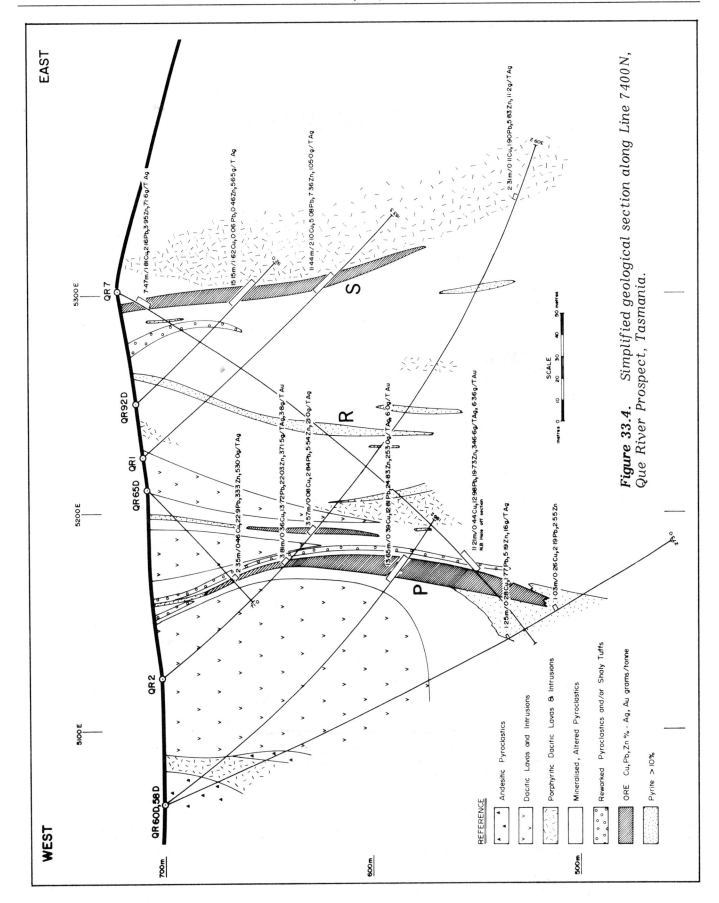

Figure 33.4. Simplified geological section along Line 7400N, Que River Prospect, Tasmania.

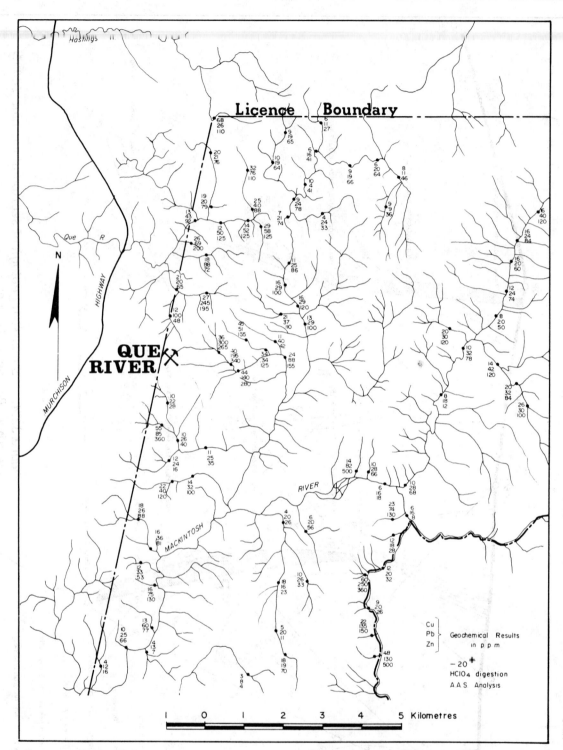

Figure 33.5. Geochemical results for the 1970-71 stream sediment survey, Que River Prospect, Tasmania.

Orange to brown clays with clasts of thoroughly weathered rock lie above weathered andesites. The C horizon may be 50 cm to 3 m thick and the andesites beneath, although compact, are weathered to approximately 20 m with kernels of fresher rock.

The C horizon results for lead (immobile) and zinc (mobile) are illustrated by Figures 33.9 and 33.10. Contour levels were selected by inspection, with the assistance of cumulative frequency plots. The absence of direct correlation between geochemical responses and EM conductors is apparent, however, the marked geochemical relief suggests that the C horizon data is reflecting sulphide occurrences. The sample spacing of 50 m by 50 m was considered too broad to consistently identify a narrow source.

Iron values (Fig. 33.12) in excess of 10 per cent broadly correlate with lead-zinc anomalies and an area of high background values for iron and zinc in the southeastern sector of the grid was identified by mapping as outcropping and

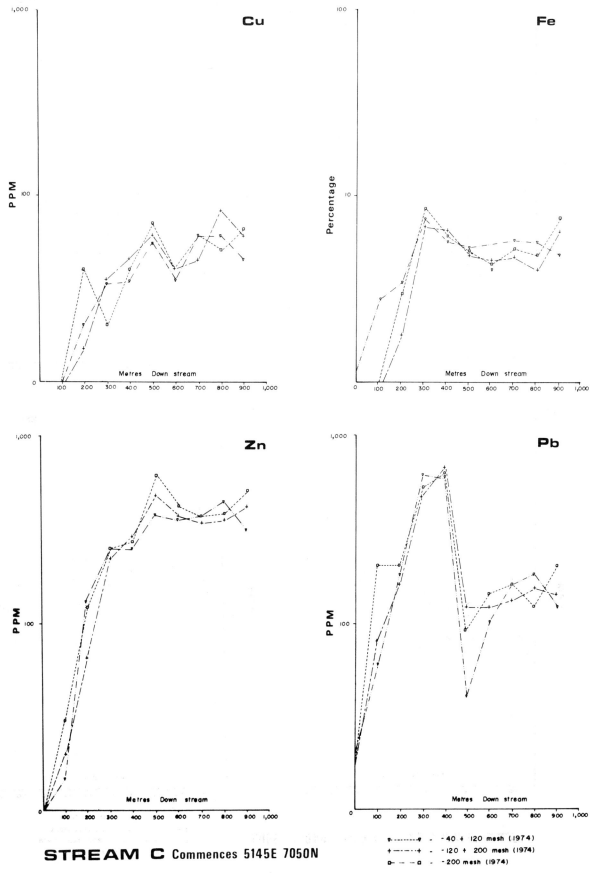

Figure 33.6. *Stream sediment geochemical results for Cu, Fe, Zn, and Pb from Stream C. Que River Prospect, Tasmania.*

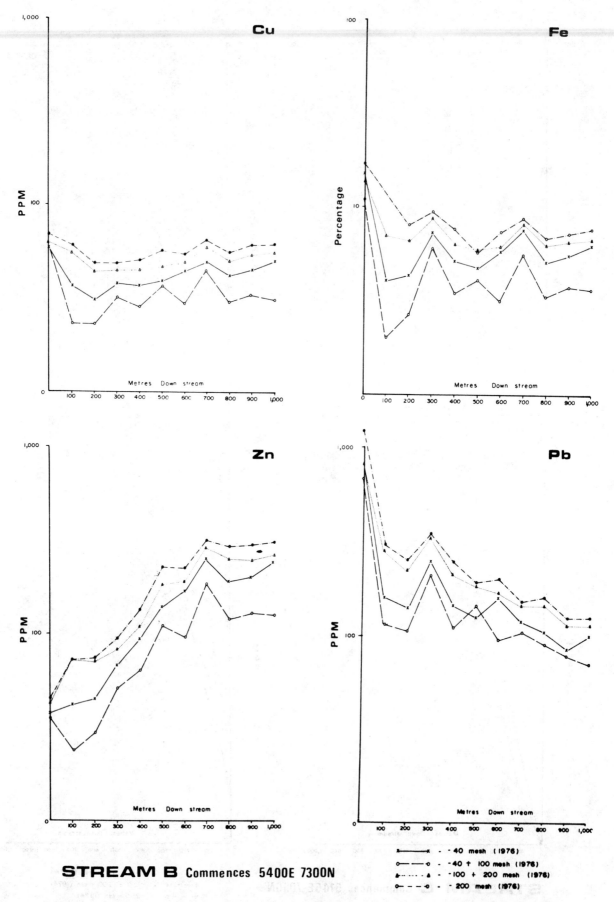

Figure 33.7. Stream sediment geochemical results for Cu, Fe, Zn, and Pb from Stream B. Que River Prospect, Tasmania.

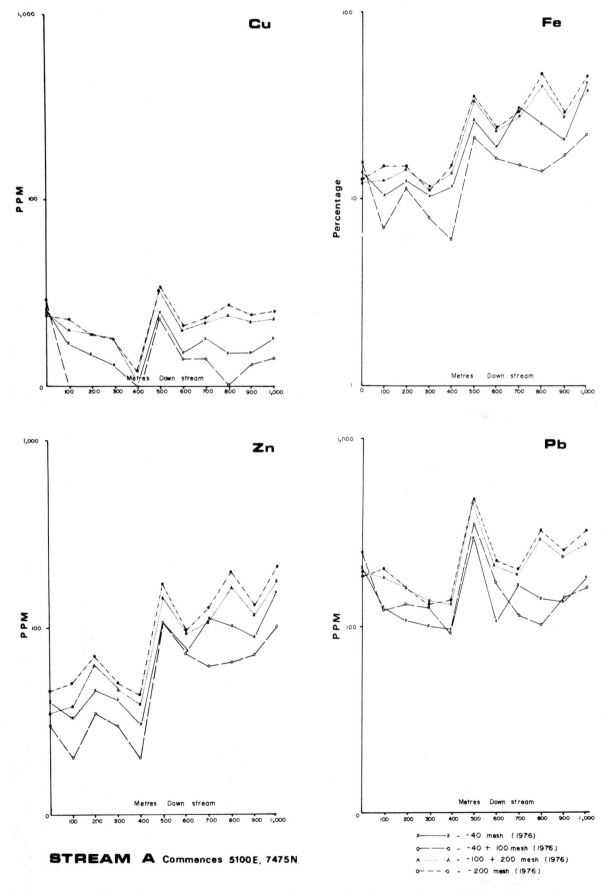

Figure 33.8. Stream sediment geochemical results for Cu, Fe, Zn, and Pb from Stream A. Que River Prospect, Tasmania.

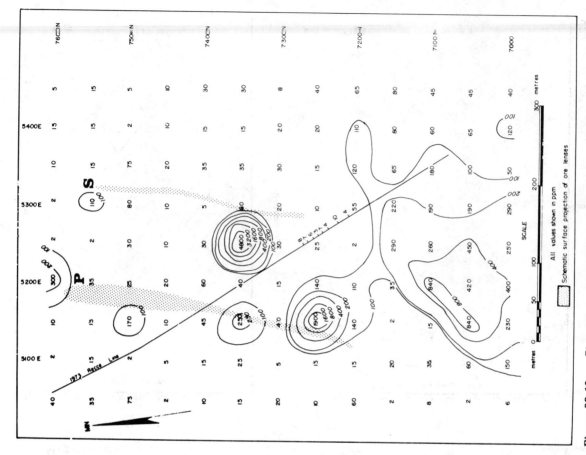

Figure 33.10. Zinc values in C horizon soils using a 50 × 50 m sampling grid. Que River Prospect, Tasmania.

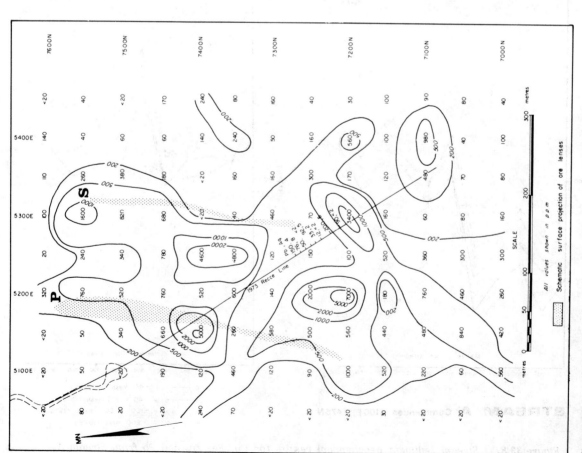

Figure 33.9. Lead values in C horizon soils using a 50 × 50 m sampling grid. Que River Prospect, Tasmania.

subcropping andesitic agglomerates with trace sulphides. The sharp termination of this geochemical zone to the northwest was inferred to be a fault of strike 045° grid, as shown in Figure 33.3.

All A horizon metal values were substantially lower than equivalent C horizon values.

Distribution patterns are broadly similar to those for the C horizon and delineate southeastern andesites as well as producing anomalies in the general vicinity of the ore lenses. However, linear trends were less evident compared with C horizon data (with the exception of iron over the western lenses). For this reason and because of the greater contrast between anomaly and background, the presumably reduced importance of hydromorphic transportation, and ease in avoiding the collection of depleted clays at the A-C horizon interface, only C horizon samples were used in follow-up.

During the 50 m sampling program, gossanous fragments and fresh sulphides in some auger samples prompted a pitting program. Sampling of soils form pit walls, (digestion and analysis of the minus 80 mesh fraction as previously described), demonstrated that an impoverished zone occurs at the base of the A horizon. Metal values then increase progressively with depth. At some locations iron-rich cellular gossans and mineralized bedrock were encountered. Best gossan values were 1100 ppm Cu, 3400 ppm Pb, 800 ppm Zn and 50% Fe and rock values attained 420 ppm Cu, 1025 ppm Pb, 10 500 ppm Zn and 15.5% Fe.

Auger samples collected from weathered bedrock had values up to 600 ppm Cu, >10 000 ppm Pb and 3400 ppm Zn which subsequently were related to the subcrop of the eastern lens as defined by drilling. This work confirmed the attractiveness of C horizon sampling in the attempt to define a linear anomaly associated with the EM conductor and 10 m spaced sampling was initiated.

As shown in Figures 33.11 and 33.12, several linear geochemical trends were indicated but these were only partly coincident with the EM conductor. Zinc, with iron, continued to define the andesites but elsewhere was less than 100 ppm except for extremely localized anomalous values in excess of 1000 ppm, which were later found to correlate with faults.

Comparison of the geochemical trends for copper, lead and iron with the position of EM conductors and subcropping ore (as subsequently determined by drilling) revealed local correlations. Elsewhere the anomalies relate to accumulations of metal after modern hydromorphic transport and to traces of base metal in pyritic zones.

Closer spaced sampling was not considered a practical exploration technique compared to geophysical methods, but one metre-spaced samples were collected on line 7400N after drilling, for research purposes. Figure 33.13 shows stacked geochemical profiles of C horizon soil values for a portion of line 7400N in the form of bar charts where each bar represents the arithmetic average of 5 point samples collected at one metre intervals.

Copper, lead and iron data show broadly coincident maxima related to mineralized pyroclastics separated by sulphide poor massive dacites. The greater level of copper values in the eastern zone (5225E to 5270E) is attributed to secondary supply of metal from the relatively copper-rich eastern lens. Iron is also greater in this zone due to numerous veins of massive pyrite within the pyroclastics.

The erratic lead response and the displacement of the trough between the major maxima, relative to copper and iron, is due to the relative immobility of this metal. Trace amounts of galena in this environment may cause soil anomalies as strong as those caused by subcropping ore.

Although varying from metal to metal, a narrow anomaly is evident over, or adjacent to, the eastern lens. That this anomaly is not in proportion to the grade of subcropping ore, relative to the dominant anomaly, must be due to secondary dispersive effects. Zinc in particular shows a narrow anomaly succeeded westwards by a depleted zone through which the metal passes before reaching the stagnant swamp environment.

Clearly, detailed soil geochemistry at Que River is not an adequate tool alone for the identification of drill targets. The role of geochemistry in this environment is in the selection of zones for geophysical appraisal.

AIRBORNE GEOPHYSICAL SURVEY PROGRAM

In February 1972, a combined aeromagnetic/electromagnetic survey of a 400 km^2 area was flown by McPhar Geophysics Pty. Ltd. using a 320 m. The total magnetic field was measured by a Barringer proton precession magnetometer with a noise envelope of 5 gammas. The electromagnetic measurements were made using a McPhar H400, two frequency (340 Hz and 1070 Hz) quadrature system utilizing a large transmitter (horizontal dipole) to receiver (vertical dipole) separation of 130 m, mounted in a helicopter (Jet Ranger 206B). This system and its installation is described in more detail by Fountain and Bottos (1970).

The data from the two systems were recorded in analogue form as shown in Figure 33.14, which illustrates the discovery data from the 1972 survey for traverse line 43A.

Normal qualitative interpretation of H-400 data, as outlined by Fountain and Bottos (1970), is performed by assessing three anomaly characteristics; the amplitude and shape of the anomaly and the ratio of the low frequency response to the high frequency response. If this ratio is less than 0.5 the conductor is rated as "poor"; between 0.5 and 0.75 the conductor is "fair"; and between 0.75 and 1.00 the conductor is considered "good". The shape of the anomaly is rated from A for a steep-sided, bell-shaped pattern through to D for a broad, flat-topped curve. The anomaly pattern in Figure 33.14 over the Que River deposit has an amplitude of 4 parts per thousand (ppt), is rated as an A shape and exhibits a relative amplitude ratio of 0.5 which indicates the response of a shallow tabular source of "fair" conductivity.

Quantitative interpretation of the data at fiducial 1622 on line 43A, using the charts of Ghosh (1972), indicates a conductivity – thickness (σt) parameter of 2.3 mhos, not allowing for the effects of finite length and depth extent. This estimation is indicative of a "fair" conductor, according to the classification outlined in Table 33.1.

The flight path recovery map for this survey (Fig. 33.15) shows that poor survey control in the vicinity of the ore lenses resulted in only one line crossing the conductor. Due to this survey deficiency and the acquisition of additional tenure to the west, a second survey was flown by Geoex Pty. Ltd. in 1975 with 160 m line spacing. This detailed survey, utilizing an improved version of the H-400 system, obtained a three-line anomaly over the Que River orebody. Altitude attenuation tests over the orebody are illustrated in Figure 33.16 and indicate there is a recognizable response up to 245 m terrain clearance, which confairms the scale model results of Ward (1969), and the conclusions of Seiberl (1975).

In 1975, Comstaff Pty. Ltd. conducted an INPUT airborne electromagnetic survey in the district and extended several traverses to cover the Que River orebody. The INPUT response over the deposit, illustrated in Figure 33.17, comprises a four channel response indicative of a fair conductor. A similar response is observed over the more extensive black shale unit to the west of the deposit.

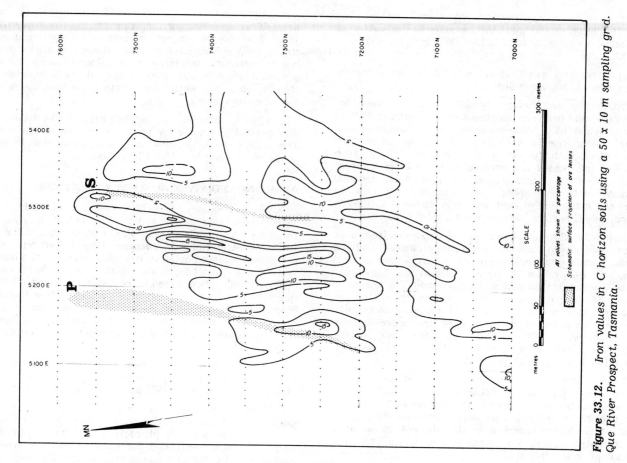

Figure 33.12. Iron values in C horizon soils using a 50 x 10 m sampling grid. Que River Prospect, Tasmania.

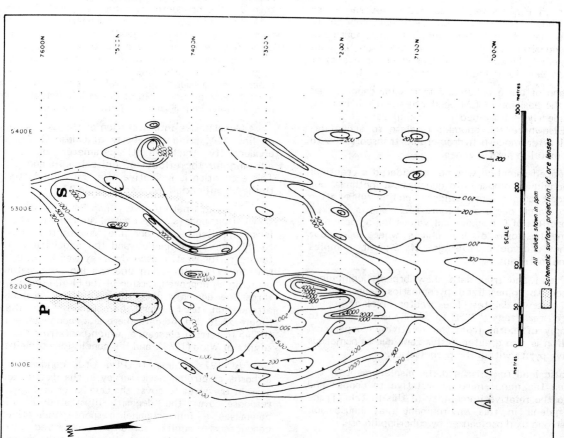

Figure 33.11. Lead values in C horizon soils using a 50 x 10 m sampling grid. Que River Prospect, Tasmania.

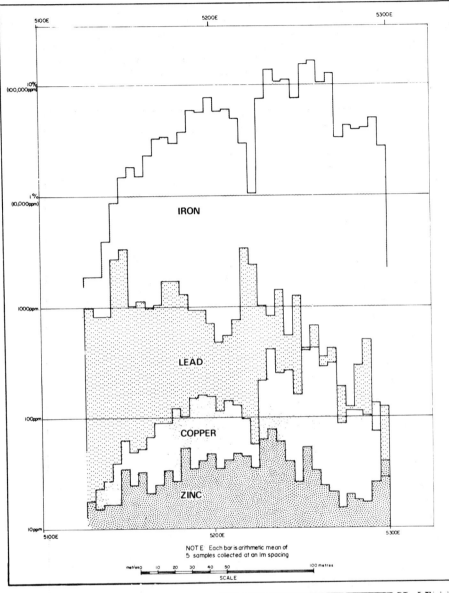

Figure 33.13.

Averaged iron, lead, copper and zinc values obtained from C horizon soil sampling at one metre intervals along Line 7400N. Que River Prospect, Tasmania.

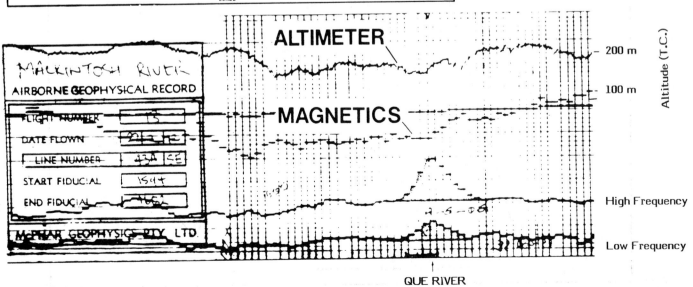

Figure 33.14. Airborne electromagnetic discovery profile across the Que River Prospect using the McPhar H400 two-frequency quadrature system (1972).

GROUND GEOPHYSICAL SURVEY PROGRAM

Ground Electromagnetic and Magnetic Surveys

To accurately locate and delineate the Que River airborne electromagnetic anomaly, a survey grid was established with cross lines, each 400 m long, cut every 50 m for a baseline length of 600 m over the position located during the 1972 survey. The grid was initially surveyed using the horizontal-loop electromagnetic method and a proton precession magnetometer. After the anomaly had been positively identified, the grid was surveyed using the vertical loop electromagnetic method in the broadside configuration, to accurately locate the axis of the anomalous source.

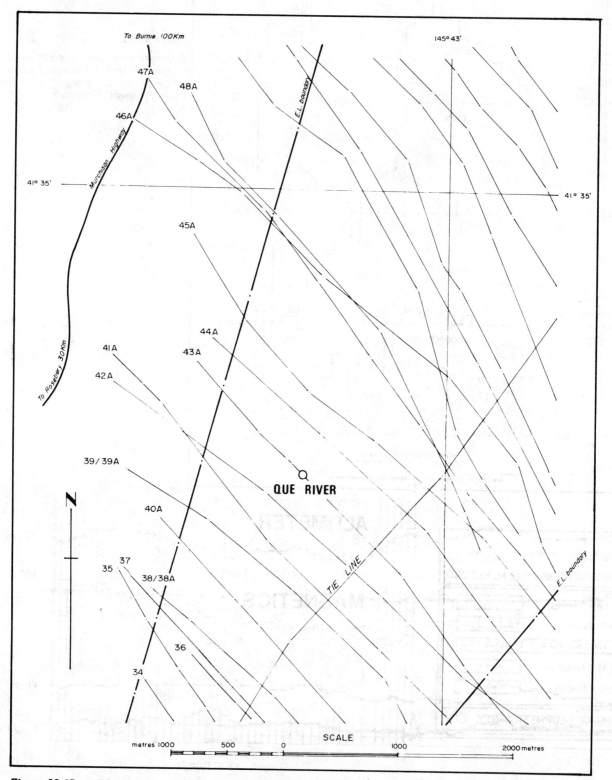

Figure 33.15. *Flight path map for the 1972 airborne EM survey by McPhar Geophysics Pty. Ltd. Que River Prospect, Tasmania.*

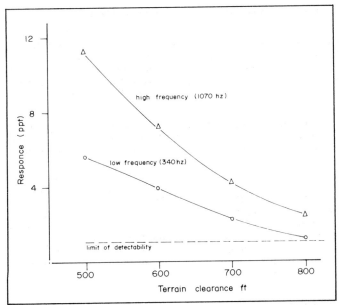

Figure 33.16. Attenuation tests over the Que River anomaly using the McPhar H-400 airborne EM system flown by Geoex Pty Ltd. in 1975.

A McPhar VHEM unit was chosen for the grid survey work because of its versatility in either horizontal or vertical loop mode of operation and its dual frequency (600 Hz and 2400 Hz) capability. A transmitter-to-receiver separation of 92 m was used for the horizontal-loop EM survey (Fig. 33.18) to achieve a reasonable depth of penetration. Some short cable (terrain) effects were expected, but these only constituted a minor problem in the area. Total-intensity magnetic field measurements were obtained with a McPhar GP 70 proton precession magnetometer which had a sensitivity of ± 1 gamma.

The vertical loop electromagnetic data were recorded by Geoex Pty. Ltd. utilizing a McPhar SS15 unit with a 5 m diameter vertical-loop which was positioned over the conductor axis and kept in maximum coupling with the receiver as the lines were traversed. The operational frequencies were 1000 Hz and 5000 Hz. The grid was surveyed with the McPhar SS15 vertical-loop EM from two transmitter locations 7150N, 5270E (Fig. 33.19) and 7400N, 5275E, thus covering the area of interest within the most effective range of the equipment.

Qualitative Analysis of Electromagnetic Data

The horizontal loop electromagnetic traverses (Fig. 33.18) showed the presence of a definite conductor from lines 7500N to 7250N with strongest response on lines 7450N and 7400N in the vicinity of 5250E to 5300E. Responses detected along strike on lines 7250N to 7350N are indicative of a poor condutor. These EM anomalies are clearly due to the eastern (S lens) mineralization.

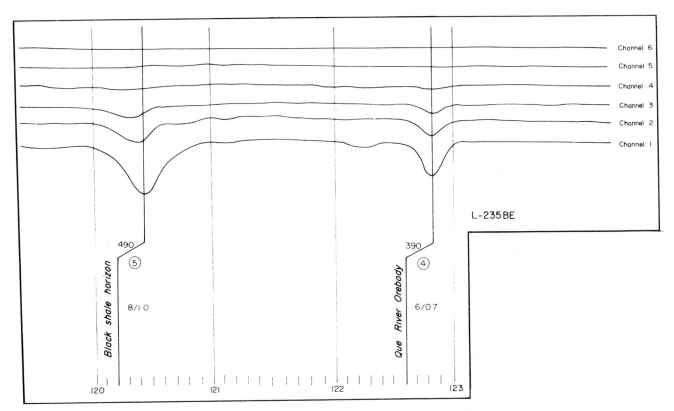

Figure 33.17. INPUT airborne EM profiles across the Que River Prospect, Tasmania.

All traverse lines crossed the western (P lens) mineralization, but no response is evident. This result is surprising, when consideration is given to coil spacing, source geometry and the resistivities of the flanking barren rock types.

A weak, but definite magnetic anomaly was detected on lines 7350N to 7550N, with a maximum relief of 200 gammas on line 7500N in the vicinity of 5150E. The magnetic data, however, proved to be of no value in this environment, due to the lack of magnetic minerals in the ore and related rock units, and the results are not included in this paper.

For ease of presentation, the vertical-loop electromagnetic (VEM) data have been transformed to their first derivatives by the procedure of Fraser (1969) and plotted in contour form in Figure 33.20. This procedure results in anomaly axes being located along contour highs, instead of at cross-over points. The VEM data show the presence of a conductor between lines 7250N and 7500N in close proximity to the base line, i.e. 5300E. The presence of a weak second conductor is readily observed at 5250E on lines 7350N and 7300N, this response is due to a barren pyrite lens, known as R lens (see Fig. 33.4). The main conductor can be classified as "strong" from 7350N to 7450N whilst the western flanking conductor can be classified as "poor". The proximity of this second "poor" conductor probably explains the only "fair" overall horizontal-loop electromagnetic (HEM) response on line 7350N. The two conductors are observed to merge on line 7400N. There are again no significant responses over the (P) western lens system.

Quantitative Analysis of Electromagnetic Data

The horizontal-loop electromagnetic data recorded at Que River have been interpreted, according to the charts of Strangway (1967), to determine conductivity-thickness (σt) parameters and source depths for classification purposes. These parameters are only approximate, as readings were taken every 50 m along traverses, i.e. at approximately half-loop separation.

Table 33.1 lists the results of this analysis, and the classification of the conductivity-thickness parameters according to the system outlined in Table 33.1. Each anomaly gave an apparent depth to source value of less than 10 metres, i.e. 0.1 times the coil separation, the limit of resolution for this technique. The interpretation curves used in this analysis were those computed for a vertically dipping source, which was assumed appropriate from geological consideration and inferred from the near symmetry of EM data. Minor asymmetries in the HEM curve shapes are probably due to the multiple sources indicated on several lines by the VEM data. These limitations were not expected to be a source of major error in the results.

Self-potential Survey

A self-potential survey conducted over the original grid produced a strong anomaly of the order of -200 to -300 millivolts over the electromagnetic conductor (Fig. 33.21). This sharp anomaly was superimposed on a broad anomaly of -20 to -30 mv, which appears to delineate the pyritic suite. A weak northeast-southwest gradient crosses 5300E on line 7200N and marks the fault contact between mineralized pyroclastics and nonmineralized andesite.

Table 33.1

Classification of conductors detected by the horizontal-loop electromagnetic survey. σt is the conductivity-thickness product

Line	Anomaly	σt (mhos) High Frequency	Abminco Classification	σt (mhos) Low Frequency	Abminco Classification	Remarks
7500N	5310E	1.2	Poor	-		
7450N	5290E	11.5	Good	11.5	Good	
7400N	5285E	11.5	Good	9.1	Good/Fair	
7350N	5285E 5250E	2.9	Fair	3.9	Fair	Double Conductor
7300N	5285E 5250E	2.9	Fair	-		Double Conductor
7250N	5270E	0.9	Poor	-		
7200N	5240E	N.D.	Probably Poor			

N.B. Abminco Classification of EM conductors.

Classification	σt mhos
Excellent	>15.0
Good	6.0 – 15.0
Fair	1.5 – 6.0
Poor	<1.5

DRILLING PROGRAM – FIRST PHASE

A seven-hole diamond drilling program was designed to evaluate the prospect. The first hole encountered 11.4 m of sulphide mineralization which assayed 2.10% Cu, 5.08% Pb, 7.86% Zn and 105 grams/tonne Ag. The second hole was sited as a deep test of this zone and as an evaluation of a broad soil geochemical anomaly. A second mineralized zone was intersected over 3.81 m which averaged 0.86% Cu, 13.72% Pb, 22.03% Zn, 371 grams/tonne Ag and 3.8 grams/tonne Au.

GROUND GEOPHYSICAL SURVEY PROGRAM – SECOND PHASE

Mise-a-la-masse Survey

As intersections of conductive mineralization were anticipated in the drilling program, provisions were made to survey the prospect with the mise-à-la-masse technique. The objective of this work was to attempt to ascertain the strike length of the eastern mineralization and its electrical conductivity, by placing a current electrode in drillhole QR 1 adjacent to the mineralization.

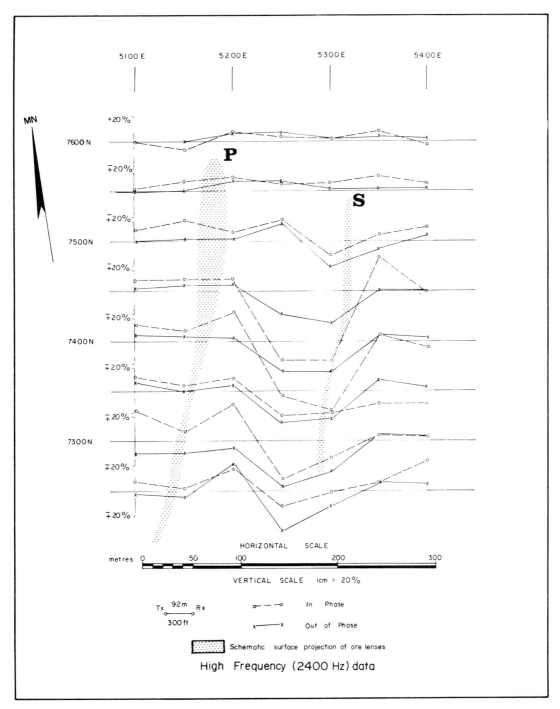

Figure 33.18. *Horizontal-loop electromagnetic profiles using McPhar VHEM equipment operating at 2400 Hz. Separation of transmitter/receiver was 92 metres. Que River Prospect, Tasmania.*

The surface potential mapped when this electrode was energized is shown in Figure 33.24 and indicates electrical continuity within the eastern mineralization between 7300N and 7550N, with possible continuity to 7200N, which was confirmed by drill results. The asymmetrical pattern is due to the effects of the far current electrode, at about 4300E on line 7500N, plus asymmetry of the host rock conductivity along strike relative to the conductivity normal to strike. The combination of these effects precludes quantitative estimation of the conductivity of the mineralization.

An attempt was made to energize the western mineralization via an electrode in QR 2, however, the resulting surface potential pattern suggested that the host pyroclastic unit was being energized in preference to the mineralization.

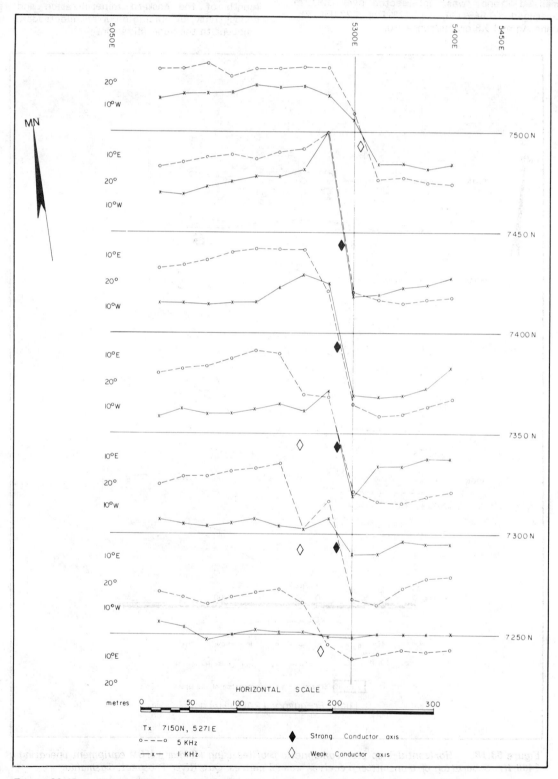

Figure 33.19. Vertical-loop electromagnetic profiles obtained using the McPhar SS15 system. Transmitter located at 7150N, 5270E. Que River Prospect, Tasmania.

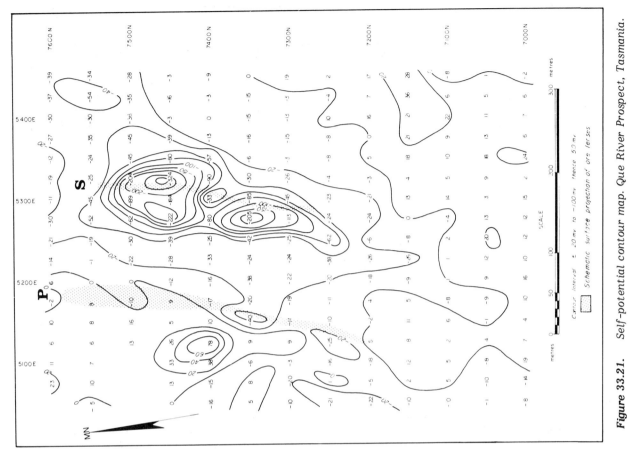

Figure 33.21. Self-potential contour map. Que River Prospect, Tasmania.

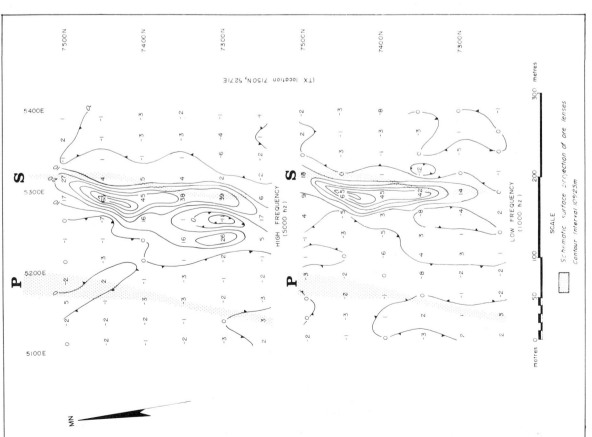

Figure 33.20. First horizontal derivative map of vertical-loop EM data obtained using Fraser (1969) procedure. Contour interval 10°/25 m. Que River Prospect, Tasmania.

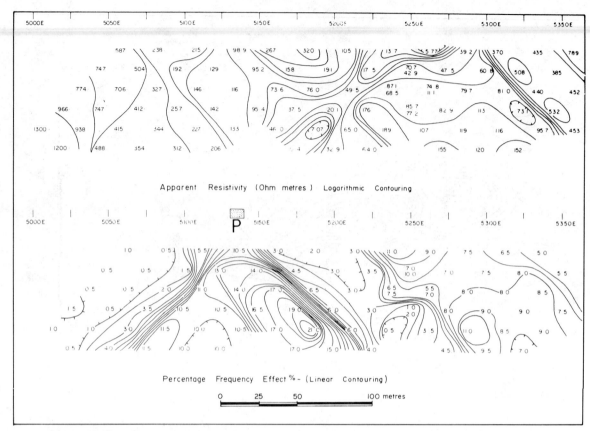

Figure 33.22a. Resistivity and frequency domain IP results along Line 7250N using a dipole-dipole array. Frequencies used ranged from 0.3 to 2.5 Hz. Que River Prospect, Tasmania.

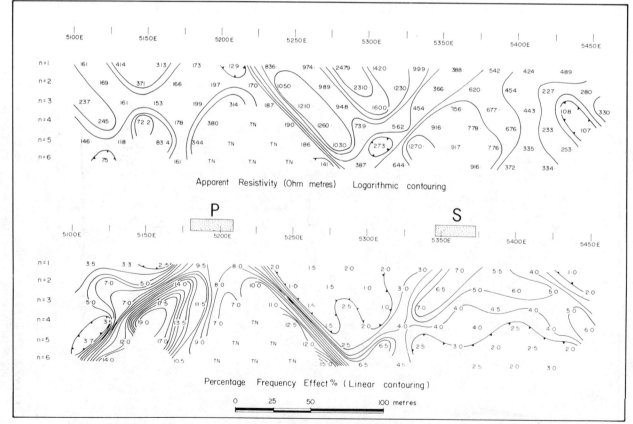

Figure 33.22b. Resistivity and frequency-domain IP results along Line 7800N. Frequencies used range from 0.3 to 2.5 Hz. Que River Prospect, Tasmania.

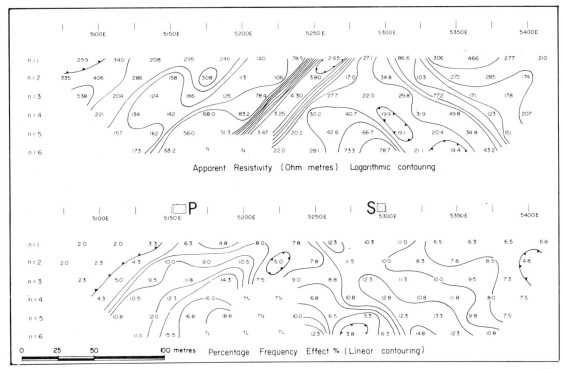

Figure 33.23. Resistivity and frequency-domain IP results along Line 7350N obtained using a 25 m dipole-dipole array. Frequencies used ranged from 0.3 to 2.5 Hz. TN and TL indicates that readings were either too noisy or the voltage was too low for measurement. Que River Prospect, Tasmania.

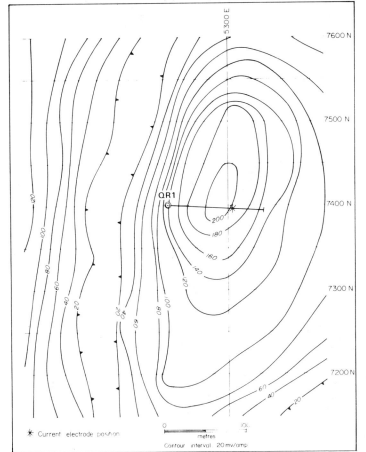

Figure 33.24. Mise-à-la-masse map survey results using a current electrode located in the eastern mineralization in drillhole QR1. Que River Prospect, Tasmania.

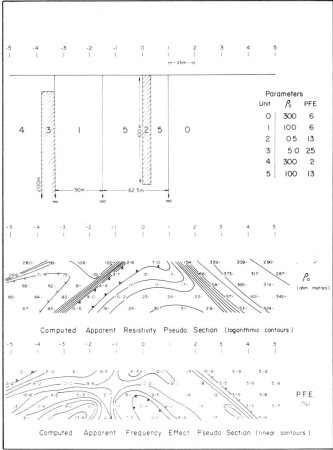

Figure 33.25. Model results derived from the dipole-dipole array IP results along Line 7350N.

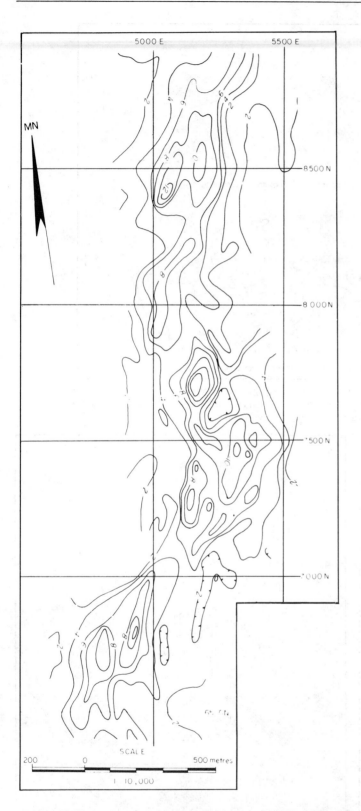

Figure 33.26. Average percentage frequency effect (PFE) map. Contour interval 2%. Que River Prospect, Tasmania.

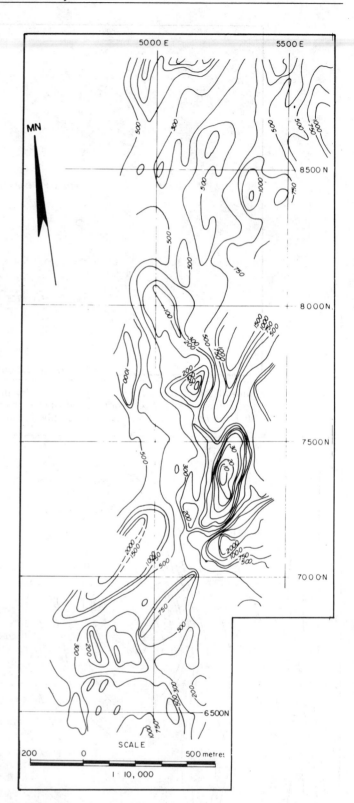

Figure 33.27. Averaged apparent resistivity in ohm-metres using a logarithmic contour interval. Que River Prospect, Tasmania.

Downhole electrical logging of the mineralization and lithologies could not be undertaken due to poor ground conditions which prevented the holes from remaining accessible.

Induced Polarization Survey

Following the discovery of the western (P) ore lens in drillhole QR 2, a test induced polarization (IP) survey was undertaken to ascertain if this technique could detect the apparently nonconductive mineralization and possibly distinguish between the two ore types. A dipole-dipole frequency-domain survey using Geoscience equipment and frequencies of 0.3 and 3 hertz, was initially undertaken with array spacings of 25 m and 50 m on nine lines spaced 50 m apart.

The resistivity and frequency effect data for line 7350N, with 25 m spreads, are shown in Figure 33.23 and clearly show two types of anomalous response. The eastern conductive lens at 5275E is depicted by a strong apparent resistivity low, less than 5 ohms, in the usual "double-pants leg" pattern for a shallow tabular source. The asymmetry of the anomaly pattern is probably due to the location of electrodes relative to the conductor and the resistivity asymmetry of flanking rock-types. A broad diffuse frequency effect anomaly is evident from 5250E to 5325E, due to multiple sources and disseminated sulphides in the host pyroclastics.

The strong frequency effect anomaly deep beneath 5175E is inferred to represent the relatively non-conductive western mineralization, which produces only a minor inflection in the resistivity gradients in this vicinity. This composite anomaly pattern is evident on lines 7300N to 7600N inclusive, beyond which the eastern anomaly becomes subordinate to the western mineralization, as is evident in data for line 7250N (Fig. 33.22a) and line 7800N (Fig. 33.22b).

An important feature of the dipole-dipole pseudo-section is the apparent depth control which indicates that the eastern mineralizaiton outcrops whereas the western zone is "topped off" on nearly all lines. This interpretation is confirmed by the drilling results which show the western lenses only partially coming to surface, and may explain the lack of EM response over this lens position.

The geology for section 7350N was computer modelled for IP response, by the procedure of Dodds (1976), to confirm the above interpretation. The model results (Dodds, pers. comm.) shown in Figure 33.25, are in close agreement with the observed data, considering that the computer model is two dimensional. The cost of the computer modelling, to test 13 models to obtain the best fit, was equivalent to the cost of acquiring the data. The success of the computer modelling exercise, in duplicating a real, complex IP pseudo-section, illustrates the need for readily available, inexpensive programs to facilitate the use of this procedure on a routine basis prior to drilling.

The IP coverage was later extended to the north (8900N) and south (6400E) in an attempt to assist the siting of development drilling plus locate targets for exploratory drilling. This coverage was completed with a 25 m dipole separation on lines spaced 100 m apart, compared with the discovery grid coverage of lines spaced 50 m apart. Plan presentation of these data (Fig. 33.26, 33.27) was accomplished by averaging the data for the first three dipole separations at each receiver position. Three dipole separations were averaged to remove the noise often evident in n=1 data plus obtaining some response from "topped-off" anomalies.

A qualitative interpretation of these results clearly shows the mineralized "host" dacite over 2500 m strike length flanked by resistive barren andesite and dacite units. The barren dacite unit between the two ore lenses is clearly evident. The structural displacements which bound the ore lenses are also well illustrated, indicating the use of geophysics for post discovery geological purposes. The resistivity low over the eastern lens indicates the short strike length (300 m) of the conductor which was detected in the airborne surveys relative to the strike length of the western mineralization (700 m) which was not detected.

DRILLING PROGRAM – SECOND PHASE

An ore delineation drill program of 108 drillholes totalling 25 500 m, based on the results of this exploration, enabled an ore reserve estimate of 6 million tonnes containing 800 000 tonnes of lead and zinc, to be calculated. Of this total reserve, the eastern (S) mineralization comprises only 750 000 tonnes, the balance being in the western lens system.

CONCLUSIONS

The discovery of the Que River deposit resulted from the implementation of an integrated multitechnique exploration program. The ability to focus on a specific target, as defined by an airborne electromagnetic response coincident with a zone of anomalous stream sediment geochemistry, considerably reduced exploration expenditure. The exploration program demonstrated that broad spaced stream sediment geochemistry was a valid reconnaissance technique in the northwest Tasmania drainage environment, and for the first time in Australia (to the authors' knowledge) the airborne electromagnetic method was successful as the prime focusing technique. Soil geochemistry alone was found not to be generally acceptable for the selection of drilling targets due to secondary dispersion effects and the intermittent subcrop of ore lenses.

The application of several electrical and electromagnetic ground techniques failed to indicate the presence of the significant western sulphide mineralization in the Que River deposit. That these ore lenses did not outcrop may partially explain the lack of responses, but effectively the mineralization is nonconductive. The same mineralization was however strongly responsive to the induced polarization technique.

It is therefore concluded that in this environment, a necessary criterion is that drill targets should exhibit both a soil geochemical anomaly and an induced polarization anomaly in close proximity.

ACKNOWLEDGMENTS

The authors wish to thank the management of Abminco N.L. for permission to publish this paper, and for their support in the preparation of the text and plates. Special acknowledgment goes to Dr. Max Richards and Mr. Lindsay Gentle who conceived and encouraged much of the exploration effort. Many staff members contributed to the overall program and to the generation of new concepts in a difficult area. We thank those people who constructively reviewed the manuscript and the Abminco drafting section whose creativity contributed to the overall presentation.

REFERENCES

Corbett, K.D., Reid, K.O., Corbett, E.B., Green, G.R., Wells, K., and Sheppard, N.W.
1974: The Mt. Read Volcanics and Cambro-Ordovician relationships at Queenstown, Tasmania; J. Geol. Soc. Aust., v. 21 (2), p. 173-186.

Dodds, A.R.
1976: A parametric study of induced polarisation models; Commonwealth Scientific and Industrial Research Organisation, Investigation Report No. 118.

Fountain, D.K. and Bottos, F.B.
1970: The McPhar 400 series dual frequency airborne EM system; McPhar Geophysics Ltd., Publication, 30 p.

Fraser, D.C.
1969: Contouring of VLF-EM data; Geophysics, v. 34 (6), p. 958-967.

Gee, C.E., Jago, J.B., and Quilty, P.G.
1970: The age of the Mt. Read Volcanics in the Que River area, western Tasmania; J. Geol. Soc. Aust., v. 16 (2), p. 761-763.

Ghosh, M.K.
1972: Interpretation of airborne EM measurements based on thin sheet models; unpubl Ph.D. thesis Univ. of Toronto, Toronto, Canada.

Seiberl, W.
1975: The F400 series quadrature component airborne EM system; Geoexploration, v. 13, p. 99-115.

Strangway, D.W.
1966: Electromagnetic parameters of some sulphide ore bodies; Soc. Explor. Geophys., Min. Geophys., v. 1, p. 227-242.

Ward, S.
1969: A model study of the McPhar F400 airborne EM method; Paper Presented 39th SEG Annual Meeting, Calgary.

GEOPHYSICS AND GEOCHEMISTRY IN THE DISCOVERY AND DEVELOPMENT OF THE LA CARIDAD PORPHYRY COPPER DEPOSIT, SONORA, MEXICO

D.F. Coolbaugh
Consultant Industrias Peñoles, Mexico

Coolbaugh, D.F., Geophysics and Geochemistry in the discovery and development of the La Caridad porphyry copper deposit, Sonora, Mexico; in Geophysics and Geochemistry in the Search for Metallic Ores; Peter J. Hood, editor; Geological Survey of Canada, Economic Geology Report 31, p. 721-725, 1979.

Abstract

The discovery of La Caridad porphyry copper deposit in Sonora, Mexico was the result of planned mineral exploration which was carried out over a large potential area employing reconnaissance exploration methods of photogeology and geochemistry followed by semidetailed and detailed exploration methods of geophysics, geochemistry and geology. Intense hydrothermal alteration exposures in the district were visually observed from fixed-wing aircraft but before this area could be field checked anomalous copper and molybdenum values in stream sediments were found 18 km downstream in the drainage from the deposit. These anomalous values were followed upstream to the area of the La Caridad deposit. The stream sediment sampling was followed by more detailed soil and rock-chip sampling and reconnaissance geological mapping. Induced polarization surveys, geochemical studies and geological mapping were used to delimit the La Caridad deposit. Very low induced polarization and geochemical values were noted over the central part of the deposit due to deep leaching. Alteration and mineralization exhibit classic porphyry copper patterns. As a result of this exploration a deposit was delineated which contains in excess of 700 million metric tons with an average grade of 0.72% copper.

Résumé

La découverte du gisement du minerai porphyrique de cuivre de la Caridad à Sonora au Mexique, a été le résultat d'un projet d'exploration minière, réalisé dans une région ayant un grand potentiel minier. Les méthodes d'exploration de reconnaissance utilisées ont été la photogéologie et la géochimie, suivies d'une exploration géophysique, géochimique, et géologique détaillée et semidétaillée. Dans le secteur on a observé d'un avion, à l'oeil nu, des affleurements présentant une altération hydrothermale intense; mais, avant d'en faire l'exploration au sol, on a trouvé des anomalies en cuivre et molybdène dans des dépôts fluviatiles à 18 km en aval du gisement. On a suivi ces anomalies en amont jusqu'à la région du gisement de la Caridad. L'échantillonnage des dépôts alluvionnaires a été suivi d'un échantillonnage plus détaillé des sols et de fragments de roches, et d'un levé géologique de reconnaissance. On a fait des levés de polarisation provoquée, des études géochimiques, et des levés géologiques pour délimiter le gisement de la Caridad. On a relevé de très faibles valeurs géochimiques et de polarisation provoquée au-dessus de la partie centrale du gisement à cause du profond lessivage. L'altération et la minéralisation sont typiques des minerais porphyriques cuprifères. Cette exploration a permis de délimiter un gisement contenant plus de 700 millions de tonnes métriques ayant une teneur moyenne en cuivre de 0.72%.

INTRODUCTION

Successful mine exploration requires sound planning and modern exploration techniques but it also requires perseverance and often the good fortune of being in the right place at the right time. The La Caridad mineral area had been known for over 70 years but, as with many mines, it was not 'discovered' until viewed under different economic conditions and different exploration philosophies. The La Caridad deposit is located in the State of Sonora, Mexico, approximately 120 km south of the international boundary between Mexico and the United States, at Douglas, Arizona (Fig. 34.1). The two nearest large copper deposits are Cananea, 105 km to the northwest and Bisbee, 128 km to the north. The Pilares mine, formerly operated by the Moctezuma Copper Company of Phelps Dodge Corporation, is located approximately 10 km to the west. Nacozari, Sonora, the nearest village is 15 km northwest of La Caridad.

The deposit is situated approximately 1825 m above sea level. The climate is semiarid with an annual rainfall of 38 cm. The topography is rugged and diversified.

The Cananea and Pilares copper mines were in operation over 80 years ago. In 1925, Alfred Wandke gave a description of the La Caridad Vieja mine which is located about 3 km northeast of the La Caridad porphyry deposit. Mineralization in the Vieja deposit occurs along a shear zone in quartz porphyry; a very small mining operation recovered some copper and a little silver (Wandke, 1925).

The Guadalupe mine, located in a pegmatite body on the flank of La Caridad, was exploited between 1954 and 1959. During this period 116 000 tons of ore containing 3.44% Cu and 0.43% Mo were mined.

During the period 1962-1967, the Mexican Government in conjunction with the United Nations undertook a joint mineral survey of selected zones in Mexico. The largest of these zones covered a 50 000 km^2 area in northern Sonora which included Cananea, Mexico's only large copper-producing mine, and the La Caridad deposit, which is the subject of this paper.

Subsequently a great amount of geochemical and geophysical work was carried out by the United Nations on the La Caridad deposit, and an excellent geological study of

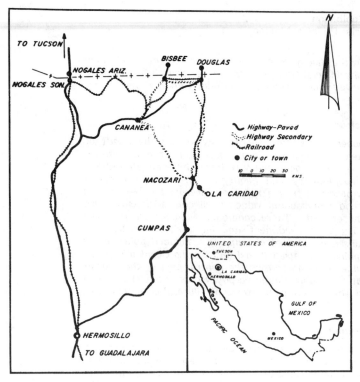

Figure 34.1. Location of La Caridad Deposit.

the deposit has been presented by Saegert et al. (1974). This paper, a discovery case history, will be confined primarily to the discovery methods, procedures and results and will not attempt to be a complete treatise.

METHOD OF EXPLORATION

Area Selection

Effective exploration is actually a process of selection and elimination. First the desired material must be selected and then the general area such as country or countries in which to explore must first be chosen. A selection of the area for reconnaissance surveys is made through library and similar studies by eliminating the areas with least potential. Reconnaissance surveys are carried out to eliminate the majority of the area and leave zones of greater potential to be studied by semidetailed and later detailed surveys.

The selection of the area for exploration that led to the La Caridad discovery was based on geological and economic decisions. The decision to explore for copper had been taken; also, it had been decided that the large bulk low-grade deposits were to be the exploration target.

Mexico is a mineral-rich country; however, in 1963 it only had one large copper-producing mine located at Cananea, Sonora. Immediately to the north of Cananea in the state of Arizona, U.S.A. are located the important copper mines of Bisbee, Pima, Mission, Esperanza, Twin Buttes and Ajo. Therefore, from an economic standpoint northern Sonora was a favourable area for exploration.

Northeastern Sonora is located along the north-south trending Wasatch-Jerome crustal lineament at or near its intersection with the northwest-trending Texas lineament. Correlation between these lineaments and the distribution of bulk low-grade copper deposits has been remarkable. Thus the foregoing favourable structural conditions and the fact that this area is known to contain many acid intrusive rocks of Laramide age, and the presence of existing mines, made the selection of northeastern Sonora as an area to explore for bulk low-grade copper deposits an easy decision.

Reconnaissance Exploration

The reconnaissance methods of exploration carried out in 1963-1965 included photogeology with aerial overflights by trained geologists and stream-sediment geochemical surveys.

The photogeological work was begun at about the same time as the other reconnaissance surveys. The results of this work could have been used more effectively had this work been completed before the other reconnaissance surveys were begun.

By using aerial photographs and topographic maps, aerial observation flights were planned and were flown along the sides of the mountain ranges in a fixed-wing aircraft with one photogeologist and two geologists observing geological features of interest such as alteration, mineralization, rock type and structure. These observations were plotted on photographs and topographic maps and were the basis, along with the photogeology and reconnaissance geochemistry, for the follow-up ground investigations.

Regional geochemical reconnaissance was carried out over the area utilizing stream-sediment sampling in tributaries to all major drainages and analyzing these samples for copper and for molybdenum by the cold extractable method (Hawkes, and Webb, 1962) which was used throughout this reconnaissance survey. Additional samples were taken where anomalous conditions were noted.

Geochemical Exploration

In late 1963 and early 1964, stream-sediment samples were taken from the arroyos (streams) entering the main drainage systems of northern Sonora within the zone of exploration of the Mexican-UN joint survey. This was a reconnaissance survey to locate areas for further exploration. Previously, geochemical stream sediment, soil and rock chip sampling over known copper deposits had demonstrated the feasibility of this type of survey in the area.

In early 1964, during the sampling of the arroyos emptying into the Rio Bavispe, anomalous values were noted in the arroyo Cruz de Cañada and in a few other streams. The method applied was to take three samples about 50 m apart where the arroyo emptied into the main drainage. The reason for three samples was that one sample might not prove reliable due to erratic results; however, three samples would constitute reasonable control. The values at the mouth

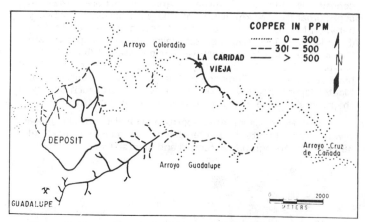

Figure 34.2. Results of stream sediment sampling for copper, La Caridad, Sonora, Mexico (Lee and Osoria, 1964).

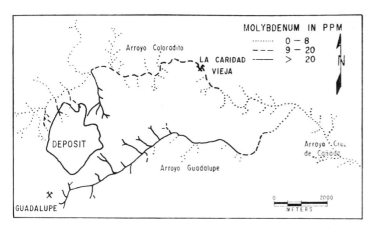

Figure 34.3. Results of stream sediment sampling for molybdenum, La Caridad, Sonora, Mexico (Lee and Osoria, 1964).

of the Cruz de Cañada arroyo ranged from 24 to 90 ppm copper and 2 to 15 ppm molybdenum in an area with a copper background of 20 ppm and a molybdenum background of 1.5 ppm (Lee and Osoria, 1965).

During this reconnaissance stream-sediment survey in a number of arroyos, of which Cruz de Cañada was only one, anomalous values were found. Although Cruz de Cañada gave one of the strongest anomalies, follow-up sampling was not undertaken for about four months. The follow-up consisted of sampling the arroyo upstream for approximately 3 km. The laboratory results of these samples showed them to be consistently anomalous at between three and five times background and these anomalous values were confined only to the main Cruz de Cañada drainage. Since the values were still anomalous, the field crew returned and sampled the Cruz de Cañada and the drainage entering it for another 7 km. Again values of copper and molybdenum gave values of between three and six times background with a few erratic higher values.

Since the arroyo Cruz de Cañada had proven to be anomalous over 10 km, the geochemical survey was quickly resumed. For the next 4 km upstream the values slowly increased to the junction of the arroyos Guadalupe and Coloradito (see Fig. 34.2). Both arroyos were anomalous and values increased more rapidly upstream. Following the arroyo Guadalupe, the values increased (to values more than 20 times background for copper and molybdenum) for another 7 km where the small abandoned copper-molybdenum workings of the Guadalupe Mine were found. This mine dump was the source of much of the downstream contamination; however, the drainage from the north into the arroyo Guadalupe had shown values of 5 to 30 times background in copper and molybdenum and the drainage from the south showed anomalous values in molybdenum up to 20 times background.

Values increased rapidly upstream in the arroyo Coloradito (to more than 20 times background for copper and molybdenum) for 4 km where the abandoned workings of the La Caridad Vieja mine, consisting of a few superficial workings, was found. Enargite, tetrahedrite and chalcocite were found. This mine had given rise to the heavy contamination downstream of copper and molybdenum. However, sampling was continued upstream from the workings and anomalous values equal to or sometimes even higher were encountered in the next 2 km.

The stream sediment sampling definitely located a large anomalous area that was not just related to the contamination of the two, previously unknown, abandoned small mine workings. Whether anomalous values would have been detected 18 km downstream from the now known La Caridad deposit without the contamination from these two old workings might be questionable, but this writer believes that anomalous values would have been encountered.

In July 1973, this drainage was resampled (Osoria, 1973) and the results were similar except higher values were obtained as the analyses were made by the atomic absorption method. The pH of the waters in the arroyos was between 5 and 6.

In order to maintain chronological sequence it is noted here that, while the reconnaissance geochemical studies were being conducted, this area was overflown and the prominent red hills lying approximately 3 km to the east of the La Caridad deposit were noted and plotted on aerial reconnaissance maps for later ground checking.

Induced polarization geophysical surveys were begun in late 1964 as soon as the La Caridad Vieja mine had been located and this work continued at the same time as the more detailed geochemical surveys were being made. The results of these surveys will be presented later. With the favourable geochemical results of stream sediments known, additional detailed stream sediment sampling in the local area was carried out leading to the closer determination of the location of the mineralized body (Figs. 34.2, 34.3).

Soil sample and rock chip sample surveys were also conducted during 1965-1966 to assist in determining the exact location of the deposit. Although this area is very steep and there is much talus on the slopes making true outcrops difficult to determine, and the soils were not well developed, samples were nevertheless taken of this material. Figures 34.4 and 34.5 show the results of the soil sample study. In Figure 34.4 it is easily seen that the surface over the mineralized zone gave much lower copper values than over the periphery; in fact, in the local region these values

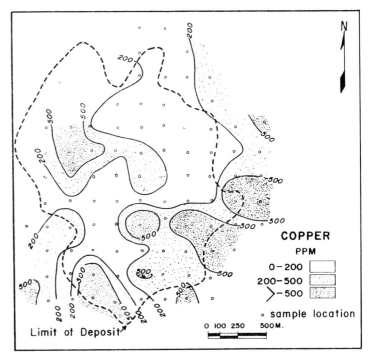

Figure 34.4. Geochemical soil survey results for copper in parts per million, La Caridad deposit, Sonora, Mexico.

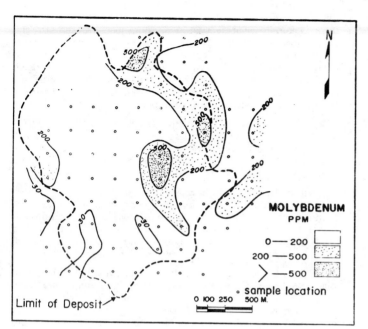

Figure 34.5. Geochemical soil survey results for molybdenum in parts per million, La Caridad deposit, Sonora, Mexico.

show as an anomalous low. This is easily explained since the rocks directly over the deposit have been leached of their copper which has been deposited in the supergene zone.

In Figure 34.5, higher values of molybdenum occur over parts of the deposit than in the surrounding area. This is due to the fact that the molybdenum has been retained and concentrated as molybdite in the leached capping. The anomaly indicated the higher molybdenum values in the deposit to be in the eastern part. The soil sample analyses were made by using the atomic absorption technique.

In 1976, the Mexican Geological Survey (Consejo de Recursos Minerales) in conjunction with the U.S. Geological Survey resampled the La Caridad drainage by collecting stream sediments and analyzed these using the cold extractable method. Their results correlated very well with the original survey. In this study zinc, silver and tungsten were also analyzed. The conclusions (Lee et al., 1976) were that copper and tungsten values were anomalous downstream to the Rio Bavispe; that molybdenum values were anomalous to the Rio Bavispe and as far as 13 km downstream in that river; that silver values were anomalous 10 km below the deposit and that zinc could not be considered a pathfinder element for the La Caridad deposit.

Geophysical Exploration

Induced polarization surveys were employed as a follow-up of the favourable stream sediment geochemical survey and were first conducted over the La Caridad Vieja mine area. Although widespread sulphide mineralization was not indicated, a shear zone in quartz monzonite gave values that indicate sulphides within a narrow zone. This zone was considered of minor importance and no drilling was planned.

An extension of the induced polarization survey lines to the south gave rather high readings over a granodiorite. These data suggested near-surface primary sulphides. A close examination of the surface of this granodiorite and an examination of a few prospect pits showed an abundance of pyrite, sufficient to account for the induced polarization anomaly. As the induced polarization survey was continued up the arroyo Guadalupe, anomalous values were recorded in the areas where the highest geochemical anomalies had been found. Here pyrite, chalcopyrite, chalcocite and molybdenite were sometimes observed at the surface. Anomalously low induced polarization values, that corresponded to the background or lower geochemical values for copper, were found over what is now known to be the leached zone above the La Caridad orebody. The area surrounding this leached zone contained the highest induced-polarization values.

This anomalous effect is very easy to understand if it is recognized that the deepest leaching penetrates to a depth of up to 225 m directly below the highest elevations. However, the thickness of this leached capping decreases on the flanks of the deposit, which are at a lower elevation, and sometimes sulphides outcrop at the lower elevations (Fig. 34.6). In the capping, the copper sulphides have been almost completely leached and, therefore, the geochemical values are low. The total volume of sulphides within the induced polarization area of influence is also decreased as the result of removal of sulphides from the capping. In addition, there is a pyrite halo around the deposit which enhanced the induced polarization values.

The induced polarization surveys were carried out using Seigel Associates Mark VG, 7KW, time domain equipment. The three electrode array was used with spacings of 100, 200 and in places, 300 m. Wenner expanding electrode arrays of up to 500 m spacing were also used in a few selected locations.

DESCRIPTION OF LA CARIDAD PORPHYRY COPPER DEPOSIT

Although this paper is a case history of the discovery and development of the La Caridad porphyry copper deposit, the account would not be complete without a description of the deposit.

The first hole was drilled in 1967 at La Caridad, and it encountered supergene mineralization with interesting copper and molybdenum values. The geochemical and geophysical studies had indicated a high probability for the occurrence of a bulk low-grade copper deposit and the first drillhole confirmed mineralization. Later drilling, based primarily on detailed geological studies, confirmed the existence of a major copper deposit.

There are no features of the La Caridad deposit which make it greatly different from other porphyry deposits. The ore is confined to the supergene enrichment zone where the predominant ore mineral is chalcocite. The underlying protore averages less than 0.25% Cu. The supergene blanket underlies a leached capping that varies in thickness from 0 to 255 m. No oxidized copper minerals are found in the capping.

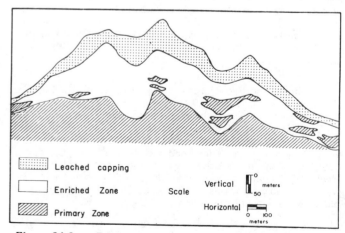

Figure 34.6. Typical section through La Caridad deposit.

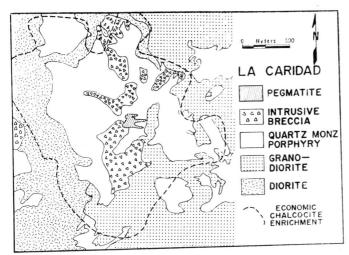

Figure 34.7. Generalized bedrock geology (after – Saegart, Sell, Kilpatrick).

Pre-mineral rock types found within the altered and mineralized area are diorite, granodiorite, quartz monzonite porphyry and pegmatites. Supergene copper mineralization is found in all these rock types (Fig. 34.7). Alteration in the central part of the deposit is phyllic with abundant sericite. This zone is surrounded by an irregular argillitic band which, in turn, is surrounded by a propylitic zone. Hypogene mineralization consists of pyrite, chalcopyrite and molybdenite. Supergene mineralization consists primarily of chalcocite. The chalcocite blanket has an average diameter of 1700 m and an average thickness of 90 m (Fig. 34.6). Both hypogene and supergene copper and molybdenum are disseminated in the igneous rocks, in fractures, in the cement of breccias or disseminated in pegmatites. Within both the breccias and pegmatites, the copper values are appreciably higher than in the surrounding rocks. A fairly large breccia pipe has been observed near the centre of the deposit. Small, randomly-spaced satellitic pipes are close to the main pipe, but the pipes are not restricted to any one rock type.

SUMMARY

A favourable bulk low-grade copper district was located in Sonora, Mexico in 1964. During the next two years, the most favourable area was investigated with the assistance of geochemical, geophysical and geological studies and, in 1967, the first hole drilled encountered very favourable mineralization and confirmed a porphyry copper discovery. During the next year limited additional drilling indicated a large deposit of favourable grade copper with values in molybdenum. In 1968, Asarco Mexicana, S.A. obtained exploration rights and drilling began which led to the delineation of an orebody in excess of 700 000 000 metric tons with an average of 0.72% copper (using a 0.4% copper cutoff) plus values in molybdenum. La Caridad is now owned by Mexicana de Cobre and production will begin in 1978 at the rate of 72 000 metric tons per day.

Although this discovery was the joint effort of various exploration disciplines, it can be stated that the reconnaissance geochemical survey was the first to pinpoint the favourable area. Both geochemical and induced polarization surveys were used to delineate the most favourable zone and were instrumental in choosing the location of the first drillhole which confirmed a discovery. Later development drilling was guided by information obtained by geological investigations.

A major orebody was discovered. How was it found? 1) First it was through planned regional reconnaissance exploration in a geologically favourable area, 2) through the use of modern exploration techniques, 3) through team effort of a dedicated group of exploration personnel who knew what they were looking for, and 4) a little bit of luck.

ACKNOWLEDGMENTS

The discovery and exploration of the La Caridad deposit have been by team effort. Special acknowledgment, however, should be given to the Mexican Government and the United Nations Development Program for the vision in initiating the project, to the dedicated Mexican engineers of the Consejo de Recursos Minerales, to the UN team of international experts, to the American Smelting and Refining Company consultants and to the management of Asarco Mexicana, S.A. and Mexicana de Cobre, S.A. for their vigorous approach to exploration and development.

REFERENCES

Chaffee, M.A., Lee Moreno, J.L., Caire, L.F., Mosier, E.L., and Frisken, J.G.
    1976: Results of geochemical investigations comparing samples of stream sediment, panned concentrate, and vegetation in the vicinity of the Caridad porphyry copper deposit, northern Sonora, Mexico; U.S. Geol. Surv., Open File Rept. 76-559, 34 p.

Hawkes, H.E. and Webb, J.S.
    1962: Geochemistry in mineral exploration; New York, Harper and Row.

Lee Moreno, J.L., Chafee, M.A., Mosier, E.L., and Frisken, J.G.
    1976: Resultados de Investigaciones Geoquímicas en la Vecindad del Pórfido Cuprifero de La caridad, en el noreste de Sonora, Mexico, Comparando Muestras de Sedimentos de Arroyo, Concentrados de Batea y Vegetación; Geomimet No. 84, Nov./Dic.. 1976.

Lee Moreno, J.L., y Osoria H.
    1965: Informe Preliminar sobre la Anomalía Geoquímica Cruz de Cañada; Informe CRNNR.

Osoria H.
    1973: Exploración Geoquímica en La Caridad, Nacozari, Sonora, Mexico. Paper presented at the X Convention of the AIMMGM in Chihuahua, Chihuahua.

Saegart, W.E., Sell, J.D., Kilpatrick, B.E.
    1974: Geology and mineralization of La Caridad Porphyry Copper Deposit, Sonora, Mexico; Econ. Geol., v. 67(7), p. 1069-1077.

Wandke, A.
    1925: The Caridad mine, Sonora, Mexico; Econ. Geol., v. 20, p. 311-318.

EXPLORATION OF THE REAL DE ANGELES SILVER-LEAD-ZINC SULPHIDE DEPOSIT, ZACATECAS, MEXICO

Lee R. Stoiser
Esso Eastern Inc., Houston, Texas, U.S.A.

José Bravo Nieto
Cía. Industrias Peñoles, San Luis Potosí, México

Stoiser, Lee R. and Bravo Nieto, José, Exploration of the Real de Angeles silver-lead-zinc sulphide deposit, Zacatecas, Mexico; in Geophysics and Geochemistry in the Search for Metallic Ores; Peter J. Hood, editor; Geological Survey of Canada, Economic Geology Report 31, p. 727-734, 1979.

Abstract

In the Real de Angeles district of Zacatecas, Mexico, galena, sphalerite, and silver ore minerals occur with pyrrhotite, pyrite and arsenopyrite in narrow fault veins and fractures, as disseminated grains, and along bedding planes. The asymmetrical, funnel-shaped body measures 500 x 400 m in plan near surface and extends to a known depth of 300 m. The host rock is composed of moderately warped, interbedded and gradational carbonaceous greywacke, sandstone, and shale of the Upper Cretaceous Caracol Formation. The mineral deposit outcrops in places but is largely overlain by a cover of caliche, numerous mine dumps, and the stone ruins of the village of the Real de Angeles. Conventional ground geophysical methods, including induced polarization, magnetic and resistivity surveys, were employed during the exploration diamond drilling program as a means to better determine the lateral and depth limits of the large deposit. Generally, the ore minerals are not chargeable nor are they naturally magnetic. However, the associated gangue sulphide minerals are chargeable and/or magnetic. The body could therefore be outlined by its high chargeability, low resistivity, and high magnetism, thereby reducing the number of drillholes required in the exploration program.

Résumé

Dans le district de Real de Angeles à Zacatecas, au Mexique, des minerais de galène, sphalérite et argent accompagnent la pyrrhotine, la pyrite et l'arsénopyrite dans des veines faillées étroites et dans des fractures, sous forme de particules disséminées, ainsi que le long des plans de stratification. Le corps asymétrique, de forme conique, a pour dimensions 500 m par 400 m dans un plan proche de la surface, et se prolonge jusqu'à une profondeur connue de 300 m. La roche favorable est composée d'une grauwacke charbonneuse, de grès, et d'une argile litée de la formation de caracol du Crétacé supérieur; ces roches sont modérément plissées, sont stratifiées en alternance répétée, et granoclassées. Le gîte minéral affleure en certains endroits, mais est en grande partie recouvert d'une croûte calcaire, de nombreux déblais de mine, et des pierres provenant des ruines du village de Real de Angeles. Les méthodes géophysiques au sol conventionnelles, en particulier les levés de polarisation induite, magnétiques et de résistivité, ont été employées au cours du programme d'exploration par forage au diamant, pour mieux déterminer les limites latérales et en profondeur du vaste gisement. Généralement, le minérai n'est ni chargeable ni naturellement magnétique. Cependant, les minéraux sulfureux associés qui constituent la gangue sont chargeables ou magnétiques, ou bien les deux à la fois. Par conséquent, on pourrait délimiter le corps grâce à sa forte capacité de polarisation, sa faible résistivité et son magnétisme élevé, ce qui permettrait de réduire ainsi le nombre de trous de forage nécessaires au programme d'exploration.

LOCATION

The Real de Angeles silver-lead-zinc mineral district is located on a regional, semiarid central plateau near the geographic centre of Mexico, in the southeast part of the State of Zacatecas, about midway between the cities of Zacatecas and San Luis Potosí (Fig. 35.1), and lies at an elevation of 2300 m above sea level. The mines and mining village of Real de Angeles, whose ruins cover the top of the mineralized hill, were reportedly established by the Spanish late in the 16th century.

THE PURPOSE OF THE GEOPHYSICAL STUDIES

Explomin S.A. de C.V. obtained the mineral rights over the Real de Angeles mineral district between late 1973 and mid-1974. Geophysical surveys were conducted over the district and a large surrounding area for the purpose of better defining the limits of the Real deposit in three dimensions, to search for extensions thereof, and to investigate the ore potential under the nearby proposed millsite, mine dump and tailings disposal areas. This work was carried out in 1975 while surface diamond core drilling and geological mapping were in progress. The geophysical work consisted of ground magnetometer, resistivity, and induced polarization (IP) surveys. A VLF-EM survey was attempted but proved ineffective.

The known portion of the Real deposit is largely covered by rock debris of numerous small shaft mine workings, the stone ruins of the village, and a caliche capping several metres thick. Other than by blind drilling and/or dewatering more than 30 very old mine shafts and drifts, the lateral and vertical extent of the mineral body and the surrounding mineral potential could not be determined easily. It was hoped that the geophysical responses of the gangue sulphide minerals closely associated with the ore minerals, could lead indirectly to a better definition of the limits of the silver-lead-zinc mineral body, prospect for blind, unknown orebodies and provide guidance to the drilling program.

Figure 35.1. Location map of the Real de Angeles mining district, Zacatecas, Mexico.

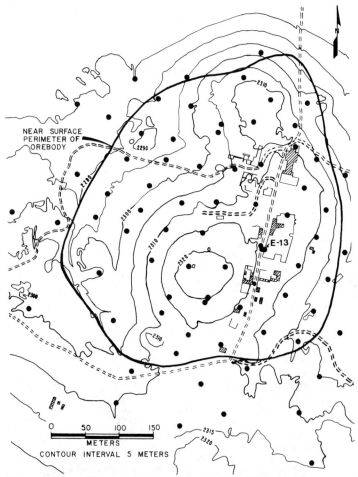

Figure 35.2. Topographic map of the Real de Angeles mineral area, orebody outline, drillhole location, and village ruins. Note location of DDH E-13.

GEOLOGICAL STRUCTURE OF THE MINERAL DEPOSIT

The Real sulphide deposit has sharply defined lateral limits, beyond which both ore and barren sulphides diminish rapidly. The mineral deposit near the ground surface measures 500 m by 400 m (Fig. 35.2). The body is asymmetrically funnel-shaped and plunges steeply to the southwest. Continuous sulphide mineralization is indicated in one sulphide drillhole to a depth of 362 m or a vertical depth of 332 m. Overburden consists of up to 15 m of mine waste rock, rock rubble of the village, and a caliche layer which is in turn successively underlain by a leached capping 5 to 10 m thick and a transition zone of mixed sulphides and oxides 25 m thick.

The sulphide body, as shown in Figure 35.3, is hosted by fine grained shallow-marine clastic rocks of the Caracol Formation of Late Cretaceous age, which consists of thin bedded, lenticular, interbedded, and intercalated sandstone, greywacke, siltstone, shale and argillite. This flysch-type stratigraphy is characteristic of the host rock. Crossbedding and slumping are observed.

Two older calcareous units, the Cuesta del Cura and Indidura formations, are exposed in the small hill immediately south of the mineral deposit, where they are overthrust above the Caracol Formation. Based on top determinations the Caracol Formation is right side up in the immediate vicinity of the mineral deposit.

MINERALOGY

Ore, associated gangue sulphide, and nonmetallic hypogene minerals occur generally as separate, discrete, small- to medium-size grains disseminated in the clayey matrix of the sedimentary rocks, as fillings and aggregates in fractures, joints, and bedding planes, and as vein matter along narrow, discontinuous faults. Galena, the iron-bearing variety of sphalerite (marmatite), and freibergite are the ore minerals. Galena and freibergite carry important amounts of silver. Pyrite, low-temperature pyrrhotite, arsenopyrite and marcasite are the gangue sulphides. They are closely associated spatially with the silver-lead-zinc mineralization.

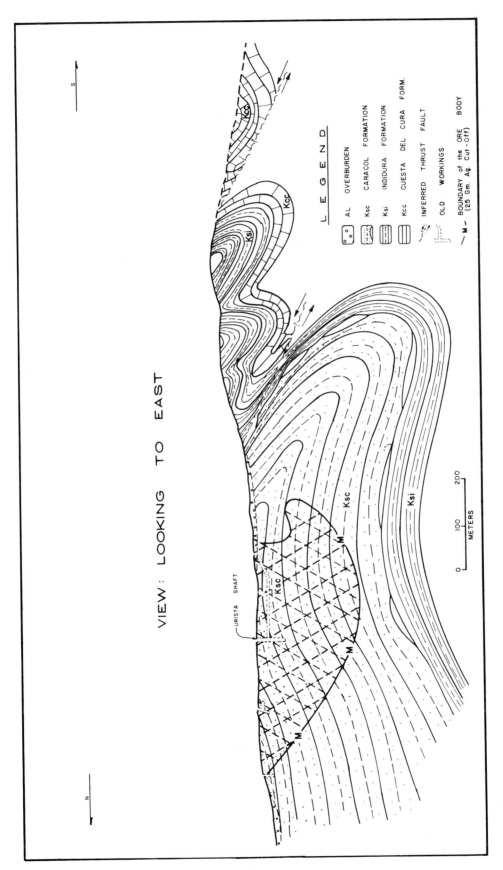

Figure 35.3. North-south geological cross-section through Real orebody.

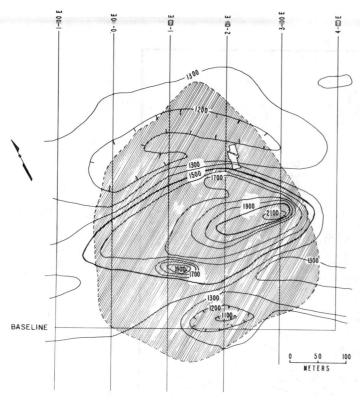

Figure 35.4. Ground vertical-field magnetic anomaly map; contour interval is 100 gamma. Real orebody is indicated by shaded area.

that cut the disseminated body. The high concentration of pyrrhotite contained in those veins, therefore, is most likely the cause for the anomaly. The fact that this central core also defines the zone where the mineral body lies closest to the surface is undoubtedly an important factor.

The well-developed arcuate 250 gamma low located along the northern margin of the Real deposit is a direct consequence of the relatively shallow inclination of the earth's magnetic field (I = 50°N) intersecting a fairly steeply dipping magnetic body. The characteristics of the anomaly in profile change along strike. The shape is smooth and broad at the west end, suggesting a deeper source. At the east end, the anomaly is characterized by a sharp, positive peak, bounded by two sharp but weaker positive lows. Actually the ore zone has limited depth extent and is centred directly under the positive magnetic high.

The magnetic data obtained from this survey show that the magnetic response from pyrrhotite could be used to determine the location and approximate areal extent of the Real silver-lead-zinc orebody. It should be mentioned that no magnetic anomaly was indicated by a ground survey over an area of anomalous electrical conductivity encountered about 1200 m northeast of the Real deposit and another located 1000 m southwest. These areas will be referred to in more detail under Induced Polarization Survey.

Induced Polarization Survey

A portable time-domain unit (Scintrex IPC-7 and IPR-8) and a three electrode (pole-dipole) array were used in the induced polarization survey. The transmitter had a maximum output of 5000 volts and five amperes, and the receiver was set for a pulse duration of 1.5 s and an integration time of 0.3 s, which conforms to the mean value of the curve. Seven parallel lines spaced 300 m apart and one transverse line, totalling in all 15 km of traverse, were surveyed in an area of about 2.2 x 2.0 km. Readings were taken along the lines using electrode spacings of 50, 100, and 200 m, and in selected segments at 300 and 400 m spacings. The remote electrode in each case was placed off to the south side of the line (see Fig. 35.5, 35.6, 35.7, and 35.8).

Because caliche is a very poor electrical conductor, electrical contact problems were expected. Fortunately, these did not materialize and currents sufficiently high to generate voltages above the strong SP noise level were achieved in all but a few cases. However, some deviation from the regular electrode spacings had to be made when rock dumps and rock rubble and buildings of the village were encountered.

Figure 35.5 shows the chargeability and apparent resistivity pseudosections obtained on traverse line 20N heading S 62° E looking northeast, and Figure 35.6 is a line (23E) at right angles to the former. Location of the traverse lines is shown on Figures 35.7 and 35.8. The relative position of the orebody on each traverse line is also shown.

The field work and data interpretation were done by the Mexican Government Consejo de Recursos Minerales. In both cases, the IP data are plotted on vertical pseudosections according to the Consejo de Recursos Minerales practice of using the centre of the three-electrode array as the data point co-ordinate, vertically below which are plotted the IP readings at scaled distances equal to the electrode spacing and then contoured. This representation should not be taken as an electrical section of the underlying ground; rather, it is a convenient schematic plot of results from various electrode spacings on a simple visual format. The Consejo has found this schematic procedure satisfactory for their purposes in Mexico.

Syngenetic, diagenetic and hydrothermal processes have been proposed by various geologists for the origin of the Real de Angeles deposit. In our opinion, evidence for hydrothermal deposition prevails. No intrusive rocks are known in the vicinity. Dr. G.L. Cumming of the Department of Physics at the University of Alberta (Canada) (pers. comm.) has determined isotopically that the galena in the deposit is probably less than 25 million years old.

Pyrrhotite is distributed through the whole mineral body, averaging about 5% by volume but locally may exceed 10%. No magnetite was detected in the ore. Total sulphide content is about 5-15% by volume, of which less than 3% corresponds to the ore minerals galena and sphalerite. The volume of sulphides decreases rapidly outside the deposit to one per cent or less with pyrite becoming the dominant sulphide mineral.

GROUND GEOPHYSICAL SURVEYS

Ground Magnetometer Survey

A hand-held, vertical-field magnetometer was used in an attempt to determine whether the magnetic pyrrhotite associated with the ore minerals would give a detectable, meaningful magnetic response. A favourable response was obtained on the initial test line across the Real deposit, so five additional lines spaced 100 m apart were surveyed. Readings were taken at 20 m intervals on each line. The results contoured in Figure 35.4 show a good correlation of the magnetic anomaly with the sulphide body. The anomaly consists of an 800 gamma high flanked on the north by a 150 gamma low and trends west-northwesterly within the mineral area. This direction conforms fairly well to the strike orientation and position of the major set of fault veins

It can be readily seen in Figure 35.5 that the chargeability within the orebody of as much as seven times background; i.e., 20 vs. 3 millivolts, particularly near the surface, and that the apparent resistivity low of 20 ohm-metres, or less than one tenth of background resistivity, are spatially related to the orebody.

Figure 35.6 shows an IP and apparent resistivity pseudosection on line 23E, as viewed looking N 62° W. Again, the coincident chargeability high and resistivity low are directly related to the sulphide body.

The strength of these electrical responses, in particular, the resistivity anomaly, is quite surprising given the low overall metallic sulphide content in the orebody. Some observers were skeptical about the results obtained by the IP survey, thinking that graphite shale layers might have been responsible, in spite of the fact that the graphite seems to have been leached out by the hydrothermal solutions and subsequent emplacement of the sulphide minerals. This uncertainty was settled by the results of laboratory tests on mineralized core of drillhole E-13, located just east of the centre of the mineralized area (Fig. 35.2). The results are shown in Table 35.1. There is a strong IP effect in the stockwork mineralized sample at 100.5 m, which has only 4% sulphide by volume. Evidently a very high degree of interconnected sulphide stringers reduces the resistivity of these rocks. Such stringers are quite common in the Real de Angeles ore.

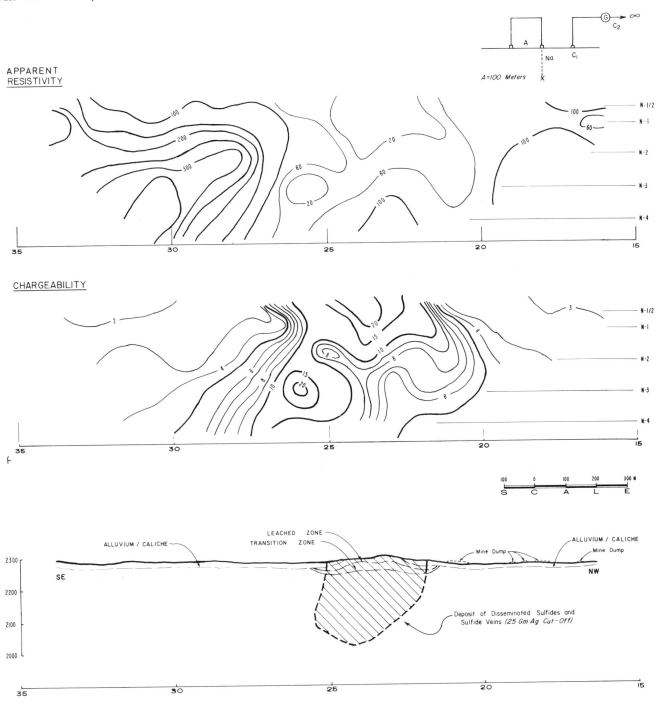

Figure 35.5. Vertical IP and apparent resistivity pseudo-sections on line 20 N oriented N28°E.

Figure 35.7 shows a resistivity contour plan produced from readings taken of the same area surveyed by IP at 100 m electrode spacings; the contour interval is 20 ohm-metres. The orebody outline is shown superimposed on the resistivity results for comparison purposes. The low resistivity anomaly coincides with the known location and nearsurface outline of the mineral body.

A chargeability contour map (Fig. 35.8) based on readings of the same 100-metre electrode spacings as before shows a similar coincidence of a strong chargeability anomaly high with the Real orebody. Another anomaly with moderate chargeabilities of 8 to 10 milliseconds was encountered about 1200 m northeast of the Real de Angeles body. Unlike the Real deposit, this later anomaly is not accompanied by a resistivity low or magnetic response. Several holes drilled in this anomaly to depths of 150 m encountered only weakly pyritized Caracol rocks. The small chargeability anomaly 1000 m southwest of the Real orebody has no corresponding apparent resistivity or magnetic anomaly. This anomaly was not drilled (Fig. 35.8).

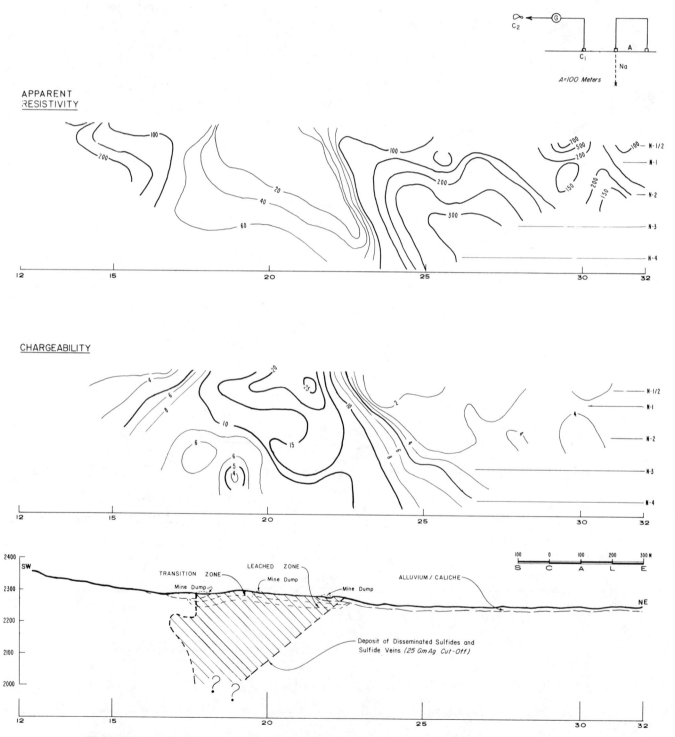

Figure 35.6. IP and apparent resistivity pseudo-sections on line 23 E oriented N62°W.

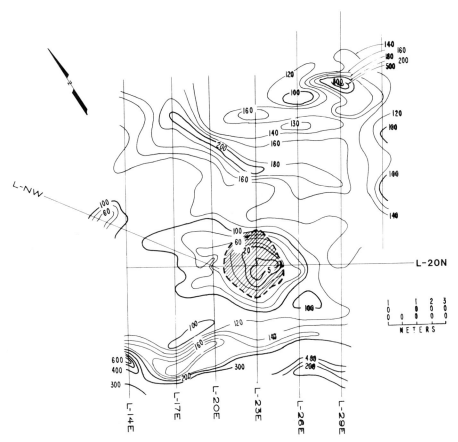

Figure 35.7. Apparent resistivity contour map; electrode spacing is 100 m; contour units are millivolts; Real orebody is shown by shaded area.

Table 35.1

Chargeability and mineralogical description of samples in drillhole E-13, Real de Angeles orebody

Specimen Depth (M)	% PFE	Chargeability Milliseconds	ASSAYS			Hole Interval (M)	Description
			%Pb	%Zn	Gm.Ag.		
88.5	12.7	50	1.95	1.40	180	87-90	Diss Py, Po, Sl, Gn, in grey 1 mm sandy matrix. 4% vol. sulphides. Spec. quite magnetic.
100.5	48	200	4.0	2.50	167	99-102	Same as above but network of 2-5 mm Gn, Po, Sl grains. Numerous hairlike veinlets. Whole rock wkly. magnetic, but Po sections strongly magnetic.
139.5	50	200	1.95	1.35	77	138-141	15 mm wide Po, Sl, Gn, $CaCO_3$ veinlet in fine grain arenite. No diss. sulphides. Only large magnetic Po grains.

(Abbrev.: Py (Pyrite), Po (Pyrrhotite), Sl (Sphalerite), Gn (Galena)).

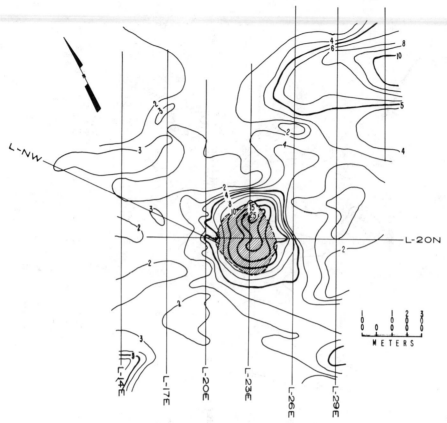

Figure 35.8. Chargeability contour map; electrode spacing is 100 m; contour units are ohm-metres; Real orebody is shown by shaded area.

Ground Electromagnetic Survey

A ground VLF-electromagnetic survey was attempted in the hope that massive sulphide veins and bedded layers of massive sulphides could be detected. The VLF signals from the transmitters at Seattle and Washington, USA, and Balboa, Panama were utilized. The survey proved ineffective because the orebody is not conductive enough to give a readable response.

DIAMOND DRILLING

EXPLOMIN and a previous exploration company together drilled a total of 13 123 m in 78 surface holes, a large majority of which were oriented S 28°W at angles of minus 45 to 75°. Hole depths ranged from 41 to 366 m, and, in the latter instance continuous sulphide mineralization was encountered to a hole depth of 362 m. Core size was principally NXWL. Also, two holes totalling 254 m were drilled to test the satellite geophysical anomaly located northeast of the Real de Angeles deposit.

MINERAL RESERVES

As a result of the foregoing development and exploration work, the Real de Angeles silver-lead-zinc deposit appears amenable to low-cost open pit mining methods because of its physical shape, low waste-to-ore ratio, and the uniform distribution of ore minerals therein. Ore reserves have been calculated at 51.1 million metric tons grading 78.4 g (2.5 oz.) silver per ton, and about 1% each of lead and zinc, using a 25 g silver cut-off. Recoverable amounts of cadmium and copper are also present. At the time of writing (September, 1977), a final decision to bring the deposit into production was awaiting further economic evaluations.

CONCLUSIONS

Ground geophysical surveys co-ordinated with core drilling and geological mapping were carried out in an attempt to use the magnetic and electrical properties of the more abundant gangue sulphide minerals as a guide to determine indirectly the limits of the Real de Angeles silver-lead-zinc orebody and to prospect for other blind orebodies under a caliche cover. The geophysical survey was a qualified success in correctly delineating extent of the Real de Angeles orebody. No additional orebodies were discovered as a result of this survey.

ACKNOWLEDGMENTS

The authors wish to express their appreciation to the Banco de Comercio, Comision de Fomento Minero, and Placer Development Ltd., the shareholders of Explomin S.A. de C.V. for permission to present this paper on the Real de Angeles mining property. The data used in this case history were drawn from technical reports written by staff geophysicists of the Mexican Consejo de Recursos Minerales while under contract to Explomin, and from technical input of the exploration staffs of both Explomin and Placer. The efforts of our colleagues and the Mexican Government geophysicists are gratefully acknowledged. Also, we wish to thank R.A. Rivera and R.K. Warren for their critical review of the technical data covered in this paper.

APPLICATION OF X-RAY DIFFRACTION ALTERATION AND GEOCHEMICAL TECHNIQUES AT SAN MANUEL, ARIZONA

D.M. Hausen
Newmont Exploration Limited, Danbury, Connecticut

Hausen, D.M., Application of X-ray diffraction alteration and geochemical techniques at San Manuel, Arizona; in Geophysics and Geochemistry in the Search for Metallic Ores; Peter J. Hood, editor; Geological Survey of Canada, Economic Geology Report 31, p. 735-744, 1979.

Abstract

Alteration zoning at the San Manuel-Kalamazoo porphyry deposit near San Manuel, Arizona has been evaluated by the X-ray diffraction method of monomineralic contouring. Over 350 drill and draft samples from the San Manuel segment have been analyzed by X-ray diffraction for major alteration minerals, including sericite, quartz, pyrite, K-feldspar, plagioclase, chlorite, etc., followed by plotting and contouring of XRD data.

Contouring of alteration data indicates that the San Manuel orebody is bound by concentric zones of alteration assemblages that outline the general shape of the ore zone. Alteration zones include: (a) external zones that surround the orebody as concentric outer envelopes of sericitization, silicification and pyritization, and (b) internal zones of high plagioclase-low K-feldspar (locally high biotite) that make up the porphyry core of the deposit, and (c) K-feldspar-rich zones that usually coincide with the orebody, but are occasionally discordant to the ore and alteration zones.

Monomineralic contouring for each type of alteration corroborates the general conclusion that the San Manuel orebody is terminated circumferentially and at the east end by closed contours, but remains open to the west and on top, where the Kalamazoo segment is believed to have been removed.

Résumé

On a étudié les auréoles d'altération du gisement porphyrique de San Manuel-Kalamazoo, situé près de San Manuel (Arizona), par la méthode consistant à tracer les contours lithogéochimiques de chaque élément identifié par diffraction X. On a analysé plus de 350 échantillons de surface ou recueillis par forage dans le secteur de San Manuel, par la méthode de diffraction X, afin d'identifier les principaux minéraux d'altération: séricite, quartz, pyrite, K-feldspath, plagioclase, chlorite, etc.; on a ensuite reporté les résultats et tracé les contours lithogéochimiques qui correspondent aux minéraux d'altération.

Le tracé des contours montre que le corps minéralisé de San Manuel est délimité par des auréoles concentriques d'altération, qui donnent la configuration générale de la zone minéralisée. Les auréoles d'altération sont constituées (a) des zones externes qui enveloppent le corps minéralisé, à savoir les auréoles concentriques, de séricitisation, de silicification et de pyritisation et (b) les zones internes riches en plagioclase et pauvres en K-feldspath (avec localement une forte concentration de biotite), qui forment le noyau porphyrique du gisement et (c) les zones riches en K-feldspath, qui coïncident habituellement avec le corps minéralisé, mais ne concordent pas toujours avec les zones minéralisées ni avec les auréoles d'altération.

Le tracé géolithochimique correspondant à chaque minéral indicateur pour toutes les phases d'altération, confirme la conclusion générale à laquelle on était parvenu à savoir que le corps minéralisé de San Manuel est bordé à l'est par des contours fermés, mais des contours ouverts à l'ouest et à son sommet, le segment de Kalamazoo ayant probablement été érodé.

INTRODUCTION

In recent years, Newmont Exploration Limited (NEL) has developed a quantitative method for the measurement of alteration minerals in altered wall rock. This method, termed monomineralic contouring, has been successfully demonstrated in the evaluation of various mineralized prospects (Hausen and Kerr, 1971; Hausen, 1973), and applied to the detailed investigation of alteration trends at the San Manuel porphyry copper deposit, San Manuel, Arizona.

This alteration study correlates pervasive wall rock alteration with copper mineralization at San Manuel, and provides a test case for the contouring of XRD alteration data around mineralized centres in a large porphyry copper deposit. Techniques of monomineralic contouring are described from a case history aspect at San Manuel, and as a state-of-the-art development in mineral exploration, supplementing conventional methods of geochemistry and geophysics.

HISTORY AND DEVELOPMENT

Magma Copper Company, a wholly owned subsidiary of Newmont Mining Corporation, operates one of the world's largest underground copper production facilities at San Manuel, Arizona. The San Manuel copper deposit is in the desert valley of the San Pedro River between the Santa Catalina and Galiuro mountains, about 35 miles northeast of Tucson, in the Old Hat District, Pinal County, Arizona (Fig. 36.1).

Early exploration at San Manuel included drilling by the U.S. Bureau of Mines in 1944, which established the existence of the copper deposit (Schwartz, 1953). The Bureau discontinued work in 1945 when the Magma Copper Company obtained an option on the property. Development churn drilling was continued by Magma from 1945 to 1947, conforming to the original Bureau of Mines' drill pattern. By 1947, approximately 120 million tons of ore averaging 0.8 per

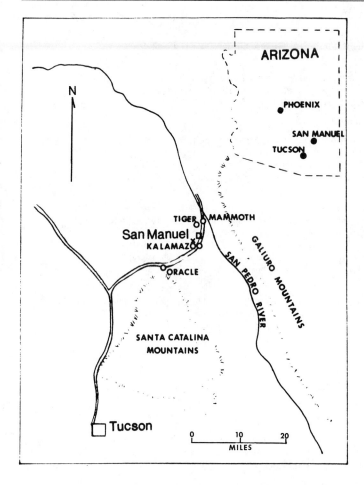

Figure 36.1. Location map of San Manuel area, Arizona.

cent copper (Steele and Rubly, 1947) were blocked out. This resulted in the development of the San Manuel underground block-caving mine which has a current production capacity of about 60 000 tons per day.

An exploration project west of the San Manuel deposit was initiated by Quintana Minerals Corporation in 1967, resulting in the discovery of the down-faulted segment of the San Manuel orebody. The project was based on the concept by J.D. Lowell (Lowell, 1968) that the original orebody was cylindrical with concentric alteration zoning, the orebody being first tilted approximately 70°, then bisected by the flat San Manuel normal fault into the lower plate San Manuel segment and the upper plate Kalamazoo segment. Reserves of the combined San Manuel-Kalamazoo orebody are reported at approximately a billion tons of 0.75 per cent copper (Dayton, 1972).

According to Lowell (op. cit.), the Kalamazoo orebody has the shape of an overturned canoe, complementing the upturned San Manuel ore shell to the east. Development drilling was guided by geology and mineral zone symmetry extrapolated from the San Manuel deposit.

Recent geochemical investigations by Chaffee (1976) on drill samples from Kalamazoo indicated that sulphur and selenium offer considerable promise as pathfinder elements. According to L.A. Thomas (pers. comm.), the inner edge of the selenium anomaly at San Manuel and Kalamazoo is essentially coincident with the inner edge of the ore shell. The outer edge of anomalous selenium is significantly outside of the ore zone, extending for distances up to 1000 feet (305 m), or more, into the outer pyritic-sericitic zones of altered wall rock.

GEOLOGIC BACKGROUND

Most of the San Manuel area is underlain by Gila Conglomerate, consisting of coarse boulder conglomerate to fine marly silt, representing late fanglomerates and playa (or lake) deposits, respectively.

The Cloudburst Formation of late Cretaceous or early Tertiary age consists of more than 6000 feet (1800 m) of intercalated fanglomerates and propylitized latite flows which unconformably underlie the Gila Conglomerate, and overlie the Precambrian granitic rocks of the area.

A large area of volcanic rocks (basalt flows, flow breccias, and agglomerate) occurs north and northwest of Tiger; these appear to lie below the main beds of the Gila Conglomerate. Dykes and irregular masses of rhyolite, which cut all formations older than the Gila, are also found over a wide area.

Precambrian basement rocks comprise mostly Oracle quartz monzonite and an older granodiorite. Leucocratic rocks (aplite and alaskite) appear to be related but slightly later than the Oracle quartz monzonite. The crosscutting relations of late diabase dykes suggest partly Precambrian and partly Cretaceous age (Creasey, 1965).

Laramide porphyry intrusives ranging from dacite, granodiorite, and latite to monzonite occur as small irregularly-shaped masses and dykes in the Oracle quartz monzonite, and are the principal host rock for the disseminated copper ores of the San Manuel deposit.

The San Manuel deposit is situated within a central concentration of Laramide intrusives. Much of the orebody lies within the porphyry intrusives and along highly brecciated contacts with the Oracle quartz monzonite. The deposit is roughly 2000 to 3000 feet (600 to 900 m) in overall diameter, more than 6000 feet (1800 m) in length, and is developed in multiple, rod-shaped injections of porphyry that interfinger with lenticular slices or fragments of Oracle quartz monzonite. Most of the elongate Oracle fragments and porphyry plugs have a similar orientation and are parallel to the cylindrical axis of the ore shell.

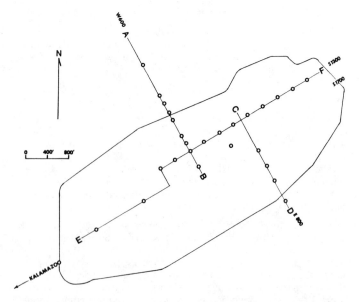

Figure 36.2. San Manuel plan map, sections AB, CD, and EF.

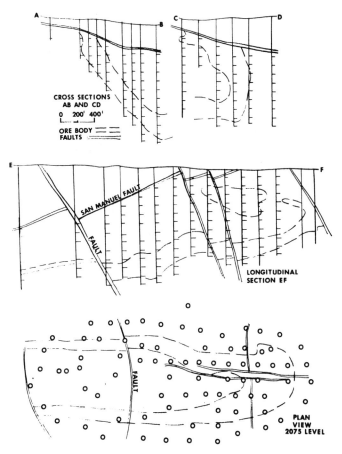

Figure 36.3. Sampling patterns along sections in San Manuel deposit.

Rock types in the ore zone are commonly heterogeneous and hybrid in appearance, often consisting of mixtures of porphyry and Oracle quartz monzonite. Such mixtures have been highly sheared, K-feldspathized and recrystallized, and appear to represent hydrothermally-healed shear zones. Hybrid zones, consisting mostly of Oracle quartz monzonite, appear to have been remobilized during shearing and multi-injections of porphyry; these zones probably represent major avenues of hydrothermal solution movement during the period of mineralization.

LOCATION AND DESCRIPTION OF SAMPLING

Whole rock samples from the San Manuel orebody were obtained along several sections, including two cross-sections, one longitudinal section, and one horizontal section through the 2075 level of the mine (Fig. 36.2). Most of the samples were collected from churn drill cuttings obtained in the 1940s and retained in storage for over 30 years. The irreplaceable nature of these samples cannot be overstated, since they provide the means for reconstructing the original alteration zoning of the deposit. The deposit has since been block-caved to a major degree and is inaccessible to sampling.

Ten-foot (3 m) samples were composited every 200 feet (60 m) along equivalent horizons (elevations) from each drillhole through the longitudinal and cross-sections (Fig. 36.3). Samples from the Plan view (2075 level) were also selected from churn drill cuttings, and supplemented by additional chip samples collected by hand from locations in haulage drifts at this level where no previous churn drilling existed (Fig. 36.3). The grid spacing of sampling at this level was between 200-500 feet (60-150 m).

Most of the churn drill pulps were received in a sufficiently pulverized condition for direct analysis by X-ray diffraction.

FEATURES OF MINERALIZATION

The semi-elliptical shape of the San Manuel orebody in cross-section is outlined by the 0.6 per cent Cu contour. The 0.3 and 0.1 per cent Cu contours show the distribution of lower grade copper mineralization around the orebody.

The orebody outlined in cross-sections (Fig. 36.4) is distributed about equally between the Oracle quartz monzonite on the west and the Laramide porphyry on the east.

If the two halves were brought into juxtaposition, the orebody would appear as a tilted U-shaped configuration, with the open portion of the "U" oriented upwards toward the west.

Contours of copper mineralization are generally smooth and uniformly spaced, indicative of gradational decreases away from the ore zone (Fig. 36.5). The eastern perimeter of the deposit shows a relatively wide zone of low grade copper mineralization (0.1-0.2 per cent) persisting for an indeterminate distance. This greater thickness of low grade copper dissemination is related in part to the sharp flexure (greatest curvature) of the ore shell, and in part to the porphyry host rock which commonly carries low grade copper mineralization.

The orebody in longitudinal section (defined by the 0.6 per cent Cu contour) appears for the most part as a planar-shaped layer (Fig. 36.4), plunging gently towards the southwest, locally segmented by steeply-dipping normal faults to the east, and largely terminated by the Hangover Fault on the west. Copper mineralization continues on the west side

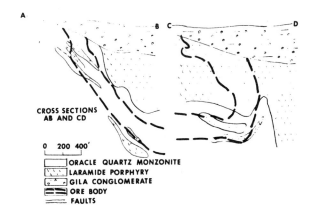

Figure 36.4. Geologic cross-sections of San Manuel deposit.

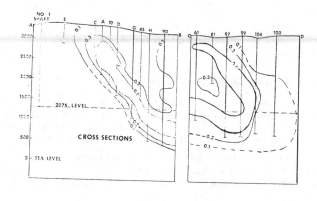

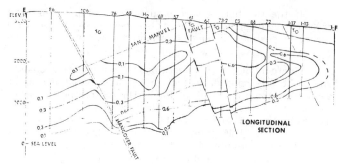

Figure 36.5. Per cent copper distribution in San Manuel sections.

of the fault, but at lower grades and at higher elevations. The orebody throughout much of its extent follows the contact between the Oracle quartz monzonite and the Laramide porphyry.

A major flexure in the orebody occurs at the eastern end, where the zone of high copper values (>0.6 per cent Cu) swings upward and rolls back towards the east at higher elevations (Fig. 36.4, 36.5). This upper ore zone is largely oxidized and variable in thickness and continuity.

The San Manuel orebody follows an irregular inter-tongued contact in the horizontal section through the 2075 level of the mine (Fig. 36.6). The orebody at this level (defined by the 0.6 per cent contour) has a horseshoe-shaped configuration, with the open end oriented toward the Kalamazoo Block to the southwest. Copper contours are uniformly spaced, indicative of gradational changes in copper values away from the ore zone.

TECHNIQUES OF X-RAY DIFFRACTION ANALYSIS

X-ray diffraction analysis provides a direct measure of most major alteration features, including sericitization, K-feldspathization, silicification, and pyritization, and more qualitatively of chloritization, kaolinization, and carbonatization. The use of such facies terms as "potassic" and "propylitic" are thus supplemented by mineralogic terms such as "K-feldspathic", "sericitic", "chloritic", "pyritic", etc., which relate directly to the distribution of their mineralogic equivalents, e.g., K-feldspar, sericite, chlorite, pyrite, etc., as alteration replacements in wall rocks.

A transistorized Norelco X-ray diffractometer with a wide range goniometer, a curved graphite crystal monochromator, and a transistorized Honeywell recorder was utilized in obtaining X-ray diffraction patterns for this investigation. Pulp samples were pulverized to nominal 200 mesh at San Manuel, and scanned from 2 through 40 degrees, 2θ, at a scanning speed of 2 degrees per minute, using Cu Kα radiation. Samples were continuously scanned

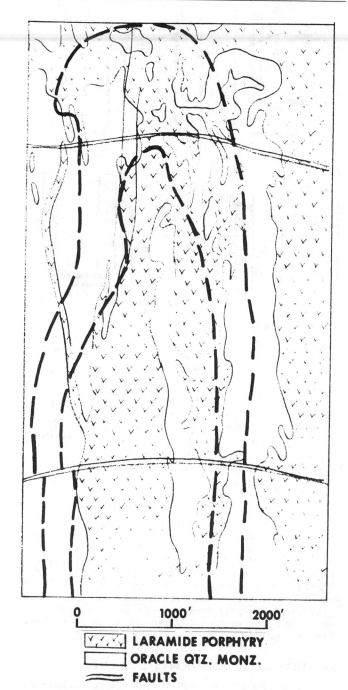

Figure 36.6. Geologic plan of 2075 level.

and rotated at an operating voltage of 40 kv and 25 ma at a sensitivity of 1000 cps. Sensitivity and alignment were checked periodically, using a standard quartz mount, to insure minimum instrumental deviation.

Special care was taken to prepare sample surfaces that were relatively reproducible under X-ray analysis. Preferred orientation of mineral grains is the source of largest error, and is minimized by fine grinding and random packing. Random orientation is approached by impressing the surface with a grid design similar to that described by Peters (1970), which is called simply a Peters grid.

Sample holders were one-inch (2.54 cm) in diameter and recessed 0.020 inch (0.051 cm) with a 0.020 inch (0.051 cm) thick rim to retain about 1/2 gram of pulverized sample.

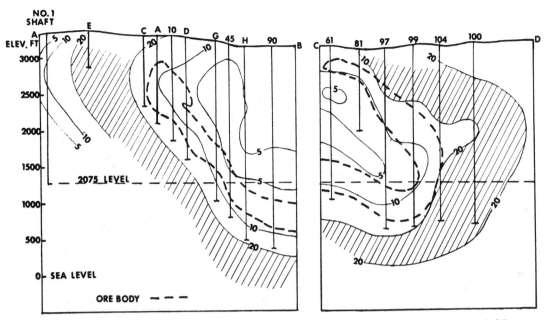

Figure 36.7. Sericitization: per cent sericite distribution in cross-sections AB and CD.

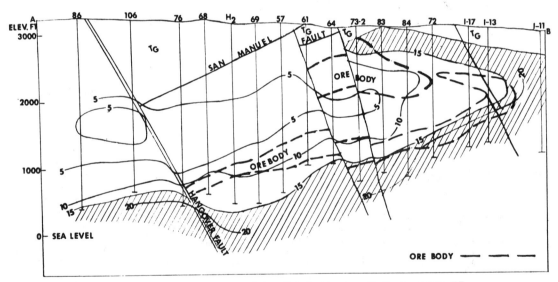

Figure 36.8. Per cent sericite distribution longitudinal section EF.

The sample, when ready for scanning, was placed in a Philips rotating flat specimen holder, Type No. 52413, and rotated continuously during analysis.

Peak intensities for characteristic reflections of each mineral were calculated into weight percentages by means of computer processing, using standard curves for each mineral and normalizing to 100 per cent. The mineralogic composition for each sample was calculated in the same manner, to provide semiquantitative weight percentages for quartz, plagioclase, K-feldspar, sericite, chlorite, calcite, pyrite, amphiboles, etc.

For quartz, peak intensities for only the 100 (4.26Å) reflection were measured as a guide to silicification. Only one peak was required for reproducible analyses, because of the refinements in use of a rotating sample holder and Peters grid.

For pyrite, peak intensities for the 200 (2.71Å) reflection were measured as a guide to pyritization. This peak has no apparent interferences, other than the 104 spacing for hematite near 2.68Å, and shows good reproducibility using a Peters grid and a rotating sample holder.

For sericite, peak intensities for two reflections, including the 004 (4.97Å), and 110 (4.47Å) spacings, corresponding to $2M_1$ muscovite, were measured, integrated, and converted into estimated percentages as a guide to sericitic alteration. Coarse grained micas including biotite were estimated largely from the 001 (10.0Å) spacing, to which some intensities are contributed by sericite and illite.

In the evaluation of K-feldspathic alteration, a ratio method was used, whereby the intensity of the 050 spacing for K-feldspar (3.24Å) is divided by that of plagioclase (3.18Å), and expressed as a quotient.

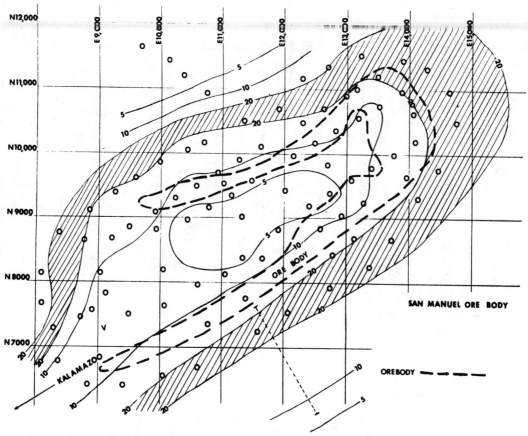

Figure 36.9. Sericitization: per cent sericite distribution horizontal section 2075 level.

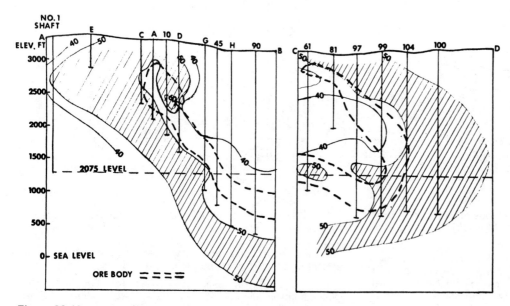

Figure 36.10. Silicification: per cent quartz distribution in cross-sections AB and CD.

Coefficients of variation have been compared for deviations of 1, 2, and 3 sigma, representing, respectively, the 65, 95, and 99.7 percentage confidence levels. Of the minerals tested, quartz shows the lowest variation, ranging between 5 and 10 per cent for one sigma, and averaging near 7 per cent for all of the sets (n = 60). Coarse micas show the highest variation, averaging near 15.6 per cent for all of the sets. Plagioclase shows relatively close precision (averaging near 8.5 per cent for one sigma), whereas sericite, chlorite, and calcite tend to group between 13 to 15 per cent.

The occasional large deviations in XRD analysis are usually attributed to preferred orientation of cleavage flakes, a feature that is often a problem in analyzing feldspars and sericitic micas. The problem is more serious using a flat pack instead of a gridded pack. However, major alteration trends are usually discernible in spite of higher deviations, because of the multiplicity of data points in contouring.

On the basis of repetitive mineral analyses, the XRD data obtained by the methods described are considered sufficiently precise (although semi-quantitative) for contouring alteration data and evaluating mineralizing trends in mineral exploration.

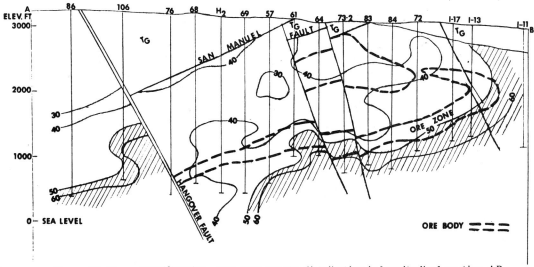

Figure 36.11. Silicification: per cent quartz distribution in longitudinal section AB.

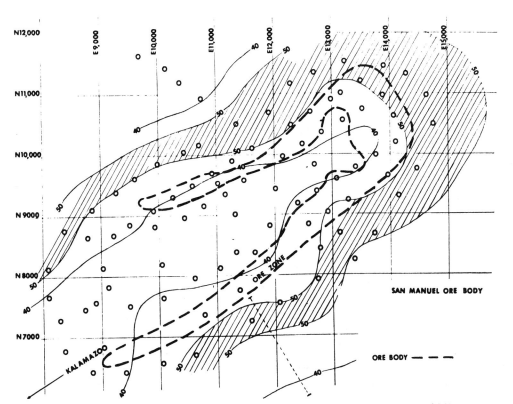

Figure 36.12. Silicification: per cent quartz distribution horizontal section 2075 level.

FEATURES OF ALTERATION

Relevant alteration minerals, including sericite, quartz, feldspars, pyrite, etc., were analyzed by X-ray diffraction, as described above. Semi-quantitative values for each alteration mineral are plotted and contoured in sections, analogous to copper values, to outline the trends of wall rock alteration.

Sericitization

Estimated percentages of sericite are plotted and contoured in cross-section in Figure 36.7, showing a conformable halo of sericitization surrounding the orebody. Sericite values range from about 5 per cent or less in samples from the inner porphyry core and outermost zones away from ore, to over 20 per cent immediately outside the ore zone. A zone of high sericite follows the outer perimeter of the ore zone, thickening along the eastern margin. The 5 and 10 per cent sericite contours within the inner porphyry core are conformable to the orebody and to the sericite high around the exterior margins of the ore shell, and reflect the overall symmetry of the deposit. Contours are closed at both east and west ends of the deposit and appear to swing around the top of the deposit, intersecting the surface.

In longitudinal section (Fig. 36.8), the zone of intense sericitic alteration occurs as a blanket layer, conformable to the lower margins of the primary ore zone and the contact zone between the porphyry and underlying Oracle quartz monzonite. Contours of sericitization (5, 10, 15, and 20 per cent) are conformable to the general outline of copper mineralization and appear to swing around the top of the upper oxidized orebody on the east, and to intersect the surface in the vicinity of the San Manuel fault.

In horizontal section (Fig. 36.9), sericitization occurs as a conformable halo, surrounding the orebody and outlining its general U-shaped configuration. The zone of highest sericite values (20 per cent or more) averages about 500 feet (150 m) in thickness, becoming thicker along the major flexure at the northeast end of the orebody. Sericite diminishes inside and outside of this curvilinear blanket around the ore shell, decreasing to 10-15 per cent within the ore shell, and to less than 10 per cent within the central porphyry core. Sericite also decreases along the exterior margins, although more gradually on the north and east sides within porphyry. Contours remain open to the southwest towards the Kalamazoo segment.

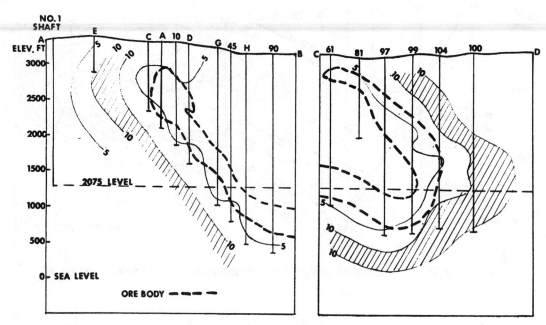

Figure 36.13. Pyritization: per cent pyrite distribution cross-sections AB and CD.

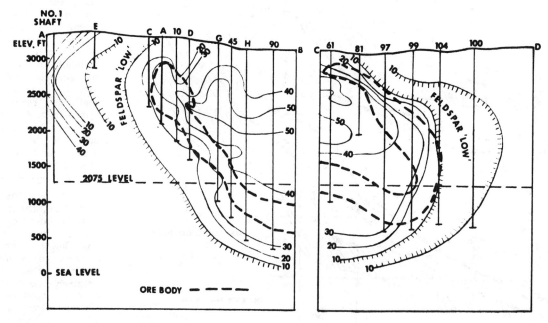

Figure 36.14. Percentages of total feldspar in cross-sections AB and CD.

Silicification

Most features of the silicification in cross-section (Fig. 36.10) are conformable to the orebody and similar in shape and dimensions to the sericite halo. Anomalous quartz values, defined by the 50 per cent contour, occur as a crescent-shaped zone along the perimeter of the orebody, extending away from the ore for distances up to 1000 feet (300 m) or more. Thickening of this zone occurs along the eastern margin.

Contours of decreasing silicification (40 and 30 per cent quartz) show conformable relations to the orebody along the inner and exterior margins of the deposit. Contours are closed inside the orebody, but open upward toward the surface where they lap over at both ends in a recumbent manner.

The features of the silicification in longitudinal section are also conformable (Fig. 36.11). High quartz values occur as a peripheral zone of silicification underlying the primary ore zone. The thickness of this exterior fringe cannot be determined because none of the holes penetrated through this zone. Within the ore shell contours of decreasing silicification are concentrically distributed, showing conformable geometry to the orebody and extending around the eastern nose of the deposit.

In horizontal section (Fig. 36.12), high quartz values form a continuously concordant zone paralleling the ore shell and overlapping the sericite blanket outside of the ore zone. The silicification halo has similar dimensions and shape as the sericitization halo, and shows similar decreases in intensity both inside and outside of the ore shell. Contours are open at the southwest end toward the Kalamazoo segment.

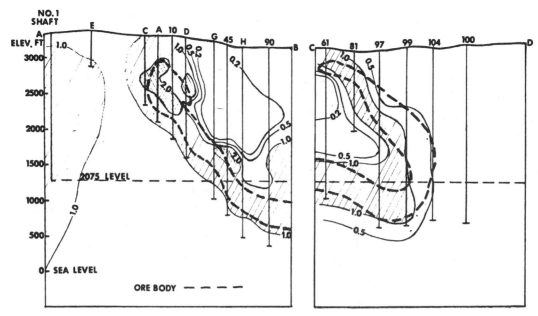

Figure 36.15. K-feldspathization: ratios of K-feldspar/plagioclase in cross-sections AB and CD.

Pyritization

In cross-section a zone of pyritization, defined by the 10 per cent contour, follows the perimeter of the outer ore boundary, surrounding the orebody as a crescent-shaped halo (Fig. 36.13). Contours are generally closed around the ore shell, except toward the surface.

In longitudinal section, pyritization is limited mostly to the underside of the primary ore zone, where pyrite values commonly range up to 8 to 10 per cent. No attempt was made to extrapolate the zone of high pyrite around the eastern closure of the orebody, because of the oxidized condition of samples from the upper ore zone.

In horizontal section, a similar zone of moderate pyritization (5-10 per cent pyrite) surrounds the ore shell, defining its general shape, and is similar in form to the zones of sericitization and silicification. Contours are generally open towards the Kalamazoo orebody to the southwest.

Feldspar Distributions

Two parameters were chosen to define the changes in feldspar distribution through the San Manuel orebody, (1) percentages of total feldspars (plagioclase and K-feldspars), and (2) ratios of K-feldspar/plagioclase.

Contours of percentages of total feldspars in cross-section (Fig. 36.14) reveal a strong feldspar low, lying just outside of the ore zone, overlapping zones of sericitization, silicification, and pyritization. Total feldspars comprise less than 10 per cent of the rock in this zone, and apparently have been replaced in most part by sericite, quartz, and pyrite. Contouring of percentage feldspar lows may thus be used to considerable advantage, since they magnify the cumulative effects of various types of alteration.

Amounts of feldspar increase inward through the orebody, and into the inner porphyry core, as defined by the contours 20, 30, 40, and 50 per cent feldspar. Each contour is conformable to the ore zone and may be used to delineate roughly the general shape of the orebody.

Total percentage of feldspars also increases sharply on the external side of the feldspar low (away from the deposit). Contours generally enclose the entire orebody, except at surface, where they remain largely open.

In longitudinal section, feldspar "lows", defined by the 10 per cent contour, follow continuously the lower boundary of the ore zone, locally offset by normal faults. Contour closures around the ore flexure at the east end are inferred for the 30 and 20 per cent contours. Contours are open above most of the orebody to the west, where the inner core is exposed. This is the portion of the orebody from which the Kalamazoo segment is assumed to have been removed.

Segments of the San Manuel orebody and associated alteration zones are in apparent agreement with concept of Lowell (1968) of a faulted cylindrical ore shell for the two segments.

The distribution of total feldspar in horizontal section reflects the concentric horseshoe-shaped outline of the orebody. A feldspar low surrounds most of the ore shell, and generally overlaps zones of high sericite, quartz, and pyrite. Contours of increasing feldspars are well defined inside the barren core, as well as exterior to the deposit. Contours are open to the southwest towards Kalamazoo, as in the case of other forms of alteration.

Ratios of K-feldspar/plagioclase were contoured in cross-section in Figure 36.15 to outline the zones of K-feldspathization. Zones of anomalous K-feldspar follow the inner part of the ore shell throughout the deposit. This feature is also confirmed by the contoured distributions in longitudinal and horizontal sections. This zone of high K-feldspathization (as defined by the 0.1 contour) lies just inside the zone of sericitization within the ore shell, and extends for several hundred feet into the barren core. However, the inner edge grades sharply into relatively unaltered (slightly propylitized) porphyry with low K-feldspar and high plagioclase.

PRELIMINARY COMPARISON OF SAN MANUEL SEGMENTS WITH LOWELL'S MODEL

Sufficient alteration data are available from cross-sections of the San Manuel orebody for limited comparison with the cylindrical model proposed by J.D. Lowell (1968) for the Kalamazoo and San Manuel segments. According to Lowell, the two orebodies represent one cylindrical-shaped orebody with concentric alteration zoning that was tilted and bisected by the San Manuel fault.

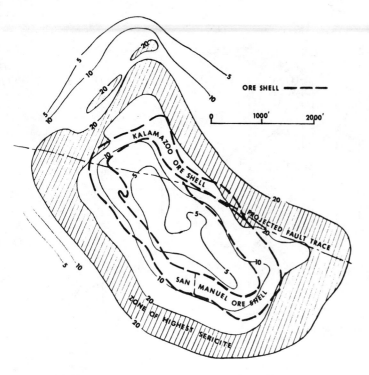

Figure 36.16. Sericite halo around composite San Manuel-Kalamazoo ore shell.

Samples available from churn drilling during the mid-1940s in the San Manuel segment indicate a U-shaped orebody with concentric zones of wall rock alteration, similar to the Kalamazoo segment (unpublished NEL Report), except that the zoning is inverted, following the underside of the San Manuel orebody compared with the topside at Kalamazoo.

Alteration contours are open above most of the San Manuel segment to the west, where the inner core is exposed along the San Manuel fault. This portion of the deposit provides a reasonably good geometric fit with the Kalamazoo segment, where contours of alteration and mineralization are open (unpublished NEL Report) and are believed to connect between the two ore segments (Fig. 36.16).

This evidence strongly supports the conceptual cylindrical model proposed by Lowell (1968). The repetitive patterns for each type of alteration (sericitization, silicification, pyritization, etc.) all show similar zonal distributions that reach maximum intensities along the outside perimeter of the ore zone and display textbook concentric symmetry around the ore. On the basis of this supportive evidence, a distinction can be made with some certainty between the outside and inside boundaries of the ore shell from any of its segments.

GENERAL CONCLUSIONS

X-ray diffraction analysis of alteration minerals has been found useful and practical in delineating trends in wall rock alteration in the San Manuel deposit. The relationship of alteration to mineralization is readily apparent after comparison of monomineralic contouring with that of copper.

Alteration contouring should find increasingly wider application in the evaluation of wall rock alteration, supplementing the interpretation and use of such classification terms as propylitic, phyllic, potassic, argillic, etc., which depend mostly on the qualitative and deductive judgment and experience of the observer.

On the basis of alteration contouring to date, major mineralized trends at San Manuel are clearly defined by intense zones of alteration. Chances for discovery of additional significant tonnages of ore do not appear favourable without the association of these zones of intense alteration.

ACKNOWLEDGMENTS

Acknowledgments are extended to Mr. L.A. Thomas, Chief Planning and Geological Engineer, and his mine staff at San Manuel, for their excellent assistance and considerable efforts in providing representative whole rock sampling, supplemented by detailed maps, assays, and descriptive field data, essential for this type of study.

REFERENCES

Chaffee, M.A.
 1976: The zonal distribution of selected elements above the Kalamazoo porphyry copper deposit, San Manuel District, Pinal County, Arizona; J. Geochem. Explor., v. 5, p. 145-165.

Creasey, S.C.
 1965: Geology of the San Manuel Area, Pinal County, Arizona; U.S. Geol. Surv., Prof. Paper 471.

Dayton, S.
 1972: Magma closes the mine to market gap; Eng. Min. J., April, p. 73-83.

Hausen, D.M.
 1973: Application of quantitative mineralogy by X-ray diffraction to problems in mineral exploration; in Quantitative Mineral Exploration, Colo. Sch. Mines Q., v. 68 (1), p. 61-85.

Hausen, D.M. and Kerr, P.F.
 1971: X-ray diffraction methods of evaluating potassium silicate alteration in porphyry mineralization; in Geochemical Exploration, Editors R.W. Boyle and J.I. McGerrigle, Can. Inst. Min. Met., Spec. Vol. 11, p. 334-340.

Lowell, J.D.
 1968: Geology of the Kalamazoo orebody, San Manuel District, Arizona; Econ. Geol., v. 63, p. 645-654.

Peters, T.
 1970: A simple device to avoid orientation effects in X-ray diffraction samples; Norelco Reporter, v. 17 (2), Dec., p. 23-24.

Schwartz, G.M.
 1953: Geology of the San Manuel copper deposit, Arizona; U.S. Geol. Surv. Prof. Paper 256.

Steele, J.H. and Rubly, G.R.
 1947: San Manuel prospect; Am. Inst. Min. Eng., Tech. Publ. 2255.

EXPLORATION FOR MASSIVE SULPHIDES IN DESERT AREAS USING THE GROUND PULSE ELECTROMAGNETIC METHOD

Duncan Crone
Crone Geophysics Limited, Mississauga, Ontario

Crone, Duncan, Exploration for massive sulphides in desert areas using the ground pulse electromagnetic method; in Geophysics and Geochemistry in the Search for Metallic Ores; Peter J. Hood, editor; Geological Survey of Canada, Economic Geology Report 31, p. 745-755, 1979.

Abstract

Examples of ground Pulse electromagnetic surveys from Arizona, the Sultanate of Oman, and Australia, show that both massive and fracture-filling sulphide bodies can be detected in desert conditions. The wide frequency spectrum of Pulse EM equipment, and its capability of defining the shape of the conductor, enable the method to differentiate between oxidized sulphides and conductive surficial layers, even though the conductivity contrast is slight. It is important for exploration purposes to retain the high frequency portion of the Pulse electromagnetic spectrum. Low sulphide content marker horizons and narrow oxidized sulphide zones are usually detectable only at high frequencies. New detailed and deep penetration methods utilizing the ground Pulse EM are being developed to locate accurately the position of the sulphide body; as exploration methods reach greater depths of penetration, this becomes increasingly important. Borehole Pulse electromagnetic equipment has been built that will detect sulphide bodies 100 m to the side of a borehole. The capabilities of Borehole Pulse EM surveys in detecting and defining the position of sulphide bodies, should encourage deep exploration in the vicinity of known mineral deposits.

Résumé

Les exemples d'études électromagnétiques au sol en Arizona, dans le Sultanat d'Oman et en Australie, montrent que les corps sulfurés massifs et ceux remplissant les fractures peuvent être décelés dans les milieux désertiques. La vaste gamme de fréquences de l'appareillage et la capacité de celui-ci à définir la forme du conducteur, permettent de distinguer les sulfures oxydés des couches superficielles conductrices, même quand le contraste de conductivité est faible. Il est important pour l'exploration de garder seulement la gamme de hautes fréquences. Les horizons repères à faible taux de sulfures et les étroites zones de sulfures oxydés sont normalement détectés seulement aux hautes fréquences. On a mis au point de nouvelles méthodes détaillées, de prospection profonde utilisant les ondes électromagnétiques au sol, qui permettent de localiser avec plus de précision les minerais sulfurés; cet aspect de la prospection prend de l'importance, à mesure que les méthodes d'exploration permettent une exploration plus profonde du sous-sol. On a construit un appareil de prospection EM utilisé dans les forages, pour repérer les minerais sulfurés dans un rayon de 100 m à partir du sondage. Les possibilités offertes par la méthode de levés EM à partir de forages pour déceler et délimiter les sulfures, devraient encourager l'exploration profonde à proximité des gîtes minéraux connus.

EXPLORATION OBJECTIVES AND PROBLEMS ENCOUNTERED IN DESERT AREAS

A ground geophysical survey for mineral exploration normally has three specific objectives:

— to locate an airborne electromagnetic anomaly on the ground, or to discover an anomaly that could be an orebody;
— to provide sufficient information to permit an evaluation of the anomaly in comparison with other anomalies;
— to obtain results that allow determination of the dip, depth and width of the target with sufficient accuracy to position an exploration drillhole.

When an exploration program for massive sulphide deposits is carried out in a desert region, attaining these objectives becomes increasingly difficult. The primary difficulty is surface weathering that gradually reduces the conducting sulphides to resistive oxides. This is a highly variable process whose effect may range from a few metres to 200 metres in depth from surface. The oxidation weathering process also tends to break down the inter-crystal electronic connection within a sulphide body. The presence of even minor oxidation can drastically reduce the conductivity of a massive sulphide body. Targets in desert areas are therefore usually deeper and are weaker conductors than those encountered in unweathered areas. The desert climate also produces large areas of high surficial conductivity that may consist of brackish groundwater or conductive rock formations such as conglomerates or limestones. This surficial conductivity reduces the penetration of electrical and electromagnetic (EM) methods and creates a background of confusing spurious anomalies.

Induced polarization has traditionally been the most effective geophysical method in the exploration for sulphides in desert areas. The method is suited primarily for the detection of large disseminated deposits, but it is not effective in the exploration for smaller, massive sulphide bodies (Dolan, 1967). Most conventional EM systems have been designed for use in resistive environments and can be usefully applied, only in nonconductive desert areas. Thus suitable ground geophysical equipment that would be effective in the search for massive sulphides in desert areas has, in the past, not been available.

DEVELOPMENT OF THE GROUND PULSE EM METHOD

The Pulse EM system was selected by Crone Geophysics Ltd. as an exploration tool since it appeared to have the most likely capability of providing conductivity, depth, dip and width information for subsurface conductors. The wide frequency spectrum of measurement of a Pulse EM system is capable of resolving the variance of conductivity encountered under desert conditions, and the low frequency portion of the spectrum can penetrate through the surficial conductive layer. Crone Geophysics Limited initiated a Pulse EM development program in 1972 with the co-operation of Newmont Mining Corporation. Newmont held the original Pulse EM patents (Wait, 1956) and had developed a large Pulse EM instrument which had been used successfully in Cyprus (Dolan, 1967). The Crone equipment (Crone, 1975) consists of a moving horizontal loop system; two persons operate the transmitter and one the receiver. The transmitter-receiver coil separation is 50 to 150 m. The transmitter is a multiturn loop of wire 6 to 15 m in diameter laid out in a rough circle on the ground. The current waveform is 10.8 ms on, 10.8 ms off with a 1.4 ms ramp shut-off. Eight delay time-windows, or channels, of the secondary field are sampled after the current shut-off at 0.15, 0.30, 0.55, 0.90, 1.45, 2.40, 4.00, and 6.40 milliseconds to the centre of the sample. The sample amplitude is normalized by setting to 1000, a sample taken of the maximum shut-off voltage amplitude measured at the receiver. The sample measurements are therefore without dimensions. The first sample (0.15 ms) is in units of 1/1000 of the shut-off sample, the eighth sample in units of 1/10 000 of the shut-off sample, with a logarithmic dispersion in between. Unlike conventional horizontal loop EM surveys variance in coil separation and elevation effects are not critical with this time domain method.

Field Examples of Ground Pulse EM Surveys of Poor Conductors

The following are case histories and recent developments in the application of a ground Pulse electromagnetic system designed for exploration in desert areas. Figure 37.1 is a typical example of the response from a narrow (width less than 3 m), weathered, massive sulphide zone using the Crone Pulse EM technique. This profile was obtained over the Ghayth copper-zinc showing in the Sultanate of Oman. Narrow, massive sulphide zones such as this tend to weather to considerable depths. This zone is therefore a poor conductor and is detected by the first Pulse EM sample only. Figure 37.2, also from the Sultanate of Oman, represents the type of anomaly obtained from a zone of disseminated sulphides. In this case, the weathering is shallow (only 30 m), with the lack of conductivity being caused by a low sulphide content of approximately 5 per cent. The important factor illustrated by these two examples is the necessity of retaining the high frequency information generated by a Pulse EM system. This information enables the operator to detect and trace out narrow or weakly mineralized zones that are favourable geological horizons. These zones may expand into larger, more massive bodies along strike.

The conductivity-thickness (σt) of these sulphide zones often is the same order as that of the conductive surficial layer. The sulphide zone is detected only because of the geometrical presence of both vertical and horizontal conductive sheets. The importance of detecting weak conductors is illustrated in Figure 37.3, which shows the discovery Pulse EM profile over the Bayda copper-zinc orebody in Oman. The Bayda showing consisted of an ancient exploratory pit in a small gossan zone at the side of a hill. This showing produces a weak two-sample anomaly that was traced downhill until it strengthened to a six-sample anomaly on line 1+00N. Drilling this section intersected high grade massive mineralization.

Response of Wide Massive Sulphide Conductors Using the Pulse EM Technique

One supportive aspect of desert weathering, we soon discovered, is that wide (greater than 10 m) massive sulphide bodies often self-seal themselves against further oxidation.

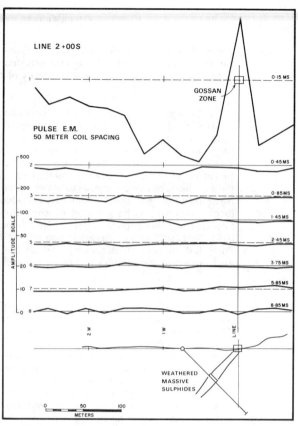

Figure 37.1. Pulse EM profile, 50 m coil spacing, moving coils method, Ghayth showing, Sultanate of Oman.

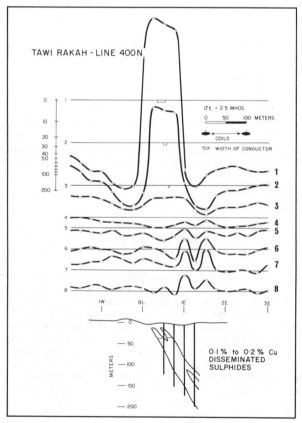

Figure 37.2. Pulse EM profile, 100 m coil spacing, moving coils method, Tawi Rakah, Sultanate of Oman.

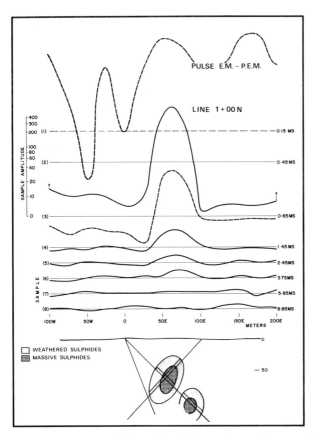

Figure 37.3. Pulse EM profile, 50 m coil spacing, moving coils method, Bayda showing, Sultanate of Oman.

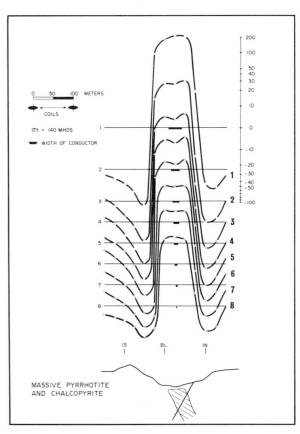

Figure 37.5. Pulse EM profile, 100 m coil spacing, moving coils method, Maydan deposit, Sultanate of Oman.

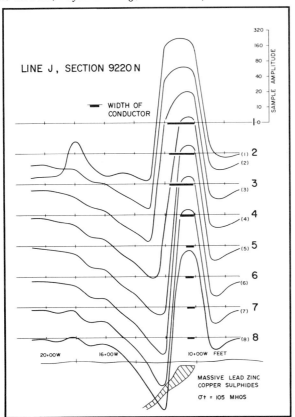

Figure 37.4. Pulse EM profile, 200 ft coil spacing, moving coils method, Jododex, Woodlawn deposit, Australia.

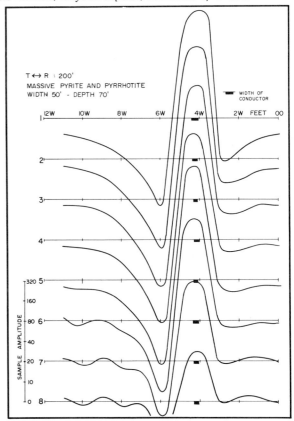

Figure 37.6. Pulse EM profile 200 ft coil spacing, moving coils method, Massive Sulphide Body, Arizona.

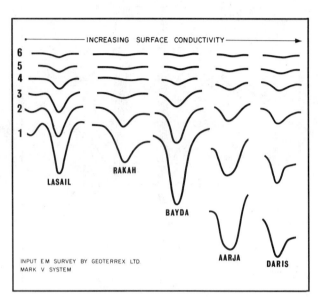

Figure 37.7. Airborne Input EM Anomalies over Massive sulphide orebodies, Sultanate of Oman (Geoterrex Ltd., Ottawa).

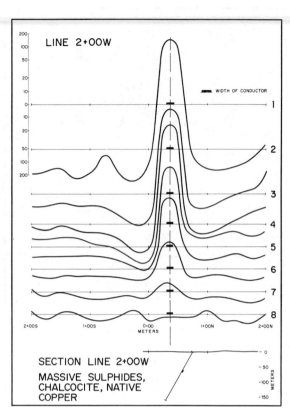

Figure 37.9. Pulse EM profile, 50 m coil spacing, moving coils method, Daris deposit, Sultanate of Oman.

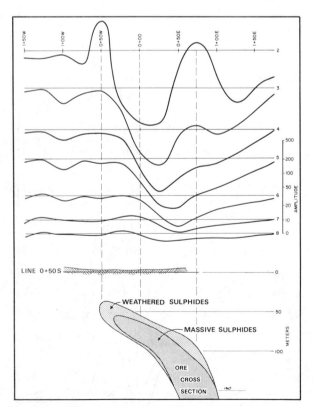

Figure 37.8. Pulse EM profile, 50 m coil spacing, moving coils method, Lasail Orebody, Sultanate of Oman.

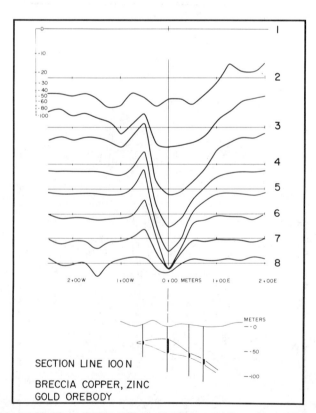

Figure 37.10. Pulse EM profile, 50 m coil spacing, moving coils method, section line 100N, Rakah orebody, Sultanate of Oman.

This halting of the oxidation process leaves the central core of the body as fresh sulphides of high conductivity. An excellent example of this phenomenon occurs in the Woodlawn lead-zinc-copper orebody in New South Wales, Australia (Fig. 37.4). Weathering of this 50 m-wide orebody stops at a depth of 12 m. The near-surface, fresh, massive sulphides of the Woodlawn orebody were at first considered unusual for Australian climatic conditions. We have since encountered several similar occurrences (without the Woodlawn grade) in Australia and other desert areas. Two examples are shown; the first example presented in Figure 37.5 is a Pulse EM profile across the Maydan massive sulphide body in the mountains of Oman, which is 40 m wide and oxidized to 10 m. Figure 37.6 presents a Pulse EM profile across a massive sulphide body in the Precambrian of Arizona, which is 15 m wide and is weathered to a depth of approximately 20 m. All three bodies occur in areas where oxidation of narrow sulphide zones extends down to at least 40 m. The occurrence of a strong, isolated zone of very high conductivity in a desert area can usually be attributed to a wide, massive sulphide body. Such anomalies are uncommon, but are important exploration targets.

Field Examples from the Sultanate of Oman

Since 1973, Crone Geophysics Limited, has been consulting for Prospection Limited of Toronto, who are managing an exploration program in the Sultanate of Oman. A large area of Oman was flown by Geoterrex, with the Input Mark V airborne EM system. Approximately 70 per cent of this area is covered by surficial conductors that produce a background response down to the lowest Input EM channel. In Figure 37.7, the Input airborne EM response from five massive sulphide ore deposits is shown with background conductivities, varying from low at the Lasail orebody, to high around the Daris deposit. All five orebodies are clearly detected by the airborne EM survey with their signature superimposed on the background surficial conductivity response. The ground Pulse EM profile over the Lasail copper orebody is shown in Figure 37.8. The background surficial response at Lasail is again almost zero. In this section the upper surface of the Lasail orebody is wider than the 50 m coil spacing. Both edges of the body are shown as positive-trending peaks with a large negative response being produced when both coils are located directly over the wide conductor. Figure 37.9 is a ground Pulse EM profile over the Daris deposit which is located on the outwash flats between the coastal mountains and the Gulf of Oman. The surficial conductive overburden in that area produces a large negative anomaly of minus 200 (normalized) divisions on the first sample. With both the airborne and ground EM anomalies, the surficial response in the Daris area in the early samples, exceeds in magnitude the peak response from the sulphides of the Lasail orebody. In both techniques, the Daris body is clearly detected with the sulphide anomaly being superimposed on the surficial anomaly.

The discovery Pulse EM profile over the Rakah orebody, is shown in Figure 37.10. It is a typical response of a flat, dipping conductor with a conductivity thickness of 45 Mhos. The test borehole for such an anomaly was spotted as a vertical hole at the maximum negative response; it intersected 30 m of 3% copper in a breccia sulphide zone. The Rakah body occurs at the foot of a mountain ridge and dips under the mountain. A Pulse EM profile along the steep side slope of the mountain is shown in Figure 37.11. The ore zone under this profile is at a depth of 100 m and is barely discernible with the 50 m spacing Pulse EM profile. This type of anomaly may weakly represent a conductor at this depth, but it lacks the definition required to accurately locate a test drillhole.

Detailed Pulse EM Methods

When a marginal anomaly is detected by a ground Pulse EM survey, a detailed survey is required to first establish if the anomaly is actual or noise, and secondly to provide the additional information required to locate a test drillhole. The deeper the conductive target, the greater the need for accurate position, depth, dip and width information.

The Pulse electromagnetic method has no geometrical restrictions and therefore detailed measurements can be obtained with various transmit-receive coil configurations. The method also has the advantage of being able to measure the secondary electromagnetic fields directly rather than as a resultant field reading in the presence of a primary transmitted field. One of the most effective detailed methods is to measure both horizontal and vertical field components and from these determine the direction of the secondary field. From this information measured at several stations, the induced eddy current paths can be located. The induced eddy current paths within a conductor are shown by Lamontagne and West (1971) to be spaced apart. The distance of separation is dependent on the conductivity-thickness of the body and the frequency of the induced field. For measurement points directly over the induced current path, the field is approximately circular and the eddy current position is located at the intersection point of lines drawn at right angles to the secondary field direction. Figure 37.12, illustrates the eight induced eddy current paths that would occur in a weathered, massive sulphide body. The high-frequency, early-sample current paths will flow along the outside of the body and also within the poorly conductive weathered sulphide and gossan areas. The late-sample eddy current paths will confine themselves to the highly conductive, inner core of the massive body. As an example, a standard, moving horizontal coil Pulse EM profile over a known massive orebody located in Western Australia is presented in the upper part of Figure 37.13. The depth to the top of the body is 100 m. The response of the standard method is within the noise level so that the observed anomaly would normally be considered questionable. A detailed survey was therefore carried out which used a vertical transmitter loop oriented coplanar with the strike direction on a line 120 m from the transmitter loop. The Pulse EM Sample 1 and 2 field directions and orthogonal lines are shown in the bottom part of Figure 37.13. The eddy current path for Sample 1 is located at a depth of 90 m. The slight scatter of the intersection points is caused by surficial conductivity effects. The Sample 2 current position is accurately defined at 96 m depth. The vertical loop survey was used in this case to selectively energize one edge of the sulphide body and to define the location and depth of one side of the body.

THE DEEPEM TECHNIQUE

An alternative detailed method utilizes a 100 m square, single-turn transmit loop out on surface. This has the advantage of a much stronger transmitted field providing greater penetration, consequently the technique is called the Deepem method. The transmitter loop is laid out on one side of the area to be detailed and the survey lines extend away from the loop, starting 50 m from one wire out to a distance of 350 m. Both horizontal and vertical components are measured and the induced current paths are determined as before.

An example of the use of the Deepem technique from a test survey over the Flying Doctor prospect, North Broken Hill area of Australia, is shown in Figure 37.14. In this case the surficial conductivity which is caused by brackish groundwater, has a conductivity of 4.2 ohm-metres to a depth of 7 m and a conductivity-thickness of 1.5 mhos. The massive sulphide body consists almost entirely of galena and

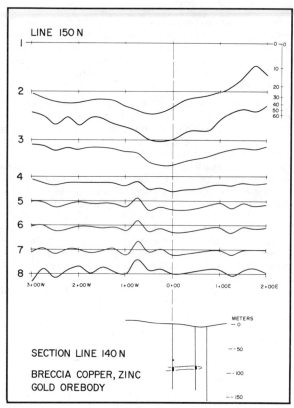

Figure 37.11. Pulse EM profile, 50 m coil spacing, moving coils method, section line 140N, Rakah orebody, Sultanate of Oman.

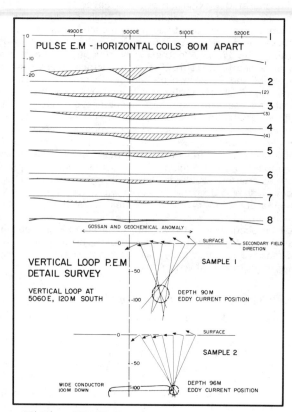

Figure 37.13. PULSE EM surveys moving coils method and vertical loop detail method. Massive sulphide body, western Australia.

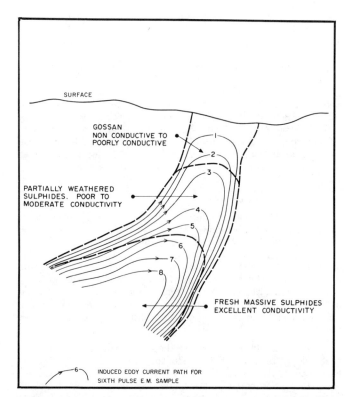

Figure 37.12. Induced eddy currents in a weathered massive sulphide body from the PULSE EM transmitter.

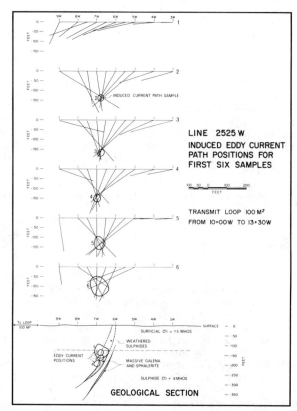

Figure 37.14. PULSE EM survey, detail vertical loop (Deepem) method, induced eddy current paths. Flying Doctor prospect, North Broken Hill, Australia.

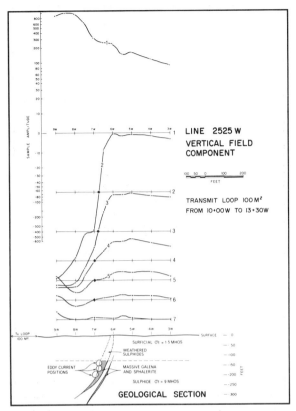

Figure 37.15. Pulse EM survey, detail vertical loop (Deepem) method, vertical component, Flying Doctor prospect, North Broken Hill, Australia.

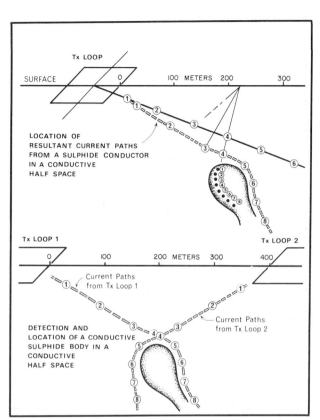

Figure 37.17. Pulse EM Deepem detail method, induced current paths in conductive half space containing a massive sulphide body and the application of two transmit loops.

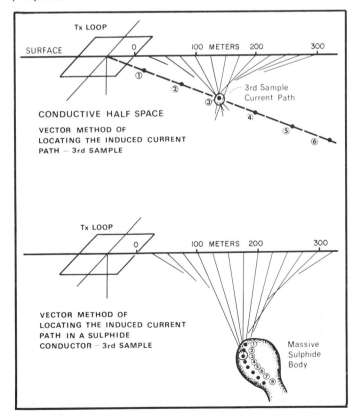

Figure 37.16. PULSE EM Deepem detail method, Induced current paths in a conductive half space and massive sulphide body.

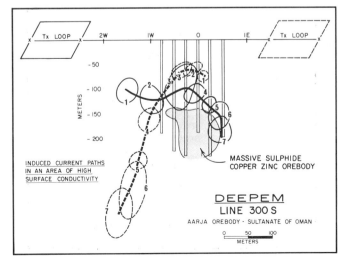

Figure 37.18. Pulse EM Deepem detail method over Aarja Orebody, Sultanate of Oman.

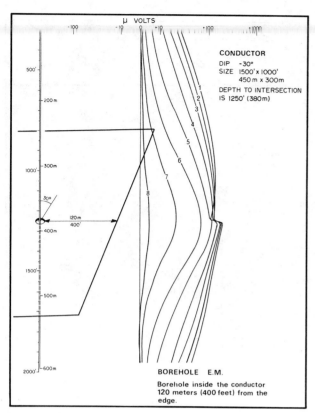

Figure 37.19. Pulse EM borehole method, model study result from a conductive sheet, borehole intersecting the sheet 400 ft from its edge.

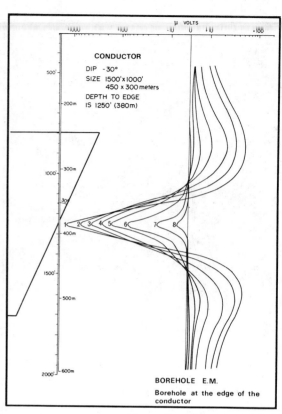

Figure 37.21. Borehole pulse EM method, model study result from a conductive sheet, borehole just outside edge of sheet.

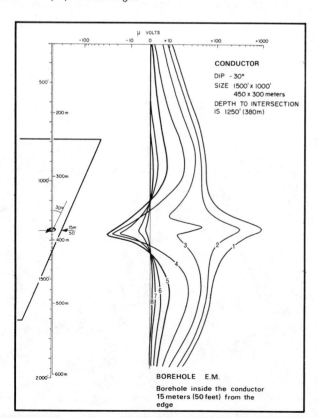

Figure 37.20. Borehole pulse EM method, model study result from a conductive sheet, borehole intersecting the sheet 50 ft from its edge.

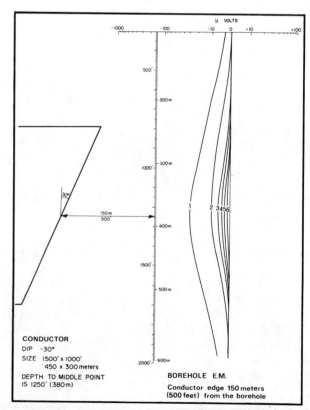

Figure 37.22. Borehole pulse EM method, model study result from a conductive sheet, borehole 500 ft and outside edge of sheet.

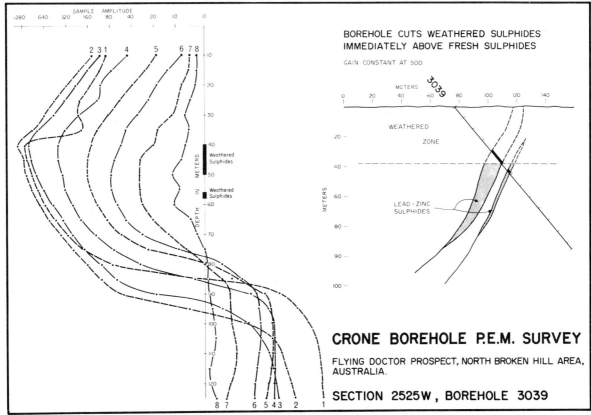

Figure 37.23. Borehole pulse EM survey, borehole just outside the upper edge of Flying Doctor lead-zinc body, North Broken Hill, Australia.

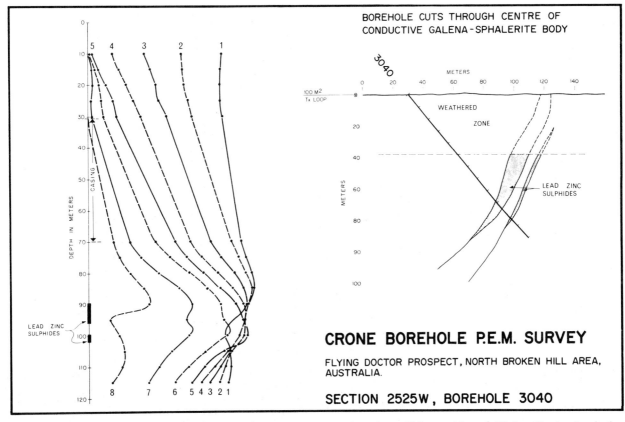

Figure 37.24. Borehole pulse EM survey, borehole intersecting the middle portion of Flying Doctor lead-zinc body, North Broken Hill, Australia.

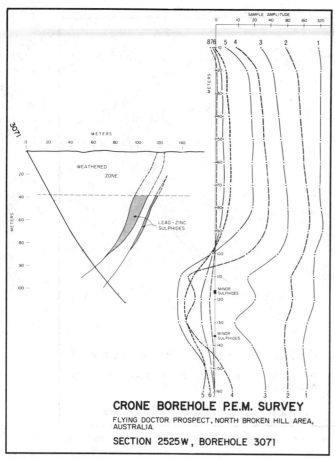

Figure 37.25. Borehole pulse EM survey 20 metres away from the lower edge of Flying Doctor lead-zinc body, North Broken Hill, Australia.

sphalerite with a calculated conductivity-thickness from the Pulse survey of 9 Mhos. The body is weathered to a depth of 40 m. The first Pulse EM sample is dominated by the surficial conductivity and does not form an eddy current path position. The second, third and fourth samples produce eddy current path positions along the contact of the sulphide lens that faces the transmitter loop. Because the sulphide lens has a low conductivity, the response of the fifth and sixth samples are weak and the eddy current paths are not accurately defined. In order to show the amplitudes of the responses measured, the vertical component is usually plotted as shown in Figure 37.15. The conductor is located below the cross-over position.

Limitations of the Deepem Detail Method in Areas of High Surficial Conductivity

The presence of a conductive half space below the transmitter loop results in eddy currents that flow in concentric rings around and outside the loop. The first sample eddy current flows close to the loop, with the later samples spaced farther out as shown in the upper portion of Figure 37.16. The interval between the eddy current paths decreases as the conductivity-thickness of the half space increases. The lower portion of Figure 37.16, shows the current paths induced in a massive sulphide body without the presence of a surficial conductive zone. When the sulphide body occurs in a conductive half space, then both surficial and sulphide eddy currents are present as shown in the upper portion of Figure 37.17. In this case, a resultant current path will be detected by the receiver that is shown as a dashed line. This resultant current path for the early or high frequency samples, will be dominated by the surficial conductivity response. The later sample or low frequency resultant response, will be influenced to a greater degree by the more conductive sulphide body. The net effect is for the resultant current path to form a line that lies between the surficial and sulphide current paths. If transmitter loops are employed on either side of the target area, then the approximate position of the sulphide zone can be determined as the area enclosed by the two resultant current paths, as shown in the lower portion of Figure 37.17.

The two-transmitter Deepem procedure was first tested in a profile over the Aarja massive sulphide body in the Sultanate of Oman. The body approximates a cylinder some 50 to 100 m in diameter of massive sulphides that has a shallow plunge of 20° from the horizontal. The test section has a depth to the top of the sulphide zone of 150 m. The surficial conductivity in this area is 9 ohm m to a depth of approximately 30 or 40 m. The resultant induced current paths are shown in Figure 37.18; the eddy current paths from the eastern transmit loop as dashed circles, the paths from the western loop as solid circles with the orebody occurring between.

Further tests with the Deepem method in Australia, indicate that the induced current path method does not outline deep (100 m plus) massive sulphide conductors, when the surficial conductivity-thickness product is of the order of 10 Mhos. In this case the anomalous information is available in the measured readings but the eddy current path method lacks the sensitivity to unlock the sulphide response from the strong surficial response. Computer processing of the observed data to strip off the surficial conductivity background effect is now being investigated.

BOREHOLE PULSE EM METHOD

Crone Geophysics developed in 1976, a Borehole Pulse EM system for the Geological Survey of Canada with depth capabilities in excess of 1000 m. The method uses a large single-turn transmitter loop laid out on the ground and a receiver probe sent down the borehole. The advantages of the Borehole Pulse EM system are; (1) Since the method is free of geometrical effects, anomalies are not caused by variances in the straightness of the borehole. (2) The measurement of the secondary fields directly provides accurate interpretative information. (3) The wide frequency spectrum enables the method to separate effects from weakly mineralized sulphide zones intersected within the hole from large massive bodies located outside the hole.

A model study was carried out (Woods, 1975) as an aid to the interpretation of Borehole Pulse EM data. This study illustrates that Borehole Pulse EM is effective in detecting massive sulphide bodies and also provides an idea of the size, shape and position of the sulphide body. Figures 37.19 to 37.22 illustrate the change in the response pattern obtained when the borehole is moved from an intersection from 120 m inside a conductive sheet to 150 m outside. The response unit in this model study are microvolts of signal received at the downhole probe and are not normalized.

Field results are shown from a survey of three holes in a section from the Flying Doctor prospect. This is the same section that was detailed with the Deepem method. The three holes were all surveyed from a 100 m square, single turn, transmitter loop located immediately west of the collar of borehole 3071. In Figure 37.23, the Borehole Pulse EM results from borehole 3039 are shown. This hole intersects the upper weathered portion of the sulphide zone. The survey curves match those of Figure 37.21 of the model study. This would indicate that the weathered sulphides intersected in the hole are nonconductive, but the hole is just outside the conductive sulphide zone. In Figure 37.24, the downhole results from hole 3040 are shown to produce the typical positive response obtained when a borehole intersects a conductive sulphide zone, with the intersection point located towards the central part of the body (model study Fig. 37.19). Casing in this hole blocked out readings between 30 and 65 m. The survey of borehole 3071 (Fig. 37.25) shows that the minor sulphide zones intersected at 117 and 136 m are both nonconductive but that two conductive bodies are located some 10 to 20 m from the hole. The upward displacement of the negative response peaks from the intersection of minor sulphides, indicates that the massive sulphides occur up dip from the intersection.

CONCLUSIONS

The ground Pulse electromagnetic method is an effective exploration tool in the search for massive sulphide bodies in desert areas. The method is very flexible as far as coil configurations and field component measurements are concerned. A large number of field measurements can be obtained at each observation point. Advancements in the computer processing of such a large data base, should lead to further increased depths of penetration and accuracy of the Pulse EM method. Developments in the Borehole Pulse EM technique have expanded the radius of detection of mineralization from a drillhole from a few centimetres of core up to 150 m. Thus this capability of the Borehole Pulse EM technique will permit exploration at depth in the vicinity of known ore deposits.

REFERENCES

Crone, J.D.
 1975: Pulse Electromagnetic – PEM ground method and equipment as applied in mineral exploration; AIME Annual Meeting, New York, 1975, Preprint 75-L-93, 8 p.

Dolan, W.M.
 1967: Geophysical detection of deeply buried sulphide bodies in weathered regions; in Mining and Groundwater Geophysics, Geol. Surv. Can., Econ. Geol. Rept. 26, p. 336-344.

Lamontagne, Y.I. and West, G.F.
 1971: EM response of a rectangular thin plate; in Research in Applied Geophysics, v. 2, Geophysics Lab., Univ. of Toronto.

Wait, J.R.
 1956: Method of geophysical exploration; U.S. Patent Office 2,735,980.

Woods, D.V.
 1975: A model study of the Crone Borehole Pulse Electromagnetic (PEM) system; unpubl. M.A. thesis, Dept. of Geological Sciences, Queen's Univ., Kingston, Ontario.

THE APPLICATION OF AIRBORNE AND GROUND GEOPHYSICAL TECHNIQUES TO THE SEARCH FOR MAGNETITE-QUARTZITE ASSOCIATED BASE-METAL DEPOSITS IN SOUTHERN AFRICA

Geoff Campbell and R. Mason
Johannesburg Consolidated Investment Company Ltd.
Johannesburg, South Africa

Campbell, Geoff and Mason, R., The application of airborne and ground geophysical techniques to the search for magnetite-quartzite associated base-metal deposits in Southern Africa; in Geophysics and Geochemistry in the Search for Metallic Ores; Peter J. Hood, editor; Geological Survey of Canada, Economic Geology Report 31, p. 757-777, 1979.

Abstract

Within Southern Africa, much of the recent prospecting of Precambrian high grade metamorphic terranes has concentrated on the search for Cu-Pb-Zn deposits associated with highly aluminous host rocks and spatially related magnetite-quartzites (or banded iron formations). These orebodies comprise stratabound massive or (more commonly) disseminated sulphides which are essentially nonmagnetic, but are usually highly polarizable and in some cases highly conductive. The related magnetite-quartzites are of special prospecting significance in that they act as unique marker horizons for such mineralization, their strike extent appears to partly control the distribution of mineralization, and they yield readily identifiable magnetic responses.

During the period 1973 to 1975 the Johannesburg Consolidated Investment Company Ltd. conducted major exploration programs to locate deposits of this type, within the Damara Sequence of South West Africa and the Bushmanland Sequence of South Africa. Given the extensive overburden cover and lack of geological control, effective prospecting of these areas demanded a combined geophysical and geochemical approach with later geological mapping and percussion/diamond drilling.

The geophysical signatures of known deposits in the Bushmanland and Damara sequences were assessed, and were utilized in devising a search target model whose criteria were applied throughout the exploration program. As base-metal deposits were found to be characterized by fairly unique magnetic signals, the aeromagnetic technique was used as the prime reconnaissance prospecting tool in overburden-covered areas. Aeromagnetic anomalies designated as "significant" were ground checked by magnetometer surveys aimed at delineating potential magnetite-quartzite horizons, geochemical soil sampling being used initially to assess the base-metal potential of each magnetite-quartzite occurrence. Where preliminary results were encouraging, these were followed up by further ground geophysical surveys (i.e. EM, IP) to delineate possible sulphide zones, the resulting anomalies being first investigated by percussion drilling.

By late 1975, this exploration program had delineated several small orebodies, two of which reflected original grass roots discoveries.

Résumé

La plupart des travaux récents de prospection exécutés dans le sud de l'Afrique en terrains précambriens fortement métamorphisés ont consisté en bonne partie à chercher des gîtes de Cu-Pb-Zn associés à des roches encaissantes à forte teneur en alumine ainsi qu'à des quartzites à magnétite (ou formations ferrifères rubanées). Ces gisements comprennent des sulfures stratifiés ou (plus communément) disséminés qui sont essentiellement non magnétiques mais habituellement très polarisables et dans certains cas très conducteurs. Les quartzites à magnétite associés présentent un intérêt tout particulier pour les travaux de prospection puisqu'ils constituent des horizons repères uniques pour une telle minéralisation. L'étendue de leur direction semble contrôler en partie la répartition de la minéralisation et ils produisent des réactions magnétiques facilement identifiables.

De 1973 à 1975, la Johannesburg Consolidated Investment Company Ltd. a exécuté de grands programmes d'exploration afin de détecter des gisements de ce genre dans la séquence de Damara dans le sud-ouest africain et dans la séquence de Bushmanland en Afrique du Sud. Compte tenu de l'épaisse couche de morts-terrains et du manque de données géologiques, il a fallu, pour mener à bon terme les travaux de prospection, faire appel à des techniques géochimiques et géophysiques puis dresser des cartes géologiques et enfin effectuer des forages par percussion et au diamant.

Les sismogrammes de gisements connus dans les séquences de Bushmanland et de Damara ont été évalués et ont servi à établir un modèle de cible de recherche dont les critères ont été appliqués à tout le programme d'exploration. Étant donné que les gisements de métaux non précieux se sont caractérisés par l'émission de signaux magnétiques à peu près uniques, la technique aéromagnétique a été utilisée comme principal outil de prospection préliminaire dans les régions recouvertes de morts-terrains. Les anomalies aéromagnétiques considérées importantes ont fait l'objet de vérifications au sol au moyen de levés par magnétomètre afin de *localiser les horizons possibles de quartzite à magnétite; l'échantillonnage géochimique du sol a servi à évaluer initialement le potentiel de métaux non précieux de chaque venue de quartzite à magnétite. Lorsque les résultats préliminaires s'avéraient encourageants, ils étaient suivis d'autres levés géophysiques (EM, IP) sur le terrain pour déterminer les zones possibles de minéraux sulfurés et les anomalies relevées étaient d'abord étudiées au moyen de forages par percussion.*

Vers la fin de 1975, le programme avait permis de localiser plusieurs petits gisements de minerai dont deux reflétaient les découvertes initiales en surface.

INTRODUCTION

Major base-metal discoveries during the early nineteen-seventies of the Aggenys Pb-Zn-Cu-Ag deposit (Phelps Dodge) and the Gamsberg Zn deposit (Newmont) in the Northwest Cape Province of South Africa, and of the Otjihase Cu deposit (Johannesburg Consolidated Investment Company) near Windhoek in South West Africa, gave renewed impetus to the base-metal exploration activity initiated in Southern Africa following the Prieska Cu-Zn discovery by Anglovaal in 1969. All these discoveries were made following diamond drilling of prominent surface gossan outcrops discovered by conventional prospecting. It was surprising to many explorationists outside Southern Africa that such prominent surface expressions of major mineralization had laid unrecognized for so long. The stratabound Aggenys – Gamsberg – Otjihase deposits are all closely associated with magnetite-quartzites (banded iron formations), and it is interesting to note that similar associations had already been recognized as exploration targets for a decade or more in Canada and Australia.

During the period 1973 to 1975, the Geological Department of Johannesburg Consolidated Investment Company Ltd (JCI) conducted a major exploration program aimed at locating deposits of this type within the Bushmanland Sequence of South Africa and the Damara Sequence of South West Africa. Extensive overburden cover and a consequent lack of geological control, resulted in the use of geophysical and geochemical surveys as the major prospecting techniques, with later back-up from percussion and diamond drilling. This paper discusses the contributions made to the exploration program by geophysical surveys over the two most significant (as subsequently determined) prospecting areas, namely the Pofadder East Block of the Northwest Cape Province, South Africa, and the Gorob Prospect of South West Africa (see Fig. 38.1).

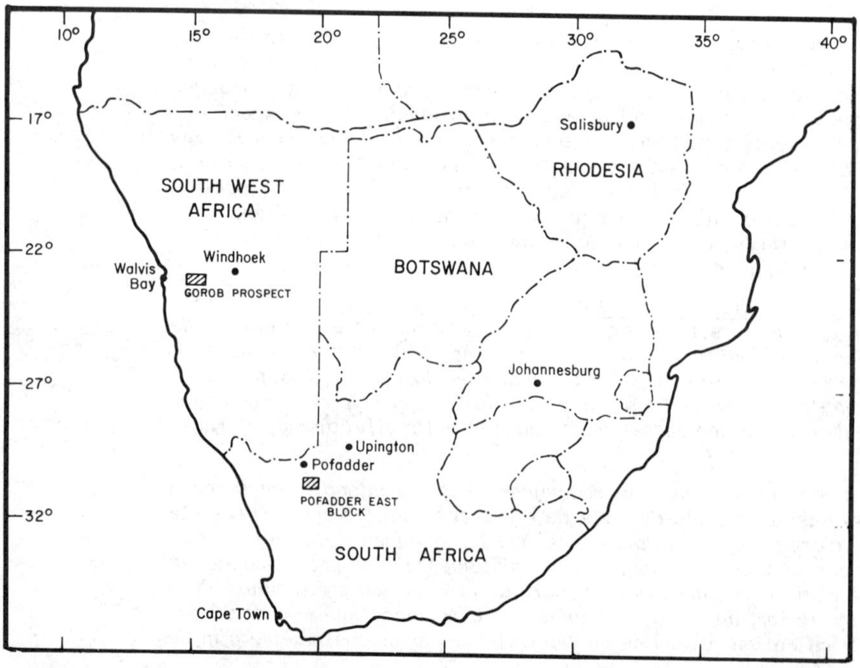

The Pofadder East Block, South Africa

The Pofadder East Block occupies a rectangular-shaped area some 17 km (N-S) by 31 km (E-W), the centre of which is approximately 40 km southeast of Pofadder and 90 km east of the Aggenys and Gamsberg base-metal deposits. The region is sparsely vegetated, semi-desert scrubland characterized by low relief. Outcrop is largely obscured by the presence of extensive sand and calcrete cover, maximum overburden thickness being approximately 15 m but more typically 2 to 7 m.

The ground was taken under option by JCI following an aerial reconnaissance exercise carried out to trace the Aggenys-Gamsberg lithological sequence eastwards towards Pofadder. This exercise was facilitated by tracing prominent metaquartzite ridges which occur close to the mineralized parts of the Bushmanland Sequence. Prior to this program, there were no reported instances of base-metal mineralization within the area. The so-called Putsberg copper deposit was subsequently located during the ground follow-up of aeromagnetic anomalies within the Pofadder East Block during 1974.

The Gorob Prospect, South West Africa

The Gorob Prospect falls within the Namib Desert of southwest Africa, and comprises a roughly rectangular-shaped block measuring some 20 km (NW-SE) by 80 km (NE-SW). The area lies some 70 km east of Walvis Bay and 200 km west of the Otjihase Copper mine near Windhoek. The prospect area is of minimal relief and has extensive alluvium and calcrete cover, increasing from approximately 2 m thickness in the southwest up to a maximum of 40 m thickness in the northeast.

Copper mineralization within the area was noted at the turn of the century, with subsequent small scale (and short-lived) workings being concentrated at the outcropping, so-called Gorob Mine. The latter and its associated deposits along strike, had previously been drilled by Rand Mines Limited and investigated in more detail by Penarroya, prior to the subsequent exploration by JCI. JCI's interest in the Gorob area was stimulated by the realization that its mineralization was related to a narrow belt of metavolcanic rocks called the Matchless Amphibolite Belt,

Figure 38.1. *Location map, Pofadder East Block and Gorob Prospect, Southern Africa.*

with which were also associated the Otjihase – Ongeama – Ongombo copper deposits (discovered in 1970-71 by JCI. Follow-up work by JCI over major aeromagnetic anomalies, delineated two hitherto unknown deposits of cupreous pyrites.

Exploration Program

Aeromagnetic surveying was used as the major reconnaissance tool in prospecting these overburden covered areas, with the prime aim of defining the locale of magnetite-quartzite horizons, and hence, of potentially economic base-metal deposits occurring in close association.

Aeromagnetic anomalies deemed as significant were detailed by ground magnetometer surveys, and, where thick sand and/or calcrete cover was absent, geochemical soil sampling surveys were used to initially assess the mineralization potential of each "magnetite-quartzite" occurrence.

Where preliminary results were encouraging, EM or IP surveys were executed over and in the locale of the magnetic horizon, the resulting anomalies being first investigated by percussion drilling. The philosophy underlying the geophysical prospecting program is discussed later.

OUTLINE OF GEOLOGY

Pofadder East Block, South Africa

The Pofadder East Block forms part of the high grade Namaqualand metamorphic complex which straddles the lower parts of the Orange River valley in the Northwest Cape and in southern South West Africa. Throughout this metamorphic terrane a 100 to 500 m thick mid-Precambrian Sequence of gneisses, schists, metaquartzites and amphibolites (the Bushmanland Sequence) overlies a granitoid basement (Joubert, 1971). The Bushmanland Sequence is in recumbent fold complexes. Basal quartzofeldspathic gneisses are followed upwards by heterogeneous gneisses which may or may not be aluminous and/or siliceous; amphibolites of distinctive volcanic and hypabyssal types and aluminous schists with associated magnetite-quartzite horizons occur therein. This heterogeneous part of the Bushmanland Sequence contains all the significant base-metal mineralization discovered to date (Aggenys, Gamsberg etc.) and is overlain by prominent metaquartzite horizons, which become progressively more feldspathic in the areas east of Pofadder.

The Putsberg copper deposit occurs in a complex synformal zone of aluminous and graphitic schists, metaquartzites and amphibolites, which overlies the basal gneisses of the Bushmanland Sequence. The mineralization is intimately associated with thin siliceous horizons within biotite-sillimanite schists, and consists of chalcopyrite, minor pyrite, occasional sphalerite and rare galena. The mineralized horizon has been stretched and dismembered within the schists, and thicker intersections of it have been shown to be related to tectonic thickening in fold hinge zones (Paizes, 1975).

The resulting lack of coherency of the mineralization has made it difficult to evaluate, and it is currently considered subeconomic by the holding company in present circumstances.

Immediately below the mineralized zone at Putsberg there is a sequence of partly aluminous quartzo-feldspathic gneisses between 100 and 300 m thick which in turn overlie (structurally) a prominent zone of magnetite-bearing sillimanite gneisses with intercalated magnetite-quartzites. These are best developed in a position corresponding to the mineralization above.

Comparison has been made between the geology and mineralization of the Bushmanland Sequence and that of the Broken Hill area in Australia; the similarities are remarkable (R.L. Stanton, pers. comm.). It appears probable that the search model developed in this paper could be successfully applied to many other early to mid-Precambrian high grade metamorphic terranes elsewhere in the world.

Gorob Prospect Area, South West Africa

The cupreous pyrite deposits of the Gorob area are associated with a narrow belt of metavolcanic rocks situated within a flysch trough (the Khomas trough) of late Precambrian age, which forms part of the Damara mobile belt in South West Africa. Other cupreous pyrite deposits associated with the metavolcanic belt (the so-called Matchless Amphibolite Belt) include the Matchless Mine (Tsumeb Corporation), the Otjihase Mine, and the Ongeama, Ongombo and Kupferberg deposits near Windhoek (JCI Ltd.).

The Khomas Trough is intruded by a major granite batholith (the Donkerhoek granite) which has been partly unroofed by erosion along the northern edge of the Gorob Prospect area (Martin, 1965).

The Gorob deposits are situated around the rim of a major synformal structure which closes westwards near the Hope Mine (Fig. 38.8). The Hope – Anomaly deposits on the northern limb of the synform have been subjected to high grade contact metamorphism resulting from the intrusion of the granite to the north. Thus the predominant iron sulphide is pyrrhotite rather than pyrite, the mineralization is coarse grained rather than fine- to medium-grained, and the enclosing rocks contain upper-amphibolite facies mineral assemblages (the mineralization is confined within drag-folded sections of magnetite-quartzite horizons). The deposits on the southern limb of the synform (Gorob, Vendome and Luigi) are developed in magnetite-bearing quartz-sericite schists adjacent to barren magnetite-quartzite horizons. Ore shoots are contained within drag fold closures and the mineralization is predominantly pyritic with subordinate chalcopyrite, silver and minor sphalerite. All the orebodies at Gorob are characterized by well-developed pyritic-chloritic-aluminous lenses adjacent to the quartzitic rocks. While ore-grade deposits have been delineated within the Prospect, these are currently considered subeconomic by the holding company.

The geological setting and characteristics of the Gorob deposits are similar to those of many other late Precambrian-Phanerozoic pyritic ore deposits, and exploration for these could again follow the search procedure outlined in this paper.

EXPLORATION STRATEGY AND GEOPHYSICAL METHODS

The Search Target Model and Exploration Philosophy

Test aeromagetic surveys over economic orebodies associated with magnetite-quartzites, confirmed that their common denominator of well-developed, sympathetically correlating magnetite-quartzite horizons, yielded readily identifiable aeromagnetic anomalies of limited strike extent. Data released subsequent to the initiation of the exploration program have confirmed the results of the test surveys, and isomagnetic contour maps covering the Aggenys deposit in the Bushmanland Sequence and the Otjihase deposit in the Damara Sequence, are shown in Figures 38.2 and 38.3, respectively.

The magnetic contour map of Figure 38.2 is taken from a regional aeromagnetic survey (1976) of the Northwest Cape Province, executed on behalf of the South African Geological Survey. The survey was flown using a Geometrics

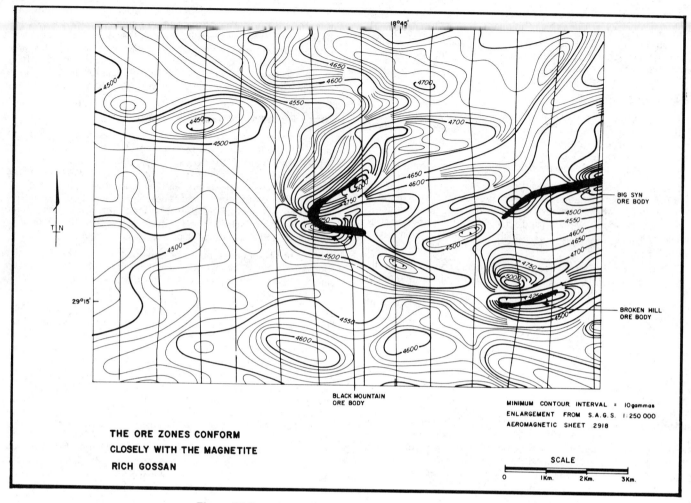

Figure 38.2. Aeromagnetic contour map, Aggenys area.

G803 proton — precession magnetometer at a mean terrain clearance of 150 m, along north-south flight lines having a separation of 1 km. The Aggenys ore zones are distributed around the closure of a synformal structure, the limbs of which can be readily traced via the magnetic responses of a major magnetite-quartzite horizon contiguous with the mineralization. Against a background of weakly magnetic schists and gneisses, the approximately 10 m thick magnetite-quartzite unit yielded an anomaly (maximum response $\Delta T=550$ gammas) extending for 2-6 km from the nose. The predominantly positive anomaly along the northern limb and the low along the southern limb are consistent with tilted sheets magnetized by induction only in an ambient magnetic field of inclination $I=-60°$.

Figure 38.3 summarizes the results of an in-house JCI aeromagnetic survey (1974) over the Otjihase area; this survey was flown using a Geometrics G803 proton-precession magnetometer at a terrain clearance of 70 m, along north-northwest-south-southeast flight lines having a separation of 400 m. The ore-zone is characterized by a 3 km long, linear magnetic anomaly produced by the 3 m thick magnetite-quartzite horizon associated with the major sulphide zone. The predominantly negative response (maximum $\Delta T=100$ gammas) is consistent with a sheet dipping approximately 20° to the north and magnetized by induction only.

From the results of the test surveys, it was concluded that orebodies of the search target type were characterized by narrow, linear magnetic anomalies having strike-lengths in excess of 1.5 km, but probably not greater than 5 km, with the anomalous horizon showing well-developed continuity along strike. Available evidence showed that such horizons had (i) magnetically identifiable widths in the range 10 to 50 m, but were generally not greater than 20 m, and (ii) a bulk magnetite content varying from 5 to 40 per cent by volume. While the strike-trace of the magnetite-quartzite units generally correlated well with the mineralized trace, inferring either contiguous horizons or an intimate association of both, one instance of a considerable separation (100 to 200 m) between ore horizon and magnetite-quartzite was noted at the Rozyn-Bosh Prospect, Northwest Cape Province. A discrete difference in stratigraphic levels between the sulphide mineralization and the magnetite-quartzites is a common feature of these types of deposits (Stanton, R.L. pers. comm.), and the phenomenon was incorporated into the final search model.

Whilst recognizing that the magnetite-quartzites do not always constitute the major ore-bearing horizons, their association with mineralization and their geophysical responses, were such as to render these units of prime prospecting significance in the search for Cu-Pb-Zn and cupreous pyrite deposits. In general, the magnetite-quartzites were assumed to act as unique marker horizons for such mineralization, and as discrete units their strike extent appeared to exert considerable control (albeit indirectly) on the spatial development of sympathetically correlating sulphides.

Given the resultant magnetic signal responses of the search target mineralization, and the limited number of magnetite-bearing geological units within both prospect areas, the aeromagnetic technique recommended itself as the prime reconnaissance prospecting tool, both in terms of its cost-effectiveness and its fast-target generation capabilities. Furthermore, this strategy eliminated large areas with no economic potential (i.e. areas of little or no magnetic relief) at the outset. Thus, the initial prospecting approach was an indirect one, with areas of possible mineralization potential being selected on the basis of their proximity to interpreted magnetite-quartzite horizons.

At this stage of the program, little attempt was made to utilize any of the conventional geophysical prospecting techniques in defining the target sulphide zones themselves. Within the Northwest Cape area, the generally disseminated nature of the sulphide zones precluded their detection by airborne electromagnetic methods, whereas portions of the overburden-covered areas of the Gorob Prospect were known, from a previous Input EM survey executed on behalf of Penarroya, to be highly conductive, thereby negating the effectiveness of AEM surveys over a large portion of the area.

Regional geochemical sampling was not employed in the reconnaissance phase of the Northwest Cape program, due to doubts (since dispelled) as to its effectiveness in calcrete-covered areas, and its inferior cost-effectiveness/rate of coverage when compared to the aeromagnetic method. Geochemical surveys were utilized in the Gorob area, but proved ineffective in the deep calcrete-covered zones.

Selection of Potential Magnetite-Quartzite Horizons from Aeromagnetic Data

Because of their geometry, magnetite-quartzite horizons produce an anomalous aeromagnetic profile having the characteristic thin-dyke (i.e. sensor ground clearance >> dyke width) waveform (Gay, 1963; Reford, 1964). While this waveform is an important aeromagnetic parameter in identifying potential magnetite-quartzite units on the basis of their geometrical configuration i.e. differentiating between thick/thin dykes, plugs etc., it cannot be taken as uniquely diagnostic of the causative source. Other magnetite-bearing units (e.g. amphibolites) of long strike extent and widths not greater than the sensor ground clearance (i.e. about 100 m), may be expected to produce similar magnetic targets. The only additional criteria that can be utilized in separating significant horizons from those of lesser significance, short of ground checking, is an assessment of anomaly amplitudes within any one prospect area. However the amplitude parameter must be used with caution, as discussed below.

The magnetic response (peak-to-peak amplitude) of a magnetite-quartzite unit is not in itself significant in terms of (a) its identity and (b) the existence or otherwise of sympathetically correlating sulphide mineralization.

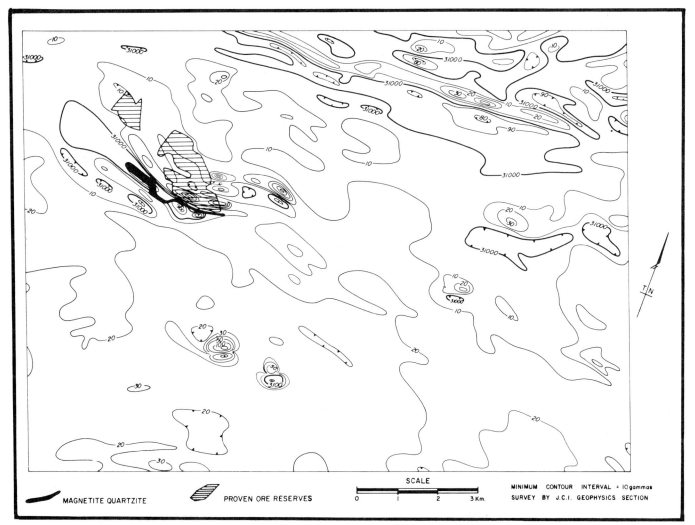

Figure 38.3. Aeromagnetic contour map, Otjihase area.

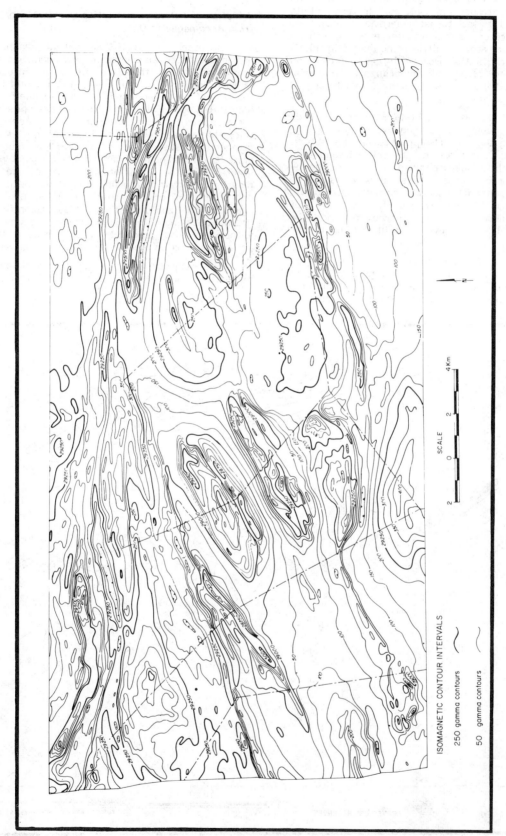

Figure 38.4. Aeromagnetic contour map, *Pofadder East Block.*

Neglecting remanence, the amplitude of the magnetic response is mainly dependant on the product of the width and magnetite content of the horizon, and may vary by an order of magnitude, as shown by the test surveys over Aggenys and Otjihase, which demonstrated magnetite content (by volume) x width characteristics of the order of 300 m% and 20 m% respectively. The development of magnetite-quartzite units is of greatest significance, rather than their width or magnetite content, although the major orebodies do seem to be associated with higher magnetite contents in most cases.

In addition, localized tectonic thickening of the magnetite-quartzite unit e.g. drag folding, which in itself is a favourable indicator of possible contiguous sulphide deposition, will generally result in an increase in the apparent intensity of magnetization of this unit, producing an amplitude-enhanced magnetic response. Thus, given other favourable parameters, it remains a valid approach to assign high priorities to those horizons, or discrete strike sections, showing the greatest magnetic activity within an area.

Reconnaissance Phase – Ground Magnetometer Surveys

Ground magnetometer surveys were executed over selected airborne anomalies with the purpose of (a) accurately delineating the anomaly on the ground and (b) resolving ambiguities inherent in the selection of thin-dyke anomalies from the airborne data. Based on waveform analysis, (Martin, 1966; Koulomzine et al., 1970, and Am, 1972), the superior resolving power of the ground surveys permitted the rejection of heterogeneous magnetite-bearing units which had appeared as composite units from the air, plus those units which on inspection proved to have widths greater than 40-50 m. In outcropping areas, nonsignificant anomalies were rejected from geological considerations.

Follow-up Exploration Phase

Based on prior empirical observations, the search model adopted allowed for possible spatial separation between the magnetite-quartzite horizon and its associated mineralization. Thus the search for sulphide zones was not only restricted to the delineation of possible ore-bearing horizons contiguous with the pre-delineated magnetite-quartzite horizon, but recognized that such zones, while paralleling the former, might occur up to (say) 200 m away from it.

Magnetic horizons conforming to the search target type were accurately located on the ground, and were covered by detailed geochemical surveys aimed at assessing the base-metal potential of the horizon and adjacent area, up to 500 m on either side of the horizon. Soil and/or other surface material samples were taken at 10 m intervals along line, with the minus 80 mesh fraction being treated by total acid extraction and analyzed for Cu, Zn, and Pb on atomic absorption analytical equipment.

Within the Northwest Cape, geochemical sampling was followed by IP surveys over the same grids, with the aim of determining the sulphide potential of (a) prominent base-metal geochemical anomalies or (b) the entire grid, where reasonable doubt existed as to the effectiveness of the geochemical technique in any one area. Under favourable circumstances, the IP technique is capable of yielding polarization responses from sulphide zones having as little as 2 per cent sulphides by volume, and thus lent itself ideally to the search for what, in the main, were expected to be disseminated (about 5 per cent by volume) sulphides.

As the Gorob Prospect area was known to be typified by contiguous magnetite-quartzite and massive pyrite sulphide horizons, the search for the latter was initiated using the Turam electromagnetic technique. Interpretation problems arising from the use of Turam led to its early rejection, and except where geological problems dictated otherwise (see later), investigation of potential sulphide-bearing magnetite-quartzite horizons was undertaken by systematic percussion drilling.

GEOPHYSICAL RESULTS

The Pofadder East Block, South Africa

Aeromagnetic Survey

The aeromagnetic survey of the Pofadder East Block was flown using a Scintrex Map-2 proton-precession magnetometer at a terrain clearance of 70 m, along north-south flight lines having a separation of 300 m. An isomagnetic contour map covering the major portion of this block is shown in Figure 38.4. The original contour interval of 10 gammas has been coarsened in Figure 38.4 to 50 gammas for the sake of visual clarity. The corresponding magnetic interpretation map is shown in Figure 38.5.

Some 40 per cent of the total survey area is underlain by weak to moderately magnetic rock units, which show up as composite assemblages comprising closely-spaced, subparallel magnetic anomalies. The short wavelength nature of the latter indicates a near-surface origin for their causative sources, calculated depths of burial being in the range 0-20 m i.e. well within the depth penetration capabilities of the IP technique.

Major magnetic discontinuities, other than those due to contact-type sources, are not common within the area, and this implied scarcity of faults/shears has since been substantiated by later geological mapping (guided by the aeromagnetic data). While the major structure of this Block conforms to that of an easterly-plunging synform, the presence of a rapidly alternating series of subparallel antiformal and synformal fold axes is readily detectable from the data. Where such fold axes have been interpreted, the selection of an antiform or synform was based on the assumption of induced magnetization of the outer magnetic horizons.

Four major, magnetically identifiable rock units coded A to D (see Fig. 38.5) have been defined within the survey area, and are discussed below.

Unit A, which occupies the central portion of the survey area, is correlated with nonmagnetic biotite gneiss assemblages. Although these gneisses can be geologically subdivided, no such distinction is possible from the aeromagnetic data.

Unit B occupies some 40 per cent of the total survey area, and correlates with areas underlain by basal quartzofeldspathic gneisses which occupy interpreted antiformal structures. These weakly magnetic units generally exhibit a domal structure, and their nonconformable nature is apparent in the central portion of the survey area.

Moderately magnetic zones interpreted as Unit C occur within the central and eastern areas, and are correlated with discrete calc-silicate units falling within the topmost part of the local sequence. This is consistent with an interpretation of their association with east-west oriented synformal structures, and isolated units of this type presumably represent erosional remnants.

Unit D encompasses a fairly large zone in the northwest and western portions of the area, where it is geologically correlated with pink gneisses and biotite-schist units, which have associated magnetite-quartzites, metaquartzites and amphibolites. This moderately magnetic unit is the potentially mineralized portion of the Bushmanland Sequence in the Northwest Cape.

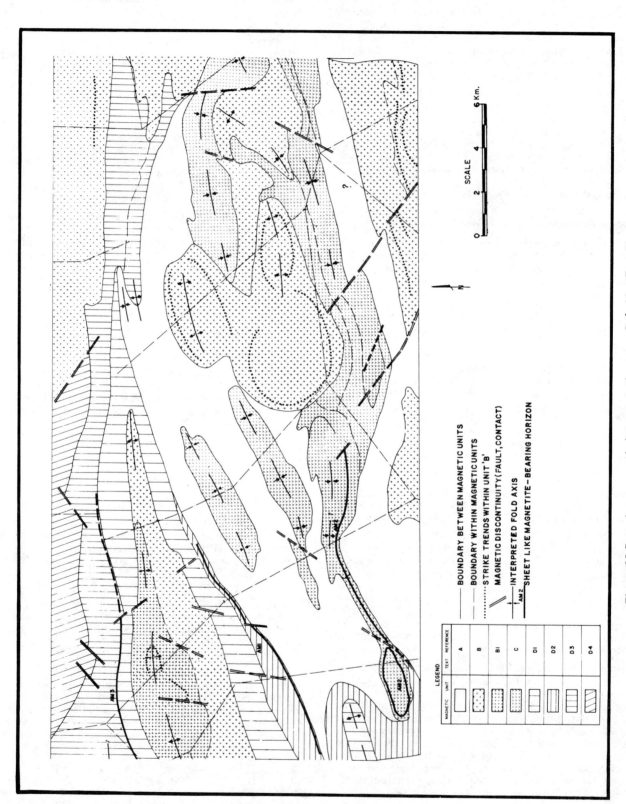

Figure 38.5. Aeromagnetic interpretation map, Pofadder East Block.

The weakly magnetic characteristics of units A and B were taken as precluding the presence of significant magnetite-quartzites therein. Based on the search target model, this automatically excluded some 50-60 per cent of the present area as having little or no economic potential, which subsequent ground checking proved to be correct.

The high density of magnetic horizons within units C and D, resulted in the selection therein of 18 priority areas covering aeromagnetic anomalies thought likely to reflect magnetite-quartzite horizons. True magnetite-quartzite units are now known to lie only within unit D, where they are identified by aeromagnetic anomalies AM1 and AM3 (see Fig. 38.5). Elsewhere, major magnetic responses are attributable to magnetite-enriched lenses within amphibolitic-type rocks, with the exception of AM2, which reflects magnetite layers in an intrusive metagabbroic sill within biotite gneiss country rocks.

AM1 and AM3 appear as predominantly negative anomalies which, given the confirmed near-vertical dip of enclosing strata, implies that remanent rather than induced magnetism must be taken as the dominant magnetization contributor, with the remanent magnetization vector being anti-parallel to the present earth's field inclination. As both the unit C and D rocks are characterized by normal induced responses only, it would appear these remanently magnetized magnetite-quartzite horizons must occupy a unique position within the local Bushmanland stratigraphy. AM3 was subsequently proven to have no base-metal sulphide associations.

The northeast-striking magnetic anomaly AM1 extends for some 10 km, and for a considerable portion of its length is spatially associated with a mineralized horizon at Putsberg. The Putsberg mineralized stratigraphic horizon parallels the magnetite-quartzite unit which has produced AM1 and occurs some 100-250 m to the south, where it lies close to an interpreted contact between units B and D. It appears significant in terms of mineralization control, that the aeromagnetic anomaly attains its greatest amplitude adjacent to the mineralized zone, indicating a marked increase in its width and/or magnetite content within this section. In fact it is a combination of increased width and increased magnetite content which produces the major anomaly. Magnetic discontinuities are apparent at the extremity of the highly anomalous section of the anomaly, and are related to local terminations of the magnetite-quartzite horizon.

Given its exploration significance, the magnetization characteristics of this anomaly may assume special importance. Insufficient evidence exists to establish any direct relationship between areas of possible mineralization and their association with remanently magnetized units, but it may be significant that the magnetite-quartzite at the Gamsberg zinc deposit is also reported to show a large negative remanent component.

Ground Surveys

The ground follow-up program over 18 selected aeromagnetic targets comprised ground magnetometer, geochemical soil sampling and induced polarization surveys carried out during the period August 1973 to July 1974. Magnetometer surveys were conducted at 20 m intervals along traverse lines having a spacing of 200 m, total coverage being 700 line kilometres. The instruments used were Scintrex MF-2 fluxgate magnetometers and Geometrics G816 proton-precession magnetometers. Induced polarization surveys totalling 100 line kilometres were carried out using a McPhar P660 frequency-domain unit, the preferred electrode configuration being dipole-dipole with an 'a' spacing of 25 m or 50 m (prior investigation had shown the thin overburden cover to be moderately resistive (about 300 Ωm) and the upper weathered layer to be less than 10 m thick).

Systematic deployment of the above techniques over and adjacent to selected aeromagnetic anomalies, resulted in the delineation of a zone of copper mineralization on the Putsberg farm. The ground magnetometer coverage of the significant portion of aeromagnetic anomaly AM1 (which first drew attention to the area) is shown in isomagnetic contour form in Figure 38.6b, along with the subsequently determined strike trace of the mineralized horizon.

The magnetic zone identified with AM1 extends for a distance of 6.5 km, from line 9000 in the southwest to line 15 500 in the northeast, and is flanked by nonmagnetic rock assemblages (with the exception of a narrow amphibolite horizon to the south). For the greater part of its length (lines 9000 – 12 800) its major component is a sharply peaking magnetic low (ΔT=1000-2000 gammas), which reflects a magnetite-quartzite horizon having a width in the range 15-25 m, and an average "depth to top" of 10 m. This unit is intercalated within a magnetite-rich sillimanite gneiss which averages about 60 m in thickness. Both causative sources exhibit remanent magnetization, with the polarization vector being anti-parallel to the earth's present field inclination (I Ω -60°).

Lying some 100-250 m south of, and paralleling aeromagnetic anomaly AM1, the metalliferous horizon was delineated by detailed IP surveys carried out over and along the projected strike trace of weak, discontinuous Cu-Zn geochemical anomalies. The copper anomaly is subdued and values between two and three times a background of 25 ppm Cu are the norm, with a maximum spot value of 260 ppm cu. The zinc is more erratic but values four to five times a background of about 80 ppm are common, and zinc values are sympathetic to the copper values.

Figure 38.6a shows two of the original magnetic/geochemical/IP discovery traverses over the mineralized zone. The 4 km-long IP anomaly reflects a narrow (about 20 m wide) steeply-dipping causative source having a shallow depth (< 15 m) to top, which was characterized by strongly persistent responses in the range 3 to 4.5 per cent PFE. The polarizable unit shows no unique correlating resistivity low, thereby largely precluding massive (conducting-type) sulphide mineralization. The hanging-wall biotite-gneisses/schists are characterized by resistivities of 200 Ω m, while the footwall grey/pink gneisses show values in the range 1000 to 3000 Ω m, thereby readily permitting mapping of their contact by the resistivity method.

Contemporaneous percussion drilling guided on site by the IP survey results, showed that the polarizable unit reflected a 4 km long, persistently mineralized horizon lying under 1-5 m of overburden, and containing erratic copper values of up to 4 per cent

Subsequent work has shown the polarizable horizon to contain finely disseminated pyrite (between 1 per cent and rarely 10 per cent by volume) within a 10 m thick assemblage of heterogeneous schists and thin siliceous beds. The major base-metal mineral is chalcopyrite which is present in sufficient quantity to generate subeconomic mineralized bodies at three sections along the mineralized horizon, between lines 10300-10500, lines 1100-11500 and lines 12100-12500. Because these ore zone sections are not characterised by a significant build up in sulphide content as such, the IP technique proved incapable of defining them uniquely within the boundaries of the mineralized horizon.

Geophysical Test Surveys over Putsberg Mineralized Horizons

In order to provide a more complete geo-electrical "finger-print" of the Putsberg mineralized horizon, test surveys were carried out over selected sections using the

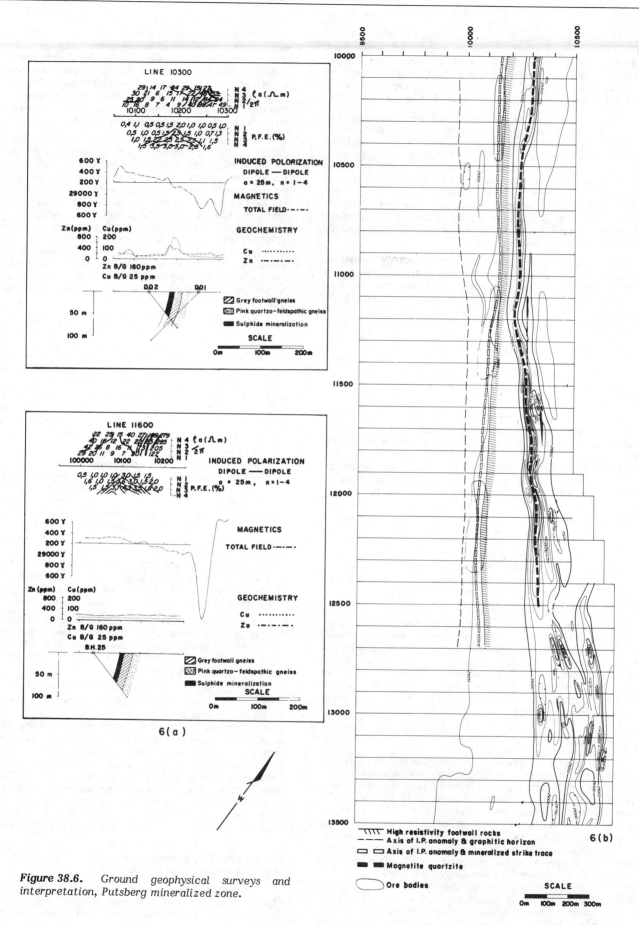

Figure 38.6. Ground geophysical surveys and interpretation, Putsberg mineralized zone.

self-potential, electromagnetic and time-domain IP methods, and an example of the results obtained is shown in Figure 38.7.

While consistently significant anomalies were obtained by both frequency and time-domain IP systems using a variety of electrode configurations, it is noteworthy (although not unexpected) that the sulphide horizon failed to generate distinct horizontal-loop EM and self-potential anomalies. Given the near-surface location (about 5 m depth) of the sulphides, this is of course directly attributable to the lack of massive sulphides in situ, thereby precluding any significant degree of electrical interconnection between sulphide grains.

Gorob Prospect Area, South West Africa

Aeromagnetic Survey

Figure 38.8 shows the results of an in-house JCI aeromagnetic survey over the major portion of the Prospect area, the original minimum contour interval of 2.5 gammas having been coarsened to 10 gammas for the sake of visual clarity. The survey was flown using a Geometrics G803 proton-precession magnetometer at a mean terrain clearance of 100 m, along northwest-southeast traverse lines having a separation of 400 m.

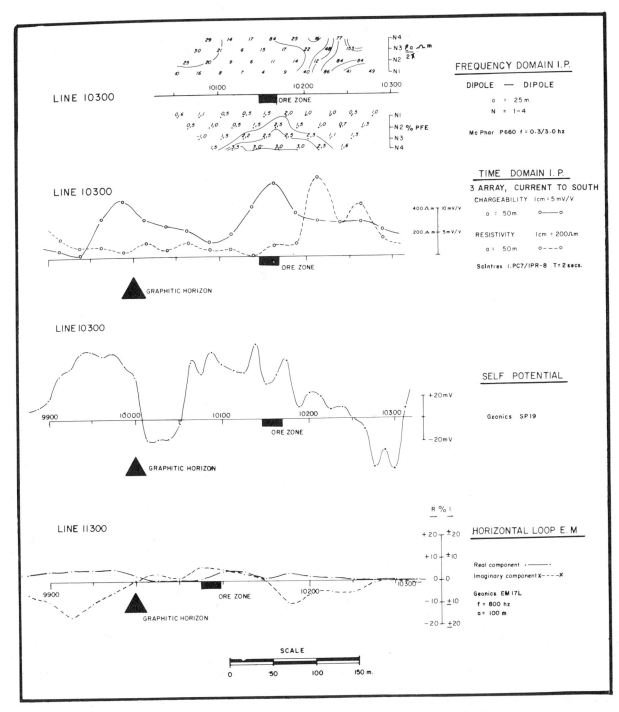

Figure 38.7. Ground geophysical test surveys, Putsberg mineralized zone.

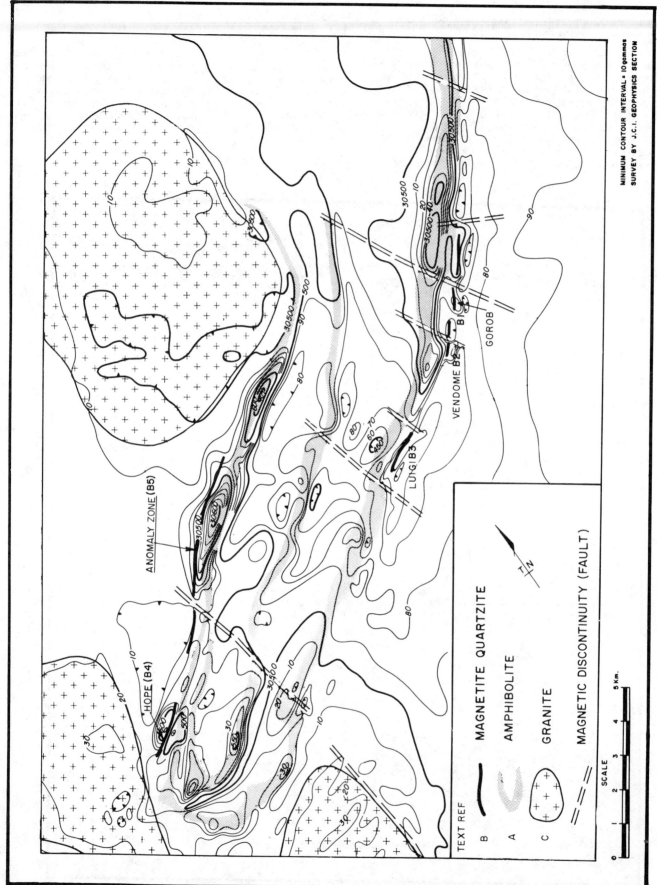

Figure 38.8. Aeromagnetic contour and interpretation map, Gorob prospect.

Magnetic activity within the area is largely restricted to magnetite-bearing amphibolites of the so-called Amphibolite Belt, which fall within essentially nonmagnetic schists and gneisses. The magnetite-bearing nature of these amphibolites is in sharp contrast to those occurring in the vicinity of the Otjihase Mine, which contain little or no magnetite.

Major magnetic discontinuities trending north-northwest are apparent from the aeromagnetic map, and in most cases have been correlated with faults/shear-zones identified from photo-interpretation and geological mapping. The short wavelength nature of most magnetic sources indicates shallow to moderate depths of burial, ranging from at or near-surface in the south of the area, up to 50 m in the north.

Unit A reflects a nearly continuous and for the most part linear amphibolite horizon, whose strike trace delineates a major synformal structure in the western sector of the area. The amphibolites vary in thickness from 50 to 300 m, and have magnetite contents up to 5 per cent by volume. Their attitudes, as interpreted from the magnetic data and confirmed in the field, are consistent with a synclinal structure. Along the southern limb the amphibolites form an outcropping or sub-outcropping unit, whereas along the northern limb there is a marked variation in the "depth to top", ranging from sub-outcrop near the nose of the fold, up to a maximum to 50 m in the eastern extremity of the area, reflecting the increasing thickness of calcrete/alluvium cover in this direction.

The subdivided unit B reflects discrete magnetite-quartzite horizons lying structurally below the amphibolite unit, and separated from it (in plan position) by distances varying from 100 to 300 m. The magnetite-quartzites B1, B2, and B3 falling along the southern limb of the syncline, correlate with short-strike length (< 200 m) mineralized horizons delineated prior to the survey. B1 reflects a 500 m-long magnetite-quartzite horizon (maximum response $\Delta T = 50$ gammas) contiguous to the 150 m strike length pyrite orebody at the so-called Gorob Mine.

B4 reflects a 1.5 km-long magnetite-quartzite assemblage (maximum response $\Delta T = 150$ gammas) correlating, in part at least, with the oxidized outcrop of the mineralized zone of the "Hope Mine". The magnetite-quartzite associated with the mineralization is complexly drag folded in a tight synformal structure (plunging shallowly to the east-northeast) which, prior to the survey, had only been traced via sub-outcrop for a distance of about 300 m from the nose of the syncline. Subsequent work along that portion of the ore-carrying structure revealed by the magnetic survey results, has proved the existence of substantial sulphide mineralization along the down-plunge axis.

Unit B5, since renamed Anomaly Zone, reflects an approximately 4.5 km-long magnetite-quartzite horizon, paralleling and lying some 200-350 m north of the major amphibolite unit defining the northern limb of the syncline. Its aeromagnetic anomaly overlaps with that of the amphibolite horizon, such that it is only clearly distinguishable as a unique horizon from the original profile data. The magnetite-quartzite unit is now known to contain ore-grade copper mineralization within three discrete, short strike-length (about 500 m) sections. For the entirety of its strike-length, the Anomaly Zone magnetite-quartzite is overlain by some 2-40 m of surficial calcrete and gravels.

Unit C reflects weakly magnetic granitic intrusions, which have removed the eastward strike extension of the northern limb of the syncline in the northeast parts of the area.

Ground Surveys – "Anomaly Zone"

Ground magnetometer surveys along a 10 m by 50 m grid were carried out over the Anomaly Zone area, using a Geometrics G816 proton-precession magnetometer. Survey results were used to guide contemporaneous percussion drilling of the 4.5 km-long buried magnetite-quartzite horizon, and resulted in the early delineation of intercalated, narrow (less than 2 m wide) massive sulphide sections carrying traces of copper. In an attempt to define the mineralized horizons more precisely, and in particular to isolate those sections exhibiting the highest sulphide content, ground electromagnetic suveys were carried out over the magnetometer grid. These surveys utilized a Scintrex SE-71 Turam EM unit operating at a frequency of 400 Hz, loop dimensions of 1000 x 1000 m, and a receiver coil spacing of 25 m.

Magnetometer survey results over the entire Anomaly Zone grid, and representative Turam EM results from the grid are shown in Figure 38.9, along with relevant geological information, including the position of significantly mineralized zones proved by limited drilling.

The magnetic data indicate that the magnetite-quartzite unit shows considerable pinching and swelling along strike, with a maximum thickness of about 20 m. Drilling has shown mineralized sulphide zones of low-grade copper in three places, namely Anomaly West, Central and East. (In fact, Anomaly West and Central are contiguous and form the hinge zone of an easterly-plunging drag fold). It is significant that it is within these sections that the magnetite-quartzite unit attains its greatest thickness and/or magnetite content. Using the magnetic width(t) x susceptibility (k) product as an index, these 3 sections are characterized by k*t values in excess of 150 cgs units, vs an "average" value of about 70 cgs units (Koulomzine et al., 1970; McGrath and Hood, 1970; Paterson et al., 1975); the contribution from pyrrhotite may be significant here. In general the magnetic data show the magnetite-quartzite unit to have a near vertical to steep southerly dip, although in the locality of Anomaly West/Central this is reversed and a steep northerly dip is indicated (later verified by diamond drilling). Interpreted depths of burial range from 10-50 m.

The broad deep-seated anomaly to the west of Anomaly Central has proved to be related to an amphibolite lens detached from the main zone to the south.

The Turam EM survey delineated a continous 4 km-long conducting horizon, with an axis parallel and some 10-20 m north of the magnetite-quartzite horizon. The causative unit in general displayed a monotonously regular conductivity – thickness value averaging some 10 mhos, and, assuming a thin dyke source (Bosschart, 1964) an average depth of burial of 40 to 50 m. The electromagnetic data were disappointing in that no unique responses were obtained in the vicinity of the mineralized sections, which are now known to contain up to 40 per cent by volume of sulphides.

Given that massive sulphides are not present along the entire strike extent of the magnetite-quartzite horizon, the strike persistency and regular stratigraphic position of the conducting horizon, indicate that its major causative source must be attributed to conducting minerals other than sulphides in a zone flanking the magnetite-quartzite unit. In this instance, it thus appears that the EM results are not diagnostic of buried massive sulphides, but relate more to a wide lithological unit, probably graphitic schists which have been intersected in one of the few diamond-drill holes in this sector.

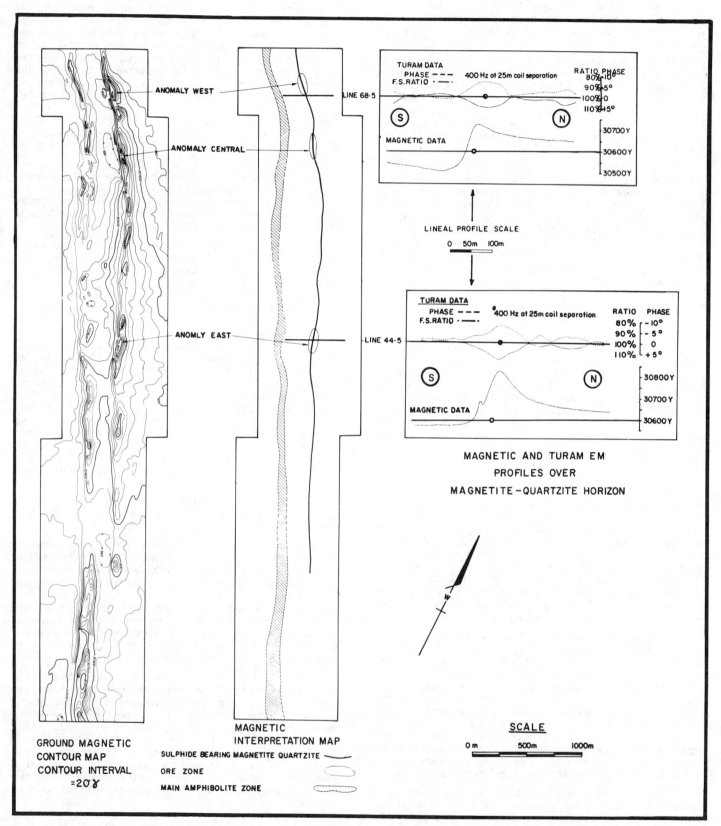

Figure 38.9. Ground geophysical surveys and interpretation, Anomaly Zone area.

Ground Surveys — "Hope Mine" Zone

Ground magnetometer data for this area are summarized in the isomagnetic contour map of Figure 38.10. Survey instrumentation and grid configuration were the same as for the Anomaly Zone area. The magnetite-quartzite horizon generates a maximum anomalous response of $\Delta T=3000$ gammas in the locality of the synformal closure and diminishes eastwards in accordance with the shallow plunge in this direction. Individual anomalous centres relating to the northern and southern limbs, are clearly identified on the profile data for a distance of 200 m down-plunge from the fold closure. Thereafter, the increasing depth of burial of the magnetic units, and their close separation (about 40 m), combine to produce a single broad anomalous response which can be traced, albeit with some difficulty, for a distance of up to 1500 m along strike from the fold closure.

Early drilling along the synclinal axis (as interpreted from the magnetic data) indicated that ore-grade massive sulphide mineralization was distributed erratically throughout the refolded keel of the syncline, where it conformed to a near cylindrical mass with a diameter in the range of 30-50 m. Typical magnetic and geological cross-sections are shown in Figure 38.10b.

Magnetic profile data were interpreted using both the thick-dyke and ribbon models (Martin, 1966; Paterson et al., 1975), although it was recognized that the geometrical complexity of the magnetite-bearing units hardly accorded with either of these simple models. Theoretically the ribbon model should give the best approximation for a synformal structure of the type developed at Hope, but it was found in practice that the thick-dyke model yielded (a) the more internally consistent results and (b) depth and dip values closer to those revealed by later drilling. In all cases, induced magnetization was assumed, and simple field tests have confirmed a magnetization vector within 10° of the earth's present field inclination of -60°. In practice, the thick dyke model was used for determining the axis of the synclinal structure, which, when combined with the extrapolated plunge of the mineralized zone, allowed the designation of subsurface drilling targets.

Taking the magnetite-quartzite limbs as one composite magnetic source, the thick dyke interpretation showed the synclinal structure as having a shallow easterly plunge (<12°) and a steeply-dipping axial plane (>70°N/S). The thick dyke and ribbon interpretations were largely in agreement to the east of line 115W, with the latter generating a ribbon length greater than 1000 m in this sector i.e. the magnetite-quartzites did not appear to be significantly depth limited. West of line 115 the ribbon model inevitably exhibited greater depths of burial (by 20-50 m), with the ribbon at times (fortuitously?) intersecting the mineralized zone. Lack of practical necessity has precluded further investigation of the appropriateness of either model in such a complex geological situation, especially given that the magnetite distribution within the synclinal unit is known to vary considerably in both the horizontal and vertical planes.

In an attempt to define the highly localized ore-zone more closely, recourse was made to electrical survey techniques after core tests had shown the sulphide mineralization to be electrically conducting. A down-hole resistivity survey in drillhole Hope 5 (see Fig. 38.11) showed the upper 70 m of strata to have a resistivity of about 20 Ωm, resistivities down to 150 m being of the order of approximately 40 Ωm. These depressed resistivity values were attributed to heavy rains having generated highly saline groundwater, although similar values had not been observed in the adjacent Anomaly Zone. Given that the upper layer thus demonstrated a minimum conductivity x thickness value of 3 mhos, surface EM techniques were initially discounted, especially given their lack of diagnostic responses in the Anomaly Zone. The large depth to mineralization and limited dimensions of the sulphide zone, also militated against the use of the IP technique, leading finally to the employment of the mise-à-la-masse method (Parasnis, 1967).

Figure 38.11 shows the results of such surveys along the strike trace of the magnetic unit, utilizing energizing current electrodes in drillholes Hope 5, 9 and 17 respectively. Infinite current electrodes were set out 2 km to the north and south of the mineralized zone, and surface voltage potentials read (with respect to an "infinite" potential electrode) at intervals of 12.5 m along line. Instrumentation consisted of a Huntec 7.5 kw, time-domain Ip transmitter and a Scintrex IPR-8 receiver unit.

The isopotential data from Hope 5 and 9 (only discrete sections of which are shown in Fig. 38.11) confirmed the presence of a continuous, moderately conducting ribbon extending for a minimum distance of 700 m. Of particular note is the strike trace flexure in the locality of lines 115-114.5W. The relatively low values for the normalized potential field response over the mineralization and the lack of a sharply peaking response, are due to both the considerable depth of burial of the mineralization, and the thick, upper conducting layer above it, which acts to short-circuit and attenuate a major portion of the subsurface current flow.

The mineralized zone intersected by Hope 9 appears to terminate in the locality of line 114. Mineralization intersected by Hope 17 does not appear to be continuous with that intersected in Hope 9, nor exhibit significant longitudinal dimensions.

The results of the mise-à-la-masse survey are summarized, along with the magnetic data, in the geophysical interpretation maps of Figure 38.12. Results from a Pulse EM survey utilizing a Crone Pulse EM unit (see below) are also presented thereon.

An interpretation of the magnetic data using the thick dyke model shows the synclinal fold axis to be arcuate in outline and steeply dipping. The width of the structure determined is up to 60 m greater than that indicated by drilling, which is due to the presence of magnetite in the schists adjacent to the quartzite. Based on the "depth-to-top" of the magnetite-quartzites, the topmost portion of the synclinal structure shows a plunge of 4° along its western section, which steepens east of line 115W to an angle of 12°. The marked discontinuity in apparent plunge angles is most probably due to faulting believed to occur between lines 115 and 114W.

For the greater part of its known strike length and for depths of burial of up to 125 m, the mineralized zone is well delineated by both the Pulse EM and mise-à-la-masse data. Only the latter, however, provides information on the known northern flexure of the zone at its eastern extremity. The lack of continuity of the mise-à-la-masse conductive axis to the east of line 114.5W, presumably ties in with the fault indicated as occurring at this locality (from both magnetic and drilling information).

The contributions made by both magnetic and electrical methods to this particular search problem, comprising as they do a mixture of quantitative and qualitative data interpretations, are an effective testimony to the value of multi-technique geophysical surveys in assisting drilling programs aimed at assessing relatively small, discrete, and therefore difficult drilling targets at moderate depths of burial.

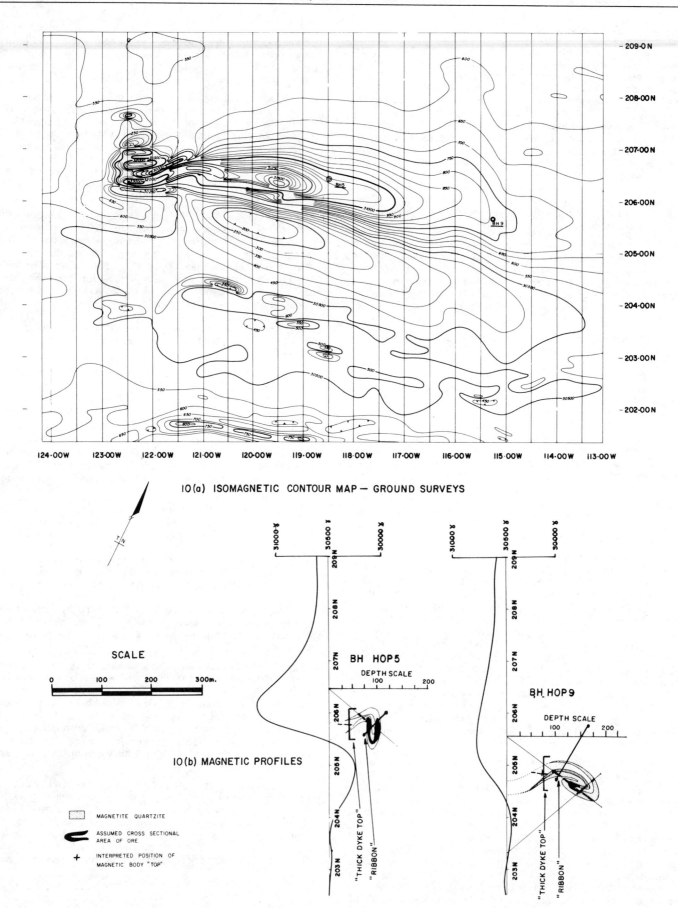

Figure 38.10. Ground magnetometer surveys, Hope Mine Zone.

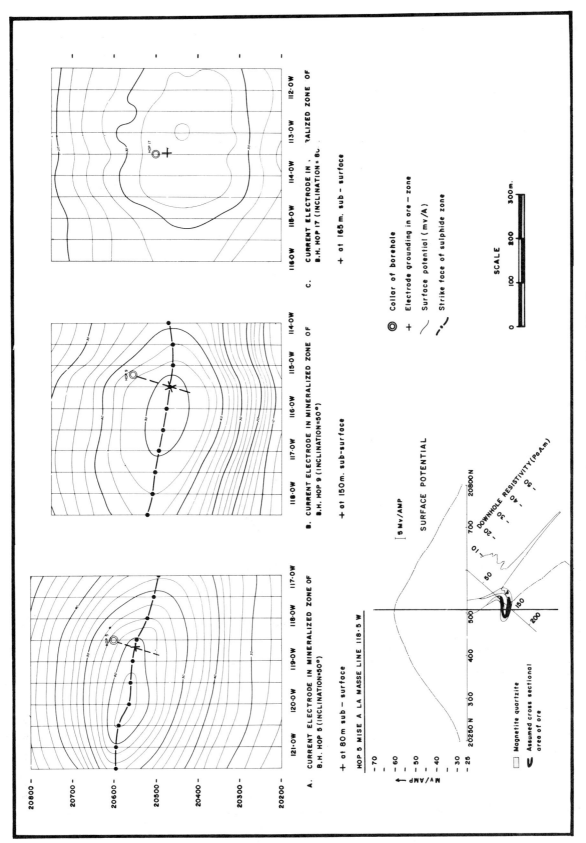

Figure 38.11. Mise-à-la-masse surveys, Hope Mine Zone.

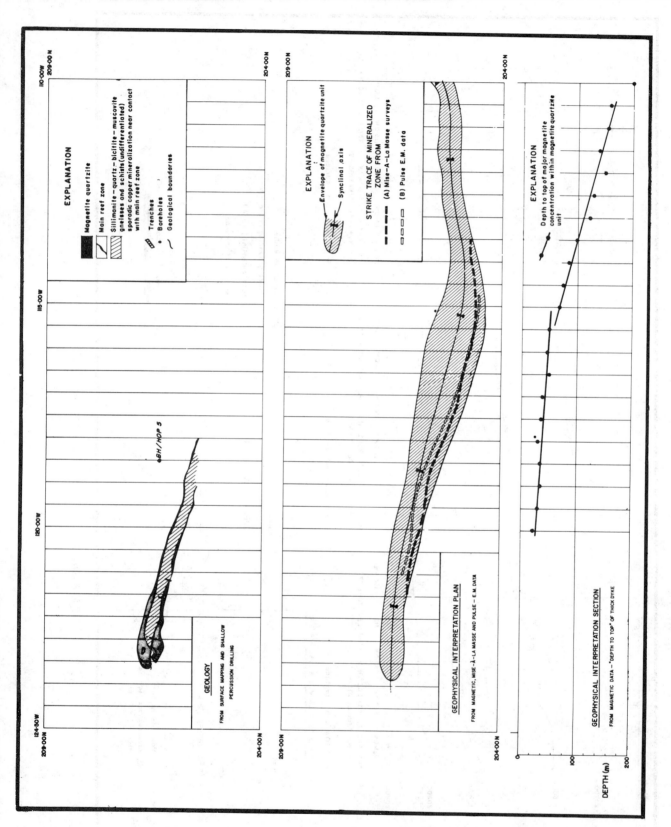

Figure 38.12. Geophysical ground surveys interpretation map, Hope Mine Zone.

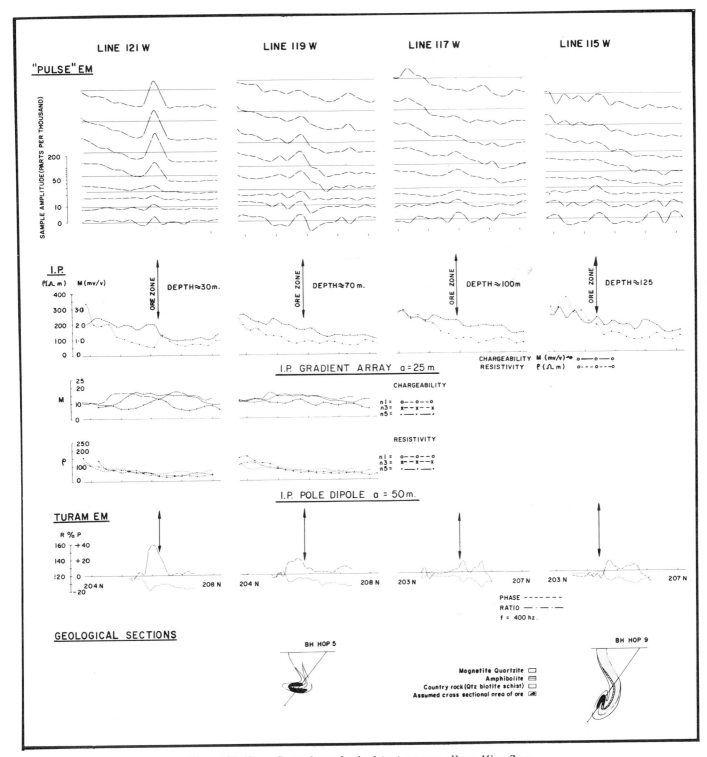

Figure 38.13. Ground geophysical test surveys, Hope Mine Zone.

Test Geophysical Surveys – Hope Mine Ore-Zone

Because of its limited lateral dimensions, large depth of burial, and relatively low conductivity (about 1mho/m), detection of this sulphide ore-zone presents a challenging geophysical search problem, especially given the conducting nature of the local country rocks. The postdrilling availability of a Crone Pulse EM system (through Geoterrex Ltd.) prompted the execution of a small test survey aimed at determining the time-domain EM response of the ore-zone. This was followed by a series of comparative surveys employing the IP, Turam EM and horizontal-loop EM (Geonics EM17L) methods, sample results from all except the latter being shown in Figure 38.13. The 600 m strike extent covered in these surveys reflected ore-zone depths of burial from 30 m (L121W) to 125 m (L115W).

The Crone Pulse EM equipment (Crone, 1976) comprised horizontal Tx/Rx dipoles having a separation of 75 m. Data are presented in the conventional manner as a logarithmic response functions (signal amplitude in ppt) for the 7 receiver sample channels, with channel 1 reflecting the shortest and channel 7 the longest time delays i.e. anomalies in the higher numbered channels reflect a progressively increasing conductivity of the causative source and/or more deep-seated responses.

The lack of a comprehensive interpretation scheme (at the time of the survey) precluded more than an assessment of the presence or otherwise of readily identifiable channel anomalies. While only line 121W (depth-30 m) yielded an anomaly with a good signal-to-noise ratio, it is possible to trace the ore-zone responses through, on progressively higher channel numbers, to the vicinity of line 115W (depth about 125m). However, the poor signal-to-noise ratios from line 117W eastwards, would most probably have precluded confident anomaly identification had the presence of a conducting zone not already been known. It is interesting to note that the horizontal-loop EM results (a=200 m, f=800 Hz), while being of inferior quality to the Pulse EM data, permitted delineation of the ore-zone up to line 117W.

In contrast, the Turam survey results failed to delineate the ore-zone past line 119W, and even on the latter did not conclusively resolve the nature of the causative source. The highly conducting surface layer proved a serious drawback in the field utilization of this method. While the nature of the search problem demanded small loop sizes (say 500 m x 500 m) and low frequencies (200 Hz on the Scintrex SE-71) (West, pers. comm., 1975), attenuation of the primary and secondary magnetic fields via the conducting layer, resulted in extremely poor signal-to-noise ratios, and almost complete loss of signal some 300 m from the leading edge of the loop. Field operations in this case were only practicable using a 1000 x 1000 m loop and a frequency of 400 Hz, with a coil spacing a = 25 m. The surprisingly large field-strength ratio (about 40 per cent) anomaly on line 121W may, in part, be due to the current-gathering anomaly effect (due to host rock conductivity) described by Lamontagne (1970), Lajoie (1973) and West (pers. comm., 1975). Elsewhere, minor anomalous features have been attributed to the heterogeneous surficial conducting layer.

Time-domain IP surveys were carried out using a Huntec 7.5 kw transmitter (2 sec. on/off) and a Scintrex IPR-8 receiver. Given the search target dimensions, the lack of significant anomalous IP responses along any of the sections traversed is not unexpected. The 5 mV/V IP anomaly on line 121W (gradient array + pole-dipole, n=1) presumably reflects the narrow, polarizable magnetite-quartzites. Host rock resistivities in the vicinity of the ore-zone are generally about 50 Ωm, confirming (in a qualitative sense) the results of the down-hole resistivity survey. The disparity in value (20 Ωm vs 50 Ωm) between the surface and down-hole surveys may be due to the combination of a thin, surface resistive layer, and the presence of conducting muds etc. in the borehole.

Given the discrete nature of a sulphide target, and its conductive geoelectrical environment, it would thus appear that its delineation by surface geophysical surveys can only be done by utilizing close-coupled EM techniques.

SUMMARY AND CONCLUSIONS

1. While no major base-metal discoveries resulted from the present exploration program, adoption of a pragmatic search target model did result in the discovery of several small orebodies, in particular the Putsberg grassroots discovery in the Pofadder East Block, South Africa, plus the Hope Mine and Anomaly Zone orebodies within the Gorob Prospect, South West Africa. At present all these deposits are considered subeconomic by the holding company, and thus are not being exploited.

2. Within the Southern African context, there is a valid correlation between magnetite-quartzites and spatially correlating base-metal orebodies of both the complex Pb-Zn-Cu-Ag and cupreous pyrite type, in the cover rocks of many metamorphic terranes.

3. Base-metal sulphides may be directly associated with the magnetite-quartzite horizons or may be spatially separated (up to 300 m) from these horizons.

4. Detailed aeromagnetic surveys are invaluable in the search for magnetite-quartzite horizons, especially in geological environments where magnetic activity is restricted to a limited number of lithological units. Cost-effective benefits include rapid target generation, the rejection of obviously nonpotential areas, and the building up of a semi-regional geological framework.

5. Interpretation of the aeromagnetic data should emphasize the selection of anomalies reflecting thin-dyke causative sources and should be based on interpretation of profiles rather than contour maps. Priority targets are those showing a structural discontinuity or an enhanced magnetic response resulting from tectonic thickening and/or an increased magnetite content.

6. In the first instance it is the presence of the magnetite-quartzite itself and its relative (and not absolute) magnetic response, which are significant in the search for associated base metal sulphides.

7. Ground magnetic surveys as a follow-up to aeromagnetic surveys serve (a) to accurately delineate the magnetic unit on the ground, (b) to reject those horizons which do not accord with the thin-dyke model, (c) to permit an accurate interpretation of the geometrical configuration of metalliferous zones in cases where the latter are intercalated with magnetite-quartzites and lie under substantial overburden cover, and (d) to provide an invaluable aid to interpretation of the geology of an area.

8. Having located the target using magnetics and geology (where possible), the next logical step is to assess the target geochemically where overburden is not an inhibiting factor, using a grid which permits coverage up to 500 m on either side of the axis of the magnetic target.

9. Thereafter significant geochemical anomalies should be followed-up using IP/EM techniques, bearing in mind that possible ambiguities may arise where sulphide mineralization is directly associated with the magnetite-quartzite.

10. Exploration of long strike-extent magnetite-quartzite zones is most cost effective when inexpensive percussion drilling can be used to probe the anomaly (i.e. where the depth to fresh sulphides is generally less than 40 m). In cases where IP anomalies are generated in areas covered by overburden and close to the search target-type of magnetic anomaly, percusison drilling will most cost effectively indicate whether significant mineralization is developed there or not.

The exploration model adopted here, and the geophysical techniques and interpretation schemes utilized, are deceptively simple. However, in the South African context they have proved very successful in exploration for the various types of mineralization which can occur in association with magnetite-quartzites. Close liason must be maintained between the geologist and geophysicist in order to control excesses which inevitably develop where one or the other adopts too prejudiced and/or dominant a role. There is no doubt in our minds that detailed, carefully interpreted aeromagnetic data over potentially mineralized areas still provide the most reliable guide to significant targets. Thereafter, carefully-planned exploration programs utilizing a variety of techniques to suit particular situations, can be adopted with a good expectation of success.

ACKNOWLEDGMENTS

The authors would like to thank the management of Johannesburg Consolidated Investment Company Ltd., for permission to use their data in this publication, and to the exploration staff of the company in South Africa and South West Africa who provided them with assistance.

REFERENCES

Am, K.
　1972: The arbitrarily magnetized dyke: interpretation by characteristics; Geophys. Explor., v. 10(2), p. 63-90.

Bosschart, Robert A.
　1964: Analytical interpretation of fixed source electromagnetic prospecting data; Ph.D. Thesis, Technische Hogeschool, Delft, The Netherlands, 103 p.

Crone, J.D.
　1976: Ground Pulse EM – Examples of survey results in the search for massive sulphides and new developments with the Crone PEM equipment; 25th Int. Geol. Cong., Sydney, Australia (Abs.).

　1979: Exploration for massive sulphides in desert areas; in Geophysics and Geochemistry in the Search for Metallic Ores; Geol. Surv. Can., Econ. Geol. Rept. 31, Paper 37.

Gay, S.P. Jr.
　1963: Standard curves for interpretation of magnetic anomalies over long tabular bodies; Geophysics, v. 28(2), p. 161-200.

Joubert, P.
　1971: The regional tectonism of the gneisses of Part of Namaqualand; P.R.U. Bulletin No. 10, University of Cape Town, South Africa.

Koulomzine, T., Lamontagne, Y., and Nadeau, A.
　1970: New methods for the direct interpretation of magnetic anomalies caused by inclined dikes of infinite length; Geophysics, v. 35(5), p. 812-830.

Lajoie, J.J.
　1973: The electromagnetic response of a conductive inhomogeneity in a layered earth; Ph.D. Thesis, University of Toronto, Canada.

Lamontagne, Y.
　1970: Model studies of the Turam electromagnetic method; M.Sc. Thesis, University of Toronto, Canada.

Martin, Luciano
　1966: Manual of magnetic interpretation; Computer Applications and Systems Engineering, Toronto, Canada.

McGrath, P.H. and Hood, P.J.
　1970: The dipping dyke case: a computer curve matching method of magnetic interpretation; Geophysics, v. 35(5), p. 831-848.

Paizes, P.E.
　1975: The geology of an area between Vaalkop and Aggenys in the vicinity of Pofadder, Northwest Cape Province; unpubl. M.Sc. Thesis, University of the Witwatersrand, South Africa.

Parasnis, D.S.
　1967: Three-dimensional electric mise-à-la-masse survey of an irregular lead-zinc-copper deposit in Central Sweden; Geophys. Prosp., v. 15 p. 407-432.

Paterson, Grant and Watson Limited
　1975: Magmod-magnetic model-fitting programs for the dipping ribbon, the sloping step, the infinite dyke and the prism. Iterative curve fitting computer programmes prepared for J.C.I. Ltd., by Paterson Grant and Watson Limited, Consulting Geophysicists, Toronto, Canada.

Reford, M.S.
　1964: Magnetic anomalies over thin sheets; Geophysics v. 29(4), p. 532-536.

GEOPHYSICAL AND GEOCHEMICAL METHODS FOR MAPPING GOLD-BEARING STRUCTURES IN NICARAGUA

R.S. Middleton
Rosario Resources Corp., Toronto, Ontario

E.E. Campbell
Rosario Mining of Nicaragua

Middleton, R.S., Campbell, E.E., Geophysical and geochemical methods for mapping gold-bearing structures in Nicaragua; in Geophysics and Geochemistry in the search for Metallic Ores; Peter J. Hood, editor; Geological Survey of Canada, Economic Geology Report 31, p. 779-798, 1979.

Abstract

Gold deposits in northeastern Nicaragua which are associated with Tertiary volcanic centres occur as fracture-controlled epithermal vein systems or disseminated sulphide zones within areas of deep tropical weathering volcanic breccias.

Time-domain induced polarization and magnetic surveys were used to outline a gold-bearing disseminated pyrite-sphalerite deposit hosted in a volcanic breccia. This deposit, known as Coco Mina, is located near the Honduras-Nicaragua border. Anomalous gold, copper, zinc, lead and silver geochemical values over the volcanic breccia defined the main portion of the zinc-bearing volcanic breccia, however a broader gold anomaly at threshold levels (0.03 ppm) outlined a much larger area which subsequently was defined by IP as an extensive disseminated pyrite-kaolinite-sericite alteration zone. The dispersion of zinc was greater than lead or copper, and the geochemical patterns for Cu, Pb, Zn appeared to be highly affected by drainage patterns and water courses along faults whereas the gold patterns were controlled by the lithology and the alteration zone, and correlated best with the IP – resistivity and magnetic anomalies.

VLF electromagnetic surveys were also carried out by Rosario Resources Corp. to map the fractures and alteration zones associated with two quartz vein deposits at Guapinol and LaLuna. Tropical weathering appears to enhance the conductivity of the controlling fractures, making them ideal VLF-EM targets. By filtering the in-phase response, the veins are delineated by the resultant contoured data. Soil geochemical surveys over the vein deposits revealed anomalous gold values (0.3 – 0.5 ppm), in disconnected patterns along the trend of the veins, however, the VLF-EM results were of considerable value in interpreting the geochemical patterns and when taken together define much better the various veins in the system.

Résumé

Les gisements aurifères du nord-est du Nicaragua, qui sont liés à des centres volcaniques du Tertiaire, se présentent en réseaux de filons épithermaux à fracture contrôlée ou en zones disséminées de sulfure à l'intérieur de régions de profondes brèches volcaniques à altération météorologique d'influence tropicale.

Des levés magnétiques et des levés par polarisation provoquée en régime transitoire ont servi à délimiter un gisement disséminé de pyrite-sphalérite, contenant de l'or, dans une brèche volcanique. Ce gisement, connu sous le nom de Coco Mina, est situé près de la frontière du Honduras et du Nicaragua. L'or, le cuivre, le zinc, le plomb et l'argent présentent des anomalies géochimiques au-dessus de la brèche volcanique et définissent la partie principale de la brèche volcanique à teneur en zinc, tandis qu'une anomalie aurifère plus importante à des niveaux limites (0.03 ppm) a défini une zone beaucoup plus grande qui, par la suite, a été reconnue, par la méthode PP, comme étant une vaste zone d'altération disséminée de pyrite-kaolinite-séricite. La dispersion du zinc était supérieure à celle du plomb ou du cuivre et les structures géochimiques du Cu, du Pb et du Zn ont semblé fortement altérées par les structures d'écoulement et les cours d'eau le long des failles alors que les structures aurifères étaient contrôlées par la lithologie et la zone d'altération, et correspondaient davantage aux anomalies de résistivité des levés par la méthode PP et aux anomalies magnétiques.

Des levés électromagnétiques à très basse fréquence ont été aussi effectués par la Rosario Resources Corp. dans le but de cartographier les zones de fractures et d'altération reliées à deux gisements filoniens de quartz à Guapinol et LaLuna. Des altérations météorologiques d'influence tropicale semblent accroître la conductivité des fractures de contrôle, faisant d'elles des objectifs de levés électromagnétiques à très basse fréquence. En filtrant la réponse en phase, on peut délimiter les filons à partir des données profilées qui en résultent. Des levés géochimiques du sol au-dessus des gisements filoniens ont révélé des teneurs en or anormales (0.3 à 0.5 ppm) dans les structures disjointes le long de la direction des filons, tandis que les résultats de LE-TBF (levés électromagnétiques à très basse fréquence) étaient d'une importance considérable pour l'analyse des structures géochimiques, quand ils étaient pris ensemble, ces résultats définissaient beaucoup mieux les filons du réseau.

INTRODUCTION

Gold exploration and mining was carried out in Nicaragua before the turn of the century and numerous small quartz lode prospects were mined in northeastern Nicaragua between 1900 and 1928. The history of some operations, including the skarn gold-copper type deposits typified by the La Luz mine at Siuna and the Rosita copper mine, have been described by Plecash and Hopper (1963) and Beven (1973). Present day operations in Nicaragua consist of Sententrion's (Noranda subsidiary) mine near Leon on the west coast, the Neptune Mining Company (Asarco and Rosario) operations at Bonanza 30 km north of Siuna in northeastern Nicaragua, and the Rosario Mining of Nicaragua mines at Riscos de Oro, Blag and La Luna, 25 km northeast of Rosita. All of these ongoing operations are in epithermal quartz vein structures with widths of 2-10 m which occurs in Tertiary basaltic lavas proximal to circular-shaped caldera structures.

Three case histories are presented in this paper illustrating the application of combined geophysical and geochemical methods in outlining gold-bearing epithermal veins (La Luna and Guapinol) as well as a low-grade disseminated sulphide gold-zinc deposit called Coco Mina which is hosted in a caldera-related volcanic breccia.

In the past 10 years geochemical surveys for gold have become more of a reality with the refinement of atomic absorption analytical methods which allow for detection limits below 30 ppb Au. VLF-EM methods have become accepted for mapping structural geological features such as faults, shears, fold axes, conductive contacts etc. Multielement geochemistry is an aid to defining geological-mineralogical settings. These approaches, along with conventional time-domain IP, resistivity and magnetic surveys, are demonstrated in this paper as being effective methods for mapping gold-bearing structures in areas of deep tropical weathering.

COCO MINA DEPOSIT

History and Geology

The Coco Mina gold-zinc deposit is located near the Honduras-Nicaragua border as shown on Figure 39.1. The deposit is situated within a small caldera-like structure in the shape of a cone-shaped mountain approximately 2 km in diameter, which interrupts a folded belt of Mesozoic sediments.

The Coco Mina deposit is contained in a zone of altered (kaolinized-sericitized-pyritized) intermediate volcanic breccia (andesite-dacite) and consists of black sphalerite filling the voids (matrix) of the breccia. The volcanic fragments are usually 10-25 cm in size. Minor chalcopyrite occurs within the sphalerite. Pyrite, which averages 7-10 per cent of the rock, occurs as crystalline masses 2-5 cm in size within the matrix as well as finely disseminated crystals and grains throughout the alteration zone. Isolated arsenopyrite crystals have been observed but are rare. Gold occurs in association with the pyrite and sphalerite and it is speculated from preliminary metallurgical tests that fine free gold also occurs in the upper portions of the deposit. Quartz veining has not veen observed although numerous dykelets of unaltered light grey green andesite cut the altered breccia and sulphides. The altered breccia also occurs around the central lake (lago in Spanish on Fig. 39.2 to 39.8) immediately west of the deposit. The lake itself is situated in a basin with a circular ridge composed of unaltered basalt-andesite surrounding the basin (see height of land apparent from the location of the headwaters of small creeks flowing into the lake in Fig. 39.4 to 39.8).

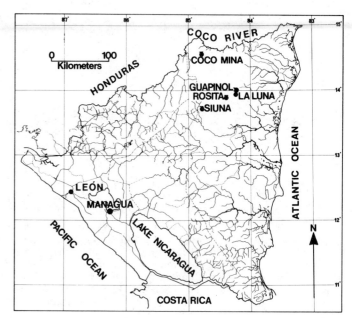

Figure 39.1. Location map showing the position of the Coco Mina, Guapinol and La Luna gold deposits in Nicaragua, central America.

The area surrounding the deposit has been under study by Rosario and Fresnillo since 1974. The placers surrounding the deposit were first worked by a Mr. Mueller in 1900-1944 and then by the Texana Mining Co. (1942-1945), who worked the gold-bearing laterite soil from the east flank of the deposit. Stream geochemical surveys surrounding the deposit and surface sampling were carried out in 1974 followed by the cutting of a survey grid for soil sampling, magnetometer surveys and initial drilling in 1975. An IP-resistivity survey and more drilling was done in 1976. Four more holes were completed in 1977 along with drifting and raising to take a metallurgical sample and verify drillhole assay results.

Geochemical Surveys

Placer operations yielding rice-grain sized gold granules having been carried out for a number of years downstream from the Coco Mina hill, testifying to a source of gold in the hill. Stream sediments show anomalous values in zinc downstream from the deposit; however, the gold appears to show a greater (mechanical and chemical) dispersion than zinc as is evident by comparing Figures 39.2 and 39.3.

A soil-sampling geochemical program was carried out by Rosario Resources Corp. to define the mineralized areas in Coco Mina Hill. Samples were taken at 30 m intervals on lines spaced at 60 m intervals. Samples of the hard clay-like laterite were obtained by auger from a depth of 0.5 m. Complete pulverization of the soil sample followed by screening to -80 mesh was done in the field. Analysis for Au, Ag, Pb, Zn, Cu was carried out by atomic absorption analysis. Gold was extracted into solution by placing the sample in hot aqua regia for 90 minutes and a detection limit of 30 ppb was achieved.

The results of the soil geochemical program are presented in Figures 39.4 to 39.8. The silver anomaly (Fig. 39.4) is confined to the deposit whereas the zinc (Fig. 39.5) is highly dispersed down drainage and around the base of the hill containing the deposit. The laterite soils immediately overlying the deposit are depleted in zinc and an annular-shaped anomaly results (see Fig. 39.9 also).

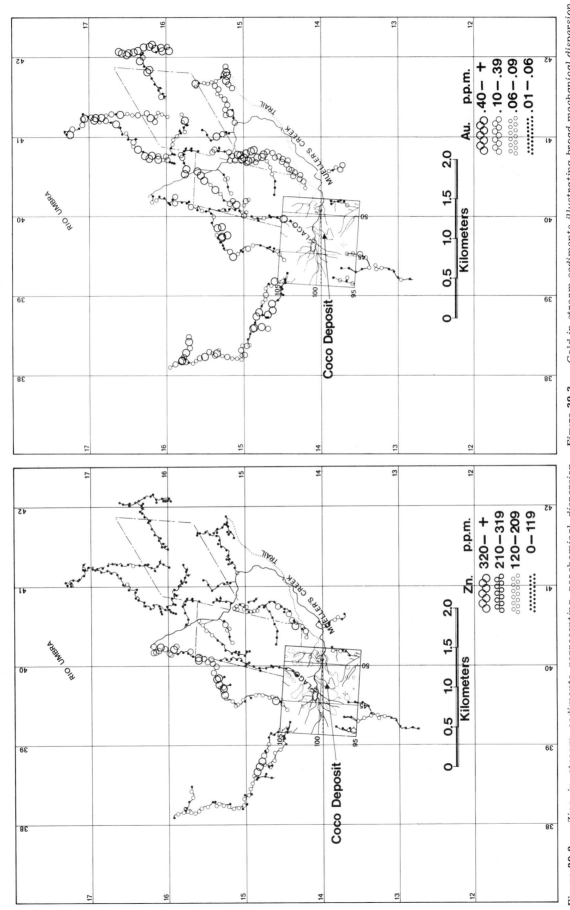

Figure 39.3. Gold in stream sediments illustrating broad mechanical dispersion from the Coco Mina deposit.

Figure 39.2. Zinc in stream sediments representing geochemical dispersion from the Coco Mina Deposit.

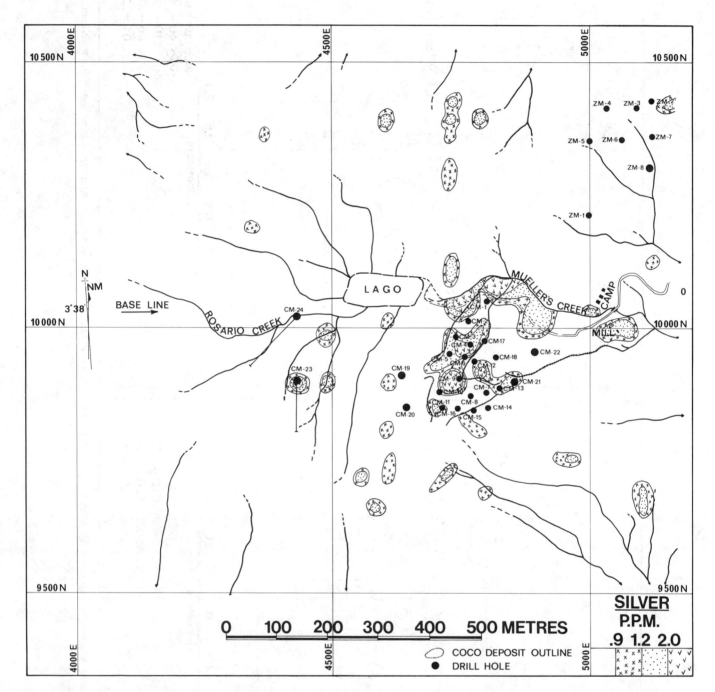

Figure 39.4. Silver soil-geochemical results over Coco Mina.

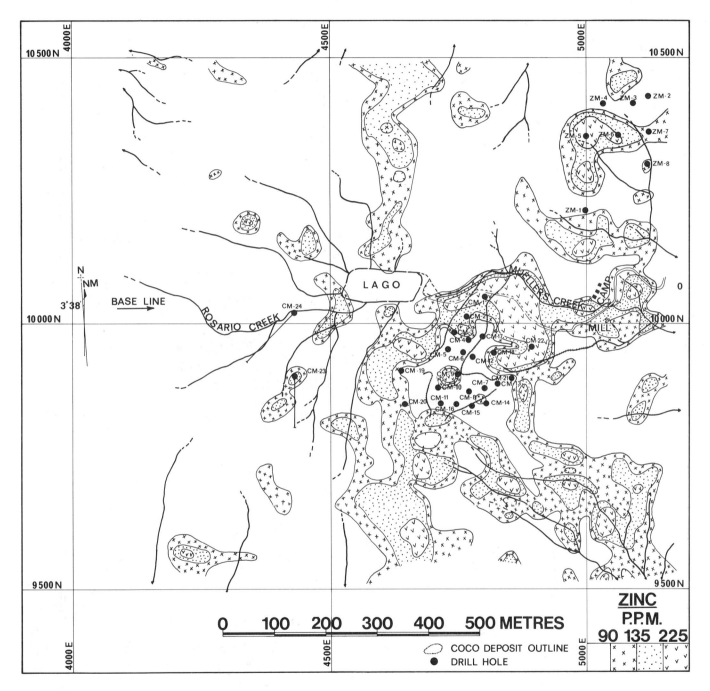

Figure 39.5.   *Zinc soil-geochemical results over Coco Mina.*

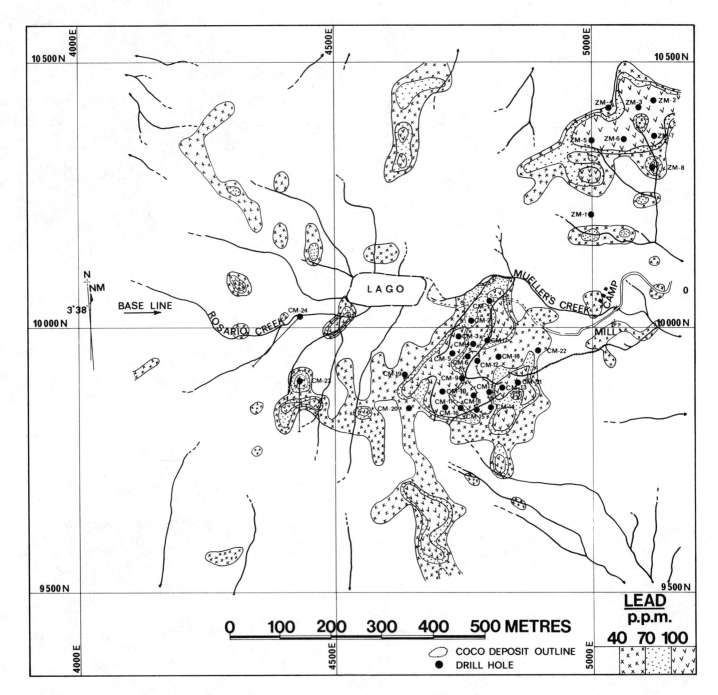

Figure 39.6. Lead soil-geochemical results over Coco Mina.

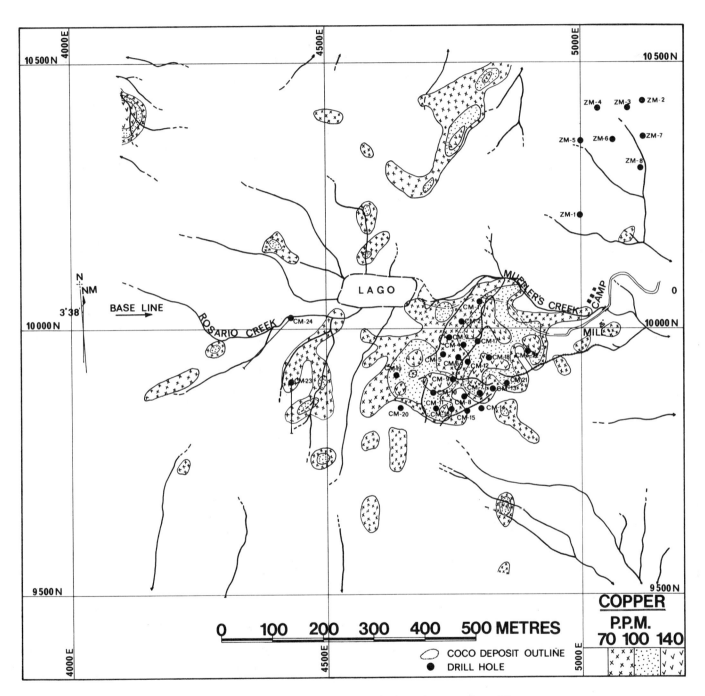

Figure 39.7.   Copper soil-geochemical results over Coco Mina.

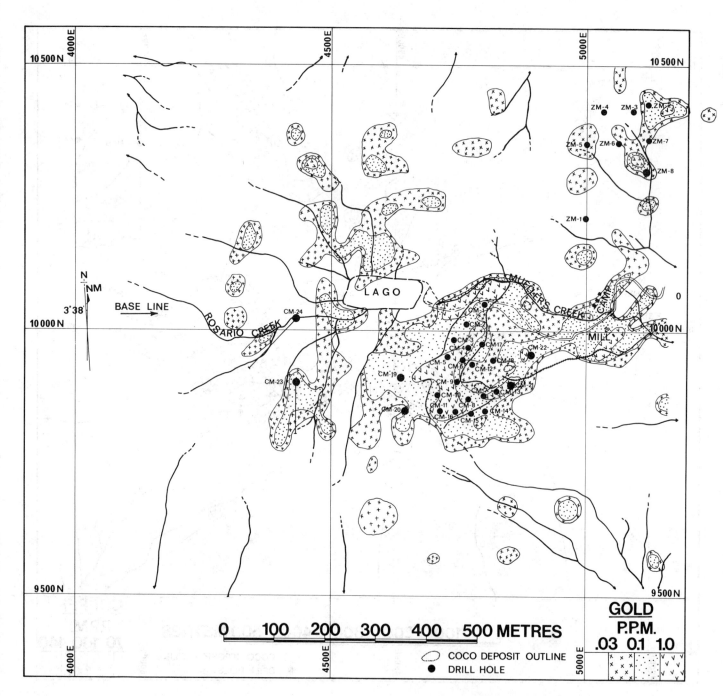

Figure 39.8. Gold soil-geochemical results over Coco Mina.

Lead (Fig. 39.6) is more closely confined to the deposit area but also forms a large anomaly 400 m northeast of the deposit which is known as Zona Margarita (ZM on the figures). Drilling and time-domain IP work showed the Zona Margarita to be narrow and limited in extent and that the results of the soil geochemistry survey have exaggerated the apparent dimensions of the mineralization.

Copper (Fig. 39.7) is also confined to the Coco Mina zone. Anomalous gold values (Fig. 39.8) occur directly over the deposit and also form a series of smaller anomalies around the lake. This gold halo corresponds to areas of altered volcanic breccia and, as will be seen from the following geophysical results, the gold pattern corresponds to a magnetic low, resistivity low and chargeability high.

A north-south soil geochemical profile across the main portion of the Coco deposit is presented in Figure 39.4. Samples of the upper 2 cm of soil were taken and analyzed with an RF plasma unit by Barringer Research in Toronto, Canada to compare with the 0.5 m depth soil-sampling geochemical results analyzed by the atomic absorption technique. Essentially the same pattern of geochemical anomalies was obtained with a slight increase of the metal values occurring in the near-surface material.

Geophysical Surveys

A total field magnetometer survey was carried out in October, 1975 on the geochemical survey grid and the results are given in Figure 39.10. Areas of unaltered basalt (high magnetic susceptibility rocks) are defined essentially by the 41 000 gamma contour and it can be seen that a large ring-shaped pattern encloses the lake, with a branch magnetic high occurring 100 m south of the lake. This branch corresponds roughly to a chargeability low shown on Figure 39.11 and is interpreted as an area of unaltered andesite. The results of a detailed time-domain IP-resistivity survey are given in Figures 39.11 to 39.15. North-south lines were run at 60 and 120 m line spacings with east-west lines being run at 120 m line spacings.

A dipole-dipole array with an "a" spacing of 60 m was used with readings being taken at n=1 to n=5. The instrumentation consisted of a Scintrex IPR-8 receiver, IPC-7 transmitter and a 2.5 Kw motor generator system. Chargeability values on the maps represent the 650-1160 millisecond interval after the shut-off of the pulse. A 2 second on 2 second off square wave was transmitted and the chargeability values were averaged over 2 cycles (4 pulses).

The resulting pattern in plan for n=2 (90 m) (Fig. 39.13) shows the actual extent of the pyritized host volcanic breccia beyond the zinc-gold concentrations of the main Coco deposit. A basin-shaped structure, pyritized host is suggested by the IP results with dips toward the lake; however, the actual main zone on the east side of the lake is steeply-dipping to the east within the altered breccia. A chargeability low immediately west of the lake is interpreted to be a thin capping (sill?) of unaltered andesite since high chargeabilities are seen below this area on n=4 (150 m). The area to the northeast of the Coco Mina deposit known as Zona Margarita (ZM) has only moderate chargeabilities and a narrow northeast-trending pattern. High chargeabilities are somewhat contracted at n=4, again suggesting a dip toward the centre of the lake.

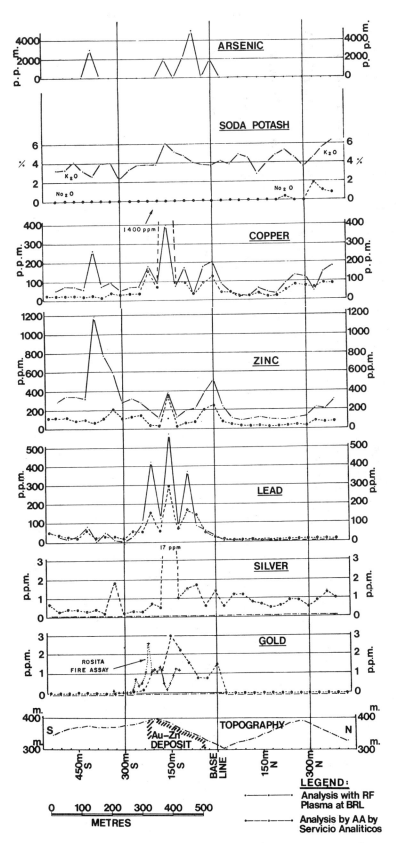

Figure 39.9. North-south geochemical profile through the Coco Mina deposit showing atomic absorption results for soil samples taken from 0.5 m depth with surficial samples taken from the upper 2 cm analyzed in a RF plasma.

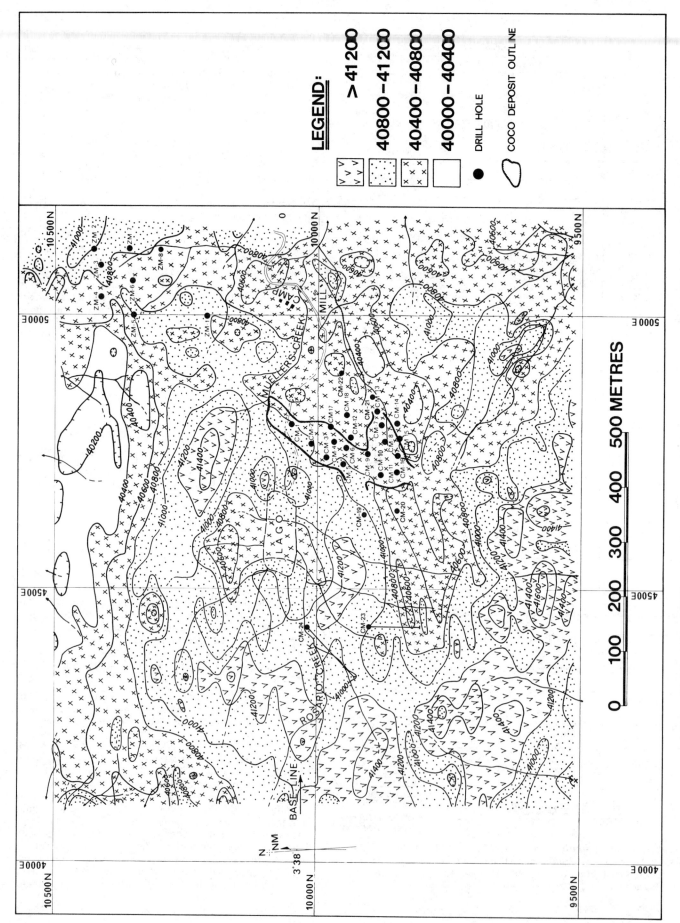

Figure 39.10. *Total field magnetic survey over Coco Mina showing ring-shaped pattern produced by unaltered basalt-andesite.*

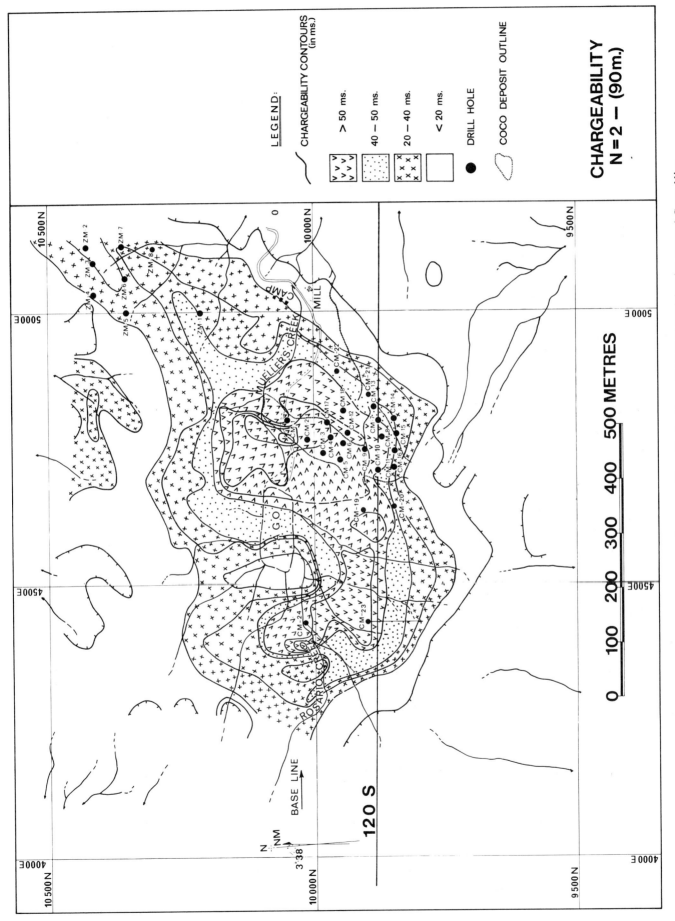

Figure 39.11. IP chargeability values for upper 90 m illustrating widespread pyritization around Coco Mina.

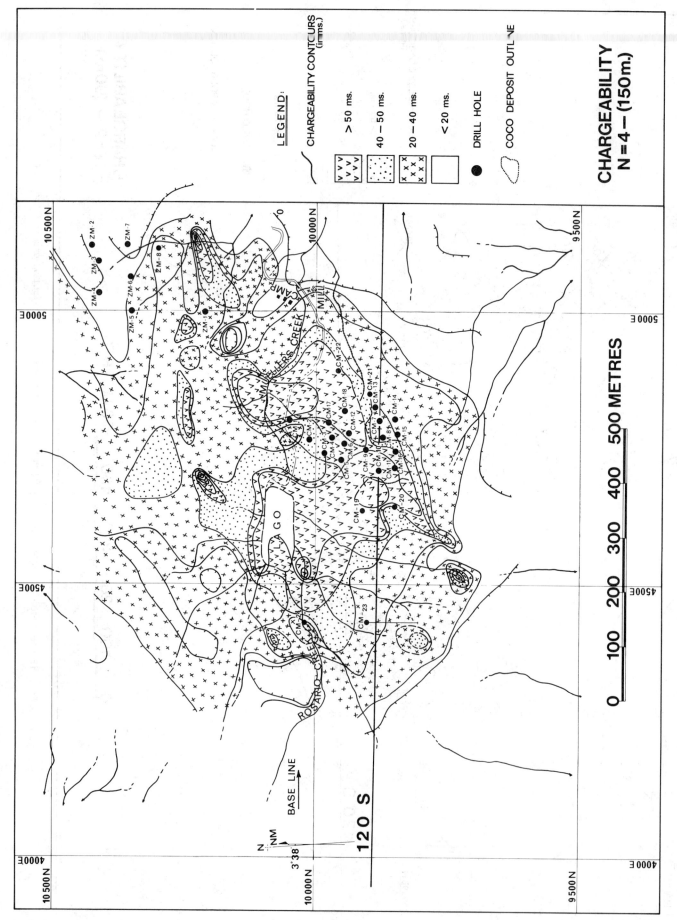

Figure 39.12. IP chargeability values for upper 150 m Coco Mina.

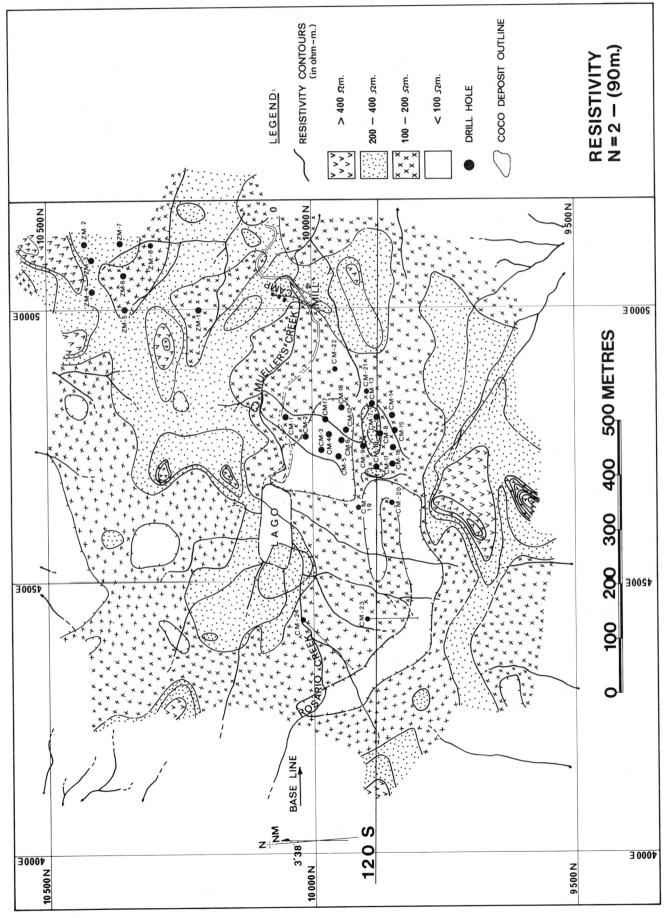

Figure 39.13. Apparent resistivity values of the upper 90 m over Coco Mina.

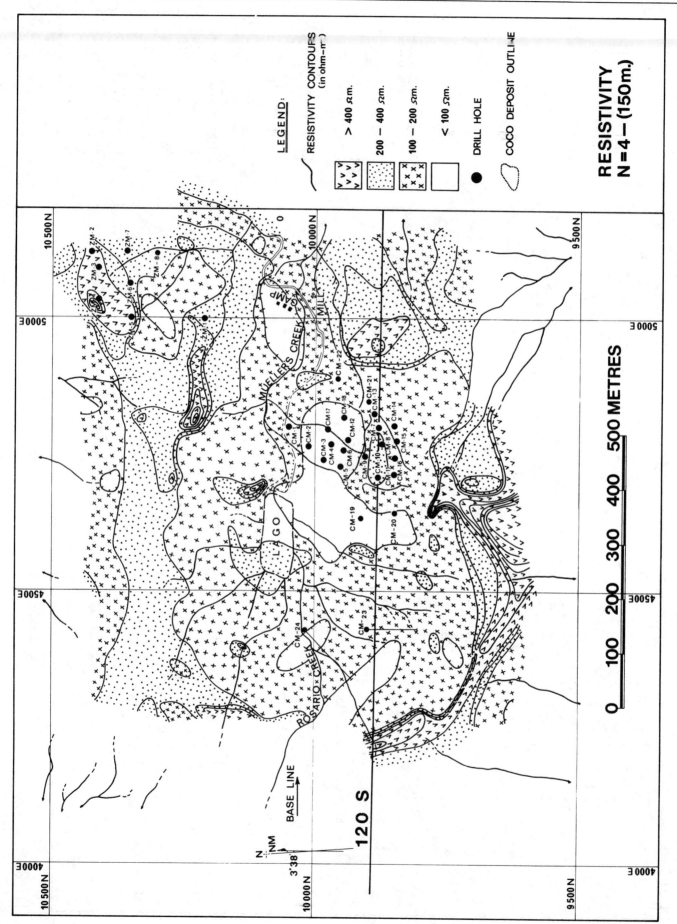

Figure 39.14. Apparent resistivity values of the upper 150 m over Coco Mina.

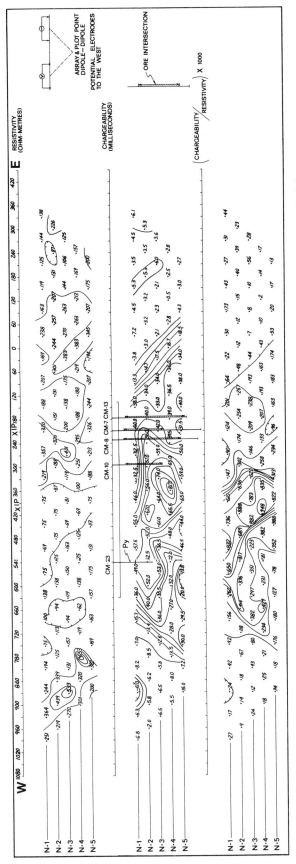

Figure 39.15. East-west section along Line 120S of Coco Mina showing apparent resistivity, chargeability and metal factor relative to some drill intersections.

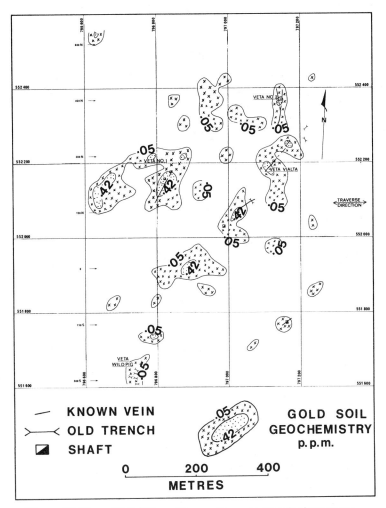

Figure 39.16. VLF-EM profile results over Guapinol prospect.

The section along 120 S given in Figure 39.15 shows that chargeability backgrounds of 2 to 4 milliseconds occur and anomalous values over 80 milliseconds occur within the pyritized host breccia.

The resistivity maps presented in Figures 39.13 and 39.14 show values less than 100 ohm-metres for most of the main deposit and on Figure 39.13 the near-surface expression of the altered breccia is outlined. High resistivities (greater than 200 ohm-metres) are associated with the unaltered basalts to the south and north of the lake and these same areas correspond to the magnetic highs.

Twenty-two NQ-size vertical holes have been drilled in the main zone; two holes to the west and southwest of the lake and seven holes were completed in Zona Margarita for a total of 5140 m of drilling. Drilling so far shows that the Zn-Au mineralization is concentrated on the east side of the lake and that 11.3 million short tons of 0.05 oz Au, 3.4% Zn, 0.77 oz Ag are contained in proven and probable sulphide ore reserves. These grades are based on bulk sample results factored against drillhole results with the drill results on their own giving lower grades possibly due to recovery problems.

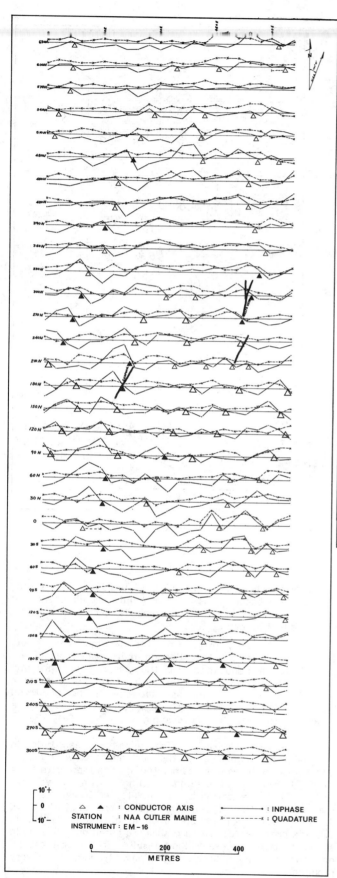

Figure 39.17. Gold soil-geochemistry map over Guapinol prospect.

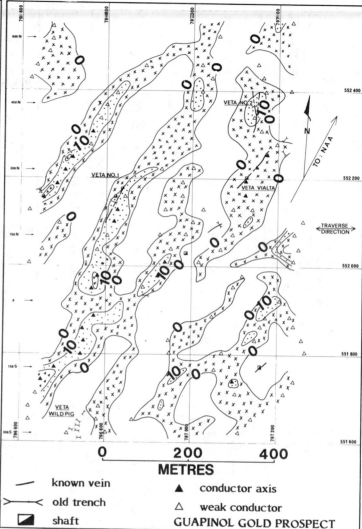

Figure 39.18. Filtered inphase VLF-EM results for Guapinol prospect showing better definition of vein structures.

GUAPINOL GOLD PROSPECT, NICARAGUA

A number of small epithermal vein occurrences are situated near the Riscos de Oro mine, 25 km northeast of Rosita at Guapinol (see Fig. 39.1). These veins are exposed in small, 10 m high hills in a broad flat jungle area. The tropical soils in the hills are residual whereas a thin layer of marine clays surrounds the hills. The exposures were first worked by the Tonopah Mining Co. in 1916-1917. Soil geochemistry sampling was done in 1975 with follow-up VLF-EM surveying in 1976 by Rosario Mining of Nicaragua. Survey stations were spaced at 30 m intervals with east-west lines at 30 m.

The gold soil-geochemical anomalies shown in Figure 39.17 outline the known prospects and hills, however, the marine clays break up the trends. The VLF-EM survey was carried out using a Geonics EM-16 to trace the possible vein extensions. The Cutler, Maine (NAA) transmitter, broadcasting at 17.8 kHz, was used because it was approximately on strike with the veins. The inphase-out-of-phase (quadrature) profiles are given in Figure 39.16 and a series of conductor axes are readily evident. Filtering was carried out on the inphase data using the mathematical operator described by Fraser (1969) in order to transform the zero-crossings to peaks for contouring purposes and reduce geological noise. The resultant contour patterns shown in Figure 39.18 outline the vein extensions and indicate the gold

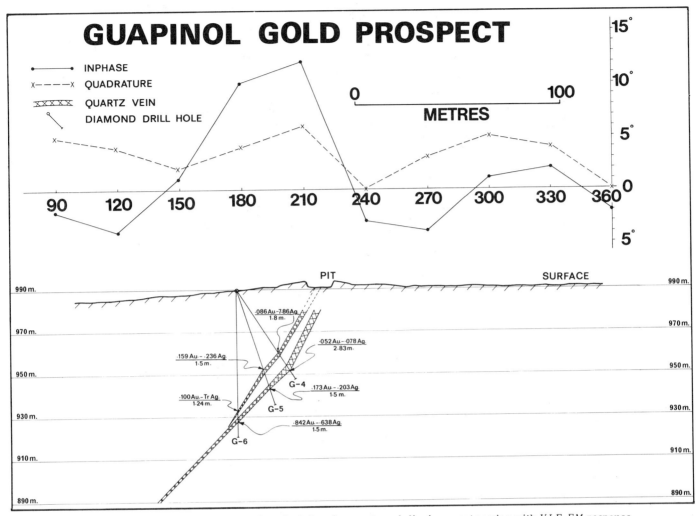

Figure 39.19. Drill intersections at Guapinol showing westward-dipping quartz veins with VLF-EM response.

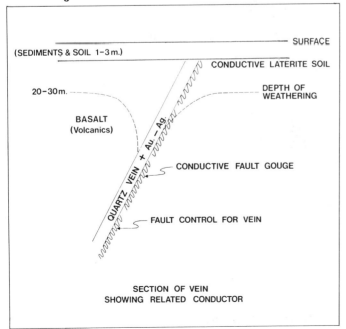

Figure 39.20. Diagrammatic section of epithermal vein setting controlled by a fault (with conductive gouge) and deeper weathering within the fault zone area all of which create a conductive target for EM surveys.

geochemical anomalies are each associated with a particular vein. The two main veins, Veta No. 2 and Veta Vialta appear to join. Drilling shows that the quartz veins have a westward dip of 50°-60° (Fig. 39.19); vein intersections of 1.5-2 m grading 0.2 oz Au and 1.2 oz Ag were obtained. The host rock is amygdaloidal basalt flows with interbedded agglomerates. The profile at the top of Figure 39.19 shows the VLF-EM response produced by the vein, and Figure 39.20 illustrates why the quartz-gold-bearing structures are EM targets. The veins are controlled by fractures which themselves are probably related to calderas. Hot waters deposited the veins and in the process some alteration of the country rock occurred. As a result the contacts of the veins consist of conductive clay (partly clay gouge and clay formed by hydrothermal alteration). In addition, deep tropical weathering along these fracture zones forms clay along the upper portions of the veins. These combined effects result in weak conductors being formed which can be traced by VLF-EM, EM or resistivity. Underground workings and open pits have exposed the veins at Riscos de Oro (1 km northeast of Guapinol and La Luna) to show the highly hematized clay zones in place.

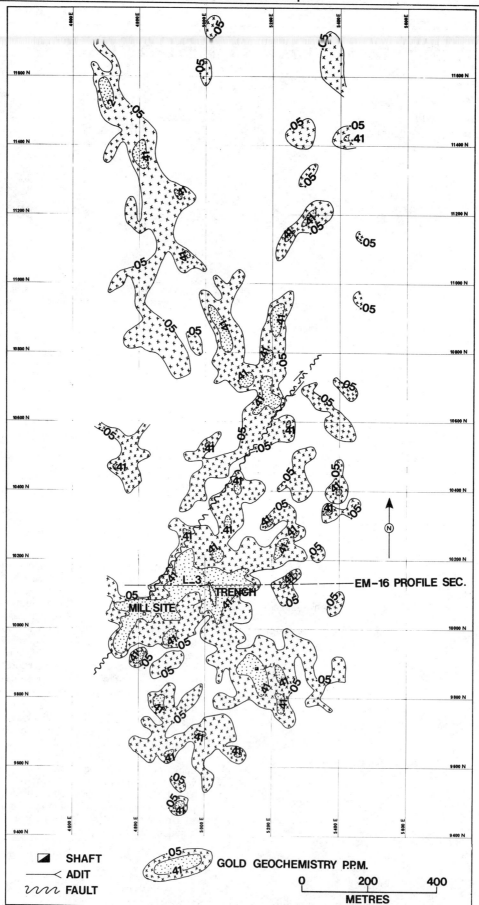

Figure 39.21. Gold soil geochemical pattern over the La Luna prospect.

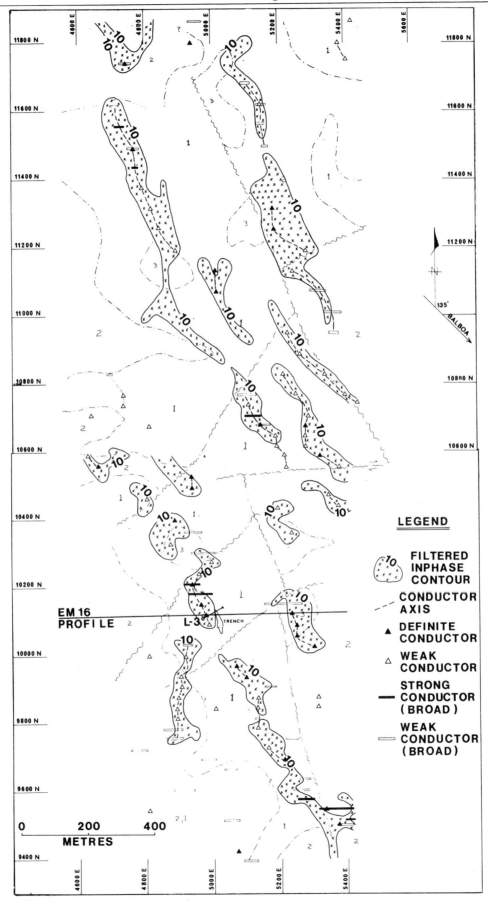

Figure 39.22. VLF-EM results for La Luna prospect showing filtered inphase values that outline vein structures.

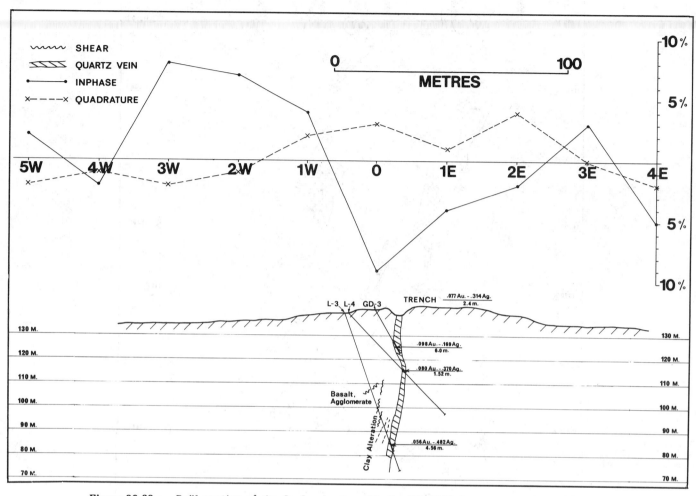

Figure 39.23. Drill section of the La Luna mine with the VLF-EM data accompanying VLF-EM profile (it should be noted that the drill section is on different azimuth than the geophysical profile creating an apparent displacement of the conductor axis).

LA LUNA MINE, NICARAGUA

The La Luna prospect (see Fig. 39.1) was worked prior to 1928 and was rediscovered in 1974 by Rosario native crews when a boiler and two Huntington mills were found in the jungle. Geochemical and VLF-EM surveys were carried out to trace the possible vein extensions from a single pit exposure. A Geonics EM-16 receiver was used for the VLF-EM survey and Balboa, Panama was used as the transmitter. The results in Figures 39.21 and 39.22 show an anomalous geochemical trend N20°W from the trench area which has a corresponding VLF-EM conductor. In addition other parallel conductors are indicted but are yet untested. Drilling (Fig. 39.23) shows a single vein and 33 000 tons of surface ore grading 0.15 oz Au/ton and 0.56 oz Ag were outlined. The mining of these reserves started in February 1977 to provide a part of the mill feed at a cyanide mill situated in Rosita. Further drilling and underground development is planned for La Luna.

CONCLUSIONS

The three examples presented in this paper show how conventional time-domain IP, resistivity, VLF-EM, geochemical, and magnetic surveys can be used to map lithology, alteration and structure which can provide a guide to outlining gold deposits in tropical areas. It was advantageous in all cases to use combined geophysical and geochemical methods which compliment each other in better defining drill targets.

ACKNOWLEDGMENTS

The writers wish to thank Rosario Resources Corporation for permission to publish this paper. Wagih Youssef assisted in the preparation of the illustrations and Jeff Robinson was the operator for the geophysical surveys at Coco Mina. A young Moskito Indian boy, Louis Lackwood, did the EM-16 surveys at La Luna and Guapinol. E.E. Campbell supervised the overall exploration on all the projects.

REFERENCES

Beven, P.A.
    1973: Rosita Mine — A brief history and geological description; Can. Inst. Min. Metall. Bull., v. 66 (73b), p. 80-84.

Fraser, D.C.
    1969: Contouring of VLF-EM data; Geophysics, v. 34, p. 958-967.

Plecash, J. and Hopper, R.
    1963: Operations at La Lus Mines and Rosita Mines, Nicaragua, Central America; Can. Inst. Min. Metall Bull., v. 56 (616), p. 624-641.

AN OUTLINE OF MINING GEOPHYSICS AND GEOCHEMISTRY IN CHINA

Delegation of geophysicists and geochemists, People's Republic of China
(Leader of delegation: Hsia Kuo-chih, National Bureau of Geology, Peking)

Hsia Kuo-chih et al., *An outline of mining geophysics and geochemistry in China*; in Geophysics and Geochemistry in the Search for Metallic Ores; Peter J. Hood, editor; Geological Survey of Canada, Economic Geology Report 31, p. 799-809, 1979.

Abstract

A great many developments and a number of outstanding achievements have occurred both in geophysics and geochemistry since the foundation of the People's Republic of China. In the field of mining exploration, geophysical techniques are used widely in both regional surveys and exploration. Under the guidance of the principle of self-reliance, a variety of geophysical instrumentation has been developed and manufactured in China, including airborne, ground and borehole geophysical instruments. The present paper provides a number of examples, illustrating some of the results obtained by Chinese geophysicists in their practical investigations through persistent summing-up of experience and raising the level of cognition. Some research programs to improve methodology and theory have been carried out to resolve problems which have originated from actual field work.

For regional surveys, a system of geochemical methods is now established and is becoming relatively mature, with a variety of terranes in consideration. Statistical and computer techniques are beginning to be widely used in the processing and assessment of geochemical prospecting data, especially those for stream-sediment surveys, which are carried out on a broad scale. The paper reviews some results from lithogeochemical surveys, with the following primary dispersion halos being discussed: 1. Anomalies having linear and complex nonlinear shapes; 2. Primary dispersion halos around a steeply dipping orebody; 3. Primary dispersion halos from a skarn-type ore deposit. The present paper also presents experimental results from a soil mercury-vapor survey carried out in a broad loess-covered area.

Résumé

De très nombreux développements et un grand nombre de réalisations importantes ont eu lieu dans les domaines de la géophysique et de la géochimie, depuis la fondation de la République populaire de Chine. Dans le domaine de la prospection minière, des techniques géophysiques sont utilisées sur une grande échelle lors des levés et de la prospection dans les régions. Dans l'esprit du principe d'autosuffisance, une gamme d'instruments de géophysique ont été mis au point et fabriqués en Chine, notamment des instruments de prospection par sondage. Le présent document fournit un certain nombre d'exemples, en expliquant quelques-uns des résultats obtenus par les géophysiciens chinois dans leurs recherches appliquées, grâce à la revue constante des progrès accomplis et au perfectionnement continuel des connaissances. Quelques programmes de recherche visant à améliorer les côtés pratique et théorique ont été mis en oeuvre, afin de résoudre les problèmes qui ont surgi lors des travaux sur le terrain.

En ce qui concerne les levés de régions, il existe maintenant un système de méthodes géochimiques qui est en train de devenir assez perfectionné, en ce qu'il tient compte d'une grande variété de terrains. Des techniques d'étude statistique et informatique commencent à être utilisées sur une grande échelle pour le traitement et l'évaluation de données de prospection géochimique, particulièrement les données relatives aux levés de sédiments fluviaux, levés qui sont effectués sur une grande échelle. Le document analyse quelques résultats obtenus à partir de levés lithogéochimiques, notamment les auréoles de dispersions primaires suivantes: 1) anomalies ayant des fa onnements linéaires et non linéaires complexes; 2) auréoles de dispersion primaire autour d'un gisement en amas à inclinaison raide; 3) auréoles de dispersion primaire à partir d'un gisement de type skarn. Le présent document contient aussi des résultats d'expériences à partir d'un levé de la vapeur de mercure du sol effectué dans une vaste zone couverte de loess.

INTRODUCTION

The history of the development of geophysics in China can be traced far back to ancient times. According to early records, it was in the third century B.C. that the phenomenon that "magnet attracts iron" was first observed, and in the early years of the first century A.D. the polarity of a magnet was known. The compass had been widely used by the beginning of the twelfth century A.D.

With respect to geochemistry, over 2000 years ago scientific ideas of applying geochemistry in ore prospecting were plainly put forward in China in the Spring and Autumn Period (770-475 B.C.), as described in the works Kuan Tze "where there is red sand (i.e. rusty quartz) above, there is gold below, where there are magnetic stones (i.e. magnetite) above, there are copper and gold below, where there is ochre (i.e. gossans) above, there is iron ore below, and where there is lead above, there is silver below". But owing to the long period of feudal domination and semi-feudal and semi-colonial rule, exploration geophysics and geochemistry, like other disciplines of science and technology did not progress in China. As a result before liberation in 1949 there were only a few mining geophysicists engaged in a very limited amount of experimental investigations. In geochemistry the situation was so bad that until the eve of liberation there was not even a single specialist in this field.

For more than twenty years since the founding of the People's Republic of China on October 1, 1949, under the leadership of the great leader Chairman Mao and the Communist Party of China, the geophysical and geochemical population of China has expanded from a small to a large force, with a total staff presently numbering tens of thousands; thus geophysical and geochemical methods have found a wide application in mining exploration in China; at the same time geophysical methods also have been used widely in the fields of petroleum and coal, as well as in hydrogeology and engineering geology. Under the guidance of dialectical materialism, Chinese geophysicists and geochemists have made their work more effective. In recent years most of the magnetic iron-ore deposits found have been discoveries resulting from geophysical exploration. Successful results have been also obtained from geophysical and geochemical prospecting for such mineral resources as copper, lead, zinc, chromium, nickel, molybdenum, vanadium, and radioactive elements. In order to support the nation-wide 1:200 000 scale geological mapping and areal airborne magnetic survey program and a regional geochemical investigation have been carried out, which have provided a large amount of fundamental information. Consequently mining geophysics and geochemistry have been taken more seriously in geological investigations with every passing day. A brief review of the application of geophysics and geochemistry to search for metallic ores in China follows.

GEOPHYSICAL PROSPECTING IN CHINA

The socialist system has provided a broad possibility for the rapid development of geophysical prospecting in China using the principles of independence, initiative and self-reliance. Geological instrument factories have been built-up step by step and research institutions have also been established. In addition to the foregoing, research projects of methodology and instrumentation have also been undertaken by the various universities and colleges concerned as well as by some of the geophysical field parties. With respect to magnetic instrumentation, suspension-wire magnetometers, quartz magnetometers, fluxgates, proton-procession and optically-pumped magnetometers have been successfully designed and manufactured. In the field of gravimetry high-precision portable gravimeters have been developed and fabricated. We have also developed and manufactured a variety of instruments for electrical exploration and for borehole investigations. In addition a compensation-type airborne EM system has been successfully developed. In respect to the treatment and interpretation of geophysical data we have established and are continuing to establish several data-processing centres equipped with computers.

While learning advanced techniques from abroad, we are developing instruments specifically suited to our actual geological and geophysical conditions. For instance, a horizontal magnetometer was designed for measurement of the horizontal magnetic component in an arbitrary direction. Such an instrument has proven to be applicable to prospecting for magnetic orebodies at low latitudes. In support of the scientific investigation in the Mt. Jolmo Lungma region, a "Mt. Jolma Lungma"-type gravimeter was successfully manufactured and used for station measurements at an altitude of 7790 m (Fig. 40.1) To enable the induced polarization method (IP) to prospect areas rapidly, a modified IP receiver has been developed. By using this instrument, the measurement line is separated from the current line without any synchronizing unit and the signal received by the receiver at the cutoff of the primary field serves as the triggering signal. Being portable, simple to read and permitting the simultaneous operation of several measuring groups. The instrument makes possible a high working efficiency for the IP method.

Chinese geophysicists in order to obtain the most reasonable interpretation examine anomalies and sum up their collective experience. Thus a number of successful results have been obtained. For example, Figure 40.2 shows

Figure 40.1. Chinese mountaineers making measurements with the "Mt. Jolmo Lungma" type gravimeter in the Mt. Jolmo Lungma region. (GSC 203492-D)

Anomaly M 85 indicated by airborne magnetic survey on the prairie of northeastern China. Judging by its intensity and shape and taking the geological environment into account this anomaly was considered to be caused by an ore occurrence. The ground magnetic follow-up survey also located the anomaly and the causative orebody was interpreted to be at a depth of about 210 m. Drilling of the anomaly followed and revealed a thick bedded magnetite orebody at a depth of 184 m. Then ground magnetic, gravimetric, electromagnetic and borehole three-component magnetic measurements were conducted to furnish information on the shape, mode of occurrence, and size of the orebody to guide further exploration. Figure 40.3 shows the ore occurrence with the result of the three-component magnetic measurement and the downward continuation of the vertical field (OZ) anomaly.

It should be noted that, so far as Anomaly M 85 is concerned, from the actual discovery of this anomaly by airborne magnetic survey to the final confirmation by drilling took less than 4 months.

From many case histories we have come to realize that one should learn to see the essence through the appearance and that very often correct knowledge can be obtained only after a process of repeated practice and repeated cognition.

For example, aeromagnetic Anomaly M 19, was delineated in the lower reaches of the Yangtze River. Within this anomaly zone three local anomalies, A, B, and C, were indicated by the subsequent ground magnetic follow-up. In 1962, a number of boreholes were drilled for verification of Anomaly A (Fig. 40.4) and only a limited quantity of iron reserves were obtained. Afterwards in 1970 when making a further investigation of this area, it was noted that apart from the local anomaly there exists around it still a broad and gentle anomaly of low intensity, which had been thought in

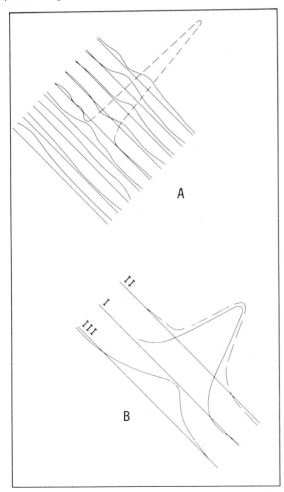

A. Aeromagnetic ΔT anomaly, 1 cm = 400γ
B. Ground magnetic ΔZ anomaly, 1 cm = 1000γ

Figure 40.2. Aeromagnetic Anomaly M-85 and ground magnetic follow-up results, northeastern China.

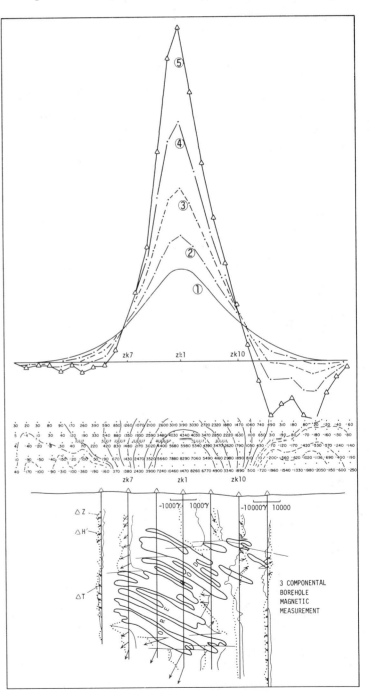

1. Measured curve
2. Continued downward to a level of 40 m below the surface
3. To a level of 80 m
4. To a level of 120 m
5. To a level of 160 m

Figure 40.3. The vertical field (ΔZ) downward continuation, 3-component borehole magnetic measurements and geological section for Anomaly M-85.

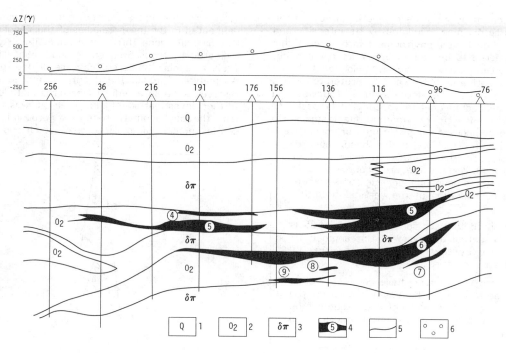

1. Quaternary deposits
2. Ordovician limestone
3. Diorite
4. Magnetite ore and orebody number
5. ΔZ curve
6. Theoretical ΔZ curve

Figure 40.4. *A composite section of Anomaly A along exploration line 3, Yangtze River.*

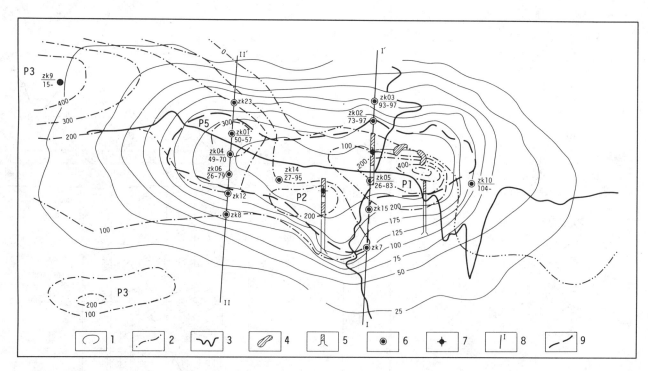

1. Equipotentials of the Mise-à-la-masse method
2. Contour map of the self-potential measurement
3. Gradient curve of the Mise-à-la-masse method
4. Outcrop of the known orebody
5. Exposure of the orebody in trench
6. Borehole
7. Charging point
8. Exploration line
9. Orebody boundary located with geophysical method

Figure 40.5. *A map showing the survey results by the self-potential and Mise-à-la-masse method over a deposit, Chinghai-Tibet Plateau.*

the past to be the effect of diorite-porphyrite. Magnetic susceptibility determination of the diorite-porphyrite, however, showed that it had too low a magnetization to give rise to such an anomaly. It was suggested that the boreholes drilled before had only intersected the upper contact orebodies, and all ended in the diorite-porphyrite. With this in consideration borehole ZK 191 was drilled and high-grade ores were encountered at a depth of 217 m underneath the diorite-porphyrite, leading to the discovery of the second and third ore beds. The theoretical curve calculated gave a fairly good coincidence with the one practically measured, indicating that the second and third ore beds appear to be the cause of the gentle low-intensity anomaly.

It is quite obvious that any single geophysical method can at best reflect only one aspect of the subsurface geology. It is for this reason that several methods reasonably integrated are usually employed in an exploration program to get a more comprehensive picture of the subsurface geology. Close cooperation of geological, geophysical and geochemical means of prospecting is also required to make up for the deficiency of any given technique. Figure 40.5 shows several self-potential anomalies indicated in reconnaissance surveys on the Chinghai-Tibet Plateau. Mineralization was revealed in all the three galleries driven initially on the self-potential anomalies P1 and P2. This mineralization, however, was then regarded as two small isolated bodies.

Sometime later when the work resumed in this area, the Mise-à-la-masse method was applied with the purpose of clarifying the relationship between these orebodies. The results obtained with all the three groundings used showed nearly the same picture, indicating that all these orebodies were connected with one another with an excellent electric conductivity, and therefore the size of the body tested by drilling appeared to be much larger than it was supposed before. Borehole ZK 14 and those on lines 1 and 2 drilled in accordance with the interpretation all intersected the body. Thus all the isolated self-potential anomalies appeared to be due to the local shallowest exposures of the body.

We have also carried out research programs into methodology so as to resolve prolems which have originated from actual field work; for example, the method of interpretation of magnetic anomalies in the case of oblique magnetization, the techniques for gravity survey in mountains with complicated terrain conditions, the potential application of IP method and the approaches for its interpretation as well as the possible application of holographic techniques to mineral exploration. In addition we are still expanding the aeromagnetic coverage of high-mountainous regions. The development and achievements of geophysical prospecting in China also should be attributed to the persistent implementation of the mass line. In addition to the specialized geophysical brigades, geophysical prospecting teams or groups have been organized in almost every geological field brigade, and the geological detachments of some counties are also equipped with magnetometers and potentiometers to do geophysical prospecting for mineral resources for the county and commune-run industries as well as groundwater investigations for the development of agriculture.

GEOCHEMICAL PROSPECTING IN CHINA

Since the late fifties, a more or less systematic investigation of the primary halos of mineral deposits has been carried out in combination with the mining geology mapping the prospecting for blind orebodies in/or around known mineral deposits. For example, a rock sampling program was carried out over a lead-zinc district of northeast China which had been exploited for some years. The primary halos controlled by tectonic fractures were found to be linearly distributed around lead-zinc bodies. However, the anomalies caused by the postmineral quartz veins appeared to greatly interfere with the resultant geochemical data because of their similarity to the ore minerals both in composition and configuration. But the Cu/Pb ratio of the anomalies was quite different. Based on a large amount of data, it was found that directly over blind Pb-Zn orebodies related to fracturing, lead was usually over 300 ppm, arsenic over 80 ppm, with the Cu/Pb ratio being less than 0.2. An attempt was made to use this criterion in an evaluation of other geochemical anomalies. As a result new blind lead-zinc orebodies of economic interest were discovered in the area (Fig. 40.6). Thus an old mine was renewed and its useful life extended. Many examples of complicated nonlinear anomalies observed during later geochemical prospecting activities on a broad scale have further enriched our knowledge of primary halos. Within these anomalies, element contents fluctuate locally over a wide range, so that mathematical methods are used to suppress noise and permit anomalies to be divided into three concentration zones: namely the inner zone, the middle zone and the outer zone, according to the geometrical series of the background values. In this way, the pattern, size, and internal structure of various nonlinear anomalies can be

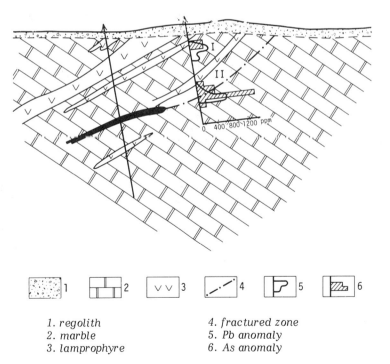

1. regolith
2. marble
3. lamprophyre
4. fractured zone
5. Pb anomaly
6. As anomaly

Figure 40.6. A blind lead orebody discovered from the Pb and As anomalies in drill cores.

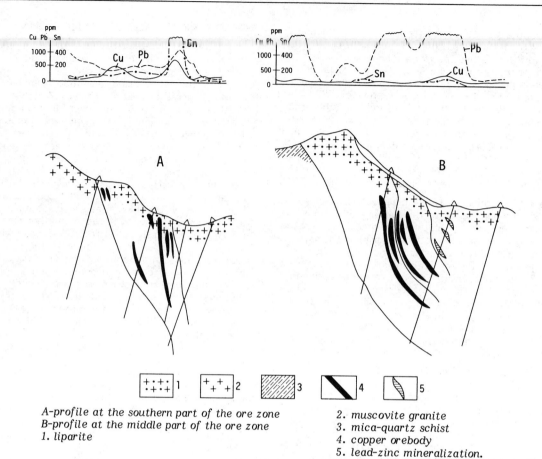

A-profile at the southern part of the ore zone
B-profile at the middle part of the ore zone
1. liparite
2. muscovite granite
3. mica-quartz schist
4. copper orebody
5. lead-zinc mineralization.

Figure 40.7. Primary anomalies over copper orebodies, southern China.

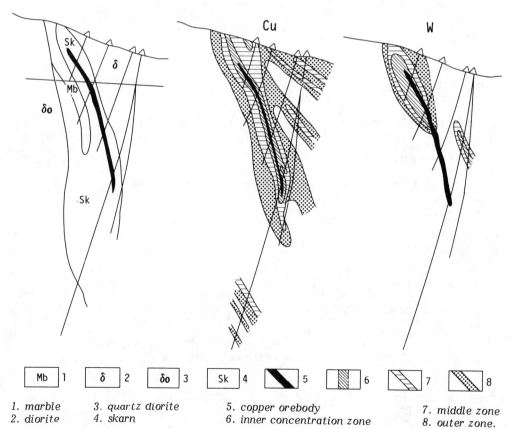

1. marble
2. diorite
3. quartz diorite
4. skarn
5. copper orebody
6. inner concentration zone
7. middle zone
8. outer zone.

Figure 40.8. Copper and tungsten halos around a skarn copper orebody.

compared. It has been proven in practice that the presence of an outer zone at the ground surface and its size may be used to evaluate blind orebodies at depth, while a middle zone of fairly large scale is an indication of a shallow-buried one.

Based on a systematic study of a vast amount of data obtained in various areas, it is considered that a primary halo surrounding a steeply-dipping orebody may be divided into three parts: a front part (or front halo), an adjacent part (or adjacent halo) and a rear part (or rear halo). There are distinct longitudinal chemical zoning at the front and rear halos. The most developed elements within the front halo of a Pb-Zn hydrothermal deposit involve Hg, Ag, Ba, As, Mn, and Pb, whereas those developed at the adjacent and rear halos are Cu, Zn, Cd, Sn, W, Mo, and Bi. By investigating the characteristic chemical composition of primary halos on the land surface, predictions of blind ores at depth and an evaluation of the erosion level of outcropping orebodies can be made.

In south China, a geophysical prospecting team discovered an anomaly zone extending over 3000 metres on either side of a fault. The anomaly to the south of the fault was characterized by a Sn-Cu-Pb association, while that on the north was mainly characterized by a relatively high content of Pb, Zn, and Ag. At first sight the mineralization was widespread in the south where ancient workings were scattered everywhere, but it was not very noticeable in the north. According to the zoning of primary halos, the prospecting team came to the conclusion, that concealed ore might be encountered in the north and that the orebody in the south was either buried at a shallow depth or was already eroded to a certain level. This supposition was proved later by drilling (Fig. 40.7).

A more complicated picture was encountered in the study of the primary halos of skarn copper deposits. The "synthesized zoning sequence" established from hydrothermal infiltration halos is inapplicable to skarn copper deposits. For example, it can be seen from Figure 40.8 that W occupying a place usually at the end of the "synthesized zoning sequence" becomes the most typical front element. To study the primary halo of this type of deposit, factor analysis was first applied to discriminate the multistages of halo-formation.

Exploration geochemistry has already received a wide application in regional geological investigations. Through the repeated practices of many years, the method and techniques

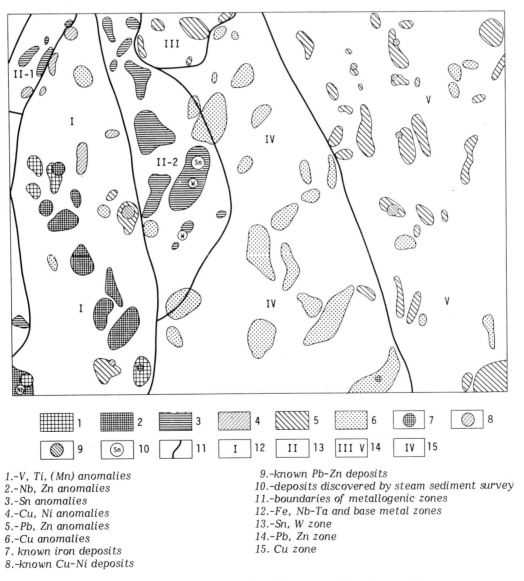

1.- V, Ti, (Mn) anomalies
2.- Nb, Zn anomalies
3.- Sn anomalies
4.- Cu, Ni anomalies
5.- Pb, Zn anomalies
6.- Cu anomalies
7. known iron deposits
8.- known Cu-Ni deposits
9.- known Pb-Zn deposits
10.- deposits discovered by steam sediment survey
11.- boundaries of metallogenic zones
12.- Fe, Nb-Ta and base metal zones
13.- Sn, W zone
14.- Pb, Zn zone
15. Cu zone

Figure 40.9. Spatial distribution of stream sediment anomalies.

1. Quaternary deposits
2. Cretaceous volcanic formations
3. Upper and Middle Cambrain carbonate rocks
4. Lower Cambrian carbonaceous rocks
5. Sinian carbonate rock, sandstone and volcanic rocks
6. Yenshan granite
7. contour of trend surface
8. positive deviation of copper (>20 ppm) from trend surface

Figure 40.10. *Fourth degree trend surface for copper.*

of regional geochemistry have been improved and are taking shape in our country. In a country like China with vast mountainous terrain, stream-sediment surveys should be carried out on a large scale to enhance efficiency and effectiveness in ore prospecting instead of low-efficiency soil surveys.

For example, in a mountainous area in southwest China, a stream-sediment survey was carried out in conjunction with 1:200,000 scale geological mapping. As a result, 86 anomalies were discovered, among which 19 anomalies were derived from known deposits, 72 being newly discovered and many of them were missed during the soil survey accompanying regional geological mapping. Five metallogenic zones were delineated according to the geological structures (Fig. 40.9), thus providing valuable information for further mineral exploration and theoretical research in this region. While making a further examination of 17 more promising anomalies, a large tin deposit of economic interest, a niobium-tantalum deposit, a tungsten deposit, two lead-zinc occurrences, and a gold prospect were discovered.

In recent years statistical methods using computers have been widely applied in China to process the vast amount of data obtained in regional geochemical surveys. The moving average method and trend surface analysis are considered as routine in compilation and map production. Cluster analysis, discriminant analysis, and factor analysis are used as tools for the evaluation of anomalies.

Figure 40.10 shows the result of trend surface analysis of copper in stream sediments (4th-degree trend surface). It can be noted clearly from the map, that there is a high-background zone with two large anomalies (A and B), superimposed on it, so it is necessary to delineate a more promising area in the high-background zone to the northeast. Figure 40.11 shows the result of the discriminant analysis. From the ΔR curve ($\Delta R = R - R_o$, where R is the discriminant score, and R_o the discriminant index), it is obvious that the background anomaly has been effectively filtered out.

Though stream-sediment survey technique has proven to be an effective prospecting method in mountainous areas, in special cases, however, it is also subject to some limitations, so one must choose the particular geochemical method with the local conditions in mind. For example, in high mountainous areas with strongly dissected terrain, fine materials cannot remain in the drainage system owing to the frequent washing by torrential waterflow, in which case stream-sediment surveys can hardly reveal any distinct anomaly; as an alternative, soil surveys would give a better result. In some arid regions subjected to severe wind erosion such as in northwest China, there is usually neither soil cover, nor a well developed drainage system. In such a case only colluvium sampling or rock sampling may be used. While conducting a soil survey in a region, where a humus layer is developed, one can succeed sometimes in discovering anomalies even by sampling from the humus layer. But in a cold climate, high-altitude grassland or tundra where only herbage with shallow root systems grow, it is necessary to take samples from the upper permafrost under the humus layer owing to the impoverishment of the humus layer in metals by leaching. In regions where ancient or recent glaciation are noticeable, samples must be taken from eluvium beneath the glacial drift.

In south China, the open ground in mountainous terrain is mostly cultivated and has become paddy fields. In the past these areas were considered as being useless for geochemical sampling. In recent years, however, it has been recognized that soils in the paddy fields are mostly of eluvial origin, or have been transported only a short distance. In such cases, anomalies could be readily found by sampling from under the cultivated layer by augering. The Kwangtung Geophysical Team has discovered a copper anomaly extending over 3000 m with a Cu content mostly of 100–200 ppm, which was accompanied by several small separate magnetic anomalies, over a large area of paddy fields in the vicinity of a long-abandoned mine; subsequently a skarn-type copper orebody over 100 m thick was revealed by drilling under the 20-30 m of overburden (Fig. 40.12).

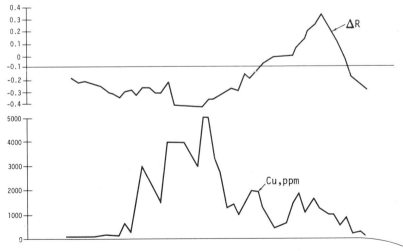

1. Quaternary deposits
2. quartz diorite
3. limestone
4. skarn

Figure 40.11

Distinction between superjacent anomaly and background anomaly by discriminant analysis.

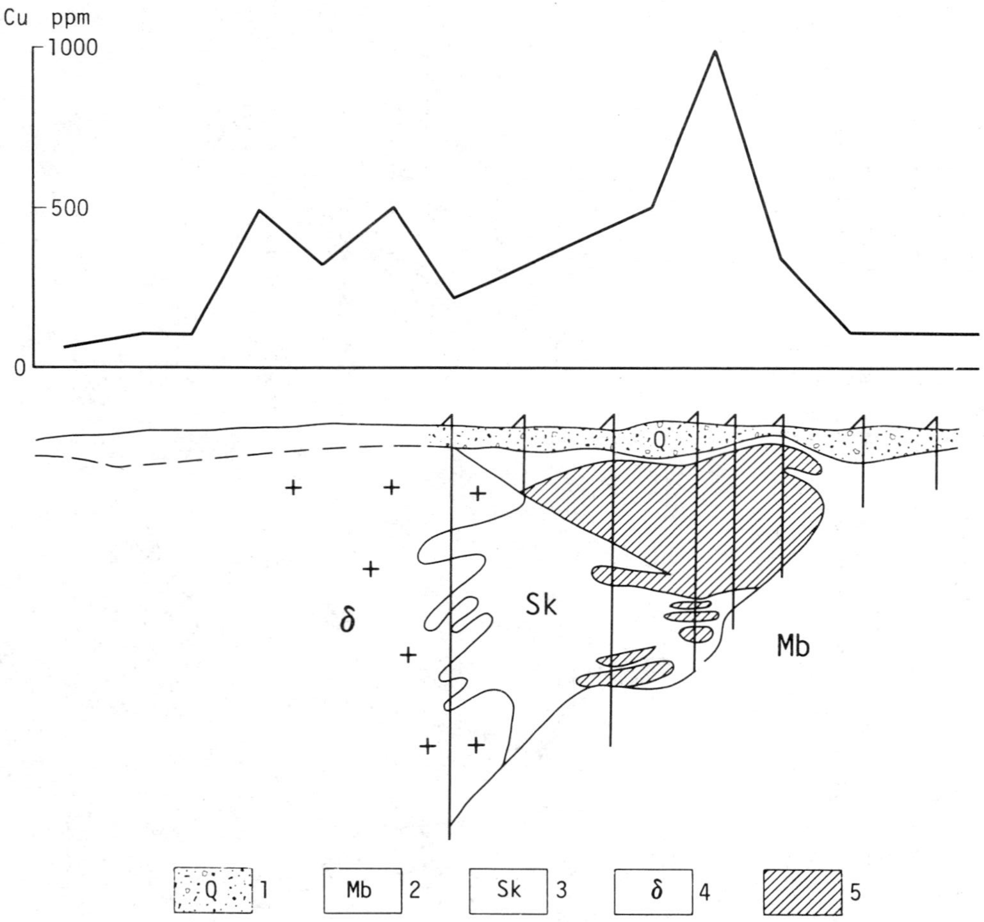

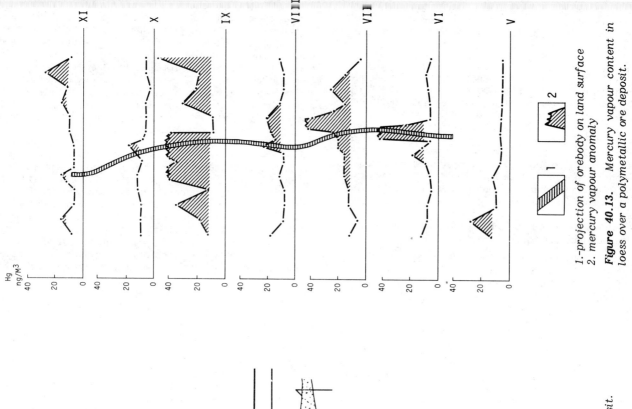

Figure 40.12. A copper anomaly in paddy field over a skarn copper deposit.

1. Quaternary deposit
2. -marble
3. -skarn
4. -quartz diorite
5. orebodies

Figure 40.13. Mercury vapour content in loess over a polymetallic ore deposit.

1.-projection of orebody on land surface
2. mercury vapour anomaly

In the northern and northwestern parts of China, where loess covers a large percentage of the area, the application of coventional geochemical methods is quite ineffective. Consequently soil gas and atmosphere mercury-vapour surveys have been attempted. Figures 40.13 and 40.14 show the result of a soil mercury-vapour survey in loess over a polymetallic ore deposit.

In order to complement the wide application of geochemical prospecting methods along with carrying out research on methodology, a great deal of attention has been paid in China to the development of geochemical instruments for analysis and other applications. Mass production of medium and large-sized quartz or glass prism spectrographs, various types of grating spectrometers, and atomic absorption spectrophotometers has been effected. In addition various kinds of mercury spectrometer, a series of portable X-ray fluorescence spectrometers suitable for base or in situ analysis, gamma-spectrometers and cold-extraction field kits for the analysis of 7 elements have also been developed.

CONCLUSION

Although a notable success has been achieved in the fields of mining geophysics and geochemistry in our country, it is, however, far from the requirements of socialist construction. We wish to bring the spirit of self-reliance and hard struggle into full play in order to make an even greater contribution to the fulfilment of the various tasks put forward by the 11th Congress of the Communist Party of China.

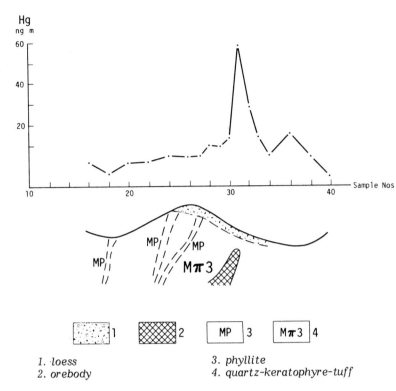

1. loess
2. orebody
3. phyllite
4. quartz-keratophyre-tuff

Figure 40.14. Variation of mercury content in soil gas profile VII of a polymetallic district.

AUTHOR INDEX

	Page
Barringer, A.R.	363
Becker, A.	33, 605
Bloom, H.	411
Boldy, J.	593
Bølviken, B.	295
Bondar, W.F.	365
Boyle, R.W.	25
Bradshaw, P.M.D.	327
Bravo Nieto, J.	727
Bristow, Q.	135
Brooks, R.R.	397
Cameron, E.M.	435
Campbell, E.E.	779
Campbell, G.	757
Cannon, H.L.	385
Coker, W.B.	435
Coolbaugh, D.F.	721
Coope, J.A.	575
Crone, D.	745
Czubek, J.A.	231
Davidson, M.J.	575
Derry, D.R.	1
Dyck, W.	489
Gibb, R.A.	105
Gleeson, C.F.	295
Glenn, W.E.	273
Govett, G.J.S.	339
Grasty, R.L.	147
Gregory, A.F.	511
Hausen, D.M.	735
Holroyd, M.T.	77
Hood, P.J.	77
Hornbrook, E.H.W.	435
Howarth, R.L.	545
Hsia Kuo-chih	799

	Page
Keller, G.V.	63
Ketola, M.	665
Killeen, P.G.	163
Klein, J.	653
Lajoie, J.J.	653
McGrath, P.H.	77
Martin, L.	545
Mason, R.	757
Meyer, W.T.	411
Middleton, R.S.	779
Miller, W.R.	479
Nelson, P.H.	273
Nichol, I.	339
Parker, W.	527
Podolsky, G.	641
Reed, L.E.	631
Seigel, H.O.	7
Skey, E.H.	697
Slankis, J.A.	641
Spector, A.	527
Stoiser, L.R.	727
Sumner, J.S.	123
Tanner, J.G.	105
Telford, W.M.	605
Theobald, P.K., Jr.	411
Thomson, I.	327
Ward, F.N.	365
Ward, S.H.	45
Webster, S.S.	697
Witherly, K.E.	685